HARNESSING AutoCAD® 2004

HARNESSING AutoCAD® 2004

THOMAS A. STELLMAN
G.V. KRISHNAN

autodesk® press

Australia • Canada • Mexico • Singapore • Spain • United Kingdom • United S

autodesk press

Harnessing AutoCAD® 2004

Thomas A. Stellman / G.V. Krishnan

Autodesk Press Staff

Vice President, Technology and Trades SBU:
Alar Elken

Editorial Director:
Sandy Clark

Senior Acquisitions Editor:
James DeVoe

Senior Development Editor:
John Fisher

Editorial Assistant:
Mary Ellen Martino

Executive Marketing Manager:
Cynthia Eichelman

Channel Manager:
Fair Huntoon

Marketing Coordinator:
Sarena Douglass

Production Director:
Mary Ellen Black

Production Manager:
Andrew Crouth

Production Editor:
Tom Stover

Art and Design Specialist:
Mary Beth Vought

Printed in Canada
1 2 3 4 5 XX 06 05 04 03

For more information contact Delmar Learning, Executive Woods, 5 Maxwell Drive, PO Box 8007, Clifton Park, NY 12065-8007

Or find us on the World Wide Web at www.delmarlearning.com or www.autodeskpress.com

Library of Congress Cataloging-in-Publication Data

Stellman, Thomas A.
Harnessing AutoCAD 2004 /
Thomas A. Stellman, G.V. Krishnan.
p. cm.
ISBN 0-4018-5079-0

Notice To The Reader

Publisher does not warrant or guarantee any of the products described herein or perform any independent analysis in connection with any of the product information contained herein. Publisher does not assume, and expressly disclaims, any obligation to obtain and include information other than that provided to it by the manufacturer.

The reader is expressly warned to consider and adopt all safety precautions that might be indicated by the activities herein and to avoid all potential hazards. By following the instructions contained herein, the reader willingly assumes all risks in connection with such instructions.

The publisher makes no representation or warranties of any kind, including but not limited to, the warranties of fitness for particular purpose or merchantability, nor are any such representations implied with respect to the material set forth herein, and the publisher takes no responsibility with respect to such material. The publisher shall not be liable for any special, consequential, or exemplary damages resulting, in whole or part, from the readers' use of, or reliance upon, this material.

Trademarks

CONTENTS

HARNESSING THE POWER OF AUTOCAD 2004

AutoCAD 2000 brought its "Heads Up" interface innovations and AutoCAD 2002 brought its interconnectivity capabilities and data management. AutoCAD 2004 has received some significant fine-tuning that makes it even more manageable than it has been, which is quite a task for a design/drafting program of its complexity.

Harnessing AutoCAD 2004 brings you comprehensive descriptions, explanations, and examples of the basics of AutoCAD 2004 and its new innovations. It has been written to be used both in the classroom as a textbook and in industry by the professional CADD designer/drafter as a reference and learning tool. The chapter on Utilities includes explicit instructions and descriptions on the new Tool Palettes Window feature. The chapter on the DesignCenter includes coverage of Searching for Content using the new DesignCenter Online feature. There is also thorough coverage of new features such as enhancements to presentation graphics (gradient fills, more colors), the new Multiline Text Editor, Revision Cloud, Password Protection and Digital Signatures. Whether you're new to AutoCAD or a seasoned user upgrading your skills, *Harnessing AutoCAD 2004* will show you how to rein in the power of AutoCAD to improve your professional skills and increase your productivity.

HIGHLIGHTS AND FEATURES OF THIS NEW EDITION

The improvements that AutoCAD 2004 brings include the following:

- **Status bar tray icons** provide access to features.
- **Tool palettes** provide access to blocks and hatch patterns.
- **DesignCenter Online** provides access to CAD libraries and product information on the Internet.
- **Gradient Fills, True Color and Color Books,** and **Shaded Plots** enhance presentation graphics.
- **Design Publisher** permits distribution of drawing sets viewable without AutoCAD.
- **Password Protection** and **Digital Signatures** provide added security.
- **Revision Cloud, Multiple Undo/Redo,** new **Multiline Text Editor** provide higher productivity.

- **External Reference Drawings (Xrefs)** can now be opened immediately in a new window. Xrefs can now be saved with their path relative to the host drawing.
- **CAD Standards** are verifiable to assure better quality.
- **License Borrowing** permits you to run AutoCAD while disconnected from the network.
- **License Timeout** sets a timeout period for all idle network licenses.

HOW TO USE THIS BOOK

OVERVIEW

The first chapter of this text provides an overview of the AutoCAD program, its interface, the commands, special features and warnings, and AutoCAD 2004 enhancements. Specific commands are described in detail throughout the book, along with lessons on how to use them.

FUNDAMENTALS

Harnessing AutoCAD contains five chapters devoted to teaching the fundamentals of AutoCAD 2004. Fundamentals I introduces some of the basic commands and concepts, and Fundamentals II through V continue to build logically on that foundation until the student has a reasonable competency in the most basic functions of AutoCAD.

INTERMEDIATE

After mastering the fundamentals, you move on to the intermediate topics, which include dimensioning, plotting and printing, hatching and boundaries, blocks and attributes, external references, and drawing environments. Other chapters teach students to make the most of AutoCAD using utility commands, scripts and slides, 3D commands, rendering, and the digitizing tablet.

ADVANCED

For the advanced AutoCAD user, this book offers a chapter on customizing AutoCAD 2004 (including toolbar customization) and Visual LISP. These two chapters teach you to make AutoCAD 2004 more individualized and powerful as you tailor them to your special needs.

APPENDICES

There are eight appendices in the back of this book. Appendix A is an introduction to hardware and software requirements of AutoCAD. Appendix B is a quick reference of AutoCAD commands with a brief description of their basic functions, and Appendix C provides a visual reference of AutoCAD toolbars.

Appendix D lists system variables, including default setting, type, whether or not it is read-only, and an explanation of the system variables. To see hatch and fill patterns, fonts, and linetypes provided with the AutoCAD program, refer to Appendices E, F, and G respectively. Appendix H lists available AutoCAD command aliases.

STYLE CONVENTIONS

In order to make this text easier for you to use, we have adopted certain conventions that are used throughout the book:

Convention	Example
Command names are in small caps	The MOVE command
Menu names appear with the first letter capitalized	Draw pulldown menu
Toolbar menu names appear with the first letter capitalized	Standard toolbar
Command sequences are indented. User inputs are indicated by boldface. Instructions are indicated by italics and are enclosed in parentheses	Command: **move** Enter variable name or [?]: **snapmode** Enter group name: *(enter group name)*

HOW TO INVOKE COMMANDS

Methods of invoking a command are summarized in a table

Standard toolbar	Choose REDO (see Figure 3–75)
Edit menu	Choose Redo
Command:	**redo** (ENTER)

DRAWING EXERCISES AND BONUS CHAPTER ON AUTOLISP

Harnessing AutoCAD 2004, like its predecessors, still offers comprehensive learning exercises associated with the lessons in each chapter. These exercises are representative of the types of discipline-related drawing problems found in the design industry today. They can be found on the CD in the back of this book and in the companion exercise book: *Harnessing AutoCAD 2004 Exercise Manual.* The CD contains PDF files of each chapter in the *Harnessing AutoCAD 2004 Exercise Manual.* These chapters correspond to the text in this book and contain a project exercise and specific exercises for the following disciplines: mechanical, architectural, civil, electrical, and piping. Exercise icons and tabs identify the discipline sections (refer to the following table of exercise icons). In addition, a bonus chapter entitled "Introduction to AutoLisp" is included on the CD in pdf format .

EXERCISE ICONS

Step-by-step Project Exercises are identified by the special icon shown in the following table. Exercises that give you practice with types of drawings that are often found in a particular discipline are identified by the icons shown in the following table.

Type of Exercise	Icon	Type of Exercise	Icon
Project Exercises		Civil	
Electrical		Mechanical	
Piping		Architectural	

HARNESSING AUTOCAD 2004 EXERCISE MANUAL

This printed exercise manual contains project exercises and discipline-specific exercises for Chapters 2 through 11 and Chapter 15 of the core text.

ISBN 1-4018-5080-4.

ONLINE COMPANION™

The Online Companion™ is your link to AutoCAD on the Internet. Updates are posted monthly, including a command of the month, tutorials, and FAQs. To access the Online Companions, go to the following URL:

http://www.autodeskpress.com/resources/olcs/index.html

E-RESOURCE

E-resource is an educational resource that creates a truly electronic Classroom. It is a CD-ROM containing tools and instructional resources that enrich your classroom and make your preparation time shorter. The elements of e-resource link directly to the text and tie together to provide a unified instructional system. Spend your time teaching, not preparing to teach.

Features contained in e-resource include:

> Syllabus: Lesson plans created by chapter that list goals and discussion topics. You have the option of using these lesson plans with your own course information.
>
> Chapter Hints: Objectives and teaching hints that provide direction on how to present the material and coordinate the subject matter with student projects.
>
> Answers to Review Questions: These solutions enable you to grade and evaluate end of chapter tests.
>
> PowerPoint® Presentation: These slides provide the basis for a lecture outline that helps you to present concepts and material. Key points and concepts can be graphically highlighted for student retention.
>
> Exam View Computerized Test Bank: Over 800 questions of varying levels of difficulty are provided in true/false and multiple-choice formats. Exams can be generated to assess student comprehension, or questions can be made available to the student for self evaluation.
>
> Animations: These .AVI files graphically depict the execution of key concepts and commands in drafting, design, and AutoCAD and let you bring multimedia presentations into the classroom.

Spend your time teaching, not preparing to teach!

ISBN 1-4018-5081-2.

ABOUT THE AUTHORS

Thomas A. Stellman received a B.A. degree in Architecture from Rice University and has over 20 years of experience in the architecture, engineering, and construction industry. He has taught at the college level for over ten years and has been teaching courses in AutoCAD since the introduction of version 1.4 in 1984. He conducts seminars covering both introductory and advanced AutoLISP. In addition, he develops and markets third-party software for AutoCAD. He currently is a CADD consultant, AutoLISP programmer, and project coordinator for Sunland Group in Austin, Texas.

G.V. Krishnan is director of the Applied Business and Technology Center, University of Houston-Downtown, a Premier Autodesk Training Center. He has used AutoCAD since the introduction of version 1.4 and writes about AutoCAD from the standpoint of a user, instructor, and general CADD consultant to area industries. Since 1985 he has taught courses ranging from basic to advanced levels of AutoCAD, including customizing, 3D AutoCAD, solid modeling, and AutoLISP programming.

ACKNOWLEDGMENTS

We would like to thank and acknowledge the many professionals who reviewed the manuscript to help us publish this *Harnessing AutoCAD 2004* text. Special thanks go to Lee Seroka, Technical Editor, Gail Taylor, Copy Editor and Ralph Grabowski for providing relevant information for Appendix I.

The authors would like to acknowledge and thank the following staff members of Delmar Thomson Learning:

Publisher: Alar Elken

Senior Acquisitions Editor: James DeVoe

Production Manager: Larry Main

Senior Developmental Editor: John Fisher

Production Editor: Tom Stover

Art & Design Coordinator: Mary Beth Vought

Editorial Assistant: Mary Ellen Martino

The authors also would like to acknowledge and thank the following:

Composition: John Shanley and Phoenix Creative Graphics

Getting Started

INTRODUCTION

The key phrase for AutoCAD 2004 is "better and easier to use." The heads-up scheme has been improved with new and enhanced palettes. Presentations benefit from gradient fills and a wider range of available colors. Security is more secure with password and digital signature upgrade. Productivity and quality assurance is higher with improved drafting features and access to product information and CAD libraries.

Advancements from AutoCAD 2002 to AutoCAD 2004 include the following:

- **Status bar tray icons** provide access to features.
- **Tool palettes** provide access to blocks and hatch patterns.
- **DesignCenter Online** provides access to CAD libraries and product information on the Internet.
- **Gradient Fills, True Color and Color Books**, and **Shaded Plots** enhance presentation graphics.
- **Design Publisher** permits distribution of drawing sets viewable without AutoCAD.
- **Password Protection** and **Digital Signatures** provide added security.
- **Revision Cloud, Multiple Undo/Redo**, new **Multiline Text Editor** provide higher productivity.
- **External Reference Drawings (Xrefs)** can now be opened immediately in a new window. Xrefs can now be saved with their path relative to the host drawing.
- **CAD Standards** are verifiable to assure better quality.
- **License Borrowing** permits you to run AutoCAD while disconnected from the network.
- **License Timeout** sets a timeout period for all idle network licenses.

STARTING AUTOCAD

Design/drafting is what AutoCAD (and this book) is all about. So how do you get into AutoCAD? Choose the Start button (Windows 2000/ME, Windows NT 4.0 and XP operating systems), select the AutoCAD 2004 program group, and then select the AutoCAD 2004 program.

THE STARTUP WINDOW

First Time Startup

By default, when AutoCAD is started, it displays a blank drawing window surrounded by menus and toolbars, as shown in Figure 1–1.

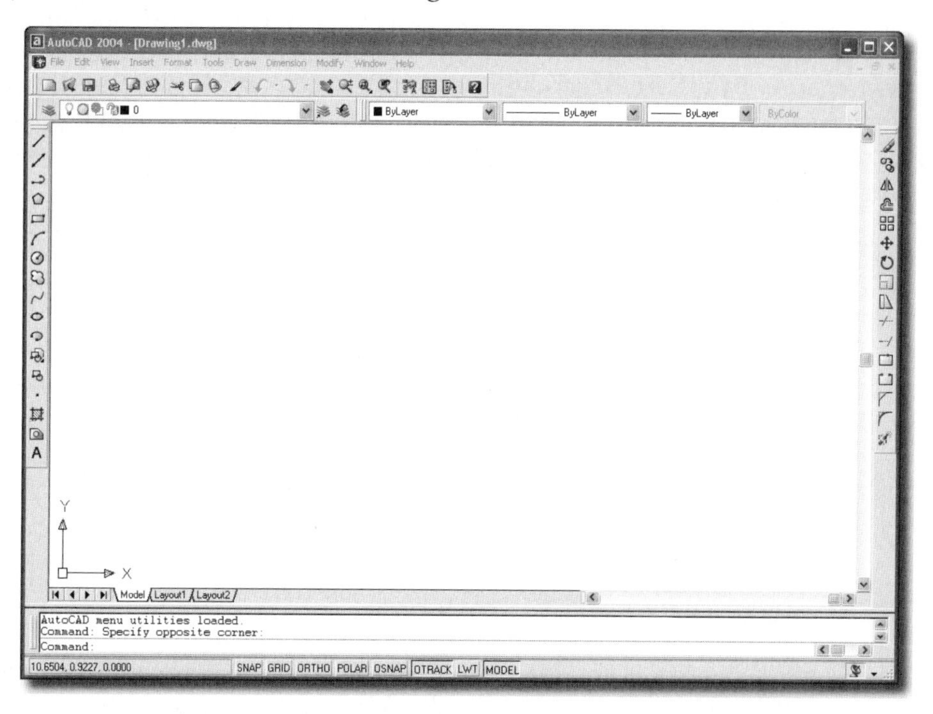

Figure 1–1 *AutoCAD 2004 OOTB (Out of the Box) Startup window*

The window shown in Figure 1–1 is one of the possible windows that might appear when AutoCAD is opened. This one appears when you start the program for the first time. The layout in the graphics area conforms to a particular set of drawing parameters. Other startup windows and layouts are possible, depending on how AutoCAD has been configured or if you have double-clicked on an AutoCAD drawing file icon in the Windows Explorer window. As covered later in this chapter, it is possible to configure AutoCAD to open with a Startup dialog box to assist you in setting up parameters different from the default opening layout.

Within the AutoCAD 2004 program window you can create drawings for viewing, printing (referred to as plotting in the trade), solving geometry and engineering problems, accumulating data, creating three-dimensional views of objects, and various other design, graphics and engineering applications. Whatever your objective is, you will very likely have to make changes in the layout and drawing parameters, or you can configure the startup configuration to suit your needs.

The Drawing Layout

In this chapter, the significance of the width and height of the graphics area on the Startup window and how they correlate to a final plotted sheet will be dealt with. The three key elements of drafting are location, direction, and distance. When you start out to draw something on a paper sheet, you have in mind a starting point on the drawing sheet for locating a point on an object to be drawn, and an orientation for the object. You must also consider the measurements of the object. The graphics area of the AutoCAD screen operates like a zoom lens on a camera through which you are looking at an imaginary drawing sheet. The imaginary drawing area itself is limitless (although you can place limits on inputting points). The AutoCAD graphics area on the screen has dimensions relative to dimensions on the imaginary drawing sheet, depending on the "zoom" factor in effect.

The initial AutoCAD startup graphics area is a full view of a 12 unit wide by 9 unit high drawing sheet. The features and commands in AutoCAD permit you to move your view around the drawing area and zoom in for a closer look or zoom out to see a broader area.

Note: Do not automatically assume that because the screen drawing area on the monitor is approximately 12" wide by 9" high, the units of measurement (12 units by 9 units) in the drawing must be inches. As described later, the units can be whatever distance of measurement you need, perhaps millimeters, or even miles.

Startup Dialog Box

There is a system variable called STARTUP whose value can be changed, causing the AutoCAD program to start up with a dialog box. It allows you to make changes in the drawing parameters to suit the size and type of drawing you wish to create. The setting of the system variable and use of the dialog box will be discussed later in this chapter.

Startup with an Existing Drawing

You can start the AutoCAD program by choosing a drawing file (one with the extension of .DWG) from the Windows Explorer window and double-clicking on its icon or filename. The AutoCAD program will be started. This is similar to the way other Windows-based programs are started by double-clicking on one of the types of files that is created and edited by that particular program. When AutoCAD is started in this manner (double-clicking on a .DWG file), the initial screen will normally display the drawing that was double-clicked using the view in which it was last saved.

AutoCAD SCREEN

The AutoCAD screen (see Figure 1–2) consists of the following elements: the graphics window, status bar and tool tray, title bar, toolbars, menu bar, model tab/layout tabs, and the command window.

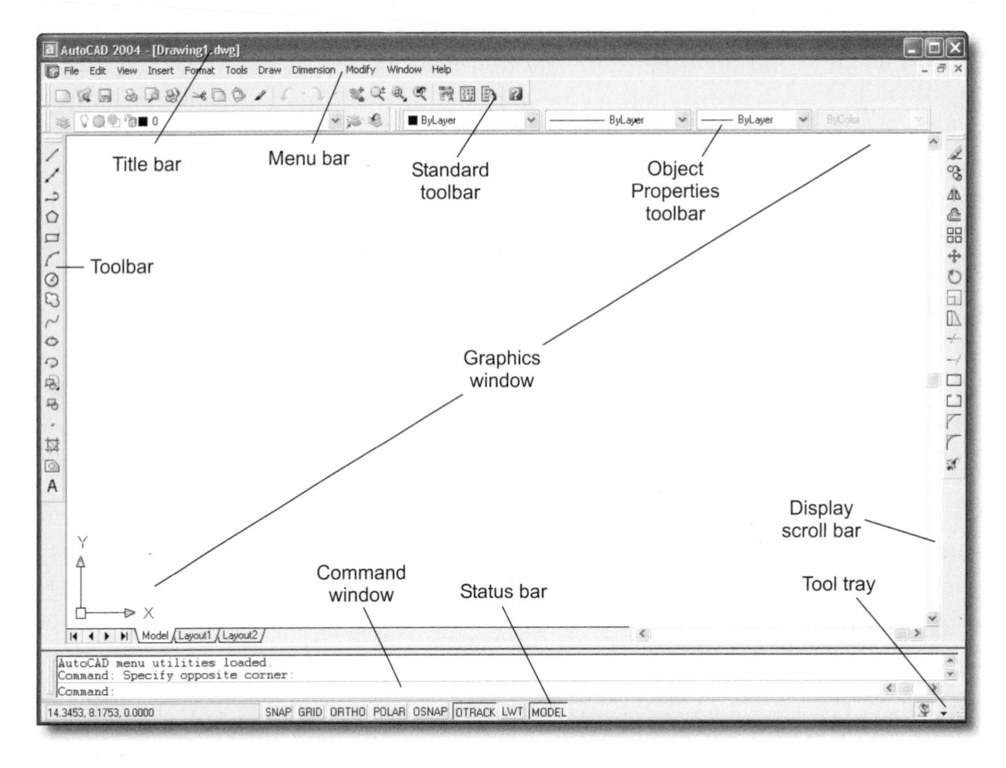

Figure 1–2 *The AutoCAD screen*

GRAPHICS WINDOW

The graphics window is where you can view the objects you create. In this window, AutoCAD displays the cursor, indicating your current working point. As you move your pointing device (usually a mouse or puck) around on a digitizing tablet, mouse pad, or other suitable surface, the cursor mimics your movements on the screen. When AutoCAD prompts you to select a point, the cursor is in the form of crosshairs. It changes to a small pick box when you are required to select an object on the screen. AutoCAD uses combinations of crosshairs, boxes, dashed rectangles, and arrows in various situations so you can quickly see what type of selection or pick mode to use.

Note: It is possible to enter coordinates outside the viewing area for AutoCAD to use for creating objects. As you become more adept at AutoCAD, you may find a need to do this. Until then, working within the viewing area is recommended.

STATUS BAR AND TOOL TRAY

The status bar at the bottom of the screen displays the cursor's coordinates and important information on the status of various modes. On the right end of the status bar is the tool tray, which contains icons for quick access to the Communications Center, Xref Manager, CAD Standards alert, and Digital Signature authenticator, as shown in Figure 1–3.

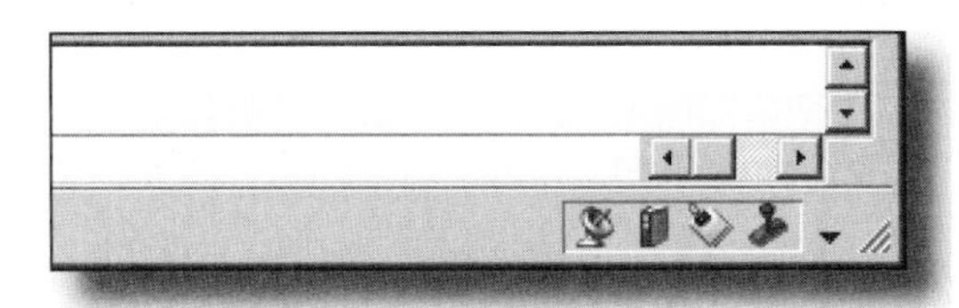

Figure 1–3 *Status bar with tool tray icons*

Communications Center

The Communications Center icon, when selected, causes the Communication Center dialog box to be displayed. A Welcome message appears stating "The Communication Center is your direct connection to the latest software updates, product support announcements and more." From here you specify your country and preferred update frequency, connect to the Internet and download available information, and specify which information channels you wish to view.

External Reference Manager

The Xref manager icon appears when your drawing has an external drawing attached. A message appears when an Xref needs to be reloaded or resolved. Chapter 11 explains using External References.

CAD Standards

The CAD Standards icon appears when there is a standards file associated with your drawing. A message appears when a standards violation occurs. Chapter 13 explains using CAD Standards.

Digital Signatures

The Validate Digital Signatures icon appears when the drawing has a digital signature. Select the icon to validate a digital signature. Chapter 13 explains using Digital Signatures.

TITLE BAR

The title bar displays the current drawing name for the AutoCAD application window.

TOOLBARS

The toolbars contain tools, represented by icons, from which you can invoke commands. Click a toolbar button to invoke a command, and then select options from a dialog box or respond to the prompts on the command line. If you position your pointer over a toolbar button and wait a moment, the name of the tool is displayed, as shown in Figure 1–4. This is called the ToolTip. In addition to the ToolTip, AutoCAD displays a very brief explanation of the function of the command on the status bar.

Figure 1–4 *Toolbar with a ToolTip displayed*

Some of the toolbar buttons have a small triangular symbol in the lower right corner of the button indicating that there are *flyout* buttons underneath that contain subcommands. Figure 1–5 shows the ZOOM command flyout located on the Standard toolbar. When you pick a flyout option, it remains on top to become the default option.

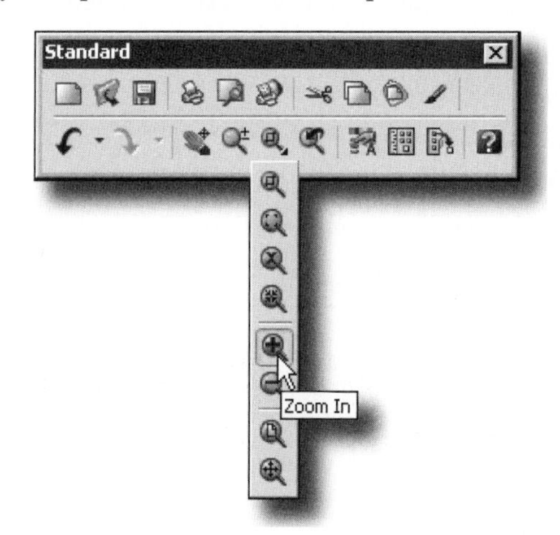

Figure 1–5 *Display of the Zoom flyout located on the Standard toolbar*

You can display multiple toolbars on screen at once, change their contents, resize them, and dock or float them. A *docked* toolbar attaches to any edge of the graphics window. A *floating* toolbar can lie anywhere on the screen and can be resized.

Docking and Undocking a Toolbar

To dock a toolbar, position the cursor on the caption, and press the pick button on the pointing device. Drag the toolbar to a dock location to the top, bottom, or either side of the graphics window. When the outline of the toolbar appears in the docking area, release the pick button. To undock a toolbar, position the cursor on the left end (for horizontal toolbars) or the top end (for vertical toolbars) of the toolbar and drag and drop it outside the docking regions. To place a toolbar in a docking region without docking it, hold down CTRL as you drag. By default, the Standard toolbar and the Properties toolbar are docked at the top of the graphics window (see Figure 1–2). Figure 1–6 shows the Standard toolbar and the Properties toolbar docked at the top of the graphics window, and the Draw and Modify toolbars docked on the left side of the graphics window.

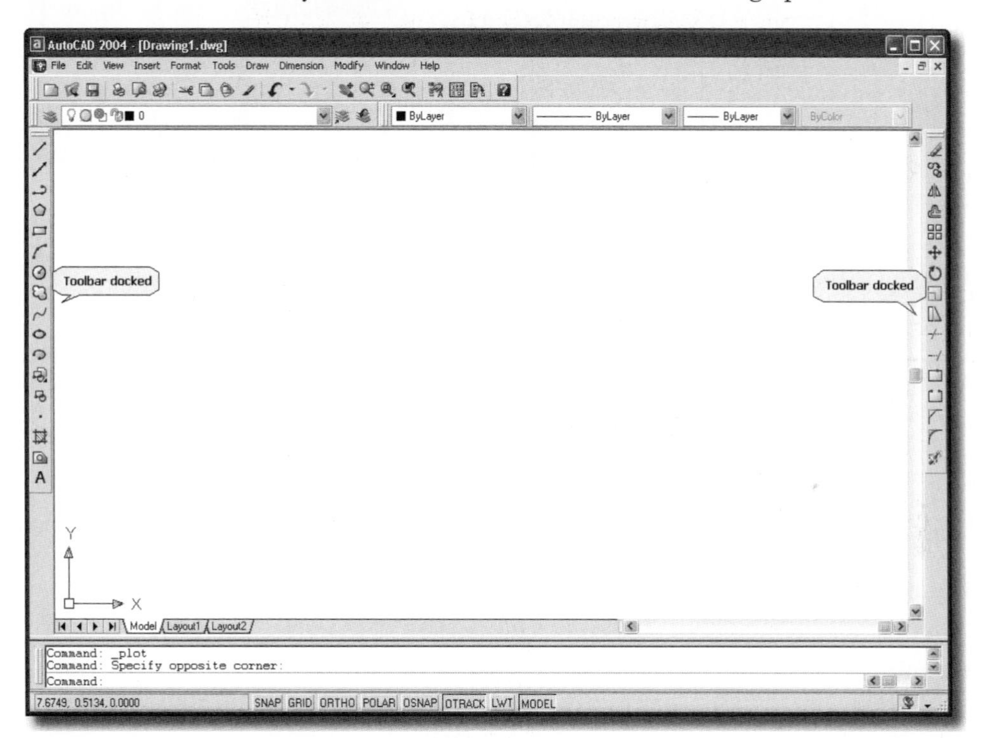

Figure 1–6 *Docking of toolbars in the graphics window*

Resizing a Floating Toolbar

If necessary, you can resize a floating toolbar. To resize a floating toolbar, position the cursor anywhere on the border of the toolbar, and drag it in the direction you want to resize. Figure 1–7 shows different combinations of resizing the Draw toolbar.

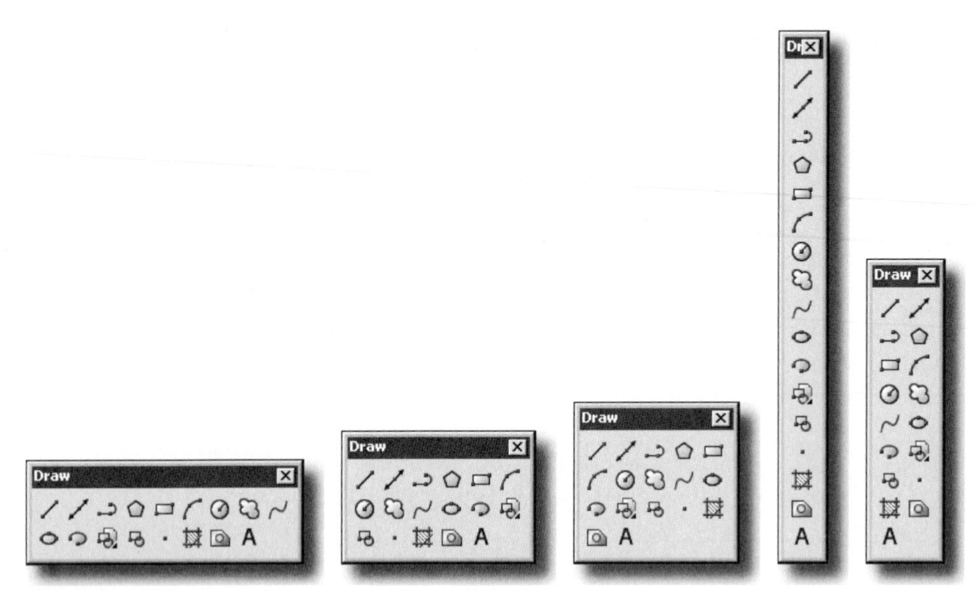

Figure 1–7 *Draw toolbar in different resizing positions*

Closing a Toolbar

To close a toolbar, position the cursor on the X located in the upper right corner of the toolbar, as shown in Figure 1–8, and press the pick button on your pointing device. The toolbar will disappear from the graphics window.

Figure 1–8 *Positioning the cursor to close a toolbar*

Opening a Toolbar

AutoCAD 2004 comes with 31 toolbars. To open any of the available toolbars, invoke the TOOLBAR command:

View menu	Choose Toolbars
Command: prompt	**toolbar** (ENTER)

If you choose **Toolbars** from the **View** menu or enter **toolbar** at the Command: prompt, AutoCAD displays the Customize dialog box with the **Toolbars** tab selected, as shown in Figure 1–9.

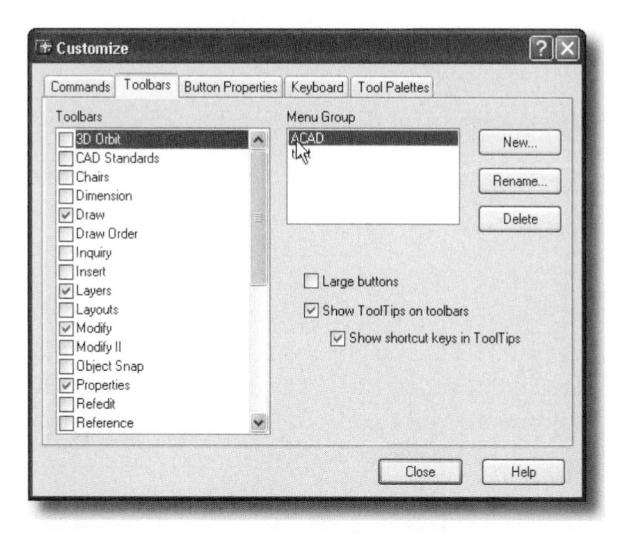

Figure 1–9 *Customize dialog box with the Toolbars tab selected*

The Customize dialog box has five tabs: **Commands, Toolbars, Properties, Keyboard,** and **Tool Palettes.** The **Commands, Properties,** and **Keyboard** tabs are explained in Chapter 17, "Customizing AutoCAD."

To open a toolbar, first select the check box for the toolbar from the **Toolbars:** list box on the Toolbars tab of the Customize dialog box. AutoCAD displays the selected toolbar. The **Show ToolTips on toolbars** check box controls whether or not to display the ToolTips. The **Show shortcut keys in ToolTips** check box controls whether or not the Tool Tips include shortcut keys. For example, with this check box checked, the Tool Tip for the CUT TO CLIPBOARD command will read "Cut to Clipboard (Ctrl + X)". The **Large Buttons** checkbox controls the size (large vs. regular) at which buttons are displayed. The default display size is 16 x 15 pixels. Setting the Large Buttons checkbox to ON displays the icons at 24 x 22 pixels. The Menu Group section will list the ACAD menu group that comes standard with AutoCAD. Other menu groups, along with the **New, Rename,** and **Delete** buttons are explained in Chapter 17. Figure 1–10 shows the command icons available on the Standard toolbar; Figure 1–11 shows the commands available on the Properties toolbar. Appendix C lists all the toolbars available in AutoCAD.

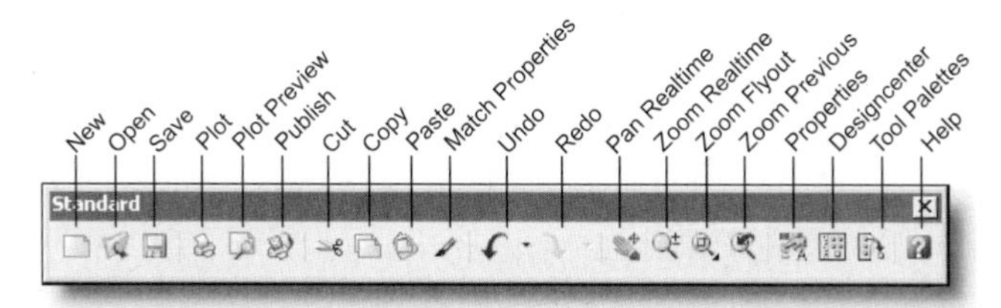

Figure 1–10 *Standard toolbar*

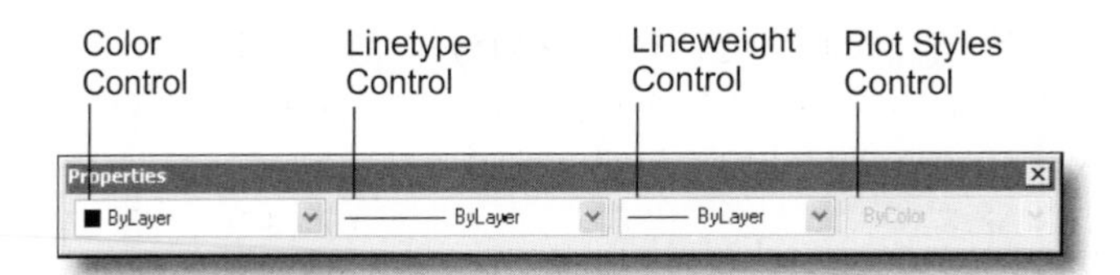

Figure 1–11 *Properties toolbar*

You can also open toolbars from the shortcut menu. Place your cursor anywhere on a toolbar that is displayed in the graphic window, click your right-button on your pointing device, and a shortcut menu appears, listing all the available toolbars (see Figure 1–12). Select the toolbar you want to open. You can also close a toolbar by removing the check mark next to the name of the toolbar.

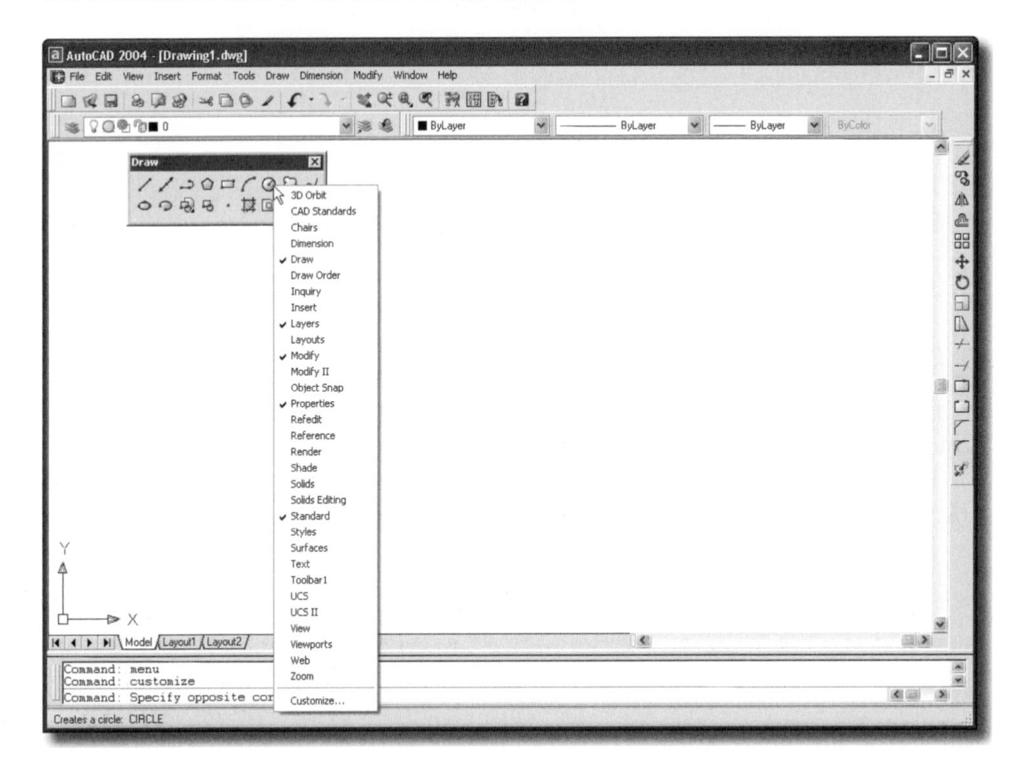

Figure 1–12 *Shortcut menu with listing of toolbars*

MENU BARS

The menus are available from the menu bar at the top of the screen. To select any of the available commands, move the crosshairs cursor into the menu bar area and press the pick button on your pointing device, which pops that menu bar onto the screen (see Figure 1–13). Selecting from the list is a simple matter of moving the cursor until the desired item is highlighted and then pressing the designated pick button on the pointing device. If a menu item has an arrow to the right, it has a cascading submenu. To display the submenu, move the pointer over the item and press the pick button. Menu items that include ellipses (...) display dialog boxes. To select one of these, just pick that menu item.

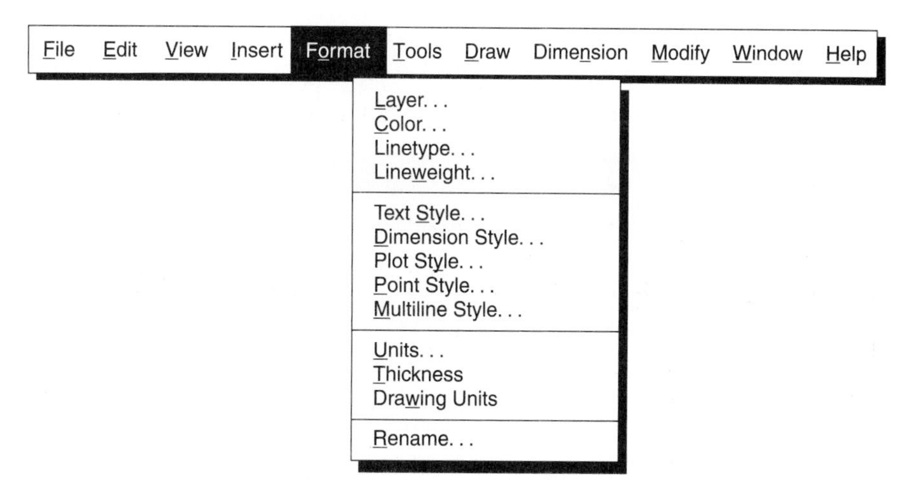

Figure 1–13 *Example of a menu bar*

MODEL TAB/LAYOUT TABS

AutoCAD allows you to switch your drawing between model (drawing) space and paper (layout) space. You generally create your designs in model space, and then create layouts to plot your drawing in paper space. Refer to Chapter 8 for a detailed explanation on plotting from layouts.

COMMAND WINDOW

The command window is a window in which you enter commands and in which AutoCAD displays prompts and messages. The command window can be a floating window with a caption and frame. You can move the floating command window anywhere on the screen and resize its width and height by dragging a side, bottom, or corner of the window.

There are two components to the command window: the single command line where AutoCAD prompts for input and you see your input echoed back, as shown in Figure 1–14, and the command history area, which shows what has transpired in the current drawing session. One display of the single command line remains at the bottom of the screen.

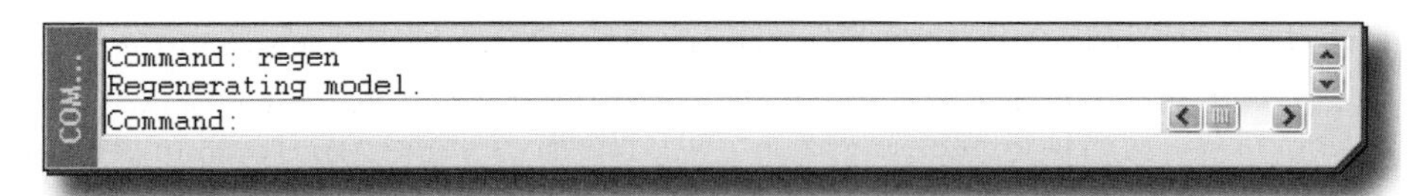

Figure 1–14 *Command window*

The command history area can be enlarged like other windows by picking the top edge and dragging it to a new size. You can also scroll inside the enlarged area to see previous command activity by using the scroll bars (see Figure 1–15).

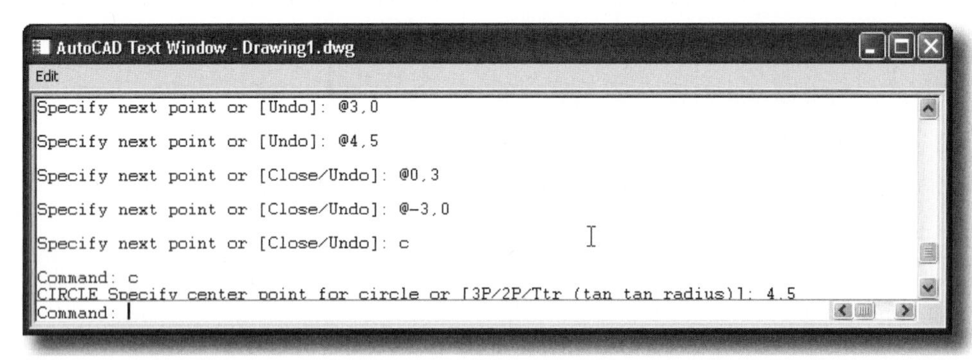

Figure 1–15 *Command history*

When you press F2, the command history text window switches between being displayed and being hidden. When the text window is displayed, you can scroll through the command history.

When you see "Command:" displayed in the Command window, it signals that AutoCAD is ready to accept a command. After you enter a command name and press ENTER or select a command from one of the menus or toolbars, the prompt area continues to inform you of the type of response(s) that you must furnish, until the command is either completed or terminated. For example, when you pick the LINE command, the prompt displays "Specify first point:"; after selecting a starting point by appropriate means, you will see "Specify next point or [Undo]:" asking for the endpoint of the line.

Each command has its own series of prompts. The prompts that appear when a particular command is used in one situation may differ from the prompts or sequence of prompts when invoked in another situation. You will become familiar with the prompts as you learn to use each command.

When you enter the command name or give any other response by typing from the keyboard, make sure to press ENTER or SPACEBAR. Pressing ENTER sends the input to the program for processing. For example, after you enter **line**, you must press ENTER or SPACEBAR in order for AutoCAD to start the part of the program that lets you draw lines. If you type **lin** and press ENTER or SPACEBAR, you will get an error message, unless someone has customized the program and created a command alias or command named "lin." Likewise, typing **linez** and pressing ENTER or SPACEBAR is not a standard AutoCAD command.

Pressing SPACEBAR has the same function as ENTER except when entering strings of words, letters, or numbers in response to the TEXT and MTEXT commands.

To repeat the previous command, you can press ENTER or SPACEBAR at the "Command:" prompt. A few commands skip some of their normal prompts and assume default settings when repeated in this manner.

Terminating a Command

There are three ways to terminate a command:

- Complete the command sequence and return to the "Command:" prompt.
- Press ESC to terminate the command before it is completed.
- Invoke another command from one of the menus, which automatically cancels any command in progress.

AUTOCAD COMMANDS AND INPUT METHODS

AUTOCAD COMMANDS

As much as possible, AutoCAD divides commands into related categories. For example, **Draw** is not a command, but a category of commands used for creating primary objects such as lines, circles, arcs, text (lettering), and other useful objects that are visible on the screen. Categories include **Modify**, **View**, and another group listed under the **Options** section of the **Tools** menu for controlling the electronic drawing environment. The commands under **Format** are also referred to as drawing aids and utility commands throughout the book. Learning the program can progress at a better pace if the concepts and commands are mentally grouped into their proper categories. This not only helps you find them when you need them, but also helps you grasp the fundamentals of computer-aided drafting more quickly.

INPUT METHODS

There are several ways to input an AutoCAD command: from the keyboard, toolbars, menu bars, the side screen menu, dialog boxes, the shortcut menu, or the digitizing tablet.

Keyboard

To invoke a command from the keyboard, simply type the command name at the "Command:" prompt and then press ENTER or SPACEBAR (ENTER and SPACEBAR are interchangeable except when entering a space in a text string).

If at the "Command:" prompt you want to repeat a command you have just used, press ENTER, SPACEBAR, or right-click on your pointing device. Right-clicking causes a shortcut menu to appear on the screen, from which you can choose the Repeat <*last command*> option. If the Default Mode in the Right-Click Customization subdialog box (available on the **User Preferences** tab of the Options dialog box) has been set up to **Repeat Last Command**, right-clicking will cause the last command to automatically repeat without displaying a shortcut menu. You can also repeat a command by using the up-arrow and down-arrow keys to display the commands you previously entered from the keyboard. Use the up-arrow key to display the previous line in the command history; use the down-arrow key to display the next line in the command history. Depending on the buffer size, AutoCAD stores all the information you entered from the keyboard in the current session.

AutoCAD also allows you to use certain commands transparently, which means they can be entered on the command line while you are using another command. Transparent commands frequently are commands that change drawing settings or drawing tools, such as GRID, SNAP, and ZOOM. To invoke a command transparently, enter an apostrophe (') before the command name while you are using another command. After you complete the transparent command, the original command resumes.

Toolbars

The toolbars contain tools that represent commands. Click a toolbar button to invoke the command, and then select options from a dialog box or follow the prompts on the command line.

Menu Bar

The menus are available from the menu bar at the top of the screen. You can invoke almost all of the available commands from the menu bar. You can choose menu options in one of the following ways:

- First select the menu name to display a list of available commands, and then select the appropriate command.
- Press and hold down ALT and then enter the underlined letter in the menu name. For example, to invoke the LINE command, first hold down ALT then press **D** (that is, press ALT + D) to open the Draw menu, and then press **L**.

The default menu file is *ACAD.MNU*. You can load a different menu file by invoking the MENU command.

Side Screen Menu

The side screen menu provides another, traditional way to enter AutoCAD commands. By default, the side screen menu is turned off in AutoCAD 2004. While this book does not refer to the side screen menu, traditional DOS users of AutoCAD may be more comfortable using it. To display the side screen menu, type **options** at the "Command:" prompt and press ENTER or SPACEBAR. AutoCAD displays the Options dialog box, as shown in Figure 1–16.

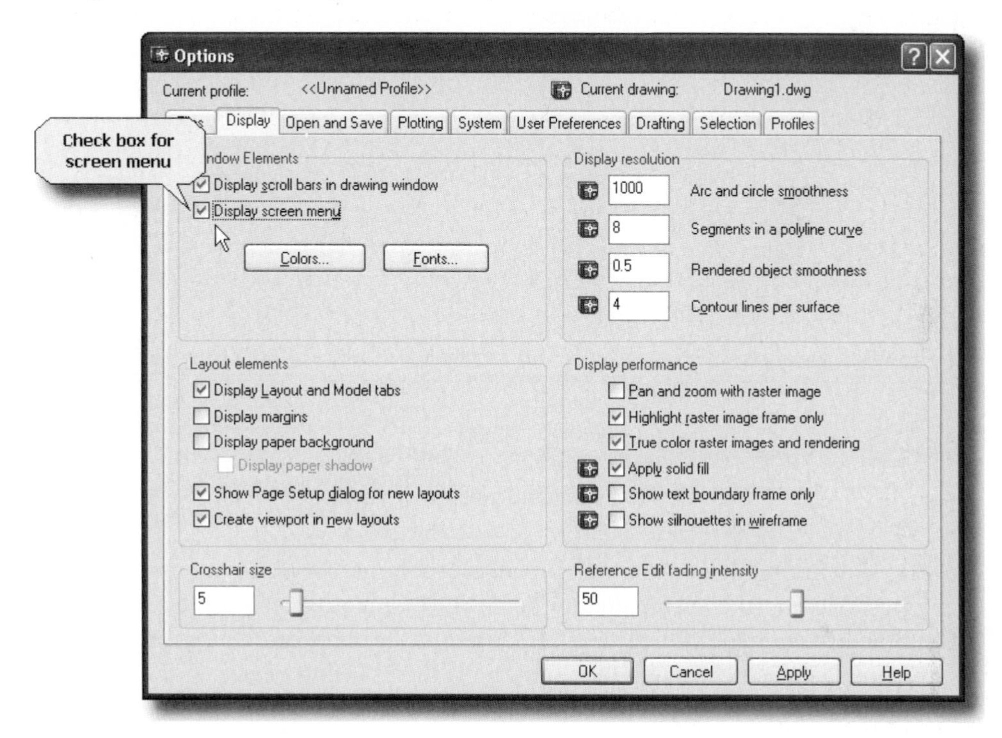

Figure 1–16 *Options dialog box*

Select the **Display** tab, and set the checkbox for **Display screen menu** to ON located in the **Window Elements** section of the dialog box. Choose the **OK** button to close the dialog box and save the settings. AutoCAD displays the side screen menu, as shown in Figure 1–17.

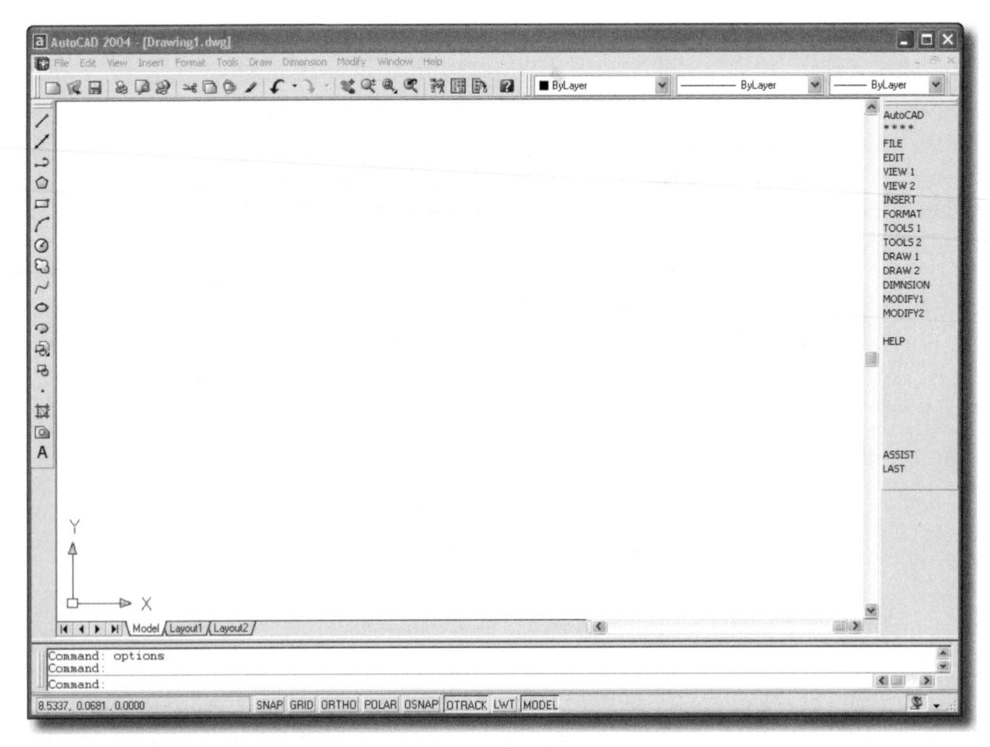

Figure 1–17 *AutoCAD screen window with the side screen menu*

Moving the pointing device to the right will cause the cursor to move into the screen menu area. Moving the cursor up and down in the menu area will cause selectable items to be highlighted. When the desired item is highlighted, you can choose that item by pressing the designated pick button on the pointing device. If the item is a command, either it will be put into action or the menu area will be changed to a list of actions that are options of that command. The screen menu is made up of menus and submenus. At the top of every screen menu is the word *AutoCAD*. When selected, it will return you to what is called the *root menu*. The root menu is the menu that is displayed when you first enter AutoCAD. It lists the primary classifications of commands or functions available.

Dialog Boxes

Many commands, when invoked, cause a dialog box to appear unless you prefix the command with a hyphen. For example, entering **insert** causes the dialog box to be displayed as shown in Figure 1–18, and entering **-insert** causes responses to be displayed in the Command: prompt area. Dialog boxes display the lists and descriptions of options, long rectangles for receiving your input data, and, in general, are the more convenient and user-friendly method of communicating with the AutoCAD program for that particular command.

The commands listed in the menu bar that include ellipses (...), such as **Plot**... and **Hatch**..., display dialog boxes when selected. For a detailed discussion of different dialog box components, see the section on "Using Dialog Boxes."

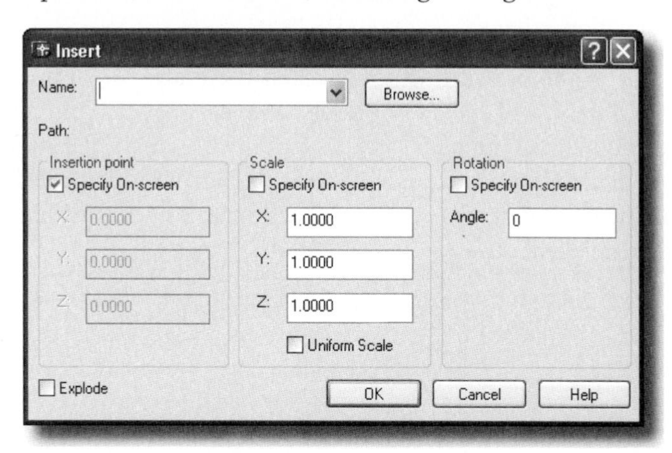

Figure 1–18 *Dialog box invoked from the* INSERT *command*

Cursor Menu

The AutoCAD cursor menu (see Figure 1–19) appears at the location of the cursor by pressing the middle button on a three-or-more-button mouse. On a two-button mouse you can invoke this feature by pressing SHIFT and right clicking. On a two-button mouse, the right button usually causes the shortcut menu to appear. The cursor menu (different from the shortcut menu) includes the handy Object Snap mode options along with the *X,Y,Z* filters. The reason that the Object Snap modes and Tracking are in such ready access will become evident when you learn the significance of these functions.

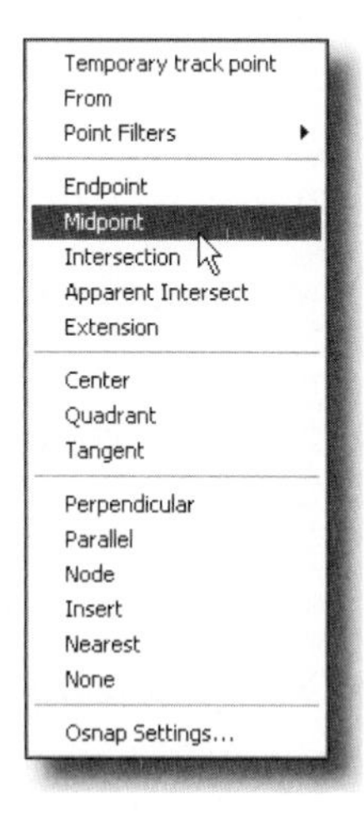

Figure 1–19 *Cursor menu*

Shortcut Menu

The AutoCAD shortcut menu appears at the location of the cursor when you press the right button (right-click) on the pointing device. The contents of the shortcut menu depend on the situation at hand.

If you right-click in the drawing window when there are no commands in effect, then the shortcut menu will include options to repeat the last command, a section for editing objects such as **Cut** and **Copy**, a section with **Undo**, **Pan**, and **Zoom**, and a section with **Quick Select**, **Find**, and **Options**, as shown in Figure 1–20. Selections that cannot be invoked under the current situation will appear in lighter text than those that can be invoked.

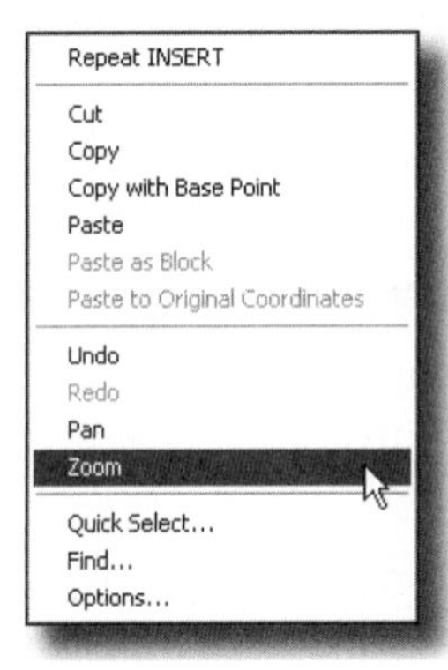

Figure 1–20 *Shortcut menu when no command is in effect*

If you select one or more objects (system variable PICKFIRST and/or GRIPS set to ON), and no commands are in effect, then right-click, and the shortcut menu will include some of the editing commands, as shown in Figure 1–21.

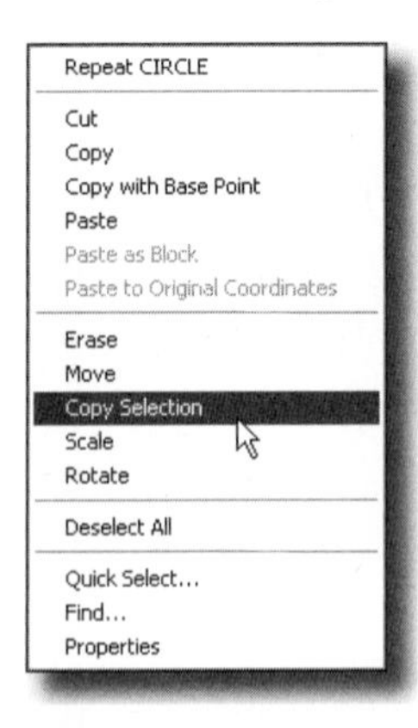

Figure 1–21 *Shortcut menu with one or more objects selected when no command is in effect*

Whenever you have entered a command and do not wish to proceed with the default option, you can invoke the shortcut menu and select the desired option with the mouse. For example, if instead of the default center-radius method of drawing a circle, you wished to use the TTR (tangent-tangent-radius), 2P (two-point), or 3P (three-point) option, you can select one of them from the shortcut menu, as shown in Figure 1–22. You can also select the PAN and ZOOM commands (transparently) or cancel the command. If pressing ENTER is required, that is also available.

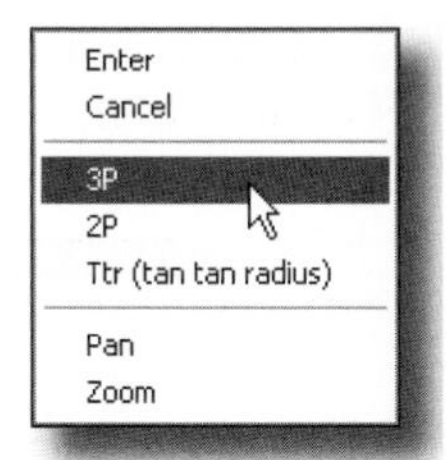

Figure 1–22 *Shortcut menu when the* CIRCLE *command is in effect*

Right-click anywhere in the Command window and the shortcut menu provides an access to the six most recently used commands (see Figure 1–23).

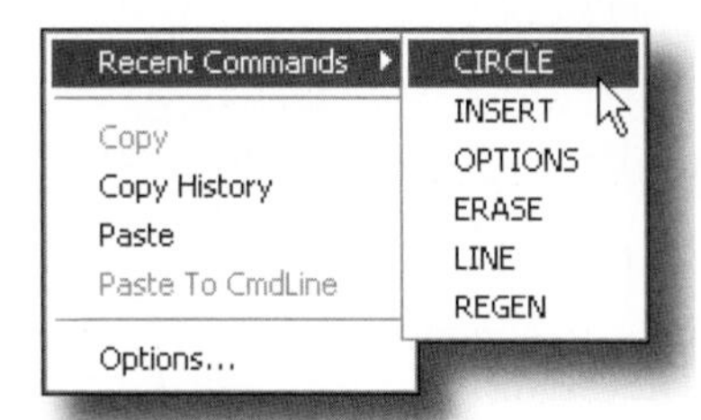

Figure 1–23 *Shortcut menu while the cursor is in the Command window*

Right-click on any of the buttons in the status bar, and the shortcut menu provides toggle options for drawing tools and a means to modify their settings.

Right-click on the **Model** tab or **Layout** tabs in the lower left corner of the drawing area, and the shortcut menu provides display plotting, page setup, and various layout options.

Right-click on any of the open AutoCAD dialog boxes and windows, and the shortcut menu provides context-specific options. Figure 1–24 shows an example for the Layer Properties Manager dialog box with a shortcut menu.

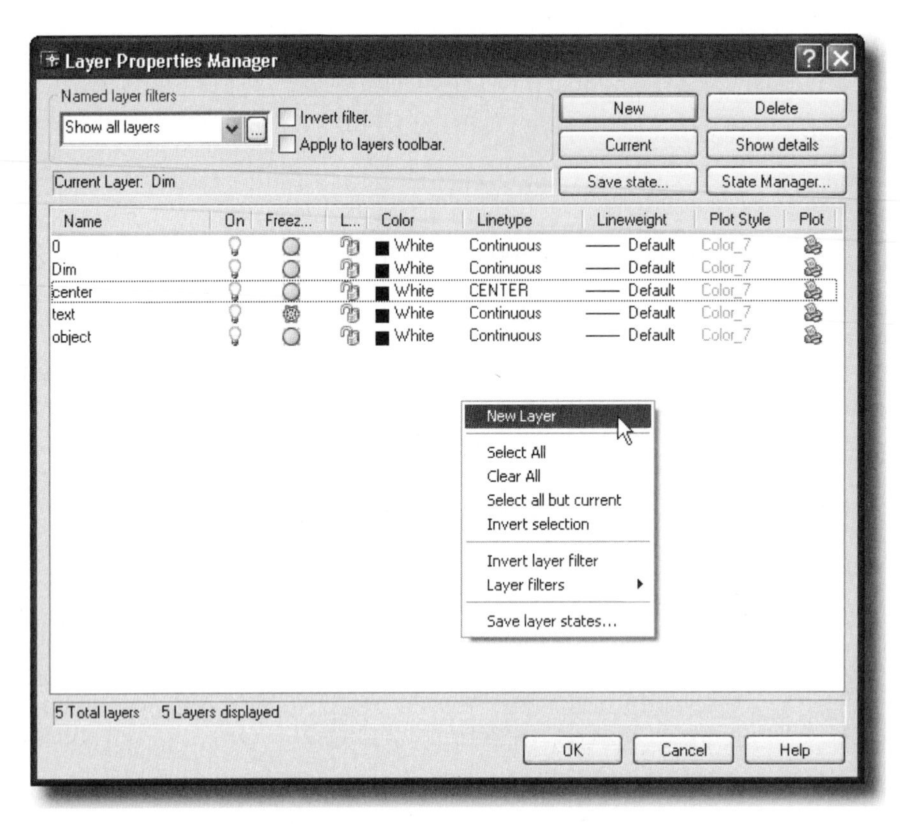

Figure 1–24 *Shortcut menu in the Layer Properties Manager dialog box*

Digitizing Tablet

The most common input device after the mouse is the digitizing tablet. It combines the screen cursor control of a mouse with its own printed menu areas for selecting items. However, with the new Heads-up features and customizability in AutoCAD since release 2000, the tablet overlay with command entries is becoming obsolete. One powerful feature of the tablet (not related to entering commands) is that it allows you to lay a map or other picture on the tablet and trace over it with the puck (the specific pointing device for a digitizing tablet), thereby transferring the objects to the AutoCAD drawing. The new interfaces with other platforms that allow you to insert a picture of an aerial photograph, for example, make tablet digitizing less of a demand.

USING DIALOG BOXES

When a dialog box appears, the crosshairs cursor changes to an arrow, pointing up and to the left. You can use the arrow keys on your keyboard to make selections in the dialog box, but it is much easier to use your pointing device. Another way to make selections in a dialog box is to use keyboard equivalents. You can move the cursor from one field to another by pressing TAB when the cursor is not in the text box.

TEXT BOX

A text box is an area that accepts one line of text entry. It is normally used to specify a name, such as a layer name or even a file name, including the drive and/or folder. Text boxes often function as an alternative to selecting from a list of names when the desired name is not in the list box. Once the correct text is keyed in, enter it by pressing ENTER.

Moving the pointer into the text box causes the text cursor to appear in a manner similar to the cursor in a word processor. The text cursor, in combination with special editing keys, can be used to make changes to the text. You can see both the text cursor and the pointer at the same time, making it possible to click the pointer at a character in the text box and relocate the text cursor to that character.

Right and Left Arrows > < keys

These move the cursor right or left (respectively) across text without having any effect on the text.

Backspace key

This deletes the character to the left of the cursor and moves the cursor to the space previously occupied by the deleted character.

Delete key

This deletes the character at the location of the cursor, causing any text to the right to move one space to the left.

BUTTONS

Actions are immediately initiated when you select one of the dialog box buttons.

Default Buttons

If a heavy line surrounds a button (like the **OK** button in most cases), then it is the default button and pressing ENTER is the same as clicking that button.

Buttons with Ellipses (...)

Buttons with ellipses display a second dialog box, called a subdialog or child dialog box.

Note: When a subdialog box appears, you must respond to the options in the subdialog box before the underlying dialog box can continue.

Screen Action Buttons

Buttons that are followed by an arrow (<) require a graphic response, such as selecting an object on the screen or picking/specifying coordinates.

Disabled Buttons

Buttons with actions that are not currently acceptable will be disabled. They appear grayed out.

Character Equivalents

A button with a label that has an underlined character in it can be activated by holding down ALT + pressing the underlined character.

Radio Buttons

Radio buttons are options when only one of two or more selections can be active at a time (see Figure 1–25). Selecting one radio button will deactivate any other in the group, like selecting a station button on the radio.

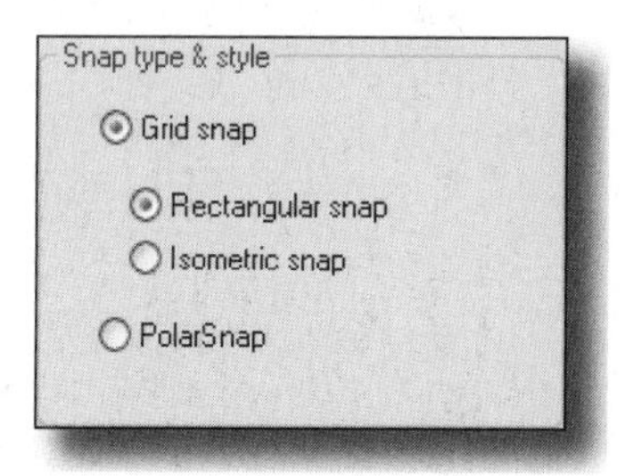

Figure 1–25 *Radio buttons*

CHECK BOXES

A check box acts as a toggle. When selected, the check box switches the named setting between ON and OFF, as shown in Figure 1–26. A check or X in the check box means the option is set to ON; no check or X (an empty box) means the option is set to OFF.

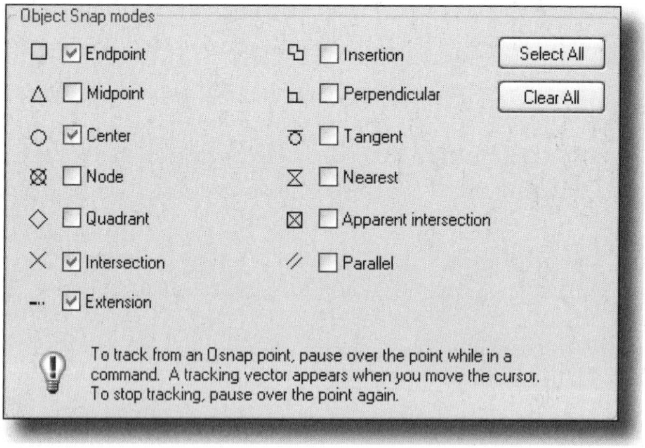

Figure 1–26 *Check boxes*

LIST BOXES AND SCROLL BARS

List boxes make it easy to view, select, and enter a name from a list of existing items, such as file names and fonts. Move the pointer to highlight the desired selection. When you click on the item, it appears in the text box. You accept this item by selecting **OK** or by double-clicking on the item. For example, Figure 1–27 shows the list box from the standard Select File dialog box.

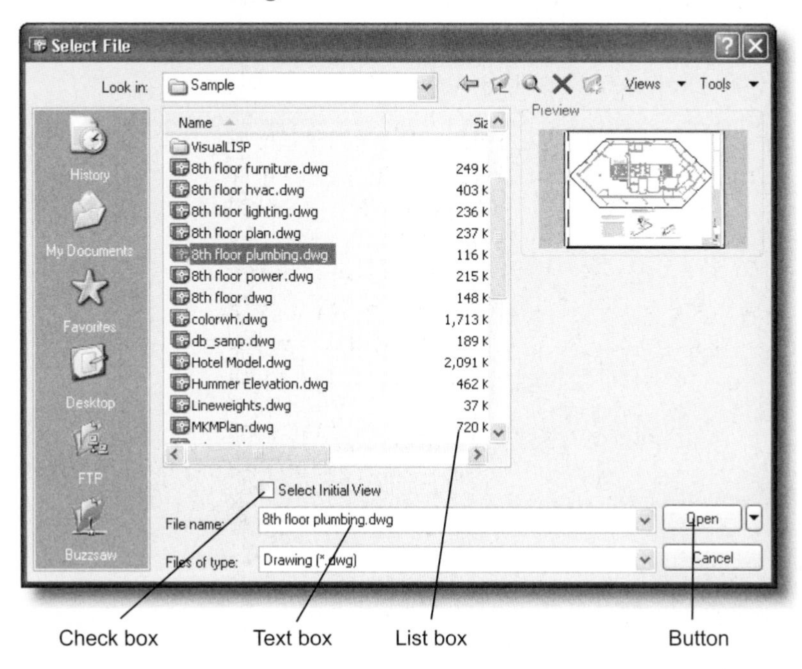

Figure 1–27 *List box from the standard Select File dialog box*

List boxes are accompanied by scroll bars to facilitate moving long lists left to right and right to left in the list box.

Some boxes that have multiple options available are really just unexpanded list boxes. They will have a down arrow on the right side. Selecting the down arrow displays the expanded list.

Note: Many dialog boxes have a Help button. If you are not sure how to use the feature in the dialog box, choose the Help button for a brief explanation of the dialog box.

TOOL PALETTES WINDOW

Tool Palettes

Tool palettes (which are separate tabbed areas within the Tool Palettes window) are introduced in AutoCAD 2004 and make it quicker and easier to insert blocks, draw hatch patterns and implement custom tools developed by a third party. Blocks (See Chapter 10) and Hatch Patterns (See Chapter 9) are the primary tools that are managed with tool palettes. The Tool Palettes feature allows blocks and hatch patterns of similar usage and type to be grouped in their own tool palette. For example, one tool palette can be named Plumbing Fixtures and contain, of course, blocks representing plumbing fixtures as shown in Figure 1–28.

Figure 1–28 *The Tool Palettes window in the docked position with the Plumbing Fixtures tab displayed*

Figure 1–28 shows the default Tool Palettes window that comes with AutoCAD 2004. Other tabs attached to the Tool Palettes window are the Sample Office Project tab, the Imperial Hatches tab, and the ISO Hatches tab, all of which contain icons representing blocks, hatch patterns or both.

The TOOLPALETTES command causes the Tool Palettes window to be displayed.

Docking and Undocking the Tool Palettes Window

The default position for the Tool Palettes window is docked on the right side of the screen. Its position can be changed by placing the cursor over the double line bar at the top of the window and either double-clicking or dragging the window into the screen area (or across to a docking position on the left side of the screen). Double-clicking causes the Tool Palettes window to become undocked and to float in the drawing area as shown in Figure 1–29. When the Tool Palettes window is undocked, it can be docked by double-clicking in the title bar (which may be on the left or right side of the window) or by placing the cursor over the title bar and dragging the window all the way to the side where you wish to dock it.

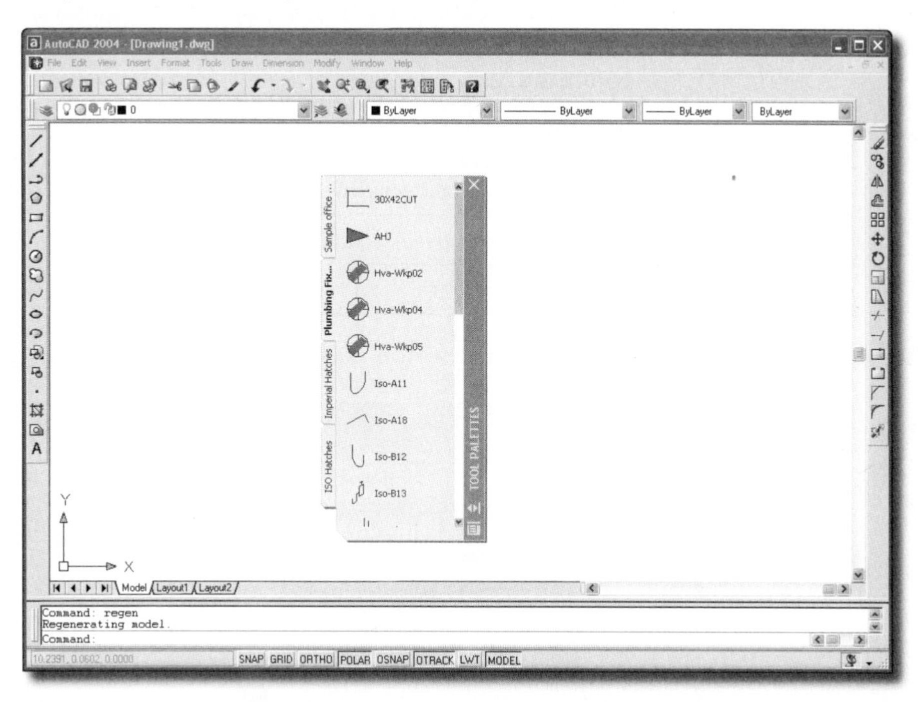

Figure 1–29 *The Tool Palettes window in the floating position*

Inserting Blocks and Hatch Patterns from a Tool Palette

To insert a block (see Chapter 10) from a tool palette simply place the cursor on the block symbol in the tool palette, press the pick button, and drag the symbol into the drawing area. The block will be inserted at the point where the cursor is located when the pick button is released. This procedure is best implemented by using the appropriate OSNAP mode (see chapter 3 for Object Snap modes). Another method of inserting a block from a tool palette is to select the block symbol in the tool palette and then select a point in the drawing area for the insertion point.

To draw a hatch pattern (see Chapter 9) that is a tool in a tool palette, place the cursor on the hatch pattern symbol in the tool palette, press the pick button, and drag the symbol into the boundary to receive the hatch pattern and release the pick button. Another method of drawing a hatch pattern that is a tool in a tool palette is to select the hatch pattern symbol in the tool palette and then select a point within a boundary in the drawing area.

GETTING HELP

ACTIVE ASSISTANCE

AutoCAD provides automatic or on-demand context-sensitive help in the form of an Active Assistance window. The Active Assistance window can be configured to be displayed automatically when you begin an AutoCAD session or to not be displayed until you invoke the ASSIST command.

Whenever you need AutoCAD's Active Assistance, invoke the ASSIST command:

Windows system tray	Right-click the Active Assistance icon
Help menu	Choose Active Assistance
Command: prompt	**assist** (ENTER)

AutoCAD displays the **Active Assistance window**, as shown in Figure 1–30.

Figure 1–30 *Active Assistance window*

By dragging a side, bottom or corner of the Active Assistance window you can enlarge it to display more (sometimes all) of the information in the window.

While the Active Assistance window is being displayed, and you invoke a command, the Active Assistance window will display information about the command just invoked. For example, when you invoke the CIRCLE command, Help topics about the CIRCLE command will be displayed as shown in Figure 1–31. When you select one of the Help topics, the AutoCAD Help window is displayed with information about the topic selected.

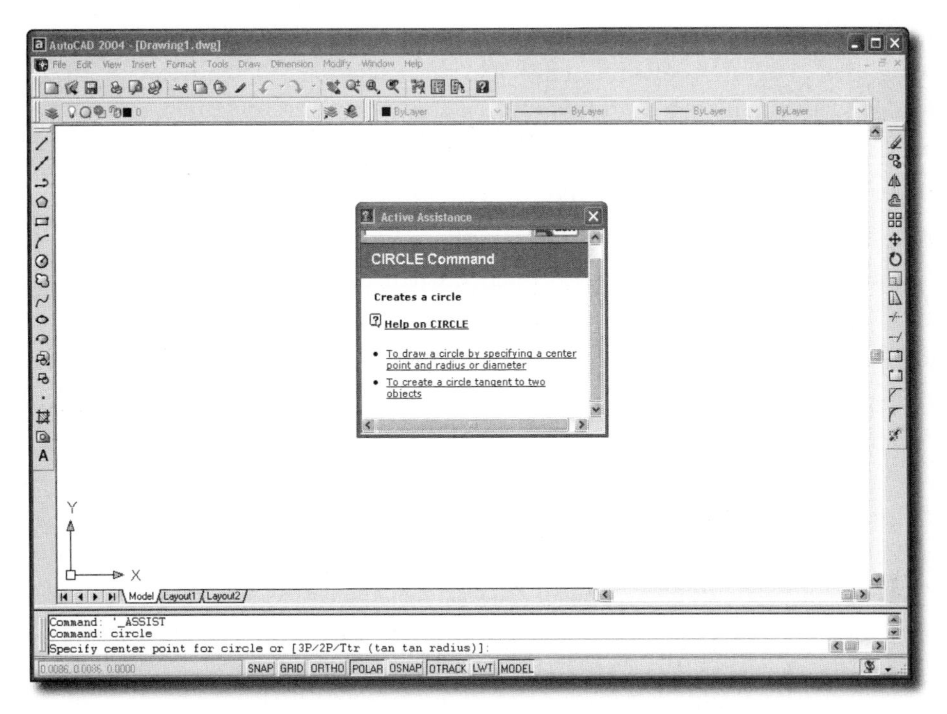

Figure 1–31 *The Active Assistance window automatically displaying Help topics about the* CIRCLE *command*

Active Assistance Window Settings

The settings of the Active Assistance feature can be accessed by right-clicking in the Active Assistance window and then selecting **Settings**. AutoCAD will display the Active Assistance Settings dialog box as shown in Figure 1–32.

If the **Show on start** check box is checked, AutoCAD will automatically display the Active Assistance window when you begin a new session. It will not automatically display the window if the **Show on start** check box is not checked.

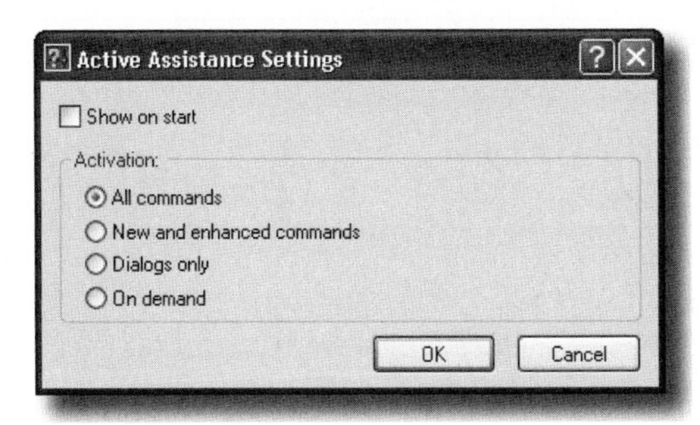

Figure 1–32 *The Active Assistance dialog box*

In the Activation section of the Active Assistance Settings dialog box, there are four options that determine how the Active Assistance window will be initiated; **All commands, New and enhanced commands, Dialogs only**, and **On demand.**

When the **All commands** radio button is selected, the Active Assistance window is automatically opened when any command is invoked.

When the **New and enhanced commands** radio button is selected, the Active Assistance window is automatically opened when any new or enhanced commands are invoked.

When the **Dialogs only** radio button is selected, the Active Assistance window is automatically opened when a dialog box is displayed and cannot be closed while the dialog box is being displayed.

When the **On demand** radio button is selected, the Active Assistance window is opened only by double-clicking the Active Assistance icon in the system tray, right-clicking the icon and choosing Show Active Assistance, or entering **assist** at the Command: prompt.

In the **Ask** text box, enter a question (or even a single word) and click **Ask**, and the AutoCAD Help: User Documentation dialog box will be displayed with a list of topics in accordance with the parameter(s) specified in the **Ask** text box.

TRADITIONAL HELP

When you are in the graphics window, AutoCAD provides a context-sensitive help facility to list its commands and what they do. The HELP command provides online assistance within AutoCAD. When an invalid command is entered, AutoCAD displays a message to remind you of the availability of the help facility.

Whenever you need AutoCAD help, invoke the HELP command:

Standard toolbar	Choose the HELP command (see Figure 1–33)
Help menu	Choose AutoCAD Help Topics
Command: prompt	**?** or HELP (ENTER)

Figure 1–33 *Invoking the* HELP *command from the Standard toolbar*

AutoCAD displays the AutoCAD Help:User Documentation dialog box, as shown in Figure 1–34.

The HELP command can be invoked while you are in the middle of another command. This is referred to as a *transparent command*. To invoke a command transparently (if it is one of those that can be used that way), simply prefix the command name with an apostrophe. For example, to use HELP transparently, enter **'help** or '? in response to any prompt that is not asking for a text string. AutoCAD displays help for the current command. Often the help is general in nature, but sometimes it is specific to the command's current prompt.

As an alternative, press the function key F1 to invoke the HELP command. Or you can invoke help by choosing the Help menu from the menu bar at the top of your screen.

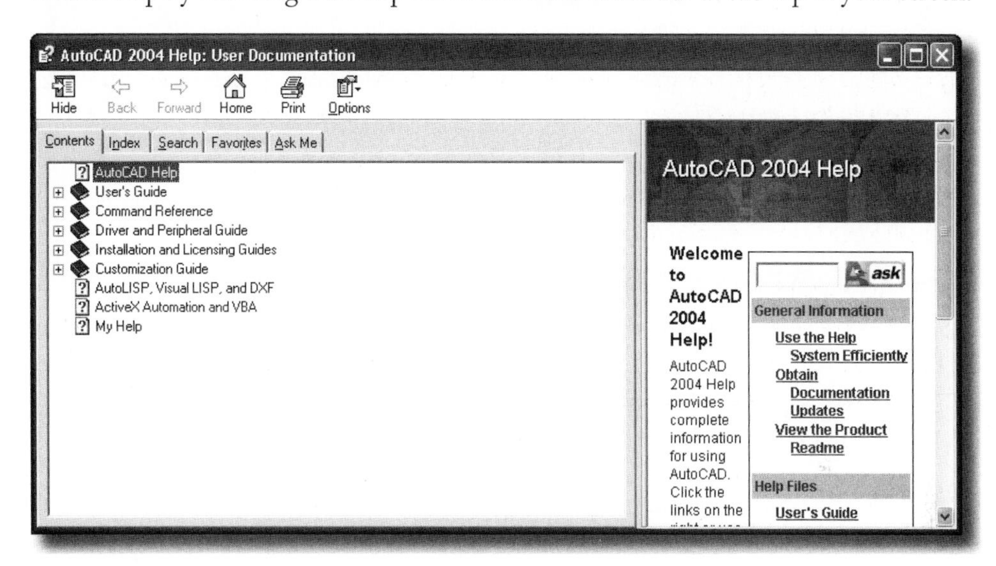

Figure 1–34 *AutoCAD Help: User Documentation dialog box*

Help switches to an independent window, so when you are through with the help utility you will need to switch to the AutoCAD program window to continue drawing. You do not have to close the AutoCAD Help: User Documentation window to work in an AutoCAD drawing session.

At the top of the AutoCAD Help:User Documentation window is a button that will let you hide the tab (left) side, therefore shrinking the size of the window with the **Hide** button or, if it has been shrunk, enlarge the window and show the tabs with the **Show** button. The **Back** button returns you to the previous screen when possible. The **Forward** button reverses the previous action of the **Back** button. The information (right) side is the instruction or description area where information about the selected subject or command is displayed. The tab side has text boxes and list boxes to aid in getting the help you need on all subjects and commands in AutoCAD. The **Home** button will take you to the home page of the User documentation. The **Print** button sends the contents of the information area to the printer. The **Options** button displays a menu with the options shown in Figure 1–35.

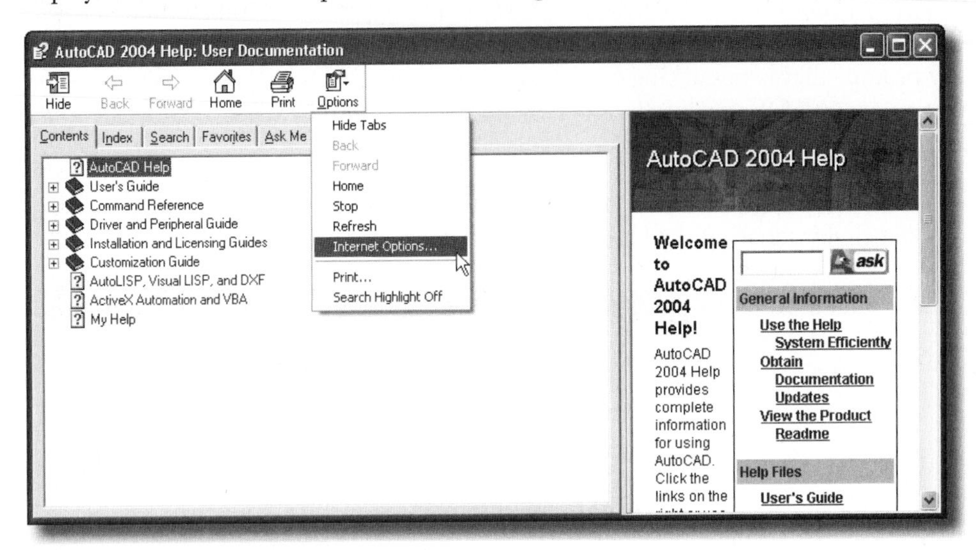

Figure 1–35 *AutoCAD Help:User Documentation dialog box with Options button selection menu*

The **Hide Tabs** and **Show Tabs** options hide or display respectively the tab (left) side of the AutoCAD Help:User Documentation dialog box.

The **Back** and **Forward** options cause previous or next screens respectively to be displayed in the information area on the right side of the AutoCAD Help: User Documentation dialog box.

The **Stop, Refresh,** and **Internet Options...** options are used in conjunction with connectivity with other sites and are explained in Chapter 14.

The **Contents** tab of the AutoCAD Help: User Documentation dialog box (see Figure 1–36) has a list of subjects prefixed by a closed-book icon. You may scroll through the list, pressing ENTER when the item on which you need help is highlighted, or you may double-click that item in the text box. This causes the subject's icon to become an open book, and below the selected subject, a sublist will be displayed, indented, containing items that may be prefixed by a closed book or a ? button, as shown in Figure 1–36. Items with closed-book icons lead to sublists, and ? buttons display help available on the highlighted item in the information (right) side of the dialog box.

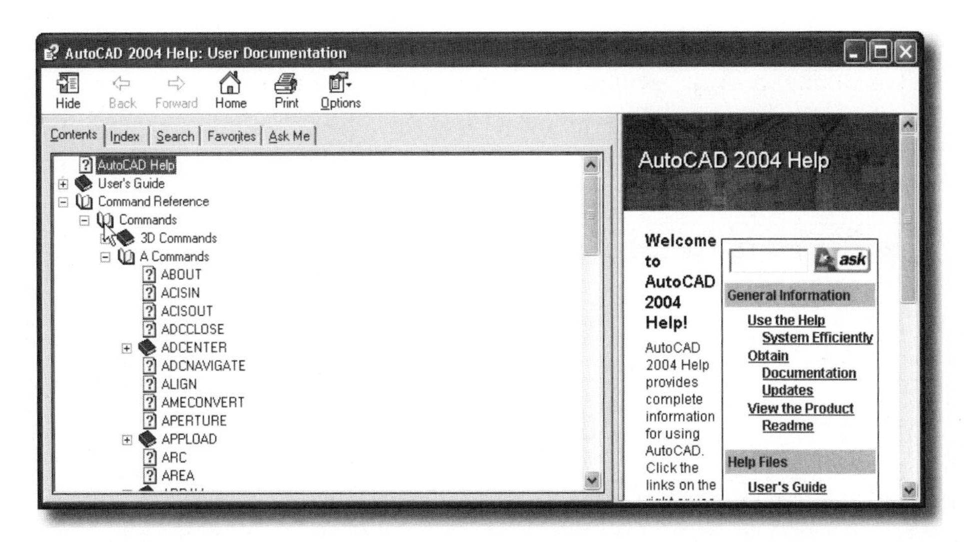

Figure 1–36 *Contents tab of the AutoCAD Help:User Documentation dialog box*

The **Index** tab provides help on a particular item. First begin typing the desired topic in the top text box, as shown in Figure 1–37. These items might be a command name; a concept, such as *color*; or maybe just a symbol, such as an asterisk or a backslash. The initial character(s) typed in will highlight the topic in the lower text box whose initial characters match the string of characters. You may scroll through the list, choosing the **Display** button when the item on which you need help is highlighted, or you may double-click that item in the text box. AutoCAD then displays a window containing the help available on the highlighted item in the narrative (right) side of the dialog box.

The **Search** tab in the AutoCAD Help: User Documentation window has a text box labeled **Type in the word(s) to search for:**, as shown in Figure 1–38. When you have typed in one or more words, choose the **List Topics** button, and a list of topics relative to the words typed in the above list box will be displayed in the **Select topic**: list box. The topics listed will be the result of a search done in accordance with the selected options at the bottom of the dialog box; **Search previous results, Match similar words,** and **Search titles only.** You may scroll through the list, pressing ENTER when the word on which you need help is highlighted, or you may select that item in the information (right) side of the window.

The **Favorites** tab (see Figure 1–39) lets you assemble a list of favorite topics for quick access. Highlighted topics can be removed from the list by selecting the **Remove** button or displayed in the information section of the dialog box by selecting the Display button. If the current topic (shown in the **Current topic:** text box) is not already in the favorites list, it can be added to the list by selecting the **Add** button.

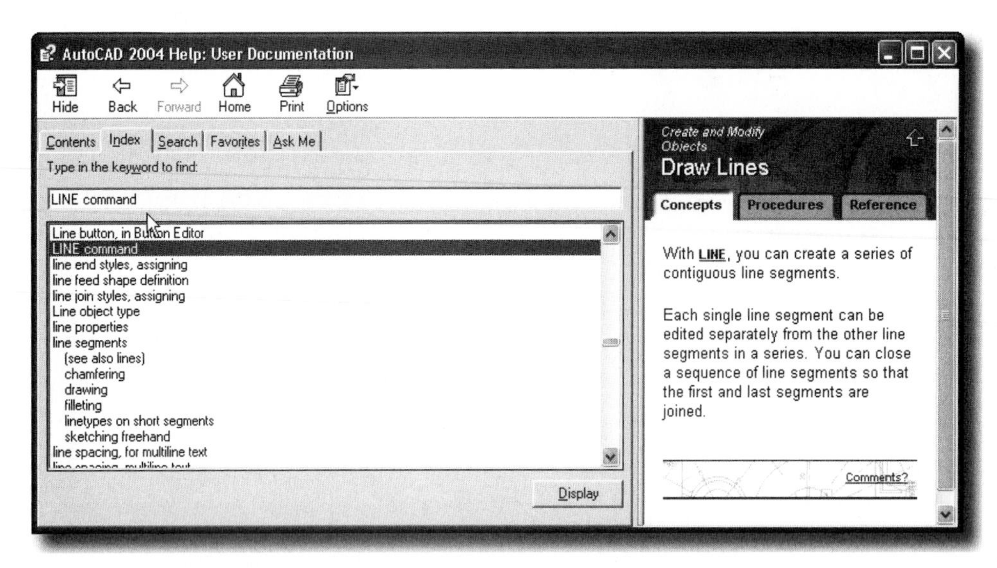

Figure 1–37 *Index tab of the AutoCAD Help:User Documentation dialog box*

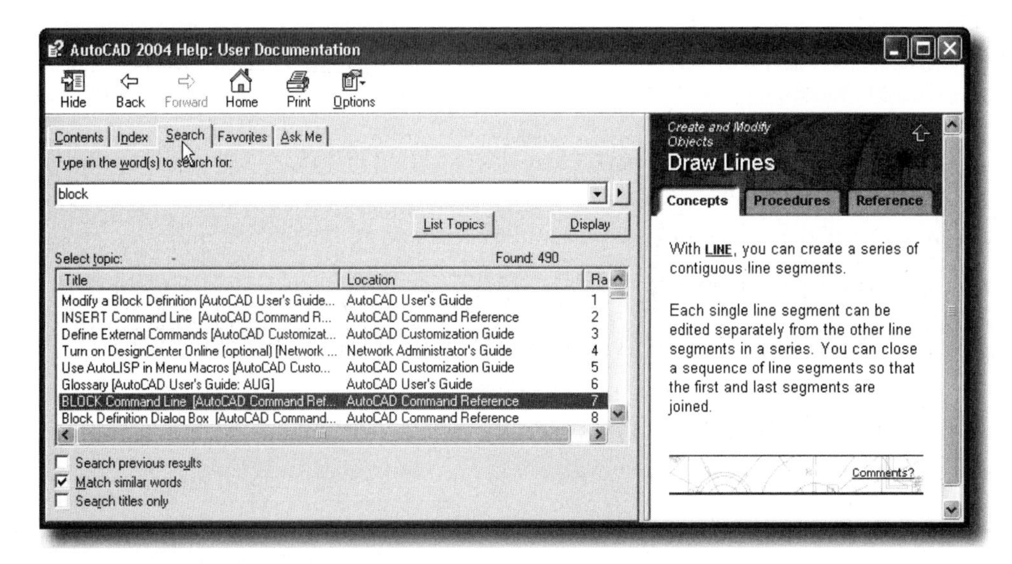

Figure 1–38 *Search tab of the AutoCAD Help:User Documentation dialog box*

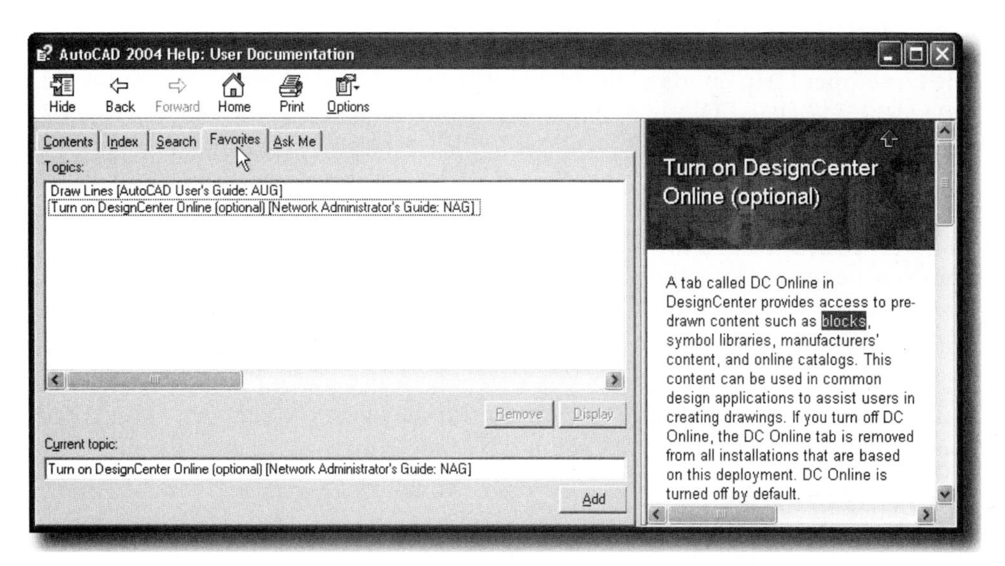

Figure 1–39 *Favorites tab of the AutoCAD Help: User Documentation dialog box*

The **Ask Me** tab (see Figure 1–40) lets you enter a question (or even a single word) in the **Type in a question and press Enter** text box and AutoCAD will list topics in accordance with the parameter selected in the **List of components to search:** text box.

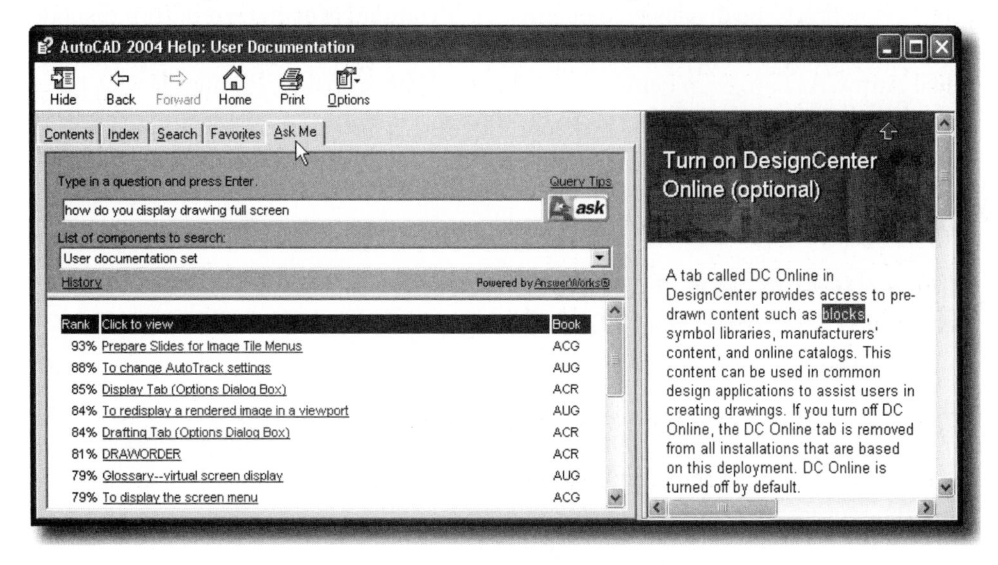

Figure 1–40 *Ask Me tab of the AutoCAD Help: User Documentation dialog box*

DEVELOPER HELP

The **Developer Help** option of the Help menu, when selected, causes the AutoCAD 2004 Help: Developer Documentation dialog box to be displayed. The topics in this dialog box have to do with subjects important to third-party developers such as customization, AutoLISP, DXF, ActiveX, and VBA applications.

NEW FEATURES WORKSHOP

The **New Features Workshop** option of the Help menu, when selected, causes the New Features Workshop dialog box to be displayed. New features introduced in AutoCAD 2004 are described and explained. Topics included on this dialog box are New Features Workshop Interface, Workspace Enhancements, Presentation Graphics, Design Publishing, i-drop, Drafting and Productivity Tools, External Reference (xref) Management, CAD Standards, and Network Improvements.

ONLINE RESOURCES

The **Online Resources** option of the Help menu, when selected, displays a submenu with options for Product Support, Training, Customization, and Autodesk User Group International. Each of these options, when selected, launches your Internet browser and links to the Autodesk address associated with the option selected.

ABOUT

The **About** option of the Help menu, when selected, causes the About AutoCAD 2004 dialog box to be displayed. Selecting the **Product Information** button displays the Product Information dialog box, which contains information about the individual AutoCAD program being run, including Product name, Product service pack, License type, Product release version, Product serial number, and License expiration date. Selecting the **License Agreement** button displays the License agreement in the default word processor. Selecting the **Save as Text File** button allows you to save the Product Information Properties to a text file.

BEGINNING A NEW DRAWING WITH THE NEW COMMAND (STARTUP SYSTEM VARIABLE SET TO 0)

Drawing1.dwg

When the first new drawing is started in an AutoCAD drawing session, it is given the temporary name *DRAWING1.DWG.* It will not be saved with a name of your choice until you use a form of the SAVE command. The second new drawing in a session is given the temporary name *DRAWING2.DWG* (and so on).

The setting of the system variable STARTUP affects what you see on the screen when an AutoCAD drawing session is begun. It also controls the type of dialog box that is displayed when the NEW command is invoked. Here we discuss using the NEW command when AutoCAD is configured with the system variable STARTUP set to 0 (default setting).

The New command

The NEW command allows you to begin a new drawing. In this case, with the system variable STARTUP set to 0, AutoCAD has started from "scratch" using the *ACAD.DWT* template file to put you right into a drawing session with the temporary name *DRAWING1.DWG*. Therefore, when you started AutoCAD, it automatically invoked the NEW command once. If you invoke the NEW command in this situation, it is really for the second time in the session. AutoCAD prompts for the selection of a template and will utilize the temporary name *DRAWING2.DWG*.

Invoke the NEW command:

Standard toolbar	Choose the QNEW command (see Figure 1–41)
File menu	Choose New
Command: prompt	**new** (ENTER)

Figure 1–41 *Invoking the NEW command from the Standard toolbar*

The NEW command, with the system variable STARTUP set to 0, displays the Select Template dialog box as shown in Figure 1–42.

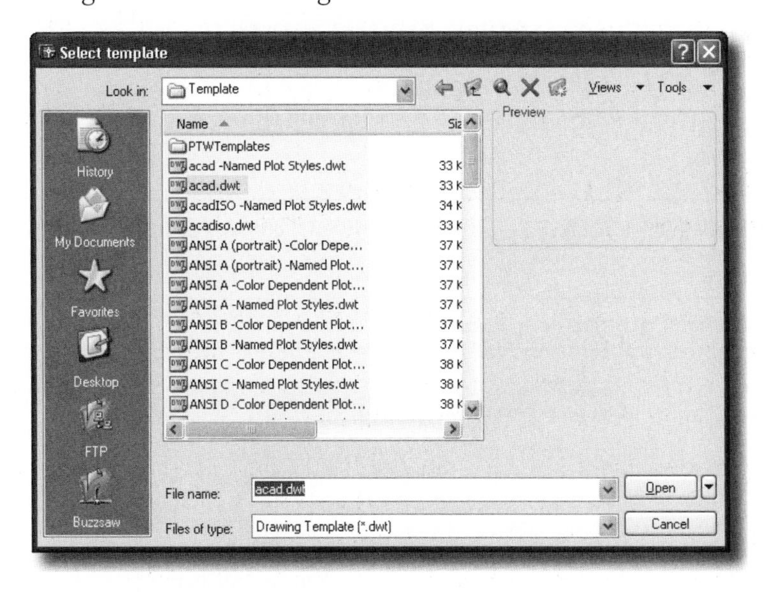

Figure 1–42 *Select Template dialog box*

The Select Template dialog box operates in a manner similar to Windows file management dialog boxes. It contains a **Preview** window that will show a thumbnail sketch of the template file selected if it is available.

Note: The NEW command causes one dialog box to be displayed when the STARTUP system variable is set to 0 and a different dialog box to be displayed when STARTUP is set to 1 (see the explanation in the following section of this chapter)

Drawing Template File

A Drawing Template file is a drawing file with selected parameters already preset to meet certain requirements, so that you do not have to go through the process of setting them up each time you wish to begin drawing with those parameters. A template drawing might have the imaginary drawing sheet dimensions preset, or the units of measurement preset, or it could contain objects already drawn on the drawing. In many cases a blank title block has already been created on a standard sheet size. Or the type of coordinate system that is needed to make the drawing could already be set up with the origin (x,y,z-coordinates of 0,0,0) located where needed relative to the edges of the envisioned drawing sheet.

The AutoCAD program files contain over 60 templates for drawings of various standard sizes containing predrawn title block conforming to standards such as ANSI, DIN, Gb, ISO, and JIS. You can create templates by making a drawing with the desired preset parameters and predrawn objects and then saving the drawing as a template file with the extension of DWT.

OPENING AN EXISTING DRAWING WITH THE OPEN COMMAND

The OPEN command allows you to open an existing drawing.

Invoke the OPEN command:

Standard toolbar	Choose OPEN (see Figure 1–43)
File menu	Choose Open
Command: prompt	**open** (ENTER)

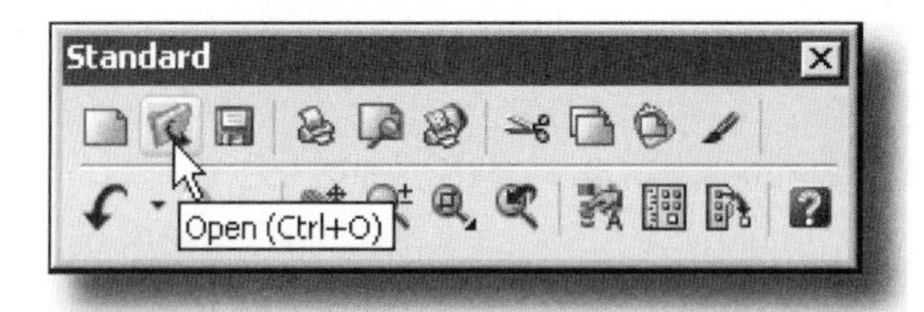

Figure 1–43 *Invoking the OPEN command from the Standard toolbar*

AutoCAD displays the Select File dialog box, as shown in Figure 1–44. The dialog box is similar to the standard file selection dialog box, except that it includes options for selecting an initial view and for setting **Open Read-Only, Partial Open,** and **Partial Open Read-Only** modes. In addition, when you click on the file name, AutoCAD displays a bitmap image in the **Preview** section. And there is a window on the left side of the dialog box displaying quick access icons to folders on your computer: Desktop and My Documents; icons for History (recently opened drawings) and Favorites; and locations on the Internet: Buzzsaw, and FTP.

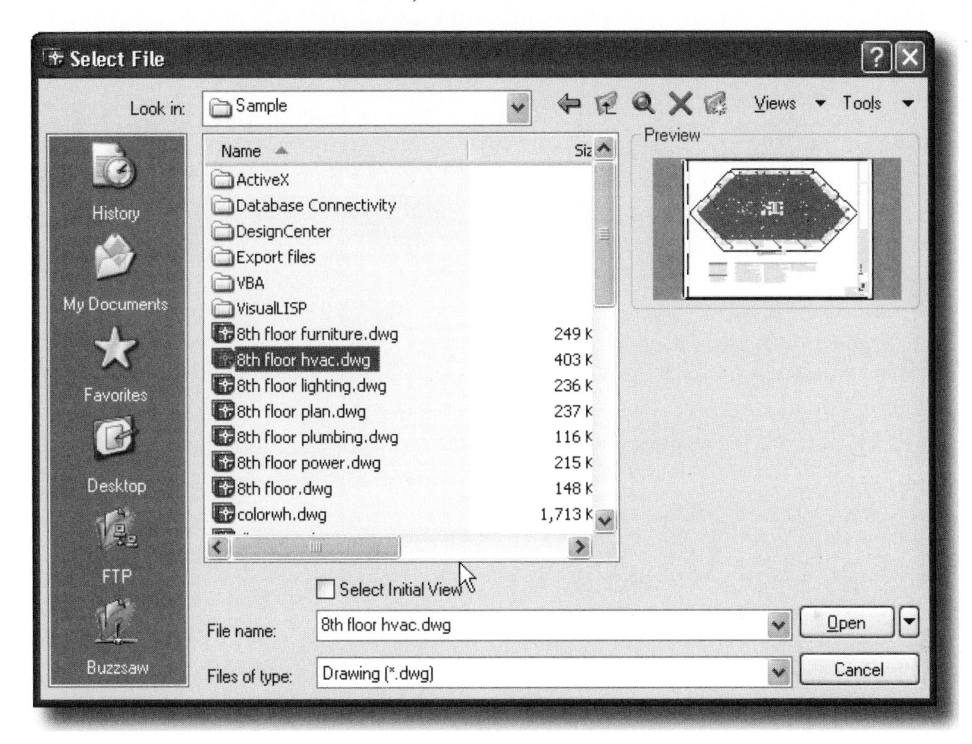

Figure 1–44 *Select File dialog box*

The **Select Initial View** check box permits you to specify a view name in the named drawing to be the startup view. If there are named views in the drawing, an M or P beside their name will tell if the view is model or paper space, respectively.

Buttons to the right of the **Look in:** list box are **Back to <the last folder>, Up one level, Search the Web, Delete, Create New Folder, Views,** and **Tools**. The **Back to** <the last folder>, **Up one level, Search the Web, Delete,** and **Create New Folder** options are similar to most file handling dialog boxes for reaching the location of the drawing you wish to open.

The **Views** button displays a pulldown menu with the options of **list** or **details** to determine how folders and files are displayed and a preview option that opens a Preview window to show a thumbnail sketch of drawing selected.

The **Tools** button displays a pulldown menu with the options of **Find, Locate, Add/Modify FTP Locations, Add current folder to Places,** and **Add to Favorites.**

Choosing the **Find** option causes the **Find** dialog box to be displayed. Various drives and folders are searched using search criteria. The Find dialog box combines the usual Windows file/path search of files by name and location in the **Name & Location** tab and by date ranges in the **Date Modified** tab.

The **Name & Location** tab, shown in Figure 1–45, permits you to search a specific path or paths that meet the search.

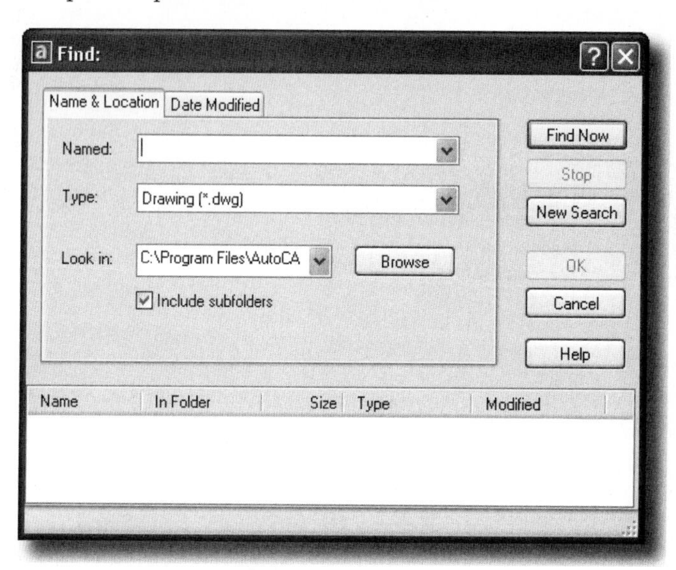

Figure 1–45 *Name and Location tab of the Find dialog box*

Named: permits you to read a list of files that have been searched. Here is where you can enter a name to search.

Type: permits you to specify the file type.

Look in: lets you specify the drive/path to search.

The **Browse** button displays the tree form of drives and folders available for searching.

The search will include subfolders if the **Include subfolders** check box is checked.

The **Date Modified** tab, shown in Figure 1–46, permits you specify dates and/or date ranges as criteria to search for drawings.

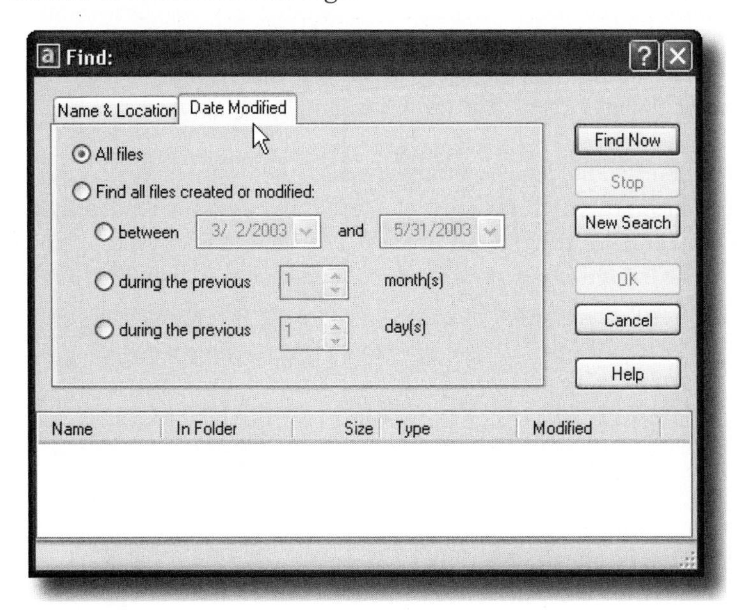

Figure 1–46 *Date Modified tab of the Find dialog box*

All Files causes the search to display the name of all files meeting the criteria of the **Name & Location** tab parameters selected. If the **Named:** text box is blank, then AutoCAD will include drawings of all names.

The **Find all files created or modified:** section has options for date ranges that are: **between** (text box for date) **and** (text box for date), **during the previous** (text box for number of months) **month(s)**, and **during the previous** (text box for number of days) **day(s).** The **between** (text box for date) **and** (text box for date) option has pulldown calendars to allow you to select the month and day with the mouse.

The **Find Now** button starts the search. The **Stop** button will stop the search. The **New Search** button clears the Named: text box for a new search. Once one or more drawing(s) that match the name and criteria has caused one or more drawings to be listed in the list box at the bottom of the dialog box, you can highlight one and it will be opened for editing when you choose the **OK** button. The **Cancel** button closes the Find dialog box. The **Help** button opens the AutoCAD Help: User Documentation dialog box.

Note: You can open and edit an AutoCAD Release 12, 13, or 14 drawing. If necessary, you can save the drawing in other formats by using the SAVEAS command. Possible formats include 2004 Drawing [*.dwg], 2000/LT 2000 Drawings [*dwg], Drawing Standards [*.dws], Template [*.dwt], 2004 DXF [*.dxf], 2000/LT 2000 DXF [*.dxf], and R12/LT 2 DXF [*.dxf]. Certain limitations apply when doing this, which are explained in the saveas section later in this chapter.

A dropdown menu is displayed when you select the down arrow to the right of the **Open** button. From this menu you may open a drawing in the read-only mode, which permits you to view the drawing but not save it with its current name. You can open a drawing with the **Partial Open** option which when selected displays the **Partial Open** dialog box as shown in Figure 1-47. You can also select the **Partial Open and Read-Only** option. However, you can edit and save it under a different name by invoking the SAVEAS command.

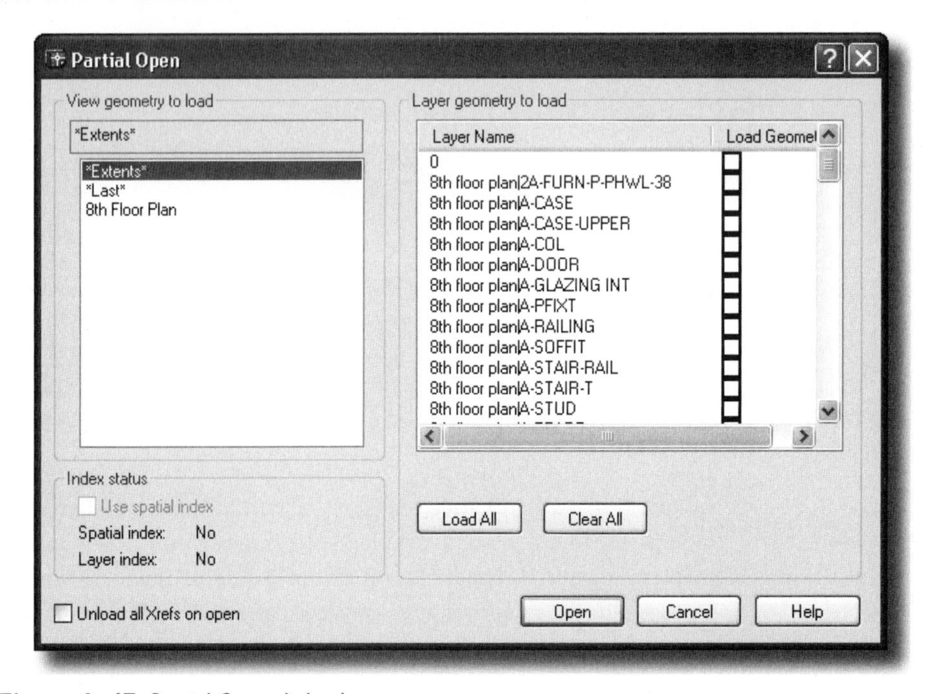

Figure 1–47 *Partial Open dialog box*

In the **View geometry to load** section of the **Partial Open** dialog box, a list of options is displayed including saved views and *Extents* and *Last*. Only model space views are available for loading. Paper space geometry can be loaded by loading the layer on which the paper space geometry is drawn. AutoCAD will open the drawing in the view selected and with the layers on that are checked in the **Layer geometry to load** section.

In the **Layer geometry to load** section available layers are listed. You can select the **Load All** button to have all of the layers loaded, or select the **Clear All** button to clear the checks from the check boxes and then select only the layers whose geometry you wish to have loaded.

Note: Even though a drawing is partially open, all objects are still loaded. All layers are available, but only the specified layers will have their geometry appear in the drawing when it is opened.

In the **Index status** section there are options to turn on or off the **Spatial index** and the **Layer index** if the selected drawing contains spatial or layer indexes.

The **Spatial index** item shows whether the selected drawing file contains a spatial index. A spatial index arranges objects based on their location in space. AutoCAD uses a spatial index when partially opening a drawing to determine what portion of the drawing is read, which makes opening the drawing take less time.

The **Layer index** item shows whether the selected drawing file contains a layer index. A layer index is a list showing the objects that are on each layer. AutoCAD uses a layer index when partially opening a drawing to determine which layers of the drawing are read, which makes opening the drawing take less time.

The **Unload all Xrefs on open** check box, when checked, causes all external references to be loaded when the drawing is opened. If a partially opened drawing that contains a bound xref, only the portion of the xref that is loaded (defined by the selected view) is bound to the partially open drawing.

The **Open** button opens the drawing file, loading only combined geometry from the selected view and layers in accordance with spatial and layer indexes.

STARTING AUTOCAD WITH A DIALOG BOX (STARTUP SYSTEM VARIABLE SET TO 1)

AutoCAD can be configured to start by displaying a Startup dialog box as shown in Figure 1-48. In order to set up AutoCAD to start with the Startup dialog box, you must set the system variable named STARTUP to the value of 1. AutoCAD as it is shipped the system variable is set to 0. Following is the command sequence to change the value of STARTUP system variable from 0 to 1.

Command: **startup** (ENTER)
Enter the new value for STARTUP <0>: **1** (ENTER)

You can also change the value by right-clicking in the drawing area, select **Options...** from the shortcut menu and then, on the **System** tab of the Options dialog box, select **Show Startup dialog box** from the **Startup:** list box in the **General Options** section.

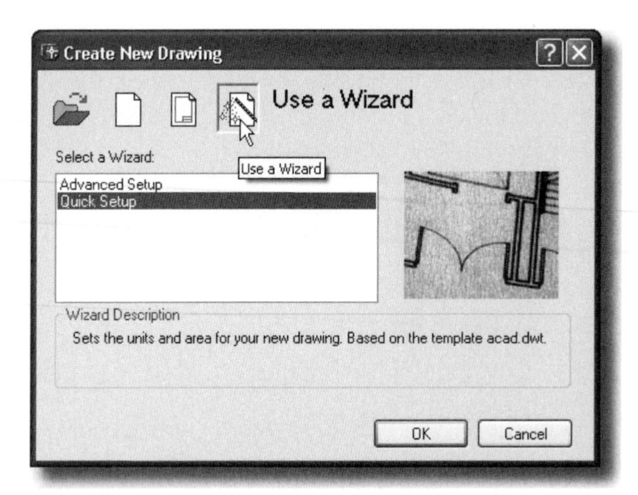

Figure 1–48 *Startup dialog box*

Whenever you begin a new drawing, whether by means of one of the two available wizards or one of the available templates or by starting from scratch, AutoCAD creates a new drawing called *DRAWING1.DWG*. You can begin working immediately and save the drawing to a file name later, using the SAVE or SAVEAS command. During any one session of AutoCAD, subsequent new drawings will be named *DRAWING2.DWG*, *DRAWING3.DWG*,... until each is either saved to a name you choose or terminated without saving.

STARTING A NEW DRAWING WITH THE WIZARDS

If from one of the Startup dialog boxes you select the **Use a Wizard** option as shown in Figure 1–48, AutoCAD leads you through the basic steps of setting up a drawing, using either the **Quick Setup** or the **Advanced Setup**. The initial drawing settings correspond to those in either the template *ACAD.DWT* (English units) or the template *ACADISO.DWT* (metric units), based on the current setting of the MEASUREINIT system variable in the registry. When MEASUREINIT is set to 0, the drawing settings are based on the template *ACAD.DWT*; when it is set to 1, the drawing settings are based on the template *ACADISO.DWT*. Depending on which wizard you select, you can then set the values of such variables as limits, units, and angle direction.

Quick Setup

The QuickSetup dialog box has two pages: the **Units** page (see Figure 1–49) and the **Area** page.

The **Units** page (Figure 1–49) applies to how the linear units of the drawing are entered. They also determine how the linear units are reported in the status bar. When you select one of the radio buttons, an example of how linear units will be written is displayed to the right of the radio buttons. AutoCAD allows you to choose from several formats for the display and entry of the coordinates and distances. For example, you can choose feet and fractional inches for architectural drafting. Other options include scientific notation and engineering formats. Selection of **Decimal** units allows you to draw in inches, feet, millimeters, or whatever units you require. This enables you to draw with real-world values and eliminates the possibility of scaling errors. Once the drawing is complete, you can plot it at whatever scale you like. As mentioned earlier, drawing to real-world size is an advantage of AutoCAD that some overlook. You can plot a drawing at several different scales, thereby eliminating the need for separate drawings at different scales.

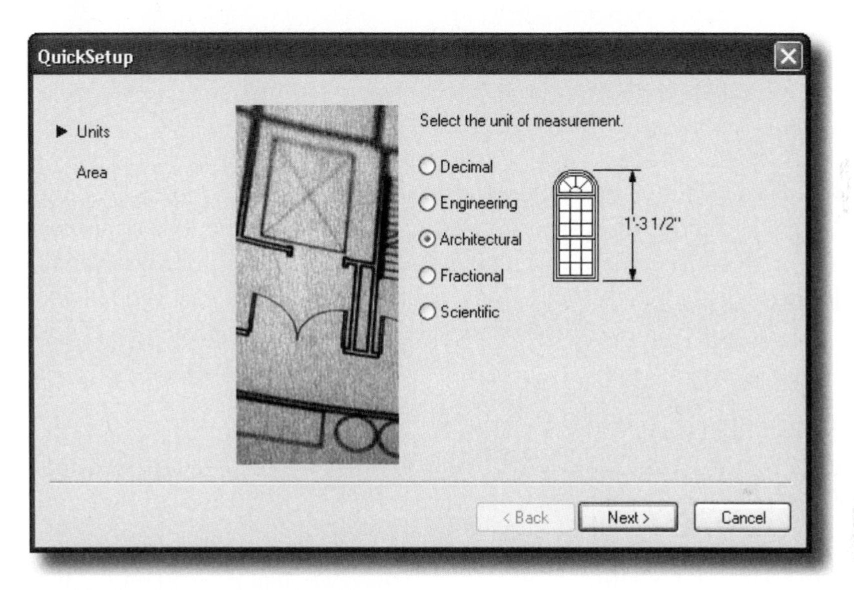

Figure 1–49 *QuickSetup dialog box: Units page*

After specifying the units, choose the **Next** > button. AutoCAD displays the **Area** page.

The **Area** page (see Figure 1–50) allows you to set your drawing area's width (left-to-right dimension) and length (bottom-to-top dimension), also known as the *limits*. You may set the limits to accommodate your drawing. For example, if you are drawing a printed circuit board that is 8 inches wide by 6 inches long, you can choose a decimal drawing unit and set the width to 8 and length to 6. If the drawing exceeds your original plans or the drawing limits become too restrictive, you can change the drawing limits. A detailed description of how to set up limits appears later in the chapter.

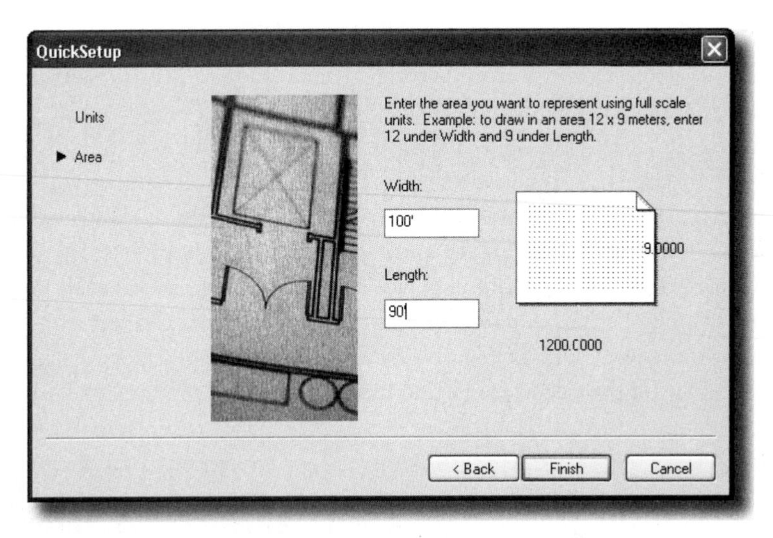

Figure 1–50 *QuickSetup dialog box: Area page*

Choose the **Finish** button to close the QuickSetup dialog box. AutoCAD automatically adjusts the scale factors for dimension settings and text height. The adjusted settings are based on the full-scale size of the objects you draw. The dimension variables that are adjusted include DIMASZ, DIMCEN, DIMDLI, DIMEXE, DIMEXO, DIMGAP, and DIMTXT. Refer to Chapter 7, "Dimensioning," for a detailed discussion of dimension variable settings. In addition, AutoCAD adjusts the linetype scale and hatch pattern scale.

Advanced Setup

The Advanced Setup dialog box has five pages: **Units**, **Angle**, **Angle Measure**, **Angle Direction**, and **Area**.

The **Units** page and **Area** page settings are the same as for the **Quick Setup** wizard. In the **Units** page, the **Precision:** text box lets you set the number of decimal places or the fractional precision of the denominator to which coordinates and linear distances are reported.

In the **Angle** page (see Figure 1–51) you can set the type of units in which angular input and reporting is given. When you select one of the radio buttons, an example of how the angular units will be written is displayed to the right of the radio buttons. You can select the format used for the display and entry of angles. Degrees in decimal form are a common choice. However, you might also select gradient, radians, degrees/minutes/seconds, or surveyor's units. The **Precision:** text box lets you set the number of decimal places or the degrees, minutes, seconds, or decimal precision of seconds to which angles are reported. After specifying the angle, choose the **Next >** button. AutoCAD displays the **Angle Measure** page.

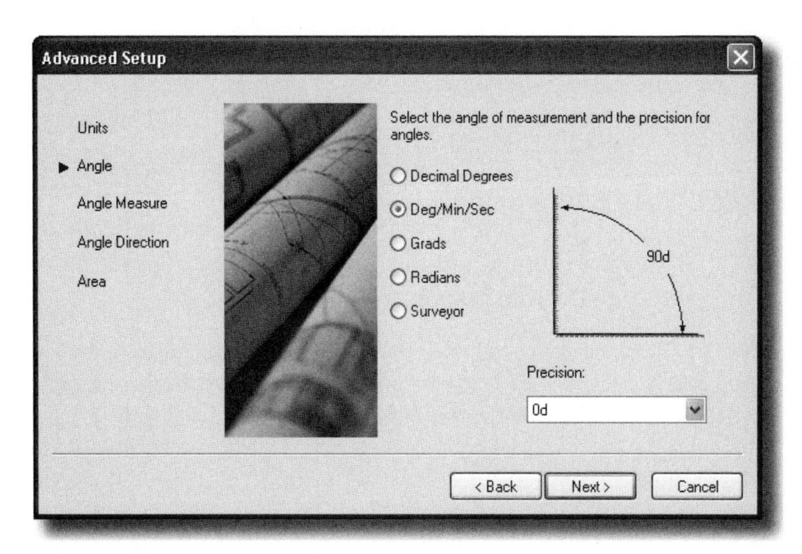

Figure 1–51 *Advanced Setup dialog box: Angle page*

The **Angle Measure** page (see Figure 1–52) lets you set zero degrees for your drawing relative to the universally accepted map compass, where North is up on the drawing and East is 90 degrees clockwise from North. You can select from the **East, North, West**, or **South** radio buttons or the **Other** button, in which you can enter the direction in the text box. After specifying the angle measure, choose the **Next >** button. AutoCAD displays the **Angle Direction** page.

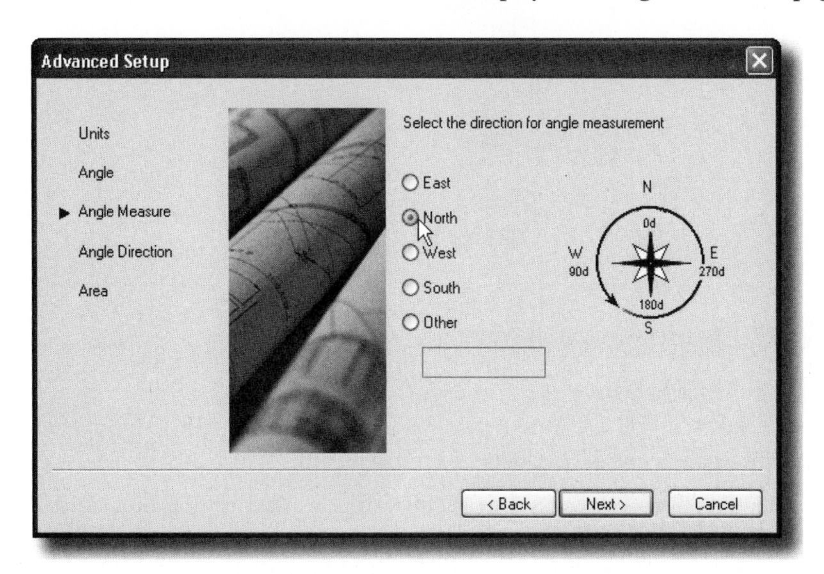

Figure 1–52 *Advanced Setup dialog box: Angle Measure page*

The **Angle Direction** page (see Figure 1–53) lets you set the direction (clockwise or counterclockwise) in which angle values increase. After specifying the angle direction, choose the **Next** > button. AutoCAD displays the **Area** page.

Choose the **Finish** button after specifying the area to close the Advanced Setup dialog box. As with the **Quick Setup** wizard, AutoCAD automatically adjusts the scale factors for dimension settings and text height. The adjusted settings are based on the full-scale size of the objects you draw.

Note: The wizards let you set up the drawing parameters that most commonly vary from one drawing to the next. It is convenient to have those variables accessible in a single place, and it lessens the chance of forgetting one. You should note, however, that the wizards' **Area** page does not permit you to use a point other than 0,0 as a lower left corner of the limits.

Any settings you changed in the wizard can be changed again later by invoking the UNITS and LIMITS commands. Detailed discussions of the UNITS and LIMITS commands are provided later in the chapter.

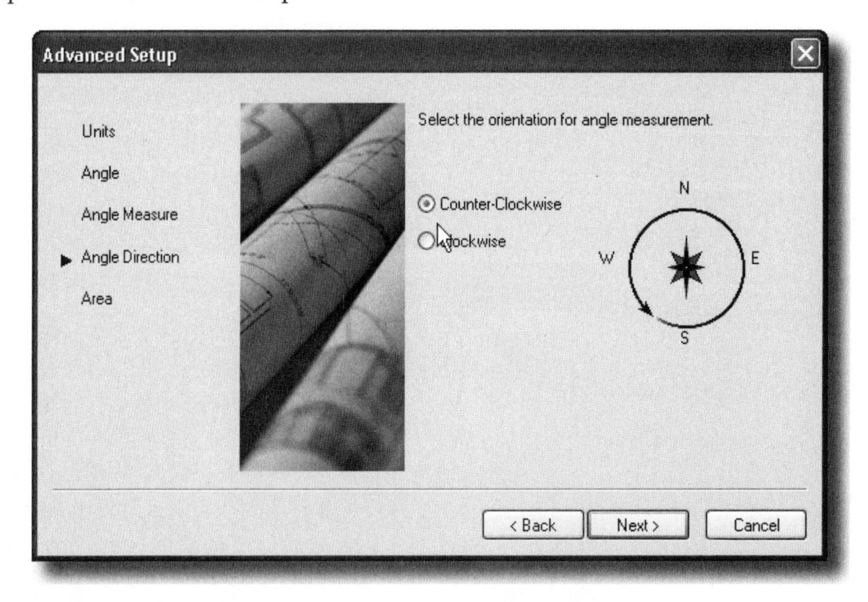

Figure 1–53 *Advanced Setup dialog box: Angle Direction page*

STARTING A NEW DRAWING WITH A TEMPLATE

If AutoCAD has been configured to start with a Startup dialog box, then from the Startup dialog box you can select the **Use a Template** option, as shown in Figure 1–54. In the **Select a Template:** section AutoCAD lists the available templates. An

AutoCAD template is a drawing file with a file extension of .DWT instead of .DWG. AutoCAD does not actually let you make changes to the template, but lets you begin with a drawing that is exactly like the template, only it is named *DRAWINGn.DWG*. This is useful to be able to have the drawing parameters such as the paper size, units of measure, and layers already set up and also have a border and title block already drawn, ready to fill in the blanks.

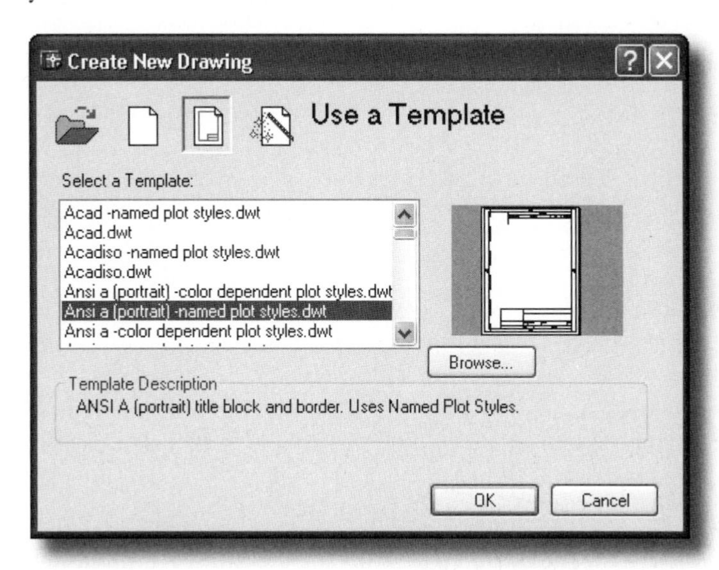

Figure 1–54 *Startup dialog box with the Use a Template option selection*

The preview graphics box on the right of the **Select a Template:** list box lets you see a small view of the template that is highlighted in the **Select a Template:** list box. If a description has been saved with the selected template, it will be displayed in the **Template Description** section. To begin your drawing in the identical setup as the selected template, either choose **OK** or double-click the highlighted name of the template in the **Select a Template:** list box. You can also choose the **Browse...** button to invoke the Select Template File dialog box to select a template file from a different folder, drive or Web site as shown in Figure 1–55. If AutoCAD has been configured to open without a Startup dialog box, and you select the **New** option or invoke the NEW command, the Select a Template File dialog box will be displayed as shown in Figure 1–55.

Figure 1–55 *Select a Template File dialog box.*

When you use a template to create a new drawing, AutoCAD copies all the information from the template drawing to the new drawing. The most common use of drawing templates is to enable you to start with a border and title block already drawn, layers and styles already created, and the system variables set to values that suit the drawing you intend to make. For example, a template for a house floor plan might have the limits set for drawing the floor plan to draw at a full scale and to plot at a scale of 1/4" = 1'-0" on a 24" x 18" sheet, the units set to architectural, and the border and title block drawn to full scale. Dimension variables might be saved in the desired style, and separate layers might be set for drawing walls, doors, windows, cabinetry, plumbing, dimensions, text, and any other object you wish to have created on its own layer. You can even use an existing .DWG drawing file as a template to create a new drawing.

STARTING A NEW DRAWING FROM SCRATCH

If you choose the **Start from Scratch** option in the Startup dialog box, choose from the two options: **Imperial (feet and inches)** and **Metric** as shown in Figure 1–56.

Note: When you choose **Start from Scratch**, it is the same as using the Template option with the drawing named *ACAD.DWT* as the template. The *ACAD.DWT* drawing comes with drawing environmental variables (called SYSTEM VARIABLES) set "at the factory" so to speak. Any and all of these SYSTEM VARIABLES can be changed. Also, there are no objects drawn in the *ACAD.DWT* template. It is advisable to NOT replace *ACAD.DWT* with a drawing having settings of the SYSTEM VARIABLES different from the ones in the original template unless you know exactly how you want the settings of all of the SYSTEM VARIABLES to be each time you choose **Start from Scratch**. Also it is not advisable to have objects already drawn in the New Drawing started from scratch.

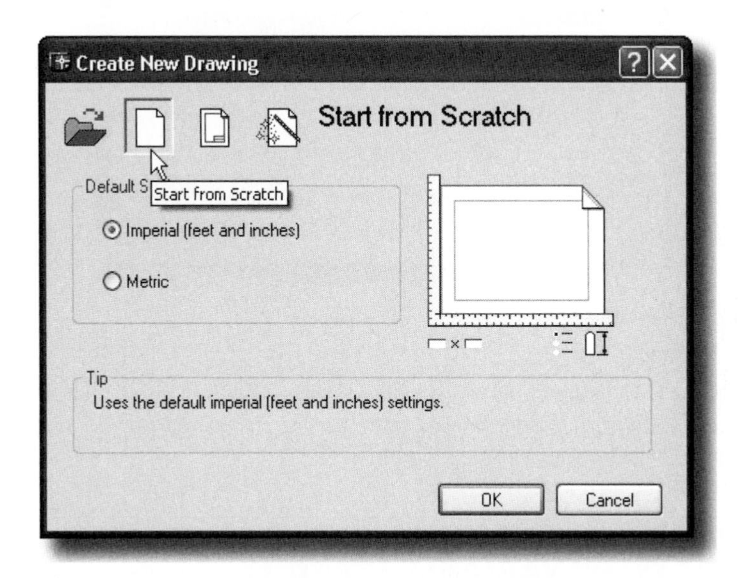

Figure 1–56 *Startup dialog box with the Start from Scratch option selection*

The **Imperial (feet and inches)** option sets the units to feet and inches settings; the **Metric** option sets the units to metric settings. By default the limits are set to 0,0 for the lower left corner and 12,9 for the upper right corner. If necessary, you can change the settings at the beginning of the drawing session or at any time during the drawing session.

OPENING AN EXISTING DRAWING

If, from the traditional Startup dialog box, you select the **Open a Drawing** option, as shown in Figure 1–57, AutoCAD lists the drawing files available from the current folder. If you need to open a drawing from a different folder, then choose the **Browse...** button. AutoCAD displays a Select File dialog box. Select the appropriate folder and drawing to open. Choose the **OK** button to open the selected drawing, and AutoCAD closes the Startup dialog box and displays the AutoCAD screen.

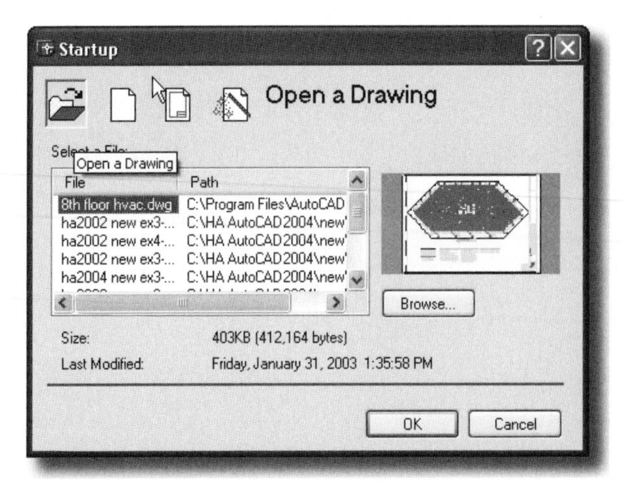

Figure 1–57 *Startup dialog box: Open a Drawing option*

Note: The **Open a Drawing** option is available only in the Startup dialog box. If you need to open an existing drawing after starting a drawing, then you have to invoke the OPEN command to open an existing drawing.

BEGINNING A NEW DRAWING WITH THE NEW COMMAND (STARTUP SYSTEM VARIABLE SET TO 1)

The NEW command, with the system variable STARTUP set to 1, displays the Create New Drawing dialog box, as shown in Figure 1-58. Except for the name, the Create New Drawing dialog box is exactly the same as the Startup dialog box described in the previous section.

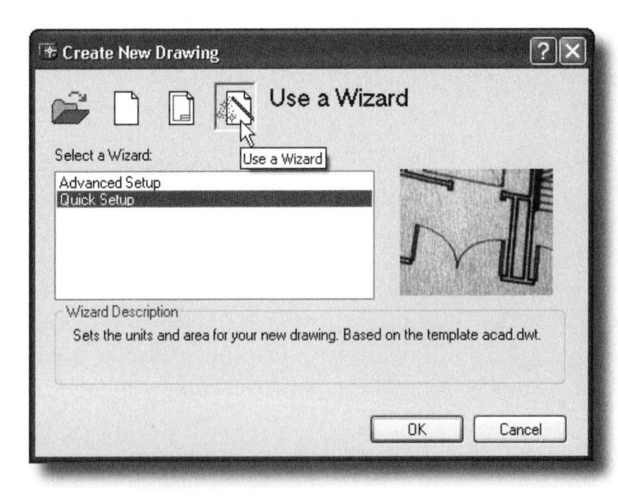

Figure 1–58 *Create New Drawing dialog box*

CHANGING UNITS

The UNITS command lets you change the linear and angular units by means of the Drawing Units dialog box. In addition, it lets you set the display format measurement and precision of your drawing units. You can change any or all of the following:

- Unit display format
- Unit display precision
- Angle display format
- Angle display precision
- Angle base
- Angle direction

Invoke the UNITS command:

Format menu	Choose Units
Command: prompt	**units** (ENTER)

AutoCAD displays the Drawing Units dialog box, as shown in Figure 1–59.

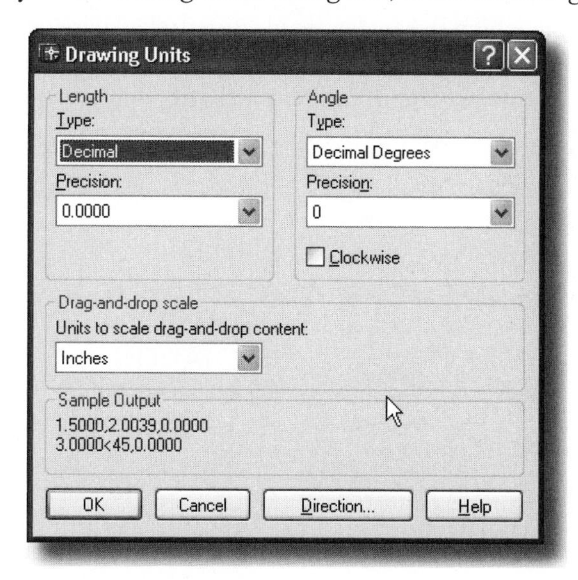

Figure 1–59 *The Drawing Units dialog box*

Length

The **Length** section of the Drawing Units dialog box allows you to change the units of linear measurement. From the **Type:** list box select one of the five types of report formats you prefer. For the selected report format, choose precision from the **Precision:** list box.

The engineering and architectural report formats produce feet-and-inches displays. These formats assume each drawing unit represents 1 inch. The other formats (scientific, decimal, and fractional) make no such assumptions, and they can represent whatever real-world units you like.

Drawing a 150-ft-long object might, however, differ depending on the units chosen. For example, if you use decimal units and decide that 1 unit = 1 foot, then the 150-ft-long object will be 150 units long. If you decide that 1 unit = 1 inch, then the 150-ft-long object will be drawn 150 x 12 = 1,800 units long. In architectural and engineering units modes, the unit automatically equals 1 inch. You may then give the length of the 150-ft-long object as 150' or 1,800" or simply 1,800.

Angles

The **Angle** section of the Drawing Units dialog box allows you to set the drawing's angle measurement. From the **Type:** list box select one of the five types of report formats you prefer. For the selected format, choose precision from the **Precision:** list.

Select the direction in which the angles are measured, clockwise or counterclockwise. If the **Clockwise** check box is set to ON, then the angles will increase in value in the clockwise direction. If it is set to OFF, the angles will increase in value in the counterclockwise direction. (see Figure 1–60).

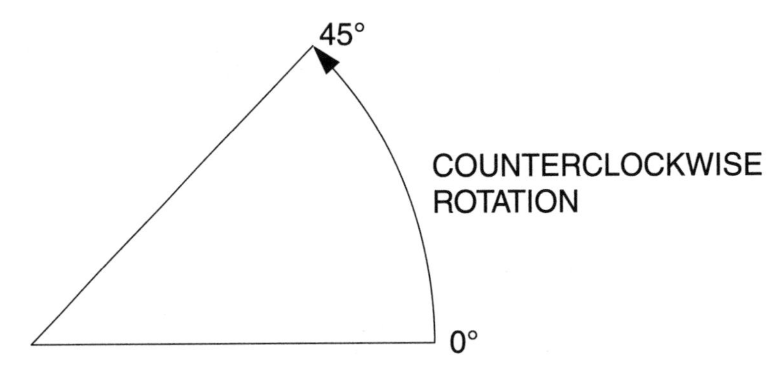

Figure 1–60 *The default, counterclockwise direction of angle measurement*

Drag-and-Drop Scale

The units that you select from the **Units to scale drag-and-drop content:** list box determines the unit of measure used for block insertions from AutoCAD DesignCenter, tool palettes or i-drop. If a block is created in units different from the units specified in the list box, they will be inserted and scaled in the specified units. If you select **Unitless**, the block will be inserted as is, and the scale will not be adjusted to match the specified units.

Note: Source content units and **Target drawing units** settings on the **User Preferences** tab of the Options dialog box are used when Insert units are not defined. See Figure 1–61.

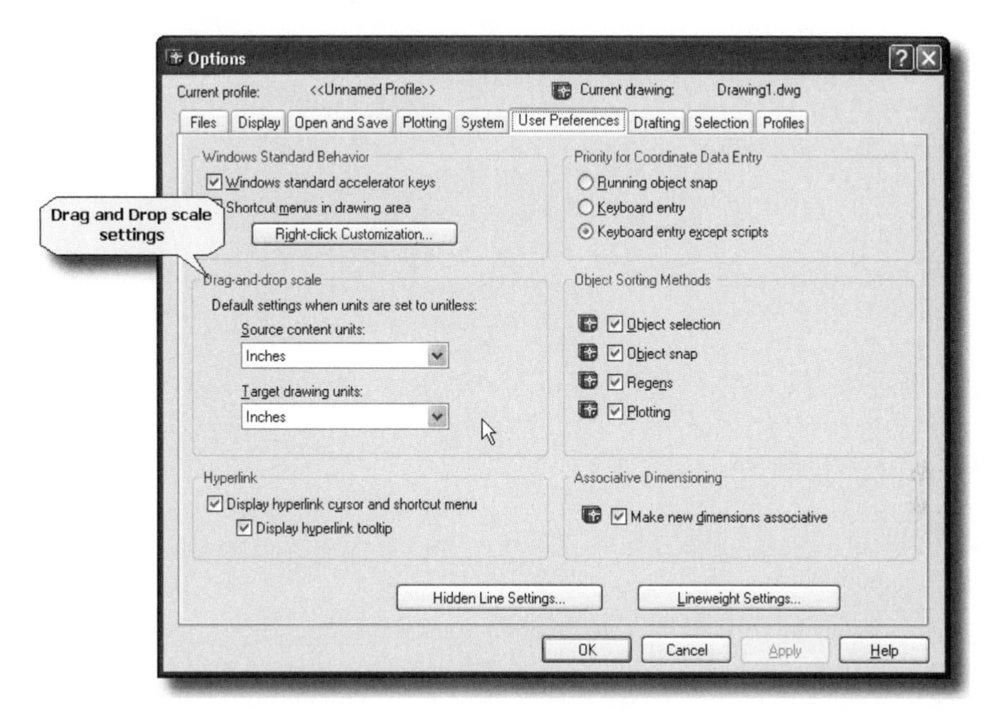

Figure 1–61 *The User Preferences tab of the Options dialog box*

Direction

To control the direction of angles, choose the **Direction...** button; the Direction Control subdialog box appears, as shown in Figure 1–62.

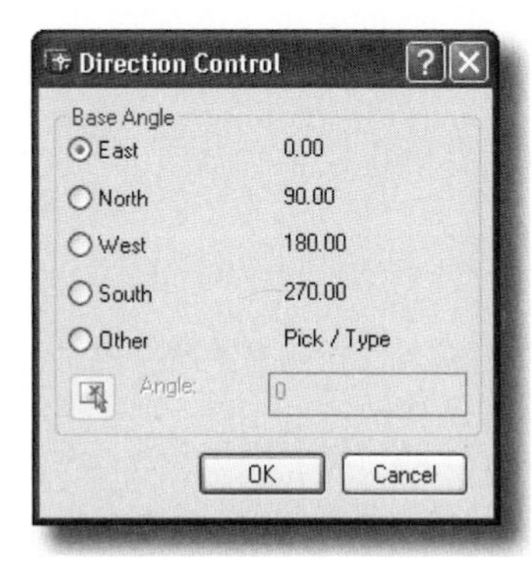

Figure 1–62 *Direction Control subdialog box*

AutoCAD, by its default setting, assumes that 0 degrees is to the right (east, or 3 o'clock) (see Figure 1–63), and that angles increase in the counterclockwise direction.

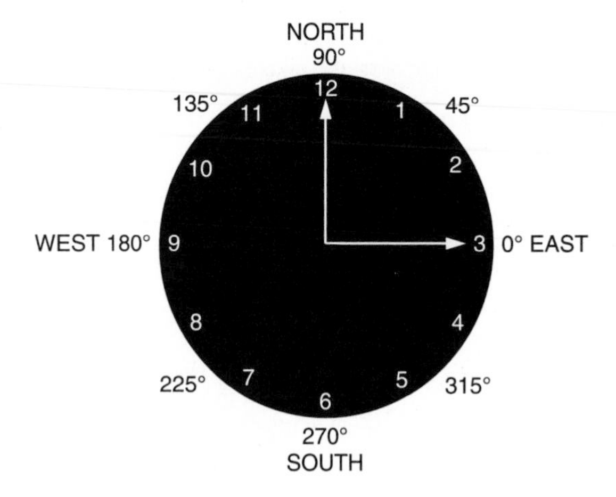

Figure 1–63 *Default angle setting direction*

You can change measuring angles to start at any compass point by selecting one of the five radio buttons.

You can also show AutoCAD the direction you want for angle 0 by specifying two points. This can be done by selecting the radio button for **Other** and choosing the **Pick an angle** button. AutoCAD prompts you for two points and sets the direction for angle 0. Choose the **OK** button to close the Direction Control subdialog box.

Once you are satisfied with all of the settings in the **Drawing Units** dialog box, choose the **OK** button to set the appropriate settings to the current working drawing and close the dialog box.

Note: When AutoCAD prompts for a distance, displacement, spacing, or coordinates, you can always reply with numbers in integer, decimal, scientific, or fractional format. If engineering or architectural report format is in effect, you can also input feet, inches, or a combination of feet and inches. However, feet-and-inches input format differs slightly from the report format because it cannot contain a blank. For example, a distance of 75.5 inches can be entered in the feet/inches/fractions format as 6'3-1/2". Note the absence of spaces and the hyphen in the unconventional location between the inches and the fraction. Normally, it will be displayed in the status area as 6'-3 1/2.

If you wish, you can use the SETVAR command to set the UNITMODE system variable to 1 (default UNITMODE setting is 0) to display feet-and-inches output in the accepted format. For example, if you set UNITMODE to 1, AutoCAD displays the fractional value of 45 1/4 as you enter it: 45-1/4. The feet input should be followed by an apostrophe (') and inches with a trailing double quote (").

When engineering or architectural report format is in effect, the drawing unit equals 1 inch, so you can omit the trailing double quote (") if you like. When you enter feet-and-inches values combined, the inch values should immediately follow the apostrophe, without an intervening space. Distance input does not permit spaces because, except when entering text, pressing SPACEBAR functions the same as pressing ENTER.

CHANGING THE LIMITS

The LIMITS command allows you to place an imaginary rectangular drawing sheet in the CAD drawing space. But, unlike the limitations of the drawing sheet of the board drafter, you can move or enlarge the CAD electronic sheet (the limits) after you have started your drawing. The LIMITS command does not affect the current display on the screen. The defined area determined by the limits governs the portion of the drawing indicated by the visible grid (for more on the GRID command, see Chapter 3). Limits are also factors that determine how much of the drawing is displayed by the ZOOM ALL command (for more on the ZOOM ALL command, see Chapter 3).

The limits are expressed as a pair of *2D* points in the World Coordinate System, a lower left and an upper right limit. For example, to set limits for an A-size sheet, set lower left as 0,0 and upper right as 11,8.5 or 12,9; for a B-size sheet set lower left as 0,0 and upper right as 17,11 or 18,12. Most architectural floor plans are drawn at a scale of 1/4" = 1'-0". To set limits to plot on a C-size (22" x 17") paper at 1/4" = 1'-0", the limits are set lower left as 0,0 and upper right as 88',68' (4 x 22, 4 x 17).

Invoke the LIMITS command:

Format menu	Choose Drawing Limits
Command: prompt	**limits** (ENTER)

AutoCAD prompts:

Command: **limits** (ENTER)
Specify lower left corner or [ON/OFF] <current>: *(press* ENTER *to accept the current setting, specify the lower left corner, or right-click for shortcut menu and select one of the available options)*
Specify upper right corner <current>: *(press* ENTER *to accept the current setting, or specify the upper right corner)*

The response you give for the upper right corner gives the location of the upper right corner of the imaginary rectangular drawing sheet.

There are two additional options available for the LIMITS command. When AutoCAD prompts for the lower left corner, you may respond to the ON or OFF options. The ON/OFF options determine whether or not you can specify a point outside the limits when prompted to do so.

When you select the ON option, limits checking is turned on and you cannot start or end an object outside the limits, nor can you specify displacement points required by the MOVE or COPY command outside the limits. You can, however, specify two points (center and point on circle) that draw a circle, part of which might be outside the limits. The limits check is simply an aid to help you avoid drawing off the imaginary rectangular drawing sheet. Leaving the limits checking ON is a sort of safety net to keep you from inadvertently specifying a point outside the limits. On the other hand, limits checking is a hindrance if you need to specify such a point.

When you select the OFF option (default), AutoCAD disables limits checking, allowing you to draw the objects and specify points outside the limits.

Whenever you change the limits, you will not see any change on the screen unless you use the All option of the ZOOM command. ZOOM ALL lets you see entire newly set limits on the screen. For example, if your current limits are 12 by 9 (lower left corner 0,0 and upper right corner 12,9) and you change the limits to 42 by 36 (lower left corner 0,0 and upper right corner 42,36), you still see the 12 by 9 area. You can draw the objects anywhere on the limits 42 by 36 area, but you will see on the screen the objects that are drawn only in the 12 by 9 area. To see the entire limits, invoke the ZOOM command using the All option.

Invoke the ZOOM ALL command:

View menu	Choose Zoom > All
Command: prompt	**zoom** (ENTER)

AutoCAD prompts:

Command: **zoom** (ENTER)
Specify corner of window, enter a scale factor (nX or nXP), or [All/Center/Dynamic/Extents/Previous/Scale/Window] <real time>: **all** (ENTER)

You see the entire limits or current extents (whichever is greater) on the screen. If objects are drawn outside the limits, ZOOM ALL displays all objects. (For a detailed explanation of the ZOOM command, see Chapter 3).

Whenever you change the limits, you should always invoke ZOOM ALL to see the entire limits or current extents on the screen.

For example, the following command sequence shows steps to change limits for an existing drawing (see Figures 1–64 and 1–65).

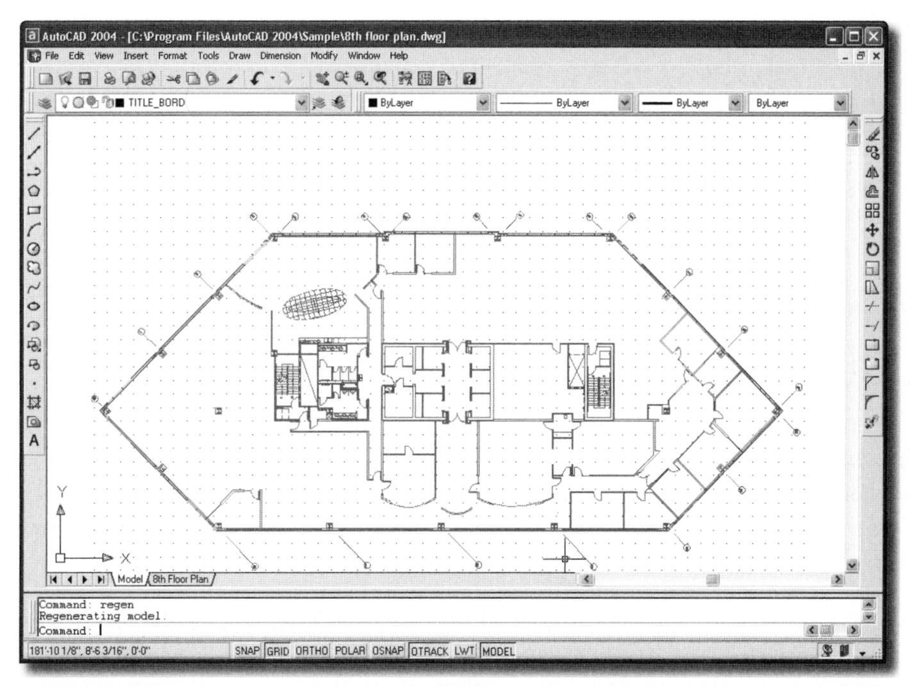

Figure 1–64 *The limits of an existing drawing before being changed by the* LIMITS *command*

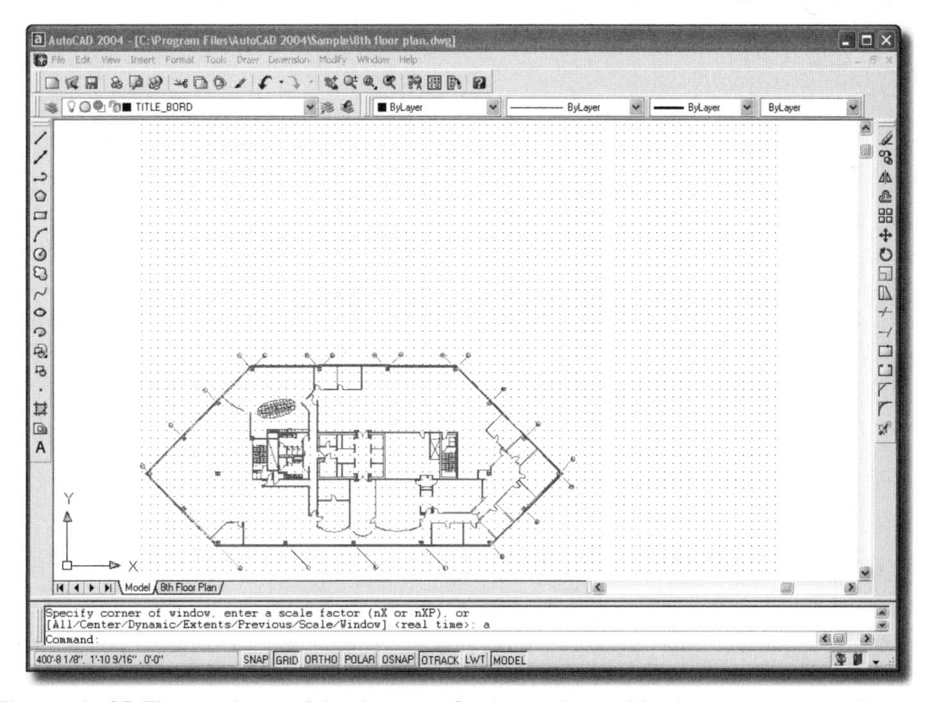

Figure 1–65 *The new limits of the drawing, after being changed by the* LIMITS *command*

Command: **limits** (ENTER)
ON/OFF/<lower left corner><0.00,0.00>: (ENTER)
Upper right corner <22.00,17.00>: **42,36** (ENTER)

Command: **zoom** (ENTER)
Specify corner of window, enter a scale factor (nX or nXP), or [All/Center/Dynamic/Extents/Previous/Scale/Window] <real time>: **all** (ENTER)

WORKING WITH MULTIPLE DRAWINGS

AutoCAD 2004 allows you to work in more than one drawing in a single AutoCAD session (Multiple Document). When multiple drawings are open, you can switch between the drawings, by selecting the appropriate name of the drawing from the Window menu. If the drawing is arranged in Tile or Cascade mode, then simply click anywhere in the drawing to make it active. You can also use CTRL + F6 or CTRL + TAB to switch between open drawings. However, you cannot switch between drawings during certain long operations such as regenerating the drawing.

You can copy and paste objects between drawings, and use the Properties palette or DesignCeNter to transfer properties from objects in one drawing to objects in another drawing. You can also use AutoCAD object snaps, the Copy with Base point command, and the Paste to Original Coordinates command to ensure accurate placement, especially when copying objects from one drawing to another.

If you want to turn off the Multiple Document mode, open the Options dialog box:

Tools menu	Choose Options
Command: prompt	**options** (ENTER)

AutoCAD displays the Options dialog box. Select the **System** tab, and under **General Options** set the **Single-drawing compatibility mode** to ON (see Figure 1–66). Choose the **OK** button to save the changes. AutoCAD will allow you to open only one drawing at a time (similar to AutoCAD Release 14).

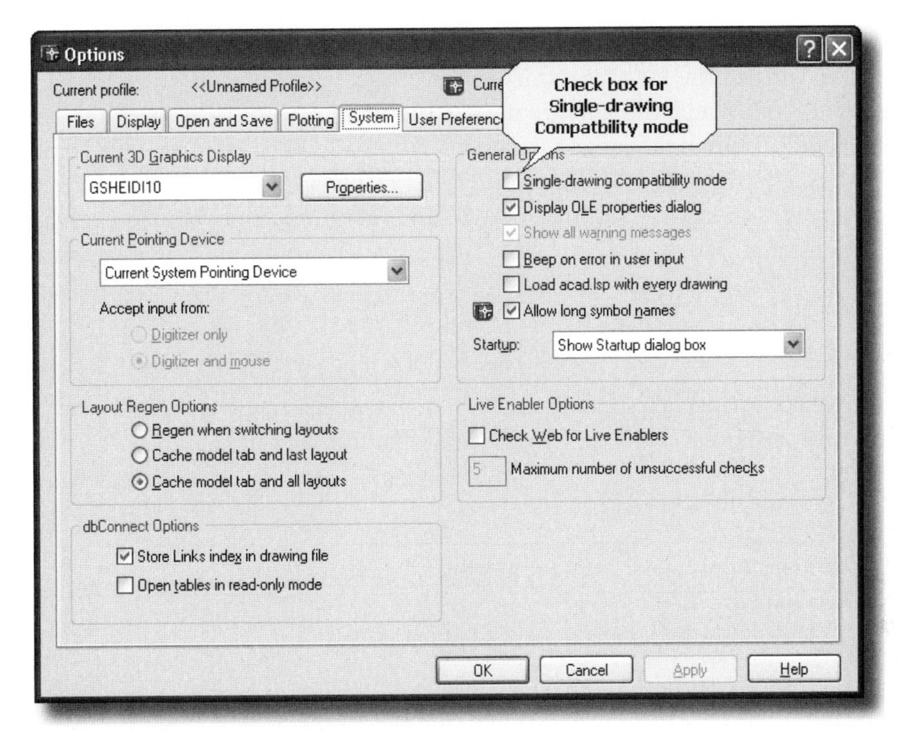

Figure 1–66 *Options dialog box – System tab*

SAVING A DRAWING

While working in AutoCAD, you should save your drawing once every 10 to 15 minutes without exiting AutoCAD. By saving your work periodically, you are protecting your work from possible power failures, editing errors, and other disasters. This can be done automatically by setting the SAVETIME system variable to a specific interval (in minutes). In addition, you can also manually save by using the SAVE, SAVEAS, and QSAVE commands.

The SAVE command saves an unnamed drawing with a file name. If the drawing is already named, then it works like the SAVEAS command, when the SAVE command is invoked from the Command: prompt.

The SAVEAS command saves an unnamed drawing with a file name or renames the current drawing. If the current drawing is already named, then AutoCAD saves the drawing to the current drawing name, or prompts for a new file name, and sets the current drawing to the new file name you specified. If the current drawing is already named and you accept the current default file name, AutoCAD saves the current drawing and continues to work on the updated drawing. If you specify a file name that already exists in the current folder, AutoCAD displays a message warning you that you are about to overwrite another drawing file. If you do not want to overwrite it, specify a different file name. The SAVEAS command also allows you to save in various formats. Possible formats include 2004 Drawing [*.dwg], 2000/LT 2000 Drawings [*dwg], Drawing Standards [*.dws], Template [*.dwt], 2004 DXF [*.dxf], 2000/LT 2000 DXF [*.dxf], R12/LT 2 DXF [*.dxf], and as a drawing template.

The QSAVE command saves an unnamed drawing with a file name. If the drawing is named, AutoCAD saves the drawing without requesting a file name.

Invoke the SAVE command:

Command: prompt	**save** (ENTER)

AutoCAD displays the Save Drawing As dialog box.

Select the appropriate folder in which to save the file, and type the name of the file in the **File name:** text field. There is a window on the left side of the dialog box displaying quick access icons to folders on your computer: Desktop and My Documents; icons for History (recently opened drawings) and Favorites; and locations on the Internet: Buzzsaw and FTP. This lets you quickly specify where to save the drawing. The file name can contain up to 255 characters including embedded spaces and punctuation. File names cannot include any of the following characters: forward slash (/), backslash (\), greater than sign (>), less than sign (<), asterisk (*), question mark (?), quotation mark ("), pipe symbol (|), colon (:), or semicolon (;). Following are the examples of valid filenames:

this is my first drawing

first house

machine part one

AutoCAD automatically appends DWG as a file extension. If you save it as a template file, then AutoCAD appends DWT as a file extension.

Invoke the SAVEAS command:

File menu	Choose Save As
Command: prompt	**saveas** (ENTER)

AutoCAD displays the Save Drawing As dialog box. Select the appropriate folder in which to save the file, and type the name of the file in the **File name:** text field. The file name can contain up to 255 characters including embedded spaces and punctuation. File names cannot include any of the following characters: forward slash (/), backslash (\), greater than sign (>), less than sign (<), asterisk (*), question mark (?), quotation mark ("), pipe symbol (|), colon (:), or semicolon (;). Select the format type you want to save to from the **Files of type:** option list.

If you want to save the current drawing to the given file name, then select the **Save** button without changing the file name. AutoCAD displays a message to warn you that you are about to overwrite another drawing file. Select **Yes** to save to the current file name and **No** to not replace the older file with the current file name.

Invoke the QSAVE command:

Standard toolbar	Select the SAVE command (see Figure 1–67)
File menu	Choose Save
Command: prompt	**qsave** (ENTER)

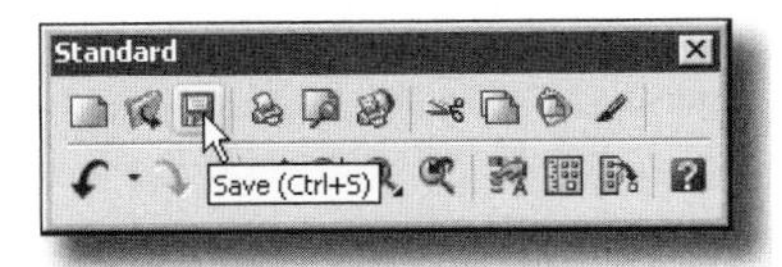

Figure 1–67 *Invoking the* QSAVE *command from the Standard toolbar*

If the drawing is unnamed, then AutoCAD displays the Save Drawing As dialog box. Select the appropriate folder in which to save the file, and type the name of the file in the **File name:** text field. The file name can contain up to 255 characters including embedded spaces and punctuation. File names cannot include any of the following characters: forward slash (/), backslash (\), greater than sign (>), less than sign (<), asterisk (*), question mark (?), quotation mark ("), pipe symbol (|), colon (:), or semicolon (;). If the drawing is named, AutoCAD saves the drawing without requesting a file name.

If you are working in multiple drawings, the CLOSE command closes the active drawing. Invoke the CLOSE command:

Window menu	Choose the CLOSE command
Command: prompt	**close** (ENTER)

AutoCAD closes the active drawing. If you have not saved the drawing, since the last change, AutoCAD displays the AutoCAD alert box – Save changes to Filename.DWG. If you select No then AutoCAD closes the drawing. If you select Yes, then AutoCAD displays the Save Drawing As dialog box. Select the appropriate folder in which to save the file, and type the name of the file in the **File name:** text field.

Note: The CLOSE command is not available when AutoCAD is in Single Document mode.

If you are working in multiple drawings, the CLOSEALL command closes all the open drawings. Invoke the CLOSEALL command:

Window menu	Choose the Close All command
Command: prompt	**closeall** (ENTER)

AutoCAD closes all the open drawings. AutoCAD displays a message for any unsaved drawing, in which you can save any changes (since the last SAVE) to the drawing before closing it.

Note: The CLOSEALL command is not available when AutoCAD is in Single Document mode.

EXITING AUTOCAD

The EXIT or QUIT command allows you to exit AutoCAD. The EXIT or QUIT command exits the current drawing if there have been no changes since the drawing was last saved. If the drawing has been modified, AutoCAD displays the Drawing Modification dialog box to prompt you to save or discard the changes before quitting.

Invoke the EXIT command:

File menu	Choose the Exit command
Command: prompt	**exit** (ENTER)

If the drawing has been modified, AutoCAD displays the Drawing Modification dialog box to prompt you to save or discard the changes before exiting.

Invoke the QUIT command:

Command: prompt	**quit** (ENTER)

If the drawing has been modified, AutoCAD displays the Drawing Modification dialog box to prompt you to save or discard the changes before quitting.

REVIEW QUESTIONS

1. If you executed the following commands in order: LINE, CIRCLE, ARC, and ERASE, what would you need to do to re-execute the CIRCLE command?

   a. press PGUP, PGUP, PGUP, ENTER

   b. press DOWNARROW, DOWNARROW, DOWNARROW, ENTER

   c. press UPARROW, UPARROW, UPARROW, ENTER

   d. press PGDN, PGDN, PGDN, ENTER

   e. press LEFTARROW, LEFTARROW, LEFTARROW, ENTER

2. In a dialog box, when there is a set of mutually exclusive options (a list of several from which you must select exactly one), these are called:

   a. text box

   b. check boxes

   c. radio buttons

   d. scroll bars

   e. list boxes

3. If you lost all pulldown menus, what command could you use to load the standard menu?

   a. LOAD
   b. OPEN
   c. NEW
   d. MENU
   e. PULL

4. What is the extension used by AutoCAD for template drawing files used by the setup wizard?

   a. .DWG
   b. .DWT
   c. .DWK
   d. .TEM
   e. .WIZ

5. If you performed a ZOOM ALL and your drawing shrunk to a small portion of the screen, one possible problem might be:

   a. out of computer memory

   b. misplaced drawing object

   c. Grid and Snap set incorrectly

   d. this should never happen in AutoCAD

   e. Limits are set much larger than current drawing objects.

6. What is an external tablet used to input absolute coordinate addresses to AutoCAD by means of a puck or stylus called?
   a. digitizer
   b. input pad
   c. coordinate tablet
   d. touch screen
7. The menu that can be made to appear at the location of the crosshairs is called:
   a. mouse menu
   b. cross-hair menu
   c. cursor menu
   d. none of the above, there is no such menu
8. In order to save basic setup parameters (such as snap, grid, etc.) for future drawings, you should:
   a. create an AutoCAD Macro
   b. create a prototype drawing template file
   c. create a new configuration file
   d. modify the ACAD.INI file
9. To cancel an AutoCAD command, press:
   a. CTRL + A
   b. CTRL + X
   c. ALT + A
   d. ESC
   e. CTRL + ENTER
10. The SAVE command:
    a. saves your work
    b. does not exit AutoCAD
    c. is a valuable feature for periodically storing information to disk
    d. all of the above

Fundamentals I

INTRODUCTION

This chapter introduces some of the basic commands and concepts in AutoCAD that can be used to complete a simple drawing. The project drawing used in this chapter is relatively uncomplicated, but for the newcomer to AutoCAD it presents ample challenge. It has fundamental problems that provide useful material for lessons in drawing setup, and in creating and editing objects. When you learn how to access and use the commands, how to find your way around the screen, and how AutoCAD makes use of coordinate geometry, you can apply these skills to the chapters containing more advanced drawings and projects.

After completing this chapter, you will be able to do the following:

- Construct geometric figures with LINE, RECTANGLE, CIRCLE, and ARC commands
- Use coordinate systems
- Use various object selection methods
- Use the ERASE command

CONSTRUCTING GEOMETRIC FIGURES

AutoCAD gives you an ample variety of drawing elements, called objects. It also provides you with many ways to generate each object in your drawing. You will learn about the properties of these objects as you progress in this text. It is important to keep in mind that the examples in this text of how to generate the various lines, circles, arcs, and other objects are not always the only methods available. You are invited, even challenged, to find other more expedient methods to perform tasks demonstrated in the lessons. You will progress at a better rate if you make an effort to learn as much as possible as soon as possible about the descriptive properties of the individual objects. When you become familiar with how the CAD program creates, manipulates, and stores the data that describes the objects, you are then able to create drawings more effectively.

DRAWING LINES

The primary drawing object is the line. A series of connected straight line segments can be drawn by invoking the LINE command and then selecting the proper sequence of endpoints. AutoCAD connects the points with a series of lines. The LINE command is one of the few AutoCAD commands that automatically repeats in this fashion. It uses the ending point of one line as the starting point of the next, continuing to prompt you for each subsequent ending point. To terminate this continuing feature you must give a null response (press ENTER or right-click and choose Enter from the shortcut menu). Even though a series of lines is drawn using a single LINE command, each line is a separate object, as though it had been drawn with a separate LINE command.

You can specify the endpoints using either two dimensional (x,y) or three dimensional (x,y,z) coordinates, or a combination of the two. If you enter 2D coordinates, AutoCAD uses the current elevation as the Z element of the point (zero is the default). This chapter is concerned only with 2D points whose elevation is zero. (3D concepts and nonzero elevations are covered in later chapters.)

Invoke the LINE command:

Draw toolbar	Choose the LINE command (see Figure 2–1)
Draw menu	Choose Line
Command: prompt	**line** (ENTER)

Figure 2–1 *Invoking the LINE command from the Draw toolbar*

AutoCAD prompts:

Command: **line** (ENTER)
Specify first point:

Where to Start

The first point of the first object in a drawing normally establishes where all of the points of other objects must be placed. It is like the cornerstone of a building. Careful thought should go into locating the first point.

You can specify the starting point of the line by entering absolute coordinates (see the section on Coordinate Systems for a detailed explanation) or by using your pointing device (mouse or puck). After specifying the first point, AutoCAD prompts:

Specify next point or [Undo]:

Where To from Here?

In addition to the first point's being the cornerstone, the direction of the first object is also critical to where all other points of other objects are located with respect to one another.

You can specify the end of the line by means of absolute coordinates, relative coordinates, or by using your pointing device to specify the end of the line on the screen. Again, AutoCAD repeats the prompt:

Specify next point or [Undo]:

You can enter a series of connected lines. To save time, the LINE command remains active and prompts "Specify next point:" after each point you specify. When you have finished entering a connected series of lines, give a null reply (press ENTER) to terminate the LINE command.

If you are placing points with a cursor instead of specifying coordinates, a rubber-band preview line is displayed between the starting point and the crosshairs. This helps you see where the resulting line will go. In Figure 2–2 the dotted lines represent previous cursor positions.

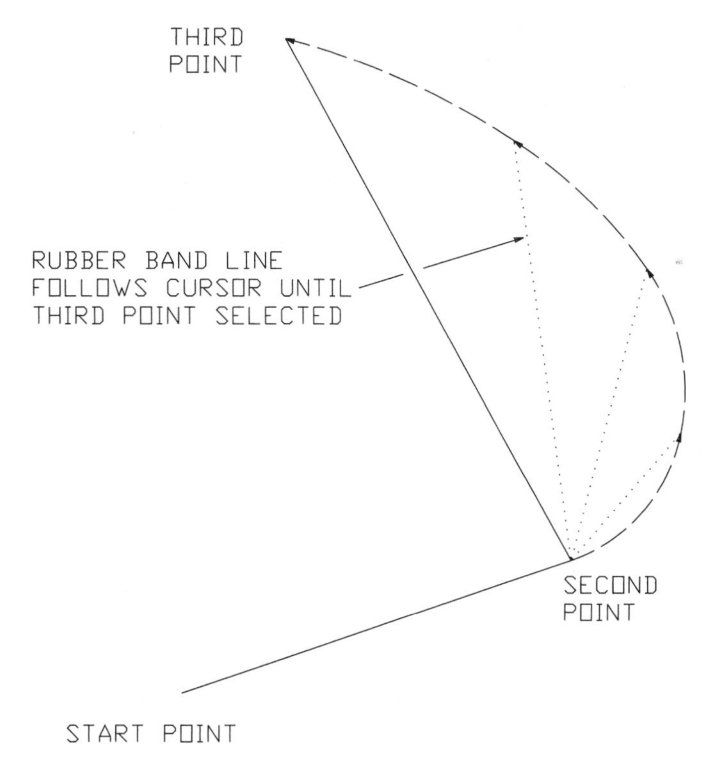

Figure 2–2 *Placing points with the cursor rather than with keyboard coordinates input*

Most of the AutoCAD commands have a variety of options. For the LINE command, three options are available: Continue, Close, and Undo. The available options can be selected from the shortcut menu that appears when you right-click on your pointing device after you invoke the LINE command.

Continue Option

When you invoke the LINE command and respond to the "Specify first point:" prompt by pressing ENTER, AutoCAD automatically sets the start of the line to the end of the most recently drawn line or arc. This provides a simple method for constructing a tangentially connected line in an arc-line continuation.

The subsequent prompt sequence depends on whether a line or arc was more recently drawn. If the line is more recent, the starting point of the new line will be set as the ending point of that most recent line, and the "Specify next point:" prompt appears as usual. If an arc is more recent, its end defines the starting point and the direction of the new line. AutoCAD prompts for:

```
Length of line:
```

Specify the length of the line to be drawn, and then AutoCAD continues with the normal "Specify next point:" prompt.

The following command sequence shows an example using the Continue option (see Figure 2–3).

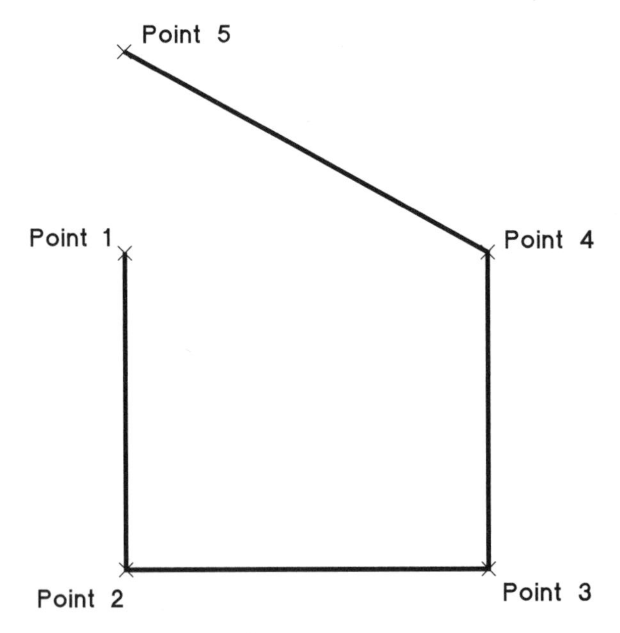

Figure 2–3 *Using the LINE command's Continue option*

Command: **line** (ENTER)
Specify first point: *(specify Point 1)*
Specify next point or [Undo]: *(specify Point 2)*
Specify next point or [Undo]: *(specify Point 3)*
Specify next point or [Close/Undo]: (ENTER)
Command: *(right-click and choose the* LINE *command again)*
Specify first point: *(to continue the line from Point 3, press* ENTER *or the* SPACEBAR*)*
Specify next point or [Undo]: *(specify Point 4)*
Specify next point or [Undo]: *(specify Point 5)*
Specify next point or [Close/Undo]: (ENTER)

Close Option

If you are drawing a sequence of lines to form a polygon, then you can use the Close option to join the last and first points automatically. AutoCAD draws the closing line segment if you respond to the "Specify next point:" prompt by right-clicking and choosing CLOSE from the shortcut menu. AutoCAD performs two steps when you choose the Close option. The first step closes the polygon, and the second step terminates the LINE command (equivalent to a null response) and returns you to the "Command:" prompt.

The following command sequence shows an example of using the Close option (see Figure 2–4).

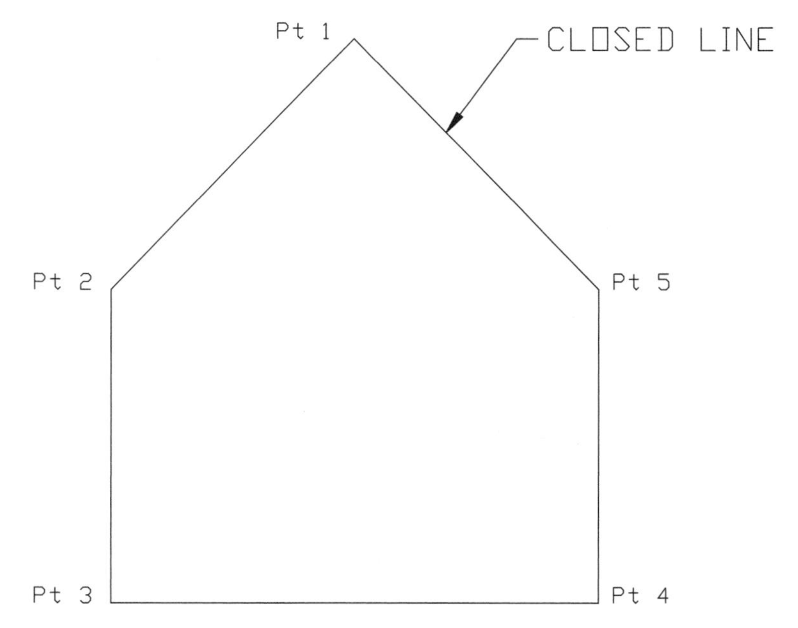

Figure 2–4 *Using the* LINE *command's Close option*

Command: **line** (ENTER)
Specify first point: *(specify Pt 1)*
Specify next point or [Undo]: *(specify Pt 2)*
Specify next point or [Undo]: *(specify Pt 3)*
Specify next point or [Close/Undo]: *(specify Pt 4)*
Specify next point or [Close/Undo]: *(specify Pt 5)*
Specify next point or [Close/Undo]: *(right-click and choose* CLOSE *option from the shortcut menu)*

Undo Option

While drawing a series of connected lines, you may wish to erase the most recent line segment and continue from the end of the previous line segment. You can do so and remain in the LINE command without exiting by using the Undo option. Whenever you wish to erase the most recent line segment, at the "Specify next point:" prompt, choose the UNDO option from the shortcut menu. If necessary, you can select multiple Undos; this will erase the most recent line segment one at a time. Once you are out of the LINE command, it is too late to use the Undo option of the LINE command to erase the most recent line segment.

The following command sequence shows an example using the Undo option (see Figure 2–5).

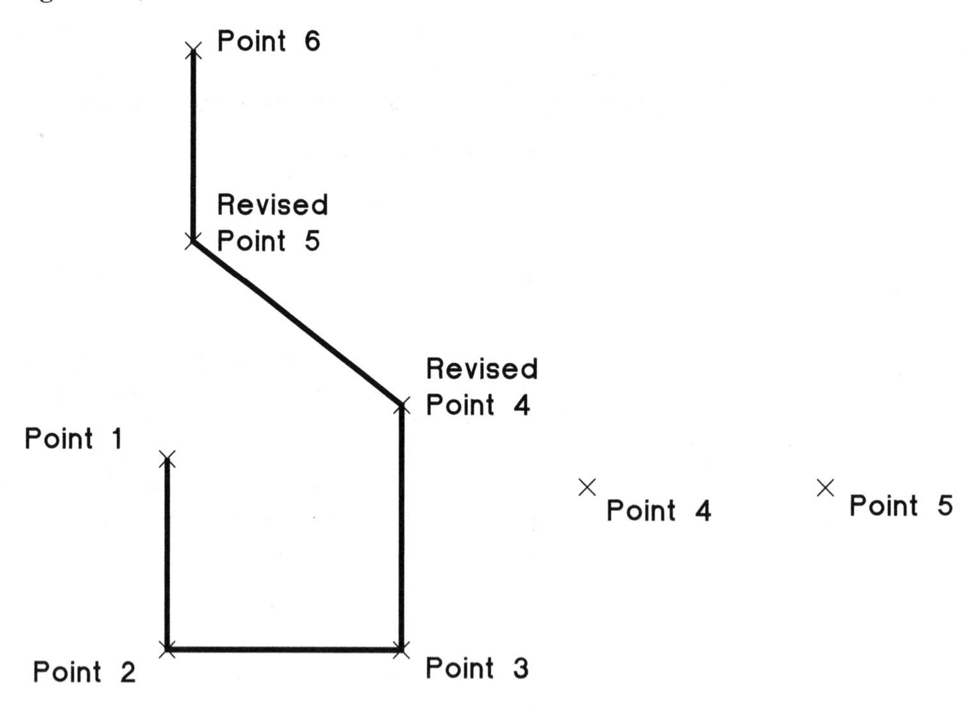

Figure 2–5 *Using the* LINE *command's Undo option*

Command: **line** (ENTER)
Specify first point: *(specify Point 1)*
Specify next point or [Undo]: *(specify Point 2)*
Specify next point or [Undo]: *(specify Point 3)*
Specify next point or [Close/Undo]: *(specify Point 4)*
Specify next point or [Close/Undo]: *(specify Point 5)*
Specify next point or [Close/Undo]: *(right-click and choose Undo option from the shortcut menu)*
Specify next point or [Close/Undo]: *(right-click and choose Undo option from the shortcut menu)*
Specify next point or [Close/Undo]: *(specify Revised Point 4)*
Specify next point or [Close/Undo]: *(specify Revised Point 5)*
Specify next point or [Close/Undo]: *(specify Point 6)*
Specify next point or [Close/Undo]: (ENTER)

DRAWING RECTANGLES

When it is necessary to create a rectangular box, you can use the RECTANGLE command.

Invoke the RECTANGLE command:

Draw toolbar	Choose the RECTANGLE command (see Figure 2–6)
Draw menu	Choose Rectangle
Command: prompt	**rectangle** (ENTER)

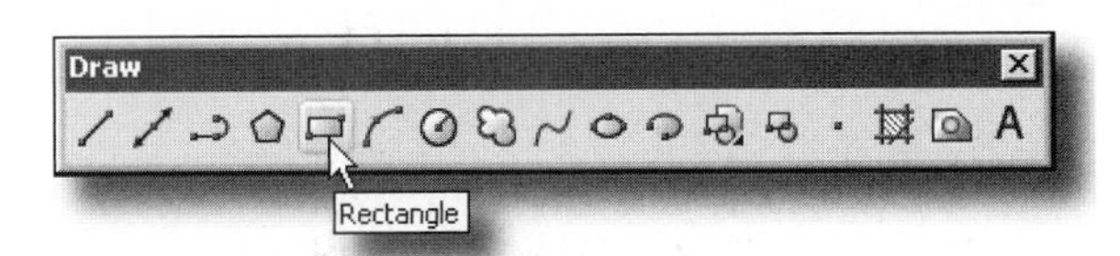

Figure 2–6 *Invoking the* RECTANGLE *command from the Draw toolbar*

AutoCAD prompts:

Command: **rectangle** (ENTER)
Specify first corner point or [Chamfer/Elevation/Fillet/Thickness/Width]: *(specify first corner point to define the start of the rectangle or right-click for shortcut menu and choose one of the available options)*
Specify other corner point or [Dimensions]: *(specify a point to define the opposite corner of the rectangle or right-click for shortcut menu and choose* DIMENSIONS*)*

Chamfer Option

The Chamfer option sets the chamfer distance for the rectangle to be drawn. Refer to Chapter 4 for a detailed explanation on the usage of the CHAMFER command and its available settings.

Elevation Option

The Elevation option specifies the elevation of the rectangle to be drawn. Refer to Chapter 15 for a detailed explanation on the usage of the Elevation setting.

Fillet Option

The Fillet option sets the fillet radius for the rectangle to be drawn. Refer to Chapter 4 for a detailed explanation on the usage of the FILLET command and its available settings.

Thickness Option

The Thickness option specifies the thickness of the rectangle to be drawn. Refer to Chapter 15 for a detailed explanation on the Thickness setting.

Width Option

The Width option selection allows you set the line width for the rectangle to be drawn. The default width is set to 0.0.

Dimensions

The Dimensions option allows you create a rectangle using length and width values. When you choose the Dimensions option, AutoCAD prompts:

Specify length for rectangles <default>: *(specify length of the rectangle)*
Specify width for rectangles <default>: *(specify width of the rectangle)*
Specify other corner point or [Dimensions]: *(specify a point by moving the cursor to one of the four possible locations for the diagonally opposite corner of rectangle)*

DRAWING WIDE LINES

When it is necessary to draw thick lines, the TRACE command may be used instead of the LINE command. Traces are entered just like lines except that the line width is set first. To specify the width, you can specify a distance or select two points and let AutoCAD use the measured distance between them. When you draw using the TRACE command, the previous trace segment is not drawn until the next endpoint is specified.

Invoke the TRACE command:

Command: prompt	**trace** (ENTER)

AutoCAD prompts:

Command: **trace** (ENTER)
Specify trace width <current>: *(specify the trace width and press ENTER)*
Specify start point: *(specify the starting point of the trace)*

Specify a series of connected trace lines. When you have finished specifying a connected series of lines, give a null reply (press ENTER) to terminate the TRACE command.

For example, the following command sequence shows the placement of connected lines using the TRACE command (see Figure 2–7).

Command: **trace** (ENTER)
Specify trace width <current>: .05 (ENTER)
Specify start point: *(specify Point 1)*
Specify next point: *(specify Point 2)*
Specify next point: (specify *Point 3)*
Specify next point: (specify *Point 4)*
Specify next point: (ENTER)

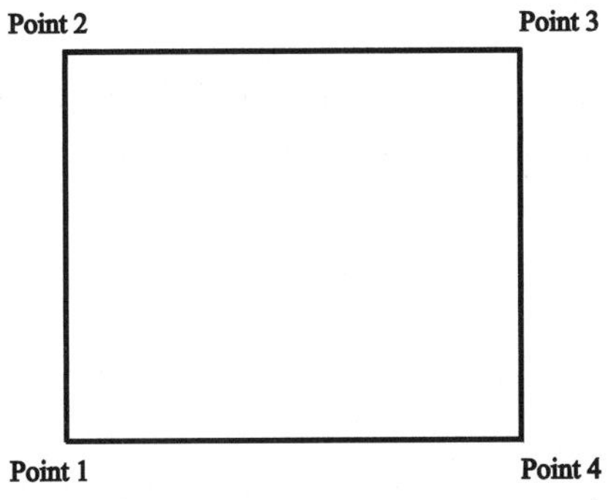

Figure 2–7 *Placing wide connected lines using the* TRACE *command*

COORDINATE SYSTEMS

In accordance with the conventions of the Cartesian coordinate system, horizontal distances increase in the positive *X* direction, toward the right, and vertical distances increase in the positive *Y* direction, upward. Distances perpendicular to the *XY* plane that you are viewing increase toward you in the positive *Z* direction. This set of axes defines the *World Coordinate System*, abbreviated as WCS.

The significance of the WCS is that it is always in your drawing; it cannot be altered. An infinite number of other coordinate systems can be established relative to it. These others are called *User Coordinate Systems* (UCSs) and can be created with the UCS command. Refer to Chapter 15 for a detailed explanation of creating and modifying User Coordinate Systems. Even though the WCS is fixed, you can view it from any angle, side, or rotation without changing to another coordinate system.

AutoCAD provides what is called a *coordinate system icon* to help you keep your bearings among different coordinate systems in a drawing. The icon will show you the orientation of your current UCS by indicating the positive directions of the *X* and *Y* axes. Figure 2–8 shows some examples of coordinate system icons.

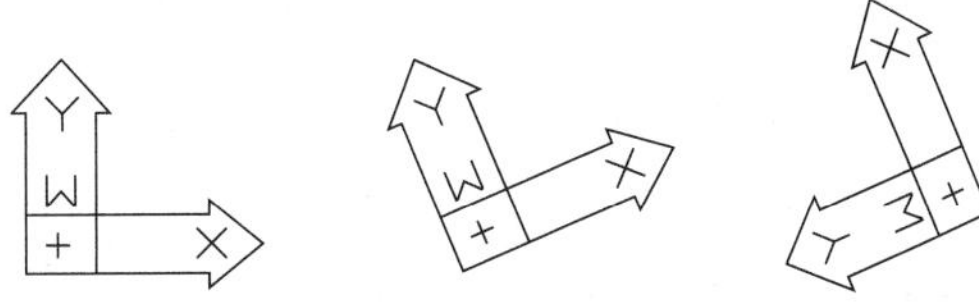

Figure 2–8 *Examples of the UCS icons*

Computer-aided drafting permits you always to draw an object at its true size and then make the border, title block, and other non-object associated features fit the object. The completed combination is reduced (or increased) to fit the plotted sheet size you require when you plot.

A more complicated situation is when you wish to draw objects at different scales on the same drawing. This can be handled easily by one of several methods with the more advanced features and commands provided in AutoCAD.

Drawing a schematic that is not to scale is one situation where the graphics and computing power are hardly used to their potential. But even though the symbols and the distances between them have no relationship to any real-life dimensions, the sheet size, text size, line widths, and other visible characteristics of the drawing must be considered in order to give your schematic the readability you desire. Some planning, including sizing, needs to be applied to all drawings.

When AutoCAD prompts for the location of a point, you can use one of several available point entry techniques, including absolute rectangular coordinates, relative rectangular coordinates, relative polar coordinates, spherical coordinates, and cylindrical coordinates.

ABSOLUTE RECTANGULAR COORDINATES

The rectangular coordinates method is based on specifying the location of a point by providing its distances from two intersecting perpendicular axes in a 2D plane or from three intersecting perpendicular planes for 3D space. Each point's distance is measured along the *X* axis (horizontal), *Y* axis (vertical), and *Z* axis (toward or away from the viewer). The intersection of the axes, called the origin (*X*,*Y*,*Z* = 0,0,0) divides the coordinates into four quadrants for 2D or eight sections for 3D (see Figure 2–9).

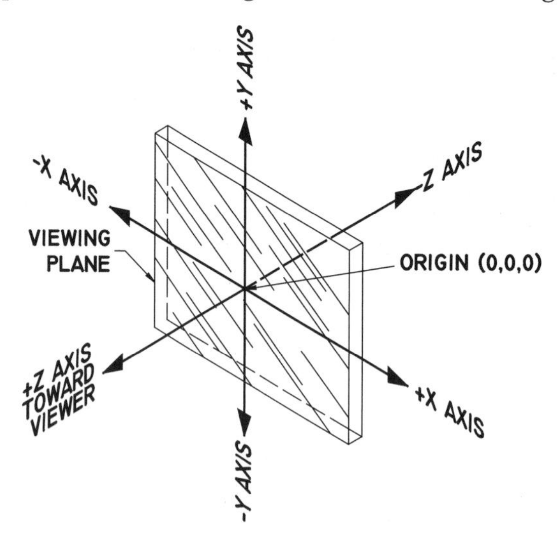

Figure 2–9 *Specifying rectangular coordinates using the intersections of the X, Y, and Z axes*

Points are located by absolute rectangular coordinates in relation to the origin. You specify the reference to the WCS origin or UCS origin. In AutoCAD, by default the origin (0,0) is located at the lower left corner of the grid display, as shown in Figure 2–10.

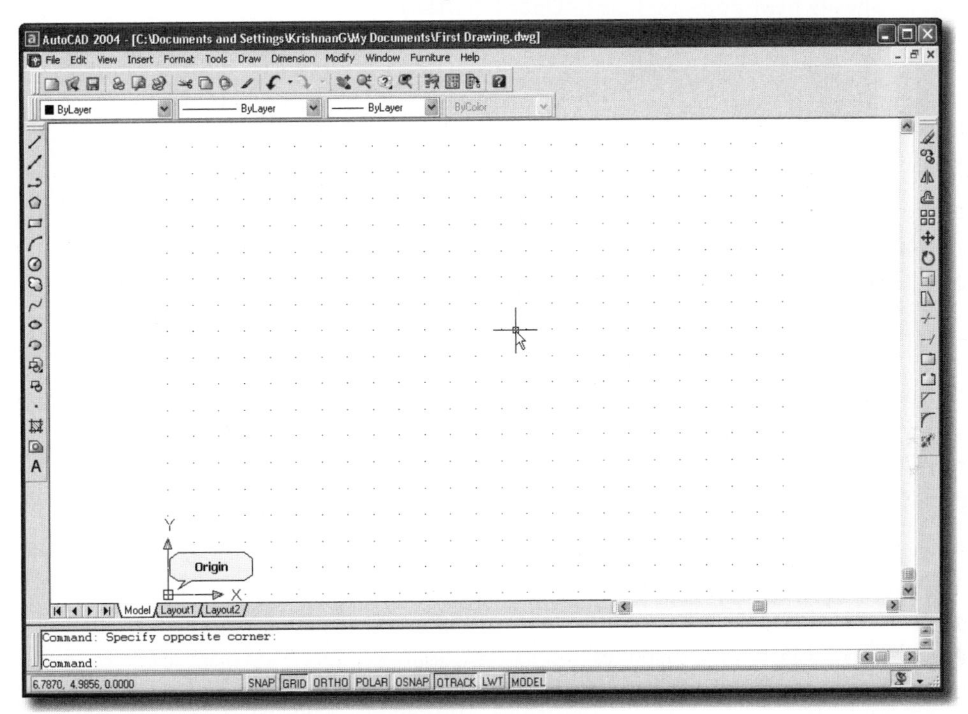

Figure 2–10 *Default location of the AutoCAD origin*

As mentioned earlier, the horizontal distance increases in the positive *X* direction from the origin, and the vertical distance increases in the positive *Y* direction from the origin. You specify a point by entering its *X,Y,Z* coordinates in decimal, fractional, or scientific notation separated by commas.

For example, the following command sequence shows placement of connected lines as shown in Figure 2–11 by absolute coordinates (see Figure 2–12):

Command: **line** (ENTER)
Specify first point: **2,2** (ENTER)
Specify next point or [Undo]: **2,4** (ENTER)
Specify next point or [Undo]: **3,5** (ENTER)
Specify next point or [Close/Undo]: **5,5** (ENTER)
Specify next point or [Close/Undo]: **5,7** (ENTER)
Specify next point or [Close/Undo]: **7,9** (ENTER)
Specify next point or [Close/Undo]: **10,9** (ENTER)

Specify next point or [Close/Undo]: **13,4** (ENTER)
Specify next point or [Close/Undo]: **13,2** (ENTER)
Specify next point or [Close/Undo]: **2,2** (ENTER)
Specify next point or [Close/Undo]: (ENTER)
Command:

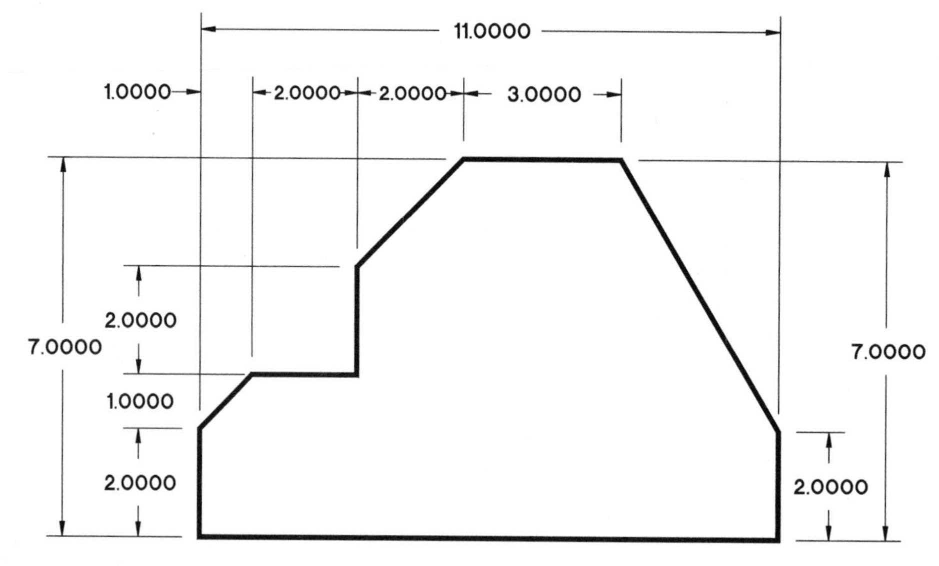

Figure 2–11 *Placing connected lines*

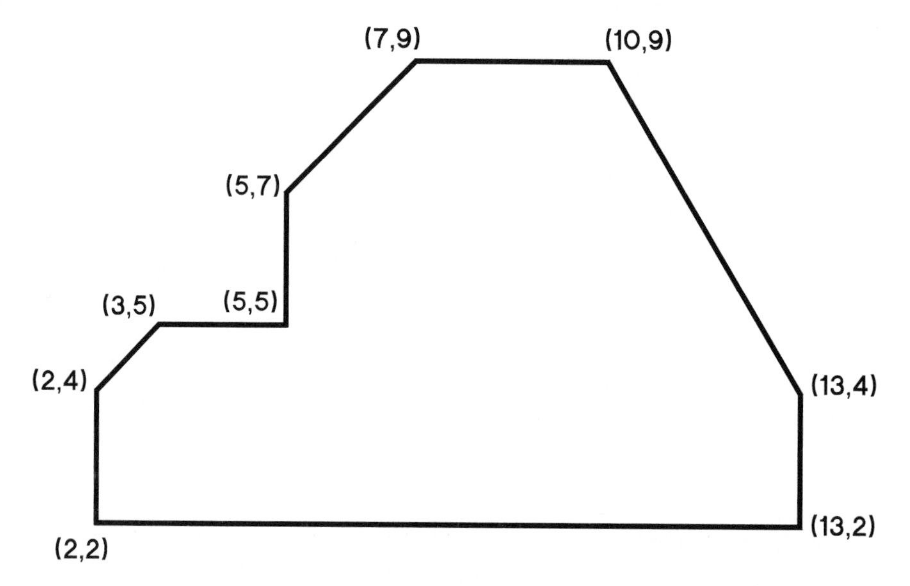

Figure 2–12 *Placing connected lines using absolute coordinates*

RELATIVE RECTANGULAR COORDINATES

Points are located by relative rectangular coordinates in relation to the last specified position or point, rather than the origin. This is like specifying a point as an offset from the last point you entered. In AutoCAD, whenever you specify relative coordinates, the @ ("at" symbol) must precede your entry. This symbol is selected by holding the SHIFT key and simultaneously pressing the key for the number 2 at the top of the keyboard.

The following command sequence shows placement of connected lines as shown in Figure 2–11 by relative rectangular coordinates (see Figure 2–13):

Command: **line** (ENTER)
Specify first point: **2,2** (ENTER)
Specify next point or [Undo]: **@0,2** (ENTER)
Specify next point or [Undo]: **@1,1** (ENTER)
Specify next point or [Close/Undo]: **@2,0** (ENTER)
Specify next point or [Close/Undo]: **@0,2** (ENTER)
Specify next point or [Close/Undo]: **@2,2** (ENTER)
Specify next point or [Close/Undo]: **@3,0** (ENTER)
Specify next point or [Close/Undo]: **@3,–5** (ENTER)
Specify next point or [Close/Undo]: **@0,–2** (ENTER)
Specify next point or [Close/Undo]: **@–11,0** (ENTER)
Specify next point or [Close/Undo]: (ENTER)
Command:

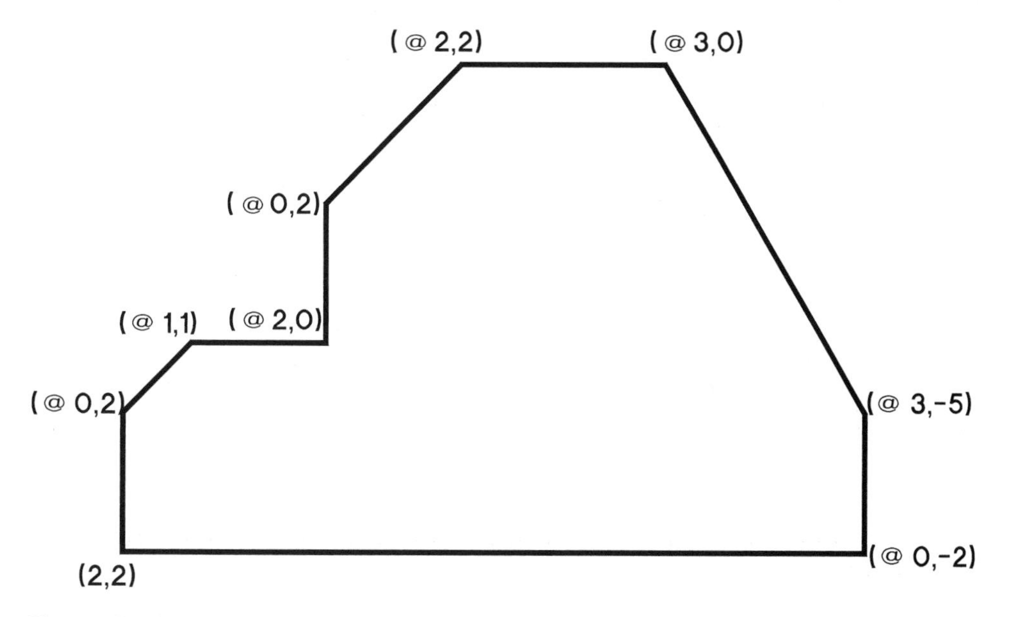

Figure 2–13 *Placing connected lines using relative rectangular coordinates*

RELATIVE POLAR COORDINATES

Polar coordinates are based on a distance from a fixed point at a given angle. In AutoCAD, a polar coordinate point is determined by the distance from a previous point and angle measured from the zero degree, rad, or gradient. In AutoCAD, by default the angle is measured in the counterclockwise direction. It is important to remember that for points located using polar coordinates they are to be positioned relative to the previous point and not the origin (0,0). You can specify a point by entering its distance from the previous point and its angle in the *XY* plane, separated by < (not a comma). This symbol is selected by holding the SHIFT key and simultaneously pressing the comma (",") key at the bottom of the keyboard. Failure to use the @ symbol will cause the point to be located relative to the origin (0,0).

The following command sequence shows the placement of the connected lines shown in Figure 2–11 by using a combination of polar and rectangular coordinates (see Figure 2–14).

Command: **line** (ENTER)
Specify first point: **2,2** (ENTER)
Specify next point or [Undo]: **@2<90** (ENTER)
Specify next point or [Undo]: **@1,1** (ENTER)
Specify next point or [Close/Undo]: **@2<0** (ENTER)
Specify next point or [Close/Undo]: **@2<90** (ENTER)
Specify next point or [Close/Undo]: **@2,2** (ENTER)
Specify next point or [Close/Undo]: **@3,0** (ENTER)
Specify next point or [Close/Undo]: **@3,–5** (ENTER)
Specify next point or [Close/Undo]: **@2<270** (ENTER)
Specify next point or [Close/Undo]: **@11<180** (ENTER)
Specify next point or [Close/Undo]: (ENTER)
Command:

If you are working in a UCS and would like to enter points with reference to the WCS, enter coordinates preceded by an asterisk (*). For example, to specify the point with an *X* coordinate of 3.5 and a *Y* coordinate of 2.57 with reference to the WCS, regardless of the current UCS, enter:

***3.5,2.57**

In the case of relative coordinates, the asterisk will be preceded by the @ symbol. For example:

@*4,5

This represents an offset of 4,5 from the previous point with reference to the WCS.

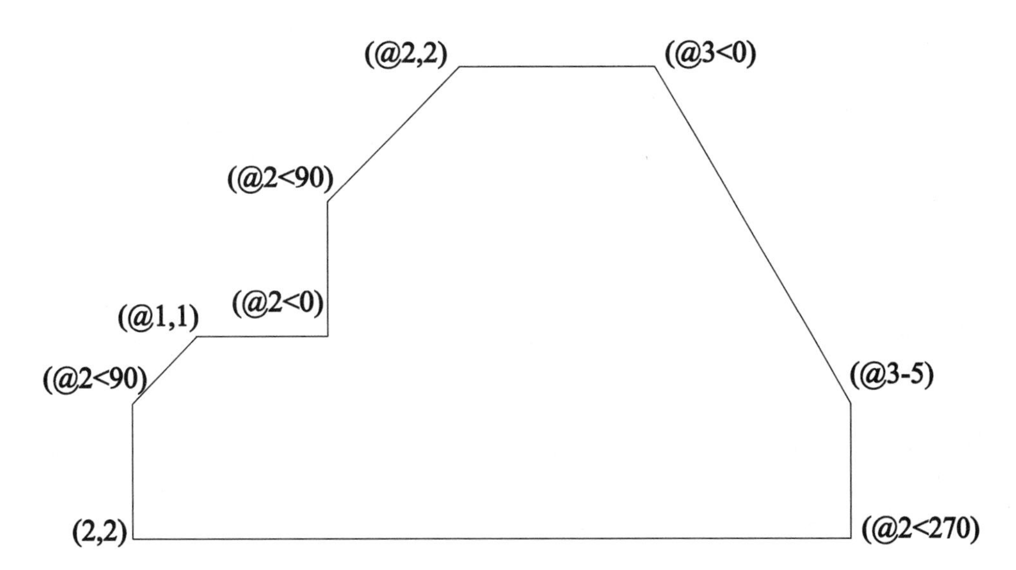

Figure 2–14 *Placing connected lines using a combination of polar and rectangular coordinates*

COORDINATE DISPLAY

The Coordinates Display is a report in the status bar of the cursor coordinates at the bottom of the screen. It has three settings. On most systems the F6 function toggles between the three settings. The three settings are as follows:

1. This setting causes the display to report the location of the cursor when the prompt is in the "Command:" status or when you are being prompted to select the first point selection of a command. It then changes to a relative polar mode when you are prompted for a second point that could be specified relative to the previous point. In this case the report is in the form of the direction/distance. The direction is given in terms of the current angular units setting and the distance in terms of the current linear units setting.

2. This setting is similar to the previous one, except that the display for the second location is given in terms of its absolute coordinates, rather than relative to the previous point.

3. This setting is used to save either the location in the display at the time you toggle to this setting or the last point entered. It does not change dynamically with the movement of the cursor.

DRAWING CIRCLES

The CIRCLE command offers five different options for drawing circles: Center-Radius (default), Center-Diameter, 2 Point, 3 Point, and Tangent, Tangent, Radius (TTR).

Center-Radius

The Center-Radius option draws a circle based on a center point and a radius. Invoke the CIRCLE command:

Draw toolbar	Choose the CIRCLE command (see Figure 2–15)
Draw menu	Choose Circle > Center, Radius
Command: prompt	**circle** (ENTER)

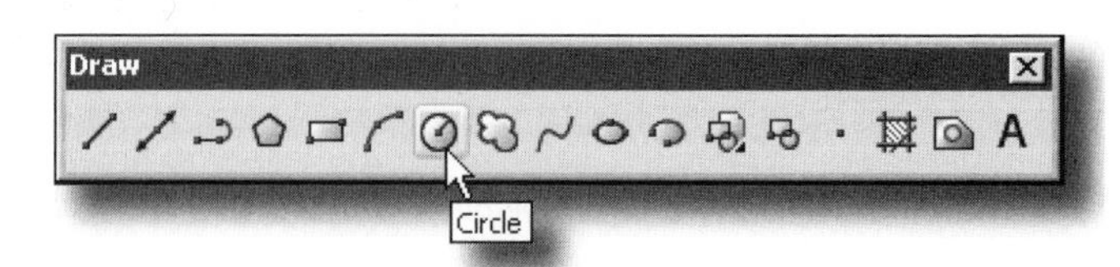

Figure 2–15 *Invoking the* CIRCLE *command from the Draw toolbar*

AutoCAD prompts:

Command: **circle** (ENTER)
Specify center point for circle or [3P/2P/Ttr (tan tan radius)]: *(specify center point to define the center of the circle)*
Specify radius of circle or [Diameter]: *(specify the radius of the circle)*

The following command sequence shows an example (see Figure 2–16):

Command: **circle** (ENTER)
Specify center point for circle or [3P/2P/Ttr (tan tan radius)]: **2,2** (ENTER)
Specify radius of circle or [Diameter]: **1** (ENTER)

The same circle can be drawn as follows (see Figure 2–17):

Command: **circle** (ENTER)
Specify center point for circle or [3P/2P/Ttr (tan tan radius)]: **2,2** (ENTER)
Specify radius of circle or [Diameter]: **3,2** (ENTER)

In the last example, AutoCAD used the distance between the center point and the second point given as the value for the radius of the circle.

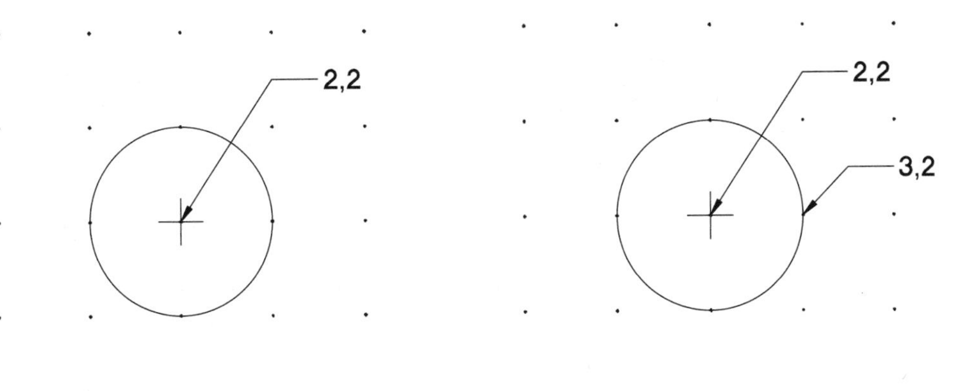

Figure 2–16 *A circle drawn with the* CIRCLE *command's default options: Center, Radius*

Figure 2–17 *A circle drawn with the Center-Radius option by specifying the coordinates*

Center-Diameter

The Center-Diameter option draws a circle based on a center point and a diameter. Invoke the Center-Diameter option:

Draw menu	Choose Circle > Center, Diameter
Command: prompt	**circle** (ENTER)

AutoCAD prompts:

Command: **circle** (ENTER)
Specify center point for circle or [3P/2P/Ttr (tan tan radius)]: *(specify center point to locate the center of the circle)*
Specify radius of circle or [Diameter]: **d** (ENTER)
Specify diameter of circle: *(specify the diameter of the circle)*

The following command sequence shows an example:

Command: **circle** (ENTER)
Specify center point for circle or [3P/2P/Ttr (tan tan radius)]: **2,2** (ENTER)
Specify radius of circle or [Diameter]: **d** (ENTER)
Specify diameter of circle: **2** (ENTER)

The same circle can be generated as follows (see Figure 2–18):

Command: **circle** (ENTER)
Specify center point for circle or [3P/2P/Ttr (tan tan radius)]: **2,2** (ENTER)
Specify radius of circle or [Diameter]: **d** (ENTER)
Specify diameter of circle: **4,2** (ENTER)

Note: Specifying a point causes AutoCAD to use the distance to the point specified from the previously selected center as the value for the diameter of the circle to be drawn.

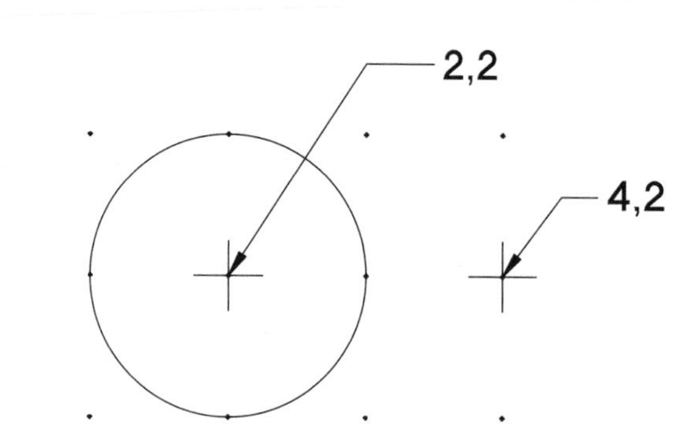

Figure 2–18 *A circle drawn using the Center-Diameter option*

Three-Point Circle

The Three-Point circle option draws a circle based on three points on the circumference.

Invoke the Three-Point circle option:

Draw menu	Choose Circle > 3 Points
Command: prompt	**circle** (ENTER)

AutoCAD prompts:

Command: **circle** (ENTER)
Specify center point for circle or [3P/2P/Ttr (tan tan radius)]: **3p** (ENTER)
Specify first point on circle: *(specify a point or a coordinate)*
Specify second point on circle: *(specify a point or a coordinate)*
Specify third point on circle: *(specify a point or a coordinate)*

The following command sequence shows an example (see Figure 2–19).

Command: **circle** (ENTER)
Specify center point for circle or [3P/2P/Ttr (tan tan radius)]: **3P** (ENTER)
Specify first point on circle: **2,1** (ENTER)
Specify second point on circle: **3,2** (ENTER)
Specify third point on circle: **2,3** (ENTER)

Note: The 3P response allows you to override the Center Point default. The 3 Point option can also be selected from the shortcut menu.

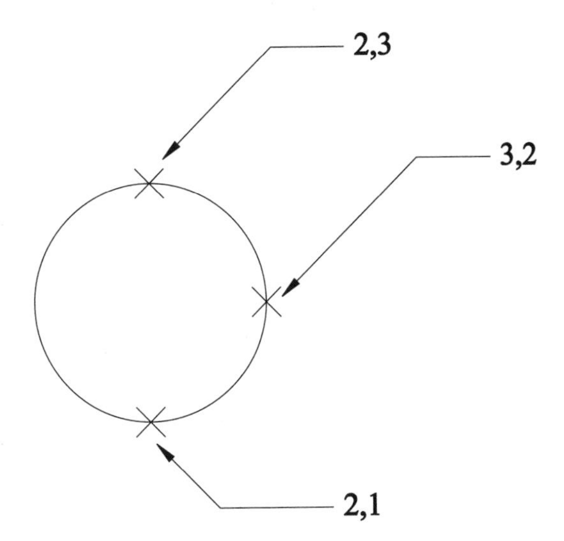

Figure 2–19 *A circle drawn with the Three-Point option*

Two-Point Circle

The Two-Point Circle option draws a circle based on two endpoints of the diameter.

Invoke the Two-Point circle option:

Draw menu	Choose Circle > 2 Points
Command: prompt	**circle** (ENTER)

AutoCAD prompts:

Command: **circle** (ENTER)
Specify center point for circle or [3P/2P/Ttr (tan tan radius)]: **2p** (ENTER)
Specify first end point of circle's diameter: *(specify a point or a coordinate)*
Specify second end point of circle's diameter: *(specify a point or a coordinate)*

The following command sequence shows an example (see Figure 2–20).

Command: **circle** (ENTER)
Specify center point for circle or [3P/2P/Ttr (tan tan radius)]: **2P** (ENTER)
Specify first end point of circle's diameter: **1,2** (ENTER)
Specify second end point of circle's diameter: **3,2** (ENTER)

Note: The 2P response overrides the Center Point default. The 2 Point option can also be selected from the shortcut menu.

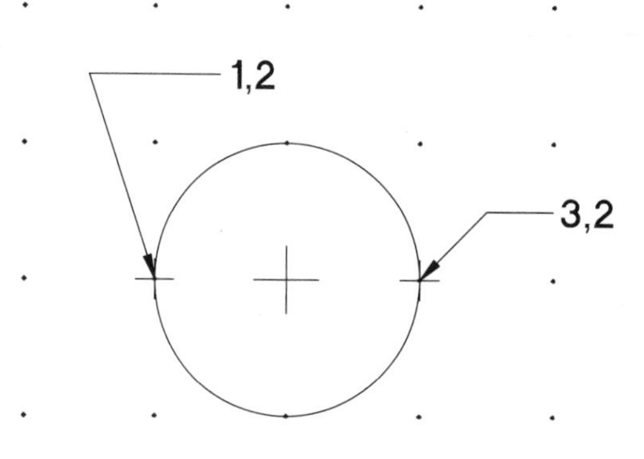

Figure 2–20 *A circle drawn with the Two-Point option*

Tangent, Tangent, Radius (TTR)

The Tangent, Tangent, Radius option draws a circle tangent to two objects (either lines, arcs, or circles) with a specified radius.

Invoke the Tangent, Tangent, Radius option:

Draw menu	Choose Circle > Tan, Tan, Radius
Command: prompt	**circle** (ENTER)

AutoCAD prompts:

Command: **circle** (ENTER)
Specify center point for circle or [3P/2P/Ttr (tan tan radius)]: **ttr** (ENTER)
Specify point on object for first tangent of circle: *(specify a point on an object for first tangent of circle)*
Specify point on object for second tangent of circle: *(specify a point on an object for second tangent of circle)*
Specify radius of circle: *(specify radius)*

For specifying the "tangent-to" objects, it normally does not matter where on the objects you make your selection. However, if more than one circle can be drawn to the specifications given, AutoCAD will draw the one whose tangent point is nearest to the selection made.

Note: Until it is changed, the radius/diameter you specify in any one of the options becomes the default setting for subsequent circles to be drawn.

DRAWING ARCS

The ARC command makes an arc and offers eleven combinations to draw an arc:

1. Three-point (3 points)
2. Start, center, end (S,C,E)
3. Start, center, included angle (S,C,A)
4. Start, center, length of the chord (S,C,L)
5. Start, end, included angle (S,E,A)
6. Start, end, direction (S,E,D)
7. Start, end, radius (S,E,R)
8. Center, start, end (C,S,E)
9. Center, start, included angle (C,S,A)
10. Center, start, length of the chord (C,S,L)
11. Continuation from line or arc (LinCont or ArcCont)

Methods 8, 9, and 10 are just rearrangements of methods 2, 3, and 4, respectively.

Three-Point Arc

The Three-Point Arc option draws an arc using three specified points on the arc's circumference. The first point specifies the start point, the second point specifies a point on the circumference of the arc, and the third point is the arc endpoint. You can specify a three-point arc either clockwise or counterclockwise.

Invoke the Three-Point arc option:

Draw toolbar	Choose the ARC command (see Figure 2–21)
Draw menu	Choose Arc > 3 Points
Command: prompt	**arc** (ENTER)

Figure 2–21 *Invoking the ARC command from the Draw toolbar*

AutoCAD prompts:

Command: **arc** (ENTER)
Specify start point of arc or [Center]: *(specify start point of arc)*
Specify second point of arc or [Center/End]: *(specify second point of arc)*
Specify end point of arc: *(specify end point of arc)*

The following command sequence shows an example (see Figure 2–22).

Command: **arc** (ENTER)
Specify start point of arc or [Center]: **1,2** (ENTER)
Specify second point of arc or [Center/End]: **2,1** (ENTER)
Specify end point of arc: **3,2** (ENTER)

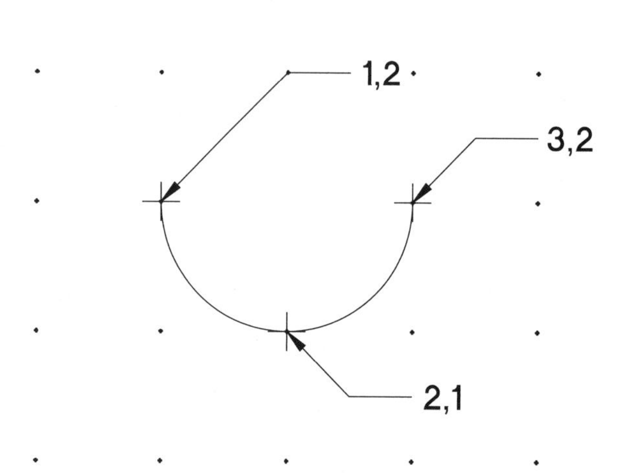

Figure 2–22 *An arc drawn with the* ARC *command's default option: 3 Points*

Start, Center, End (or S,C,E for short)

The Start, Center, End option draws an arc using three specified points. The first point specifies the start point, the second point specifies the center point of the arc to be drawn, and the third point is the arc endpoint.

Invoke the Start, Center, End option:

Draw menu	Choose Arc > Start, Center, End
Command: prompt	**arc** (ENTER)

AutoCAD prompts:

Command: **arc** (ENTER)
Specify start point of arc or [Center]: *(specify start point of arc)*
Specify second point of arc or [Center/End]: **c** (ENTER)
Specify center point of arc: (specify the center point of arc)
Specify end point of arc or [Angle/chord Length]: *(specify end point of arc)*

The following command sequence shows an example (see Figure 2–23).

Command: **arc** (ENTER)
Specify start point of arc or [Center]: **1,2** (ENTER)
Specify second point of arc or [Center/End]: **c** (ENTER)
Specify center point of arc: **2,2** (ENTER)
Specify end point of arc or [Angle/chord Length]: **2,3** (ENTER)

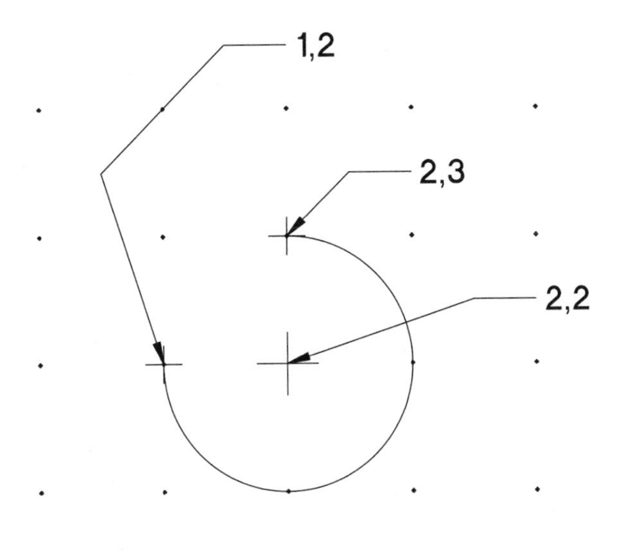

Figure 2–23 *An arc drawn with the Start, Center, End (S,C,E) option*

Arcs drawn by this method are always drawn counterclockwise from the starting point. The distance between the center point and the starting point determines the radius. Therefore, the point specified in response to "end point" needs only to be on the same radial line of the desired endpoint. For example, specifying the point 2,2.5 or 2,4 draws the same arc.

An alternative to this method is to specify the center point first, as follows (see Figure 2–24):

Command: **arc** (ENTER)
Specify start point of arc or [Center]: **c** (ENTER)
Specify center point of arc: **2,2** (ENTER)
Specify start point of arc: **1,2** (ENTER)
Specify end point of arc or [Angle/chord Length]: **2,3** (ENTER)

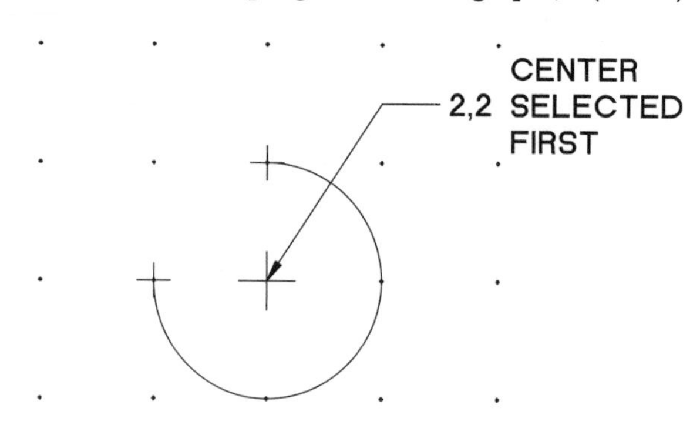

Figure 2–24 *An arc drawn by specifying the center point first*

Start, Center, Angle (or S,C,A for short)

The Start, Center, Angle option draws an arc similar to the Start, Center, End option method, but it places the endpoint on a radial line at the specified angle from the line between the center point and the start point. If you specify a positive angle as the included angle, an arc is drawn counterclockwise; for a negative angle, the arc is drawn clockwise.

Invoke the Start, Center, Angle option:

Draw menu	Choose Arc > Start, Center, Angle
Command: prompt	**arc** (ENTER)

AutoCAD prompts:

Command: **arc** (ENTER)
Specify start point of arc or [CEnter]: *(specify start point of arc)*
Specify second point of arc or [CEnter/End]: **c** (ENTER)
Specify center point of arc: *(specify the center point of arc)*
Specify end point of arc or [Angle/chord Length]: **a** (ENTER)
Specify Included Angle: *(specify the included angle of the arc to be drawn)*

The following command sequence shows an example (see Figure 2–25).

Command: **arc** (ENTER)
Specify start point of arc or [CEnter]: **1,2** (ENTER)
Specify second point of arc or [CEnter/End]: **c** (ENTER)
Specify center point of arc: **2,2** (ENTER)
Specify end point of arc or [Angle/chord Length]: **a** (ENTER)
Specify Included Angle: **270** (ENTER)

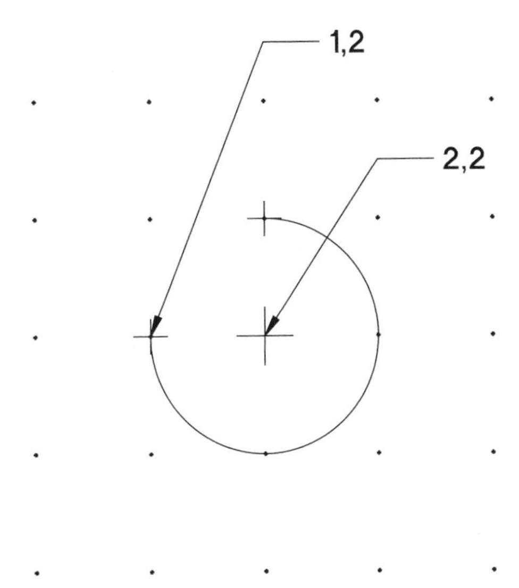

Figure 2–25 *An arc drawn with the Start, Center, Angle (S,C,A) option*

Note: If a point directly below the specified center were selected (in the previous example) in response to the "Included angle:" prompt, AutoCAD would read the angle of the line (270 degrees from zero) as the included angle for the arc. In Figure 2–25, the point selected in response to the "Included angle:" prompt causes AutoCAD to read the angle between the line it establishes from the center and the zero direction (east in the default coordinate system). It does not measure the angle between the line the point establishes from the center and the line established from the center to the start point.

Start, Center, Length of Chord (or S,C,L for short)

The Start, Center, Length of Chord option uses the specified chord length as the straight-line distance from the start point to the endpoint. With any chord length (equal to or less than the diameter length) there are four possible arcs that can be drawn: a major arc in either direction and a minor arc in either direction. Therefore, all arcs drawn by this method are counterclockwise from the start point. A positive value for the length of chord will cause AutoCAD to draw the minor arc; a negative value will result in the major arc.

Invoke the Start, Center, Length of chord option:

Draw menu	Choose Arc > Start, Center, Length
Command: prompt	**arc** (ENTER)

AutoCAD prompts:

Command: **arc** (ENTER)
Specify start point of arc or [CEnter]: *(specify start point of arc)*
Specify second point of arc or [CEnter/End]: **c** (ENTER)
Specify center point of arc: *(specify the center point of arc)*
Specify end point of arc or [Angle/chord Length]: **l** (ENTER)
Specify length of chord: *(specify the length of the chord of the arc to be drawn)*

The following command sequence shows an example of drawing a minor arc (shown in Figure 2–26).

Command: **arc** (ENTER)
Specify start point of arc or [CEnter]: **1,2** (ENTER)
Specify second point of arc or [CEnter/End]: **c** (ENTER)
Specify center point of arc: **2,2** (ENTER)
Specify end point of arc or [Angle/chord Length]: **l** (ENTER)
Specify length of chord: **1.414** (ENTER)

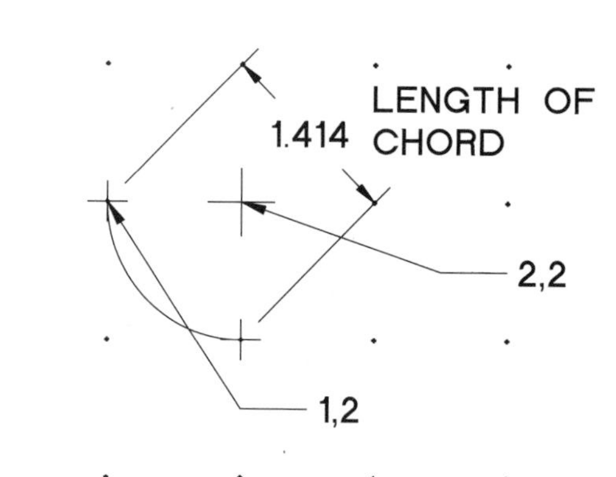

Figure 2–26 *A minor arc drawn with the Start, Center, Length of chord (S,C,L) option.*

The following command sequence shows an example of drawing a major arc (shown in Figure 2–27).

Command: **arc** (ENTER)
Specify start point of arc or [CEnter]: **1,2** (ENTER)
Specify second point of arc or [CEnter/End]: **c** (ENTER)
Specify center point of arc: **2,2** (ENTER)
Specify end point of arc or [Angle/chord Length]: **l** (ENTER)
Specify length of chord: **–1.414** (ENTER)

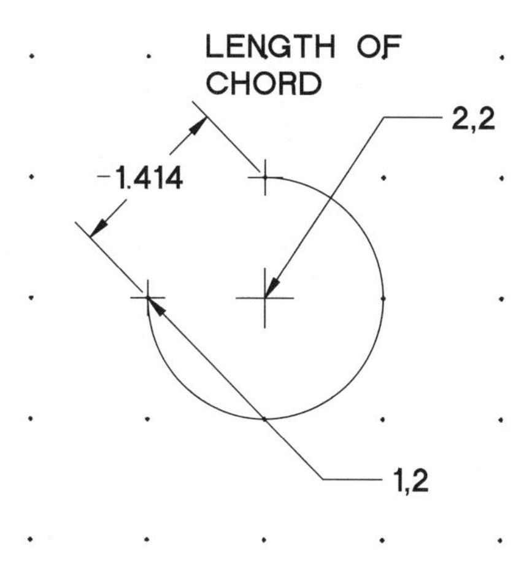

Figure 2–27 *A major arc drawn with the Start, Center, Length of chord (S,C,L) option*

Start, End, Angle (or S,E,A for short)

The Start, End, Angle option draws an arc similar to the Start, Center, Angle option method, and places the endpoint on a radial line at the specified angle from the line between the center point and the start point. If you specify a positive angle for the included angle, an arc is drawn counterclockwise; for a negative angle the arc is drawn clockwise.

Invoke the Start, End, Angle option:

Draw menu	Choose Arc > Start, Center, Angle
Command: prompt	**arc** (ENTER)

AutoCAD prompts:

Command: **arc** (ENTER)
Specify start point of arc or [CEnter]: *(specify start point of arc)*
Specify second point of arc or [CEnter/End]: **e** (ENTER)
Specify end point of arc: *(specify the end point of arc)*
Specify center point of arc or [Angle/Direction/Radius]: **a** (ENTER)
Specify included angle: *(specify the included angle of the arc to be drawn)*

The arc shown in Figure 2–28 is drawn using the following sequence:

Command: **arc** (ENTER)
Specify start point of arc or [CEnter]: **3,2** (ENTER)
Specify second point of arc or [CEnter/End]: **e** (ENTER)
Specify end point of arc: **2,3** (ENTER)
Specify center point of arc or [Angle/Direction/Radius]: **a** (ENTER)
Specify included angle: **90** (ENTER)

The arc shown in Figure 2–29 is drawn with a negative angle using the following sequence:

Command: **arc** (ENTER)
Specify start point of arc or [CEnter]: **3,2** (ENTER)
Specify second point of arc or [CEnter/End]: **e** (ENTER)
Specify end point of arc: **2,3** (ENTER)
Specify center point of arc or [Angle/Direction/Radius]: **a** (ENTER)
Specify included angle: **–270** (ENTER)

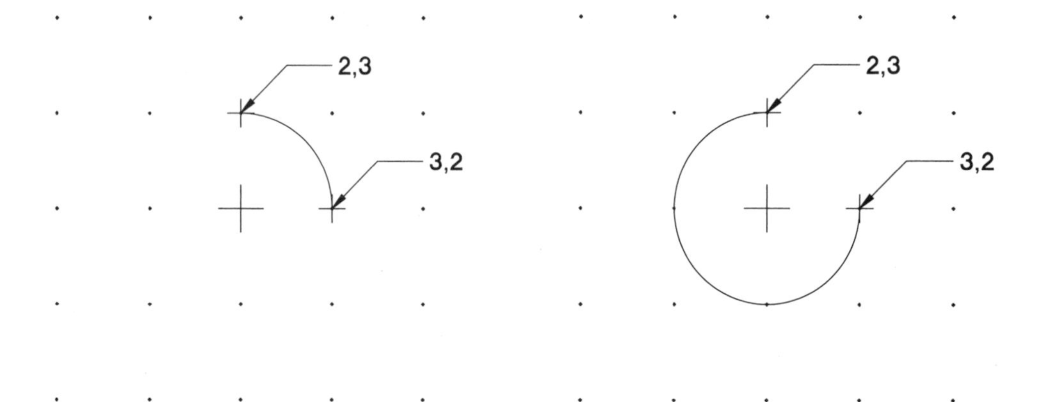

Figure 2–28 *An arc drawn counter-clockwise with the Start, End, Included Angle (S,E,A) option*

Figure 2–29 *An arc drawn clockwise with the Start, End, Included Angle (S,E,A) option*

Start, End, Direction (or S,E,D for short)

The Start, End, Direction option allows you to draw an arc between selected points by specifying a direction in which the arc will start from the selected start point. Either the direction can be keyed in or you can select a point on the screen with your pointing device.

If you select a point on the screen, AutoCAD uses the angle from the start point to the selected point as the starting direction.

Invoke the Start, End, Direction option:

Draw menu	Choose Arc > Start, End, Direction
Command: prompt	**arc** (ENTER)

AutoCAD prompts:

Command: **arc** (ENTER)
Specify start point of arc or [Center]: *(specify start point of arc)*
Specify second point of arc or [CEnter/End]: **e** (ENTER)
Specify end point of arc: *(specify end point of arc)*
Specify center point of arc or [Angle/Direction/Radius]: **d** (ENTER)
Direction from start point: *(specify the direction from the start point of the arc to be drawn)*

The arc shown in Figure 2–30 is drawn using the following sequence:

Command: **arc** (ENTER)
Specify start point of arc or [Center]: **3,2** (ENTER)
Specify second point of arc or [CEnter/End]: **e** (ENTER)
Specify end point of arc: **2,3** (ENTER)
Specify center point of arc or [Angle/Direction/Radius]: **d** (ENTER)
Direction from start point: **90** (ENTER)

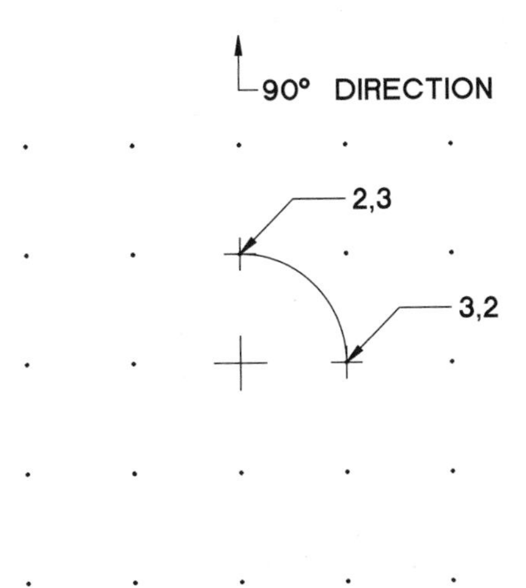

Figure 2–30 *An arc drawn with the Start, End, Direction (S,E,D) option*

Start, End, Radius (or S,E,R for short)

The Start, End, Radius option allows you to specify a radius after selecting the two endpoints of the arc. As with the Chord Length method, there are four possible arcs that can be drawn: a major arc in either direction and a minor arc in either direction.

Therefore, all arcs drawn by this method are counterclockwise from the start point. A positive value for the radius causes AutoCAD to draw the minor arc; a negative value results in the major arc.

Invoke the Start, End, Radius option:

Draw menu	Choose Arc > Start, End, Radius
Command: prompt	**arc** (ENTER)

AutoCAD prompts:

Command: **arc** (ENTER)
Specify start point of arc or [CEnter]: *(specify start point of arc)*
Specify second point of arc or [CEnter/End]: **e** (ENTER)
Specify end point of arc: *(specify the end point of arc)*
Specify center point of arc or [Angle/Direction/Radius]: **r** (ENTER)
Specify radius of arc: *(specify the radius of the arc to be drawn)*

The following command sequence shows an example of drawing a major arc, as shown in Figure 2–31.

Command: **arc** (ENTER)
Specify start point of arc or [CEnter]: **1,2** (ENTER)
Specify second point of arc or [CEnter/End]: **e** (ENTER)
Specify end point of arc: **2,3** (ENTER)
Specify center point of arc or [Angle/Direction/Radius]: **r** (ENTER)
Specify radius of arc: **-1** (ENTER)

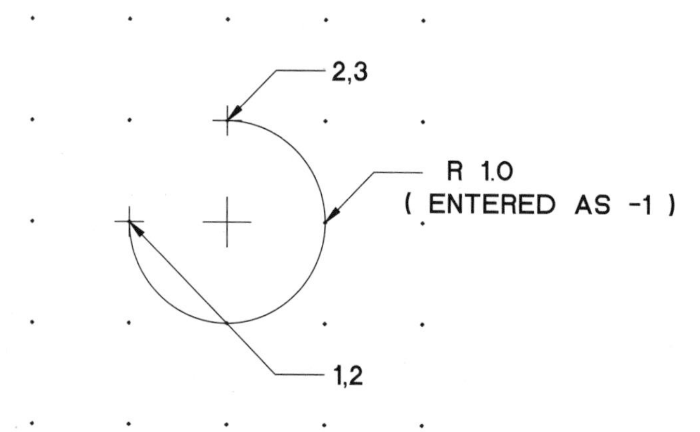

Figure 2–31 *A major arc drawn with the Start, End, Radius (S,E,R) option*

The following command sequence shows an example of drawing a minor arc, as shown in Figure 2–32.

Command: **arc** (ENTER)
Specify start point of arc or [CEnter]: **2,3** (ENTER)
Specify second point of arc or [CEnter/End]: **e** (ENTER)
Specify end point of arc: **1,2** (ENTER)
Specify center point of arc or [Angle/Direction/Radius]: **r** (ENTER)
Specify radius of arc: **1** (ENTER)

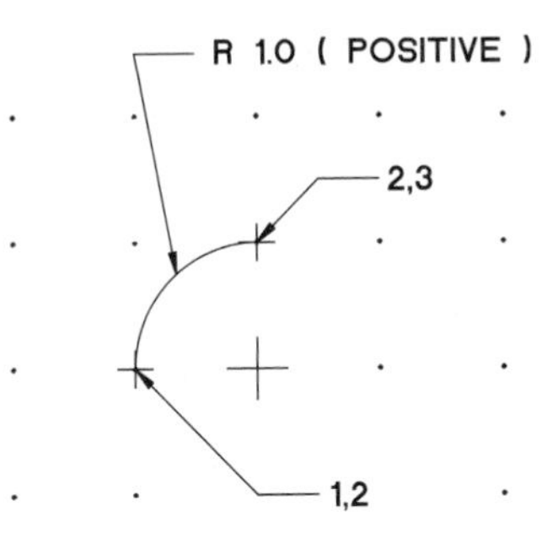

Figure 2–32 *A minor arc drawn with the Start, End, Radius (S,E,R) option*

Center, Start, End (or C,S,E for short)

The Center, Start, End option is similar to the Start, Center, End (S,C,E) method, except in this option the beginning point is the center point of the arc rather than the start point.

Center, Start, Angle (or C,S,A for short)

The Center, Start, Angle option is similar to the Start, Center, Angle (S,C,A) method, except in this option the beginning point is the center point of the arc rather than the start point.

Center, Start, Length (or C,S,L for short)

The Center, Start, Length option is similar to the Start, Center, Length (S,C,L) method, except that the beginning point is the center point of the arc rather than the start point.

Line-Arc and Arc-Arc Continuation

You can use an automatic start point, endpoint, starting direction method to draw an arc by pressing ENTER as a response to the first prompt of the ARC command. After pressing ENTER, the only other input is to select or specify the endpoint of the arc you wish to draw. AutoCAD uses the endpoint of the previous line or arc (whichever was drawn last) as the start point of the new arc. AutoCAD then uses the ending direction of that last drawn object as the starting direction of the arc. Examples are shown in the following sequences and figures.

The start point of the existing arc is 2,1 and the endpoint is 3,2 with a radius of 1. This makes the ending direction of the existing arc 90 degrees, as shown in Figure 2–33.

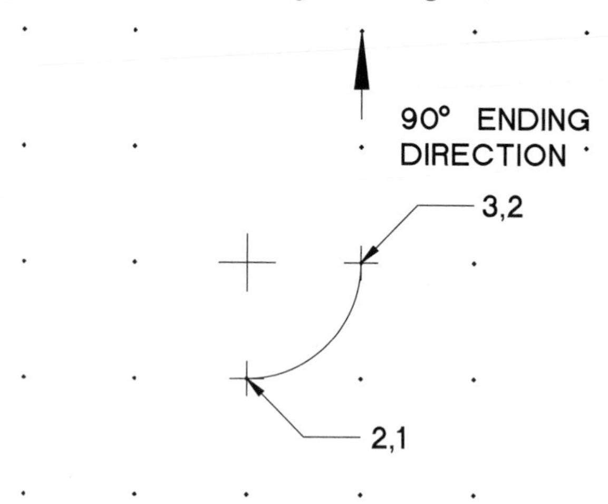

Figure 2–33 *An arc drawn with a start point (2,1), an endpoint (3,2), and a radius of 1.0*

The following command sequence continues drawing an arc, from the last-drawn arc, as shown in Figure 2–34 (arc-arc continuation).

Command: **arc** (ENTER)
Specify start point of arc or [Center]: (ENTER)
Specify end point of arc: **2,3** (ENTER)

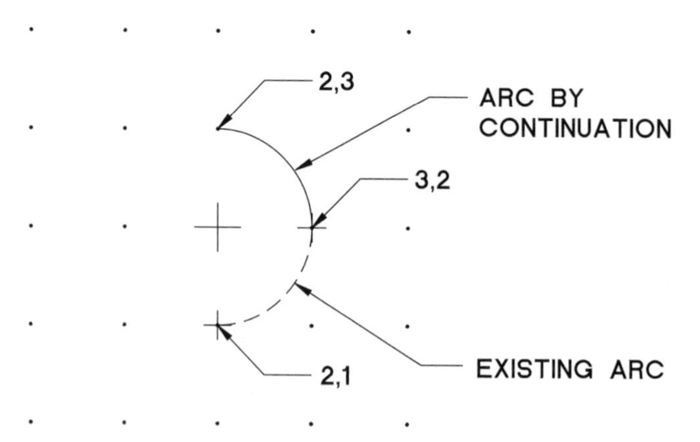

Figure 2–34 *An arc drawn by means of the Arc-Arc Continuation method*

The arc, as shown in Figure 2–33, is drawn clockwise instead, with its start point at 3,2 to an endpoint of 2,1 (see Figure 2–35).

The following command sequence will draw the automatic start point, endpoint, starting direction arc, as shown in Figure 2–36.

Command: **arc** (ENTER)
Specify start point of arc or [Center]: (ENTER)
Specify end point of arc: **2,3** (ENTER)

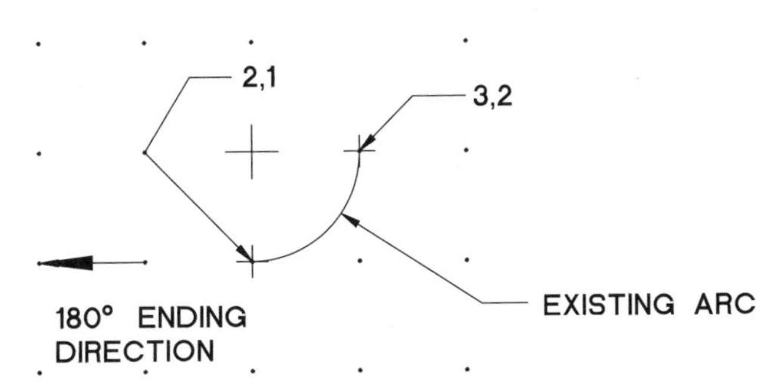

Figure 2–35 *An arc drawn clockwise with start point (3,2) and endpoint (2,1)*

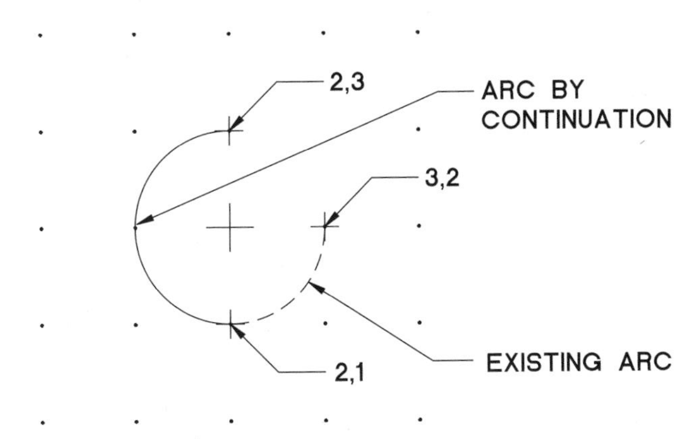

Figure 2–36 *An arc drawn by means of the automatic Start point–Endpoint-Starting direction method*

In the last case, the direction used is 180 degrees. The same arc would have been drawn if the last "line-or-arc" drawn was a line starting at 4,1 and ending at 2,1.

Note: This method uses the last drawn arc or a line. If you draw an arc, then draw a line, then draw a circle, and then use this continuation method, AutoCAD will use the line as the basis for the start point and direction. This is because the line was the last of the "line-or-arc" objects drawn.

OBJECT SELECTION

Many AutoCAD modify and construct commands prompt you to select one or more objects for manipulation. When you select one or more objects, AutoCAD highlights them by displaying them with dashed lines. The group of objects selected for the manipulation is called the selection set. There are several different ways of selecting the objects for manipulation. The selection options include Window, Window Polygon (WP), Crossing, Crossing Polygon (CP), Box, Fence, All, Last, Previous, Group Add, Remove, Single, Multiple, and Undo.

All modify and construct commands require a selection set, for which AutoCAD prompts:

Select objects:

AutoCAD replaces the screen crosshairs with a small box called the object selection target. With the target cursor, select individual objects for manipulation. Using your pointing device (or the keyboard's cursor keys), position the target box so it touches only the desired object or a visible portion of it. The object selection target helps you select the object without having to be very precise. Every time you select an object, the "Select objects:" prompt reappears. To indicate your acceptance of the selection set, give a Null reply (press ENTER) at the "Select objects:" prompt.

Sometimes, it is difficult to select objects that are close together or lie directly on top of one another. You can use the pick button to cycle through these objects, one after the other, until you reach the one you want.

To cycle through objects for selection, at the "Select Objects" prompt, hold down the CTRL button. Select a point as near as possible to the object. Press the pick button on your pointing device repeatedly until the object you want is highlighted, and press ENTER to select the object.

WINDOW

The Window option for selecting objects allows you to select all the objects contained completely within a rectangular area or dynamically manipulated window.

The window can be defined by specifying a point at the appropriate location to the "Select Objects:" prompt and moving the cursor toward the right of the first data point. AutoCAD prompts:

Specify opposite corner: *(specify opposite corner)*

You can also define a window for selection of objects, by entering **w** in response to the "Select Objects:" prompt, and AutoCAD prompts:

Specify first corner: *(specify first corner)*
Specify opposite corner: *(specify opposite corner)*

If there is an object that is partially inside the rectangular area, then that object is not included in the selection set. You can select only objects currently visible on the screen. To select a partially visible object, you must include all its visible parts within the window. See Figure 2–37, in which only the lines will be included, not the circles, because a portion of each of the circles is outside the rectangular area.

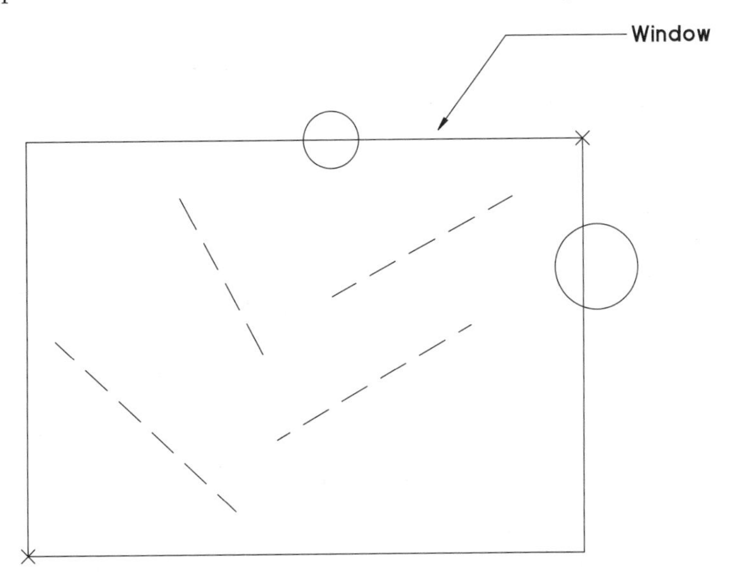

Figure 2–37 *Selecting objects by means of the Window option*

Crossing

The Crossing option for selecting objects allows you to select all the objects contained completely in a rectangular area as well as the objects that are crossing the window.

The crossing rectangle can be placed by specifying a point at the appropriate location to the "Select Objects:" prompt and moving the cursor toward the left of the first data point. AutoCAD prompts:

> Specify opposite corner: *(specify opposite corner)*

You can also define a crossing window for the selection of objects, by entering **c** to the "Select Objects:" prompt, and AutoCAD prompts:

> Specify first corner: (specify first corner)
> Specify opposite corner: (specify opposite corner)

See Figure 2–38, in which all the lines and circles are included, though parts of the circles are outside the rectangle.

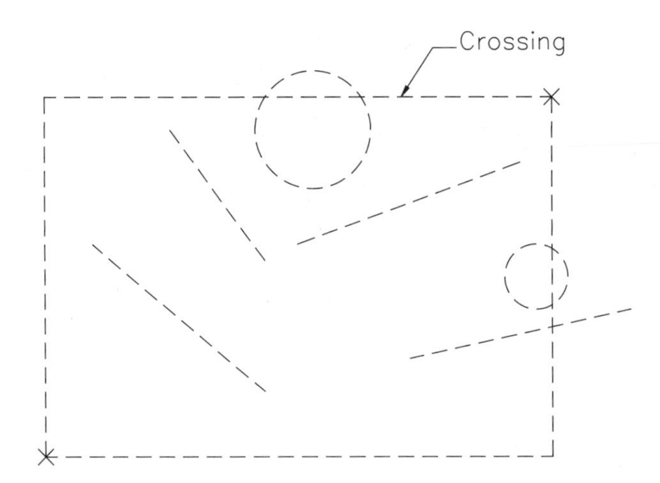

Figure 2–38 *Selecting objects by means of the Crossing option*

Previous

The Previous option enables you to perform several operations on the same object or group of objects. AutoCAD remembers the most recent selection set and allows you to reselect it with the Previous option. For example, if you moved several objects and now wish to copy them elsewhere, you can invoke the COPY command and respond to the "Select objects:" prompt by entering **p** to select the same objects again. (There is a command called SELECT that does nothing but create a selection set; you can then use the Previous option to refer to this set in subsequent commands.)

Last

The Last option is an easy way to select the most recently created object currently visible. Only one object is designated, no matter how often you use the Last option, when constructing a particular selection set. The Last option is invoked by entering **l** at the "Select Objects:" prompt.

The Wpolygon, Cpolygon, Fence, All, Group, Box, Auto, Undo, Single, Multiple, Add, and Remove options are explained in Chapter 5.

MODIFY OBJECTS

AutoCAD not only allows you to draw objects easily, but also allows you to modify the objects you have drawn. Of the many modifying commands available, the ERASE command will probably be the one you use most often. Everyone makes mistakes, but in AutoCAD it is easier to erase them. Or, if you are through with an object that you have created for aid in constructing other objects, you may wish to erase it.

ERASING OBJECTS

To erase objects from a drawing, invoke the ERASE command:

Modify toolbar	Choose the ERASE command (see Figure 2–39)
Modify menu	Choose Erase
Command: prompt	**erase** (ENTER)

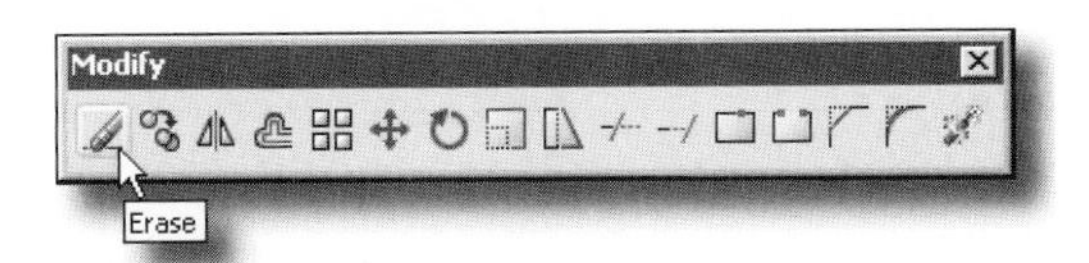

Figure 2–39 *Invoking the* ERASE *command from the Modify toolbar*

Command: **erase** (ENTER)
Select objects: (select objects to be erased, and then press the SPACEBAR or ENTER)

You can use one or more available object selection methods. After selecting the object(s), press ENTER (null response) in response to the next "Select objects:" prompt to complete the ERASE command. All the objects that were selected will disappear.

The following command sequence shows an example of erasing individual objects (shown Figure 2–40).

Command: **erase** (ENTER)
Select objects: *(select Line 2, the line is highlighted)* 1 selected, 1 found
Select objects: *(select Line 4, the line is highlighted)* 1 selected, 1 found
Select objects: (ENTER)

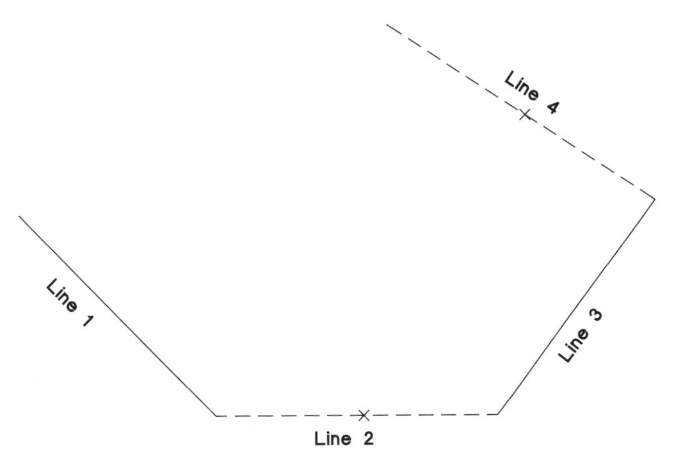

Figure 2–40 *Selection of individual objects to erase*

The following command sequence shows an example of erasing a selection of objects by the Window option (shown in Figure 2–41).

Command: **erase** (ENTER)
Select objects: (specify a point)
Specify opposite corner: *(Specify a point for the diagonally opposite corner)*
Select objects: (ENTER)

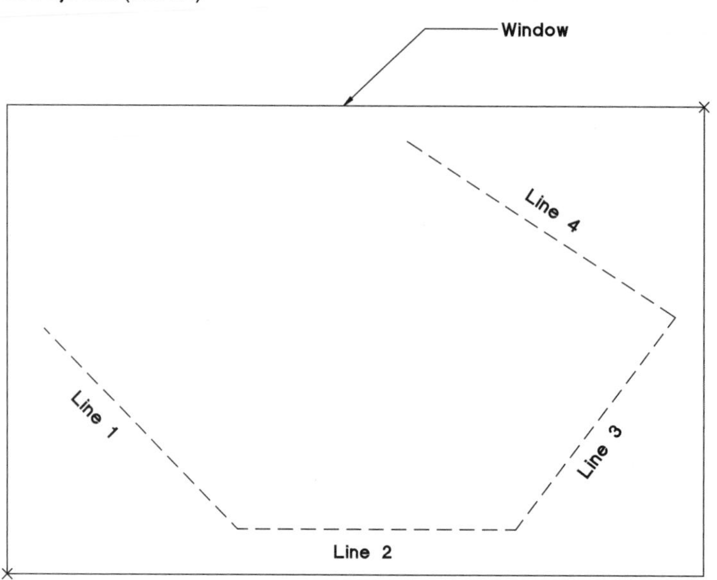

Figure 2–41 *Erasing a group of objects*

GETTING IT BACK

The OOPS command restores objects that have been unintentionally erased. Whenever the ERASE command is used, the last group of objects erased is stored in memory. The OOPS command will restore the objects; it can be used at any time. It only restores the objects erased by the most recent erase command. See Chapter 3 on the UNDO command, if you need to step back further than one ERASE command.

To restore objects erased by the last ERASE command, invoke the OOPS command:

Command: prompt	**oops** (ENTER)

AutoCAD restores the objects erased by the last erase command.

The following example shows the command sequence for using the OOPS command in conjunction with the erase command:

Command: **erase** (ENTER)
Select objects: *(specify a point)*
Other corner: *(specify a point for the diagonally opposite corner)*
Select objects: (ENTER)
Command: **oops** (ENTER) *(will restore the erased objects)*

Open the Exercise Manual PDF file for Chapter 2 on the accompanying CD for project and discipline specific exercises.

If you have the accompanying Exercise Manual, refer to Chapter 2 for project and discipline specific exercises.

REVIEW QUESTIONS

1. The RECTANGLE command requests what information when drawing a rectangle?
   a. an initial corner, the width, and the height
   b. the coordinates of the four corners of the rectangle
   c. the coordinates of diagonally opposite corners of the rectangle
   d. the coordinates of three adjacent corners of the rectangle

2. When drawing a trace line, after selecting the second point:
   a. nothing appears on the screen
   b. you are prompted for the trace width
   c. the segment is drawn and the command terminates
   d. the first segment is drawn and you are prompted for the next point

3. To draw multiple connected line segments, you must invoke the LINE command multiple times?
   a. True
   b. False

4. The file extension .BAK stands for:
   a. backup drawing file
   b. binary file
   c. binary attribute file
   d. drawing file
   e. both b and c

5. The HELP command cannot be used:
   a. while in the LINE command
   b. while in the CIRCLE command option TTR
   c. to list commands
   d. for a text string

6. Points are located by relative rectangular coordinates in relation to:

   a. the last specified point or position

   b. the global origin

   c. the lower left corner of the screen

   d. all of the above

7. Polar coordinates are based on a distance from:

   a. the global origin

   b. the last specified position at a given angle

   c. the center of the display

   d. all of the above

8. To enter a command from the keyboard, simply enter the command name at the "Command:" prompt:

   a. in lower-case letters

   b. in upper-case letters

   c. either a or b or mixed case

   d. commands cannot be entered via the keyboard

9. The "C" option used in the LINE command at the "From Point:" prompt will:

   a. Continue the line from the last line or arc that was drawn

   b. Close the previous set of line segments

   c. Displays an error message

10. A rectangle generated by the RECTANGLE command will always have horizontal and vertical sides.

    a. True

    b. False

11. Which of the following coordinates will define a point at the screen default origin?

    a. 000

    b. 00

    c. 0.0,0.0

    d. 112

    e. @0,0

12. Switching between graphics and text screen can be accomplished by:

    a. pressing (ENTER) twice

    b. entering (CTRL) and (ENTER) at the same time

    c. pressing the (ESC) key

    d. pressing the (F2) function key

    e. both b and c

13. To draw a line a length of eight feet, four and five-eights inches in the 12 o'clock direction from the last point selected, enter:

    a. @8'4-5/8<90

    b. 8'-4-5/8<90

    c. @8-45/8<90

    d. 8'-45/8<90

    e. None of the above

14. By default, what direction does a positive number indicate when specifying angles in degrees?

    a. Clockwise

    b. Counterclockwise

    c. Has no impact when specifying angles in degrees

    d. None of the above

15. When erasing objects, if you select a point which is not on any object, AutoCAD will:

    a. Terminate the ERASE command

    b. delete the selected objects and continue with the ERASE command

    c. allow you to drag a window to select multiple objects within the area

    d. ignore the selection and continue with the ERASE command

16. Regarding the ARC options, what does "S,C,E" mean?

    a. Start, Center, End

    b. Second, Continue, Extents

    c. Second, Center, End

    d. Start, Continue, End

17. The number of different methods by which a circle can be drawn is:

    a. 1

    b. 3

    c. 4

    d. 7

    e. None of the above

18. When using the ERASE command, AutoCAD deletes each object from the drawing as you select it.

    a. True

    b. False

19. Once an object is erased from a drawing, which of the following commands could restore it to the drawing?

    a. OOPS

    b. RESTORE

    c. REPLACE

    d. CANCEL

20. When drawing a circle with the two-point potion, the distance between the two points is equal to:

    a. the circumference

    b. the perimeter

    c. the shortest chord

    d. the radius

    e. the diameter

21. A circle may be created by any of the following options, except:

    a. 2P

    b. 3P

    c. 4P

    d. Cen,Rad

    e. TTR

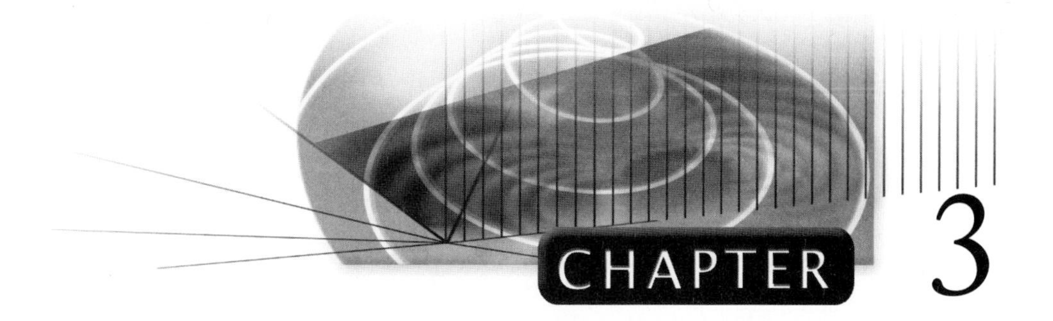

Fundamentals II

INTRODUCTION

AutoCAD provides various tools to make your drafting and design work easier. The drawing tools described in this chapter will assist you in creating drawings rapidly, while ensuring the highest degree of precision.

After completing this chapter, you will be able to do the following:

- Use and control drafting settings (e.g., GRID, SNAP, ORTHO, POLAR TRACKING, and OBJECT SNAP)
- Use Tracking and Direct Distance
- Use display control commands (e.g., ZOOM, PAN, REDRAW, and REGEN)
- Create tiled viewports
- Use layering techniques
- Use the UNDO and REDO commands

DRAFTING SETTINGS

The GRID, SNAP, ORTHO, POLAR TRACKING, and OBJECT SNAP commands do not create objects. However, they make it possible to create and modify them more easily and accurately. Each of these Drafting Settings commands can be readily toggled ON when needed and OFF when not. These commands, when turned ON, operate according to settings that can also be changed easily. When used appropriately, these commands provide the power, speed, and accuracy associated with Computer Aided Design/Drafting.

SNAP COMMAND

The SNAP command provides an invisible reference grid in the drawing area. When set to ON, the Snap feature forces the cursor to lock onto the nearest point on the specified Snap Grid. Using the SNAP command, you can specify points quickly, letting AutoCAD ensure that they are placed precisely. You can always override the snap spacing by entering absolute or relative coordinate points by means of the keyboard, or by simply turning the Snap mode OFF. When the Snap mode is set to OFF, it has no effect on the cursor. When it is set to ON, you cannot place the cursor on a point that is not on one of the specified Snap Grid locations.

Setting Snap ON and OFF

In AutoCAD, five primary methods to set the Snap ON and OFF are available. Of the five explained in this section, the first method (selecting Snap on the status bar) is the easiest and most commonly used with the pointing device. The second method (pressing the function key F9) is the most commonly used from the keyboard. The others, while not quite as convenient as the first two, might be convenient in certain situations.

1. On the status bar at the bottom of the screen, when the **SNAP** button appears to be pressed in, then the Snap is set to ON. When the **SNAP** button appears to be out, then the Snap is set to OFF. To change its setting, simply place the cursor over the snap button and press the pointing device pick button.
2. To change the Snap setting to ON when it is OFF or to OFF when it is ON, simply press the function key F9.
3. Right-click the **SNAP** button on the status bar at the bottom of the screen and from the shortcut menu, choose On to set the Snap ON or choose Off to set the Snap OFF.
4. From the Tools menu choose Drafting Settings. Then select the **Snap and Grid** tab if it is not already selected. The Snap setting is set to ON when the **Snap On (F9)** check box is checked, and is OFF when it is not checked. To change its setting, simply place the cursor over the check box as shown in Figure 3–1 and press the pointing device pick button. You can also enter **ddrmodes** at the Command: prompt to display the Drafting Settings dialog box, from which the desired tab can be selected if it is not already selected.
5. At the Command: prompt:

Command: **snap** (ENTER)
Specify snap spacing or [ON/OFF/Aspect/Rotate/Style/Type] <current>:
(enter ***on*** *to set the Snap ON and* ***off*** *to set the Snap OFF or right-click and choose ON or OFF from the shortcut menu)*

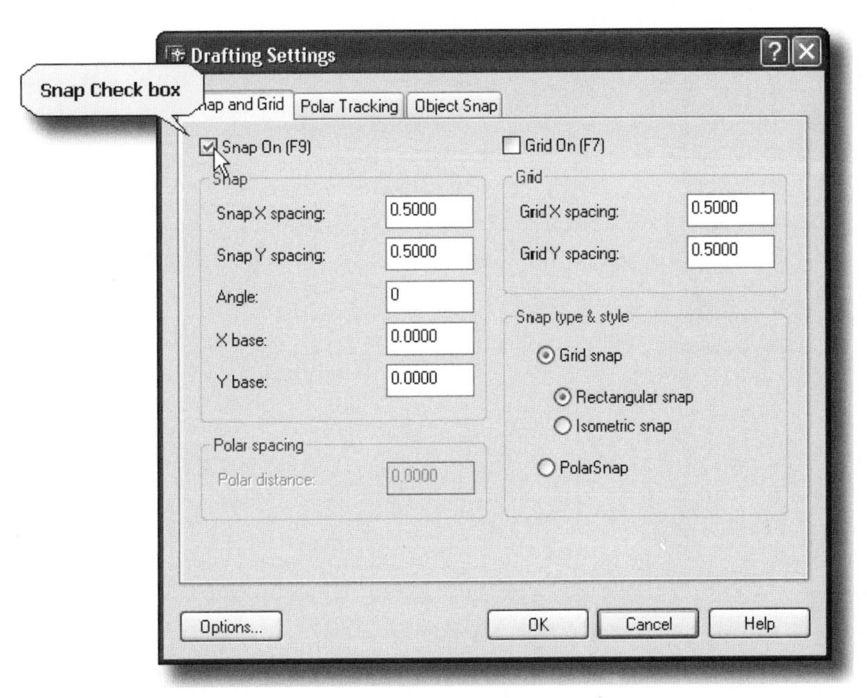

Figure 3–1 *Choosing the Snap On (F9) check box from the Snap and Grid tab of the Drafting Settings menu*

Changing Snap Spacing

In AutoCAD there are three primary methods to change the Snap spacing:

1. Right-click the **SNAP** button on the status bar at the bottom of the screen and choose SETTINGS from the shortcut menu. AutoCAD will display the Drafting Settings dialog box with the **Snap and Grid** tab selected. From the **Snap** section, you can change the settings of the X and Y spacings by entering the desired values in the **Snap X Spacing** and **Snap Y Spacing** text boxes, respectively, as shown in Figure 3–2. You can also enter **ddrmodes** at the Command: prompt to display the Drafting Settings dialog box, from which the desired tab can be selected if it is not already selected.
2. From the Tools menu, choose Drafting Settings. Then select the **Snap and Grid** tab. See the explanation in Method 1 for changing settings.
3. At the Command: prompt:

Command: **snap** (ENTER)
Specify snap spacing or [ON/OFF/Aspect/Rotate/Style/Type] <current>:
(specify distance to be used for both X and Y Snap Spacings)

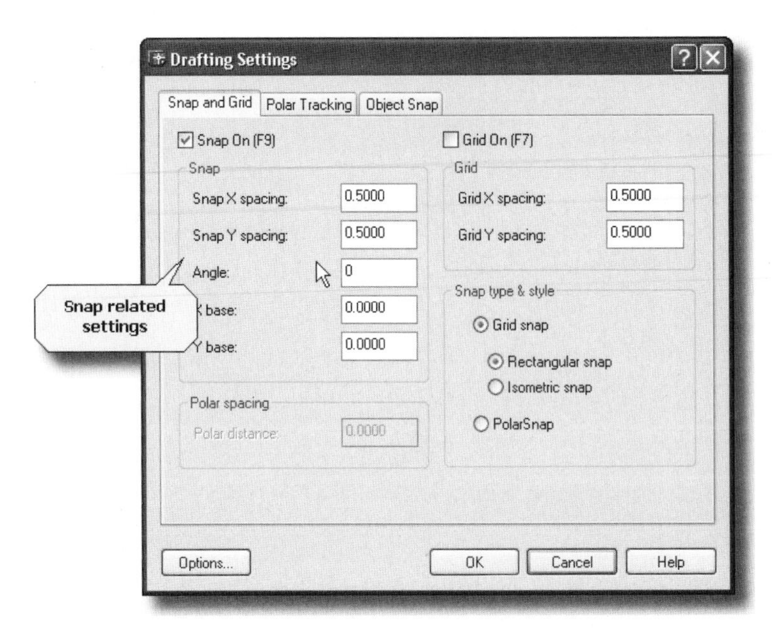

Figure 3–2 *Changing the Snap settings from the Snap and Grid section of the Drafting Settings dialog box*

Aspect

The Aspect option allows you to set the Y Snap Spacing different from the X Snap Spacing. When you select the Aspect option, AutoCAD prompts for the *X* and *Y* values independently. This is handy if the *X* and *Y* modular dimensions of your design are of unequal multiples. There are three primary methods to changing the Snap Aspect setting.

1. From the **Snap and Grid** tab of the Drafting Settings dialog box, you can set the Y spacing different from the X spacing. This is done by entering a value in the **Snap Y Spacing** text box that is different from the value in the **Snap X Spacing** text box. You can also enter **ddrmodes** at the Command: prompt to display the Drafting Settings dialog box, from which the desired tab can be selected if it is not already selected.
2. From the Tools menu, choose Drafting Settings. Then select the **Snap and Grid** tab. See the explanation in Method 1 for setting the Y spacing different from the X spacing.
3. At the Command: prompt:

Command: **snap** (ENTER)
Specify snap spacing or [ON/OFF/Aspect/Rotate/Style/Type] <1.0000>:
 a (ENTER)
Specify horizontal spacing <1.0000>: *(specify X Snap spacing)*
Specify vertical spacing <1.0000>: *(specify Y Snap spacing)*

For example, the following sequence shows how you can set a horizontal snap spacing of 0.5 and vertical snap spacing of 0.25.

Command: **snap** (ENTER)
Specify snap spacing or [ON/OFF/Aspect/Rotate/Style/Type] <1.0000>:
a (ENTER)
Specify horizontal spacing <1.0000>: **0.5** (ENTER)
Specify vertical spacing <1.0000>: **0.25** (ENTER)

Rotate

The Rotate option allows you to specify an angle to rotate both the visible Grid and the invisible Snap grid. It is a simple version of the more complicated User Coordinate System. It permits you to set up a Snap grid with an origin (*X* coordinate, *Y* coordinate of 0,0) and an angle of rotation specified with respect to the default origin and Zero-East system of direction. In conjunction with the X and Y spacing of the Snap grid, the Rotate option can make it easier to draw certain shapes.

The plot plan in Figure 3–3 is an example of where the Rotate option of the SNAP command can be applied. The property lines are drawn using surveyor's units of angular display. In this example, the decimal linear units and the surveyor's angular units are selected. The limits are set up with the lower left corner at –20',–10', and the upper right corner at 124',86'. After making sure the ORTHO mode is set to ON, the sequence for drawing the property lines is as follows:

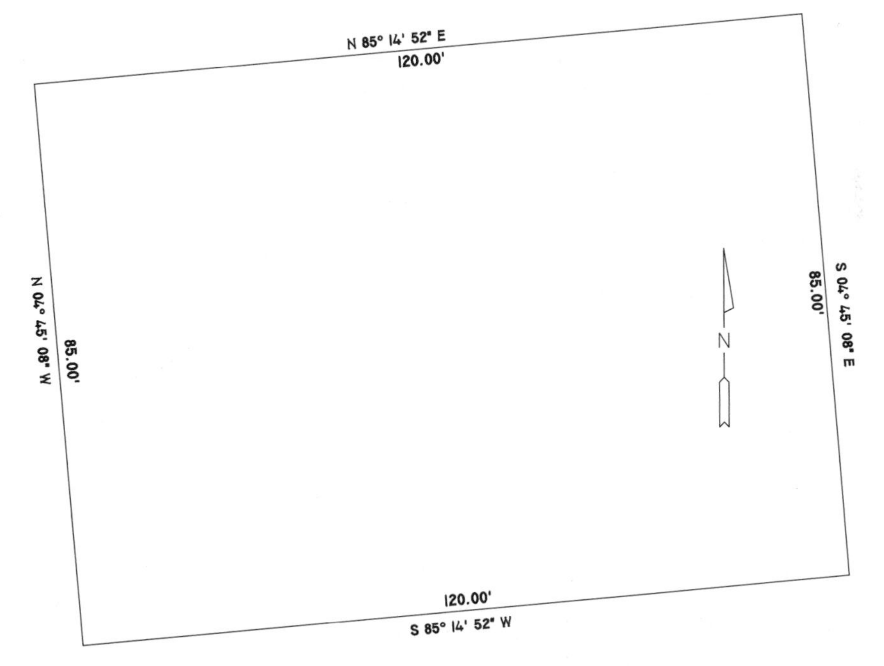

Figure 3–3 *Example drawing where the Rotate option of the* SNAP *command can be used*

Command: **snap** (ENTER)
Specify snap spacing or [ON/OFF/Aspect/Rotate/Style/Type] <current>:
r (ENTER)
Specify base point <0'-0.0",0'-0.0">: *(press* ENTER *to accept the default value)*
Specify rotation angle <0r>: **4d45'08"** (ENTER)

Command: **line** (ENTER) *(invoking the* LINE *command)*
Specify first point: **0,0** (ENTER) *(selecting the first point)*
Specify next point or [Undo]: **85'** (ENTER) *(point selected with the cursor north of the first point)*
Specify next point or [Undo]: **120'** (ENTER) *(point selected with the cursor east of the previous point)*
Specify next point or [Close/Undo]: **85'** (ENTER) *(point selected with the cursor south of the previous point)*
Specify next point or [Close/Undo]: **c** *(completes the property boundary by closing the rectangle and exiting the* LINE *command)*

Figure 3–4 shows the property boundary drawn.

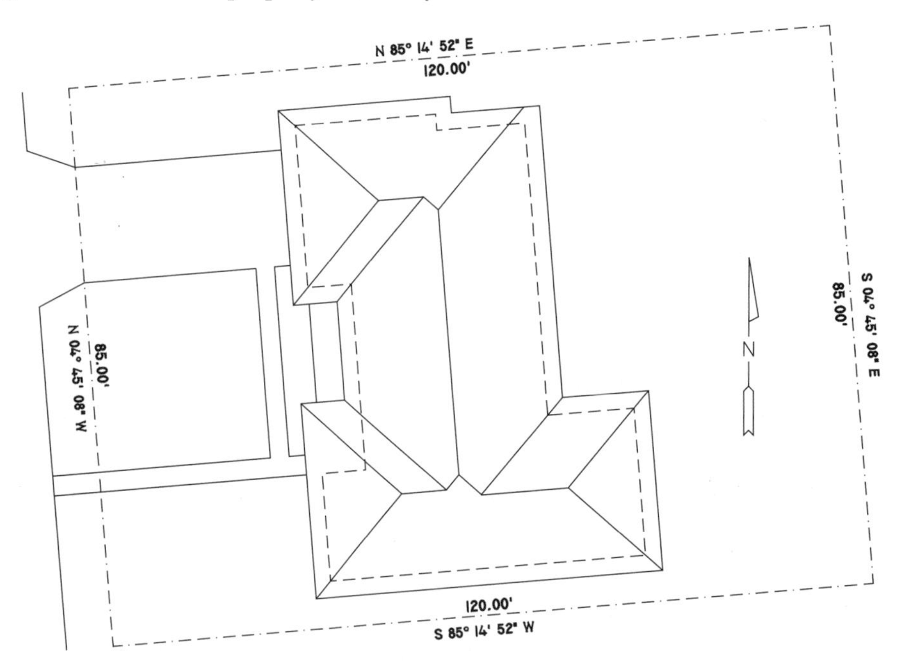

Figure 3–4 *Layout of the property boundary lines*

Style

The Style option permits you to select one of two available formats, Standard or Isometric. The Standard option refers to the normal rectangular grid (default), and the

Isometric option refers to a Grid and Snap designed for Isometric drafting purposes (see Figure 3–5).

AutoCAD prompts:

> Command: **snap** (ENTER)
> Specify snap spacing or [ON/OFF/Aspect/Rotate/Style/Type] <current>: **s** *(invoke Style option)*
> Enter snap grid style [Standard/Isometric] <S>: **i** (ENTER) *(change Style to Isometric)*
> Specify vertical spacing <1.0000>: **0.5** (ENTER) *(specify Snap spacing)*

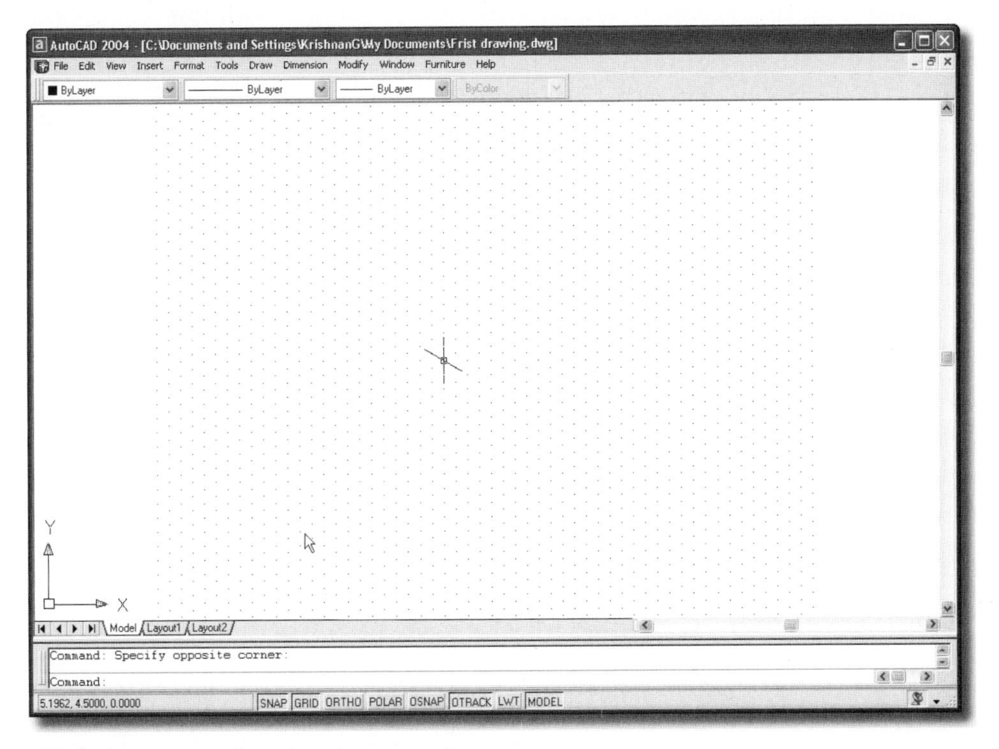

Figure 3–5 *Setting the snap for isometric drafting*

You can switch the Isoplanes between Left (90- and 150-degree angles), Top (30- and 150-degree angles), and Right (30- and 90-degree angles) by CTRL + E (the combination key strokes of holding down CTRL and then pressing **E**) or by simply press the function key F5.

Type

The Type option permits you to select one of two available Snap types, Polar or Grid. Choosing the Polar type sets the snap to Polar Tracking angles. See the explanations on Polar Tracking and ORTHO, later in this chapter. Choosing the Grid type sets the snap spacing equal to the grid spacing.

AutoCAD prompts:

Command: **snap** (ENTER)
Specify snap spacing or [ON/OFF/Aspect/Rotate/Style/Type] <current>: **t**
Enter snap type [Polar/Grid] <Grid>: **g**

GRID COMMAND

The GRID command is used to display a visible array of dots with row and column spacings that you specify. AutoCAD creates a grid that is similar to a sheet of graph paper. You can set the grid display ON and OFF, and can change the dot spacing. The grid is a drawing tool and is not part of the drawing; it is for visual reference and is never plotted. In the World Coordinate System, the grid fills the area defined by the limits.

The grid has several uses within AutoCAD. First, it shows the extent of the drawing limits. For example, if you set the limits to 42 by 36 units and grid spacing is set to 0.5 units, then each row will have 85 dots and each column will have 73 dots. This will give you a better sense of the drawing's size relative to the limits than if it were on a blank background.

Second, using the GRID command with the SNAP command is helpful when you create a design in terms of evenly spaced units. For example, if your design is in multiples of 0.5 units, then you can set grid spacing as 0.5 to facilitate point entry. You could check your drawing visually by comparing the locations of the grid dots and the crosshairs. Figure 3–6 shows a drawing with a grid spacing of 0.5 units, with limits set to 0,0 and 17,11.

Note: While the GRID, SNAP, ORTHO, POLAR TRACKING, and OBJECT SNAP commands do not create objects, they make it possible to create them more easily and accurately. Each of these drafting settings commands, when toggled ON, operates according to the value(s) to which you have set it and, when toggled OFF, has no effect. You are advised to identify and master the two skills involved in using these utilities: one is to learn how to change the settings of the utility commands, and the other is to learn how and when to best set them ON and OFF. Changing the settings of these utilities is normally done more easily by means of their associated dialog box(es). You can also set them ON and OFF from a dialog box. But, because these features are normally switched ON and OFF so frequently during a drawing session, special buttons are provided on the status bar at the bottom of the screen for this purpose.

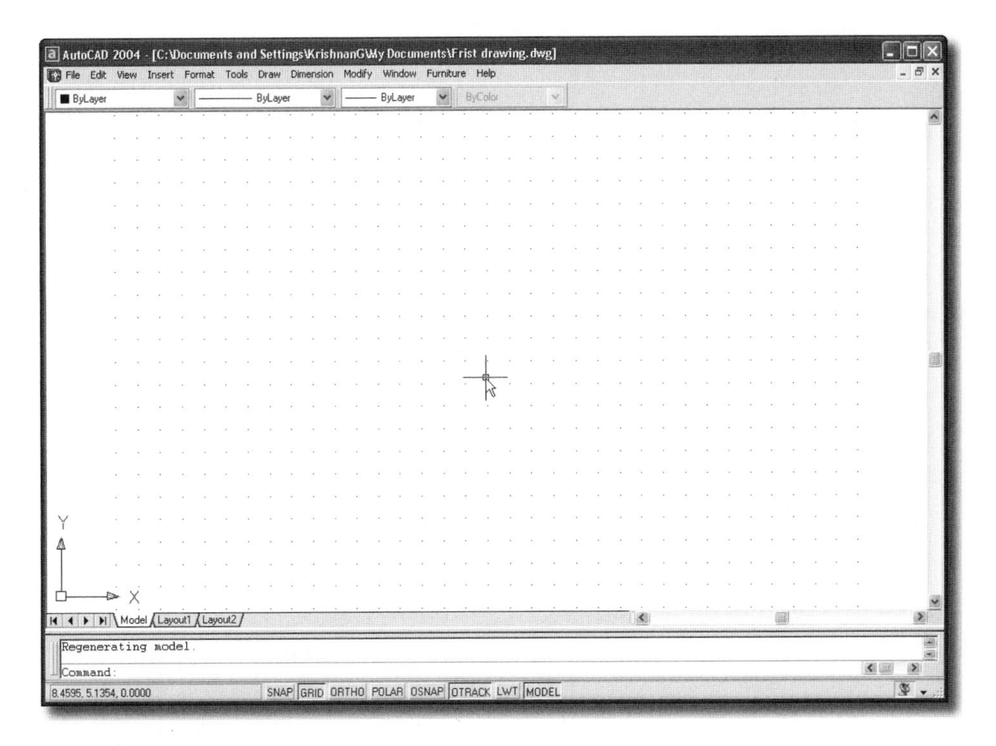

Figure 3–6 *A grid spacing of 0.5 units, with limits set to (0,0) and (17,11)*

Setting the Grid ON and OFF

In AutoCAD, five primary methods to set the Grid ON and OFF are available. Of the five explained in this section, the first method (selecting Grid on the status bar) is the easiest and most commonly used with the pointing device. The second method (pressing the function key F7) is the most commonly used from the keyboard. The others, while not quite as convenient as the first two, might be convenient in certain situations.

1. On the status bar at the bottom of the screen, when the **GRID** button appears to be pressed in, then the Grid is set to ON. When the **GRID** button appears to be out, then the Grid is set to OFF. To change its setting, simply place the cursor over the **GRID** button and press the pointing device pick button.
2. To change the Grid setting to ON when it is OFF or to OFF when it is ON, simply press the function key F7.
3. Right-click the **GRID** button on the status bar at the bottom of the screen and from the shortcut menu, choose On to set the Grid ON or choose Off to set the Grid OFF.

4. From the Tools menu, choose Drafting Settings. Then select the **Snap and Grid** tab if it is not already selected. The Grid setting is ON when the **Grid On (F7)** check box is checked, and is OFF when it is not checked. To change its setting, simply place the cursor over the check box, as shown in Figure 3–7, and press the pointing device pick button. You can also enter **ddrmodes** at the Command: prompt to display the Drafting Settings dialog box, from which the desired tab can be selected if it is not already selected.
5. At the Command: prompt:

Command: **grid** (ENTER)
Specify grid spacing(X) or [ON/OFF/Snap/Aspect] <current >: *(enter **on** to turn the Grid ON and **off** to turn the Grid OFF, or right-click and choose ON or OFF from the shortcut menu)*

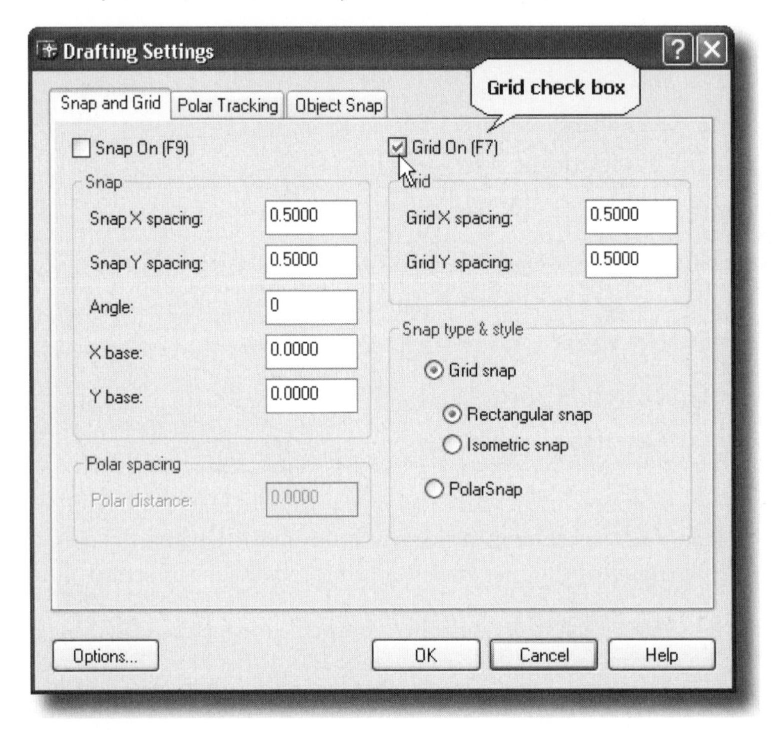

Figure 3–7 *Choosing the Grid On (F7) check box from the Snap and Grid tab of the Drafting Settings dialog box*

Changing Grid Spacing

In AutoCAD there are three primary methods to change the Grid spacing:

1. On the status bar at the bottom of the screen, right-click on the GRID button and choose Settings. AutoCAD will display the Drafting Settings dialog box with

the **Snap and Grid** tab selected. From the **Grid** section, you can change the settings of the X and Y spacings by entering the desired values in the **Grid X Spacing** and **Grid Y Spacing** text boxes, respectively, as shown in Figure 3–8.

2. From the Tools menu, choose Drafting Settings. Then select the **Snap and Grid** tab. See the explanation in Method 1 for changing settings.
3. At the Command: prompt:

Command: **grid** (ENTER)
Specify grid spacing (X) or [ON/OFF/Snap/Aspect] <current>: *(specify distance to be used for both X and Y Grid Spacings)*

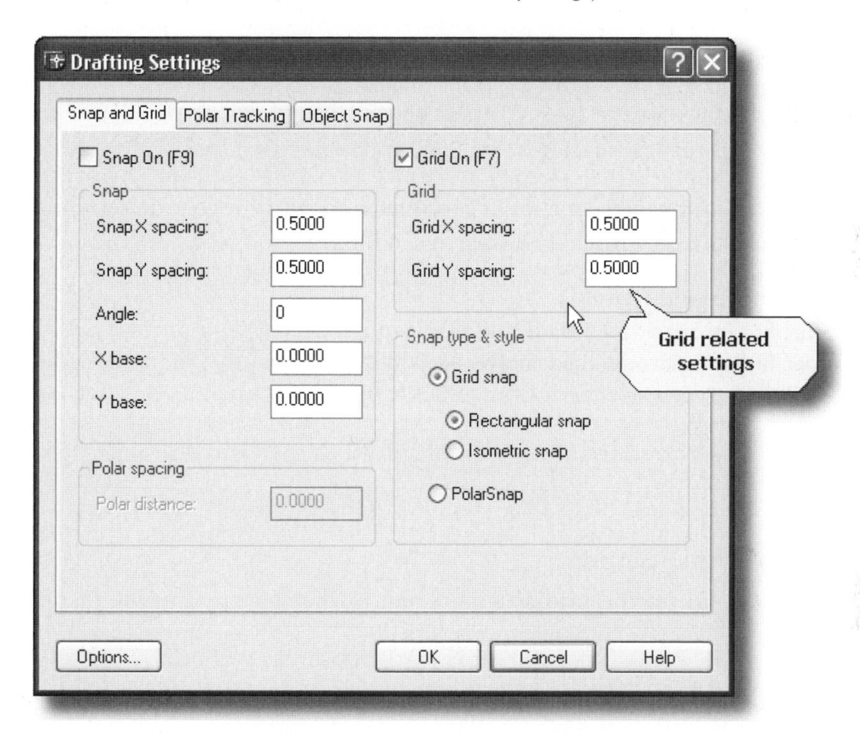

Figure 3–8 *Changing the Grid settings from the Snap and Grid section of the Drafting Settings dialog box*

Aspect

The Aspect option allows you to set the Y Grid Spacing different from the X Grid Spacing. When you select the Aspect option, AutoCAD prompts for the *X* and *Y* values independently. This is handy if the *X* and *Y* modular dimensions of your design are of unequal multiples.

1. From the **Snap and Grid** tab of the Drafting Settings dialog box, you can set the Y spacing different from the X spacing. This is done by entering a value in the **Grid Y Spacing** text box that is different from the value in the **Grid X Spacing** text box. You can also enter **ddrmodes** at the Command: prompt to display the Drafting Settings dialog box, from which the desired tab can be selected if it is not already selected.
2. From the Tools menu, choose Drafting Settings. Then select the **Snap and Grid** tab. See the explanation in Method 1 for setting the Y spacing different from the X spacing.
3. At the Command: prompt:

Command: **grid** (ENTER)
Specify grid spacing(X) or [ON/OFF/Snap/Aspect] <A>: **a** (ENTER)
Specify the horizontal spacing(X) <1.0000>: *(specify X Grid spacing)*
Specify the vertical spacing(Y) <1.0000>: *(specify Y Grid spacing)*

For example, the following sequence of commands shows how you can set a horizontal grid spacing of 0.5 and vertical grid spacing of 0.25.

Command: **grid** (ENTER)
Specify grid spacing(X) or [ON/OFF/Snap/Aspect] <A>: **a** (ENTER)
Specify the horizontal spacing(X) <1.0000>: **0.5** (ENTER)
Specify the vertical spacing(Y) <1.0000>: **0.25** (ENTER)

Applying the Aspect option in this example provides the Grid dot spacing shown in Figure 3–9.

Relationship to Snap Setting

It is often useful to set the grid spacing equal to the snap resolution, or make it a multiple of it.

To specify the grid spacing the same as the snap value, enter **s** in response to the Specify grid spacing(X) or [ON/OFF/Snap/Aspect] <A>: prompt:

Command: **grid** (ENTER)
Specify grid spacing(X) or [ON/OFF/Snap/Aspect] <current>: **s** (ENTER)

To specify the grid spacing as a multiple of the snap value, enter **x** after the value. For example, to set up the grid value as three times the current snap value (snap = 0.5 units), enter **3x** for the prompt, which is the same as setting it to 1.5 units in response to the Specify grid spacing(X) or [ON/OFF/Snap/Aspect] <A>: prompt:

Command: **grid** (ENTER)
Specify grid spacing(X) or [ON/OFF/Snap/Aspect] <current>: **3x** (ENTER)

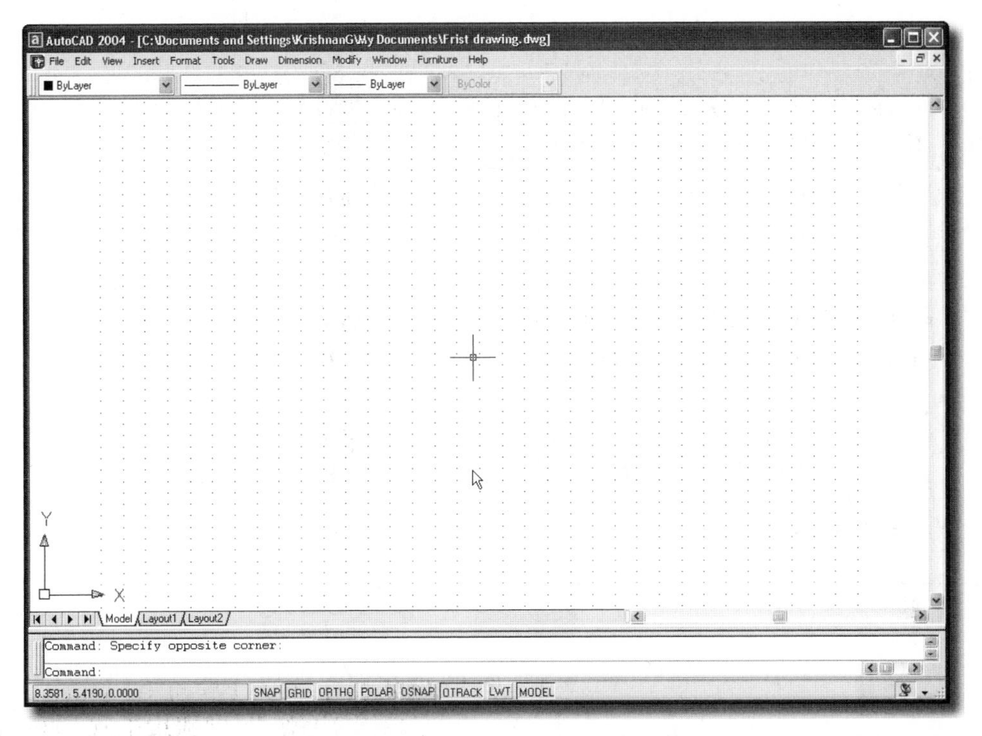

Figure 3–9 *Display after setting the grid aspect to 0.5 for horizontal and 0.25 for vertical spacing*

If the spacing of the visible grid is set too small, AutoCAD displays the following message and does not show the dots on the screen:

Grid too dense to display

To display the grid, invoke the GRID command again and specify a larger spacing.

Note: The relationship between the Grid setting and the Snap setting, when established as described in the previous section, is based on the current Snap setting. If the Snap setting is subsequently changed, the Grid setting does not change accordingly. For example, if the Snap setting is 1.00 and you enter **s** in response to the Specify grid spacing(X) or [ON/OFF/Snap/Aspect] <current>: prompt, the Grid setting becomes 1.00 and remains 1.00 even if the Snap setting is later set to something else. Likewise, if you set the Grid setting to **3x**, it becomes 3.00 and will not change with a subsequent change in the Snap setting.

ORTHO COMMAND

The ORTHO command lets you draw lines and specify point displacements that are parallel to either the *X* or *Y* axis. Lines drawn with the Ortho mode set to ON are therefore either parallel or perpendicular to each other. This mode is helpful when you need to draw lines that are exactly horizontal or vertical. Also, when the Snap Style is set to Isometric, it forces lines to be parallel to one of the three isometric axes.

Setting Ortho ON and OFF

In AutoCAD, four primary methods to set the ORTHO command ON and OFF are available. Of the four explained in this section, the first method (selecting ORTHO on the status bar) is the easiest and most commonly used with the pointing device. The second method (pressing the function key F8) is the most commonly used from the keyboard. The others, while not quite as convenient as the first two, might be convenient in certain situations.

1. On the status bar at the bottom of the screen, when the **ORTHO** button appears to be pressed in, then the Ortho is set to ON. When the **ORTHO** button appears to be out, then the Ortho is set to OFF. To change its setting, simply place the cursor over the **ORTHO** button and press the pointing device pick button.
2. To change the Ortho setting to ON when it is OFF or to OFF when it is ON, simply press the F8 function key.
3. Right-click the **ORTHO** button on the status bar at the bottom of the screen and from the shortcut menu, choose On to turn the Ortho ON or choose Off to turn the Ortho OFF.
4. At the Command: prompt:

Command: **ortho** (ENTER)
Enter mode [ON/OFF] <OFF>: *(enter **on** to set* ORTHO *ON and **off** to set* ORTHO *OFF, or right-click and choose* ON *or* OFF *from the shortcut menu)*

Note: The ORTHO and POLAR TRACKING modes (explained later in this chapter) cannot both be set to ON at the same time. They can both be set to OFF, or either one can be set to ON.

When the Ortho mode is active, you can draw lines and specify displacements only in the horizontal or vertical directions, regardless of the cursor's on-screen position. The direction in which you draw is determined by the change in the *X* value of the cursor movement compared to the change in the cursor's distance to the *Y* axis. AutoCAD allows you to draw horizontally if the distance in the *X* direction is greater than the distance in the *Y* direction; conversely, if the change in the *Y* direction is greater than the change in the *X* direction, then it forces you to draw vertically. The Ortho mode does not affect keyboard entry of points.

OBJECT SNAP

The Object Snap (or Osnap, for short) feature lets you specify points on existing objects in the drawing. For example, if you need to draw a line from an endpoint of an existing line, you can apply the Object Snap mode, called ENDpoint. In response to the "Specify first point:" prompt, place the cursor so that it touches the line nearer the desired endpoint. AutoCAD will lock onto the endpoint of the existing line when you press the pick button of your pointing device. The endpoint becomes the starting point of the new line. This feature is similar to the basic SNAP command, which locks to invisible reference grid points.

You can invoke an Object Snap mode whenever AutoCAD prompts for a point.

Object snap modes can be invoked while executing an AutoCAD command that prompts for a point, such as the LINE, CIRCLE, MOVE, and COPY commands.

Applying Object Snap Modes

Object Snap modes can be applied in either of two ways:

1. From the Object Snap toolbar, choose Object Snap Settings (see Figure 3–10), or on the status bar at the bottom of the screen, right-click on the **OSNAP** button and choose Settings. The Drafting Settings dialog box will be displayed with the **Object Snap** tab selected, as shown in Figure 3–11. An Object Snap mode has been chosen if there is a check in its associated check box. Choose the desired Object Snap mode(s) by placing the cursor over its check box and pressing the pointing device pick button. This makes it possible to use the checked mode(s) any time you are prompted to specify a point. This is referred to as the *Running Osnap* method. You can also enter **ddrmodes** at the Command: prompt to display the Drafting Settings dialog box, from which the **Object Snap** tab can be selected if it is not already selected. Running Osnap can be set to ON and OFF in the same manner as the GRID command. The most common method is to use the button on the status bar at the bottom of the screen. When the **OSNAP** button appears to be pressed in, then the Object Snap is set to ON. When the **OSNAP** button appears to be out, then the Object Snap is set to OFF. To change its setting, simply place the cursor over the **OSNAP** button and press the pointing device pick button. It is just as easy to simply press the function key F3, to change the Osnap setting to ON when it is OFF or to OFF when it is ON.

Figure 3–10 *Choosing Object Snap Settings from the Object Snap toolbar*

2. When prompted to specify a point, enter the first three letters of the name of the desired Object Snap mode, or select it from the Object Snap toolbar before specifying the point. This is a *one-time-only* Osnap method. This will override any other Running Osnap mode for this one point selection only.

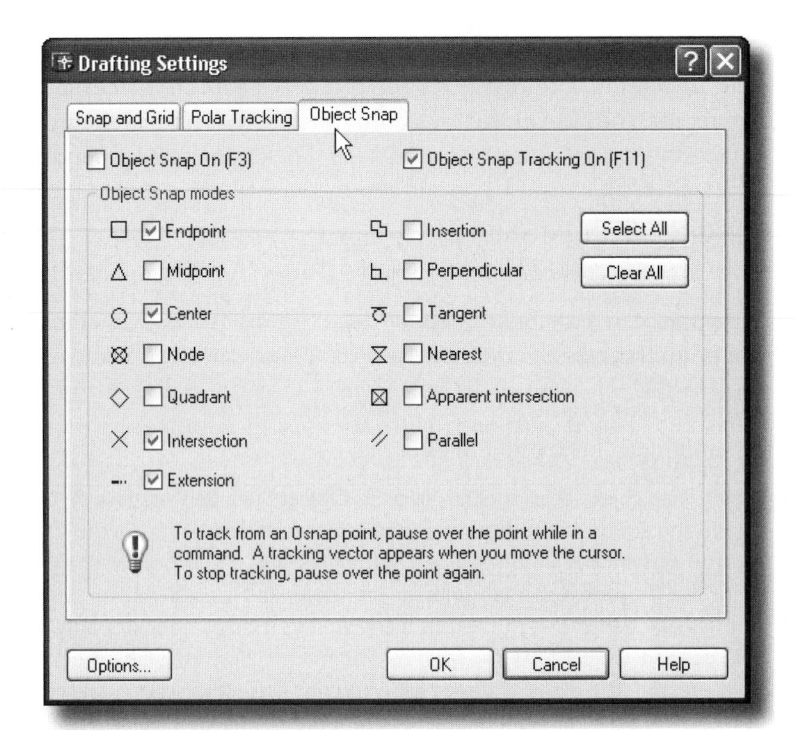

Figure 3–11 *Drafting Settings dialog box displaying the Object Snap tab*

For example, to draw a line from an end point of an existing line, invoke the LINE command, and AutoCAD prompts:

Command: **line** (ENTER)
Specify first point: **end** *(enter the first three letters of the Osnap mode endpoint)*
of *(move the cursor near the desired end of a line and press the pick button)*

The first point of the new line is drawn from the end point of the selected object.

Osnap Markers and Tooltips

Whenever one or more Object Snap modes are activated and you move the cursor target box over a snap point, AutoCAD displays a geometric shape (Marker) and Tooltip. By displaying a Marker on the snap points with a Tooltip, you can see the point that will be selected and the Object Snap mode in effect. AutoCAD displays the Marker depending on the Object Snap mode selected. In the **Object Snap** tab of the Drafting Settings dialog box, each Marker is displayed next to the name of its associated Object Snap mode.

From the **Object Snap** tab of the Drafting Settings dialog box, choose the **Options…** button. AutoCAD displays the Options dialog box with the **Drafting** tab selected. This is where the settings for displaying the Markers and Tooltips can be changed by checking or clearing the appropriate check boxes in the **AutoSnap Settings** section (see Figure 3–12).

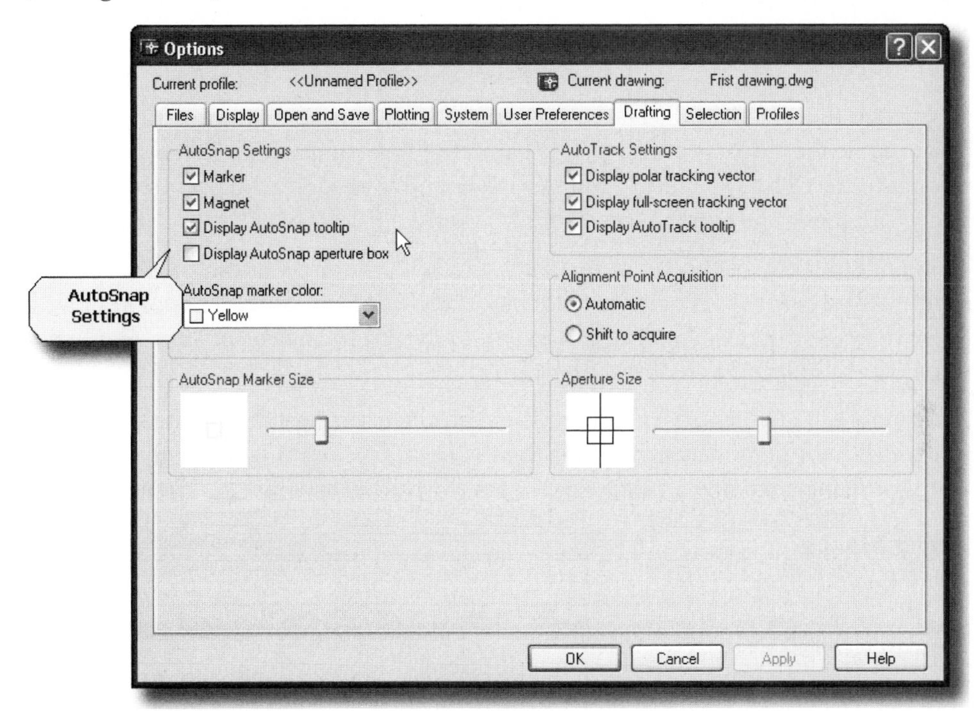

Figure 3–12 *Options dialog box with the Drafting tab selected*

In the **AutoSnap Settings** section:

The Marker setting is set to ON when the **Marker** check box is checked, and is not displayed when it is not checked. The Marker is the geometric shape that is displayed when the cursor moves over a snap point.

The Magnet setting is displayed when the **Magnet** check box is checked, and is set to OFF when it is not checked. The Magnet locks the cursor onto the snap point.

The Tooltip setting is set to ON when the **Display AutoSnap tooltip** check box is checked, and is set to OFF when it is not checked. The Tooltip is the flag that displays the name of the Object Snap mode.

The Aperture setting is set to ON when the **Display AutoSnap aperture box** check box is checked, and is set to OFF when it is not checked. The Aperture is the square target box that is displayed when AutoCAD is prompting to specify a point.

To change the ON/OFF setting of the Marker, Magnet, Tooltip, or Aperture, simply place the cursor over its check box and press the pointing device pick button. To change the color of the Marker, select the desired color from the **AutoSnap marker color** list box.

To change the size of the Marker, press and hold the pick button of the pointing device while the cursor is over the slide bar in the **Autosnap Marker Size** section and move it to the right to make the Marker larger and to the left to make it smaller. The Image tile shows the current size of the Marker.

To change the size of the Aperture, press and hold the pick button of the pointing device while the cursor is over the slide bar in the **Aperture Size** section and move it to the right to make the Aperture larger and to the left to make it smaller. The Image tile shows the current size of the Aperture.

After making the necessary changes to the AutoSnap settings, choose the **OK** button to close the Options dialog box and then choose **OK** again to close the Drafting Settings dialog box.

Osnap Modes

The following section explains the available Object Snap modes, with examples:

Object Snap—ENDpoint The ENDpoint mode allows you to snap to the closest endpoint of a line, arc, elliptical arc, multiline, polyline segment, spline, region or ray, or to the closest corner of a trace, solid, or 3Dface.

During an appropriate AutoCAD command, invoke the Object Snap—ENDpoint mode:

Object Snap toolbar	Select Snap to Endpoint (see Figure 3–13)
Command: prompt	Enter **end** and press ENTER whenever an AutoCAD command: prompt requests a point to be specified

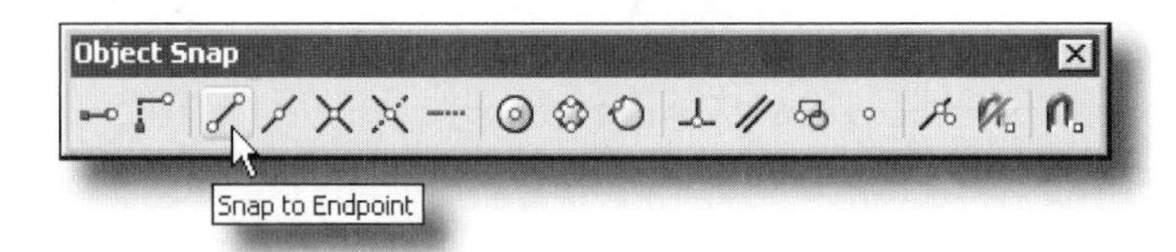

Figure 3–13 *Invoke the Object Snap ENDpoint mode from the Object Snap toolbar*

For example, to connect a line to the endpoint of an existing line, as shown in Figure 3–14, the following command sequence is used:

Command: **line** (ENTER)
Specify first point: **end** *(enter the first three letters of the Osnap mode Endpoint)*
of *(move the aperture cursor near the end of Line A and specify it)*
Specify next point or [Undo]: *(specify other end of line)*
Specify next point or [Undo]: (ENTER)

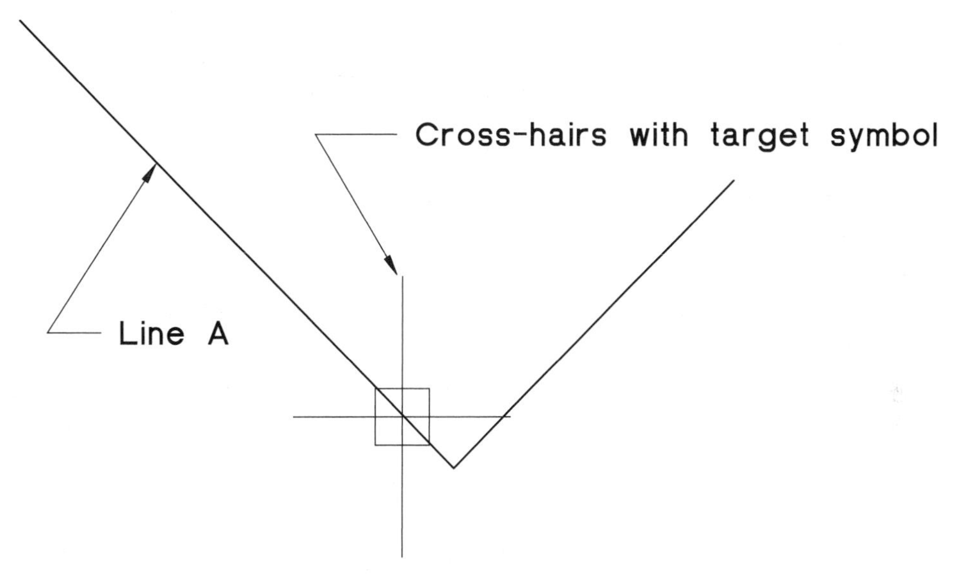

Figure 3–14 *Connecting a line to the endpoint of another line using the ENDpoint Object Snap mode*

Object Snap—MIDpoint The MIDpoint mode allows you to snap to the midpoint of a line, arc, elliptical arc, broken ellipse, multiline, polyline segment, xline, solid, or spline.

During an appropriate AutoCAD command, invoke the Object Snap—MIDpoint mode:

Object Snap toolbar	Select Snap to Midpoint (see Figure 3–15)
Command: prompt	Enter **mid** and press ENTER whenever an AutoCAD command: prompt requests a point to be specified

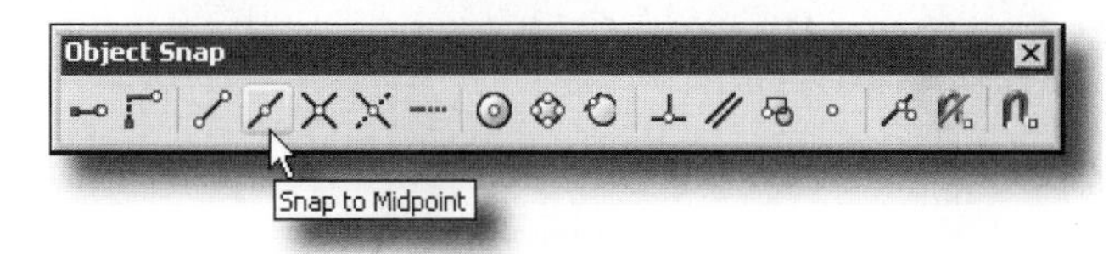

Figure 3–15 *Invoking the Object Snap—MIDpoint mode from the Object Snap toolbar*

For example, to connect a line to the midpoint of an existing line, as shown in Figure 3–16, the following command sequence is used:

Command: **line** (ENTER)
Specify first point: **mid** *(enter the first three letters of the Osnap mode Midpoint)*
of *(move the aperture cursor to anywhere on Line A and specify it)*
Specify next point or [Undo]: *(specify other end of line)*
Specify next point or [Undo]: (ENTER)

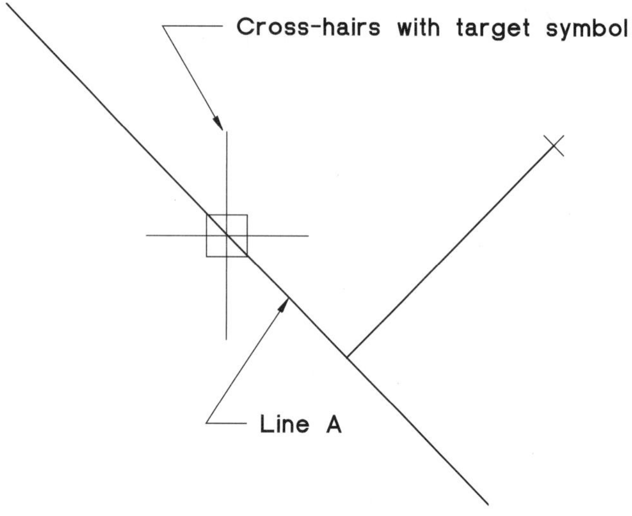

Figure 3–16 *Connecting a line to another line's midpoint using the MIDpoint Object Snap mode*

Object Snap—CENter The CENter mode allows you to snap to the center of an arc, circle, ellipse, or elliptical arc.

During an appropriate AutoCAD command, invoke the Object Snap—CENter mode:

Object Snap toolbar	Select Snap to Center (see Figure 3–17)
Command: prompt	Enter **cen** and press ENTER whenever an AutoCAD command: prompt requests a point to be specified

Figure 3–17 *Invoking the Object Snap—CENter mode from the Object Snap toolbar*

For example, to draw a line to the center of a circle, as shown in Figure 3–18, the following command sequence is used:

Command: **line** (ENTER)
Specify first point: *(specify point P)*
Specify next point or [Undo]: **cen** *(enter the first three letters of the Osnap mode Center)*
of *(specify point P1 on circle)*
Specify next point or [Undo]: (ENTER)

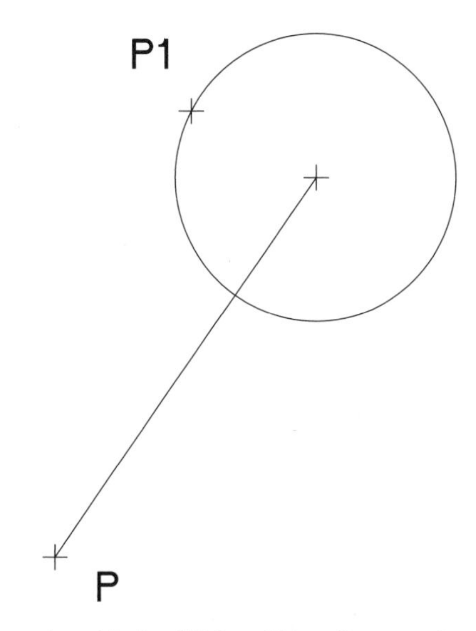

Figure 3–18 *Selecting a point with the CENter Object Snap mode*

Object Snap—NODe The NODe mode allows you to snap to a point object.

During an appropriate AutoCAD command, invoke the Object Snap—NODe mode:

Object Snap toolbar	Select Snap to Node (see Figure 3–19)
Command: prompt	Enter **nod** and press ENTER whenever an AutoCAD command: prompt requests a point to be specified

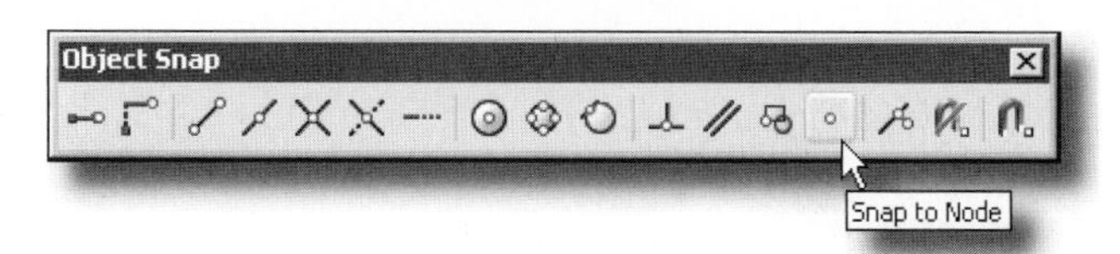

Figure 3–19 *Invoking the Object Snap NODe mode from the Object Snap toolbar*

Object Snap—QUAdrant The QUAdrant mode allows you to snap to one of the quadrant points of a circle, arc, ellipse, or elliptical arc. The quadrant points are located at 0 degrees, 90 degrees, 180 degrees, and 270 degrees from the center of the circle or arc, as shown in Figure 3–20. The quadrant points are determined by the zero degree direction of the current coordinate system.

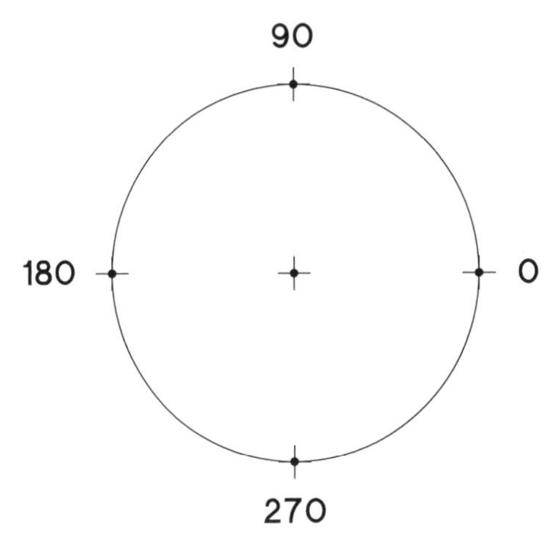

Figure 3–20 *The quadrant points recognized by the QUAdrant Object Snap mode*

During an appropriate AutoCAD command, invoke the Object Snap—QUAdrant mode:

Object Snap toolbar	Select Snap to Quadrant (see Figure 3–21)
Command: prompt	Enter **qua** and press ENTER whenever an AutoCAD command: prompt requests a point to be specified

Figure 3–21 *Invoking the Object Snap—QUAdrant mode from the Object Snap toolbar*

In Figure 3–22 a line is drawn to a quadrant as follows:

Command: **line** (ENTER)
Specify first point: *(specify point A)*
Specify next point or [Undo]: **qua** *(enter the first three letters of the Osnap mode Quadrant)*
of *(specify point A1 on circle)*
Specify next point or [Undo]: (ENTER)

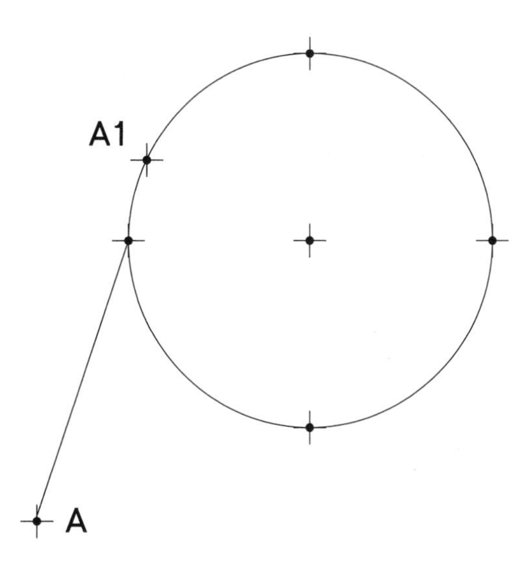

Figure 3–22 *Drawing a line to a circle's quadrant point (Method 1)*

A line can also be drawn to a quadrant as follows (see Figure 3–23):

Command: **line** (ENTER)
Specify first point: *(specify point A)*
Specify next point or [Undo]: **qua** *(enter the first three letters of the Osnap mode Quadrant)*
of *(specify point A2 on circle)*
Specify next point or [Undo]: (ENTER)

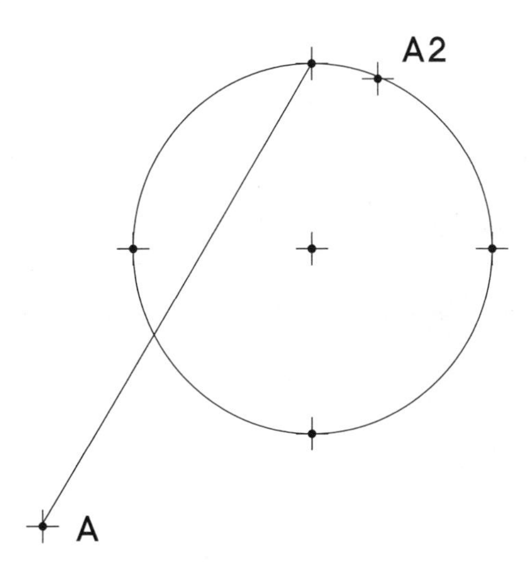

Figure 3–23 *Drawing a line to a circle's quadrant (Method 2)*

Note: Special precautions should be taken when attempting to select circles or arcs in blocks or ellipses that are rotated at an angle that is not a multiple of 90 degrees. When a circle or an arc in a block is rotated, the point of that QUAdrant Object Snap mode is also rotated. But when a circle/arc not in a block is rotated, the QUAdrant Object Snap points stay at the 0, 90, 180, or 270 degree points.

Object Snap—INTersection The INTersection mode allows you to snap to the intersection of two objects that can include arcs, circles, ellipses, elliptical arcs, lines, multilines, polylines, rays, splines, or xlines. An example with valid intersection points is shown in Figure 3–24.

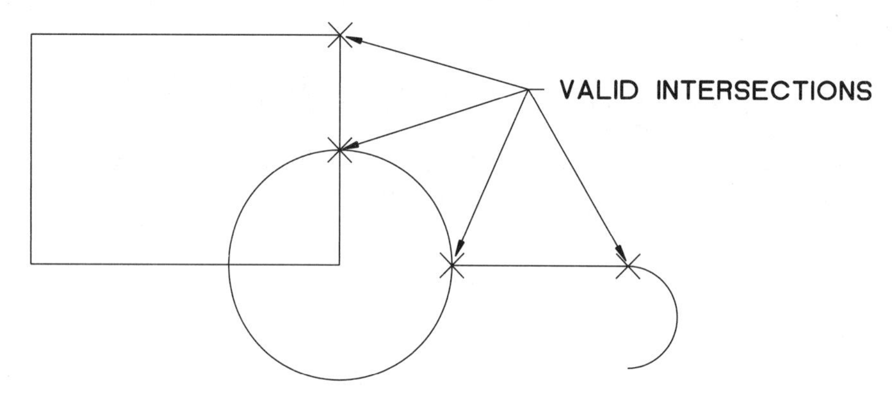

Figure 3–24 *Valid intersections that can be selected using Object Snap—INTersection mode*

During an appropriate AutoCAD command, invoke the Object Snap—INTersection mode:

Object Snap toolbar	Select Snap to Intersection (see Figure 3–25)
Command: prompt	Enter **int** and press ENTER whenever an AutoCAD command: prompt requests a point to be specified

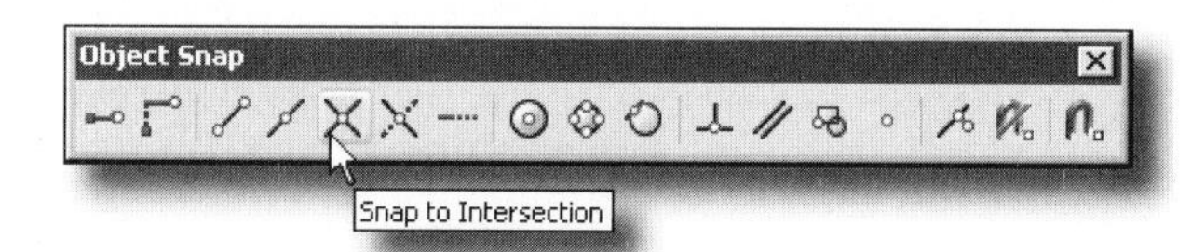

Figure 3–25 *Invoking the Object Snap—INTersection mode from the Object Snap toolbar*

AutoCAD prompts you to select two objects to establish the intersection point.

Object Snap—EXTension The EXTension mode allows you to snap to a point that is the extension of a line or an arc.

During an appropriate AutoCAD command, invoke the Object Snap - EXTension mode:

Object Snap toolbar	Select Snap to Extension (see Figure 3–26)
Command: prompt	Enter **ext** and press ENTER whenever an AutoCAD command: prompt requests a point to be specified

This example is to show how to apply the EXTension Object Snap mode to draw LINE A, as shown in Figures 3–27.

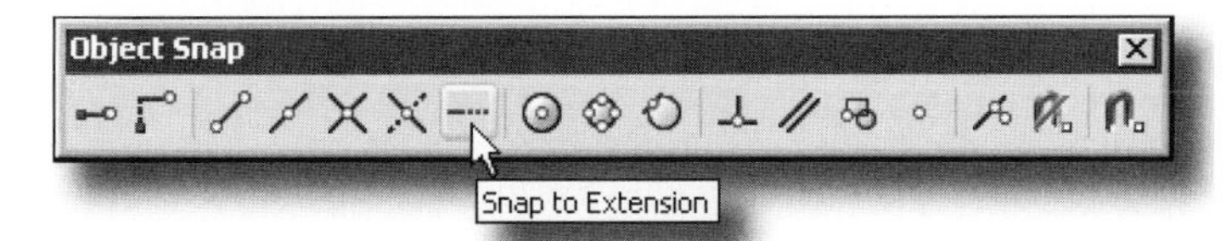

Figure 3–26 *Invoking the Object Snap EXTension mode from the Object Snap toolbar*

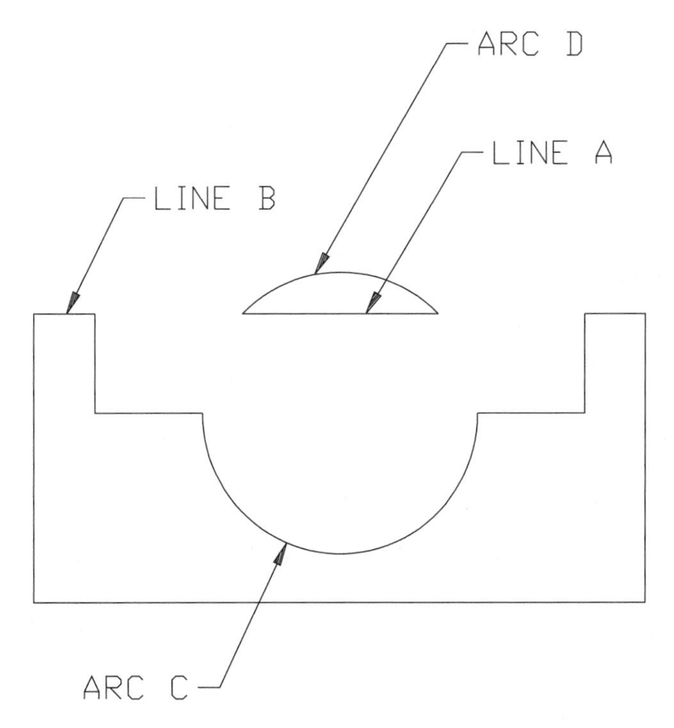

Figure 3–27 *LINE A is drawn using extensions of LINE B and ARC C*

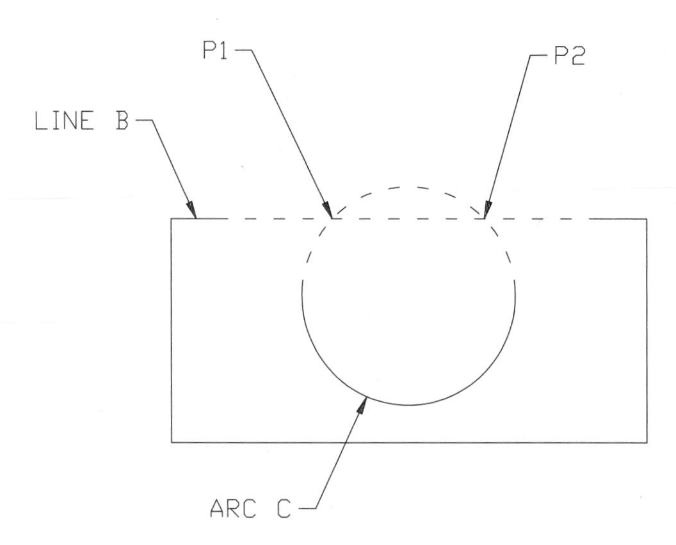

Figure 3–28 *LINE B and ARC C extensions shown with hidden lines*

LINE A is an extension of LINE B between points P1 and P2 (see Figure 3-28), where LINE B intersects the circle of which ARC C is a part, as shown in Figure 3–27. To draw LINE A, invoke the LINE command, and AutoCAD prompts:

Command: **line** (ENTER)
Specify first point: **ext** *(enter the first three letters of the Osnap mode Extension)*
of *(move the aperture cursor to the right end of LINE B and without pressing the pick button move it off of the line to the right)*

As you move the cursor away from LINE B, a dashed line follows as long as the cursor is near the extension of LINE B, as shown in Figure 3–29.

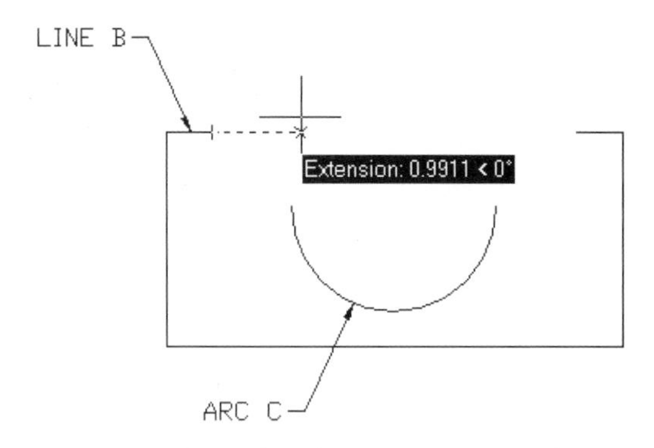

Figure 3–29 *A dashed line follows the cursor indicating the extension of LINE B*

Next, move the cursor to the left end of ARC C and then upward away from it. Another dashed arc follows until the cursor is near the extension of LINE B, as shown in Figure 3–30. At this time you can press the pick button on the pointing device and point P1 is established as the start point of LINE A.

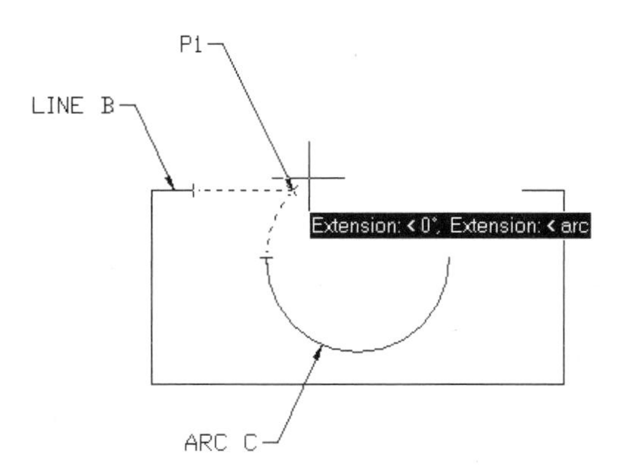

Figure 3–30 *A dashed arc follows the cursor indicating the extension of ARC C to the dashed extension of LINE B, at which point P1 is specified*

After pressing the pick button to specify point P1, move the cursor to LINE B again and then move the cursor to the right end of ARC C and then upward away from it. Another dashed arc follows until the cursor is near the extension of LINE B, as shown in Figure 3–31.

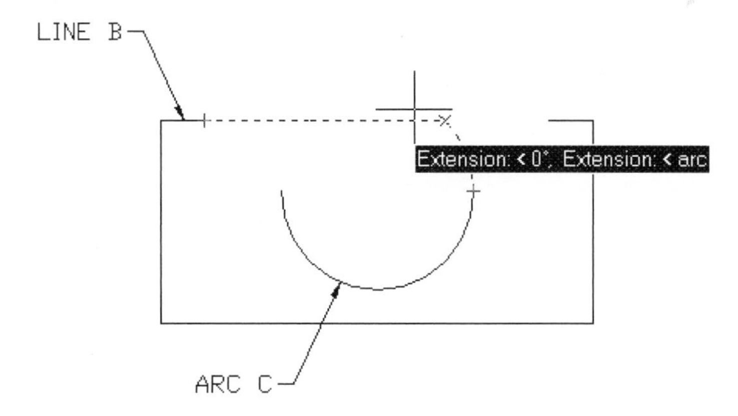

Figure 3–31 *A dashed arc follows the cursor indicating the extension of ARC C to the dashed extension of LINE B, at which point P2 is specified*

At this time, press the pick button on the pointing device and point P2, the end point of line A, is established, as shown in Figure 3–32.

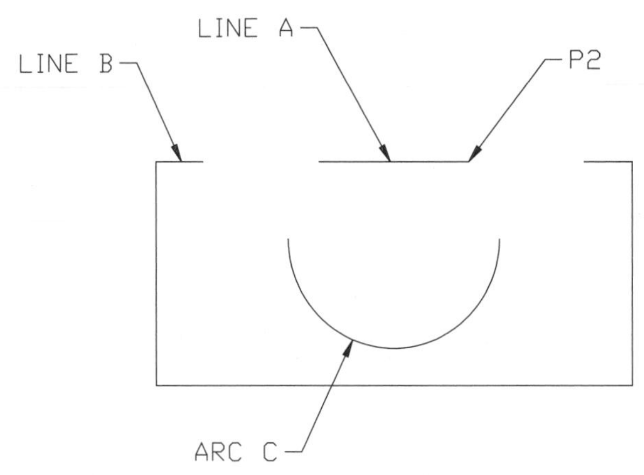

Figure 3–32 *LINE A is drawn*

After exiting the LINE command, ARC D is drawn with the Start, Center, End method (SCE). The start point and the end point of the arc is specified by using the Endpoint Object Snap mode and selecting the endpoints of LINE A. The center of the arc is specified by using the Center Object Snap mode and selecting ARC C.

Object Snap—INSertion The INSertion mode allows you to snap to the insertion point of a block, text string, attribute, or shape.

During an appropriate AutoCAD command, invoke the Object Snap—INSertion mode:

Object Snap toolbar	Select Snap to Insert (see Figure 3–33)
Command: prompt	Enter **ins** and press ENTER whenever an AutoCAD command: prompt requests a point to be specified

Figure 3–33 *Invoking the Object Snap INSert mode from the Object Snap toolbar*

Object Snap—PERpendicular The PERpendicular mode allows you to snap to a point perpendicular to a line, arc, circle, elliptical arc, multiline, polyline, ray, solid, spline, or xline. Deferred PERpendicular snap mode is automatically turned on when more than one perpendicular snap is required by the object being drawn.

During an appropriate AutoCAD command, invoke the Object Snap—PERpendicular mode:

Object Snap toolbar	Select Snap to Perpendicular (see Figure 3–34)
Command: prompt	Enter **per** and press ENTER whenever an AutoCAD command: prompt requests a point to be specified

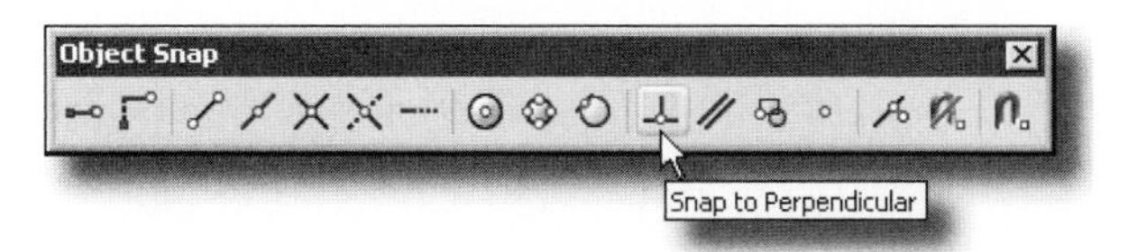

Figure 3–34 *Invoking the Object Snap—PERpendicular mode from the Object Snap toolbar*

The following command sequence demonstrates how you can draw lines using the PERpendicular Object Snap mode.

From inside the circle:

Command: **line** (ENTER)
Specify first point: *(specify point A)*
Specify next point or [Undo]: **per** *(enter the first three letters of the Osnap mode PERpendicular)*
to *(specify near-side point AN, as shown in Figure 3–35)*
Specify next point or [Undo]: (ENTER)

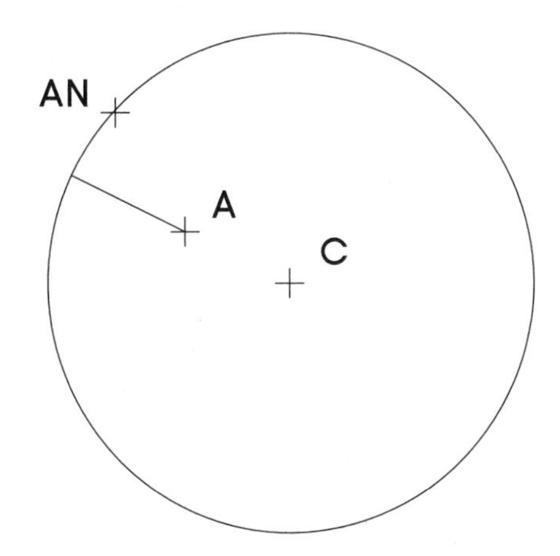

Figure 3–35 *Specifying a line's endpoint perpendicular to a circle (from inside of the circle to the near side of the circle) with the PERpendicular Object Snap mode*

Command: **line** (ENTER)
Specify first point: *(specify point A)*
Specify next point or [Undo]: **per** *(enter the first three letters of the Osnap mode PERpendicular)*
to *(specify far-side point AF, as shown in Figure 3–36)*

Specify next point or [Undo]: (ENTER)

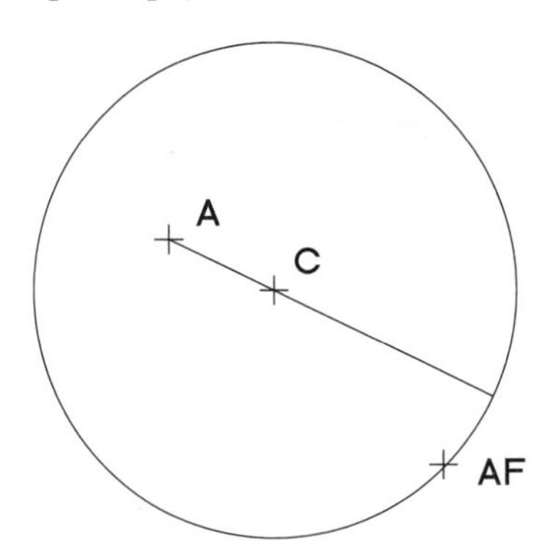

Figure 3–36 *Specifying a line's endpoint perpendicular to a circle (from inside of the circle to the far side of the circle) with the PERpendicular Object Snap mode*

From outside the circle:

Command: **line** (ENTER)
Specify first point: *(specify point B)*
Specify next point or [Undo]: **per** *(enter the first three letters of the Osnap mode PERpendicular)*
to *(specify near-side point BN, as shown in Figure 3–37)*

Specify next point or [Undo]: (ENTER)

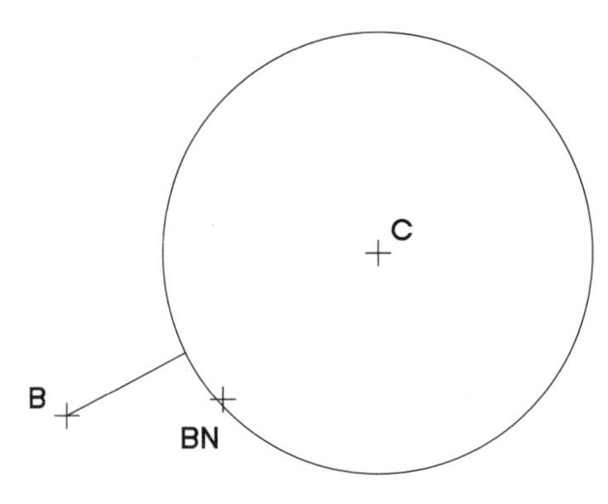

Figure 3–37 *Specifying a line's endpoint perpendicular to a circle (from outside of the circle to the near side of the circle) with the PERpendicular Object Snap mode*

Command: **line** (ENTER)
Specify first point: *(specify point B)*
Specify next point or [Undo]: **per** *(enter the first three letters of the Osnap mode PERpendicular)*
to *(specify far-side point BF, as shown in Figure 3–38)*
Specify next point or [Undo]: (ENTER)

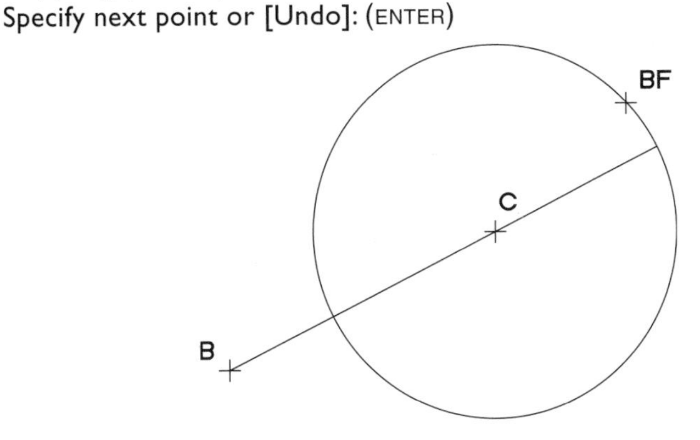

Figure 3–38 *Specifying a line's endpoint perpendicular to a circle (from outside of the circle to the far side of the circle) with the PERpendicular Object Snap mode*

When drawing a line perpendicular to another line, the point that AutoCAD establishes can be off the line selected (in response to the "perpendicular to" prompt) and the new line will still be drawn to that point. In Figure 3–39 lines from both A and B can be drawn perpendicular to line L.

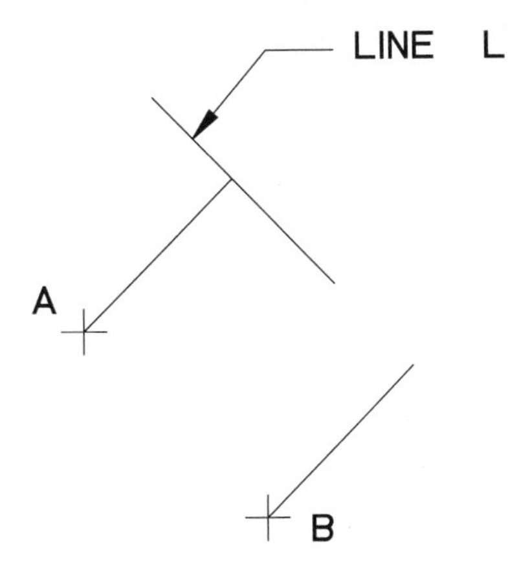

Figure 3–39 *Using the PERpendicular Osnap mode, a line is drawn perpendicular to another line*

Applying the PERpendicular Osnap mode to an arc works in a manner similar to a circle. Unlike drawing a perpendicular to a line, the point established must be on the arc.

In Figure 3–40 a line is drawn perpendicular to an arc from A but can not be drawn from B.

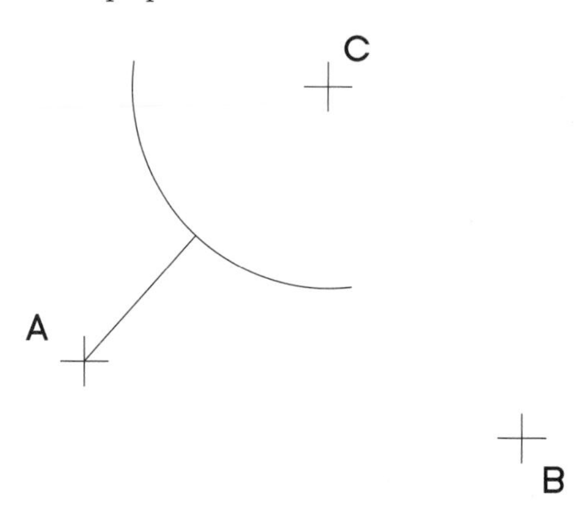

Figure 3–40 *Using the PERpendicular Osnap mode, a line is drawn perpendicular to an arc*

Object Snap–TANgent The TANgent mode allows you to snap to the tangent of an arc, circle, ellipse, or elliptical arc. Deferred TANgent snap mode is automatically turned on when more than one tangent is required by the object being drawn.

During an appropriate AutoCAD command, invoke the Object Snap—TANgent mode:

Object Snap toolbar	Select Snap to Tangent (see Figure 3–41)
Command: prompt	Enter **tan** and press ENTER whenever an AutoCAD command: prompt requests a point to be specified

Figure 3–41 *Invoking the Object Snap—TANgent mode from the Object Snap toolbar*

The following command sequence draws a line from point A tangent to a point on the circle, as shown in Figure 3–42.

Command: **line** (ENTER)
Specify first point: *(specify point A)*
Specify next point or [Undo]: **tan** *(enter the first three letters of the Osnap mode Tangent)*
to *(specify point AL toward left semicircle)*
Specify next point or [Undo]: (ENTER)

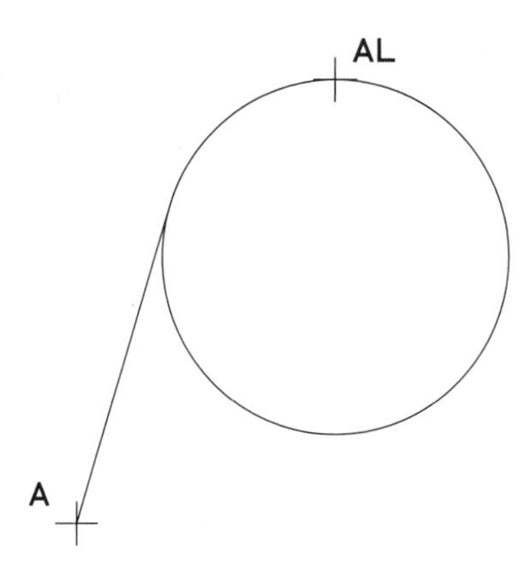

Figure 3–42 *Using the TANgent Osnap mode, a line is drawn from a point outside a circle tangent to a point on the left side of the circle*

A line can also be drawn from point A tangent to a point on the circle, as shown in Figure 3–43.

Command: **line** (ENTER)
Specify first point: *(specify point A)*
Specify next point or [Undo]: **tan** *(enter the first three letters of the Osnap mode Tangent)*
to *(specify point AR toward right semicircle)*
Specify next point or [Undo]: (ENTER)

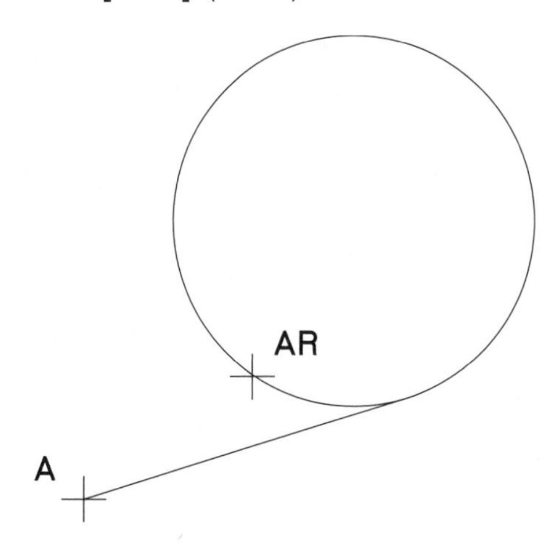

Figure 3–43 *Using the TANgent Osnap mode, a line is drawn from a point outside a circle tangent to a point on the right side of the circle*

With the TANgent Osnap mode you can also select an arc. Like the PERpendicular Osnap mode, the tangent point must be on the arc selected (see Figure 3–44).

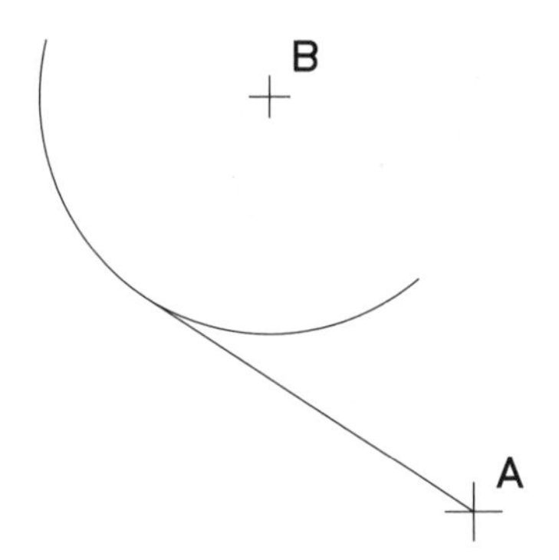

Figure 3–44 *Using the TANgent Osnap mode, a line is drawn tangent to a point on an arc*

Object Snap—NEArest The NEArest mode lets you select any object (except text and shape) in response to a prompt for a point, and AutoCAD snaps to the point on that object nearest the cursor.

During an appropriate AutoCAD command, invoke the Object Snap—NEArest mode:

Object Snap toolbar	Select Snap to Nearest (see Figure 3–45)
Command: prompt	Enter **nea** and press ENTER whenever an AutoCAD command: prompt requests a point to be specified

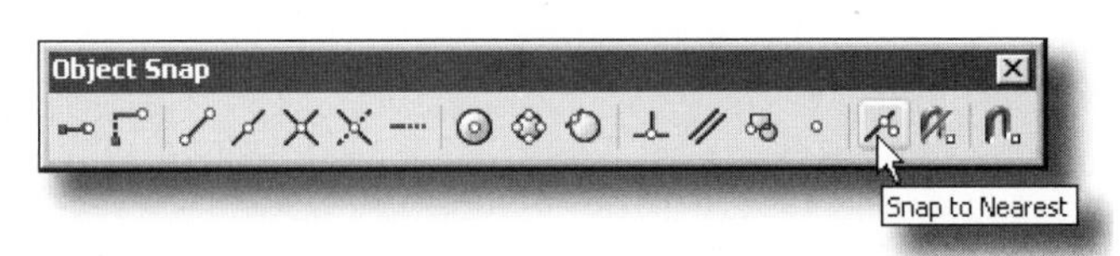

Figure 3–45 *Invoking the Object Snap NEArest mode from the Object Snap toolbar*

Object Snap—APParent intersection The APParent intersection mode allows you to snap to the apparent intersection of two objects which can include an arc, circle, ellipse, elliptical arc, line, multiline, polyline, ray, spline, or xline. These objects may or may not actually intersect but would intersect if either or both objects were extended.

During an appropriate AutoCAD command, invoke the Object Snap—APParent intersection mode:

Object Snap toolbar	Select Snap to Apparent Intersect (see Figure 3–46)
Command: prompt	Enter **app** and press ENTER whenever an AutoCAD command: prompt requests a point to be specified

Figure 3–46 *Invoking the Object Snap—APParent intersection mode from the Object Snap toolbar*

AutoCAD prompts you to select two objects to establish the apparent intersection point.

Note: The INTersection and APParent Intersection modes should not be in effect at the same time.

Object Snap—PARallel The PARallel mode allows you to draw a line that is parallel to another object. Once the first point of the line has been specified (and the Parallel Object Snap mode is selected), move the cursor over the object to which you wish to make the new line parallel. Then move the cursor near a line from the first point that is parallel to the object selected and a construction line will appear. While the construction line is visible, specify a point, and the new line will be parallel to the selected object.

During an appropriate AutoCAD command, invoke the Object Snap - PARallel mode:

Object Snap toolbar	Select Snap to Parallel (see Figure 3–47)
Command: prompt	Enter **par** and press ENTER whenever an AutoCAD command: prompt requests a point to be specified

Figure 3–47 *Invoking the Object Snap PARallel mode from the Object Snap toolbar*

TRACKING OPTION

Tracking, or moving through nonselected point(s) to a selected point, could be called a command "enhancer." It can be used whenever a command prompts for a point. If the desired point can best be specified relative to some known point(s), you can "make tracks" to the desired point by invoking the Tracking option and then specifying one or more points relative to previous point(s) "on the way to" the actual point that the command is prompting for. These intermediate tracking points are not necessarily associated with the object being created or modified by the command. The primary significance of tracking points is that they are used to establish a path to the point you wish to specify as the response to the command prompt. Some of the objects in the partial plan shown in Figure 3–48 can be drawn more easily by means of Tracking.

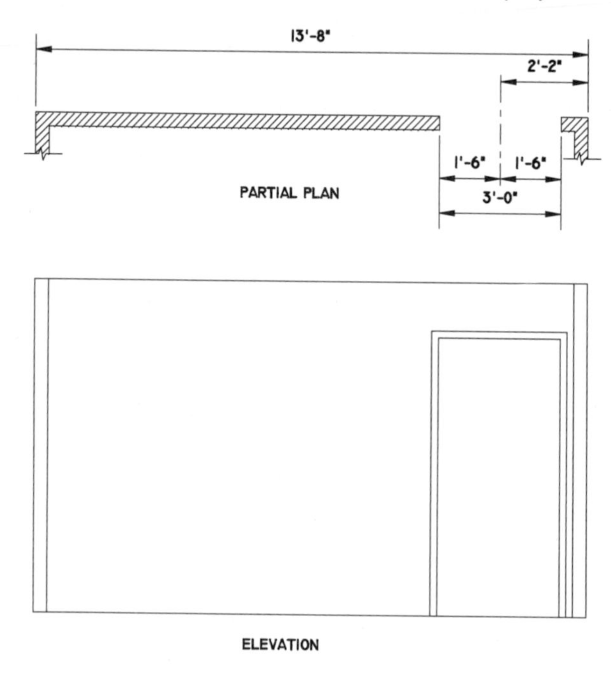

Figure 3–48 *Example of a partial plan to demonstrate the Tracking option*

The following example will use Tracking to draw lines A and B in the Partial Plan in Figure 3–49, leaving the 3'-0" door opening in the correct place. By means of Tracking, we can draw the lines with the given dimension information without having to calculate the missing information.

In Figure 3–50, line A from TK1/SP1 (tracking point 1 and starting point 1) to EP1 (ending point 1) and line B from SP2 (starting point 2) to TK2/EP2 (tracking point 2 and ending point 2) can be drawn by using the TRACKING command enhancer. Invoke the LINE command and AutoCAD prompts:

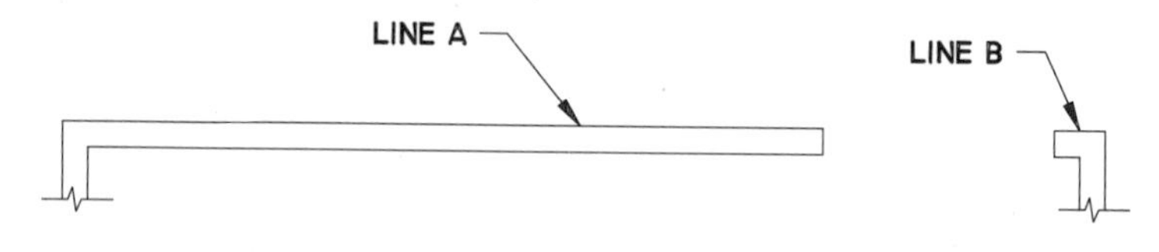

Figure 3–49 *Lines A and B to be drawn with the help of the Tracking option*

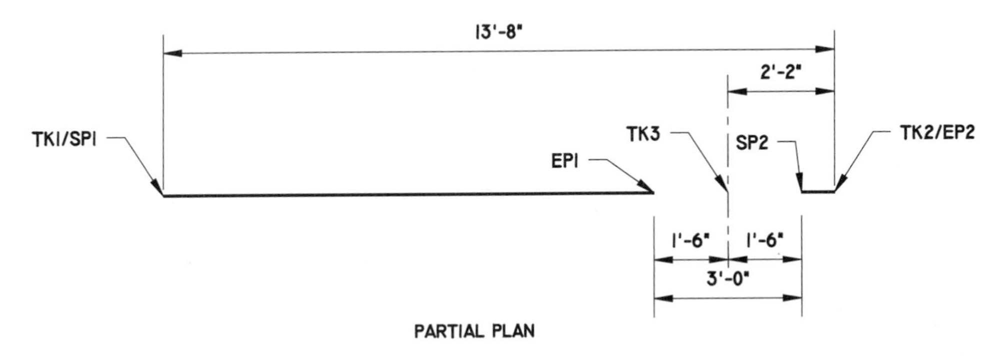

Figure 3–50 *Various points to be drawn with the help of the Tracking option*

Command: **line** (ENTER)
Specify first point: **0,12'** (ENTER) *(specify point TK1/SP1)*
Specify next point or [Undo]: **tk** (ENTER) *(invoke the Tracking feature)*
First tracking point: *(specify point SP1/TK1 again as the first tracking point)*
Next point (Press ENTER to end tracking): **@13'8,0** (ENTER) *(locates the second tracking point, TK2, as shown in Figure 3–50)*
Next point (Press ENTER to end tracking): **@2'2<180** (ENTER) *(locates the third tracking point, TK3, as shown in Figure 3–50)*
Next point (Press ENTER to end tracking): **@1'6<180** (ENTER) *(locates the fourth tracking point, EP1, as shown in Figure 3–50)*
Next point (Press ENTER to end tracking): *(press ENTER to exit Tracking; by this you are designating the point to which you have "made tracks" as the response to the prompt that was in effect when you entered Tracking)*
Specify next point or [Undo]: *(press ENTER to exit the LINE command)*

Once the Tracking option is invoked, you establish a path to EP1 by specifying the initial tracking point, TK1, and then each subsequent point relative to the previous point, that is, TK2 relative to TK1, TK3 relative to TK2, and EP1 relative to TK3. The first track in this case is specified by the Relative Rectangular method, and the next two are Relative Polar. Also, because the tracking points are all on one horizontal line, you could use the direct distance feature (explained in the next section) by placing the cursor in the correct direction (with ORTHO set to ON) and entering the distance from the keyboard.

Note: If you knew the coordinates of one of the intermediate tracking points, then it probably should be the initial tracking point. The idea behind Tracking is to establish a point by means of a path from and through other points. Thus, the shortest path is the best. If the coordinates of TK2 were known, or if you could specify it by some other method, it could become the initial tracking point. Keep this in mind as you learn to use Object Snap. You don't necessarily need to know the coordinates if you can use Object Snap to select a point from which a tracking path could be specified.

The line from SP2 to TK2/EP2 can be started and ended in a similar manner as the line from TK1/SP1 to EP1 was started, with some minor modifications. The sequence (which involves using Tracking twice) is as follows:

> Command: **line** (ENTER) *(invoke the* LINE *command)*
> Specify first point: **tk** (ENTER) *(invoke the Tracking option)*
> First tracking point: **0,12'** (ENTER) *(specify point TK1/SP1 as the first tracking point)*
> Next point (Press ENTER to end tracking): **@13'8,0** (ENTER) *(locates the second tracking point, TK2, as shown in Figure 3–50)*
> Next point (Press ENTER to end tracking): **@2'2<180** (ENTER) *(locates the third tracking point, TK3, as shown in Figure 3–50)*
> Next point (Press ENTER to end tracking): **@1'6<0** (ENTER) *(locates the fourth tracking point, SP2, as shown in Figure 3–50)*
> Next point (Press ENTER to end tracking): *(press* ENTER *to exit Tracking; AutoCAD establishes point SP2)*
> Specify next point or [Undo]: **tk** (ENTER) *(invoke the Tracking option again)*
> First tracking point: *(specify point TK1/SP1 as the first tracking point)*
> Next point (Press ENTER to end tracking): **@13'8,0** (ENTER) *(locates the second tracking point, TK2/EP2, as shown in Figure 3–50)*
> Next point (Press ENTER to end tracking): *(press* ENTER *to exit Tracking, AutoCAD establishes point EP2)*
> Specify next point or [Undo]: *(press* ENTER *to terminate the* LINE *command)*

This example shows an application of the Tracking option in which the points were established in reference to some known points.

DIRECT DISTANCE OPTION

The Direct Distance option for specifying a point relative to another point can be used with a command like LINE to permit a variation of the Relative Coordinates mode. In the case of the Direct Distance option, the distance is keyed in and the direction is determined by the current location of the cursor. This option is very useful when you know the exact distance but specifying the exact angle is not as easy as placing the cursor on a point that is at the exact angle desired.

Figure 3–51 shows a shape that can be drawn more easily by using the Direct Distance option along with setting the Snap and Ortho modes to ON and OFF at the appropriate times. Points A, B, and C are on the Snap grid (*X,Y* coordinates 3,3 for A, 3,6 for B, and 9,6 for C). By setting the Snap mode to ON with the value set to 1 or perhaps 0.5, the cursor can be placed on the required points to draw lines A-B and B-C. After drawing lines A-B and B-C (with Ortho set to either ON or OFF), the cursor should be placed on point A (with Ortho set to OFF).

If you have exited the LINE command after drawing line B-C, you must invoke the LINE command and specify C as the first point before moving the cursor to A. The rubber-band line indicates that the next line is drawn from C to A, as shown in Figure 3–52. However, you wish to draw a line only 2 units long but in the same direction as a line from C to A. With the cursor placed on A, enter 2 and press ENTER.

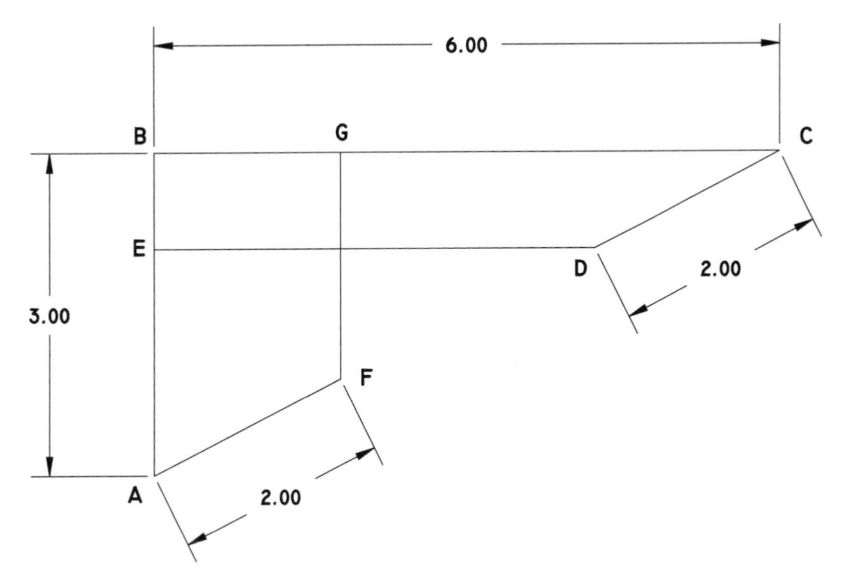

Figure 3–51 *Example of a direct distance application*

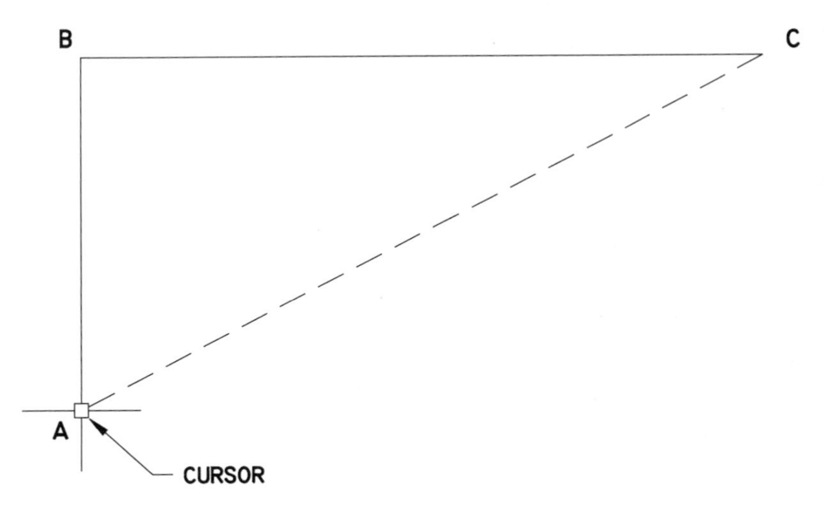

Figure 3–52 *Drawing a line from C by means of the Direct Distance option*

AutoCAD draws the line of 2 units, and Point D is established. To draw a line from D to E without exiting the LINE command, first set Ortho and Snap to ON, and then place the cursor on line A-B. The rubber-band line will indicate line D-E, as shown in Figure 3–53. In this case, you do not know the distance, but you do know that the line terminates on line A-B. Therefore, simply press the pick button on the cursor, and line D-E is drawn. You do not have to specify point D as the starting point of the line. Using the line-line continuation after drawing line C-D does that for you.

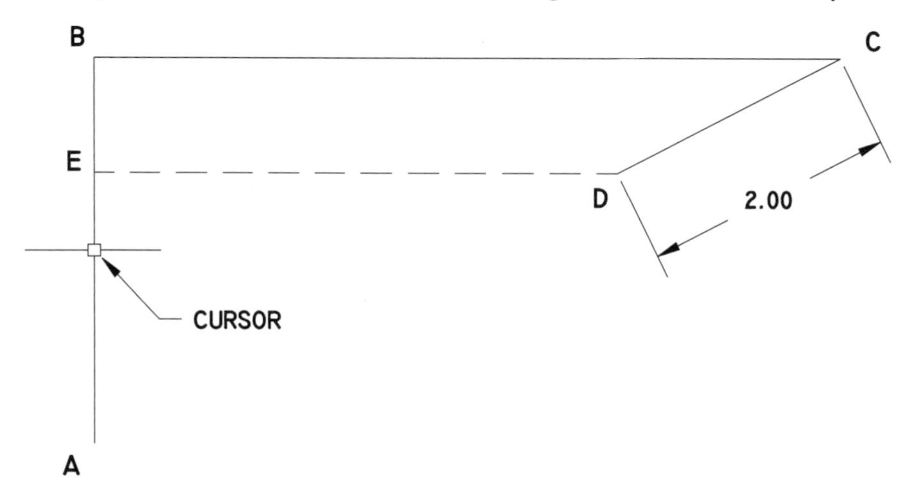

Figure 3–53 *Drawing a line from D to E*

Line A-F can be drawn in a similar manner as line C-D was drawn. Unlike line C-D, line A-F is not a continuation of another line. However, the starting point, A, can be selected with the Snap set to ON. After starting point A has been specified, place the cursor on point C and enter 2. From line A-F, line F-G can be drawn in a similar manner as line D-E was drawn. This example shows an application of the Direct Distance option in which the distance was entered in and the direction was controlled by the cursor.

POLAR TRACKING AND OBJECT SNAP TRACKING

The Polar Tracking feature lets you draw lines and specify point displacements in directions that are multiples of a specified increment angle. You can specify whether the increment angles are measured from the current coordinate system or from the angle of a previous object. You can also add up to 10 additional angles to which a line or displacement can be diverted from a base direction. The Object Snap Tracking feature lets you track your cursor from a strategic (Osnap) point on an object in a specified direction, either orthogonal or preset polar angles.

Turning Polar Tracking ON and OFF

In AutoCAD, four primary methods are available to turn Polar Tracking ON and OFF:

1. On the status bar at the bottom of the screen, when the **POLAR** button appears to be pressed in, then Polar Tracking is set to ON. When the **POLAR** button appears to be out, then Polar Tracking is set to OFF. To change its setting, simply place the cursor over the **POLAR** button and press the pointing device pick button.
2. To change the Polar Tracking to ON when it is OFF or to OFF when it is ON, simply press the function key F10.
3. Right-click the **POLAR** button on the status bar at the bottom of the screen and from the shortcut menu, choose Settings. The Drafting Settings dialog box will be displayed with the **Polar Tracking** tab selected. Polar Tracking is set to ON when the **Polar Tracking On (F10)** check box is checked, and is set to OFF when it is not checked. To change its setting, simply place the cursor over the check box and press the pointing device pick button. You can also enter **ddrmodes** at the Command: prompt to display the Drafting Settings dialog box from which the **Polar Tracking** tab can be selected if it is not already selected.
4. Right-click the **POLAR** button on the status bar at the bottom of the screen and from the shortcut menu, and choose ON to set Polar Tracking ON or choose OFF to set Polar Tracking OFF.

When the Polar Tracking mode is set to ON, it allows you to specify a displacement or draw in a direction that is a multiple of a specified increment angle. But even with Polar Tracking set to ON, you can still specify a "next point" of a line or displacement with the cursor that is not at one of the multiples of the specified Polar Tracking increment angle. This is different from the ORTHO and SNAP features where you are forced (when specifying a point with the cursor) to accept a point orthogonally or on the Snap grid when the respective mode is set to ON. You will notice, however, that when the Polar Tracking mode is ON and you move the cursor near one of the radials (multiple of increment angle), the cursor will snap to that radial and the radial will appear as a dotted construction line indicating the direction of the line that will be drawn if you specify that point by pressing the pick button on the pointing device.

Changing Polar Tracking Increment & Additional Angles

In AutoCAD there are two primary methods to change the Polar Tracking increment angle and add additional angles:

1. From the status bar at the bottom of the screen, right-click on the **POLAR** button and choose Settings. AutoCAD displays the Drafting Settings dialog box with the **Polar Tracking** tab selected. From the **Polar Angle Settings** section, enter the desired increment angle in the **Increment angle:** text box. This is the base increment angle whose multiples are used by Polar Tracking. Up to 10 additional angles can be added in the text box under the **Additional Angles** check box. Simply select the **New** button and then enter the desired additional angle(s). If there is no check in the **Additional Angles** check box, the additional angles are not available for use when Polar Tracking is set to ON.

To make the additional angle(s) available when the Polar Tracking mode is set to ON, place the cursor over the **Additional Angles** check box and press the pointing device pick button (see Figure 3–54).

2. From the Tools menu, choose Drafting Settings. Then select the **Polar Tracking** tab. See the explanation in Method 1 for changing settings.

Note: The Polar Tracking and the Ortho modes cannot both be set to ON simultaneously. They can both be set to OFF, or either one can be set to ON.

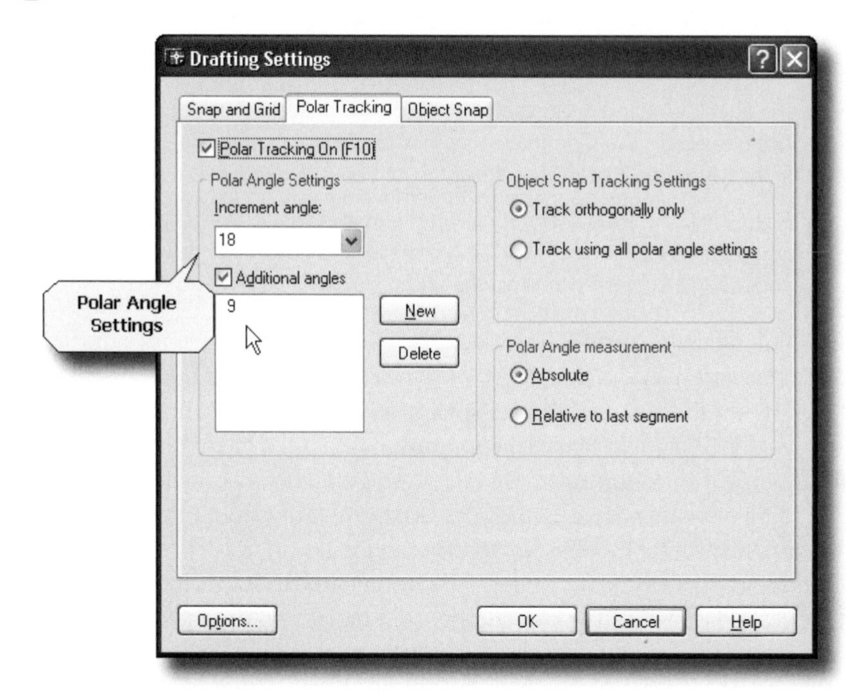

Figure 3–54 *Selecting the Polar Tracking On (F10) check box and setting the increment angle to 18° and one additional angle to 9°*

With the 18° increment angle and the **Polar Angle measurement** set to **Absolute**, you can draw lines and specify displacements at 18°, 36°, 54°, 72°, 90°, 108°… …324°, 342°, 360° around the compass by just snapping to the construction lines when they appear. The accompanying AutoTrack Tooltip will appear when you have snapped to a multiple of the increment angle if the **Display AutoTrack Tooltip** check box is set to ON on the **Drafting** tab of the Options dialog box. The Tooltip displays the cursor's distance and direction from the first specified point. You can, with the cursor snapped to one of the Polar Tracking angles, use the Direct Distance option and enter a distance from the keyboard and press ENTER. AutoCAD will draw the line or apply the displacement in accordance with the distance entered and the direction set by the cursor. For example, in Figure 3–55, the line from P1 to P2 was drawn 2 units long at 0°. This could have been done with Direct Distance and either ORTHO set to ON or with Polar Tracking. But with P2 to P3 being at 18° and from P3 to P4 being at 36°, it is easier to use Polar Tracking with the settings shown in the example in Figure 3–54. Figure 3–56 shows the three line segments drawn using Polar Tracking with the increment angle set to 18° and the **Polar Angle measurement** set to **Absolute**. Figure 3–57 shows how you can use the **Additional angles** setting of 9° to draw a 1-unit-long line at 9° from absolute zero.

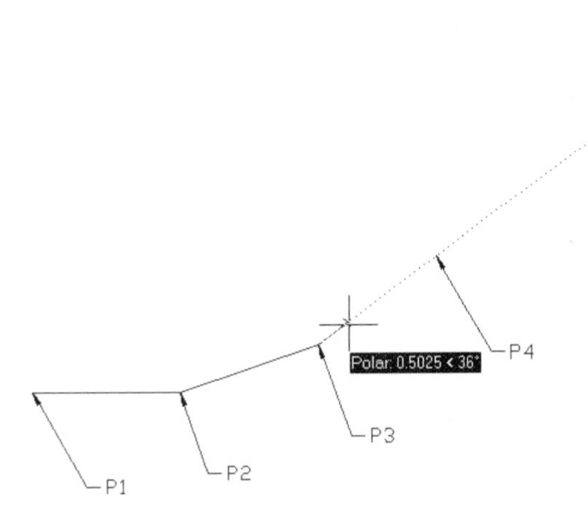

Figure 3–55 *Drawing line segments at multiples of the 18° increment angle*

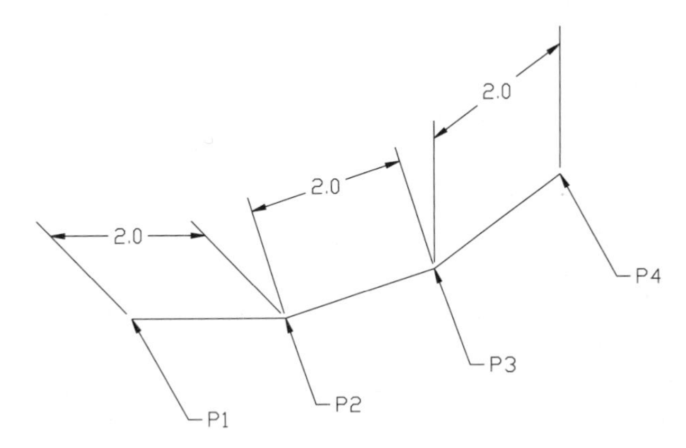

Figure 3–56 *Three line segments drawn using Polar Tracking with the increment angle set to 18° and the Polar Angle measurement set to Absolute*

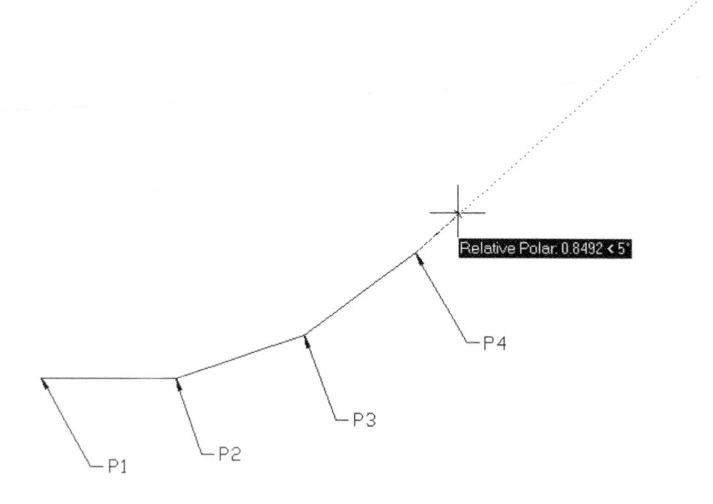

Figure 3–57 *Four line segments drawn using Polar Tracking with the increment angle set to 18° and at one additional angle at 9° from Polar Angle measurement set to Absolute*

After completing the three segments drawn at multiples of 18°, instead of adding the segment at 9°, let's change the increment angle to 5° and the **Polar Angle measurement** set to Relative To Last Segment. Continuing with three more segments, each 1 unit long, the first increment angle will be measured from the last segment, which is at 36°. The Tooltip displays "Relative Polar (distance) < 5°" but the resulting angle is 41°. Figure 3–58 shows the three additional line segments drawn using Polar Tracking with the increment angle set to 5° and the **Polar Angle measurement** set to Relative to last segment. The last two segments will actually be at 46° and 51°.

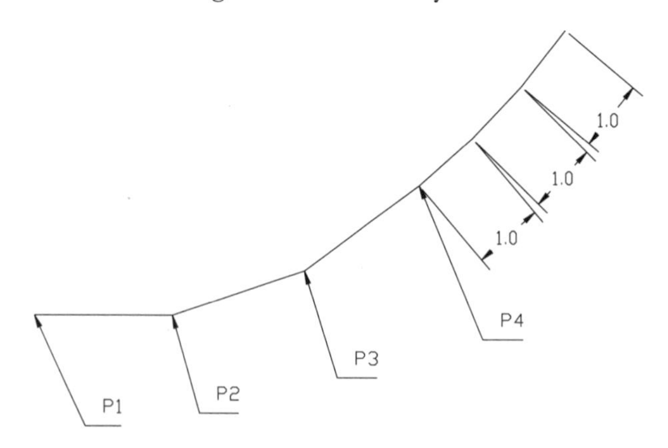

Figure 3–58 *Three line segments drawn using Polar Tracking with the increment angle set to 5° and the Polar Angle measurement set to Relative to last segment*

Polar Snap

The Polar Snap feature can be used to make drawing the 2-unit-long and 1-unit-long segments in the previous examples even easier. On the **Snap and Grid** tab of the Drafting Settings dialog box, choose the **Polar Snap** radio button in the **Snap type & style** section. Then in the **Polar distance** text box of the **Polar spacing** section, enter the desired distance. In the case of the above mentioned example, the 2-unit-long segments at 0°, 18°, and 36°, you can enter 2.0 for the distance. For the 41°, 46°, and 51° segment (additions of 5° each in the **Relative to last segment** mode), you can use a distance of 1.0.

Object Snap Tracking

Polar Tracking can be used, making one of an object's strategic points as the base point for tracking, by using the Object Snap Tracking feature, along with the Object Snap mode, that will allow acquiring the particular type of strategic point you wish such as endpoint, midpoint, or center. If you wish to start a line 1 unit and 18° from an object's endpoint, simply turn on the Object Snap Tracking by setting the **Object Snap Tracking ON (F11)** check box on the **Object Snap** tab of the Drafting Settings dialog box (or press the function key F11). Be sure the Polar Tracking angle and Polar Tracking Distance are set to 18° and 1.0, respectively. Then invoke the Object Snap mode Endpoint and acquire the desired endpoint by moving the cursor over the object's end temporarily before moving it to the second (end of tracking) point.

This next example is to show how to apply Polar Tracking and Object Snap Tracking to draw the vertical line B and the horizontal line C from existing line A and arc D, as shown in Figures 3–59 and 3–60.

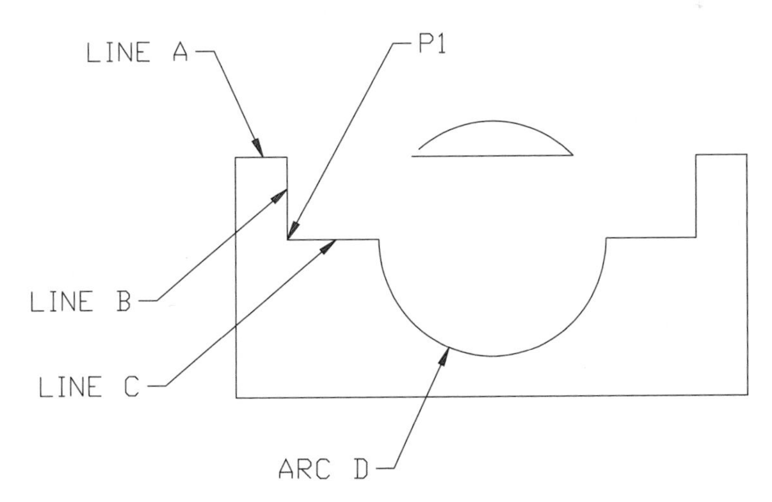

Figure 3–59 *Completed object including lines A, B, and C, and arc D*

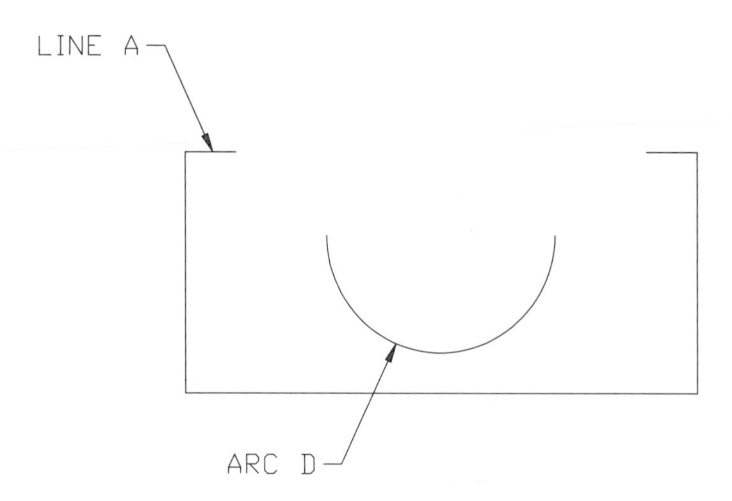

Figure 3–60 *Line A and arc D from which lines B and C will be drawn using Polar Tracking*

From the status bar at the bottom of the screen, right-click on the **POLAR** button and choose Settings. The Drafting Settings dialog box is displayed with the Polar Tracking tab selected. From the **Object Snap Tracking Settings** section choose the **Track orthoganol only** radio button. Check to be sure that Polar Tracking is set to ON. Invoke the LINE command:

> Command: **line** (ENTER)
> Specify first point: **end** *(enter the first three letters of the Osnap mode Endpoint and then select the right end of LINE A to establish the start point of LINE B)*
> Specify next point or [Undo]: *(move the aperture cursor to the right end of LINE A, keep it there until the Endpoint marker appears and without pressing a button move it downward from the line)*

As you move the cursor downward from line A, a dashed construction line will follow as long as the line between the end of line A and the cursor is vertical or horizontal, as shown in Figure 3–61.

Next, move the cursor to the left end of arc D, keep it there until the Endpoint marker appears and then move it away from it toward the left, again without pressing the pick button. A horizontal construction line will follow until the cursor is near the vertical construction line through the end of line A as shown in Figure 3–62. At this time, with both the vertical and horizontal construction lines being displayed, you can press the pick button on the pointing device and point P1, the end point of line B will be established.

After pressing the pick button to specify point P1 and complete line B, move the cursor to the left end of arc D, invoke the Endpoint Osnap mode, and select arc D. This will complete line C, as shown in Figure 3–63.

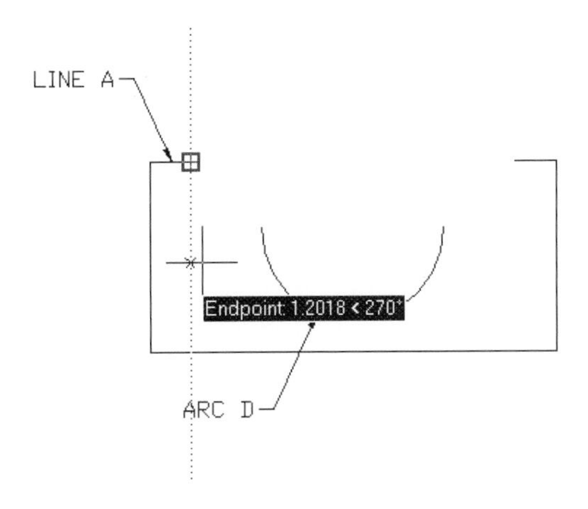

Figure 3–61 *A vertical construction line passes through the cursor and the end of LINE A*

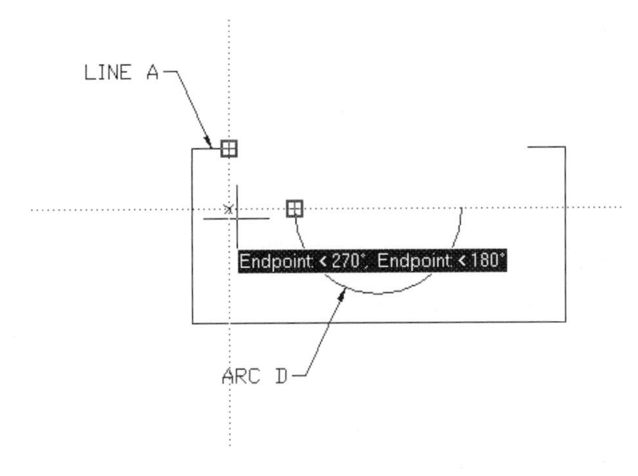

Figure 3–62 *A dashed arc follows the cursor indicating the extension of ARC D to the dashed extension of LINE B, at which point P1 is specified*

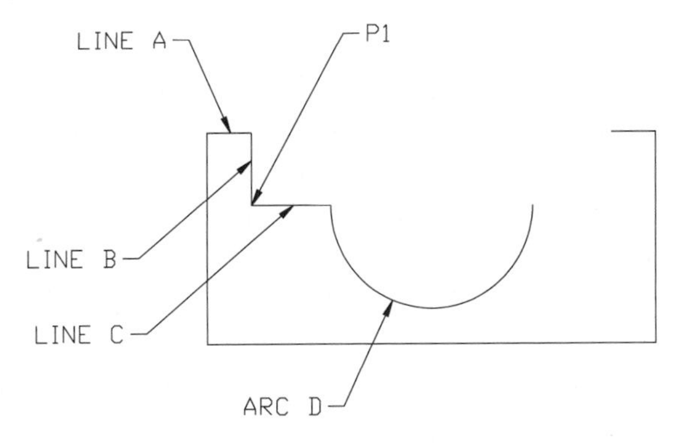

Figure 3–63 *Lines B and C are drawn with point P1 at their intersection*

DISPLAY CONTROL

There are many ways to view a drawing in AutoCAD. These viewing options vary from on-screen viewing to hard-copy plots. The hard-copy options are discussed in Chapter 8. Using the display commands, you can select the portion of the drawing to be displayed, establish 3D perspective views, and much more. By being able to view the drawing in different ways, AutoCAD provides a way to draw faster. The commands that are explained in this section are like utility commands. They make your job easier and help you to draw more accurately.

ZOOM COMMAND

The ZOOM command is like a zoom lens on a camera. You can increase or decrease the viewing area, although the actual size of objects remains constant. As you increase the visible size of objects, you view a smaller area of the drawing in greater detail as though you were closer. As you decrease the visible size of objects, you view a larger area as though you were farther away. This ability provides a close-up view for better accuracy and detail or a distant view to get the whole picture.

Invoke the ZOOM command:

Zoom toolbar	Choose one of the available options (see Figure 3–64)
View menu	Zoom > Choose one of the available options
Command: prompt	**zoom** (ENTER)

Figure 3–64 *Zoom toolbar*

AutoCAD prompts:

Command: **zoom**
Specify corner of window, enter a scale factor (nX or nXP), or [All/Center/Dynamic/Extents/Previous/Scale/Window] <real time>: *(right-click and choose one of the available options from the shortcut menu)*

Default

The default option of the ZOOM command is a zoom window. After the ZOOM command displays its list of options, simply specify two points on the screen that represent diagonally opposite corners of a rectangle. The view in the rectangle is enlarged to fill the drawing area.

Realtime

The Zoom Real-time option lets you zoom interactively to a logical extent. Once you invoke the command with the Realtime option, the cursor changes to a magnifying glass with a "±" symbol. To zoom in closer, hold the pick button and move the cursor vertically toward the top of the window. To zoom out further, hold the pick button and move the cursor vertically toward the bottom of the window. To discontinue the zooming, release the pick button. To exit the Zoom Realtime option, press ENTER, ESC, or from shortcut menu select ENTER.

The current drawing window is used to determine the zooming factor. If the cursor is moved by holding the pick button from the bottom of the window to the top of the window vertically, the zoom-in factor would be 200%. Conversely, when holding the pick button from the top of the window and moving vertically to the bottom of the window, the zoom-out factor would be 200%.

When you reach the zoom-out limit, the "-" symbol on the cursor disappears while attempting to zoom-out indicating that you can no longer zoom out. Similarly, when you reach the zoom-in limit, the "+" symbol on the cursor disappears while attempting to zoom in indicating that you can no longer zoom in.

Invoke the Zoom Realtime option:

Standard toolbar	Choose Zoom Realtime (see Figure 3–65)
View menu	Zoom > Realtime
Command: prompt	**zoom** (ENTER)

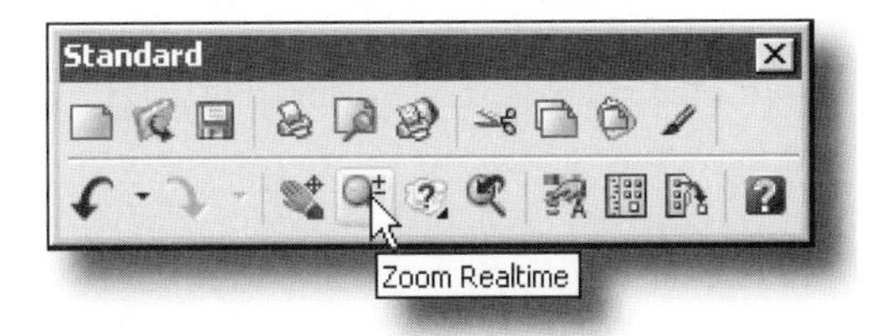

Figure 3–65 *Invoking the Zoom Realtime option from the Standard toolbar*

AutoCAD prompts:

Command: **zoom** (ENTER)
Specify corner of window, enter a scale factor (nX or nXP), or
[All/Center/Dynamic/Extents/Previous/Scale/Window] <real time>: (ENTER)
Press ESC or ENTER to exit, or right-click to activate the shortcut menu.

Control the display by moving the cursor appropriately; to exit the Zoom Realtime option, press the ESC or ENTER key. You can also exit by selecting EXIT from the shortcut menu that is displayed when you right-click with your pointing device. In addition, you can perform other operations related to ZOOM and PAN by selecting appropriate commands from the shortcut menu.

Scale

The Zoom Scale option lets you enter a display scale (or magnification) factor. The scale factor, when entered as a number (it must be a numerical value and must not be expressed in units of measure), is applied to the area covered by the drawing limits. For example, if you enter a scale value of 3, each object appears three times as large as in the Zoom All view. A scale factor of 1 displays the entire drawing (the full view), which is defined by the established limits. If you enter a value less than 1, AutoCAD decreases the magnification of the full view. For example, if you enter a scale of 0.5, each object appears half its size in the full view while the viewing area is twice the size in horizontal and vertical dimensions. When you use this option, the object in the center of the screen remains centered.

See Figures 3–66 and 3–67 for the difference between a full view and a 0.5 zoom.

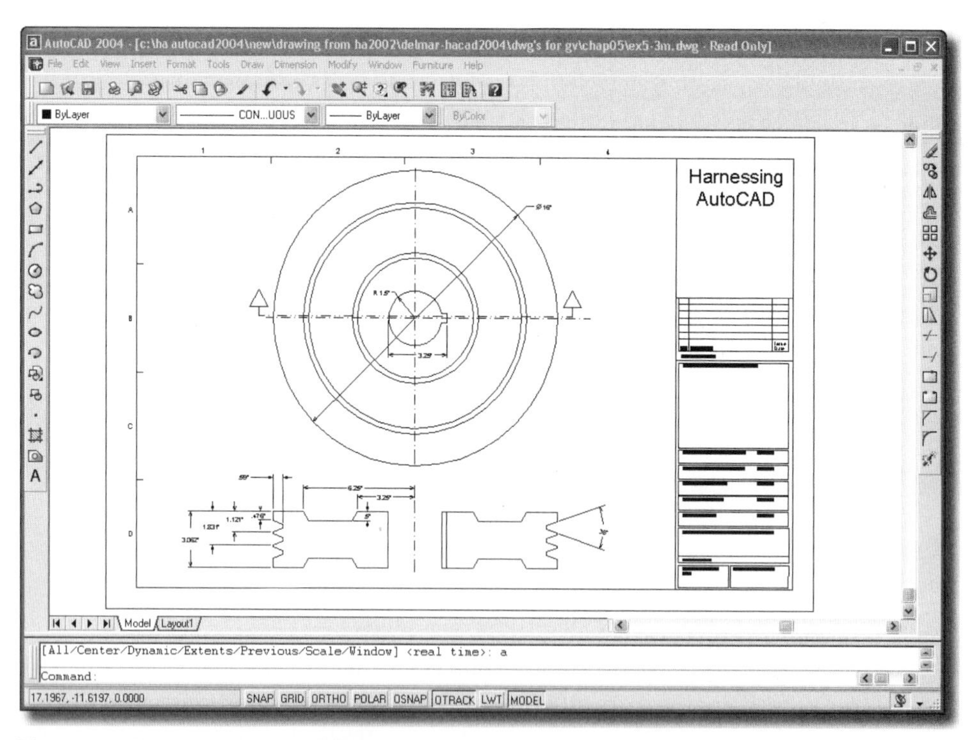

Figure 3–66 *A drawing at ZOOM All (full view)*

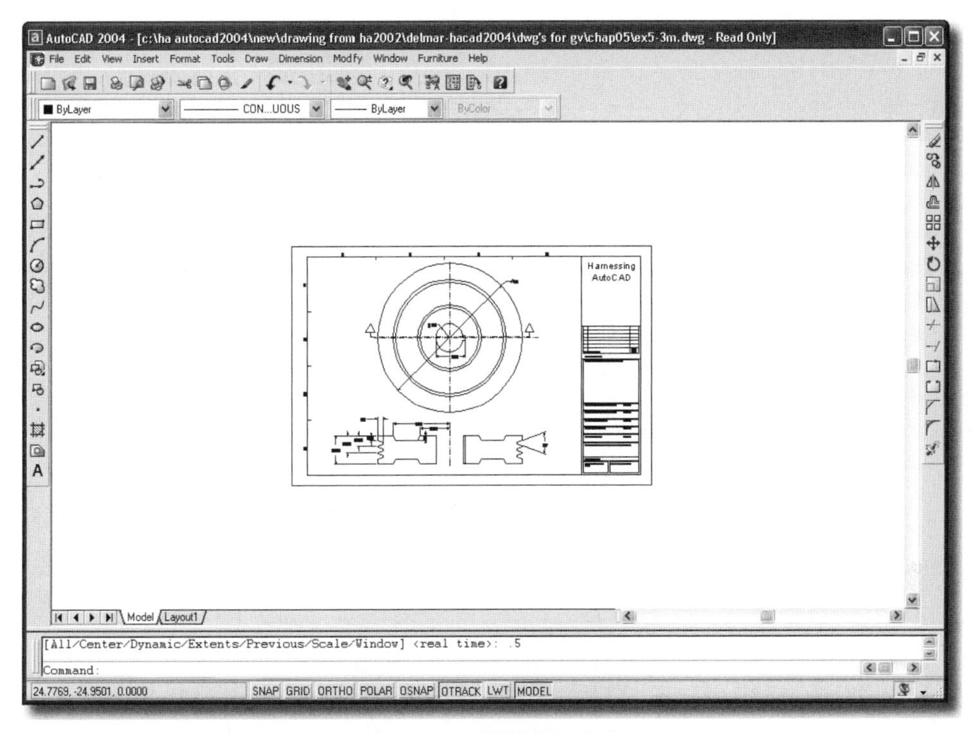

Figure 3–67 *The same drawing after setting ZOOM to 0.5*

Invoke the Zoom Scale option:

Zoom toolbar	Choose Zoom Scale (see Figure 3–68)
View menu	Zoom > Scale
Command: prompt	**zoom** (ENTER)

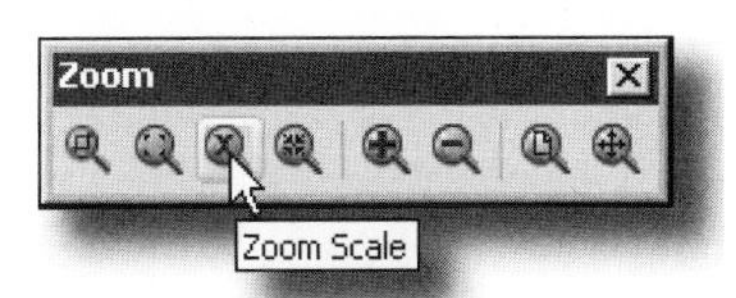

Figure 3–68 *Invoking the Zoom Scale option from the Zoom toolbar*

AutoCAD prompts:

Command: **zoom** (ENTER)
Specify corner of window, enter a scale factor (nX or nXP), or
[All/Center/Dynamic/Extents/Previous/Scale/Window] <real time>: **s**
Enter a scale factor (nX or nXP): *(specify a scale factor)*

If you enter a number followed by X, the scale is determined relative to the current view. For instance, entering 2X causes each object to be displayed two times its current size on the screen.

The scale factor XP option, related to the layout of the drawing, is explained in Chapter 7.

All

The Zoom All option lets you see the entire drawing. In a plan view, it zooms to the drawing's limits or current extents, whichever is larger. If the drawing extends outside the drawing limits, the display shows all objects in the drawing.

Invoke the ZOOM ALL command:

Zoom toolbar	Choose Zoom All (see Figure 3–69)
View menu	Zoom > All
Command: prompt	**zoom** (ENTER)

Figure 3–69 *Invoking the* Zoom All *option from the Zoom toolbar*

AutoCAD prompts:

Command: **zoom** (ENTER)
Specify corner of window, enter a scale factor (nX or nXP), or [All/Center/Dynamic/Extents/Previous/Scale/Window] <real time>:
a (ENTER)

Center

The Zoom Center option lets you select a new view by specifying its center point and the magnification value or height of the view in current units. A smaller value for the height increases the magnification; a larger value decreases the magnification.

Invoke the Zoom Center option:

Zoom toolbar	Choose Zoom Center (see Figure 3–70)
View menu	Zoom > Center
Command: prompt	**zoom** (ENTER)

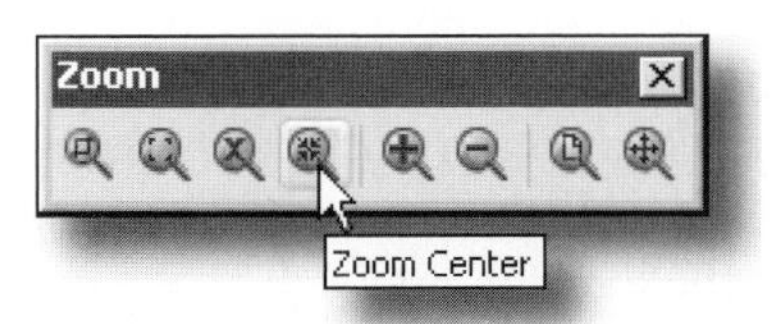

Figure 3–70 *Invoking the Zoom Center option from the Zoom toolbar*

AutoCAD prompts:

Command: **zoom** (ENTER)
Specify corner of window, enter a scale factor (nX or nXP), or
[All/Center/Dynamic/Extents/Previous/Scale/Window] <real time>:
  c (ENTER)
Center point: *(specify the center point)*
Magnification or Height <current height>: *(specify the magnification or height)*

The following command sequence produces the example of the Zoom Center option shown in Figure 3–71.

Command: **zoom** (ENTER)
Specify corner of window, enter a scale factor (nX or nXP), or
[All/Center/Dynamic/Extents/Previous/Scale/Window] <real time>:
  c (ENTER)
Specify center point: **8,6** (ENTER)
Enter magnification or height <current height>: **4**

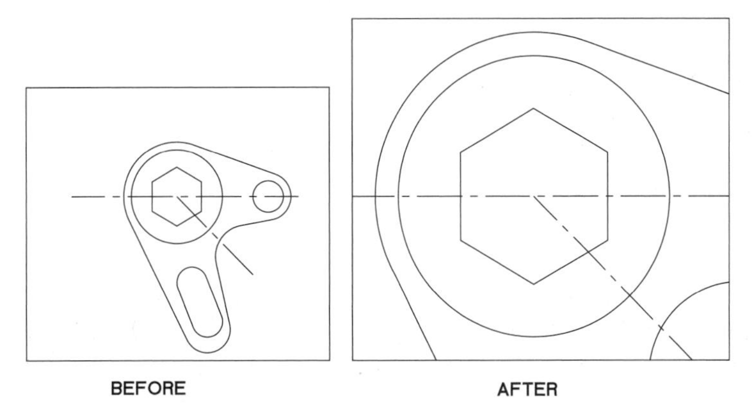

Figure 3–71 *Using the Zoom Center option*

In addition to providing coordinates for center point, you can also specify the center point by placing a point on the view window. The height can also be specified in terms of the current view height by specifying the magnification value followed by an X. A response of 3X will make the new view height three times as large as the current height.

Dynamic

The AutoCAD Zoom Dynamic option provides a quick and easy method to move to another view of the drawing. With Zoom Dynamic, you can see the entire drawing and then simply select the location and size of the next view by means of cursor manipulations. Using Zoom Dynamic is one means by which you can visually select a new display area that is not entirely within the current display.

Invoke the Zoom Dynamic option:

Zoom toolbar	Choose Zoom Dynamic (see Figure 3–72)
View menu	Zoom > Dynamic
Command: prompt	**zoom** (ENTER)

Figure 3–72 *Invoking the* Zoom Dynamic *option from the Zoom toolbar*

AutoCAD prompts:

Command: **zoom** (ENTER)
Specify corner of window, enter a scale factor (nX or nXP), or [All/Center/Dynamic/Extents/Previous/Scale/Window] <real time>:
d (ENTER)

The current viewport is then transformed into a selecting view that displays the drawing extents, as shown in the example in Figure 3–73.

When the selecting view is displayed, you see the drawing extents marked by a white or black box, the current display marked by a blue or magenta dotted box. A new view box, the same size as the current display, appears. Its location is controlled by the movement of the pointing device. Its size is controlled by a combination of the pick button and cursor movement. When the new view box has an X in the center, the box pans around the drawing in response to cursor movement. After you press the pick button on the pointing device, the X disappears and an arrow appears at the right edge of the box. The new view box is now in Zoom mode. While the arrow is in the box, moving the cursor left decreases the box size; moving the cursor right increases the size.

When the desired size has been chosen, press the pick button again to pan, or press ENTER to accept the view defined by the location/size of the new view box. Pressing ESC cancels the Zoom Dynamic and returns you to the current view.

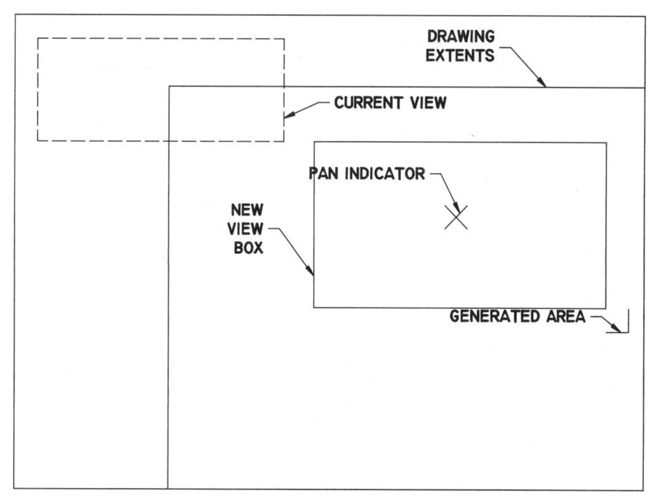

Figure 3–73 *Using the* ZOOM DYNAMIC *command to display the drawing extents*

Extents

The Zoom Extents option lets you see the entire drawing on screen. Unlike the All Option, the Extents option uses only the drawing extents and not the drawing limits. See Figures 3–74 and 3–75, which illustrate the difference between the options All and Extents.

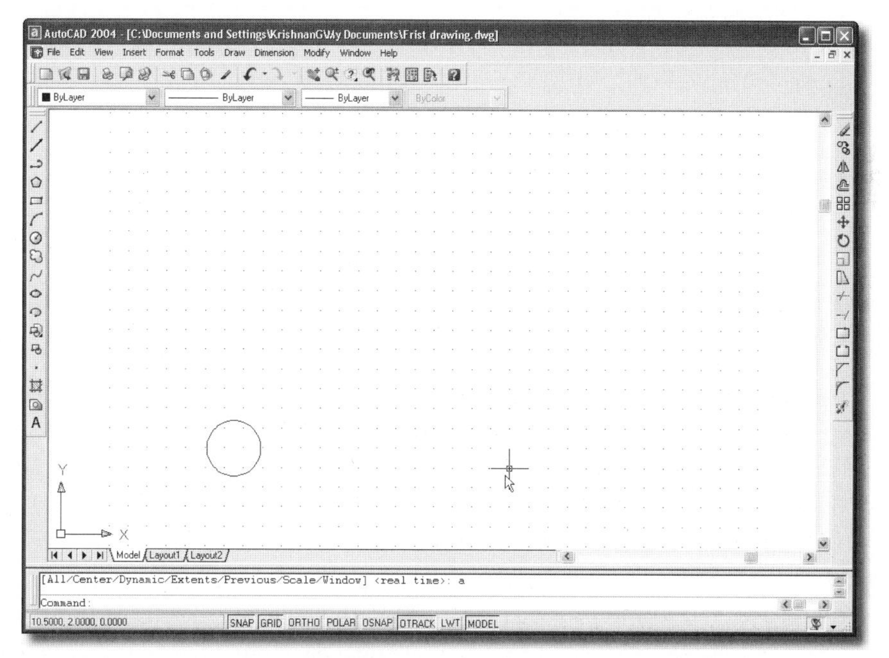

Figure 3–74 *A drawing after a ZOOM All*

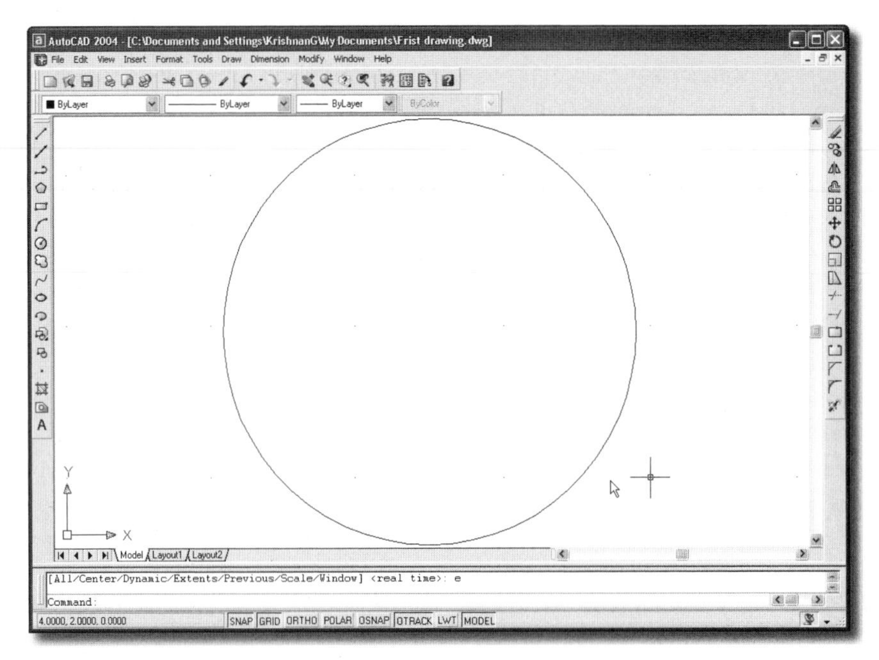

Figure 3–75 *The same drawing after a ZOOM Extents*

Invoke the Zoom Extents option:

Zoom toolbar	Choose Zoom Extents (see Figure 3–76)
View menu	Zoom > Extents
Command: prompt	**zoom** (ENTER)

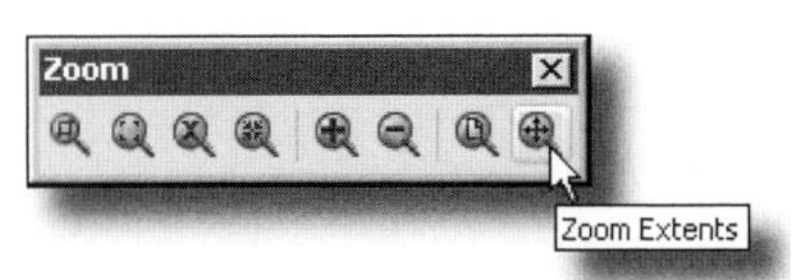

Figure 3–76 *Invoking the Zoom Extents option from the Zoom toolbar*

AutoCAD prompts:

Command: **zoom** (ENTER)
Specify corner of window, enter a scale factor (nX or nXP), or
[All/Center/Dynamic/Extents/Previous/Scale/Window] <real time>: **e** (ENTER)

Previous

The Zoom Previous option displays the last displayed view. While editing or creating a drawing, you may wish to zoom into a small area, back out to view the larger area, and then zoom into another small area. To do this, AutoCAD saves the coordinates of the current view whenever it is being changed by any of the zoom options or other view commands, so you can return to the previous view by entering the Previous option, which can restore the previous 10 views.

Invoke the Zoom Previous option:

View menu	Zoom > Previous
Command: prompt	**zoom** (ENTER)

AutoCAD prompts:

Command: **zoom** (ENTER)
Specify corner of window, enter a scale factor (nX or nXP), or
[All/Center/Dynamic/Extents/Previous/Scale/Window] <real time>: **p** (ENTER)

Window

The Zoom Window option lets you specify a smaller area of the part of the drawing being currently displayed and have that portion fill the drawing area. This is done by specifying two diagonally opposite corners of a rectangle. The center of the area selected becomes the new display center, and the area inside the window is enlarged to fill the drawing area as completely as possible.

You can enter two opposite corner points to specify an area by means of coordinates or the pointing device (see Figures 3–77 and 3–78).

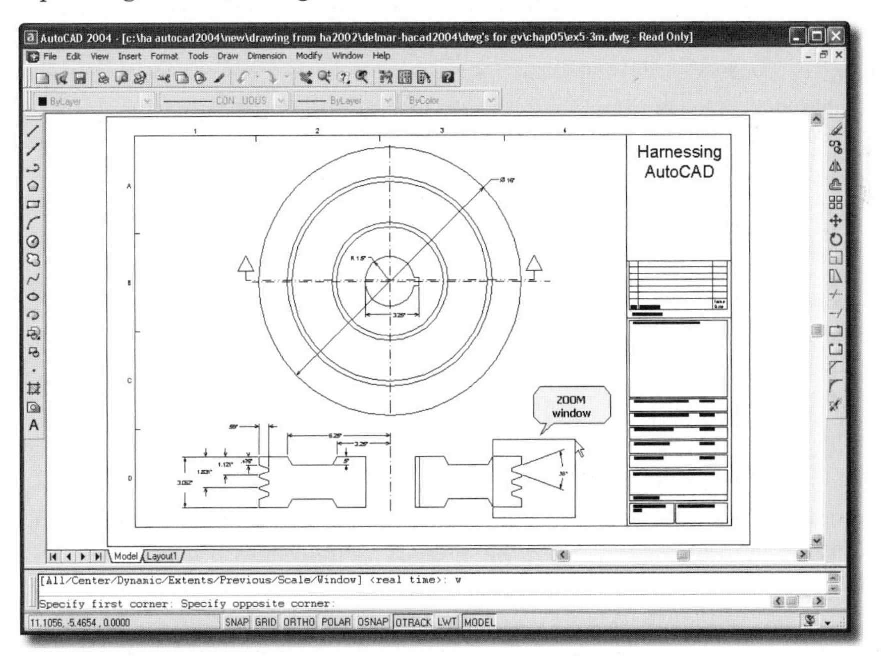

Figure 3–77 *Specifying a ZOOM Window area*

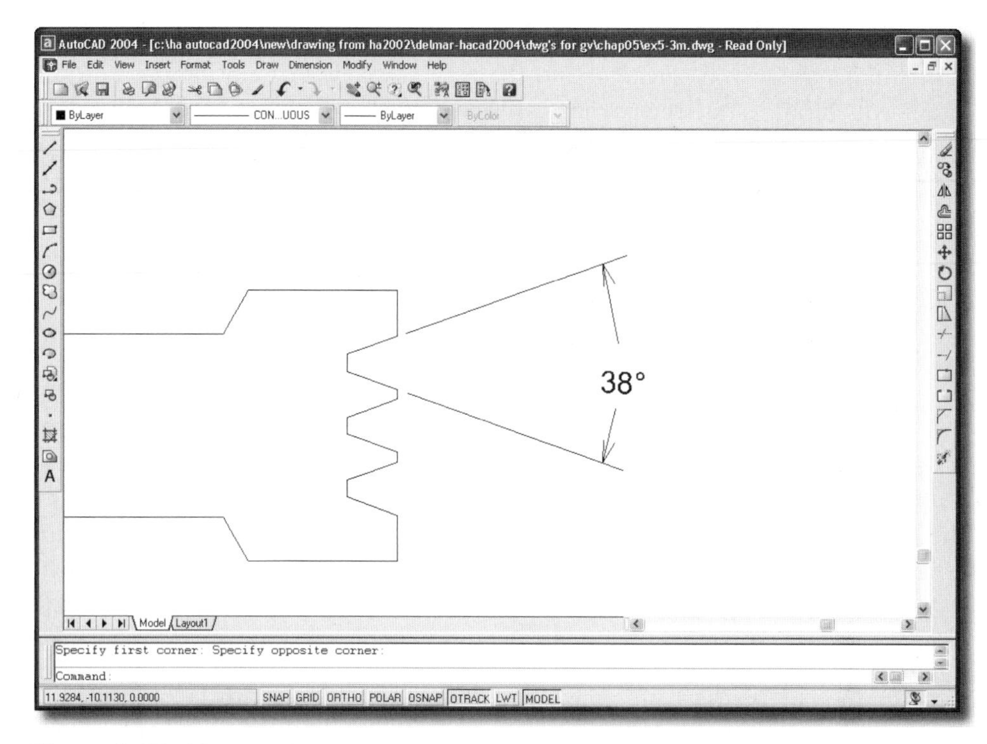

Figure 3–78 *After specifying a window area*

Invoke the Zoom Window option:

Zoom toolbar	Choose Zoom Window (see Figure 3–79)
View menu	Zoom > Window
Command: prompt	**zoom** (ENTER)

Figure 3–79 *Invoking the Zoom Window option from the Zoom toolbar*

AutoCAD prompts:

Command: **zoom** (ENTER)
Specify corner of window, enter a scale factor (nX or nXP), or

[All/Center/Dynamic/Extents/Previous/Scale/Window] <real time>:
  w (ENTER)
Specify first corner: *(specify a point to define the first corner of the window)*
Specify opposite corner: *(specify a point to define the diagonally opposite corner of the window)*

PAN COMMAND

AutoCAD lets you view a different portion of the drawing in the current view without changing the magnification. You can move your viewing area to see details that are currently off screen. Imagine that you are looking at your drawing through the display window and that you can slide the drawing left, right, up, and down without moving the window.

The Pan Realtime option lets you pan interactively to the logical extent (edge of the drawing space). Once you invoke the command, the cursor changes to a hand cursor. To pan, hold the pick button on the pointing device to lock the cursor to its current location relative to the viewport coordinate system, and move the cursor in any direction. Graphics within the window are moved in the same direction as the cursor. To discontinue the panning, release the pick button.

When you reach a logical extent (edge of the drawing space), a line-bar is displayed on the hand cursor on the side. The line-bar is displayed at the top, bottom, or left or right side of the drawing, depending upon whether the logical extent is at the top, bottom, or side of the drawing.

Invoke the Pan Realtime command:

Standard toolbar	Choose Pan Realtime (see Figure 3–80)
View menu	Pan > Real Time
Command: prompt	**pan** (ENTER)

Figure 3–80 *Invoking the* PAN REALTIME *command from the Standard toolbar*

AutoCAD prompts:

Command: **pan** (ENTER)
Press Esc or Enter to exit, or right-click to display shortcut menu.

To exit Pan Realtime, press ESC or ENTER. You can also exit by selecting EXIT from the shortcut menu that is displayed when you right-click with your pointing device. In addition, you can perform other operations related to zoom and pan by selecting the appropriate commands from the shortcut menu.

You can specify two points by entering the **–pan** command at the Command: prompt, in which case AutoCAD computes the displacement from the first point to the second.

For example, the following command sequence views a different portion of the drawing by placing two data points as shown in Figure 3–81.

Command: **-pan** (ENTER)
Specify base point or displacement: (Specify first point of displacement)
Specify second point: *(Specify second point of displacement)*

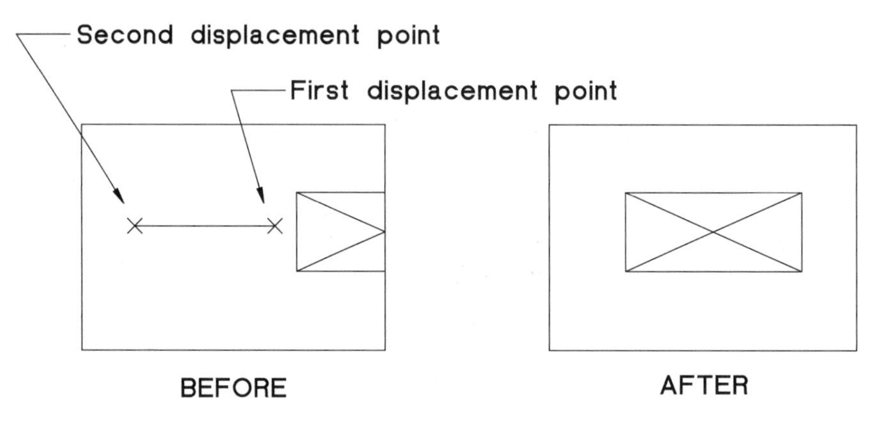

Figure 3–81 *Moving the view with the PAN command by specifying two displacement points*

AutoCAD calculates the distance and direction between the two points and pans the drawing accordingly. Sometimes it is useful to pan in exactly the horizontal or the vertical direction. In that case, set ORTHO mode to ON before invoking the PAN command to constrain cursor movement to the *X* or *Y* axis directions. You can enter a single coordinate pair indicating the relative displacement of the drawing with respect to the screen. If you give a null response to the "Second point" prompt, you are indicating that the coordinates provided represent the displacement of the drawing with respect to the origin. If you provide the coordinates for the second point instead of giving a null response, then AutoCAD computes the displacement from the first point to the second.

For example, the following command sequence views a different portion of the drawing 2 units to the left and 0.75 units up, as shown in Figure 3–82.

Command: **-pan** (ENTER)
Specify base point or displacement: **-2,.75** (ENTER)
Specify second point: (ENTER)

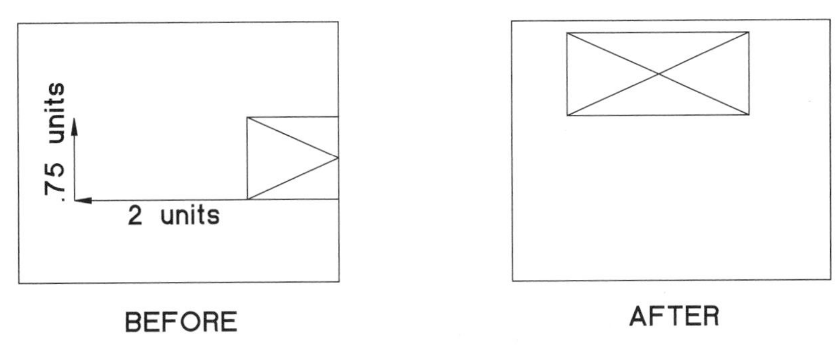

Figure 3–82 *Moving the view with the* PAN *command by specifying a pair of coordinates*

Pan Left, Right, Up, or Down

By choosing one of the **Left**, **Right**, **Up**, or **Down** options from the **View > Pan** flyout menu, it causes AutoCAD to pan the view left, right, up, or down accordingly.

AERIAL VIEW

The DSVIEWER command is used to activate the Aerial View, which provides a quick method of visually panning and zooming. By default, AutoCAD displays the Aerial View window, with the entire drawing displayed in the window, as shown in Figure 3–83. You can select any portion of the drawing in the Aerial View window by visually panning and zooming; in turn, AutoCAD displays the selected portion in the view window (current viewport).

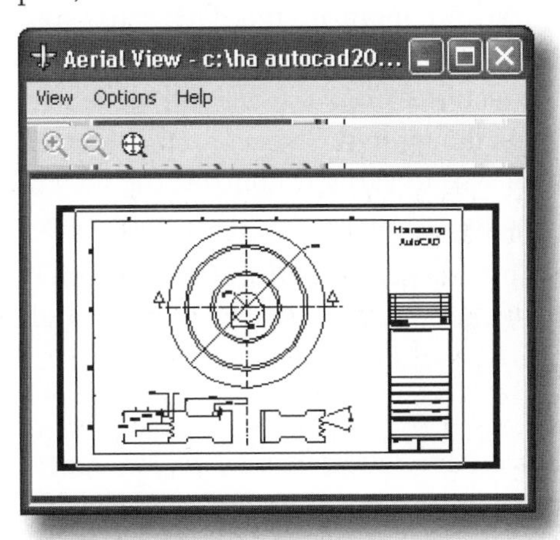

Figure 3–83 *The Aerial View dialog window*

Invoke the DSVIEWER command:

View menu	Aerial View
Command: prompt	**dsviewer** (ENTER)

AutoCAD displays the Aerial View window. Two option menus are provided to pan and zoom visually.

View Menu

The **View** menu in the Aerial View window has three options. The zoom in option causes the view to appear closer, enlarging the details of objects, but covering a smaller area. The zoom out causes the view to appear farther away, decreasing the size of objects, but covering a larger area. The global option causes the entire drawing to be viewable in the Aerial View window. You can also select the three options from the toolbar provided in the Aerial View window.

Options Menu

The **Options** menu in the Aerial View window has three options. Auto Viewport causes the active viewport to be displayed in model space. Dynamic Update toggles whether the view is updated or not in response to editing. The Realtime Zoom controls whether or not the AutoCAD window updates in real time when you zoom using the Aerial View.

CONTROLLING DISPLAY WITH INTELLIMOUSE

AutoCAD also allows you to control the display of the drawing with the small wheel provided with the IntelliMouse (two-button mouse). You can use the wheel to zoom and pan in your drawing any time without using any AutoCAD commands. By rotating the wheel forward you can zoom in and backwards you can zoom out. When double-clicking the wheel button, AutoCAD displays the drawing to the extent of the view window. To pan the display of the drawing, press the wheel button and drag the mouse. By default, each increment in the wheel rotation changes the zoom level by 10 percent. The ZOOMFACTOR system variable controls the incremental change, whether forward or backward. The higher the setting, the smaller the change.

AutoCAD also allows you to display the Object Snap shortcut menu when you click the wheel button. To do so, set the MBUTTONPAN system variable to 0. By default, the MBUTTONPAN is set to 1.

REDRAW COMMAND

The REDRAW command is used to refresh the on-screen image. You can use this command whenever you see an incomplete image of your drawing. If you draw two lines in the same place and erase one of the lines, it appears as if both the lines are erased. By invoking the REDRAW command, the second line will reappear. Also, you can use the REDRAW command to remove the blip marks on the screen. A redraw is considered a screen refresh as opposed to database regeneration.

Invoke the REDRAW command:

View menu	Choose Redraw
Command: prompt	**redraw** (ENTER)

AutoCAD prompts:

Command: **redraw** (ENTER)

AutoCAD does not provide any options for the REDRAW command.

REGEN COMMAND

The REGEN command is used to regenerate the drawing's data on the screen. In general, you should use the REGEN command if the image presented by REDRAW does not correctly reflect your drawing. REGEN goes through the drawing's entire database and projects the most up-to-date information on the screen; this command will give you the most accurate image possible. Because of the manner in which it functions, a REGEN takes significantly longer than a REDRAW.

There are certain AutoCAD commands for which REGEN takes place automatically unless REGENAUTO is set to OFF.

Invoke the REGEN command:

View menu	Choose Regen
Command: prompt	**regen** (ENTER)

AutoCAD prompts:

Command: **regen** (ENTER)

AutoCAD does not provide any options for the REGEN command.

SETTING MULTIPLE VIEWPORTS

The ability to divide the display into two or more separate viewports is one of the most useful features of AutoCAD. Multiple viewports divide your drawing screen into rectangles, permitting several different areas for drawing instead of just one. It is like having multiple zoom lens cameras, with each camera being used to look at a different portion of the drawing. You retain your menus and "Command:" prompt area.

Each viewport maintains a display of the current drawing independent of the display shown by other viewports. You can simultaneously display a viewport showing the entire drawing, and another viewport showing a closeup of part of the drawing in greater detail. A view in one viewport can be from a different point of view than those in other viewports. You can begin drawing (or modifying) an object in one viewport and complete it in another viewport. For example, three viewports could be used in a 2D drawing, two of them to zoom in on two separate parts of the drawing, showing two widely separated features in great detail on the screen simultaneously, and the third to show the entire drawing, as in Figure 3–84. In a 3D drawing, four viewports could be used to display simultaneously four views of a wireframe model: top, front, right side, and isometric, as in Figure 3–85.

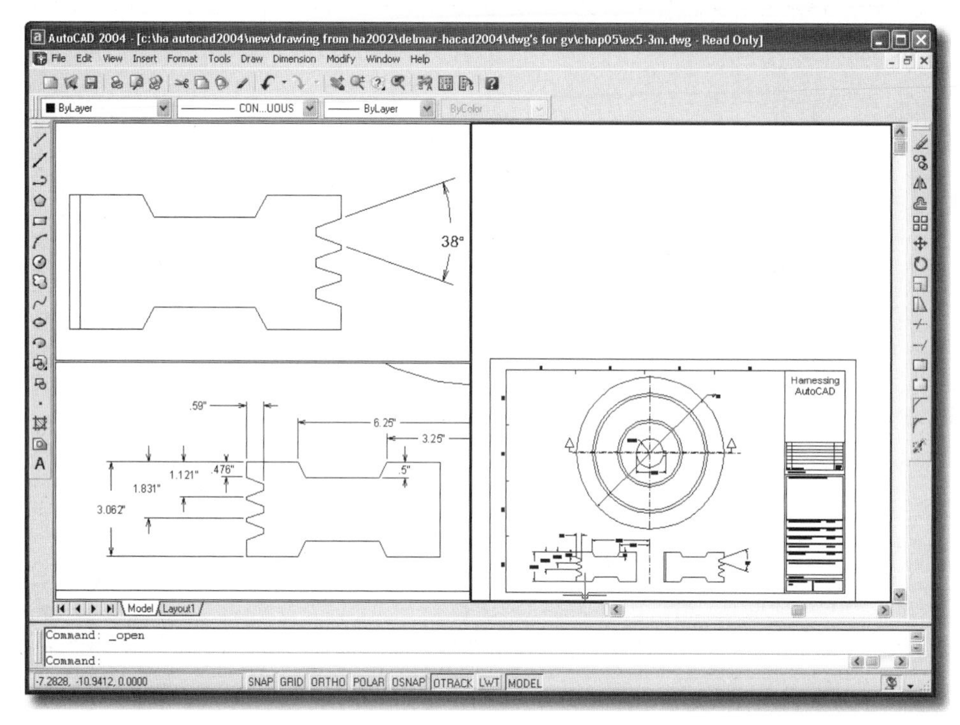

Figure 3–84 *Multiple viewports show different parts of the same 2D drawing*

AutoCAD allows you to divide the graphics area of your display screen into multiple, non-overlapping (tiled) viewports, as in Figures 3–84 and 3–85, only when you are in model space (system variable TILEMODE set to 1). The maximum number of active tiled viewports that you can have is set by the system variable MAXACTVP, and the default is 64. In addition you can also create multiple overlapping (floating) viewports when the system variable TILEMODE is set to 0. For a detailed explanation on how to create floating viewports, refer to Chapter 8.

You can work in only one viewport at a time. It is considered the current viewport. A viewport is set to current by moving the cursor into it with your pointing device and then pressing the pick button. You can even switch viewports in midcommand (except during some of the display commands). For example, to draw a line using two viewports, you must start the line in the current viewport, make another viewport current by clicking in it, and then specify the endpoint of the line in the second viewport. When a viewport is current, its border will be thicker than the other viewport borders. The cursor is active for specifying points or selecting objects only in the current viewport; when you move your pointing device outside the current viewport, the cursor appears as an arrow pointer.

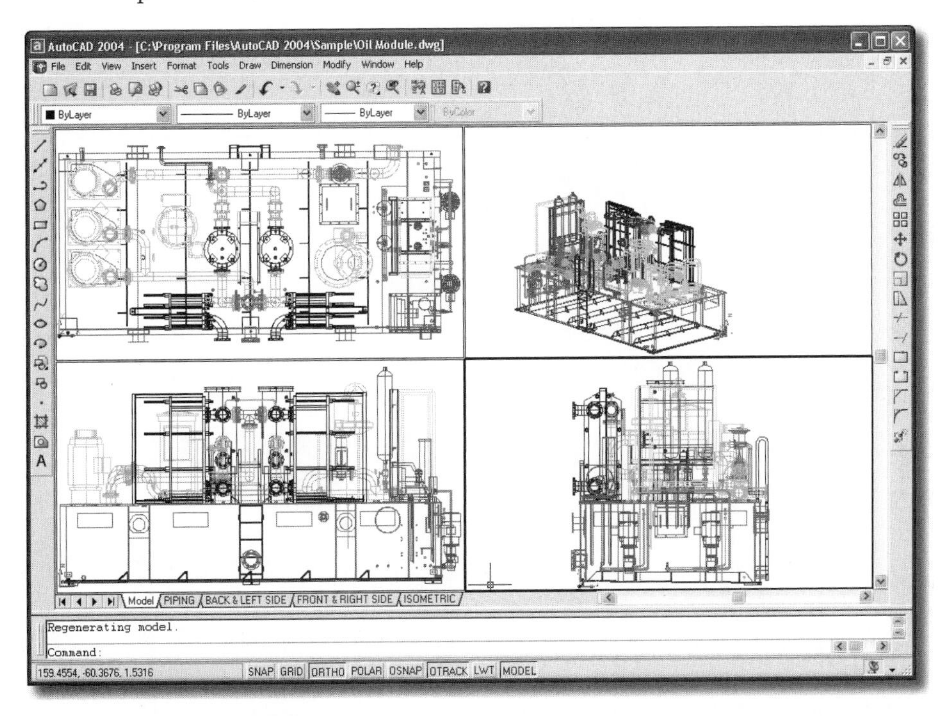

Figure 3–85 *Using viewports to show four views simultaneously for a 3D wireframe model*

Display commands like ZOOM and PAN and drawing tools like GRID, SNAP, ORTHO, and UCS icon modes are set independently in each viewport. The most important thing to remember is that the images shown in multiple viewports are all of the same drawing. An object added to or modified in one viewport will affect its image in the other viewports. You are not making copies of your drawing, just viewing its image in different viewports.

When you are working in tiled viewports, visibility of the layers is controlled globally in all the viewports. If you turn off a layer, AutoCAD turns it off in all viewports.

CREATING TILED VIEWPORTS

AutoCAD allows you to display tiled viewports in various configurations. Display of the viewports depends on the number and size of the views you need to see. By default, whenever you start a new drawing, AutoCAD displays a single viewport that fills the entire drawing area.

To create multiple viewports, open the Viewports dialog box:

Layouts toolbar	Choose Display Viewports dialog (see Figure 3–86)
View menu	Choose Viewports > New Viewports...
Command: prompt	**vports** (ENTER)

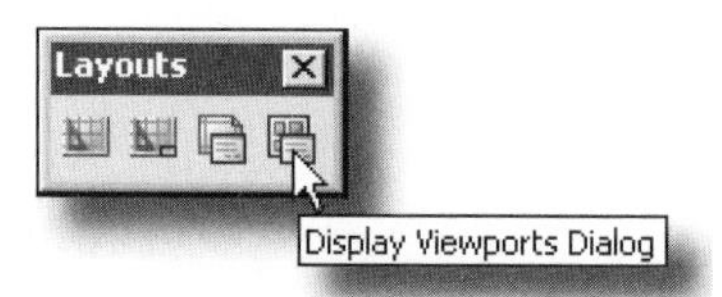

Figure 3–86 *Invoking the* VPORTS *command from the Layouts toolbar*

AutoCAD displays a Viewports dialog box similar to Figure 3–87.

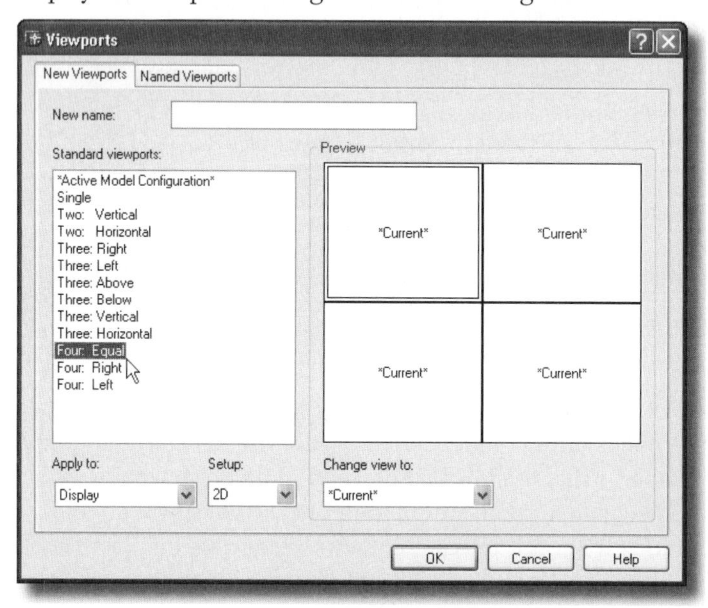

Figure 3–87 *Viewports dialog box*

Choose the name of the configuration you want to use from the **Standard viewports** list. AutoCAD displays the corresponding configuration in the preview window. If necessary, you can save the selected configuration by providing a name in the **New name:** text box. Select Display option from the **Apply to:** menu and select 2D from the **Setup:** menu for 2D viewport setup or select 3D for 3D viewport setup. Choose the **OK** button to create the selected viewport configuration. If you need additional viewports other than the standard configurations, you can subdivide a selected viewport. First select the viewport which you want to subdivide, then open the Viewports dialog box, select the appropriate configuration, and select Current Viewport from the **Apply to:** menu.

The **Named Viewports** tab lists all saved viewport configurations. At any time you can restore one of the saved viewport configurations.

You can also create tiled viewports from the command version of the VPORTS command:

Command: prompt	**-vports** (ENTER)

AutoCAD prompts:

Command: **-vports** (ENTER)
Enter an option [Save/Restore/Delete/Join/SIngle/?/2/3/4]<3>: *(select one of the available options)*

Save

The Save option allows you to save the current viewport configuration. The configuration includes the number and placement of active viewports and their associated settings. You can save any number of configurations with the drawing, to be recalled at any time. When you select this option, AutoCAD prompts:

Enter name for new viewport configuration or [?]:

You can use the same naming conventions for your configuration as you can in named objects. Instead of providing the name, you can respond with ? to request a list of saved viewport configurations.

Restore

The Restore option allows you to display a saved viewport configuration. When you select this option, AutoCAD prompts:

Enter name of viewport configuration to restore or [?]: *(Specify the name of the viewport configuration to restore)*

Delete

The Delete option deletes a named viewport configuration. When you select this option, AutoCAD prompts:

Enter name(s) of viewport configuration to delete<none>: *(Specify the name(s) of the viewport configuration you want to delete)*

Join

The Join option combines two adjoining viewports into a single viewport. The view for the resulting viewport is inherited from the dominant viewport. When you select this option, AutoCAD prompts:

Select dominant viewport <current viewport>:

You can give a null response to show the current viewport as the dominant viewport, or you can move the cursor to the desired viewport and press the pick button. Once you identify the dominant viewport, then AutoCAD prompts:

Select viewport to join:

Move the cursor to the desired viewport to join, and press the pick button. If the two viewports selected are not adjacent or do not form a rectangle, AutoCAD displays an error message and reissues the prompts.

SIngle

The SIngle option allows you to make the current viewport the single viewport.

?

The ? option displays the identification numbers and screen positions of the active viewports. When you select this option, AutoCAD prompts:

Enter name(s) of viewport configuration(s) to list <*>:

To list all saved configurations, give a null response. You also can use wild cards to list saved viewport names. All viewports are given an identification number by AutoCAD. This number is independent of any name you might give the viewport configuration. Each viewport is given a coordinate location, with respect to 0.0000,0.0000 as the lower left corner of the graphics area and 1.0000,1.0000 as the upper right corner.

2

The 2 option splits the current viewport in half. When you select this option, AutoCAD prompts:

Enter a configuration option [Horizontal/Vertical] <Vertical>:

You can select a horizontal or a vertical split, as shown in Figure 3–88. Vertical is the default.

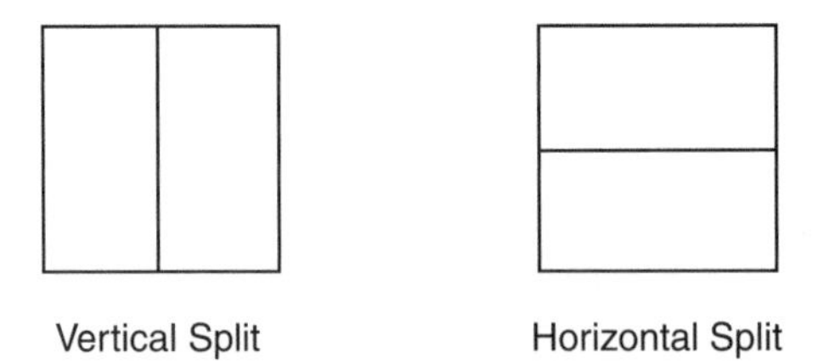

Figure 3–88 *Using the 2 option to split a current viewport in half by vertical or horizontal division*

3

The 3 option divides the current viewport into three viewports. This is the default option. When you select this option, AutoCAD prompts:

Enter a configuration option [Horizontal/Vertical/Above/Below/Left/Right] <Right>:

You can select the Horizontal or Vertical option to split the current viewport into thirds by horizontal or vertical division, as shown in Figure 3–89. The other options let you split into two small ones and one large one, specifying whether the large is to be placed above, below, or to the left or right (see Figure 3–89).

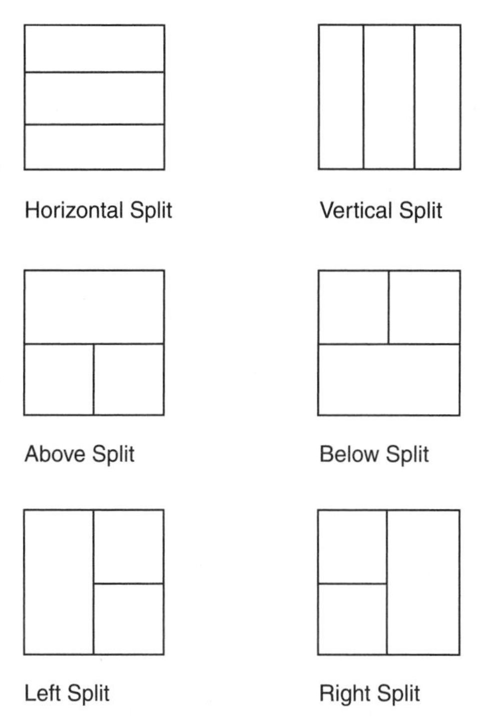

Figure 3–89 *Using the 3 option to split a current viewport into various three-way divisions*

4

The 4 option divides the current viewport into four viewports of equal size both horizontally and vertically, as shown in Figure 3–90.

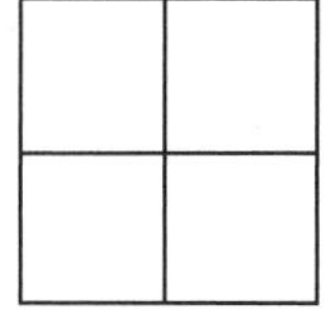

Figure 3–90 *Using the 4 option to split a current viewport into four viewports of equal size*

When you are working in multiple viewports, the REDRAW and REGEN commands will affect only the current viewport. To redraw or regenerate all the viewports simultaneously, use the REDRAWALL or REGENALL commands, respectively.

CREATING AND MODIFYING LAYER SYSTEM

AutoCAD offers a means of grouping objects in layers in a manner similar to the manual drafter's separating complex drawings into simpler ones on individual transparent sheets superimposed in a single stack. Under these conditions, the manual drafter would be able to draw on the top sheet only. Likewise, in AutoCAD you can draw only on the current layer. However, AutoCAD permits you to transfer selected objects from one layer to another (neither of which needs to be the current layer) with commands called CHANGE, CHPROP, PROPERTIES, and several others. (Let's see the manual drafter try that!)

A common application of the layer feature is to use one layer for construction (or layout) lines. You can create geometric constructions with objects, such as lines, circles, and arcs. These generate intersections, endpoints, centers, points of tangency, midpoints, and other useful data that might take the manual drafter considerable time to calculate with a calculator or to hand-measure on the board. From these you can create other objects using intersections or other data generated from the layout. Then the layout layer can be turned off (making it no longer visible) or set the layout not to plot. The layer is not lost, but can be recalled (set to ON) for viewing later as required. The same drawing limits, coordinate system, and zoom factors apply to all layers in a drawing.

To draw an object on a particular layer, first make sure that that layer is set as the "current layer." There is one, and only one, current layer. Whatever you draw will be placed on the current layer. The current layer can be compared to the manual drafter's top sheet on the stack of transparencies. To draw an object on a particular layer, that layer must first have been created; if it is not the current layer, you must make it the current layer.

You can always move, copy, or rotate any object, whether it is on the current layer or not. When you copy an object that is not on the current layer, the copy is placed on the layer that the original object is on. This is also true with the mirror or an array of an object or group of objects.

A layer can be visible (ON) or invisible (OFF). Only visible layers are displayed or plotted. If necessary, AutoCAD allows you to set a visible layer not to plot. The layer(s) that are visible and set not to plot, will not be plotted. Invisible layers are still part of the drawing; they are just not displayed or plotted. You can turn layers on and off at will, in any combination. It is possible to turn off the current layer. If this happens and you draw an object, it will not appear on the screen; it will be placed on the current layer and will appear on the screen when that layer is set to ON (provided you are viewing the area in which the object was drawn). This is not a common occurrence, but it can cause concern to both the novice and the more experienced operator who has not faced the problem before. Do not turn OFF the current layer; the results can be very confusing. When the TILEMODE system variable is set to OFF, you can make specified layers visible only in certain viewports.

Each layer in a drawing has an associated name, color, lineweight, and linetype. The name of a layer may be up to 255 characters long. It may contain letters, digits, and the special characters dollar ($), hyphen (-), underscore (_), and spaces. Always give a descriptive name appropriate to your application, such as floor-plan or plumbing. The first several characters of the current layer's name are displayed in the layer list box located on the Layers toolbar (see Figure 3–91). You can change the name of a layer any time you wish, and you can delete unused layers except layer 0.

Color

AutoCAD allows you to assign a color to a layer in a drawing. See the description later in this section of how to assign colors to layers from the range and types of colors now available in AutoCAD 2004. If necessary, you can assign the same color to more than one layer. If your graphics monitor is monochrome, all color numbers will produce the same visual effect. Even in this case, color numbers are useful, because they can be assigned to different pens on a pen plotter to plot in different line weights and colors. This works even for single-pen plotters; you can instruct AutoCAD to pause for pen changes.

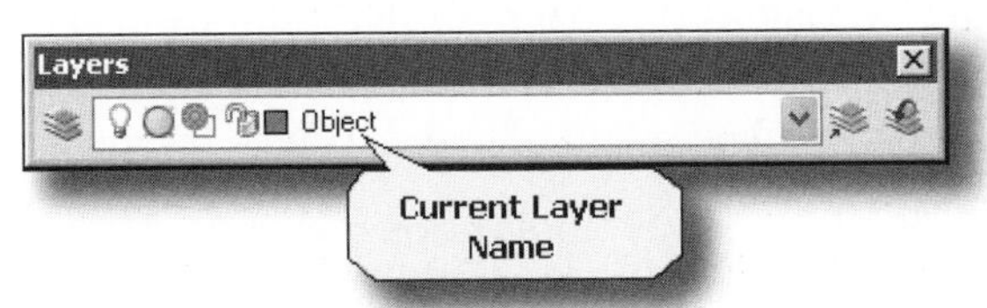

Figure 3–91 *The current layer name is displayed in the list box located on the Layers toolbar*

Lineweight

Similar to assigning color, you can also assign a specific lineweight to a layer. Lineweights add width to your objects, both on screen and on paper. Using lineweights, you can create heavy and thin lines to show varying object thicknesses in details. For example, by assigning varying lineweights to different layers, you can differentiate between new, existing, and demolition construction.

AutoCAD allows you to assign a lineweight to a specific layer or an object in either inches or millimeters, with millimeters being the default. If necessary, you can change the default setting by invoking the LINEWEIGHT command.

A lineweight value of 0 is displayed as one pixel in model space and plots at the thinnest lineweight available on the specified plotting device. Appendix G lists all the available lineweights as well as associated industry standards.

Lineweights are displayed differently in model space than in a paper space layout (refer to Chapter 8 for a detailed explanation on layouts). In model space, lineweights are displayed in relation to pixels. In a paper space layout, lineweights are displayed in the exact plotting width. You can recognize that an object has a thick or thin lineweight in model space but the lineweight does not represent an object's real-world width. A lineweight of 0 will always be displayed on screen with the minimum display width of one pixel. All other lineweights are displayed using a pixel width in proportion to its real-world unit value. A lineweight displayed in model space does not change with the zoom factor. For example, a lineweight value that is represented by a width of four pixels is always displayed using four pixels regardless of how close you zoom in to your drawing.

If necessary, you can change the display scale of lineweights of the objects in the model space to appear thicker or thinner. Changing the display scale does not affect the lineweight plotting value. However, AutoCAD regeneration time increases with lineweights that are represented by more than one pixel. By default, all lineweight values that are less than or equal to 0.01 in or 0.25 mm are displayed at one pixel and do not slow down performance in AutoCAD. If you want to optimize AutoCAD performance when working in the Model Space, set the lineweight display scale to the minimum value. The LINEWEIGHT command allows you to change the display scale of lineweights.

In model space, AutoCAD allows you to turn OFF the display of lineweights by toggling the LWT toggle button on the status bar. With the LWT toggle button OFF, AutoCAD displays all the objects with a lineweight of 0 and reduces regeneration time.

When exporting drawings to other applications or cutting/copying objects to the Clipboard, objects retain lineweight information. The system variables LWUNITS and LWDEFAULT set the units in which lineweight is displayed and applied to newly created objects and layers, and set the default value for lineweights, respectively.

Linetype

A linetype is a repeating pattern of dashes, dots, and blank spaces. AutoCAD adds the capability of including the repeated objects in the custom linetypes. The assigned linetype is normally used to draw all objects on the layer unless you set the current linetype to some other linetype.

The following are some of the linetypes that are provided in AutoCAD in a library file called ACAD.LIN:

Border	Dashdot	Dot
Center	Dashed	Hidden
Continuous	Divide	Phantom

See Appendix G for examples of each of these linetypes. Linetypes are another means of conveying visual information. You can assign the same linetype to any number of layers. In some drafting disciplines, conventions have been established giving specific meanings to particular dash-dot patterns. If a line is too short to hold even one dash-dot sequence, AutoCAD draws a continuous line between the endpoints. When you are working on large drawings, you may not see the gap between dash-dot patterns in a linetype, unless the scaling for the linetype is set for a large value. This can be done by means of the LTSCALE command. This command is discussed in more detail later in this chapter.

Every drawing will have a layer named 0 (zero). By default, layer 0 is set to ON and assigned the color white, the default lineweight, and the linetype continuous. Layer 0 cannot be renamed or deleted.

If you need additional layers, you must create them. By default, each new layer is assigned the properties of the layer directly above it in the table. If necessary, you can always reassign the color, lineweight, and linetype of the newly created layer.

CREATING AND MANAGING LAYERS WITH THE LAYER PROPERTIES MANAGER DIALOG BOX

The Layer Properties Manager dialog box or command line version of the LAYER command can be used to set up and control layers.

To open the Layer Properties Manager dialog box, invoke the LAYER command:

Layers toolbar	Choose Layers Properties manager (see Figure 3–92)
Format menu	Choose Layer...
Command: prompt	**layer** (ENTER)

Figure 3–92 *Invoking the LAYER command from the Layers toolbar*

AutoCAD displays the Layer Properties Manager dialog box, as shown in Figure 3–93.

AutoCAD lists the available layer names in the Layer Properties Manager dialog box. By default, AutoCAD provides one layer called 0 that is set to ON and is assigned the color white, the lineweight set to default, and the linetype continuous.

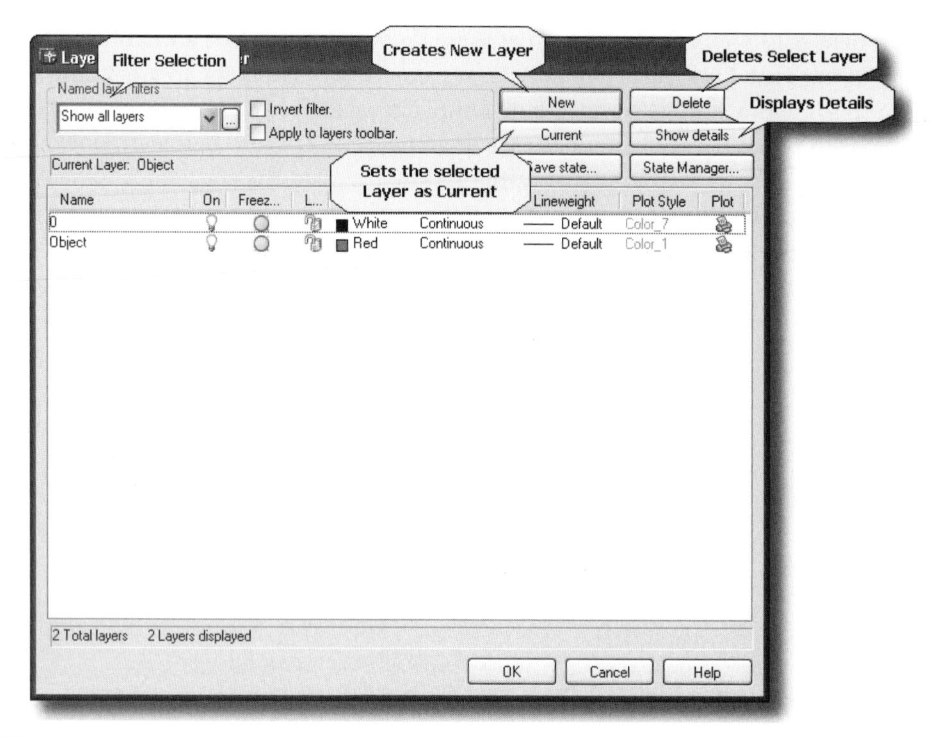

Figure 3–93 *The Layer Properties Manager dialog box*

Creating New Layers

To create new layer(s), choose the **New** button. AutoCAD then creates a new layer by assigning the name "Layer1" and setting it to ON, with the properties of the layer directly above it in the table. When first listed in the layer list box, the name Layer1 is highlighted and ready to be edited. Just enter the desired name for the new layer. If you accept the name Layer1 you can still change it later. To rename the layer, click anywhere on the line where the layer is listed and the whole line is highlighted. Then double-click on the layer name, enter a new name, and press ENTER.

Note: The layer name cannot contain wild-card characters (such as * and ?). You cannot duplicate existing names.

Making a Layer Current

To make a layer current, first choose the layer name from the layer list box in the Layer Properties Manager dialog box, and then choose the **Current** button located on the top right side of the dialog box. You can also make a layer current by double-clicking on the name. AutoCAD displays the name of the current layer just above the listing of layers.

Filters for Showing/Not Showing Layers

In the **Named layer filters** section of the Layer Properties Manager dialog box, the list box at the left shows the sets of filters by which layers are displayed in the layer list box. Choosing the down-arrow button displays the available options. The default options include *Show all layers, Show all used layers,* and *Show all Xref dependent layers.* The Show all layers selection lists all the layers created in the current drawing, the Show all used layers selection lists all the layers that have objects drawn, and the Show all Xref dependent layers selection lists all the layers that belong to external reference drawings (see Chapter 11 for a detailed explanation of external references). You can create and name other sets of filters to add to this list. Choosing the ellipsis button (...) causes the Named Layer Filters dialog box to be displayed, as shown in Figure 3–94. From this dialog box you can create and name a set of filters that cause AutoCAD to filter layers with respect to their name, color, lineweight, linetype, plot style, and whether they are ON/OFF, frozen/thawed, locked/unlocked (including viewports), and plotted/not plotted.

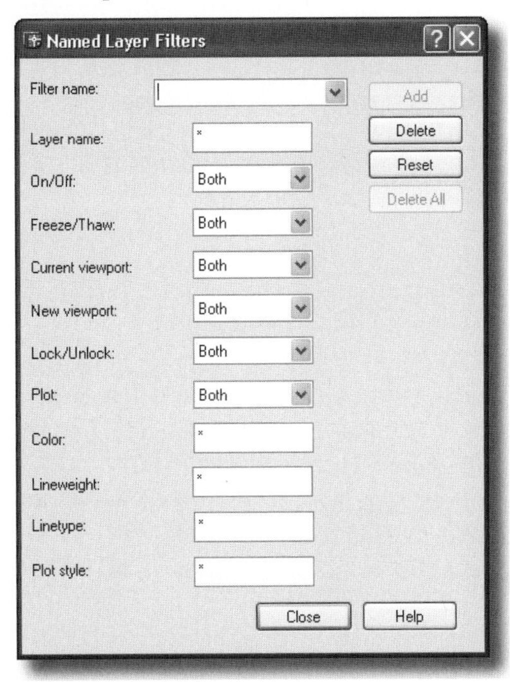

Figure 3–94 *Named Layer Filters dialog box*

When the **Invert filter** check box has no check in it, then the layers that are shown in the list are those that have all the matching characteristic(s) of the specified filter. For example, if you were to create and name a set of filters that specified the color yellow and the linetype dashed, then the list would include only layers whose color was yellow and linetype was dashed. If you put a check in the **Invert filter** check box, then all layers would be listed *except* those with both the color yellow and the linetype dashed.

When the **Apply to layers toolbar** check box has no check in it, then the layers that are filtered (or Invert filtered) are shown in the list box of the Layer Properties Manager dialog box only. When there is a check in the **Apply to layers toolbar** check box, then the list in the text box for layers in the Layers toolbar will also be filtered (or Invert filtered), depending on the status of the **Invert filter** check box.

Show Details

The **Details** section of the Layer Properties Manager dialog box is displayed by choosing the **Show details** button, as shown in Figure 3–95.

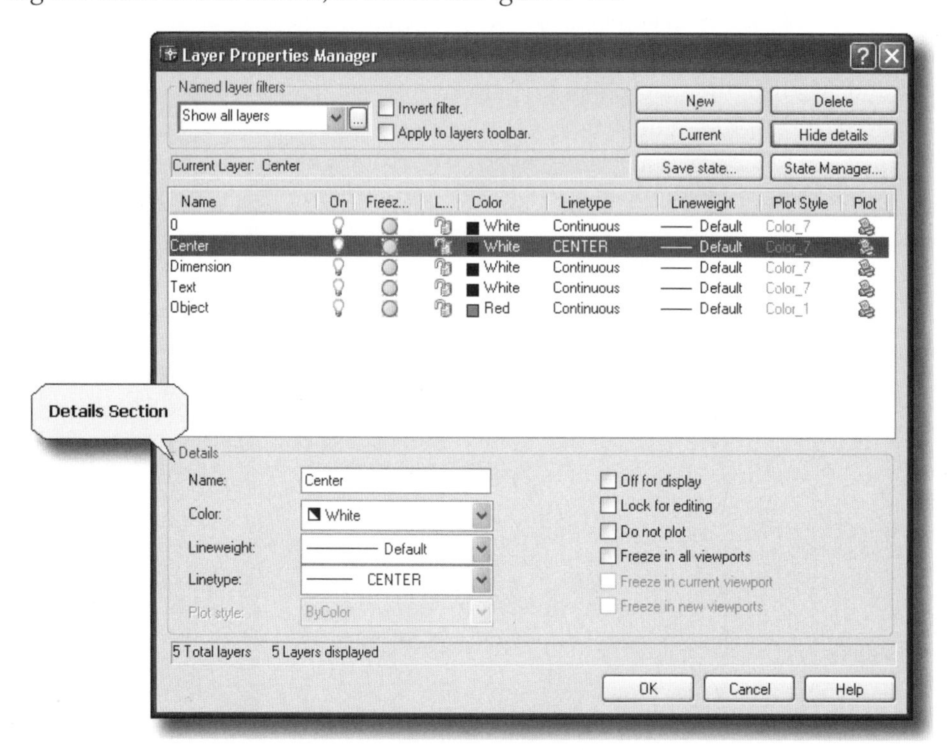

Figure 3–95 *The Layer Properties Manager dialog box with the Details section showing*

When the **Details** section is displayed, the **Show details** button becomes the **Hide details** button which, when chosen, causes the **Details** section to be hidden. The **Details** section has text boxes and check boxes showing the status of the layer that is highlighted in the list box above it. The text boxes include the **Name, Color, Lineweight, Linetype**, and **Plot style** of the highlighted layer. The check boxes include **Off for display, Lock for editing, Do not plot, Freeze in all viewports, Freeze in cur-**

rent viewport, and **Freeze in new viewport**. The Details section provides a means of changing the properties and status of layers quickly and easily.

Note: You cannot delete layer 0 or layers that contain objects. To select more than one layer, hold down CTRL and select the layer(s) from the list box.

Layer States

At any time during a drawing session, the collective status of all layer's properties settings is known as the layer state. This state can be saved and given a name by which it can be recalled later, thus having every setting of selected properties of every layer revert to what they were when that particular layer state was named and saved. Layer states are saved in files with the extension of **LAS.**

To create a new layer state, from the Layer Properties Manager dialog box choose the **Save state** button, which causes the Save Layer States dialog box to be displayed, as shown in Figure 3–96. From this dialog box you can name and save a layer state based on settings of selected properties. Once the desired properties have been selected, enter the name you wish to give the layer state to be created and choose the **OK** button. For example, you may wish to save and name a layer state based only on the visibility of the layers. Therefore you would select the **On/Off** check box and when the named layer state is restored, all of the layers will revert to the visibility status at which they were when the layer state was created. Other properties would not be changed.

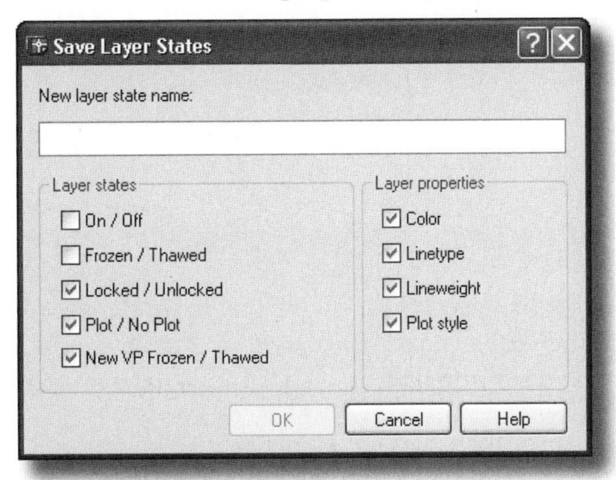

Figure 3–96 *The Save Layer States dialog box*

A saved layer state can be restored. From the Layer Properties Manager dialog box choose the **State Manager...** button, which causes the Layer States Manager dialog box to be displayed, as shown in Figure 3–97. From this dialog box you can select a saved layer state.

Figure 3–97 *The Layer States Manager dialog box*

To restore a saved layer state, select its name in the Layer States Manager dialog box and then choose the **Restore** button.

To edit a saved layer state, select its name in the Layer States Manager dialog box and then choose the **Edit** button. This causes the Edit Layer State dialog box to be displayed. The procedure is the same as in the Save Layer States dialog box except the layer state has already been named.

To delete a named and saved layer state, from the Layer States Manager dialog box choose the **Delete** button when the layer state you wish to delete is highlighted in the **Layer states** list.

To import one or more layer states into your drawing, from the Layer States Manager dialog box choose the **Import** button. This causes the Import layer state dialog box to be displayed. This dialog box is similar to other Windows file seeking and handling dialog boxes, as shown in Figure 3–98. From this dialog box you can select a saved layer state to import. Layer states are saved in files with the extension of **LAS.**

To export one or more layer states from your drawing, from the Layer States Manager dialog box choose the **Export** button. This causes the Export layer state dialog box to be displayed. This dialog box is similar to other Windows file seeking and handling dialog boxes, as shown in Figure 3–99. From this dialog box you can select a saved layer state to export. Layer states are saved in files with the extension of **LAS.**

When the layer state handling needs are complete, the Layer States Manager dialog box can be exited by choosing the **Close** button.

Figure 3–98 *The Import layer states dialog box*

Figure 3–99 *The Export layer states dialog box*

Visibility of Layers

When you turn a layer OFF, the objects on that layer are not displayed in the drawing area and they are not plotted. The objects are still in the drawing, but they are not visible on the screen. And they are still calculated during regeneration of the drawing, even though they are invisible.

To change the setting for the visibility of layer(s), select the icon corresponding to the layer name located under the **On** column (second column from the left) in the Layer Properties Manager dialog box. The icon is a toggle for ON/OFF of layers.

Freezing and Thawing Layers

In addition to turning the layers OFF, you can freeze layers. The layers that are frozen will not be visible in the view window, nor will be they be plotted. In this respect, frozen layers are similar to layers that are OFF. However, layers that are simply turned OFF still go through a screen regeneration each time the system regenerates your drawing, whereas the layers that are frozen are not considered during a screen regeneration. If you want to see the frozen layer later, you simply thaw it, and automatic regeneration of the drawing area takes place.

To change the setting for the visibility of layer(s) by freezing/thawing, select the icon corresponding to the layer name located under the **Freeze in All Viewports** column (third column from the left). The icon is a toggle for FREEZE/THAW of layers.

Assigning Plot Style to Layers

Plot styles are a collection of property settings (such as color, linetype, and lineweight) that can be assigned to a layer or to individual objects. These property settings are contained in a named plot style table (for a detailed explanation on creating plot styles, refer to Chapter 8). When applied, the plot style can affect the appearance of the plotted drawing. By default, AutoCAD assigns the plot style to Normal to a newly created layer if the Default plot style behavior is set to Named plot styles. To change the assigned plot style, choose the plot style name corresponding to the layer name located under the **Plot Style** column (eighth column from the left). AutoCAD displays the Select Plot Style dialog box, which allows you to change the Plot Style of the selected layer(s). Select the appropriate plot style from the list box, and choose the **OK** button to accept the plot style selection. To cancel the selection, choose the **Cancel** button. A layer that is assigned a Normal plot style assumes the properties that have already been assigned to that layer. You can create new plot styles, name them, and assign them to individual layers. You cannot change the plot style if the Default plot style behavior is set to Color Dependent plot styles.

Plotting Layers

AutoCAD allows you to turn plotting ON or OFF for visible layers. For example, if a layer contains construction lines that need not be plotted, you can specify that the layer is not plotted. If you turn OFF plotting for a layer, the layer is displayed but is not plotted. At plot time, you do not have to turn OFF the layer before you plot the drawing.

To change the setting for the plotting of layers, select the icon corresponding to the layer name located under the **Plot** column (ninth column from the left). The icon is a toggle for plotting or not plotting layers.

Locking and Unlocking Layers

Objects on locked layers are visible in the view window but cannot be modified by means of the modifying commands. However, it is still possible to draw on a locked layer by making it the current layer, changing the linetypes, lineweights, and colors, freezing them, and using any of the inquiry commands and Object Snap modes on them.

To change the setting for the Lock/Unlock of layer(s), select the icon corresponding to the layer name located under the **Lock** column (fourth column from the left). The icon is a toggle for LOCK/UNLOCK of layers.

Changing the Color of Layers

By default, AutoCAD assigns the color of the layer directly above it to a newly created layer. To change the assigned color, choose the corresponding icon to the layer name located under the **Color** column (fifth column from the left). AutoCAD displays the Select Color dialog box (see Figure 3–100 with the **Index Color** tab displayed), which allows you to change the color of the selected layer(s). You can select one of the 256 colors. Use the cursor to select the color you want, or enter its name or number in the **Color:** text box. Choose the **OK** button to accept the color selection. To cancel the selection, choose the **Cancel** button. AutoCAD 2004 introduces new color capabilities. In the Select Color dialog box there are two new tabs: **True Color** (see Figure 3-101) and **Color Books** (see Figure 3–102). As well as choosing from 256 standard colors, you can choose colors from the True Color graphic interface with its controls for Hue, Saturation, Luminance, and Color Model or from standard Color Books (such as Pantone). **True Color** and **Color Books** makes it easier to match colors in your drawing with colors of actual materials.

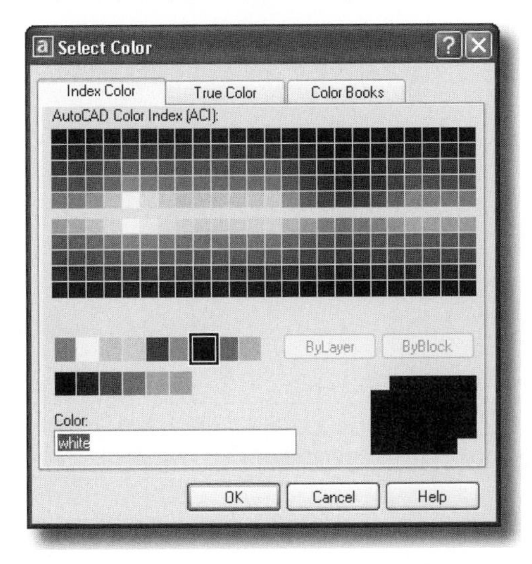

Figure 3–100 *Select Color dialog box with Index Color tab displayed*

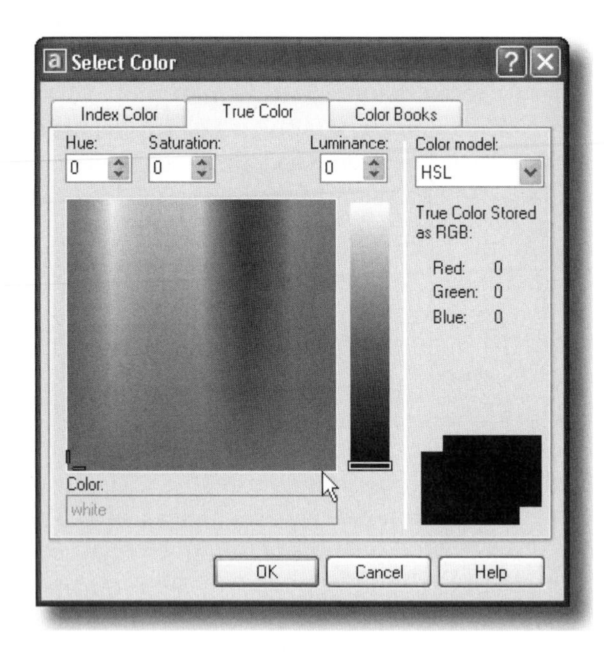

Figure 13–101 *Select Color dialog box with True Color tab displayed*

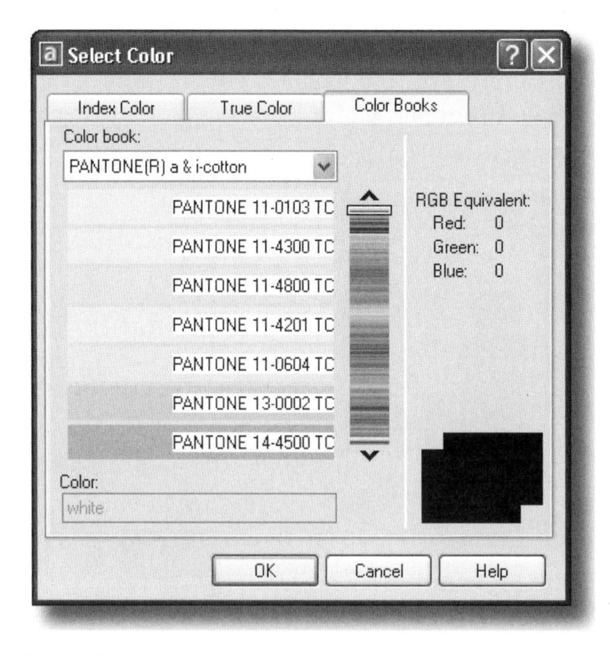

Figure 13–102 *Select Color dialog box with Color Books tab displayed*

Changing the Linetype of Layers

By default, AutoCAD assigns the linetype of the layer directly above it to a newly created layer. To change the assigned linetype, choose the linetype name corresponding to the layer name located under the **Linetype** column (sixth column from the left). AutoCAD displays the Select Linetype dialog box, which allows you to change the linetype of the selected layer(s). Select the appropriate linetype from the list box, and choose the **OK** button to accept the linetype selection. To cancel the selection, choose the **Cancel** button.

Note: AutoCAD lists only the linetypes that are loaded in the current drawing in the Select Linetype dialog box. To load additional linetypes in the current drawing, choose the **Load...** button, and AutoCAD displays the Load or Reload Linetypes dialog box. AutoCAD lists the available linetypes from the default linetype file ACAD.LIN. Select all the linetypes that need to be loaded, and select the **OK** button to load the linetypes into the current drawing. If necessary, you can change the default linetype file ACAD.LIN to another file by choosing the **File...** button and selecting the appropriate linetype file. You can also load the linetypes into the current drawing from the **Linetype** tab in the Layer and Linetype Properties dialog box.

Changing the Lineweight of Layers

By default, AutoCAD assigns the lineweight of the layer directly above it to a newly created layer. To change the assigned lineweight, choose the lineweight name corresponding to the layer name located under the **Lineweight** column (seventh column from the left). AutoCAD displays the Lineweight dialog box, which allows you to change the lineweight of the selected layer(s). Select the appropriate lineweight from the list box, and choose the **OK** button to accept the lineweight selection. To cancel the selection, choose the **Cancel** button.

If necessary, you can drag the widths of the column headings to see additional characters of the layer name, full legend for each symbol and color name, or number in the list box. If you want to sort the order in which layers are displayed in the list box, choose the column headings. The first selection lists the layers in descending order (Z to A, then numbers), and a second selection lists the layers in ascending order (numbers, A to Z). Choosing the status column headers lists the layers by the property in the list.

You can also select New Layers, Select All, Clear All, Select All But Current, Invert Selection, Invert Layer Filter, and select one of the available options from LAYER FILTERS from the shortcut menu (see Figure 3–103) that is displayed by right-clicking on your pointing device in the Layer Properties Manager dialog box.

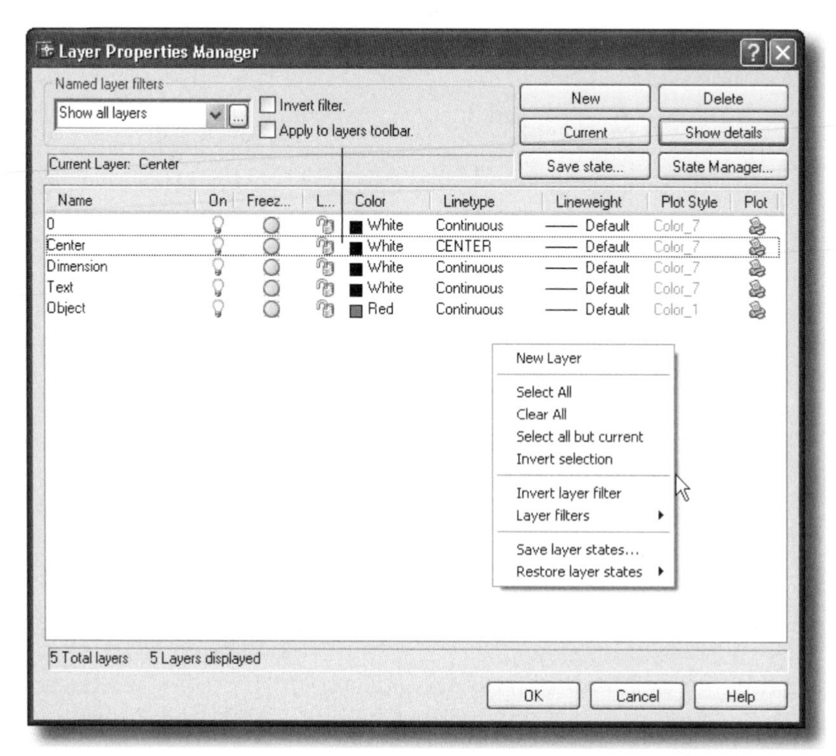

Figure 3–103 *The shortcut menu displayed in the Layer Properties Manager dialog box*

After making the necessary changes, choose the **OK** button to keep the changes and close the dialog box. To discard the changes, choose the **Cancel** button and close the dialog box.

CHANGING LAYER PROPERTIES FROM THE LAYERS TOOLBAR

You can also toggle on/off, freeze/thaw, lock/unlock, or plot/no plot, in addition to making a layer current from the layer list box provided in the Layers toolbar. Select the appropriate icon next to the layer name you wish to toggle, as shown in Figure 3–104.

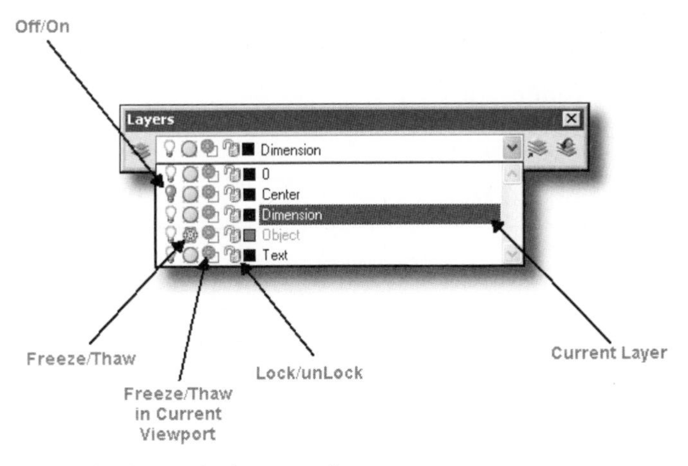

Figure 3–104 *Layer list box in the Layers toolbar*

MAKING AN OBJECT'S LAYER CURRENT

AutoCAD allows you to select an object in the drawing to make its layer the current layer. To do so, invoke the command:

Layers toolbar	Choose Make Object's Layer Current (See Figure 3–105)

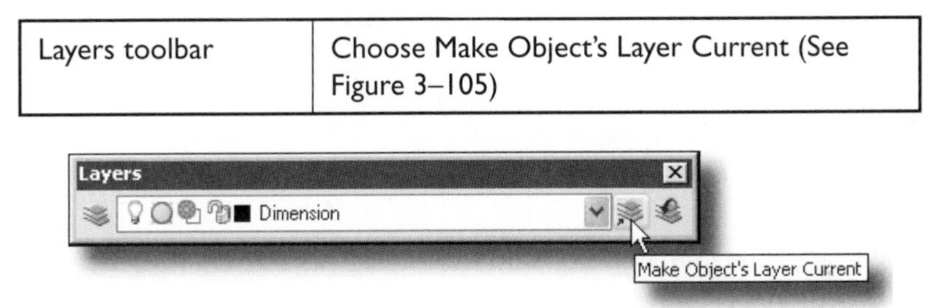

Figure 3–105 *Invoking the* MAKE OBJECT'S LAYER CURRENT *command from the Layers toolbar*

AutoCAD prompts:

> Select object whose layer will become current: *(select the object in the drawing to make its layer current)*

AutoCAD makes the selected object's layer current.

UNDOING LAYER SETTINGS

AutoCAD allows you to undo the last change or set of changes made to layer settings. To do so, invoke the command:

Layers toolbar	Choose Layer Previous (see Figure 3–106)
Command: prompt	**layerp** (ENTER)

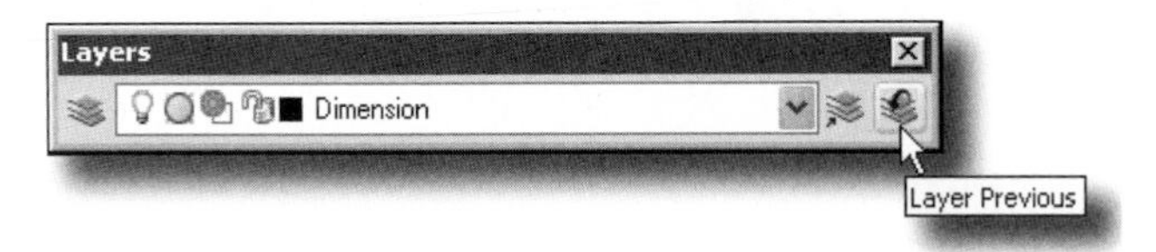

Figure 3–106 *Invoking the* LAYERP *from the Layers toolbar*

This command undoes changes you have made to layer settings such as color or linetype. If settings are restored, AutoCAD displays the message "Restored previous layer states." The LAYER PREVIOUS command restores the original properties but not the original name if the layer is renamed, and it also does not restore if you delete or purge a layer.

SETTING UP AND CONTROLLING LAYERS BY MEANS OF THE COMMAND VERSION OF THE LAYER COMMAND

Invoke the command version of the LAYER command:

Command: prompt	**-layer** (ENTER)

AutoCAD prompts:

Command: **-layer** (ENTER)
Enter an option
[?/Make/Set/New/ON/OFF/Color/Ltype/LWeight/Plot/Freeze/Thaw/LOck/
Unlock/stAte]: *(select one of the available options)*

? Option

The ? option lists the names of layers defined in the current drawing, showing their on/off, freeze/thaw, lock/unlock, lweight, plot status, color, and linetypes. When you select this option, AutoCAD prompts:

Enter layer name(s) to list <*>:

You can enter a list of layer names, using wild cards if you wish, or press the SPACEBAR or ENTER to accept the default to list all layer names.

Make

The Make option effectively does three things. When this option is invoked, AutoCAD prompts you for a layer name. Once you enter a name, Make does the following:

1. Searches for the layer to determine whether it exists.
2. Creates it if the layer does not exist, and assigns the color white and the linetype continuous.
3. Sets the newly made layer to be the current layer.

When you choose this option, AutoCAD prompts:

```
Enter name for new layer (becomes the current layer) <0>:
```

If the layer you selected exists and is presently turned OFF, AutoCAD turns it ON automatically, using the color and linetype previously assigned to that layer, and makes it the current layer.

Set

The Set option tells AutoCAD which layer you want to draw on (making it the current layer). When you choose this option, AutoCAD prompts:

```
Enter layer name to make current <0>:
```

Enter the layer's name (it must be an existing layer) and press the SPACEBAR or ENTER. This layer becomes the current layer. The first several characters of the layer name are displayed in the Layers toolbox layer list box.

New

The New option allows you to create new layers. When you choose this option, AutoCAD prompts:

```
Enter name list for new layer(s):
```

You can enter more than one name at a time by separating the names with a comma. If you need to separate characters in the name, use the underscore (_) instead of pressing the SPACEBAR (in AutoCAD, pressing the SPACEBAR is the same as pressing ENTER). Each layer thus created is automatically set to ON and assigned the color white and linetype continuous.

OFF

The OFF option allows you to turn OFF selected layers. When you choose this option, AutoCAD prompts:

```
Enter name list of layer(s) to turn off:
```

The list should contain only existing layer names. The names may include wild-card characters. When you turn OFF a layer, its associated objects will not be displayed and they are not plotted. The objects still exist in the drawing; they are just invisible. And they are still calculated during the regeneration of the drawing, even though they are not visible.

ON

The ON option allows you to turn ON layers that have been turned OFF. When you choose this option, AutoCAD prompts:

```
Enter name list of layer(s) to turn on:
```

This list should contain only existing layer names. The name may include wild-card characters. Each designated layer is turned on using the color and linetype previously associated with it. Turning a layer ON does not cause it to be the current layer.

Color

The Color option allows you to change the color associated with that specific layer. When you choose this option, AutoCAD first prompts:

```
Enter color name or number (1-255):
```

Reply with one of the standard color names or with a legal color number from 1 to 255 (for example, 2 for yellow). After you specify the color, AutoCAD prompts:

```
Enter name list of layer(s) for color 2 (yellow) <0>:
```

Reply with the names of existing layers, separated by commas. The names may include wild-card characters. The specified layers are given the color you designated and are then automatically set to ON if they are set to OFF. If you would prefer to assign the color but to turn the layers OFF, precede the color with a minus sign (–).

Ltype

The Ltype option allows you to change the linetype associated with a specific layer. When you choose this option, AutoCAD prompts:

```
Enter loaded linetype name or [?] <Continuous>:
```

Reply with the name of an existing defined linetype. AutoCAD then asks for a list of layer names to which the linetype should be applied. For example, if you had replied to the first prompt with the linetype named HIDDEN, the next prompt would be:

```
Enter name list of layer(s) for linetype "hidden" <0>:
```

Reply with the names of existing layers, separated by commas. The names may include wild-card characters.

LWeight

The LWeight option allows you to change the lineweight associated with a specific layer. When you choose this option, AutoCAD prompts:

```
Enter lineweight (0.0mm - 2.11mm):
```

Enter a numeric value for the lineweight. AutoCAD then asks for a list of layer names to which the lineweight should be applied. For example, if you had replied to the first prompt with the lineweight of 0.1, the next prompt would be:

```
Lineweight rounded to nearest valid value of 0.09 mm
Enter name list of layers(s) for lineweight 0.09mm <default>:
```

Reply with the names of existing layers, separated by commas.

Freeze and Thaw

The layers that are frozen will not be visible on the display, nor will they be plotted. In this respect, Freeze is similar to OFF. However, layers that are simply turned OFF still go through a screen regeneration each time the system regenerates your drawing. If you want to see the frozen layer later, you simply Thaw it, and automatic regeneration of the screen takes place. The layers that are frozen will not be regenerated.

When you choose the Freeze option, AutoCAD prompts:

```
Enter name list of layer(s) to freeze:
```

You can enter more than one name at a time by separating the names with a comma.

When you choose the Thaw option, AutoCAD prompts:

```
Enter name list of layer(s) to thaw:
```

You can enter more than one name at a time by separating the names with a comma.

LOck and Unlock

Objects on locked layers are visible on the display but cannot be modified with the modify commands. When you choose the LOck option, AutoCAD prompts:

```
Enter name list of layer(s) to lock:
```

You can enter more than one name at a time by separating the names with a comma.

When you choose the Unlock option, AutoCAD prompts:

```
Enter name list of layer(s) to unlock:
```

You can enter more than one name at a time by separating the names with a comma.

Plot and No Plot

Objects on layers whose Plot Status is Plot will be plotted. Objects on layers whose Plot Status is No Plot will not be plotted. When you choose the Plot option, AutoCAD prompts:

```
Enter a plotting preference [Plot/No plot] <Plot>: p
Enter layer name(s) for this plot preference <0>:
```

You can enter more than one name at a time by separating the names with a comma.

When you choose the No Plot option, AutoCAD prompts:

```
Enter a plotting preference [Plot/No plot] <Plot>: n
Enter layer name(s) for this plot preference <0>:
```

You can enter more than one name at a time by separating the names with a comma. Whenever you give a null response to the LAYER command, it returns you to the "Command:" prompt.

State

The state can be saved and given a name by which it can be recalled later, thus having every setting of selected properties of every layer revert to what they were when that particular layer state was named and saved. Layer states are saved in files with the extension of LAS.

When you choose the stAte option, AutoCAD prompts:

```
Enter an option [?/Save/Restore/Edit/Name/Delete/Import/EXport]:
```

When you choose the ? option, AutoCAD lists the currently saved layer states. For example, if layer states "test01" and "test02" haved been saved in the drawing AutoCAD's response to the ? option would be to display:

```
Layer state
----------------------------------------
test01
test02
```

When you choose the Save option, AutoCAD prompts:

```
Enter an option [?/Save/Restore/Edit/Name/Delete/Import/EXport]: s
```

AutoCAD prompts for the new layer state name:

```
Enter new layer state name:
```

Enter a name for the new layer state and press ENTER. AutoCAD displays the AutoCAD Text Window with each line of information shows one of the layer properties along with its current status.

After you enter a name, AutoCAD prompts:

```
Enter states to change [On/Frozen/Lock/Plot/Newvpfreeze/Color/lineType/
   lineWeight/plotStyle]:
```

When the property name or keyletter is entered a new list is displayed with the selected property status changed and AutoCAD continues to prompt you enter states to change. To terminate the states option respond with ENTER:

Enter states to change [On/Frozen/Lock/Plot/Newvpfreeze/Color/lineType/
lineWeight/plotStyle]:(ENTER)

AutoCAD returns you to the layer state prompt:

Enter an option [?/Save/Restore/Edit/Name/Delete/Import/EXport]:

When you choose the Restore option, AutoCAD prompts: for the name of the layer state to restore. For example, if layer states "test01" and "test02" haved been saved in the drawing AutoCAD's response to the ? option would be to display

Layer state
----
test01
test02
Enter name of layer state to restore or [?]:

Enter the name of the desired layer state.

When you choose the Edit option, AutoCAD prompts:

Enter name of layer state to edit or [?]

Enter a name for the layer state to edit and press (ENTER). AutoCAD displays the AutoCAD Text Window with each line shows one of the layer properties along with its current status.

After you enter a name, AutoCAD prompts:

Enter states to change [On/Frozen/Lock/Plot/Newvpfreeze/Color/lineType/
lineWeight/plotStyle]:

When the property name or keyletter is entered a new list is displayed with the selected property status changed and AutoCAD continues to prompt you enter states to change. To terminate the states option respond with ENTER:

Enter states to change [On/Frozen/Lock/Plot/Newvpfreeze/Color/lineType/
lineWeight/plotStyle]:(ENTER)

AutoCAD returns you to the layer state prompt:

Enter an option [?/Save/Restore/Edit/Name/Delete/Import/EXport]:

When you choose the Name option, AutoCAD prompts:

Enter name of layer state to rename or [?]:

Enter a name for the layer state to rename and press (ENTER)

When you choose the Delete option, AutoCAD prompts:

Enter name of layer state to delete or [?]:

Enter a name for the layer state to delete and press (ENTER)

When you choose the Import option, AutoCAD prompts:

Enter name of layer state to import or [?]:

Enter a name for the layer state to import and press (ENTER)

When you choose the Export option, AutoCAD prompts:

Enter name of lay'er state to export or [?]:

Enter a name for the layer state to export and press (ENTER)

SETTING THE LINETYPE SCALE FACTOR

The linetype scale factor allows you to change the relative lengths of dashes and spaces between dashes and dots linetypes per drawing unit. The definition of the linetype instructs AutoCAD on how many units long to make dashes and the spaces between dashes and dots. As long as the linetype scale is set to 1.0, the displayed length of dashes and spaces coincides with the definition of the linetype. The LTSCALE command allows you to set the linetype scale factor.

Invoke the LTSCALE command:

Command: prompt	**ltscale** (ENTER)

AutoCAD prompts:

Command: **ltscale** (ENTER)
Enter new linetype scale factor <1.0000>: *(specify the scale factor)*

Changing the linetype scale affects all linetypes in the drawing. If you want dashes that have been defined as 0.5 units long in the dashed linetype to be displayed as 10 units long, you set the linetype scale factor to 20. This also makes the dashes that were defined as 1.25 units long in the center linetype display as 25 units long, and the short dashes (defined as 0.25 units long) display as 5 units long. Note that the 1.25-unit-long dash in the center linetype is 2.5 times longer than the 0.5-unit-long dash in the dashed linetype. This ratio will always remain the same, no matter what the setting of LTSCALE. So if you wish to have some other ratio of dash and space lengths between different linetypes, you will have to change the definition of one of the linetypes in the ACAD.LIN file.

Remember that linetypes are for visual effect. The actual lengths of dashes and spaces are bound more to how they should look on the final plotted sheet than to distances or sizes of any objects on the drawing. An object plotted full size can probably use an LTSCALE setting of 1.0. A 50'-long object plotted on an 18" x 24" sheet might be plotted at a 1/4" = 1'-0" scale factor. This would equate to 1 = 48. An LTSCALE setting of 48 would make dashes and spaces plot to the same lengths as the full-size plot with a setting of 1.0. Changing the linetype scale factor causes the drawing to regenerate.

WILD CARDS AND NAMED OBJECTS

AutoCAD lets you use a variety of wild cards for specifying selected groups of named objects when responding to prompts during commands. By placing one or more of these wild cards in the string (your response), you can specify a group that includes (or excludes) all of the objects with certain combinations or patterns of characters.

The types of objects associated with a drawing that are referred to by name include blocks, layers, linetypes, text styles, dimension styles, named User Coordinate Systems, named views, shapes, plot styles and named viewport configurations.

The wild-card characters include the two commonly used in DOS (* and ?), as well as eight more that come from the UNIX operating system. Here is a table listing the wild cards and their uses.

Wild Card	Use
# (pound)	Matches any numeric digit
@ (at)	Matches any alphanumeric character
. (period)	Matches any character except alphanumeric
* (asterisk)	Matches any string. It can be used anywhere in the search pattern: the beginning, middle, or end of the string.
? (question mark)	Matches any single character
~ (tilde)	Matches anything but the pattern
[...]	Matches any one of the characters enclosed
[~...]	Matches any character not enclosed
[-] (hyphen)	Specifies single-character range
‘ (reverse quote)	Reads characters literally

The following table shows some examples of wild card patterns.

Pattern	Will match or include ...	But not ...
ABC	Only ABC	
~ABC	Anything but ABC	
?BC	ABC through ZBC	AB, BC, ABCD, XXBC
A?C	AAC through AZC	AC, ABCD, AXXC, ABCX
AB?	ABA through ABZ	AB, ABCE, XAB
A*	Anything starting with A	XAAA
A*C	Anything starting with A and ending with C	XA, ABCDE
*AB	Anything ending with AB	ABCX, ABX
AB	AB anywhere in string	AXXXB
~*AB*	All strings without AB	AB, ABX, XAB, XABX
[AB]C	AC or BC	ABC, XAC
[A-K]D	AD, BD, through KD	ABC, AKC, KD

U, UNDO, AND REDO COMMANDS

The UNDO command undoes the effects of the previous command or group of commands, depending on the option employed. The U command reverses the most recent operation, and the REDO command is a one-time reversal of the effects of the previous U and UNDO commands.

U COMMAND

The U command undoes the effects of the previous command by displaying the name of that command. Pressing ENTER after using the U command undoes the next-previous command, and continues stepping back with each repetition until it reaches the state of the drawing at the beginning of the current editing session.

When an operation cannot be undone, AutoCAD displays the command name but performs no action. An operation external to the current drawing, such as plotting or writing to a file, cannot be undone.

Invoke the U command:

Standard toolbar	Choose UNDO (see Figure 3–107)
Pull-down menu	Edit > Undo
Command: prompt	**u** (ENTER)

Figure 3–107 *Invoking the* U *command from the Standard toolbar*

AutoCAD reverses the most recent operation. For example, if the previous command sequences drew a circle and then copied it, two U commands would undo the two previous commands in sequence, as follows:

Command: **u** (ENTER)
COPY
Command: (ENTER)
CIRCLE

Using the U command after commands that involve transparent commands or subcommands causes the entire sequence to be undone. For example, when you set a dimension variable and then perform a dimension command, a subsequent U command nullifies the dimension drawn and the change in the setting of the dimension variable.

UNDO COMMAND

The UNDO command permits you to select a specified number or marked group of prior commands for undoing.

Invoke the UNDO command:

Command: prompt	**undo** (ENTER)

AutoCAD prompts:

Command: **undo** (ENTER)
Enter the number of operations to undo or [Auto/Control/BEgin/End/Mark/Back]: *(specify number of undo operations to undo or select one of the available options)*

Control

The Control option controls the number of available options. By limiting the number of options, you can free up the memory and disk space that is otherwise being used to save undoing operation information. AutoCAD prompts as follows when the Control option is selected:

Command: **undo** (ENTER)
Enter the number of operations to undo or [Auto/Control/BEgin/End/Mark/Back]: **c** (ENTER)
Enter an UNDO control option [All/None/One]<All>: *(select one of the three available options)*

Selection of the All option (the default setting of the UNDO command) enables all Undo options. The None option disables the U and UNDO commands but not the Control option of the UNDO command that re-enables various options. And the One option sets the UNDO command to work like the U command.

Mark and Back

If you are at a point in the editing session at which you would like to experiment but would still like the option of undoing the experiment, you can mark that point. An example of the use of the Mark and Back options is as follows:

Command: **line** (ENTER) *(draw a line)*
Command: **circle** (ENTER) *(draw a circle)*
Command: **undo** (ENTER)
Auto/Control/Begin/End/Mark/Back/<Number>: **m** (ENTER)
Command: **text** (ENTER) *(enter text)*
Command: **arc** (ENTER) *(draw an arc)*
Command: **undo** (ENTER)
Auto/Control/Begin/End/Mark/Back/<Number>: **b** (ENTER)

The Back option returns you to the state of the drawing that has the line and the circle. Following this UNDO Back with a U removes the circle. Another U removes the line. Another U displays the following prompt:

Everything has been undone

Using the Back option when no Mark has been established will prompt:

This will undo everything. OK? <Y>

Responding Y undoes everything done since the current editing session was begun or since the last SAVE command.

Note: The default is Y; think twice before pressing ENTER in response to the "This will undo everything" prompt.

Begin and End

AutoCAD's U and UNDO commands treat the operations between an UNDO Begin and an UNDO End as one command. A Begin option entered after another Begin option (before an UNDO End) will automatically invoke an UNDO End option, thereby grouping the operations since that prior Begin option. If the UNDO Control has been set to None or One, the Begin option will not work. Using the U command is permissible after Begin and before a U End, to undo operations, but only back to the UNDO Begin.

The Begin and End options are normally intended for use in strings of menu commands where a menu pick involves several operations.

Auto

The Auto option causes multiple operations invoked by a single menu pick to be treated as one command by the U or UNDO command. UNDO Begin should be placed at the beginning of a menu string, with UNDO End at the end of the string. It has no effect if the UNDO Control has been set to None or One, however.

The effects of the following commands cannot be undone:

AREA, ATTEXT, DBLIST, DELAY, DIST, DXFOUT, END, FILES, FILMROLL, GRAPHSCR, HELP, HIDE, ID, IGESOUT, LIST, MSLIDE, PLOT, PRPLOT, QUIT, REDRAW, REDRAWALL, REGENALL, RESUME, SAVE, SHADE, SHELL, STATUS, and TEXTSCR.

Undo History Arrow

When the down arrow to the right of the UNDO command icon on the Standard menu is selected, a drop-down menu is displayed with a list of commands that can be undone by using the UNDO command. They are listed from the most recent to the earliest. You can select any command on the list and AutoCAD will undo all of the commands back to and including the command selected.

REDO COMMAND

The REDO command permits reversal of prior U or UNDO commands. Prior versions were limited to only one REDO operation, but withAutoCAD 2004, all UNDO commands can be reversed with the REDO command. In order to function, the REDO command must be used following the U or UNDO command and prior to any other action.

Invoke the REDO command:

Standard toolbar	Choose REDO (see Figure 3–105)
Edit menu	Choose Redo
Command: prompt	**redo** (ENTER)

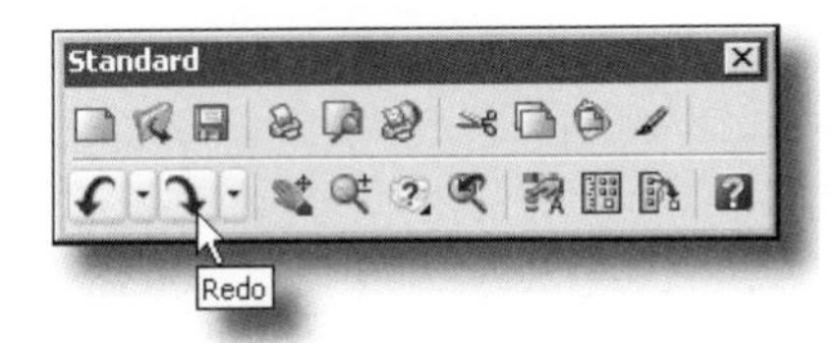

Figure 3–108 *Invoking the REDO command from the Standard toolbar*

AutoCAD reverses the prior U or UNDO command.

The REDO command does not have any options.

Open the Exercise Manual PDF file for Chapter 3 on the accompanying CD for project and discipline specific exercises.

If you have the accompanying Exercise Manual, refer to Chapter 3 for project and discipline specific exercises.

REVIEW QUESTIONS

1. The invisible grid, which the crosshairs lock onto, is called:

   a. SNAP
   b. GRID
   c. ORTHO
   d. Cursor Lock

2. To globally change the sizes of the dashes for all dashed lines, you should adjust:

   a. Line Scale
   b. LTSCALE
   c. SCALE
   d. Layer Scale

3. To reverse the effect of the last 11 commands, you could:

   a. Use the UNDO command
   b. Use the U command multiple times
   c. Either A or B
   d. It is not possible, AutoCAD only retains the last 10 commands

4. The following are AutoCAD tools available, except:

   a. GRID
   b. SNAP
   c. ORTHO
   d. TSNAP
   e. OSNAP

5. If you just used the U command, what command would restore the drawing to the state before the U command?

   a. RESTORE
   b. U
   c. REDO
   d. OOPS

6. Which of the following cannot be modified in the Drafting Settings dialog box?

   a. Snap
   b. Grid
   c. Ortho
   d. Limits
   e. All of the above

7. If the spacing of the visible grid is set too small, AutoCAD responds as follows:
   a. does not accept the command
   b. produces a "Grid too dense to display" messages
   c. produces a display that is distorted
   d. automatically adjusts the size of the grid so it will display
   e. displays the grid anyway

8. The smallest number that can be displayed in the denominator when setting units to architectural units is:
   a. 8
   b. 16
   c. 64
   d. 128
   e. none of the above

9. All of the following are considered valid options of the -LAYER command, except:
   a. On
   b. Use
   c. Make
   d. Lock
   e. Set

10. Which of the following is not a valid option of the Layer Properties Manager dialog box?
    a. Close
    b. Lock
    c. On
    d. Freeze
    e. Color

11. After having drawn a 3-point circle, you want to begin a line at the exact center of the circle. What tool in AutoCAD would you use?
    a. Snap
    b. Object Snap
    c. Entity Snap
    d. Geometric Calculator

12. How many previous zooms are available with the previous option of the ZOOM command?
    a. 4
    b. 6
    c. 8
    d. 10

13. The assignment of a specific color to a specific layer is permitted by the -LAYER command option:

a. Set
b. New
c. Color
d. Make
e. Both C and D

14. In general, a REDRAW is quicker than a REGEN.

a. True
b. False

15. When a layer is ON and THAWed:

a. the objects on that layer are visible on the monitor
b. the objects on that layer are not visible on the monitor
c. the objects on that layer are ignored by a REGEN
d. the drawing REDRAW time is reduced
e. the objects on that layer cannot be selected

16. To ensure the entire limits of the drawing are visible on the display, you should perform a ZOOM-

a. All
b. Previous
c. Extents
d. Limits

17. A layer where objects may not be edited or deleted, but are still visible on the screen and may be OSNAPed to is considered:

a. Frozen
b. Locked
c. On
d. unSet
e. fiXed

18. The -LAYER command will allow you to:

a. assign colors
b. assign linetypes
c. list previously created layers
d. selectively turn ON and OFF the layers
e. all of the above

19. Which Osnap option allows you to select the closest endpoint of a line, arc, or polyline segment?

    a. ENDpoint

    b. MIDpoint

    c. CENter

    d. INSertion point

    e. PERpendicular

20. Which Osnap option allows you to select the point in the exact center of a line?

    a. ENDpoint

    b. MIDpoint

    c. CENter

    d. INSertion point

    e. PERpendicular

21. When drawing a line which Osnap option allows you to select the point on a line or polyline segment where the angle formed with the line is a 90 degree angle?

    a. ENDpoint

    b. MIDpoint

    c. CENter

    d. INSertion point

    e. PERpendicular

22. Which Osnap option allows you to select the location where two lines, arcs, or polyline segments cross each other?

    a. NODe

    b. QUADrant

    c. TANgent

    d. NEArest

    e. INTersection

23. Which Osnap option allows you to select the location where a "point" has been established?

    a. NODe

    b. QUADrant

    c. TANgent

    d. NEArest

    e. INTersection

24. Which Osnap option allows you to select a point on a circle that is 0, 90, 180, or 270 degrees from the circle's center?

    a. NODe

    b. QUADrant

    c. TANgent

    d. NEArest

    e. INTersection

25. Which Osnap option allows you to select a point on any object, except text, that is closest to the cursor's position?

    a. NODe

    b. QUADrant

    c. TANgent

    d. NEArest

    e. INTersection

26. Which Osnap option allows you to select the location where a two lines or arcs may or may not cross each other in 3D space?

    a. QUADrant

    b. APParent intersection

    c. NEArest

    d. INTersection

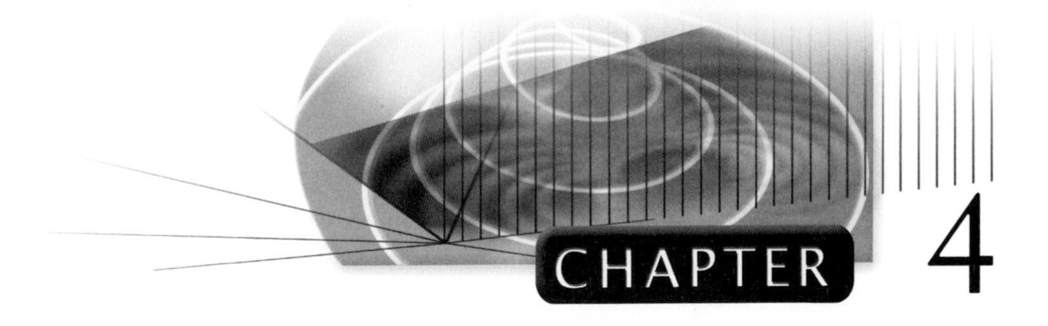

Fundamentals III

INTRODUCTION

After completing this chapter, you will be able to do the following:

- Draw construction lines using the XLINE and RAY commands
- Construct geometric figures with polygons, ellipses, and polylines
- Create single line text and multi-line text using appropriate styles and sizes, to annotate drawings
- Use the construct commands: COPY, ARRAY, OFFSET, MIRROR, FILLET, and CHAMFER
- Use the modify commands MOVE, TRIM, BREAK, and EXTEND

DRAWING CONSTRUCTION LINES

AutoCAD provides a powerful tool for drawing lines (XLINE and RAY commands) that extend infinitely in one or both directions. These lines have no effect, however, on the ZOOM EXTENTS command. They can be moved, copied, and rotated like any other objects. If necessary, you can trim the lines, break them anywhere with the BREAK command, draw an arc between two non-parallel construction lines with the FILLET command, and draw a chamfer between two non-parallel construction lines with the CHAMFER command.

The construction lines can be used as reference lines for creating other objects. To keep the construction lines from being plotted, you can draw them on a layer that can be turned off or set not to plot at the same time the objects are visible on the drawing.

XLINE COMMAND

The XLINE command allows you to draw lines that extend infinitely in both directions from the point selected when being created.

Invoke the XLINE command:

Draw toolbar	Choose the Construction Line command (see Figure 4–1)
Draw menu	Choose Construction Line
Command: prompt	**xline** (ENTER)

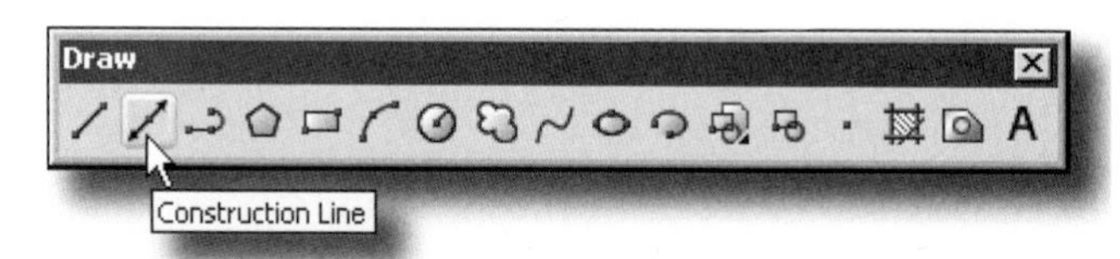

Figure 4–1 *Invoking the XLINE command from the Draw toolbar*

AutoCAD prompts:

Command: **xline** (ENTER)
Specify a point or [Hor/Ver/Ang/Bisect/Offset]: *(specify a point or right-click for shortcut menu and choose one of the available options)*

When you specify a point to define the root of the construction line, this point becomes the conceptual midpoint of the construction line. AutoCAD prompts:

Specify through point: *(specify a point through which the construction line should pass)*

AutoCAD draws a line that passes through two points and extends infinitely. AutoCAD continues to prompt for additional points to draw construction lines. To terminate the command sequence, press ENTER or the SPACEBAR.

Horizontal

The Horizontal option allows you to draw a construction line through a point that you specify and is parallel to the *X* axis of the current UCS.

Vertical

The Vertical option allows you to draw a construction line through a point that you specify and is parallel to the *Y* axis of the current UCS.

Angle

The Angle option allows you to draw a construction line at a specified angle. AutoCAD prompts:

Enter angle of xline or [Reference]>: *(specify an angle at which to place the construction line)*
Specify through point: *(specify a point through which the construction line should pass)*

AutoCAD draws the construction line through the specified point, using the specified angle.

The Reference option allows you to draw a construction line at a specific angle for a selected reference line. The angle is measured counterclockwise from the reference line.

Bisect

The Bisect option allows you to draw a construction line through the first point bisecting the angle determined by the second and third points, with the first point being the vertex. AutoCAD prompts:

Specify angle vertex point: *(specify a point for the vertex of an angle to be bisected and through which the construction line will be drawn)*
Specify angle start point: *(specify a point to determine one boundary line of an angle)*
Specify angle endpoint: *(specify a point to determine second boundary line of angle)*

The construction line lies in the plane determined by the three points.

Offset

The Offset option allows you to draw a construction line parallel to and at the specified distance from the line object selected and on the side selected. AutoCAD prompts:

Specify offset distance or [Through]: *(specify an offset distance or right-click for shortcut menu and choose one of the available options)*
Select a line object: *(select a line, pline, ray, or xline)*
Specify side to offset? *(specify a point to draw a construction line parallel to the selected object)*

The Through option allows you to specify a point through which a construction line is drawn to the line object selected.

RAY COMMAND

The RAY command allows you to draw lines that extend infinitely in one direction from the point selected when the line is being created.

Invoke the RAY command:

Draw menu	Choose RAY
Command: prompt	**ray** (ENTER)

AutoCAD prompts:

Command: **ray** (ENTER)
Specify start point: *(specify the start point to draw the ray)*
Specify through point: *(specify a point through which you want the ray to pass)*
Specify through point: *(specify a point to draw additional rays or press* ENTER *to terminate the command sequence)*

The ray is drawn starting at the first point and extending infinitely in one direction through the second point. AutoCAD continues to prompt for through points until you provide a null response to terminate the command sequence.

DRAWING POLYGONS

AutoCAD allows you to draw 2D polygons (edges with equal length) with the POLYGON command. The number of sides can be anywhere from 3 (which forms an equilateral triangle) to 1024. AutoCAD draws a polygon as a polyline with zero width and no tangent information. If necessary, with the help of the PEDIT command you can modify the polygon, such as changing its width.

Invoke the POLYGON command:

Draw toolbar	Choose the POLYGON command (see Figure 4–2)
Draw menu	Choose Polygon
Command: prompt	**polygon** (ENTER)

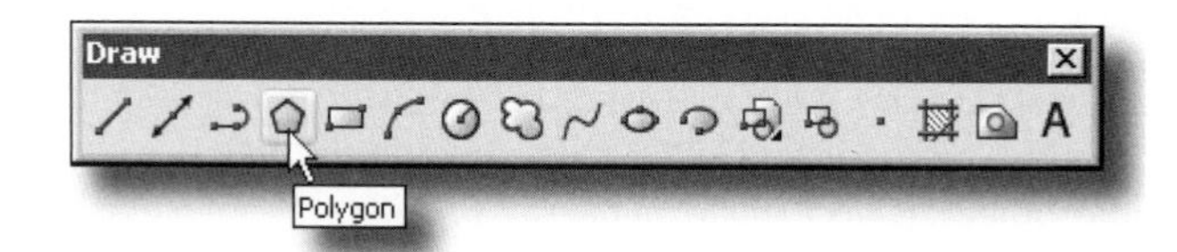

Figure 4–2 *Invoking the* POLYGON *command from the Draw toolbar*

AutoCAD prompts:

Command: **polygon** (ENTER)
Enter number of sides<default>: *(specify the number of sides for the polygon to be drawn)*
Specify center of polygon or [Edge]: *(specify center point for polygon or right-click for shortcut menu and choose Edge option)*

After specifying the center point of polygon, AutoCAD provides two options to draw the polygon: Inscribed in circle and Circumscribed about circle.

The Inscribed in circle option selection draws the polygon of equal length for all sides inscribed inside an imaginary circle (see Figure 4–3) having the same diameter as the distance across opposite polygon corners (for an even number of sides).

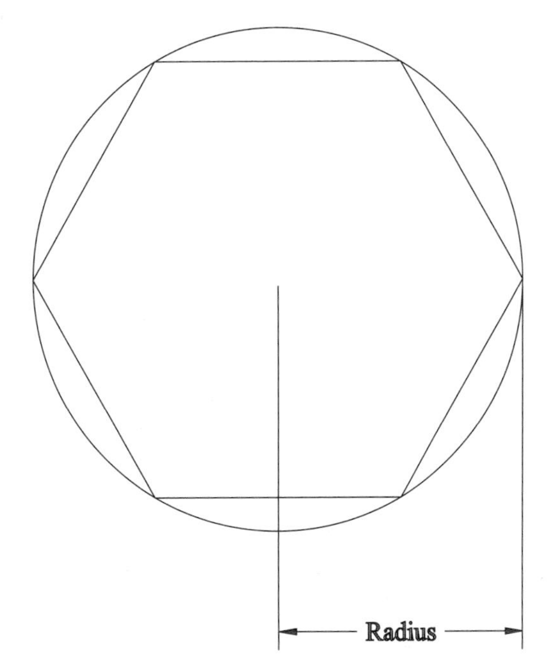

Figure 4–3 *Example of displaying a polygon inscribed inside an imaginary circle*

For example, the following command sequence shows steps in drawing a six-sided polygon (see Figure 4–4).

Command: **polygon** (ENTER)
Enter number of sides<default>: **6** (ENTER)
Specify center of polygon or [Edge]: **3,3** (ENTER)
Enter an option [Inscribed in circle/Circumscribed about circle]: **I** (ENTER)
Specify radius of circle: **2** (ENTER)

AutoCAD draws a polygon with six sides, centered at 3,3, whose edge vertices are 2 units from the center of the polygon.

Specifying the radius with a specific value draws the bottom edge of the polygon at the current snap rotation angle. If instead, you specify the radius with your pointing device or by means of coordinates, AutoCAD determines the rotation and size of the polygon.

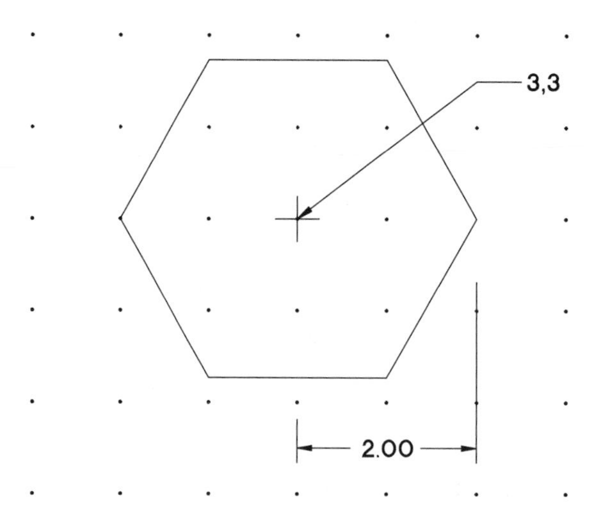

Figure 4–4 *Polygon drawn with six sides by selecting the Inscribed in circle option*

The following command sequence is an example of drawing a polygon by specifying the radius with relative coordinates.

Command: **polygon** (ENTER)
Enter number of sides<default>: **6** (ENTER)
Specify center of polygon or [Edge]: **3,3** (ENTER)
Enter an option [Inscribed in circle/Circumscribed about circle]: **I** (ENTER)
Specify radius of circle: **@2<90** (ENTER)

The Circumscribed about circle option selection draws a polygon circumscribed around the outside of an imaginary circle having the same diameter as the distance across the opposite polygon sides (for an even number of sides) (see Figure 4–5).

For example, the following command sequence shows steps in drawing an eight-sided polygon with the selection of the Circumscribed about circle option (see Figure 4–6).

Command: **polygon** (ENTER)
Enter number of sides<default>: **8** (ENTER)
Specify center of polygon or [Edge]: **3,3** (ENTER)
Enter an option [Inscribed in circle/Circumscribed about circle]: **c** (ENTER)
Specify radius of circle: **2** (ENTER)

AutoCAD draws a polygon with eight sides, centered at 3,3, whose edge vertices are 2 units from the center of the polygon.

Specifying the radius with a value draws the bottom edge of the polygon at the current snap rotation angle. If instead, you specify the radius with your pointing device or by means of coordinates, AutoCAD determines the rotation and size of the polygon.

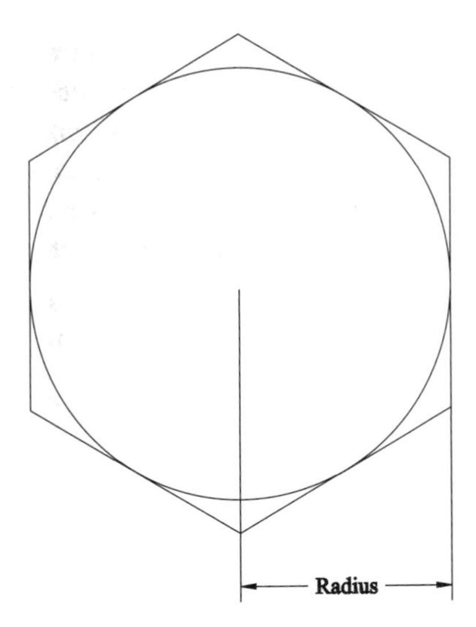

Figure 4–5 *Example of displaying a polygon circumscribed around the outside an imaginary circle*

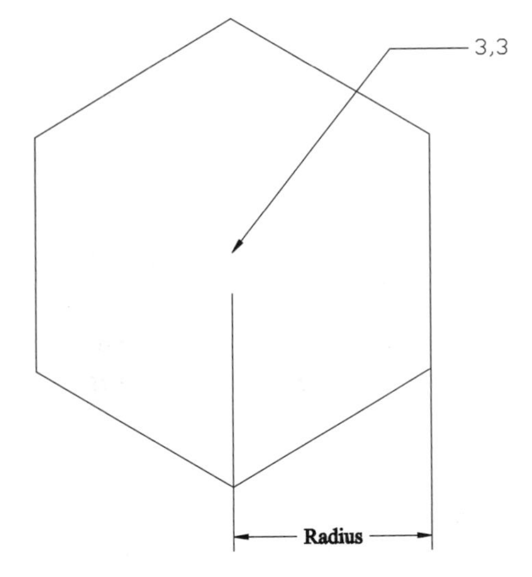

Figure 4–6 *Polygon drawn with eight sides by selecting the Circumscribed about circle option*

Edge

The Edge option allows you to draw a polygon by specifying the endpoints of the first edge.

For example, the following command sequence shows steps in drawing a seven-sided polygon (see Figure 4–7).

Command: **polygon** (ENTER)
Enter number of sides<default>: **7** (ENTER)
Specify center of polygon or [Edge]: **e** (ENTER)
Specify first endpoint of edge: **1,1** (ENTER)
Specify second endpoint of edge: **3,1** (ENTER)

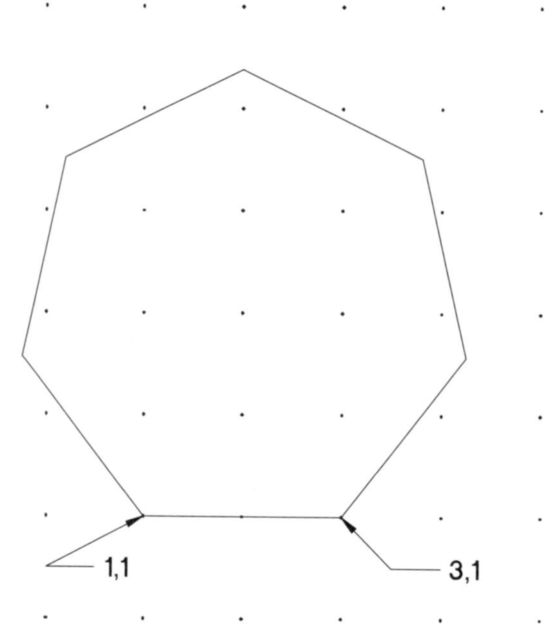

Figure 4–7 *Polygon drawn with seven sides by selecting the Edge option*

DRAWING ELLIPSES

AutoCAD allows you to draw an ellipse or an elliptical arc with the ELLIPSE command. Invoke the ELLIPSE command:

Draw toolbar	Choose the ELLIPSE command (see Figure 4–8)
Draw menu	Choose Ellipse
Command: prompt	**ellipse** (ENTER)

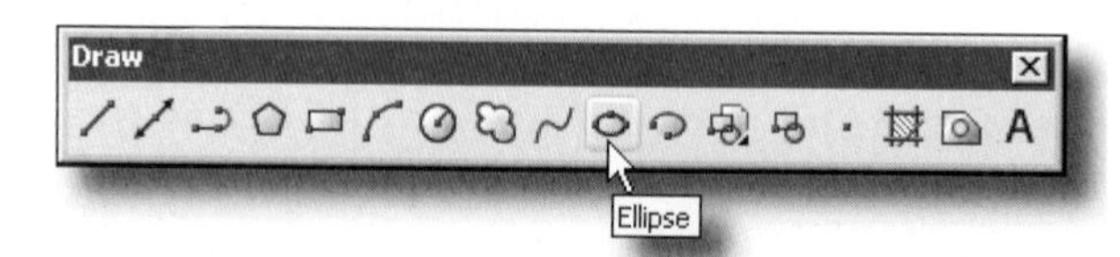

Figure 4–8 *Invoking the ELLIPSE command from the Draw toolbar*

AutoCAD prompts:

Command: **ellipse** (ENTER)
Specify axis endpoint of ellipse or [Arc/Center]: *(specify axis endpoint of the ellipse to be drawn or right-click for shortcut menu and choose one of the available options)*

Drawing an Ellipse by Defining Axis Endpoints

This option allows you to draw an ellipse by specifying the endpoints of the axes. AutoCAD prompts for two endpoints of the first axis. The first axis can define either the major or the minor axis of the ellipse. Then AutoCAD prompts for an endpoint of the second axis as the distance from the midpoint of the first axis to the specified point.

For example, the following command sequence shows steps in drawing an ellipse by defining axis endpoints (see Figure 4–9).

Command: **ellipse** (ENTER)
Specify axis endpoint of ellipse or [Arc/Center]: **1,1** (ENTER)
Specify other endpoint of axis: **5,1** (ENTER)
Specify distance to other axis or [Rotation]: **3,2** (ENTER)

AutoCAD draws an ellipse whose major axis is 4.0 units long in a horizontal direction and whose minor axis is 2.0 units long in a vertical direction, as shown in Figure 4–9.

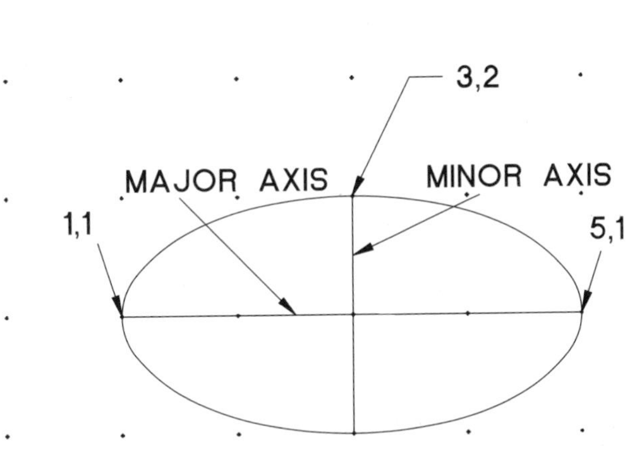

Figure 4–9 *An ellipse drawn by specifying the major and minor axes*

Drawing an Ellipse by Defining the Center of the Ellipse

This option allows you to draw an ellipse by defining the center point and axis endpoints. First, AutoCAD prompts for the ellipse center point. Then AutoCAD prompts for an endpoint of an axis as the distance from the center of the ellipse to the specified point. The first axis can define either the major or the minor axis of the ellipse. Then AutoCAD prompts for an endpoint of the second axis as the distance from the center of the ellipse to the specified point.

For example, the following command sequence shows steps in drawing an ellipse by defining the ellipse center point.

Command: **ellipse** (ENTER)
Specify axis endpoint of ellipse or [Arc/Center]: **c** (ENTER)
Specify center of ellipse: **3,1** (ENTER)
Specify endpoint of axis: **1,1** (ENTER)
Specify distance to other axis or [Rotation]: **3,2** (ENTER)

AutoCAD draws an ellipse similar to the previous example, with a major axis 4.0 units long in a horizontal direction and a minor axis 2.0 units long in a vertical direction.

Drawing an Ellipse by Specifying the Rotation Angle

AutoCAD allows you to draw an ellipse by specifying a rotation angle after defining two endpoints of one of the two axes. The rotation angle defines the major-axis-to-minor-axis ratio of the ellipse by rotating a circle about the first axis. The greater the rotation angle value, the greater the ratio of major to minor axes. AutoCAD draws a circle if you set the rotation angle to 0 degrees.

For example, the following command sequence shows steps in drawing an ellipse by specifying the rotation angle.

Command: **ellipse** (ENTER)
Specify axis endpoint of ellipse or [Arc/Center]: **3,–1** (ENTER)
Specify other endpoint of axis: **3,3** (ENTER)
Specify distance to other axis or [Rotation]: **r** (ENTER)
Specify rotation around major axis: **30** (ENTER)

See Figure 4–10 for examples of ellipses with various rotation angles.

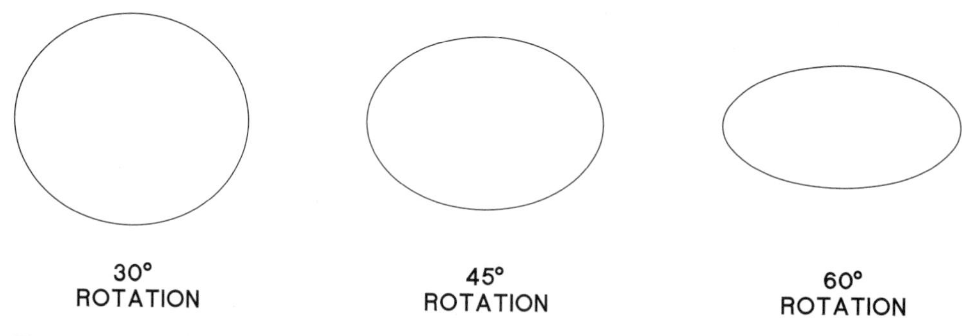

Figure 4–10 *Ellipses drawn with different rotation angles*

Drawing an Elliptical Arc with the Arc option

The Arc option allows you to draw an elliptical arc. After you specify the major and minor axis endpoints, AutoCAD prompts for the start and end angle points for the elliptical arc to be drawn. Instead of specifying the start angle or the end angle, you can toggle to the Parameter option, which prompts for the Start parameter and End parameter point locations. AutoCAD creates the elliptical arc using the following parametric vector equation:

$$p(u) = c + a^{x} + \cos(u) + b^{x} \sin(u)$$

where c is the center of the ellipse, and a and b are its major and minor axes, respectively. Instead of specifying the end angle, you can also specify the included angle of the elliptical arc to be drawn.

The ELLIPSE command, with the Arc option, is accessible directly from the Draw toolbar or you can select the Arc option while in the ELLIPSE command.

For example, the following command sequence shows steps in drawing an elliptical arc.

```
Command: ellipse (ENTER)
Specify axis endpoint of ellipse or [Arc/Center]: a (ENTER)
Specify axis endpoint of elliptical arc or [Center]: 1,1 (ENTER)
Specify other endpoint of axis: 5,1 (ENTER)
Specify distance to other axis or [Rotation]: 3,2 (ENTER)
Specify start angle or [Parameter]: 3,2 (ENTER)
Specify end angle or [Parameter/Included angle]:
Arc/Center/<Axis endpoint 1>: 1,1 (ENTER)
```

AutoCAD draws an elliptical arc with the start angle at 3,2 and the ending angle at 1,1.

Isometric Circles (or Isocircles)

By definition, Isometric Planes (*iso* meaning "same" and *metric* meaning "measure") are all being viewed at the same angle of rotation (see Figure 4–11). The angle is approximately set to 54.7356 degrees. AutoCAD uses this angle of rotation automatically when you wish to represent circles in one of the isoplanes by drawing ellipses with the Isocircle option.

Normally, a circle 1 unit in diameter being viewed in one of the isoplanes will project a short axis dimension of 0.577350 units. One of its diameters parallel to an isoaxis will project a dimension of 0.816497 units. A line drawn in isometric that is parallel to one of the three main axes will also project a dimension of 0.816497 units. We would like these lines and circle diameters to project a dimension of exactly 1.0 unit. Therefore, you automatically increase the entire projection by a fudge factor of 1.22474 (the reciprocal of 0.816497) in order to be able to use true dimensioning parallel to one of the isometric axes.

This means that circles 1 unit in diameter will be measured along one of their isometric diameters rather than along their long axis. This facilitates using true lengths as the lengths of distances projected from lines parallel to one of the isometric axes. So a 1-unit-diameter isocircle will project a long axis that is 1.224744871 units and a short axis that is 0.707107 (0.577350 x 1.22474) units. These "fudge" factors are built into AutoCAD isocircles.

The Isometric Circle method is available as one of the options of the ELLIPSE command when you are in the isometric Snap mode.

Command: **ellipse** (ENTER)
Specify axis endpoint of ellipse or [Arc/Center/Isocircle]: **i** (ENTER)
Specify center of isocircle: *(select the center of the isometric circle)*
Specify radius of isocircle or [Diameter]: *(specify the radius or right-click for the shortcut menu and choose one of the available options)*

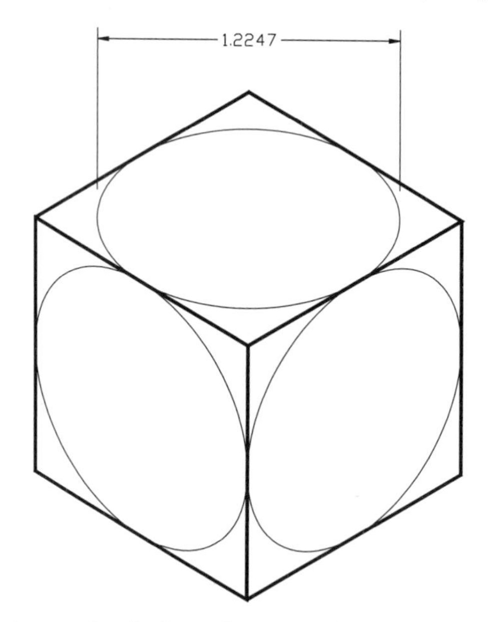

Figure 4–11 *Ellipses drawn using the Isocircle option of the* ELLIPSE *command*

If you override the last prompt default by typing d, the following prompt will appear:

Circle diameter: *(specify the desired diameter)*

Note: The Iso and Diameter options will work only when you are in the isometric Snap mode.

DRAWING POLYLINES

The poly in polyline refers to a single object with multiple connected straight-line and/or arc segments. The polyline is drawn by invoking the PLINE command and then selecting a series of points. In this respect, PLINE functions much like the LINE command. However, when completed, the segments act like a single object when operated on by modify commands. You specify the endpoints using only 2D (*X,Y*) coordinates.

The versatile PLINE command also draws lines and arcs of different widths, linetypes, tapered lines, and a filled circle. The area and perimeter of a 2D Polyline can be calculated.

By default, polylines are drawn as optimized polylines. The optimized polyline provides most of the functionality of 2D polylines but with much improved performance and reduced drawing file size. The vertices are stored as an array of information on one object. When you use the PEDIT command to edit the polyline to spline fitting or curve fitting, the polyline loses its optimization feature and vertices are stored as separate entities, but it still behaves as a single object when operated on by modify commands.

Note: If you open an AutoCAD Release 13 drawing containing polylines, they are transparently converted to optimized polylines, unless the 2D polyline is a curve-fit or splined polyline.

Invoke the PLINE command:

Draw toolbar	Choose the POLYLINE command (see Figure 4–12)
Draw menu	Choose Polyline
Command: prompt	**pline** (ENTER)

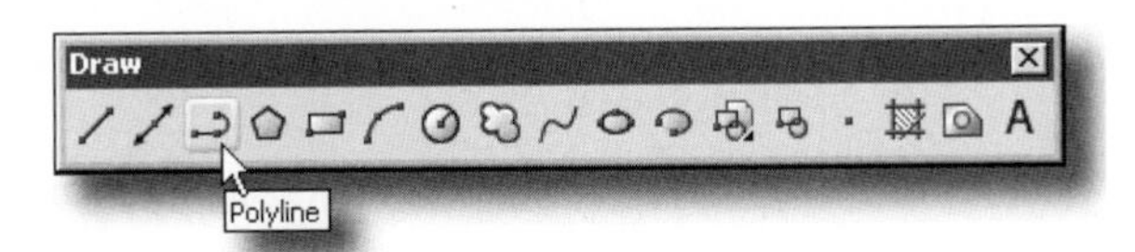

Figure 4–12 *Invoking the* PLINE *command from the Draw toolbar*

AutoCAD prompts:

Command: **pline** (ENTER)
Specify start point: *(specify the start point of the polyline)*
Specify next point or [Arc/Halfwidth/Length/Undo/Width]: *(specify next point or right-click for shortcut menu and choose one of the available options)*

You can specify the end of the line by means of absolute coordinates, relative coordinates, or by using your pointing device to specify the end of the line on the screen. After you do, AutoCAD repeats the prompt:

Specify next point or [Arc/Close/Halfwidth/Length/Undo/Width]:

Having drawn a connected series of lines, you can give a null reply (press ENTER) to terminate the PLINE command. The resulting figure is recognized by AutoCAD modify and construct commands as a single object.

The following command sequence presents an example of connected lines, as shown in Figure 4–13, drawn by means of the PLINE command.

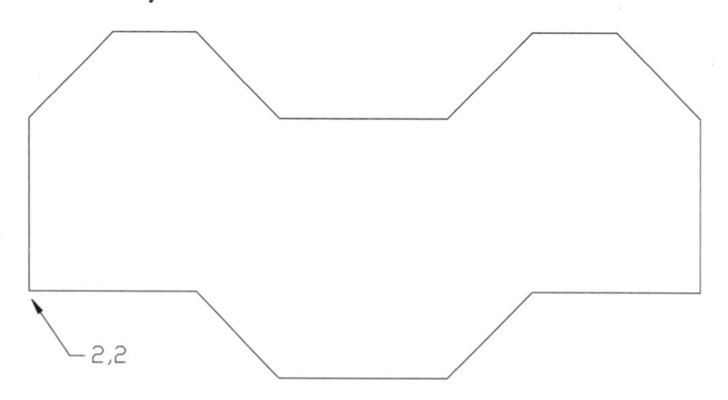

Figure 4–13 *Example of connected line segments drawn by means of the* PLINE *command*

Command: **pline** (ENTER)
Specify start point: **2,2** (ENTER)
Current line-width is 0.0000
Specify next point or [Arc/Halfwidth/Length/Undo/Width]: **4,2** (ENTER)
Specify next point or [Arc/Close/Halfwidth/Length/Undo/Width]: **5,1** (ENTER)
Specify next point or [Arc/Close/Halfwidth/Length/Undo/Width]: **7,1** (ENTER)
Specify next point or [Arc/Close/Halfwidth/Length/Undo/Width]: **8,2** (ENTER)
Specify next point or [Arc/Close/Halfwidth/Length/Undo/Width]: **10,2** (ENTER)
Specify next point or [Arc/Close/Halfwidth/Length/Undo/Width]: **10,4** (ENTER)
Specify next point or [Arc/Close/Halfwidth/Length/Undo/Width]: **9,5** (ENTER)
Specify next point or [Arc/Close/Halfwidth/Length/Undo/Width]: **8,5** (ENTER)
Specify next point or [Arc/Close/Halfwidth/Length/Undo/Width]: **7,4** (ENTER)
Specify next point or [Arc/Close/Halfwidth/Length/Undo/Width]: **5,4** (ENTER)
Specify next point or [Arc/Close/Halfwidth/Length/Undo/Width]: **4,5** (ENTER)

Specify next point or [Arc/Close/Halfwidth/Length/Undo/Width]: **3,5** (ENTER)
Specify next point or [Arc/Close/Halfwidth/Length/Undo/Width]: **2,4** (ENTER)
Specify next point or [Arc/Close/Halfwidth/Length/Undo/Width]: **c** (ENTER)

Close and Undo

The Close and Undo options work similarly to the corresponding options in the LINE command.

Width

After selecting a starting point, you may enter **w** or right-click for the shortcut menu and choose the width option to specify a starting and an ending width for a wide segment. When you select this option, AutoCAD prompts:

Specify starting width <default>: *(specify starting width)*
Specify ending width <default>: *(specify ending width)*

You can specify a width by entering a value at the prompt or by selecting a width determining points on the screen. When you specify points on the screen, AutoCAD uses the distance from the starting point of the polyline to the point selected as the starting width. You can accept the default value for the starting width by providing a null response (press ENTER), or enter in a new value. The starting width you enter becomes the default for the ending width. If necessary, you can change the ending width to another width, which results in a tapered segment or an arrow. The ending width, in turn, becomes the uniform width for all subsequent segments until you change the width again.

The following command sequence presents an example of connected lines with tapered width, as shown in Figure 4–14, drawn by means of the PLINE command.

Command: **pline** (ENTER)
Specify start point: **2,2** (ENTER)
Current line-width is 0.0000
Specify next point or [Arc/Halfwidth/Length/Undo/Width]: **w** (ENTER)
Specify starting width <default>: **0** (ENTER)
Specify ending width <default>: **.25** (ENTER)
Specify next point or [Arc/Halfwidth/Length/Undo/Width]: **2,2.5** (ENTER)
Specify next point or [Arc/Close/Halfwidth/Length/Undo/Width]: **2,3** (ENTER)
Specify next point or [Arc/Close/Halfwidth/Length/Undo/Width]: **w** (ENTER)
Specify starting width <0.2500>: (ENTER)
Specify ending width <0.2500>: **0** (ENTER)

Specify next point or [Arc/Close/Halfwidth/Length/Undo/Width]: **2,3.5** (ENTER)
Specify next point or [Arc/Close/Halfwidth/Length/Undo/Width]: **w** (ENTER)
Specify starting width <0.0000>: (ENTER)
Specify ending width <0.0000>: **.25** (ENTER)
Specify next point or [Arc/Close/Halfwidth/Length/Undo/Width]: **2.5,3.5** (ENTER)
Specify next point or [Arc/Close/Halfwidth/Length/Undo/Width]: **3,3.5** (ENTER)
Specify next point or [Arc/Close/Halfwidth/Length/Undo/Width]: **w** (ENTER)
Specify starting width <0.2500>: (ENTER)
Specify ending width <0.2500>: **0** (ENTER)
Specify next point or [Arc/Close/Halfwidth/Length/Undo/Width]: **3.5,3.5** (ENTER)
Specify next point or [Arc/Close/Halfwidth/Length/Undo/Width]: **w** (ENTER)
Specify starting width <0.0000>: (ENTER)
Specify ending width <0.0000>: **.25** (ENTER)
Specify next point or [Arc/Close/Halfwidth/Length/Undo/Width]: **3.5,3** (ENTER)
Specify next point or [Arc/Close/Halfwidth/Length/Undo/Width]: **3.5,2.5** (ENTER)
Specify next point or [Arc/Close/Halfwidth/Length/Undo/Width]: **w** (ENTER)
Specify starting width <0.2500>: (ENTER)
Specify ending width <0.2500>: **0** (ENTER)
Specify next point or [Arc/Close/Halfwidth/Length/Undo/Width]: **3.5,2** (ENTER)
Specify next point or [Arc/Close/Halfwidth/Length/Undo/Width]: **w** (ENTER)
Specify starting width <0.0000>: (ENTER)
Specify ending width <0.0000>: **.25** (ENTER)
Specify next point or [Arc/Close/Halfwidth/Length/Undo/Width]: **3,2** (ENTER)
Specify next point or [Arc/Close/Halfwidth/Length/Undo/Width]: **2.5,2** (ENTER)
Specify next point or [Arc/Close/Halfwidth/Length/Undo/Width]: **w** (ENTER)
Specify starting width <0.2500>: (ENTER)
Specify ending width <0.2500>: **0** (ENTER)
Specify next point or [Arc/Close/Halfwidth/Length/Undo/Width]: **c** (ENTER)

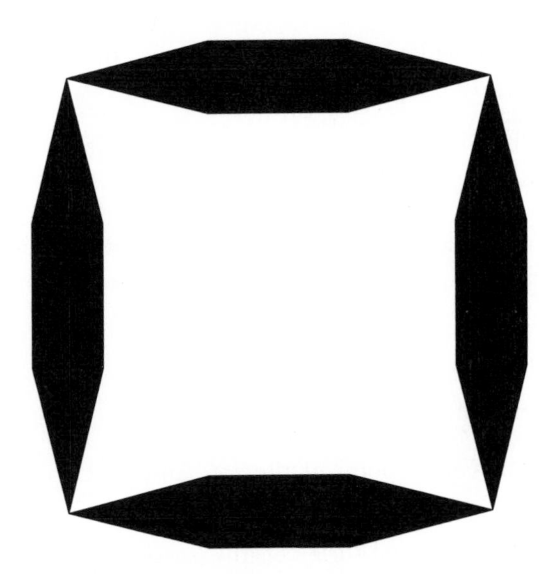

Figure 4–14 *Example of connected line segments with tapered width drawn by means of the* PLINE *command*

Halfwidth

The Halfwidth option is similar to the width option, including the prompts, except it lets you specify the width from the center of a wide polyline to one of its edges. In other words, you specify half of the total width. For example, it is easier to input 1.021756 as the halfwidth than to figure out the total width by doubling. You can specify a halfwidth by selecting points on the screen in the same manner used to specify the full width.

Arc

The Arc option allows you to draw a polyline arc. When you select the Arc option, AutoCAD displays another submenu:

> Specify endpoint of arc or [Angle/CEnter/CLose/Direction/Halfwidth/Line/Radius/Second pt/Undo/Width]: *(specify end point of arc or right-click for shortcut menu and choose one of the available options)*

If you respond with a point, it is interpreted as the endpoint of the arc. The endpoint of the previous segment is the starting point of the arc, and the starting direction of the new arc will be the ending direction of the previous segment (whether the previous segment is a line or an arc). This resembles the ARC command's Start, End, Direction (S,E,D) option, but requires only the endpoints to be specified or selected on the screen.

The CLose, Width, Halfwidth, and Undo options are similar to the corresponding options for the straight-line segments described earlier.

The Angle option lets you specify the included angle by prompting:

> Specify included angle: *(specify an angle)*

The arc is drawn counterclockwise if the value is positive, clockwise if it is negative. After the angle is specified, AutoCAD prompts for the endpoint of the arc.

The CEnter option lets you override with the location of the center of the arc and AutoCAD prompts:

> Specify center point: *(specify center point)*

When you provide the center point of the arc, AutoCAD prompts for additional information:

> Specify endpoint of arc or [Angle/Length]: *(specify end point of arc or right-click for shortcut menu and choose one of the available options)*

If you respond with a point, it is interpreted as the endpoint of the arc. Selecting Angle or Length allows you to specify the arc's included angle or chord length.

The Direction option lets you override the direction of the last segment, and AutoCAD prompts:

> Specify the tangent direction for the start point of arc: *(specify the direction)*

If you respond with a point, it is interpreted as the starting point of the direction and AutoCAD prompts for the endpoint for the arc.

The Line option reverts to drawing straight-line segments.

The Radius option allows you to specify the radius by prompting:

> Specify radius of arc: *(specify the radius of the arc)*

After the radius is specified, you are prompted for the endpoint of the arc.

The Second point option causes AutoCAD to use the three-point method of drawing an arc by prompting:

> Specify second point on arc: *(specify second point)*

If you respond with a point, it is interpreted as the second point and then you are prompted for the endpoint of the arc. This resembles the ARC command's Three-point option.

Length

The Length option continues the polyline in the same direction as the last segment for a specified distance.

DRAWING TEXT

You have learned how to draw some of the geometric shapes that make up your design. Now it is time to learn how to annotate your design. When you draw on paper, adding descriptions of the design components and the necessary shop and fabrication notes is a tedious, time-consuming process. AutoCAD provides several text commands and tools (including a spell checker) that greatly reduce the tedium of text placement and the time it takes.

Text is used to label the various components of your drawing and to create the necessary shop or field notes needed for fabrication and construction of your design. AutoCAD includes a large number of text fonts. Text can be stretched, compressed, obliqued, mirrored, or drawn in a vertical column by applying a style. Each text string can be sized, rotated, and justified to meet your drawing needs. You should be aware that AutoCAD considers a text string (all the characters that comprise the line of text) as one object.

Note: If you do not know how to type, you can place text quickly and easily after a period of learning the keyboard and developing typing skills. If you create drawings that require a lot of text entry, it may be worth your time to learn to type with all ten fingers ("touch-typing"). There are several computer programs that can help you teach yourself to type, and almost all colleges offer typing classes. If you do not have time to learn touch-typing, there is no need to worry, many "two-finger" typists productively place text in their drawings.

AutoCAD has a toolbar just for Text-related commands. Included are MULTILINE TEXT, SINGLE LINE TEXT, EDIT TEXT, FIND AND REPLACE, TEXT STYLE, SCALE TEXT, JUSTIFY TEXT, and CONVERT DISTANCE BETWEEN SPACES, as shown in Figure 4–15.

Figure 4–15 *Text toolbar*

CREATING A SINGLE LINE OF TEXT

The TEXT command allows you to create several lines of text in the current style. If necessary, you can change the current style. To modify a style or to create a new style, refer to the section Creating and Modifying Text Styles, later in this chapter.

Invoke the TEXT command:

Text toolbar	Choose Single Line Text
Command: prompt	**text** (ENTER)

AutoCAD prompts:

Command: **text** (ENTER)
Current text style: "Standard" Text height: 0.2000
Specify start point of text or [Justify/Style]: *(specify start point of text or right-click for shortcut menu and choose one of the available options)*

The start point indicates the lower left corner of the text. If necessary, you can change the location of the justification point. You can specify the starting point in absolute coordinates or by using your pointing device. After you specify the starting point, AutoCAD prompts:

Specify height <default>: *(specify the text height)*

This allows you to select the text height. You can accept the default text height by giving a null response, by using your pointing device, or you can enter the appropriate text height. Next, AutoCAD prompts:

Specify rotation angle of text <default>: *(specify the rotation angle)*

This allows you to place the text at any angle in reference to 0 degrees (default is 3 o'clock, or east, measured in the counterclockwise direction). The default value of the rotation angle is 0 degrees, and the text is placed horizontally at the specified start point. The last prompt is:

Enter text: *(type the desired text and press* ENTER)

A cursor appears on the screen at the starting point you have selected. After you enter the first line of text and press ENTER, you will notice the cursor drop down to the next line, anticipating that you wish to enter more text. If this is the case, enter the next line of text string; when you are through with typing text strings, press ENTER at the "Enter Text:" prompt to terminate the command sequence.

If you are in the TEXT command and notice a mistake (or simply want to change a character or word), backspace to the text you want to change. This, however, deletes all of the text you backspaced over to get to the point you want to change. If this involves erasing several lines of text, it may be faster to use the CHANGE or DDEDIT command to make changes to the text string.

One feature of TEXT that will speed up text entry on your drawing is the ability to move the crosshairs cursor to a new point on the drawing while staying in the TEXT command. As you move the crosshairs cursor and specify a new point, you will notice the cursor moves to this new point, allowing you to enter a new string of text. However, you must remember to give a null response (press ENTER) to the "Enter text:" prompt to terminate the command.

For example, the following command sequence shows placement of left-justified text by providing the starting point of the text, as shown in Figure 4–16.

Command: **text** (ENTER)
Specify start point of text or [Justify/Style]: *(specify start point of text as shown in Figure 4–16)*
Specify height <default>: **.25** (ENTER)
Specify rotation angle of text: (ENTER)
Enter text: **Sample Text Left Justified** (ENTER)
Enter text: (ENTER)

Sample Text Left Justified

Start Point

Figure 4–16 *Using the* TEXT *command to place left-justified text by specifying a start point*

Justify

The Justify option allows you to place text in one of the 14 available justification points. When you select this option, AutoCAD prompts:

Enter an option [Align/Fit/Center/Middle/Right/TL/TC/TR/ML/MC/MR/BL/BC/BR]: *(select one of the available options or right-click for shortcut menu and choose one of the available options)*

The Center option allows you to select the center point for the baseline of the text. Baseline refers to the line along which the bases of the capital letters lie. Letters with descenders, such as g, q, and y, dip below the baseline. After providing the center point, enter the text height and rotation angle. The justification does not occur on the screen until ENTER key is pressed to exit the TEXT command.

For example, the following command sequence shows placement of center-justified text, by providing the center point of the text, as shown in Figure 4–17.

Command: **text** (ENTER)
Current text style: "Standard" Text height: 0.2000
Specify start point of text or [Justify/Style]: **j** (ENTER)
Enter an option [Align/Fit/Center/Middle/Right/TL/TC/TR/ML/MC/MR/BL/BC/BR]: **c** (ENTER)
Specify height <default>: **.25** (ENTER)
Specify rotation angle of text:<0>:÷ **0** (ENTER)
Enter text: **Sample Text Center Justified** (ENTER)

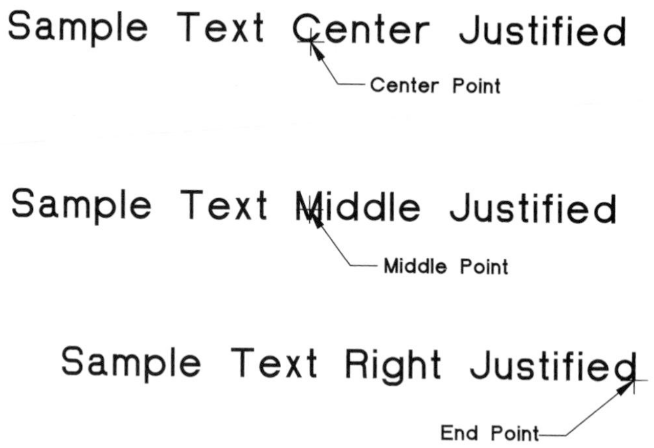

Sample Text Center Justified

Center Point

Sample Text Middle Justified

Middle Point

Sample Text Right Justified

End Point

Figure 4–17 *Using the* TEXT *command to place text by specifying a center point (center justified), a middle point (middle justified), or an endpoint (right justified)*

The Middle option allows you to center the text both horizontally and vertically at a given point. After providing the middle point, enter the text height and rotation angle.

For example, the following command sequence shows placement of middle-justified text by providing the middle point of the text, as shown in Figure 4–17.

Command: **text** (ENTER)
Current text style: "Standard" Text height: 0.2500
Specify start point of text or [Justify/Style]: **j** (ENTER)
Enter an option [Align/Fit/Center/Middle/Right/TL/TC/TR/ML/MC/MR/BL/BC/BR]: **m** (ENTER)
Specify height <default>: **.25** (ENTER)
Specify rotation angle of text: **0** (ENTER)
Enter text: **Sample Text Middle Justified** (ENTER)

The Right option allows you to place the text in reference to its lower right corner (right justified). Here, you provide the point where the text will end. After providing the right point, enter the text height and rotation angle.

For example, the following command sequence shows placement of right-justified text, as shown in Figure 4–17.

Command: **text** (ENTER)
Specify start point of text or [Justify/Style]: **j** (ENTER)
Enter an option [Align/Fit/Center/Middle/Right/TL/TC/TR/ML/MC/MR/BL/BC/BR]: **r** (ENTER)
Specify height <default>: **.25** (ENTER)
Specify rotation angle of text: 0 (ENTER)
Enter text: **Sample Text Right Justified** (ENTER)

Other options are combinations of the previously mentioned options:

TL top left
TC top center
TR top right
ML middle left
MC middle center
MR middle right
BL bottom left
BC bottom center
BR bottom right

The Align option allows you to place the text by designating the endpoints of the baseline. AutoCAD computes the text height and orientation so that the text just fits proportionately between two points. The overall character size adjusts in proportion to the height. The height and width of the character will be the same.

For example, the following command sequence shows placement of text using the Align option as shown in Figure 4–18.

Command: **text** (ENTER)
Specify start point of text or [Justify/Style]: **j** (ENTER)
Enter an option [Align/Fit/Center/Middle/Right/TL/TC/TR/ML/MC/MR/BL/
BC/BR]: **a** (ENTER)
Specify first endpoint of text baseline: *(specify the first point)*
Specify second endpoint of text baseline: *(specify the second point)*
Enter text: **Sample Text Aligned** (ENTER)

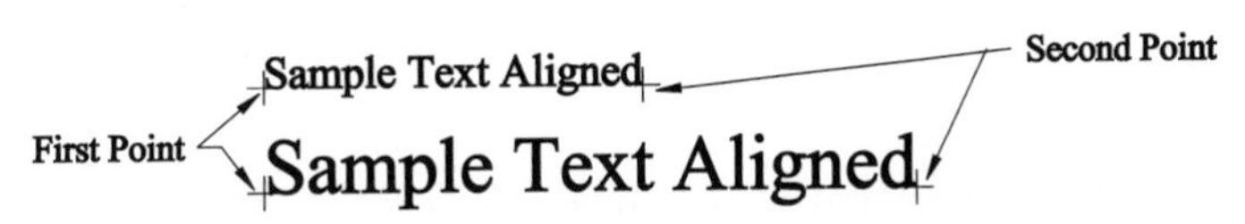

Figure 4–18 *Using the Align and Fit option of the* TEXT *command to place text*

The Fit option is similar to the Align option, but in the case of the Fit option AutoCAD uses the current text height and adjusts only the text's width, expanding or contracting it to fit between the points you specify.

For example, the following command sequence shows placement of text using the Fit option as shown in Figure 4–18.

Command: **text** (ENTER)
Specify start point of text or [Justify/Style]: **j** (ENTER)
Enter an option [Align/Fit/Center/Middle/Right/TL/TC/TR/ML/MC/MR/BL/BC/BR]: **f** (ENTER)
Specify first endpoint of text baseline: *(specify the first point)*
Specify second endpoint of text baseline: *(specify the second point)*
Specify height <default>: **0.25** (ENTER)
Enter text: **Sample Text Fit** (ENTER)

Style

The Style option allows you to select one of the available styles in the current drawing. To modify a style or to create a new style, refer to the section "Creating and Modifying Text Styles," later in this chapter.

CREATING MULTILINE TEXT

AutoCAD offers the MTEXT command for drawing text by "processing" the words in paragraph form; the width of the paragraph is determined by the user-specified rectangular boundary. It is an easy way to have your text automatically formatted as a multiline group, with left, right, or center justification as a group. Each multiline text object is a single object, regardless of the number of lines it contains. The text boundary remains part of the object's framework, although it is not plotted or printed. New features introduced in AutoCAD 2004 include indents and tabs, making it easier to correctly align your text for tables and numbered lists. At the top of the user-specified rectangle is a ruler, as shown in Figure 4–20, similar to those in word processors. Individual characters in the Multiline Text Editor can be selected for applying formatting styles such as bold, underline, and italics.

Invoke the MTEXT command:

Draw toolbar	Choose the Multiline Text command (see Figure 4–19)
Text toolbar	Choose Multiline Text...
Command: prompt	**mtext** (ENTER)

Figure 4–19 *Invoking the MTEXT command from the Draw toolbar*

AutoCAD prompts:

Command: **mtext** (ENTER)
Current text style: "Standard" Text height: 0.2500
Specify first corner: *(specify the first corner of the rectangular boundary)*
Specify opposite corner or [Height/Justify/Line spacing/Rotation/Style/Width]: *(specify the opposite corner of the rectangular boundary, or right-click for shortcut menu and choose one of the available options)*

When you drag the cursor after specifying the first corner of the rectangular boundary, referred to as the bounding box, AutoCAD displays an arrow within the rectangle to indicate the direction of the paragraph's text flow. After you specify the opposite corner of the bounding box, AutoCAD displays the Text Formatting toolbar and input/editing area, shown in Figure 4–20.

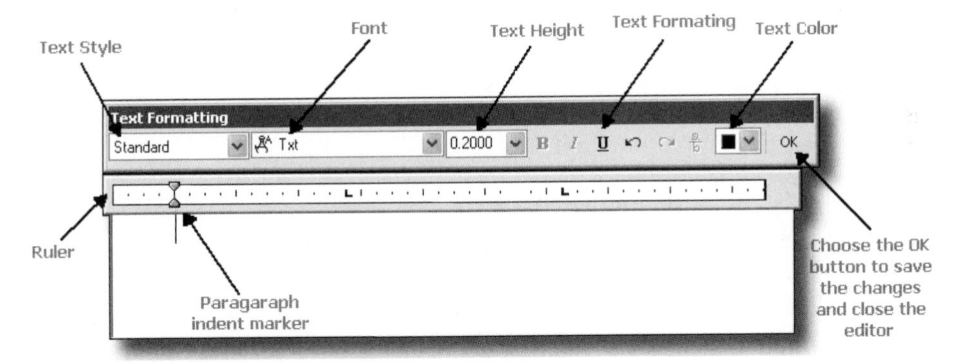

Figure 4–20 *Text Formatting toolbar, ruler and multiline text input/editing bounding box*

The bounding box part of the Multiline Text Editor can be configured to envelop the text in its actual size and shape on the screen (unless it is considered too small or too large), or to become a amply sized word processing page covering a major portion of the screen. These options are set by the MTEXTFIXED system variable. To cause the bounding box to fit around the text in its true size and shape, set MTEXTFIXED to **0** (default). To cause it to be displayed as a larger page, set MTEXTFIXED to **1.** When MTEXTFIXED is set to **0**, it can be made transparent by using the Transparency dialog box. This dialog box is displayed when you select the **Transparency** option on the shortcut menu that appears when you right-click on the Toolbars palette. The fit and transparency of the boundary box apply only when the Multiline Text object is not rotated from the angle of 0 degrees. When the Multiline Text object is rotated, the bounding box part of the Multiline Text Editor becomes an opaque word processing page. When the bounding box is transparent, it can be made opaque by clicking on the lower edge of the ruler.

Enter the text in the text input/editing bounding box. If you right-click in the bounding box, AutoCAD displays a shortcut menu with options that include **Undo, Redo, Cut, Copy, Paste, Indents and Tabs, Justification, Find and Replace, Select All, Change Case, Autocaps, Remove Formatting, Combine Paragraphs, Symbol, Import Text**, and **Help**.

The text can be highlighted in the following three ways: holding down the pick button while dragging across the selected text, double-clicking to select the entire word, or triple-clicking to select the entire line of text.

Text Formatting Toolbar

The options provided on the Text Formatting toolbar (see Figure 4–20) include character formatting: **Text Style, Font, Height, Bold, Italic, Underline, Undo, Redo, Stack, Text Color**, and **OK**.

The **Text Style** option allows you to apply an existing style to new text or selected text. If you apply a new style to an existing text object, AutoCAD overrides character formatting such as font, height, bold, and italic attributes. Styles that have backwards or upside-down effects are not applied.

You can also select the text style by means of the Style option when AutoCAD prompts for the opposite corner of the rectangular boundary. Once the text style is set, AutoCAD returns to the previous prompt until the opposite corner of the rectangular boundary is specified

The **Font** option allows you to specify a font for new text or changes the font of selected text. All of the available TrueType fonts and SHX fonts are listed in the drop-down list box.

The **Height** option sets the character height in drawing units. The default value for the height is based on the current style. (A detailed discussion is provided later in the chapter on creating or modifying a text style.) If the current style is set to 0, then the value of the height is based on the value stored in the TEXTSIZE system variable. Each multiline text object can contain a text string of varying text size. When you highlight the text string in the dialog box, AutoCAD displays the selected text height in the list box. If necessary, you can specify a new height in addition to those listed.

You can also set the text height by selecting the **Height** option when AutoCAD prompts for the opposite corner of the rectangular boundary. Once the text height is specified, AutoCAD returns to the previous prompt until the opposite corner of the rectangular boundary is specified.

The **Bold** option allows you to turn ON and OFF bold formatting for new text or selected text. The Bold option is available only to the characters that belong to the TrueType font.

The **Italic** option allows you to turn ON and OFF italic formatting for new text or selected text. The Italic option is available only to the characters that belong to the TrueType font.

The **Underline** option allows you to turn ON and OFF underlining for new text or selected text.

The **Undo** option undoes the last edit action in the Multiline Text Editor dialog box that includes changes in the content of the text string or formatting.

The **Redo** option cancels the previous Undo.

The **Stack** option is used to place one part of a selected group of text over the remaining part. Before using the **Stack** option, the selected text must contain a forward slash (/) to separate the top part (to the left of the /) from the bottom part (to the right of the /). The slash will cause the horizontal bar to be drawn between the upper and lower parts, necessary for fractions with center justification. Instead of a slash (/), you can use the caret (^) symbol. In this case, AutoCAD will not draw a horizontal bar between the upper and lower parts with left justification, which is useful for placing tolerance values.

The **Color** option sets the color for new text or changes it for the selected text. You can assign the color by BYLAYER, BYBLOCK, or one of the available colors.

The **OK** option, when selected, accepts the text in the boundary box and terminates the MLINE command.

Text Editor Shortcut Menu

Undo, Redo, Cut, Copy, and **Paste** operate in the same manner as similar commands in other Windows-based word processors and text handling programs.

The **Indents and Tabs...** option, when selected, causes the Indents and Tabs dialog box to be displayed as shown in Figure 6-21. From this dialog box you can set indentation for paragraphs and the first lines of paragraphs and set tab stops.

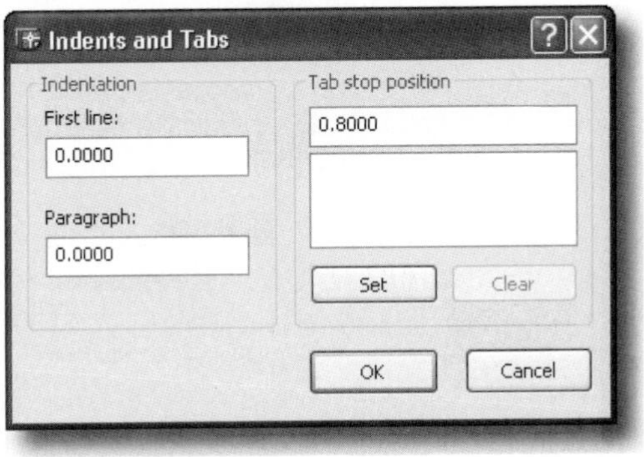

Figure 4–21 *The Indents and Tabs dialog box*

In the **Indentation** section of the Indents and Tabs dialog box, the **First line:** text box lets you set the indentation of the first line of the current or selected paragraphs. The **Paragraph:** text box lets you set the indentation for the current or selected paragraph. In the **Tab stop position** section, the upper text box lets you set tab positions, which are then recorded in the lower text box. The **Set** button, when selected, causes the values in the upper text box to be copied to and listed in the lower text box. The **Clear** button, when selected, causes the selected tab stop to be removed from the list.

The **Justification** option, when selected, causes a shortcut menu to be displayed from which the customary AutoCAD Text justification and alignment for new or selected text options can be selected. Text is center, left, or right justified with respect to the left and right text boundaries, and aligned from the middle, top, or bottom of the paragraph with respect to the top and bottom text boundaries. If the selected justification is one of the "top" options (**Top Left, Top Center, Top Right**), excess text will "spill" out of the bottom of the specified boundary box. If the selected justification is one of the "bottom" options (**Bottom Left, Bottom Center, Bottom Right**), excess text will "spill" out of the top of the specified boundary box.

You can also set the justification by selecting the Justify option when AutoCAD prompts for the opposite corner of the rectangular boundary. The available justification options are the same as in the TEXT command. Once justification is specified, AutoCAD returns to the previous prompt until the opposite corner of the rectangular boundary is specified.

The **Find and Replace...** option, when selected, causes the Replace dialog box to be displayed (see Figure 4–22), which includes the **Find what** and **Replace with** text boxes, and **Match Case**, and **Match whole word only** check boxes. These are used to search for specified text strings and replace them with new text.

Enter the text string to be searched for in the **Find what** text box, and then choose the **Find Next** button to start the search. AutoCAD highlights the appropriate text string in the bounding box. To continue the search, choose the **Find Next** button again.

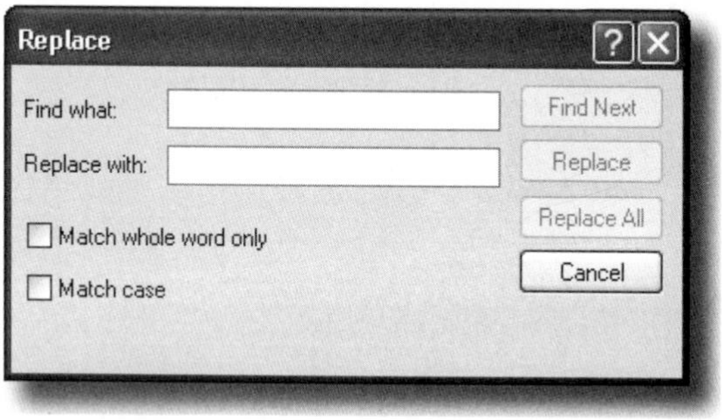

Figure 4–22 *Replace dialog box*

In the **Replace with** text box, type the text string that you want as replacement for the text string in the **Find what** text box. Then choose the **Replace** button to replace the highlighted text with the text in the **Replace with** text box. If you choose the **Replace All** button, all instances of the specified text will be replaced.

When the **Match Case** check box is selected, AutoCAD finds text only if the case of all characters in the text object is matched to that of the text characters in the **Find what** text box. When it is not selected, AutoCAD finds a match for the specified text string regardless of the case of the characters.

When the **Match whole word only** check box is selected, AutoCAD finds text only if the text string is a single word. If the text is part of another text string, it is ignored. When it is not selected, AutoCAD finds a match for the specified text string whether it is a single word or part of another word.

The **Select All** option, when selected, causes all the text in the Multiline text object to be selected and highlighted.

The **Change Case** option, when selected, causes a shortcut menu to be displayed with options for **Uppercase** and **Lowercase.** Selecting **Uppercase** causes any selected and highlighted text to be uppercase. Selecting **Lowercase** causes any selected and highlighted text to be lowercase.

The **AutoCAPS** option, when selected, operates in a manner similar to pressing the CAPS LOCK key, toggling the CAPS LOCK key on and off.

The **Remove Formatting** option, when selected, removes any bold, italics, or underlining formatting to selected text.

The **Combine Paragraphs** option, when selected, combines selected paragraphs into a single paragraph and replaces each paragraph return with a space.

The **Symbol** option, when selected, causes a shortcut menu to be displayed from which symbols for **Degrees, Plus/Minus,** and **Diameter** can be selected along with the **Non-Breaking Space.** Also on the shortcut menu is the **Other...** option, which causes the Character Map dialog box to be displayed. The Character Map dialog box operates in the same manner as the similar dialog box in other Windows-based word processors and text handling programs.

In addition to the options provided in the Multiline Text Editor dialog box for drawing special characters, you can draw them by means of the control characters. The control characters for a symbol begin with a double percent sign (%%). The next character you enter represents the symbol. The control sequences defined by AutoCAD are presented in Table 4–1.

Table 4–1 Control Character Sequences for Drawing Special Characters and Symbols

Special Character or Symbol	Control Character Sequence	Example	
		Text String	Control Character Sequence
° (degree symbol)	%%d	104.5°F	104.5%%dF
± (plus/minus tolerance symbol)	%%p	34.5±3	34.5%%p3
∅ (diameter symbol)	%%c	56.06∅	56.06%%c
% (single percent sign; necessary only when it must precede another control sequence	%%%	34.67%±1.5	34.67%%%%%P1.5
Special coded symbols (where nnn stands for a three-digit code)	%%nnn	@	%%064

The **Import Text** option, when selected, causes the Select File dialog box to be displayed from which a file of ASCII or RTF format can be imported. The file imported is limited to 32K.

Property settings accessible after the first corner of the boundary box has been specified include the Height, Justify, Line Spacing, Rotation, Style and Width. The Height, Style, and Justification options have been discussed previously in this section.

The Width option sets the paragraph width for new or selected text. If it is set to the No Wrap option, the resulting multiline text object appears on a single line.

You can also set the width by selecting the Width option when AutoCAD prompts for the opposite corner of the rectangular boundary. Once the width is specified, AutoCAD returns to the previous prompt until the opposite corner of the rectangular boundary is specified.

The Rotation option sets the rotation angle for new or selected text, in the current unit of angle measurement.

You can also set the rotation by selecting the Rotation option when AutoCAD prompts for the opposite corner of the rectangular boundary. Once the rotation is specified, AutoCAD returns to the previous prompt until the opposite corner of the rectangular boundary is specified.

Line Spacing Options

The options that are provided specify line spacing for the text objects. Two options are provided for line spacing: At Least and Exactly.

The At Least selection adjusts lines of text automatically based on the height of the largest character in the line. When At Least is selected, lines of text with taller characters have added space between lines.

The Exactly selection forces the line spacing to be the same for all lines of text in the mtext object. Spacing is based on the text height of the object or text style.

Once the line spacing is specified, AutoCAD returns to the previous prompt until the opposite corner of the rectangular boundary is specified.

Note: Exact spacing is recommended when you use MTEXT to create a table.

EDITING TEXT

The DDEDIT command allows you to edit text and attributes. An attribute is informational text associated with a block. See Chapter 10 for a detailed discussion of blocks and attributes.

Invoke the DDEDIT command:

Text toolbar	Choose Edit Text (see Figure 4–23)
Command: prompt	**ddedit** (ENTER)

Figure 4–23 *Invoking the DDEDIT command from the Text toolbar*

AutoCAD prompts:

Command: **ddedit** (ENTER)
Select an annotation object or [Undo]: *(select the text or attribute definition, or right-click for shortcut menu and choose one of the available options)*

If you select a text string created by means of a TEXT command, AutoCAD displays the Edit Text dialog box, as shown in Figure 4–24. Make the necessary changes in the text string and choose the **OK** button to keep the changes.

If instead you select text created by means of the MTEXT command, AutoCAD displays the Multiline Text Editor dialog box shown in Figure 4–25. Make the necessary changes in the text string, and click the **OK** button to keep the changes.

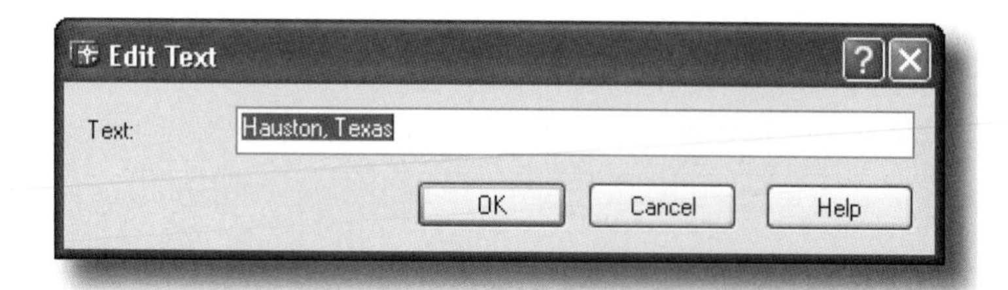

Figure 4–24 *Edit Text dialog box*

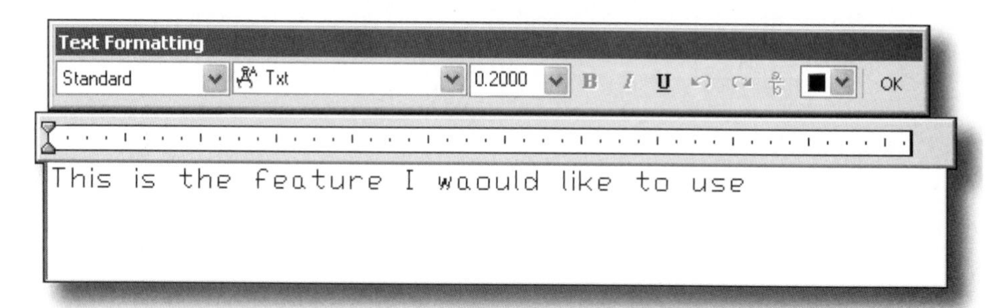

Figure 4–25 *Multiline Text Formatting toolbar, Ruler, and Bounding box*

AutoCAD continues to prompt you to select a new text string to edit, or you can enter U to undo the last change made to the text. To terminate the command sequence, give a null response (press ENTER).

FIND AND REPLACE

The FIND command is used to find a string of specified text and replace it with another string of specified text.

Invoke the FIND command:

Text toolbar	Choose Find and Replace (see Figure 4–26)
Command: prompt	**find** (ENTER)

Figure 4–26 *Invoking the FIND command from the Text toolbar*

AutoCAD displays the Find and Replace dialog box, similar to Figure 4–27.

In the **Find text string:** text box, enter the string that you wish to replace. In the **Replace with:** text box, enter the new text string. Select the **Find** button and an instance of the string to be replaced is displayed in the **Context** area of the **Search results** section of the dialog box. Select the **Replace** button to replace this instance with the string in the **Replace with:** text box. To skip over this instance without replacing it, select the **Find** button. To replace all instances of the string entered in **the Find text string:** text box, select the **Replace All** button.

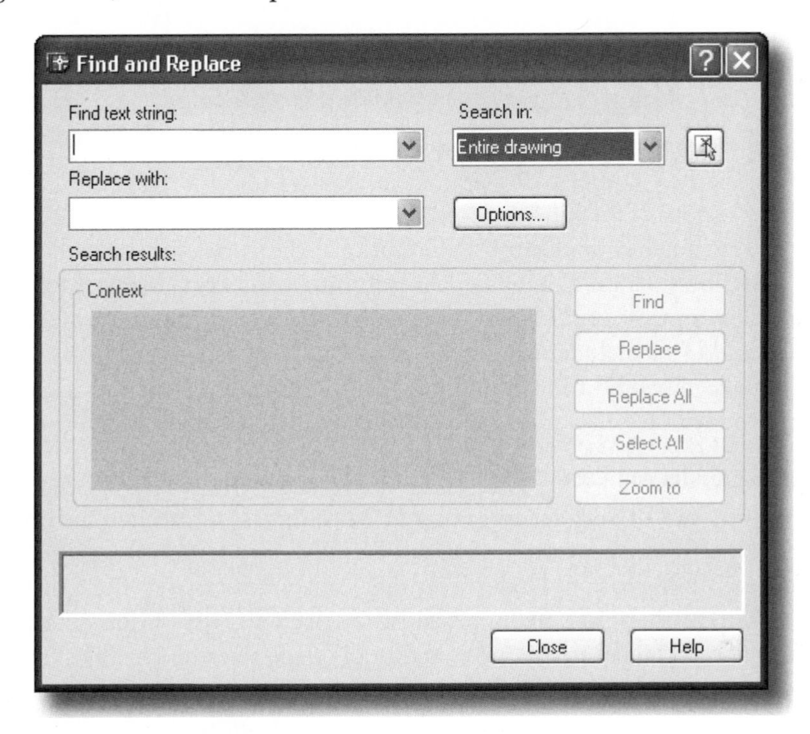

Figure 4–27 *Find and Replace dialog box*

In the **Search in:** text box, you can direct AutoCAD to search either in the **Entire drawing**, or the **Current selection**, for the string to be replaced. The **Select objects** button returns you to the graphics screen to select text objects. Then AutoCAD searches for the string to be replaced. Selecting the **Options** button causes the Find and Replace Options dialog box to appear, similar to Figure 4–28.

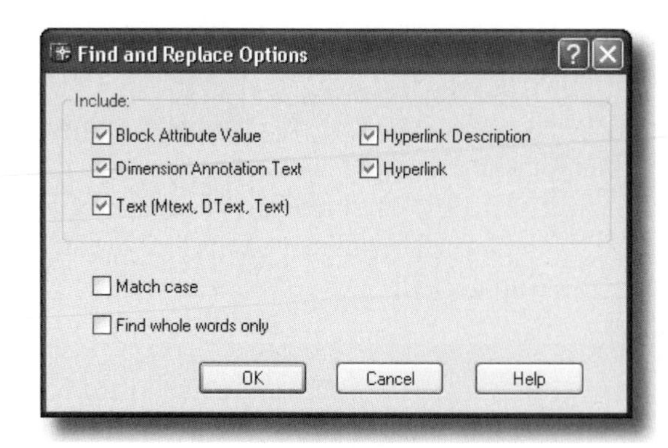

Figure 4–28 *Find and Replace Options dialog box*

The **Include:** section of the Find and Replace Options dialog box lets you filter the type of text to be included in the Find search. Optional categories for text to be included are **Block Attribute Value, Dimension Annotation Text, Text (Mtext, Dtext, Text), Hyperlink Description**, and **Hyperlink.** The **Match case** check box, when checked, causes AutoCAD to include text strings whose case matches that of the specified text string to search. The **Find whole words only** check box, when checked, causes AutoCAD to include only whole words that match the specified text string.

SCALING TEXT IN AN OLE OBJECT

To scale text in an OLE object according to its font, first right-click the object and select Properties from the shortcut menu. The OLE Properties dialog box displays the current size information. Under **Text Size**, select a font from the list. The font list contains all of the fonts that appear in the selected OLE object. Select a point size from the list. The point size list contains all of the sizes for the selected font. In the box that follows the equal sign, enter a value in drawing units. This value is the height for text in the selected font and point size. For example, if you select Arial and 10 points, and then enter .5 after the equal sign, all text in the selected OLE object that is currently 10-point Arial, changes to 0.5 drawing units in height. All other text in the object changes size in relation to the selected font. The boundary box adjusts to accommodate the new text sizes.

JUSTIFYING TEXT

The JUSTIFYTEXT command lets you change the justification point of a text string without having to change its location.

Invoke the JUSTIFYTEXT command:

Text toolbar	Choose Justify Text (see Figure 4–29)
Command: prompt	**justifytext** (ENTER)

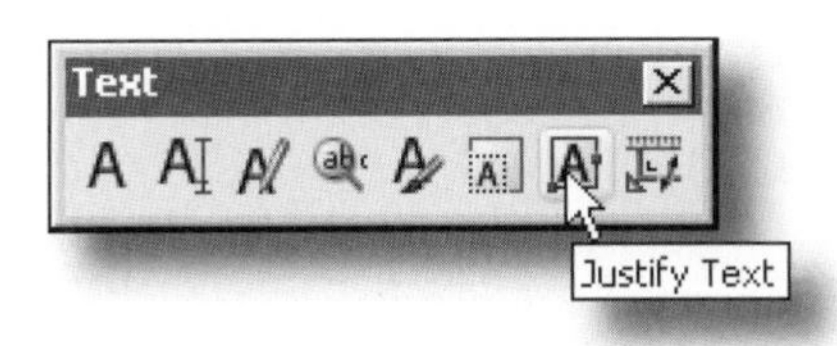

Figure 4–29 *Invoking the* JUSTIFYTEXT *command from the Text toolbar*

AutoCAD prompts:

Command: **justifytext** (ENTER)
Select objects: *(select a text object)*
Select objects: (ENTER)
Enter a justification option
[Left/Align/Fit/Center/Middle/Right/TL/TC/TR/ML/MC/MR/BL/BC/BR]
<Left>: *(select one of the available options to change the justification)*

CHANGING TEXT FROM ONE SPACE TO ANOTHER

The SPACETRANS command converts distances between model space units and paper space units. By using SPACETRANS transparently, you can provide commands with distance entries relative to another space. For example, if you wish to create a text object in model space that matches the height of other text in a layout, enter the following from model space:

Command: **text**
Specify start point of text or [Justify/Style]: **6.5,2.75**
Specify height <0.20>: **'spacetrans**
>>Specify paper space distance <1.000>: **1/2**
Resuming TEXT command
Specify height <0.20>: **0.875**

When the command is complete, a text object is created in model space with a height of 0.875, which appears as 1/2 when viewed from a layout.

SPELL-CHECKING

The SPELL command is used to correct the spelling of text objects created with the TEXT or MTEXT command in addition to attribute values in Blocks.

Invoke the SPELL command:

Tools menu	Choose Spelling
Command: prompt	**spell** (ENTER)

AutoCAD prompts:

> Command: **spell** (ENTER)
> Select objects: *(select one or more text strings, and press* ENTER *to terminate object selection)*

AutoCAD displays the Check Spelling dialog box, similar to Figure 4–30, only when it finds a dubious word in the selected text objects.

AutoCAD displays the name of the current dictionary in the top of the Check Spelling dialog box. If necessary, you can change to a different dictionary by clicking the **Change Dictionaries...** button and selecting the appropriate dictionary from the Change Dictionaries dialog box.

AutoCAD displays each misspelled word in the **Current Word** section and lists the suggested alternate spellings in the **Suggestions:** list box. Click the **Change** button to replace the current word with the selected suggested work, or click the **Change All** button to replace all instances of the current word. Alternatively, click the **Ignore** button to skip the current word, or click the **Ignore All** button to ignore all subsequent entries of the current word.

The **Add** button allows you to include the current word (up to 63 characters) in the current or custom dictionary. The **Lookup** button allows you to check the spelling of the word in the suggestions box.

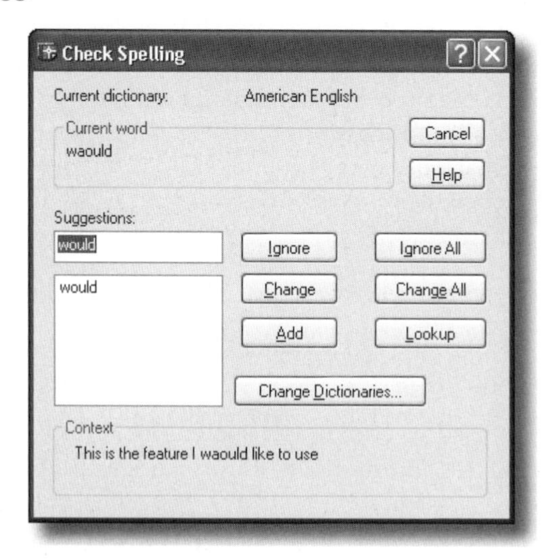

Figure 4–30 *Check Spelling dialog box*

After completion of the spelling check, AutoCAD displays an AutoCAD message informing you that the spelling check is complete.

CONTROLLING THE DISPLAY OF TEXT

The QTEXT command is a utility command for TEXT and MTEXT that is designed to reduce the redraw and regeneration time of a drawing. Regeneration time becomes a significant factor if the drawing contains a great amount of text and attribute information and/or if a fancy text font is used. Using QTEXT, the text is replaced with rectangular boxes of a height corresponding to the text height. These boxes are regenerated in a fraction of the time required for the actual text.

If a drawing contains many text and attribute items, it is advisable to set QTEXT to ON. However, before plotting the final drawing, or inspection of text details, the QTEXT command is set to OFF and is followed by the REGEN command.

Invoke the QTEXT command:

Command: prompt	**qtext** (ENTER)

AutoCAD prompts:

Command: **qtext** (ENTER)
Enter mode ON/OFF <current>: *(right-click and select one of the available options from the shortcut menu and choose one of the available options)*

CREATING AND MODIFYING TEXT STYLES

The Style option of the TEXT and MTEXT commands (in conjunction with the STYLE command) lets you determine how text characters and symbols appear, other than adjusting the usual height, slant, and angle of rotation. To specify a text style from the Style option of the TEXT and MTEXT commands, it must have been defined by using the STYLE command. In other words, the STYLE command creates a new style or modifies an existing style. The Style option under the TEXT or MTEXT command allows you to choose a specific style from the styles available.

There are three things to consider when using the STYLE command.

First, you must name the newly defined style. Style names may contain up to 255 characters, numbers, and special characters ($, –, and _). Names like "title block," "notes," and "bill of materials" can remind you of the purpose for which the particular style was designed.

Second, you may apply a particular font to a style. The font that AutoCAD uses as a default is called TXT. It has blocky looking characters, which are economical to store in memory. But the TXT.SHX font, made up entirely of straight-line (noncurved) segments, is not considered as attractive or readable. Other fonts offer many variations in characters, including those for foreign languages. All fonts are stored for use in files of their font name with an extension of .SHX. The most effective way to get a distinctive appearance in text strings is to use a specially designed font. You can also use TrueType fonts and Type 2 postscript. See Appendix F for a list of fonts that come with AutoCAD. If necessary, you can buy additional fonts from third-party vendors. AutoCAD can also read hundreds of PostScript fonts available in the marketplace.

The third consideration of the STYLE command is in how AutoCAD treats general physical properties of the characters, regardless of the font that is selected. These properties are the height, width-to-height ratio, obliquing angle, backwards, upside-down, and orientation (horizontal/vertical) options.

Invoke the STYLE command:

Text toolbar	Choose Text Style
Command: prompt	**style** (ENTER)

AutoCAD displays the Text Style dialog box, similar to Figure 4–31.

Choose the **New...** button to create a new style. AutoCAD displays the New Text Style dialog box shown in Figure 4–32. Enter the appropriate name for the text style and choose the **OK** button to create the new style.

To rename an existing style, first select the style from the **Style Name** list box in the New Text Style dialog box, and then click the **Rename...** button. AutoCAD displays the Rename Text Style dialog box. Make the necessary changes in the name of the style, and choose the **OK** button to rename the text style.

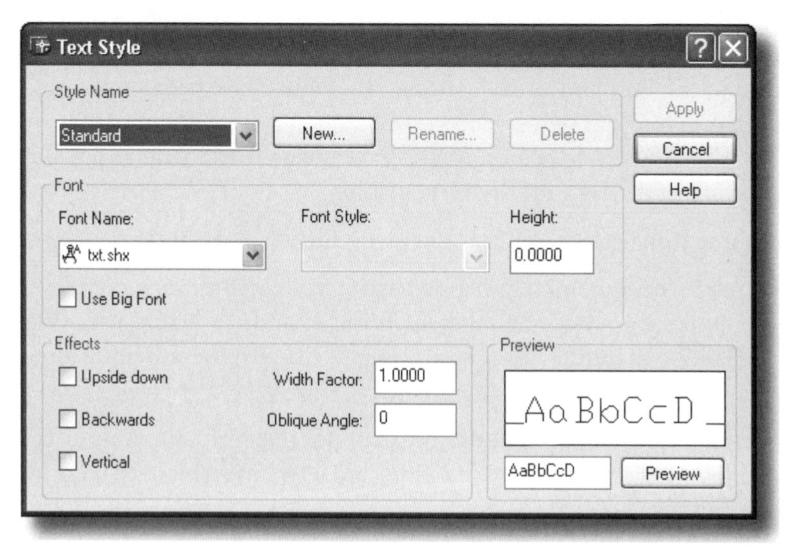

Figure 4–31 *Text Style dialog box*

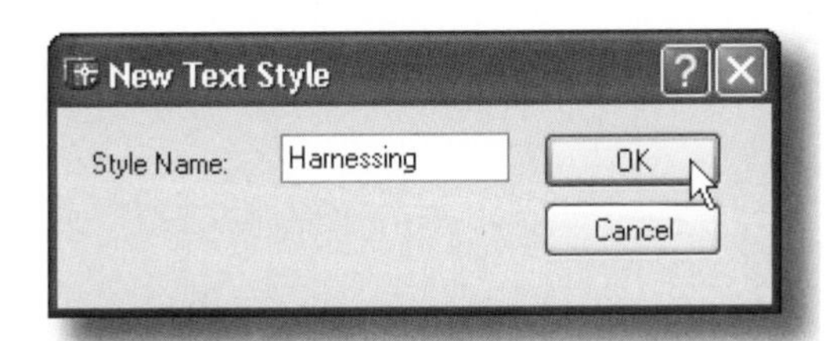

Figure 4–32 *New Text Style dialog box*

To delete an existing style, first select the style from the **Style Name** list box in the Text Style dialog box, and then click the **Delete** button. AutoCAD displays the AutoCAD Alert dialog box to confirm the deletion of the selected style. Click the **Yes** button to confirm the deletion or the **No** button to cancel the deletion of the selected style.

To assign a font to the selected text style, select the appropriate font from the **Font Name:** list box. Similarly, select a font style from the **Font Style:** list box. The font style specifies font character formatting, such as italic, bold, or regular.

The **Height:** text box sets the text height to the value you enter. If you set the height to 0 (zero), then when you use this style in the TEXT or MTEXT command, you are given an opportunity to change the text height with each occurrence of the command. If you set it to any other value, then that value will be used for this style and you will not be allowed to change the text height.

The **Upside down** and **Backwards** check boxes control whether the text is drawn right to left (with the characters backward) or upside down (left to right), respectively. See Figure 4–33.

THIS IS AN EXAMPLE OF UPSIDE-DOWN TEXT

Figure 4–33 *Examples of backward and upside-down text*

The **Vertical** check box controls the display of the characters aligned vertically. The Vertical option is available only if the selected font supports dual orientation. See Figure 4–34 for an example of vertically oriented text.

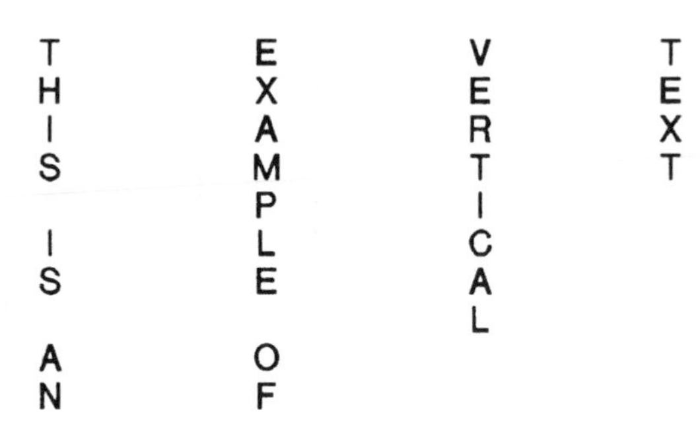

Figure 4–34 *Example of vertically oriented text*

The **Width Factor:** text box sets the character width relative to text height. If it is set to more than 1.0, the text widens; if set to less than 1.0, it narrows.

The **Oblique Angle:** text box sets the obliquing angle of the text. If it is set to 0 (zero) degrees, the text is drawn upright (or in AutoCAD, 90 degrees). A positive value slants the top of the characters toward the right, or in the clockwise direction. A negative value slants the characters in the counterclockwise direction. See Figure 4–35 for examples of oblique angle settings applied to a text string.

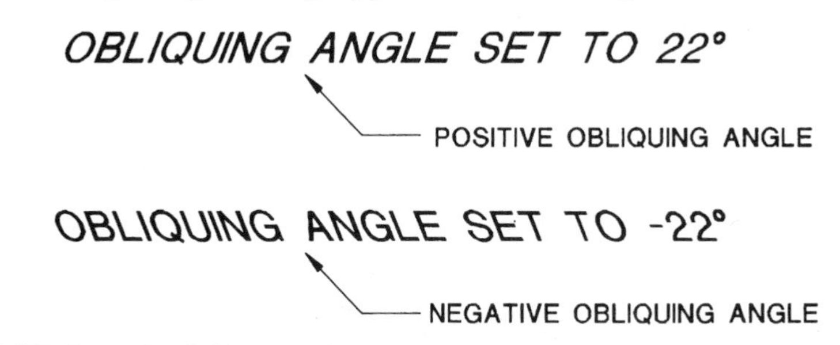

Figure 4–35 *Example of oblique angle settings applied to a text string*

The **Preview** section of the Text Style dialog box displays sample text that changes dynamically as you change fonts and modify the effects. To change the sample text, enter characters in the box below the larger preview image.

After making the necessary changes in the Text Style dialog box, choose the **Apply** button to apply the changes. Click the **Close** button to close the Text Style dialog box.

CREATING OBJECTS FROM EXISTING OBJECTS

AutoCAD not only allows you to draw objects easily, but also allows you to create additional objects from existing objects. This section discusses six important commands that will make your job easier: COPY, ARRAY, OFFSET, MIRROR, FILLET, and CHAMFER.

COPYING OBJECTS

The COPY command places copies of the selected objects at the specified displacement, leaving the original objects intact. The copies are oriented and scaled the same as the original. If necessary, you can make multiple copies of selected objects. Each resulting copy is completely independent of the original and can be edited and manipulated like any other object.

Invoke the COPY command:

Modify toolbar	Choose the Copy Object command (see Figure 4–36)
Modify menu	Choose Copy
Command: prompt	**copy** (ENTER)

Figure 4–36 *Invoking the* COPY *command from the Modify toolbar*

AutoCAD prompts:

Command: **copy** (ENTER)
Select objects: *(select the objects and press* ENTER *to complete the selection)*
Specify base point or displacement, or [Multiple]: *(specify base point or right-click for shortcut menu and choose one of the available options)*
Specify second point of displacement or <use first point as displacement>: *(specify a point for displacement, or press* ENTER *to use first point as displacement)*

You can use one or more object selection methods to select the objects. If you specify two data points, AutoCAD computes the displacement and places a copy accordingly. If you provide a null response to the second point of displacement, AutoCAD considers the point provided as the second point of a displacement vector with the origin (0,0,0) as the first point, indicating how far to copy the objects and in what direction.

The following command sequence shows an example of copying a group of objects selected by means of the Window option, as shown in Figure 4–37, by placing two data points:

Command: **copy** (ENTER)
Select objects: *(specify a point to place one corner of a window)*
Other corner: *(specify a point to place the opposite corner of the window)*

Select objects: (ENTER)
Specify base point or displacement, or [Multiple]: *(specify the base point as shown in Figure 4–37)*
Specify second point of displacement or <use first point as displacement>: *(specify the second point as shown in Figure 4–37)*

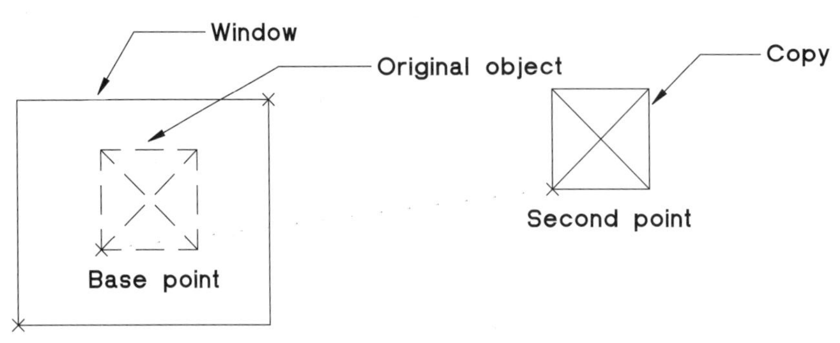

Figure 4–37 *Using the COPY Window option to copy a group of objects by specifying two data points*

Creating Multiple Copies

To make multiple copies, first invoke the COPY command, and respond to the "Specify second point of displacement or <use first point as displacement>:" prompt by entering **m**, for multiple, or right-click and choose **Multiple** from the shortcut menu. The "Specify base point" prompt then reappears, followed by repeated "Specify second point of displacement or <use first point as displacement>:" prompts, and a copy of the selected objects is made at a location determined by each displacement you enter. Each displacement is relative to the original base point. When you have made all the copies you need, give a null response to the "Specify second point of displacement or <use first point as displacement>:" prompt to terminate the command sequence.

The following command sequence shows an example of placing multiple copies of a group of objects selected by means of the Window option, as shown in Figure 4–38.

Command: **copy** (ENTER)
Select objects: *(specify a point to place one corner of the window)*
Other corner: *(specify a point to place the opposite corner of the window)*
Select objects: (ENTER)
Specify base point or displacement, or [Multiple]: (right-click and choose MULTIPLE from shortcut menu and choose one of the available options)
Specify base point: *(select base point as shown in Figure 4–38)*
Specify second point of displacement or <use first point as displacement>: *(specify the second point for Copy 1 as shown in Figure 4–38)*
Specify second point of displacement or <use first point as displacement>: *(specify the second point for Copy 2 as shown in Figure 4–38)*

Specify second point of displacement or <use first point as displacement>:
(specify the second point for Copy 3 as shown in Figure 4–38)
Specify second point of displacement or <use first point as displacement>:
(ENTER)

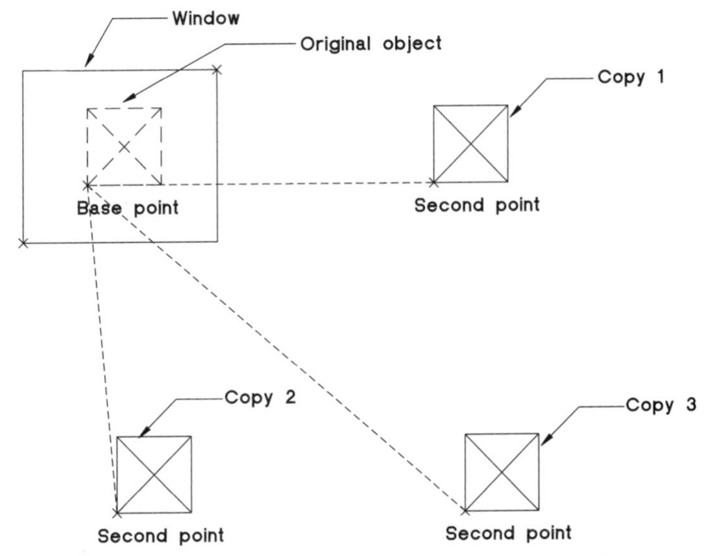

Figure 4–38 *Using the COPY Window option to make multiple copies of a group of objects*

CREATING A PATTERN OF COPIES

The ARRAY command is used to make multiple copies of selected objects in either rectangular or polar arrays (patterns). In the rectangular array, you can specify the number of rows, the number of columns, and the spacing between rows and columns (row and column spacing may differ). The whole rectangular array can be rotated at a selected angle. In the polar array, you can specify the angular intervals, the number of copies, the angle that the group covers, and whether or not the objects maintains their orientation as they are arrayed.

Array

Invoke the ARRAY command:

Modify toolbar	Choose the ARRAY command (see Figure 4–39)
Modify menu	Choose Array
Command: prompt	**array** (ENTER)

Figure 4–39 *Invoking the ARRAY command from the Modify toolbar*

AutoCAD displays the Array dialog box as shown in Figure 4–40.

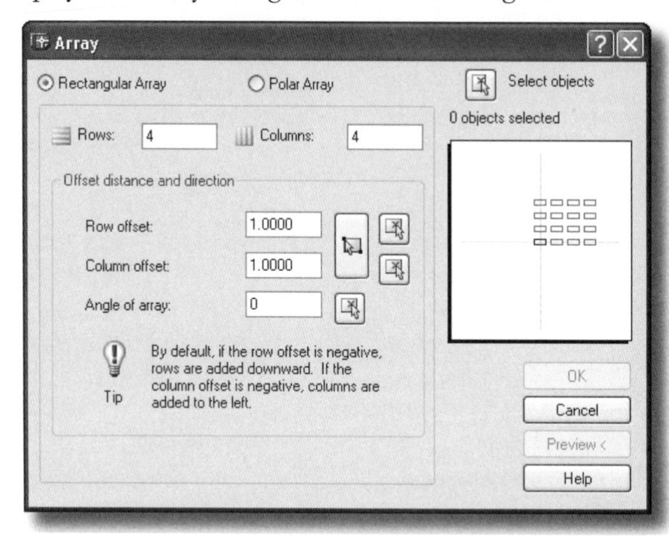

Figure 4–40 *Array dialog box*

In the Array dialog box, you can select the type of array (rectangular or polar). For the rectangular type of array you can select the number and spacings of rows and columns of the array and the angle of the array. For the polar type of array you can select the center of the array, the angle through which objects are arrayed, the number of objects in the array and whether the arrayed objects are rotated as they are arrayed or they keep the orientation of the original object when arrayed.

Rectangular Array

In the Array dialog box, select **Rectangular Array** as shown in Figure 4–40. Enter the numbers for rows and columns in their respective text boxes. At least one of them must be equal to two or greater.

Note: The numbers for rows and columns you enter include the original object. For an array of objects that will be four rows high and three columns wide the numbers four and three must be entered in the **Rows** and **Columns** text boxes. A positive number for the column and row spacing causes the elements to array toward the right and upward, respectively. Negative numbers for the column and row spacing cause the elements to array toward the left and downward, respectively.

In the **Offset distance and direction** section of the Array dialog box, enter the value for the spacing between objects in rows in the **Row offset** text box. Or you can select the button to the far right of the **Row offset:** text box (with a single arrow and crossmark icon). This will allow you to specify the row offset (spacing) by selecting two points on the screen with the cursor. The two points can be in any direction. AutoCAD will use the distance between them for the row offset.

Enter the value for the spacing between objects in columns in the **Column offset:** text box. Or you can select the button to the far right of the **Column offset:** text box (with a single arrow and crossmark icon). This will allow you to specify the column offset (spacing) by selecting two points on the screen with the cursor. The two points can be in any direction. AutoCAD will use the distance between them for the row column.

Note: The offset you specify determines the distance between corresponding points of adjacent objects and not the space between adjacent objects. For example, if you arrayed 3" diameter circles with 2" offsets, the adjacent circles would overlap 1".

You can also specify the row and column offsets in one maneuver by first selecting the large button just to the right of the **Row offset** and **Column offset** text boxes. Then select two points on the screen with the cursor that specify the opposite corners of a rectangle called a unit cell. AutoCAD uses the width of the unit cell as the horizontal distance(s) between columns and the height as the vertical distance(s) between rows.

The array can be rotated by entering the angle in the **Angle of array**: text box. Or you can select the button to the right of the **Angle of array:** text box (with a single arrow and crossmark icon). This will allow you to specify the rotation angle by selecting two points on the screen with the cursor. The two points can be in anywhere on the screen. AutoCAD will use the direction between them (measured from the first point selected to the second) for the angle that the array will be rotated.

The array sample window on the right of the dialog box shows the rows and columns selected. Choose the **Select objects** button to return to the drawing screen in order to select the objects to be arrayed. Once one or more objects are selected, the number of objects is shown above the array sample window and you can preview the resulting array on the screen by choosing the **Preview<** button.

Once the settings have been satisfactorily specified, choose the **OK** button.

Polar Array

In the Array dialog box, select **Polar Array** as shown in Figure 4–41. Next to **Center point:**, enter the X and Y coordinates for the center of the polar array in their respective text boxes. Or you can choose the button with the arrow and crossmark, which will return you to the drawing screen to specify the center point with your cursor.

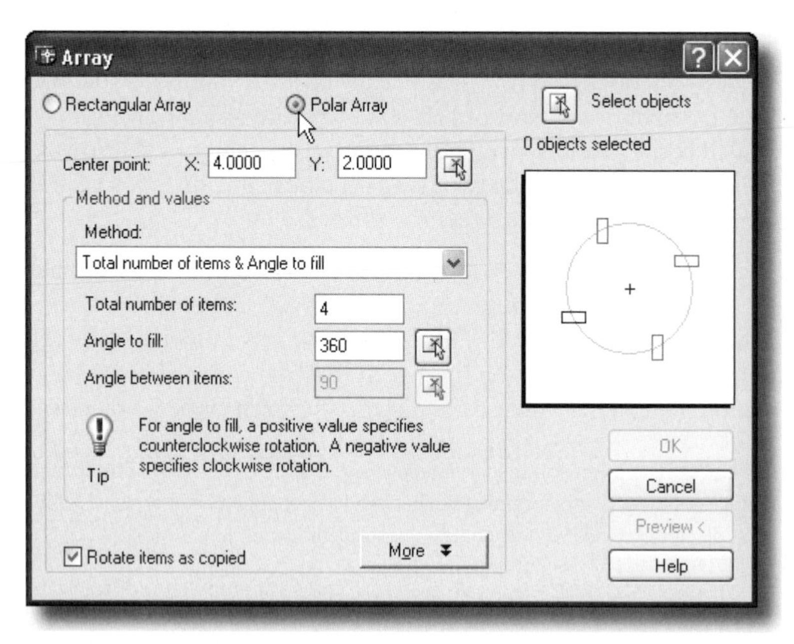

Figure 4–41 *Array dialog box—Polar array selection*

In the **Method and values** section, there are three methods from which to choose; **Total number of items & Angle to fill**, **Total number of items & Angle between items**, and **Angle to fill & Angle between items**. The method you select will determine which two the three text boxes will be active of the **Total number of items:** text box, the **Angle to fill:** text box, and the **Angle between items:** text box.

The total number of items includes the original item, just as in the rectangular array. Note that a positive angle causes items to array in a counterclockwise direction. A negative angle causes them to array in a clockwise direction.

The array sample window on the right of the dialog box shows a representative example of the number items selected, the total angle of the array and the angle between items. Choose the **Select objects** button to return to the drawing screen in order to select the objects to be arrayed. Once one or more objects are selected, the number of objects is shown above the array sample window and you can preview the resulting array on the screen by choosing the Preview button..

Check the **Rotate items as copied** check box to cause the items to each be rotated from the original item's orientation the same angle that particular the item is arrayed.

Choose the **More** button to display more options in the dialog box. If the **Rotate items as copied** check box in not checked, then you can specify a new reference (base) point relative to the selected objects that will remain at a constant distance from the center point of the array as the objects are arrayed. AutoCAD uses the distance from the array's center point to a base point on the last object selected. The point used is determined by the type of object, as shown in the following table.

Type of Object	Base Point
Arc, circle, ellipse	Center point
Polygon, rectangle	First corner
donut, line, polyline, 3D polyline, ray, spline	Starting point
Block, paragraph text, single-line text	Insertion point
Construction lines	Midpoint
Region	Grip point

If the **Set to Object's Default** check box is checked, AutoCAD uses the default base point of the object to position the arrayed object. To specify a different base point, clear this check box.

You can use the **Base Point:** text boxes to set new *X* and *Y* base point coordinates. Or you can choose **Pick Base Point** to temporarily close the dialog box and specify a point with your cursor. After you specify a point, the Array dialog box is redisplayed.

Once the settings have been satisfactorily specified, choose the **OK** button.

You can also create rectangular/polar arrays by using the command line version of the ARRAY command.

AutoCAD prompts:

Command: **-array** (ENTER)
Select objects: *(select the objects and then give a null response to complete the selection)*
Enter the type of array [Rectangular/Polar] <R>: *(choose* **r** *for rectangular array or right-click for shortcut menu and choose one of the available options)*
Enter the number of rows (—) <1>: *(specify a nonzero integer for number of rows)*
Enter the number of columns (|||) <1> *(specify a nonzero integer for number of columns)*
Enter the distance between rows or specify unit cell (—):*(specify distance between rows, or specify two points to measure the distance between rows)*
Specify the distance between columns (|||):*(specify a distance between columns, or specify two points to measure the distance between columns)*

Any combination of whole numbers of rows and columns may be entered (except both 1 row and 1 column, which would not create any copies). AutoCAD includes the original object in the number you enter. Row and column spaces can be different from each other. They can be entered separately when prompted, or you can select two points that specify the opposite corners of a rectangle called a unit cell. AutoCAD uses the width of the unit cell as the horizontal distance(s) between columns and the height as the vertical distance(s) between rows.

A positive number for the column and row spacing causes the elements to array toward the right and upward, respectively. Negative numbers for the column and row spacing cause the elements to array toward the left and downward, respectively.

AutoCAD creates rectangular arrays along a baseline defined by the current snap rotation. By default, the snap rotation is set to 0 degrees, so that rows and columns are orthogonal with respect to the *X* and *Y* drawing axes. The **Rotate** option of the SNAP command allows you to change the rotation angle and creates a rotated array.

The following command sequence shows an example of placing a rectangular array with 4 rows and 6 columns, as shown in Figure 4–42.

Command: **array** (ENTER)
Select objects: *(select objects)*
Enter the type of array [Rectangular/Polar] <R>: **r** (ENTER)
Enter the number of rows (—) <1>: **4** (ENTER)
Enter the number of columns (|||) <1> **6** (ENTER)
Enter the distance between rows or specify unit cell (—): **1** (ENTER)
Specify the distance between columns (|||): **1.5** (ENTER)

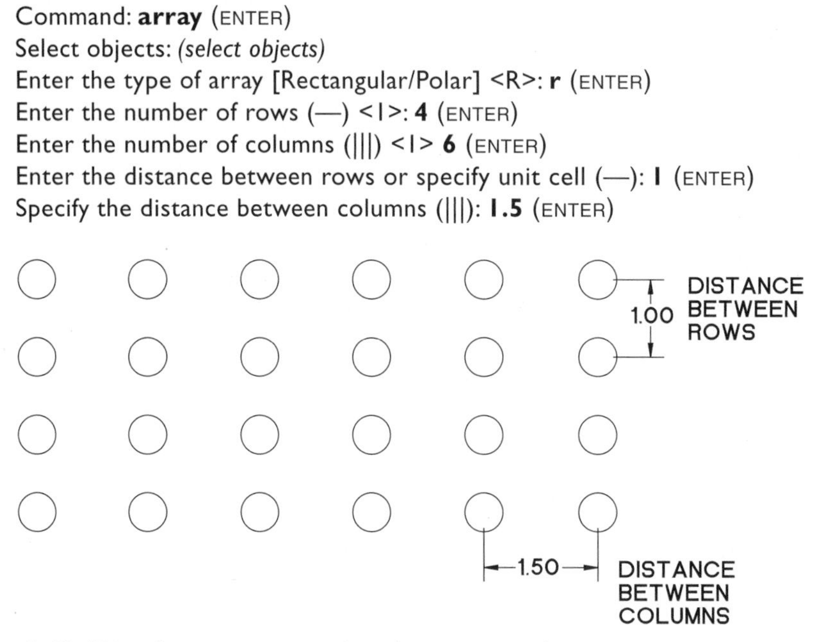

Figure 4–42 *Using the* ARRAY *command to place a rectangular array*

You can also create Polar arrays by using the command version of the ARRAY command:

AutoCAD prompts:

Command: **-array** (ENTER)
Select objects: *(select the objects and then give a null response to complete the selection)*
Enter the type of array [Rectangular/Polar] <R>: *(choose* ***p*** *for polar array or right-click for shortcut menu and choose one of the available options)*

Specify center point of array or [Base]: *(specify a point around which you want the array to form)*
Enter the number of items in the array: *(specify a positive integer for number of itmes in the array, include the original item or press* ENTER*)*
Specify the angle to fill (+=ccw, -=cw) <360>: *(specify an angle to fill or press* ENTER*)*
Specify Angle between items (+=CCW, -=CW): *(specify an angle between items; this prompt appears only if a null response is provided for Angle to fill)*
Rotate arrayed objects? [Yes/No] <Y>: *(choose* **y**, *for yes to Rotate as they are copied, or* **n**, *for no to Rotate as they are copied)*

If you specify the number of items for an array, you must specify either the angle to fill or the angle between items. If you provide a null response to the number of items, then you must specify the angle to fill and the angle between items.

The following command sequence presents an example of placing a rotated polar array as shown in Figure 4–43.

Command: **array** (ENTER)
Select objects: *(select objects)*
Enter the type of array [Rectangular/Polar] <R>: **p** (ENTER)
Specify center point of array or [Base]: *(specify the center point as shown in Figure 4–43)*
Enter the number of items in the array: **8** (ENTER)
Specify the angle to fill (+=ccw, -=cw) <360>: **360** (ENTER)
Rotate arrayed objects? [Yes/No] <Y>: (ENTER)

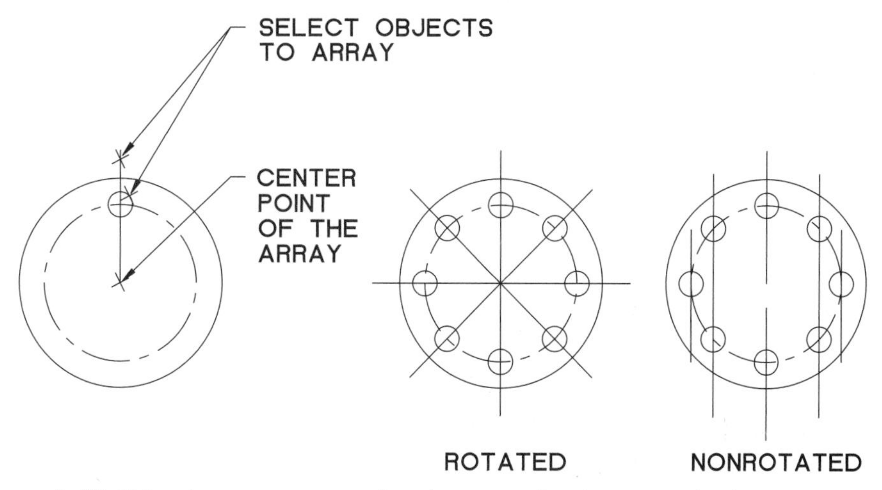

Figure 4–43 *Using the* ARRAY *command to place rotated and nonrotated polar arrays*

Figure 4–43 shows both non-rotated and rotated polar arrays.

CREATING PARALLEL LINES, PARALLEL CURVES, AND CONCENTRIC CIRCLES

The OFFSET command creates parallel lines, parallel curves, and concentric circles relative to existing objects, as shown in Figure 4–44. Special precautions must be taken when using the OFFSET command to prevent unpredictable results from occurring when using the command on arbitrary curve/line combinations in polylines.

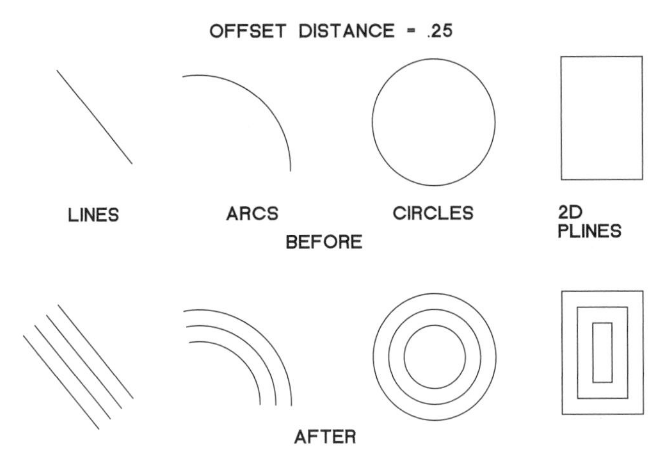

Figure 4–44 *Examples created using the* OFFSET *command*

Invoke the OFFSET command:

Modify toolbar	Choose the OFFSET command (see Figure 4–45)
Modify menu	Choose Offset
Command: prompt	**offset** (ENTER)

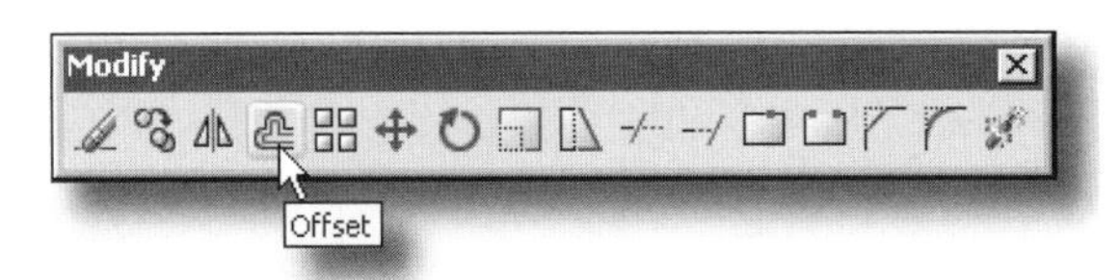

Figure 4–45 *Invoking the* OFFSET *command from the Modify toolbar*

AutoCAD prompts:

Command: **offset** (ENTER)
Specify offset distance or [Through] <1.0000>: *(specify offset distance, or right-click for shortcut menu and choose one of the available options)*
Select object to offset or <exit>: *(select an object to offset)*
Specify point on side to offset: *(specify a point to one side of the object to offset)*

Select object to offset or <exit>: *(continue selecting additional objects for offset, and specify the side of the object to offset, or press* ENTER *to terminate the command sequence)*

Instead of specifying the offset distance, select the Through option, and AutoCAD prompts for a through point. Specify a point, and AutoCAD creates an object passing through the specified point.

Valid Objects to Offset

Valid objects include the line, spline curve, arc, circle, and 2D polyline. If you select another type of object, such as text, you will get the following error message:

Cannot offset that object.

The object selected for offsetting must be in a plane parallel to the current coordinate system. Otherwise you will get the following error message:

Object not parallel with UCS.

Offsetting Miters and Tangencies

The OFFSET command affects single objects in a manner different from a polyline made up of the same objects. Polylines whose arcs join lines and other arcs in a tangent manner are affected differently than polylines with nontangent connecting points. For example, in Figure 4–46 the seven lines are separate objects. When you specify side to offset as shown, there are gaps and overlaps at the ends of the newly created lines.

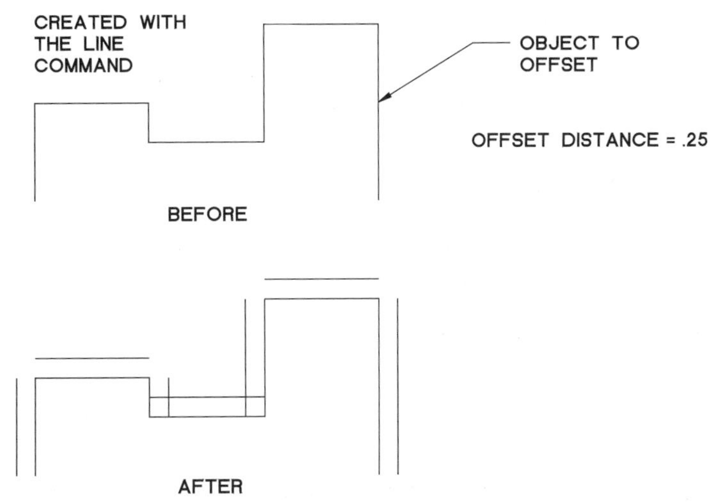

Figure 4–46 *Using the* OFFSET *command with single objects*

In Figure 4–47, the lines have been joined together (see PEDIT in Chapter 5) as a single polyline. See how the OFFSET command affects the corners where the new polyline segments join.

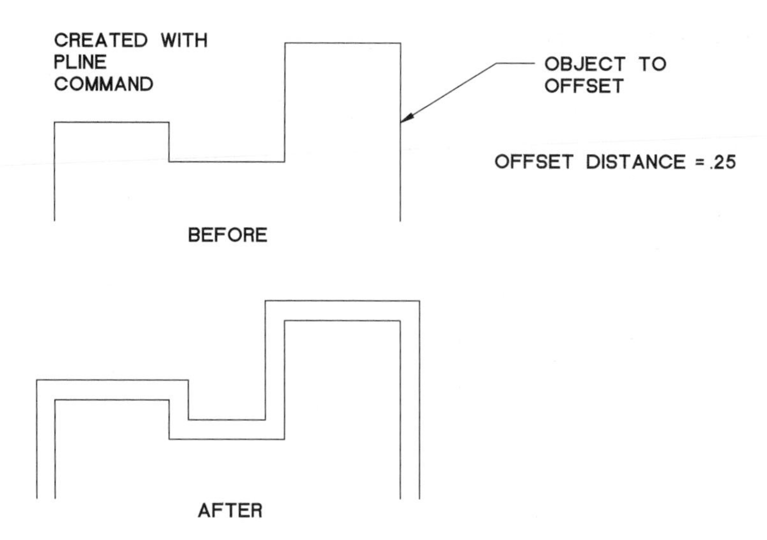

Figure 4–47 *Using the* OFFSET *command with polylines*

Note: The results of offsetting polylines with arc segments that connect other arc segments and/or line segments in dissimilar (nontangent) directions might be unpredictable. Examples of offsetting such polylines are shown in Figure 4–48.

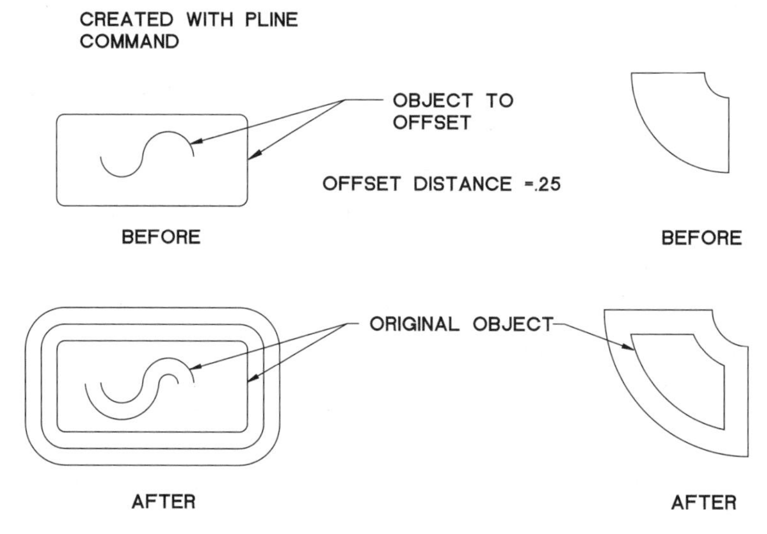

Figure 4–48 *Using the* OFFSET *command with nontangent arc and/or line segments*

If you are not satisfied with the resulting new polyline configuration, you can use the PEDIT command to edit it. Or, you can explode the polyline and edit the individual segments.

CREATING A MIRROR COPY OF OBJECTS

The MIRROR command creates a copy of selected objects in reverse, that is, mirrored about a specified line. Invoke the MIRROR command:

Modify toolbar	Choose the MIRROR command (see Figure 4–49)
Modify menu	Choose Mirror
Command: prompt	**mirror** (ENTER)

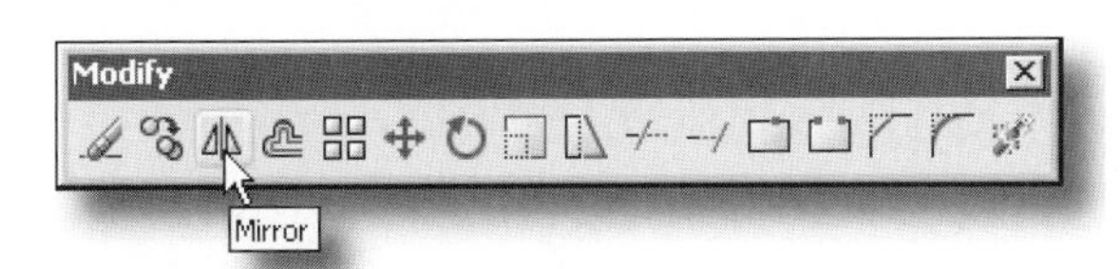

Figure 4–49 *Invoking the MIRROR command from the Modify toolbar*

AutoCAD prompts:

Command: **mirror** (ENTER)
Select objects: *(select the objects and then give a null response to complete the selection)*
Specify first point of mirror line: *(specify a point to define the first point of the mirror line)*
Specify second point of mirror line: *(specify a point to define the second point of the mirror line)*
Delete source objects? [Yes/No] <N>: *(enter **y**, for yes to delete the original objects, or **n**, not to delete the original objects, that is, to retain them)*

The first and second points of the mirror line become the endpoints of an invisible line about which the selected objects will be mirrored.

The following command sequence shows an example of mirroring a group of selected objects by means of the Window option, as shown in Figure 4–50:

Command: **mirror** (ENTER)
Select objects: *(specify POINT 1 to place one corner of a window)*
Second corner: *(specify POINT 2 to place the opposite corner of the window)*
Select objects: (ENTER)
Specify first point of mirror line: *(specify POINT 3, as shown in Figure 4–50)*
Specify second point of mirror line: *(specify POINT 4, as shown in Figure 4–50)*
Delete source objects? [Yes/No] <N>: (ENTER)

The text as mirrored is located relative to other objects within the selected group. But the text will or will not retain its original orientation, depending on the setting of the system variable called MIRRTEXT. If the value of MIRRTEXT is set to 1, then text items in the selected group will have their orientations and location mirrored. That is, if their characters were normal and they read left to right in the original group, in the mirrored copy they will read right to left and the characters will be backwards. If MIRRTEXT is set to 0 (zero), then the text strings in the group will have their locations mirrored, but the individual text strings will retain their normal, left-to-right, character appearance. The MIRRTEXT system variable, like other system variables, is changed by the SETVAR command or by typing MIRRTEXT at the "Command:" prompt, as follows:

Command: **mirrtext** (ENTER)
Enter New Value for MIRRTEXT <1>: **0** (ENTER)

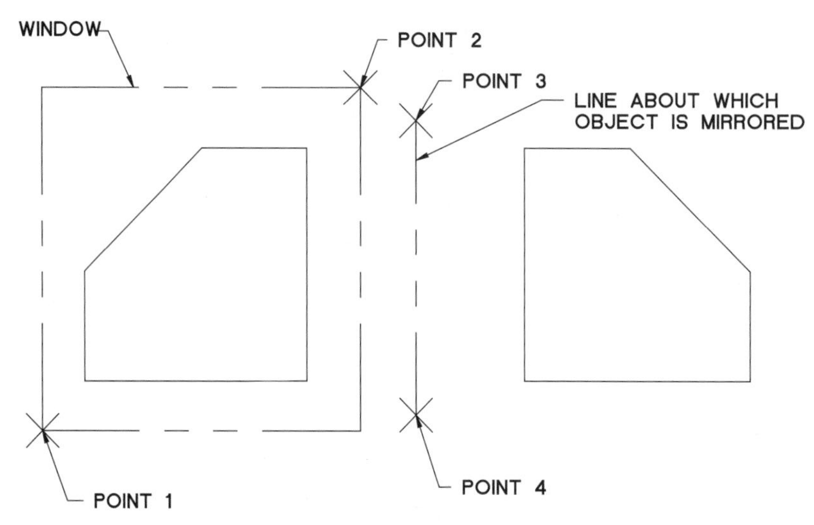

Figure 4–50 *Mirroring a group of objects selected by means of the Window option*

This setting causes mirrored text to retain its readability. Figures 4–51 and 4–52 show the result of the MIRROR command when the MIRRTEXT variable is set to 1 and 0, respectively.

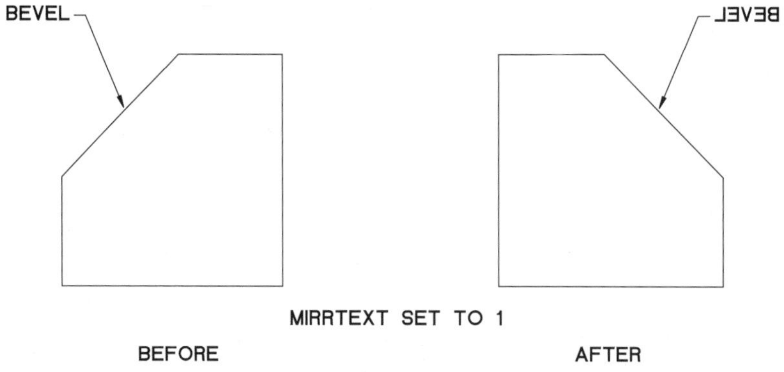

Figure 4–51 *The MIRROR command with the MIRRTEXT variable set to 1*

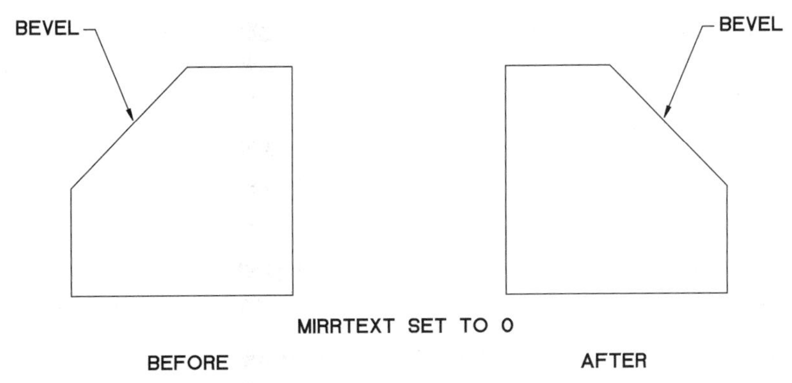

Figure 4–52 *The* MIRROR *command with the* MIRRTEXT *variable set to 0*

CREATING A FILLET BETWEEN TWO OBJECTS

The FILLET command fillets (rounds) the intersecting ends of two arcs, circles, lines, elliptical arcs, polylines, rays, xlines, or splines with an arc of a specified radius.

If the TRIMMODE system variable is set to 1 (default), then the FILLET command trims the intersecting lines to the endpoints of the fillet arc. And if TRIMMODE is set to 0 (zero), then the FILLET command leaves the intersecting lines at the endpoints of the fillet arc.

Invoke the FILLET command:

Modify toolbar	Choose the FILLET command (see Figure 4–53)
Modify menu	Choose Fillet
Command: prompt	**fillet** (ENTER)

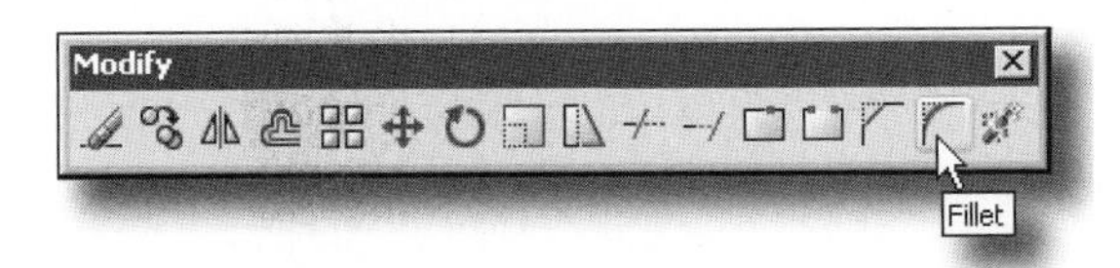

Figure 4–53 *Invoking the* FILLET *command from the Modify toolbar*

AutoCAD prompts:

Command: **fillet** (ENTER)
Current settings: Mode = TRIM, Radius = <current>
Select first object or [Polyline/Radius/Trim/mUltiple]: *(select one of the two objects to fillet, or right-click for the shortcut menu and choose one of the available options)*

By default, AutoCAD prompts you to select an object. If you select an object to fillet, then AutoCAD prompts:

Select second object: *(select the second object to fillet)*

AutoCAD joins the two objects with an arc having the specified radius. If the objects selected to be filleted are on the same layer, AutoCAD creates the fillet arc on the same layer. If not, AutoCAD creates the fillet arc on the current layer.

AutoCAD allows you to draw a fillet between parallel lines, xlines, and rays. The first selected object must be a line or ray, but the second object can be a line, xline, or ray. The diameter of the fillet arc is always equal to the distance between the lines. The current fillet radius is ignored and remains unchanged.

Radius

The Radius option allows you to change the current fillet radius. The following command sequence sets the fillet radius to 0.5 and draws the fillet between two lines, as shown in Figure 4–54.

Command: **fillet** (ENTER)
Current settings: Mode = TRIM, Radius = <current>
Select first object or [Polyline/Radius/Trim/mUltiple]: **r** (ENTER)
Specify fillet radius <current>: 0.25 (ENTER)
Select first object or [Polyline/Radius/Trim/mUltiple]: *(select one of the lines, as shown in Figure 4–54)*
Select second object: *(select the other line to fillet, as shown in Figure 4–54)*

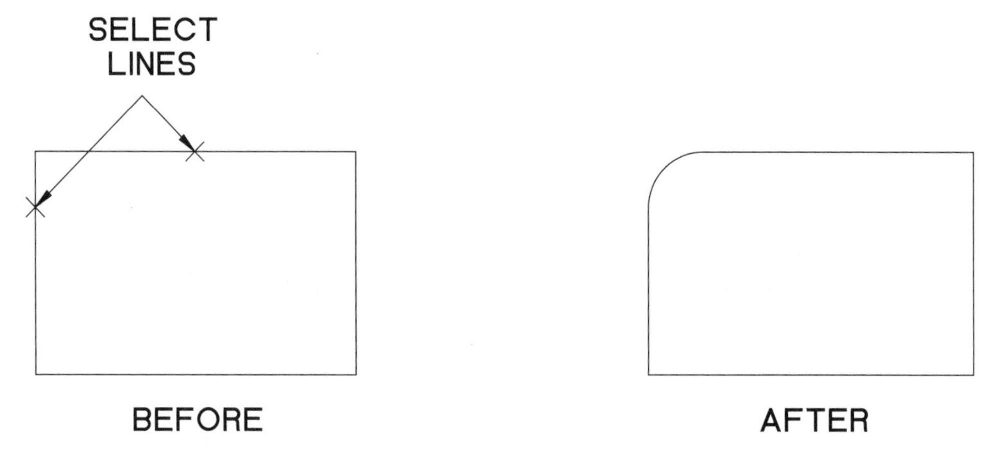

Figure 4–54 *Fillet drawn with a radius of 0.25*

If you select lines or arcs, AutoCAD extends these lines or arcs until they intersect or trims them at the intersection, keeping the selected segments if they cross. The following command sequence sets the fillet radius to 0 and draws the fillet between two lines as shown in Figure 4–55.

Command: **fillet** (ENTER)
Current settings: Mode = TRIM, Radius = <current>
Select first object or [Polyline/Radius/Trim/mUltiple]: **r** (ENTER)
Specify fillet radius <current>: 0 (ENTER)
Select first object or [Polyline/Radius/Trim/mUltiple]: *(select the first object, as shown in Figure 4–55)*
Select second object: *(select the second object to fillet as shown in Figure 4–55)*

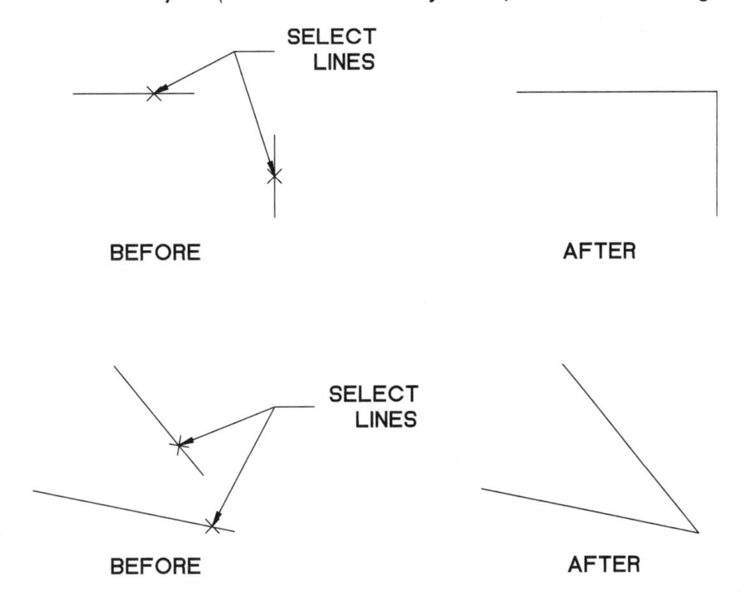

Figure 4–55 *Fillet drawn with a radius of 0*

Polyline

With the Polyline option, AutoCAD draws fillet arcs at each vertex of a 2D polyline where two line segments meet. The following command sequence sets the fillet radius to 0.5 and draws the fillet at each vertex of a 2D polyline, as shown in Figure 4–56:

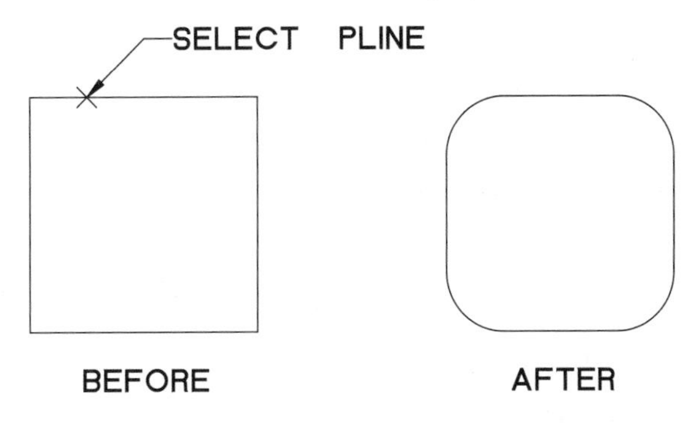

Figure 4–56 *Fillet with a radius of 0.5 drawn to a polyline*

Command: **fillet** (ENTER)
Current settings: Mode = TRIM, Radius = <current>
Select first object or [Polyline/Radius/Trim/mUltiple]: **r** (ENTER)
Specify fillet radius <current>: 0.5 (ENTER)
Select first object or [Polyline/Radius/Trim/mUltiple]: **p** (ENTER)
Select 2D polyline: *(select the polyline as shown in Figure 4–56)*

Trim

The Trim option (Trim/No Trim) controls whether or not AutoCAD trims the selected edges to the fillet arc endpoints. This option is similar to setting the TRIMMODE system variable from 1 to 0 or 0 to 1.

Multiple

The Multiple option, when selected, allows you to specify multiple pairs of objects to be filleted without exiting the FILLET command. To exit the command, select ENTER or CANCEL from the shortcut menu, press ESC, or invoke another command.

CREATING A CHAMFER BETWEEN TWO OBJECTS

The CHAMFER command allows you to draw an angled corner between two lines. The size of the chamfer is determined by the settings of the first and the second chamfer distances. If it is to be a 45-degree chamfer for perpendicular lines, then the two distances are set to the same value.

If the TRIMMODE system variable is set to 1 (default), then the CHAMFER command trims the intersecting lines to the endpoints of the chamfer line. And if TRIMMODE is set to 0 (zero), then the CHAMFER command leaves the intersecting lines at the endpoints of the chamfer line.

Invoke the CHAMFER command:

Modify toolbar	Choose the CHAMFER command (see Figure 4–57)
Modify menu	Choose Chamfer
Command: prompt	**chamfer** (ENTER)

Figure 4–57 *Invoking the CHAMFER command from the Modify toolbar*

AutoCAD prompts:

Command: **chamfer** (ENTER)
(TRIM mode) Current chamfer Dist1 = <current>, Dist2 = <current>

Select first line or [Polyline/Distance/Angle/Trim/Method/mUltiple]: *(select one of the two lines to chamfer, or right-click for the shortcut menu and choose one of the available options)*

By default, AutoCAD prompts you to select the first line to chamfer. If you select a line to chamfer, then AutoCAD prompts:

Select second line: *(select the second line to chamfer)*

AutoCAD draws a chamfer to the selected lines. If the selected lines to be chamfered are on the same layer, AutoCAD creates the chamfer on the same layer. If not, AutoCAD creates the chamfer on the current layer.

Distance

The Distance option allows you to set the first and second chamfer distances. The following command sequence sets the first chamfer and second chamfer distance to 0.5 and 1.0, respectively, and draws the chamfer between two lines as shown in Figure 4–58.

Command: **chamfer** (ENTER)
(TRIM mode) Current chamfer Dist1 = <current>, Dist2 = <current>
Select first line or [Polyline/Distance/Angle/Trim/Method/mUltiple]: **d** (ENTER)
Specify first chamfer distance <0.5000>: **0.5** (ENTER)
Specify second chamfer distance <0.5000>: **1.0** (ENTER)
Select first line or [Polyline/Distance/Angle/Trim/Method/mUltiple]: *(select the first line, as shown in Figure 4–58)*
Select second line: *(select the second line, as shown in Figure 4–58)*

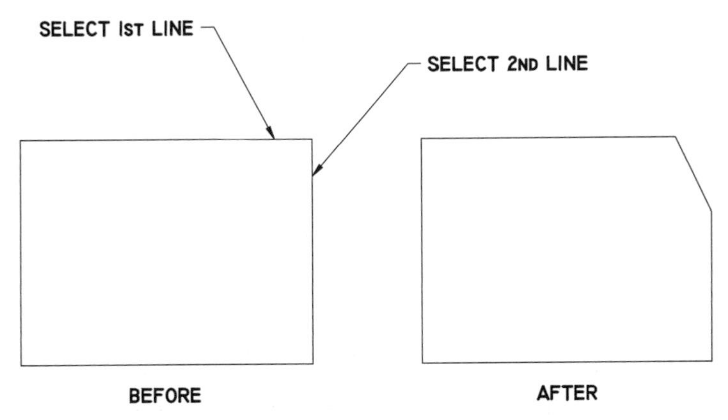

Figure 4–58 *Chamfer drawn with distances of 0.5 and 1.0*

Polyline

With the Polyline option, AutoCAD draws chamfers at each vertex of a 2D polyline where two line segments meet. The following command sequence sets the chamfer distances to 0.5 and draws the chamfer at each vertex of a 2D polyline, as shown in Figure 4–59.

Command: **chamfer** (ENTER)
(TRIM mode) Current chamfer Dist1 = <current>, Dist2 = <current>
Select first line or [Polyline/Distance/Angle/Trim/Method/mUltiple]: **d** (ENTER)
Specify first chamfer distance <0.5000>: **0.5** (ENTER)
Specify second chamfer distance <0.5000>: **0.5** (ENTER)
Select first line or [Polyline/Distance/Angle/Trim/Method/mUltiple]: **p** (ENTER)
Select 2D polyline: *(select the polyline, as shown in Figure 4–59)*

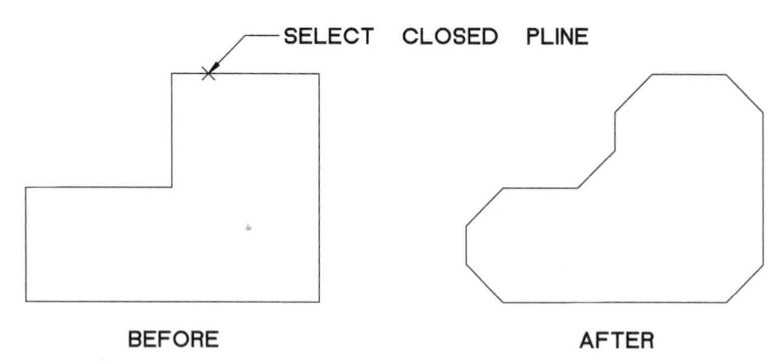

Figure 4–59 *Chamfer with distances of 0.5 drawn on a polyline*

Angle

The Angle option is similar to the Distance option, but instead of prompting for the first and second chamfer distances, AutoCAD prompts for the first chamfer distance and an angle from the first line. This is another method by which to create the chamfer line.

Method

The Method option controls whether AutoCAD uses two distances or a distance and an angle to create the chamfer line.

Trim

The Trim option (Trim/No Trim) controls whether or not AutoCAD trims the selected edges to the chamfer line endpoints. This option is similar to setting the TRIMMODE system variable from 1 to 0 or from 0 to 1.

Note: The CHAMFER command set to zero distance operates the same way the FILLET command operates set to zero radius.

Multiple

The Multiple option, when selected, allows you to specify multiple pairs of objects to be chamfered without exiting the CHAMFER command. To exit the command, select ENTER or CANCEL from the shortcut menu, press ESC, or invoke another command.

MODIFYING OBJECTS

In this section, four additional MODIFY commands are explained: MOVE, TRIM, BREAK, and EXTEND. (The ERASE command was explained in Chapter 2.)

MOVING OBJECTS

The MOVE command lets you move one or more objects from their present location to a new one without changing orientation or size.

Invoke the MOVE command:

Modify toolbar	Choose the MOVE command (see Figure 4–60)
Modify menu	Choose Move
Command: prompt	**move** (ENTER)

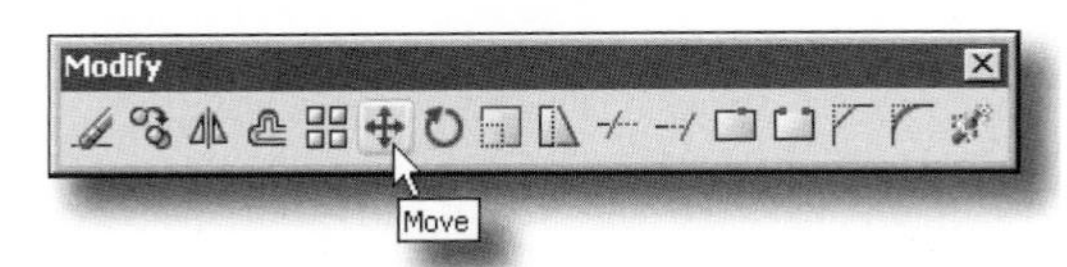

Figure 4–60 *Invoking the* MOVE *command from the Modify toolbar*

AutoCAD prompts:

Command: **move** (ENTER)
Select objects: *(select the objects)*
Specify base point or displacement: *(specify a point)*
Specify second point of displacement: *(specify a point for displacement, or press* ENTER*)*

You can use one or more object selection methods to select the objects. If you specify two data points, AutoCAD computes the displacement and moves the selected objects accordingly. If you specify the points on the screen, AutoCAD assists you in visualizing the displacement by drawing a rubber-band line from the first point as you move the crosshairs to the second point. If you provide a null response by pressing ENTER to the second point of displacement, then AutoCAD interprets the base point as relative *X,Y,Z* displacement.

The following command sequence shows an example of moving a group of objects, selected by means of the Window option, by relative displacement, as shown in Figure 4–61.

Command: **move** (ENTER)
Select objects: *(specify Point 1 to place one corner of a window)*
Other corner: *(specify Point 2 to place the opposite corner of the window)*
Select objects: (ENTER)
Specify base point or displacement: **2,3** (ENTER)
Specify second point of displacement: (ENTER)

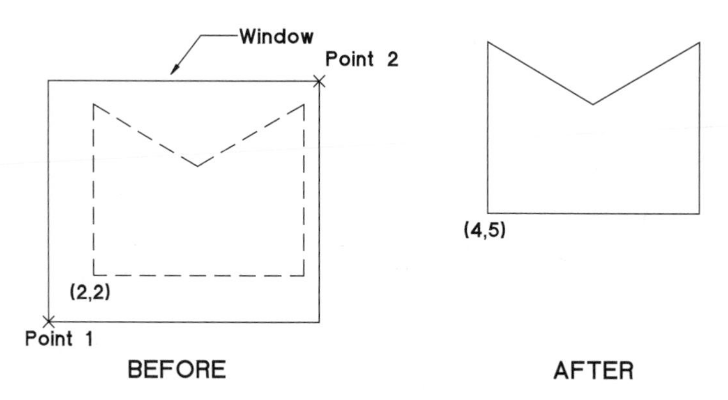

Figure 4–61 *Using the Window option of the* MOVE *command to move a group of objects by means of vector displacement*

The following command sequence shows an example of moving a group of objects selected by the Window option and moving the objects by specifying two data points, as shown in Figure 4–62.

> Command: **move** (ENTER)
> Select objects: *(pick Point 1 to place one corner for a window)*
> Other corner: *(pick Point 2 to place the opposite corner of the window)*
> Select objects: (ENTER)
> Specify base point or displacement: *(specify base point)*
> Specify second point of displacement: *(specify second point)*

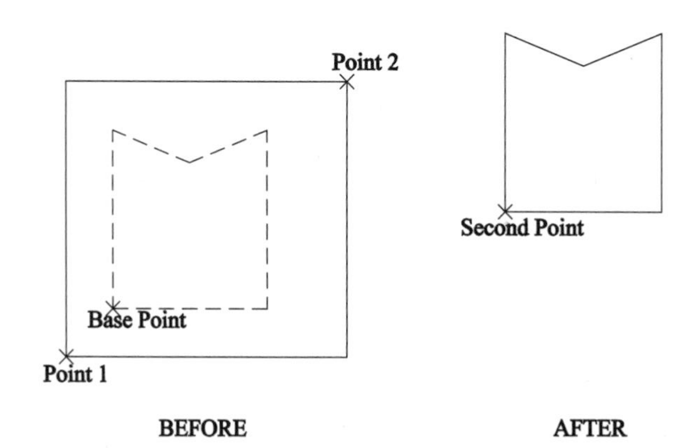

Figure 4–62 *Using the Window option of the* MOVE *command to move a group of objects by specifying two data points*

TRIMMING OBJECTS

The TRIM command is used to trim the portion of the object(s) that is drawn past a cutting edge or from an implied intersection defined by other objects. Objects that can be trimmed include lines, arcs, elliptical arcs, circles, 2D and 3D polylines, xlines, rays, and splines. Valid cutting edge objects include lines, arcs, circles, ellipses, 2D and 3D polylines, floating viewports, xlines, rays, regions, splines, and text.

Invoke the TRIM command:

Modify toolbar	Choose the TRIM command (see Figure 4–63)
Modify menu	Choose Trim
Command: prompt	**trim** (ENTER)

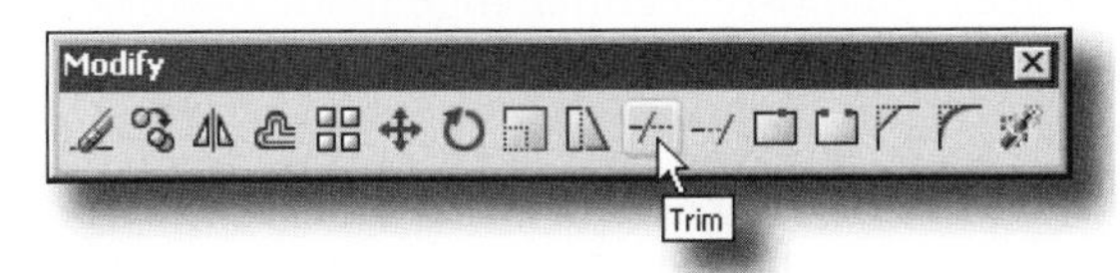

Figure 4–63 *Invoking the TRIM command from the Modify toolbar*

AutoCAD prompts:

Command: **trim** (ENTER)
Current settings: Projection=UCS, Edge=None
Select cutting edge(s)...
Select objects: *(select the objects and press ENTER to terminate the selection of the cutting edges)*
Select object to trim or shift-select to extend or [Project/Edge/Undo]: *(select object(s) to trim or right-click for shortcut menu and choose one of the available options)*

The TRIM command initially prompts you to "Select cutting edge(s):". After selecting one or more cutting edges to trim, press ENTER. You are then prompted to "Select object to trim:". Select one or more objects to trim and then press ENTER to terminate the command. If you press ENTER in response to "Select cutting edge(s)" prompt with out selecting any objects, by default AutoCAD selects all the objects. You can switch to the EXTEND command by holding the SHIFT key while selecting the objects which will be extended to the specified cutting edge instead of being trimmed. See the section on the EXTEND command in this chapter.

Note: Don't forget to press ENTER after selecting the cutting edge(s). Otherwise, the program will not respond as expected. In fact, TRIM continues expecting more cutting edges until you terminate the cutting edge selecting mode.

Edge

The Edge option determines whether objects that extend past a selected cutting edge or to an implied intersection are trimmed. AutoCAD prompts as follows when the Edge option is selected:

Enter an implied edge extension mode [Extend/No extend] <current>:
(right-click for the shortcut menu and select one of the two available options)

The Extend selection extends the cutting edge along its natural path to intersect an object in 3D (implied intersection).

The No Extend selection specifies that the object is to be trimmed only at a cutting edge that intersects it in 3D space.

Undo

The Undo option reverses the most recent change made by TRIM command.

Project

The Project option specifies the projection mode and coordinate system AutoCAD uses when trimming objects. By default, it is set to the current UCS.

Figure 4–64 shows examples of using the TRIM command.

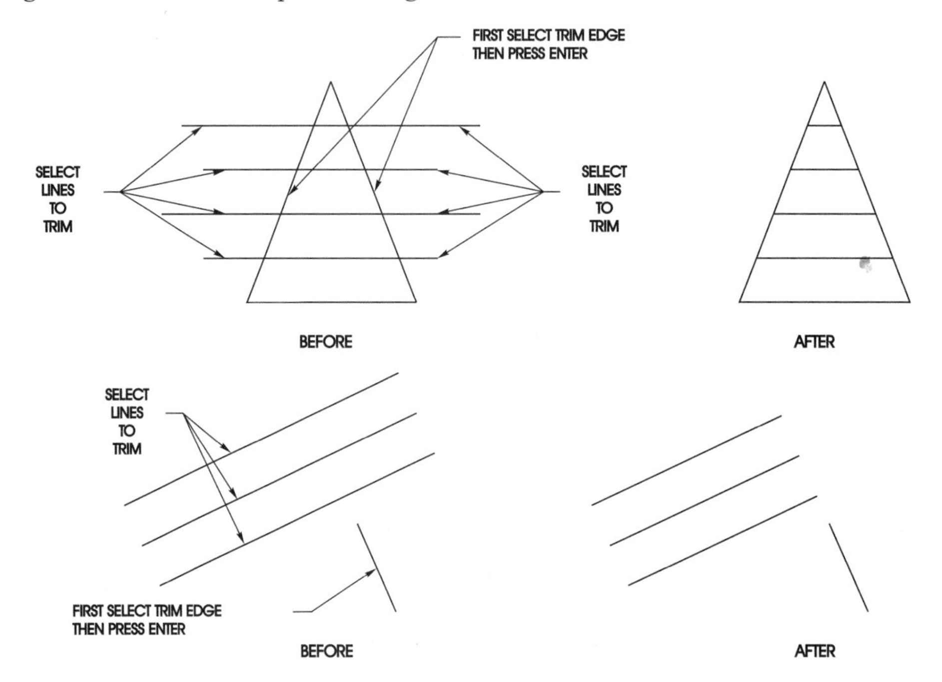

Figure 4–64 *Examples of using the* TRIM *command*

ERASING PARTS OF OBJECTS

The BREAK command is used to remove parts of objects or to split an object in two parts, and it can be used on lines, xlines, rays, arcs, circles, ellipses, splines, donuts, traces, and 2D and 3D polylines.

Invoke the BREAK command:

Modify toolbar	Choose the BREAK command (see Figure 4–65)
Modify menu	Choose Break
Command: prompt	**break** (ENTER)

Figure 4–65 *Invoking the* BREAK *command from the Modify toolbar*

AutoCAD prompts:

Command: **break** (ENTER)
Select object: *(select an object)*
Specify second break point or [First point]: *(specify the second break point, or right-click for shortcut menu and choose one of the available options)*

See Figure 4–66 for examples of applications of the BREAK command.

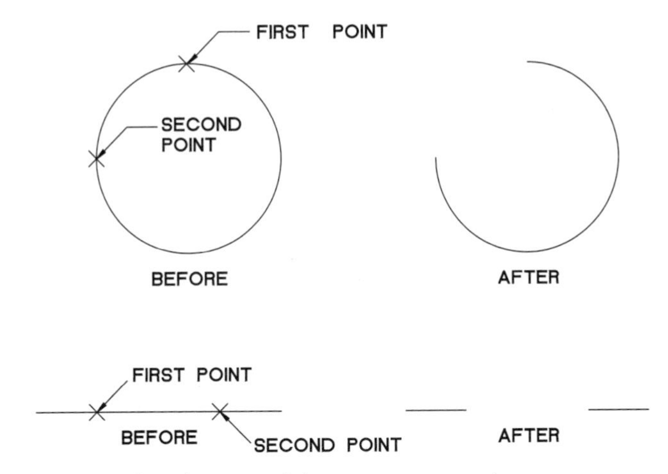

Figure 4–66 *Examples of applications of the* BREAK *command*

AutoCAD erases the portion of the object between the first point (the point where the object was selected) and second point. If the second point is not on the object, then AutoCAD selects the nearest point on the object. If you need to erase an object to one end of a line, arc, or polyline, then specify the second point beyond the end to be removed.

If instead of specifying the second point, you choose First point from the shortcut menu, AutoCAD prompts for the first point and then for the second point.

An object can be split into two parts without removing any portion of the object by selecting the same point as the first and second points. You can do so by entering @ to specify the second point.

If you select a circle, then AutoCAD converts it to an arc by erasing a piece, moving counterclockwise from the first point to the second point. For a closed polyline, the part is removed between two selected points, moving in direction from the first to the last vertex. And in the case of 2D polylines and traces with width, the BREAK command will produce square ends at the break points.

EXTENDING OBJECTS TO MEET ANOTHER OBJECT

The EXTEND command is used to change one or both endpoints of selected lines, arcs, elliptical arcs, open 2D and 3D polylines, and rays to extend to lines, arcs, elliptical arcs, circles, ellipses, 2D and 3D polylines, rays, xlines, regions, splines, text string, or floating viewports.

Invoke the EXTEND command:

Modify toolbar	Choose the EXTEND command (see Figure 4–67)
Modify menu	Choose Extend
Command: prompt	**extend** (ENTER)

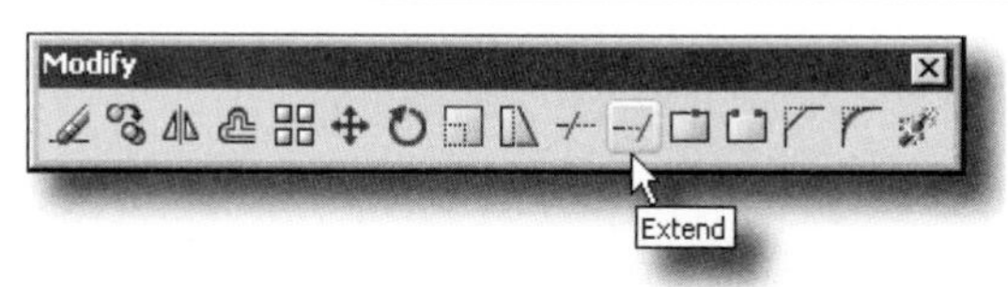

Figure 4–67 *Invoking the EXTEND command from the Modify toolbar*

AutoCAD prompts:

Command: **extend** (ENTER)
Select boundary edges...
Select objects: *(select the objects and then press ENTER to complete the selection)*
Select object to extend or shift-select to trim or [Project/Edge/Undo]: *(select the object(s) to extend and press ENTER to terminate the selection process or right-click for shortcut menu and choose one of the available options)*

The EXTEND command initially prompts you to "Select boundary edges:". After selecting one or more boundary edges, press ENTER to terminate the selection process. Then AutoCAD prompts you to "Select object to extend:" (default option). Select one or more objects to extend to the selected boundary edges. After selecting the required objects to extend, press ENTER to complete the selection process. You can switch to the TRIM command by holding the SHIFT key while selecting the objects which will be trimmed to the specified cutting edge instead of being extended. See the section on the TRIM command in this chapter.

The EXTEND and TRIM commands are very similar in this method of selecting. With EXTEND you are prompted to select the boundary edge to extend to; with TRIM you are prompted to select a cutting edge. If you press ENTER in response to "Select boundary edge(s)" prompt with out selecting any objects, by default AutoCAD selects all the objects.

Edge

The Edge option determines whether objects are extended past a selected boundary or to an implied edge. AutoCAD prompts as follows when the Edge option is selected:

Enter an implied edge extension mode [Extend/No extend] <current>: *(right-click for the shortcut menu and select one of the two available options)*

The Extend selection extends the boundary object along its natural path to intersect another object in 3D space (implied edge).

The No Extend selection specifies that the object is to extend only to a boundary object that actually intersects it in 3D space.

Undo

The Undo option reverses the most recent change made by the EXTEND command.

Project

The Project option specifies the projection and AutoCAD uses when trimming objects. By default, it is set to the current UCS.

Figure 4–68 shows examples of the use of the EXTEND command.

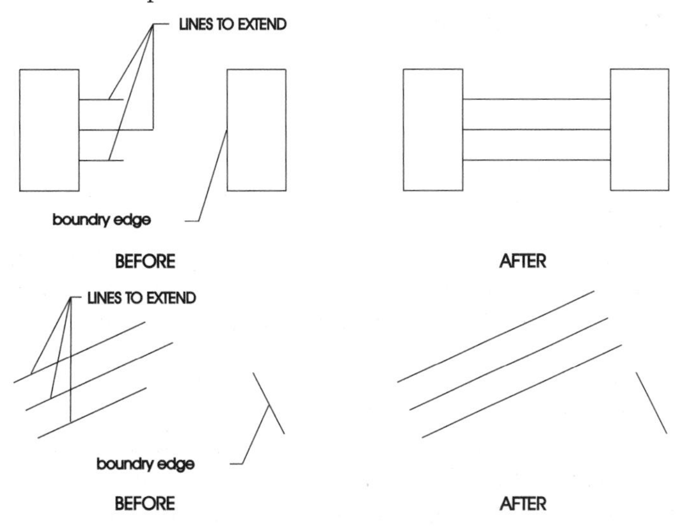

Figure 4–68 *Examples of applications of the* EXTEND *command*

Open the Exercise Manual PDF file for Chapter 4 on the accompanying CD for project and discipline specific exercises.

If you have the accompanying Exercise Manual, refer to Chapter 7 for project and discipline specific exercises.

REVIEW QUESTIONS

1. In order to draw two rays with different starting points, you must use the RAY command twice.

   a. True b. False

2. When you place xlines on a drawing, they:

   a. may effect the limits of the drawing

   b. may effect the extents of the drawing

   c. always appear as construction lines on layer 0

   d. can be constructed as offsets to an existing line

   e. none of the above

3. Filleting two non-parallel, non-intersecting line segments with a zero radius will:

   a. return an error message

   b. have no effect

   c. create a sharp corner

   d. convert the lines to rays

4. Objects can be trimmed at the points where they intersect existing objects.

   a. True b. False

5. The default justification for text is:

   a. TL d. BR

   b. BL e. None of the above

   c. MC

6. Which command allows you to change the location of the objects and allows a duplicate to remain intact?

   a. CHANGE c. COPY

   b. MOVE d. MIRROR

7. The maximum number of sides accepted by the POLYGON command is:

   a. 8 d. 1024

   b. 32 e. infinite (limited by computer memory, but VERY large)

   c. 128

8. Portions of objects can be erased or removed by using the command:

   a. ERASE d. EDIT

   b. REMOVE e. PARERASE

   c. BREAK

9. Ellipses are drawn by specifying:

   a. The major and minor axes

   b. The major axis and a rotation angle

   c. Any three points on the ellipse

   d. Any of the above

   e. Both a and b

10. Which of the following commands can be used to place text on a drawing?

    a. TEXT d. Both a and c

    b. CTEXT e. None of the above

    c. MTEXT

11. Two lines are drawn. Then the first line is erased and the second line is moved. Executing the OOPS command at the "Command:" prompt will:

    a. restore the erased line

    b. execute the LINE command automatically

    c. replace the second line at its original position

    d. restore both lines to their original positions

    e. none of the above

12. To efficiently move multiple objects, which option would be more efficient?

    a. Objects

    b. Last

    c. Window

    d. Add

    e. Undo

13. While using the TEXT command, AutoCAD will display the text you are typing:
    a. in the command prompt area
    b. in the drawing screen area
    c. both a and b
    d. neither a or b

14. While using the TEXT command, AutoCAD will display the text you are typing:
    a. in the command prompt area
    b. in the drawing screen area
    c. both a and b
    d. neither a or b

15. The MOVE command allows you to:
    a. move objects to new locations on the screen
    b. dynamically drag objects on the screen
    c. move only the objects that are on the current layer
    d. move an object from one layer to another
    e. both a and b

16. To create a rectangular array of objects, you must specify:
    a. the number of items and the distance between them
    b. the number of rows, the number of items, and the unit cell size
    c. the number of rows, the number of columns, and the unit cell size
    d. none of the above

17. A polyline:
    a. can have width
    b. can be exploded
    c. is one object
    d. all of the above

18. Polylines are:
    a. made up of line and arc segments, each of which is treated as an individual object
    b. are connected sequences of lines and arcs
    c. both a and b
    d. none of the above

19. To create an arc which is concentric with an existing arc, you could use what command?

    a. ARRAY

    b. COPY

    c. OFFSET

    d. MIRROR

20. The following are all options of the PLINE command except:

    a. Undo
    b. Halfwidth
    c. Arc
    d. Ltype
    e. Width

Directions: Answer the following questions based on Figure QX4–1.

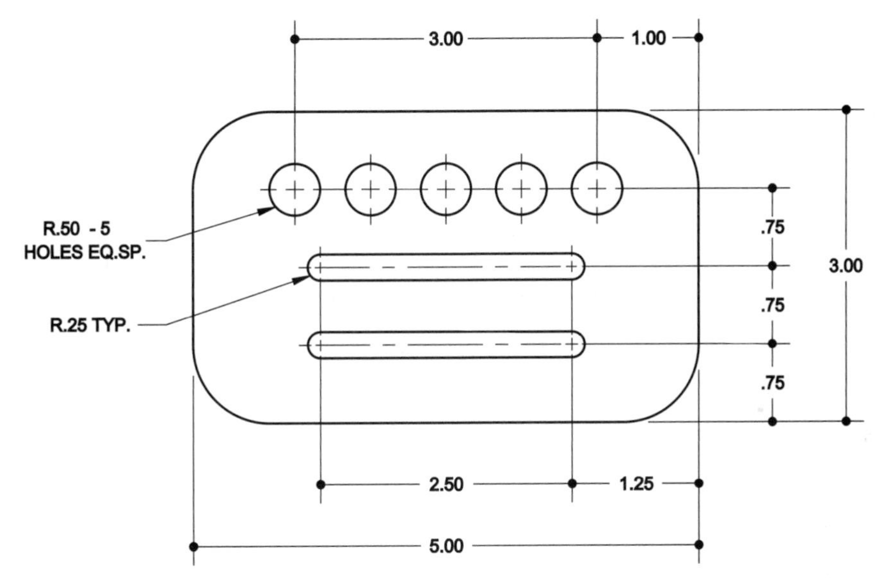

Figure QX4–1 *Mechanical part*

21. If the RECTANGLE command were used to create the object outline with the first corner at 0,0 what "*specify other corner point:*" coordinate location input would be required to complete the rectangle shape?

    a. @3.0,3.0

    b. 5.0,3.0

    c. 3.0,5.0

    d. @3.0,5.0

22. Which command would be the easiest and quickest to create rounded corners on the object?

    a. ARC

    b. ROUND

    c. FILLET

    d. CHAMFER

23. Creating the five holes equally spaced can be best achieved with which command?

    a. MIRROR

    b. OFFSET

    c. COPY

    d. ARRAY

24. Which ARRAY type would the five equally spaced holes are created with?

    a. POLAR

    b. RECTANGULAR

    c. COPY

    d. ARRAY

25. If the first corner of the object outline is at 0,0 what coordinate location input is required to position the center of the circle on the left?

    a. @1.25,2.5

    b. -2.25,1.25

    c. @1.00,2.25

    d. 1.25,2.25

26. What is the *distance between cells* (holes) value required when using the ARRAY command to locate the five holes?

    a. 3.00 c. 0.50

    b. 0.75 d. 0.25

27. If the top horizontal line of the long slot is drawn with the LINE command, which command can easily establish the location of the remaining horizontal lines for the slots?

    a. XLINE

    b. MIRROR

    c. OFFSET

    d. MOVE

Fundamentals IV

INTRODUCTION

After completing this chapter, you will be able to do the following:

- Construct geometric figures with the DONUT, SOLID, and POINT commands
- Create freehand line segments with the SKETCH command
- Use advanced object selection methods and modes to modify objects
- Use the modify commands: LENGTHEN, STRETCH, ROTATE, SCALE, PEDIT (edit Polyline), and MATCHPROP

CONSTRUCTING GEOMETRIC FIGURES

DRAWING SOLID-FILLED CIRCLES

The DOUGHNUT (or DONUT) command lets you draw solid-filled circles and rings by specifying outer and inner diameters of the filled area. The fill display depends on the setting of the FILLMODE system variable.

Invoke the DONUT command:

Draw menu	Choose Donut
Command: prompt	**donut** (ENTER)

AutoCAD prompts:

Command: **donut** (ENTER)
Specify inside diameter of donut <current>: *(specify a distance, or press* ENTER *to accept the current setting)*
Specify outside diameter of donut <current>: *(specify a distance, or press* ENTER *to accept the current setting)*
Specify center of donut or <exit>: *(specify a point to draw the donut)*

You may specify the inside and outside diameters of the donut to be drawn by specifying two points at the appropriate distance apart on the screen for either or both diameters, and AutoCAD will use the measured distance for the diameter(s). Or you may enter the distances from the keyboard.

You can select the center point by entering its coordinates or by selecting it with your pointing device. After you specify the center point, AutoCAD prompts for the center of the next donut and continues prompting for subsequent center points. To terminate the command, enter a null response.

Note: Be sure the FILLMODE system variable is set to ON (a value of 1). Check at the Command: prompt by entering **fill** and pressing ENTER. If FILLMODE is set to OFF (value of 0), then the PLINE, TRACE, DONUT, and SOLID commands display only the outline of the shapes. With FILLMODE set to ON, the shapes you create with these commands appear solid. If FILLMODE is reset to ON after it has been set to OFF, you must use the REGEN command in order for the screen to display as filled any unfilled shapes created by these commands. Switching between ON and OFF affects only the appearance of shapes created with the PLINE, TRACE, DONUT, and SOLID commands. Solids can be selected or identified by specifying the outlines only. The interior of a solid area is not recognized as an object when specified with the pointing device.

For example, the following command sequence shows placement of a solid-filled circle, as shown in Figure 5–1, by use of the DONUT command.

Command: **donut** (ENTER)
Specify inside diameter of donut <0.5000>: **0** (ENTER)
Specify outside diameter of donut <1.0000>: **1** (ENTER)
Specify center of donut or <exit>: **3,2** (ENTER)
Specify center of donut or <exit>: (ENTER)

Figure 5–1 *Using the DONUT command to place a solid-filled circle and a filled circular shape*

The following command sequence shows placement of a filled circular shape, as shown in Figure 5–1, by use of the DONUT command.

Command: **donut** (ENTER)
Specify inside diameter of donut <0.0000>: **0.5** (ENTER)

Specify outside diameter of donut <1.0000>: **1** (ENTER)
Specify center of donut or <exit>: **6,2** (ENTER)
Specify center of donut or <exit>: (ENTER)

DRAWING SOLID-FILLED POLYGONS

The SOLID command creates a solid-filled straight-sided area whose outline is determined by points you specify on the screen. Two important factors should be kept in mind when using the SOLID command: (1) the points must be selected in a specified order or else the four corners generate a bowtie instead of a rectangle; and (2) the polygon generated has straight sides. (Closer study reveals that even filled donuts and PLINE-generated curved areas are actually straight-sided, just as arcs and circles generate as straight-line segments of small enough length to appear smooth.)

Invoke the SOLID command:

Surfaces toolbar	Choose the SOLID command (see Figure 5–2)
Draw menu	Choose > Surfaces > 2D Solid
Command: prompt	**solid** (ENTER)

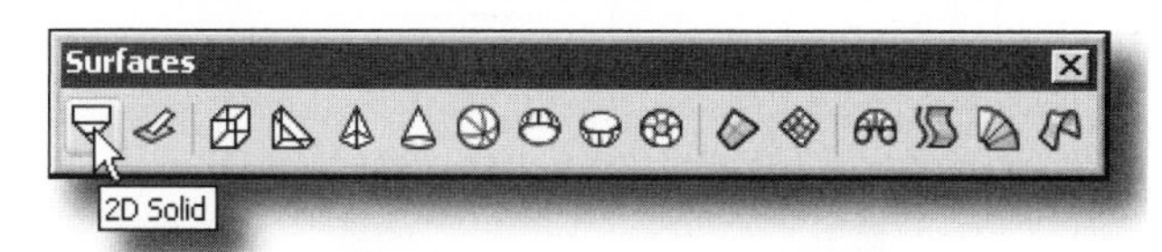

Figure 5–2 *Invoking the SOLID command from the Surfaces toolbar*

AutoCAD prompts:

Command: **solid** (ENTER)
Specify first point: *(specify a first point)*
Specify second point: *(specify a second point)*
Specify third point: *(specify a third point diagonally opposite the second point)*
Specify fourth point or <exit>: *(specify a fourth point, or press ENTER to exit)*

When you specify a fourth point, AutoCAD draws a quadrilateral area. If instead you press ENTER, AutoCAD creates a filled triangle.

For example, the following command sequence shows how to draw a quadrilateral area, as shown in Figure 5–3.

Command: **solid** (ENTER)
Specify first point: *(pick PT. 1)*
Specify second point: *(pick PT. 2)*
Specify third point: *(pick PT. 3)*
Specify fourth point or <exit>: *(pick PT. 4)*
Specify third point: *(press ENTER to exit)*

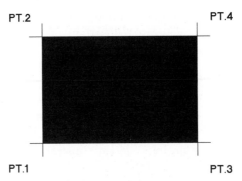

Figure 5–3 *The pick order for creating a quadrilateral area with the* SOLID *command*

To create the solid shape, the odd-numbered picks must be on one side and the even-numbered picks on the other side. If not, you get an effect such as shown in Figure 5–4.

You can use the SOLID command to create an arrowhead or triangle, such as the one shown in Figure 5–5. Polygon shapes can be created with the SOLID command by keeping the odd picks along one side and the even picks along the other side of the object, as shown in Figure 5–6.

Figure 5–4 *Results of using the* SOLID *command when odd/even points are not specified correctly*

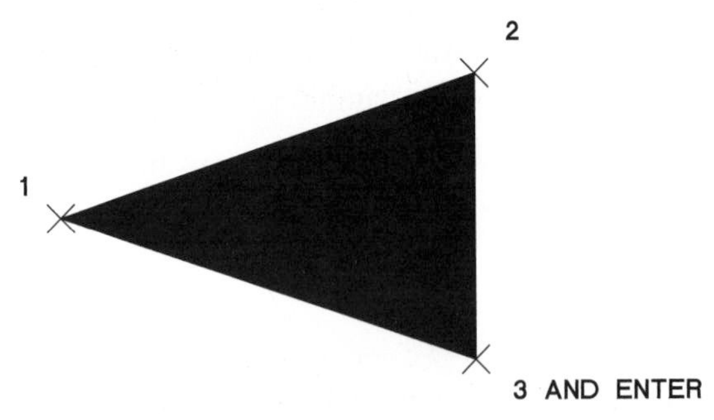

Figure 5–5 *Using the* SOLID *command to create a solid triangular shape*

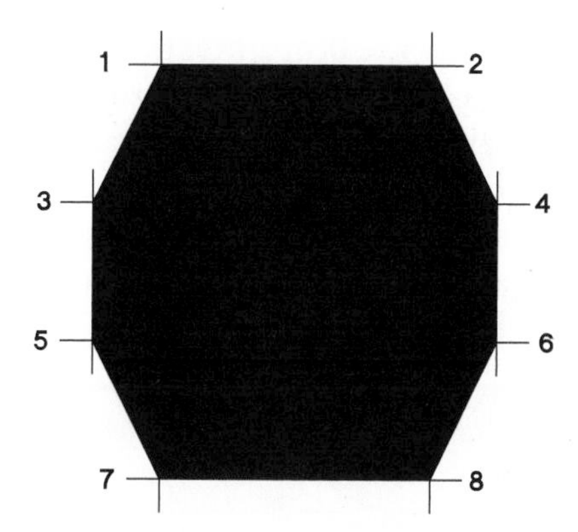

Figure 5–6 *Using the* SOLID *command to create a polygonal shape*

DRAWING POINT OBJECTS

The POINT command draws points on the drawing, and these points are drawn on the plotted drawing sheet with a single "pen down." You can enter such points to be used as reference points for object snapping when necessary. When the drawing is finished, simply erase them from the drawing or freeze their layer. Points are entered by specifying 2D or 3D coordinates or with the pointing device.

Invoke the POINT command:

Draw toolbar	Choose the POINT command (see Figure 5–7)
Draw menu	Choose Point > Single Point
Command: prompt	**point** (ENTER)

AutoCAD prompts:

Command: **point** (ENTER)
Point: *(specify a point)*

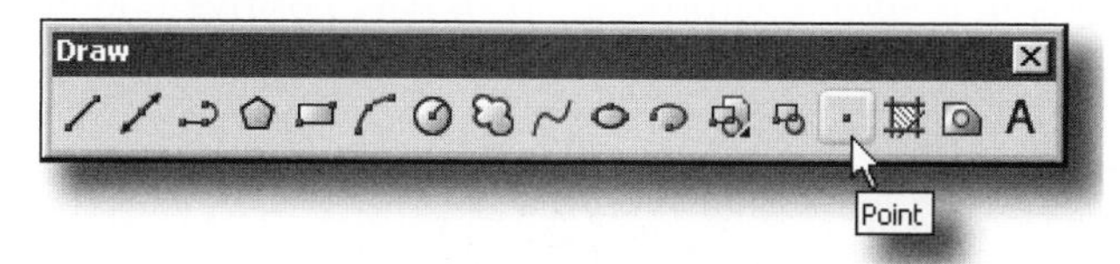

Figure 5–7 *Invoking the* POINT *command from the Draw toolbar*

Point Modes

When you draw the point, it appears on the display as a blip (+) if the BLIPMODE system variable is set to ON (default is OFF). After a REDRAW command, it appears as a dot (.).

You can make the point appear as a +, x, 0, or any of the available symbols by changing the PDMODE system variable. This can be done by entering PDMODE at the "Command: " prompt and entering the appropriate value. You can also change the PDMODE value by using the icon menu, as shown in Figure 5–8, invoked by typing DDPTYPE at the Command: prompt and pressing ENTER. The default value of PDMODE is zero, which means the point appears as a dot. If PDMODE is changed, all previous points drawn are replaced with the current setting.

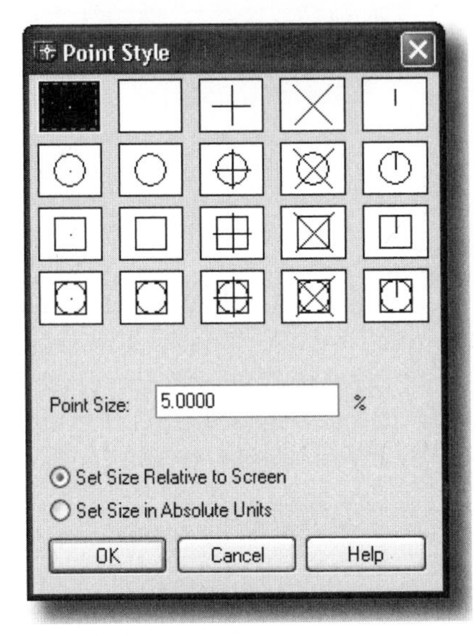

Figure 5–8 *The Point Style icon menu lets you select the shape and size of the point object*

Point Size

The size that the point appears on the screen depends on the value to which the PDSIZE system variable is set. If necessary, you can change the size via the PDSIZE command. The default for PDSIZE is zero (one pixel in size). Any positive value larger than this will increase the size of the point accordingly.

DRAWING SKETCH LINE SEGMENTS

The SKETCH command creates a series of freehand line segments. It is useful for freehand drawings, contour mapping, and signatures. The sketched lines are not added to the drawing until they are recorded.

Invoke the SKETCH command:

Command: prompt	**sketch** (ENTER)

AutoCAD prompts:

Command: **sketch** (ENTER)
Record increment <current>: *(specify a distance, or press* ENTER *to accept the current value)*

You may also respond by specifying two points, either keyed in or specified on the screen, causing AutoCAD to use the distance between the points as the record increment distance. Once a record increment is specified, AutoCAD displays the following list of options:

Sketch Pen eXit Quit Record Erase Connect

You can use any of the available options while you are in the SKETCH command. They are accessible either as single-key entries or as a mouse/puck button, provided your mouse/puck has the number of buttons corresponding to the option. The following table shows the optional subcommands, and their key, button number, and function. Normal button functions are not available while in the SKETCH mode.

Command Character	Pointer Button	Function
P	Pick	Raise/lower pen
.(period)	1	Line to point
R	2	Record lines
X, Spacebar, or ENTER	3	Record lines and exit
Q, or ESC	4	Discard lines and exit
E	5	Erase
C	6	Connect

P (Pen Up and Down)

An imaginary pen follows the cursor movement. When the pen is down, AutoCAD sketches a connected segment whenever the cursor moves the specified increment distance from the previously sketched segment. When the pen is up, the pen follows the cursor movement without drawing.

The pen is raised (up) and lowered (down) either by pressing the pick button on the mouse/puck or by pressing P on the keyboard. When you invoke a PEN UP, the current location of the pen will be the endpoint of the last segment drawn, which will be shorter than a standard increment length.

A PEN UP does not take you out of the SKETCH mode. Nor does a PEN UP permanently record the lines drawn during the current SKETCH session.

. (Period; Line-to-Point)

While the pen is up, you cause AutoCAD to draw a straight line from the last segment to the current cursor location and return to the PEN UP status by entering . (period) from the keyboard. This is convenient for long, straight lines that might occur in the middle of irregular shapes.

R (Record)

Lines being displayed while the cursor is moved (with the pen down) are temporary. They will appear green (or red if the current color for that layer or object is green) until they are permanently recorded. These temporary segments are subject to being modified with special Sketch options until you press **R** to record the latest lines. These may include several groups of connected lines drawn during PEN DOWN sequences separated by PEN UPs. When the Record option is invoked by entering either **R** ENTER or the third mouse/puck button, the total number of recorded segments is reported as follows:

```
nnn lines recorded
```

The nnn indicates the number of recorded segments.

E (Erase)

Prior to any group(s) of connected lines being recorded with the Record option, you may use the **E** (Erase) option to remove any or all of the lines from the last segment back to the first. The sequence of prompts is as follows:

```
Erase:
Select end of delete
```

The pen is automatically set to UP, and you may then use the cursor to remove segments, starting from the last segment. Press **P** or the pick button to accept the selected segments to delete. To abort the deletion of the selected segments and return to the SKETCH mode, press **E** again (or any other option), and the following message will be displayed:

```
Erase aborted
```

C (Connect)

Whenever a disconnect occurs (PEN UP or ERASE), you can reconnect and continue sketching from the point of the last disconnect as long as you have not exited the SKETCH command. The sequence is as follows:

```
Connect:
Move to endpoint of line.
```

At this prompt you can move the cursor near the end of the last segment. When you are within a specified increment length, sketching begins, connected to that last endpoint. This option is meaningless if invoked during PEN DOWN. A message also tells you:

No last point known

if no last point exists. The Connect option can be canceled by pressing **C** a second time.

X (Record and Exit)

The X option exits SKETCH mode after recording all temporary lines. This can also be accomplished by pressing either ENTER or the SPACEBAR.

Q (Quit)

The Q option exits SKETCH mode without recording any temporary lines. It is the same as pressing ESC.

OBJECT SELECTION

As mentioned earlier, all the modify commands initially prompt you to select objects. In most modify commands, the prompt allows you to select any number of objects. In some of the modify commands, however, AutoCAD limits your selection to only one object, for instance, the BREAK, DIVIDE, and MEASURE commands. In the case of the FILLET and CHAMFER commands, AutoCAD requires you to select two objects. And, whereas in the DIST and ID commands AutoCAD requires you to select a point, in the AREA command AutoCAD permits selection of either a series of points or an object.

Compared to the basic object selection option, the options covered in this section give you more flexibility and greater ease of use when you are prompted to select objects for use by the modify commands. The options that are explained in this section include WPolygon (WP), CPolygon (CP), Fence, All, Multiple, Box, Auto, Undo, Add, Remove, and Single.

WPOLYGON (WP)

The WPolygon option is similar to the Window option, but it allows you to define a polygon-shaped window rather than a rectangular area. You define the selection area as you specify the points about the objects you want to select. The polygon is formed as you select the points. The polygon can be of any shape but may not intersect itself. The polygon is formed as you specify points and includes rubber-band lines to the graphics cursor. When the selected points define the desired polygon, press ENTER. Only those objects that are totally inside the polygon shape are selected.

To select the WPolygon option, type **WP** and press ENTER at the "Select objects:" prompt. The UNDO option lets you undo the most recent polygon pick point.

CPOLYGON (CP)

The CPolygon option is similar to the WPolygon option, but it selects all objects within or crossing the polygon boundary. If there is an object that is partially inside the polygon area, then the whole object is included in the selection set.

To select the CPolygon option, type **CP** and press ENTER at the "Select objects:" prompt. The UNDO option lets you undo the most recent polygon pick point.

FENCE (F)

The Fence option is similar to the CPolygon option, except you do not close the last vector of the polygon shape. The selection fence selects only those objects it crosses or intersects. Unlike the WPolygons and CPolygon, the fence can cross over and intersect itself.

To select the Fence option type **F** and press ENTER at the "Select objects:" prompt.

ALL

The All option selects all the objects in the drawing, including objects on frozen or locked layers. After selecting all the objects, you may use the REMOVE (R) option to remove some of the objects from the selection set.

The All option must be spelled out in full (All) and not applied as an abbreviation as you may do with the other options.

MULTIPLE

The Multiple option helps you overcome the limitations of the Point, Window, and Crossing options. The Point option is time-consuming for use in selecting many objects. AutoCAD does a complete scan of the screen each time a point is picked. By using the Multiple modifier option, you can pick many points without delay and when you press ENTER, AutoCAD applies all of the points during one scan.

Selecting one or more objects from a crowded group of objects is sometimes difficult with the Point option. It is often impossible with the Window option. For example, if two objects are very close together and you wish to point to select them both, AutoCAD normally selects only one no matter how many times you select a point that touches them both. By using the Multiple option, AutoCAD excludes an object from being selected once it has been included in the selection set. As an alternative, use the Crossing option to cover both objects. If this is not feasible, then the Multiple modifier may be the best choice.

BOX

The Box option is usually employed in a menu macro to give the user a double option of window and Crossing, depending on how and where the picks are made on the screen. It must be spelled out in full (Box) and not applied as an abbreviation as you may do with the others, for example, W for Window. Because of this, you normally would not go to the trouble to use the Box option from the keyboard when a single key (W or C) provides a decidedly faster selection.

When the Box modifier is invoked as a response to a prompt to select an object, the options are applied as follows:

> If the picks are made left to right (the first point is to the left of the second), then the two points become diagonally opposite corners of a rectangle that is used as a Window option. That is, all visible objects totally within the rectangle are part of that selection.

If the picks are right to left, then the selection rectangle becomes the Crossing option. That is, all visible objects that are within or partially within the rectangle are part of that selection.

AUTO

The Auto option is actually a triple option. It includes the Point option with the two Box options. If the target box touches an object, then that object is selected as you would in using the Point option. If the target box does not touch an object, then the selection becomes either a Window or Crossing option, depending on where the second point is picked in relation to the first.

Note: The Auto option is the default option when you are prompted to "Select objects:".

UNDO

The Undo option allows you to remove the last item(s) selected from the selection set without aborting the "Select objects:" prompt and then to continue adding to the selection set. It should be noted that if the last option to the selection process includes more than one object, the Undo option will remove all the objects from the selection set that were selected by that last option.

ADD

The Add option lets you switch back from the Remove mode in order to continue adding objects to the selection set by however many options you wish to use.

REMOVE

The Remove option lets you remove objects from the selection set. The "Select objects:" prompt always starts in the Add mode. The Remove mode is a switch from the Add mode, not a standard option. Once invoked, the objects selected by whatever and however many options you use will be removed from the selection set. It will be in effect until reversed by the Add option.

SINGLE

The Single option causes the object selection to terminate and the command in progress to proceed after you use only one object selection option. It does not matter if one object is selected or a group is selected with that option. If no object is selected and the point selected cannot be the first point of a Window or Crossing rectangle, AutoCAD will not abort the command in progress; however, once there is a successful selection, the command proceeds.

OBJECT SELECTION MODES

AutoCAD provides six selection modes that will enhance object selection. You can toggle on/off one or more object selection modes from the Selection tab of the Options dialog box. By having the appropriate selection mode set to ON, you have various methods that give you more flexibility and greater ease of use in the selection of the objects.

To open the Selection Modes settings box, invoke the DDSELECT command:

Tools menu	Choose Options *(then select the Selection tab)*
Command: prompt	**ddselect or options** (ENTER)

AutoCAD displays the Options dialog box and select the **Selection** tab, as shown in Figure 5–9.

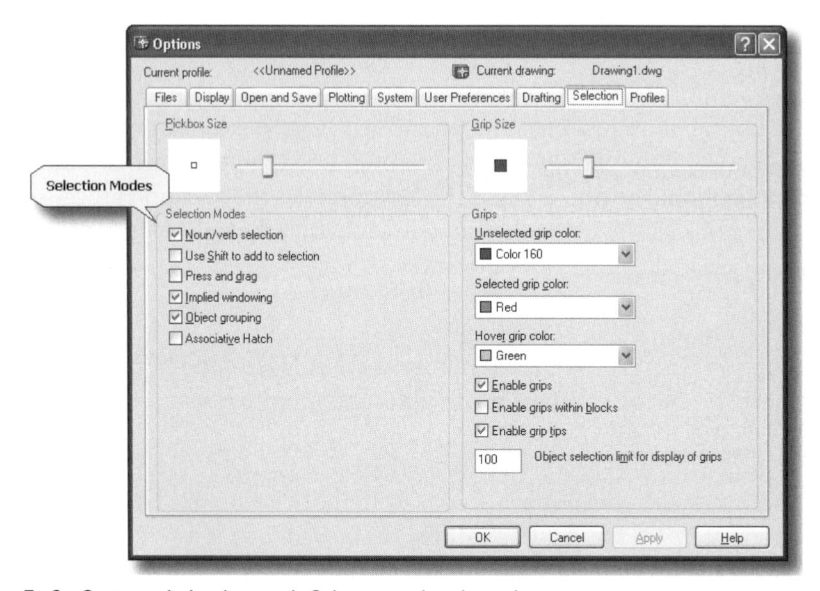

Figure 5–9 *Options dialog box with Selection tab selected*

You can toggle any one or more combinations of the settings provided under the selection modes. **Noun/verb selection, Implied windowing**, and **Object grouping** are the defaults.

NOUN/VERB SELECTION

The **Noun/verb selection** feature allows the traditional verb-noun command syntax to be reversed for most modifying commands. When the **Noun/verb selection** is set to ON, you can select the objects first at the Command: prompt and then invoke the appropriate modifying command you want to use on the selection set. For example, instead of invoking the copy command followed by selecting the objects to be copied, with **Noun/verb selection** set to ON, you can select the objects first and then invoke the COPY command, and AutoCAD skips the object selection prompt.

When **Noun/verb selection** is set to ON, the cursor at the Command: prompt changes to resemble a Running-Osnap cursor. Whenever you want to use the **Noun/verb selection** feature, first create a selection set at the Command: prompt. Subsequent modify commands you invoke execute by using the objects in the current selection set without prompting for object selection. To clear the current selection set, press ESC key at the "Command:" prompt. This clears the selection set, so any subsequent editing command will once again prompt for object selection.

Note: Another way to turn ON/OFF **Noun/verb selection** is to use the PICKFIRST system variable. TRIM, EXTEND, BREAK, CHAMFER, and FILLET are the commands not supported by the Noun/Verb selection feature.

USE SHIFT TO ADD

The **Use Shift to add to selection** feature controls how you add objects to an existing selection set. When **Use Shift to add to selection** is set to ON, it activates an additive selection mode in which SHIFT must be held down while adding more objects to the selection set. For example, if you first pick an object, it is highlighted. If you pick another object, it is highlighted and the first object is no longer highlighted. The only way you can add objects to the selection set is to select objects by holding down the SHIFT key. Similarly, the way to remove objects from the selection set is to select the objects by holding down the SHIFT key.

When **Use Shift to add to selection** is set to OFF (default), objects are added to the selection set by just picking them individually or by using one of the selection options; AutoCAD adds the objects to the selection set.

Note: Another way to select objects with this method is to set the PICKADD system variable to 0.

PRESS AND DRAG

The **Press and drag** feature controls the manner by which you draw the selection window with your pointing device. When **Press and drag** is set to ON, you can create a selection window by holding down the pick button and dragging the cursor diagonally while you create the window.

When **Press and drag** is set to OFF (default), you need to use two separate picks of the pointing device to create the selection window. In other words, you need to pick once to define one corner of the selection window, and pick a second time to define its diagonal corner.

Note: Another way to control how selection windows are drawn is to set the PICKDRAG system variable to 0 to draw the selection window using two points and to 1 to draw the selection window using dragging.

IMPLIED WINDOWING

The **Implied windowing** feature allows you to create a selection window automatically when the "Select objects:" prompt appears. When **Implied windowing** is set to ON (default), it works like the box option, explained earlier. If **Implied windowing** is set to OFF, you can create a selection window by using the Window or Crossing selection set methods.

Note: Another way to control the **Implied windowing** option is to set the PICKAUTO system variable to 0 to turn PICKAUTO OFF and to 1 to turn PICKAUTO ON.

OBJECT GROUPING

The **Object grouping** feature controls the automatic group selection. If **Object grouping** is set to ON, then selecting an object that is a member of a group selects the whole group. Refer to Chapter 6 for a detailed description of how to create groups.

ASSOCIATIVE HATCH

The **Associative Hatch** feature controls which objects will be selected when you select an associative hatch. If **Associative Hatch** is set to ON, then selecting an associative hatch also selects the boundary objects. Refer to Chapter 9 for a detailed description of hatching.

PICKBOX SIZE

AutoCAD allows you to adjust the size of the pickbox using the **Pickbox Size** slider bar provided in the dialog box.

MODIFYING OBJECTS

In this section, six modify commands are described: LENGTHEN, STRETCH, ROTATE, SCALE, PEDIT, and MATCH PROPERTIES. Additional modify commands were explained in Chapters 2 and 4.

LENGTHENING OBJECTS

The LENGTHEN command is used to increase or decrease the length of line objects or the included angle of an arc.

Invoke the LENGTHEN command:

Modify menu	Choose Lengthen
Command: prompt	**lengthen** (ENTER)

AutoCAD prompts:

Command: **lengthen** (ENTER)
Select an object or [DElta/Percent/Total/DYnamic]: *(select an object or right-click for shortcut menu and choose one of the available options)*

When you select an object, AutoCAD displays its length in the current units display and, where applicable, the included angle of the selected object.

DElta

The DElta option changes the length or, where applicable, the included angle from the endpoint of the selected object closest to the pick point. A positive value results in an increase in extension; a negative value results in a trim. When you select the **Delta** option, AutoCAD prompts:

Select an object or [DElta/Percent/Total/DYnamic]: **de** (ENTER)
Enter delta length or [Angle] (current)>: *(specify positive or negative value)*
Select an object to change or [Undo]: *(select an object, and its length is changed on the end nearest the selection point)*
Select an object to change or [Undo]: *(select additional objects; when done, press* ENTER *to exit the command sequence)*

Instead of specifying the delta length, if you select the Angle option, AutoCAD prompts:

Enter delta length or [Angle] <current>: **a** (ENTER)
Enter delta angle <current>: *(specify positive or negative angle)*
Select an object to change or [Undo]: *(select an object, and its included angle is changed on the end nearest the selection point)*
Select an object to change or [Undo]: *(select additional objects; when done, press* ENTER *to exit the command sequence)*

The Undo option reverses the most recent change made by the LENGTHEN command.

Percent

The Percent option sets the length of an object by a specified percentage of its total length. It will increase the length/angle for values greater than 100 and decrease them for values less than 100. For example, a 12-unit-long line will be changed to 15 units by using a value of 125. A 12-unit-long line will be changed to 9 units by using a value of 75. When you select the **Percent** option, AutoCAD prompts:

Enter delta length or [Angle] <current>: **p** (ENTER)
Enter percentage length <current>: (specify positive nonzero value and press ENTER)
Select an object to change or [Undo]: *(select an object, and its length is changed on the end nearest the selection point)*
Select an object to change or [Undo]: *(select additional objects; when done, press* ENTER *to exit the command sequence)*

Total

The Total option changes the length/angle of an object to the value specified. When you select the **Total** option, AutoCAD prompts:

> Select an object or [DElta/Percent/Total/DYnamic]: **t** (ENTER)
> Specify total length or [Angle] <current)>: (specify distance, or enter **a,** for angle, and then specify an angle for change)
> Select an object to change or [Undo]: *(select an object, and its length is changed to the specified distance on the end nearest the selection point)*
> Select an object to change or [Undo]: *(select additional objects; when done, press* ENTER *to exit the command sequence)*

DYnamic

The DYnamic option changes the length/angle of an object in response to the cursor's final location, relative to the endpoint nearest to where the object is selected. When you select the DYnamic option, AutoCAD prompts:

> Select an object or [DElta/Percent/Total/DYnamic]: **dy** (ENTER)
> Select an object to change or [Undo]: *(select an object to change the endpoint)*
> Specify new endpoint: *(specify new endpoint for the selected object)*
> Select an object to change or [Undo]:*(select additional objects; when done, press* ENTER *to exit the command sequence).*

STRETCHING OBJECTS

The STRETCH command allows you to stretch the shape of an object without affecting other crucial parts that remain unchanged. A common example is to stretch a square into a rectangle. The length is changed while the width remains the same.

AutoCAD stretches lines, polyline segments, rays, arcs, elliptical arcs, and splines that cross the selection window. The STRETCH command moves the endpoints that lie inside the window, leaving those outside the window unchanged. If the entire object is inside the window, then the STRETCH command operates like the MOVE command.

Invoke the STRETCH command:

Modify toolbar	Choose the STRETCH command (see Figure 5–10)
Modify menu	Choose Stretch
Command: prompt	**stretch** (ENTER)

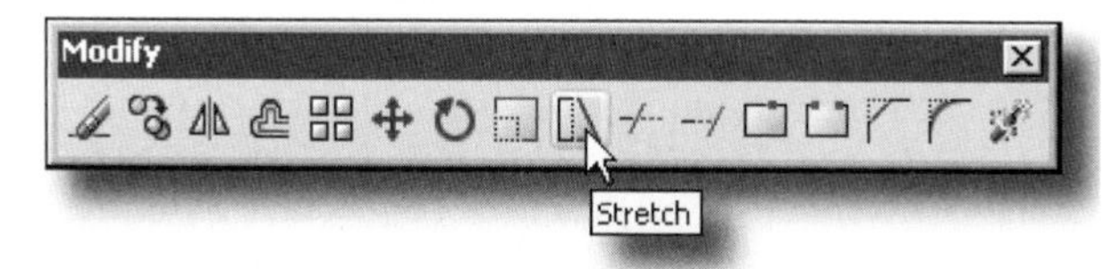

Figure 5–10 *Invoking the* STRETCH *command from the Modify toolbar*

AutoCAD prompts:

Command: **stretch** (ENTER)
Select objects to stretch by crossing-window or crossing-polygon...
Select objects: *(select the first corner of a crossing window or polygon)*
Specify opposite corner: *(select the opposite corner of the crossing window or polygon)*
Select objects: *(select additional objects; when done, press* ENTER *to exit the command sequence)*
Specify base point or displacement: *(specify a base point or press* ENTER*)*
Specify second point of displacement or <use first point as displacement>: *(specify the second point of displacement or press* ENTER *to use first point as displacement)*

If you provide the base point and second point of displacement, AutoCAD stretches the selected objects the vector distance from the base point to the second point. If you press ENTER at the prompt for the second point of displacement, then AutoCAD considers the first point as the X,Y displacement value.

Figure 5–11 shows some examples of using the STRETCH command.

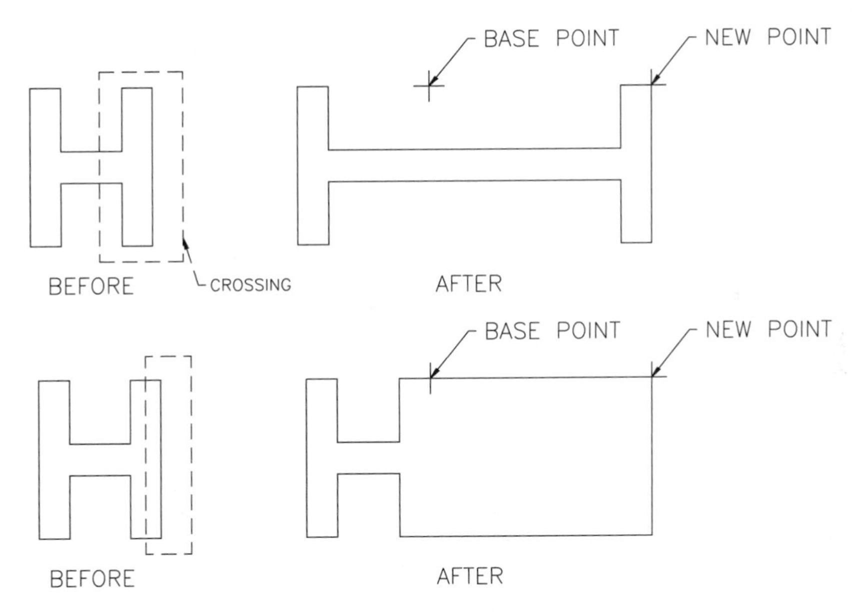

Figure 5–11 *Examples of using the* STRETCH *command*

ROTATING OBJECTS

The ROTATE command changes the orientation of existing objects by rotating them about a specified point, labeled as the base point. Design changes often require that an object, feature, or view be rotated. By default, a positive angle rotates the object in the counterclockwise direction, and a negative angle rotates in the clockwise direction.

Invoke the ROTATE command:

Modify toolbar	Choose the ROTATE command (see Figure 5–12)
Modify menu	Choose Rotate
Command: prompt	**rotate** (ENTER)

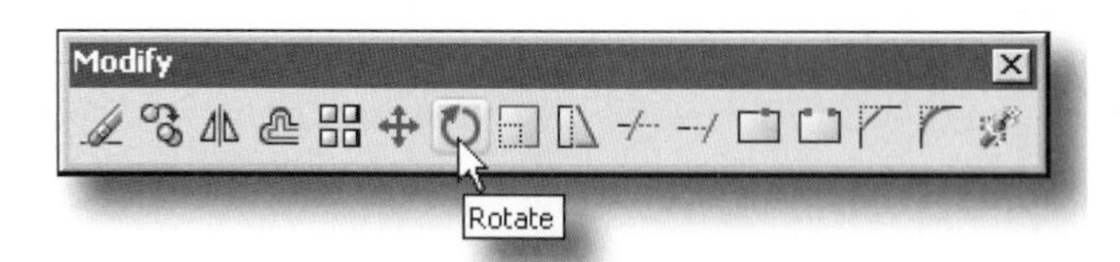

Figure 5–12 *Invoking the* ROTATE *command from the Modify toolbar*

AutoCAD prompts:

Command: **rotate** (ENTER)
Select objects: *(select the objects to rotate and press* ENTER *to complete the selection)*
Specify base point: *(specify a base point about which selected objects are to be rotated)*
Specify rotation angle or [Reference]: *(specify a positive or negative rotation angle, or enter* ***r*** *to select the Reference option)*

The base point can be anywhere in the drawing. If the base point selected is on the selected object itself, then the selected base point becomes an anchor point for rotation. The rotation angle determines how far an object rotates around the base point.

The following command sequence shows an example of rotating a group of objects selected by the Window option, as shown in Figure 5–13.

Command: **rotate** (ENTER)
Select objects: *(select the objects to rotate as shown in Figure 5–13 and press* ENTER *to complete the selection)*
Specify base point: *(pick the base point as shown in Figure 5–13)*
Specify rotation angle or [Reference]: **45** (ENTER)

AutoCAD rotates the selected objects by 45 degrees, as shown in Figure 5–13.

Reference Angle

If an object has to be rotated in reference to the current orientation, you can use the **Reference** option to do that. Specify the current orientation as referenced by the angle, or show AutoCAD the angle by pointing to the two endpoints of a line to be rotated, and specify the desired new rotation. AutoCAD automatically calculates the rotation angle and rotates the object appropriately. This method of rotation is very useful when you want to straighten an object or align it with other features in a drawing.

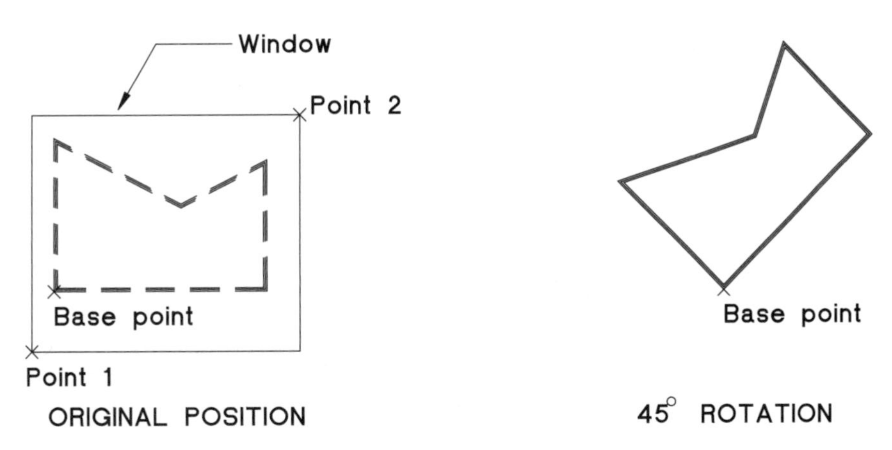

Figure 5–13 *Rotating a group of objects by means of the* ROTATE *command*

The following command sequence shows an example of rotating a group of objects selected by the Window option in reference to the current orientation, as shown in Figure 5–14.

Command: **rotate** (ENTER)
Select objects: *(select the objects to rotate, as shown in Figure 5–14)*
Specify base point: *(pick the base point, as shown in Figure 5–14)*
Specify rotation angle or [Reference]: **r** (ENTER)
Specify the reference angle <0>: **60** (ENTER)
Specify the new angle: **30** (ENTER)

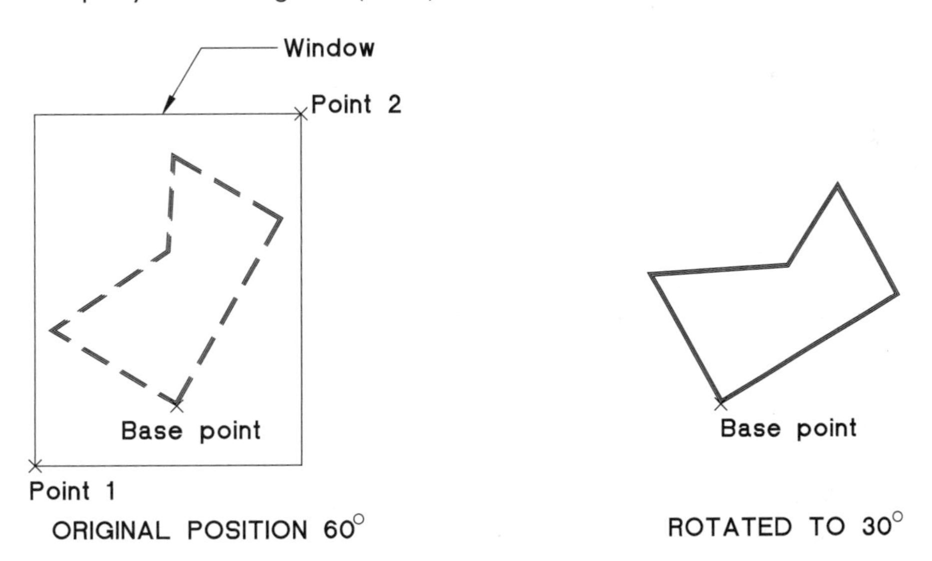

Figure 5–14 *Rotating a group of objects by means of the* ROTATE *command in reference to the current orientation*

SCALING OBJECTS

The SCALE command lets you change the size of selected objects or the complete drawing. Objects are made larger or smaller; the same scale factor is applied to the X, Y, and Z directions. To enlarge an object, specify a scale factor greater than 1. For example, a scale factor of 3 makes the selected objects 3 times larger. To reduce the size of an object, use a scale factor between 0 and 1. Do not specify a negative scale factor. For example, a scale factor of 0.75 would reduce the selected objects to three-quarters of their current size.

Invoke the SCALE command:

Modify toolbar	Choose the SCALE command (see Figure 5–15)
Modify menu	Choose Scale
Command: prompt	**scale** (ENTER)

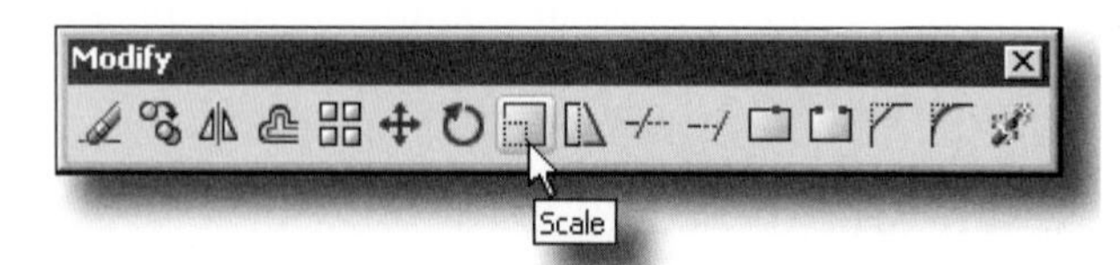

Figure 5–15 *Invoking the SCALE command from the Modify toolbar*

AutoCAD prompts:

Command: **scale** (ENTER)
Select objects: *(select the objects to scale and press ENTER to complete the selection)*
Specify base point: *(specify a base point about which selected objects are to be scaled)*
Specify scale factor or [Reference]: *(specify a scale factor, or enter **r** to select the Reference option)*

The base point can be anywhere in the drawing. If the base point selected is on the selected object itself, then the selected base point becomes an anchor point for scaling. The scale factor multiplies the dimensions of the selected objects by the specified scale.

The following command sequence shows an example of enlarging a group of objects by a scale factor of 3, as shown in Figure 5–16.

Command: **scale** (ENTER)
Select objects: *(select the objects to enlarge, as shown in Figure 5–16, and press ENTER to complete the selection)*
Specify base point: *(pick the base point, as shown in Figure 5–16)*
Specify scale factor or [Reference]: **3** (ENTER)

AutoCAD enlarges the selected objects by a factor of 3, as shown in Figure 5–16.

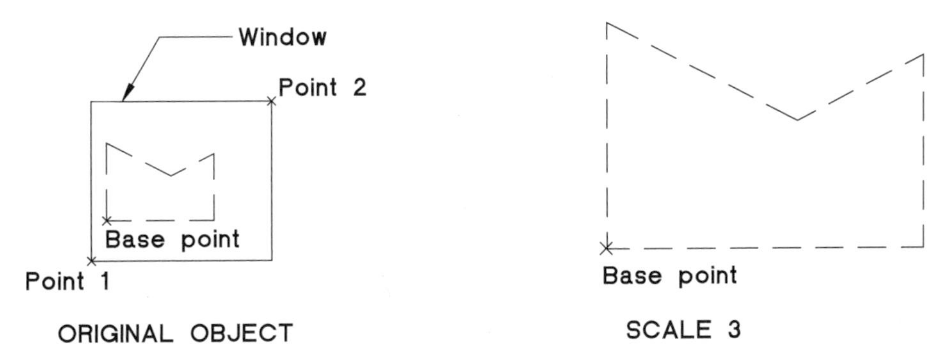

Figure 5–16 *Enlarging a group of objects by means of the* SCALE *command*

Reference

The Reference option is used to scale objects relative to the current dimension. Specify the current dimension as a reference length, or select two endpoints of a line to be scaled, and specify the desired new length. AutoCAD will automatically calculate the scale factor and enlarge or shrink the object appropriately.

The following command sequence shows an example of using the SCALE command to enlarge a group of objects selected by means of the Window option in reference to the current dimension, as shown in Figure 5–17.

Command: **scale** (ENTER)
Select objects: *(select the objects to enlarge, as shown in Figure 5–17)*
Specify base point:*(pick the base point, as shown in Figure 5–17)*
Specify scale factor or [Reference]: **r** (ENTER)
Specify reference length <1>:**3.8** (ENTER)
Specify new length: **4.8** (ENTER)

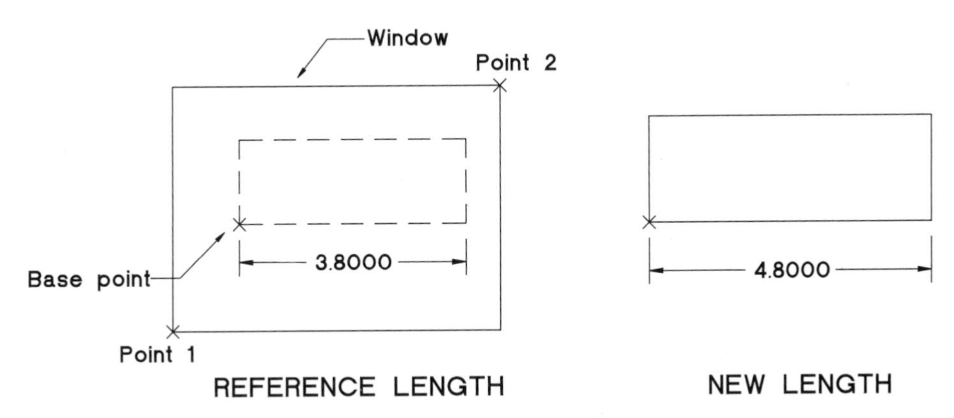

Figure 5–17 *Enlarging a group of objects by means of the* SCALE *command in reference to the current dimension*

MODIFYING POLYLINES

The PEDIT command allows you to modify polylines. In addition to using such modify commands as MOVE, COPY, BREAK, TRIM, and EXTEND, you can use the PEDIT command to modify polylines. The PEDIT command has special editing features for dealing with the unique properties of polylines and is perhaps the most complex AutoCAD command, with several multi-option submenus totaling some 70 command options.

Invoke the PEDIT command:

Modify II toolbar	Choose the PEDIT command (see Figure 5–18)
Modify menu	Choose Object > Polyline
Command: prompt	**pedit** (ENTER)

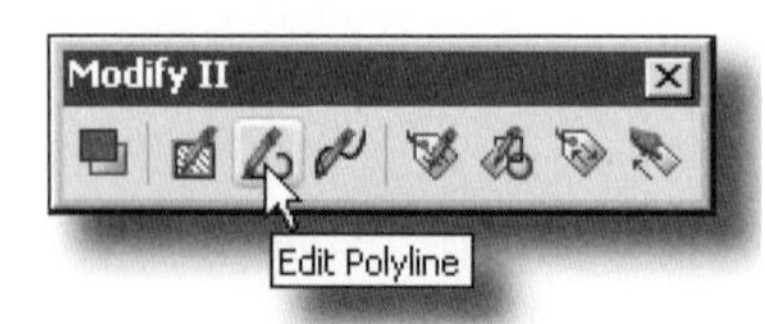

Figure 5–18 *Invoking the* EDIT POLYLINE *command from the Modify II toolbar*

AutoCAD prompts:

Command: **pedit** (ENTER)
Select polyline or [Multiple]: *(select polyline, line, or arc, or enter M for multiple option selection)*

If you select a line or an arc instead of a polyline, you are prompted as follows:

Object selected is not a polyline.
Do you want it to turn into one? <Y>

Responding **Y** or pressing ENTER turns the selected line or arc into a single-segment polyline that can then be edited. Normally this is done in order to use the **Join** option to add other connected segments that, if not polylines, will also be transformed into polylines. It should be emphasized at this time that in order to join segments together into a polyline, their endpoints must coincide. This occurs during line-line, line-arc, arc-line, and arc-arc continuation operations, or by using the endpoint Object Snap mode.

The second prompt does not appear if the first segment selected is already a polyline. It may even be a multisegment polyline. If you select multiple option, then AutoCAD allows you to select more than one polyline to modify. After the object selection process, you will be returned to the multioption prompt as follows:

Enter an option [Close/Join/Width/Edit vertex/Fit/Spline/Decurve/Ltype gen/Undo]: *(select one of the available options)*

Close

The Close option performs in a manner similar to the Close option of the LINE command. If, however, the last segment was a polyline arc, then the next segment will be similar to the arc-arc continuation, using the direction of the last polyarc as the starting direction and drawing another polyarc, with the first point of the first segment as the ending point of the closing polyarc.

Figures 5–19 and 5–20 show some examples of the application of the **Close** option.

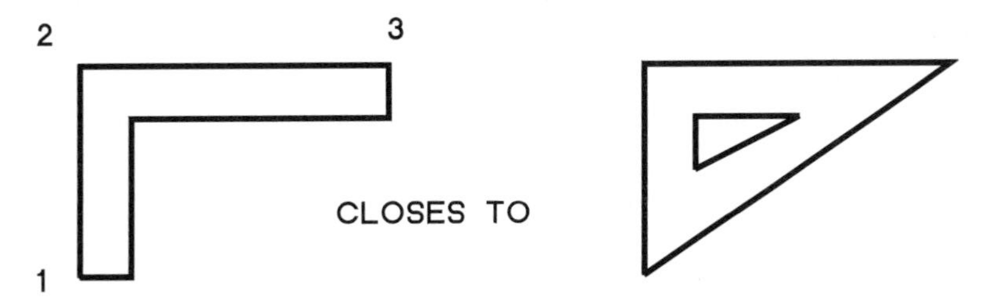

Figure 5–19 *Using the* PEDIT *command Close option with polylines*

Figure 5–20 *Using the* PEDIT *command Close option with polyarcs*

Open

The Open option deletes the segment that was drawn with the Close option. If the polyline had been closed by drawing the last segment to the first point of the first segment without using the Close option, then the Open option will not have a visible effect.

Join

The Join option takes selected lines, arcs, or polylines and combines them with a previously selected polyline into a single polyline if all segments are connected at sequential and coincidental endpoints.

Width

The Width option permits uniform or varying widths to be specified for polyline segments.

Edit Vertex

A vertex is the point where two segments join. When you select the Edit vertex option, the visible vertices are marked with an X to indicate which one is to be modified. You can modify vertices of polylines in several ways. When you select the Edit vertex option, AutoCAD prompts you with additional suboptions:

Command: **pedit** (ENTER)
Select polyline or [Multiple]: *(select a polyline)*
Enter an option [Close/Join/Width/Edit vertex/Fit/Spline/Decurve/Ltype gen/Undo]: **e** (ENTER)
[Next/Previous/Break/Insert/Move/Regen/Straighten/Tangent/Width/eXit]<N>:

Next and **Previous** The N (Next) or P (Previous) options can be used when you wish to move the mark to the next or previous vertex, whether or not you have modified the marked vertex.

Break The Break option establishes the marked vertex as one vertex for the Break option and then prompts:

Enter an option [Next/Previous/Go/eXit] <N>:

The choices of the Break option permit you to step to another vertex for the second break point, or to initialize the break, or to exit the option. If two vertices are selected, you may use the Go option to have the segment(s) between the vertices removed. If you select the endpoints of a polyline, this option will not work. If you select the Go option immediately after the Break option, the polyline will be divided into two separate polylines. Or, if it is a closed polyline, it will be opened at that point.

Insert The Insert option allows you to specify a point and have the segment between the marked vertex and the next vertex become two segments meeting at the specified point. The selected point does not have to be on the polyline segment.

For example, the following command sequence shows the application of the Insert option as shown in Figure 5–21.

Command: **pedit** (ENTER)
Select polyline or [Multiple]: *(select the polyline as shown in Figure 5–21)*
Enter an option [Close/Join/Width/Edit vertex/Fit/Spline/Decurve/Ltype gen/Undo]:**e** (ENTER)
[Next/Previous/Break/Insert/Move/Regen/Straighten/Tangent/Width/eXit]<N>: **i** (ENTER)
Specify location for new vertex: *(specify a new vertex)*

Figure 5–21 *Using the* PEDIT *command Insert option*

Move The Move option allows you to specify a point and have the marked vertex be relocated to the selected point.

For example, the following command sequence shows the application of the Move option, as shown in Figure 5–22.

Command: **pedit** (ENTER)
Select polyline or [Multiple]: *(select the polyline as shown in Figure 5–22)*
Enter an option [Close/Join/Width/Edit vertex/Fit/Spline/
Decurve/Ltype gen/Undo]: **e** (ENTER)
Enter a vertex editing option
[Next/Previous/Break/Insert/Move/Regen/Straighten/Tangent/
Width/eXit]<N>: **m** (ENTER)
Specify new location for marked vertex: *(specify the new location)*

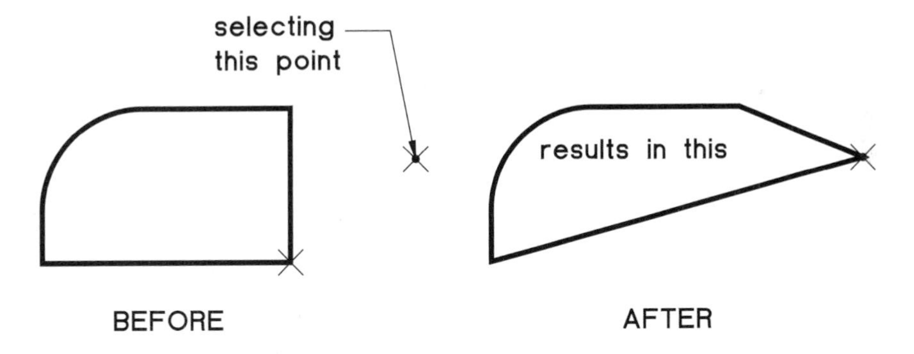

Figure 5–22 *Using the* PEDIT *command Move option*

Regen The Regen option regenerates the polyline without having to cancel the PEDIT command to invoke the REGEN command at the Command: prompt.

Straighten The Straighten option establishes the marked vertex as one vertex for the Straighten option and then prompts:

Enter an option [Next/Previous/Go/eXit]<N>:

These choices of the Straighten option permit you either to step first to another vertex for the second point or to exit the option. When the two vertices are selected, you may use the Go option to have the segment(s) between the vertices replaced with a single straight-line segment.

For example, the following command sequence shows the application of the Straighten option, as shown in Figure 5–23.

Command: **pedit** (ENTER)
Select polyline or [Multiple]: *(select the polyline as shown in Figure 5–23)*
Enter an option [Close/Join/Width/Edit vertex/Fit/Spline/
  Decurve/Ltype gen/Undo]: **e** (ENTER)
Enter a vertex editing option
[Next/Previous/Break/Insert/Move/Regen/Straighten/Tangent/
  Width/eXit]<N>: **s** (ENTER)
Enter an option [Next/Previous/Go/eXit]<N>: **n** (ENTER)
Enter an option [Next/Previous/Go/eXit]<N>: **g** (ENTER)
Enter an option [Next/Previous/Go/eXit]<N>: **x** (ENTER)

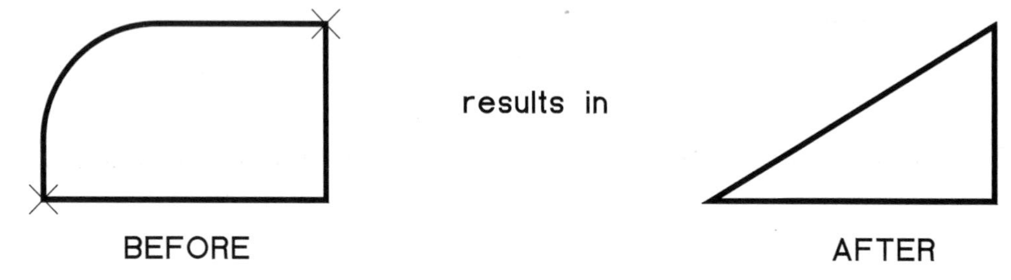

Figure 5–23 *Using the* PEDIT *command Straighten option*

Tangent The Tangent option permits you to assign to the marked vertex a tangent direction that can be used for the curve fitting option. The prompt is as follows:

Specify direction of vertex tangent:

You can either specify the direction with a point or type in the coordinates at the keyboard.

Width The Width option permits you to specify the starting and ending widths of the segment between the marked vertex and the next vertex. The prompt is as follows:

Specify new width for all segments:

For example, the following command sequence shows the application of the Width option as shown in Figure 5–24.

Command: **pedit** (ENTER)
Select polyline or [Multiple]: *(select a polyline)*
Enter a vertex editing option
[Next/Previous/Break/Insert/Move/Regen/Straighten/Tangent/
Width/eXit]<N>: **e** (ENTER)
Enter an option [Close/Join/Width/Edit vertex/Fit/Spline/
Decurve/Ltype gen/Undo]: **w** (ENTER)
Specify new width for all segments: **0.25** (ENTER)

eXit The Exit option exits from the Vertex Editing option and returns to the PEDIT multioption prompt.

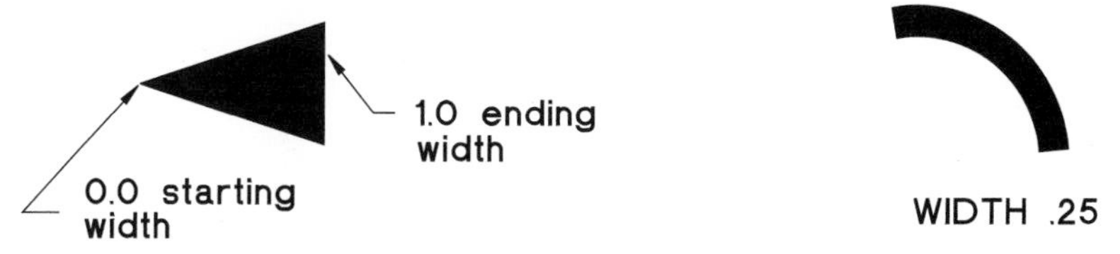

Figure 5–24 *Using the* PEDIT *command Width option*

Fit

The Fit option draws a smooth curve through the vertices, using any specified tangents.

Spline

The Spline option provides several ways to draw a curve based on the polyline being edited. These include Quadratic B-spline and Cubic B-spline curves.

Decurve

The Decurve option returns the polyline to the way it was drawn originally. See Figure 5–25 for differences between Fit Curve, Spline, and Decurve.

Ltype gen

The Ltype gen option controls the display of linetype at vertices. When it is set to ON, AutoCAD generates the linetype in a continuous pattern through the vertices of the polyline. And when it is set to OFF, AutoCAD generates the linetype starting and ending with a dash at each vertex. The Ltype gen option does not apply to polylines with tapered segments.

Undo

The Undo option reverses the latest PEDIT operation.

eXit

The eXit option exits the PEDIT command.

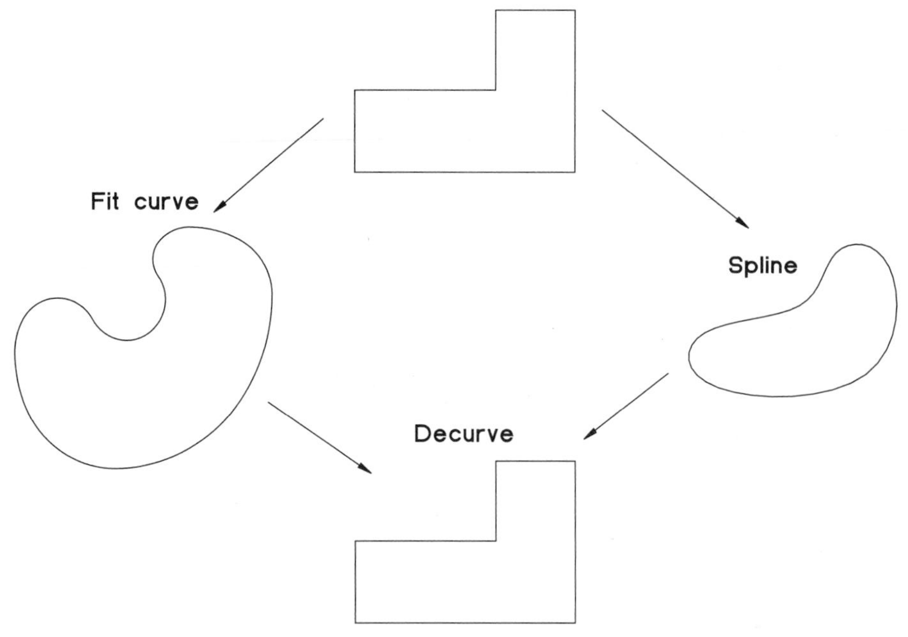

Figure 5–25 *Comparing* PEDIT *command Fit Curve, Spline Curve, and Decurve options*

MATCHING PROPERTIES

The MATCHPROP command allows you to copy selected properties from one object to one or more other objects located in the current drawing or any other drawing currently open. Properties that can be copied include color, layer, linetype, linetype scale, lineweight, thickness, plot style, and in some cases, dimension, text, polyline, viewport, and hatch.

Invoke the MATCHPROP command:

Standard toolbar	Choose the Match Properties command (see Figure 5–26)
Modify menu	Choose Match Properties
Command: prompt	**matchprop or painter** (ENTER)

Figure 5–26 *Invoking the* MATCH PROPERTIES *command from the Standard toolbar*

AutoCAD prompts:

> Command: **matchprop** (ENTER)
> Select source object: *(select the object whose properties you want to copy)*
> Select destination object(s) or [Settings]: *(select the destination objects and press* ENTER *to terminate the selection, or enter* **s** *to select the Settings option to display the Property Settings)*

Selection of the Settings option displays the Property Settings dialog box, similar to Figure 5–27. Use the Settings option to control which object properties are copied. By default, all object properties in the Property Settings dialog box are set to ON for copying.

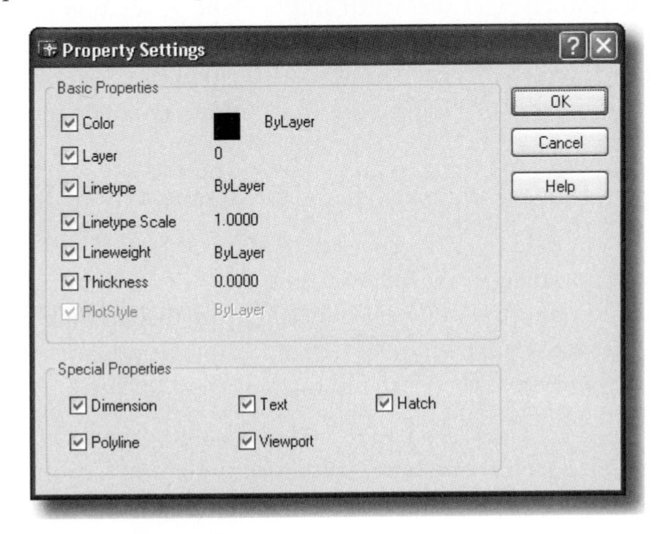

Figure 5–27 *Property Settings dialog box*

The **Color** selection changes the color of the destination object to that of the source object.

The **Layer** selection changes the layer of the destination object to that of the source object.

The **Linetype** selection changes the linetype of the destination object to that of the source object. Available for all objects except attributes, hatches, multiline text, points, and viewports.

The **Linetype Scale** selection changes the linetype scale factor of the destination object to that of the source object. Available for all objects except attributes, hatches, multiline text, points, and viewports.

The **Lineweight** selection changes the lineweight of the destination object to that of the source object.

The **Thickness** selection changes the thickness of the destination object to that of the source object. Available only for arcs, attributes, circles, lines, points, 2D polylines, regions, text, and traces.

The **PlotStyle** selection changes the plot style of the destination object to that of the source object. If you are working in color-dependent plot style mode (PSTYLEPOLICY is set to 1), this option is unavailable.

The **Dimension** selection changes the dimension style of the destination object to that of the source object. Available only for dimension, leader, and tolerance objects.

The **Polyline** selection changes the width and linetype generation properties of the destination polyline to those of the source polyline. The fit/smooth property and the elevation of the source polyline are not transferred to the destination polyline. If the source polyline has variable width, the width property is not transferred to the destination polyline.

The **Text** selection changes the text style of the destination object to that of the source object.

The **Viewport** selection changes the following properties of the destination paper space viewport to match those of the source viewport: on/off, display locking, standard or custom scale, shade plot, snap, grid, and UCS icon visibility and location. The settings for clipping and for UCS per viewport and the freeze/thaw state of the layer are not transferred to the destination object.

The **Hatch** selection changes the hatch pattern of the destination object to that of the source object.

After making the necessary changes, choose the **OK** button to close the Property Settings dialog box. AutoCAD continues with the Select Destination Object(s): prompt. Press ENTER to complete the object selection.

Open the Exercise Manual PDF file for Chapter 5 on the accompanying CD for project and discipline specific exercises.

If you have the accompanying Exercise Manual, refer to Chapter 5 for project and discipline specific exercises.

REVIEW QUESTIONS

1. The command that allows you to draw freehand lines is:

   a. SKETCH

   b. DRAW

   c. PLINE

   d. FREE

2. To create a six-sided area which would select all the objects completely within it, you should respond to the "Select Objects:" prompt with:

   a. WP

   b. CP

   c. W

   d. C

3. The MATCHPROP command does not allow you to modify an object's:

   a. linetype

   b. fillmode

   c. color

   d. layer

4. By default, most selection set prompts default to which option:

   a. All

   b. Auto

   c. Box

   d. Window

   e. Crossing

5. The multiple option selection sets allows you to:
   a. select multiple objects which lie on top of each other
   b. scan the database only once to find multiple objects
   c. use the window or crossing options
   d. both A and B
   e. both B and C
6. In regard to using the SOLID command, which of the following statements is true:
   a. The order of point selection is unimportant
   b. FILL must be turned ON in order to use the SOLID command
   c. The points must be selected on existing objects
   d. The points must be selected in a clockwise order
   e. None of the above
7. What command is commonly used to create a filled rectangle?
   a. FILL-ON
   b. PLINE
   c. RECTANGLE
   d. LINE
   e. SOLID
8. To turn a series of line segments into a polyline, one uses which option of the PEDIT command?
   a. Join
   b. Fit curve
   c. Spline
   d. Connect
   e. Line segments cannot be connected.
9. Which command allows you to change the size of an object, where the X and Y scale factors are changed equally?
   a. ROTATE
   b. SCALE
   c. SHRINK
   d. MODIFY
   e. MAGNIFY

10. The Box option for creating selection sets is most useful for:

    a. creating custom functions

    b. creating rectangular selection areas

    c. extending a selection into 3D

    d. Nothing, it is not a valid option

11. Using the SCALE command, what number would you enter to enlarge an object by 50%?

    a. 0.5

    b. 50

    c. 3

    d. 1.5

12. To avoid changing the location of an object when using the SCALE command:

    a. the reference length should be less than the limits

    b. the scale factor should be less than one

    c. the base point should be on the object

    d. the base point should be at the origin

13. The ROTATE command is used to rotate objects around:

    a. any specified point

    b. the point -1,-1 only

    c. the origin

    d. is only usable in 3D drawings

    e. none of the above

14. The Remove option for forming selection sets deletes objects from the drawing in much the same manner as the ERASE command.

    a. True

    b. False

15. Which of the following is not supported by the noun/verb feature?

    a. COPY

    b. MOVE

    c. TRIM

    d. ROTATE

    e. ERASE

16. When noun/verb selection is set to ON, you may add objects to the selection set by:

    a. picking the objects

    b. windowing the objects

    c. using SHIFT to add

    d. using CTRL to add

Directions: Use Figure CH5RQ–1 to answer the following questions.

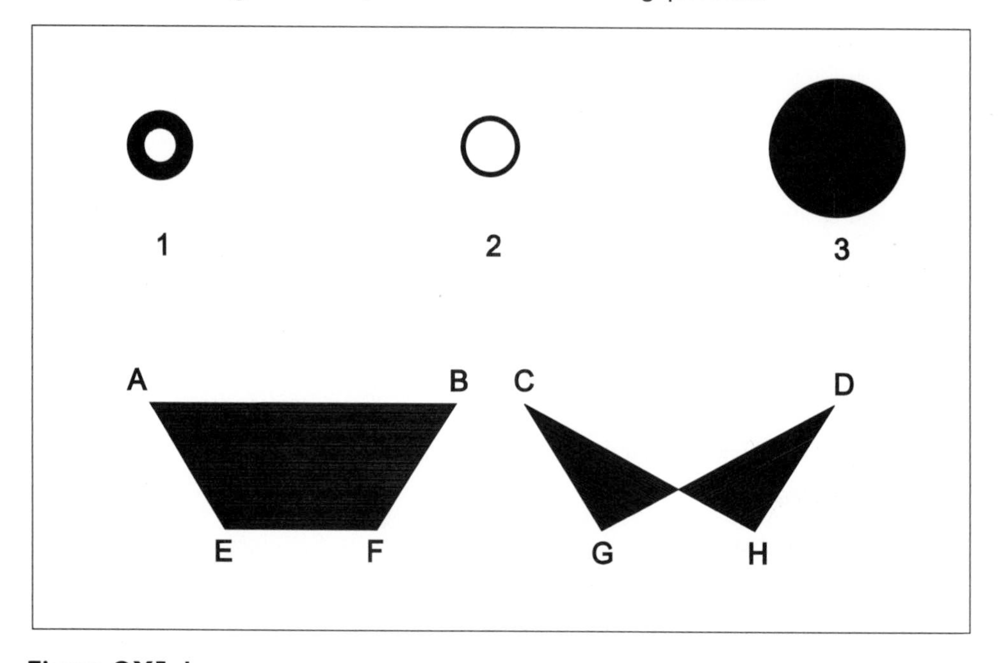

Figure QX5–1

17. Which of the figures were drawn with the DONUT command?

    a. 1

    b. 2

    c. 3

    d. all of the above

    e. 1, 2 & 3

18. Which of the DONUT objects has the smallest inside diameter?

    a. 1

    b. 2

    c. 3

19. DONUTS are positioned in a manner similar to which OSNAP option?

 a. QUADrant

 b. CENter

 c. TsANgent

 d. NODe

20. Which of the options below is the correct sequence to create the SOLID object on the left?

 a. A, B, F, E

 b. A, E, F, B

 c. A, B, E, F

 d. F, E, A, B

 e. B, F, A, E

21. Which of the options below is the correct sequence to create the SOLID object on the right?

 a. H, G, C, D

 b. G, H, D, C

 c. G, D, H, C

 d. D, C, H, G

 e. C, G, H, D

22. If the space between the two SOLID shapes were to be filled what sequence of points would completely fill this space?

 a. G, B, F, C

 b. B, G, F, C

 c. C, B, F, G

 d. F, G, C, B

 e. B, F, C, G

Fundamentals V

INTRODUCTION

After completing this chapter, you will be able to do the following:

- Construct geometric figures by means of the MULTILINE (MLINE), SPLINE, WIPEOUT and REVCLOUD commands
- Modify multilines and splines
- Create or modify multiline styles
- Use grips to modify objects, group objects, and filter certain types of objects by Quick Select for modification
- Use inquiry commands
- Change the settings of system variables

MULTILINES

AutoCAD allows you to draw multiple parallel line segment with the MLINE command. In addition, you can also modify the intersections of two or more multilines or cut gaps with the MLEDIT command. AutoCAD also allows you to create a new multiline style or edit an existing one comprised of up to 16 lines, called elements, with the MLSTYLE command.

DRAWING MULTIPLE PARALLEL LINES

The MLINE command allows you to draw multiple parallel line segments, similar to polyline segments that have been offset one or more times. Examples of applying the MLINE command are shown in Figure 6–1.

Figure 6–1 *Examples of multilines*

The properties of each element of a multiline are determined by the style that is current when the multiline is drawn. The properties of the multiline that can be determined by the style include whether to display the line at the joints (miters) and ends and whether to close the ends with a variety of half circles, connecting inner and/or outer elements. In addition, the style controls the element properties, such as color, linetype, and offset distance between two parallel lines.

Invoke the MLINE command:

Draw menu	Choose Multiline
Command: prompt	**mline** (ENTER)

AutoCAD prompts:

Command: **mline** (ENTER)
Current settings: Justification = Top, Scale = 1.00, Style = STANDARD
Specify start point or [Justification/Scale/STyle]: *(specify a style or right-click for a shortcut menu and select one of the available options)*

Start Point

The Start point option (default) lets you specify the starting point of the multiline, known as its origin. Once you specify the starting point for a multiline, AutoCAD prompts:

Specify next point:

When you respond by selecting a point, the first multiline segment is drawn according to the current style. You are then prompted:

Specify next point or [Undo]:

If you specify a point, the next segment is drawn, along with segments of all other elements specified by the current style. After two segments have been drawn the prompt will include the Close option:

Specify next point or [Close/Undo]:

Choosing the Close option causes the next segment to join the starting point of the multiline, fillets all elements, and exits the command.

Selecting Undo (u) after any segment is drawn (and the MLINE command has not been terminated) causes the last segment to be erased, and you are prompted again for a point.

Justification

The Justification option determines the relationship between the elements of the multiline and the line you specify by way of the placement of the points. The justification is set by selecting one of the three available suboptions.

Enter justification type [Top/Zero/Bottom] <top>:

The Top option causes the element with the greatest offset value to be drawn on the line of selected points. All other elements will be to the right of the line of points as viewed from the starting point to the ending point of each segment. In other words, if the line is drawn left to right, the line of points (and the element with the greatest offset value) will be on top of (above) all other elements.

The Zero option causes the baseline to coincide with the line of selected points. Elements with positive offsets will be to the right of, and those with negative offsets to the left of, the line of selected points, as viewed from the starting point to the ending point of the each segment.

The Bottom option causes the element with the least offset value to be drawn on the line of selected points. All other elements will be to the left of the line of points as viewed from the starting point to the ending point of each segment. In other words, if the line is drawn left to right, the line of points (and the element with the least offset value) will be on the bottom of (below) all other elements.

Figure 6–2 shows the location for various justifications.

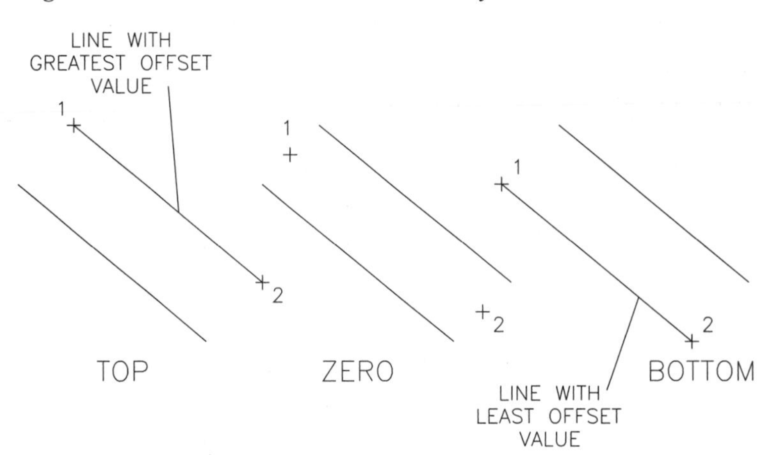

Figure 6–2 *Location of various justifications*

Scale

The Scale option determines the value used for offsetting elements when drawing them relative to the values assigned to them in the style. For instance, if the scale is changed to 3.0, elements that are assigned 0.5 and –1.5 will be drawn with offsets of 1.5 and –4.5, respectively. If a negative value is given for the scale, then the signs of the values assigned to them in the style will be changed (positive to negative and negative to positive). The value can be entered in decimal form or as a fraction. A 0 (zero) scale value produces a single line.

Style

The Style option sets the current multiline style from the available styles. AutoCAD prompts:

> Specify start point or [Justification/Scale/STyle]: **st** (ENTER)
> Enter mline style name or [?]: *(specify the name of an existing style)*

Detailed explanation is provided later in this chapter for creating or modifying multiline styles.

EDITING MULTIPLE PARALLEL LINES

The MLEDIT command helps you modify the intersections of two or more multilines or cut gaps in the lines of one multiline. The tools are available for the type of intersection operated on (cross, tee, or vertex) and if one or more elements need to be cut or welded.

Invoke the MLEDIT command:

Modify menu	Choose > Object > Multiline
Command: prompt	**mledit** (ENTER)

AutoCAD displays the Multiline Edit Tools dialog box, as shown in Figure 6–3.

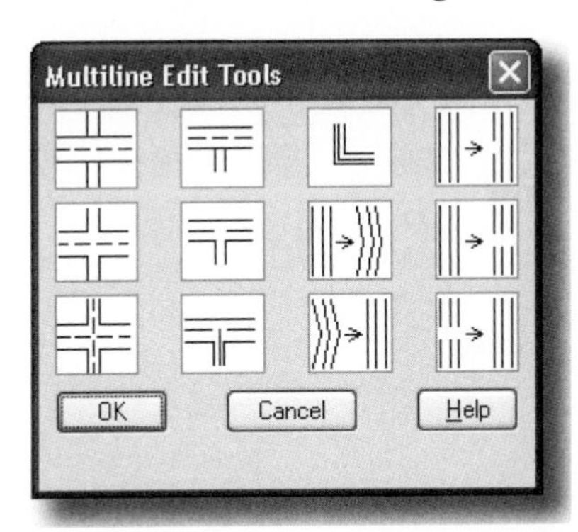

Figure 6–3 *Multiline Edit Tools dialog box*

To choose one of the available options, select one of the image tiles. AutoCAD then prompts for appropriate information.

The first column in the Multiline Edit Tools dialog box works on multilines that cross, the second works on multilines that form a tee, the third works on corner joints and vertices, and the fourth works on multilines to be cut or welded.

The Closed Cross option cuts all lines that make up the second multiline you select at the point where it crosses the first multiline, as shown in Figure 6–4. Choose the **Closed Cross** image tile, as shown in Figure 6–5, to invoke the Closed Cross option. AutoCAD then prompts:

Select first mline: *(select the first multiline, as shown in Figure 6–4)*
Select second mline: *(select the second multiline, as shown in Figure 6–4)*

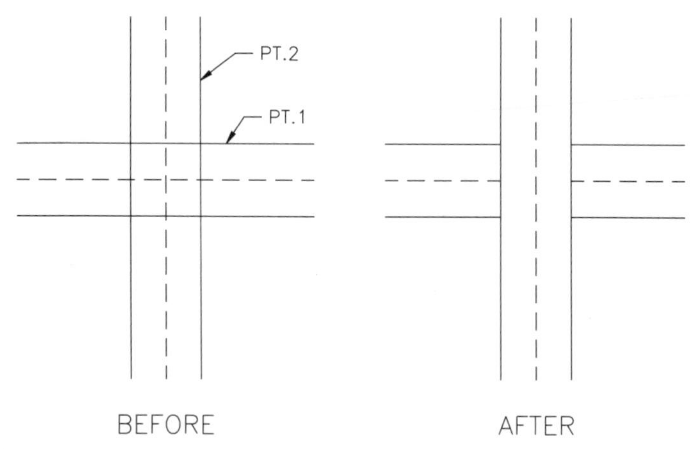

Figure 6–4 *An example of a closed cross*

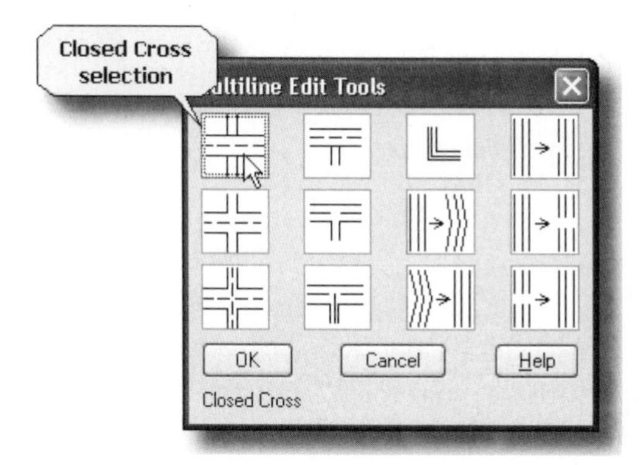

Figure 6–5 *Invoking the Closed Cross option from the Multiline Edit Tools dialog box*

After the closed cross intersection is created, AutoCAD prompts:

Select first mline (or Undo): *(select another multiline, enter **u**, or press* ENTER*)*

Selecting another multiline repeats the prompt for the second mline. Entering **u** undoes the closed cross just created.

The Open Cross option cuts all lines that make up the first multiline you select and cuts only the outside line of the second multiline, as shown in Figure 6–6. Choose the **Open Cross** image tile, as shown in Figure 6–7, to invoke the Open Cross option; AutoCAD prompts:

Select first mline: *(select a multiline, as shown in Figure 6–6)*
Select second mline: *(select a multiline that intersects the first multiline, as shown in Figure 6–6)*

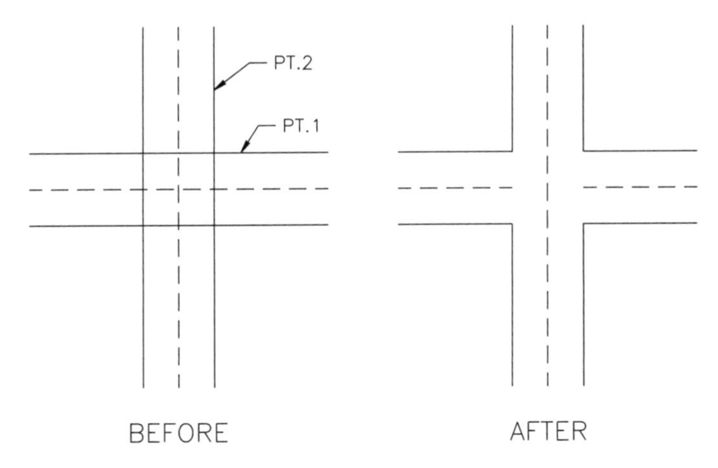

Figure 6–6 *An example of an open cross*

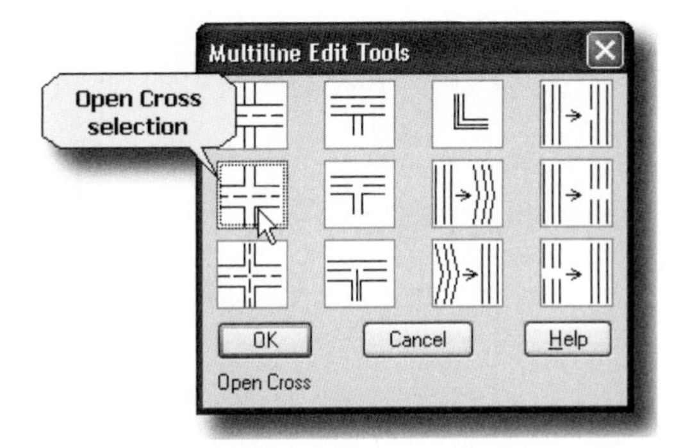

Figure 6–7 *Invoking the Open Cross option from the Multiline Edit Tools dialog box*

After the open cross intersection is created, AutoCAD prompts:

> Select first mline (or Undo): *(select another multiline, enter* **u**, *or press* ENTER*)*

Selecting another multiline repeats the prompt for the second mline. Entering **u** undoes the open cross just created.

The Merged Cross option cuts all lines that make up the intersecting multiline you select except the centerlines, as shown in Figure 6–8. Choose the **Merged Cross** image tile, as shown in Figure 6–9, to invoke the Merged Cross option; AutoCAD prompts:

> Select first mline: *(select a multiline, as shown in Figure 6–8)*
> Select second mline: *(select a multiline that intersects the first multiline, as shown in Figure 6–8)*

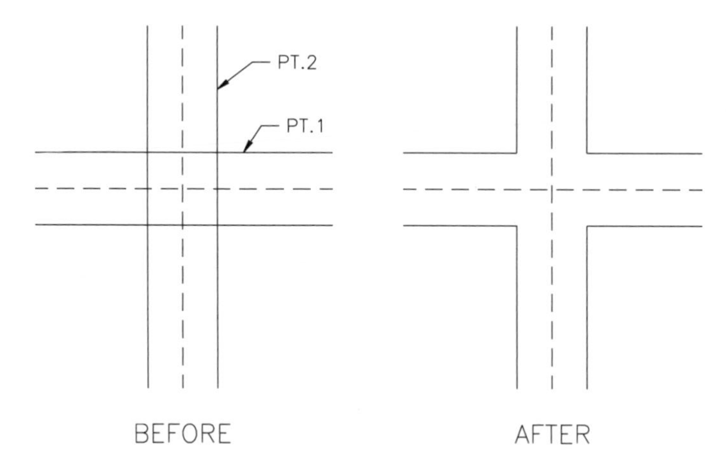

Figure 6–8 *An example of a merged cross*

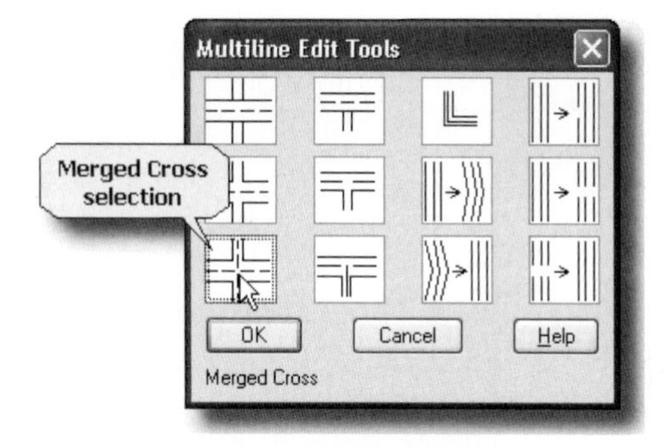

Figure 6–9 *Invoking the Merged Cross option from the Multiline Edit Tools dialog box*

After the merged cross intersection is created, AutoCAD prompts:

Select first mline (or Undo): *(select another multiline, enter **u**, or press* ENTER*)*

Selecting another multiline repeats the prompt for the second mline. Entering **u** undoes the merged cross just created. The order in which the multilines are selected is not important.

The Closed Tee option extends or shortens the first multiline you identify to its intersection with the second multiline, as shown in Figure 6–10. Choose the **Closed Tee** image tile, as shown in Figure 6–11, to invoke the Closed Tee option; AutoCAD prompts:

Select first mline: *(select the multiline to trim or extend, as shown in Figure 6–10)*
Select second mline: *(select the intersecting multiline, as shown in Figure 6–10)*

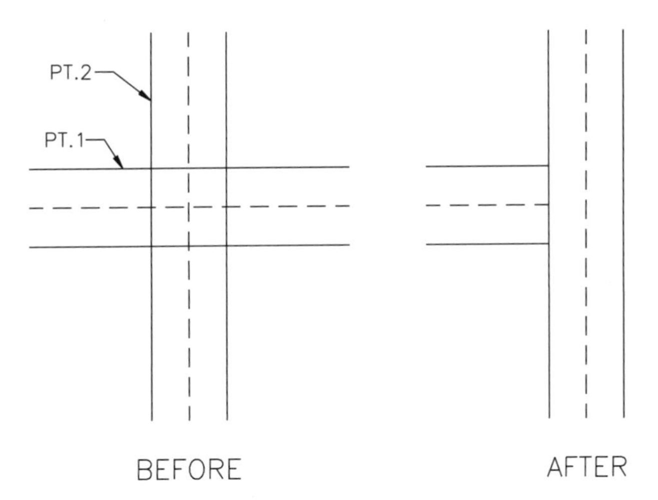

Figure 6–10 *An example of a closed tee*

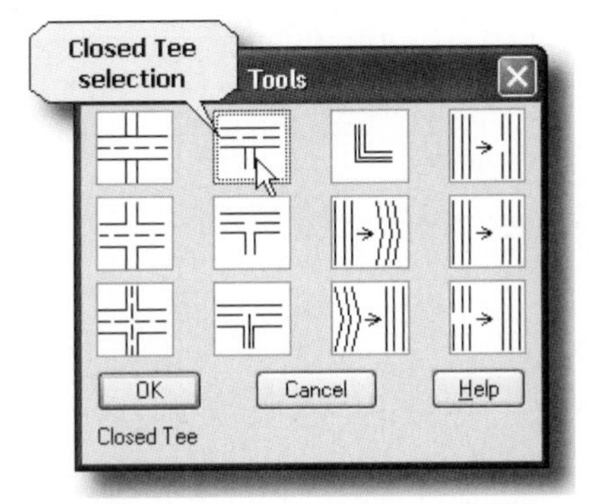

Figure 6–11 *Invoking the Closed Tee option from the Multiline Edit Tools dialog box*

After the closed tee intersection is created, AutoCAD prompts:

Select first mline (or Undo): *(select another multiline, enter **u**, or press* ENTER*)*

Selecting another multiline repeats the prompt for the second mline. Entering **u** undoes the closed tee just created.

The Open Tee option is similar to the Closed Tee option, except that it leaves an open end at the intersecting multiline, as shown in Figure 6–12. Choose the **Open Tee** image tile, as shown in Figure 6–13, to invoke the Open Tee option; AutoCAD prompts:

Select first mline: *(select the multiline to trim or extend, as shown in Figure 6–12)*
Select second mline: *(select the intersecting multiline, as shown in Figure 6–12)*

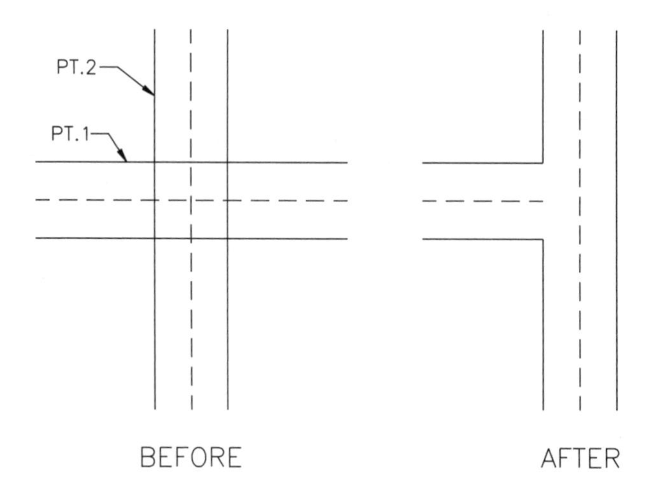

Figure 6–12 *An example of an open tee*

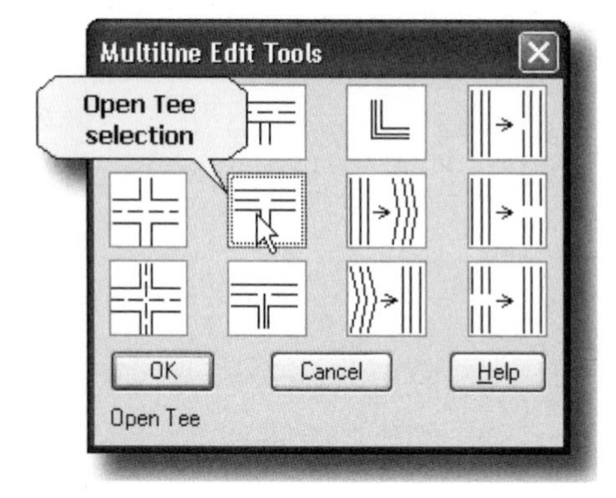

Figure 6–13 *Invoking the Open Tee option from the Multiline Edit Tools dialog box*

After the open tee intersection is created, AutoCAD prompts:

Select first mline (or Undo): *(select another multiline, enter* ***u****, or press* ENTER*)*

Selecting another multiline repeats the prompt for the second mline. Entering **u** undoes the open tee just created.

The Merged Tee option is similar to the Open Tee option, except that the centerline of the first multiline is extended to the center of the intersecting multiline, as shown in Figure 6–14. Choose the **Merged Tee** image tile, as shown in Figure 6–15, to invoke the Merged Tee option; AutoCAD prompts:

Select first mline: *(select the multiline to trim or extend, as shown in Figure 6–14)*
Select second mline: *(select the intersecting multiline, as shown in Figure 6–14)*

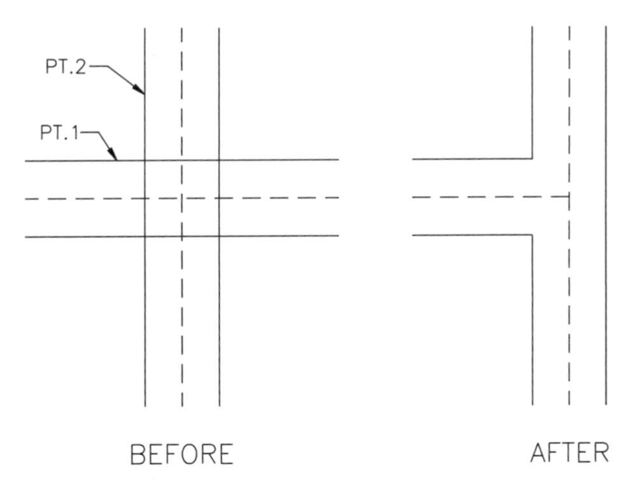

Figure 6–14 *An example of a merged tee*

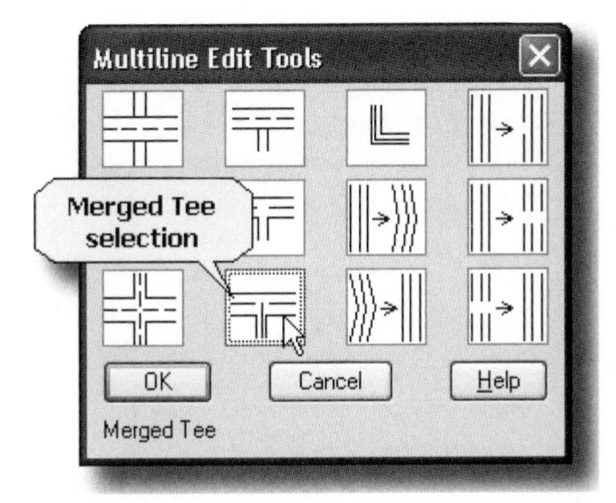

Figure 6–15 *Invoking the Merged Tee option from the Multiline Edit Tools dialog box*

After the merged tee intersection is created, AutoCAD prompts:

> Select first mline (or Undo): *(select another multiline, enter* ***u****, or press* ENTER*)*

Selecting another multiline repeats the prompt for the second mline. Entering **u** undoes the merged tee just created.

The Corner Joint option lengthens or shortens each of the two multilines you select as necessary to create a clean intersection, as shown in Figure 6–16. Choose the **Corner Joint** image tile, as shown in Figure 6–17, to invoke the Corner Joint option; AutoCAD prompts:

> Select first mline: *(select the multiline to trim or extend, as shown in Figure 6–16)*
> Select second mline: *(select the intersecting multiline, as shown in Figure 6–16)*

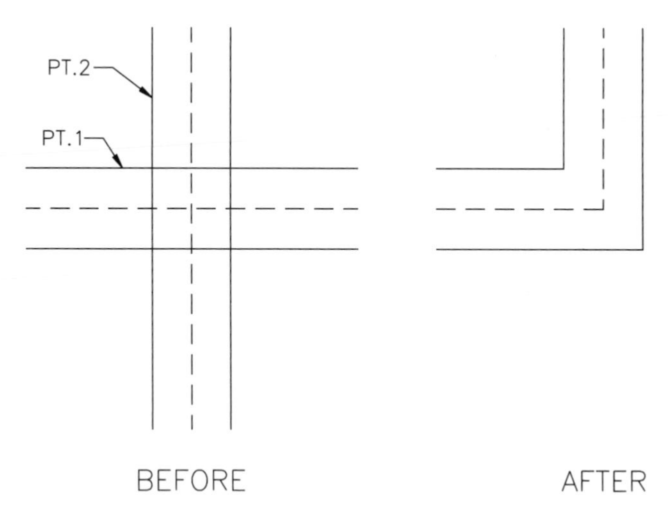

Figure 6–16 *An example of a corner joint*

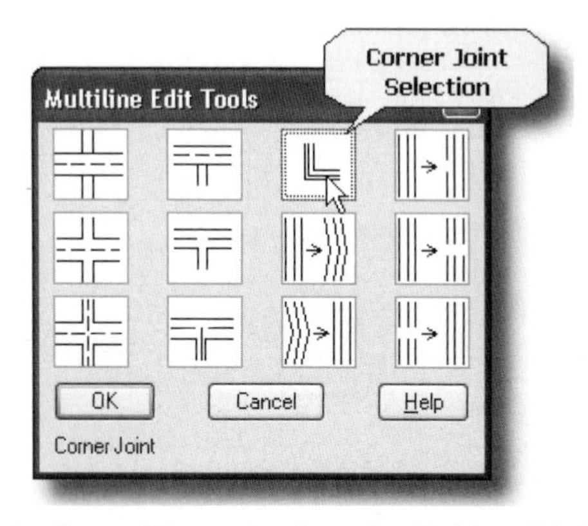

Figure 6–17 *Invoking the Corner Joint option from the Multiline Edit Tools dialog box*

After the corner joint intersection is created, AutoCAD prompts:

Select first mline (or Undo): *(select another multiline, enter **u**, or press* ENTER*)*

Selecting another multiline repeats the prompt for the second mline. Entering **u** undoes the corner joint just created.

The Add Vertex option adds a vertex to a multiline, as shown in Figure 6–18. Choose the **Add Vertex** image tile, as shown in Figure 6–19, to invoke the Add Vertex option; AutoCAD prompts:

Select mline: *(select a multiline to add a vertex, as shown in Figure 6–18)*

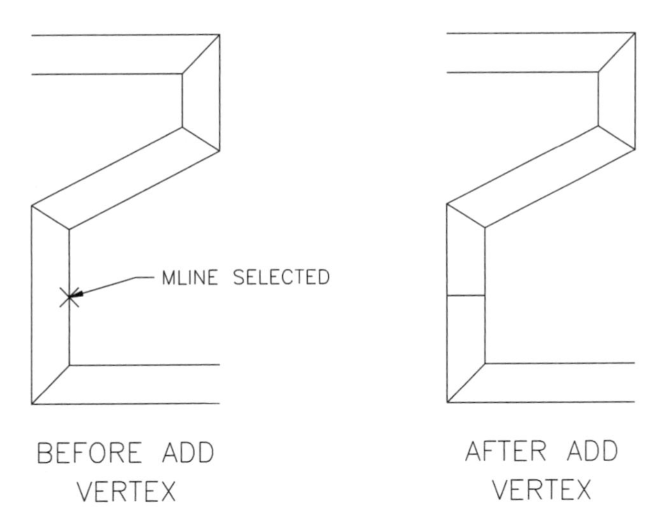

Figure 6–18 *An example of adding a vertex*

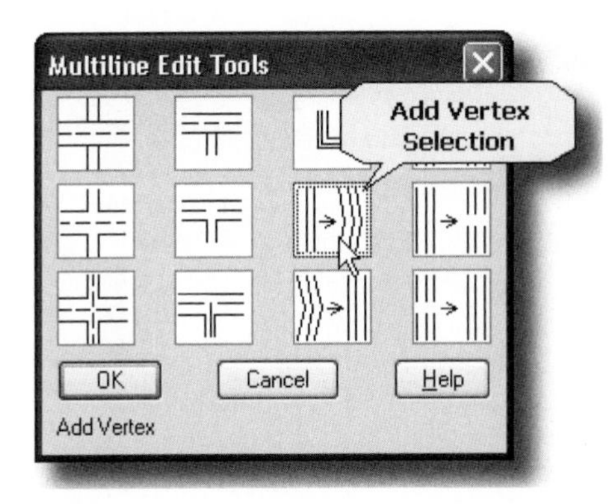

Figure 6–19 *Invoking the Add Vertex option from the Multiline Edit Tools dialog box*

AutoCAD adds the vertex at the selected point and prompts:

Select mline (or Undo): *(select another multiline, enter **u**, or press* ENTER*)*

Selecting another multiline allows you to add another vertex. Entering **u** undoes the vertex just created.

The Delete Vertex option deletes a vertex from a multiline, as shown in Figure 6–20. Choose the **Delete Vertex** image tile, as shown in Figure 6–21, to invoke the Delete Vertex option; AutoCAD prompts:

Select mline: *(select a multiline to delete a vertex, as shown in Figure 6–20)*

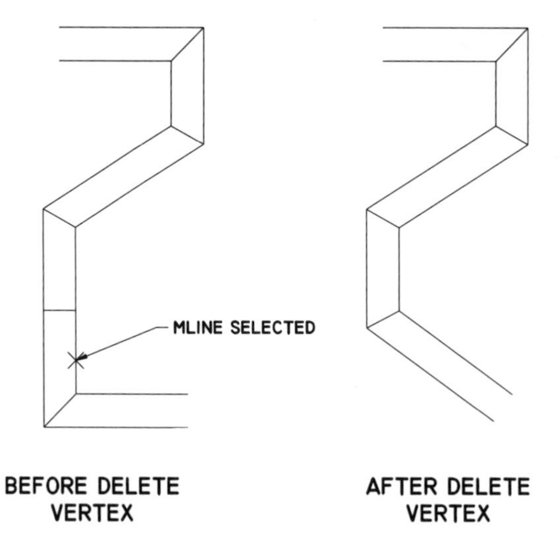

Figure 6–20 *An example of deleting a vertex*

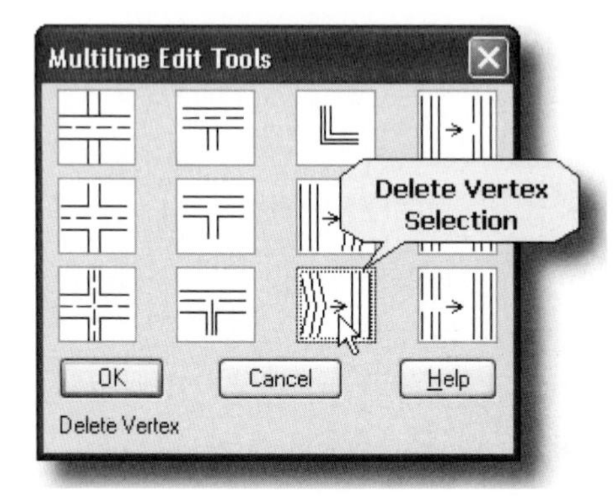

Figure 6–21 *Invoking the Delete Vertex option from the Multiline Edit Tools dialog box*

AutoCAD deletes the vertex at the selected point and prompts:

Select mline (or Undo): *(select another multiline, enter **u**, or press* ENTER*)*

Selecting another multiline allows you to delete another vertex. Entering **u** undoes the operation and displays the "Select mline:" prompt.

The Cut Single option cuts a selected element of a multiline between two cut points, as shown in Figure 6–22. Choose the Cut Single image tile, as shown in Figure 6–23, to invoke the Cut Single option; AutoCAD prompts:

Select mline: *(select a multiline and the selected point becomes the first cut point, as shown in Figure 6–22)*

Select second point: *(select the second cut point on the multiline, as shown in Figure 6–22)*

Figure 6–22 *An example of removing a selected element of a multiline between two cut points*

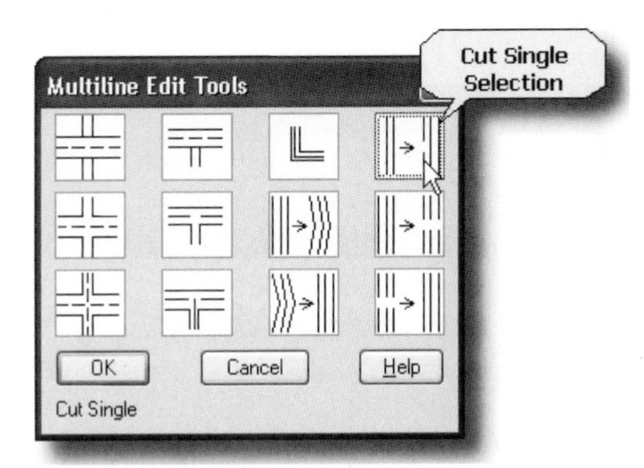

Figure 6–23 *Invoking the Cut Single option from the Multiline Edit Tools dialog box*

AutoCAD cuts the multiline and prompts:

Select mline (or Undo): *(select another multiline, enter **u**, or press* ENTER*)*

Select another multiline to continue, or enter **u** to undo the operation and display the Select mline: prompt.

The Cut All option removes a portion of the multiline you select between two cut points, as shown in Figure 6–24. Choose the **Cut All** image tile, as shown in Figure 6–25, to invoke the Cut All option; AutoCAD prompts:

Select mline: *(select a multiline and the selected point becomes first cut point, as shown in Figure 6–24)*
Select second point: *(select the second cut point on the multiline, as shown in Figure 6–24)*

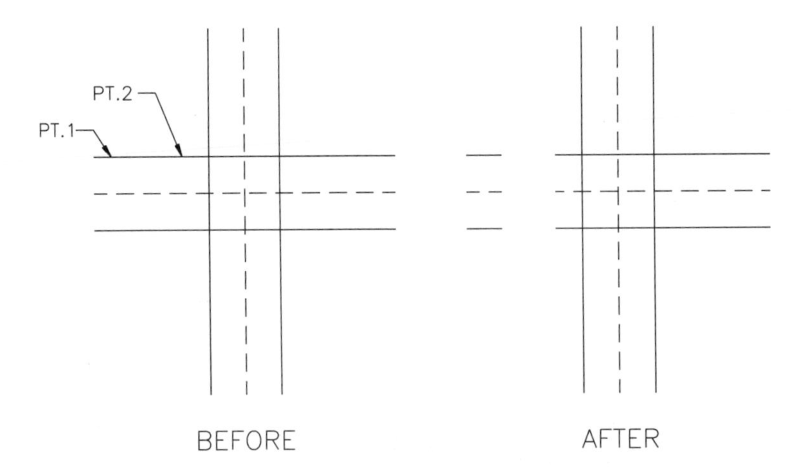

Figure 6–24 *An example of removing a portion of a multiline between two cut points*

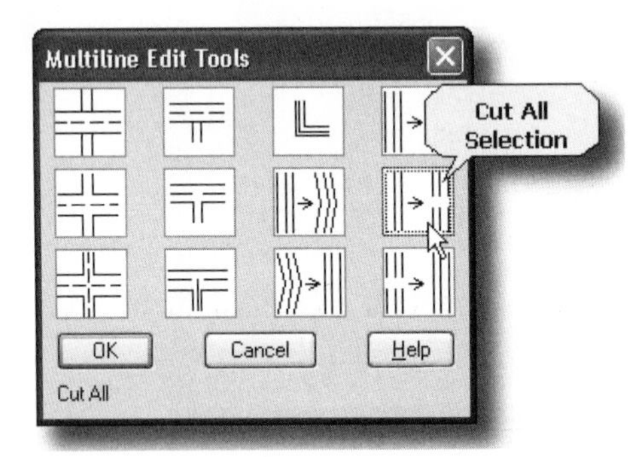

Figure 6–25 *Invoking the Cut All option from the Multiline Edit Tools dialog box*

AutoCAD cuts all the elements of the multiline and prompts:

> Select mline (or Undo): *(select another multiline, enter* ***u****, or press* ENTER*)*

Select another multiline to continue, or enter **u** to undo the operation and display the Select mline: prompt.

The Weld All option rejoins multiline segments that have been cut, as shown in Figure 6–26. Choose the **Weld All** image tile, as shown in Figure 6–27, to invoke the Weld All option; AutoCAD prompts:

> Select mline: *(select the multiline, as shown in Figure 6–26)*
> Select second point: *(select the endpoint on the multiline to be joined, as shown in Figure 6–26)*

Figure 6–26 *An example of rejoining multiline segments that have been cut*

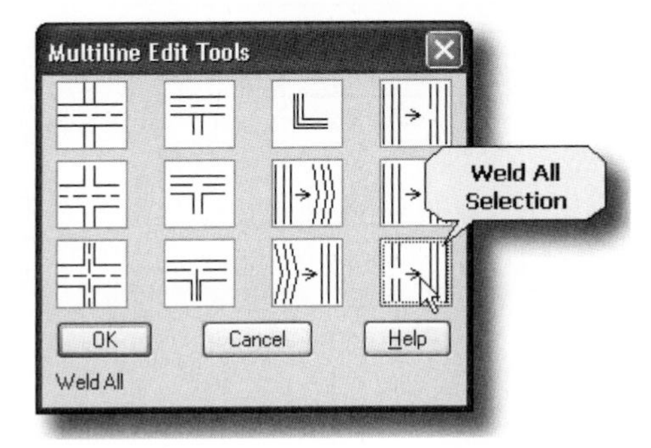

Figure 6–27 *Invoking the Weld All option from the Multiline Edit Tools dialog box*

AutoCAD joins the multiline and prompts:

Select mline (or Undo): *(select another multiline, enter **u**, or press* ENTER*)*

Select another multiline to continue, or enter **u** to undo the operation and display the Select mline: prompt.

CREATING AND MODIFYING MULTILINE STYLES

The MLSTYLE command is used to create a new Multiline Style or edit an existing one. You can define a multiline style comprised of up to 16 lines, called elements. The style controls the number of elements and the properties of each element. In addition, you can specify the background color and the end caps of each multiline.

Invoke the MLSTYLE command:

Format menu	Choose Multiline Style
Command: prompt	**mlstyle** (ENTER)

AutoCAD displays the Multiline Styles dialog box, as shown in Figure 6–28.

Figure 6–28 *Multiline Styles dialog box*

The **Current:** list box allows you to choose from the available multiline styles loaded in the current drawing. To make a specific multiline style current, first choose the appropriate multiline style from the list box and then choose **OK** to close the dialog box. If necessary, you can load additional multiline styles into the current drawing by means of the Load option.

The **Name:** text box allows you to specify a name for a new Multiline Style or rename an existing one.

To create a new Multiline Style, first define the element and multiline properties, then enter a name for the newly created Multiline Style in the **Name:** text box. Choose the **Save...** button, and AutoCAD displays the Save Multiline Style dialog box shown in Figure 6–29. By default, AutoCAD saves the Multiline Style definition in the ACAD.MLN library file. If necessary, you can select another file or provide a new file name with .MLN as the file extension. Choose the **Save** button to save the multiline .MLN library file.

To make the newly created Multiline Style current, choose the **Add** button. AutoCAD then adds the newly created multiline style to the **Current:** list box and makes it the current Multiline Style.

The **Description:** text box allows you to add a description of up to 255 characters, including spaces.

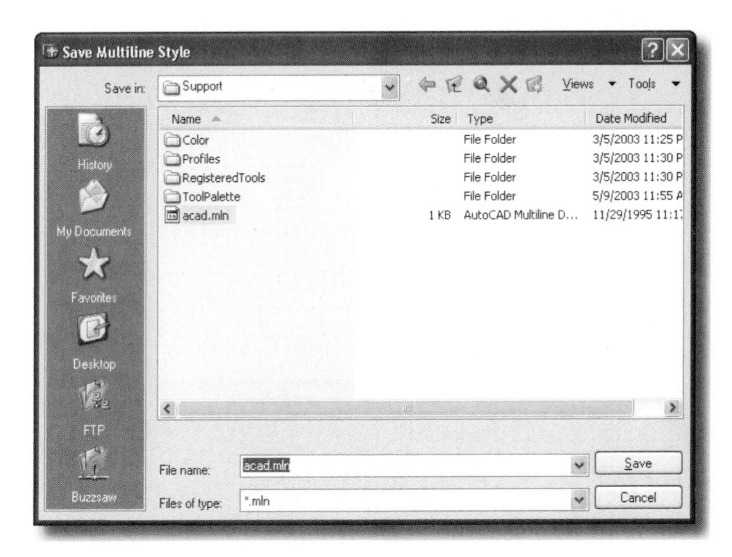

Figure 6–29 *Save Multiline Style dialog box*

The **Load...** button allows you to load a Multiline Style from a multiline library file into the current drawing. To load a Multiline Style into the current drawing, choose the **Load...** button. AutoCAD then displays the Load Multiline Styles dialog box, as shown in Figure 6–30. Select one of the available Multiline Styles from the list box. If you need to load a Multiline Style from a different library file, then choose the **File...** button. AutoCAD lists the available multiline library files. Select the appropriate library file, and AutoCAD lists the Multiline Styles available from the library file selected. After Choosing the Multiline Style, choose the **OK** button to close the dialog box.

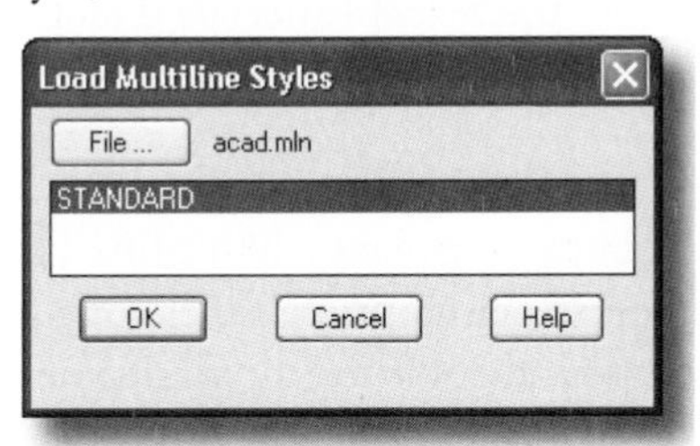

Figure 6–30 *Load Multiline Styles dialog box*

The **Add** button, as mentioned earlier, allows you to add the newly created Multiline Style to the Current: list box.

ELEMENT PROPERTIES

The **Element Properties...** button displays an Element Properties dialog box similar to the one shown in Figure 6–31, with options to set or change the offset, number, line type, or color of the multiline elements.

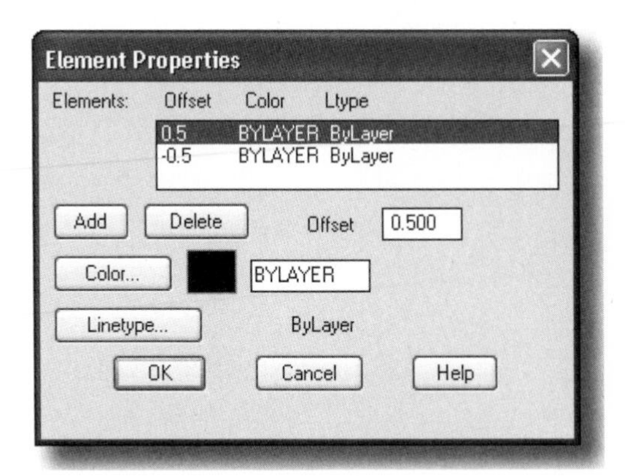

Figure 6–31 *Element Properties dialog box*

The **Elements:** list box displays existing elements along with the color and linetype of each.

Choose the **Add** or **Delete** button to add a line element to the multiline style or delete a line element from the multiline style, respectively.

The **Offset** text box sets the distance from the baseline for the elements selected from the **Elements:** list box. AutoCAD will use this distance to offset the element when the multiline is drawn and the scale factor is set to 1.0. Otherwise, the offset will be a ratio determined by the scale value.

The **Color...** button displays the Select Color dialog box for setting the color for line elements (selected from **Elements:** list box) in the multiline style.

The **Linetype...** button displays the Select Linetype dialog box for setting the linetype for line elements (selected from the **Elements:** list box) in the multiline style.

Multiline Properties

The **Multiline Properties...** button displays a Multiline Properties dialog box similar to the one shown in Figure 6–32, with options to change the type of joints, end caps (and their angles), and background color for the multiline.

The **Display joints** toggle button controls the display of the joints (miter) at the vertices of each multiline segment, as shown in Figure 6–33.

The **Caps** section has four options to specify the appearance of multiline start and end caps.

The **Line** toggle button controls the display of the start and end caps by a straight line, as shown in Figure 6–34.

The **Outer arc** toggle button controls the display of the start and end caps by connecting the ends of the outermost elements with a semicircular arc, as shown in Figure 6–35.

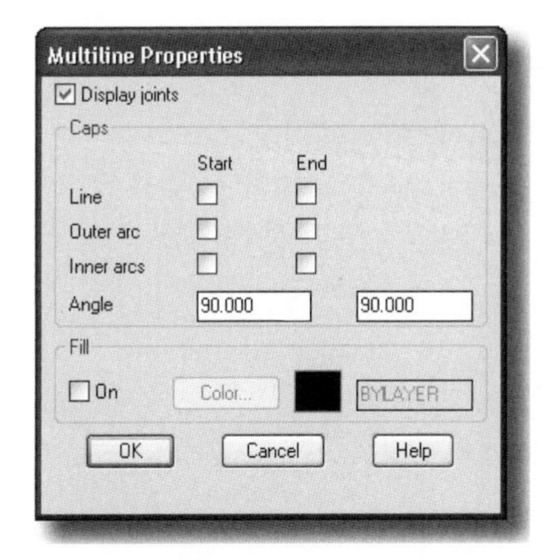

Figure 6–32 *Multiline Properties dialog box*

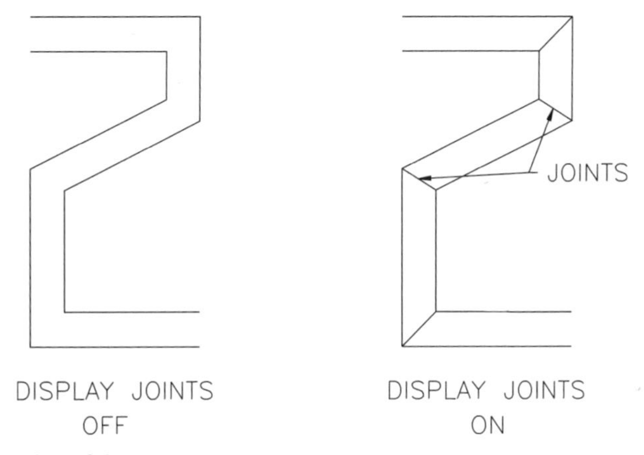

Figure 6–33 *Display of the joints*

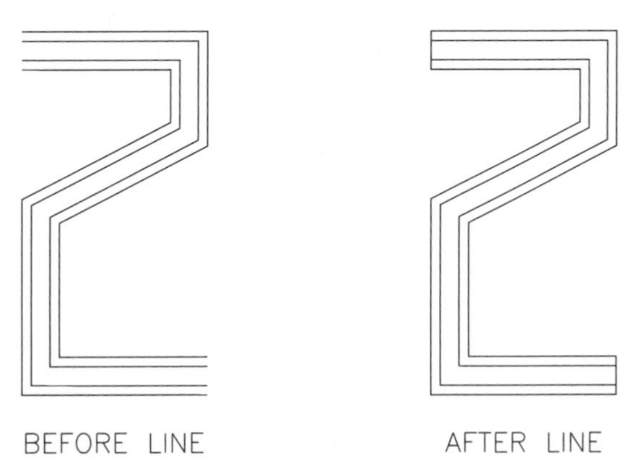

Figure 6–34 *Display of the line for start and end caps*

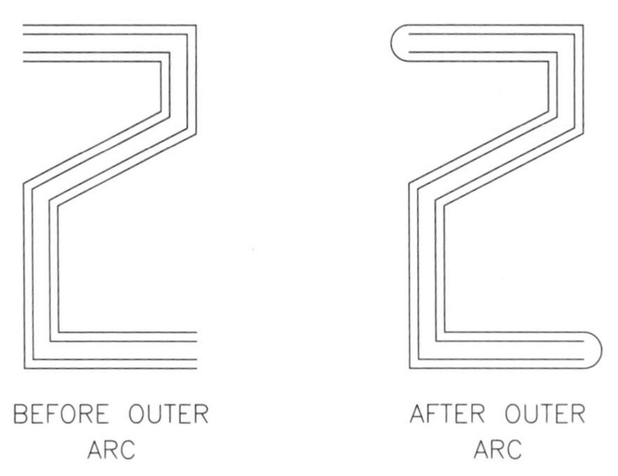

Figure 6–35 *Display of the outer arc for start end caps*

The **Inner arcs** toggle button controls the display of the start and end caps by connecting the ends of the innermost elements with a semicircular arc, as shown in Figure 6–36. For a multiline with an odd number of elements, the center element is not connected. For an even number of elements, connected elements are paired with elements that are the same number from each edge. For example, the second element from the left outer will be connected to the second element from the right outer, the third to the third, and so forth.

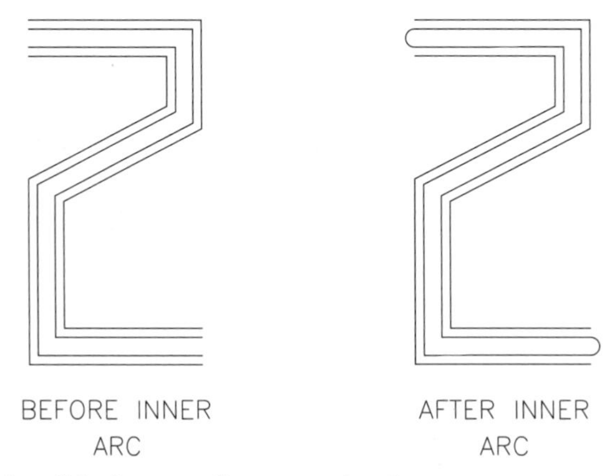

Figure 6–36 *Display of the inner arc for start and end caps*

The **Angle** edit field sets the angle of endcaps, as shown in Figure 6–37.

The **Fill** toggle button controls the background fill of the multiline.

The **Color** button displays the Select Color dialog box and sets the color of the background fill.

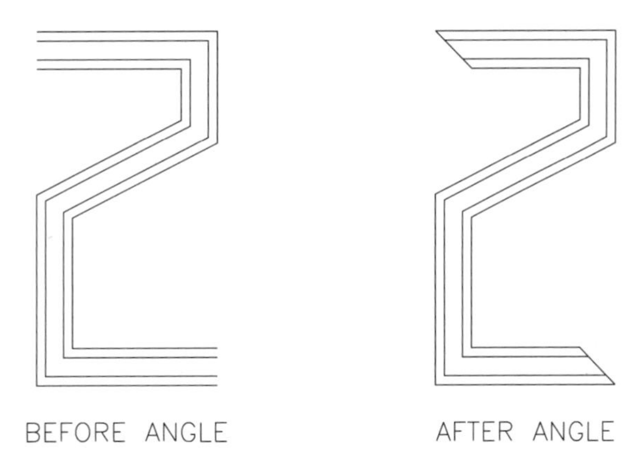

Figure 6–37 *Display of the end caps with an angular cap*

Once you have set the appropriate element properties and multiline properties for a new multiline style, enter the name and description in the **Name:** and **Description:** edit fields, respectively. Choose the **Save** button to save the newly created multiline style.

SPLINE CURVES

DRAWING SPLINE CURVES

The SPLINE command is used to draw a curve through or near points in a series. The type of curve is a nonuniform rational B-spline (NURBS). This type is used for drawing curves with irregularly varying radii, such as topographical contour lines.

The spline curve is drawn through a series of two or more points, with options either to specify end tangents or to use Close to join the last segment to the first. Another option lets you specify a tolerance, which determines how close to the selected points the curve is drawn.

Invoke the SPLINE command:

Draw toolbar	Choose the SPLINE command (see Figure 6–38)
Draw menu	Choose Spline
Command: prompt	**spline** (ENTER)

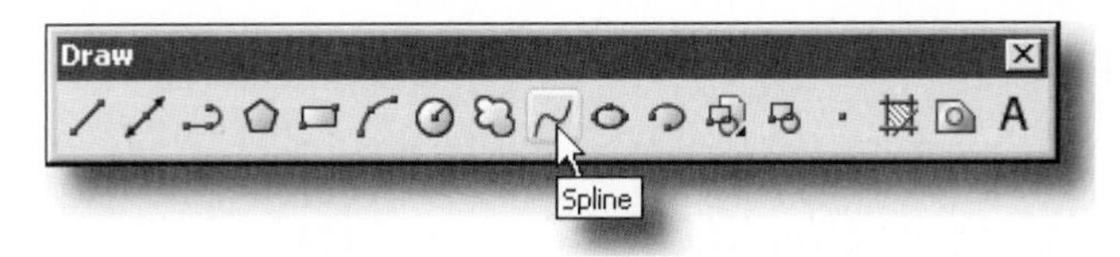

Figure 6–38 *Invoking the SPLINE command from the Draw toolbar*

AutoCAD prompts:

Command: **spline** (ENTER)
Specify first point or [Object]: *(specify a point or enter* **o** *for the Object option)*

The default option lets you specify the point from which the spline starts and to which it can be closed. After entering the first point, a preview line appears. You will then be prompted:

Specify next point:

When you respond by specifying a point, the spline segments are displayed as a preview spline, curving from the first point, through the second point, and ending at the cursor. You are then prompted:

Specify next point or [Close/Fit tolerance] <start tangent>:

If you specify a point, the next segment is added to the spline. This will occur with each additional point specified until you use the Close option or enter a null response by pressing ENTER.

Choosing a null response terminates the specifying of segment determining points. You are then prompted for a start tangent determining point as follows:

Specify start tangent:

If you specify a point (for tangency), its direction from the start point determines the start tangent. If you press ENTER, the direction from the first point to the second point determines the tangency. After the start tangency is established, you are prompted:

Specify end tangent:

If you specify a point (for tangency), its direction from the endpoint determines the end tangent. If you press ENTER, the direction from the last point to the previous point determines the tangency.

The Close option uses the original starting point of the first spline segment as the endpoint of the last segment and terminates segment placing. You are then prompted:

Specify tangent:

You can specify a point to determine the tangency at the connection of the first and last segments. If you press ENTER, AutoCAD calculates the tangency and draws the spline accordingly. You can also use the Perp or Tan options to cause the tangency of the spline to be perpendicular or tangent to a selected object.

The Fit tolerance option lets you vary how the spline is drawn relative to the selected points. You are then prompted:

Specify fit tolerance <current>:

Entering 0 (zero) causes the spline to pass through the specified points. A positive value causes the spline to pass within the specified value of the points.

Object

The Object option is used to change spline-fit polylines into splines. This can be used for 2D or 3D polylines, which will be deleted depending on the setting of the DELOBJ system variable.

EDITING SPLINE CURVES

Splines created by means of the SPLINE command have numerous characteristics that can be changed with the SPLINEDIT command. These include quantity and location of fit points, end characteristics such as open/close and tangencies, and tolerance of the spline (how near the spline is drawn to fit points).

SPLINEDIT operations on control points (which are different than fit points) of the selected spline include adding control points (with the Add or the Order option) and changing the weight of individual control points, which determines how close the spline is drawn to individual control points.

Invoke the SPLINEDIT command:

Modify II toolbar	Choose the Edit Spline command (see Figure 6–39)
Modify menu	Choose Object > Spline
Command: prompt	**splinedit** (ENTER)

Figure 6–39 *Invoking the* EDIT SPLINE *command from the Modify II toolbar*

AutoCAD prompts:

Command: **splinedit** (ENTER)
Select Spline: *(specify a spline curve)*
Enter an option [Fit data/Open/Move vertex/Refine/rEverse/Undo]: *(right-click to choose one of the available options from the shortcut menu)*

Control points appear in the grip color, and, if the spline has fit data, fit points also appear in the grip color. If you select a spline whose fit data is deleted, then the Fit Data option is not available. A spline can lose its fit data if you use the purge option while editing fit data, refine the spline, move its control vertices, fit the spline to a tolerance, or open or close the spline.

The Open option will replace Close if you select a closed spline, and vice versa.

Fit Data

The Fit data option allows you to edit the spline by providing the following suboptions:

Enter a fit data option [Add/Open/Delete/Move/Purge/Tangents/toLerance/eXit]<eXit>:

The Add suboption allows you to add fit points to the selected spline. AutoCAD prompts:

Specify control point <exit>: *(select a fit point)*

After you select one of the fit points, AutoCAD highlights it and you are prompted for the next point:

Specify new point <exit>: *(specify a point)*
Specify new point <exit>: *(specify another point or press* ENTER*)*

Selecting a point places a new fit point between the highlighted ones.

The Close suboption closes an open spline smoothly with a segment or smoothes a spline with coincidental starting and ending points.

The Open suboption opens a closed spline, disconnecting it and changing the starting and ending points.

The Delete suboption deletes a selected fit point.

The Move suboption moves fit options to a new location by prompting:

Specify new location or [Next/Previous/Select point/eXit]<N>:

The Next option steps forward through fit points.

The Previous option steps backwards through fit points.

The Select point option permits you to select a fit point.

The eXit option exits this set of suboptions.

The Enter New Location option moves the highlighted point to a specified point.

The Purge suboption deletes fit data for the selected spline.

The Tangents suboption edits the start and end tangents of a spline by prompting:

> Specify start tangent or [System default]: *(specify a point, enter an option, or press* ENTER*)*
> Specify end tangent or [System default]: *(specify a point, enter an option, or press* ENTER*)*

For a closed spline, the prompt is:

> Specify tangent or [System default]:

If you choose the System Default, AutoCAD calculates the tangents. You can choose Tan or Perp and select an object for the spline tangent to be tangent to or perpendicular to the object selected.

The toLerance suboption refits the spline to the existing points with new tolerance values by prompting:

> Enter fit tolerance <current>: *(enter a value or press* ENTER*)*

The value you enter determines how near the spline will be fit to the points.

The eXit suboption exits the Fit Data options and returns to the main prompt.

Close

The Close option of the SPLINEDIT command causes the spline to be joined smoothly at its start point.

Open

The Open option opens a closed spline. Previously open splines with coincidental starting and ending points will lose their tangency. Others will be restored to a previous state.

Move Vertex

The Move vertex option relocates a spline's control vertices by providing the following suboptions:

> Specify new location or [Next/Previous/Select point/eXit]<N>:

The Specify New Location suboption moves the highlighted point to a specified point.

The Next suboption steps forward through fit points.

The Previous suboption steps backward through fit points.

The Select point suboption permits you to select a fit point.

The eXit suboption exits this set of suboptions.

Refine

The Refine option of the SPLINEDIT command allows you to fine-tune a spline definition by providing the following suboptions:

```
Enter a refine option [Add control point/Elevate order/Weight/
  eXit]<eXit>:
```

The Add control point suboption increases the number of control points that control a portion of a spline.

The Elevate order suboption increases the order of the spline. You can increase the current order of a spline up to 26 (the default is 4), causing an increase in the number of control points.

The Weight suboption changes the weight at various spline control points by providing the following sub-options:

```
Enter new weight (current = 1.0000) or [Next/Previous/Select
  point/eXit]<N>:
```

The Next suboption steps forward through fit points.

The Previous suboption steps backward through fit points.

The Select point suboption permits you to select a fit point.

The eXit suboption exits this set of suboptions and returns you to the Refine prompt.

The default weight value for a control point is 1.0. Increasing it causes the spline to be drawn near the selected point. A negative or zero value is not valid.

From the Refine prompt the **eXit** suboption returns you to the main prompt.

rEverse

The rEverse option of the SPLINEDIT command reverses the direction of the spline. Reversing the spline does not delete the Fit Data.

Undo

The Undo option undoes the effects of the last subcommand.

eXit

The eXit option terminates the SPLINEDIT command.

WIPEOUT

The WIPEOUT command allows you to create an area on the screen that obscures previously drawn objects within its boundary. These areas can be displayed with or without a visible boundary (called a frame).

Invoke the WIPEOUT command:

Draw menu	Choose Wipeout
Command: prompt	**wipeout** (ENTER)

AutoCAD prompts:

Specify start point: *(specify the start point of the wipeout area)*
Specify next point or [Undo]: *(specify the second point or right-click for shortcut menu and choose one of the available options)*
Specify next point or [Close/Undo]: *(specify the third point or right-click for shortcut menu and choose one of the available options)*

Having drawn a connected series of lines, you can give a null reply (press ENTER) to terminate the WIPEOUT command. The series of responses determines the polygonal boundary of the wipeout object from a series of points. The shortcut menu displayed when you right-click at the prompt for the first point includes options for changing the display (ZOOM & PAN) and the Frames and Polyline options.

The Frames option determines whether the edges of all wipeout objects are displayed or hidden.

AutoCAD prompts

Enter mode ON/OFF <existing setting>: *(Enter* ***on*** *or* ***off****)*

Enter ON to display all wipeout frames. Enter OFF to suppress the display of all wipeout frames.

The Polyline option allows you to select a polyline, which determines the polygonal boundary of the wipeout area.

AutoCAD prompts:

Select a closed polyline: *(Use an object selection method to select a closed polyline)*
Erase polyline? [Yes/No]<No>: *(Enter* ***y*** *or* ***n****)*

Enter **y** to erase the polyline that was used to create the wipeout object. Enter **n** to retain the polyline. The shortcut menu displayed when you right-click at the prompt for the third or subsequent point includes options for changing the display (ZOOM & PAN) and the Close and Undo options. These operate in the same way as in the LINE and POLYLINE commands. Also included are the ENTER and CANCEL options.

Note: The area and its frame created by the WIPEOUT command will cover existing objects. Objects drawn after and covering the area will not be hidden by the area. If one or more of the objects being covered is modified, by the MOVE command for example, then it will no longer be covered. Likewise, if the wipeout area is modified, it will then cover all objects that overlap it.

REVISION CLOUD

The REVCLOUD command allows you to draw a connected series of arcs encircling objects in a drawing to signify an area on the drawing that has been revised as shown in Figure 6–40.

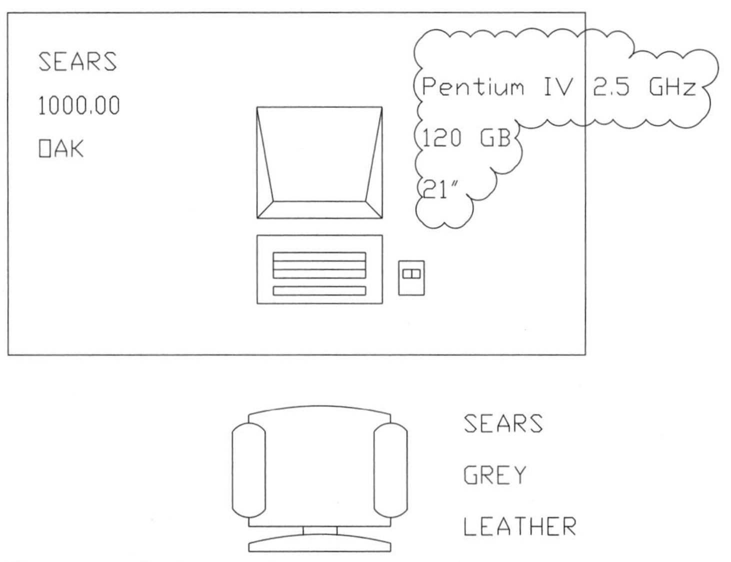

Figure 6–40 *An example of usage of REVCLOUD command*

Invoke the REVCLOUD command:

Draw toolbar	Choose the REVCLOUD command (see Figure 6–41)
Draw menu	Choose Revision Cloud
Command: prompt	**revcloud** (ENTER)

Figure 6–41 *Invoking the* REVCLOUD *command from the Draw toolbar*

AutoCAD prompts:

Minimum arc length: 0.5000 Maximum arc length: 0.5000
Specify start point or [Arc length/Object] <Object>:*(specify the start point of the revision cloud)*
Guide crosshairs along cloud path... *(move the cursor crosshairs along the path of the desired revision cloud)*

When the revision cloud is almost complete and the cursor crosshairs approach the starting point, the cloud is automatically closed without requiring any additional action or input.

AutoCAD prompts:

Revision cloud finished.

The size range of the arcs can be set for the desired appearance when prompted for minimum and maximum arc lengths. This value is then multiplied by the value of the dimension variable DIMSCALE to compensate for drawings with different scale factors.The Object option allows you to select a closed shape (polyline, rectangle, circle, etc.) from which AutoCAD creates a revision cloud. You are then prompted:

Reverse direction [Yes/No] <No>: *(Enter **y** or **n**)*

Selecting Yes causes the arcs to be redrawn on the opposite side of the line/arc defining the cloud. Selecting No causes the revision cloud to remain as drawn.

EDITING WITH GRIPS

The grips feature allows you to edit AutoCAD drawings in an entirely different way than using the traditional AutoCAD modify commands. With grips you can move, stretch, rotate, copy, scale, and mirror selected objects without invoking one of the regular AutoCAD modify commands. When you select an object with grips enabled, small squares appear at strategic points on the object that enable you to edit the selected objects.

To enable the grips feature, first open the Options dialog box:

Tools menu	Choose Options
Command: prompt	**options** (ENTER)

AutoCAD displays the Options dialog box, as shown in Figure 6–42.

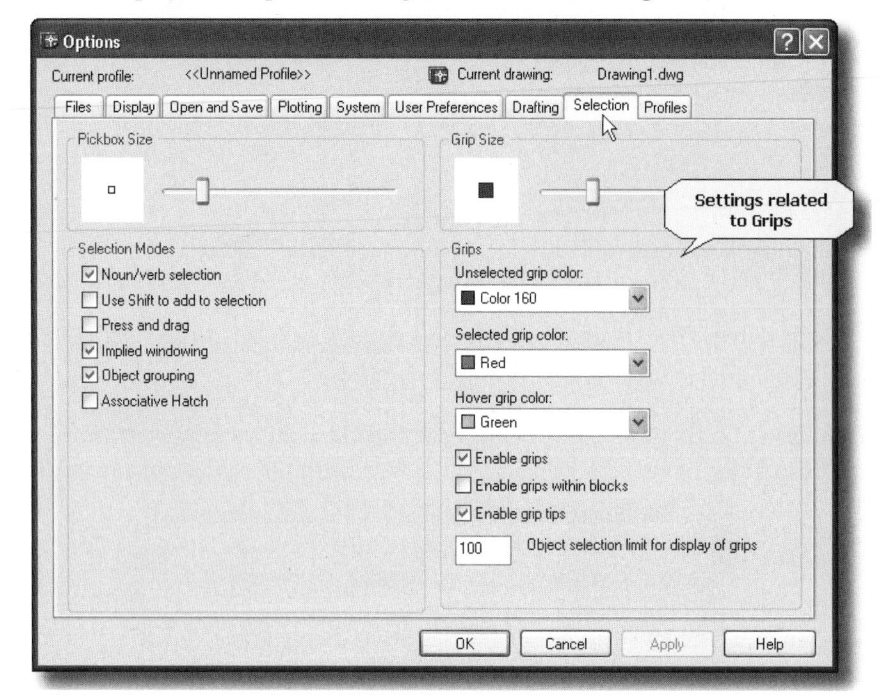

Figure 6–42 *Options dialog box*

Choose the **Selection** tab, and AutoCAD displays various options related to Grips, as shown in Figure 6–42.

The **Enable grips** toggle button controls the display of the grips. If it is set to ON, the grips display is enabled; if it is set OFF, the grips display is disabled.

You can also enable the grips feature by setting the system variable GRIPS to 1.

The **Enable grips within blocks** toggle button controls the display of grips on objects within blocks. If it is set to ON, the grips are displayed on all objects within the block; if it is set to OFF, a grip is displayed only on the insertion point of the block.

The **Enable grip tips** toggle button, when checked, causes a grip-specific tip to be displayed when the cursor hovers over a grip on a custom object that supports grip tips. Standard AutoCAD objects are not affected by this option.

The **Unselected grip color** list box allows you to change the color to the unselected grips.

The **Selected grip color** list box allows you to change the color to the selected grips.

The **Hover grip color** list box allows you to change the color to the grips in the hover mode.

The **Object select limit for display of grips** text box allows you to specify the number which limits the display of grips when the initial selection set includes more than the specified number of objects. The valid range is 1 to 32,767. The default setting is 20.

The **Grip Size** slider bar allows you to change the size of the grips. To adjust the size of grips, move the slider box left or right. As you move the slider, the size is illustrated to the right of the slider.

After making the necessary settings, choose the **OK** button to keep the changes and close the Options dialog box.

AutoCAD gives you a visual cue when grips are enabled by displaying a pick box at the intersection of the crosshairs, even when you are at the Command: prompt, as shown in Figure 6–43.

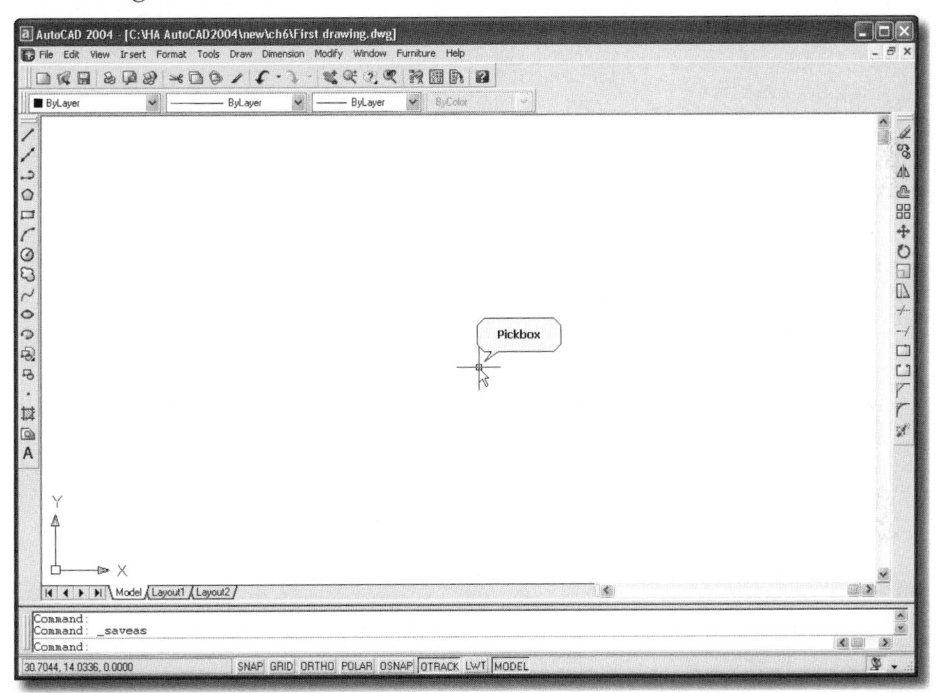

Figure 6–43 *The pick box displayed at the intersection of the crosshairs at the Command: prompt*

Note: The pick box is also displayed on the crosshairs when the PICKFIRST (Noun/Verb selection) system variable is set to ON.

To place grips directly from the Command: prompt, select one or more objects you wish to manipulate.

Note: To place grips, you can either select objects individually or select multiple objects by placing two points to specify the diagonally opposite corners of a rectangle.

Grips appear on the endpoints and midpoint of lines and arcs, on the vertices and endpoints of polylines, on quadrants and the center of circles, on dimensions, text, solids, 3dfaces, 3dmeshes, and viewports, and on the insertion point of a block. Figure 6–44 shows location of the grips on some of the commonly used objects.

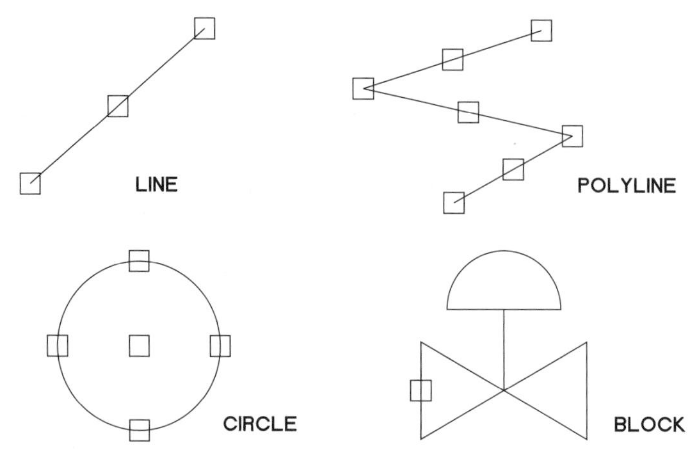

Figure 6–44 *Locations of grips on commonly used objects*

USING GRIPS

This section explains how to utilize grips in modifying your drawing. Learning to use grips speeds up the editing of your drawing while at the same time it maintains the accuracy of your work.

Snapping to Grips

When you move your cursor over a grip, it automatically snaps to the grip point. This allows you to specify exact locations in the drawing without having to use grid, snap, ortho, object snap, or coordinate entry tools.

Condition of Grips

Grips are categorized as being hot, warm, or cold, depending on their use.

A grip becomes hot when it is selected with your cursor. It has a solid-filled color and is the base point unless another grip is selected. You can make more than one grip hot. Hold down the SHIFT key while selecting the grips.

A grip is said to be warm if it is on an object in the current selection set that you haven't selected with the cursor. The object(s) are highlighted to indicate that they are in the selection set.

A grip is said to be cold if it is on an object that is not in the current selection set. Although the objects with cold grips are not highlighted, they will look identical to those with warm grips, and you can still use the grips to snap to. Figure 6–45 shows examples of hot, warm, and cold grips.

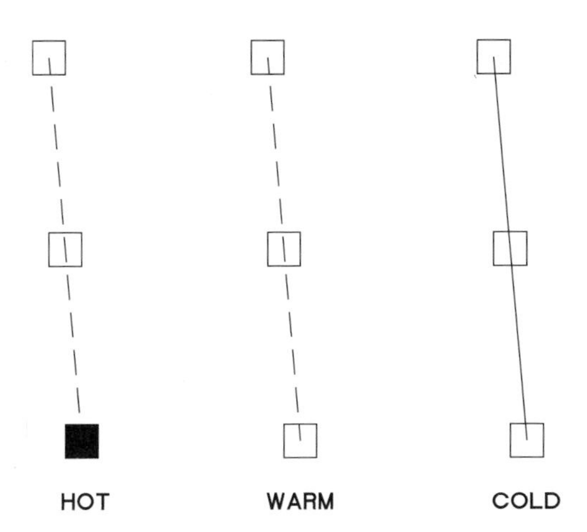

Figure 6–45 *Examples of hot, warm, and cold grips*

Clearing Grips

To clear grips from a selection set, press the ESC key. The grips on the selected objects will be cleared. When you invoke a non-modifying AutoCAD command, such as LINE or CIRCLE, AutoCAD clears the grips from a selection set.

To use grips to edit the selected objects, specify a grip to act as the base point for the editing operation at the "Command: prompt. Specifying a grip starts the Grip modes, which includes STRETCH, MOVE, ROTATE, SCALE, and MIRROR. You can cycle through the Grip modes by pressing the SPACEBAR, ENTER, entering a keyboard shortcut, or selecting from the right-click shortcut menu. To cancel Grip mode, enter x (for the mode's eXit option); AutoCAD returns to the Command: prompt. You can also use a combination of the current Grip mode and a multiple copy operation on the selection set.

As mentioned earlier, you can also select the available Grip modes from the shortcut menu when you specify a grip at the Command: prompt. Right-click after selecting a grip, and AutoCAD displays the shortcut menu, listing of the Grip modes, as shown in Figure 6–46.

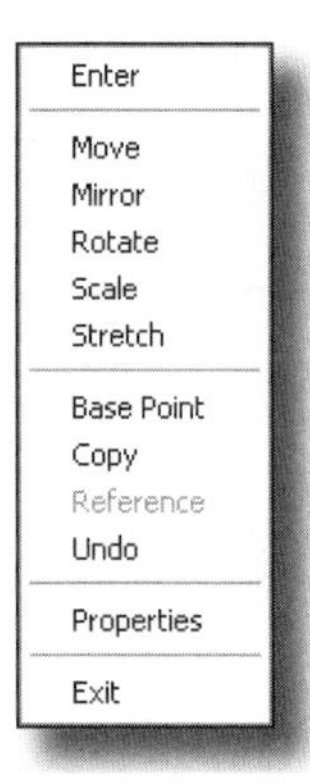

Figure 6–46 *Shortcut menu displaying the Grip modes*

Stretch Mode

The Stretch mode works similarly to the STRETCH command. It allows you to stretch the shape of an object without affecting other crucial parts that remain unchanged. When you are in the Stretch Mode, the following prompt appears:

> **STRETCH**
> Specify stretch point or [Base point/Copy/Undo/eXit]:

The Specify stretch point option (the default)refers to the stretch displacement point. As you move the cursor, you see that the shape of the object is stretched dynamically from the base point. You can specify the new point with the cursor or by entering coordinates. The displacement is applied to all selected hot grips.

If necessary, you can change the base point to be other than the base grip, by selecting Base point from the shortcut menu. Then specify the point with the cursor, or enter the coordinates.

To make multiple copies while stretching objects, choose COPY from the shortcut menu. Then specify destination copy points with the cursor, or enter their coordinates.

If you press the SHIFT key while selecting multiple stretch points, the Copy mode is activated and for subsequent copies the cursor snaps to offset points based on the first two points selected. When the first offset is specified while holding the SHIFT key, then the subsequent second, third and fourth offsets can be specified by moving the cursor until it locks onto the points that have been preset to the same offset as the first one. It you release the SHIFT key, then you can place the stretch point at any location.

Move Mode

The Move mode works similarly to the MOVE command. It allows you to move one or more objects from their present location to a new one without changing their orientation or size. In addition, you can make copies of the selected objects at the specified displacement, leaving the original objects intact. To invoke the Move mode, cycle through the modes by entering a null response until it takes you to the Move mode, or choose MOVE from the shortcut menu. When you are in the Move mode, the following prompt appears:

> **MOVE**
> Specify move point or [Base point/Copy/Undo/eXit]:

The Specify move point option refers to the move displacement point. As you move the cursor, AutoCAD moves all the objects in the current selection set to a new point relative to the base point. You can specify the new point with the cursor or by entering the coordinates.

If necessary, you can change the base point to be other than the base grip by selecting BASE POINT from the shortcut menu. Then specify the point with the cursor or enter the coordinates.

To make multiple copies while moving objects, choose COPY from the shortcut menu and then specify the destination copy points with the cursor, or enter their coordinates.

If you press the SHIFT key while moving the object, the Copy mode is activated and for subsequent copies the cursor snaps to offset points based on the first two points selected. When the first offset is specified while holding the SHIFT key, then the subsequent second, third and fourth offsets can be specified by moving the cursor until it locks onto the points that have been preset to the same offset as the first one. It you release the SHIFT key, then you can place the copy at any location.

Rotate Mode

The Rotate mode works similarly to the ROTATE command. It allows you to change the orientation of objects by rotating them about a specified base point. In addition, you can make copies of the selected objects and at the same time rotate them about a specified base point.

To invoke the Rotate mode, cycle through the modes by entering a null response until it takes you to the Rotate mode, or choose ROTATE from the shortcut menu. When you are in the Rotate mode, the following prompt appears:

> **ROTATE**
> Specify rotation angle or [Base point/Copy/Undo/Reference/eXit]:

The Specify rotation angle option refers to the rotation angle to which objects are rotated. As you move the cursor, AutoCAD allows you to drag the rotation angle, to position all the objects in the current selection set at the desired orientation. You can specify the new orientation with the cursor or from the keyboard. If you specify an angle by entering a value from the keyboard, this is taken as the angle that the objects should be rotated from their current orientation, around the base point. A positive angle rotates counterclockwise, and a negative angle rotates clockwise. Similar to the ROTATE command, you can use the Reference option to specify the current rotation and the desired new rotation.

If necessary, you can change the base point to be other than the base grip by choosing Base Point from the shortcut menu. Then specify the point with the cursor, or enter the coordinates.

To make multiple copies while rotating objects, choose COPY from the shortcut menu. Then specify destination copy points with the cursor, or enter the coordinates.

If you press the SHIFT key while rotating the object, the Copy mode is activated and for subsequent copies the cursor snaps to offset points based on the first two points selected. When the first offset is specified while holding the SHIFT key, then the subsequent second, third and fourth offsets can be specified by moving the cursor until it locks onto the points that have been preset to the same offset as the first one. It you release the SHIFT key, then you can place the copy at any location.

Scale Mode

The Scale mode works similarly to the SCALE command. It allows you to change the size of objects about a specified base point. In addition, you can make copies of the selected objects and at the same time change the size about a specified base point. To invoke the Scale mode, cycle through the modes by entering a null response until it takes you to the Scale mode, or choose SCALE from the shortcut menu. When you are in the Scale mode, the following prompt appears:

```
**SCALE***
Specify scale factor or [Base point/Copy/Undo/Reference/eXit]:
```

The Specify scale factor option refers to the scale factor to which objects are made larger or smaller. As you move the cursor, AutoCAD allows you to drag the scale factor, to change all the objects in the current selection set to the desired size. You can specify the new scale factor with the cursor or from the keyboard. If you specify the scale factor by entering a value from the keyboard, this is taken as a relative scale factor by which all dimensions of the objects in the current selection set are to be multiplied. To enlarge an object, enter a scale factor greater than 1; to shrink an object, use a scale factor between 0 and 1. Similar to the SCALE command, you can use the Reference option to specify the current length and the desired new length.

If necessary, you can change the base point to be other than the base grip, by choosing Base Point from the shortcut menu. Then specify the point with the cursor, or enter the coordinates.

To make multiple copies while scaling objects, choose COPY from the shortcut menu, and then specify the destination copy points with the cursor, or enter their coordinates.

If you press the SHIFT key while scaling the object, the Copy mode is activated and for subsequent copies the cursor snaps to offset points based on the first two points selected. When the first offset is specified while holding the SHIFT key, then the subsequent second, third and fourth offsets can be specified by moving the cursor until it locks onto the points that have been preset to the same offset as the first one. It you release the SHIFT key, then you can place the copy at any location.

Mirror Mode

The Mirror mode works similarly to the MIRROR command. It allows you to make mirror images of existing objects. To invoke the Mirror mode, cycle through the modes by entering a null response until it takes you to the Mirror mode, or choose MIRROR from the shortcut menu. When you are in the Mirror Mode, the following prompt appears:

> **MIRROR**
> Specify second point or [Base point/Copy/Undo/eXit]:

Two points are required in AutoCAD to define a line about which the selected objects are mirrored. AutoCAD considers the base grip point as the first point; the second point is the one you specify or enter in response to the Second point.

If necessary, you can change the base point to be other than the base grip, by choosing BASE POINT from the shortcut menu. Then specify the point with the cursor, or enter the coordinates.

To make multiple copies while retaining original objects, choose COPY from the shortcut menu. Then specify mirror points by specifying the point(s) with the cursor, or enter the coordinates.

If you press the SHIFT key while mirroring the object, the Copy mode is activated and for subsequent copies the cursor snaps to offset points based on the first two points selected. When the first offset is specified while holding the SHIFT key, then the subsequent second, third and fourth offsets can be specified by moving the cursor until it locks onto the points that have been preset to the same offset as the first one. It you release the SHIFT key, then you can place the copy at any location.

SELECTING OBJECTS BY QUICK SELECT

The QSELECT command is used to create selection sets based on objects that have or do not have similar characteristics or properties as determined by filters that you specify. For example, you can create a selection set of all lines that are equal to or less than 2.5 units long. Or you can create a selection set of all objects that are not text objects on one certain layer. The combinations of possible filters are almost limitless. You can cause the created selection set to replace the current selection set or be appended to it.

To create a filtered selection set, invoke the QSELECT command:

Tools menu	Choose Quick Select
Command: prompt	**qselect** (ENTER)

You can also invoke QSELECT command from the shortcut menu any time when you are at the Command: prompt. AutoCAD displays the Quick Select dialog box, similar to shown in Figure 6–47.

The **Apply to:** text box lets you select whether to apply the specified filters to the Current selection or to the Entire drawing. If there is a current selection, then Current selection is the default, otherwise, Entire drawing is the default. If the **Append to current selection set** check box is checked, then Current selection is not an option. If you wish to create a selection set, choose the **Select Objects** button (located next to the **Apply to:** text box). The **Select Objects** button is available only when **Append to current selection set** is not checked.

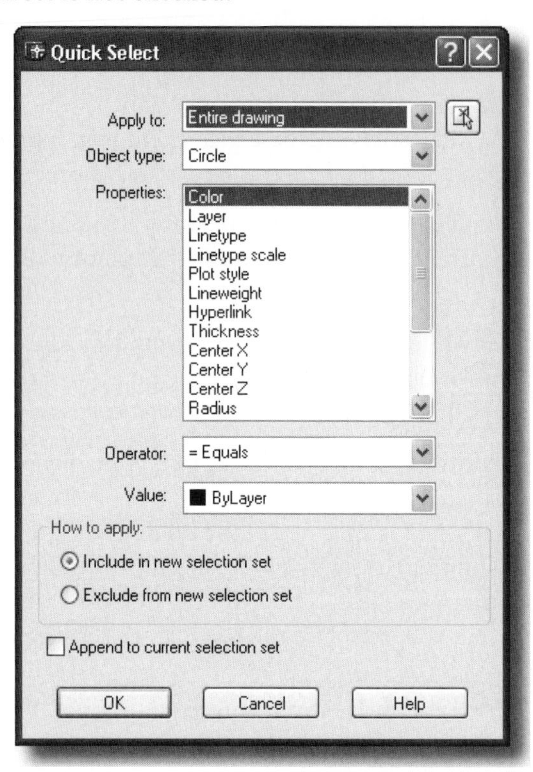

Figure 6–47 *Quick Select dialog box*

The **Select Objects** button returns you to the drawing screen so you can select objects. After selecting the objects to be included in the selection set, press ENTER.

The **Object type:** text box lets you select whether to include or exclude only certain types of objects or multiple objects.

The **Properties** section lists the properties that can be used for filters for the type of object(s) specified in the Object Type text box.

The **Operator** text box lets you apply logical operators to values. These include Greater Than, Less Than, Equal To, Not Equal To, and Wildcard Match. Greater

Than and Less Than apply primarily to numeric values, and Wildcard Match applies to text strings.

The **Value** text box lets you specify a value to which the operator applies. If you specify the Greater Than Operator to the Value of 1.0 for the Length of lines, then the lines with lengths greater than 1.0 will be filtered to be either included in or excluded from the selection set, depending on which radio button in the **How to apply:** section has been chosen.

The **How to apply:** section lets you specify whether the filtered objects will be included in or excluded from the selection set.

The **Append to current selection set** check box lets you specify whether the filtered selection set replaces the current selection or is appended to it.

Once you have all the required selection criteria set, choose the **OK** button to close the dialog box. AutoCAD displays grips on the selected objects. You can proceed with the appropriate modification required for the selected objects.

SELECTION SET BY FILTER TOOL

The FILTER command displays a dialog box that lets you create filter lists that you can apply to the selection set. This is another method of selecting objects. With the FILTER command, you can select objects based on object properties, such as location, object type, color, linetype, layer, block name, text style, and thickness. For example, you could use the FILTER command to select all the blue lines and arcs with a radius of 2.0 units. You can even name filter lists and save them to a file.

The new selection set that is created by the FILTER command can be used as the Previous option at the next Select object: prompt. If you use the FILTER command transparently, then AutoCAD passes the new selection set directly to the command in operation. This will save you a considerable amount of time.

Invoke the FILTER command:

Command: **filter** (ENTER)

AutoCAD displays the Object Selection Filters dialog box shown in Figure 6–48.

The list box displays the filters currently being used as a selection set. The first time you use the FILTER command in the current drawing, the list box is empty.

The **Select Filter** section lets you add filters to the list box based on object properties. Select the object or logical operator from the list box. You can use the grouping operators AND, OR, XOR, and NOT from the list box. The grouping operators must be paired and balanced correctly in the filter list. For example, each Begin OR operator must have a matching End OR operator. If you select more than one filter, AutoCAD by default uses an AND as a grouping operator between each filter.

Figure 6–48 *Object Selection Filters dialog box*

The **Select...** button displays a dialog box that lists all items of the specified type within the drawing. From the list, you can select as many items as you want to filter. This process saves you from typing the specific filter parameters.

The **Add to List** button lets you add the current **Select Filter** selection to the filter list.

The **Add Selected Object<** button allows you to select an object from the drawing and add it to the filter list.

The **Substitute** button replaces the selected filter with the current one in the Select Filter box.

The **Edit Item** button moves the selected filter into the **Select Filter** area for editing. First select the filter from the filter list box and then choose the **Edit Item** button. Make the necessary changes in the **Select Filter** section and choose the **Substitute** button. The edited filter replaces the selected filter.

The **Delete** button allows you to delete the selected filter in the filter list box.

The **Clear List** button allows you to clear the filter list from the filter list box.

To save the current filter list, enter the name for the filter list in the **Save As:** edit field, and choose the **Save As:** button to save the list in the given name.

The **Current:** list box displays the saved filter lists. Select a list to make it current.

The **Delete Current Filter List** button allows you to delete filter lists from the default filter file.

The **Apply** button lets you close the dialog box; AutoCAD displays the Select Objects: prompt, where you create a selection set. AutoCAD uses the filter list on the objects you select.

CHANGING PROPERTIES OF SELECTED OBJECTS

The PROPERTIES command is used to manage and change properties of selected objects by means of the Properties window.

To manage or change the objects in a selection set, invoke the PROPERTIES command:

Standard toolbar	Choose the PROPERTIES command (see Figure 6–49)
Tools menu	Choose Properties
Command: prompt	**properties** (ENTER)

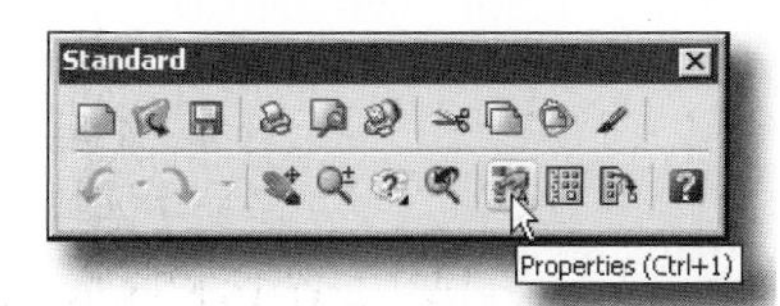

Figure 6–49 *Invoking the* PROPERTIES *command from the Standard toolbar*

AutoCAD displays the Properties Window, as shown in Figure 6–50.

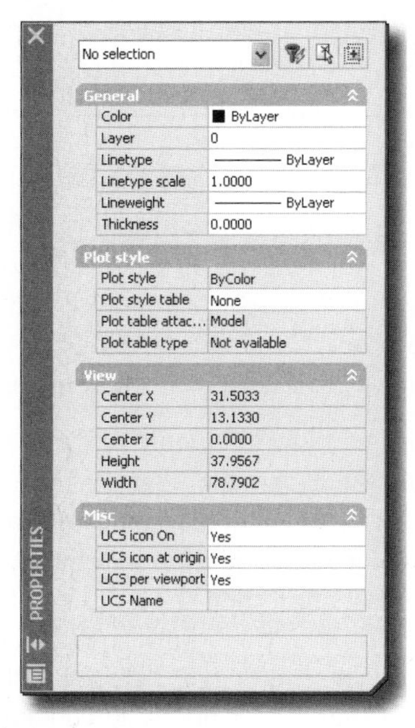

Figure 6–50 *Properties window*

The Properties window can be displayed by right-clicking when one or more objects have been preselected and are highlighted on the screen.

Options available for displaying the Properties window are accessible by right-clicking anywhere on the Properties window except the title bar. Available options include **Close, Undo**, and **Description**. Choosing **Close** causes the Properties Window to be hidden. Checking **Description** causes a description section to be displayed at the bottom of the Properties window describing a property when it has been selected. The **Undo** option undoes the last operation. The default position for the Properties window is docked on the left side of the screen. Its position can be changed by placing the cursor over the double line bar at the top of the window and either double-clicking or dragging the window into the screen area (or across to a docking position on the right side of the screen). Double-clicking causes the Properties window to become undocked and to float in the drawing area. When the Properties window in undocked, it can be docked by double-clicking in the title bar (which may be on the left or right side of the window) or by placing the cursor over the title bar and dragging the window all the way to the side you wish to dock it.

The Properties window is called a window instead of a dialog box because you can draw and modify objects in the drawing area while it is being displayed. When the Properties window is being displayed and one object, say a circle, is selected (make sure PICKFIRST and/or GRIPS system variable is set to ON), then the Properties window lists all the properties of that circle. Not only are properties like color, linetype and layer listed, but the center *X*, *Y*, and *Z* coordinates, radius, diameter, circumference, and area also listed. However, if two circles are selected, only the values of properties that are the same for both circles are listed. Other properties that are not the same but are common properties are left blank. If you enter a value in the blank area, the new value will be applied to both circles. For example, if the centers of the two circles are different and you enter *X* and *Y* coordinates in the Center X and Center Y edit fields respectively, both circles will be moved to have their centers coincide with the specified center. If different types of objects are in the selection set, then the properties listed in the Properties window will include only those common to the selected objects such as color, layer and linetype. Whenever you change the properties of selected objects in the Properties window, AutoCAD reflects the changes immediately in the drawing window.

You can also open Properties window by double-clicking on an object that has to be modified when the PICKFIRST system variable is set to ON. You can also open the Quick Select dialog box by choosing the **Quick Select** button located top right side of the Properties window to select objects. If you need to select the objects by the traditional method (by window and/or crossing), then choose the **Select Objects** button located on the top right side of the Properties window. In addition, you can change PICKADD system variable value by choosing the Toggle Value of **PICKADD** sysvar button located on the top right side of the Properties window. The PICKADD system variable setting controls how you add objects to an existing selection set.

You can also change properties of the selected objects by invoking the DDCHPROP or CHANGE commands.

The DDCHPROP command also displays the Properties window, as shown in Figure 6–50.

The CHANGE command also allows you to change properties for selected objects such as their color, lineweight, linetype, lineweight, layer, elevation, thickness, and plotstyle. In addition, the CHANGE command lets you modify some of the characteristics of lines, circles, text, and blocks. Invoke the CHANGE command:

> Command: **change** (ENTER)
> Select objects: *(select objects and press* ENTER *to complete the selection)*
> Specify change point or [Properties]: *(specify change point for the selected objects, or choose properties to change one of the properties of the selected object)*

The Change point option allows you to modify some of the characteristics of lines, circles, text, and blocks. If you select one or more lines, the closest endpoint(s) are moved to the new change point. If you select a circle, change point allows you to change the radius of the circle. If you selected more than one circle, AutoCAD moves on to the next circle and repeats the prompt. If the selected object is a text string, then AutoCAD allows you to change one or more parameters such as text height, text style, text rotation angle and text string. And if the selected object is a block, then specifying a new location repositions the block.

Instead of changing a point, you can change the properties of selected objects by choosing the Properties option. The properties that can be changed include color, lineweight, linetype, lineweight, layer, elevation, thickness, and plotstyle.

GROUPING OBJECTS

The GROUP command adds flexibility in modifying a group of objects. It allows you to name a selection set. Naming a selection set combines two powerful AutoCAD drawing features. One is being able to modify a group of unrelated objects as a group. It is similar to using Previous to select the last selection set when prompted to Select object: for a modify command. The advantage of using GROUP instead of Previous is that you are not restricted to only the last selection set. The other feature combined in the GROUP command is that of giving a name to a selected group of objects for recall later by the name of the group. This is similar to the BLOCK command. The advantage of using GROUP instead of BLOCK is that the GROUP command's "selectable" switch can be set to OFF for modifying an individual member without losing its "membership" in the group. Also, named groups, like blocks, are saved with the drawing.

A named group can be selected for modifying as a group only when its "selectable" switch is set to ON. Figure 6–51 shows the result of trimming an object with the selectable switch set to ON and to OFF. Modifying objects (such as with the MOVE or COPY command) that belong to a group can be selected by two methods. One is to select one of its members. The other method is to select the Group option by typing g at the Select Objects: prompt. AutoCAD prompts for the group's name. Enter the group name and press ENTER or the SPACEBAR. AutoCAD highlights the objects that belong to the selected group.

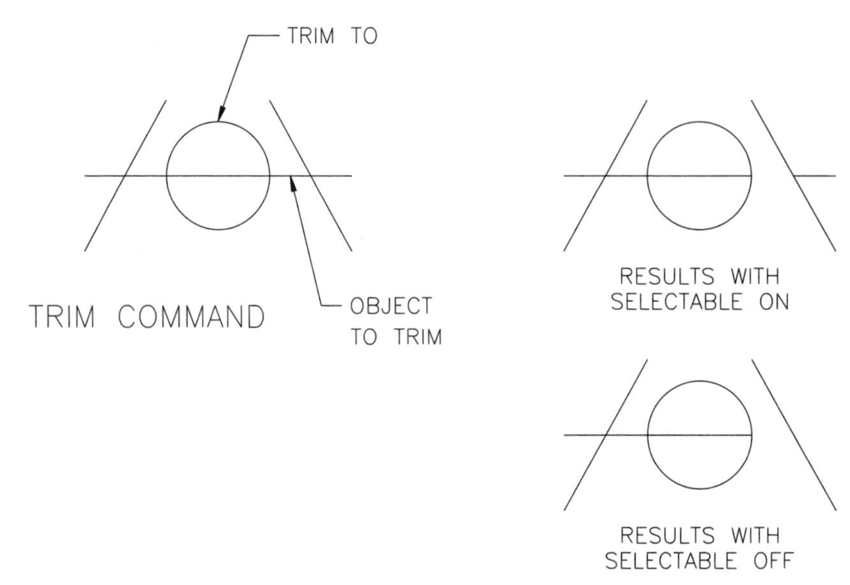

Figure 6–51 *Trimming an object with the selectable switch set to ON and to OFF*

To create a new group or edit an existing group, invoke the GROUP command:

Command: prompt	**group** (ENTER)

AutoCAD displays the Object Grouping dialog box, similar to Figure 6–52.

The dialog box is divided into four sections:

Group Name list box

Group Identification

Create Group

Change Group

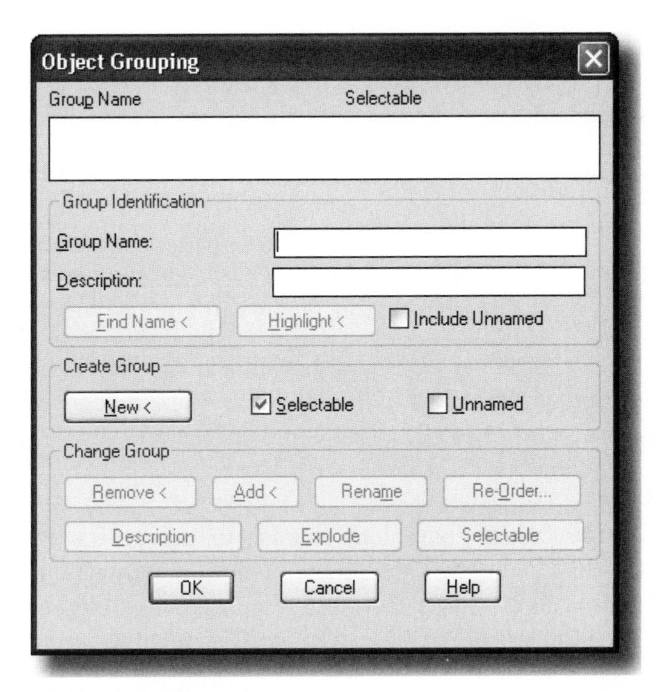

Figure 6–52 *Object Grouping dialog box*

GROUP NAME LIST BOX

The **Group Name** list box lists the names of the existing groups defined in the current drawing. The Selectable column indicates whether a group is selectable. If it is listed as selectable, then selecting a single group member selects all the members except those on locked layers. If it is listed as unselectable, then selecting a single group member selects only that object.

GROUP IDENTIFICATION

The **Group Identification** section is where AutoCAD displays the group name and a description when a group is selected in the **Group Name** list.

The **Find Name<** button lists the groups to which an object belongs. AutoCAD prompts for the selection of an object and displays the Group Member List dialog box, which lists the group or groups to which the selected object belongs.

The **Highlight<** button lets you see the members of the selected group from the **Group Name** list box.

The **Include Unnamed** toggle box controls the listing of the unnamed groups in the **Group Name** list box.

CREATE GROUP

The **Create Group** section is used for creating a new group with or without a group name. In addition, you can set whether or not it is initially selectable.

To create a new group, enter the group name and description in the **Group Name:** and **Description:** text boxes. Group names can be up to 31 characters long and can include letters, numbers, and the special characters $ and _. To create an unnamed group, set the **Unnamed** toggle box to ON. AutoCAD assigns a default name, *An, to unnamed groups. The n represents a number that increases with each new group.

Set the **Selectable** box to ON or OFF, and then choose the **New<** button. AutoCAD prompts for the selection of objects. Select all the objects to be included in the new group, and press ENTER or the SPACEBAR to complete the selection.

CHANGE GROUP

The **Change Group** section is for making changes to individual members of a group or to the group itself. The buttons are disabled until a group name is selected in the **Group Name:** list box.

The **Remove<** button lets you remove selected objects from the selected group. To remove objects from the selected group, choose the **Remove<** button; AutoCAD prompts:

Remove objects: *(select objects that are to be removed from the selected group and press* ENTER *or the* SPACEBAR*)*

AutoCAD redisplays the Object Grouping dialog box.

The **Add<** button allows you to add the selected objects to the selected group. To add objects from the selected group, choose the **Add<** button; AutoCAD prompts:

Select objects: *(select objects that are to be added to the selected group and press* ENTER *or the* SPACEBAR*)*

AutoCAD redisplays the Object Grouping dialog box.

The **Rename** button allows you to change the name of the selected group to the name entered in the **Group Name:** text box in the **Group Identification** section.

The **Re-order...** option allows you to change the numerical order of objects within the selected group. Initially, the objects are numbered in the order in which they were selected to form a group. Reordering is useful when creating tool paths. Choose the **Re-order...** button; AutoCAD displays the Order Group dialog box shown in Figure 6–53.

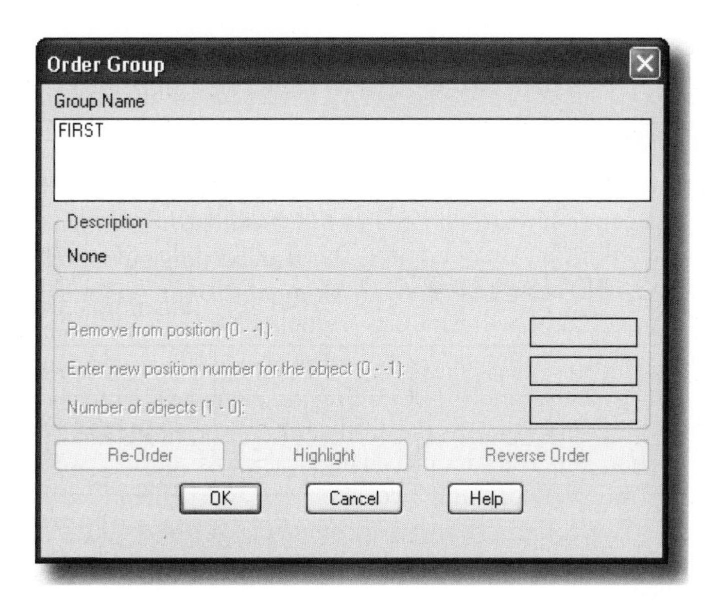

Figure 6–53 *Order Group dialog box*

The **Group Name** list box gives the names of the groups defined in the current drawing. Members of a group are numbered sequentially starting with 0 (zero).

Remove from position (0 – *n*): identifies the position number of an object.

Replace at position (0 – *n*): identifies the new position number of the object.

Number of objects (1 – *n*): identifies the number/range of objects to reorder.

The **Re-Order** and **Reverse Order** buttons allow you to change the numerical order of objects as specified and reverses the order of all members, respectively.

The **Highlight** button allows AutoCAD to display the members of the selected group in the graphics area.

The **Explode** button in the **Change Group** section of the Object Grouping dialog box deletes the selected group from the current drawing. Thus, the group no longer exists as a group. The members remain in the drawing and in any other group(s) of which they are members.

The **Description** button allows you to change the description of the selected group.

The **Selectable** button controls whether the group is selectable.

After making the necessary changes to the Object Grouping dialog box, choose the **OK** button to keep the changes and close the dialog box.

INFORMATION ABOUT OBJECTS

AutoCAD provides several commands for displaying useful information about the objects in the drawing. These commands do not create anything, nor do they modify or have any effect on the drawing or objects therein. The only effect on the AutoCAD editor is that on single-screen systems, the screen switches to the AutoCAD Text window (not to be confused with the TEXT command) and the information requested by the particular inquiry command is then displayed on the screen. If you are new to AutoCAD, it is helpful to know the FLIP SCREEN feature that returns you to the graphics screen so you can continue with your drawing. On most systems this is accomplished with the F2 function key. You can also change back and forth between graphic and text screens with the GRAPHSCR and TEXTSCR commands, entered at the Command: prompt, respectively. The inquiry commands include LIST, AREA, ID, DBLIST, and DIST.

LIST COMMAND

The LIST command displays information about individual objects stored by AutoCAD in the drawing database. The information includes:

The location, layer, object type, and space (model or paper) of the selected object as well as the color, lineweight, and linetype if not set to BYLAYER or BYBLOCK.

The distance in the main axes between the endpoints of a line, that is, the delta-*X*, delta-*Y*, and delta-*Z*.

The area and circumference of a circle or the area of a closed polyline.

The insertion point, height, angle of rotation, style, font, obliquing angle, width factor, and actual character string of a selected text object.

The object handle, reported in hexadecimal.

Invoke the LIST command:

Inquiry toolbar	Choose the LIST command (see Figure 6–54)
Tools menu	Choose Inquiry > List
Command: prompt	**list** (ENTER)

Figure 6–54 *Invoking the LIST command from the Inquiry toolbar*

AutoCAD prompts:

Command: **list** (ENTER)
Select objects: *(select the objects and press* ENTER *to terminate object selection)*

AutoCAD lists the information about the selected objects.

DBLIST COMMAND

The DBLIST command lists the data about all of the objects in the drawing. It can take a long time to scroll through all the data in a large drawing. DBLIST can, like other commands, be terminated by canceling by pressing ESC key.

Invoke the DBLIST command:

Command: prompt	**dblist** (ENTER)

AutoCAD lists the information about all the objects in the drawing.

AREA COMMAND

The AREA command is used to report the area (in square units) and perimeter of a selected closed geometric figure on the screen, such as a circle, polygon, or closed polyline. You may also specify a series of points that AutoCAD considers a closed polygon, compute, and report the area.

Invoke the AREA command:

Inquiry toolbar	Choose the AREA command (see Figure 6–55)
Tools menu	Choose Inquiry > Area
Command: prompt	**area** (ENTER)

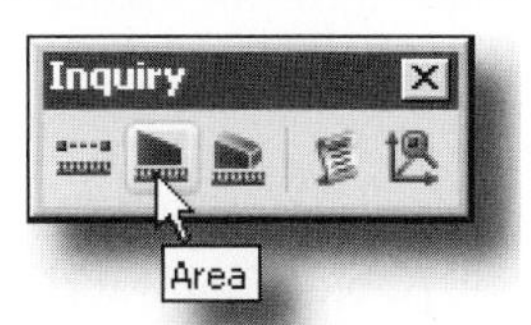

Figure 6–55 *Invoking the* AREA *command from the Inquiry toolbar*

AutoCAD prompts:

Command: **area** (ENTER)
Specify first corner point or [Object/Add/Subtract]: *(specify a point or select one of the available options from the shortcut menu)*

The default option calculates the area when you select the vertices of the objects. If you want to know the area of a specific object such as a circle, polygon, or closed polyline, select the Object option.

The following command sequence is an example of finding the area of a polygon using the Object option, as shown in Figure 6–56.

Command: **area** (ENTER)
Specify first corner point or [Object/Add/Subtract]: **o** (ENTER)
Select objects: *(select an object, as shown in Figure 6–56)*
Area = 12.21 Perimeter = 13.79

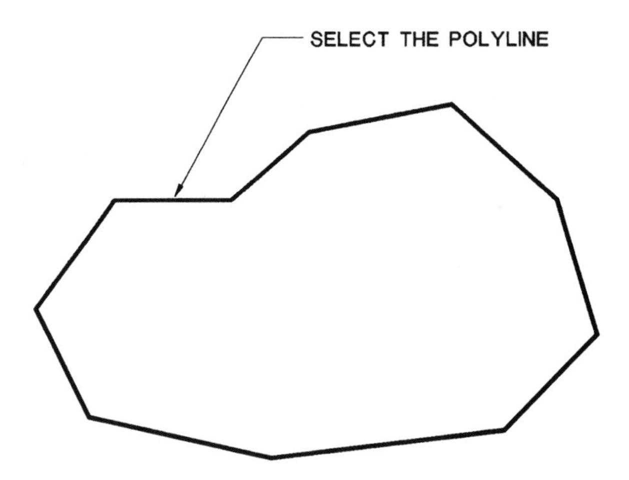

Figure 6–56 *Finding the area of a polygon using the* AREA *command Object option*

The Add option allows you to add selected objects to form a total area; then you can use the Subtract option to remove selected objects from the running total.

The following example demonstrates the application of the Add and Subtract options. In this example, the area is determined for the closed shape after subtracting the area of the four circles, as shown in Figure 6–57.

Command: **area** (ENTER)
Specify first corner point or [Object/Add/Subtract]: **a** (ENTER)
Specify first corner point or [Object/Subtract]: **o** (ENTER)
(ADD mode) Select objects: *(select the polyline, as shown in Figure 6–57)*
Area = 12.9096, Perimeter = 15.1486
Total area = 12.9096
(ADD mode) Select objects: (ENTER)
Specify first corner point or [Object/Add/Subtract]: **s** (ENTER)
Specify first corner point or [Object/Add]: **o** (ENTER)
(SUBTRACT mode) Select objects: *(select circle A, as shown in Figure 6–57)*
Area = 0.7125, Circumference = 2.9992
Total area = 12.1971
(SUBTRACT mode) Select objects: *(select circle B, as shown in Figure 6–57)*
Area = 0.5452, Circumference = 2.6175

Total area = 11.6179
(SUBTRACT mode) Select objects: *(select circle C, as shown in Figure 6–57)*
Area = 0.7125, Circumference = 2.9922
Total area = 10.9394
(SUBTRACT mode) Select objects:*(select circle D, as shown in Figure 6–57)*
Area = 0.5452, Circumference = 2.6175
Total area = 10.3942
(SUBTRACT mode) Select objects: (ENTER)
Specify first corner point or [Object/Add]: (ENTER)

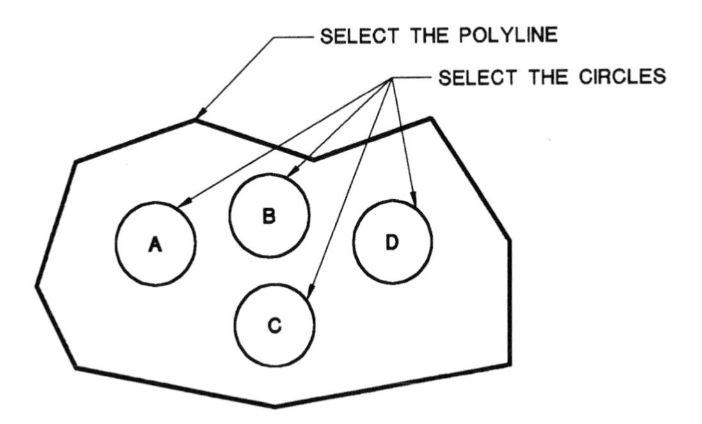

Figure 6–57 *Using the Add and Subtract options of the* AREA *command*

ID COMMAND

The ID command is used to obtain the coordinates of a selected point. If you do not use an Object Snap mode to select a point that is not in the current construction plane, AutoCAD assigns the current elevation as the *Z* coordinate of the point selected.

Invoke the ID command:

Inquiry toolbar	Choose the Locate Point command (see Figure 6–58)
Tools menu	Choose Inquiry > ID Point
Command: prompt	**id** (ENTER)

Figure 6–58 *Invoking the* LOCATE POINT *command from the Inquiry toolbar*

AutoCAD prompts:

Command: **id** (ENTER)
Specify point: *(select a point)*

AutoCAD displays the information about the selected point.

If the BLIPMODE system variable is set to ON (the default), a blip appears on the screen at the specified point, provided it is in the viewing area.

DIST COMMAND

The DIST command prints out the distance, in the current units, between two points, either selected on the screen or keyed in from the keyboard. Included in the report are the horizontal and vertical distances (delta-X and delta-Y, respectively) between the points and the angles in and from the *XY* plane.

Invoke the DIST command:

Inquiry toolbar	Choose the DISTANCE command (see Figure 6–59)
Tools menu	Choose Inquiry > Distance
Command: prompt	**dist** (ENTER)

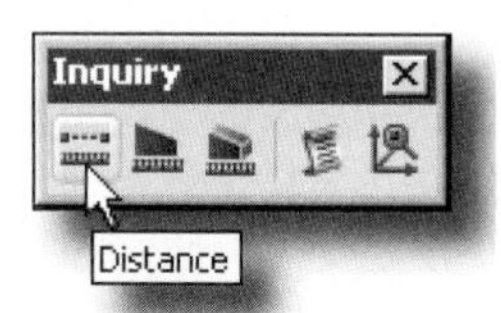

Figure 6–59 *Invoking the DIST command from the Inquiry toolbar*

AutoCAD prompts:

Command: **dist** (ENTER)
Specify first point: *(specify the first point to measure from)*
Specify second point: *(specify the endpoint to measure to)*

AutoCAD displays the distance between two selected points.

SYSTEM VARIABLES

AutoCAD stores the settings (or values) for its operating environment and some of its commands in system variables. Each system variable has an associated type: integer (for switching), integer (for numerical value), real, point, or text string. Unless they are read-only, you can examine and change these variables at the Command: prompt by typing the name of the system variable, or you can change them by means of the SETVAR command.

Integers (for Switching) System variables that have limited nonnumerical settings can be switched by setting them to the appropriate integer value. For example, the snap can be either ON or OFF. The purpose of the SNAPMODE system variable is to turn the snap ON or OFF by using the AutoCAD SETVAR command or the AutoLISP (setvar) function.

Turning the snap ON or OFF is demonstrated in the following example by changing the value of its SNAPMODE system variable. First, its current value is set to "0", which is OFF.

Command: **setvar** (ENTER)
Enter variable name or [?]: **snapmode** (ENTER)
Enter new value for SNAPMODE <0>: **1** (ENTER)

This sequence may seem unnecessary, because the snap mode is so easily toggled with a press of a function key. Changing the snap with the SETVAR command is inconvenient, but doing so does allow you to view the results immediately.

For any system variable whose status is associated with an integer, the method of changing the status is just like the preceding example. In the case of SNAPMODE, "0" turns it OFF and "1" turns it ON. In a similar manner, you can use SNAPISOPAIR to switch one isoplane to another by setting the system variable to one of three integers: 0 is the left isoplane, 1 is the top, and 2 is the right isoplane.

It should be noted that the settings for the Osnap system variable named OSMODE are members of the binomial sequence. The integers are 1, 2, 4, . . . , 512, 1024, 2048, 4096, 8192. See Table 6–1 for the meaning of OSMODE values. While the settings are switches, they are more than just ON and OFF. There may be several Object Snap modes active at one time. It is important to note that the value of an integer (switching) has nothing to do with its numerical value.

Table 6–1 Values for the OSMODE System Variable

NONe	0
ENDpoint	1
MIDpoint	2
CENter	4
NODe	8
QUAdrant	16
INTersection	32
INSertion	64
PERpendicular	128
TANgent	256

Table 6–1 Values for the OSMODE System Variable (continued)

NEArest	512
QUIck	1024
APP int	2048
EXTension	4096
PARallel	8192

Integers (for Numerical Value) System variables such as APERTURE and AUPREC are changed by using an integer whose value is applied numerically in some way to the setting, rather than just as a switch. For instance, the size of the aperture (the target box that appears for selecting Osnap points) is set in pixels (picture elements) according to the integer value entered in the SETVAR command. For example, setting the value of APERTURE to 9 should render a target box that is three times larger than setting it to 3.

AUPREC is the variable that sets the precision of the angular units in decimal places. The value of the setting is the number of decimal places; therefore, it is considered a numerical integer setting.

Real System variables that have a real number for a setting, such as VIEWSIZE, are called real.

Point (X Coordinate,Y Coordinate) LIMMIN, LIMMAX, and VIEWCTR are examples of system variables whose settings are points in the form of the *X* coordinate and *Y* coordinate.

Point (Distance,Distance) Some system variables, whose type is point, are primarily for setting spaces rather than a particular point in the coordinate system. For instance, the SNAPUNIT system variable, though called a point type, uses its X and Y distances from (0,0) to establish the snap X and Y resolution, respectively.

String These variables have names like CLAYER, for the current layer name, and DWG-NAME, for the drawing name.

Open the Exercise Manual PDF file for Chapter 6 on the accompanying CD for project and discipline specific exercises.

If you have the accompanying Exercise Manual, refer to Chapter 6 for project and discipline specific exercises.

REVIEW QUESTIONS

1. The elements of a multiline can have different colors.

   a. True

   b. False

2. A SPLINE object:

   a. does not actually pass through the control points

   b. is always shown as a continuous line

   c. requires you to specify tangent information for each control point

   d. requires you to specify tangent information for the first and last points only

3. When spell-checking a drawing, AutoCAD automatically checks the entire drawing.

   a. True

   b. False

4. To use incline lettering so that the vertical portions of the characters point to 2 o'clock for a horizontal line of text, the obliquing angle should be:

   a. –60

   b. –30

   c. 0

   d. 30

   e. 60

5. Once objects are grouped together, they can be ungrouped by:

   a. using the EXPLODE command

   b. using the UNGROUP command

   c. using the GROUPcommand

   d. they cannot be ungrouped

6. To insert the diameter symbol into a string of text, you should enter:

   a. %%d

   b. %%c

   c. %%phi

   d. %%dia

7. It is possible to force all text on a drawing to display as an open rectangle in order to speed up the redisplay of the drawing by using the:

   a. TEXT command

   b. RTEXT command

   c. QTEXT command

   d. DTEXT command

8. In order to select a different font for use in the TEXT command, a text style must be created.

   a. True

   b. False

9. The grips dialog box allows you to change all of the following, except:

   a. grip size

   b. grip color

   c. toggle grips system variable ON/OFF

   d. specify the location of the grips

10. Grips do not allow you to _______ an object.

   a. trim

   b. erase

   c. mirror

   d. move

   e. stretch

11. To edit a word in the middle of a string of text, a reasonable command to use would be:

   a. DDEDIT

   b. CHPROP

   c. DDCHPROP

   d. MODIFY

   e. CHANGE

12. Multiline styles can be saved to an external file, thus allowing their use in multiple drawings.

    a. True   b. False

13. The ID command will:

    a. display the serial number of the AutoCAD program

    b. display the X, Y, and Z coordinates of a selected point

    c. allow you to password protect a drawing

    d. none of the above

14. A quick way to find the length of a line is to use what command?

    a. LIST   d. LENGTH

    b. DISTANCE   e. COORDINATES

    c. DIMENSION

15. To obtain a full listing of all the objects contained in the current drawing, you should use:

    a. LIST   d. DBDUMP

    b. DBLIST   e. DUMP

    c. LISTALL

16. When filleting multilines, you must specify:

    a. the radius of the innermost fillet

    b. the radius of the center line

    c. the radius of the outermost fillet

    d. multilines cannot be filleted

17. Which of the following are not valid options when creating a text style?

    a. width factor

    b. upside down

    c. vertical

    d. backwards

    e. none of the above (all answers are valid)

18. You can change the setting of a system variable by using the command:

    a. SYSVAR   d. VARSYS

    b. SETVAR   e. none of the above

    c. VARSET

19. Which of the following is not a valid type of system variable?
    a. real
    b. integer
    c. point
    d. string
    e. double

20. To locate text such that it is centered exactly within a circle, a reasonable justification would be:
    a. Center
    b. Middle
    c. Full
    d. Both A and B
    e. none, simply zoom in and approximate it

Directions: Use the following X and Y locations to construct figure Qx6–1. Your completed drawing will be used to answer the remaining questions.

Point	X value	Y value
A	1.5000	1.0000
B	2.1717	1.0000
C	2.7074	1.7500
D	4.5000	1.7500
E	4.8200	1.0000
F	5.5000	1.0000
G	4.1838	4.0711
H	3.5071	3.8100
J	3.7974	3.3861
K	3.6104	3.3139
L	2.8125	2.2500
M	4.2843	2.2500

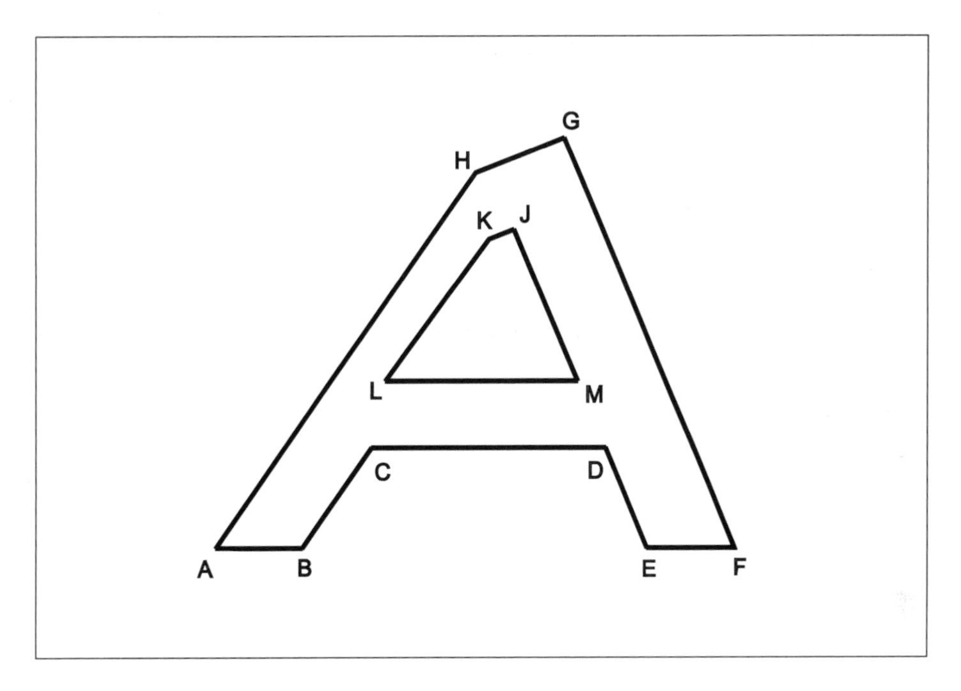

Figure Qx6–1

21. Using the appropriate AutoCAD command what is the length of Line AH?
    a. 3.5432
    b. 3.4523
    c. 3.4532
22. Using the appropriate AutoCAD command what is the angle of Line AH?
    a. 324 degrees
    b. 234 degrees
    c. 432 degrees
23. Using the appropriate AutoCAD command what is the length of Line CD?
    a. 1.7962
    b. 1.7926
    c. 1.7629

24. Using the appropriate AutoCAD command what is the length of Line FG?
    a. 3.3143
    b. 3.3314
    c. 3.3413
25. Using the appropriate AutoCAD command what is the angle of Line FG?
    a. 131 degrees
    b. 311 degrees
    c. 113 degrees
26. Using the appropriate AutoCAD command what is the length of Line GH?
    a. 0.7253
    b. 0.7532
    c. 0.7325
27. Using the appropriate AutoCAD command what is the angle of Line GH?
    a. 210 degrees
    b. 102 degrees
    c. 201 degrees
28. What is the AREA of shape JKLM?
    a. .8833
    b. .8383
    c. .8838
29. What is the perimeter of shape JKLM?
    a. 4.1281
    b. 4.1812
    c. 4.1182
30. What is the linear distance from point E to point J?
    a. 2.2589
    b. 2.5859
    c. 2.5958

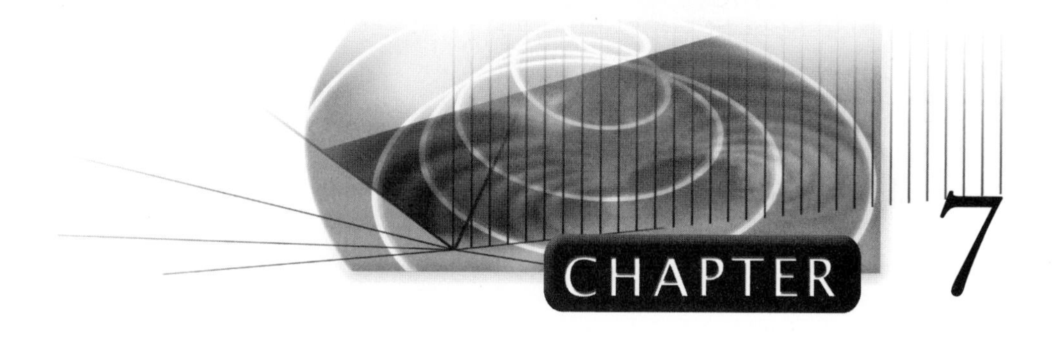

Dimensioning

INTRODUCTION

AutoCAD provides a full range of dimensioning commands and utilities to enable the drafter to comply with the conventions of most disciplines, including architecture and civil, electrical, and mechanical engineering, among many others.

After completing this chapter, you will be able to do the following:

- Draw linear dimensioning
- Draw aligned dimensioning
- Draw angular dimensioning
- Draw diameter and radius dimensioning
- Draw leaders with annotation and geometric tolerance
- Draw ordinate dimensioning
- Draw baseline and continue dimensioning
- Edit dimension text
- Create and modify dimension styles

AutoCAD makes drawing dimensions easy. For example, the width of the rectangle shown in Figure 7–1 can be dimensioned by specifying the two endpoints of the top corners (using an OSNAP mode such as Indpoint or Intersection) and then specifying a point to determine the location of the Dimension Line. AutoCAD drags a preview image of the dimension to indicate how it will look while you move the cursor to specify the location of the Dimension Line.

Once Linear Dimensioning is mastered, other types of dimensions, such as Diameter, Radius, Angular, Baseline, and Ordinate Dimensioning, can be drawn quickly and accurately.

The dimension types available are Linear, Angular, Diameter, Radius, and Ordinate. There are primary and secondary commands available for each of these types. There are also other general utility, editing, and style-related commands and subcommands that assist you in drawing the correct dimensions quickly and with accuracy.

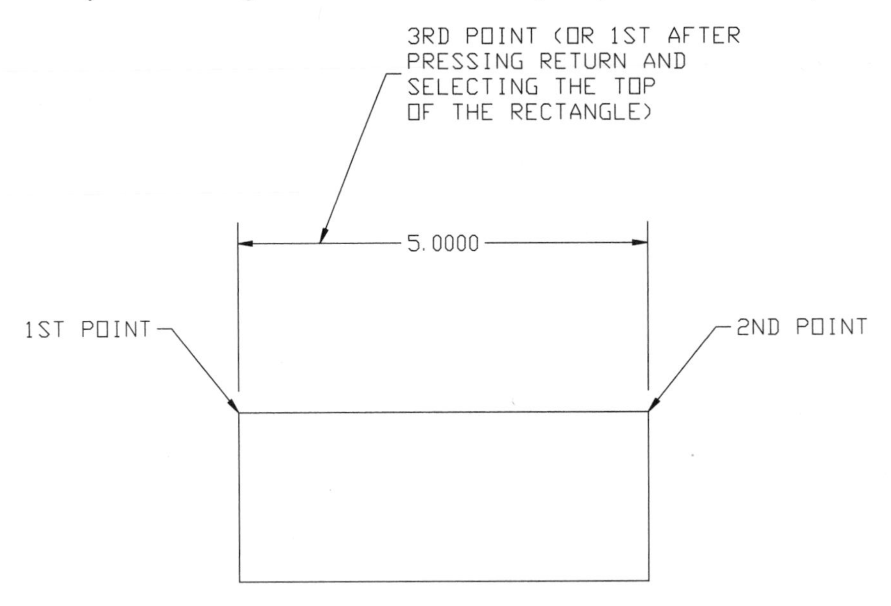

Figure 7–1 *An example of linear dimensioning*

The Linear Dimensioning command options include Horizontal, Vertical, Aligned, and Rotated. Angular dimensioning is covered by the ANGULAR command. Diameter, radius, and ordinate dimensioning are covered by the DIAMETER, RADIUS, and ORDINATE commands, respectively.

Approximately 60 Dimensioning System Variables are available specifically for dimensioning. Most of these have names that begin with "DIM." They are used for such purposes as determining the size of the gap between the extension line and the point specified on an object or whether one or both of the extension lines will be drawn or suppressed. It is combinations of these variable settings that can be named and saved as Dimension Styles and later recalled for applying when needed. See Appendix C for the listing of available Dimensioning System Variables.

Dimension variables change when their associated settings are changed in the Modify Dimension Style dialog box. For example, the value of the DIMEXO (extension line offset) dimensioning system variable is established by the number in the **Offset from origin** text box in the Modify Dimension Style dialog box. See the section later in this chapter on "Dimension Styles."

Dimension utilities include Override, Center, Leader, Baseline, Continue, and Feature Control Frames for adding tolerancing information.

The dimension editing command options include Hometext, Newtext, Oblique, Tedit, and Trotate.

DIMENSION TERMINOLOGY

Following are the terms for the different parts that make up dimensions in AutoCAD.

Figure 7–2 shows the different components of a typical dimension.

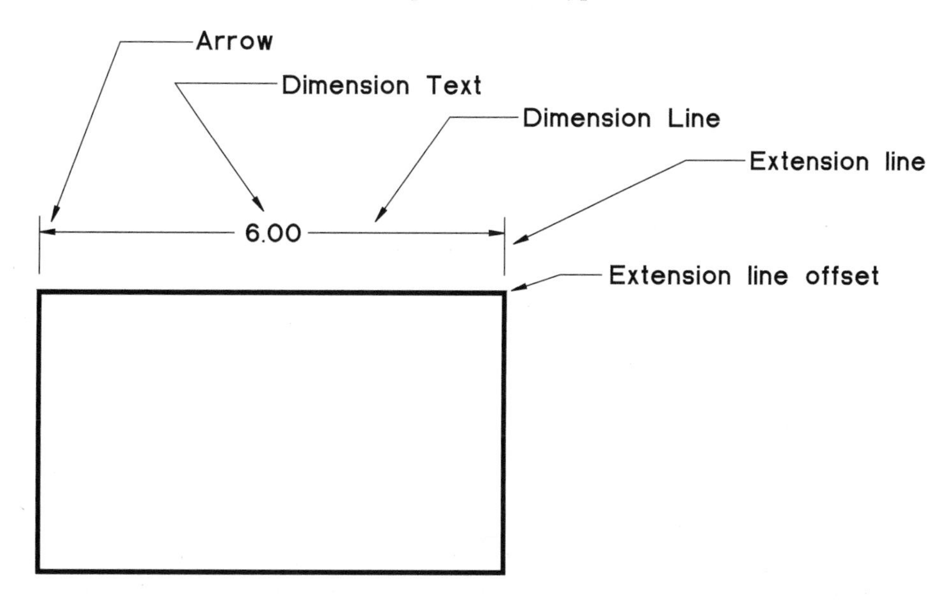

Figure 7–2 *The different components of a typical dimension*

DIMENSION LINE

The Dimension Line is offset from the measured feature. The Dimension Line (sometimes drawn as two segments outside the Extension Lines if a single line with its related text will not fit between the extension lines) indicates the direction and length of the measured distance. It is normally offset for clarity. If the dimension is measured between parallel lines of one or two objects, then it may not be offset, but is drawn on the object or between two objects. Dimension Lines are usually terminated with markers such as arrows or ticks (short slanted lines). Angular Dimension Lines become arcs whose centers are at the vertex of the angle.

ARROWHEAD

The arrowhead is a mark at the end of the Dimension Line to indicate its termination. Shapes other than arrows are used in some styles.

EXTENSION LINE

When the Dimension Line is offset from the measured feature, the Extension Lines (sometimes referred to as witness lines) indicate such offset. Unless you have invoked the Oblique option, the Extension Lines will be perpendicular to the direction of the measurement.

DIMENSION TEXT

The Dimension Text consists of numbers, words, characters, and symbols used to indicate the measured value and type of dimension. Unless the text has been altered in the standard Dimension Style, the number/symbol format is decimal. This conforms to the same linear and angular units as the default settings of the units. The text style usually conforms to that of the current text style.

LEADER

The leader is a radial line used to point from the Dimension Text to the circle or arc whose diameter or radius is being dimensioned. A leader can also be used for general annotation.

CENTER MARK

The center mark is made up of lines or a series of lines that cross in the center of a circle for the purpose of marking its center.

ASSOCIATIVE/NON-ASSOCIATIVE/EXPLODED DIMENSIONS

Dimensions in AutoCAD can be drawn as associative, non-associative, or exploded, depending on the setting of the DIMASSOC dimensioning system variable. The setting for associative dimensions is 2 (default), non-associative is 1, and exploded dimensions is 0.

ASSOCIATIVE

An associative dimension becomes associated with an object by selecting points on the object (using an Object Snap mode) when prompted to do so during a dimensioning command. If the object is subsequently modified in a manner that changes the location(s) of one or both of the selected points, the associated dimension is automatically updated to correctly indicate the new distance or angle. Associative dimensioning does not support multilines. The associativity between a dimension and a block is lost when the block is redefined.

An associative dimension drawn with the DIMASSOC dimensioning system variable set to 2 has all of its separate parts become members of a single object. Therefore, if any one of its members is selected for modifying, all members are highlighted and subject to being modified. This is similar to the manner in which member objects of a block reference are treated. In addition to the customary visible parts, AutoCAD draws point objects at the ends, where the measurement actually occurs on the object. If you have dimensioned the width of a rectangle with an associative dimension and then specified one end of the rectangle to stretch, the dimension will also be stretched, and the Dimension Text will be changed to correspond to the new measurement.

The associativity of associative dimensions created in AutoCAD 2004 is usually maintained when saved to a previous release and then reopened in the current release. However, if you modify dimensioned objects (forming new objects) using an earlier version of AutoCAD, the dimension associations change when the drawing is loaded into the current release. For example, if a circle that was dimensioned is broken so that portions of the circle are removed, two arc objects result and the associated dimension applies to only one of the arc objects.

NON-ASSOCIATIVE

The non-associative dimensions are drawn while the DIMASSOC dimensioning system variable is set to 1. The separate parts of the non-associative dimension are, like those of the associative dimensions, considered members of a single object when any of them are selected for modifying. However, if the object that was dimensioned is selected for modifying without selecting the dimension, the dimension itself will remain unchanged.

EXPLODED

Exploded dimensions are drawn while the DIMASSOC dimensioning system variable is set to 0, and the members are drawn as separate objects. If one of the components of the dimension is selected for modifying, that component will be the only one modified.

Note: An associative dimension can be converted to an exploded dimension with the EXPLODE command. Once the dimension is exploded, you cannot recombine the separate parts back into the associative dimension from which they were exploded (except by means of the UNDO command if feasible). Note that when you explode an associative dimension, the measurement-determining points (nodes) remain in the drawing as point objects.

DIMENSIONING COMMANDS

In AutoCAD at least five ways to invoke dimensioning commands (DIMALIGNED in the following examples) are available:

1. At the "Command:" prompt, enter **dimaligned**.
2. From the Dimension toolbar (see Figure 7–3), choose the dimensioning command button labeled Aligned Dimension.
3. From the Dimension menu (see Figure 7–4), choose **Aligned**.
4. At the "Command:" prompt, enter **dim.** From the Dimensioning mode enter **ali.**
5. At the "Command:" prompt, enter **dim1.** From the Dimensioning mode enter **ali.**

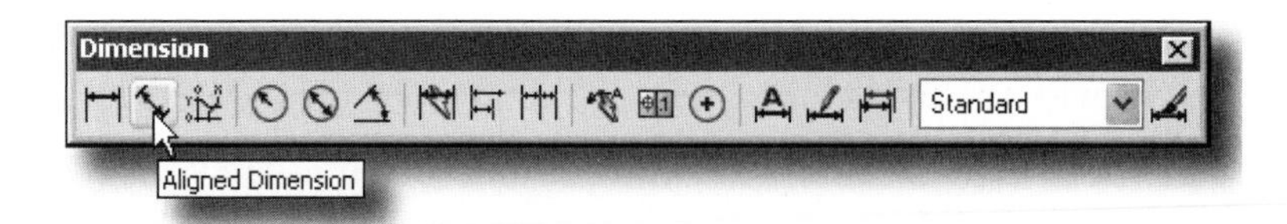

Figure 7–3 *Choosing the* ALIGNED DIMENSION *command from the Dimension toolbar*

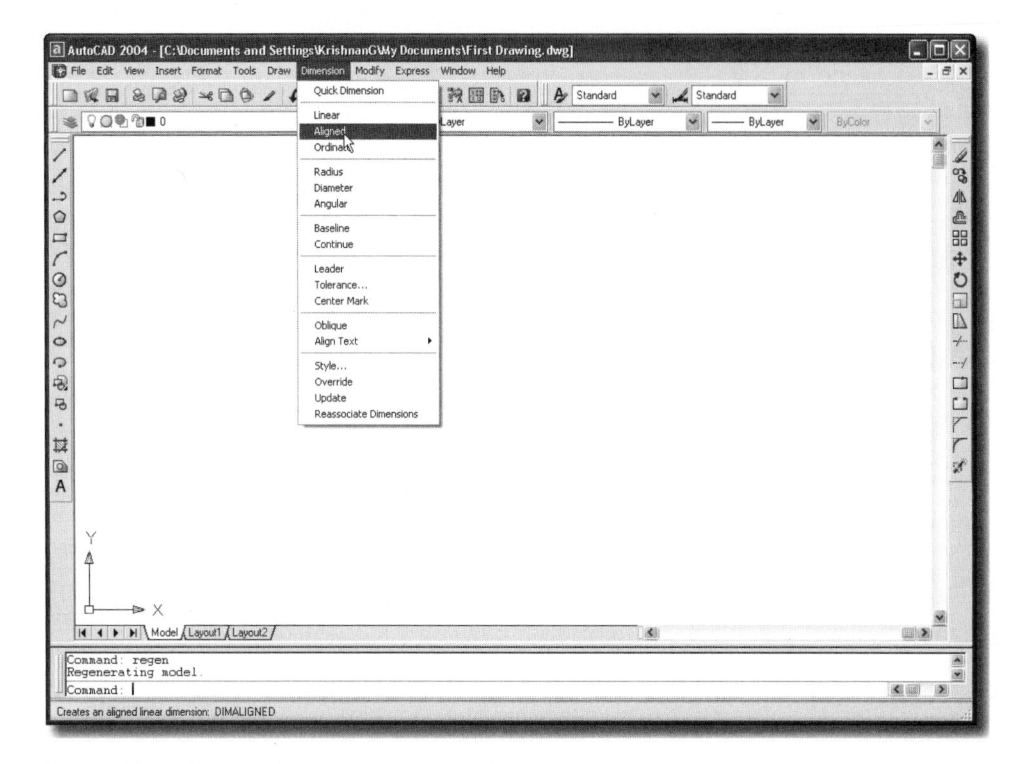

Figure 7–4 *Choosing the* ALIGNED *command from the Dimension menu*

DIM AND DIM1

Entering **dim** or **dim1** at the "Command:" prompt causes AutoCAD to change to the Dimensioning mode. AutoCAD prompts:

Dim: *(enter an abbreviated dimension command)*

Note: Dimensioning commands entered at the "Command:" prompt differ from dimensioning commands entered at the "Dim:" prompt (in Dimensioning mode) as explained in the following paragraph and illustrated in Appendix B.

Invoking a dimensioning command from the "Dim:" prompt requires entering the name of the command without the "dim" prefix. For example, **dimradius** is entered as **radius** and can even be shortened to **rad.** However, **dimlinear** cannot be entered as **linear.** You must enter either **horizontal** or **vertical** (**hor** or **ver**) and will not be able to change dynamically between the two as when entered as **dimlinear** at the "Command:" prompt or invoked from the Dimension toolbar or Dimension menu. Other "Command:" prompt dimensioning commands and their corresponding "Dim:" prompt commands are shown in Appendix B.

The **dim** command changes to the Dimensioning mode, and AutoCAD continues in the Dimensioning mode after a dimensioning command is utilized, repeating the "Dim:" prompt. You can exit the Dimensioning mode by invoking another (non-dimensioning) command or you can return to the "Command:" prompt by pressing ESC.

The **dim1** command also changes to the Dimensioning mode, but AutoCAD returns to the "Command:" prompt after a dimensioning command is utilized, or you can return to the "Command:" prompt by pressing ESC without having utilized a dimensioning command.

Dimensioning from the Dimensioning mode is not a common method. It does offer a difference that, however minor, might be useful in special circumstances and would be worth practicing in case the difference will make certain dimensions easier to draw. Because dimensioning from the Dimensioning mode is not commonly used, explanations and examples of dimensioning commands in this chapter will be only for dimensioning commands entered from the "Command:" prompt or selected from the Dimension toolbar or Dimension menu. The fundamentals will be the same when dimensioning commands are invoked from the "Dim:" prompt.

LINEAR DIMENSIONING

Invoke the LINEAR DIMENSION command:

Dimension toolbar	Choose the Linear Dimension command (see Figure 7–5)
Dimension menu	Choose Linear
Command: prompt	**dimlinear** (ENTER)

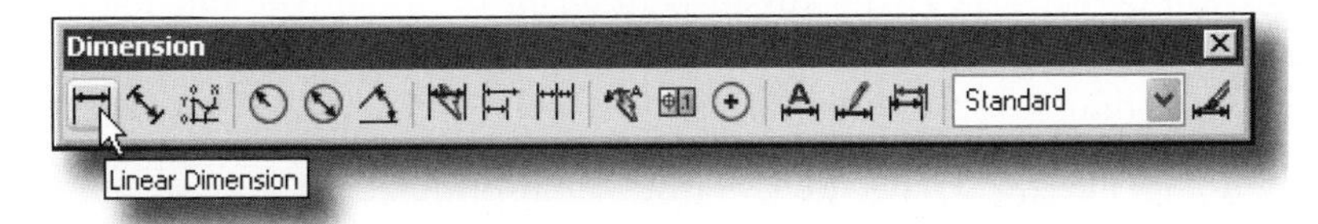

Figure 7–5 *Invoking the* DIMINEAR *command from the Dimension toolbar*

AutoCAD prompts:

Command: **dimlinear** (ENTER)
Specify first extension line origin or <select object>: *(specify a point)*

AutoCAD uses it as the start point (origin) for the first extension line. This point can be the endpoint of a line, the intersection of objects, the center point of a circle, or even the insertion point of a text object. You can specify a point on the object itself. AutoCAD provides a gap between the object and the extension line that is equal to the value of the DIMEXO dimensioning system variable, which you can change at any time. After you specify the start point (origin), AutoCAD prompts:

Specify second extension line origin: *(specify a point at which the second extension line should start)*

If neither of the points was specified by using an OSNAP mode or selecting a object, then AutoCAD prompts:

Non-associative dimension created.

Dynamic Horizontal/Vertical Dimensioning

Dynamic horizontal/vertical dimensioning is an option after you have specified two points in response to the DIMLINEAR command. If you specify two points on the same horizontal line, moving the cursor above or below the line causes the preview image of the dimension to appear. AutoCAD assumes you wish to draw a horizontal dimension. Likewise, AutoCAD assumes a vertical dimension if the specified points are on the same vertical line.

Dynamic Drag (switching between horizontal and vertical) is more applicable when the two points specified are not on the same horizontal or vertical line. That is, they can be considered diagonally opposite corners of an imaginary rectangle with both width and height. After the two points are specified, you are prompted to specify the location of the dimension line. You will also be shown a preview image of where the dimension will be located, by where the cursor is located relative to the imaginary rectangle formed by the two points. If the cursor is above the top line or below the bottom line of the rectangle, then the dimension will be horizontal. If the cursor is to the right of the right side or to the left of the left side of the rectangle, then the dimension will be vertical. If the cursor is dragged to one of the outside quadrants or inside of the rectangle, it will maintain the type of dimension in effect before the cursor was moved.

After two points have been specified, AutoCAD prompts:

Specify dimension line location or [Mtext/Text/Angle/Horizontal/Vertical/Rotated]: *(specify location for dimension line or right-click for the shortcut menu and select one of the available options)*

AutoCAD displays the value of the dimension:

Dimension text = *(measured dimension)*

Drawing The Dimension Line

After specifying the two points (or using the right-click shortcut menu to change the dimension text or type), specify a point through which the dimension line is to be drawn. AutoCAD draws the line on which the dimension text is drawn. If there is enough room between the extension lines, the dimension text will be centered in line with or above this line. If the dimension text is to be in line with the dimension line, then the dimension line will be broken to allow room for the text. However, if the dimension line, arrows, and text do not fit between the extension lines, they are drawn outside. The text will be drawn near the second extension line.

Changing Dimension Text

The Mtext and Text options allow you to change the measured dimension text. The Angle option allows you to change the rotation angle of the dimension text. After responding appropriately for Text or Angle, AutoCAD repeats the prompts for the dimension line location.

Changing Dimension Type

The Horizontal option forces a horizontal dimension to be drawn (even when the dynamic drag switching calls for a vertical dimension). Likewise, the Vertical option forces a vertical dimension to be drawn (even when the dynamic drag switching calls for a horizontal dimension).

The Rotated option allows you to draw the dimension at a specified angle that is neither horizontal nor vertical. Nor is the desired angle at the angle determined by the first two points specified as it is in the case of aligned dimensioning (see the section on Aligned Dimensioning in this chapter). Figure 7–6 shows a situation where Rotated Dimension is applicable. Here is a case where the desired dimension is from point A to point B. However, the desired angle of dimensioning is the angle created by a line from point B to point C. The group of dimensions can be started by choosing the Rotated option, after specifying points A and B. Then points B and C are specified to determine the angle.

Right-click and choose **Rotated** from the shortcut menu, and then AutoCAD prompts:

Specify angle of dimension line <0>: *(Specify the dimension line angle)*

In lieu of specifying an angle, you can specify two points to determine the angle to rotate the dimension line. The points specified do not have to be parallel to the direction of the dimensioned distance.

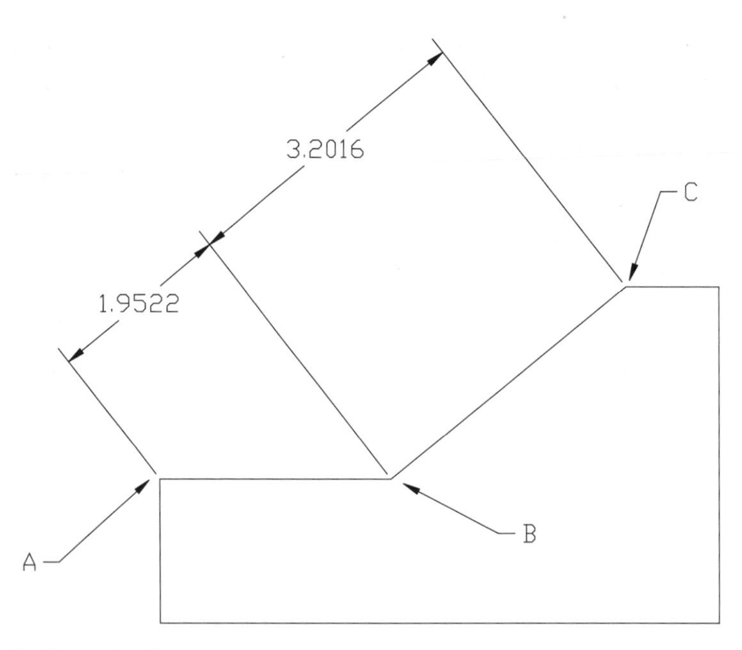

Figure 7–6 *Example of using rotated dimensioning*

The following command sequence is an example of drawing linear dimensioning for a horizontal line by providing two data points for the first and second line origins, respectively, as shown in Figure 7–7.

Command: **dimlinear** (ENTER)
Specify first extension line origin or <select object>: *(specify the origin of the first extension line)*
Specify second extension line origin: *(specify the origin of the second extension line)*
Specify dimension line location or [Mtext/Text/Angle/Horizontal/Vertical/Rotated]: *(specify the location for the dimension line)*
Dimension text = *(measured dimension)*

The following command sequence is an example of drawing a linear dimensioning for a vertical line by providing two data points for the first and second line origins, respectively, as shown in Figure 7–8.

Command: **dimlinear** (ENTER)
Specify first extension line origin or <select object>: *(specify the origin of the first extension line)*
Specify second extension line origin: *(specify the origin of the second extension line)*

Specify dimension line location or [Mtext/Text/Angle/Horizontal/Vertical/Rotated]: *(specify the location for the dimension line)*

Note: You can dynamically switch between the angle of the dimension being parallel to the angle determined by the two points specified and an angle of the dimension that is perpendicular to the angle determined by the two points specified. This is similar to dynamically switching between horizontal and vertical dimensioning for normal linear dimensioning that has not been rotated.

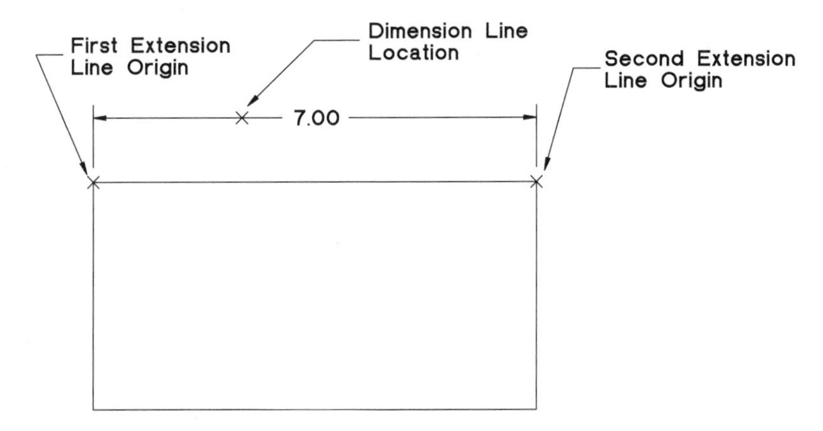

Figure 7–7 *Drawing a linear dimensioning for a horizontal line*

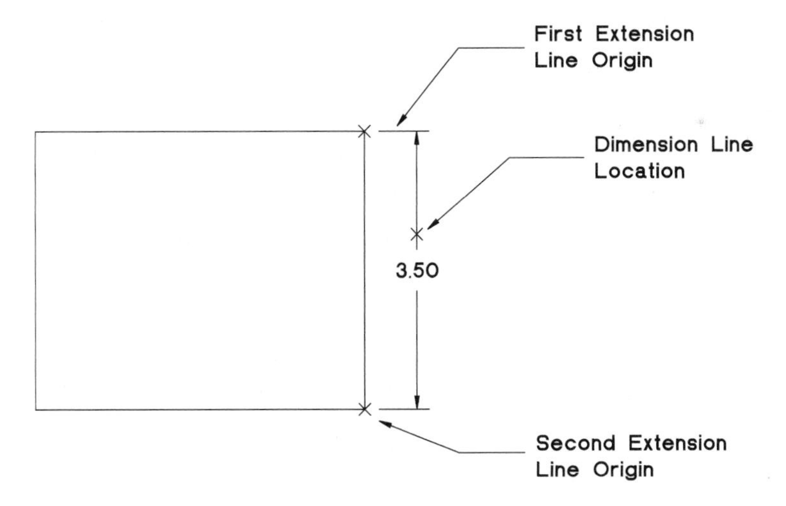

Figure 7–8 *Drawing linear dimensioning for a vertical line*

LINEAR DIMENSIONING BY SELECTING AN OBJECT

If the dimension you wish to draw is between endpoints of a line, arc, or circle diameter, AutoCAD allows you to bypass having to specify those endpoints separately. This speeds up drawing the dimension, especially if you have to invoke the endpoint object snap feature. When you have invoked the DIMLINEAR command and you press ENTER in response to the "Specify first extension line origin or <select object>:" prompt, AutoCAD prompts:

> Select object to dimension: *(select line, polyline segment, arc, or circle object)*
> Specify dimension line location or[Mtext/Text/Angle/Horizontal/Vertical/ Rotated]: *(specify location for dimension line or right-click for the shortcut menu and select one of the available options)*

AutoCAD displays the value of the dimension:

> Dimension text = (measured dimension)

If you select a line object, AutoCAD automatically uses the line's endpoints as the first and second points to determine the distance to measure. You will be prompted to specify the location of the dimension line. If you right-click and choose the **Horizontal** mode, a horizontal dimension is drawn accordingly. In a similar manner, if you choose the **Vertical** mode, a vertical dimension is drawn. If you do not choose the type of dimension to be drawn, then a horizontal dimension is drawn if the specified point for the dimension line is above or below the line object, and a vertical dimension is drawn if the specified point for the dimension line is to the right or left of the line object. If you right-click and choose the **Rotated** mode from the shortcut menu, AutoCAD will use the endpoints of the line for the distance reference points and then use the next two points specified to determine the direction of the dimension and measure the distance between the two endpoints of the line object in that direction. The dimension line passes through the last point specified.

The following command sequence shows an example of drawing a linear dimension applied to a single object, as shown in Figure 7–9.

> Command: **dimlinear** (ENTER)
> Specify first extension line origin or <select object>: (ENTER)
> Select object to dimension: *(select the line)*
> Specify dimension line location or [Mtext/Text/Angle/Horizontal/Vertical/ Rotated]: *(specify the location for the dimension line)*
> Dimension text = *(measured dimension)*

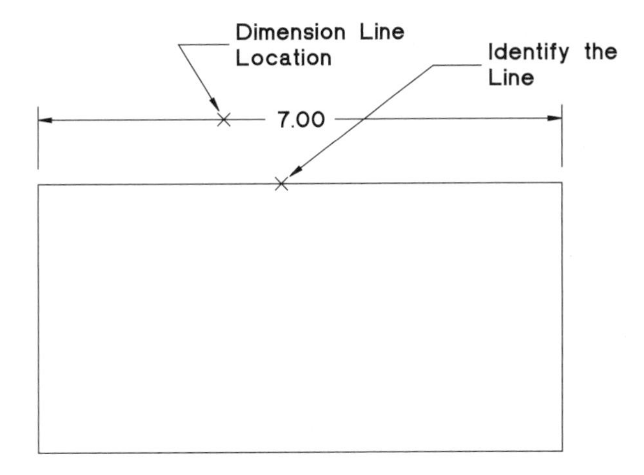

Figure 7–9 *Drawing a linear dimension for a single object*

If the object is a circle, AutoCAD automatically uses the diameter of the circle as the distance to measure and the point at which you selected the circle as one diameter endpoint for one end of the measured direction. If you are in the Horizontal mode, a horizontal dimension is drawn using the endpoints of the horizontal diameter. Likewise, with the Vertical mode the endpoints of the vertical diameter is used. If you are in the Rotated mode, AutoCAD uses the first two points specified to determine the direction of the dimension and to measure the distance between the two endpoints of a diameter in that direction. The dimension line passes through the last point specified.

If the object is an arc, AutoCAD automatically uses the arc's endpoints for the first and second points in determining the distance to measure. You will be prompted to specify the location of the dimension line. If you are in the Horizontal mode, a horizontal dimension is drawn accordingly. Likewise with the Vertical mode. If you have not yet determined the type of dimension to be drawn, then a horizontal dimension will be drawn if the specified point for the dimension line is above or below the arc, and a vertical dimension is drawn if the specified point for the dimension line is to the right or left of the arc. If you are in the Aligned mode, AutoCAD uses the endpoints of the selected arc as the first and second points and the direction from one endpoint to the other endpoint as the direction to measure. If you are in the Rotated mode, AutoCAD uses the first two points specified to determine the direction of the dimension and to measure the distance between the two endpoints of the arc in that direction. The dimension line passes through the last point specified.

ALIGNED DIMENSIONING

When dimensioning a line drawn at an angle, it may be necessary to align the dimension line with an object line.

Invoke the ALIGNED DIMENSION command:

Dimension toolbar	Choose the Aligned Dimension command (see Figure 7–10)
Dimension menu	Choose Aligned
Command: prompt	**dimaligned** (ENTER)

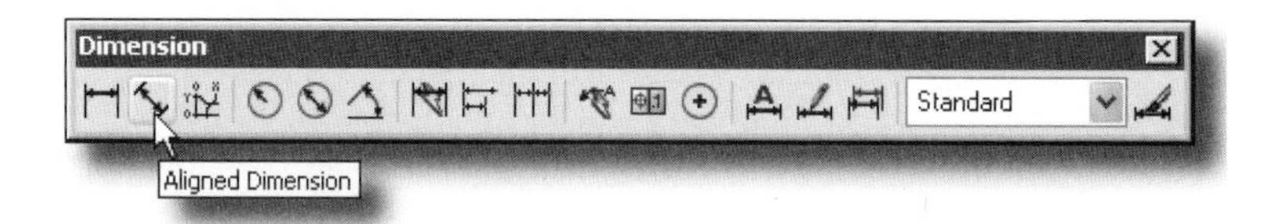

Figure 7–10 *Invoking the* DIMALIGNED *command from the Dimension toolbar*

AutoCAD prompts:

Command: **dimaligned** (ENTER)
Specify first extension line origin or <select object>: *(specify the origin of the first extension line)*
Specify second extension line origin: *(specify the origin of the second extension line)*

After two points have been specified, AutoCAD prompts:

Specify dimension line location or [Mtext/Text/Angle]: *(specify the location for the dimension line or right-click for the shortcut menu and select one of the available options)*

Specify a point where the dimension line is to be drawn or choose one of the available options. The Mtext, Text, and Angle options are the same as in Linear Dimensioning, explained earlier in this section.

The following command sequence shows an example of drawing aligned dimensioning for an angular line by providing two data points for the first and second line origins, respectively, as shown in Figure 7–11.

Command: **dimaligned** (ENTER)
Specify first extension line origin or <select object>: *(specify the origin of the first extension line)*
Specify second extension line origin: *(specify the origin of the second extension line)*

Specify dimension line location or [Mtext/Text/Angle]: *(specify the location of the dimension line)*
Dimension text = *(measured dimension)*

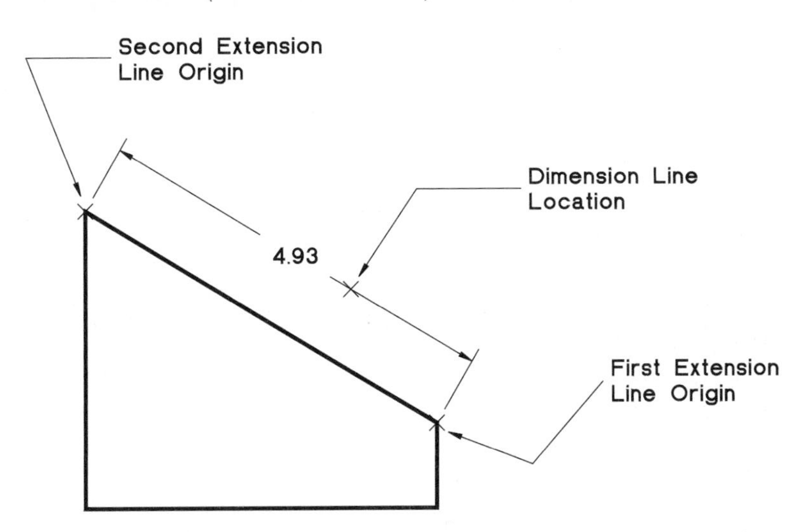

Figure 7–11 *Drawing aligned dimensioning for a line drawn at an angle*

ORDINATE DIMENSIONING

AutoCAD uses the mutually perpendicular *X* and *Y* axes of the World Coordinate System or current User Coordinate System as the reference lines from which to base the *X* or *Y* coordinate displayed in an ordinate dimension (sometimes referred to as a datum dimension). In the following examples, Figure 7–13 is valid when the base of the rectangle lies on the *X* axis, giving it a *Y* value of 0.0000 and Figure 7–14 is valid when the left side of the rectangle lies on the *Y* axis, giving it an *X* value of 0.0000. If this were not the case, where the objects are located in the drawing and you still wished to have their values to be 0.0000, a new coordinate system would have to be created (even if temporarily for drawing the dimensions). See Chapter 15 for a detailed discussion on creating a User Coordinate System.

Invoke the ORDINATE DIMENSION command:

Dimension toolbar	Choose the Ordinate Dimension command (see Figure 7–12)
Dimension menu	Choose Ordinate
Command: prompt	**dimordinate** (ENTER)

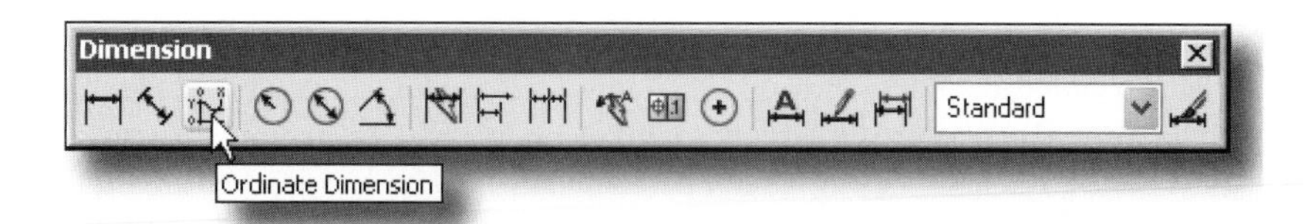

Figure 7–12 *Invoking the* DIMORDINATE *command from the Dimension toolbar*

AutoCAD prompts:

Command: **dimordinate** (ENTER)
Specify feature location: *(specify a point)*

Although the default prompt is "Specify feature location:", AutoCAD is actually looking for a point that is significant in locating a feature point on an object, such as the endpoint/intersection where planes meet or the center of a circle representing a hole or shaft. Therefore, an Object Snap mode, such as endpoint, intersection, quadrant, or center, will normally need to be invoked when responding to the "Specify feature location:" prompt. Specifying a point determines the origin of a single orthogonal leader that will point to the feature when the dimension is drawn. AutoCAD prompts:

Specify leader endpoint or [Xdatum/Ydatum/Mtext/Text/Angle]: *(specify a point or right-click for the shortcut menu and select one of the available options)*
Dimension text = *(measured dimension)*

If the Ortho mode is set to ON, the leader will be a single horizontal line for a Ydatum ordinate dimension, as shown in Figure 7–13, or a single vertical line for an Xdatum ordinate dimension, as shown in Figure 7–14.

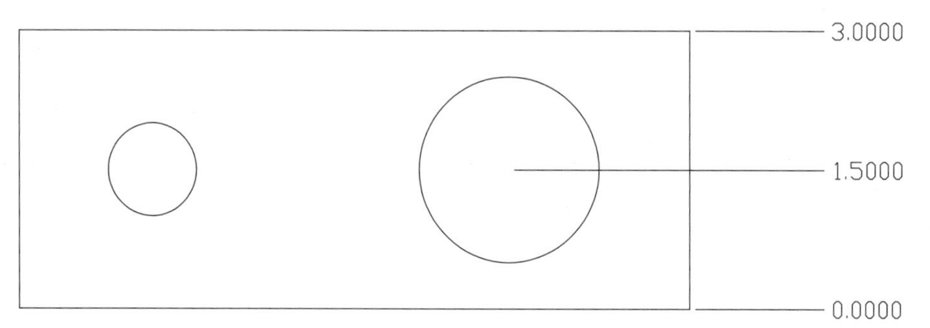

Figure 7–13 *Ydatum dimension*

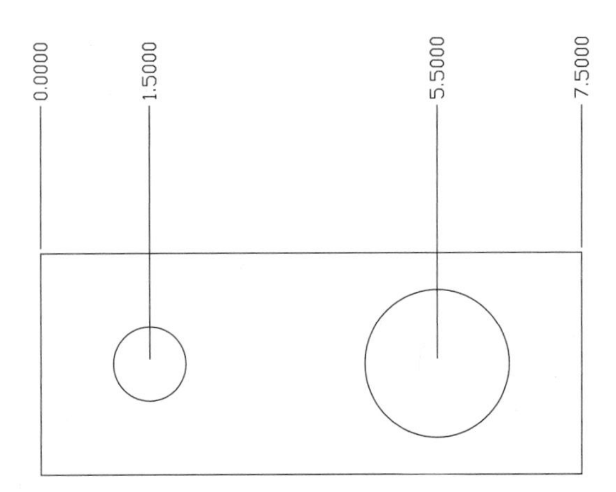

Figure 7–14 *Xdatum dimension*

If the Ortho mode is set to OFF, the leader will be a three part line consisting of orthogonal lines on each end joined by a diagonal line in the middle. It may be necessary to use the non-orthogonal leader if the text has to be offset to keep from interfering with other objects in the drawing. The type of dimension drawn (Ydatum or Xdatum) depends on which is greater of the horizontal and vertical distances between the specified "feature location" point and the "leader endpoint" point. A preview image of the dimension is displayed during specification of the "leader endpoint".

Right-click and choose Xdatum or Ydatum from the shortcut menu, and AutoCAD then draws an Xdatum dimension or Ydatum dimension, respectively, regardless of the location of the "leader endpoint" point relative to the "feature location" point.

RADIUS DIMENSIONING

The radius dimensioning feature provides commands to create radius dimensions, as shown in Figure 7–15 for a circle and an arc. The type of dimensions that AutoCAD utilizes depends on the Dimensioning System Variable settings (see the section on "Dimension styles," later in this chapter, for how to change the Dimensioning System Variable settings to draw appropriate radius dimensioning).

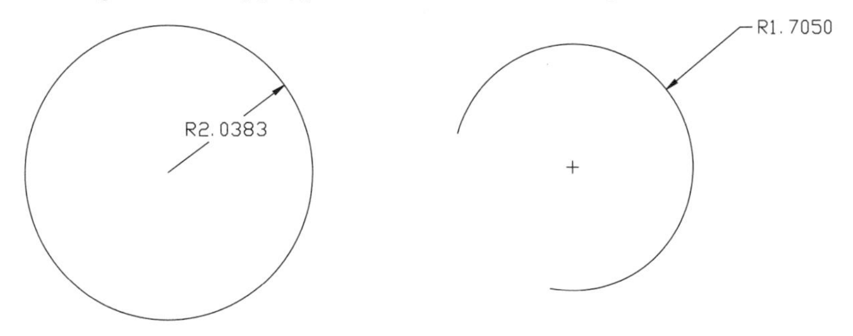

Figure 7–15 *Radius dimensioning of a circle and an arc*

Invoke the RADIUS DIMENSION command:

Dimension toolbar	Choose the Radius Dimension command (see Figure 7–16)
Dimension menu	Choose Radius
Command: prompt	**dimradius** (ENTER)

Figure 7–16 *Invoking the* DIMRADIUS *command from the Dimension toolbar*

AutoCAD prompts:

Command: **dimradius** (ENTER)
Select arc or circle: *(select an arc or a circle to dimension)*
Dimension text = *(measured dimension)*
Specify dimension line location or [Mtext/Text/Angle]: *(specify the location for the dimension leader line or right-click for the shortcut menu and select one of the available options)*

The Mtext, Text, and Angle options are the same as in Linear Dimensioning, explained earlier in this section. Dimension text for radius dimensioning is preceded by the letter R.

The following command sequence shows an example of drawing radius dimensioning for a circle, as shown in Figure 7–17.

Command: **dimradius** (ENTER)
Select arc or circle: *(select the circle object)*
Dimension text = *(measured dimension)*
Specify dimension line location or [Mtext/Text/Angle]: *(specify a point to draw the dimension leader line or right-click for the shortcut menu and select one of the available options)*

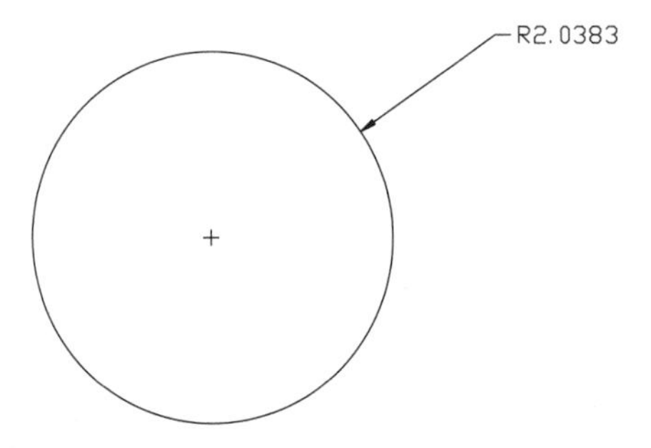

Figure 7–17 *Radius dimensioning of a circle*

DIAMETER DIMENSIONING

The diameter dimensioning feature provides commands to create diameter dimensions, as shown in Figure 7–18 for a circle and an arc. The type of dimensions that AutoCAD utilizes depends on the Dimensioning System Variable settings (see the section on "Dimension Styles," later in this chapter, for how to change the Dimensioning System Variable settings to draw an appropriate diameter dimensioning).

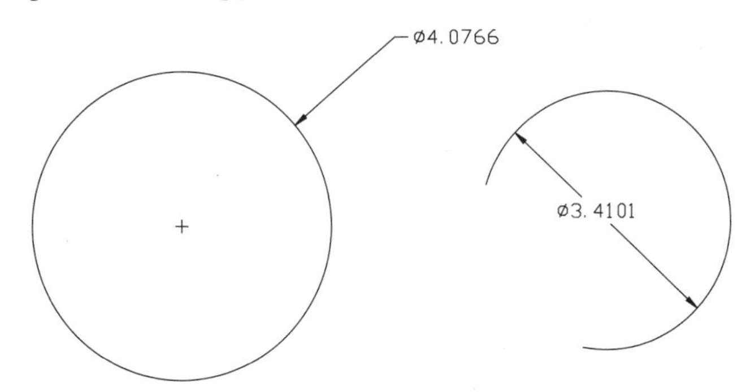

Figure 7–18 *Diameter dimensioning of a circle and an arc*

Invoke the DIAMETER DIMENSION command:

Dimension toolbar	Choose the Diameter Dimension command (see Figure 7–19)
Dimension menu	Choose Diameter
Command: prompt	**dimdiameter** (ENTER)

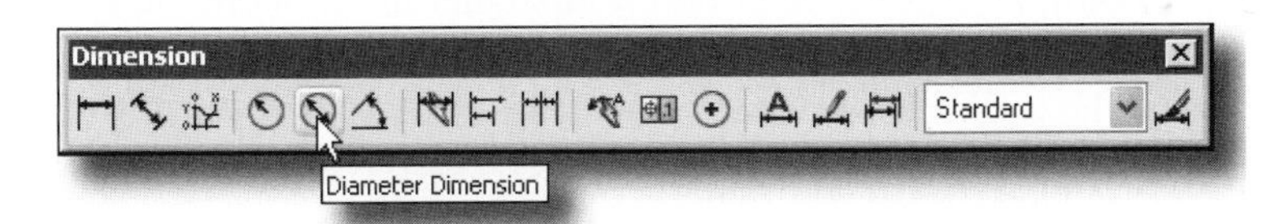

Figure 7–19 *Invoking the* DIMDIAMETER *command from the Dimension toolbar*

AutoCAD prompts:

Command: **dimdiameter** (ENTER)
Select arc or circle: *(select an arc or a circle to dimension)*
Dimension text = *(measured dimension)*
Specify dimension line location or [Mtext/Text/Angle]: *(specify the location for the dimension line or right-click for the shortcut menu and select one of the available options)*

The Mtext, Text, and Angle options are the same as in Linear Dimensioning, explained earlier in this section.

Specifying a point determines the location of the diameter dimension. Note that the dimension for an arc of less than 180 degrees cannot be forced to where neither end of the dimension is on the arc.

The following command sequence shows an example of drawing diameter dimensioning for a circle, as shown in Figure 7–20.

Command: **dimdiameter** (ENTER)
Select arc or circle: *(select an arc or a circle to dimension)*
Dimension text = *(measured dimension)*
Specify dimension line location or [Mtext/Text/Angle]: *(specify a point to draw the dimension)*

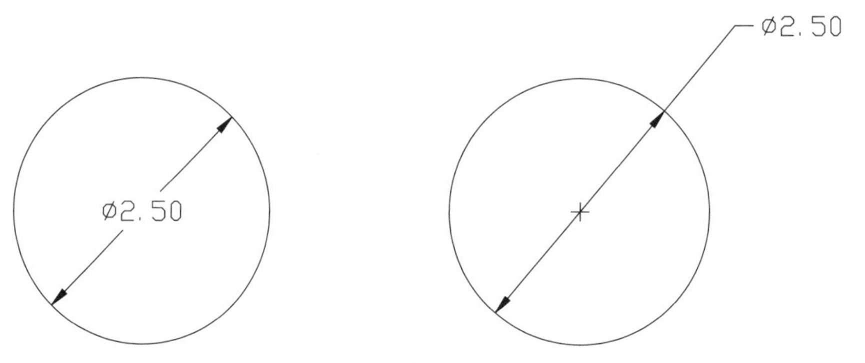

Figure 7–20 *Diameter dimensioning of a circle*

ANGULAR DIMENSIONING

The ANGULAR DIMENSION command allows you to draw angular dimensions using three points (vertex, point, point), between two nonparallel lines, on an arc (between the two endpoints of the arc, with the center as the vertex), and on a circle (between two points on the circle, with the center as the vertex).

Invoke the ANGULAR DIMENSION command:

Dimension toolbar	Choose the Angular Dimension command (see Figure 7–21)
Dimension menu	Choose Angular
Command: prompt	**dimangular** (ENTER)

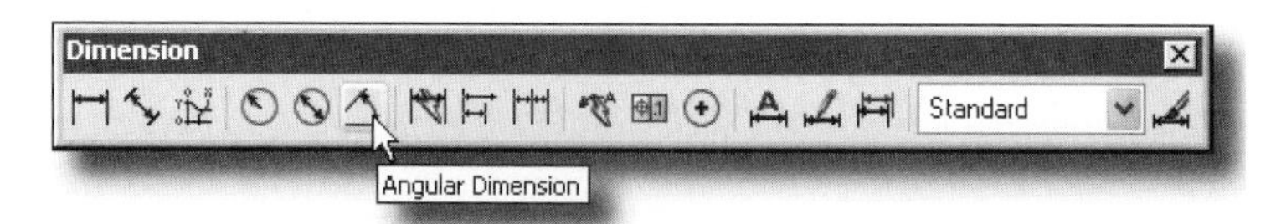

Figure 7–21 *Invoking the DIMANGULAR command from the Dimension toolbar*

AutoCAD prompts:

> Command: **dimangular** (ENTER)
> Select arc, circle, line, or <specify vertex>: *(select an object or press* ENTER *to specify a vertex)*

The default method of angular dimensioning is to select an object.

If the object selected is an arc, as shown in Figure 7–22, AutoCAD automatically uses the arc's center as the vertex and its endpoints for the first angle endpoint and second angle endpoint to determine the three points of a Vertex/Endpoint/Endpoint angular dimension. AutoCAD prompts:

> Specify dimension arc line location or [Mtext/Text/Angle]: *(specify the location for the dimension arc line or right-click for the shortcut menu and select one of the available options)*
> Dimension text = *(measured angle)*

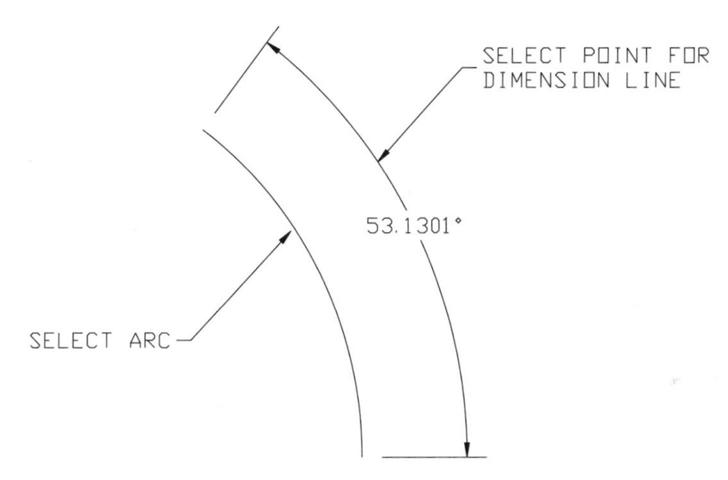

Figure 7–22 *Angular dimensioning of an arc*

The Mtext, Text, and Angle options are the same as in Linear Dimensioning, explained earlier in this section. Specify a point for the location of the dimension. AutoCAD automatically draws radial extension lines.

If the object selected is a circle, as shown in Figure 7–23, AutoCAD automatically uses the circle's center as the vertex and the point at which you selected the circle as the endpoint for the first angle endpoint, and prompts:

> Specify second angle endpoint: *(specify a point to determine the second angle endpoint)*

Specify a point, and AutoCAD makes this point the second angle endpoint to use along with the previous two points as the three points of a Vertex/Endpoint/Endpoint angular dimension. Note that the last point does not have to be on the circle; however, it does determine the origin for the second extension line. Then, AutoCAD prompts:

> Specify dimension arc line location or [Mtext/Text/Angle]: *(specify the location for the dimension arc line or right-click for the shortcut menu and select one of the available options)*
> Dimension text = *(measured angle)*

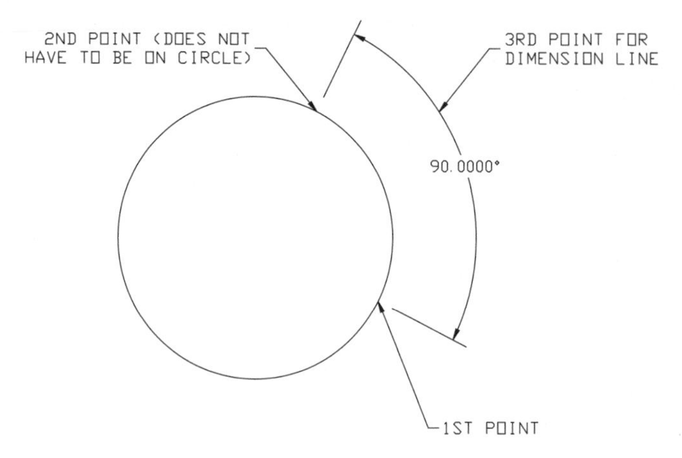

Figure 7–23 *Angular dimensioning of a circle*

The Mtext, Text, and Angle options are the same as in Linear Dimensioning, explained earlier in this section.

If you specify a point for the location of the dimension arc line, AutoCAD automatically draws radial extension lines and draws either a minor or a major angular dimension, depending on whether the point used to specify the location of the dimension arc line is in the minor or major projected sector.

If the object selected is a line, as shown in Figure 7–24, then AutoCAD prompts:

> Select second line: *(select a line object)*

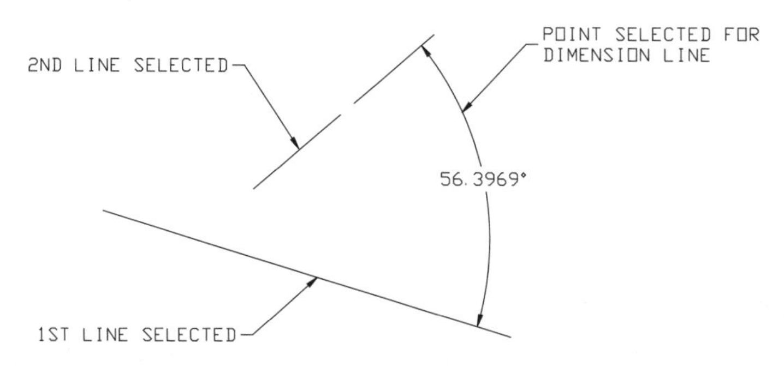

Figure 7–24 *Angular dimensioning of a line*

Select another line, and AutoCAD uses the apparent intersection of the two lines as the vertex for drawing a Vertex/Vector/Vector angular dimension. You are then prompted to specify the location of the dimension arc, which will always be less than 180 degrees.

If the dimension arc is beyond the end of either line, AutoCAD adds the necessary radial extension line(s). Then AutoCAD prompts:

> Specify dimension arc line location or [Mtext/Text/Angle]: *(specify the location for the dimension arc line or right-click for the shortcut menu and select one of the available options)*
> Dimension text = *(measured angle)*

The Mtext, Text, and Angle options are the same as in Linear Dimensioning, explained earlier in this section.

If you specify a point for the location of the dimension arc, AutoCAD automatically draws extension lines and draws the dimension text.

If, instead of selecting an arc, a circle, or two lines for angular dimensioning, you press ENTER, AutoCAD allows you to do three-point angular dimensioning. The following command sequence shows an example of drawing angular dimensioning by providing three data points, as shown in Figure 7–25.

> Command: **dimangular** (ENTER)
> Select arc, circle, line, or <specify vertex>: (ENTER)
> Specify angle vertex: *(specify a point)*
> Specify first angle endpoint: *(specify a point)*
> Specify second angle endpoint: *(specify a point)*
> Specify dimension arc line location or [Mtext/Text/Angle]: *(specify the dimension arc line location or right-click for the shortcut menu and select one of the available options)*
> Dimension text = *(measured angle)*

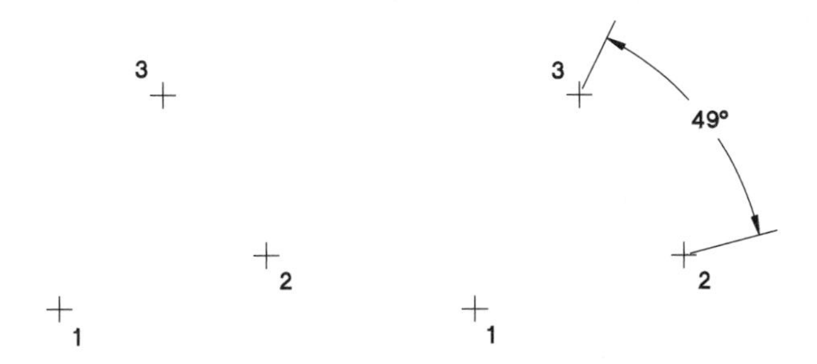

Figure 7–25 *Using three points for angular dimensioning*

BASELINE DIMENSIONING

Baseline dimensioning (sometimes referred to as parallel dimensioning) is used to draw dimensions to multiple points from a single datum baseline, as shown in Figure

7–26. The first extension line origin of the initial dimension (it can be a linear, angular, or ordinate dimension) establishes the base from which the baseline dimensions are drawn. That is, all of the dimensions in the series of baseline dimensions will share a common first extension line origin. AutoCAD automatically draws a dimension line/arc beyond the initial (or previous baseline) dimension line/arc. The location of the new dimension line/arc is an offset distance established by the DIMDLI (for dimension line increment) dimensioning variable.

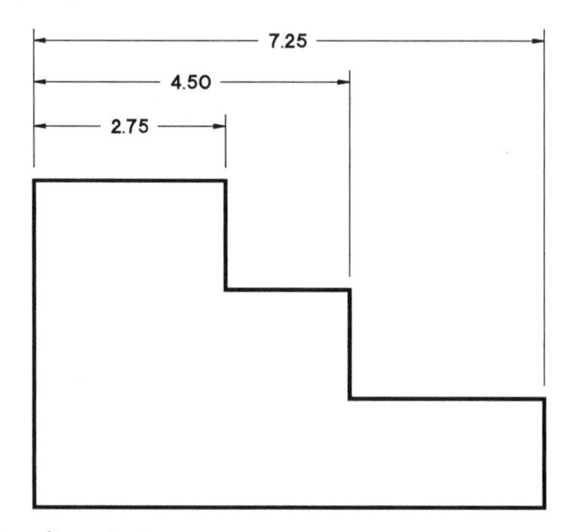

Figure 7–26 *Baseline dimensioning*

Invoke the BASELINE DIMENSION command:

Dimension toolbar	Choose the Baseline Dimension command (see Figure 7–27)
Dimension menu	Choose Baseline
Command: prompt	**dimbaseline** (ENTER)

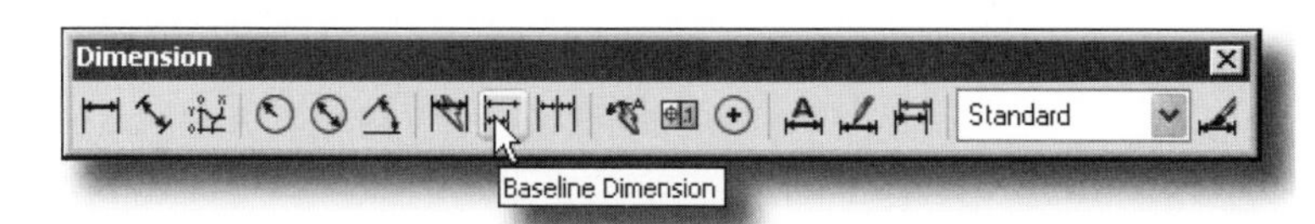

Figure 7–27 *Invoking the* DIMBASELINE *command from the Dimension toolbar*

AutoCAD prompts:

Command: **dimbaseline** (ENTER)
Specify a second extension line origin or [Undo/Select>]: *(specify a point or right-click for the shortcut menu and select one of the available options)*

After you specify a point for the second extension line origin, AutoCAD will use the first extension line origin of the previous linear, angular, or ordinate dimension as the first extension line origin for the new dimension, and the prompt is repeated. To exit the command, right-click and choose ENTER from the shortcut menu.

For the BASELINE DIMENSION command to be valid, there must be an existing linear, angular, or ordinate dimension. If the previous dimension was not a linear, angular, or ordinate dimension, or if you press ENTER without providing the second extension line origin, AutoCAD prompts:

Select base dimension: *(select a dimension object)*

You may select the base dimension, with the baseline being the extension line nearest to where you select the dimension.

The following command sequence shows an example of drawing baseline dimensioning for a circular object to an existing angular dimension, as shown in Figure 7–28.

Command: **dimbaseline** (ENTER)
Second extension line origin or [Undo<Select>]: *(specify a point)*
Dimension text = *(measured angle or dimension)*
Second extension line origin or [Undo<Select>]: *(specify a point)*
Dimension text = *(measured angle or dimension)*
Second extension line origin or [Undo<Select>]: *(press ESC or right-click and choose ENTER from the shortcut menu)*
Dimension text = *(measured angle or dimension)*

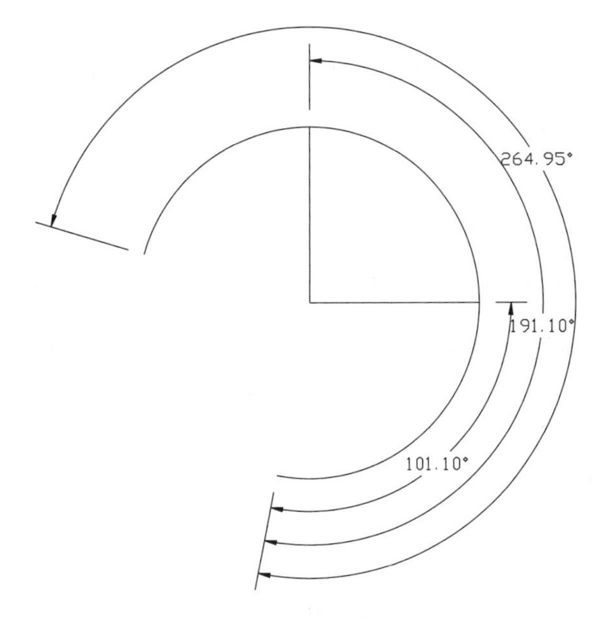

Figure 7–28 *Baseline dimensioning of a circular object*

CONTINUE DIMENSIONING

Continue dimensioning, shown in Figure 7–29, is used for drawing a string of dimensions, each of whose second extension line origin coincides with the next dimension's first extension line origin.

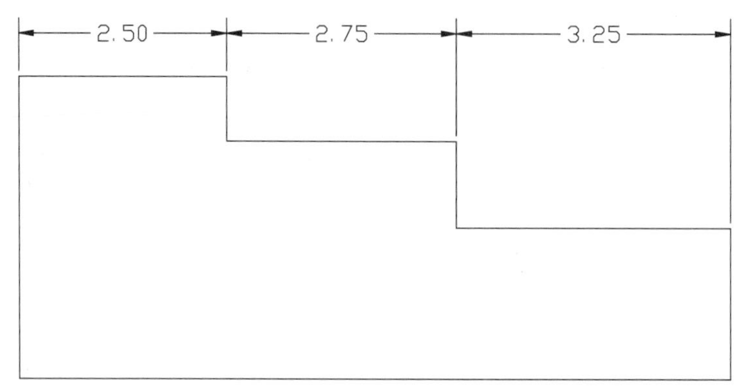

Figure 7–29 *Continue dimensioning*

Invoke the CONTINUE DIMENSION command:

Dimension toolbar	Choose the Continue Dimension command (see Figure 7–30)
Dimension menu	Choose Continue
Command: prompt	**dimcontinue** (ENTER)

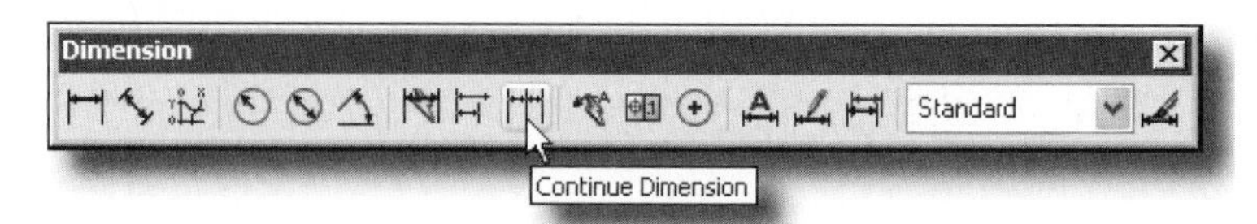

Figure 7–30 *Invoking the* DIMCONTINUE *command from the Dimension toolbar*

AutoCAD prompts:

Command: **dimcontinue** (ENTER)
Specify a second extension line origin or [Undo<Select>]: *(specify a point or right-click for the shortcut menu and select one of the available options)*
Dimension text = *(measured angle or dimension)*

After you specify a point for the second extension line origin, AutoCAD will use the second extension line origin of the previous linear, angular, or ordinate dimension as the first extension line origin for the new dimension, and the prompt is repeated. To exit the command, right-click and select ENTER from the shortcut menu.

For the CONTINUE DIMENSION command to be valid, there must be an existing linear, angular, or ordinate dimension. If the previous dimension was not a linear, angular, or ordinate dimension, or if you press ENTER without providing the second extension line origin, AutoCAD prompts:

Select continued dimension: *(select a dimension object)*

You may select the continued dimension, with the coincidental extension line origin being the one nearest to where the existing dimension is selected.

The following command sequence shows an example of drawing continue dimensioning for a linear object to an existing linear dimension, as shown in Figure 7–31.

Command: **dimcontinue** (ENTER)
Specify a second extension line origin or [Undo<Select>]: *(specify the right end of the 3.60-unit line)*
Dimension text = *(measured angle or dimension)*
Specify a second extension line origin or [Undo<Select>]: *(specify the right end of the right 1.80-unit line)*
Dimension text = *(measured angle or dimension)*
Specify a second extension line origin or [Undo<Select>]: *(press* ESC *or right-click and choose* ENTER *from the shortcut menu)*
Dimension text = *(measured angle or dimension)*

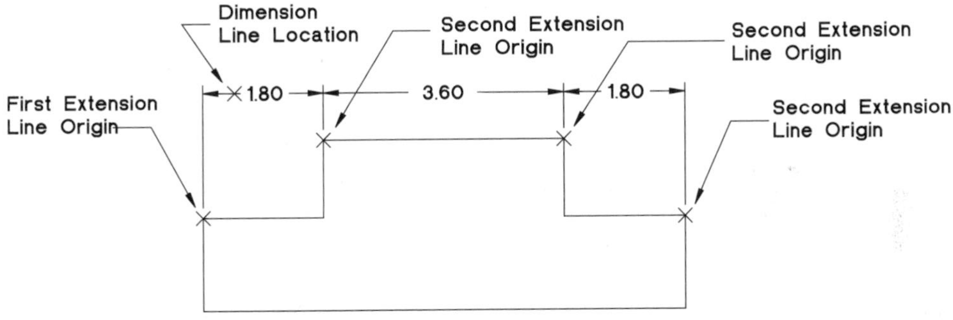

Figure 7–31 *Continue dimensioning of a linear object*

QUICK DIMENSIONING

Quick Dimensioning or the QDIM command is used to draw a string of dimensions between all of the end and center points of the selected object(s).

Invoke the QUICK DIMENSION command:

Dimension toolbar	Choose the QDIM command (see Figure 7–32)
Dimension menu	Choose Quick Dimension
Command: prompt	**qdim** (ENTER)

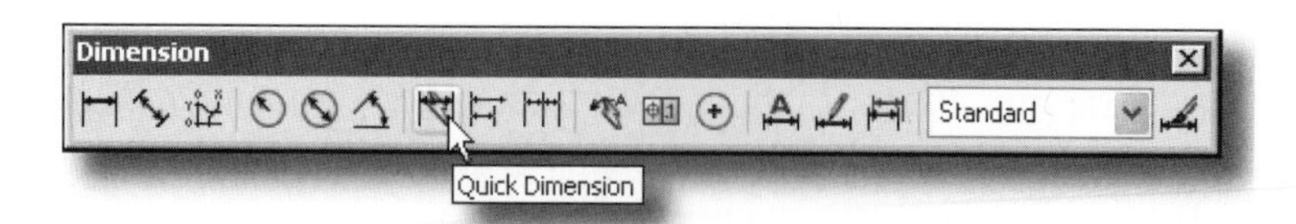

Figure 7–32 *Invoking the* QDIM *command from the Dimension toolbar*

AutoCAD prompts:

Command: **qdim** (ENTER)
Associative dimension priority = Endpoint
Select geometry to dimension: *(select one or more objects and then press* ENTER*)*
Specify dimension line position, or [Continuous/Staggered/Baseline/
Ordinate/Radius/Diameter/datumPoint/Edit] <Continuous>: *(specify location for dimension line or right-click for the shortcut menu and select one of the available options)*

If you specify a location for a dimension line, AutoCAD will draw continuous dimensioning between all end or center points of the objects selected horizontally or vertically, depending on where the dimension line location is specified (see Figure 7–33 for an example). You can right-click and select one of the options that will draw the type of dimension chosen.

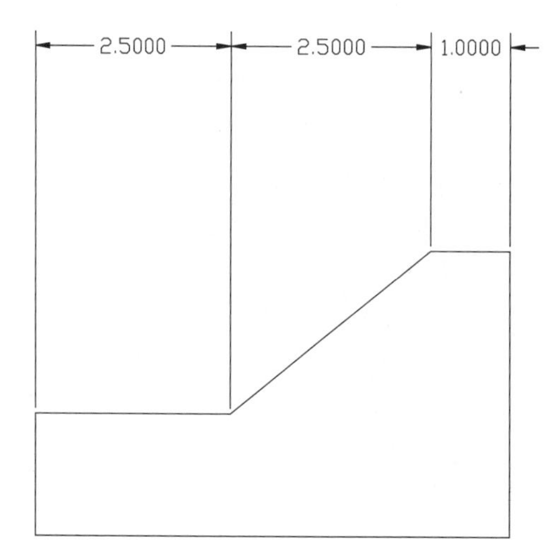

Figure 7–33 *An example of Quick Dimensioning*

QLEADER AND LEADER

The QLEADER command (the default version of the LEADER command) assumes that you wish to enter multiline text that is top left justified. The QLEADER and LEADER commands minimize the steps required to draw the text with the line(s) and arrowhead pointing to an object (or feature on an object) for annotations and callouts used to describe them.

Invoke the LEADER command:

Command: prompt	**qleader** (ENTER)

Invoke the QLEADER command:

Dimension toolbar	Choose the QUICK LEADER command (see Figure 7–34)
Dimension menu	Choose Leader

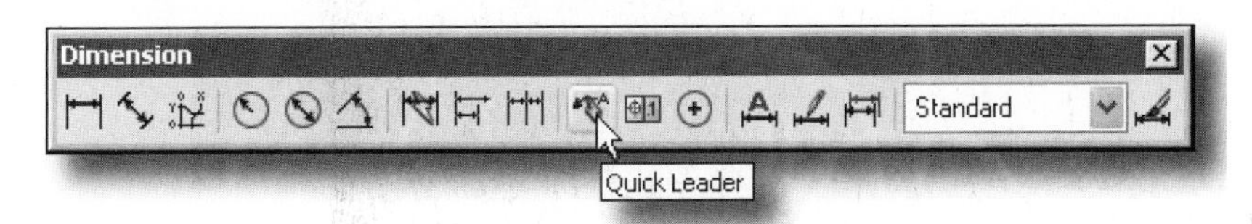

Figure 7–34 *Invoking the* QUICK LEADER *command from the Dimension toolbar*

Qleader

The LEADER command, when invoked from the Dimension toolbar or the Dimension menu, invokes QLEADER, and AutoCAD prompts:

Command: **_qleader** (ENTER)
Specify first leader point, or [Settings]: *(specify a point for the arrowhead end of the leader line or right-click for the shortcut menu)*
Specify next point: *(specify another point for the end of the first leader segment opposite the arrowhead)*
Specify text width <0.0000>: *(specify a point to determine the maximum width of the multiline text)*
Enter first line of annotation text <Mtext>: *(enter the first line of annotation text or right-click for the Multiline Text Editor dialog box. See Figure 7-39)*
Enter next line of annotation text: *(enter the next line of annotation text or* ENTER *to complete the* QLEADER *command)*

Note that the text width that is specified causes the text to wrap to the next line when the text on the current line exceeds the specified width. If no width is specified, then the entire text is drawn on one line.

If you respond when prompted to specify the first leader point by right-clicking and selecting **Settings** from the shortcut menu, AutoCAD displays the Leader Settings dialog box, shown in Figure 7-35.

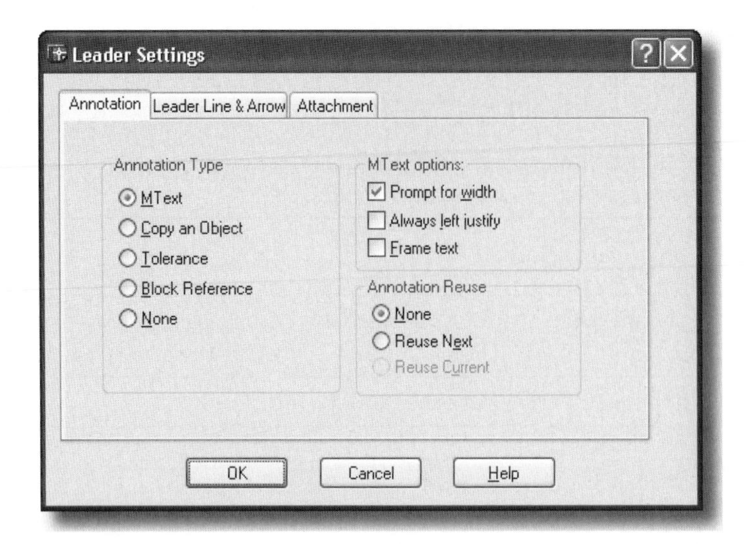

Figure 7–35 *Leader Settings dialog box with Annotation tab chosen*

Annotation

The **Annotation** tab allows you to set the **Annotation Type**, select **Mtext options**, and determine **Annotation Reuse**.

In the **Annotation Type** section there are five radio buttons that let you choose the type of annotation that will be used. The options are as follows:

Choosing the **Mtext** button causes AutoCAD to use the MTEXT command when the annotation is called for. See Figure 7–39 and the section on the MTEXT command in Chapter 4.

Choosing the **Copy an Object** button causes AutoCAD to prompt you to "Select an object to copy:" when the annotation is called for. If you have not chosen a text object, AutoCAD will prompt: "Please select an mtext, text, block reference, or tolerance object." The text object you select will become the text for leader annotation.

Choosing the **Tolerance** button causes the Tolerance option to be used (as described in the following section) when the annotation is called for.

Choosing the **Block Reference** button causes AutoCAD to use the INSERT command (for inserting a block) when the annotation is called for.

Choosing the **None** button causes AutoCAD to draw the leader and exit the LEADER command without prompting for or placing text or other object.

In the **MText options:** section there are three check boxes that let you specify how MText will be formatted. The **MText options:** section is active for use only if the **Mtext** button has been selected in the **Annotation Type** section. The options are as follows:

Checking the **Prompt for width** check box causes the causes AutoCAD to prompt you for the width of the MText box when the annotation is called for.

Checking the **Always left justify** check box causes the annotation text to be left justified whether the leader is to the left or the right of the annotation. When this box is checked, the **Prompt for width** check box becomes disabled.

Checking the **Frame text** check box causes AutoCAD to draw a rectangle around the annotation text.

In the **Annotation Reuse** section there are three radio buttons that let you specify whether or not AutoCAD reuses the next or the current annotation text for subsequent leader annotation text.

Choosing the **None** button causes AutoCAD to not reuse annotation text and to prompt you for annotation text to be used.

Choosing the **Reuse Next** button causes AutoCAD to reuse the next annotation text used for subsequent annotation text.

Choosing the **Reuse Current** button causes AutoCAD to reuse the current annotation text for subsequent annotation text.

Leader Line & Arrow Tab

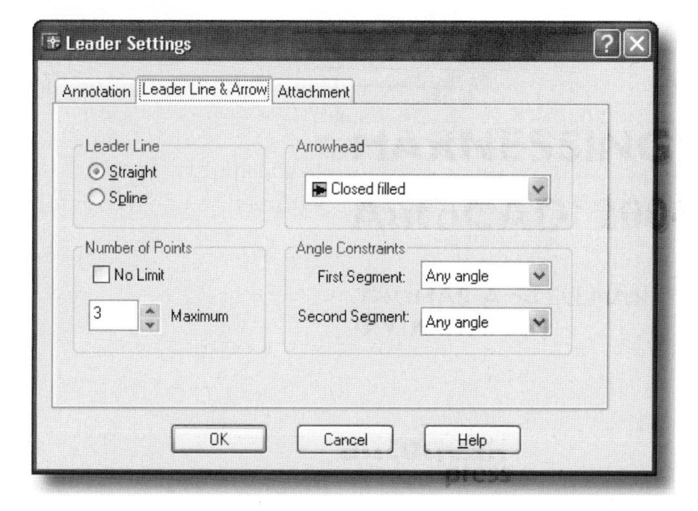

Figure 7–36 *Leader Settings dialog box with Leader Line & Arrow tab chosen*

The **Leader Line & Arrow** tab as shown in Figure 7–36 allows you to modify the geometry of the lines and arrow heads that make up the leader, along with specifying the number of points permitted (which limits the number leader segments that may be drawn) and the angle constraints of the first and second leader lines. The options are as follows:

In the **Leader Line** section:

Choosing the **Straight** radio button causes leaders to be drawn with straight segments.

Choosing the **Spline** radio button causes leaders to be drawn with curved segments using the same input as in the SPLINE command.

In the **Number of Points** section:

Checking the **No Limit** check box lets you put in any number of leader segments.

If the **No Limit** check box is not checked, the number of points you are can specify to draw leader segments is limited to the number (between 2 and 999) in the **Maximum** text box. The number of segments will be one less than the number of points specified.

In the **Arrowhead** section:

From the list box you can select one of the many available arrowhead shapes. Select **None** to have no arrowhead or select **User Arrow...** and name a user-defined block to be used as an arrowhead.

In the **Angle Constraints** section:

The **First Segment:** list box lets you specify if the first segment can be drawn at any angle or will be constrained to one of the following angles: 90, 45, 30, or 15 degrees.

The **Second Segment:** list box lets you specify if the first segment can be drawn at any angle or will be constrained to one of the following angles: 90, 45, 30, or 15 degrees.

Attachment Tab

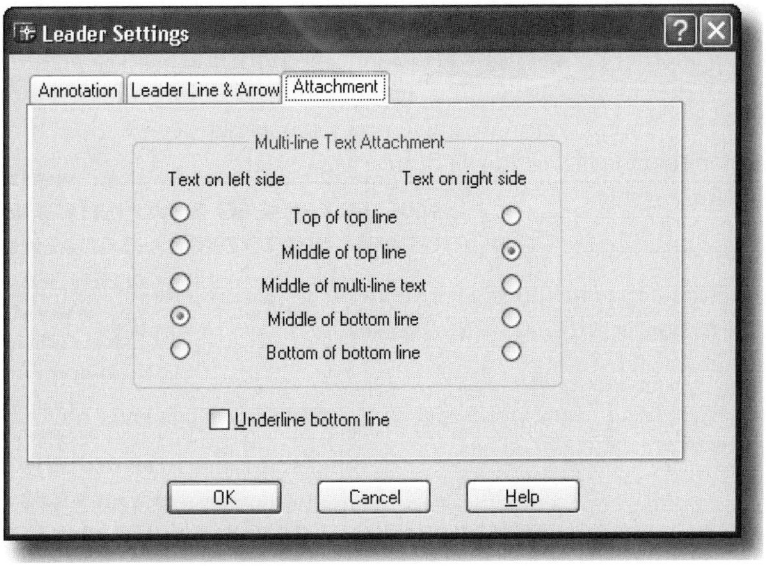

Figure 7–37 *Leader Settings dialog box with Attachment tab chosen*

The **Attachment** tab as shown in Figure 7–37 allows you to specify which part of Multi-line text will be lined up with the annotation end of the leader. The attachment can be specified to line up one way when the text is to the left of the leader and a different way when it is to the right of the leader. An option in the **Multi-line Text Attachment** section is specified by selecting its radio button. The five options are as follows: **Top of top line, Middle of top line, Middle of multi-line text, Middle of bottom line,** and **Bottom of bottom line.**

The **Attachment** tab has a check box that, when checked, causes the leader to terminate at the bottom line of text and continue as an underline of the text. This overrides any of the first four attachment options mentioned above.

Leader

Invoke the LEADER command:

Command: prompt	**leader** (ENTER)

When you invoke the LEADER command at the Command: prompt, AutoCAD prompts:

Command: **leader** (ENTER)
Specify leader start point: *(specify a point for the arrowhead end of the leader)*
Specify next point: *(specify another point for the end of the first leader segment opposite the arrowhead)*

> Specify next point or [Annotation/Format/Undo] <Annotation>: *(specify a point or right-click for the shortcut menu)*

If you specify another point, AutoCAD connects another leader segment to the previous one and repeats the last prompt. If the leader radial direction is more than 15 degrees from horizontal, AutoCAD will add a horizontal segment pointing toward the annotation that is equal in length to an arrow, as determined by the DIMASZ dimensioning system variable.

Annotation

The Annotation option allows you to draw alphanumerics or symbols as text for the annotation at the end of the last leader segment opposite the arrowhead. AutoCAD prompts:

> Enter first line of annotation text or <options>: *(you can enter characters now or press* ENTER*)*

Enter the alphanumerics or symbols that will be used as text for the annotation at the end of the leader line opposite the arrowhead. If you press ENTER instead, AutoCAD prompts:

> Enter an annotation option [Tolerance/Copy/Block/None/Mtext] <Mtext>: *(right-click and choose one of the available options from the shortcut menu)*

Choosing the Mtext option displays the Mtext dialog box, shown in Figure 7–38, for creating or editing single-line or multiline text for annotation.

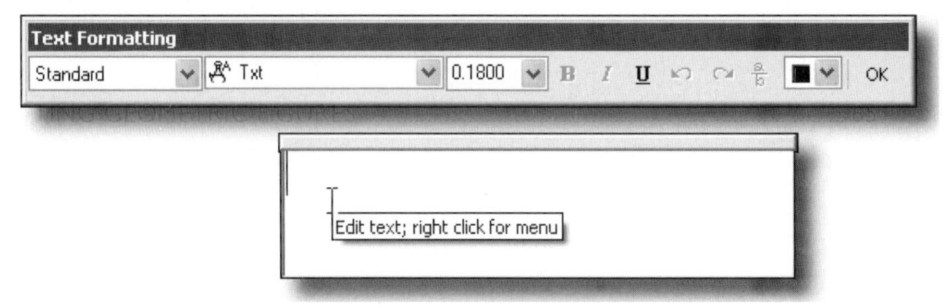

Figure 7–38 *Multiline Text Editor dialog box*

AutoCAD draws the primary dimension where it finds the < > (left arrow/right arrow) in a string of characters in the text box of the Mtext dialog box. If the measured dimension is 1/2" and **COPE** < > is displayed in the text box, the dimension would be drawn as **COPE 1/2"**. In a similar manner, [ ] (left bracket/right bracket) causes the secondary dimension to be drawn.

The Tolerance option allows you to draw a Feature Control Frame for use in describing standard tolerances by means of the Geometric Tolerance dialog box, as described in the section on "Tolerances" in this chapter.

The Copy option allows you to copy Text, Feature Control Frames, Mtext, or Block Reference.

The Block option allows you to insert a named Block Reference. (Refer to Chapter 10 for a discussion on Blocks.)

The None option allows you to draw the Leader without the annotation.

Format

The Format option is used to determine the appearance of the Leader and the arrowhead. AutoCAD prompts:

Spline/STraight/Arrow/None/<Exit>: *(select one of the available options)*

The Spline option allows you to draw a leader in a manner similar to the SPLINE command.

The STraight option allows you to draw a leader with straight-line segments.

The Arrow option allows you to draw an arrowhead at the first specified point of the leader.

The None option allows you to draw a leader without an arrowhead.

The Exit option exits the Format option.

Undo

The Undo option undoes the last segment of the leader line.

TOLERANCES

Tolerance symbology and text can be included with the dimensions that you draw in AutoCAD. AutoCAD provides a special set of subcommands for the two major methods of specifying tolerances. One set is for lateral tolerances; the other is for geometric tolerances.

Lateral tolerances are the traditional tolerances. Though they are often easier to apply, they do not always satisfy tolerancing in all directions (circularly and cylindrically), and are more subject to misinterpretations, especially in the international community.

Geometric tolerance values are the maximum allowable distances that a form or its position may vary from the stated dimension(s).

Note: Lateral tolerance symbols and text are accessible from the Dimension Style dialog box. Geometric tolerance symbols and text are accessible from the Dimension toolbar, from the Dimension menu, and at the "Command:" prompt.

Lateral Tolerance

Lateral tolerancing draws the traditional symbols and text for Limit, Plus or Minus (unilateral and bilateral), Single Limit, and Angular tolerance dimensioning.

Lateral tolerance is the range from the smallest to the greatest that a dimension is allowed to deviate and still be acceptable. For example, if a dimension is called out as 2.50 ± 0.05, then the tolerance is 0.1 and the feature being dimensioned may be anywhere between 2.45 and 2.55. This is the symmetrical plus-or-minus convention. If the dimension is called out as 2.50 $^{+0.10}_{-0.00}$, then the feature may be 0.1 greater than 2.50, but may not be smaller than 2.50. This is referred to as unilateral. The Limits tolerance dimension may also be shown as $^{2.55}_{2.45}$.

To set the tolerance method, first open the Dimension Style Manager dialog box (see Figure 7–51. Choose **Modify**, and AutoCAD displays the Modify Dimension Style dialog box. Select the **Tolerances** tab and select the tolerance method from the **Method:** option menu. Figure 7–39 shows all the options available in the **Tolerance Format** section and the **Alternate Unit Tolerance** section when the **Tolerances** tab is chosen in the Modify Dimension Style dialog box. Selecting a method (other than **None**) in the **Method:** text box applies only to lateral tolerancing.

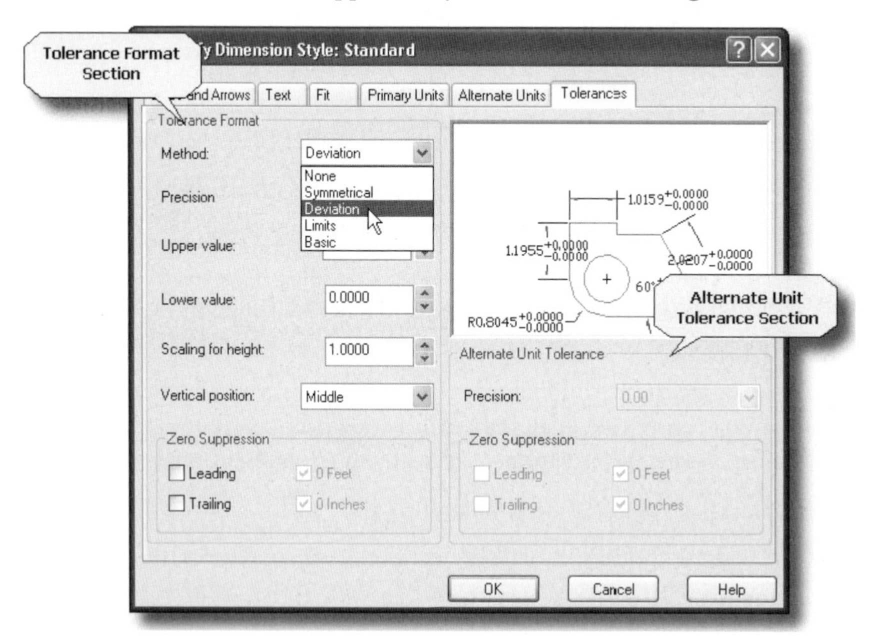

Figure 7–39 *Modify Dimension Style dialog box with the Tolerances tab selected.*

Choosing the **Method:** list box in the **Tolerance Format** section of the **Tolerances** tab displays the following options: **None, Symmetrical, Deviation, Limits,** and **Basic**. See Table 7–1 for examples of the various options.

Table 7–1 Examples of Lateral Tolerance Methods

Tolerance Method	Description	Example
Symmetrical	Only the Upper Value box is usable. Only one value is required.	1.00 ± 0.05
Deviation	Both Upper Value and Lower Value boxes are active. A 0.00 value may be entered in either box, indicating that the variation is allowed in only one direction.	$1.00\,^{+0.07}_{-0.03}$
Limits	Both Upper and Lower Value boxes are active. In this case, there is no base dimension. The Justification box is not active, because the values are not tolerance values but are actual dimensions to be drawn in full height -text. Using an Upper Value of 0.0500 and a Lower Value of –0.0250 will cause annotation of a 1.000 basic dimension to be $^{0.0500}_{0.0250}$.	1.070 0.930
Basic	The dimension value is drawn in a box, indicating that it is the base value from which a general tolerance is allowed. The general tolerance is usually given in other notes or specifications on the drawing or other documents.	1.00

Note: To get a preview of how your tolerance will appear in relation to the dimension, the format is set in the **Primary Units** tab of the Modify Dimension Style dialog box.

In the **Tolerance Format** and **Alternate Unit Tolerance** sections, the **Precision:** text box lets you determine how many decimals the text will be shown in decimal units. This value is recorded in DIMDEC. The **Upper value**: and **Lower value**: text boxes allow you to preset the values where they apply. The **Scaling for height**: text box lets you set the height of the tolerance value text. The **Vertical position**: selection box determines if the tolerance value that is drawn after the dimension will be at the top, middle, or bottom of the text space.

In the **Tolerance Format** and **Alternate Unit Tolerance** sections, the **Zero Suppression** subsection lets you specify whether or not zeros are displayed, as follows:

Selecting the **Leading** check box causes zeros ahead of the decimal point to be suppressed. *Example:* .700 is displayed instead of 0.700.

Selecting the **Trailing** check box causes zeros behind the decimal point to be suppressed. *Example:* 7 or 7.25 is displayed instead of 7.000 or 7.250, respectively.

Selecting the **0 Feet** check box causes zeros representing feet to be suppressed if the dimension text represents inches and/or fractions only. *Example:* 7" or 7 1/4" is displayed instead of 0'-7" or 0'-7 1/4".

Selecting the **0 Inches** check box causes zeros representing inches to be suppressed if the dimension text represents feet only. *Example:* 7' is displayed instead of 7'-0".

Geometric Tolerance

Geometric tolerancing draws a Feature Control Frame for use in describing standard tolerances according to the geometric tolerance conventions. Geometric tolerancing is applied to forms, profiles, orientations, locations, and runouts. Forms include squares, polygons, planes, cylinders, and cones.

Invoke the Geometric TOLERANCE command:

Dimension toolbar	Choose the TOLERANCE command (see Figure 7–40)
Dimension menu	Choose Tolerance...
Command: prompt	**tolerance** (ENTER)

Figure 7–40 *Invoking the Geometric* TOLERANCE *command from the Dimension toolbar*

AutoCAD displays the Geometric Tolerance dialog box, shown in Figure 7–41.

Figure 7–41 *Geometric Tolerance dialog box*

The conventional method of expressing a geometric tolerance for a single dimensioned feature is in a Feature Control Frame, which includes all necessary tolerance information for that particular dimension. A Feature Control Frame has the Geometric Characteristic Symbol box and the Tolerance Value box. Datum reference/material condition datum boxes may be added where needed. The Feature Control Frame is shown in Figure 7–42. An explanation of the Characteristic Symbols is given in Figure 7–43. The supplementary material conditions of datum symbols are shown in Figure 7–44.

Once the symbol button located in the **Sym** column in the Geometric Tolerance dialog box is chosen, AutoCAD displays the Symbol dialog box, as shown in Figure 7–45. Select one of the available symbols and specify appropriate tolerance values in the **Tolerance** field.

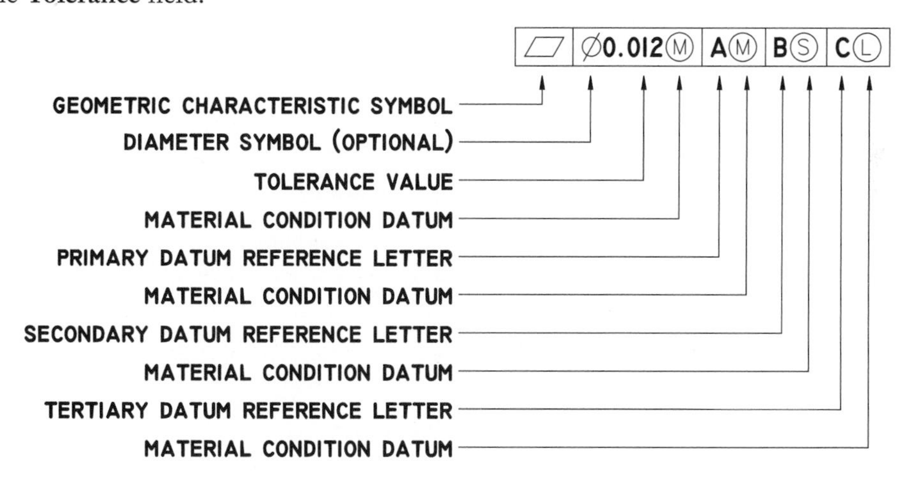

Figure 7–42 *Feature control frame*

Maximum Materials Condition (MMC) means a feature contains the maximum material permitted within the tolerance dimension; that is, the minimum hole size, maximum shaft size.

Least Material Condition (LMC) means a feature contains the least material permitted within the tolerance dimension, that is maximum hole size, minimum shaft size.

Regardless of Feature Size (RFS) applies to any size of the feature within its tolerance. This is more restrictive. For example, RFS does not allow the tolerance of the center-to-center dimension of a pair of pegs fitting into a pair of holes greater leeway if the peg diameters are smaller and/or the holes are bigger, whereas MMC does.

The diameter symbol, Ø, is used in lieu of the abbreviation DIA.

CHARACTERISTIC SYMBOLS		
FORM	⏥	FLATNESS
	⏤	STRAIGHTNESS
	○	ROUNDNESS
	⌭	CYLINDRICITY
	⌒	LINE PROFILE
	⌓	SURFACE PROFILE
	∠	ANGULARITY
	//	PAREALLELISM
	⊥	PERPENDICULARITY
LOCATION	◎	CONCENTRICITY
	⌖	POSITION
	⌯	SYMMETRY
RUNOUT	↗	CIRCULAR RUNOUT
	⌰	TOTAL RUNOUT

Figure 7–43 *Geometric characteristics symbols*

MATERIAL CONDITIONS OF DATUM SYMBOLS	
Ⓜ	MAXIMUM MATERIAL CONDITIONS (MMC)
Ⓢ	REGARDLESS OF FEATURE SIZE
Ø	DIAMETER

Figure 7–44 *Material conditions of datum symbols*

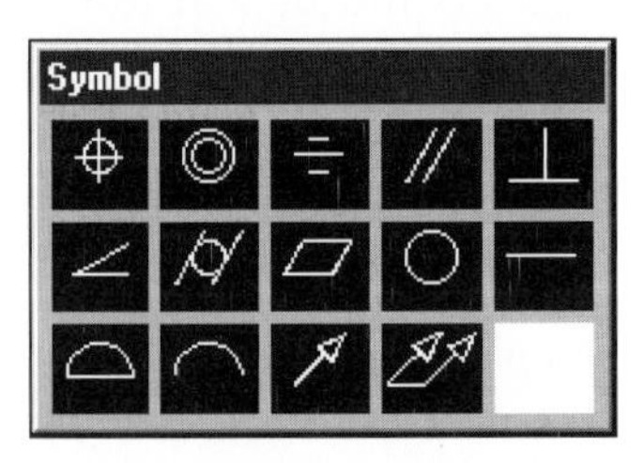

Figure 7–45 *Symbol dialog box*

Geometric tolerancing is becoming widely accepted. It is highly recommended that you study the latest drafting texts concerning the significance of the various symbols, so that you will be able to apply geometric tolerancing properly.

DRAWING CROSS MARKS FOR ARCS OR CIRCLES

As shown in Figure 7–46, the DIMCENTER command is used to draw the cross marks that indicate the center of an arc or circle.

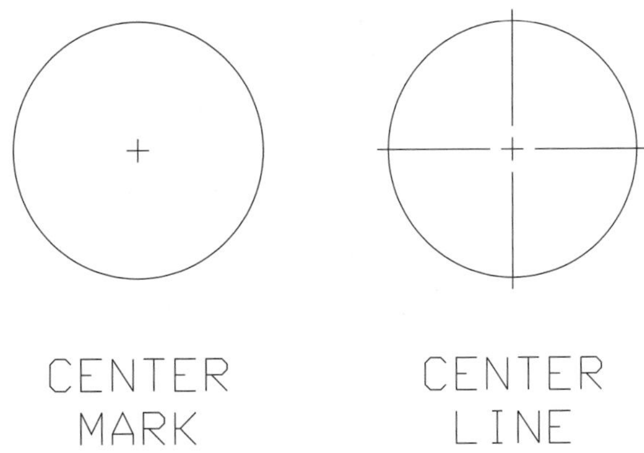

Figure 7–46 *Circles with center cross marks*

Invoke the CENTER MARK command:

Dimension toolbar	Choose the Center Mark command (see Figure 7–47)
Dimension menu	Choose Center Mark
Command: prompt	**dimcenter** (ENTER)

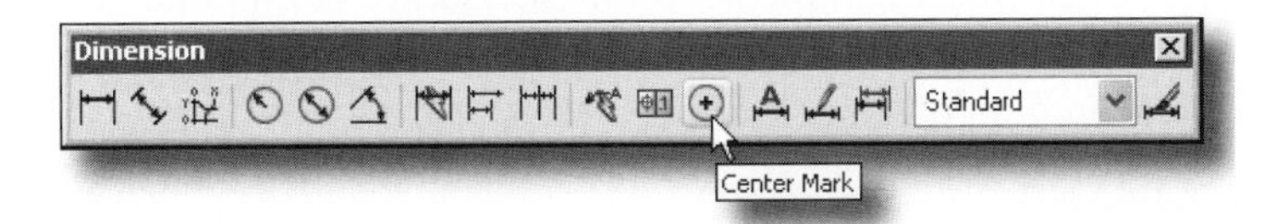

Figure 7–47 *Invoking the DIMCENTER command from the Dimension toolbar*

AutoCAD prompts:

Command: **dimcenter** (ENTER)
Select arc or circle:

Select an arc or circle, and AutoCAD draws the cross marks in accordance with the setting of the DIMCEN dimensioning variable.

OBLIQUE DIMENSIONING

The OBLIQUE command allows you to slant the extension lines of a linear dimension to a specified angle. The dimension line will follow the extension lines, retaining its original direction. This is useful for having the dimension stay clear of other dimensions or objects in your drawing. It is also a conventional method of dimensioning isometric drawings.

Invoke the OBLIQUE command:

Dimension menu	Choose Oblique

AutoCAD prompts:

Command: **_dimedit**
Enter type of dimension editing [Home/New/Rotate/Oblique] <Home>: **_o**
Select objects: *(select the dimension(s) for obliquing)*
Enter obliquing angle (press ENTER for none): *(specify an angle or press* ENTER*)*

Note: The OBLIQUE command is actually the Dimension Style Edit command in which AutoCAD automatically chooses the oblique option for you. All you have to do is select a dimension.

The extension lines of the selected dimension(s) will be slanted at the specified angle.

From the "Command:" prompt you can invoke the OBLIQUE command by first changing to the Dimensioning mode and then entering **oblique**:

Command: **dim**
Dim: **oblique**
Select objects: *(select the dimension(s) for obliquing)*
Enter obliquing angle (press ENTER for none): *(specify an angle or press* ENTER*)*

EDITING DIMENSION TEXT

AutoCAD allows you to edit dimensions with Modify commands and grip editing modes. Also, AutoCAD provides two additional Modify commands specifically designed to work on dimension text objects, DIMEDIT and DIMTEDIT.

DIMEDIT COMMAND

The options that are available in the DIMEDIT command allow you to replace the dimension text with new text, rotate the existing text, move the text to a new location, and if necessary, restore the text back to its home position, which is the position defined by the current style. In addition, these options allow you to change the angle of the extension lines (normally perpendicular) relative to the direction of the dimension line (by means of the Oblique option).

Invoke the DIMEDIT command:

Dimension toolbar	Choose the Dimension Edit command (see Figure 7–48)
Command: prompt	**dimedit** (ENTER)

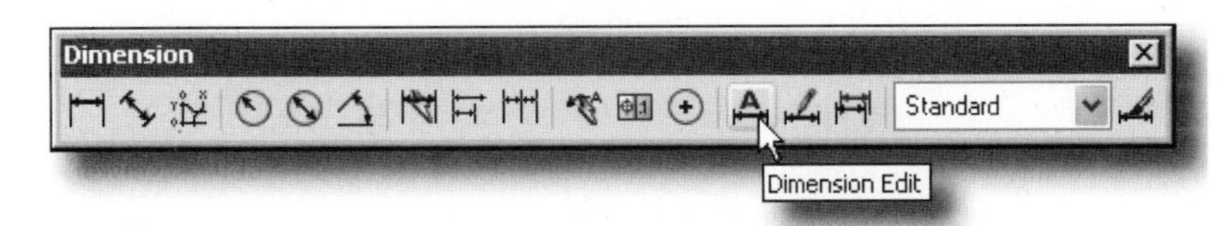

Figure 7–48 *Invoking the* DIMENSION EDIT *command from the Dimension toolbar*

AutoCAD prompts:

Command: **dimedit** (ENTER)
Enter type of dimension editing [Home/New/Rotate/Oblique]<Home>:
(press ENTER *for the Home option or right-click and choose one of the available options from the shortcut menu)*

Home

The Home option returns the dimension text to its default position. AutoCAD prompts:

Select objects: *(select the dimension objects and press* ENTER*)*

New

The New option allows you to change the original dimension text to the new text. AutoCAD opens the Multiline Text Editor dialog box. Enter the new text and choose the **OK** button. AutoCAD then prompts:

Select objects: *(select the dimension objects for which the existing text will be replaced by the new text)*

Rotate

The Rotate option allows you to change the angle of the dimension text. AutoCAD prompts:

Specify angle for dimension text: *(specify the rotation angle for text)*
Select objects: *(select the dimension objects for which the dimension text has to be rotated)*

Oblique

The Oblique option adjusts the obliquing angle of the extension lines for linear dimensions. This is useful to keep the dimension parts from interfering with other objects in the drawing. Also, it is an easy method by which to generate the slanted dimensions used in isometric drawings. AutoCAD prompts:

Select objects: *(select the dimension objects)*
Enter obliquing angle (press ENTER for none): *(specify the angle or press ENTER)*

DIMTEDIT COMMAND

The DIMTEDIT command is used to change the location of dimension text (with the Left/Right/Center/Home options) along the dimension line and its angle (with the Rotate option).

Invoke the DIMTEDIT command:

Dimension toolbar	Choose the Dimension Text Edit command (see Figure 7–49)
Command: prompt	**dimtedit** (ENTER)

Figure 7–49 *Invoking the* DIMENSION TEXT EDIT *command from the Dimension toolbar*

AutoCAD prompts:

Command: **dimtedit** (ENTER)
Select dimension: *(select the dimension object to modify)*

A preview image of the dimension selected is displayed on the screen, with the text located at the cursor. You will be prompted:

Specify new location for dimension text or [Left/Right/Center/Home/Angle]: *(specify a new location for the dimension text or right-click for the shortcut menu and select one of the available options)*

By default, AutoCAD allows you to position the dimension text with the cursor, and the dimension updates dynamically as it drags.

Center

The Center option will cause the text to be drawn at the center of the dimension line.

Left

The Left option will cause the text to be drawn toward the left extension line.

Right

The Right option will cause the text to be drawn toward the right extension line.

Home

The Home option returns the dimension text to its default position.

Angle

The Angle option changes the angle of the dimension text. AutoCAD prompts:

Specify angle for dimension text: *(specify the angle)*

The angle specified becomes the new angle for the dimension text.

EDITING DIMENSIONS WITH GRIPS

If the Grips feature is set to ON, you can select an associative dimension object and its grips will be displayed at strategic points. The grips will be located at the object ends of the extension lines, the intersections of the dimension and extension lines, and at the insertion point of the dimension text. In addition to the normal grip editing of the dimension as a group (rotate, move, copy, etc.), each grip can be selected for editing the configuration of the dimension as follows: Moving the object end grip of an extension line will move that specified point, making the value change accordingly. Horizontal and vertical dimensions remain horizontal and vertical. Aligned dimensions follow the alignment of the relocated point. Moving the grip at the intersection of the dimension line and one of the extension lines causes the dimension line to be nearer to or farther from the object dimensioned. Moving the grip at the insertion point of the text does the same as the intersection grip and also permits you to move the text back and forth along the dimension line.

DIMDISASSOCIATE COMMAND

The DIMDISASSOCIATE command removes associativity from selected associative dimensions.

DIMREASSOCIATE COMMAND

The DIMREASSOCIATE command associates selected dimensions to geometric objects. When prompted, select dimension(s) to be reassociated to object(s) and then step through the point selection process (using Object Snap mode) when each extension line point is marked with an X. If the X is in a box, then that point is already associated with a point on an object. You may skip this point by pressing ENTER.

DIMREGEN COMMAND

The DIMREGEN command updates the locations of all associative dimensions. This is sometimes necessary after panning or zooming with a wheel mouse, after opening a drawing that was modified in an earlier version of AutoCAD, or after opening a drawing that contains external references that have been modified.

DIMENSION STYLES

Each time a dimension is drawn it conforms to the settings of the Dimensioning System Variables in effect at the time. The entire set of Dimensioning System Variable settings can be saved in their respective states as a Dimension Style, with a name by which it can be recalled for application to a dimension later in the drawing session or in a subsequent session. Some Dimensioning System Variables affect every dimension. For example, every time a dimension is drawn, the DIMSCALE setting determines the relative size of the dimension. But DIMDLI, the variable that determines the offsets for Baseline Dimensions, comes into effect only when a Baseline Dimension is drawn. However, when a Dimension Style is created and named, all Dimensioning System Variable settings (except DIMASO and DIMSHO) are recorded in that Dimension Style, whether or not they will have an effect. The DIMASO and DIMSHO dimensioning variable settings are saved in the drawing separate from the Dimension Styles. As stated at the beginning of this chapter: "Dimension variables change when their associated settings are changed in the Modify Dimension Style dialog box. For example, the value of the DIMEXO (extension line offset) dimensioning system variable is established by the number in the **Offset from origin** text box in the Modify Dimension Style dialog box."

DIMENSION STYLE MANAGER DIALOG BOX

AutoCAD provides a comprehensive set of dialog boxes accessible through the Dimension Style Manager dialog box for creating new Dimension Styles and managing existing ones. In turn, these dialog boxes compile and store Dimensioning System Variable settings. Creating Dimension Styles through use of the DIMSTYLE command's dialog boxes allows you to make the desired changes to the appearance of dimensions without having to search for or memorize the names of the Dimensioning System Variables in order to change the settings directly.

Invoke the DIMSTYLE command:

Dimension toolbar	Choose the DIMSTYLE command (see Figure 7–50)
Dimension menu	Choose Style
Command: prompt	**dimstyle** (ENTER)

Figure 7–50 *Invoking the DIMSTYLE command from the Dimension toolbar*

AutoCAD displays the Dimension Style Manager dialog box, shown in Figure 7–51.

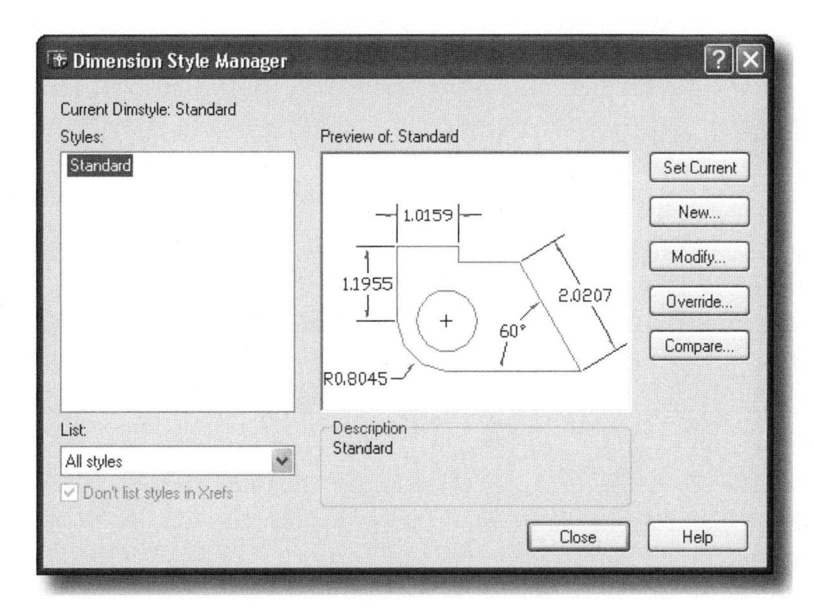

Figure 7–51 *Dimension Style Manager dialog box*

Name of the Current Dimension Style:

The **Current Dimstyle** heading shows the name of the Dimension Style to which the next drawn dimension will conform. The name of the **Current Dimstyle** is recorded as the value of the DIMSTYLE dimensioning system variable.

Listing of the Available Dimension Styles:

The **Styles:** section displays the name(s) of the current style and any other style(s) available depending on the option chosen in the **List:** selection box. If you choose one of the styles in the list, it is shown in white characters on a blue background and will be the Dimstyle acted upon when one of the buttons (**Set Current, New, Modify, Override,** or **Compare**) is chosen. It will also be the Dimstyle whose appearance is shown in the **Preview of:** viewing area.

Selection for listing the Dimension Styles:

The options in the **List:** selection box allows you to list all of the styles available in the drawing displayed or only those in use by choosing the appropriate option from this selection box. The Standard style cannot be deleted and is always available to be listed. Styles from externally referenced drawings can be listed, though they are not changeable in the current drawing.

Preview:

The view below the **Preview of:** heading shows how dimensions will appear when drawn using the current Dimstyle.

Description

The **Description** section shows the difference(s) between the Dimstyle chosen in the **Styles:** section and the current Dimstyle. For example, if the current Dimstyle is Standard and the Dimstyle chosen in the **Styles:** section is named Harnessing, then the **Description** section might say "Standard + Angle format = 1, Fraction format = 1, Length units = 4" to indicate that the Angle format is in Degrees/Minutes/Seconds, the Fraction format is in Diagonal, and the Length units is in Architectural, these settings being different from the corresponding settings for the Standard Dimstyle.

Set Current

Choosing the **Set Current** button causes the Dimstyle chosen in the **Styles**: section to become the current Dimstyle.

New

Choosing the **New** button causes the Create New Dimension Style dialog box to appear (see Figure 7–52).

The Create New Dimension Style dialog box allows you to create and name a new Dimstyle.

The **New Style Name** text box is where you enter the name of the new Dimstyle you wish to create.

The **Start With:** list box lets you choose the existing Dimstyle that you would like to start with in creating your new Dimstyle. The Dimensioning System Variables will be the same as those in the **Start With:** list box until you change their settings. In many cases you may wish to change only a few Dimensioning System Variable settings.

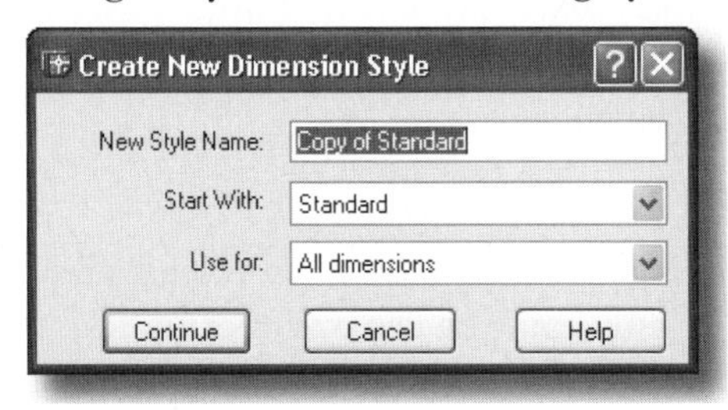

Figure 7–52 *Create New Dimension Style dialog box*

The **Use For:** list box lets you choose the type(s) of dimensions for which to apply the new Dimstyle you will be creating.

Modify

Choosing the **Modify** button in the Dimension Style Manager dialog box causes the Modify Dimension Style dialog box to appear (see Figure 7–53). There are six tabs to choose from: **Lines and Arrows, Text, Fit, Primary Units, Alternate Units,** and **Tolerances.**

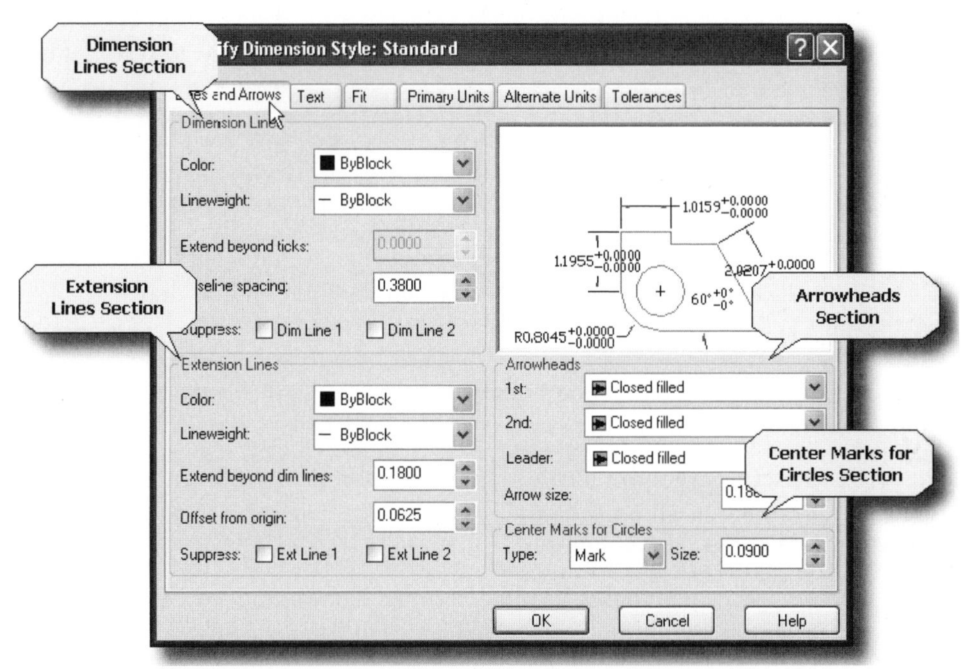

Figure 7–53 *Modify Dimension Style dialog box with Lines and Arrows tab chosen*

The **Lines and Arrows** tab allows you to modify the geometry of the various elements that make up the dimensions, such as Dimension Lines, Extension Lines, Arrowheads, and Center Marks for circles. There is a preview showing how dimensions will appear when drawn using the settings as you change them.

In both the **Dimension Line** and **Extension Line** sections:

> **Color:** list boxes let you choose how the color of Dimension Lines and Extension Lines will be determined. By Layer causes the color of the line to match that of its layer. By Block causes the color of the line to match that of its block reference if it is part of a block reference. You can also choose one of the standard colors or an Other color for the line color to match.
>
> **Lineweight:** list boxes lets you choose how the lineweight of Dimension and Extension Lines will be determined. By Layer causes the lineweight

of the line to match that of its layer. By Block causes the lineweight of the line to match that of its block reference if it is part of a block reference. You can also choose one of the standard lineweights or enter a value.

In the **Dimension Line** section only:

Extend beyond ticks: text box lets you enter the distance that Dimension Lines will be drawn past Extension Lines if the arrowhead type is a tick (small diagonal line).

Baseline spacing: text box lets you enter the distance that AutoCAD uses between Dimension Lines when they are drawn using the BASELINE DIMENSION command.

Suppress: check boxes let you determine if one or more of the Dimension or Extension Lines is suppressed when the dimension is drawn.

In the **Extension Line** section only:

Extend beyond dim lines: text box lets you enter the distance that Extension Lines will be drawn past Dimension Lines.

Offset from origin: text box lets you enter the distance that AutoCAD uses between the Extension Line and the origin point specified when drawing an extension line.

Suppress: check boxes let you determine if one or more of the Dimension or Extension Lines is suppressed when the dimension is drawn.

In the **Arrowhead** section:

1st, 2nd, and **Leader** list boxes let you determine if and how arrowheads will be drawn at the terminations of the Extension Lines or the start point of a Leader. Unless you specify a different type for the second line termination, it will be the same type as the first line termination. The type of arrowheads is recorded in DIMBLK if both the first and second are the same or in DIMBLK1 and DIMBLK2 if they are different..

The **Arrow size:** text box lets you enter the distance that AutoCAD uses for the length of an arrowhead when drawn in a dimension or leader.

Some of the arrowheads types included in the standard library are **None, Closed, Dot, Closed Filled, Oblique, Open, Origin Indication**, and **Right-Angle**. Figure 7–54 presents an example of each.

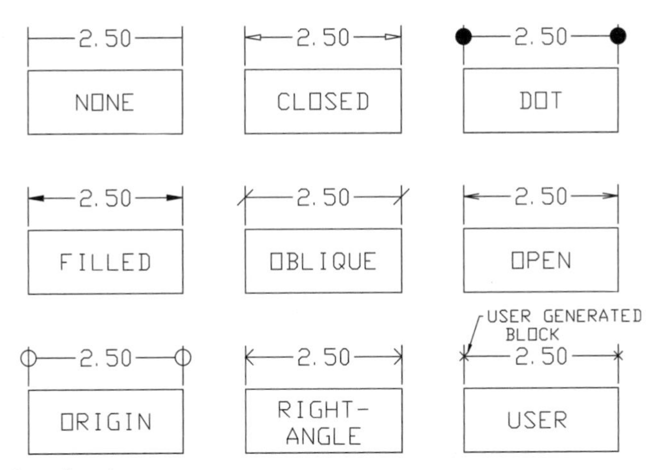

Figure 7–54 *Arrowhead types*

The **User Arrow** selection lets you use a previously saved block definition by entering its name. Figure 7–55 shows the related dialog box. The block definition should be created as though it were drawn for the right end of a horizontal Dimension Line, with the insertion point at the intersection of the Extension Line and the Dimension Line.

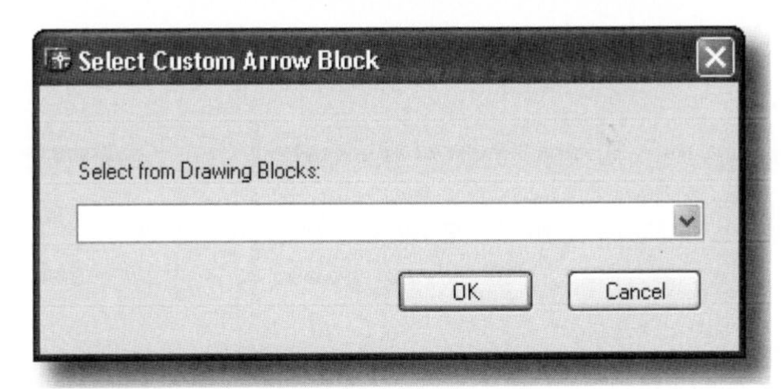

Figure 7–55 *Select Custom Arrow dialog box*

In the **Center Marks for Circles** section:

The **Type:** list box lets you choose if and what type of center mark is drawn when a Diameter Dimension or Radius Dimension is drawn. You may choose from **None, Mark,** and **Line.** If **Mark** is chosen, a Center Mark will be drawn with its length determined by the value in the **Size:** text box. If **Line** is chosen, the Center Marks will be drawn as lines that extend outside the circle a distance determined by the value in the **Size:** text box.

The **Text** tab lets you modify the appearance, location, and alignment of Dimension Text that is included when a dimension is drawn. There is a preview showing how dimensions will appear when drawn using the settings as you change them. (see Figure 7–56)

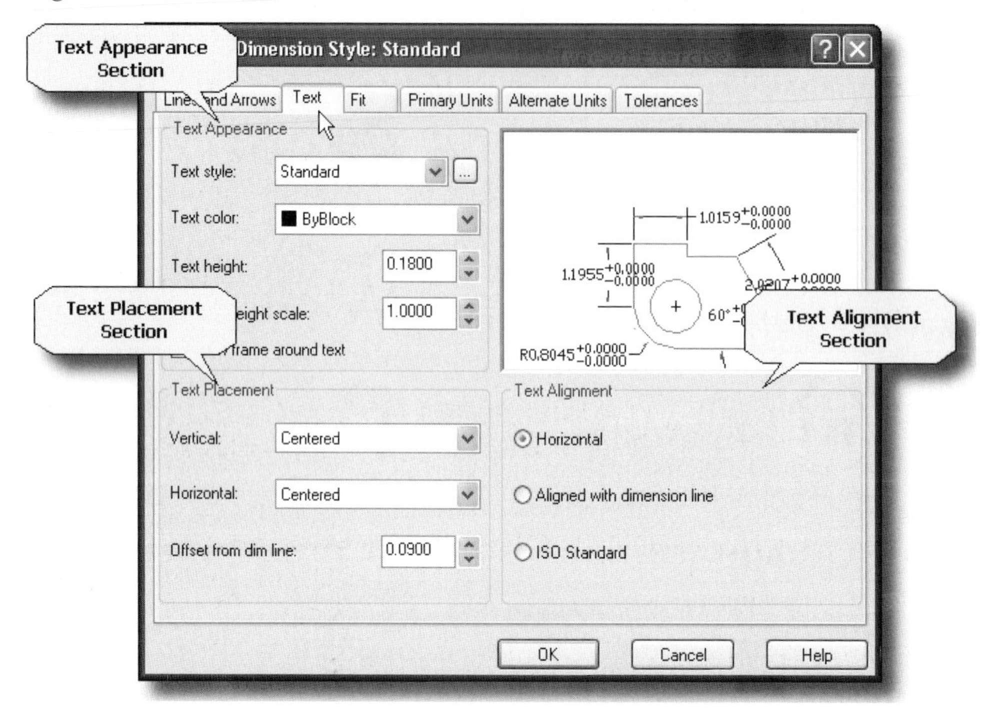

Figure 7–56 *Modify Dimension Style dialog box with Text tab chosen*

In the **Text Appearance** section:

The **Text style:** list box lets you choose the text style to which the Dimension Text will conform.

Note: Do not confuse Text Style with Dimension Style. Dimensions are drawn in accordance with the current Dimstyle, which has as part of its configuration a Text Style to which Dimension Text will conform.

The **Text color:** list box lets you choose how the color of Dimension Text will be determined. By Layer causes the color of the text to match that of its layer. By Block causes the color of the text to match that of its block reference if it is part of a block reference. You can also choose one of the standard colors for the text color to match.

The **Text height:** text box lets you enter the distance that AutoCAD uses for text height when drawing Dimension Text as part of the dimension.

The **Fraction height scale:** text box lets you enter a scale which determines the distance that AutoCAD uses for text height when drawing Dimension Text for fractions as part of the Dimension Text. This scale is the ratio of the text height for normal Dimension Text to the height of the fraction text.

The **Draw frame around text** check box, when checked, causes AutoCAD to draw the Dimension Text inside a rectangular frame.

In the **Text Placement** section:

The **Vertical:** list box lets you choose how the Dimension Text will be drawn in relation to the Dimension Line. The options include **Centered, Above, Outside,** and **JIS**.

The **Horizontal:** list box lets you choose how the Dimension Text will be drawn in relation to the Extension Lines. The options include **Centered, At Ext Line 1, At Ext Line 2, Over Ext Line 1,** and **Over Ext Line 2**.

The **Offset from dim line:** text box lets you enter the distance from the Dimension Line that AutoCAD uses when drawing Dimension Text as part of the dimension.

In the **Text Alignment** section, there are three radio buttons that let you choose how the text will be aligned in relation to the Dimension Lines. The options include **Horizontal, Aligned with dimension line,** and **ISO Standard**.

The **Fit** tab lets you to determine the arrangement of the various elements of dimensions when a dimension is drawn. There is a preview showing how dimensions will appear when drawn using the settings as you change them (see Figure 7–57).

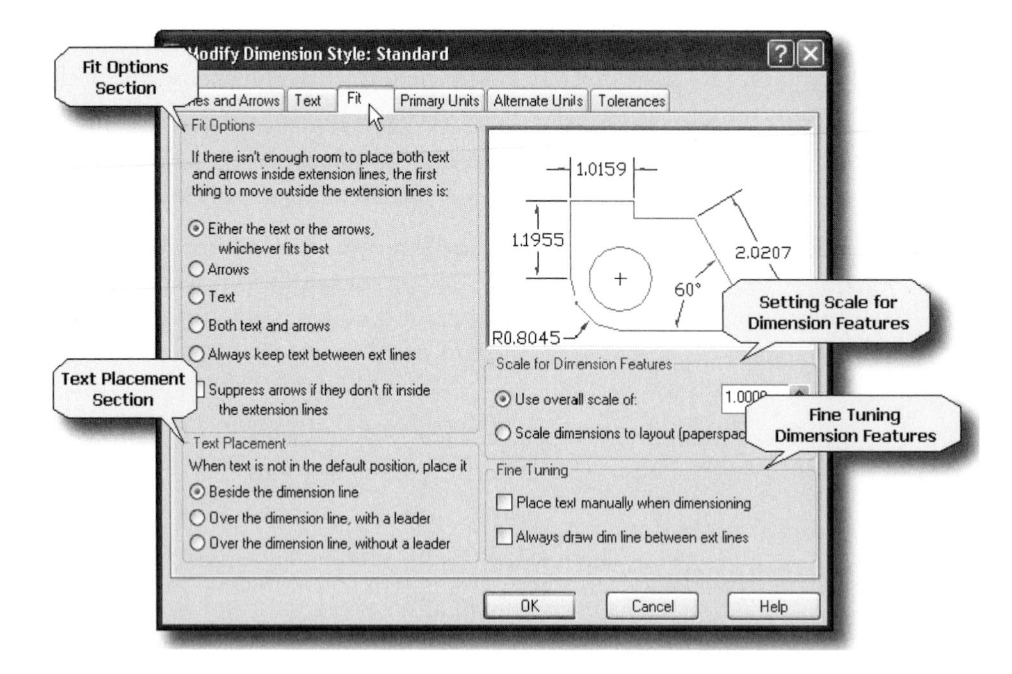

Figure 7–57 *Modify Dimension Style dialog box with Fit tab chosen*

In the **Fit Options** section, there are five radio buttons that let you choose which of the text or arrowheads will be drawn between the extension lines. The options include:

Choosing the **Either the text or the arrows, whichever fits best** button causes the text to be drawn outside and the arrows inside the extension lines if there is room for the arrows only.

Choosing the **Arrows** button causes the text to be drawn outside and the text inside the extension lines if there is room for the arrows only.

Choosing the **Text** button causes the arrows to be drawn outside and the text inside the extension lines if there is room for the text only.

Choosing the **Both text and arrows** button causes both the text and arrows to be drawn outside if the Dimension Lines are forced outside and both to be drawn inside if space is available.

Choosing the **Always keep text between ext lines** button causes text to always be drawn between the Extension Lines.

The **Suppress arrows if they don't fit inside the extension lines** check box, when checked, causes AutoCAD to not draw the arrowheads between extension lines if they do not fit.

In the **Text Placement** section there are three radio buttons that let you choose how the text will be placed when text is not in the default position. The options include: **Beside the dimension line, Over the dimension line, with a leader,** and **Over the dimension line, without a leader**.

In the **Scale for Dimension Features** section:

> The **Use overall scale of:** radio button activates the text box that lets you enter the number AutoCAD uses as a ratio of the true scale to drawn dimensions. This is useful when you wish to create different parts of the drawing at different scales but have the dimension elements uniform in size. This does not include distances, coordinates, angles, or tolerances.
>
> The **Scale of Dimension Features** section lets you increase or decrease the size of all of the dimensioning components uniformly by the factor entered. For example, say you have set the sizes of the components to plot at a scale factor of 1/4" = 1'-0" (1:48) and wish to draw a detail to true size (including the dimensions) and then scale it up to 3/4" = 1'-0" (1:16). You can set the dimensioning scale to 1/3 so that the components, when scaled up by a factor of 3, will plot at the same size as those not scaled up.
>
> Choosing the **Scale dimensions to layout (paperspace)** radio button causes dimensions to be drawn with elements scaled to layout.

In the **Fine Tuning** section:

> Choosing the **Place text manually when dimensioning** check box allows you to dynamically specify where Dimension Text is placed when the dimension is drawn.
>
> Choosing the **Always draw dim line between ext lines** button causes a Dimension Line to be drawn between the extension lines regardless of the distance between Extension Lines.

The **Primary Units** tab lets you to determine the appearance and format of numerical values of distances and angles when they are drawn with dimensions. There is a preview showing how dimensions will appear when drawn using the settings as you change them (see Figure 7–58).

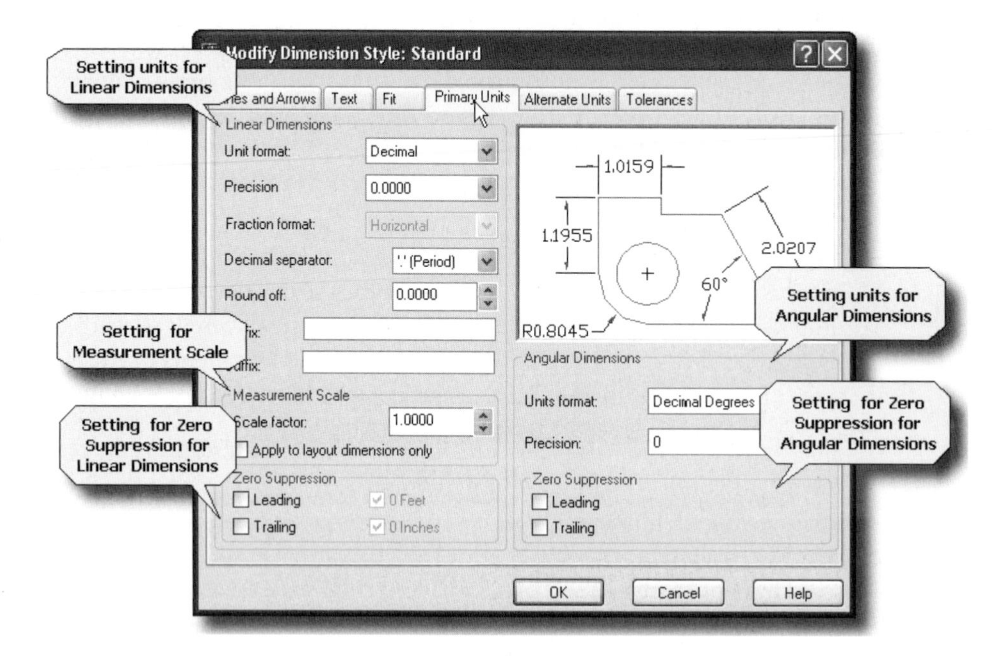

Figure 7–58 *Modify Dimension Style dialog box with Primary Units tab chosen*

In the **Linear Dimensions** section:

The **Unit format:** list box lets you choose the format of the units that AutoCAD uses when drawing Dimension Text as part of the dimension. The options include **Scientific**, **Decimal**, **Engineering**, **Architectural**, **Fractional**, and **Windows Desktop**. (see Figure 7–59). The value is recorded in DIMUNIT.

The **Precision:** list box lets you determine to how many decimals the text will be shown if the format is in Scientific, Decimal, or Windows Desktop Units. It will show to how many decimals the inches will be shown if the format is in Engineering Units. It will show how small the fraction will be shown in Architectural or Fractional Units, and the options in the **Fractional format:** shows whether the fractions will be shown **Horizontal**, **Diagonal**, or **Not Stacked**. This value is recorded in DIMDEC.

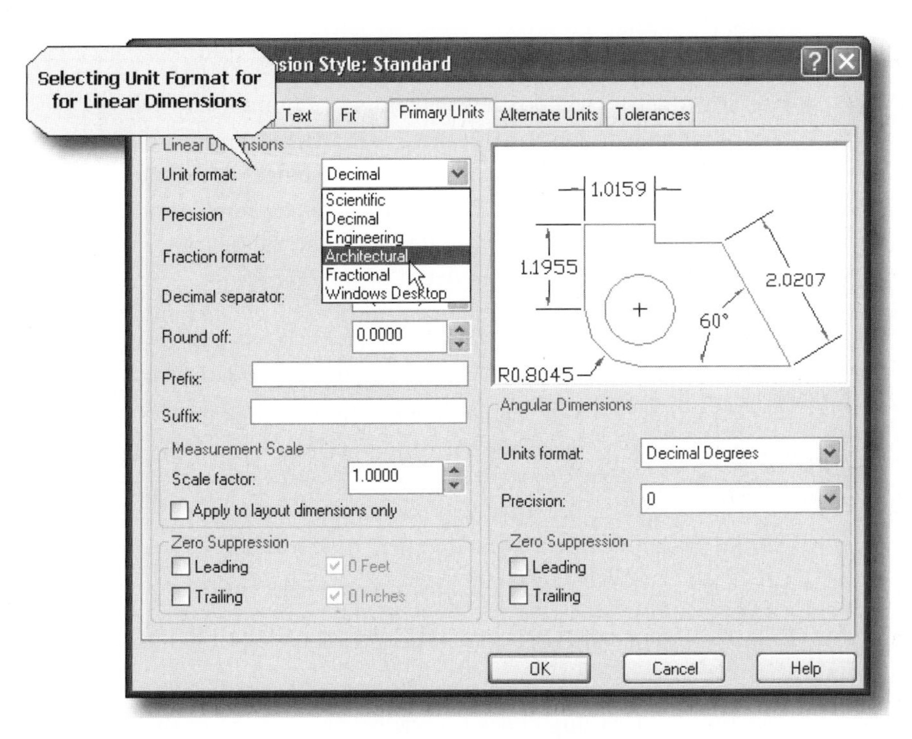

Figure 7–59 *Unit Format options in the Primary Units tab*

The **Decimal separator:** list box lets you choose whether the decimal will be separated by a **Period, Comma,** or a **Space** if the format is in Scientific, Decimal, or Windows Desktop Units.

The **Round off:** text box lets you enter the value to which distances will be rounded off. For example, a value of 0.5 causes dimensions to be rounded to the nearest 0.5 units.

The **Prefix:** text box allows you to include a prefix in the Dimension Text. The prefix text will override any default prefixes, such as those used in radius (R) dimensioning. The value is recorded in DIMPOST.

The **Suffix:** text box allows you to include a suffix in the Dimension Text. If you specify tolerances, AutoCAD includes the suffix in the tolerances, as well as in the main dimension. The value is recorded in DIMPOST.

In the **Measurement Scale** section:

The **Scale factor:** text box lets you enter the number AutoCAD uses as a ratio of the true dimension distances to drawn dimensions distances, or the linear scale factor for the linear measured distances of a dimension without affecting the components, angles, or tolerance values. For example, you are drawing with the intention of plotting at the quarter-size scale, or 3" = 1'-0". You have scaled up a detail by a factor of 4 so that it will plot to full scale. If you wish to dimension it after it has been enlarged, you can set the linear scale factor in this section of the Primary Units dialog box to .25 so that dimensioned distances will represent the dimension of the object features before it was scaled up. This method keeps the components at the same size as dimensions created without a scale change. The value is recorded in DIMLFAC. This is useful when you wish to create different parts of the drawing at different scales but have the dimension elements uniform in size. Selecting the **Apply to layout dimensions only** check box causes the ratio to be applied only to layout dimensions.

In the **Zero Suppression** section you can specify whether or not zeros are displayed, as follows:

Selecting the **Leading** check box causes zeros ahead of the decimal point to be suppressed. For example: .700 is displayed instead of 0.700.

Selecting the **Trailing** check box causes zeros behind the decimal point to be suppressed. For example: 7 or 7.25 is displayed instead of 7.000 or 7.250, respectively.

Selecting the **0 Feet** check box causes zeros representing feet to be suppressed if the Dimension Text represents inches and/or fractions only. For example: 7" or 7 1/4" is displayed instead of 0'-7" or 0'-7 1/4".

Selecting the **0 Inches** check box causes zeros representing inches to be suppressed if the Dimension Text represents feet only. For example: 7' is displayed instead of 7'-0".

In the **Angular Dimensions** section:

The **Units format:** list box lets you choose the format of the units for the angular Dimension Text. Options include **Decimal Degrees, Degrees/Minutes/Seconds, Grads,** and **Radians.** The value is recorded in DIMAUNIT.

In the **Zero Suppression** section you can specify whether or not **Leading** or **Trailing** zeros are displayed.

The **Alternate Units** tab (see Figure 7–60) is similar to the **Primary Units** tab, except for the **Display alternate units** check box, which lets you enable (DIMALT set to 1) or disable (DIMALT set to 0) alternate units, and there is no **Angular Dimensions** section. There is a **Placement** section with **After primary value** and **Before primary value** radio buttons that let you determine where to place the Alternate Units.

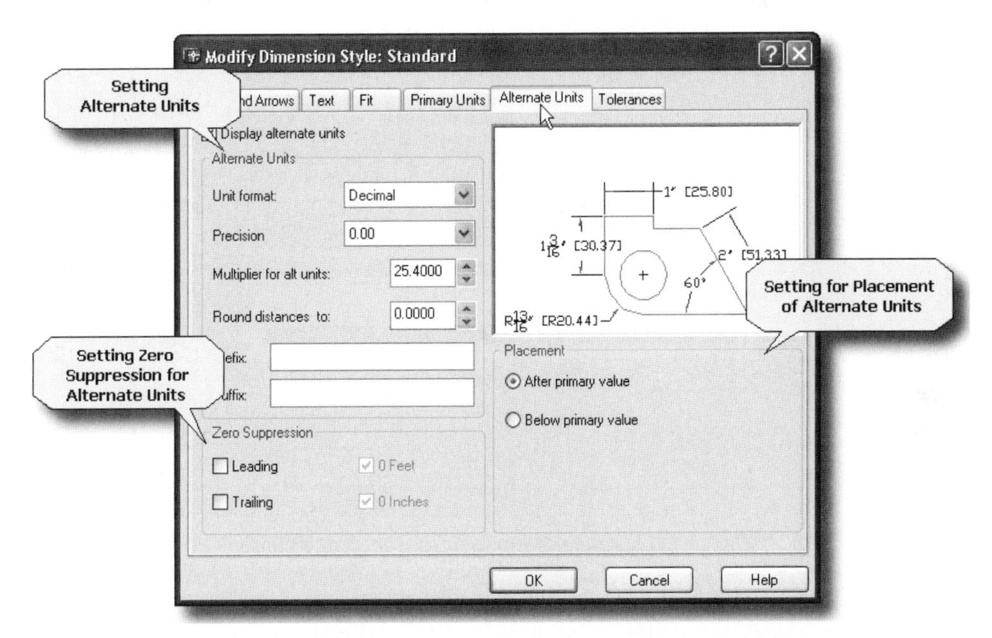

Figure 7–60 *Modify Dimension Style dialog box with Alternate Units tab chosen*

The **Tolerances** tab (see Figure 7–61) allows you to set the tolerance method and appropriate settings. Detailed explanation is provided in the "Tolerances" section, earlier in this chapter.

Once the necessary changes are made for the appropriate settings, choose the **OK** button to close Modify Dimension Style dialog box and then choose the **Close** button to close the Dimension Style Manager dialog box.

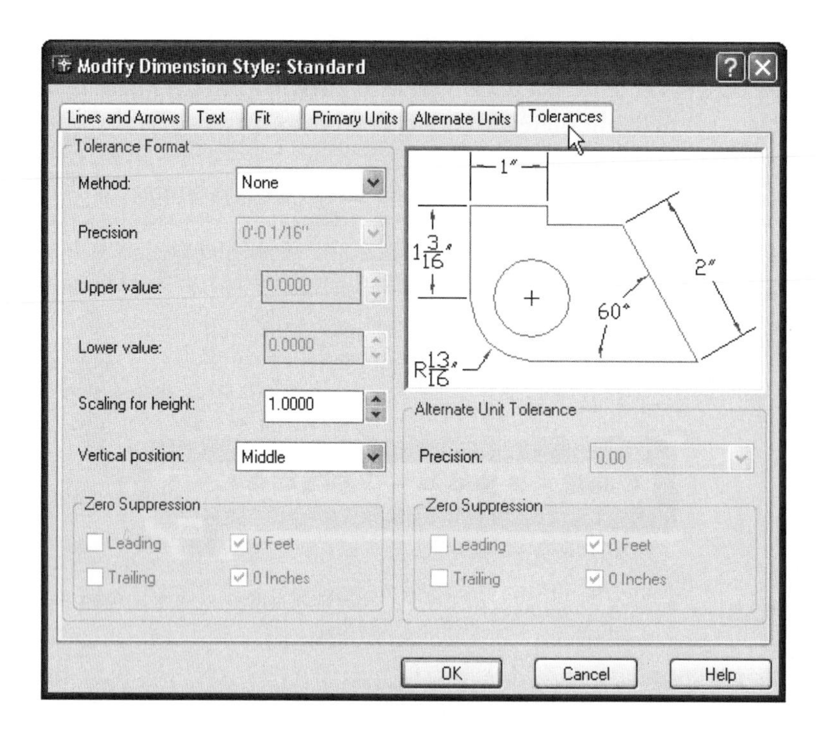

Figure 7–61 *Modify Dimension Style dialog box with Tolerances tab chosen*

EDITING DIMENSIONS USING THE SHORTCUT MENU

Dimensions already drawn can be edited easily by using the right-click shortcut menu in the following manner.

First, GRIPS must be enabled by choosing **Options...** from the Tools menu and then selecting the **Enable Grips** check box from the **Selection** section. Once the grips have been made available, select the dimension(s) that you wish to edit. You will see the grips at the strategic points of the elements of the dimension(s) you have selected. Now simply right-click (with the cursor anywhere in the drawing area) and the shortcut menu will include a section for editing dimensions (see Figure 7–62).

The **Dim Text Position** flyout includes options for **Above Dim Line, Centered, Home Text, Move Text Alone, Move With Leader,** and **Move With Dim Line.** Choosing one of these options causes the text in the selected dimension(s) to conform to the position option chosen.

The **Precision** flyout includes a range of 0.0 to 0.000000 for Decimal dimensions and a range of 0'-0 to 0'-0 1/128 for Architectural and Fractional dimensions. Choosing one of the decimal or fractional options causes the distances/angles in the selected dimension(s) to conform to the level of precision chosen.

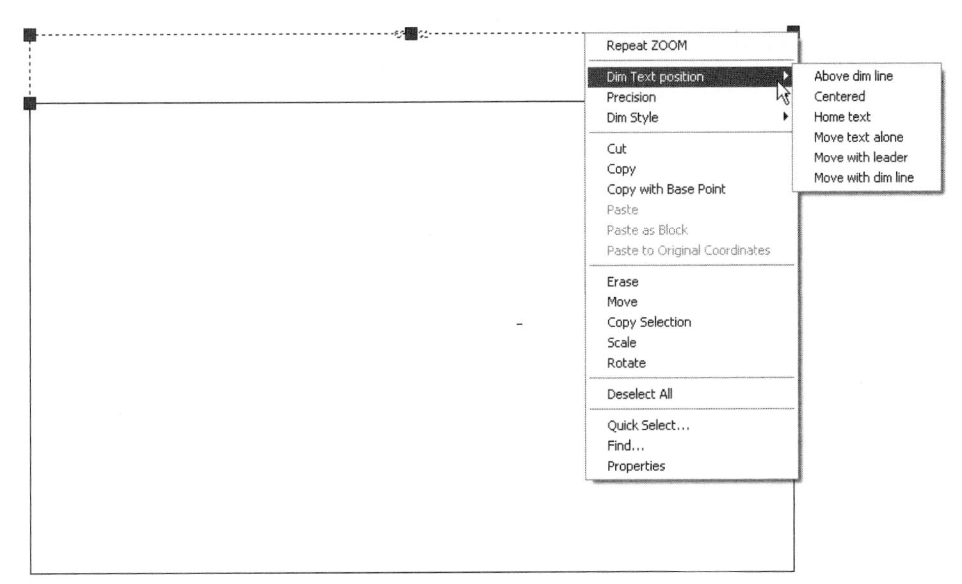

Figure 7–62 *Shortcut Menu with dimension editing section*

Note: If the dimensions selected (and identified by visible grips) include styles with both decimal and fractional units, the PRECISION option is not available in the shortcut menu.

The **Dim Style** flyout includes an option for **Save As New Style...**, and a list of available Dimstyles to choose from. If you have made some changes to settings of individual variables in a particular dimension, you can choose the **Save As New Style...** option to create and name a new, unique Dimstyle conforming to the selected dimension's variable settings.

OVERRIDING THE DIMENSION FEATURE

The DIMOVERRIDE command allows you to change one of the features in a dimension without having to change its dimension style or create a new dimension style. For example, you may wish to have one leader with the text centered at the ending horizontal line, rather than over the ending horizontal line in the manner that the current dimension style may call for. By invoking the DIMOVERRIDE command, you can respond to the prompt with the name of the Dimensioning System Variable (DIMTAD in this case) and set the value to 0 rather than 1. Then you can select the dimension leader you wish to have overridden.

Invoke the DIMOVERRIDE command:

Command: prompt	**dimoverride** (ENTER)

AutoCAD prompts:

Command: **dimoverride** (ENTER)
Enter dimension variable name to override or [Clear overrides]: *(specify the Dimensioning System Variable name or right-click and choose* CLEAR *to clear overrides)*

If you specify a Dimensioning System Variable, you are prompted:

Enter new value for dimension variable <current>: *(specify the new value)*
Enter dimension variable name to override: *(specify another Dimension System Variable or press* ENTER*)*

If you press ENTER, you are prompted:

Select objects: *(select the dimension objects)*

The dimensions selected will have the Dimensioning System Variable settings overridden in accordance with the value specified.

If you choose Clear, you are prompted:

Select objects: *(select the dimension to clear the overrides)*

The dimensions selected will have the Dimensioning System Variable setting overrides cleared.

UPDATING DIMENSIONS

The DIMENSION UPDATE command permits you to make selected existing dimension(s) conform to the settings of the current dimension style.

Invoke the DIMENSION UPDATE command:

Dimension toolbar	Choose the DIMENSION UPDATE command (see Figure 7–63)
Dimension menu	Choose Update

Figure 7–63 *Invoking the* DIMENSION UPDATE *command from the Dimension toolbar*

AutoCAD prompts:

Enter a dimension style option
[Save/Restore/STatus/Variables/Apply/?] <Restore>: apply
Select objects: *(select any dimension(s) whose settings you wish to have updated to conform to the current dimension style).*

Note: Note that the UPDATE command is actually the DIMENSION STYLE EDIT command in which AutoCAD automatically chooses the Apply option for you. All you have to do is select a dimension.

Open the Exercise Manual PDF file for Chapter 7 on the accompanying CD for project and discipline specific exercises.

If you have the accompanying Exercise Manual, refer to Chapter 7 for project and discipline specific exercises.

REVIEW QUESTIONS

1. Dimension types available in AutoCAD include:

   a. Linear d. Radius

   b. Angular e. All of the above

   c. Diameter

2. The associative dimension drawn with the DIMASO variable set to ON has all of its separate parts drawn as separate objects.

   a. True b. False

3. The LINEAR DIMENSIONING command allows you to draw horizontal, vertical, and aligned dimensions.

   a. True b. False

4. To draw a linear dimension you must (1) specify the first extension line origin, (2) locate the dimension line, and then (3) specify the second extension line.

   a. True b. False

5. The ANGULAR DIMENSION command allows you to draw angular dimensions between two parallel lines.

   a. True b. False

6. By default, the dimension text for a radius dimension is preceded by:

   a. Radius

   b. Rad

   c. R

7. Using PROPERTIES (modify properties) command will allow you to override the Dimensioning System Variable settings for a single dimension, without modifying the base dimension style.

   a. True b. False

8. The BASELINE DIMENSION command is used to draw dimensions from a single datum baseline.

   a. True

   b. False

9. The DIMCENTER command allows you to draw center cross marks or centerlines in a circle.

   a. True

   b. False

10. You must explode a dimension before you can use the DIMTEDIT command to edit the dimension text.

    a. True

    b. False

11. The suppress option in the **Lines and Arrows** tab of the Modify Dimension Style dialog box allows you to suppress only one extension line at a time.

    a. True

    b. False

12. The arrowhead section in **Lines and Arrows** tab of the Modify Dimension Style dialog box allows you to change the size and style of your arrowheads.

    a. True

    b. False

13. You must use the UNITS command to determine how many decimal places will be displayed in the dimension text.

    a. True

    b. False

Directions: For each of the following questions, select which of the dimensioning tabs the option can be found on in the following list.

a. Lines and Arrows tab

b. Text tab

c. Fit tab

d. Primary Units tab

e. cannot be found in the dimensioning dialog boxes

14. Arrowhead size
15. Associative/Non-associative setting
16. Text height
17. Arrowhead style
18. Text location
19. Suppressing of extension lines
20. Linear tolerance settings
21. Number of decimal places for a linear dimension

Directions: Answer the following questions based on Figure QX7–1.

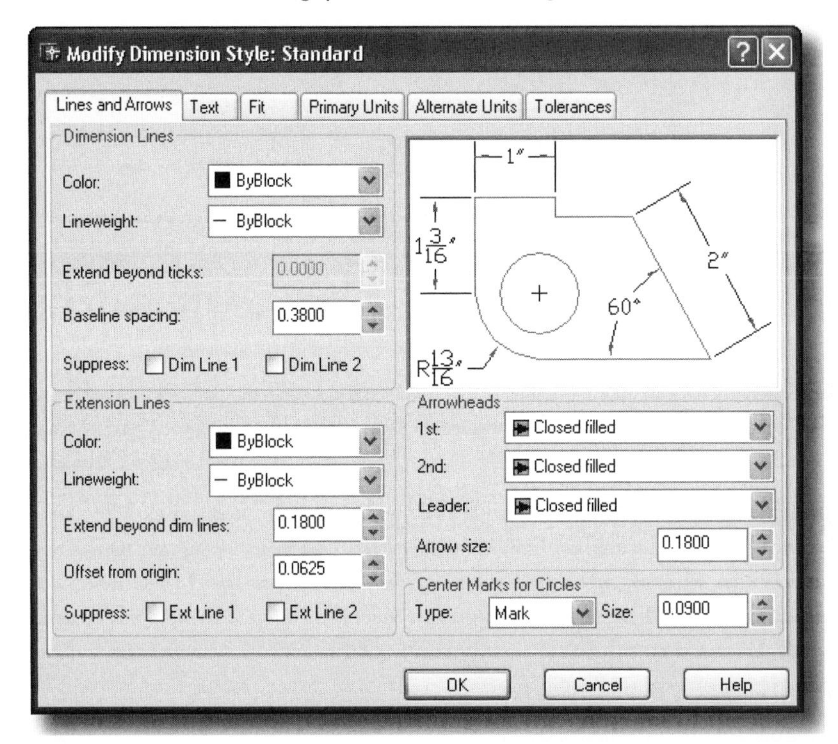

Figure QX7–1 *Modify Dimension Style dialog box with Lines and Arrows tab selected*

22. Which Dimension Variable must be changed to modify the size of Arrowheads?

 a. DIMARO
 b. DIMASZ
 c. DIMTSZ
 d. DIMDLI

23. Which Dimension Variable must be changed to modify Arrowhead types?

    a. DIMASZ c. DIMBLK

    b. DIMTSZ d. DIMEXE

24. Which Dimension Variable must be modified to change the color of dimension lines?

    a. DIMCLRE c. DIMCLRL

    b. DIMCLRT d. DIMCLRD

25. Which Dimension Variable is used to hide the first extension line?

    a. DIMEXL c. DIMSE1

    b. DIMOFF d. DIMEL1

26. Which Dimension Variable establishes the distance from the object to the start of the extension line?

    a. DIMEXO c. DIMGAP

    b. DIMSEL d. DIMSPC

27. Which Dimension Variable establishes the distance from the first dimension line to the second dimension line when using Baseline Dimensions?

    a. DIMD12 c. DIMDLI

    b. DIMSPC2 d. DIMDSPC

28. Which Dimension Variable is used to modify Architectural tick terminators?

    a. DIMARC c. DIMTIC

    b. DIMART d. DIMTSZ

29. Which Dimension Variable is used to hide the second dimension line?

    a. DIMSDL c. DIM2OFF

    b. DIMHD2 d. DIMSD2

30. When one desires for the extension lines to be drawn past the dimension lines which Dimension Variable is used to perform this function?

    a. DIMEXE c. DIMEXT

    b. DIMELD d. DIMDLE

31. Which Dimension Variable is used to have a 2nd arrowhead that is different from the 1st?

    a. DIMARW2 c. DIMBLK2

    b. DIMABLK d. dimaro2

Plotting/Printing

INTRODUCTION

One task has not changed much in the transition from board drafting to CAD: obtaining a hard copy. The term "hard copy" describes a tangible reproduction of a screen image. The hard copy is usually a reproducible medium from which prints are made and can take many forms, including slides, videotape, prints, or plots. This chapter describes the most commonly used processes for getting a hard copy: plotting/printing.

In manual drafting, if you need your drawing to be done in two different scales, you physically draw the drawing for two different scales. In CAD, with minor modifications, you plot or print the same drawing in different scale factors on different sizes of paper. In AutoCAD, you can even compose your drawing in Paper Space with limits that equal the sheet size and plot it at 1:1 scale.

After completing this chapter, you will be able to do the following:

- Plan the plotted sheet
- Plot from model space
- Set up a layout
- Create and modify a layout
- Create floating viewports
- Scale views relative to paper space
- Control the visibility of layers within viewports
- Plot from Layout (WYSIWYG)
- Create and modify plot style tables
- Change the Plot Style Property for an Object or Layer
- Configure plotters

PLANNING THE PLOTTED SHEET

Planning ahead is still required in laying out the objects to be drawn on the final sheet. The objects drawn on the plotted sheet must be arranged. At least in CAD, with its true-size capability, an object can be started without first laying out a plotted sheet.

But eventually, limits, or at least a displayed area, must be determined. For schematics, diagrams, and graphs, plotted scale is of little concern. But for architectural, civil, and mechanical drawings, plotting to a conventional scale is a professionally accepted practice that should not be abandoned just because it can be circumvented.

When setting up the drawing limits, you must take the plotted sheet into consideration to get the entire view of the object(s) on the sheet. So, even with all the power of the CAD system, some thought must still be given to the concept of scale, which is the ratio of true size to the size plotted. In other words, before you start drawing, you should have an idea at what scale the final drawing will be plotted or printed on a given size of paper.

The limits should correspond to some factor of the plotted sheet. If the objects will fit on a 24" x 18" sheet at full size with room for a border, title block, bill of materials, dimensioning, and general notes, then set up your limits to (0,0) (lower left corner) and (24,18) (upper right corner). This can be plotted or printed at 1:1 scale, that is, one object unit equals one plotted unit.

Plot scales can be expressed in several formats. Each of the following five plot scales is exactly the same; only the display formats differ.

1/4" = 1'-0"

1" = 4'

1 = 48

1:48

1/48

A plot scale of 1:48 means that a line 48 units long in AutoCAD will plot with a length of 1 unit. The units can be any measurement system, including inches, feet, millimeters, nautical miles, chains, angstroms, and light-years, but, by default, plotting units in AutoCAD are inches.

There are four variables that control the relationship between the size of objects in the AutoCAD drawing and their sizes on a sheet of paper produced by an AutoCAD plot:

- Size of the object in AutoCAD. For simplification it will be referred to as ACAD_size.
- Size of the object on the plot. For simplification it will be referred to as ACAD_plot.
- Maximum available plot area for a given sheet of paper. For simplification it will be referred to as ACAD_max_plot.
- Plot scale. For simplification it will be referred as to ACAD_scale.

The relationship between the variables can be described by the following three algebraic formulas:

ACAD_scale	= ACAD_plot / ACAD_size
ACAD_plot	= ACAD_size x ACAD_scale
ACAD_size	= ACAD_plot / ACAD_scale

EXAMPLE OF COMPUTING PLOT SCALE, PLOT SIZE, AND LIMITS

An architectural elevation of a building 48' wide and 24' high must be plotted on a 36" x 24" sheet. First, you determine the plotter's maximum available plot area for the given sheet size. This depends on the model of plotter you use.

In the case of an HP plotter, the available area for 36" x 24" is 33.5" x 21.5". Next, you determine the area needed for the title block, general notes, and other items, such as an area for revision notes and a list of reference drawings. For the given example, let's say that an area of 27" x 16" is available for the drawing.

The objective is to arrive at one of the standard architectural scales in the form of x in. = 1 ft. The usual range is from 1/16" = 1'-0" for plans of large structures to 3" = 1'-0" for small details. To determine the plot scale, substitute these values for the appropriate variables in the formula:

ACAD_scale	= ACAD_plot/ACAD_size
ACAD_scale	= 27"/48' for *X* axis
	= 0.5625"/1'-0" or 0.5625"=1'-0"

The closest standard architectural scale that can be used in the given situation is 1/2" = 1'-0" (0.5" = 1'-0", 1/24 or 1:24).

To determine the size of the object on the plot, substitute these values for the appropriate variables in the formula:

ACAD_plot	= ACAD_size x ACAD_scale
ACAD_plot	= 48' x (0.5"/1') for *X* axis
	= 24" (less than the 27" maximum allowable space on the paper)
ACAD_plot	= 24' x (0.5"/1') for *Y* axis
	= 12" (less than the 16" maximum allowable space on the paper)

If instead of 1/2" = 1'-0" scale, you wish to use a scale of 3/4" = 1'-0", then the size of the object on the plot will be 48' x (0.75"/1') = 36" for the *X* axis. This is more than the available space on the given paper, so the drawing will not fit on the given paper size. You must select a larger paper size.

Once the plot scale is determined and you have verified that the drawing fits on the given paper size, you can then determine the drawing limits for the plotted sheet size of 33.5" x 21.5".

To determine the limits for the *X* and *Y* axes, substitute the appropriate values in the formula:

ACAD_limits (*X* axis)	= ACAD_max_plot/ACAD_scale
	= 33.5"/(0.5"/1'-0")
	= 67'
ACAD_limits (*Y* axis)	= 21.5"/(0.5"/1'-0")
	= 43'

Appropriate limits settings in AutoCAD for a 36" x 24" sheet with a maximum available plot area of 33.5" x 21.5" at a plot scale of 0.5" = 1'-0" would be:

```
lower left corner: 0,0
upper right corner: 67',43'
```

Another consideration in setting up a drawing for user convenience is to have the (0,0) coordinates at some point other than the lower left corner of the drawing sheet. Many objects have a reference point from which other parts of the object are dimensioned. Being able to set that reference point to (0,0) is very helpful. In many cases, the location of (0,0) is optional. In other cases, the coordinates should coincide with real coordinates, such as those on an industrial plant area block. In still other cases, only one set of coordinates might be a governing factor.

In this example, the 48' wide x 24' high front elevation of the building is to be plotted on a 36" x 24" sheet at a scale of 1/2" = 1'-0". It has been determined that (0,0) should be at the lower left corner of the front elevation view, as shown in Figure 8–1.

Centering the view on the sheet requires a few minutes of layout time. Several approaches allow the drafter to arrive at the location of (0,0) relative to the lower left corner of the plotted sheet or limits. Having computed the limits to be 67' wide x 43' high, the half-width and half-height (dimensions from the center) of the sheet are 33.5' and 21.5' to scale, respectively. Subtracting the half-width of the building from the half-width of the limits will set the *X* coordinate of the lower left corner at

–9.5' (from the equation, 24' – 33.5'). The same is done for the *Y* coordinate –9.5' (12' – 21.5'). Therefore, the lower left corner of the limits are at (–9.5',–9.5').

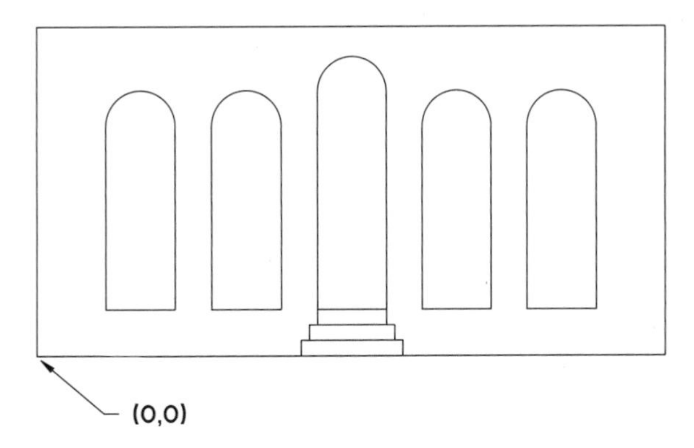

Figure 8–1 *Setting the reference point to the origin (0,0) in a location other than the lower left corner*

Appropriate limits settings in AutoCAD for a 36" x 24" sheet with a maximum available plot area of 33.5" x 21.5" by centering the view at a plot scale of 0.5" = 1'-0" (see Figure 8–2) is:

lower left corner: –9.5',–9.5'
upper right corner: 57.5',33.5'

Note: The absolute *X* coordinates, when added (57.5' + 9.5',) equal 67', which is the width of the limits, and the absolute *Y* coordinates (33.5' + 9.5') equal 43', which is the height of the limits.

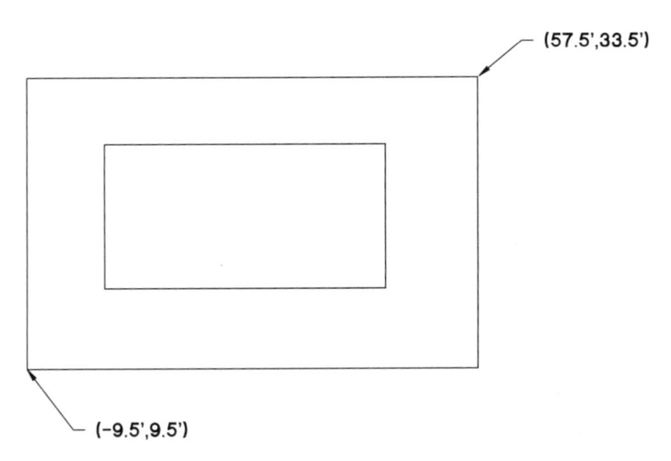

Figure 8–2 *Setting the limits to the maximum available plot area by centering the view*

SETTING FOR LTSCALE

As explained in Chapter 3, the LTSCALE system variable provides a method of adjusting the linetypes to a meaningful scale for the drawing. This sets the length of dashes and spaces in linetypes. When the value of LTSCALE is set to the reciprocal of the plot scale, the linetypes provided with AutoCAD plot out on paper at the sizes they are defined in ACAD.LIN.

LTSCALE = 1/ACAD_scale

SETTING FOR DIMSCALE

As explained in Chapter 7, AutoCAD provides a set of dimensioning variables that control the way dimensions are drawn. The DIMSCALE dimension variable is applied globally to all dimensions that are applied to sizes or distances, as an overall scaling factor. The default DIMSCALE value is set to 1. When DIMSCALE is set to the reciprocal of the plot scale, it applies globally to all dimension variables for the plot scale factor.

DIMSCALE = 1/ACAD_scale

If necessary, you can set individual dimensioning variables to the size that you want the dimension to appear on the paper by substituting the appropriate values in the following formula:

size of the plotted dimvars_value = dimvars_value x ACAD_scale x DIMSCALE

As an example, here's how to determine the arrow size DIMASZ for a plot scale of 1/2" = 1'-0", DIMSCALE set to 1, and default DIMASZ of 0.18":

size of the plotted arrow = 0.18" x 24 x 1

= 4.32"

SCALING ANNOTATIONS AND SYMBOLS

How can you determine the size at which text and symbols (blocks) will plot? As mentioned earlier, you almost always draw objects in their actual size, that is, to real-world dimensions. Even in the case of text and blocks, you place them at the real-world dimensions. In the previous example, the architectural elevation of a building 48' x 24' is drawn to actual size and plotted to a scale of 1/2" = 1'-0'. Let's say you wanted your text to plot at 1/4" high. If you were to create your text and annotations at 1/4", they would be so small relative to the elevation drawing itself that you could not read the characters.

Before you begin placing the text, you need to know the scale at which you will eventually plot the drawing. In the previous example of an architectural elevation, the plot scale is 1/2" = 1'-0" and you want the text to plot 1/4" high. You need to find a relationship between 1/4" on the paper and the size of the text for the real-world dimensions in the drawing. If 1/2" on the paper equals 12" in the model, then 1/4"-high text on the paper equals 6", so text and annotations should be drawn at 6" high in the drawing to plot at 1/4" high at this scale of 1/2" = 1'-0". Similarly, you can calculate the various text sizes for a given plot scale. Table 8–1 shows the model text size needed to achieve a specific plotted text height at some common scales.

Table 8–1 *Text Size Corresponding to Specific Plotted Text Height at Various Scales*

		PLOTTED TEXT HEIGHT								
SCALE	**FACTOR**	**1/16"**	**3/32"**	**1/8"**	**3/16"**	**1/4"**	**5/16"**	**3/8"**	**1/2"**	**5/8"**
1/16" = 1'-0"	192	12"	18"	24"	36"	48"	60"	66"	96"	120"
1/8" = 1'-0"	96	6"	9"	12"	18"	24"	30"	36"	48"	60"
3/16" = 1'-0"	64	4"	6"	8"	12"	16"	20"	24"	32"	40"
1/4" = 1'-0"	48	3"	4.5"	6"	9"	12"	15"	18"	24"	30"
3/8" = 1'-0"	32	2"	3"	4"	6"	8"	10"	12"	16"	20"
1/2" = 1'-0"	24	1.5"	2.25"	3"	4.5"	6"	7.5"	9"	12"	15"
3/4" = 1'-0"	16	1"	1.5"	2"	3"	4"	5"	6"	8"	10"
1" = 1'-0"	12	0.75"	1.13"	1.5"	2.25"	3"	3.75"	4.5"	6"	7.5"
1 1/2" = 1'-0"	8	0.5"	.75"	1"	1.5"	2"	2.5"	3"	4"	5"
3" = 1'-0"	4	0.25"	.375"	0.5"	0.75"	1"	1.25"	1.5"	2"	2.5"
1" = 10'	120	7.5"	11.25"	15"	22.5"	30"	37.5"	45"	60"	75"
1" = 20'	240	15"	22.5"	30"	45"	60"	75"	90"	120"	150"
1" = 30'	360	22.5"	33.75"	45"	67.5"	90"	112.5"	135"	180"	225"
1" = 40'	480	30"	45"	60"	90"	120"	150"	180"	240"	300"
1" = 50'	600	37.5"	56.25"	75"	112.5"	150"	187.5"	225"	300"	375"
1" = 60'	720	45"	67.5"	90"	135"	180"	225"	270"	360"	450"
1" = 70'	840	52.5"	78.75"	105"	157.5"	210"	262.5"	315"	420"	525"
1" = 80'	960	60"	90"	120"	180"	240"	300"	360"	480"	600"
1" = 90'	1080	67.5"	101.25"	135"	202.5"	270"	337.5"	405"	540"	675"
1" = 100'	1200	75"	112.5"	150"	225"	300"	375"	450"	600"	750"

WORKING IN MODEL SPACE AND PAPER SPACE

One of the useful features of AutoCAD is the option to work on your drawing in two different environments, model space and paper space. AutoCAD allows you to plot a drawing from model space as well as paper space, also referred to as Layout. Most of the drafting and design work is created in the *3D* environment of model space, even though your objects may have been drawn only in a *2D* plane. You can plot the drawing from model space to any scale by specifying the plot scale in the plot dialog box.

Paper space is a *2D* environment used for arranging various views (floating viewports) of what was drawn in model space. It represents the paper on which you arrange the drawing prior to plotting. With AutoCAD, single or multiple paper space environments (layouts) can be easily designed and manipulated. Each layout represents an individual plot output sheet, or an individual sheet in a drawing project. You can apply different scales to each floating viewport and specify different visibility for layers in the views. After arranging the views and scaling appropriately, you can plot the drawing from layout at 1:1 (full scale). Paper space allows you to plot the drawing in WYSIWYG (What You See Is What You Get) mode.

You can switch between model space and paper space by selecting the appropriate tabs provided at the bottom of the drawing window. Model space can be accessed from the Model tab and selecting one of the available Layout tabs can access paper space. You can also switch between model space and paper space by changing the value of the system variable TILEMODE. When TILEMODE is set to 1 (default), you will be working in model space; when it is set to 0, you will be working in paper space.

When you are working in a layout, you can access model space by double-clicking inside one of the floating viewports. If you double-click anywhere outside the viewport, AutoCAD will switch you to paper space. You must have at least one floating viewport in paper space to see the model.

By default, AutoCAD creates two layout tabs called Layout1 and Layout2. If necessary, you can create additional layouts. A detailed explanation is provided later in the chapter for creating and modifying layouts.

PLOTTING FROM MODEL SPACE

To plot/print the current drawing from the model space, invoke the PLOT command:

Standard toolbar	Choose the PLOT command (see Figure 8–3)
File menu	Choose Plot
Command: prompt	**plot** (ENTER)

Figure 8–3 *Invoking the* PLOT *command from the Standard toolbar*

AutoCAD displays the Plot dialog box, as shown in the Figure 8–4.

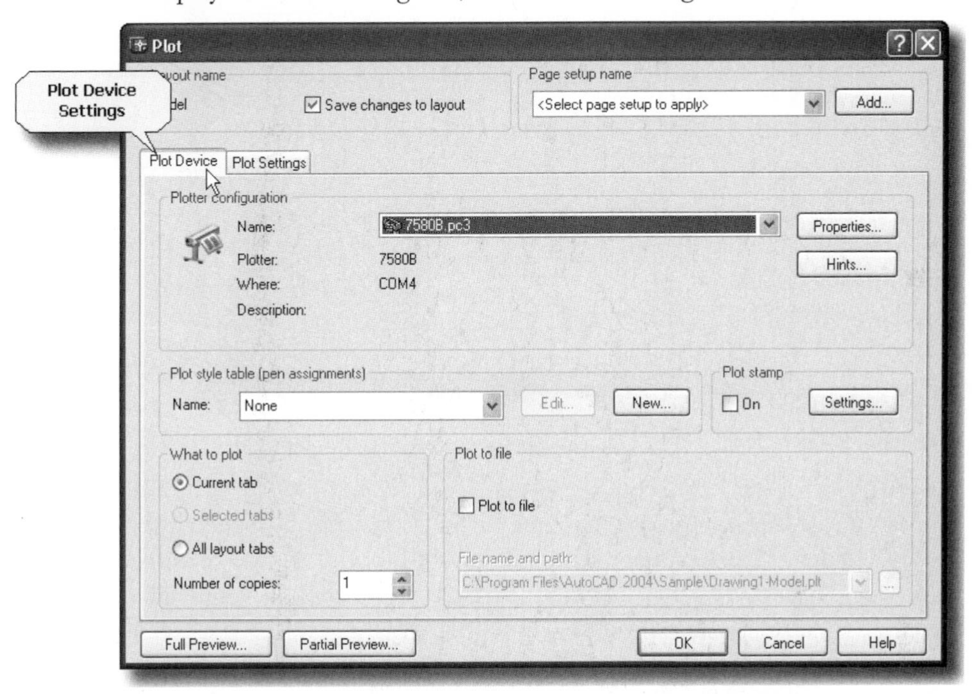

Figure 8–4 *Plot dialog box*

AutoCAD displays the current layout name or displays "Selected layouts" if multiple tabs are selected. If the Model tab is current when you choose Plot, the Layout Name shows "Model." Set the **Save changes to layout** toggle button to ON to save changes you make in the Plot dialog box to the layout. This option is not available if multiple layouts are selected.

The **Page setup name** list box displays a list of any saved page setups. You can choose from one of the saved page setups to restore the plot settings. Or you can save the current settings to a named page setup and it can be applied later to plot any drawing from model space. To save the current settings, choose the **Add...** button, and AutoCAD displays the User Defined Page Setups dialog box, as shown in Figure 8–5.

Figure 8–5 *User Defined Page Setups dialog box*

You can create, delete, or rename named page setups. To save the current settings, specify a name in the **New page setup name:** text box and choose the **OK** button. To delete the currently selected user-defined page setup, choose the **Delete** button; to rename the currently selected user-defined page setup, choose the **Rename** button and change the name. Choose the **Import...** button to import a user-defined page setup from another drawing.

PLOT DEVICE SETTINGS

The **Plot Device** tab of the Plot dialog box specifies the plotter to use, a plot style table, what to plot, plot stamp, and information about plotting to a file.

Plotter Configuration

The **Plotter configuration** section (see Figure 8–6) displays the currently configured plotting device, the port to which it's connected, or its network location. If necessary, you can plot the drawing to a different plotting device from the list of available system plotters displayed in the **Name:** list box. The default plotter is specified in the Options dialog box. To add a new plotter to the **Name:** list box, refer to the section on "Configuring Plotters," later in this chapter.

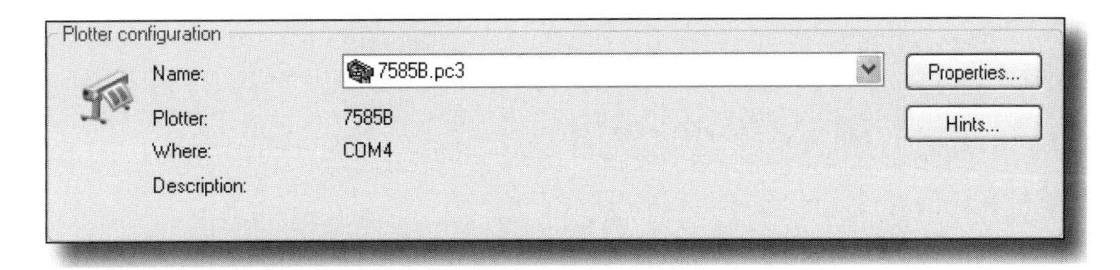

Figure 8–6 *Plot dialog box – Plotter configuration section*

Choose the **Properties…** button to change the Plotter Configuration of the currently configured plotting device. Refer to the section on "Configuring Plotters," later in this chapter, for a detailed explanation on plotter configuration.

Choose the **Hints…** button to display information of the currently configured plotting device.

Plot Style Table

The **Plot style table** section (see Figure 8–7) sets the plot style table, edits the plot style table, or creates a new plot style table. Plot style tables are settings that give you control over how objects in your drawing are plotted into hard copy plots. You can control how colors are translated into line weight and how area fills are converted into shades of gray or screened colors, as well as many other output options. You can control how the plotter treats each individual object in a drawing. Refer to the section on "Creating and Modifying Plot Style Tables," later in this chapter, for a detailed explanation on creating and modifying a plot style table.

Figure 8–7 *Plot dialog box – Plot style table and stamp sections*

Plot Stamp

The **Plot stamp** section (see Figure 8-7) controls whether or not to place a plot stamp on a specified corner of each drawing and if necessary, logs the information to a file. Plot stamp information includes: drawing name, layout name, date and time, login name, plot device name, paper size, plot scale and user defined fields, if any. Once you set Plot stamp to ON in the Plot dialog box, it remains active with whatever settings have been most recently entered until you specifically turn it OFF.

AutoCAD creates a plot stamp at the time the drawing is being plotted and it is not saved with the drawing. Before you plot the drawing, you can preview the position of the plot stamp (not the contents) in the Plot Stamp dialog box. The plot stamp can be set to plot at one of the four drawing corners and can print up to two lines.

Note: Plot stamp information is plotted with pen number 7 or the highest numbered available pen if the plotter doesn't hold 7 pens. If you are using a non-pen (raster) device, color 7 is always used for plot stamping.

To specify the information you want to be plotted as part of the plot stamp, Choose the **Settings...** button in the **Plot stamp** section. AutoCAD displays the Plot Stamp dialog box as shown in Figure 8–8.

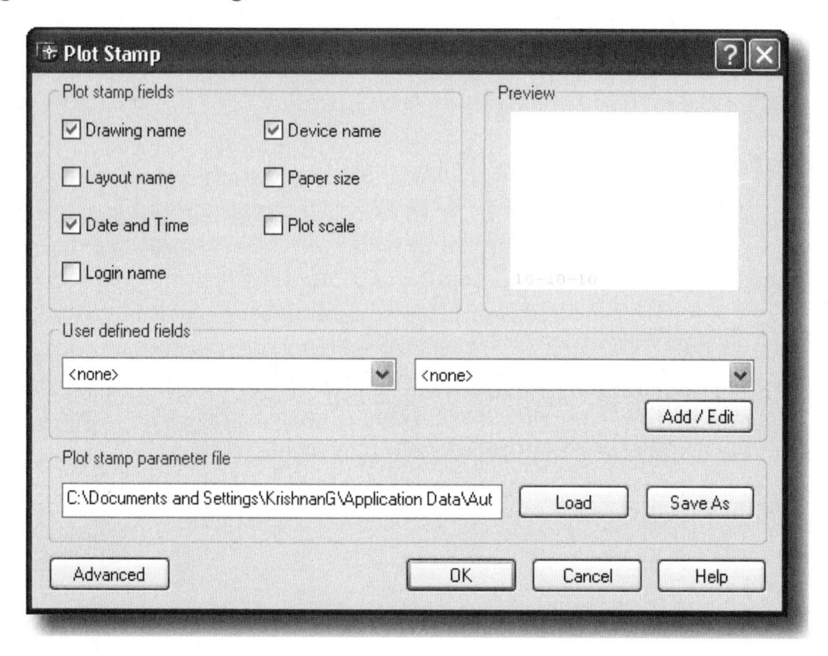

Figure 8–8 *Plot Stamp dialog box*

The **Drawing name** toggle button controls whether or not to include the drawing name and path in the plot stamp information.

The **Layout name** toggle button controls whether or not to include the name of the layout in the plot stamp information.

The **Date and Time** toggle button controls whether or not to include the date and time in the plot stamp information.

The **Login name** toggle button controls whether or not to include the Windows Login name in the plot stamp information.

The **Device name** toggle button controls whether or not to include the current plotting device name in the plot stamp information.

The **Paper size** toggle button controls whether or not to include the paper size for the currently configured plotted device in the plot stamp information.

The **Plot scale** toggle button controls whether or not to include the plot scale in the plot stamp information.

The **Preview** section of the Plot Stamp dialog box provides a visual display of the plot stamp location based on the location and rotation values specified in the Advanced Options dialog box.

Note: The preview displayed in the **Preview** section of the Plot Stamp dialog box is visual display of the plot stamp location not a preview of the plot stamp contents.

The **User defined fields** section provides text that can optionally be plotted. You can choose one or both user-defined fields for the plot stamp information. If the user-defined value is set to <none>, then no user-defined information is plotted. To add, edit, or delete user-defined fields, choose the **Add/Edit** button. AutoCAD displays the User Defined Fields dialog box as shown in Figure 8–9.

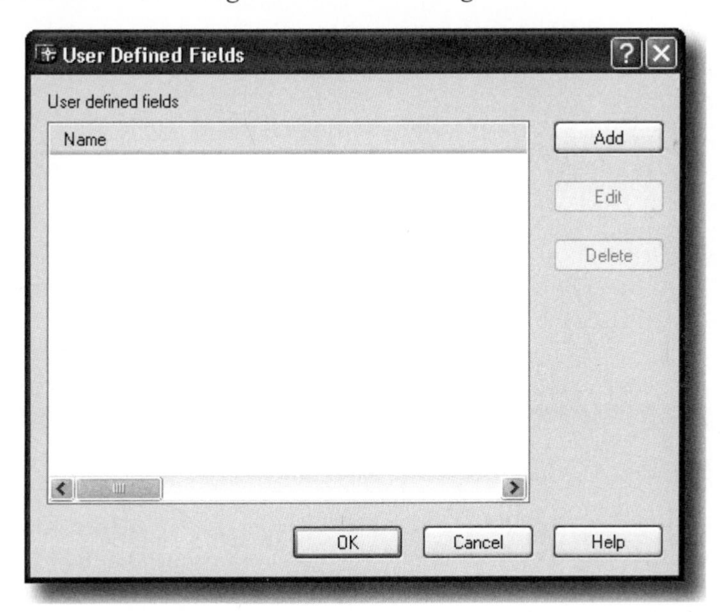

Figure 8–9 *User Defined Fields dialog box*

Choose the **Add** button to add an editable user-defined field to the bottom of the list, choose the **Edit** button to edit the selected user-defined field and choose the **Delete** button to delete the selected user-defined field. Choose the **OK** button to save the changes and close the User Defined Fields dialog box. Choose the **Cancel** button to discard the changes and close the User Defined Fields dialog box.

The **Plot stamp parameter file** section displays the name of the file in which the plot stamp settings are stored. If necessary, you can save the current plot stamp settings to a new file by choosing the **Save As** button and providing appropriate file name. AutoCAD stores plot stamp information in a file with a .PSS extension. If you need to load a different parameter file, choose the **Load** button, AutoCAD displays a standard file selection dialog box, in which you can specify the location of the parameter file you want to use.

To set the location, text properties, and units of the plot stamp, choose the **Advanced** button. AutoCAD displays Advanced Options dialog box as shown in Figure 8–10.

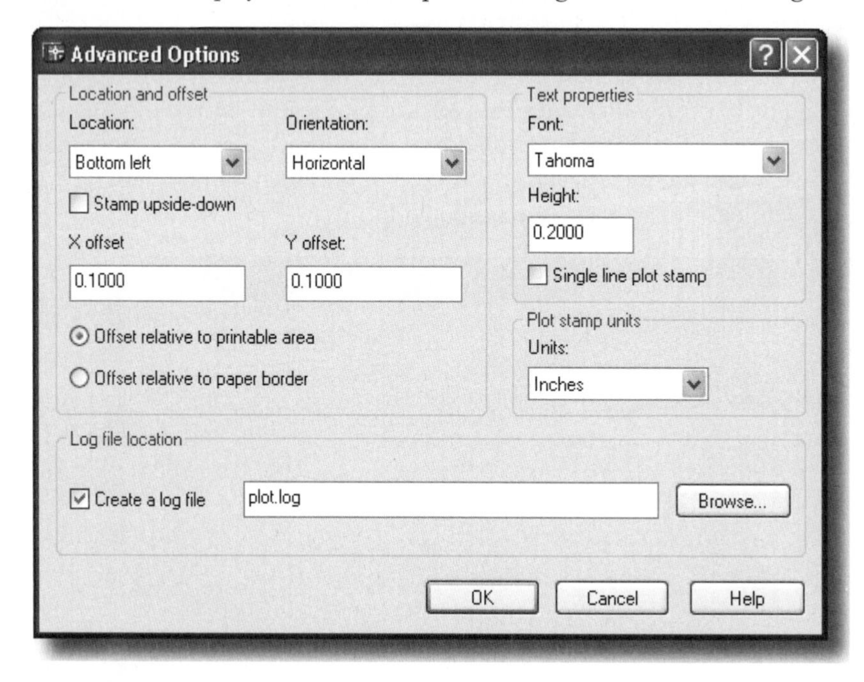

Figure 8–10 *Advanced Options (Plot Stamp) dialog box*

The **Location:** list box allows you to select the area where you want to place the plot stamp. The options include: **Top Left**, **Bottom Left** (default), **Bottom Right**, and **Top Right**. The location is relative to the image orientation of the drawing on the page.

The **Orientation:** list box allows you to select the rotation of the plot stamp in relation to the specified page. The options include **Horizontal** and **Vertical** for each of the location.

The **Stamp upside down** toggle button controls whether to rotate the plot stamp upside down.

The **X offset** and **Y offset** text fields determines the offset distance calculated from either the corner of the paper or the corner of the printable area, depending on which setting you specify. Select one of the two radio buttons: **Offset relative to printable area** or **Offset relative to paper border** to set the reference point to measure the offset distance.

The **Text properties** section determines the font, height, and number of lines you want to apply to the plot stamp text. The **Font:** list box specifies the TrueType font you want to apply to the text used for the plot stamp information. The **Height:** text field specifies the text height you want to apply to the plot stamp information. And the **Single line plot stamp** toggle button controls whether to place the plot stamp information in a single line of text or not. The plot stamp information can consist of up to two lines of text, but the placement and offset values you specify must accommodate text wrapping and text height. If the plot stamp contains text that is longer than the printable area, the plot stamp text will be truncated. If the **Single line plot stamp** is set to OFF, then plot stamp text is wrapped after the third field.

The **Plot stamp units** section allows you to specify the units used to measure X offset, Y offset, and height. You can select from the **Units** list box from one of the available units: inches, millimeters, or pixels.

The **Log file location** specifies the name of the file to which the plot stamp information is saved instead of, or in addition to, stamping the current plot. The default log file name is plot.log, and it is located in the AutoCAD folder. Choose the **Browse...** button to specify a different file name and path. After the initial plot log file is created, the plot stamp information in each succeeding plotted drawing is added to this file. Each drawing's plot stamp information is a single line of text. And, if necessary the log file can be placed on a network drive and shared by multiple users.

Choose the **OK** button to save the changes and close the Advanced Options dialog box. Choose the **Cancel** button to discard the changes and close the Advanced Options dialog box.

Choose the **OK** button to save the changes and close the Plot Stamp dialog box. Choose the **Cancel** button to discard the changes and close the Plot Stamp dialog box.

What to Plot

The **What to plot** section (see Figure 8–11) defines what you want to plot: a Model tab or a Layout tab. The **Current tab** selection plots the current Model or Layout tab. The **Selected tabs** selection plots multiple preselected Model or Layout tabs. To select multiple tabs, hold down the CTRL key while selecting the tabs. If only one tab is selected, this option is not available. The **All layout tabs** selection plots all layout tabs, regardless of which tab is selected.

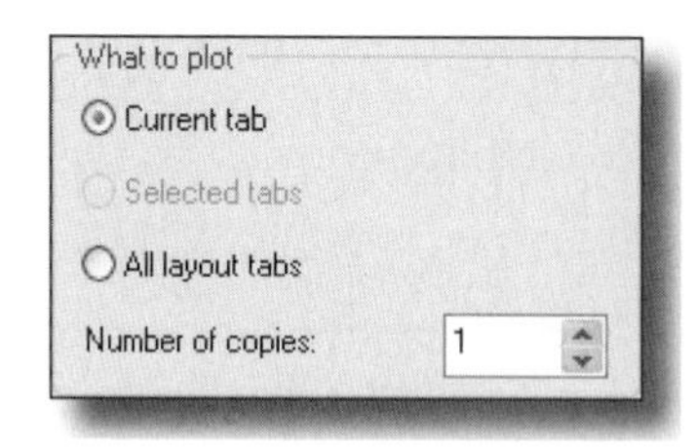

Figure 8–11 *Plot dialog box – What to plot section*

The **Number of copies:** text field specifies number of copies that are plotted.

Plot to File

The **Plot to file** section (see Figure 8–12) controls whether to create a plot file instead of using a plotter to plot the drawing. The creation of a plot file is useful if you need to use a plotter connected to another computer. You can copy the file to a floppy disk and then transfer the disk to the other computer. AutoCAD creates the plot file with .PLT as an extension to the given file name. Set the **Plot to file** toggle to ON to create a plot file. By default, the name of the plot file will be the same as the current drawing name and saved in the current folder. If you want to save it to a different file name and/or save it to a different folder, specify the file name in the **File name and path:** text field, or choose the browse button to select the location.

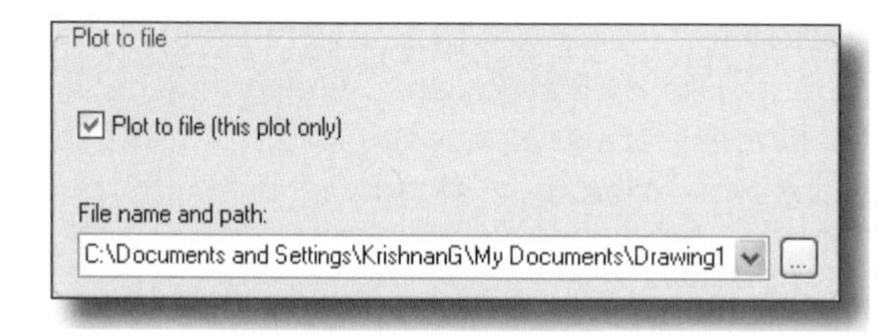

Figure 8–12 *Plot dialog box – Plot to file section*

PLOT SETTINGS

The **Plot Settings** tab (see Figure 8–13) of the Plot dialog box specifies paper size, drawing orientation, plot area and scale, plot offset, and plot options.

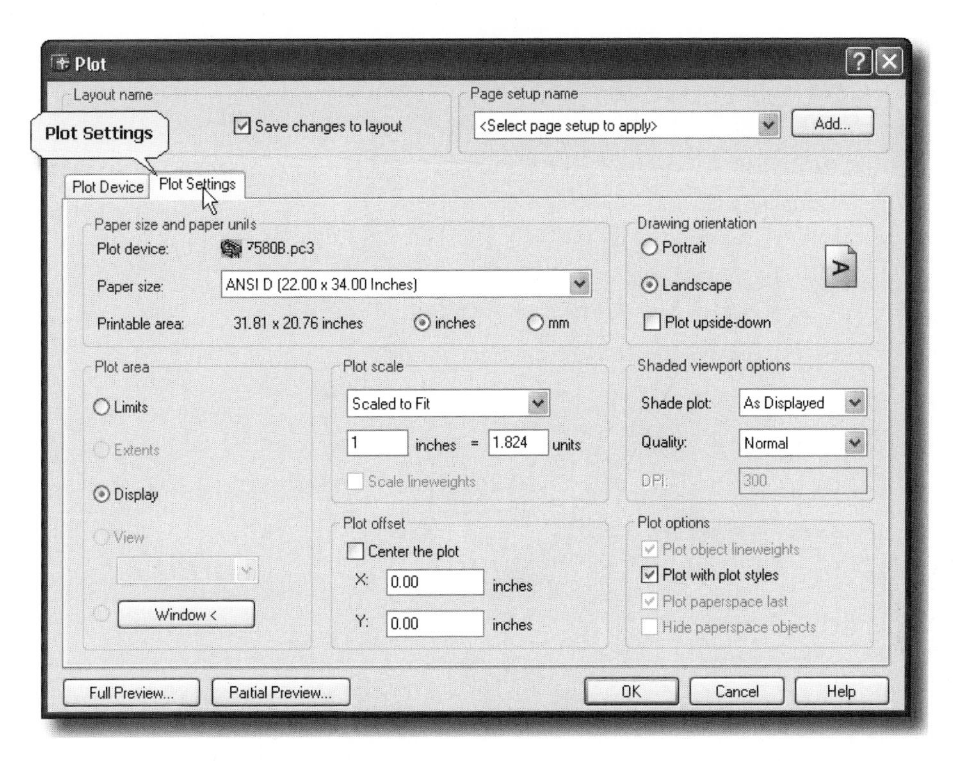

Figure 8–13 *Plot dialog box – Plot Settings tab*

Paper Size and Paper Units

The **Paper size and paper units** section allows you to select the paper size and the units to plot. Select the paper size to plot from the **Paper size:** list box for the selected plotting device. Actual paper sizes are indicated by the width (*X* axis direction) and height (*Y* axis direction). A default paper size is set for the plotting device when you create a plotter configuration file with the Add-a-Plotter wizard. For more information about the Add-a-Plotter wizard, refer to the section on "Configuring Plotters," later in this chapter.

AutoCAD displays the name of the currently selected plot device and actual area on the paper that is used for the plot based on the current paper size. Select the appropriate radio button to plot your drawing in inches or millimeters.

Drawing Orientation

The **Drawing orientation** section (see Figure 8–14) specifies the orientation of the drawing on the paper for plotters that support landscape or portrait orientation. Select the **Portrait** radio button to orient and plot the drawing so that the short edge of the paper represents the top of the page. Select the **Landscape** radio button to orient and plot the drawing so that the long edge of the paper represents the top of the page. Set the **Plot upside-down** toggle button to ON to orient and plot the drawing upside down.

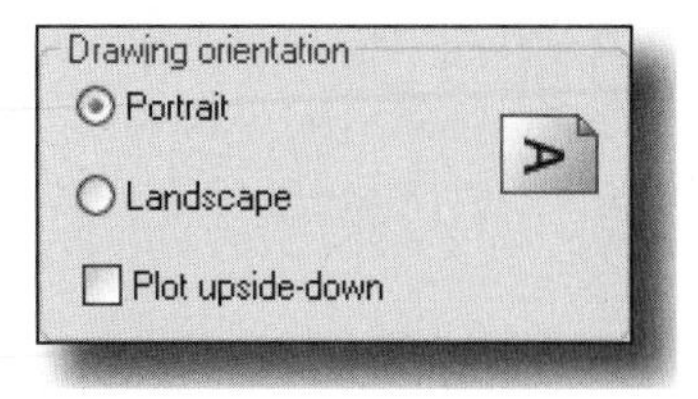

Figure 8–14 *Plot dialog box – Drawing orientation section*

Plot Area

The **Plot area** section (see Figure 8–15) specifies the portion of the drawing to be plotted.

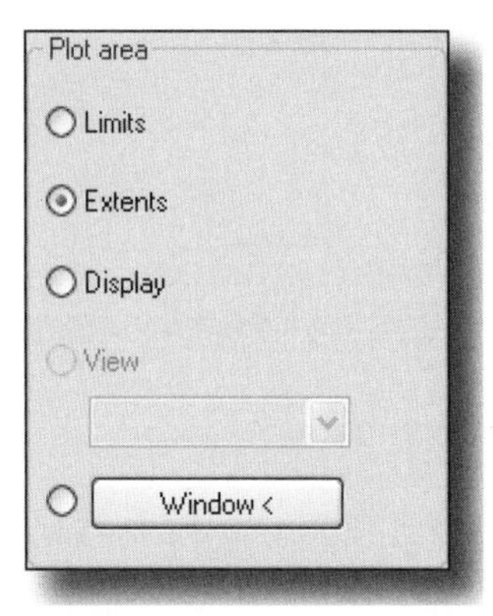

Figure 8–15 *Plot dialog box – Plot area section*

Select the **Limits** radio button to plot the drawing to its limits. In general, this makes the lower left corner of the limits of the drawing equal to the origin of the plot. This option is available only when you plot from model space.

Select the **Extents** radio button to plot the entire drawing. It forces the lower left corner of the rectangle that includes all objects and/or the current limits in the drawing, rather than the display, to become the origin of the plot. This option is similar to the ZOOM EXTENTS command and ensures that the entire drawing is plotted, including any objects drawn outside the limits.

Select the **Display** radio button to plot what is currently displayed on the screen. An important point to remember is that the lower left corner of the current display is the origin point of the plot. This option is useful if you want to plot only part of the drawing. Before you select this option, make sure the view you want to plot is displayed on the screen by using the ZOOM and PAN commands.

Select the **View** radio button to plot a previously saved view. If you plot a previously created view, the plot is identical to the screen image after the VIEW RESTORE command is used to bring the view on-screen. View plotting makes it easy to plot predefined areas of a drawing. You can select a named view from the list provided. If there are no saved views in the drawing, this option is not available.

Select the **Window** radio button to pick a window on the screen and plot the objects that are inside the window. The lower left corner of the window becomes the origin of the plot. This is similar to using the ZOOM WINDOW command to zoom into a specific portion of the drawing, and then using the Display option of the PLOT command to plot. To specify the Window, choose the **Window** < button. Specify the coordinates of the two diagonal points or designate the window by means of your pointing device.

Plot Scale

The **Plot scale** section (see Figure 8–16) controls the plot area. Select the plot scale from the list box. When plotting from model space the default setting is Scaled to Fit. When you select a standard scale, the scale is displayed in Custom text fields. You can also type scale factor in the Custom text fields to reflect the scale you want the drawing to be plotted.

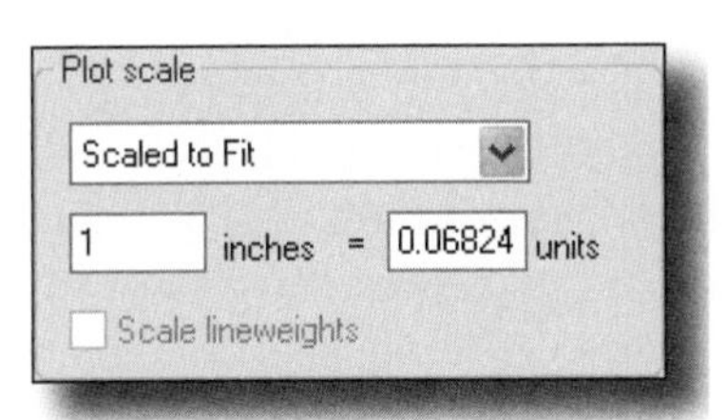

Figure 8–16 *Plot dialog box – Plot scale section*

Set the **Scale lineweights** toggle to ON to plot in proportion to the plot scale and to OFF to plot the objects with assigned lineweights. Lineweights normally specify the linewidth of printed objects and are plotted with the linewidth size regardless of the plot scale. This option is available only in a Layout (Paper Space).

Plot Offset

The **Plot offset** section (see Figure 8–17) specifies an offset of the plotting area from the lower left corner of the paper. In a layout, the lower left corner of a specified plot area is positioned at the lower left margin of the paper. You can offset the origin by entering a positive or negative value.

Set the **Center the plot** toggle button to ON to automatically center the plot on the paper.

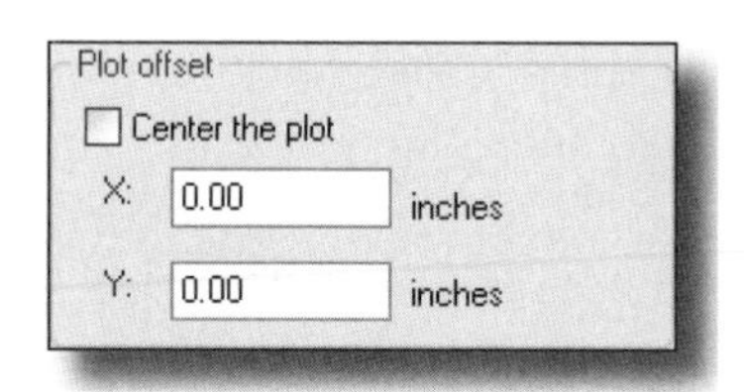

Figure 8–17 *Plot dialog box – Plot offset section*

Shaded Viewport Options

The **Shaded viewport options** section (see Figure 8–18) specifies how shaded and rendered viewports are plotted and determines their resolution level with corresponding dpi.

The **Shade plot:** list box specifies how views are plotted. **As Displayed** selection plots objects the way they are displayed on the screen. **Wireframe** selection plots objects in wireframe regardless of the way the objects are displayed on the screen. **Hidden** selection plots objects with hidden lines removed regardless of the way the objects are displayed on the screen. **Rendered** selection plots objects as rendered regardless of the way they are displayed on the screen.

The **Quality:** option menu specifies the resolution at which shaded and rendered viewports are plotted. The **Draft** selection sets rendered and shaded model space views to plot as wireframe. The **Preview** selection sets rendered and shaded model space views to plot at a maximum of 150 dpi. The **Normal** selection sets rendered and shaded model space views to plot at a maximum of 300 dpi. The **Presentation** selection sets rendered and shaded model space views to plot at the current device resolution, to a maximum of 600 dpi. The **Maximum** selection sets rendered and shaded model space views to plot at the current device resolution with no maximum. The **Custom** selection sets rendered and shaded model space views to plot at the resolution setting you specify in the **DPI:** text box, up to the current device resolution.

The **DPI:** text box specifies the dots per inch for shaded and rendered views when Custom is selected in the **Quality:** option menu, up to the maximum resolution of the current plotting device.

Plot Options

The **Plot options** section (see Figure 8–18) specifies options for lineweights, plot styles, and the current plot style table.

Set the **Plot object lineweights** toggle button to ON to plot the objects with assigned lineweights and to OFF to plot with the default lineweight. This option is not available if Plot with plot styles is set to ON.

Set the **Plot with plot styles** toggle button to ON to plot using the object plot styles that are assigned to the geometry, as defined by the plot style table. Refer to the section on "Creating and Modifying Plot Style Tables," later in this chapter, for a detailed explanation on creating and modifying a plot style table.

Set the **Plot paperspace last** toggle button to ON to plot model space geometry before paper space objects are plotted. This option is not available when plotting from model space.

Set the **Hide paperspace objects** toggle button to ON to plot layouts (paper space) with hidden lines removed. Hidden lines are those that normally would be obscured by objects placed in front of them. This option is not applicable to *2D* drawings.

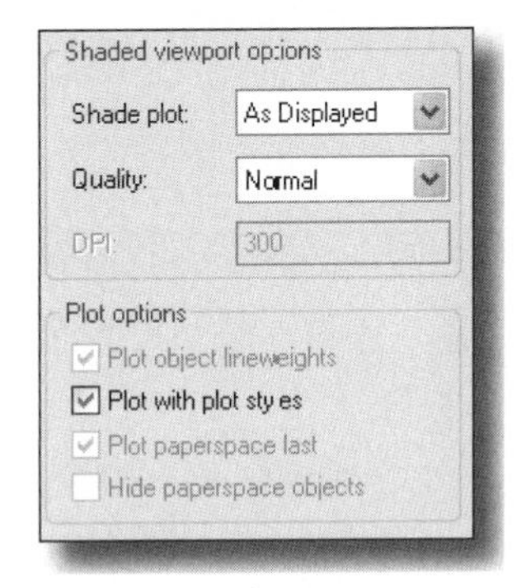

Figure 8–18 *Plot dialog box – Shaded viewport options and Plot options sections (Plotting from Model Space)*

Partial Plot Preview

Choose the **Partial Preview...** button to quickly display a partial preview of the effective plot area relative to the paper size and printable area (see Figure 8–19). In addition, AutoCAD provides information regarding paper size, printable area, and effective area. Partial plot preview also gives advance notice of any AutoCAD warnings that may be encountered when you plot the drawing.

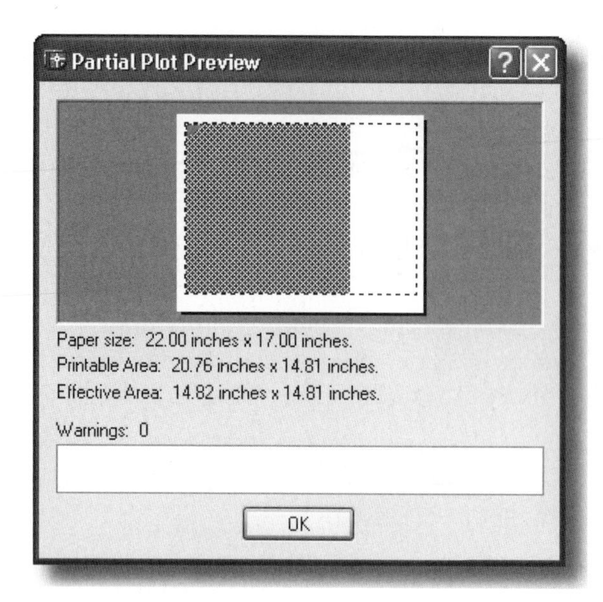

Figure 8–19 *Partial Plot Preview dialog box*

Full Preview

A full preview displays the drawing on the screen as it would appear when plotted. This takes much longer than the partial preview, but it takes less time than regular plot regeneration. This is quicker than plotting something incorrectly and having to start all over again.

To get a full preview of your plot, choose the **Full Preview...** button. AutoCAD temporarily clears the plotting dialog boxes, draws an outline of the paper size, and displays the drawing as it would appear on the paper when it is plotted (see Figure 8–20).

The cursor changes into a magnifying glass with plus and minus signs. Holding the pick button and dragging the cursor toward the top of the screen enlarges the preview image. Dragging it toward the bottom of the screen reduces the preview image. Right-click and AutoCAD displays a shortcut menu offering additional preview options: PAN, ZOOM, ZOOM WINDOW, ZOOM ORIGINAL, PLOT, and EXIT.

To end the full preview, choose the EXIT option from the shortcut menu. AutoCAD returns to the Plot dialog box.

Note: You can also access a full preview by invoking the PRINT PREVIEW command from the File menu.

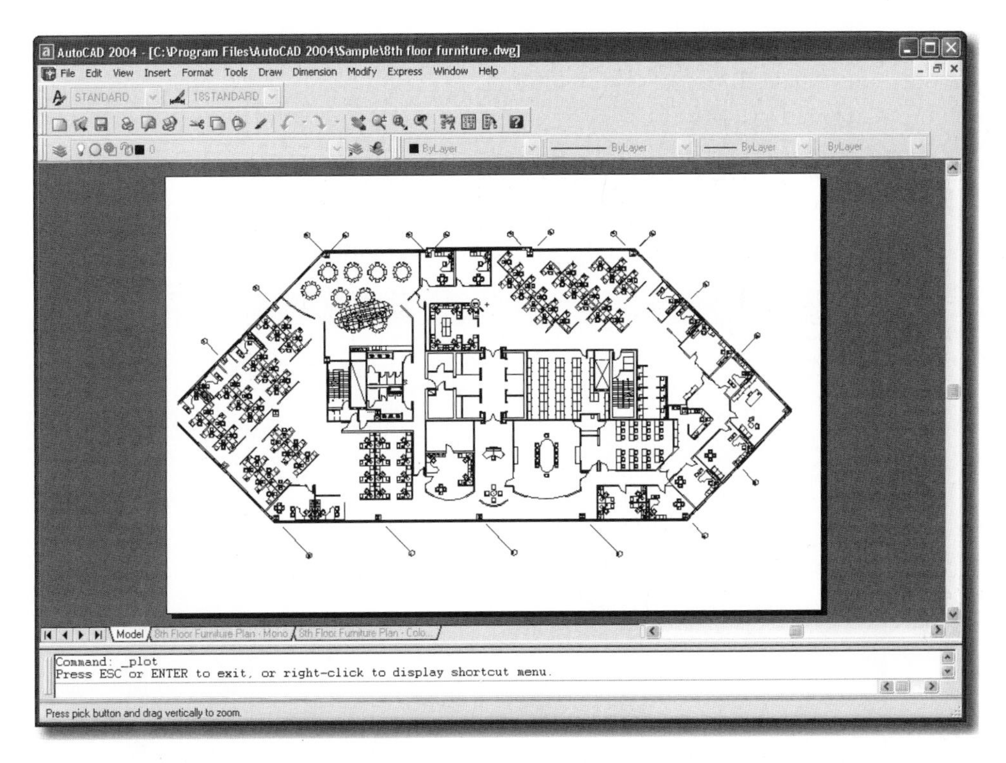

Figure 8–20 *Plot Preview dialog box*

After making the necessary changes in the plot settings, choose the **OK** button. AutoCAD starts plotting and reports its progress as it converts the drawing into the plotter's graphics language by displaying the number of vectors processed.

If something goes wrong or if you want to stop immediately, choose the **Cancel** button at any time. AutoCAD cancels the plotting.

PLOTTING FROM LAYOUT (WYSIWYG)

As mentioned earlier, AutoCAD allows you to plot a drawing from model space as well as paper space. The paper space environment allows you to set up multiple layouts. You can have as many layouts as you like, and each one can be set up for a different type of output. A layout is used to compose or layout your model drawing for plotting. A layout may consist of a title block, one or more viewports, and annotations. As you create a layout, you can design floating viewport configurations to visualize different details in your drawing. You can apply different scales to each view within the viewport, and specify different visibility for layers in the viewport. Layouts are accessible by choosing the Layout tab at the bottom of the drawing area.

By default, whenever you start a new drawing, AutoCAD creates two layouts, called Layout1 and Layout2. If you open a file created in a previous versions of AutoCAD, you only see one Layout tab.

SETTING UP A LAYOUT

To set up a layout, first select the Layout tab located at the bottom of the drawing area. AutoCAD displays the Page Setup dialog box by default, as shown in Figure 8–21.

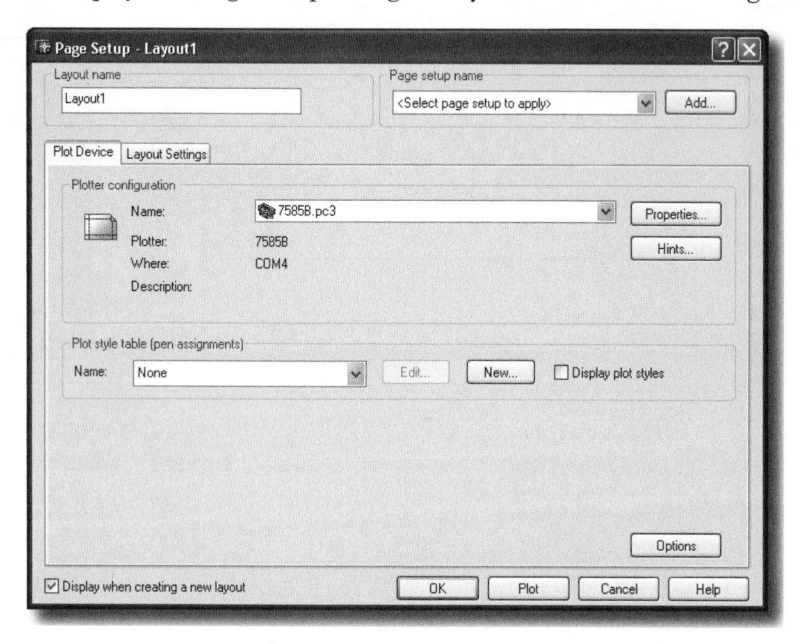

Figure 8–21 *Page Setup dialog box*

AutoCAD displays the current layout name in the **Layout name:** text field.

The **Page setup name:** list box displays a list of any saved page setups. You can choose from one of the saved page setups to restore the plot settings. Or you can save the current settings to a named page setup and it can be applied later to plot any drawing from a layout. To save the current settings, choose the **Add...** button, and AutoCAD displays the User Defined Page Setups dialog box. You can create, delete, or rename named page setups.

Plot Device Settings

The **Plot Device** tab specifies a currently configured plotting device to plot the selected layout.

The **Plotter configuration** section displays the currently configured plotting device, the port to which it's connected, or its network location. If necessary, you can plot the drawing to a different plotting device from the list of available system plotters displayed in the **Name:** list box. The default plotter is specified in the Options dialog box. To add a new plotter to the **Name:** list box, refer to the section on "Configuring Plotters," later in this chapter. Choose the **Properties...** button to change the Plotter Configuration of the currently configured plotting device. Refer to the section on "Configuring Plotters" for a detailed explanation on plotter configuration. Choose the **Hints...** button to display information on the currently configured plotting device.

The **Plot style table** section sets the plot style table, edits the plot style table, or creates a new plot style table. Refer to the section on "Creating and Modifying Plot Style Tables," later in this chapter for a detailed explanation on creating and modifying a plot style table.

Set the **Display when creating a new layout** toggle box to ON (default is set to ON) to display the Page Setup dialog box each time you select a new Layout tab to begin creating a new layout.

Choose the **Options** button and choose the **Display** tab in the Options dialog box, as shown in Figure 8–22. You can view or modify the layout elements in the Options dialog box.

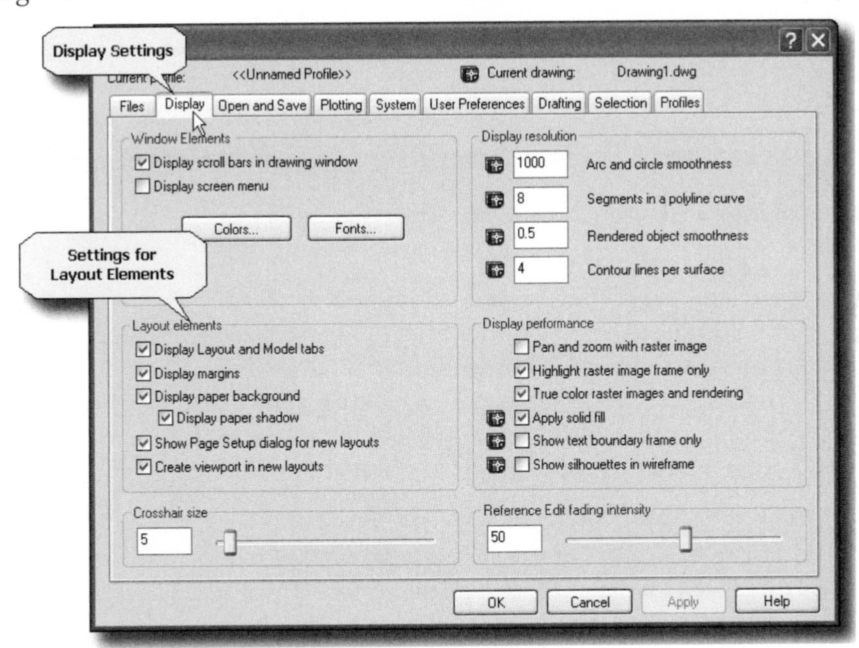

Figure 8–22 *Options dialog box with Display tab selected*

Refer to Chapter 13 for a detailed explanation of various options available in the Options dialog box.

Layout Settings

The **Layout Settings** tab (see Figure 8–23) specifies paper size, drawing orientation, plot area and plot scale, plot offset, and plot options.

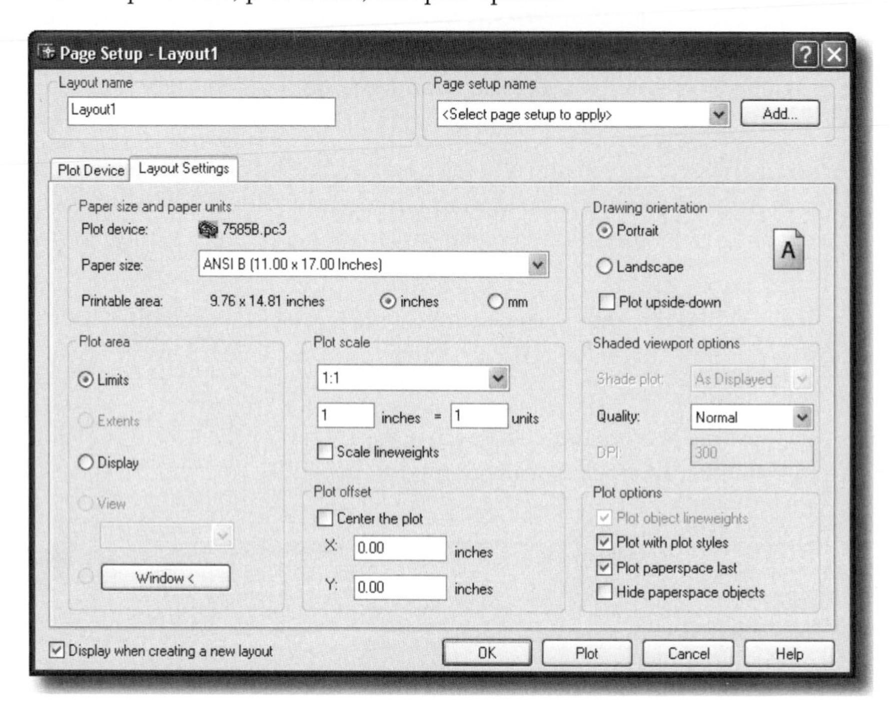

Figure 8–23 *Page Setup dialog box – Layout Settings tab*

The **Paper size and paper units** section allows you to select the paper size and the units to plot. Select the paper size to plot from the **Paper size:** list box for the selected plotting device. Actual paper sizes are indicated by the width (*X* axis direction) and height (*Y* axis direction). A default paper size is set for the plotting device when you create plotter configuration file with the Add-a-Plotter wizard. For more information about the Add-a-Plotter wizard, refer to the section on "Configuring Plotters," later in this chapter.

AutoCAD displays the name of the currently selected plot device and actual area on the paper that is used for the plot based on the current paper size. Select the appropriate radio button to plot your drawing in inches or millimeters.

The **Drawing orientation** section specifies the orientation of the drawing on the paper for plotters that support landscape or portrait orientation. Select the **Portrait** radio button to orient and plot the drawing so that the short edge of the paper represents the top of the page. Select the **Landscape** radio button to orient and plot the drawing so that the long edge of the paper represents the top of the page. Set the **Plot upside-down** toggle button to ON to orient and plot the drawing upside down.

The **Plot area** section specifies the portion of the drawing to be plotted.

Select the **Limits** radio button to plot everything within the margins of the specified paper size, with the origin calculated from 0,0 in the layout.

Select the **Extents** radio button to plot the extent of the current space of the drawing. It forces the lower left corner of the rectangle that includes all objects or the current limits in the drawing, rather than the display, to become the origin of the plot. This option is similar to the ZOOM EXTENTS command and ensures that the entire drawing is plotted, including any objects drawn outside the limits.

Select the **Display** radio button to plot what is currently displayed on the screen. An important point to remember is that the lower left corner of the current display is the origin point of the plot. This option is useful if you want to plot only part of the drawing. Before you select this option, make sure the view you want to plot is displayed on the screen by using the ZOOM and PAN commands.

Select the **View** radio button to plot a previously saved view. If you plot a previously created view, the plot is identical to the screen image after the VIEW RESTORE command is used to bring the view on-screen. View plotting makes it easy to plot predefined areas of a drawing. You can select a named view from the list provided. If there are no saved views in the drawing, this option is not available.

Select the **Window** radio button to specify a window on the screen and plot the objects that are inside the window. The lower left corner of the window becomes the origin of the plot. This is similar to using the ZOOM WINDOW command to zoom into a specific portion of the drawing, and then using the Display option of the PLOT command to plot. To specify the Window, choose the **Window** < button. Specify the coordinates of the two diagonal points or designate the window by means of your pointing device.

The **Plot scale** section controls the plot area. Select the plot scale from the **Scale:** list box. When plotting from Layout, the default setting is set to 1:1 (full scale). You can create a custom scale by entering the number of inches (or millimeters) equal to the number of drawing units in the text fields.

Set the **Scale lineweights** toggle to ON to plot in proportion to the plot scale and to OFF to plot the objects with assigned lineweights. Lineweights normally specify the linewidth of printed objects and are plotted with the linewidth size regardless of the plot scale.

The **Plot offset** section specifies an offset of the plotting area from the lower left corner of the paper. In a layout, the lower left corner of a specified plot area is positioned at the lower left margin of the paper. You can offset the origin by entering a positive or negative value. Set the **Center the plot** toggle button to ON to automatically center the plot on the paper.

The **Shaded viewport options** section specifies how shaded and rendered viewports are plotted and determines their resolution level with corresponding dpi.

The **Shade Plot:** list box displays how views are plotted and it is specified through the Properties dialog box for the selected viewport.

The **Quality:** option menu specifies the resolution at which shaded and rendered viewports are plotted. The **Draft** selection sets rendered and shaded model space views to plot as wireframe. The **Preview** selection sets rendered and shaded model space views to plot at a maximum of 150 dpi. The **Normal** selection sets rendered and shaded model space views to plot at a maximum of 300 dpi. The **Presentation** selection sets rendered and shaded model space views to plot at the current device resolution, to a maximum of 600 dpi. The **Maximum:** selection sets rendered and shaded model space views to plot at the current device resolution with no maximum. The **Custom:** selection sets rendered and shaded model space views to plot at the resolution setting you specify in the **DPI:** text box, up to the current device resolution.

The **DPI:** text box specifies the dots per inch for shaded and rendered views when **Custom** is selected in the **Quality** option menu, up to the maximum resolution of the current plotting device.

The **Plot options** section specifies options for lineweights, plot styles, and the current plot style table. Set the **Plot object lineweights** toggle button to ON to plot the objects with assigned lineweights and to OFF to plot with default lineweight. Set the **Plot with plot styles** toggle button to ON to plot using the object plot styles that are assigned to the geometry, as defined by the plot style table. Refer to the section on "Creating and Modifying Plot Style Tables," later in this chapter, for a detailed explanation on creating and modifying a plot style table. Set the **Plot paperspace last** toggle button to ON to plot model space geometry before paper space objects are plotted. Set the **Hide paperspace objects** toggle button to ON to plot layouts (paper space) with hidden lines removed from objects. Hidden lines are those that normally would be obscured by objects placed in front of them. This option is not applicable to *2D* drawings.

Choose the **Plot** button to plot the drawing from the selected Layout or choose the **OK** button to close the Page Setup dialog box.

AutoCAD sets up a layout with a rectangular outline (shadow) indicating the paper size of the currently configured plotting device (see Figure 8–24). The margins displayed (dashed lines) within the paper indicate the printable area of the paper.

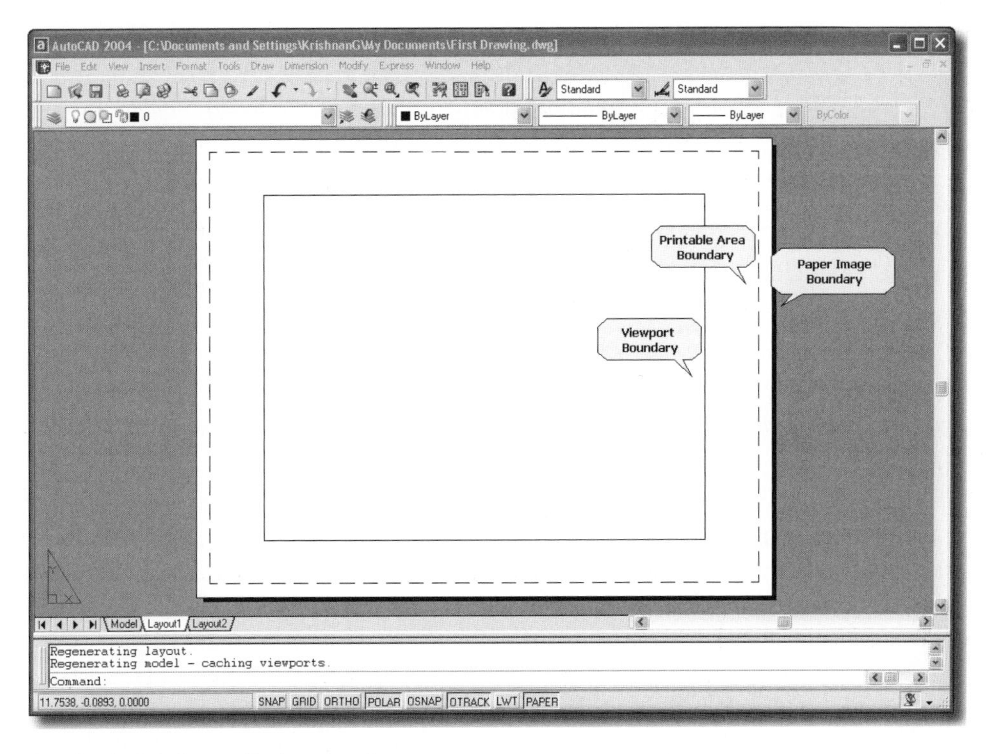

Figure 8–24 *Parts of a Layout tab*

Note: The **Display** tab of the Options dialog box controls the display of the paper background and margins.

In addition, by default, AutoCAD creates a floating viewport and displays the drawing in the viewport. Refer to the section later in this chapter on "Creating and Modifying Floating Viewports" to create and modify floating viewports.

Note: If necessary, you can make changes to the page setup anytime by invoking the PAGESETUP command.

CREATING AND MODIFYING A LAYOUT

The LAYOUT command allows you to create a new layout, copy an existing layout, rename a layout, delete a layout, save a layout, and make a layout current.

Invoke the NEW LAYOUT command:

Layouts toolbar	Choose NEW LAYOUT (see Figure 8–25)
Insert menu	Choose Layout > New Layout
Command: prompt	**layout** (ENTER)

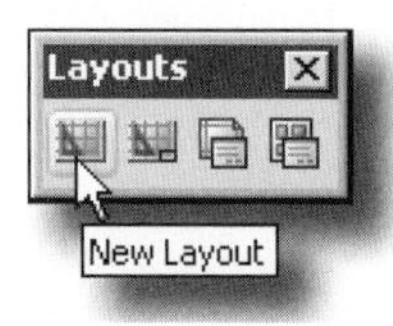

Figure 8–25 *Invoke the* NEW LAYOUT *command from the Layouts toolbar*

AutoCAD prompts:

Command: **layout** (ENTER)
Enter layout option [Copy/Delete/New/Template/Rename/SAveas/Set/?] <set>: *(specify an option or right-click for shortcut menu and select one of the options)*

New

The New option creates a new layout tab with the specified name using the default plot device. AutoCAD prompts:

Enter new Layout name <default>: *(specify a layout name or press* ENTER *to accept the default name)*

When you make the newly created layout tab current, AutoCAD displays the Page Setup dialog box by default. If necessary, you can make changes to the Plot Device settings and Layout Settings.

Note: You can also create a new layout by choosing the **New Layout** option from the shortcut menu that appears when you right-click after placing the cursor on the existing layout tab.

Copy

The Copy option creates a new layout by copying an existing layout. AutoCAD prompts:

Enter name of layout to copy <default>: *(specify the name of an existing layout name from which to copy)*
Enter layout name for copy <default>: *(specify a layout name or press* ENTER *to accept the default name)*

The new layout tab is inserted after the last current layout.

Note: You can also create a new layout by copying an existing layout. This is done by choosing the MOVE OR COPY LAYOUT option from the shortcut menu that appears when you right-click after placing the cursor on an existing layout tab.

Template

The Template option creates a new layout from an existing layout in a drawing (DWG) or template drawing (DWT) file. The layout (and all geometry from the paper space) from the specified template or drawing file is inserted into the current drawing.

When you select the Template option, AutoCAD displays the Select File dialog box. Select a drawing or template file from the appropriate directory and choose the **Open** button. AutoCAD displays the Insert Layout(s) dialog box, as shown in Figure 8–26.

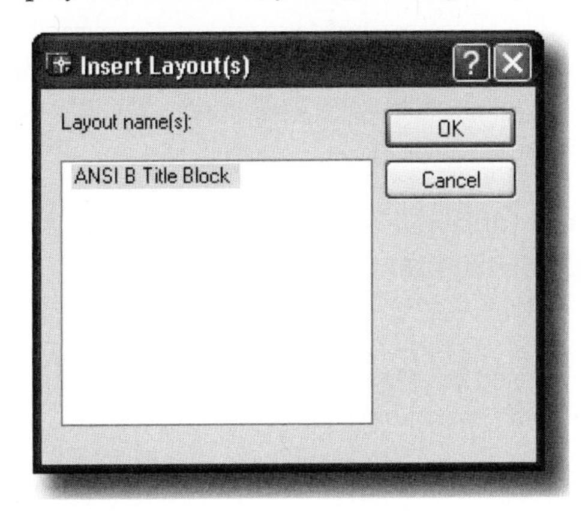

Figure 8–26 *Insert Layout(s) dialog box*

Select one or more layouts from the Insert Layout(s) dialog box and choose the **OK** button to close the dialog box.

Note: You can also create a new layout from an existing layout in a drawing or template drawing by choosing the FROM TEMPLATE option from the shortcut menu that appears when you right-click after placing the cursor on an existing layout tab.

Rename

The Rename option allows you to rename an existing layout. The last current layout is used as the default for the layout to rename. Layout names must be unique. Layout names can contain up to 255 characters, and are not case-sensitive. Only the first 32 characters are displayed in the tab.

AutoCAD prompts:

Enter layout to rename <default>: *(specify the layout name to rename)*
Enter new layout name: *(specify a new layout name)*

Note: You can also rename an existing layout by choosing the RENAME option from the shortcut menu that appears when you right-click after placing the cursor on an existing layout tab to be renamed. AutoCAD displays the Rename Layout dialog box. Specify a new layout name and choose the **OK** button.

Delete

The Delete option deletes the selected layout. The most current layout is the default. AutoCAD prompts:

Enter name of layout to delete <default>: *(press* ENTER *to accept the default layout to delete or specify the name of the layout to delete)*

Note: The Model tab cannot be deleted. To remove all the geometry from the Model tab, you must select all geometry and use the ERASE command. You can also delete layouts by choosing the DELETE option from the shortcut menu that appears when you right-click after placing the cursor on an existing layout tab to be deleted.

SAveas

The SAveas option saves the layout to a drawing template (DWT) file. The last current layout is used as the default for the layout to save.

AutoCAD prompts:

Enter layout to save as template <current>: *(press* ENTER *to accept the default or specify a layout name to save)*

AutoCAD displays the Create Drawing File dialog box. You can create a new drawing file or template drawing file with the layout. You can also save the layout to an existing drawing file or template drawing file.

Set

The Set option makes a layout current. AutoCAD prompts:

Enter layout to make current <last>: *(specify the name of the layout to make it current)*

You can also make a layout current by selecting the appropriate layout tab.

?

The ? option lists all the layouts defined in the drawing.

CREATING FLOATING VIEWPORTS

Similar to creating tiled viewports in model space (see Chapter 3 for a detailed explanation), you can also create viewports in paper space called floating viewports. In paper space, you can create multiple, overlapping, contiguous, or separated untiled viewports as shown in Figure 8–27. By default, AutoCAD sets up one floating viewport in each layout. Consider viewports as objects with a view into model space that you can move and resize. AutoCAD treats viewports in layout like any other object, such as a line, arc, or text object. You can use any of the standard AutoCAD modify commands, such as MOVE, COPY, STRETCH, SCALE, and ERASE, to manipulate the floating viewports. For example, you can use the MOVE command to grab one viewport and move it around the screen without affecting other viewports. A viewport can be of any size and can be located anywhere in the layout. You must have at least one floating viewport to view the model.

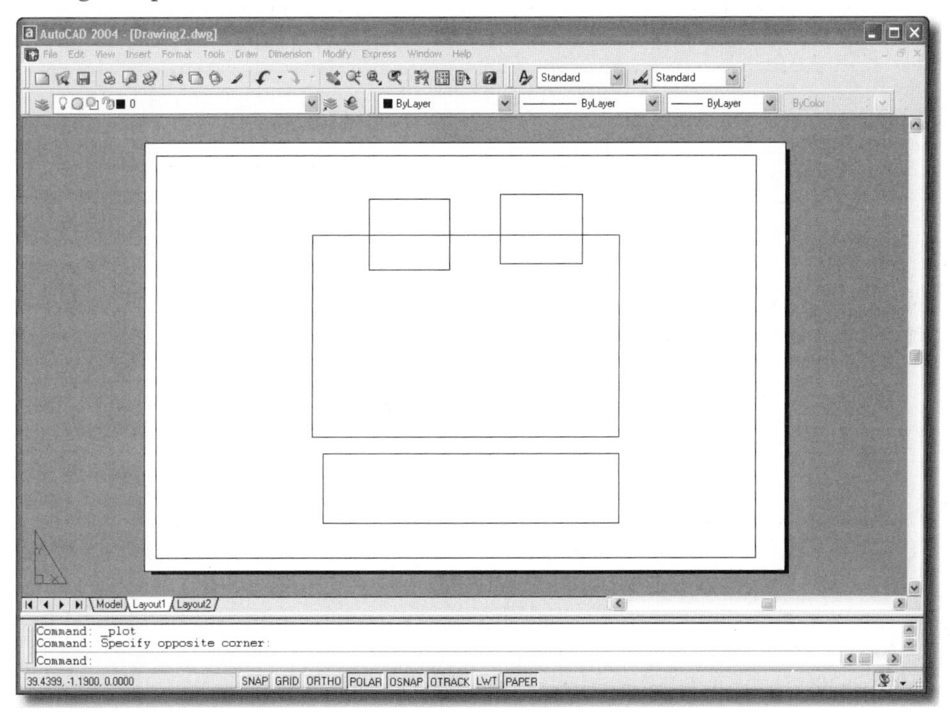

Figure 8–27 *Floating viewports in a layout*

You can create or manipulate floating viewports only when you are in the paper space. You cannot edit the model when you manipulate the floating viewports. To edit the model you must switch to model space using one of the following methods:

- Choose the Model tab.
- Double-click over the floating viewport. On the status bar, Paper changes to Model.
- Click Paper on the status bar to return to the floating viewport that was last current.

When you are working in a layout, you can switch back and forth between model space and paper space. When you make a floating viewport in a layout current, you are then working in model space in a floating viewport. Any modification to the drawing in model space is reflected in all paper space viewports as well as tiled viewports. When you double-click outside a floating viewport, AutoCAD switches the mode to paper space. In paper space you can add annotations or other graphical objects, such as title block. Objects you add in paper space do not change the model or other layouts.

To create multiple floating viewports, open the Display Viewports Dialog box:

Layouts toolbar	Choose Display Viewports Dialog (see Figure 8–28)
View menu	Choose Viewports > New Viewports...
Command: prompt	**vports** (ENTER)

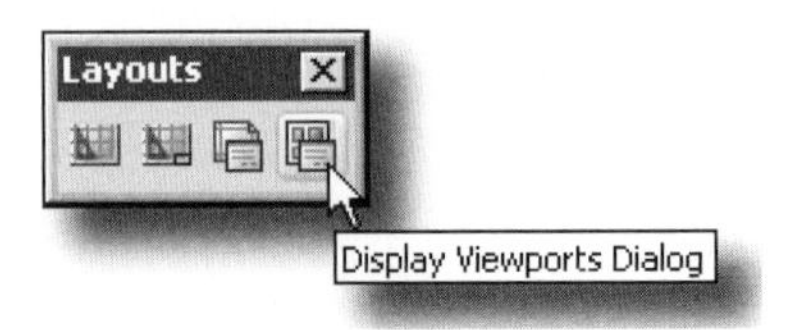

Figure 8–28 *Invoking Display Viewports Dialog from the Layouts toolbar*

AutoCAD displays the Viewports dialog box, as shown in Figure 8–29.

Choose the name of the configuration you want to use from the **Standard viewports:** list. AutoCAD displays the corresponding configuration in the preview window. Specify the spacing between the viewports if you are creating multiple viewports in the **Viewport Spacing:** text field. Select **2D** from the **Setup:** menu for *2D* viewport setup or select **3D** for *3D* viewport setup. Choose the **OK** button to create the selected viewport configuration.

AutoCAD prompts:

Specify first corner or [Fit] <Fit>: *(specify the first corner to define selected viewport configuration or right-click for shortcut menu to select one of the available options)*
Specify opposite corner: *(specify opposite corner to define selected viewport configuration)*

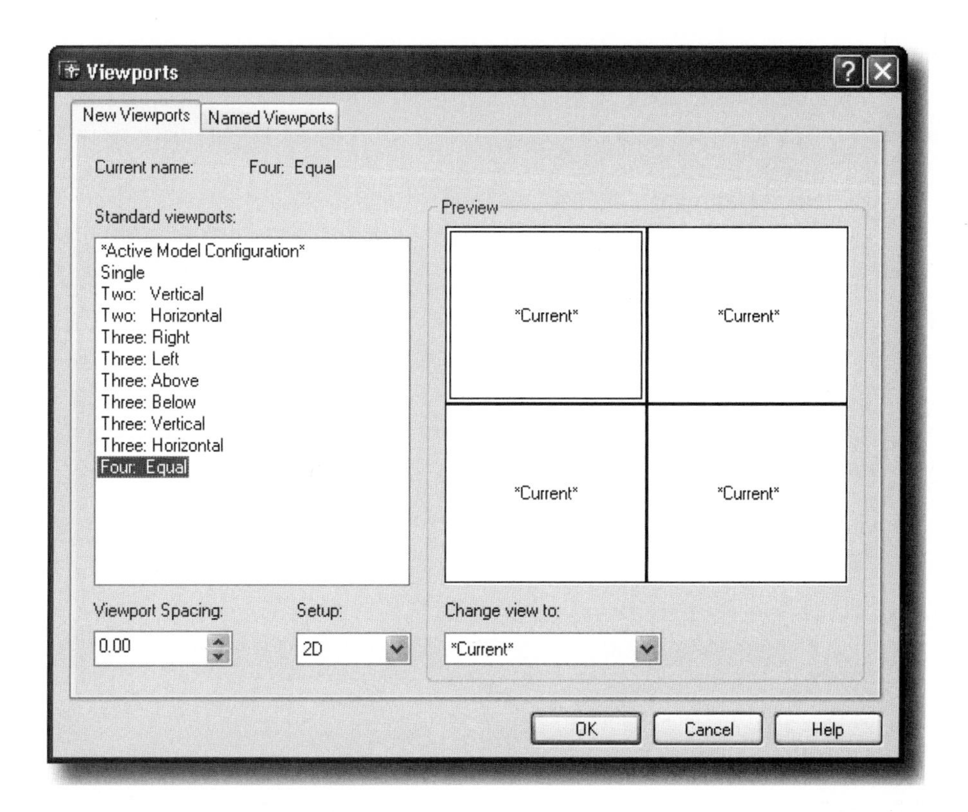

Figure 8–29 *Viewports dialog box*

Select the Fit option to create the selected viewport configuration to fit the paper size.

You can also create floating viewports from the command version of the VPORTS command:

AutoCAD prompts:

Command: **-vports** (ENTER)
Specify corner of viewport or

[ON/OFF/Fit/Hideplot/Lock/Object/Polygonal/Restore/2/3/4] <Fit>: *(specify the first corner to define a single viewport or right-click for shortcut menu to select one of the available options)*

If you specify the first corner to define a single viewport, AutoCAD prompts:

Specify opposite corner: *(specify the opposite corner to define the floating viewport)*

Fit

The Fit option (default) creates a single floating viewport to fit the paper size. The actual size of the viewport depends on the dimensions of the paper space view.

ON

The ON option turns on a viewport if it is set to OFF, making it active and making its objects visible. By default, all the new viewports are set to ON.

OFF

The OFF option turns off a viewport. When a viewport is off, its objects are not displayed, and you cannot make that viewport current. You can increase system performance by turning some viewports off or by limiting the number of active viewports. If you don't want to plot a viewport, you can turn that viewport off.

Note: You can also turn a viewport on or off from the shortcut menu that appears when you right-click after selecting the viewport, as shown in Figure 8–30.

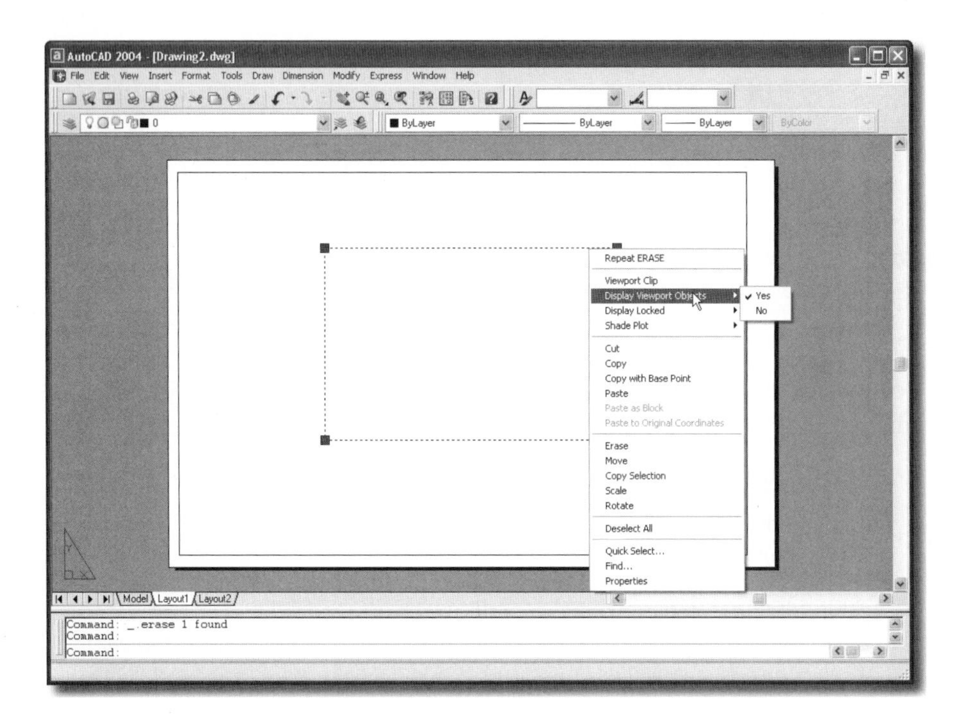

Figure 8–30 *Shortcut menu that appears after selecting a viewport*

Hideplot

The Hideplot option removes hidden lines from a viewport during plotting from paper space (layout).

Note: You can also turn the hideplot option on or off for a viewport from the shortcut menu that appears when you right-click after selecting the viewport.

Lock

The Lock option locks selected viewports. This is similar to layer locking. AutoCAD allows you to scale views relative to paper space, which establishes a consistent scale for each displayed view. To accurately scale the plotted drawing, you must scale each view relative to paper space. If you zoom in to the geometry after setting the plot scale, you are changing the viewport scale at the same time. By locking the viewport, you can zoom in without altering the viewport scale. The zoom is performed in paper space, not in the viewport.

Note: You can also turn the display lock on or off for a viewport from the shortcut menu that appears when you right-click after selecting the viewport.

Object

The Object option allows you to select a closed polyline, ellipse, spline, region, or circle to convert into a viewport. The polyline you specify must be closed and contain at least three vertices. It can be self-intersecting, and it can contain arcs as well as line segments.

When you select this option, AutoCAD prompts:

> Select object to clip viewport: *(select the object to convert into a floating viewport)*

AutoCAD creates a floating viewport from the selected object.

Polygonal

The Polygonal option creates an irregularly shaped viewport defined by specifying points.

When you select this option, AutoCAD prompts:

> Specify start point: *(specify start point to create an irregularly shaped viewport)*
> Specify next point or [Arc/Close/Length/Undo]: *(specify next point or choose an option)*

The available sub-options are the same as in the XCLIP command. Refer to Chapter 12 for a detailed explanation for all the available sub-options under the XCLIP command.

Restore

The Restore option allows you to display a saved viewport configuration. When you select this option, AutoCAD prompts:

> Enter viewport configuration name or [?]: *(Specify the name of the viewport configuration to restore)*

2

The 2 option splits the current viewport in half. When you select this option, AutoCAD prompts:

> Enter viewport arrangement [Horizontal/Vertical] <Vertical>: *(select one of the available options)*
> Specify first corner or [Fit] <Fit>: *(specify first corner to define selected viewport configuration or right-click for shortcut menu and select one of the available options)*
> Specify opposite corner: *(specify opposite corner to define selected viewport configuration)*

Select the Fit option to create the selected viewport configuration to fit the paper size.

3

The 3 option divides the current viewport into three viewports. When you select this option, AutoCAD prompts:

> Enter viewport arrangment [Horizontal/Vertical/Above/Below/Left/Right] <Right>: *(select one of the available options)*

You can select the Horizontal or Vertical option to split the current viewport into thirds by horizontal or vertical division. The other options let you split into two small ones and one large one, specifying whether the large one is to be placed above, below, or to the left or right, similar to creating tiled viewports.

4

The 4 option divides the current viewport into four viewports of equal size both horizontally and vertically. When you select this option, AutoCAD prompts:

> Specify first corner or [Fit] <Fit>: *(specify first corner to define selected viewport configuration or right-click for shortcut menu)*
> Specify opposite corner: *(specify opposite corner to define selected viewport configuration)*

Select the Fit option to create the selected viewport configuration to fit the paper size.

CREATING A LAYOUT BY LAYOUT WIZARD

The Layout Wizard lets you create a new layout (paper space) for plotting. Each wizard page instructs you to specify different layout and plot settings for the new layout you are creating. Once the layout is created using the wizard, you can modify layout settings using the Page Setup dialog box.

To open the Layout Wizard, invoke the LAYOUTWIZARD command:

Insert menu	Choose Layout > Layout wizard
Command: prompt	**layoutwizard** (ENTER)

AutoCAD displays the Begin page of the Layout Wizard, as shown in Figure 8–31.

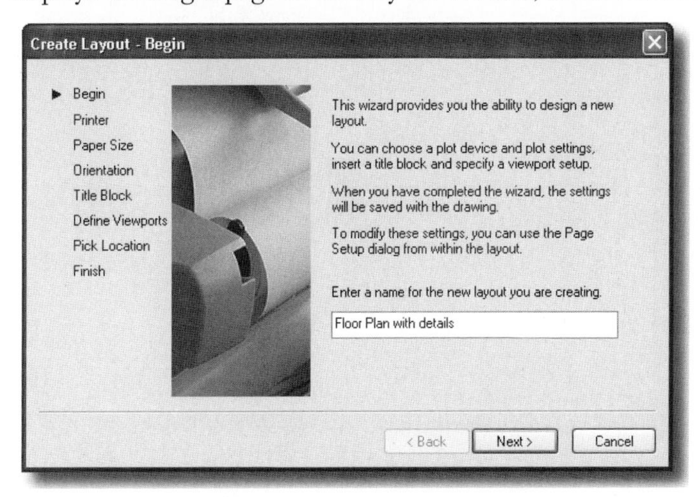

Figure 8–31 *Create Layout - Begin page*

Specify the name of the layout in the **Enter a name for the new layout you are creating** text field. Choose the **Next >** button. AutoCAD displays the **Create Layout – Printer** page of the Layout Wizard, as shown in Figure 8–32.

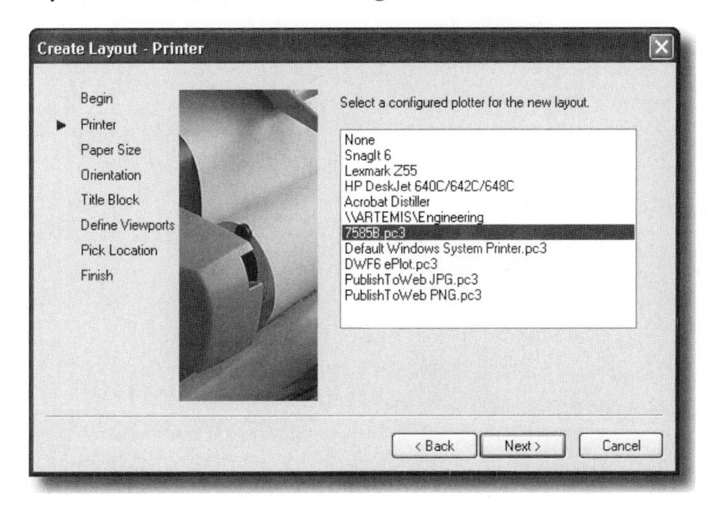

Figure 8–32 *Create Layout - Printer page*

Select a configured plotter for the new layout from the list box. If you do not see the name of the plotter to which you want to plot, refer to the section on "Configuring Plotters" to configure a plotter. Once you have selected the plotter configuration, choose the **Next >** button. AutoCAD displays the **Create Layout – Paper Size** page of the Layout Wizard, as shown in Figure 8–33.

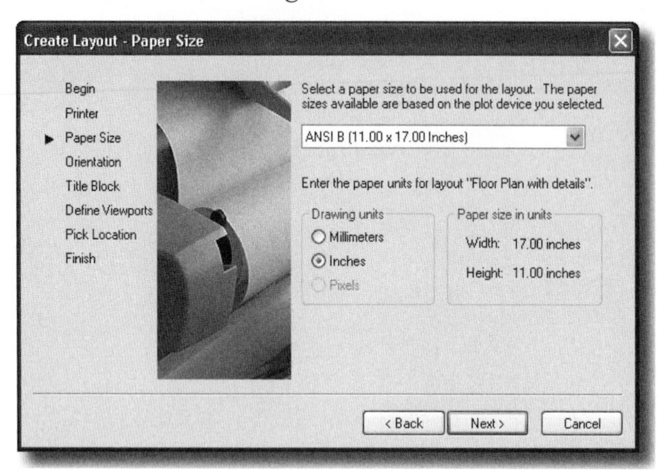

Figure 8–33 *Create Layout - Paper Size page*

Select a paper size to be used for the layout from the list box. The paper sizes available are based on the plot device you selected. Select drawing units from one of the two radio buttons located in the **Drawing units** section. Choose the **Next >** button, and AutoCAD displays the **Create Layout – Orientation** page of the Layout Wizard, as shown in Figure 8–34.

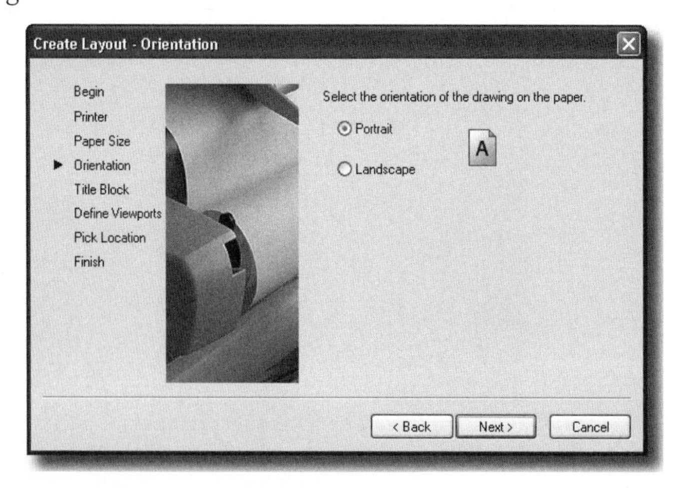

Figure 8–34 *Create Layout - Orientation page*

Select the orientation of the drawing on the paper from one of the two radio buttons: **Portrait** or **Landscape**. Choose the **Next** > button, and AutoCAD displays the **Create Layout – Title Block** page of the Layout Wizard, as shown in Figure 8–35.

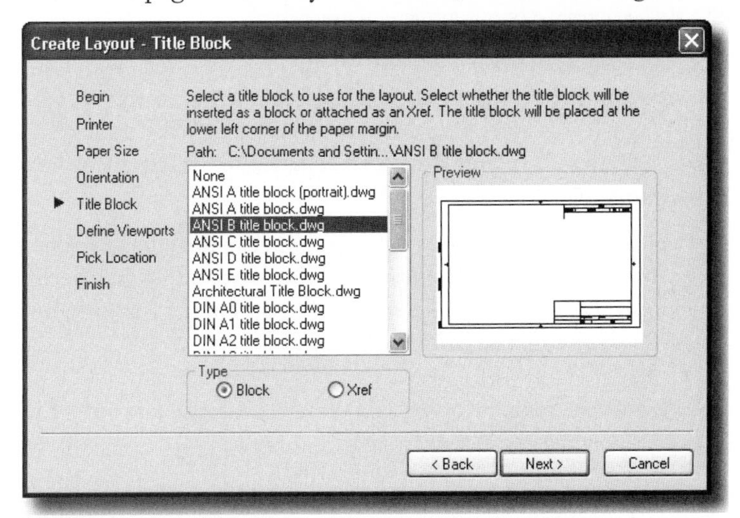

Figure 8–35 *Create Layout - Title Block page*

Select a title block from the list box to use for the layout. Select whether the title block will be inserted as a block or attached as an external reference. Choose the **Next** > button, and AutoCAD displays the **Create Layout – Define Viewports** page of the Layout Wizard, as shown in Figure 8–36.

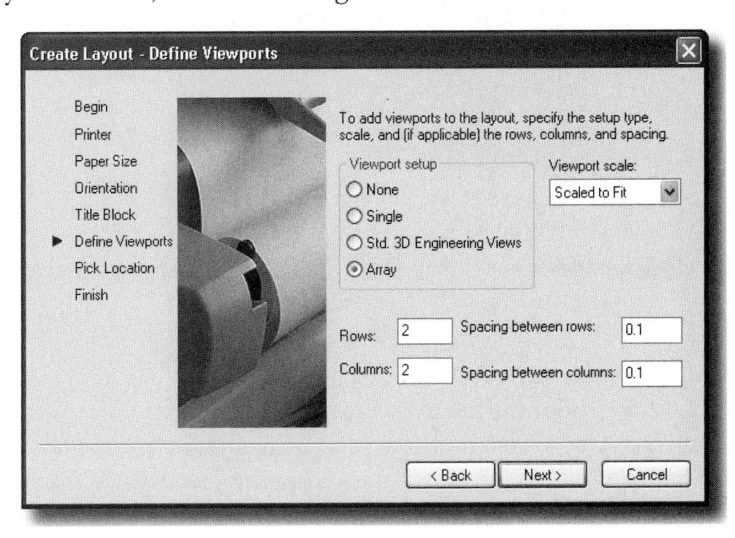

Figure 8–36 *Create Layout - Define Viewports page*

Choose one of the following four available options in the **Viewport setup** section.

- **None**—if you do not need any floating viewports.
- **Single**—to create one floating viewport.
- **Std. 3D Engineering Views**—to create four viewports with top left set for top view, top right for isometric view, bottom left for front view, and bottom right for right side view. If necessary, you can specify the distance between the viewports in the **Spacing Between Rows:** and **Spacing Between Columns:** text fields.
- **Array**—to create an array of viewports. Specify the number of viewports in rows and columns in the **Rows:** and **Columns:** text fields. If necessary, you can specify the distance between the viewports in the **Spacing between rows:** and **Spacing between columns:** text fields.

Choose the **Next>** button, and AutoCAD displays the **Create Layout – Pick Location** page of the Layout Wizard, as shown in Figure 8–37.

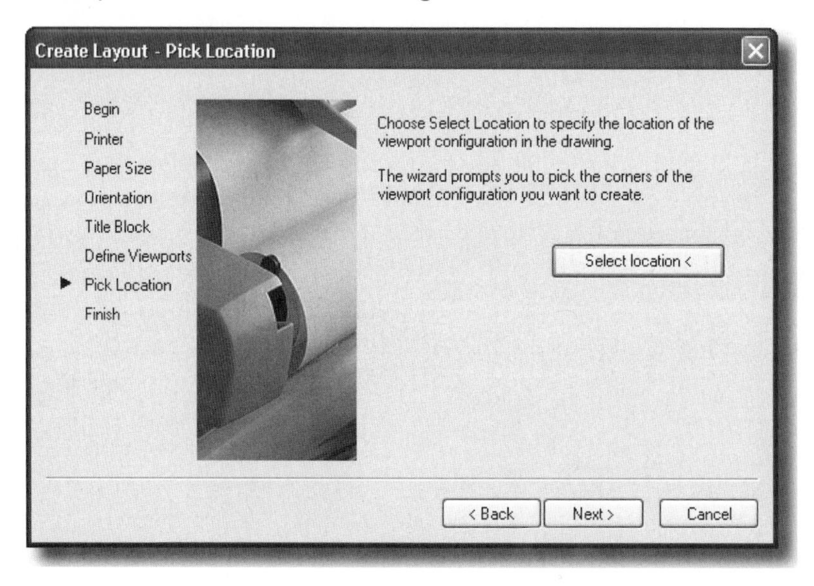

Figure 8–37 *Create Layout - Pick Location page*

Choose the **Select location<** button to specify the location of the viewport configuration in the drawing. The wizard prompts you to specify the corners of the viewport configuration you want to create. After specifying the location, AutoCAD displays the **Create Layout – Finish** page of the Layout Wizard, as shown in Figure 8–38.

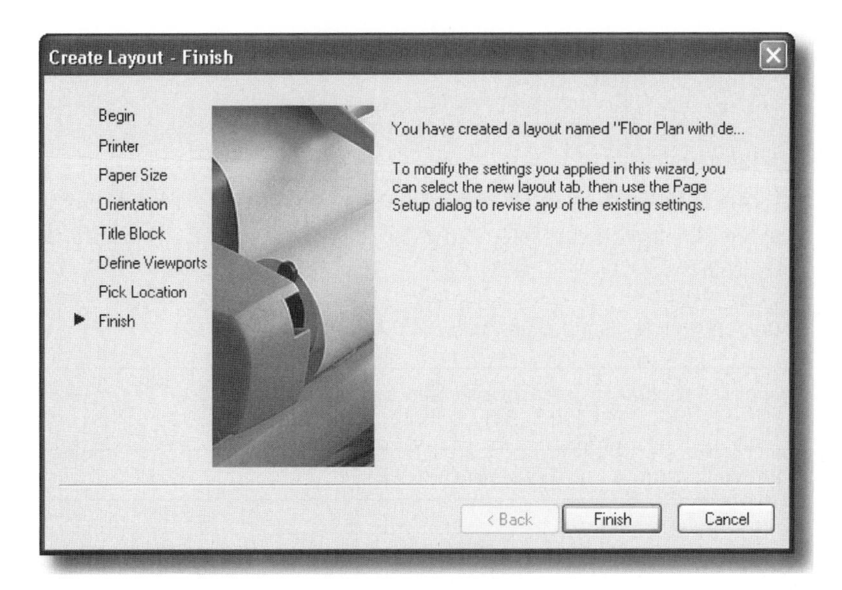

Figure 8–38 *Create Layout - Finish page*

Choose the **Finish** button to create the layout. If necessary, you can make any changes to the newly created layout by using the Page Setup dialog box.

SCALING VIEWS RELATIVE TO PAPER SPACE

AutoCAD allows you to scale views relative to paper space, which establishes a consistent scale for each displayed view. To accurately scale the plotted drawing, you must scale each view relative to paper space.

When you are working in paper space, the scale factor represents a ratio between the actual size of the model displayed in the viewports and the size of the layout. Usually the layout is plotted at a 1:1 ratio. The ratio is determined by dividing the paper space units by the model space units. The scale factor can be set with the ZOOM XP command. For example, specifying 1/24xp or 0.04167xp (1/24 = 0.01467) will display an image to a scale of 1/2" = 1'0", which is the same as 1:24 or 1/24. You can also change the plot scale of the viewport using the Viewport Scale Control on the Viewports toolbar (see Figure 8–39).

Figure 8–39 *Viewports toolbar*

Note: If necessary, after scaling the viewport you can lock the viewport with the -VPORTS LOCK command.

CLIPPING VIEWPORTS

The VPCLIP command allows you to clip a floating viewports to a user-drawn boundary. To clip a viewport, you can select an existing closed object, or specify the points of a new boundary.

To clip a floating viewport, invoke the VPCLIP command:

Viewports toolbar	Choose Clip Existing Viewport (see Figure 8–40)
Command: prompt	**vpclip** (ENTER)

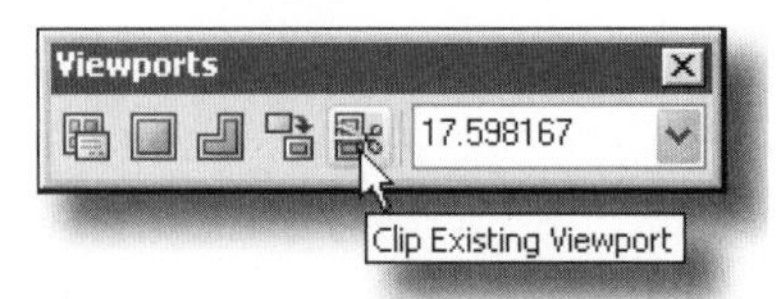

Figure 8–40 *Invoking the VPCLIP command from the Viewports toolbar*

AutoCAD prompts:

Select viewport to clip: *(select viewport to clip)*
Select clipping object or [Polygonal/Delete] <Polygonal>: *(select clipping object or select one of the available options)*

If you select an object for clipping, AutoCAD converts the object into a clipping boundary. Objects that are valid as clipping boundaries include closed polylines, circles, ellipses, closed splines, and regions.

Polygonal

The Polygonal option allows you to create a clipping boundary. When you select the Polygonal option, AutoCAD prompts:

Specify start point: *(specify start point for a clipping boundary)*

Specify next point or [Arc/Close/Length/Undo]: *(specify next point or select one of the available options)*

Arc, Close, Length, and Undo sub-options are similar to the functionality of the options available for the PLINE command. Refer to Chapter 4 for a detailed explanation for the PLINE command.

Delete

The Delete option deletes the clipping boundary of a selected viewport. This option is available only if the selected viewport has already been clipped. If you clip a viewport that has been previously clipped, the original clipping boundary is deleted and the new clipping boundary is applied.

Note: You can also use grips to create an irregularly shaped viewport just as you edit any object with grips.

CONTROLLING THE VISIBILITY OF LAYERS WITHIN VIEWPORTS

The VPLAYER (short for ViewPort LAYER) command controls the visibility of layers in a single viewport or in a set of viewports. This enables you to select a viewport and freeze a layer in it while still allowing the contents of that layer to appear in another viewport. See Figure 8–41, in which two viewports contain the same view of the drawing, but in one viewport the layer containing the dimensioning is on, and in another it is off, by using the VPLAYER command. To invoke the VPLAYER command, you must be working in layout (TILEMODE set to 0).

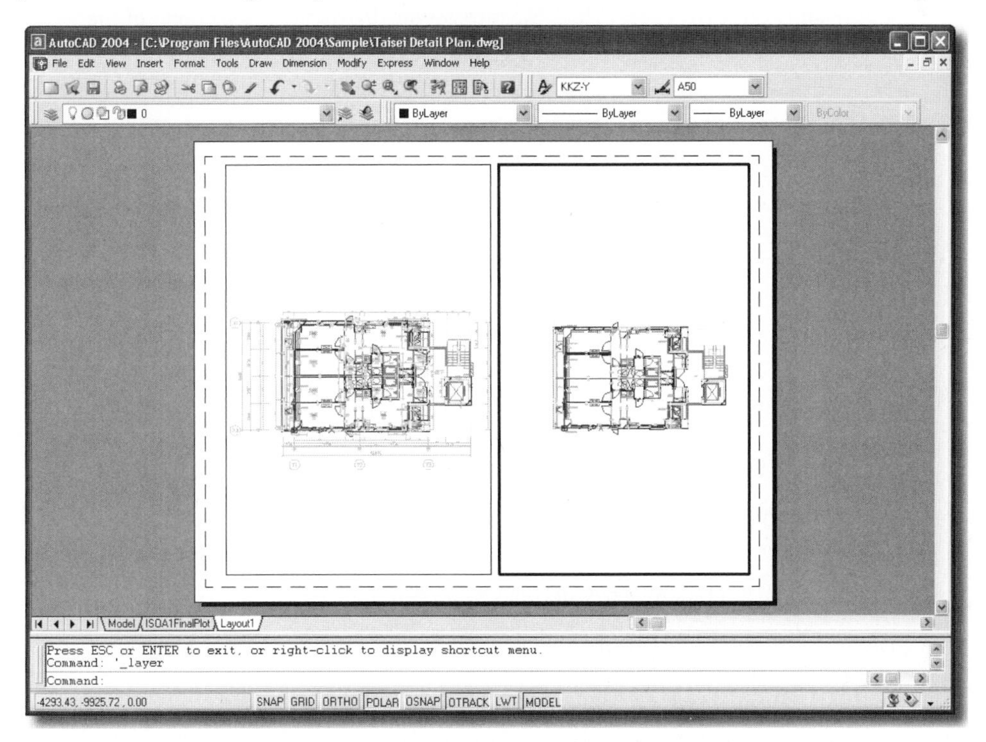

Figure 8–41 *One viewport with Dimlayer ON and another with Dimlayer OFF*

The VPLAYER command can be invoked from either model space or paper space (TILEMODE set to 0). Several options in the VPLAYER command require you to select one or more viewports in which to make your changes. AutoCAD prompts:

Enter an option [All/Select/Current] <Current>:

To accept the default option, you must be in model space; AutoCAD applies changes in the current viewport. If you choose the Select option and are in model space, AutoCAD temporarily switches to paper space so you can select a viewport. The All option applies your changes to all paper space viewports.

If you set TILEMODE to 1 (ON), the global layer settings take precedence over any VPLAYER settings.

Invoke the VPLAYER command:

Command: prompt	**vplayer** (ENTER)

AutoCAD prompts:

Command: **vplayer**
Enter an option [?/Freeze/Thaw/Reset/Newfrz/Vpvisdflt]: *(select one of the available options)*

?

The ? option displays the names of layers in a specific viewport that are frozen. When you select this option, AutoCAD prompts:

Select a viewport: *(select a viewport)*

If you are in model space, AutoCAD switches temporarily to paper space to let you select a viewport.

Freeze

The Freeze option allows you to specify one or more layers to freeze in the selected viewport. When you select this option, AutoCAD prompts:

Layer(s) to freeze:

You can respond to this prompt with a single layer name, a list of layer names separated by commas, or any valid wild-card specification. Then AutoCAD prompts:

Enter an option [All/Select/Current] <Current>:

Select the viewport(s) in which to freeze the selected layers.

Thaw

The Thaw option allows you to specify one or more layers to thaw that were frozen by the VPLAYER command in selected viewports. When you select this option, AutoCAD prompts:

Layer(s) to thaw:

You can respond to this prompt with a single layer name, a list of layer names separated by commas, or any valid wild-card specification. Then AutoCAD prompts:

Enter an option [All/Select/Current] <Current>:

Select the viewport(s) in which to thaw the selected layers.

Reset

The Reset option allows you to restore the default visibility setting for a layer in a given viewport. The default visibility is controlled by the Vpvisdflt option (explained later in this section). When you select this option, AutoCAD prompts:

Layer(s) to Reset:

You can respond to this prompt with a single layer name, a list of layer names separated by commas, or any valid wild-card specification. Then AutoCAD prompts:

Enter an option [All/Select/Current] <Current>:

Select the viewport(s) in which to reset the selected layers.

Newfrz (New Freeze)

The New Freeze option allows you to create new layers that are frozen in all viewports. If you create a new viewport, the layers that are created by the Newfrz option will be frozen by default. (The layer can be thawed in the chosen viewport by using the Thaw option.) When you select this option, AutoCAD prompts:

Enter name(s) of new layers frozen in all viewports:

You can respond to this prompt with a single layer name or a list of layer names separated by commas.

Vpvisdflt (Viewport Visibility Default)

The Viewport Visibility Default option allows you to set a default visibility for one or more existing layers. This default determines the frozen/thawed status of an existing layer in newly created viewports. When you select this option, AutoCAD prompts:

Enter layer name(s) to change viewport visibility:

You can respond to this prompt with a single layer name, a list of layer names separated by commas, or any wild-card specification. Then AutoCAD prompts:

Enter a viewport visibility option [Frozen/Thawed] <Thawed>:

You can respond to this prompt with a null response to set the default visibility to thaw, or enter F to set the default visibility to freeze.

You can also control the visibility of layers in viewports from the Layer Properties Manager dialog box. Invoke the LAYER command, and AutoCAD displays the Layer Properties Manager dialog box, similar to Figure 8–42.

The **Active VP Freeze** column (tenth column from the left) toggles between freezing and thawing of selected layer(s). Similarly, the **New VP Freeze** column (eleventh column from left) toggles between freezing and thawing of selected layer(s) for new viewports.

In Figure 8–42, the layers dim, ELEVATION, and HIDDEN are frozen in the current viewport, and OBJECT and TEXT are frozen in all the new viewports.

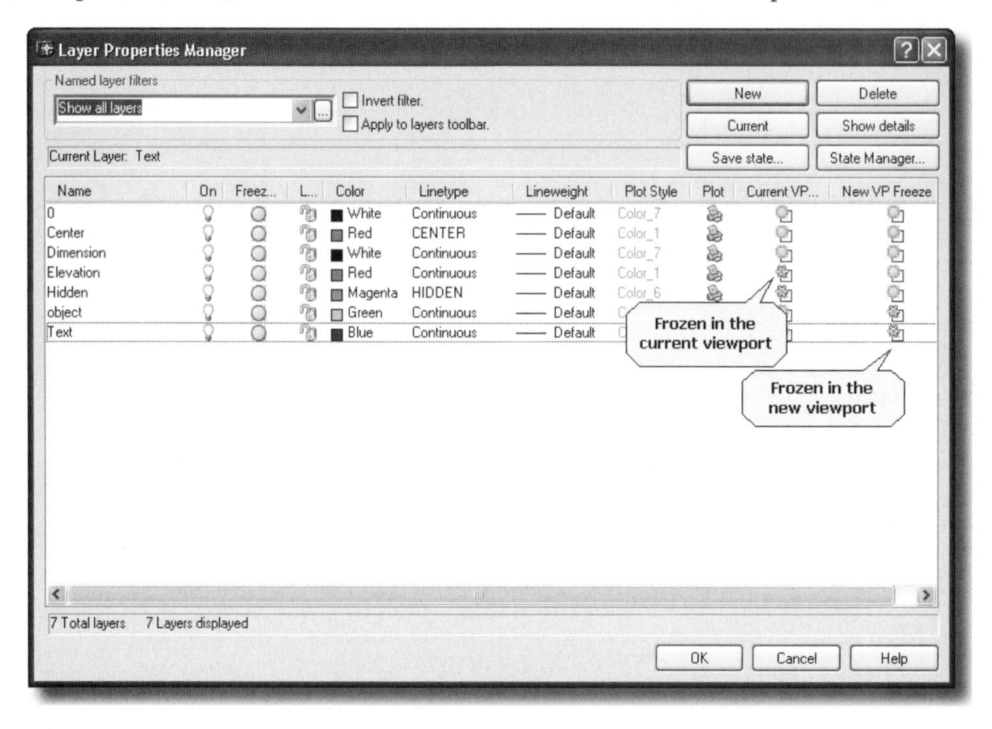

Figure 8–42 *Layer Properties Manager dialog box*

SETTING PAPER SPACE LINETYPE SCALING

Linetype dash lengths and the space lengths between dots or dashes are based on the drawing units of the model or paper space in which the objects were created. They can be scaled globally by the system variable LTSCALE factor, as explained earlier. If you want to display objects in viewports at different scales in layout (TILEMODE set to 0), the linetype objects would be scaled to model space rather than paper space by default. However, by setting paper space linetype scaling (system variable PSLTSCALE) to 1 (default), dash and space lengths are based on paper space drawing units, including the linetype objects that are drawn in model space. For example, a single linetype definition with a dash length of 0.30, displayed in several viewports with different zoom factors, would be displayed in paper space with dashes of length 0.30, regardless of the scale of the viewpoint in which it is being displayed (PSLTSCALE set to 1).

Invoke the PSLTSCALE command:

Command: prompt	**psltscale** (ENTER)

AutoCAD prompts:

Command: **psltscale**
Enter new value for PSLTSCALE <current>: *(specify the new value, or press ENTER to accept the current value)*

Note: When you change the PSLTSCALE value to 1, the linetype objects in the viewport are not automatically regenerated. Use the REGEN or REGENALL command to update the linetypes in the viewports.

DIMENSIONING IN MODEL SPACE AND PAPER SPACE

Dimensioning can be done in both model space and paper space. There are no restrictions placed on the dimensioning commands by the current mode. It is advisable to draw associative dimensions in model space, since AutoCAD places the defining points of the dimension in the space where the dimension is drawn. If the model geometry is modified with a command such as STRETCH, EXTEND, or TRIM, the dimensions are updated automatically. In contrast, if the dimensions are drawn in paper space, the paper space dimension does not change if the model geometry is modified.

For dimensioning in model space, the DIMSCALE factor should be set to 0.0. This causes AutoCAD to compute a scale factor based on the scaling between paper space and the current model space viewport. If dimensioning that describes model geometry should be created in paper space, then the DIMLFAC dimension variable scale factor should be based on the model space viewport. It is important that the length scaling be set to a value that is appropriate for the view being dimensioned.

ATTACHING PLOT STYLE TABLES TO VIEWPORTS

Plot style tables are settings that give you control over how objects in your drawing are plotted into hard copy plots. You can control how colors are translated into line weight and how area fills are converted into shades of gray or screened colors, as well as many other output options. You can control how the plotter treats each individual object in a drawing. AutoCAD allows you to assign a plot style table to a layout. In addition, you can assign a separate plot style table to floating viewports overriding the assigned plot style table to the layout. To visualize the effect of the plot styles attached to your current layout or viewport, set **Display plot styles** to ON in the Page Setup dialog box and then regenerate the drawing or preview the drawing by invoking the PRINT PREVIEW command. Refer to the section on "Creating and Modifying Plot Style Tables," later in this chapter, for a detailed explanation on creating and modifying a plot style table.

PLOTTING FROM LAYOUT

Invoke the PLOT command to plot the drawing in WYSIWYG (What You See Is What You Get). Before you plot the drawing from the layout, make sure the following tasks are completed:

Create a model drawing

Create or activate a layout

Open the page setup dialog box and set settings such as plotting device (if necessary configure plotting device), paper size, plot area, plot scale, and drawing orientation

If necessary, insert a title block or attach title block as a reference file

Create floating viewports and position them in the layout

Set the view scale of the floating viewports

Annotate or create geometry in the layout as needed.

Plot the layout

To plot the current drawing from the layout, invoke the PLOT command:

Standard toolbar	Choose the PLOT command (see Figure 8–43)
File menu	Choose Plot
Command: prompt	**plot** (ENTER)

Figure 8–43 *Invoking the PLOT command from the Standard toolbar*

AutoCAD displays the Plot dialog box, as shown in Figure 8–44.

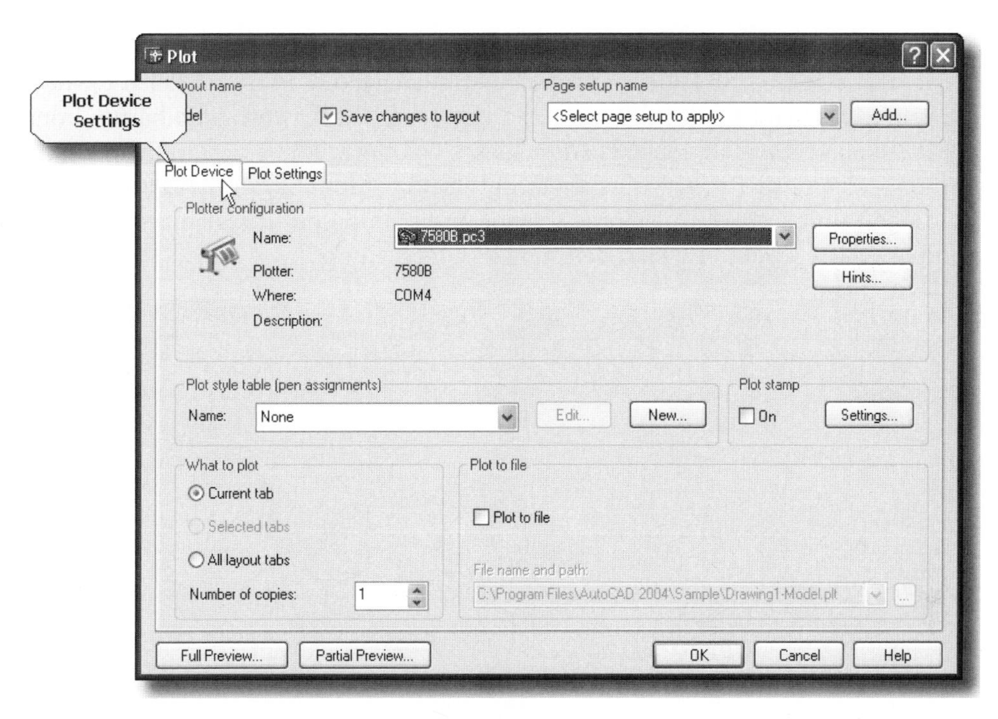

Figure 8–44 *Plot dialog box – Plot Device tab*

AutoCAD displays the current layout name or displays "Selected layouts" if multiple tabs are selected. Set the **Save changes to layout** toggle button to ON to save changes you make in the Plot dialog box to the layout. This option is not available if multiple layouts are selected.

The **Page setup name** list box displays a list of any saved page setups. Similar to plotting from model space, you can choose from one of the saved page setups to restore the plot settings. Or you can save the current settings to the named page setup by selecting the **Add...** button, and it can be applied later to plot any drawing from paper space.

Plot Device Settings

The **Plot Device** tab (see Figure 8–44) specifies the plotter to use, a plot style table, the layout or layouts to plot, and information about plotting to a file. The options that are available in the **Plot Device** tab are the same as those explained earlier in the section "Plotting From Model Space."

Plot Settings

The **Plot Settings** tab (see Figure 8–45) specifies paper size, drawing orientation, plot area and plot scale, plot offset, and plot options. The options that are available in the

Plot Settings tab are the same as those explained earlier in the section "Plotting From Model Space," except an additional option is provided in the **Plot area** called **Layout**. Whenever you want to plot the current layout, select the **Layout** radio button from the **Plot area** section. This causes AutoCAD to plot everything within the margins of the specified paper size, with the origin calculated from 0,0 in the layout. If you choose to turn off the **Display paper background** option on the **Display** tab of the Options dialog box, the **Layouts** selection becomes **Limits**.

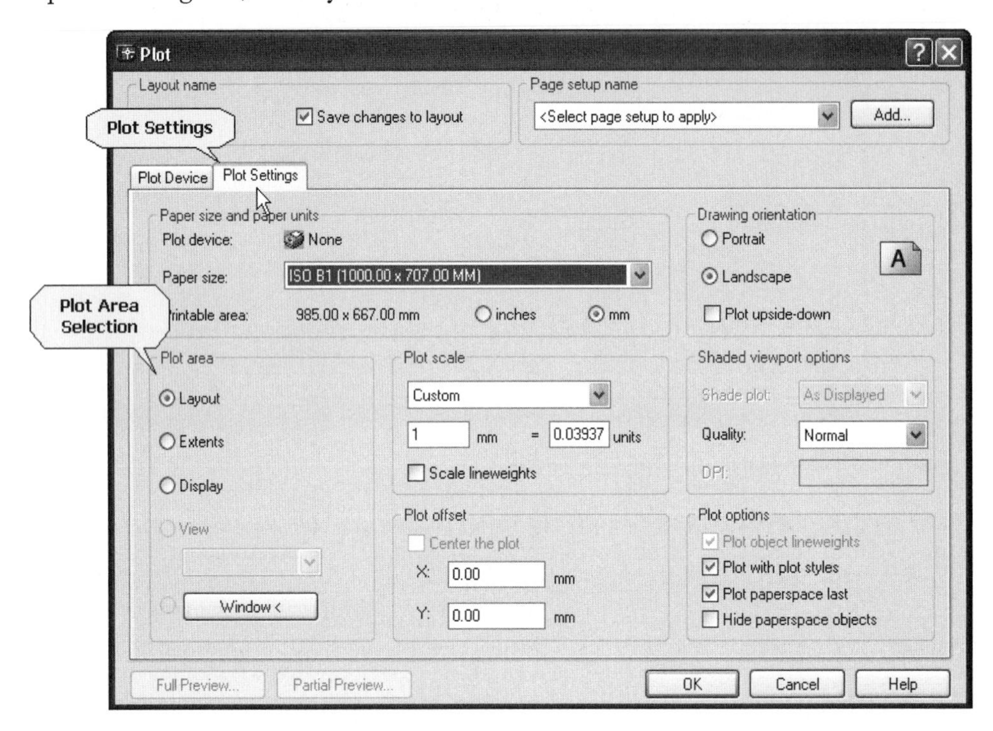

Figure 8–45 *Plot dialog box – Plot Settings tab*

Choose the **Partial Preview...** button to quickly display a partial preview of the effective plot area relative to the paper size and printable area. In addition, AutoCAD provides information regarding paper size, printable area, and effective area. Partial preview also gives advance notice of any AutoCAD warnings that may be encountered when you plot the drawing.

A full preview displays the drawing on the screen as it would appear when plotted. This takes much longer than the partial preview, but it takes less time than regular plot regeneration. This is quicker than plotting incorrectly and having to start all over again.

To get a full preview of your plot, choose the **Full Preview...** button. AutoCAD temporarily clears the plotting dialog boxes, draws an outline of the paper size, and displays the drawing as it will appear on the paper when it is plotted.

The cursor changes to a magnifying glass with plus and minus signs. Holding the pick button and dragging the cursor toward the top of the screen enlarges the preview image. Dragging it toward the bottom of the screen reduces the preview image. Right-click and AutoCAD displays a shortcut menu offering additional preview options: PAN, ZOOM, ZOOM WINDOW, ZOOM ORIGINAL, PLOT, and EXIT.

To end the full preview, choose the EXIT option from the shortcut menu. AutoCAD returns to the Plot dialog box.

Note: You can also access a full preview by invoking the PLOT PREVIEW command from File menu or the Plot Preview icon on the Standard toolbar.

After making the necessary changes in the plot settings, choose the **OK** button. AutoCAD starts plotting and reports its progress as it converts the drawing into the plotter's graphics language by displaying the number of vectors processed.

If something goes wrong or if you want to stop immediately, choose the **Cancel** button at any time. AutoCAD cancels the plotting.

CREATING AND MODIFYING PLOT STYLE TABLES

Plot style tables are settings that give you control over how objects in your drawing are plotted into hard copy plots. By modifying an object's plot style, you can override that object's color, linetype, and lineweight. You can also specify end, join, and fill styles, as well as output effects such as dithering, gray scale, pen assignment, and screening. You can use plot styles if you need to plot the same drawing in different ways.

By default, every object and layer has a plot style property. The actual characteristics of plot styles are defined in plot style tables that you can attach to a model tab and layouts within layouts. If you assign a plot style to an object, and then if you detach or delete the plot style table that defines the plot style, the plot style will not have any effect on the object.

AutoCAD provides two plot style modes: Color-Dependent and Named.

The Color-dependent plot styles are based on object color. There are 255 color-dependent plot styles. You cannot add, delete, or rename color-dependent plot styles. You can control the way all objects of the same color plot in color-dependent mode by adjusting the plot style that corresponds to that object color. Color-dependent plot style tables are stored in files with the extension .CTB.

Named plot styles work independently of object color. You can assign any plot style to any object regardless of that object's color. Named plot style tables are stored in files with the extension .STB.

By default, all the plot style table files are saved in the path that is listed in the Files section of the Options dialog box.

The default plot style mode is set in the **Plotting** tab (**Default plot style behavior for new drawings** section) of the Options dialog box, as shown in Figure 8–46.

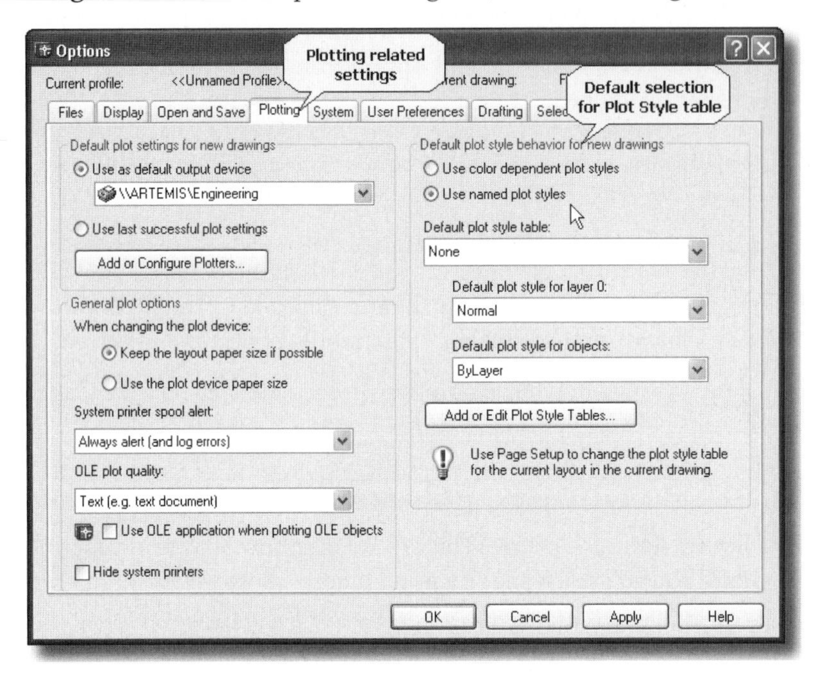

Figure 8–46 *Options dialog box (Plotting tab)*

Every time you start a new drawing in AutoCAD, the plot style mode that is set in the Options dialog box is applied. Anytime you change the mode, it is applied only for the new drawings or an open drawing that has not yet been saved in AutoCAD. You cannot change the mode for the current drawing.

CREATING A PLOT STYLE TABLE

AutoCAD allows you to create a named plot style table to utilize all the flexibility of named plot styles or a color-dependent plot style table to work in a color-based mode. The Add Plot Style Table Wizard allows you to create a plot style from scratch, modify an existing plot style table, import style properties from an acadr14.cfg file, or import style properties from an existing PCP or PC2 file.

To open the Add Plot Style Table Wizard:

Tools menu	Choose Wizards > Add Plot Style Table...

AutoCAD displays the introductory text of the Add Plot Style Table Wizard, as shown in Figure 8–47.

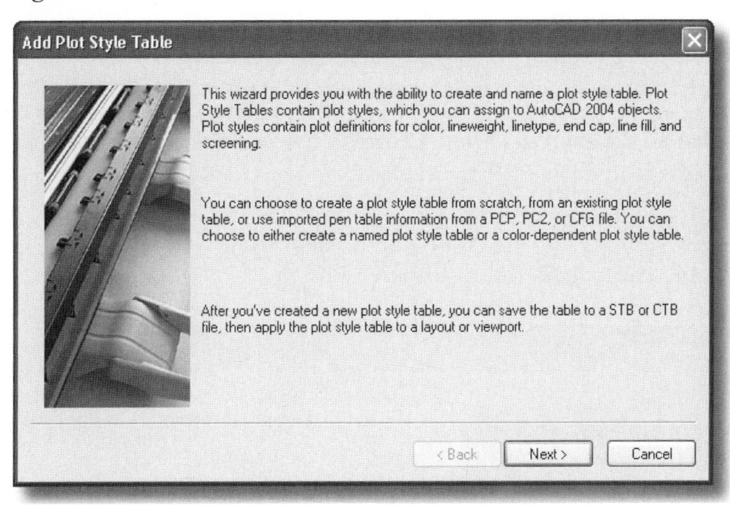

Figure 8–47 *Introductory text page of the Add Plot Style Table Wizard*

Choose the **Next>** button, and AutoCAD displays the **Add Plot Style Table – Begin** page, as shown in Figure 8–48.

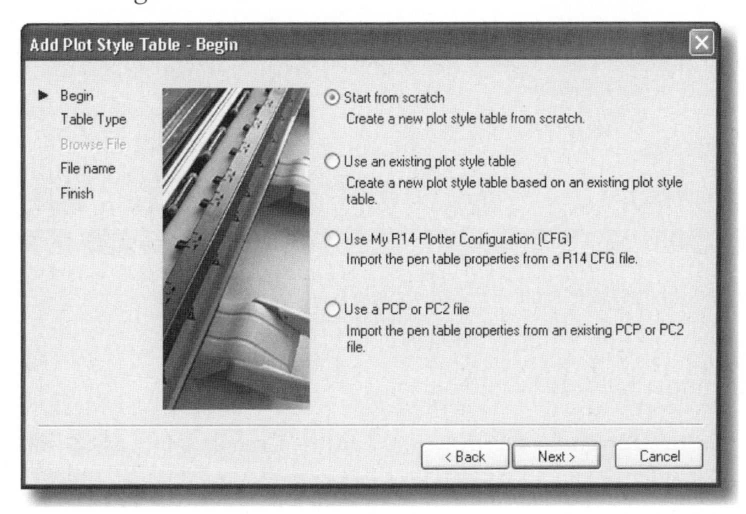

Figure 8–48 *Add Plot Style Table – Begin page*

Choose one of the following four radio buttons

- **Start from scratch**—Allows you to create a new plot style from scratch.
- **Use an existing plot style table**—Creates a new plot style using an existing plot style table.
- **Use My R14 Plotter Configuration (CFG)**—Creates a new plot style table using the pen assignments stored in the ACADR14.CFG file. Choose this option if you do not have an equivalent PCP or PC2 file.
- **Use a PCP or PC2 File**—Creates a new plot style table using pen assignments stored in a PCP or PC2 file

Choose the **Next>** button, and AutoCAD displays the **Add Plot Style Table – Pick Plot Style Table** page, as shown in Figure 8–49.

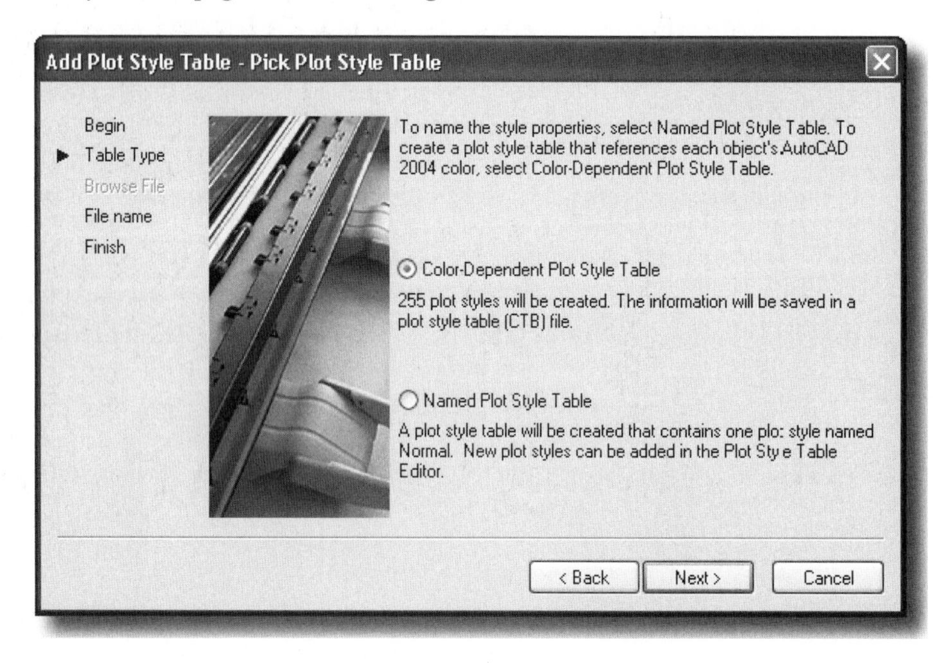

Figure 8–49 *Add Plot Style Table – Pick Plot Style Table page*

Choose one of the following radio buttons:

- **Color-Dependent Plot Style Table**—Creates a plot style table with 255 plot styles.
- **Named Plot Style Table**—Creates a named plot style table.

Note: If you selected the **Use an existing plot style table** option to create a new plot style, AutoCAD displays the **Add Plot Style Table – Browse File Name** page. Specify the plot style table file name from which to create a new plot style name.

Choose the **Next>** button, and AutoCAD displays the **Add Plot Style Table – File name** page, as shown in Figure 8–50.

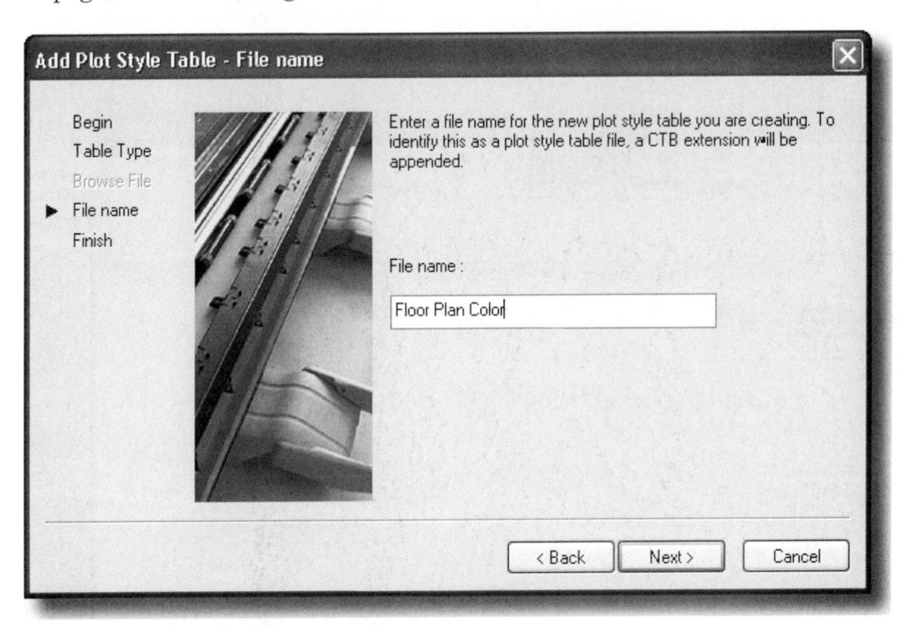

Figure 8–50 *Add Plot Style Table – File name page*

Specify the file name in the **File name:** text field. By default the new style table is saved in the path that is listed in the **Files** section of the Options dialog box.

Choose the **Next>** button, and AutoCAD displays **Add Plot Style Table – Finish** page, as shown in Figure 8–51.

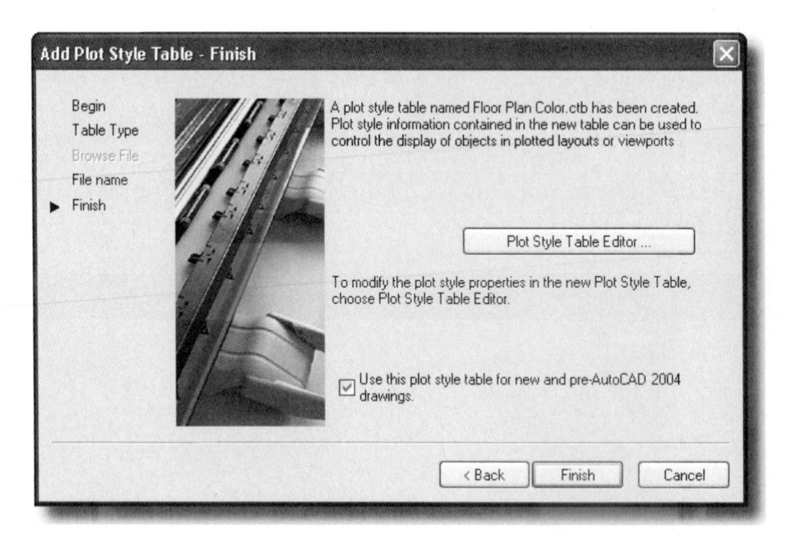

Figure 8–51 *Add Plot Style Table – Finish page*

Set the **Use this plot style table for new and pre-AutoCAD 2004 drawings** toggle button to ON to attach this plot style table to all new drawings and pre-AutoCAD 2004 drawings by default.

Choose the **Finish** button to create the plot style table and close the wizard.

MODIFYING A PLOT STYLE TABLE

AutoCAD allows you to add, delete, copy, paste, and modify plot styles in a plot style table by using the Plot Style Table Editor. You can open more than one instance of the Plot Style Table Editor at a time and copy and paste plot styles between the tables. Open the Plot Style Table Editor using any of the following methods:

- Choose the **Plot Style Table Editor** button from the Finish screen in the Add Plot Style Table Wizard.
- Open the Plot Style Manager (File menu), right-click a CTB or STB file, and then choose OPEN from the shortcut menu.
- On the **Plot Device** tab of the Plot dialog box or Page Setup dialog box, select the plot style table you want to edit from the Plot Style Table list, and then choose the **Edit** button.
- On the **Plotting** tab of the Options dialog box, choose the **Add or Edit Plot Style Tables...** button

Figure 8–52 shows an example of the Plot Style Table Editor for a named plot style table, and Figure 8–53 shows an example of the Plot Style Table Editor for a color-dependent plot style table.

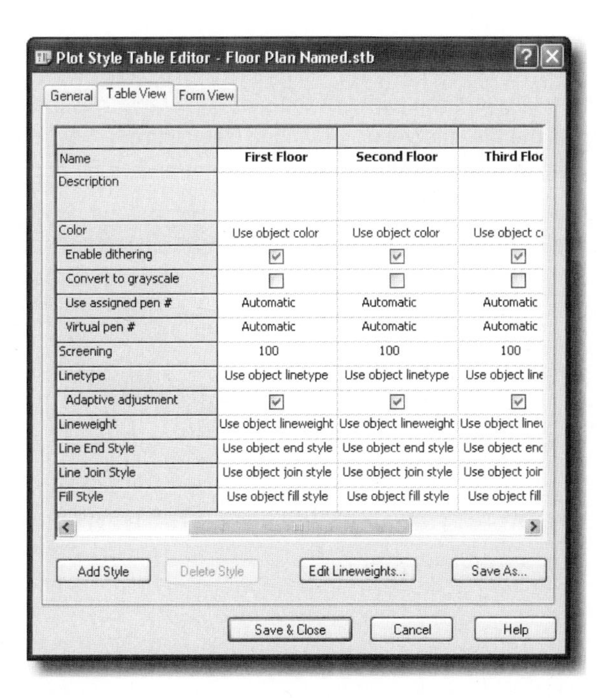

Figure 8–52 *Plot Style Table Editor for named plot style table (Table View)*

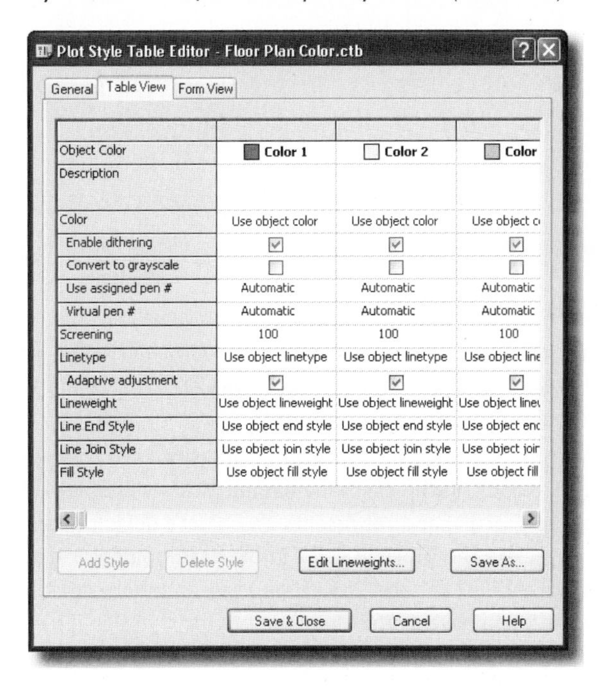

Figure 8–53 *Plot Style Table Editor for color-dependent plot style table (Table View)*

Following are the three tabs available in the Plot Style Table Editor:

- **General**—Displays the name of the plot style table, description (if any), location of the file, and version number (see Figure 8–54). You can modify the description, and apply scaling to non-ISO lines and to fill patterns.

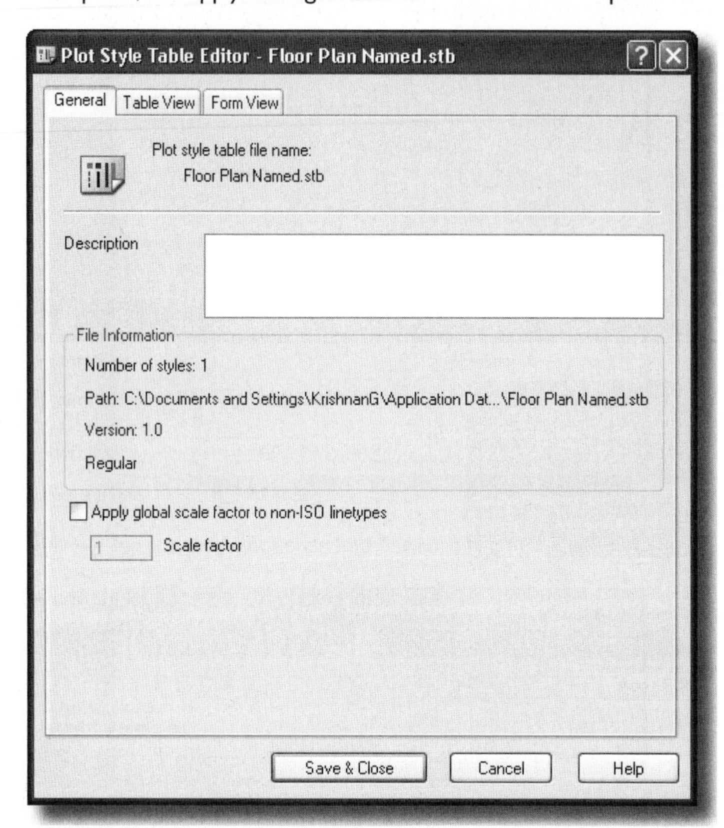

Figure 8–54 *Plot Style Table Editor – General tab*

- **Table View**—Lists entire plot styles in the plot style table and their settings in tabular form (see Figure 8–52 and Figure 8–53). The styles are displayed in columns from left to right. The setting names of each row appear at the left of the tab. By default, in the case of a named plot style table, AutoCAD sets up a style named Normal and represents an object's default properties. You cannot modify or delete the Normal style. In the case of a color-dependent plot style table, AutoCAD lists all the 255 color styles in tabular form. In general this is convenient if you have a small number of plot styles to view them in the tabular form.

- **Form View**—The plot style names are listed under the **Plot styles:** list box and the settings for the selected plot style are displayed at the right side of the dialog box (see Figure 8–55).

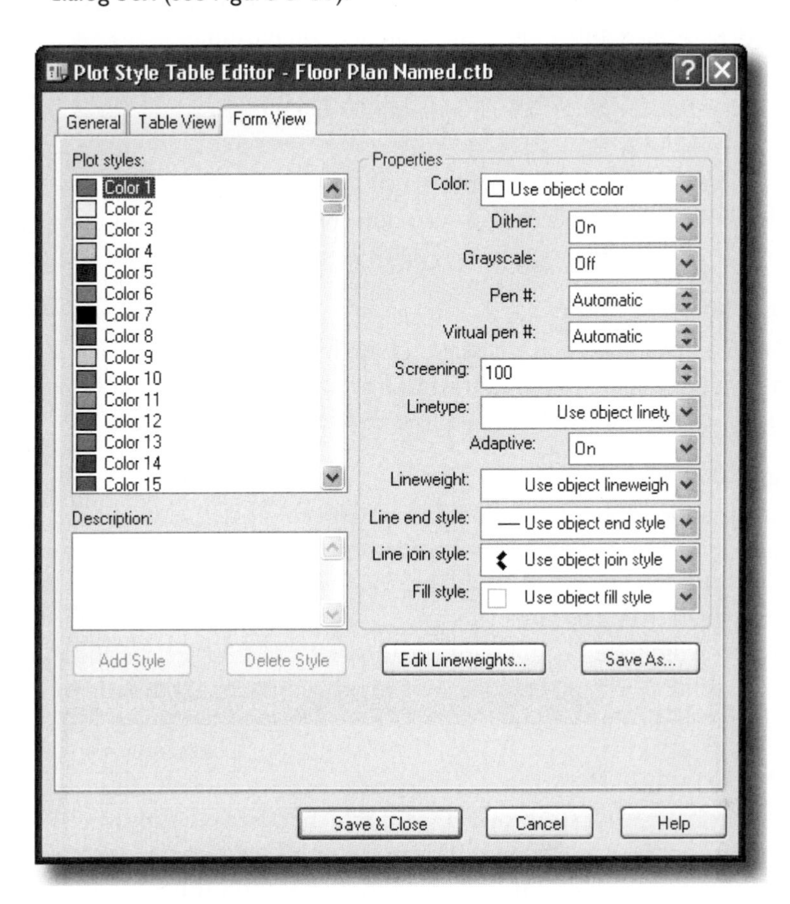

Figure 8–55 *Plot Style Table Editor – Form View*

Creating a New Plot Style

To create a new plot style, choose the **Add Style** button from the Plot Style Table Editor. AutoCAD adds a new style, and you can change the name to a descriptive name if necessary (cannot exceed 255 characters). You cannot duplicate names within the same plot style table.

Note: You cannot add or change the name for a plot style in the color-dependent style table.

Deleting a Pen Style

To delete a pen style, click the gray area above the plot style name in the Table View and choose the **Delete Style** button. In the Form View, select the style name from the **Plot styles:** list box and choose the **Delete Style** button.

Note: You cannot delete a plot style from a color-dependent style table.

Adding and Modifying a Plot Style Description

AutoCAD allows you to specify a description for plot styles and modify an existing description for a plot style if necessary. The description cannot exceed 255 characters.

Assigning Plot Style Color

The **Color:** list box allows you to assign a plot style color. If you assign a color from one of the available colors, then AutoCAD overrides the object's color at plot time. By default, all of the plot styles are set to use object color.

Enabling Dithering

If you set **Dither:** to ON, the plotter approximates colors with dot patterns, giving the impression of plotting with more colors than the ink available in the plotter. If you set the Enable Dithering to OFF, then AutoCAD maps colors to the nearest color, which limits the range of colors used for plotting. The most common reason for turning off dithering is to avoid false line typing from dithering of thin vectors and to make dim colors more visible. If the plotter does not support dithering, the dithering setting is ignored. The default setting is set to ON.

Converting to Gray Scale

If you set **Grayscale:** to ON, AutoCAD converts the object's colors to gray scale if the plotter supports gray scale. If you set **Grayscale:** to OFF, AutoCAD uses the RGB values for the object's colors. The default setting is set to OFF.

Assigning Pen Numbers

The **Pen #:** text field setting in the Plot Style Table Editor specifies which pen to use for each plot style. You can specify a pen to use in the plot style by selecting from a range of pen numbers from 1 to 32. If you specify 0, the field updates to read Automatic. AutoCAD uses the information you provided under Physical Pen Configuration in the Plotter Configuration Editor to select the pen closest in color to the object you are plotting. The default is set to Automatic.

Using Virtual Pens

Specify a virtual pen number in the **Virtual pen #:** edit field for plotters that do not use pens but can simulate the performance of a pen plotter by using virtual pens. The default is set to Automatic or 0 to specify that AutoCAD should make the virtual pen assignment from the AutoCAD Color Index. You can specify a virtual pen number between 1 and 255. The virtual pen number setting in a plot style is used only by plotters without pens and only if they are configured for virtual pens. If this is the case, all the other style settings are ignored and only the virtual pen is used. The default is set to Automatic.

Note: If a plotter without pens is not configured for virtual pens, then both the virtual and the physical pen information in the plot style are ignored and all the other settings are used.

Using Screening

The **Screening:** text field sets a color intensity setting that determines the amount of ink AutoCAD places on the paper while plotting. The valid range is 0 through 100. Selecting 0 reduces the color to white. Selecting 100 (default) displays the color at its full intensity. The default is set to 100%.

Assigning Plot Style Linetype

The **Linetype:** list box allows you to assign a plot style linetype. If you assign a linetype from one of the available linetypes, then AutoCAD overrides the object's linetype at the plot time. By default, all the plot styles are set to use object linetype.

Adaptive Adjustment

The **Adaptive:** toggle adjusts the scale of the linetype to complete the linetype pattern. Set **Adaptive:** to ON if it is more important to have complete linetype patterns than correct linetype scaling. Set **Adaptive:** to OFF if linetype scale is important. The default is set to ON.

Assigning Plot Style Lineweight

The **Lineweight:** list box allows you to assign a plot style lineweight. If you assign a lineweight from one of the available lineweights, then AutoCAD overrides the object's lineweight at the plot time. By default, all the plot styles are set to use object lineweight.

Assigning a Line End Style

The **Line end style:** list box allows you to assign a line end style. The line end style options include: Butt, Square, Round, and Diamond. If you assign a line end style from one of the available line end styles, then AutoCAD overrides the object's line end style at the plot time. By default, all the plot styles are set to use object end style.

Assigning a Line Join Style

The **Line join style:** list box allows you to assign a line end style. The line join style options include: Miter, Bevel, Round, and Diamond. If you assign line join style from one of the available line join styles, then AutoCAD overrides the object's line join style at the plot time. By default, all the plot styles are set to use object line join style.

Assigning Fill Style

The **Fill style:** list box allows you to assign a fill style. The fill style options include: Solid, Checkerboard, Crosshatch, Diamonds, Horizontal Bars, Slant Left, Slant Right, Square Dots, and Vertical Bars. The fill style applies only to solids, plines, donuts, and *3D* faces. If you assign a fill style from one of the available fill styles, then AutoCAD overrides the object's fill style at the plot time. By default, all the plot styles are set to use object fill style.

Editing Lineweights

AutoCAD allows you edit the available lineweights by choosing the **Edit Lineweights...** button. You cannot add or delete lineweights from the list.

Saving the Plot Style Table

To save the changes and close the Plot Style Table Editor, choose the **Save & Close** button. To save the changes to another plot style table, choose the **Save As** button. AutoCAD displays the Save As dialog box. Specify the file name in the **File name:** text field and choose the **Save** button to save and close the Save As dialog box.

CHANGING PLOT STYLE PROPERTY FOR AN OBJECT OR LAYER

As mentioned earlier, every object that is created in AutoCAD has a plot style property in addition to color, linetype, and lineweight. Similarly, every layer has a color, linetype, and lineweight, in addition to a plot style property. The default setting for plot styles for objects and layers are set in the Options dialog box (**Plotting** tab), as shown in Figure 8–56.

The default plot style for objects can be any of the following:

- **Normal**—Uses the object's default properties.
- **BYLAYER**—Uses the properties of the layer that contains the object.
- **BYBLOCK**—Uses the properties of the block that contains the object.
- **Named plot style**—Uses the properties of the specific named plot style defined in the plot style table.

The default plot setting for an object is BYLAYER, and the initial plot style setting for a layer is Normal. When the object is plotted it retains its original properties.

If you are working in a Named plot style mode, you can change the plot style for an object or layer at any time. If you are working in a color dependent plot style mode, you cannot change the plot style for objects or layers. By default, they are set to **By Color**.

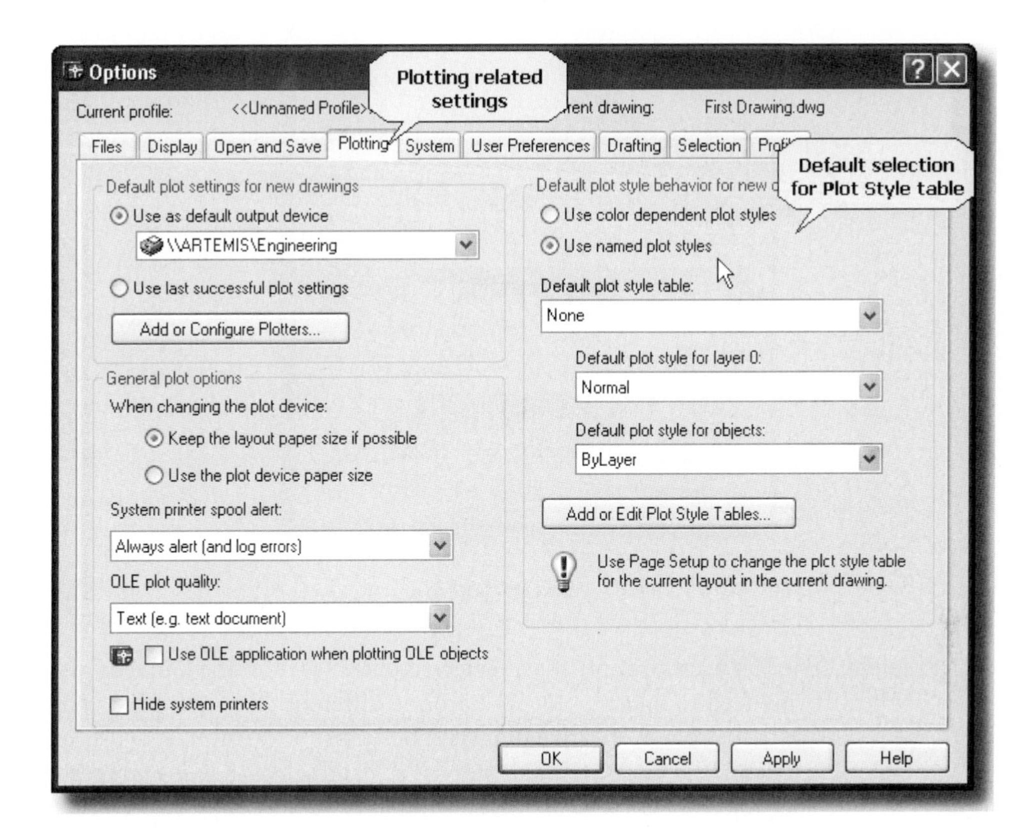

Figure 8–56 *Options dialog box (Plotting tab)*

To change the plot style for one or more objects, first select the objects (system variable PICKFIRST set to ON), and select the plot style from the **Plot style control** list box on the Properties toolbar, as shown in Figure 8–57. If the plot style does not list the one you want to select, then choose **Other**, and AutoCAD displays the Current Plot Styles dialog box, as shown in Figure 8–58.

Figure 8–57 *Properties toolbar – Plot Style control*

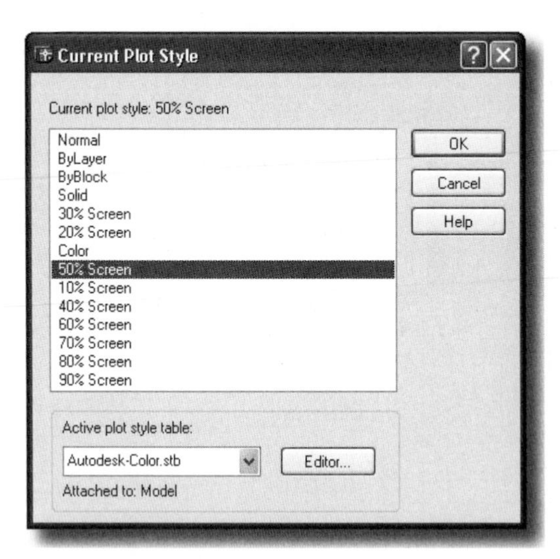

Figure 8–58 *Current Plot Style dialog box*

Select the plot style you want to apply to the selected object(s) from the **Current plot styles** list box. If you need to select a plot style from a different plot style table, then select the plot style table from the **Active plot style table:** list box. AutoCAD lists all the available plot styles in the **Current plot styles** list box and selects the one you want to apply to the select object(s). Choose the **OK** button to close the dialog box. You can also change the plot style of the selected object(s) from the properties dialog box.

To change the plot style for a layer, open the Layer Properties Manager dialog box. Select the layer you want to change and select a plot style for the selected layer similar to changing color or linetype.

CONFIGURING PLOTTERS

Autodesk® Plotter Manager allows you to configure a local or network non-system plotter. In addition, you can also configure a Windows system printer with non-default settings. AutoCAD stores information about the media and plotting device in configured plot (PC3) files. The PC3 files are stored in the path that is listed in the Files section of the Options dialog box. Plot configurations are therefore portable and can be shared in an office or on a project. If you calibrate a plotter, the calibration information is stored in a plot model parameter (PMP) file that you can attach to any PC3 files you create for the calibrated plotter.

AutoCAD allows you to configure plotters for many devices and store multiple configurations for a single device. You can create several PC3 files with different output options for the same plotter. After you create a PC3 file, it's available in the list of plotter configuration names on the **Plot Device** tab of the Plot dialog box.

Open the Autodesk® Plotter Manager:

File menu	Choose Plotter Manager
Command: prompt	**plottermanager** (ENTER)

AutoCAD displays the Plotters window explorer, shown in Figure 8–59, listing all the plotters configured.

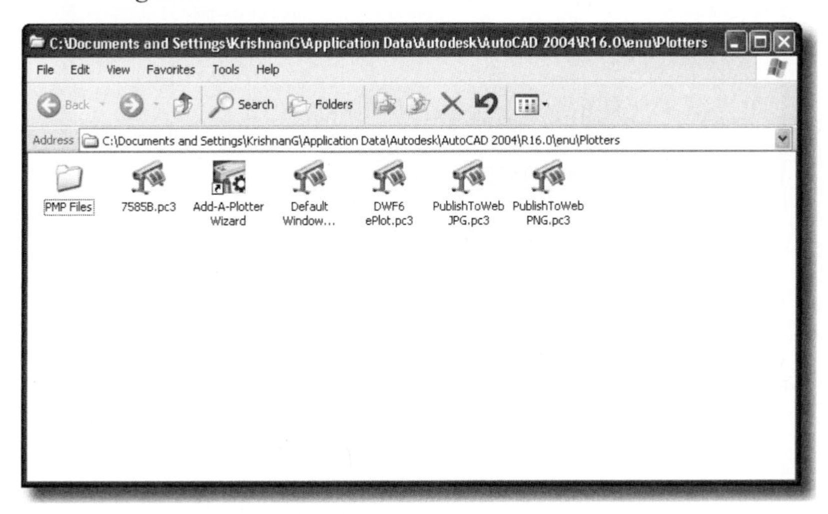

Figure 8–59 *Plotters window explorer (Windows XP version)*

Double-click **Add-A-Plotter Wizard**, and AutoCAD displays the **Add Plotter – Introduction Page**, as shown in Figure 8–60.

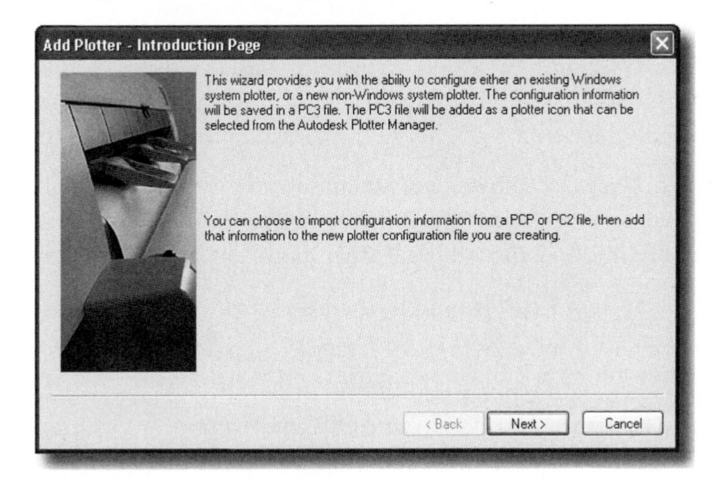

Figure 8–60 *Add Plotter – Introduction Page*

Choose the **Next** > button, and AutoCAD displays the **Add Plotter – Begin** page, as shown in Figure 8–61.

Figure 8–61 *Add Plotter – Begin page*

Choose one of the three radio buttons:

- **My Computer**—To configure a local non-system plotter.
- **Network Plotter Server**—To configure a plotter that is on the network.
- **System Printer**—To configure a Windows system printer. If you want to connect to a printer that is not in the list, you must first add the printer using the Windows Add Printer wizard in the Control Panel.

If you select the **My Computer** option, then the wizard prompts you to select a plotter manufacturer and model number, identify the port to which the plotter is connected, specify a unique plotter name, and choose the **Finish** button the close the wizard.

If you select the **Network Plotter Server** option, then the wizard prompts you to identify the network server, select the plotter manufacturer and model number, specify a unique plotter name, and choose the **Finish** button to close the wizard.

If you select the **System Printer** option, then the wizard prompts you to select one of the printers configured in the Windows operating system, specify a unique plotter name, and choose the **Finish** button to close the wizard.

AutoCAD saves the configuration file with PC3 file format with a unique given name in the path that is listed in the **Files** section of the options dialog box..

If necessary, you can edit the PC3 file using the Plotter Configuration Editor. The Plotter Configuration Editor provides options for modifying a plotter's port connections and output settings including media, graphics, physical pen configuration, custom properties, initialization strings, calibration, and user-defined paper sizes. You can drag these options from one PC3 file to another.

You can open the Plotter Configuration Editor using one of the following methods:

- From the File menu, choose **Page Setup**. Choose **Properties**.
- From the File menu, choose **Plot**. Choose **Properties**.
- Double-click a PC3 file from Windows Explorer and right-click the file and choose **Open**.
- Choose **Edit Plotter Configuration...** on the Add Plotter - Finish page in the Add-A-Plotter Wizard.

Figure 8–62 shows the Plotter Configuration Editor for an HP7580B plotter.

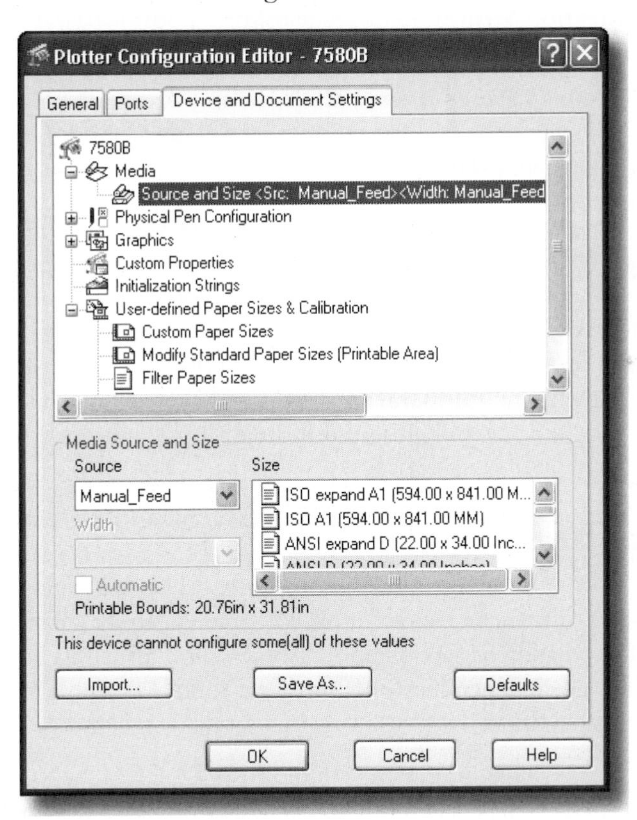

Figure 8–62 *Plotter Configuration Editor for an HP7580B plotter*

The Plotter Configuration Editor contains three tabs:

- **General** tab—Contains basic information about the configured plotter.
- **Ports** tab—Contains information about the communication between the plotting device and your computer.
- **Device and Document Settings** tab—Contains plotting options.

In the **Device and Document Settings** tab, you can change many of the settings in the configured plot (PC3) file. Following are the six areas in which you make the changes:

- **Media**—Specifies a paper source, size, type, and destination.
- **Physical Pen Configuration**—Specifies settings for pen plotters.
- **Graphics**—Specifies settings for printing vector graphics, raster graphics, and TrueType® fonts.
- **Custom Properties**—Displays settings related to the device driver.
- **Initialization Strings**—Sets pre-initialization, post-initialization, and termination printer strings.
- **User-defined Paper Sizes & Calibration**—Attaches a plot model parameter (PMP) file to the PC3 file, calibrates the plotter, and adds, deletes, or revises custom or standard paper sizes.

The areas correspond to the categories of settings in the PC3 file you're editing. Double-click any of the six categories to view and change the specific settings. When you change a setting, your changes appear in angle brackets (<>) next to the setting name unless there is too much information to display. To save the changes to another PC3 file, choose the **Save As...** button. AutoCAD displays the Save As dialog box. Specify the file name in the **File name:** text field and choose the **Save** button. To save the changes to the PC3 file and close the Plotter Configuration Editor, choose the **OK** button.

BATCH PLOTTING

In addition to plotting individual drawings, AutoCAD has a batch plotting utility that you can use to construct a list of AutoCAD drawings to be plotted. You can plot the drawings immediately or save them in a batch plot (BP3) file to plot later. The Batch Plot utility runs independently of AutoCAD and is available in the AutoCAD program folder (batchplt.exe).

Once you have created a list of drawings to plot using the Batch Plot utility, you can attach any PC3 (plot configuration) file to each drawing. Any drawing not attached to a PC3 file plots to the device that was the default before you started the Batch Plot utility.

Open the Exercise Manual PDF file for Chapter 8 on the accompanying CD for project and discipline specific exercises.

If you have the accompanying Exercise Manual, refer to Chapter 8 for project and discipline specific exercises.

REVIEW QUESTIONS

1. If you were to plot a drawing at a scale of 1"=60', what should you set LTSCALE to?

   a. 60

   b. 1/60

   c. 720

   d. 1/720

2. If you want to plot a drawing requiring multiple pens and you are using a single pen plotter, AutoCAD will:

   a. not plot the drawing at all

   b. pause when necessary to allow you to change pens

   c. invoke an error message

   d. plot all the drawing using the single pen

   e. none of the above

3. The drawing created at a scale of 1:1 and plotted to "Scaled to Fit" is plotted:

   a. at a scale of 1:1

   b. to fit the specified paper size

   c. at the prototype scale

   d. none of the above

4. To plot a full scale drawing at a scale of 1/4"=1', use a plot scale of:

   a. 0.25=12

   b. 0.25=1

   c. 48=1

   d. 12=0.25

   e. 24=1

5. What is the file extension assigned to all files created when plotting to a file?

   a. DWG

   b. DRW

   c. DRK

   d. PLO

   e. PLT

6. When plotting, pen numbers are assigned to:

   a. colors

   b. layers

   c. thickness

   d. linetypes

   e. none of the above

You need to draw three orthographic views of an airplane whose dimensions are: wingspan of 102 feet, a total length of 118 feet, and a height of 39 feet. The drawing has to be plotted on a standard 12" x 9" sheet of paper. No dimensions will be added, so you will need only 1" between the views. Answer the following five questions using the information from this drawing:

7. What would be a reasonable scale for the paper plot?

   a. 1=5'

   b. 1=15'

   c. 1=25'

   d. 1=40'

8. What would be a reasonable setting of LTSCALE ?

   a. 1

   b. 5

   c. 60

   d. 25

   e. 300

   f. 480

9. If you were plotting from paper space, what ZOOM scale factor would you use?

   a. 1/5X

   b. 1/25X

   c. 1/60X

   d. 1/300X

   e. 1/5XP

   f. 1/25XP

   g. 1/60XP

   h. 1/300XP

10. When inserting your border in paper space, what scale factor should you use?

    a. 1
    b. 5
    c. 25
    d. 60
    e. 300

11. Which of the following options will the plot preview give you?

    a. seeing what portion of your drawing will be plotted
    b. seeing the plotted size of your drawing
    c. seeing the plotted drawing relative to the page size
    d. seeing rulers around the edge of the plotted page for size comparison
    e. none of the above

12. Which of the following determine the relationship between the size of the objects in a drawing and their sizes on a plotted copy?

    a. Size of the object in the AutoCAD drawing
    b. Size of the object on the plot
    c. Maximum available plot area
    d. Plot scale
    e. All of the above

13. AutoCAD permits plotting in which of the following environment modes?

    a. Model space
    b. Paper space
    c. layout
    d. all of the above

14. Which TILEMODE system variable setting corresponds to model space?

    a. 0
    b. 1
    c. either A or B
    d. none of the above

15. When starting a new drawing, how many default plotting layouts does AutoCAD create?

    a. 0
    b. 1
    c. 2
    d. unlimited

16. Within multiple floating viewports you can establish various scale and layer visibility settings for each individual viewport.

    a. True

    b. False

17. Paper sizes are indicated by X axis direction (drawing length) and Y axis direction (drawing width)?

    a. True

    b. False

18. Floating viewports, like lines, arcs, and text, can be manipulated using AutoCAD commands such as MOVE, COPY, STRETCH, SCALE or ERASE.

    a. True

    b. False

19. While in paper space both the floating viewports and the 3D model can be modified or edited.

    a. True

    b. False

20. Which of the following can be converted into a viewport?

    a. ellipse

    b. splines

    c. circles

    d. all of the above

21. Which of the following commands allows for the control of layer visibility in a specific viewport?

    a. VPORTS

    b. VPLAYER

    c. VIEWLAYER

    d. LAYERVIS

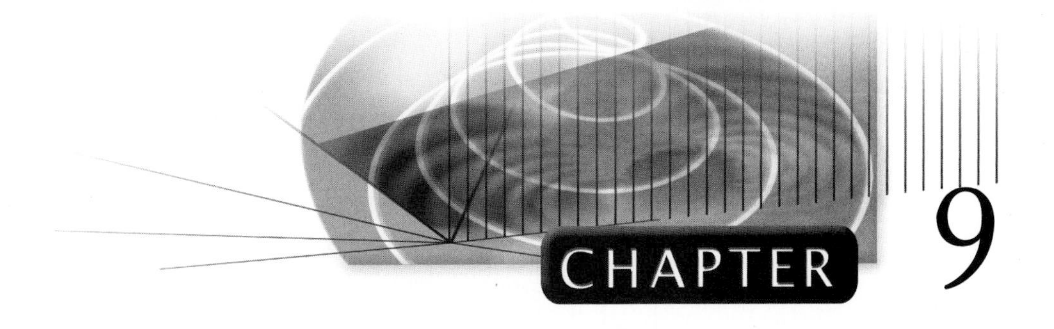

Hatching and Boundaries

INTRODUCTION

After completing this chapter, you will be able to do the following:

- Use the BHATCH and HATCH commands.
- Fine-tune the hatching boundaries by means of the **Advanced** tab of the Boundary Hatch and Fill dialog box.
- Modify hatch patterns via the HATCHEDIT command.

WHAT IS HATCHING?

Drafters and designers use repeating patterns, called hatching, to fill regions in a drawing for various purposes (see Figure 9–1). In a cutaway (cross-sectional) view, hatch patterns help the viewer differentiate between components of an assembly and indicate the material of each. In surface views, hatch patterns depict material and add to the readability of the view. In general, hatch patterns greatly help the drafter/designer achieve his or her purpose, that is, communicating information. Because drawing hatch patterns is a repetitive task, it is an ideal application of computer-aided drafting.

You can use patterns that are supplied in an AutoCAD support file called ACAD.PAT, patterns in files available from third-party custom developers, or you can create your own custom hatch patterns. See Appendix E for the list of patterns supplied with ACAD.PAT.

AutoCAD allows you to fill with a solid color in addition to a hatch pattern. AutoCAD creates an associative hatch, which updates when its boundaries are modified, or a nonassociative hatch, which is independent of its boundaries. Before AutoCAD draws the hatch pattern, it allows you to preview the hatching and to adjust the definition if necessary.

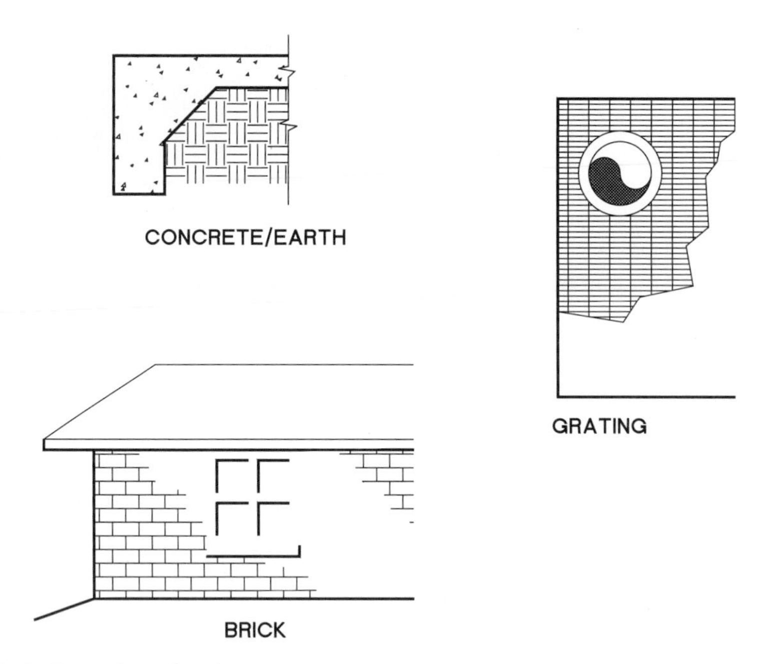

Figure 9–1 *Examples of hatch patterns*

Hatch patterns are considered as separate objects. The hatch pattern behaves as one object; if necessary, you can separate it into individual objects with the EXPLODE command. Once it is separated into individual objects, the hatch pattern will no longer be associated with the boundary object.

Hatch patterns are stored with the drawing, so they can be updated, even if the pattern file containing the hatch is not available. You can control the display of the hatch pattern with the FILLMODE system variable. If FILLMODE is set to OFF, then the patterns are not displayed, and regeneration calculates only the hatch boundaries. By default, FILLMODE is set to ON.

The hatch pattern is drawn with respect to the current coordinate system, current elevation, current layer, color, linetype and current snap origin.

DEFINING THE HATCH BOUNDARY

A region of the drawing may be filled with a hatch pattern if it is enclosed by a boundary of connecting lines, circles, or arc objects. Overlapping boundary objects can be considered as terminating at their intersections with other boundary objects. There must not be any gaps between boundary objects, however. Figure 9–2 illustrates variations of objects and the potential boundaries that might be established from them.

Note in Figure 9–2 how the enclosed regions are defined by their respective boundaries. A boundary might include all or part of one or more objects. In addition to lines, circles, and arcs, boundary objects can include 2D and 3D polylines, 3D faces, and viewports. Boundary objects should be parallel to the current UCS. You can also hatch Block References that have been inserted with unequal X and Y scale factors.

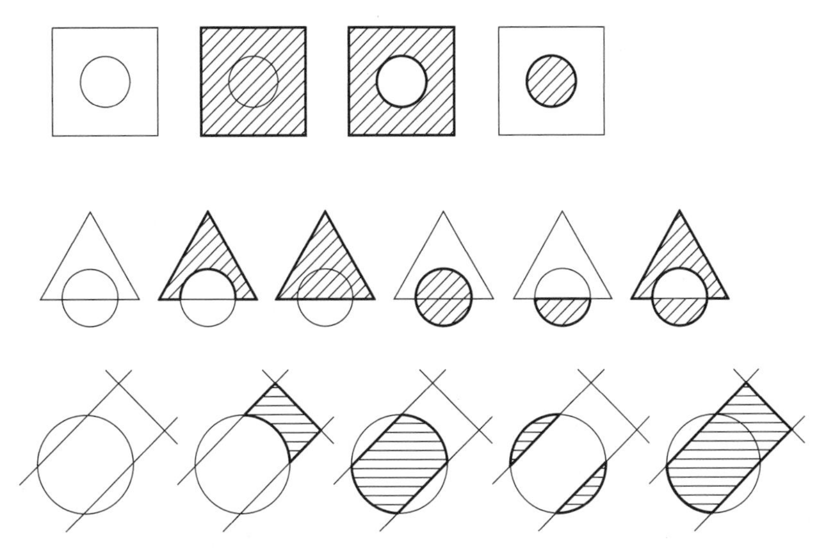

Figure 9–2 *Allowed hatching boundaries made from different objects*

BHATCH VERSUS HATCH

AutoCAD provides two commands for hatching: BHATCH and HATCH. The BHATCH command (introduced in AutoCAD Release 12) means Boundary Hatch and includes several features that greatly improve the ease of use over the HATCH command (introduced in Version 1.4). The BHATCH command automatically creates a boundary; conversely, with the HATCH command you must define a boundary or it allows you to create it. In Figure 9–3, the HATCH command selects the four lines via the Window option. These four lines comprise the hatching boundary. These four objects are valid boundary segments for use by the HATCH command. They connect at their endpoints and do not overlap. Instead of using the Window option of the object selection process, you can select the four lines individually. This may be desirable if there were unwanted objects within a window used to select them.

In Figure 9–4, the BHATCH command permits you to select a point in the region enclosed by the four lines. Then, AutoCAD creates a polyline with vertices that coincide with the intersections of the lines. There is also an option that allows you to retain or discard the boundary when the hatching is complete.

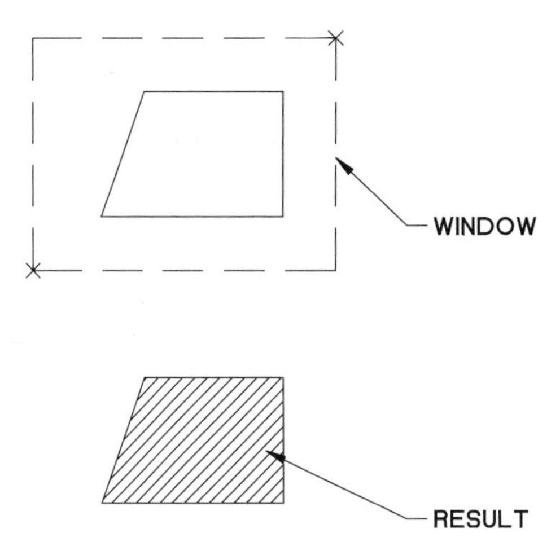

Figure 9–3 *The* HATCH *command requires the user to select objects to define the boundary*

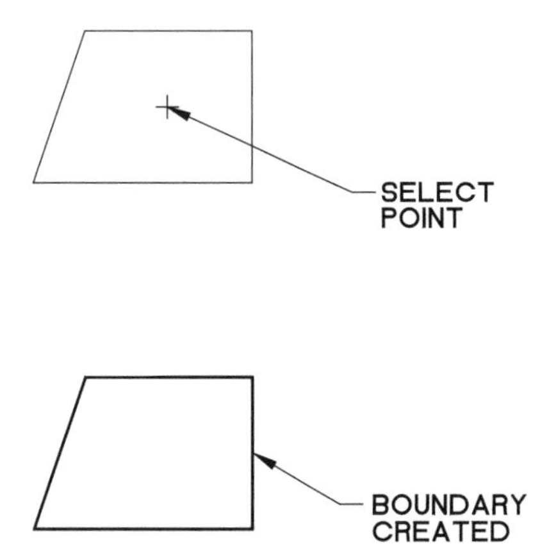

Figure 9–4 *The* BHATCH *command allows the user to pick a single point and automatically creates the boundary*

For example, if the four lines shown in Figure 9–3 had been segments of a closed polyline, then you could have selected that polyline by picking it with the cursor. Otherwise, all objects enclosing the region to be hatched must be selected when using the HATCH command, and those objects must be connected at their endpoints. For example, to use the HATCH command for the region in Figure 9–5, you would need to draw three lines (from 1 to 2, 2 to 3, and 3 to 4) and an arc from 4 to 1, select the four objects or connect them into a polyline, and select them (or it) to be the boundary.

The BHATCH command permits you to select a point in the region and have AutoCAD automatically create the needed polyline boundary. This ease of use and automation of the BHATCH command almost eliminates the need for the HATCH command except for rare, specialized applications.

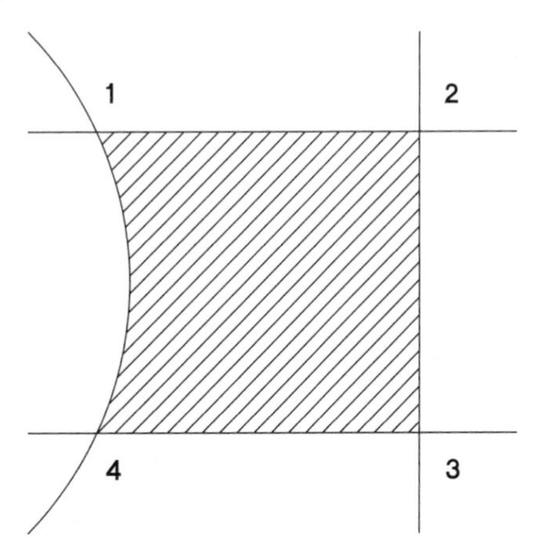

Figure 9–5 *A Region bounded by three lines and an arc*

Note: Even though the Draw menu lists "Hatch" as the item to choose and the Draw toolbar Tooltip says "Hatch", they actually invoke the BHATCH command instead of the HATCH command. To invoke the HATCH command, you must enter **hatch** at the Command: prompt.

The dialog box used by the BHATCH command provides a variety of easy-to-select options, including a means to preview the hatching before completing the command. This saves time. Consider the variety of effects possible, such as areas to be hatched, angle, spacing between segments in a pattern, and even the pattern selected. The Preview option lets you make necessary changes without having to start over.

HATCH PATTERNS WITH THE BHATCH COMMAND

Invoke the BHATCH command:

Draw toolbar	Choose the HATCH command (see Figure 9–6)
Draw menu	Choose Hatch...
Command: prompt	**bhatch** (ENTER)

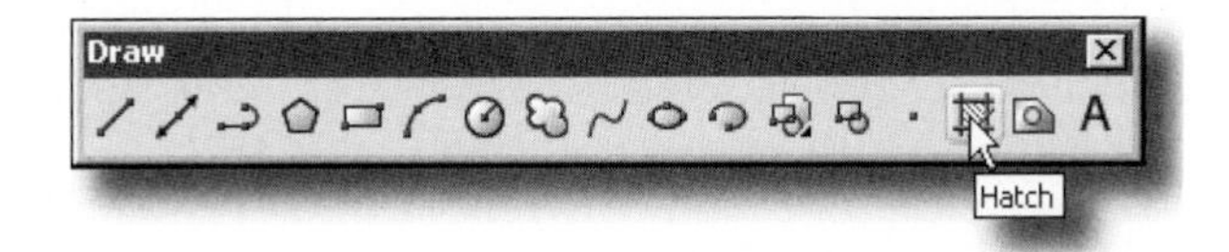

Figure 9–6 *Invoking the BHATCH command from the Draw toolbar*

AutoCAD displays the Boundary Hatch and Fill dialog box, similar to Figure 9–7.

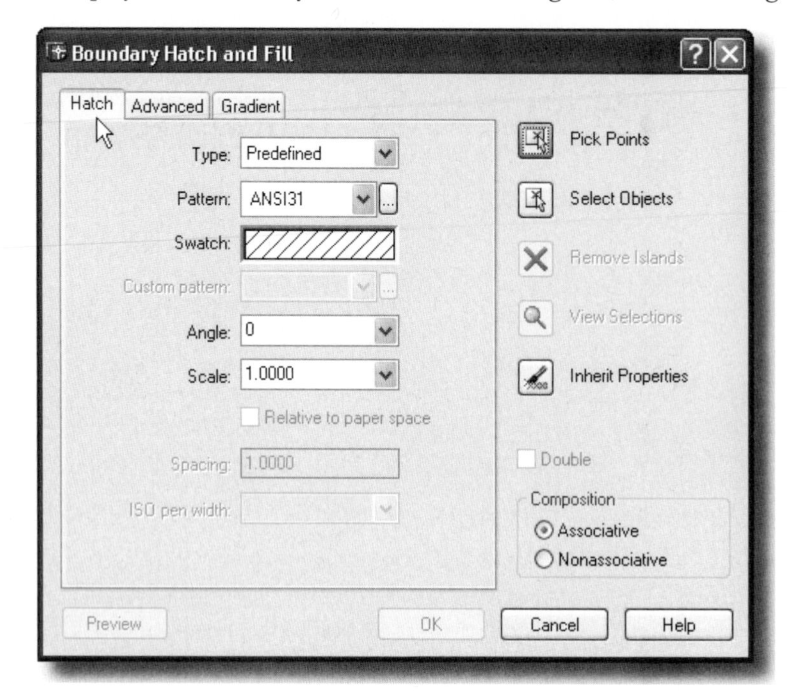

Figure 9–7 *Boundary Hatch and Fill dialog box with Hatch tab selected*

The **Type:** text box in the **Hatch** tab section of the Boundary Hatch and Fill dialog box enables you to select the type of hatch pattern to be applied. There are three choices: **Predefined, User-defined,** or **Custom.**

Predefined Pattern

When the Predefined pattern type is chosen, the **Pattern:** text box lets you select a pattern from those defined in the ACAD.PAT file. Or you can select one of the available patterns by choosing the image tile located below the **Pattern:** text box. This causes the Hatch Pattern Palette dialog box to be displayed as shown in Figure 9–8. There are four tabs from which to select predefined patterns: **ANSI, ISO, Other Predefined,** and **Custom.** Each tab displays icons representing that pattern. To select one of the patterns, choose it and then choose the **OK** button or double-click on the icon. An example of the selected pattern is displayed in the **Swatch:** box in the **Hatch** tab of the dialog box. The selected pattern becomes the value of the HPNAME system variable.

In the **Scale:** and **Angle:** text boxes, you can change the scale and angle respectively. The angle 0 (zero) corresponds to the positive *X* axis of the current UCS and refers to the hatch pattern as shown in the **Swatch** box. The default scale is set to 1 and the default angle is set to 0 degrees. These settings can be changed to suit the desired appearance, as shown in Figure 9–9.

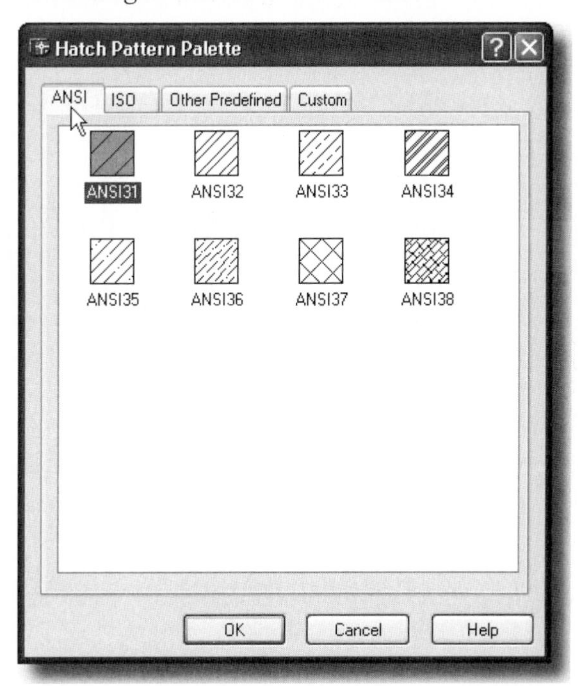

Figure 9–8 *Hatch Pattern Palette dialog box with the ANSI tab selected*

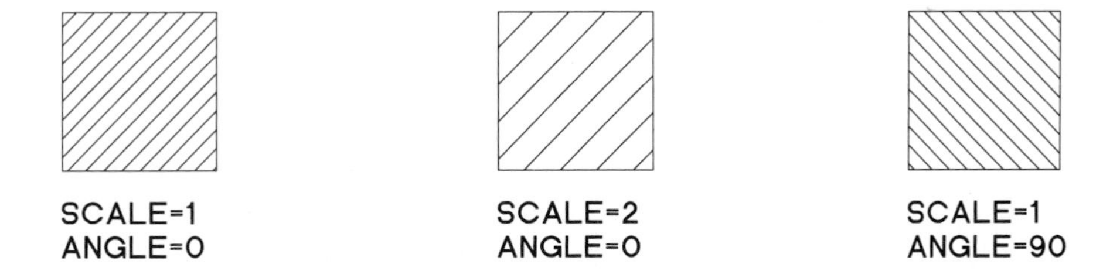

Figure 9–9 *Hatch pattern with different scale and angle values*

The **ISO pen width:** list box allows you to specify ISO-related pattern scaling based on the selected pen width. The option is available only if a predefined ISO hatch pattern is selected.

The **Relative to paper space** check box allows you to scale the hatch pattern relative to the units in Paper Space. This can only be used in layout mode.

User-Defined Pattern

The User-defined pattern type allows you to define a simple pattern using the current linetype on the fly. You can specify a simple pattern of parallel lines or two groups of parallel lines (crossing at 90 degrees) at the spacing and angle desired. Specify the angle and spacing for the user defined pattern in the **Angle:** and **Spacing:** text boxes. To draw a second set of lines at 90 degrees to the original lines, set the **Double** check box to ON.

Custom Pattern

When the Custom pattern type is chosen, the **Custom Pattern:** text box lets you select a custom pattern from a .PAT file other than the ACAD.PAT file. Or you can select one of the available patterns by choosing the image tile located below **Custom Pattern:** text box.

Solid Fill

To create a solid fill in an enclosed area, select Solid pattern. The solid fill is drawn with the current color settings, and all pattern properties are disabled, such as scale, angle, and spacing.

The Hatching Boundary

AutoCAD provides two methods by which you can select the objects, which determines the boundary for drawing hatch patterns: Pick Points and Select Objects.

The Pick Points method determines a boundary from existing objects that form an enclosed area. When you choose the **Pick Points** button to specify a point the default hatching style is set to Normal style and the default Island Detection Method is set to Flood. The available styles, Normal, Outer, and Ignore, and available detection methods, Flood and Ray Casting, are discussed later in this chapter. It is very important to make sure the appropriate hatching style and detection methods are selected.

To invoke the Pick Points method, choose the **Pick Points** button in the dialog box. AutoCAD prompts:

Select internal point: *(specify a point within the area to be hatched)*
Select internal point: *(specify a point, enter* **u** *to undo the selection, or press* ENTER *to end point specification)*

See Figure 9–10 for an example of hatching by specifying a point inside a boundary.

The **Select Objects** method allows you to select specific objects for hatching. To invoke the Select Objects option, choose the **Select Objects** button in the dialog box. AutoCAD prompts:

Select objects: *(select the object(s) by one of the standard methods, and press* ENTER *to terminate object selection)*

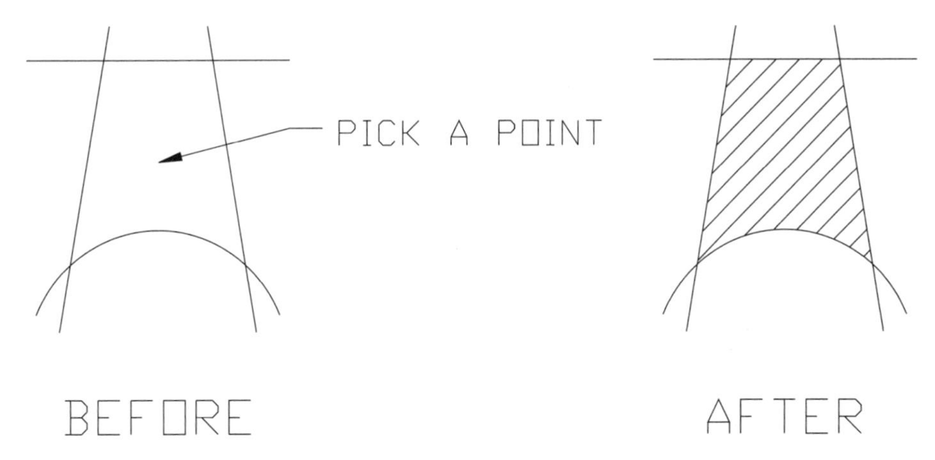

Figure 9–10 *Hatching by specifying a point*

Choosing the **View Selection** option causes AutoCAD to highlight the defined boundary set. If necessary, you can remove boundary set objects defined as a boundary set by the **Pick Points** option by choosing **Remove Islands** button. AutoCAD prompts you to select a boundary set that has to be removed from the defined boundary set. You cannot remove the outermost boundary.

The Select Objects option can be used to select an object, such as text, to cause AutoCAD not to hatch over the selected object and to leave a clear unhatched area around the text to provide better readability.

Caution must be observed when hatching over dimensioning. Dimensions are not affected by hatching as long as the DIMASO dimension variable (short for "associative dimensioning") is set to ON when the hatching is created and the dimension has not been exploded. The DIMASO system variable toggles between associative and nonassociative dimensioning. If the dimensions are drawn with DIMASO set to OFF (or exploded into individual objects), then the lines (dimension and extension) have an unpredictable (and undesirable) effect on the hatching pattern. Therefore, selecting in this case should be done by specifying the individual objects on the screen.

Blocks are hatched as though they are separate objects. Note, however, that when you select a block, all objects that make up the block are selected as part of the group to be considered for hatching.

If the selected items include text, shape, and/or attribute objects, AutoCAD does not hatch through these items if identified in the selection process. AutoCAD leaves an unhatched area around the text objects so they can be clearly viewed, as shown in Figure 9–11. Using the Ignore style will negate this feature so that the hatching is not interrupted when passing through the text, shape, and attribute objects.

Figure 9–11 *Hatching in an area where there is text*

Note: When you select objects individually (after selecting the **Select Objects** button in the Boundary Hatch and Fill dialog box), AutoCAD no longer automatically creates a closed border. Therefore, any objects selected that will be part of the desired border must be either connected at their endpoints or a closed polyline.

When a filled solid or trace with width is selected in a group to be hatched, AutoCAD does not hatch inside that solid or trace. However, the hatching stops right at the outline of the filled object, leaving no clear space around the object as it does around text, shape, and attributes.

The **Composition** section of the Boundary Hatch and Fill dialog box controls whether the hatch is associative or nonassociative. Associative selection associates the hatch pattern element with the objects they pattern. For example, if the object is stretched, the hatch pattern expands to fill the new size. Nonassociative selections do not associate hatch patterns with the object they pattern. Figure 9–12 shows an examples of associative and nonassociative hatch patterns.

Figure 9–12 *Example of the associative/nonassociative hatch pattern when object is stretched*

To apply the hatch pattern settings, such as pattern type, pattern angle, pattern scale, from an existing associative pattern to another area to be hatched, choose the **Inherit Properties** button. AutoCAD prompts:

Select associative hatch object: *(select an associative hatch pattern)*
Select internal point: *(specify a point inside a closed area to hatch)*
Select internal point: *(specify a point inside a closed area to hatch or press* ENTER *to complete the selection)*

Note: AutoCAD does not allow the properties of a nonassociative hatch pattern to be inherited.

To preview the hatch pattern for the selected objects, choose the **Preview** button. AutoCAD displays the currently defined boundaries with the current hatch settings. After previewing the hatch, press ENTER to return to the dialog box. If necessary, make any changes to the settings, choose the **OK** button to apply the hatch pattern, or choose **Cancel** to disregard the selection.

Before you choose the **OK** button, you can fine-tune additional parameters related to the boundary set if necessary; these are accessible through the **Advanced** tab of the Boundary Hatch and Fill dialog box (see Figure 9–13).

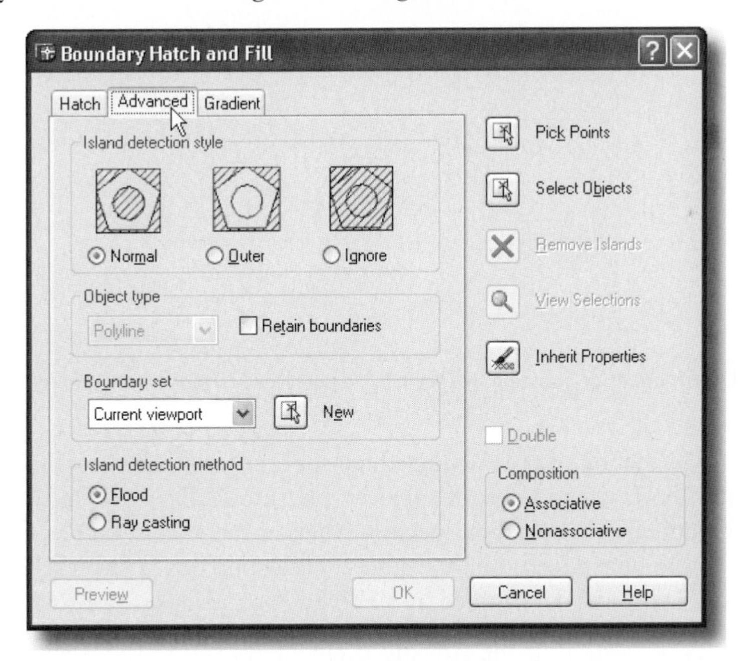

Figure 9–13 *Boundary Hatch and Fill dialog box with Advanced tab selected*

Advanced Settings

The **Island detection style** section of the **Advanced** tab specifies whether objects within the outermost boundary are used as boundary objects. These internal objects are known as islands. You can select one of the three available options: **Normal, Outer,** and **Ignore**. The **Normal** style causes AutoCAD to hatch between alternate areas, starting with the outermost area. The **Outer** style hatches only the outermost area. The **Ignore** style causes AutoCAD to hatch the entire area enclosed by the outermost boundary, regardless of how you select the object, as long as its outermost objects comprise a closed polygon and are joined at their endpoints.

For example, in Figure 9–14, specifying point P1 in response to the Pick Points option results in hatching for **Normal** style, as shown in the upper right; **Outer** style, as shown in the lower left; and **Ignore** style, as shown in the lower right.

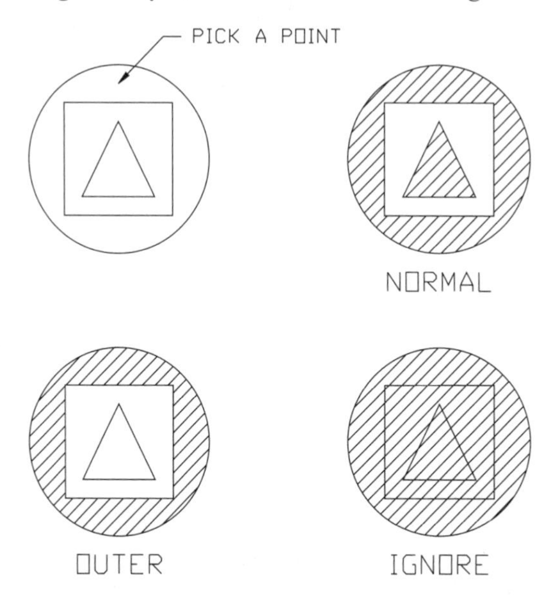

Figure 9–14 *Examples of hatching by specifying a point for Normal, Outer, and Ignore styles*

The **Object type** section of the **Advanced** tab specifies whether to retain boundaries as objects and allows you to select the object type AutoCAD applies to those objects. The **Retain boundaries** check box specifies whether the area to be hatched is to be retained. If it is to be retained then the adjacent selection box lets you select between whether the area is to be determined by a Polyline boundary or by a Region.

The **Boundary set** section of the **Advanced** tab allows you to select a set of objects (called a boundary set) that AutoCAD analyzes when defining a boundary from a specified point. The selected boundary set applies only when you use the **Pick Points** option to create a boundary to draw hatch patterns. By default, when you use **Pick Points** to define a boundary, AutoCAD analyzes all objects visible in the current viewport. To create a new boundary set, choose the **New** button in the **Boundary set** section, and AutoCAD clears the dialog box and returns you to the drawing area to select objects from which a new boundary set will be defined. AutoCAD creates a new boundary set from those objects selected that are hatchable; existing boundary sets are abandoned. If hatchable objects are selected, they remain as a boundary set until you define a new one or exit the BHATCH command. Defining a boundary set will be helpful when you are working on a drawing that has too many objects to analyze to create a boundary for hatching.

It is important to distinguish between a boundary set and a boundary. A boundary set is the group of objects from which AutoCAD creates a boundary. As explained earlier, a boundary set is defined by selecting objects in a manner similar to selecting objects for some modify commands. The objects (or parts of them) in the group are used in the subsequent boundary. A boundary is created by AutoCAD after it has analyzed the objects (the boundary set) you have selected. It is the boundary that determines where the hatching begins and ends. The boundary consists of line/arc segments, which can be considered to be a closed polygon with segments that connect at their endpoints. If objects in the boundary set overlap, then in creating the boundary AutoCAD uses only the parts of objects in the boundary set that lie between intersections with other objects in the boundary set.

The **Island detection method** section of the **Advanced** tab specifies whether to include objects (called islands) within the outermost boundary as boundary objects. Choose the **Flood** radio button to include islands within the selected boundary. The hatching is performed depending on the selection of the Island detection style (**Normal, Outer**, and **Ignore**). If the **Ray casting** radio button is chosen instead, then AutoCAD draws an imaginary line from the selected point to the nearest object and traces the boundary in the counterclockwise direction. If it cannot trace a closed boundary, AutoCAD displays an error message and will not be able to hatch.

For example, in Figure 9–15, point A is valid and point B is not when the **Flood** radio button is chosen. The object nearest to point A is the line that is part of a potential boundary (the square) of which point A is inside and AutoCAD considers the square as the hatch boundary. Conversely, point B is nearest a line that is part of a potential boundary (the triangle) of which point B is outside and AutoCAD displays an error message with the selected point as outside the boundary.

After making necessary adjustments in the **Advanced** section of the Boundary Hatch dialog box, proceed with the selection of the objects as explained earlier to create the hatch patterns.

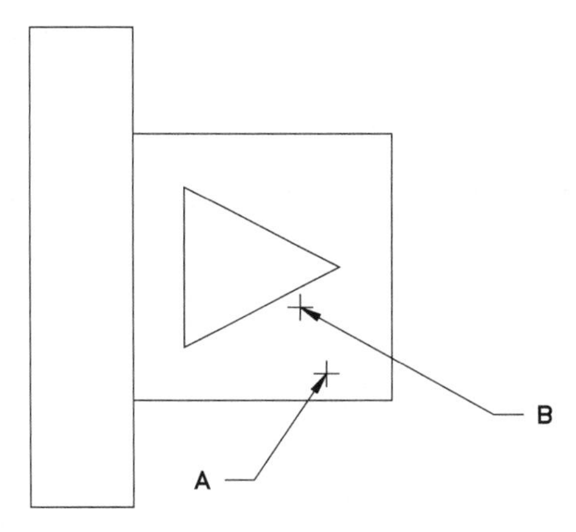

Figure 9–15 *Selecting points for hatching*

Gradient Settings

The **Gradient** tab defines the appearance of the gradient fill to be applied.

The **One color** selection specifies a fill that uses a smooth transition between darker shades and lighter tints of one color. When **One color** option is selected as shown in Figure 9–16, AutoCAD displays a color swatch with browse button and a **Shade** and **Tint** slider. The Color swatch specifies the color for the gradient fill, and the **Shade** and **Tint** slider specifies the tint (the selected color mixed with white) or shade (the selected color mixed with black) of a color to be used for a gradient fill of one color.

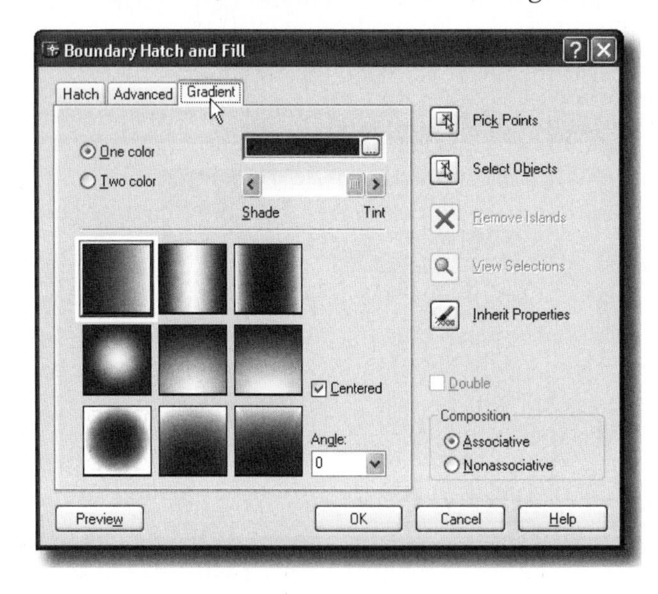

Figure 9–16 *Boundary Hatch and Fill dialog box with Gradient tab selected*

The **Two color** selection specifies a fill that uses a smooth transition between two colors. When **Two color** is selected, AutoCAD displays a color swatch with a browse button for color 1 and for color 2. The Color swatch specifies the color for the gradient fill.

The **Centered** check box specifies whether the pattern in a gradient fill is centered or is shifted up and to the left. When it is set to ON, the pattern in a gradient fill is centered and if it is set to OFF, the gradient fill is shifted up and to the left, creating the illusion of a light source to the left of the object.

The **Angle** setting specifies the angle of a gradient fill. Valid values are 0 through 360 degrees and the specified angle is relative to the current UCS. This option is independent of the angle specified for hatch patterns.

After making necessary adjustments in the Boundary Hatch and Fill dialog box, proceed with the selection of the objects as explained earlier to create the hatch patterns.

You can also hatch a closed shape by dragging a hatch pattern from a tool palette (see Figure 9–17).

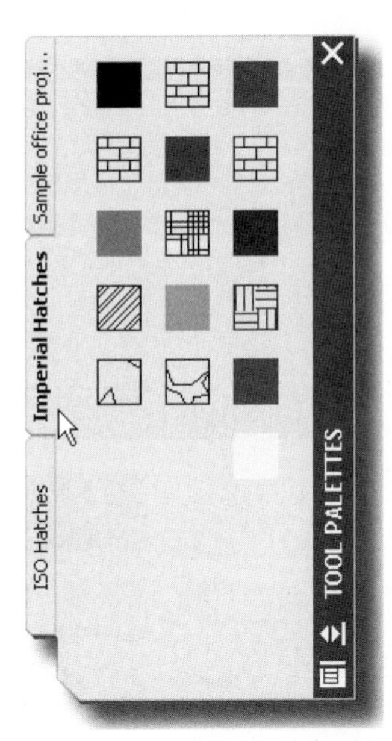

Figure 9–17 *Total Palettes window with Imperial Hatches tab selected*

Tool palettes are tabbed areas within the Tool Palettes window, and hatches that reside on a tool palette are called tools. Several tool properties including scale, rotation, and layer can be set for each tool individually. To change the tool properties, right-click a tool and select PROPERTIES... on the shortcut menu. Then you can change the tool's properties in the Tool Properties dialog box as shown in Figure 9–18. The Tool

Properties dialog box has two categories of properties— the **Pattern** properties category, which controls object-specific properties such as scale, rotation, and angle, and the **General** properties category, which overrides the current drawing property settings such as layer, color, and linetype.

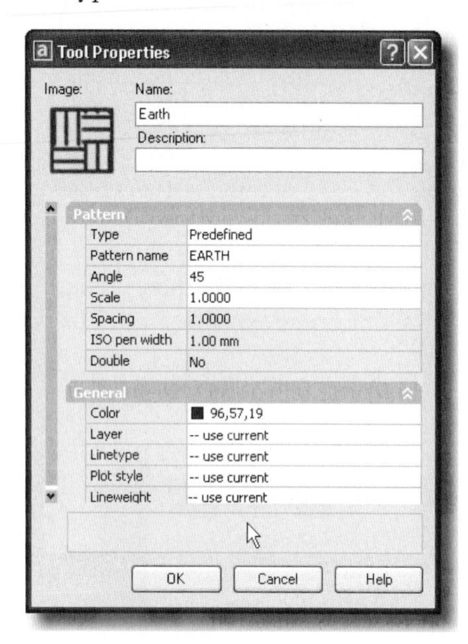

Figure 9–18 *Tool Properties of a selected tool*

You can place hatches that you use often on a tool palette by dragging hatch pattern from DesignCenter by opening ACAD.PAT file. See Chapter 12 for detailed explaining on using DesignCenter.

HATCH PATTERNS WITH THE HATCH COMMAND

The HATCH command creates a nonassociative hatch or fill. As mentioned earlier, if the boundary is made of multiple objects, then their endpoints must coincide for the hatch to be created properly with the HATCH command. AutoCAD allows you to create a polyline hatch boundary with the HATCH command if you do not have a closed boundary for drawing the hatch pattern.

Invoke the HATCH command:

Command: prompt	**hatch** (ENTER)

AutoCAD prompts:

> Enter a pattern name or [?/Solid/User defined]: *(specify a predefined or custom pattern name, or right-click for the shortcut menu and select one of the available options)*

When you specify a predefined or custom pattern name, AutoCAD prompts:

> Specify a scale for the pattern <default>: *(specify the scale for the selected pattern)*
> Specify an angle for the pattern <default>: *(specify the angle for the selected pattern)*
> Select objects to define hatch boundary or <direct hatch>: *(select the objects to draw the hatch pattern and press* ENTER *to terminate object selection)*

AutoCAD draws the hatch pattern to the selected boundary objects.

Instead of selecting the objects, if you press ENTER (null response) in response to the "Select objects:" prompt, AutoCAD allows you to draw a polyline boundary with a combination of lines and arcs for drawing the hatch pattern. The prompt sequences are as follows:

> Command: **hatch**
> Enter a pattern name or [?/Solid/User defined]: *(specify a predefined or custom pattern name)*
> Specify a scale for the pattern <default>: *(specify the scale for the selected pattern)*
> Specify an angle for the pattern <default>: *(specify the angle for the selected pattern)*
> Select objects to define hatch boundary or <direct hatch>: (ENTER)
> Retain polyline boundary? [Yes/No] <N>: *(press* ENTER *to discard the hatching boundary, or enter* **y** *to retain the hatching boundary)*
> Specify start point: *(specify a start point for the polyline boundary)*
> Specify next point or [Arc/Close/Length/Undo]: *(specify a point for the polyline boundary, or select an option)*

The options are similar to those for the PLINE command (see Chapter 4). When you've completed the polyline boundary, the HATCH command prompts you to create additional polyline boundaries. Once you have completed creating polyline boundaries, press ENTER (null response), and AutoCAD draws the selected hatch patterns.

When you specify the pattern name, you can select one of the three styles explained earlier. To invoke the Normal style, just type the name of the style. To invoke the Outermost style, type the name of the pattern followed by a comma (,) and the letter o. To invoke the Ignore style, type the name of the pattern followed by a comma (,) and the letter i.

?

The ? option lists and provides a brief description of the hatch patterns defined in the ACAD.PAT file.

Solid

The Solid option specifies a solid fill. As in drawing a hatch pattern, here also you can select objects to define a boundary or draw a polyline boundary with a combination of lines and arcs for a solid fill. The boundary of a solid fill must be closed and must not intersect itself. In addition, if the hatch area contains more than one loop, the loops must not intersect.

User-Defined

The User-defined option draws a pattern of lines using the current linetype. You can specify a simple pattern of parallel lines or two groups of parallel lines (crossing at 90 degrees) at the specified spacing and angle. The prompt sequences are as follows:

Command: **hatch**
Enter a pattern name or [?/Solid/User defined] : *(specify* **u** *for user defined)*
Specify angle for crosshatch lines <current>: *(specify the angle for crosshatch lines)*
Specify spacing between the lines <current>: *(specify the spacing for crosshatch lines)*
Double hatch area? [Yes/No} <current>: *(specify* **y** *for a second set of lines to be drawn at 90 degrees or* **n** *to draw simple pattern of parallel lines)*
Select objects to define hatch boundary or <direct hatch>: *(select objects or press* ENTER *to define a polyline boundary)*

As in drawing a hatch pattern, here also you can select objects to define a boundary or draw a polyline boundary with a combination of lines and arcs for a user-defined hatch pattern.

Note: The selection of objects for hatching must be done with awareness of how each object will affect or be affected by the HATCH command. Complex hatching of large areas can be time-consuming. Forgetting to select a vital object can change the whole effect of hatching. You can terminate hatching before it is completed by pressing the ESC key.

Hatching Base Point and Angle

Different areas hatched with the same (or a similar) pattern at the same scale and angle have corresponding lines lined up with each other in adjacent areas. This is because the families of lines were defined in the pattern(s) with the same basepoint and angle, no matter where the areas to be filled are in the drawing. This causes hatching lines to line up in adjacent hatched areas. But if you wish to offset the lines in adjacent areas, make the basepoint in one of the areas different from the basepoint in the adjacent area. Change the snap basepoint by either using the SNAP command and the Rotation option or by changing the SNAPBASE system variable. This is also useful for improving the appearance of hatching in any one area.

Changing the SNAPANG system variable or the base angle (from the SNAP/Rotate command/option) affects the angles of lines in a hatching pattern. This capability is also possible when responding to the HATCH command's "Specify angle for crosshatch lines <default>:" prompt.

Multiple Hatching

When you have finished hatching an area and have pressed ENTER to repeat the HATCH command, AutoCAD prompts only for the objects to be selected. The optional parameters of pattern, mode, scale, and angle remain unchanged. In order to change any options, you must invoke the HATCH command by entering **hatch** at the "Command:" prompt. AutoCAD resumes prompting for the hatch options.

EDITING HATCHES

The HATCHEDIT command allows you to modify hatch patterns or choose a new pattern for an existing hatch. In addition, it allows you to change the pattern style of an existing pattern.

Invoke the HATCHEDIT command:

Modify II toolbar	Choose the EDIT HATCH command (see Figure 9–19)
Modify menu	Choose Object > Hatch...
Command: prompt	**hatchedit** (ENTER)

Figure 9–19 *Invoking the HATCHEDIT command from the Modify II toolbar*

AutoCAD prompts:

Select associative hatch object: *(select a hatch object)*

AutoCAD displays the Hatch Edit dialog box, similar to Figure 9–20.

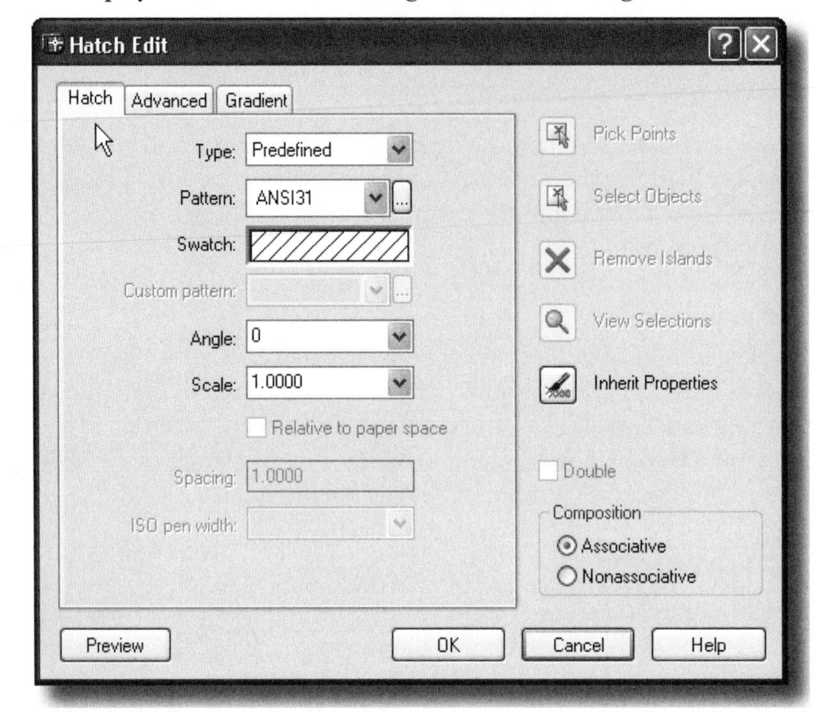

Figure 9–20 *Hatch Edit dialog box*

Select one of the three options, **Predefined, User-defined,** or **Custom,** from the **Type:** list box. If the **Predefined** option is selected, then choose one of the predefined patterns from **Pattern:** list box or choose the image tile located in the **Swatch:** text box. If User-defined option is selected, then make the necessary changes to the **Angle:** and **Scale:** settings. Instead, if Custom pattern is selected, then select one of the custom patterns from the **Custom pattern:** list box.

If necessary, you can change the Scale and Angle of the hatch pattern in the **Scale:** and **Angle:** text boxes. To inherit properties of an existing hatch pattern, select the **Inherit Properties** button, and then select an associative hatch pattern. The **Associative** check box controls whether or not selected hatching will be associative. If this option is selected, the modified hatch pattern is associative. If necessary, you can make changes in the settings in the **Advanced** and **Gradient** sections of the Hatch Edit dialog box.

Once the necessary changes are made in the Hatch Edit dialog box, choose the **OK** button to modify the selected hatch pattern.

When selecting a solid fill to modify, pick an outer edge of the hatch pattern with your pointing device or select it with the Crossing Window selection over the solid fill.

CONTROLLING THE VISIBILITY OF HATCH PATTERNS

The FILL command controls the visibility of hatch patterns in addition to the filling of multilines, traces, solids, and wide polylines.

Invoke the fill command:

Command: prompt	**fill** (ENTER)

AutoCAD prompts:

Command: **fill**
Enter mode [ON/OFF] <current>: *(specify* **ON** *to display the hatch pattern and* **OFF** *to turn off the display of hatch pattern)*

You have to invoke the REGEN command after changing the setting of fill to see the effect.

Open the Exercise Manual PDF file for Chapter 9 on the accompanying CD for project and discipline specific exercises.

If you have the accompanying Exercise Manual, refer to Chapter 9 for project and discipline specific exercises.

REVIEW QUESTIONS

1. AutoCAD will ignore text within a crosshatching boundary.

   a. True   b. False

2. The BHATCH command allows you to create an associative hatch pattern that updates when its boundaries are modified.

   a. True   b. False

3. By default, hatch patterns are drawn at a 45 degree angle.

   a. True   b. False

4. All of the following may be used as boundaries of the HATCH command, except:

   a. ARC   d. CIRCLE
   b. LINE   e. PLINE
   c. BLOCK

5. The following are all valid AutoCAD commands except:

   a. ANGLE   d. ELLIPSE
   b. POLYGON   e. MULTIPLE
   c. BHATCH

6. When using the BHATCH command with a named hatch pattern, one can change

   a. the color and scale of the pattern
   b. the angle and scale of the pattern
   c. the angle and linetype of the pattern
   d. the color and linetype of the pattern
   e. the color and angle of the pattern

7. Boundary hatch patterns inserted with an asterisk "*" preceding the name of the pattern will:

   a. exclude inside objects   d. be inserted on layer 0
   b. ignore inside objects   e. none of the above
   c. be inserted as individual objects

8. The AutoCAD hatch feature:

   a. provides a selection of numerous hatch patterns

   b. allows you to change the color and linetype

   c. hatches over the top of text when the text is contained inside the boundary

   d. all of the above

9. The BHATCH command will allow you to create a polyline around the area being hatched and to retain that polyline upon completion of the command.

   a. True b. False

10. Hatch patterns created with the BHATCH command can be nonassociative and associative.

    a. True b. False

11. The HATCH command will place a hatch pattern over any text contained within the hatch boundary.

    a. True b. False

12. If you insert a hatch pattern with an asterick (*) preceding the name, the pattern will be inserted as individual objects.

    a. True b. False

13. The ________ command automatically defines the nearest boundary surrounding a point you have specified.

    a. HATCH c. BOUNDARY

    b. BHATCH d. PTHATCH

14. The BHATCH and Hatch Edit dialog boxes look the same.

    a. True b. False

15. What default file does AutoCAD use to load hatch patterns from?

    a. ACAD.mnu c. ACAD.dwg

    b. ACAD.pat d. ACAD.hat

16. You can terminate the HATCH command before applying the hatch pattern by pressing the ESC.

    a. True b. False

17. AutoCAD allows the properties of a nonassociative hatch pattern to be inherited.

    a. True b. False

Block References and Attributes

INTRODUCTION

The AutoCAD BLOCK command feature is a powerful design/drafting tool. The BLOCK command enables a designer to create an object from one or more objects, save it under a user-specified name, and later place it back into the drawing. When block references are inserted in the drawing they can be scaled up or down in both or either of the *X* and *Y* axes. They can also be rotated as they are inserted in the drawing. Block references can best be compared with their manual drafting counterpart, the template. Even though an inserted block reference can be created from more than one object, the block reference acts as a single unit when operated on by certain modify commands, like MOVE, COPY, ERASE, ROTATE, ARRAY, and MIRROR. You can export a block reference to become a drawing file outside the current drawing and create a symbol library from which block references are inserted into other drawings. Like the plastic template, block references greatly reduce repetitious work.

The BLOCK command can save time because you don't have to draw the same object(s) more than once. Block references save computer storage because the computer needs to store the object descriptions only once. When inserting block references, you can change the scale and/or proportions of the original object(s).

After completing this chapter, you will be able to do the following:

- Create and insert block references in a drawing
- Convert individual block references into drawing files
- Define attributes, edit attributes, and control the display of attributes
- Use the DIVIDE and MEASURE commands

CREATING BLOCKS

When you invoke the BLOCK command to create a block, AutoCAD refers to this as defining the block. The resulting definition is stored in the drawing database. The same block can be inserted as a block reference as many times as needed.

Blocks may comprise one or more objects. The first step in creating blocks is to create a block definition. In order to do this, the objects that make up the block must be visible on the screen. That is, the objects that will make up the block definition must have already been drawn so you can select them when prompted to do so during the BLOCK command.

The layer the objects comprising the block are on is very important. Objects that are on layer 0 when the block is created will assume the color, linetype, and lineweight of the layer on which the block reference is inserted. Objects on any layer other than 0 when included in the block definition will retain the characteristics of that layer, even when the block reference is inserted on a different layer. See Figure 10–1 for an example.

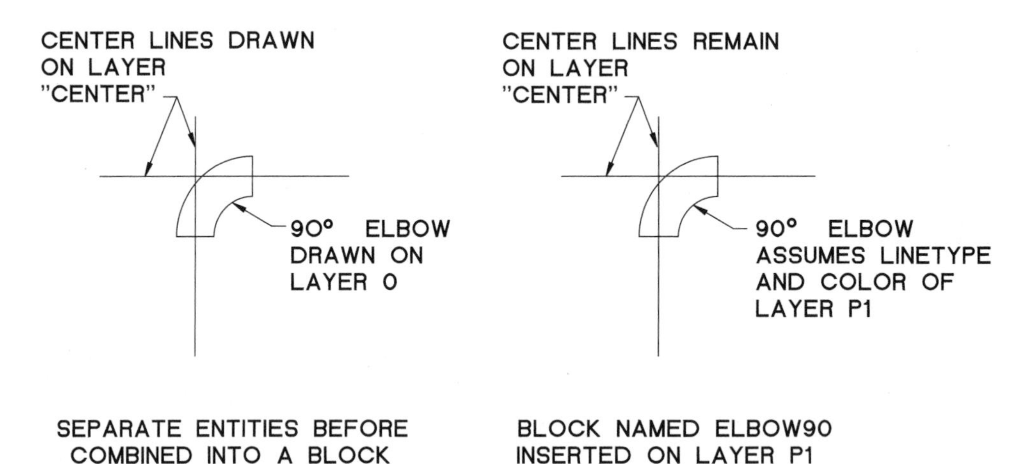

Figure 10–1 *Example of inserting block references drawn in different layers*

You should be careful when invoking the PROPERTIES command to change the color, linetype, or lineweight of elements of a block reference. It is best to keep the color, linetype, and lineweight of block references and the objects that comprise them in the BYLAYER state.

Examples of some common uses of blocks in various disciplines are shown in Figure 10–2.

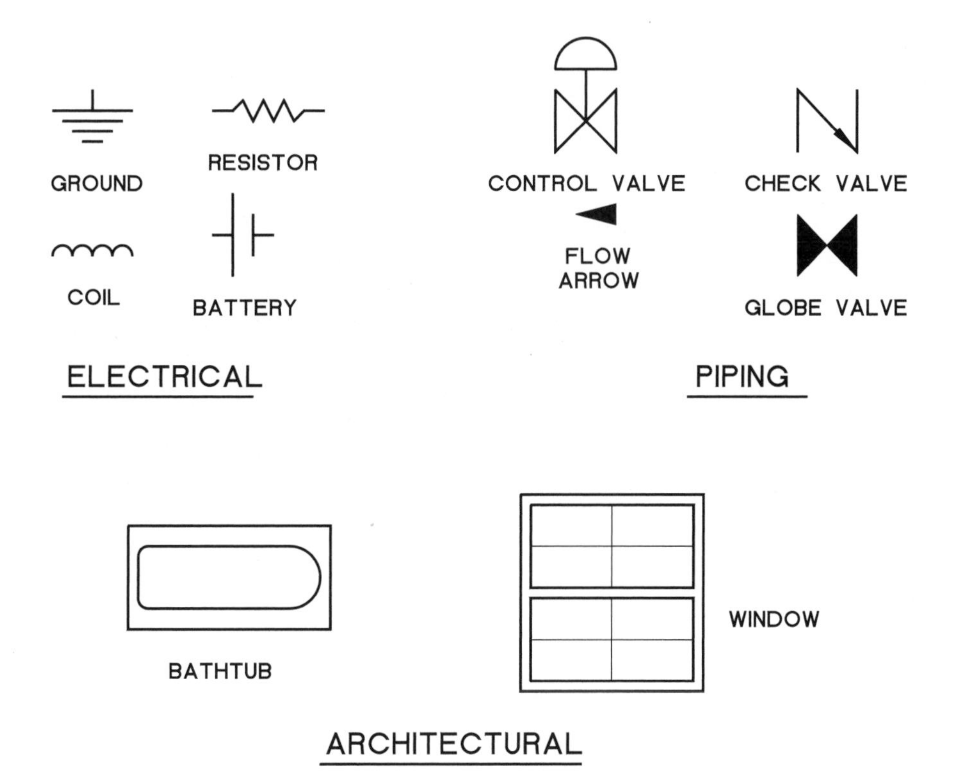

Figure 10–2 *Examples of common uses of blocks in various disciplines*

CREATING A BLOCK DEFINITION

The BLOCK command creates a block definition for selected objects. Invoke the BLOCK command:

Draw toolbar	Choose the Make Block command (see Figure 10–3)
Draw menu	Choose Block > Make
Command: prompt	**block** (ENTER)

Figure 10–3 *Invoking the BLOCK command from the Draw toolbar*

AutoCAD displays the Block Definition dialog box, similar to the Figure 10–4.

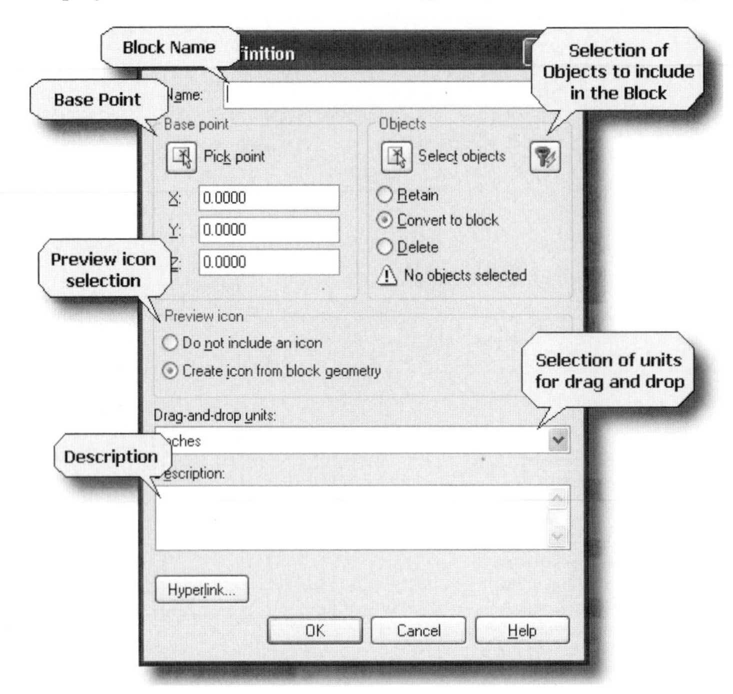

Figure 10–4 *Block Definition dialog box*

Block Name

Specify the block name in the **Name:** text box. The block name can be up to 255 characters long and may contain letters, numbers, and any special character not used by Microsoft® Windows® and AutoCAD for other purposes, if the system variable EXTNAMES is set to 1. DIRECT, LIGHT, AVE_RENDER, RM_SDB, SH_SPOT, and OVERHEAD cannot be used as block names. To list the block names in the current drawing, click the down arrow to the right of the **Name:** text box. AutoCAD lists the blocks in the current drawing.

Base Point

In the **Base point** section, you can specify the insertion point for the block.

The insertion point specified during the creation of the block becomes the basepoint for future insertions of this block as a block reference. It is also the point about which the block reference can be rotated or scaled during insertion. When determining where to locate the base insertion point it is important to consider what will be on the drawing before you insert the block reference. Therefore, you must anticipate this preinsertion state of the drawing. It is sometimes more advantageous for the insertion point to be somewhere off the object than on it.

You can specify the insertion point on the screen, or you can specify *X*, *Y*, and *Z* coordinates of the insertion point in the **X:**, **Y:**, and **Z:** text boxes, respectively, located in the **Base point** section of the Block Definition dialog box. To specify the basepoint on the screen, choose the **Pick point** button located in the Base point section of the dialog box. AutoCAD prompts:

Specify insertion point: *(specify the insertion point)*

Once you have specified the insertion point, the Block Definition dialog box reappears.

Selecting Objects

To select objects to include in the block definition, choose the **Select Objects** button located in the Objects section of the Block Definition dialog box. AutoCAD prompts:

Select objects: *(select objects using one of the AutoCAD object selection methods, and press* ENTER *to complete object selection)*

Once the objects are selected, the Block Definition dialog box reappears. You can also use the Quick Select option to define the selection of objects by clicking the icon to the right of the **Select Objects** button in the **Objects** section of the Block Definition dialog box. Choose one of the three radio buttons located in the **Objects** section to specify whether to retain or delete the selected objects or convert them to a block reference after you create the block.

If the **Retain** radio button is chosen, then the objects selected to be included in the block definition will not be deleted from the drawing after the block definition is created. They will remain in place as separate objects.

If the **Convert to Block** radio button is chosen, then the block definition created from the selected objects will become a block reference inserted into the drawing at the location that the block definition was created.

If the **Delete** radio button is chosen, then the objects selected to be included in the block definition will be deleted from the drawing after the block definition is created.

Preview Icon

In the **Preview icon** section, if the **Create icon from block geometry** radio button is chosen, then, if the AutoCAD Design Center is used to search for the block, or if it is written to a drawing and selected from the Browse feature of the INSERT command, an icon will be displayed in the Preview Icon area on the right. The icon will simulate the appearance of the block definition. If the **Do not include an icon** radio button is chosen, then no icon will be displayed.

Drag-and-drop Units

Select one of the available units from the **Drag-and-drop units:** option menu. If the block is dragged from the AutoCAD DesignCenter (for detailed explanation on DesignCenter, see Chapter 12), it will be scaled to the units in the **Drag-and-drop units** list box.

Description

The **Description** text box can be used to enter a description of the block.

Hyperlink

When you choose the **Hyperlink...** button, AutoCAD displays the Insert Hyperlink dialog box as shown in Figure 10–5.

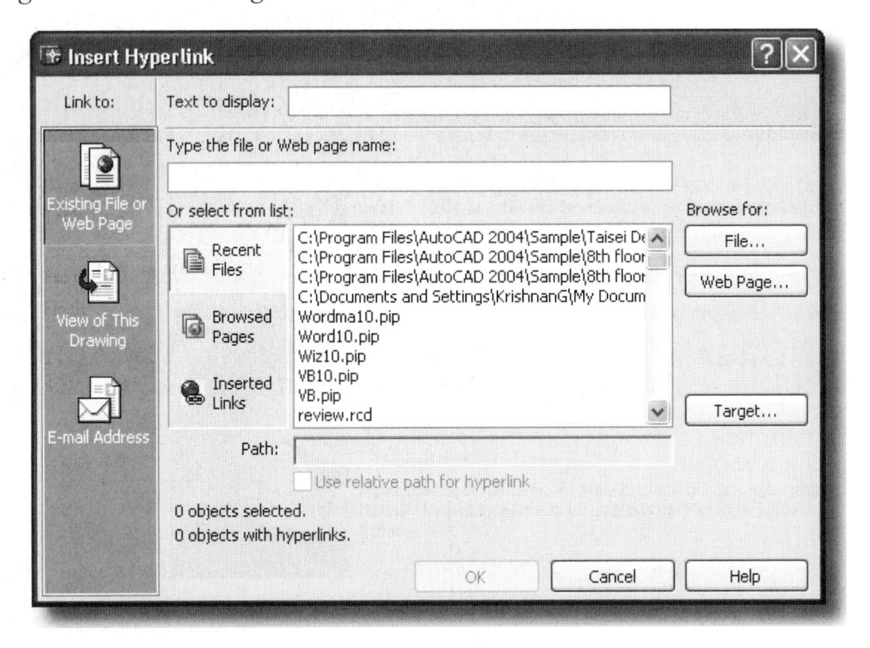

Figure 10–5 *Insert Hyperlink dialog box*

Hyperlinks are created in AutoCAD drawings as pointers to associated files. Hyperlinks can launch a word processing program and open a specific file and even point to a named location in a file. Hyperlinks can activate your Web browser and load a specified HTML page. You can specify a view in AutoCAD or a bookmark in a word processing file. Hyperlinks can be attached to a graphical object in an AutoCAD drawing.

Hyperlinks can be either **absolute** or **relative.** Absolute hyperlinks have the full path to a file location stored in them. Relative hyperlinks have only a partial path to a file location stored in them, relative to a default URL or directory you have specified by setting the HYPERLINKBASE system variable.

Hyperlinks can point to locally stored files, files on a network drive, or files on the Internet. Cursor feedback is automatically provided to indicate when the crosshairs are over a graphical object that has an attached hyperlink. You can then select the object and use the Hyperlink shortcut menu to open the file associated with the hyperlink. This hyperlink cursor and shortcut menu display can be turned off in the Options dialog box.

When a hyperlink to an AutoCAD drawing is opened that has a named view, that view is restored. This also applies to a hyperlink created with a named layout. AutoCAD opens that drawing in that layout.

If a hyperlink is created which points to an AutoCAD drawing template (DWT) files, AutoCAD will create a new drawing file based on the template. This prevents overwriting the original template.

For additional information on hyperlinks and how to create and edit them refer to Chapter 14.

Choose the **OK** button to create the block definition with the given name. If the given name is the same as an existing block in the current drawing, AutoCAD displays a warning like that shown in Figure 10–6.

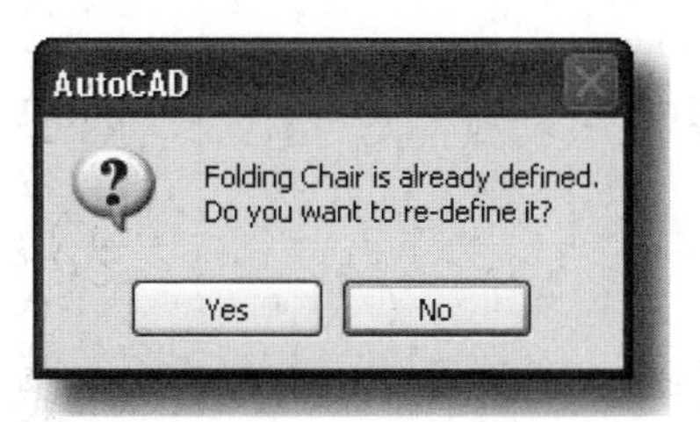

Figure 10–6 *Warning dialog box regarding block definition*

To redefine the block, choose the **Yes** button in the Warning dialog box. The block with that same name is then redefined. Once the drawing is regenerated, any insertion of this block reference already inserted in the drawing is redefined to the new block definition with this name.

Choose the **No** button in the Warning dialog box to cancel the block definition. Then, to create a new block definition, specify a different block name in the **Name:** text box of the Block Definition dialog box and choose **OK**.

If you create a block without selecting objects, AutoCAD displays a warning that nothing has been selected and provides an opportunity to select objects before the named block is created.

You can also create a block definition from selected objects by invoking the BLOCK command preceded with a hyphen from the Command: prompt, as shown in the following command sequence:

Command: **-block** (ENTER)
Enter block name or [?]: *(specify a block name, or enter* **?** *to list the block names in the current drawing)*
Specify insertion base point: *(specify the insertion base point)*
Select objects: *(select the objects to include in the block definition, and press* ENTER *to complete object selection)*

AutoCAD creates the block with the given name and then erases the objects that make up the definition from the screen. You can restore the objects by entering the OOPS command immediately after the BLOCK command.

INSERTING BLOCK REFERENCES

You can insert previously defined blocks into the current drawing by invoking the INSERT command. If there is no block definition with the specified name in the current drawing, AutoCAD searches the drives and folders on the path for a drawing of that name and inserts it instead.

Note: If blocks were created and stored in a template drawing, and you make your new drawing equal to the template, those blocks will be in the new drawing ready to insert. Any drawing inserted into the current drawing will bring with it all of its block definitions, whether they have been inserted or are only stored as definitions.

Invoke the INSERT command:

Draw toolbar	Choose the Insert Block command (see Figure 10–7)
Insert menu	Choose Block...
Command: prompt	**insert** (ENTER)

Figure 10–7 *Invoking the INSERT command from the Draw toolbar*

AutoCAD displays the Insert dialog box, similar to Figure 10–8.

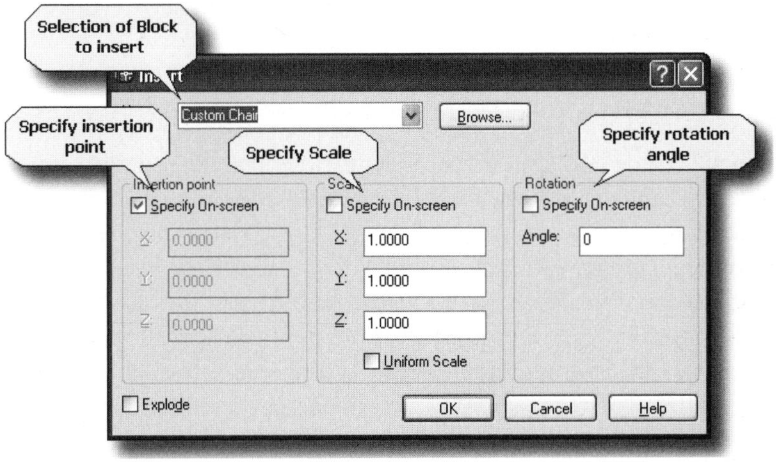

Figure 10–8 *Insert dialog box*

Specify a block name in the **Name:** text box. Or choose the down arrow to display a list of blocks defined in the current drawing (see Figure 10–9) and select the block you wish to insert.

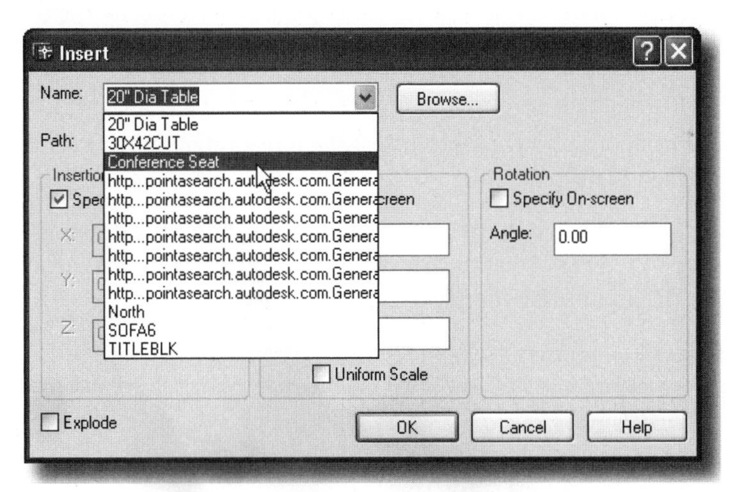

Figure 10–9 *Listing of blocks defined in the current drawing*

Insertion Point

The **Insertion point** section of the Insert dialog box allows you to specify the insertion point for inserting a copy of the block definition. You can specify the insertion point in terms of *X*, *Y*, and *Z* coordinates in the **X:**, **Y:** and **Z:** text boxes when the **Specify On-screen** check box is set to OFF. If you prefer to specify the insertion point on screen, then the **Specify On-screen** check box is set to ON.

Scale

The **Scale** section of the Insert dialog box allows you to specify the scale for the inserted block. The default scale factor is set to 1 (Full scale). You can specify a scale factor between 0 and 1 to insert the block reference smaller than the original size of the block and specify more than 1 to increase the size from the original size. If necessary, you can specify different *X* and *Y* scale factors for insertion of the block reference. If you specify a negative scale factor, then AutoCAD inserts a mirror image of the block about the insertion point. As a matter of fact, if −1 were used for both *X* and *Y* scale factors, it would "double mirror" the object, the equivalent of rotating it 180 degrees. If you prefer to specify the scale factor on screen, then set the **Specify On-screen** check box to ON. If you check the **Uniform Scale** check box, then you can enter a value only in the **X:** text box. The *Y* and *Z* scales will be the same as that entered for the *X* scale.

Rotation Angle

The **Rotation** section of the Insert dialog box allows you to specify the rotation angle for the inserted block. To rotate the block reference, specify a positive or negative angle, referencing the block in its original position. If you prefer to specify the rotation angle on screen, then set the **Specify On-screen** check box to ON.

Explode

The **Explode** check box allows you to insert the block reference as a set of individual objects rather than as a single unit.

To specify a drawing file to insert as a block definition, enter the drawing file name in the **Name:** text box. Or choose the **Browse…** button to display a standard file dialog box and select the appropriate drawing file.

Note: The name of the last block reference inserted during the current drawing session is remembered by AutoCAD. The name becomes the default for subsequent use of the INSERT command.

Choose the **OK** button to insert the selected block.

You can also insert a block by invoking the INSERT command preceded with a hyphen, as shown in the following command sequence:

Command: **-insert** (ENTER)
Enter block name or [?]: *(specify a block name, or enter* **?** *to list the block names in the current drawing)*

Specify insertion point or [Scale/X/Y/Z/Rotate/PScale/PX/PY/PZ/PRotate]: *(specify the insertion point to insert the block reference or right-click for the shortcut menu and select one of the available options)*
Enter X scale factor, specify opposite corner, or [Corner/XYZ] <1>: *(specify a scale factor for the X axis, specify a point, or press* ENTER *to accept the default scale factor)*
Enter Y scale factor <use X scale factor>: *(specify a scale factor for the Y axis, specify a point, or press* ENTER *to accept the same scale factor as for the X axis)*
Specify rotation angle <0>: *(specify a rotation angle, specify a point, or press* ENTER *to accept the default rotation angle)*

Instead of specifying the insertion point to insert the block reference, you can invoke one of the following options at the "Insertion point:" prompt to preset the scale and rotation of a block reference before you specify its position:

Scale The Scale option sets the scale factor for the *X*, *Y*, and *Z* axes.

Xscale The Xscale option sets the *X* scale factor.

Yscale The Yscale option sets the *Y* scale factor.

Zscale The Zscale option sets the *Z* scale factor.

Rotate The Rotate option sets the rotation angle.

PScale The PScale option sets the scale factor for the *X*, *Y*, and *Z* axes to control the display of the block reference as it is dragged into position.

PXscale The PXscale option sets the scale factor for the *X* axis to control the display of the block reference as it is dragged into position.

PYscale The PYscale option sets the scale factor for the *Y* axis to control the display of the block reference as it is dragged into position.

PZscale The PZscale option sets the scale factor for the *Z* axis to control the display of the block reference as it is dragged into position.

PRotate The PRotate option sets the rotation angle of the block reference as it is dragged into position.

After you provide the appropriate information to the selected option, AutoCAD returns to the "Insertion point:" prompt. AutoCAD continues with the scale factor and rotation angle prompts. A copy of the specified block is inserted, with its defined insertion point located at the specified point in the current drawing to the specified scale and rotation angle.

You can also insert a block by dragging a block from a tool palette (see Figure 10–10).

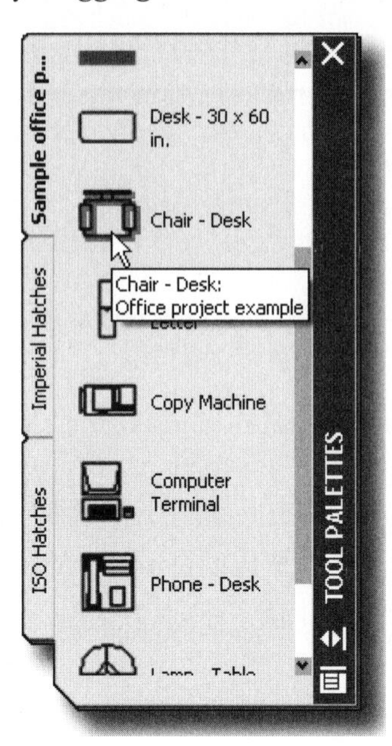

Figure 10–10 *Tool palette with Sample Office Project tab selected*

The block name will be added to the **Name:** list box of the Insert dialog box whenever a block is inserted in the current drawing by dragging from a tool palette. Tool palettes are tabbed areas within the Tool Palettes window, and blocks that reside on a tool palette are called tools. Several tool properties, including scale, rotation, and layer can be set for each tool individually. To change the tool properties, right-click a tool and select **Properties...** on the shortcut menu. Then you can change the tool's properties in the Tool Properties dialog box as shown in Figure 10–11. The Tool Properties dialog box has two categories of properties—the **Insert** properties category, which controls object-specific properties such as name of the block, name of the original source drawing name, scale, rotation, and angle, and the **General** properties category, which overrides the current drawing property settings such as layer, color, linetype. lineweight and plot style.

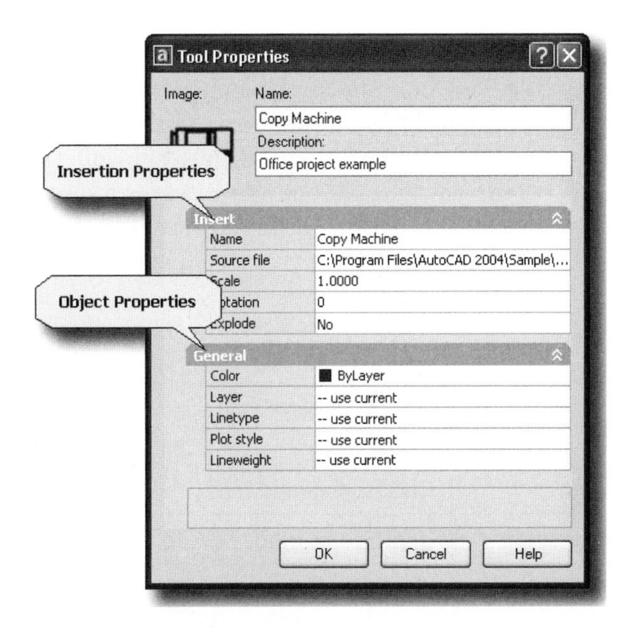

Figure 10–11 *Tool Properties of a selected tool*

You can place blocks that you use often on a tool palette by dragging blocks from DesignCenter. See Chapter 12 for details on using DesignCenter.

NESTED BLOCKS

Blocks can contain other blocks. That is, when using the BLOCK command to combine objects into a single object, one or more of the selected objects can themselves be blocks. And the blocks selected can have blocks nested within them. There is no limitation to the depth of nesting. You may not, however, use the name of any of the nested blocks as the name of the block being defined. This would mean that you were trying to redefine a block, using its old definition in the new.

Any objects within blocks (as nested blocks) that were on layer 0 when made into a block will assume the color, linetype and lineweight of the layer on which the block reference is inserted. If an object (originally on layer 0 when included in a block definition) is in a block reference that has been inserted on a layer other than layer 0, it will retain the color, linetype and lineweight of the layer it was on when its block was included in a higher-level block. For example, you draw a circle on layer 0 and include it in a block named Z1. Then, you insert Z1 on layer R, whose color is red. The circle would then assume the color of layer R (in this case it will be red). Create another block called Y3 by including the block Z1. If you insert block reference Y3 on a layer whose color is blue, the block reference Y3 will retain the current color of layer R (in this case it will be red) instead of taking up the color of blue.

EXPLODE COMMAND

The EXPLODE command causes block references, hatch patterns, and associative dimensioning to be turned into the separate objects from which they were created. It also causes polylines/polyarcs and multilines to separate into individual simple line and arc objects. The EXPLODE command causes 3D polygon meshes to become 3Dfaces, and 3D polyface meshes to become 3Dfaces and simple line and point objects. When an object is exploded, the new, separate objects are created in the space (model or paper) of the exploded objects.

Invoke the EXPLODE command:

Modify toolbar	Choose the EXPLODE command (see Figure 10–12)
Modify menu	Choose Explode
Command: prompt	**explode** (ENTER)

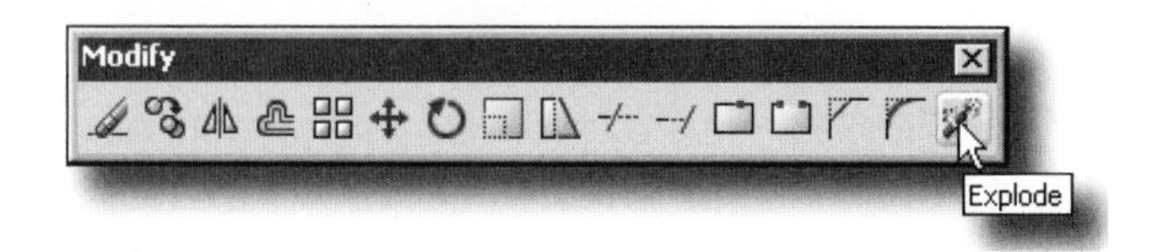

Figure 10–12 *Invoking the* EXPLODE *command from the Modify toolbar*

AutoCAD prompts:

Command: **explode**
Select objects: *(select objects to explode, and press* ENTER *to complete object selection)*

You can use one or more object selection methods. The object selected must be eligible for exploding, or an error message will appear. An eligible object may or may not change its appearance when exploded.

POSSIBLE CHANGES CAUSED BY THE EXPLODE COMMAND

A polyline segment having width will revert to a zero-width line and/or arc. Tangent information associated with individual segments is lost. If the polyline segments have width or tangent information, the EXPLODE command will be followed by the message:

Exploding this polyline has lost (width/tangent) information.
The undo command will restore it.

Individual elements within blocks that were on layer 0 when created (and whose color was BYLAYER) but were inserted on a layer with a color different than that of layer 0 will revert to the color of layer 0.

Attributes are special text objects that, when included in a block definition, take on the values (names and numbers) specified at the time the block reference is inserted. The power and usage of attributes are discussed later in this chapter. To understand the effect of the EXPLODE command on block references that include attributes, it is sufficient to know that the fundamental object from which an attribute is created is called an attribute definition. It is displayed in the form of an attribute tag before it is included in the block.

An attribute within a block will revert to the attribute definition when the block reference is exploded and will be represented on the screen by its tag. The value of the attribute specified at the time of insertion is lost. The group will revert to those elements created by the ATTDEF command prior to combining them into a block via the BLOCK command.

In brief, an attribute definition is turned into an attribute when the block in which it is a part is inserted; conversely, an attribute is turned back into an attribute definition when the block reference is exploded.

EXPLODING BLOCK REFERENCES WITH NESTED ELEMENTS

Block references containing other blocks and/or polylines are separated for one level only. That is, the highest-level block reference will be exploded, but any nested blocks or polylines will remain block references or polylines. They in turn can be exploded when they come to the highest level.

Viewport objects in a block definition cannot be turned on after being exploded unless they were inserted in paper space.

Block references with equal *X*, *Y*, and *Z* scales explode into their component objects. Block references with unequal X, *Y*, and *Z* scales (nonuniformly scaled block references) might explode into unexpected objects.

Note: Block references inserted via the MINSERT command or external references and their dependent blocks cannot be exploded.

MULTIPLE INSERTS OF BLOCK REFERENCES

The MINSERT (multiple insert) command is used to insert block references in a rectangular array. The total pattern takes the characteristics of a block, except the group cannot be exploded. This command works similar to the rectangular ARRAY command.

Invoke the MINSERT command:

Command: prompt	**minsert** (ENTER)

AutoCAD prompts:

> Command: **minsert**
> Enter block name or [?] <y>: *(enter name of the block)*
> Specify insertion point or [Scale/X/Y/Z/Rotate/PScale/PX/PY/PZ/PRotate]: *(specify the insertion point or right-click for shortcut menu and select one of the available options)*
> Enter X scale factor, specify opposite corner, or [Corner/XYZ] <1>: *(enter a number or specify a point)*
> Enter Y scale factor <use X scale factor>: *(enter a number or press* ENTER*)*
> Specify rotation angle <0>: *(enter a number or specify a point)*
> Enter number of rows (—) <default>: *(enter the number of rows)*
> Enter number of columns (|||) <default>: *(enter the number of columns)*
> Enter distance between rows or specify unit cell (—): *(specify the distance between rows)*
> Specify distance between columns (|||):*(specify the distance between columns)*

The row/column spacing can be specified by the delta-X/delta-Y distances between two points picked on the screen. For example, if, in response to the "Distance" prompt, you selected points 2,1 and 6,4 for the first and second points, respectively, the row spacing would be 3 (4 – 1) and the column spacing would be 4 (6 – 2).

INSERTING UNIT BLOCK REFERENCES

Groups of objects often need to be duplicated within a drawing. We showed earlier how the AutoCAD BLOCK command makes this task easier. The task of transferring blocks or groups of objects to another drawing is demonstrated in the section on the WBLOCK command. This section covers additional aspects of creating blocks in anticipation of inserting them later with differing scale factors (sometimes with *X* and *Y* unequal). This is referred to as a unit block.

Doors and windows are a few of the objects that can be stored as blocks in a symbol library. But blocks can be used in different ways to suit differing situations. The following examples are offered as procedures that are used without customizing menus or AutoLISP routines to enhance the process. It should be noted that these procedures can be improved either with customization or possibly with some variations involving standard commands and features. Also, the symbology and names of items are subject to variation.

You can employ a variety of symbols to represent windows in an architectural plan view. Horizontal sliding windows may need to be distinguished from single-hung windows, as shown in Figure 10–13. You may also wish to have more than one option as to how you will insert a window. You may wish to make its insertion point the center of the window sometimes or one of its edges at other times.

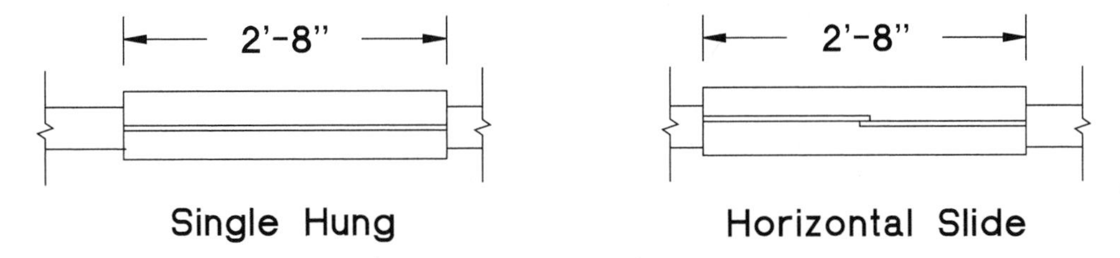

Figure 10–13 *Two different windows shown as block references*

Because there are windows of varying widths, you might have to create a separate block for each width. A group of windows might be 1'-0", 1'-6", 2'-0", 2'-8", 3'-0", 3'-4", 4'-0", 5'-4", 6'-0", and 8'-0", and some widths in between. You would have to make a block for each width in addition to the different types. The symbol for the single-hung 2'-8"–wide window could be drawn with the snap set a 0.5 (1/2") to the dimensions shown in Figure 10–14. The objects in Figure 10–14 could be saved as a block named WDW32 (for a 32"-wide window). This window could be inserted from its corner, as shown, if you have established that intersection in the wall in which it is to be drawn.

The preceding method requires a separate block for each window width. Another method is to make a drawing of a window in which the *X* and/or *Y* dimension is 1 unit, in anticipation of making the final desired dimension the *X* and/or *Y* scale factor during insertion. The following Unit Block symbol for the window can be used for a window of any width, as shown in Figure 10–15. This group of objects might be made into a block named WDW1. Then, to use it for a 2'-8" window, the INSERT command preceded with a hyphen would be:

Command: **-insert**
Enter block name or [?] <y>: **wdw1**
Specify insertion point or [Scale/X/Y/Z/Rotate/PScale/PX/PY/PZ/PRotate]:
(select insertion point)
Enter X scale factor, specify opposite corner, or [Corner/XYZ] <1>: **32**
Enter Y scale factor <use X scale factor>: **1**
Specify rotation angle <0>: (ENTER)

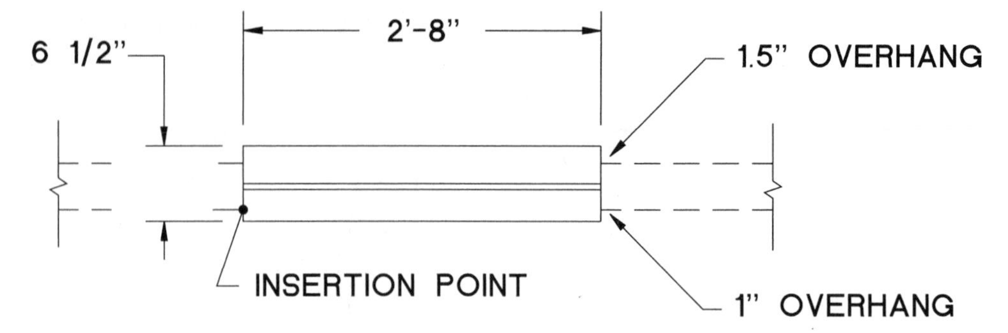

Figure 10–14 *Single-hung 2'–8"–wide window*

Note that when you specify a value for the *X* scale factor, AutoCAD assumes you wish to apply the same value to the *Y* scale factor, making the resulting shape of the object(s) proportional to the unit block from which it was generated. That is why the *Y* scale factor defaults to the *X* scale factor. Therefore, if you wish to insert the block reference with an *X* scale factor different from the *Y* scale factor, you must input a *Y* scale factor even if it is just a factor of 1. The block WDW1 can be used for a window of any width by making the desired width (in inches) the *X* scale factor and then using 1 as the *Y* scale factor. Figure 10–16 shows the window inserted in its location with the proper *X* scale factor. Also shown is the cleanup that can be done with the BREAK command.

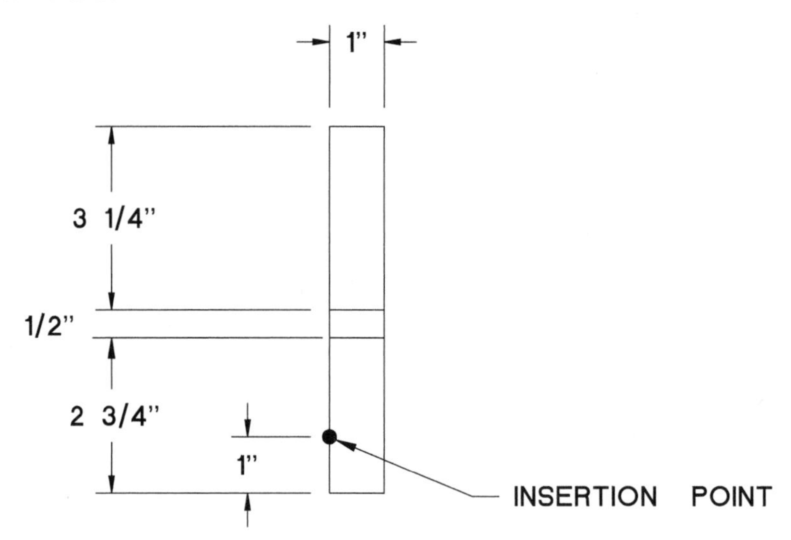

Figure 10–15 *Creating a block for a window symbol*

Another variation on the unit block WDW1 would be to use the same shape for center insertion. It would be drawn the same, but the insertion point would be chosen as shown in Figure 10–17. Figure 10–18 shows the window inserted.

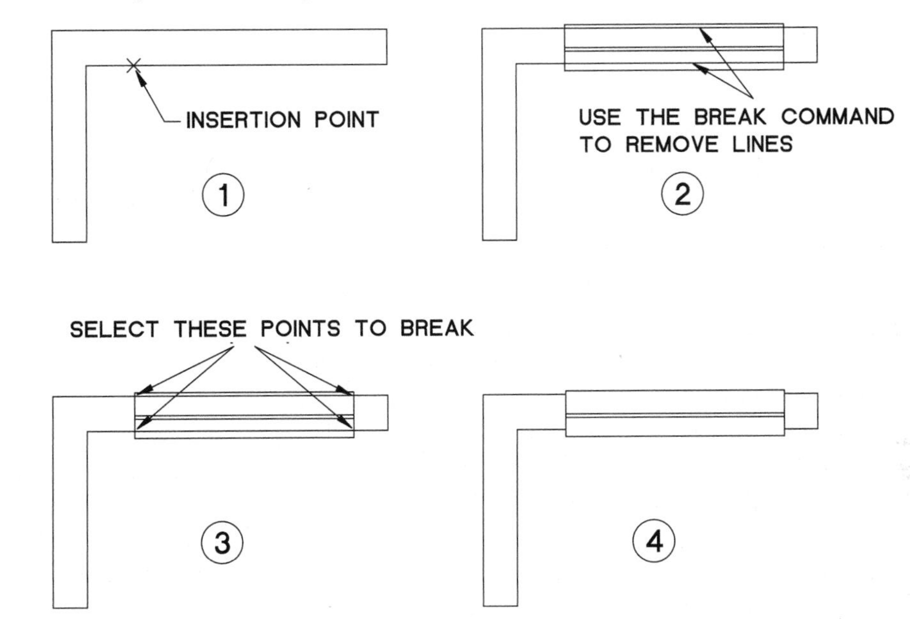

Figure 10–16 *Inserting a unit block reference with a large X scale factor*

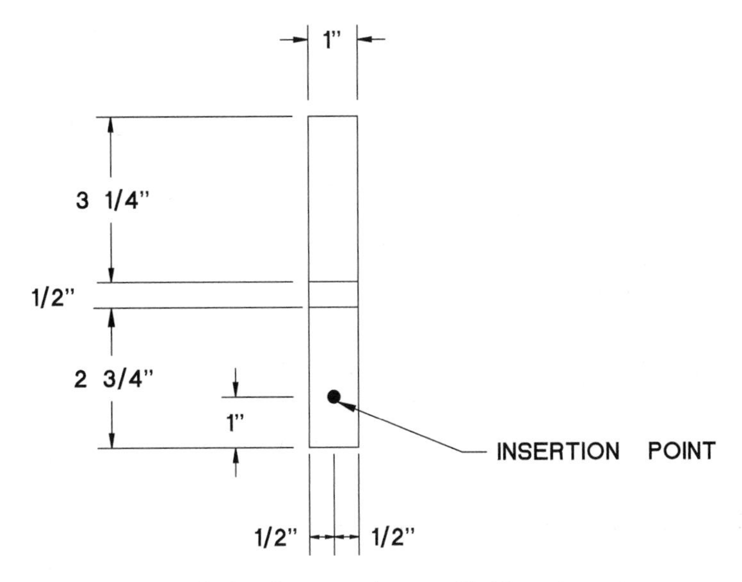

Figure 10–17 *Using a unit block reference with a specified insertion point*

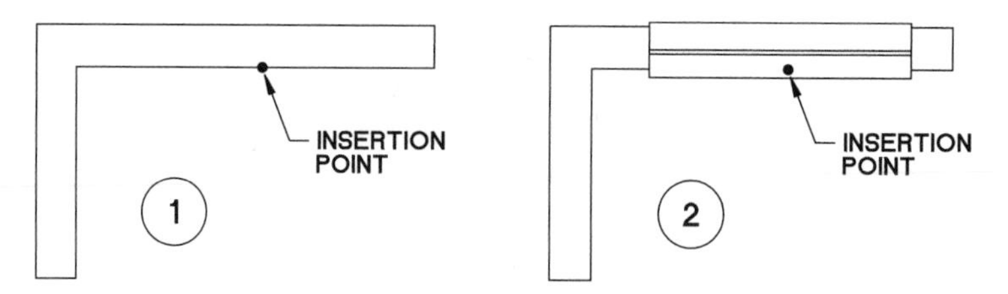

Figure 10–18 *The unit block reference inserted as a window*

Window symbols in plan view lend themselves to the application of the one-way scaling unit block reference. As demonstrated, they vary in only the *X* scale and not the *Y* scale from one size to another. Doors, however, present a special problem when we try to apply the unit block method. A symbol for a 2'-8"–wide door might look like that in Figure 10–19. The 2'-8"–wide door is drawn half open, that is, swung at 45 degrees to the wall. Therefore, if you used the same approach on the door and on the window unit block, some problems would be encountered.

First, a base block 1-unit–wide simplified door symbol would not be practical. For instance, if the unit block reference in Figure 10–20 were inserted with an *X* scale factor of 32 and a *Y* scale factor of 1, it would look like Figure 10–21. Even though the 4" lines that represent the jambs are acceptable, the "door" part of the symbol will not retain its 45-degree swing if not inserted with equal *X* and *Y* scale factors. Equal *X* and *Y* scale factors present another problem. Using a *Y* scale factor other than 1 would make the 4"-wide jamb incorrect. Therefore, combining the jambs with the door in the same symbol presents problems that might be impossible to overcome when trying to use a unit block approach to permit one block to serve for all sizes of doors.

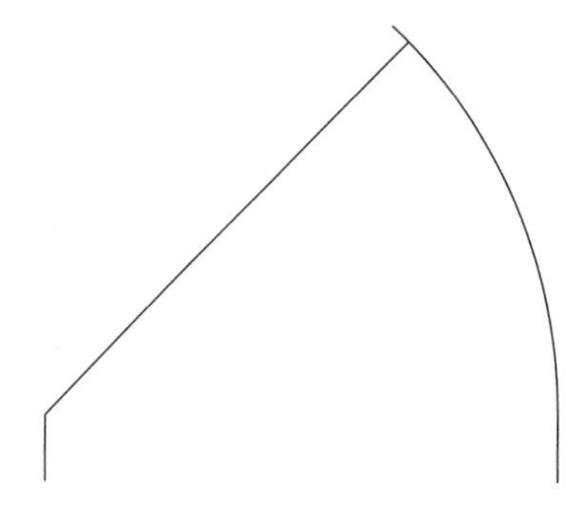

Figure 10–19 *Creating a unit block symbol for a door*

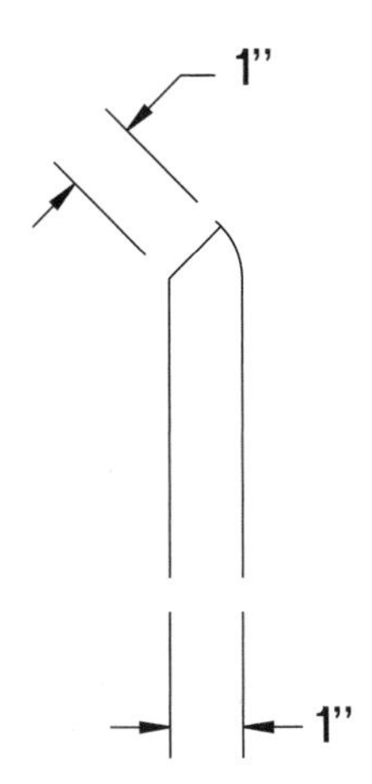

Figure 10–20 *Creating a unit block of a door symbol*

Figure 10–21 *A unit block reference of the door inserted with X and Y scale factors*

SEPARATE UNIT BLOCKS FOR ONE SYMBOL

The solution to making the jambs adaptable to using these two (jambs and door) into two block references, is shown in Figure 10–22.

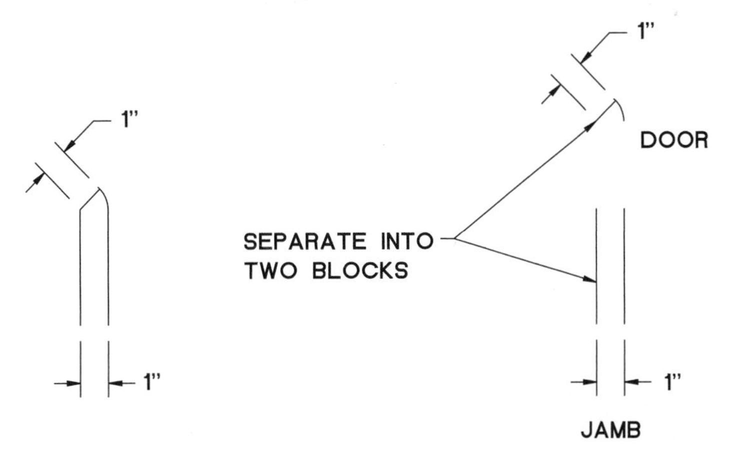

Figure 10–22 *Creating one symbol from two unit block references*

Now you can insert the block references separately, as follows (see Figure 10–23):

> Command: **-insert**
> Enter block name or [?] <y>: **jmb**
> Specify insertion point or [Scale/X/Y/Z/Rotate/PScale/PX/PY/PZ/PRotate]:
> *(specify a point)*
> Enter X scale factor, specify opposite corner, or [Corner/XYZ] <1>: **32**
> Enter Y scale factor <use X scale factor>: **1**
> Specify rotation angle <0>: (ENTER)

> Command: **-insert**
> Enter block name or [?] <y>: **dr**
> Specify insertion point or [Scale/X/Y/Z/Rotate/PScale/PX/PY/PZ/PRotate]:
> *(specify a point)*
> Enter X scale factor, specify opposite corner, or [Corner/XYZ] <1>: **32**
> Enter Y scale factor <use X scale factor>: (ENTER)
> Specify rotation angle <0>: (ENTER)

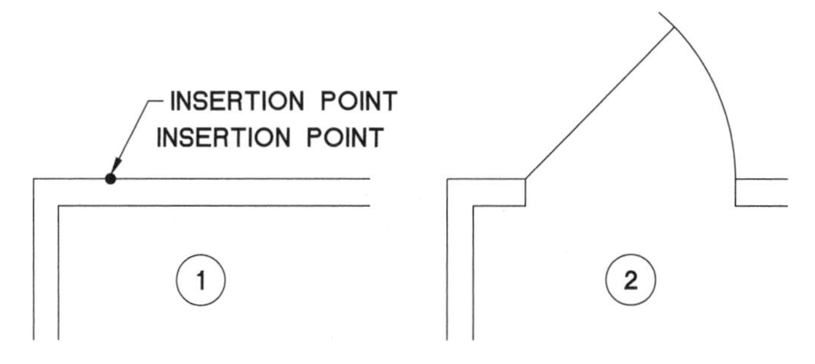

Figure 10–23 *Two unit block references inserted to create one symbol*

Note: When you intend to apply the concept of the unit block, whether or not it is to be scaled uniformly (X scale equal to Y scale), be sure that shapes and sizes of all items in the symbol will be correct at their new scale.

COLUMNS AS UNIT BLOCKS

Another unit block application is when both *X* and *Y* scale factors are changed, but not at the same factor. The column shown in Figure 10–24 is made into a block named COL1 and is inserted for a 10 x 8–wide flange symbol as follows:

> Command: **-insert**
> Block name: **col1**
> Insertion point: *(specify point)*
> X scale factor <1>: **8**
> Y scale factor (default=X>): **10**
> Rotation angle: (ENTER)

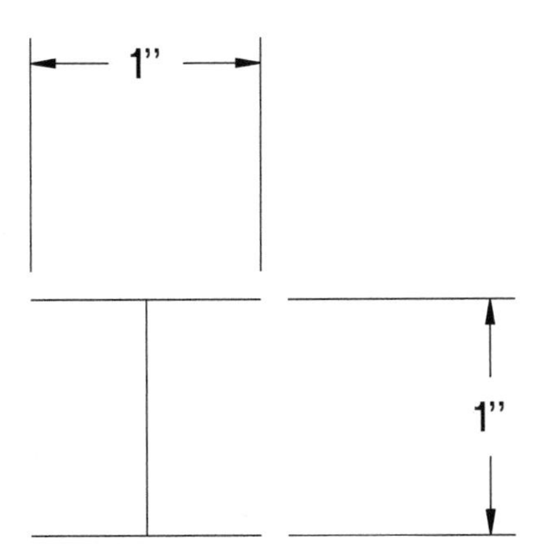

Figure 10–24 *Creating a unit block of a column symbol*

This can also be done for a 16 x 12 column rotated at 90 degrees, as shown in Figure 10–25.

Command: **-insert**
Enter block name or [?] <y>: **col1**
Specify insertion point or [Scale/X/Y/Z/Rotate/PScale/PX/PY/PZ/PRotate]:
  (specify a point)
Enter X scale factor, specify opposite corner, or [Corner/XYZ] <1>: **12**
Enter Y scale factor <use X scale factor>: **16**
Specify rotation angle <0>: **90**

Note the 90-degree rotation. But don't forget that the *X* and *Y* scale factors are applied to the block reference in the respective *X* and *Y* directions that were in effect when it was created, not to the *X* and *Y* directions after a rotated insertion.

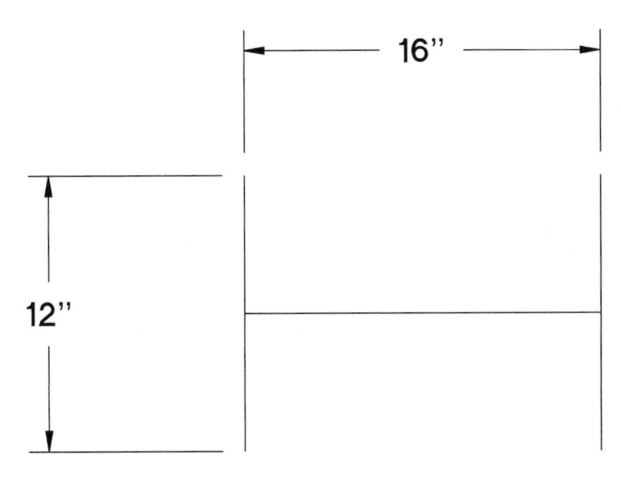

Figure 10–25 *The 16 x 12 column symbol inserted and rotated 90 degrees*

CREATING ONE DRAWING FROM ANOTHER

The WBLOCK command permits you to group objects in a manner similar to the BLOCK command. But, in addition, WBLOCK exports the group to a file, which, in fact, becomes a new and separate drawing. The new drawing (created by the WBLOCK command) might consist of a selected block reference in the current drawing. Or it might be made up of selected objects in the current drawing. You can even export the complete current drawing to the new drawing file. Whichever of these you choose to WBLOCK, the new drawing assumes the layers, linetypes, styles, and other environmental items, such as system variable settings of the current drawing.

Invoke the WBLOCK command:

Command: prompt	**wblock** (ENTER)

AutoCAD displays the Write Block dialog box, similar to Figure 10–26, where you will provide a drawing name and its path in the **File name and path:** text box of the **Destination** section of the dialog box as you do when you begin a new drawing. You should enter only the file name and not the extension. AutoCAD appends the .DWG extension automatically. The name should comply with the operating system requirements of valid characters and should be unique for drawings on the specified folder/path. Otherwise you will get the message:

Drawingname.dwg already exists. Do you want to replace it?

Choose the **Yes** button only if you wish to overwrite the existing drawing with the same name. Otherwise choose the **No** button and enter a different, unique name for the drawing.

If you wish to use an existing block, the entire drawing, or existing objects in the current drawing to create a new drawing, choose one of the radio buttons in the **Source** section of the Write Block dialog box.

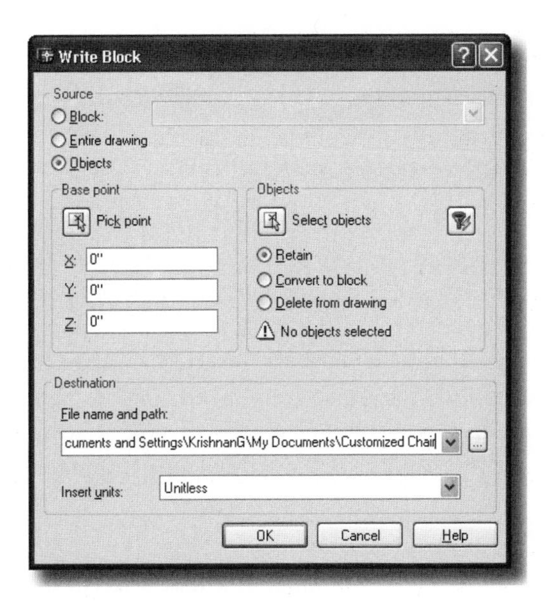

Figure 10–26 *Write Block dialog box*

Block

If you wish to create a drawing to be a duplicate of an existing block in the current drawing, then in the **Source** section of the Write Block dialog box, the **Block** radio button should be selected. Select the name of the desired block from the selection box to the right of the **Block** radio button. By default, the name of the new drawing will be the same as the block chosen. You can enter a different name for the new drawing with its corresponding path in the **File name and path:** text box.

Entire Drawing

If you wish to create a drawing to be a duplicate of the entire current drawing, then in the **Source** section of the Write Block dialog box, the **Entire drawing** radio button should be selected. Enter a name and a path for the new drawing in the **File name and path:** text box.

Objects

If you wish to create a drawing from objects to be selected in the current drawing, then in the **Source** section of the Write Block dialog box, the **Objects** radio button should be selected. You must select one or more objects to be exported as a new drawing. The object selection process is the same as when you are creating a block with the BLOCK command, described earlier in this chapter. Enter a name for the new drawing with its corresponding path in the **File name and path:** text box.

Insert Units

When the new file is inserted as a block, it will be scaled to the units in the **Insert units** list box.

Choose the **OK** button to create the drawing.

One advantage of using the WBLOCK command when the entire drawing is written to a file is that all of the unused blocks, layers, linetypes, and other unused named objects are not written. That is, the drawing is automatically purged. This means that unused items will not be written to the new drawing file. For example, unused items include block definitions that have not been inserted, noncurrent layers that have no objects drawn on them, and styles that are not being used. This can be useful if you just wish to clean up a cluttered drawing, especially one that has had other drawing files inserted into it, each bringing with it various unused named objects.

BASE COMMAND

The BASE command allows you to establish a base insertion point for the whole drawing in the same manner that you specify a base insertion point when using the BLOCK command to combine elements into a block. The purpose of establishing this basepoint is primarily so that the drawing can be inserted into another drawing by way of the INSERT command and having the specified basepoint coincide with the specified insertion point. The default basepoint is the origin (0,0,0). You can specify a 2D point, and AutoCAD will use the current elevation as the base *Z* coordinate. Or you can specify the full 3D point.

Invoke the BASE command:

Draw menu	Block > Base
Command: prompt	**base** (ENTER)

AutoCAD prompts:

Command: **base**
Enter base point <current>: *(specify a point, or press* ENTER *to accept the default)*

ATTRIBUTES

Attributes can be used for automatic annotation during insertion of a block reference. Attributes are special text objects that can be included in a block definition and must be defined beforehand and then selected when you are creating a block definition.

Attributes have two primary purposes:

> The first use of attributes is to permit annotation during insertion of the block reference to which the attributes are attached. Depending on how you define an attribute, it either appears automatically with a preset (constant) text string or it prompts you (or other users) for a string to be written as the block reference is inserted. This feature permits you to insert each block reference with a string of preset text or with its own unique string.
>
> The second (perhaps the more important) purpose of attaching attributes to a block reference is to have extractable data about each block reference stored in the drawing database file. Then, when the drawing is complete (or even before), you can invoke the ATTEXT (short for "attribute extract") command to have attribute data extracted from the drawing and written to a file in a form that database-handling programs can use. You can have as many attributes attached to a block reference as you wish. As just mentioned, the text string that makes up an attribute can be either constant or user-specified at the time of insertion.

A DEFINITION WITHIN A DEFINITION

When creating a block, you select objects to be included. Objects such as lines, circles, and arcs are drawn by means of their respective commands. Normal text is drawn with the TEXT, or MTEXT command.

As with drawing objects, attributes must be drawn before they can be included in the block. This is complicated, and it requires additional steps to place them in the drawing; AutoCAD calls this procedure defining the attribute. Therefore, an attribute definition is simply the result of defining an attribute by means of the ATTDEF command. The attribute definition is the object that is selected during the BLOCK command. Later, when the block reference is inserted, the attributes that are attached to it and the manner in which they become a part of the drawing are a result of how you created (defined) the attribute definition.

VISIBILITY AND PLOTTING

If an attribute is to be used only to store information, then you can, as part of the definition of the attribute, specify whether or not it will be visible. If you plan to use an attribute with a block as a note, label, or callout, you should consider the effect of scaling (whether equal or unequal X/Y factors) on the text that will be displayed. The scaling factor(s) on the attribute will be the same as on the block reference. Therefore, be sure that it will result in the size and proportions desired. You should also be aware of the effect of rotation on visible attribute text. Attribute text that is defined as horizontal in a block will be displayed vertically when that block reference is inserted with a 90-degree angle of rotation.

Note that, like any other object in the drawing, the attribute must be visible on the screen (or would be if the plotted view were the current display) for that object to be eligible for plotting.

ATTRIBUTE COMPONENTS

Four components associated with attributes should be understood before attempting a definition. The purpose of each is described as follows:

Tag

An attribute definition has a tag, just as a layer or a linetype has a name. The tag is the identifier of the attribute definition and is displayed where this attribute definition is located, depicting text size, style, and angle of rotation. The tag cannot contain spaces. Two attribute definitions with the same tag should not be included in the same block. Tags appear in the block definition only, not after the block reference is inserted. However, if you explode a block reference, the attribute value (described herein) changes back into the tag.

If multiple attributes are used in one block, each must have a unique tag in that block. This restriction is similar to each layer, linetype, block, and other named object having a unique name within one drawing. An attribute's tag is its identifier at the time that attribute is being defined, before it is combined with other objects and attributes by the BLOCK command.

Value

The value of an attribute is the actual string of text that appears (if the **visibility** mode is set to ON) when the block reference (of which it is a part) is inserted. Whether visible or not, the value is tied directly to the attribute, which, in turn, associates it with the block reference. It is this value that is written to the database file. It might be a door or window size or, in a piping drawing, the flange rating, weight, or cost of a valve or fitting. In an architectural drawing the value might represent the manufacturer, size, color, cost or other pertinent information attached to a block representing a desk.

Note: When an extraction of attribute data is performed, it is the value of an attribute that is written to a file, but it is the tag that directs the extraction operation to that value. This will be described in detail in the later section on "Extracting Attributes."

Prompt

The prompt is what you see when inserting a block reference with an attribute whose value is not constant or preset. During the definition of an attribute, you can specify a string of characters that will appear in the prompt area during the insertion of the block reference to prompt you to enter the appropriate value. What the prompt says to you during insertion is what you told it to say when you defined the attribute.

Default

You can specify a default value for the attribute during the definition procedure. Then, during insertion of the block reference, it will appear behind the prompt in brackets, i.e., <default>. It will automatically become the value if ENTER is pressed in response to the prompt for a value.

ATTRIBUTE COMMANDS

The four primary commands to manage Attributes are:

ATTDEF—attribute definition

ATTDISP—attribute display

ATTEDIT—attribute edit

ATTEXT—attribute extract

As explained earlier, the ATTDEF command is used to create an attribute definition. The attribute definition is the object that is selected during the BLOCK command.

The ATTDISP command controls the visibility of the attributes.

The ATTEDIT command provides a variety of ways to edit without exploding the block reference.

The ATTEXT command allows you to extract the data from the drawing and have it written to a file in a form that database-handling programs can use, as shown in the following table:

DOORS						
SIZE	**THKNS**	**CORE**	**FINISH**	**LOCKSET**	**HINGES**	**INSET**
3070	1.750	SOLID	PAINT	PASSAGE	4 x 4	3/4
3070	1.750	SOLID	VARNISH	KEYED	4 x 4	20 x 20
2868	1.375	HOLLOW	PAINT	PRIVACY	3 x 3	3/4

ROOM FINISHES					
NAME	**WALL**	**CEILING**	**FLOOR**	**BASE**	**REMARKS**
LIVING	GYPSUM	GYPSUM	CARPET	NONE	PAINT
FAMILY	PANEL	ACOUSTICAL	TILE	STAIN	STAIN
BATH	PAPER	GYPSUM	TILE	COVE	4'_CERAMIC_TILE
GARAGE	GYPSUM	GYPSUM	CONCRETE	NONE	TAPE_FLOAT_ONLY

You can include an attribute in the WDW block to record the size of the window. A suggested procedure would be to zoom in near the insertion point and create an attribute definition with a tag that reads WDW-SIZE, as shown in Figure 10–27.

If, during the insertion of the WDW block reference, you respond to the prompt for the SIZE attribute with 2054 for a 2'-0"–wide x 5'-4"–high window, the resulting block reference object would be as shown in Figure 10–28, with the normally invisible attribute value shown here for illustration purposes. Figure 10–29 shows the result of the attribute being inserted with unequal scale factors.

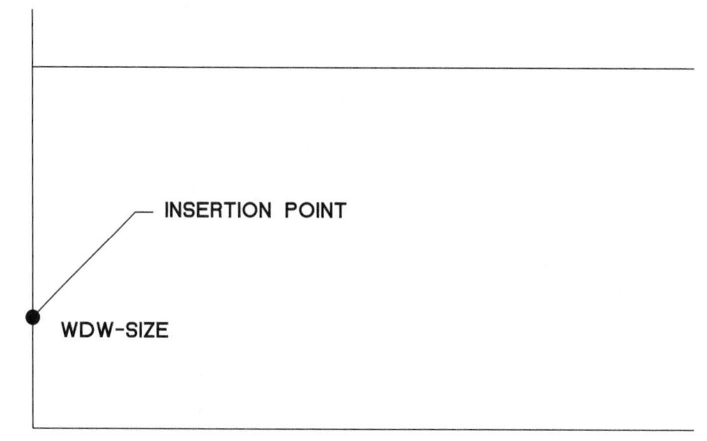

Figure 10–27 *Create the attribute definition before defining the block*

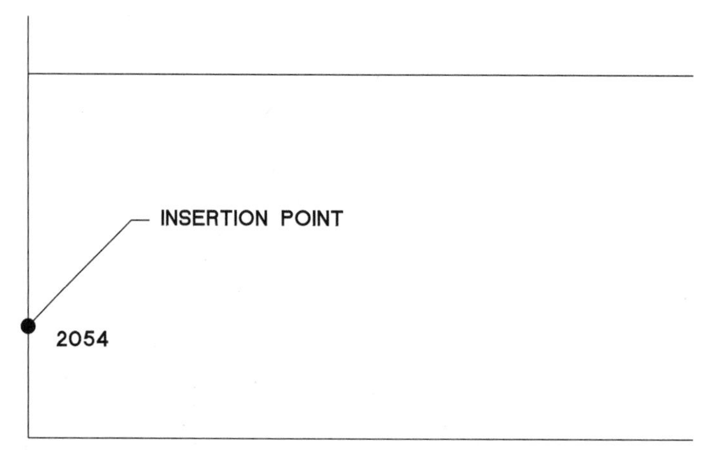

Figure 10–28 *The attribute value (2054) visible with the inserted block*

Even though the value displayed is distorted, the string is not affected when extracted to a database file for a bill of materials.

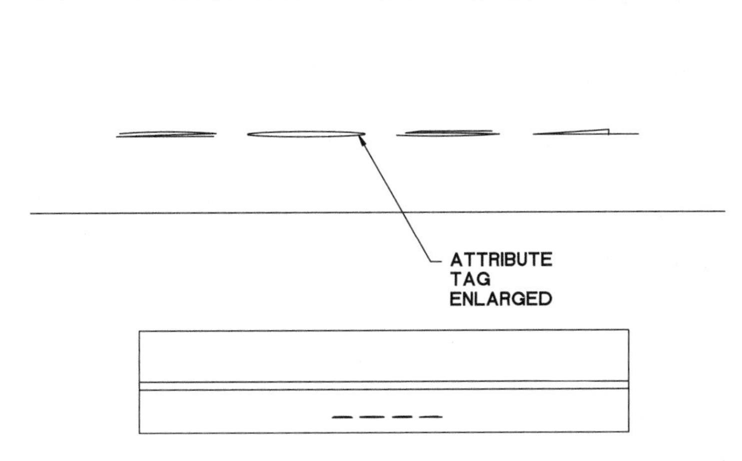

Figure 10–29 *The attribute in a block reference with exaggerated, unequal scale factors*

To solve the distortion and rotation problem, if you wish to have an attribute displayed for rotational purposes, you can create a block that contains attributes only or only one attribute. Then it can be inserted at the desired location and rotated for readability to produce the results shown in Figure 10–30.

Attribute definitions would be created as shown in Figure 10–30 and inserted as shown in Figure 10–31.

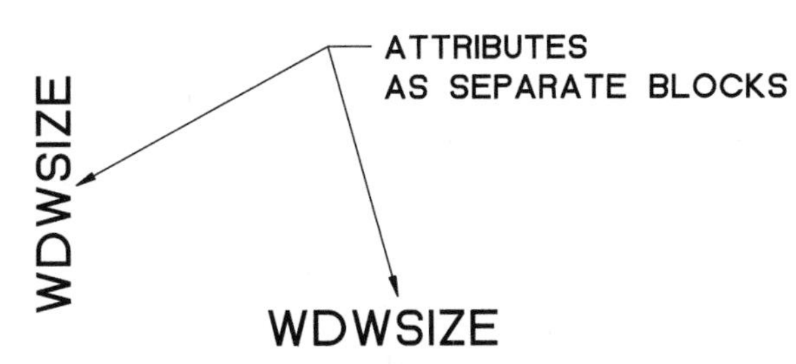

Figure 10–30 *Attributes created as separate blocks*

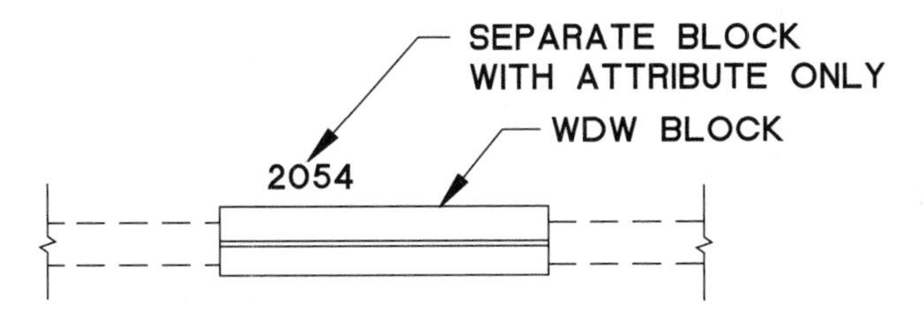

Figure 10–31 *Attribute inserted separately for the window*

The insertion points selected would correspond to the midpoint of the outside line that would result from the insertion of the WDW block reference. The SIZE attributes could be defined into blocks called WDWSIZE and inserted separately with each WDW block reference, thereby providing both annotation and data extraction.

Note: The main caution in having an attribute block reference separate from the symbol (WDW) block reference is in editing. Erasing, copying, and moving the symbol block reference without the attribute block reference could mean that the data extraction results in the wrong quantity.

There is another solution to the problem of a visible attribute not being located or rotated properly in the inserted block reference. If the attribute is not constant, you can edit it independently after the block reference has been inserted by way of the ATTEDIT command. It permits changing an attribute's height, position, angle, value, and other properties. The ATTEDIT command is covered in detail in the later section on "Editing Attributes," but it should be noted here that the height editing option applies to the *X* and *Y* scales of the text. Therefore, for text in the definition of a block that was inserted with unequal *X* and *Y* scale factors, you will not be able to edit its proportions back to equal *X* and *Y* scale factors.

CREATING AN ATTRIBUTE DEFINITION

The ATTDEF and -ATTDEF commands allow you to create an attribute definition. The ATTDEF command allows you to create an attribute definition through a dialog box; the -ATTDEF command allows you to create an attribute definition at the Command: prompt.

Invoke the ATTDEF command:

Draw menu	Block > Define Attributes
Command: prompt	**attdef** (ENTER)

AutoCAD displays the Attribute Definition dialog box, as shown in Figure 10–32.

Set one or more of the available modes to ON in the **Mode** section of the Attribute Definition dialog box.

Invisible Setting the **Invisible** mode to ON causes the attribute value *not* to be displayed when the block reference is completed. Even if visible, the value will not appear until the insertion is completed. Attributes needed only for data extraction should be invisible, to quicken regeneration and to avoid cluttering your drawing. You can use the ATTDISP command to override the **Invisible** mode setting. Setting the **Invisible** mode to Y (yes, or ON) does not affect the visibility of the tag in the attribute definition.

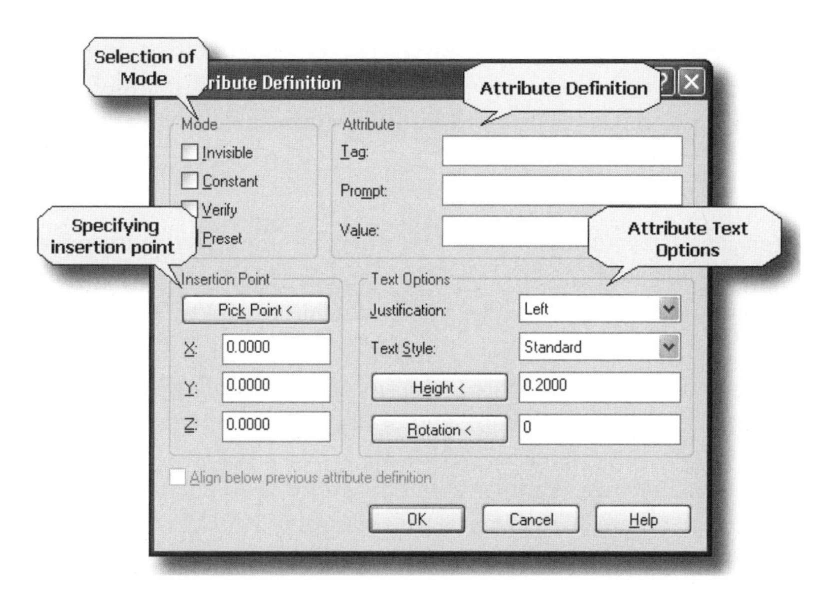

Figure 10–32 *Attribute Definition dialog box*

Constant If the **Constant** mode is set to ON, you must enter the value of the attribute while defining it. That value will be used for that attribute every time the block reference to which it is attached is inserted. There will be no prompt for the value during insertion, and you cannot change the value.

Verify If the **Verify** mode is set to ON, you will be able to verify its value when the block reference is inserted. For example, if a block reference with three (nonconstant value) attributes is inserted, once you have completed all prompt/value sequences that have displayed the original defaults, you will be prompted again, with the latest values as new defaults, giving you a second chance to be sure the values are correct before the INSERT command is completed. Even if you press ENTER to accept an original default value, it also appears as the second-chance default. If, however, you make a change during the verify sequence, you will *not* get a third chance, that is, a second verify sequence.

Preset If the **Preset** mode is set to ON, the attribute automatically takes the value of the default that was specified at the time of defining the attribute. During a normal insertion of the block reference, you will not be prompted for the value. You must be careful to specify a default during the ATTDEF command or the attribute value will be blank. A block consisting of only attributes whose defaults were blank when Preset modes were set to ON could be inserted, but it would not display anything and cannot be purged from the drawing. The only adverse effect would be that of adding to the space taken in memory. One way to get rid of a nondisplayable block reference like this is to use a visible entity to create a block with the same name, thereby redefining it to something that can be edited, that is, erased and subsequently purged.

You can duplicate an attribute definition with the COPY command, and use it for more than one block. Or you can explode a block reference and retain one or more of its attribute definitions for use in subsequent blocks.

The **Attribute** section of the Attribute Definition dialog box allows you to set attribute data. Enter the attribute's tag, prompt, and default value in the text boxes.

The attribute's tag identifies each occurrence of an attribute in the drawing. The tag can contain any characters except spaces. AutoCAD changes lowercase letters to uppercase.

The attribute's prompt appears when you insert a block reference containing the attribute definition. If you do not specify the prompt, AutoCAD uses the attribute tag as the prompt. If you turn on the **Constant** mode, the **Prompt** field is disabled.

The default **Value** specifies the default attribute value. This is optional, except if you turn on the Constant mode, for which the default value needs to be specified.

The **Insertion Point** section of the dialog box allows you to specify a coordinate location for the attribute in the drawing, either by choosing the **Pick Point** button to specify the location on the screen or by entering coordinates in the text boxes provided.

The **Text Options** section of the dialog box allows you to set the justification, text style, height, and rotation of the attribute text.

The **Align below previous attribute definition** check box allows you to place the attribute tag directly below the previously defined attribute. If you haven't previously defined an attribute definition, this option is unavailable.

Choose the **OK** button to define the attribute definition.

After you close the Attribute Definition dialog box, the attribute tag appears in the drawing. Repeat the procedure to create additional attribute definitions.

You can also define attributes by invoking ATTDEF command preceded with a hyphen from the Command: prompt, as shown in the following command sequence:

Command: **-attdef**
Current attribute modes: Invisible=N Constant=N Verify=N Preset=N
Enter an option to change [Invisible/Constant/Verify/Preset] <done>:
*(*ENTER **i**, **c**, **v**, *or* **p** *to change the current settings for attribute modes)*
Enter attribute tag name: *(specify the attribute tag name, with no spaces or exclamation marks)*
Enter attribute prompt: *(specify the text for the prompt line, or press* ENTER *to accept the attribute tag as the attribute prompt)*
Enter default attribute value: *(specify a value for the default attribute value, or press* ENTER*)*

This last prompt appears unless the **Constant** mode is set to Y (yes, or ON), in which case the prompt will be as follows:

Attribute value: *(specify a value for the attribute value)*

After the foregoing ATTDEF prompts have been answered, you will be prompted to place the tag, in the same manner as you would for placing text, except AutoCAD will create the tag in place of a text string. Subsequent attributes can be placed in a manner similar to that for placing lines of text, using the insertion point and line spacing as left justified, centered, aligned, or right justified lines of text. Simply press ENTER to invoke a repeating attribute definition.

Figure 10–33 shows a pre-drawn 2'-0"–wide door made into a block reference with attributes.

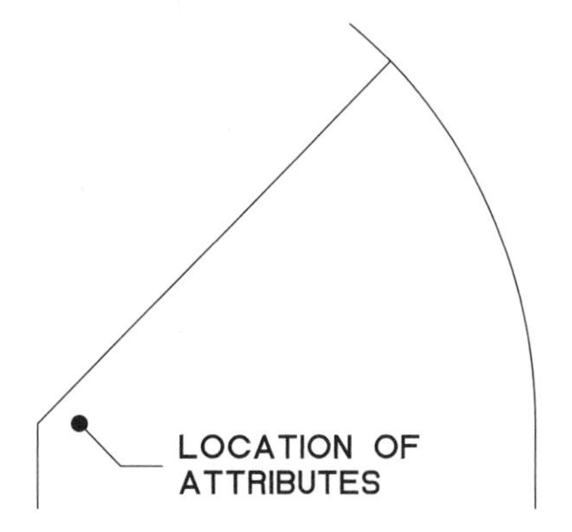

Figure 10–33 *Pre-drawn door with attributes*

Note: In this example the block is drawn to true size with jambs so it can be inserted with X and Y scale factors equal to 1 (one). But, the attributes will all be invisible, because the door might be inserted at a rotation incompatible with acceptable text orientations. The attributes can also be very small, located at a point that will be easy to find if they must be made visible in order to read (Figure 10–34).

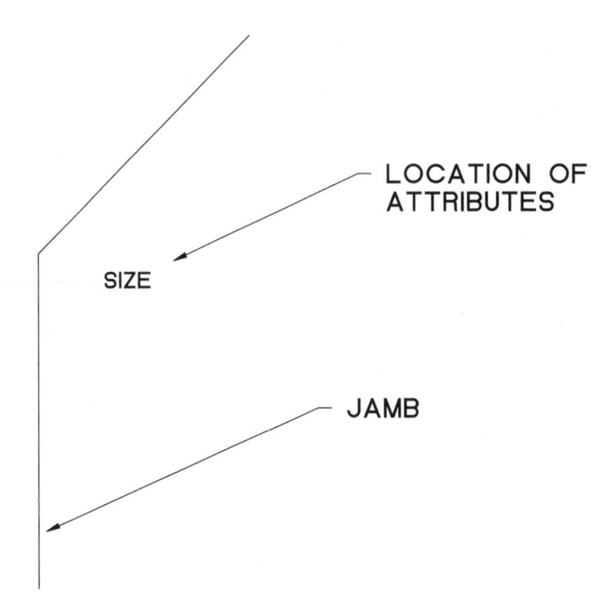

Figure 10–34 *Pre-drawn jamb block with attributes*

The Attributes are entered as follows (Figures 10–35 and 10–36):

Command: **-attdef**
Attribute modes—Invisible:N Constant:N Verify:N Preset:N
Enter an option to change [Invisible/Constant/Verify/Preset] <done>: **i**
Attribute modes—Invisible:Y Constant:N Verify:N Preset:N
Enter an option to change [Invisible/Constant/Verify/Preset] <done>: **c**
Attribute modes—Invisible:Y Constant:Y Verify:N Preset:N
Enter an option to change [Invisible/Constant/Verify/Preset] <done>: (ENTER)
Enter attribute tag name: **size**
Enter attribute value: **26681.375** *(for 2'-6" x 6'-8" x 1-3/8")*
Current text style: "Standard" Text height: 0.2000
Specify start point of text or [Justify/Style]: *(specify point)*
Specify height <0.2000>: **1/16**
Specify rotation angle of text <0>: (ENTER)

Command: (ENTER)
Attribute modes—Invisible:Y Constant:Y Verify:N Preset:N
Enter an option to change [Invisible/Constant/Verify/Preset] <done>: **c**
Attribute modes—Invisible:Y Constant:N Verify:N Preset:N
Enter an option to change [Invisible/Constant/Verify/Preset] <done>: (ENTER)
Enter attribute tag name: **matl**

Enter attribute prompt: **material**
Enter default attribute value: **mahogany**
Specify start point of text or [Justify/Style]: *(specify point)*
Command: (ENTER)

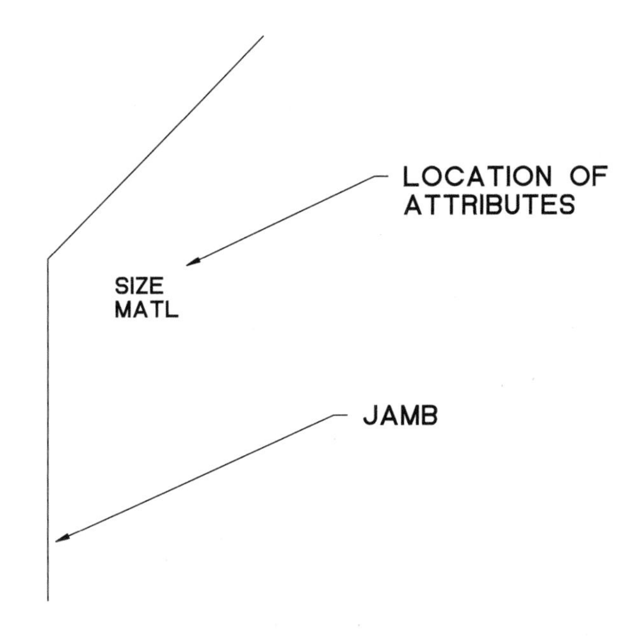

Figure 10–35 *Pre-drawn jamb block with character attributes*

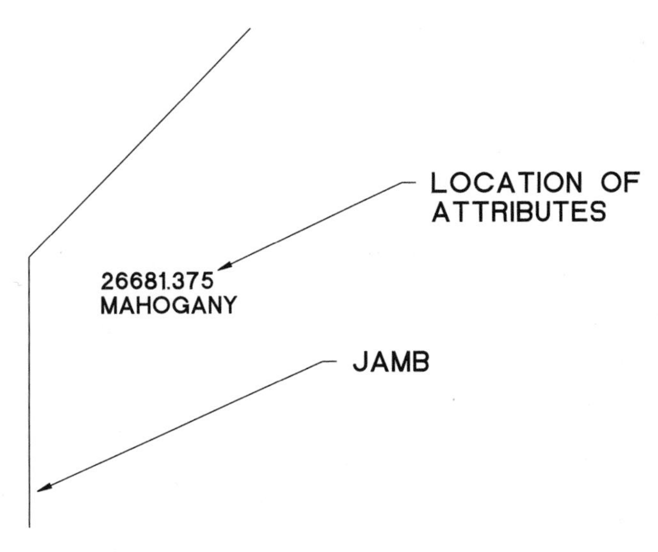

Figure 10–36 *Pre-drawn jamb block with character/numeric attributes defined*

Note: The values given for the various attributes can be written just as any string of text is written. You should note that these strings, when written to a database-handling file, might eventually need to be interpreted as numbers rather than characters. The difference is primarily of concern to the person who will use the data in a database-handling program. So, if you are not familiar with data types, such as numeric, character, and date, you might wish to consult with someone (or study a book on databases) if you are entering the values. For example, an architectural distance such as 12'-6 1/2" may need to be written in decimal feet (12.54) without the apostrophe or in decimal inches (150.5) without the inch mark if it is going to be used mathematically once extracted.

You can continue to define additional attributes in the same manner for the L.H./R.H. SWING, PAINT/VARNISH FINISH, Type of HINGE, Type of LOCKSET, and so on (see Figure 10–37).

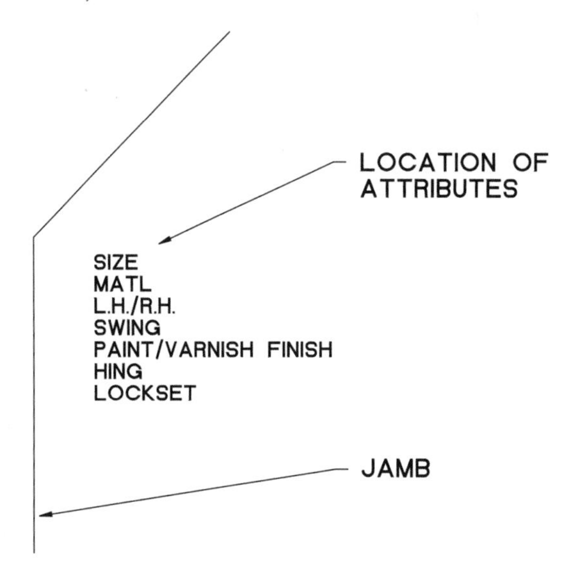

Figure 10–37 *Pre-drawn jamb with additional attributes defined*

INSERTING A BLOCK REFERENCE WITH ATTRIBUTES

Blocks with attributes may be inserted in a manner similar to that for inserting regular block references. If there are any nonconstant attributes, you will be prompted to enter the value for each. You may set the system variable called ATTREQ to 0 (zero), thereby suppressing the prompts for attribute values. In this case the values will either be blank or be set to the default values if they exist. You can later use either the DDATTE or ATTEDIT command to establish or change values.

You can set the ATTDIA system variable to a nonzero value, which will display a dialog box for attribute value input.

CONTROLLING THE DISPLAY OF ATTRIBUTES

The ATTDISP command controls the visibility of attributes. Attributes will normally be visible if the Invisible mode is set to N (normal) when they are defined.

Invoke the ATTDISP command:

View menu	Display > Attribute Display
Command: prompt	**attdisp** (ENTER)

AutoCAD prompts:

Command: **attdisp**
Enter attribute visibility setting [Normal/ON/OFF] <Normal>: *(choose one of the available options, or press* ENTER *to accept the default option)*

Responding ON makes all attributes visible; OFF makes all attributes invisible. The Normal option displays the attributes as you created them. The ATTMODE system variable is affected by the ATTDISP setting. If REGENAUTO is set to ON, changing the ATTDISP setting causes drawing regeneration.

EDITING ATTRIBUTE VALUES

Unlike other objects in an inserted block reference, attributes can be edited independently of the block reference. The EATTEDIT command allows you to change the value of attributes in blocks that have been inserted. This permits you to insert a block reference with generic attributes; that is, the default values can be used in anticipation of changing them to the desired values later. Or you can copy an existing block reference that may need only one or two attributes changed to make it correct for its new location. And, of course, there is always the chance that either an error was made in entering the value or design changes necessitate subsequent changes. Note that the ATTEDIT command can be used to change only the value of the attributes. In order to change other characteristics of attributes such as text size and font and visibility, the Block Attribute Manager, described later in this chapter, must be used.

Editing an attribute is accomplished by invoking the EATTEDIT command. The ATTEDIT command edits individual, nonconstant attribute values associated with a specific block reference, whereas the -EATTEDIT command edits both attribute values and attribute properties individually or globally, independent of the block reference.

Invoke the EATTEDIT command:

Modify II toolbar	Choose the Edit Attribute command (see Figure 10–38)
Modify menu	Choose Object > Attribute > Single
Command: prompt	**eattedit** (ENTER)

Figure 10–38 *Invoking the* EDIT ATTRIBUTE *command from the Modify II toolbar*

AutoCAD prompts:

Command: **eattedit**
Select block reference: *(select the block reference)*

AutoCAD displays the Enhanced Attribute Editor dialog box, similar to the one shown in Figure 10–39. Selecting objects that are not block references or block references that contain no attributes will cause an error message to appear.

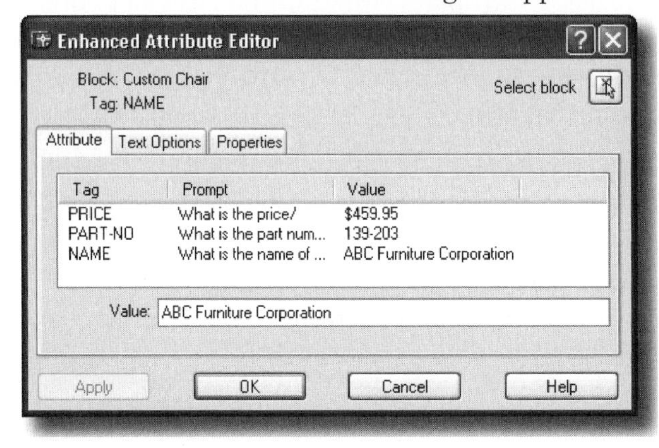

Figure 10–39 *Enhanced Attribute Editor dialog box*

The dialog box lists all the attributes defined with values for the selected block reference. Using the pointing device, you can select values to be changed in the dialog box. In addition, you can also change the text attributes and object properties of the selected attribute tag. Enter the new values and accept the changes by clicking the **OK** button or pressing ENTER. Choosing the **Cancel** button terminates the command, returning all values to their original state.

You can use the ATTEDIT command to look at the values of a selected attribute without making changes. Or you can employ repeated editing or repeated inquiry looks at attribute values by modifying the ATTEDIT command with the MULTIPLE command, as follows:

Command: **multiple**
Enter command name to repeat: **attedit**

To change attribute properties such as position, height, and style, invoke the -ATTEDIT command. The -ATTEDIT command provides a variety of ways to specify attributes to be edited. It also allows various properties of the selected attributes to be edited. It should be noted that attributes with constant values cannot be edited in this manner.

Invoke the ATTEDIT command (preceded with a hyphen):

Modify menu	Choose Object > Attribute > Global
Command: prompt	**-attedit** (ENTER)

AutoCAD prompts:

Command: **-attedit**
Edit attributes one at a time? [Yes/No] <Y>:

Responding **Y** permits you to edit visible attributes individually. You can limit those attributes eligible for selection by specifying the block name, tag, or value. In addition to an attribute's value, other properties that can be changed during this one-at-a-time mode include its position, height, and angle of rotation.

Responding **N** permits global (or mass) editing of attributes. Again, you can limit those eligible for editing to a specified block name, tag, or value. However, this mode permits editing of attribute values only and no other properties.

After responding **Y** or **N**, you will be prompted to specify eligible attributes as follows:

Enter block name specification <*>:
Enter attribute tag specification <*>:
Enter attribute value specification <*>:

For any attribute to be eligible for editing, the block name, attribute tag, and attribute value must all match the name, tag, and value specified. Use of "*" or "?" wild-card characters allows more than one string to match. Pressing ENTER in response to the "Block name specification:" prompt defaults to the asterisk, which makes all attributes eligible with regard to the block name restriction. Attributes still have to match tag and value responses in order to be eligible for editing.

Global Editing

If you responded to the "Edit attributes one at a time? <Y>:" prompt with **N** (no), thereby choosing global editing, and set the block name, tag, and value limitations, the following prompts appear:

Performing global editing of attribute values.
Edit only attributes visible on screen? [Yes/No] <Y>:

An **N** reply switches AutoCAD to the text window with the following message:

Drawing must be regenerated afterwards.

When the editing for this command is completed (and if AUTOREGEN is set to ON) a drawing regeneration occurs.

If at the "Edit only Attributes visible on screen? <Y>:" prompt you respond **Y** (or default to Y by pressing ENTER), you will be prompted as follows:

Select Attributes:

Attributes may now be selected either by specifying them on the screen, or by the window, crossing, or last method. Eligible attributes are then highlighted and the following prompts appear:

Enter string to change: *(select a string to change)*
Enter new string: *(enter a new text string)*

Note: The changes you specify will affect a group of attributes all at one time. Take care that unintended changes do not occur.

The responses to the "String to change:" prompt cause AutoCAD to search eligible selected attribute strings for matching strings. Each matching string will be changed to your response to the "New string:" prompt. For example, blocks named WDW20, having tags named SIZE and values of 2054 (for 2'-0" x 5'-4"), can be changed to 2060 by the following sequence:

Enter string to change: **54**
Enter new string: **60**

If you did not limit the attributes by block name, tag, or value, you might unintentionally change a window whose SIZE is 5440 (for 5'-4" x 4'-0") to 6040.

If you respond to "Enter string to change:" by pressing ENTER, it will cause any response to "New string:" to be placed ahead of all eligible attribute value strings. For example, if you specified only the block name WDW20 with an attribute value of 2054 to be eligible for editing, you can make an addition to the value by the following sequence:

Enter string to change: (ENTER)
Enter new string: **dbl hung**

The value 2054 will be changed to read DBL HUNG 2054. Be sure to add a space behind the G if you do not want the result to be DBL HUNG2054.

Editing Attributes One at a Time

If you responded to "Edit attributes one at a time? [Yes/No] <Y>:" by pressing ENTER (defaulting to Y), and have specified the eligible attributes through the block name, tag, and value sequences, you will be prompted:

Select Attributes:

You may select attributes by specifying them on the screen or by the window, crossing, or with the last method. The eligible selected attributes with nonconstant values will be marked sequentially with an X. The next prompt is as follows:

```
Enter an option [Value/Position/Height/Angle/Style/Layer/Color/Next]
  <N>: (choose one of the available options)
```

Angle is not an option for an attribute defined as fitted text, nor are the angle and height options for aligned attributes. Entering any option's initial letter permits a change relative to that option, followed by the repeated prompt for the list of options until you default to N for the next attribute.

The Value option, if chosen, will prompt you as follows:

```
Enter type of value modification [Change/Replace] <R>:
```

Defaulting to R causes the following prompt:

```
Enter new attribute value:
```

Any string entered will become the new value. Even a null response will become a null (blank) value. Entering C at the "Enter type of value modification [Change/Replace] <R>:" prompt will cause the following prompts to appear:

```
Enter string to change:
Enter new string:
```

Responses to these prompts follow the same rules described in the previous section on global editing. In addition to the Position/Height/Angle options normally established during attribute definition, you can change those preset properties such as Style/Layer/Color with the -ATTEDIT command.

EXTRACTING ATTRIBUTES

Extracting data from a drawing is one of the most innovative features in CAD. Paper copies of drawings have long been used to communicate more than just how objects look. In addition to dimensions, drawings tell builders or fabricators what materials to use, quantities of objects to make, manufacturers' names and models of parts in an assembly, coordinate locations of objects in a general area, and what types of finishes to apply to surfaces. But, until computers came into the picture (or pictures came into the computer), extracting data from manual drawings involved making lists (usually by hand) while studying the drawing, often checking off the data with a marker. The AutoCAD attribute feature and the ATTEXT command combine to allow complete, fast, and accurate extraction of (1) data consciously put in for the purpose of extraction, (2) data used during the drawing process, and (3) data that AutoCAD maintains about all objects (block references in this case).

The CAD drawing in Figure 10–40 shows a piping control set. The 17 valves and fittings are a fraction of those that might be on a large drawing. Each symbol is a block with attributes attached to it. Values that have been assigned to each attribute tag record the type (TEE, ELL, REDUCER, FLANGE, GATE VALVE, or CONTROL VALVE), size (3", 4", or 6"), rating (STD or 150#), weld (length of weld), and many other vital bits of specific data. Keeping track of hundreds of valves, fittings, and even cut lengths of pipe is a time-consuming task subject to omissions and errors if done manually, even if the drawing is plotted from CAD. Just as important as extracting data from the original drawing is the need to update a list of data when the drawing is changed. Few drawings, if any, remain unchanged. The AutoCAD attribute feature makes the job fast, thorough, and accurate. Examples of some of the block references with attribute definition are shown in Figures 10–41 through 10–43.

These are three examples of block references as defined and inserted, each attribute having been given a value during the INSERT command. Remember, the value that will be extracted is in accordance with how the template specifies the tags. Figure 10–44 shows the three block references inserted, with corresponding attribute values in tabular form. By using the ATTEXT command, a complete listing of all valves and fittings in the drawing can be written to a file, as shown in Table 10–1.

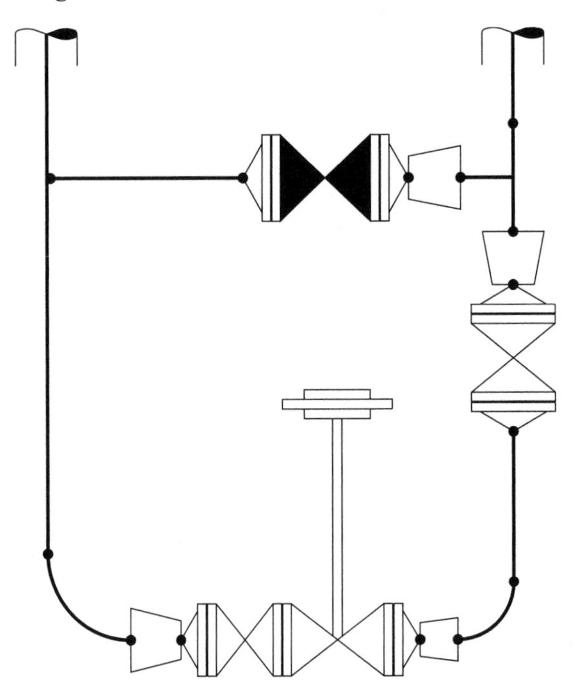

Figure 10–40 *Piping control set*

ENTITIES THAT WILL BE
INCLUDED IN EACH BLOCK

RED

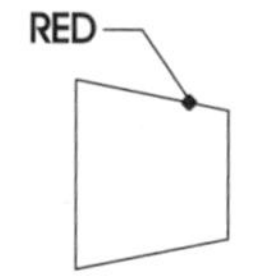

TEXT ITEMS REPRESENT
ATTRIBUTE TAGS

TYPE
SIZE
RATING
BLTQUAN
BLTSIZE
GSKETMATL
GSKETSIZE
WELD

Figure 10–41 *Reducer with attribute definition*

ENTITIES THAT WILL BE
INCLUDED IN EACH BLOCK

FLANGE

TEXT ITEMS REPRESENT
ATTRIBUTE TAGS

TYPE
SIZE
RATING
BLTQUAN
BLTSIZE
GSKETMATL
GSKETSIZE
WELD

Figure 10–42 *Flange with attribute definition*

ENTITIES THAT WILL BE
INCLUDED IN EACH BLOCK

VALVE

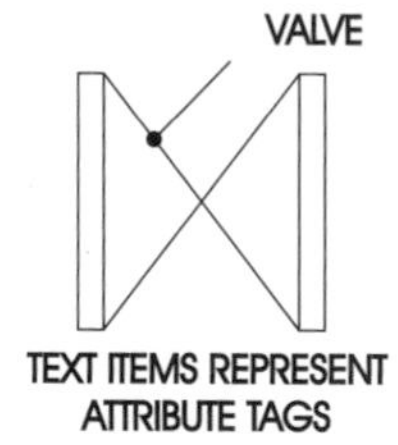

TEXT ITEMS REPRESENT
ATTRIBUTE TAGS

TYPE
SIZE
RATING
BLTQUAN
BLTSIZE
GSKETMATL
GSKETSIZE
WELD

Figure 10–43 *Valve with attribute definition*

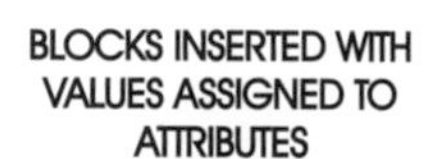

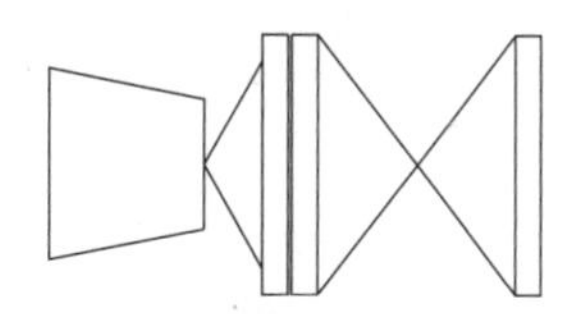

TEXT ITEMS REPRESENT
ATTRIBUTE TAGS

REDUCER	RFWN FLG	GT VLV
6X4	4"	4"
STD	150#	150#
-	8	8
-	5/8"x3"	5/8"x3"
-	COMP	COMP
-	1/8"	1/8"
34.95	14.14	-

Figure 10–44 *Inserted block references with corresponding attribute values*

Table 10–1 A List of Inventory as Specified by the ATTEXT Command and Tags

TYPE	SIZE	RATING	BLTQUAN	BLTSIZE	GSKETMATL	GSKETSIZE	WELD
TEE	6"	STD	4	5/8"X3"	COMP	1/8"	62.44
ELL	6"	STD	4	5/8"X3"	COMP	1/8"	41.63
ELL	4"	STD	4	5/8"X3"	COMP	1/8"	14.14
RED	6X4	STD	8	5/8"X3"	COMP	1/8"	34.95
RED	6X4	STD	8	5/8"X3"	COMP	1/8"	34.95
RED	6X3	STD	4	5/8"X3"	COMP	1/8"	31.81
RED	4X3	STD	8	5/8"X3"	COMP	1/8"	25.13
FLG	4"	150#	8	5/8"X3"	COMP	1/8"	14.14
FLG	4"	150#	8	5/8"X3"	COMP	1/8"	14.14
FLG	4"	150#	8	5/8"X3"	COMP	1/8"	14.14
FLG	4"	150#	8	5/8"X3"	COMP	1/8"	14.14
FLG	3"	150#	4	5/8"X3"	COMP	1/8"	11.00
FLG	3"	150#	4	5/8"X3"	COMP	1/8"	11.00
GVL	4"	150#	16	5/8"X3"	COMP	1/8"	11.00
GVL	4"	150#	16	5/8"X3"	COMP	1/8"	11.00
GVL	3"	150#	8	5/8"X3"	COMP	1/8"	11.00
CVL	3"	150#	8	5/8"X3"	COMP	1/8"	11.00

The headings above each column in Table 10–1 are for your information only. These will not be written to the extract file by the ATTEXT command. They signify the tags whose corresponding values will be extracted.

When operated on by a database program, this file can be used to sort valves and fittings by type, size, or other value. The scope of this book is too limited to cover database applications. But generating a file like Table 10–1 that a database program can use is the important linkage between computer drafting and computer management of data for inventory control, material takeoff, flow analysis, cost, maintenance, and many other applications. The CAD drafter can apply the ATTEXT feature to perform this task.

Files

Using the ATTEXT command involves concepts of computer applications other than CAD (although not necessarily more advanced). Some fundamental understanding of the computer's operating system is needed. The operating system is used to manipulate and store files.

Many different types of files or groups of files are involved in using the ATTEXT command.

EXE and *DLL:* AutoCAD program files are being used to perform the attribute extraction.

DWG: The drawing file is a data file that contains the block references and their associated attributes whose values are eligible for extraction.

TXT: A template file must be created to tell AutoCAD what type of data to extract. A line editor or word processor is used to create the TXT template file.

TXT or *DXF:* The extraction process creates a *FILENAME.TXT* file containing the data in accordance with the instructions received from the template file.

A set of database program files usually can operate on the extracted file.

Database

Extracted data can be manipulated by a database application program. The telephone directory, a database, is an alphabetical listing of names, each followed by a first name (or initial), an address, and a phone number. A listing of pipe, valves, and fittings in a piping system can be a database if each item has essential data associated with it, such as its size, flange rating, weight, material of manufacture, product that it handles, cost, and location within the system, among many others.

The two elementary terms used in a database are the record and the field. A record is like a single listing in the phone book, made up of a name and its associated first name, address, and phone number. The name Jones with its data is one record. Another Jones with a different first name (or initials) is another record. Another Jones with the same first name or initials at a different address is still another record. Each listing is a record. The types of data that may be in a record come under the heading of a field. Name is a field. All the names in the list come under the name field. Address is a field. Phone number is a field. And even though some names may have first names and some may not, first name is a field. It is possible to take the telephone directory that is listed alphabetically by name, feed it into a computer database program, and generate the same list in numerical order by phone number. You can generate a partial list of all the Jones's sorted alphabetically by the first name. Or you can generate a list of everyone who lives on Elm Street. The primary purpose of the ATTEXT command is to generate the main list that includes all of the desired objects to which the database program manipulations can be applied.

Creating a Template

The template is a file saved in ASCII format; it lists the fields that specify the tags and determine which block references will have their attribute data extracted. The template must be a file on an accessible path with the extension of .TXT. When you use a text editor or word processor in the ASCII mode, you must add this extension when you name the file.

Field Name

The field name must correspond exactly to the attribute tag if you wish for that attribute's value to be extracted. If the attribute tag is called type, then there must be a field name in the template called type. A field name in the template called ratings will not cause a tag name called rating to have its values written to the extract file.

Character-Numeric

The template tells the ATTEXT command to classify the data written to a particular field either as numeric or as character type. Characters (a, b, c, A, #, ", etc.) are always character type, but numbers do not always have to be numeric type. Characters occupy less memory space in the computer. Sometimes numbers, like an address and phone number, are better stored as characters unless they are to be operated on by mathematical functions (addition, subtraction, etc.). The only relative significance of numbers as characters is their order (1 2 3...) for sorting purposes by the database program. Characters (a b c...) have that same significance. The template contains two elements for each field. The first is the field name; the second is the field format code.

Numbers in strategic spaces in the field format code of the template file specify the number of spaces to allow for the values to be written in the extract file. Others also specify how many decimal places to carry numeric values.

The format is as follows:

fieldname	Nwwwddd	for numeric values with decimal allowance
fieldname	Nwww000	for numeric values without decimal allowance
fieldname	Cwww000	for character values

Each line in the template represents one field whose first set of characters is its field name. It could be one of the BLOCK characteristics such as BL:NAME for block name or BL:X for the X coordinate of its insertion point or it could be the Attribute tag of one of its attributes. The second set of characters (always 7 in length) is the field's field format code. The format code begins with a C or an N. The C means that the field is a character type. The N means that the field is a number type. The w's and d's are to be filled in with the necessary digits when the template file is created. Whether the field is a number (first character in the field format code is N) or a character (first character in the field format code is C), the next three characters will be the three digits of a number that establishes how many spaces in the output file will be allocated for the value of the attribute. For example, if value of the *TYPE* field might be PIPE, FLANGE, TEE, or REDUCER, then the three digits should be no less than 007, allowing seven places for the longest possible value; REDUCER. If the value of the *TYPE* field might be CONCENTRIC REDUCER, then the three digits should be no less than 018, allowing eighteen places (including one space) for the value CONCENTRIC REDUCER. If the value of the *WELD* field might be 62.44, the three digits should be 005, allowing for the four digits and the decimal point. The last three digits in the field format code determine how many decimal places to use in the value. For the value of 62.44, the last three digits should be 002. If the field is of the Character type, there is no need to establish a number for decimal places, so the last three digits will always be 000. The order of fields listed in the template does not have to coincide with the order in which attribute tags appear in a block reference. Any group of fields, in any order, is acceptable. The only requirement for a block reference to be eligible for extraction by the ATTEXT command is that there be at least one tag–field match.

A template for the example in Figure 10–43 would be written as follows:

```
TYPE          C008000
SIZE          C008000
RATING        C006000
BLTQUAN       N004000
BLTSIZE       C010000
GSKETMATL     C006000
GSKETSIZE     C006000
WELD          N006002
```

Note: Word processors add coded characters (often hidden) to files unless specifically set up to write in the ASCII format. These coded characters are not acceptable in the template. Be sure that the text editor or word processor being used is in the proper mode. Also, do not use the TAB key to line up the second column elements, but key in the necessary spaces, because the TAB key involves coded characters.

From the extracted file in Table 10–1, a database program can generate a sorted list whose items correspond to selected values. The procedure (again, to explain would take another book entirely) might list only 3" flanges, or all 3" fittings, or all reducers, or any combination of available records required corresponding to tag–field association.

BL:xxxxxx Nonattribute Fields Available

If necessary, additional data is also automatically stored with each inserted block reference. The suggested format in a template file and a description of each (in parentheses) follows:

Field Name	Field Format	Description
BL:LEVEL	Nwww000	(block nesting level)
BL:NAME	Cwww000	(block name)
BL:X	Nwwwddd	(*X* coordinate of block insertion point)
BL:Y	Nwwwddd	(*Y* coordinate)
BL:Z	Nwwwddd	(*Z* coordinate)
BL:NUMBER	Nwww000	(block counter; same for all members of a MINSERT)
BL:HANDLE	Cwww000	(block's handle; same for all members of a MINSERT)
BL:LAYER	Cwww000	(block insertion layer name)
BL:ORIENT	Nwwwddd	(block rotation angle)
BL:XSCALE	Nwwwddd	(*X* scale factor of block)
BL:YSCALE	Nwwwddd	(*Y* scale factor)
BL:ZSCALE	Nwwwddd	(*Z* scale factor)
BL:XEXTRUDE	Nwwwddd	(*X* component of block's extrusion direction)
BL:YEXTRUDE	Nwwwddd	(*Y* component)
BL:ZEXTRUDE	Nwwwddd	(*Z* component)

Note: The comments in parentheses are for your information and must not be included in the template file.

Repeating the earlier explanation, the first column lists the name of each field, for example: BL:ORIENT. The field format begins with a C or an N, denoting character or numeric data, respectively. The next three digits denote the width of the field; that is, how many spaces are to be allowed in the extract file for values to be written under this particular field. If the value to be written under this field for any record is too long for the width allowed, AutoCAD truncates the data written, proceeds with the extraction, and displays the following error message:

**Field overflow in record <record number>

The last three digits in the field format denote the number of decimal places to which numeric values will be written. Character fields should have zeros in these three places.

When fields are specified as numeric, the attribute values must be numbers, or AutoCAD will display an error message.

You can use one of the two commands available to extract attributes; EATTEXT and ATTEXT.

To extract attribute objects through the Attributes Extraction Wizard, use the EATTEXT command.

Invoke the EATTEXT command:

Modify II menu	Choose Attribute Extract
Command: prompt	**eattext** (ENTER)

AutoCAD displays the Attribute Extraction Wizard dialog box, similar to Figure 10–45.

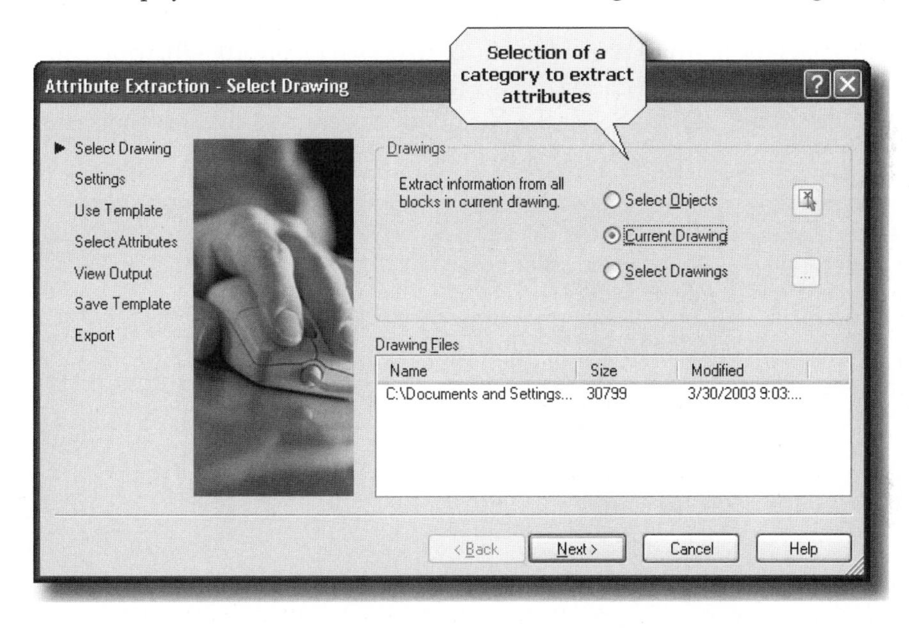

Figure 10–45 *Attribute Extraction Wizard dialog box displaying the Select Drawing page*

The Attribute Extraction Wizard dialog box has seven pages; **Select Drawing, Settings, Use Templates, Select Attributes, View Output, Save Template,** and **Export.**

The **Select Drawing** page lets you select a drawing file from which to extract information from Block Attributes or lets you specify blocks in the current drawing. In the **Drawings** section there are three radio buttons: **Select Objects, Current Drawing,** and **Select Drawings.** Choosing the **Select Objects** radio button lets you select specific blocks and Xrefs in the current drawing for extracting attribute data. The name of the current drawing is listed in the **Drawing Files** section. First, select the button to the right of the **Select Objects** radio button. AutoCAD temporarily returns to the drawing display so that you can select objects. Choosing the **Current Drawing** radio button causes all blocks and xrefs to be selected from the current drawing for extracting attribute data. The name of the current drawing is listed in the **Drawing Files** section. Choosing the **Select Drawings** radio button lets you select multiple drawing files for extracting attribute data. First, select the button to the right of the **Select Drawings** radio button. AutoCAD displays the **Select Files** dialog box. This is similar to other Windows file management dialog boxes. Specifiy the desired drawing(s) and choose **Open.** The drawing(s) will be added to the **Drawing Files** list on the **Select Drawing** page of the dialog box.

After selecting the desired objects or drawing(s), choose the **Next** button, and AutoCAD displays the **Settings** page of the dialog box (see Figure 10–46), which includes two check boxes that when checked cause the block selection process to **Include xrefs** and **Include nested blocks**. Select the **Next** button again, and the **Use Template** page of the dialog box is displayed similar to Figure 10–47.

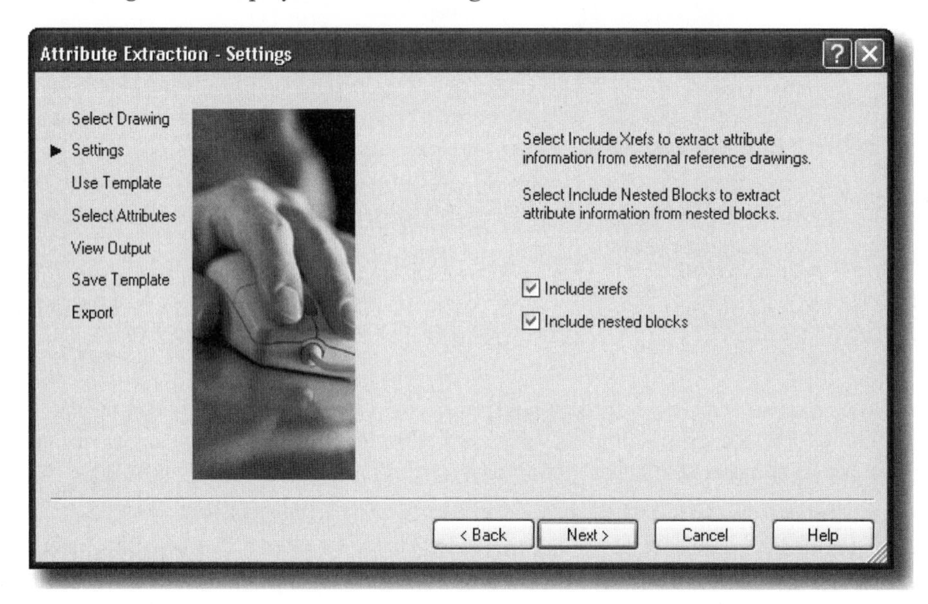

Figure 10–46 *Attribute Extraction Wizard dialog box displaying the Settings page*

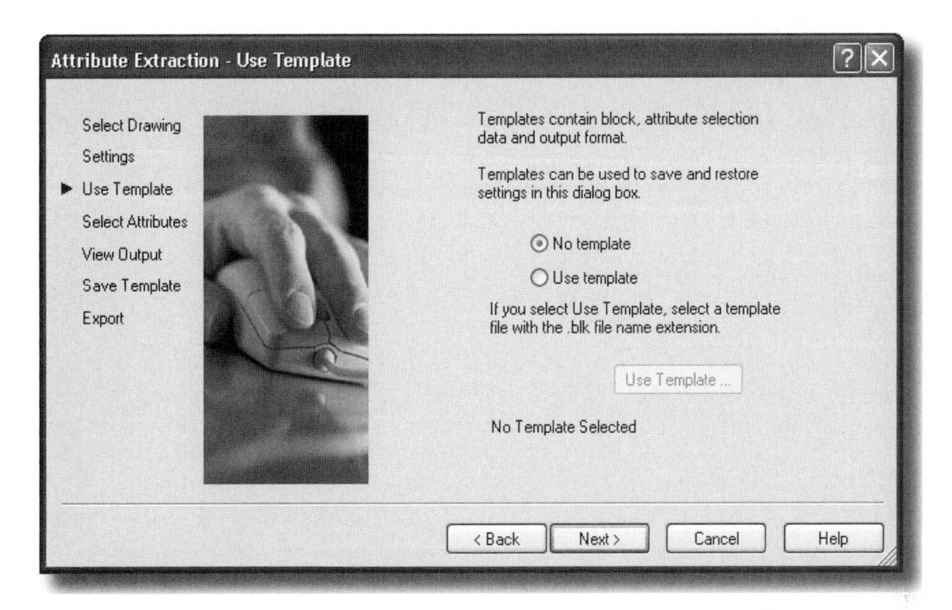

Figure 10–47 *Attribute Extraction Wizard dialog box displaying the Use Template page*

Note: Do not confuse the Attribute Extraction Template file having a file name extension of .txt with the Block Template file having a file extension of .blk. The .txt file is created in a text editor and saved with the extension of .txt. It is then used with the ATTEXT command when prompted to name a template file. The .blk may be created during the EATTEXT command and can be recalled for use from the Attribute Extraction Wizard dialog box Use Template page. Both files are used to determine what data is extracted from attributes. However, the use of the .blk file from the Wizard is more interactive during the extraction process. Also, note that the Wizard can be used without creating a template file.

The two options on the **Use Template** page are the **No template** and the **Use template** radio buttons. If you select the **Use template** radio button, the **Use Template...** button becomes active. When it is selected, AutoCAD displays a **File Open** dialog box from which you can select a template. The template file must have a .blk file name extension, Aand it must have been created during an attribute extraction process and saved while using the Attribute Extraction Wizard. After you exit the **Use Template** page of the Attribute Extraction Wizard dialog box, AutoCAD displays the **Select Attributes** page of the dialog box similar to Figure 10–48.

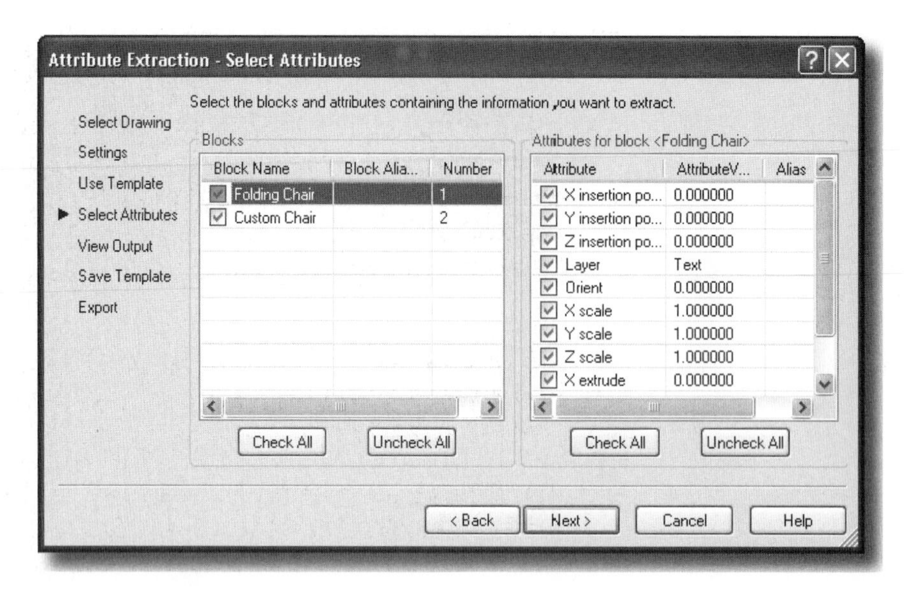

Figure 10–48 *Attribute Extraction Wizard dialog box displaying the Select Attributes page*

From the **Select Attributes** page, you can select a block from the **Blocks** section list and then select the attributes for that block reference from the **Attributes for block** section to be extracted to a data base file. Choose **Next,** and AutoCAD displays the **View Output** page, which offers two views of each block and the attribute data associated with it (see Figure 10–49).

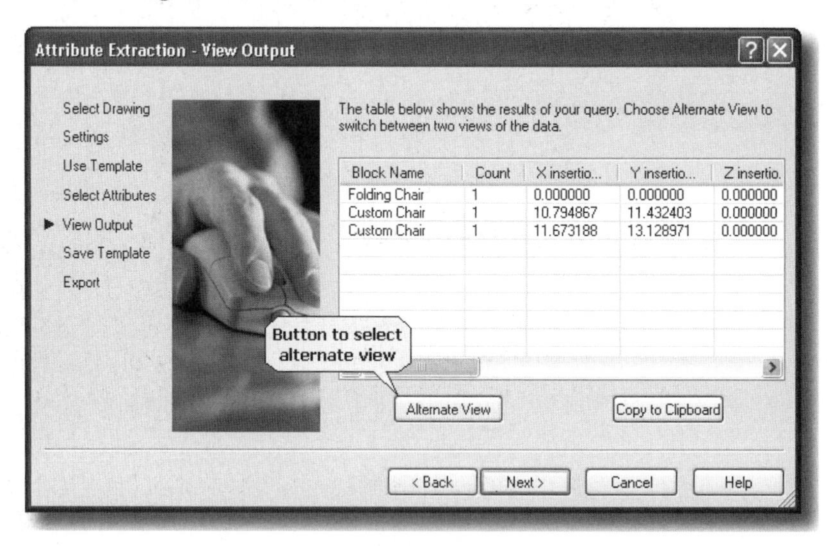

Figure 10–49 *Attribute Extraction Wizard dialog box displaying the View Output page*

Choose **Next,** and AutoCAD displays the **Save Template** page of the dialog box. Selecting the **Save Template** button causes a **Save As** dialog box to be displayed, letting you specify a name and location of a template file with a .blk extension to be created. This file can be recalled in a later attribute extraction procedure, using the same data extraction input. Choose **Next,** and AutoCAD displays the **Export** page of the dialog box similar to Figure 10–50. From this page you can specify a file name, file extension and location for the file containing the extracted data.

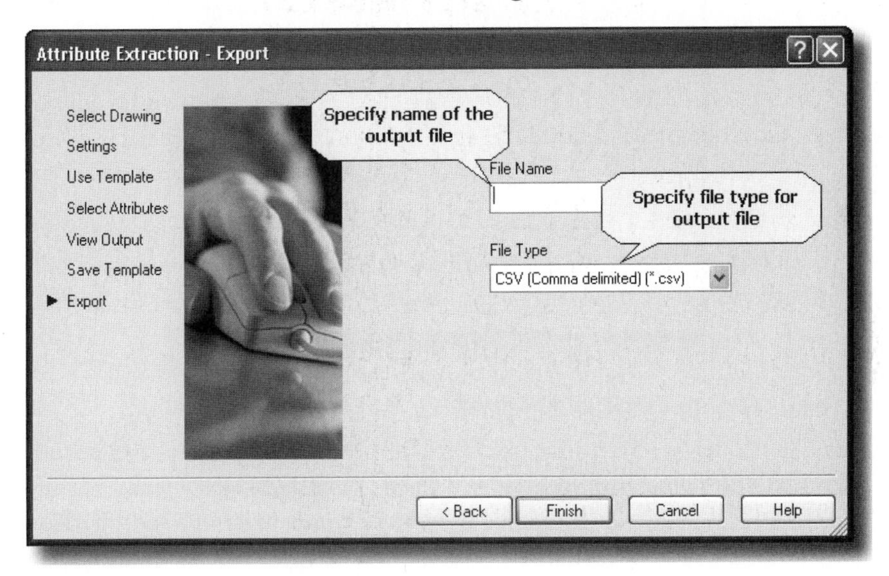

Figure 10–50 *Attribute Extraction Wizard dialog box displaying the Export page*

To extract attribute objects through the attributes extraction dialog box, use the ATTEXT command. To extract objects at the "Command:" line, use the ATTEXT command.

Invoke the ATTEXT command:

Command: prompt	**attext** (ENTER)

AutoCAD displays the Attribute Extraction dialog box, similar to Figure 10–51.

Choose one of the three radio buttons available in the **File Format** section of the Attribute Extraction dialog box.

The **Comma Delimited File (CDF)** format generates a file containing no more than one record for each block reference in the drawing. The values written under the fields in the extract files are separated by commas, with the character fields enclosed in single quotes.

The **Space Delimited File (SDF)** format writes the values lined up in the widths allowed. It is possible for adjacent mixed fields (characters and numeric) not to have spaces between them. It may be necessary to add dummy fields to provide spaces in these cases.

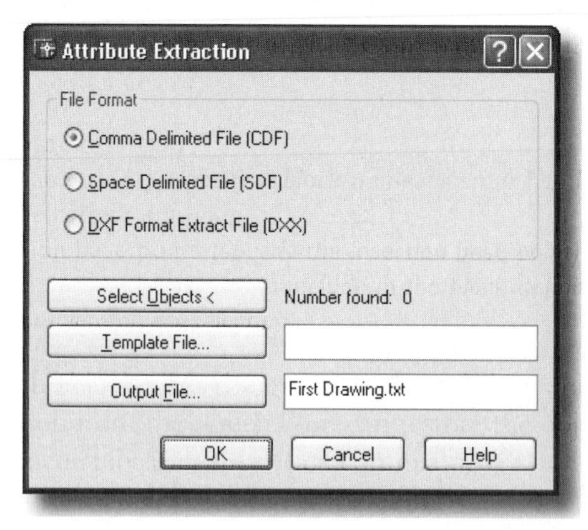

Figure 10–51 *Attribute Extraction dialog box*

The AutoCAD Drawing Interchange File format is called DXF. Unlike the DXFOUT command, extraction files generated by the ATTEXT command contain only block reference, attribute, and end-of-sequence objects.

Specify the template filename in the **Template File** edit box if you have selected either CDF or SDF file formats. AutoCAD appends the .TXT file type. If you choose the DXF file format, the **Template File...** button and edit box are disabled.

Specify the output extract file name in the **Output File...** text box. AutoCAD appends the .TXT file type for CDF and SDF file formats and DXX for DXF files. DOS permits a file name of con, for writing to the screen, or a file name of prn, for writing to a printer. In this case, a printer must be connected and ready to print.

Choose the **Select Objects** < button to select the objects to be included in the extraction of attributes. The dialog box will disappear, and AutoCAD allows you to select the attribute entities. After you have done so, the dialog box reappears, and the number of objects you selected is displayed after the **Number found:** label.

Choose the **OK** button to close the dialog box and extract the attributes from the selected block references. The extracted information is stored in the output file specified in the Attribute Extraction dialog box. You can import the output file into a database program or open the output file with the help of a text editor and analyze the extracted information.

As mentioned earlier, you can invoke the ATTEXT command preceded with a hyphen as shown in the following command sequence:

```
Command: -attext
Enter extraction type or enable object selection [Cdf/Sdf/Dxf/Objects]
  <C>:
```

Choose one of the three options: CDF, SDF, or DXF. You can also choose the Objects option, which allows you to select the objects for extraction of their attributes. Once selection is complete, the prompt reverts to the CDF, SDF, or DXF prompt without the Objects option. Then AutoCAD prompts for the template file name and extract file name, respectively.

Duplications

The example extract file listed some records whose values in every field were the same. This would serve no purpose in a telephone directory but in a list of objects in a drawing it is possible that the only difference between two objects is their location in the drawing. In a bill of materials, the purchasing agent is not concerned with where, but how many duplicate objects there are. If it were essential to distinguish every object, the BL:X, BL:Y, and BL:Z fields could be included in the template to identify each object. Multiple insertions of the same block in the same location could still be distinguished by their BL:HANDLE field if the handles system variable were set to ON, or by their BL:NUMBER if they were inserted with the MINSERT command.

Also, to the experienced pipe estimator (or astute novice) there are other duplications that are possible if not taken into account. Counting bolts and weld lengths for every fitting and valve could result in twice the quantities required. Mating flanges, each having 4 bolt holes, only require 4 bolts to assemble. An ell welded to a tee likewise requires the specified weld length only once. Therefore, Attribute value quantities associated with corresponding tags should take this and similar problems in mating assemblies into account.

If the same fittings (in the form of block references with attributes) are shown in more than one view in a drawing, some mechanism should be provided to prevent duplication of quantities. As you can see, advanced features (attributes) that provide solutions to complex problems (data extraction) often require carefully planned implementation.

Changing Objects in a Block Reference Without Losing Attribute Values

Sometimes it might be desirable to make changes to the objects within block references that have been inserted with attributes and have had values assigned to the attributes. Remember that the values of the attributes can be edited with the ATTEDIT command. But, in order to change the geometry of a block reference, it is normally required that you explode the block reference, make the necessary changes, and then redefine the block. As long as the attribute definitions keep the same tags in the new definition, the redefined blocks that have already been inserted in the drawing will retain those attributes with the original definitions.

Another method of having a new block definition applied to existing block references is to create a new drawing with the new block definition. This can be done by using the WBLOCK command to create a drawing that conforms to the old definition and then to edit that drawing. Or you can start a new drawing with the name of the block that you wish to change.

In order to apply the new definition to the existing block reference, you can call up the drawing from the OPEN command, and then use the INSERT command with the Block Name: option. For example, if you wish to change the geometry of a block named Part_1, and the changes have been made and stored in a separate drawing with the same name (Part_1), the sequence of prompts is as follows:

Command: **-insert**
Enter block name or [?] <x>: **part_1=**
Block "part_1" already exists. Redefine it? [Yes/No] <N>: **y**
Block "Part_1" redefined
Regenerating model.
Specify insertion point or [Scale/X/Y/Z/Rotate/PScale/PX/PY/PZ/PRotate]: (ENTER)

It is not necessary to specify an insertion point or to respond to scale factor and rotation angles. Simply inserting the block reference with the = (equals) behind the name will cause the definition of the drawing to become the new definition of the block with the same name residing in the current drawing.

REDEFINING A BLOCK AND ATTRIBUTES

The ATTREDEF command allows you to redefine a block reference and updates associated attributes. Invoke the ATTREDEF command:

Command: prompt	**attredef** (ENTER)

AutoCAD prompts:

Command: **attredef**
Enter name of the block you wish to redefine: *(specify the block name to redefine)*
Select objects for new Block...
Select objects: *(select objects for the block to redefine and press* ENTER*)*
Specify insertion base point of new Block: *(specify the insertion basepoint of the new block)*

New attributes assigned to existing block references are given their default values. Old attributes in the new block definition retain their old values. Old attributes not included in the new block definition are deleted from the old block references.

BLOCK ATTRIBUTE MANAGER

The BATTMAN command provides a means of managing Blocks that contain Attributes.

Invoke the BATTMAN command

Modify menu	Object > Attribute > Block Attribute Manager
Command: prompt	**battman** (ENTER)

AutoCAD displays the **Block Attribute Manager** dialog box, similar to Figure 10–52.

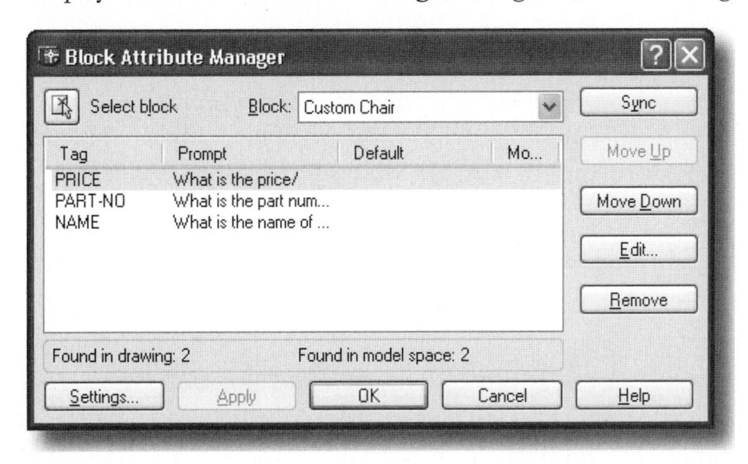

Figure 10–52 *Block Attribute Manager dialog box*

The BATTMAN command lets you edit the attribute definitions in blocks, change the order in which you are prompted for attribute values when inserting a block, and remove attributes from blocks. AutoCAD displays attributes of the selected block in the attribute list of the Block Attribute Manager dialog box. By default, the Tag, Prompt, Default, and Modes attribute properties are shown in the attribute list. You can specify which attribute properties you want displayed in the list by choosing Settings. The number of instances of the selected block is shown in a description below the attribute list.

Select Block

When you select the **Select block** button, the dialog box closes to allow you to use your pointing device to select a block from the drawing area.

Block

All block definitions in the current drawing that have attributes are listed in the **Block** text box, from which you can select the block whose attributes you want to modify.

Sync

When you select the **Sync** button, AutoCAD updates all instances of the selected block with the attribute properties currently defined.

Move Up

When the Block contains multiple attributes, selecting the **Move Up** button causes the selected attribute tag to move up earlier in the prompt sequence.

Move Down

When the Block contains multiple attributes, selecting the **Move Down** button causes the selected attribute tag to move down layer in the prompt sequence.

Edit

Selecting the **Edit…** button causes AutoCAD to display the Edit Attribute dialog box (see Figure 10-53) where you can modify attribute properties.

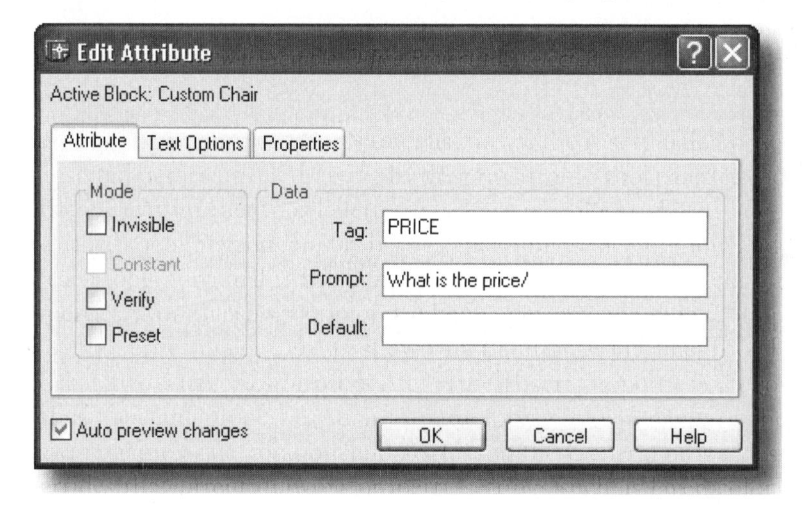

Figure 10–53 *Edit Attribute dialog box with the Attribute tab selected*

Note: This Edit Attribute dialog box is different from the Edit Attributes dialog box that is displayed in response to the ATTEDIT command.

The Edit Attribute dialog box lets you define whether or not the assigned value is visible in the drawing area, and sets the string that prompts users to enter a value. There are three tabs to choose from in the Edit Attribute dialog box; the **Attribute** tab, the **Text Options** tab, and the **Properties** tab.

Attribute Tab

The **Mode** section of the **Attribute** tab has check boxes for setting the **Invisible, Constant, Verify** and **Preset** modes. In the **Data** section there are text boxes in which you can set the tag name, prompt and default. If **Auto preview changes** check box is checked, changes to attributes are immediately visible.

Text Options Tab

The **Text Options** tab (see Figure 10–54) has text boxes to set the **Text Style, Justification, Height, Rotation, Width Factor,** and **Oblique Angle** of the Attribute text. There are check boxes for displaying the text **Backwards** or **Upside down.**

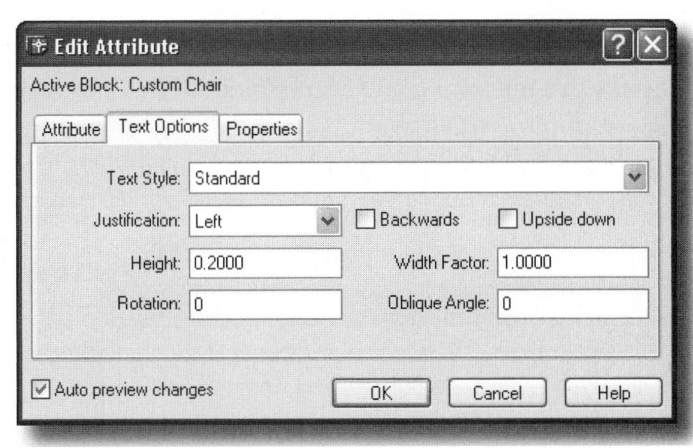

Figure 10–54 *Edit Attribute dialog box with the Text Option tab selected*

Properties Tab

The **Properties** tab (see Figure 10–55) has text boxes to set the layer that the attribute is on and the color, lineweight, and linetype for the selected attribute. If the drawing uses plot styles, you can assign a plot style to the attribute.

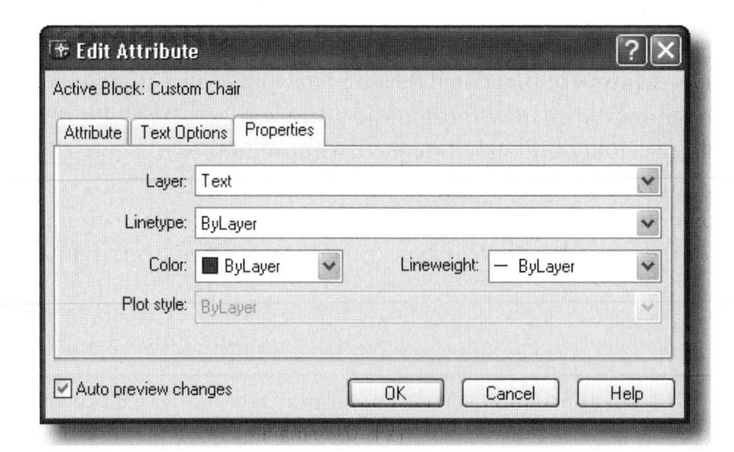

Figure 10–55 *Edit Attribute dialog box with the Properties tab selected*

Remove

The **Remove** button in the Block Attribute Manager dialog box (see Figure 10–52) removes the selected attribute from the block definition. The **Remove** button is not available for blocks with only one attribute.

Settings

Selecting the **Settings…** button causes AutoCAD to display the Settings dialog box (see Figure 10–56) where you can modify attribute properties.

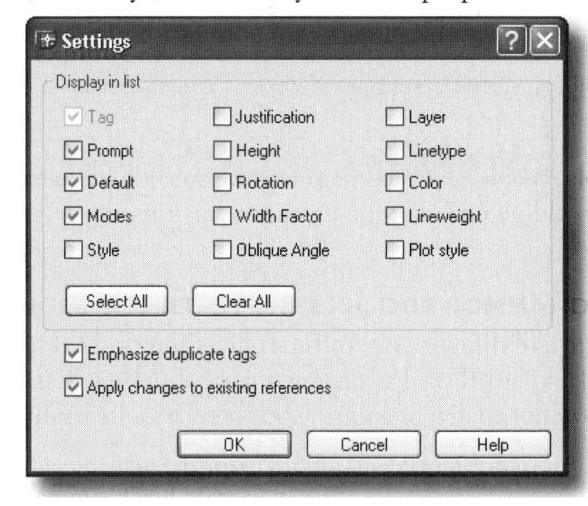

Figure 10–56 *Settings dialog box*

Display in List

There are check boxes in the **Display in list** section that, when checked, cause the checked properties to be displayed in the list in the Block Attribute Manager dialog box. The **Tag** property check box is always checked.

Select All

Selecting the **Select All** button causes all properties' check boxes to be checked.

Clear All

Selecting the **Clear All** button causes all properties' check boxes to be cleared.

Emphasize Duplicate Tags

The **Emphasize duplicate tags** check box turns duplicate tag emphasis on and off. If this option is selected, duplicate attribute tags are displayed in red type in the attribute list. If this option is cleared, duplicate tags are not emphasized in the attribute list.

Apply Changes to Existing References

The **Apply changes to existing references** check box specifies whether or not to update all existing instances of the block whose attributes you are modifying. If selected, updates all instances of the block with the new attribute definitions. If cleared, updates only new instances of the block with the new attribute definitions.

Apply

The **Apply** button in the Block Attribute Manager dialog box updates the drawing with the attribute changes you have made and leaves the Block Attribute Manager open.

DIVIDING OBJECTS

The DIVIDE command causes AutoCAD to divide an object into equal-length segments, placing markers at the dividing points. Objects eligible for application of the DIVIDE command are the line, arc, circle, ellipse, spline, and polyline. Selecting an object other than one of these will cause an error message to appear, and you will be returned to the "Command:" prompt.

Invoke the DIVIDE command:

Draw menu	Point > Divide
Command: prompt	**divide** (ENTER)

AutoCAD prompts:

Command: **divide**
Select the object to divide: *(select a line, arc, circle, ellipse, spline, or polyline)*
Enter the number of segments or [Block]: *(specify the number of segments)*

You may respond with an integer from 2 to 32767, causing points to be placed along the selected object at equal distances but not actually separating the object. The Object Snap NODe can snap at the divided points. Logically, there will be one less point placed than the number entered, except in the case of a circle. The circle will have the first point placed at the angle from the center of the current snap rotation angle. A closed polyline will have the first point placed at the first point drawn in the polyline. The total length of the polyline will be divided into the number of segments entered without regard to the length of the individual segments that make up the polyline. An example of a closed polyline is shown in Figure 10–57.

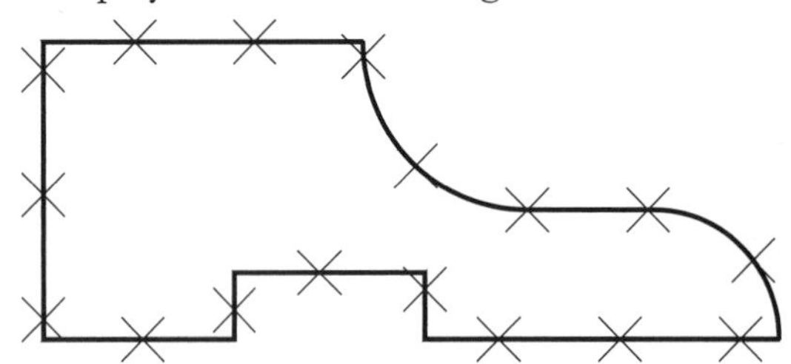

Figure 10–57 *The* DIVIDE *command as used with a closed polyline*

Note: It is advisable to set the PDSIZE and PDMODE system variables to values that will cause the points to be visible.

The Block option allows a named block reference to be placed at the dividing points instead of a point. The sequence of prompts is as follows:

Command: **divide**
Select object to divide: *(select a line, arc, circle, ellipse, spline, or polyline)*
Enter the number of segments or [Block]: **block** *(or just* **b***)*
Enter name of block to insert: *(enter the name of the block)*
Align block with object? [Yes/No] <Y>: *(press* ENTER *to align the block reference with the object, or enter* **n***, for not to align with the block reference)*
Enter the number of segments: *(specify the number of segments)*

If you respond with **No** or **N** to the "Align block with object?" prompt, all of the block references inserted will have a zero angle of rotation. If you default to **Yes**, the angle of rotation of each inserted block reference will correspond to the direction of the linear part of the object at its point of insertion or to the direction of a line tangent to a circular part of an object at the point of insertion.

MEASURING OBJECTS

The MEASURE command causes AutoCAD to divide an object into specified-length segments, placing markers at the measured points. Objects eligible for application of the MEASURE command are the line, arc, circle, ellipse, spline, and polyline. Selecting an object other than one of these will cause an error message to appear and you will be returned to the "Command:" prompt.

Invoke the MEASURE command:

Draw menu	Point > Measure
Command: prompt	**measure** (ENTER)

AutoCAD prompts:

Command: **measure**
Select object to measure: *(select a line, arc, circle, ellipse, spline, or polyline)*
Specify length of segment or [Block]: *(specify the length of the segment, or enter* **b** *for the Block option)*

If you reply with a distance, or show AutoCAD a distance by specifying two points, the object is measured into segments of the specified length, beginning with the closest endpoint from the selected point on the object. The Block option allows a named block reference to be placed at the measured point instead of a point. The sequence of prompts is as follows:

Command: **measure**
Select object to measure: *(select a line, arc, circle, ellipse, spline, or polyline)*
Specify length of segment or [Block]: **block** *(or just* **b***)*
Enter name of block to insert: *(enter the name of the block)*
Align block with object? [Yes/No] <Y>: *(press* ENTER *to align the block reference with the object, or enter* **n***, for not to align with the block reference)*
Specify length of segment: *(specify the length of segments)*

Open the Exercise Manual PDF file for Chapter 10 on the accompanying CD for project and discipline specific exercises.

If you have the accompanying Exercise Manual, refer to Chapter 10 for project and discipline specific exercises.

REVIEW QUESTIONS

1. If you used "BL??7* as the pattern for listing blocks, which of the following block names would be listed?

   a. BL7

   b. BLOCK739

   c. BLUE727

   d. BLACK

   e. none of the above

2. It is possible to force invisible attributes to display on a drawing.

   a. True

   b. False

3. Which of the following forms of attribute extraction require template files?

   a. CDF

   b. SDF

   c. DXF

   d. a and b

   e. none of the above

4. The maximum number of characters for a block name is:

   a. 8

   b. 16

   c. 23

   d. 31

   e. 255

5. A block is:

   a. a rectangular-shaped figure available for insertion into a drawing

   b. a single element found in a block formation of a building drawn with AutoCAD

   c. one or more objects stored as a single object for later retrieval and insertion

   d. none of the above

6. A block reference cannot be exploded if it:
   a. consists of other blocks (nested)
   b. has a negative scale factor
   c. has been moved
   d. has different *X* and *Y* scale factors
   e. none of the above

7. A new drawing may be created from an existing block by using the command:
   a. WBLOCK
   b. NEW
   c. BLOCK=
   d. SAVEAS
   e. none of the above

8. All of the following can be exploded, except:
   a. block references
   b. associative dimensions
   c. polylines
   d. Block references inserted with MINSERT command
   e. none of the above

9. To insert a block reference called "TABLE" and have the block reference converted to an individual object as it is inserted, what should you use for the block name?
   a. /TABLE
   b. *TABLE
   c. ?TABLE
   d. TABLE
   e. none of the above

10. The command used to write all or just part of a drawing out to a new drawing file is:
    a. SAVEAS
    b. BLOCK
    c. DXFOUT
    d. FILE
    e. WBLOCK

11. To return a block reference back to its original objects, use:
    a. EXPLODE
    b. BREAK
    c. CHANGE
    d. UNDO
    e. STRETCH

12. To identify a new insertion point for a drawing file which will be inserted into another drawing, invoke the:

    a. BASE command
    b. INSERT command
    c. BLOCK command
    d. WBLOCK command
    e. DEFINE command

13. MINSERT places multiple copies of an existing block similar to the command:

    a. ARRAY
    b. MOVE
    c. INSERT
    d. COPY
    e. MIRROR

14. If one drawing is to be inserted into another drawing and editing operations are to be performed on the inserted drawing, you must first:

    a. use the PEDIT command
    b. EXPLODE the inserted drawing
    c. UNDO the inserted drawing
    d. nothing, it can be edited directly
    e. none of the above

15. The BASE command:

    a. can be used to move a block reference
    b. is a subcommand of PEDIT
    c. will accept 3D coordinates
    d. allows one to move a dimension baseline
    e. none of the above

16. Attributes are associated with:

    a. objects
    b. block references
    c. text
    d. layers
    e. shapes

17. To merge two drawings, use:

    a. INSERT
    b. MERGE
    c. BIND
    d. BLOCK
    e. IGESIN

18. The WBLOCK command can be used to create a new block for use:
    a. in the current drawing
    b. in an existing drawing
    c. in any drawing
    d. only in a saved drawing
    e. none of the above
19. The DIVIDE command causes AutoCAD to:
    a. divide an object into equal length segments
    b. divide an object into two equal parts
    c. break an object into two objects
    d. all of the above
20. One cannot explode:
    a. polylines containing arcs
    b. blocks containing polylines
    c. dimensions incorporating leaders
    d. block references inserted with different *X*, *Y*, and *Z* scale factors
    e. none of the above
21. The DIVIDE command will:
    a. place points along a line, arc, polyline, or circle
    b. accept 1.5 as segment input
    c. place markers on the selected object and separate it into different segments
    d. divide any object into the equal number of segments
22. When using the -WBLOCK command to create a new block, one should respond to the block name with:
    a. a block name
    b. an equals sign
    c. an asterisk
    d. a file name
    e. none of the above

23. The WBLOCK command creates:
    a. a drawing file
    b. a collection of blocks
    c. a symbol library
    d. an object file
    e. none of the above

24. The MEASURE command causes AutoCAD to divide an object
    a. into specified length segments
    b. into equal length segments
    c. into two equal parts
    d. all of the above

25. A command used to edit attributes is:
    a. DDATTE
    b. EDIT
    c. EDITATT
    d. ATTFILE
    e. none of the above

26. The WBLOCK command:
    a. means "Window Block" and allows for the use of a window to define a block
    b. allows you to send a previously defined block to a file, thus creating a drawing file of the block
    c. is used in lieu of the INSERT command when the block is stored in a separate disk file
    d. none of the above

27. Attributes are defined as the:
    a. database information displayed as a result of entering the LIST command
    b. *X* and *Y* values which can be entered when inserting a block reference
    c. coordinate information of each vertex found along a SPLINE object
    d. none of the above

External References

INTRODUCTION

One of the most powerful time-saving features of AutoCAD is the ability to have one drawing become part of a second drawing while maintaining the integrity and independence of the first one. AutoCAD lets you display or view the contents of as many as 32,000 other drawing files while working in your current drawing file. This feature is provided by the XREF command, short for external reference.

After completing this chapter, you will be able to do the following:

- Attach and detach reference files
- Change the path for reference files
- Load and unload reference files
- Decide whether to attach or overlay an external reference file
- Clip external reference files
- Control dependent symbols
- Edit external references
- Manage external references
- Use the BIND command to add dependent symbols to the current drawing
- Attach and detach image files

EXTERNAL REFERENCES

Prior to Release 11, existing AutoCAD drawings could be combined in only one way, by means of the INSERT command, to insert one drawing into another. When one drawing is inserted into another, it is actually a duplicate of the inserted drawing that becomes a part of the drawing into which it is inserted. The data from the inserted drawing is added to the data of the current drawing. Once the duplicate of a drawing is inserted, no link or association remains between the original drawing from which the inserted duplicate came and the drawing it has been inserted into.

The INSERT command and the XREF command now give users a choice of two methods for combining existing drawing files. The external reference (XREF) feature does not obsolete the INSERT feature; users can decide which method is more appropriate for the current application.

When a drawing is externally referenced (instead of inserted as a block), the user can view and object snap to the external reference from the current drawing, but each drawing's data is still stored and maintained in a separate drawing file. The only information in the reference drawing that becomes a permanent part of the current drawing is the name of the reference drawing and its folder path. If necessary, externally referenced files can be scaled, moved, copied, mirrored, or rotated by using the AutoCAD modify commands. You can control the visibility, color, and linetype of the layers belonging to an external drawing file. This lets you control which portions of the external drawing file are displayed, and how. No matter how complex an external reference drawing may be, it is treated as a single object by AutoCAD. If you invoke the MOVE command and point to one line, for example, the entire object moves, not just the line you pointed to. You cannot explode the externally referenced drawing. A manipulation performed on an external reference will not affect the original drawing file, because an external reference is only an image, however scaled or rotated.

Borders are an excellent example of drawing files that are useful as external reference files. The objects that make up a border will use considerable space in a file, and commonly amount to around 20,000 bytes. If a border is drawn in each drawing file, this would waste a large amount of disk space when you multiply 20,000 bytes by 100 drawing files. If external reference files are used correctly, they can save 2 MB of disk space in this case.

Accuracy and efficient use of drawing time are other important design benefits that are enhanced through external reference files. When an addition or change is made to a drawing file that is being used as an external reference file, all drawings that use the file as an external reference will reflect the modifications. For example, if you alter the title block of a border, all the drawing files that use that border as an external reference file will automatically display the title block revisions. (Can you imagine accessing 100 drawing files to correct one small detail?) External reference files will save time and ensure the drawing accuracy required to produce a professional product. Figure 11–1 shows a drawing that externally references four other drawings to illustrate the doorbell detail.

There is a limit of 32,000 external references you can add to a drawing. In practice, this represents an unlimited number. If necessary, you can nest them so that loading one external reference automatically causes another external reference to be loaded. When you attach a drawing file as an external reference file, it is permanently attached until it is detached or bound to the current drawing. When you load the drawing with external references, AutoCAD automatically reloads each external reference drawing file; thus, each external drawing file reflects the latest state of the referenced drawing file.

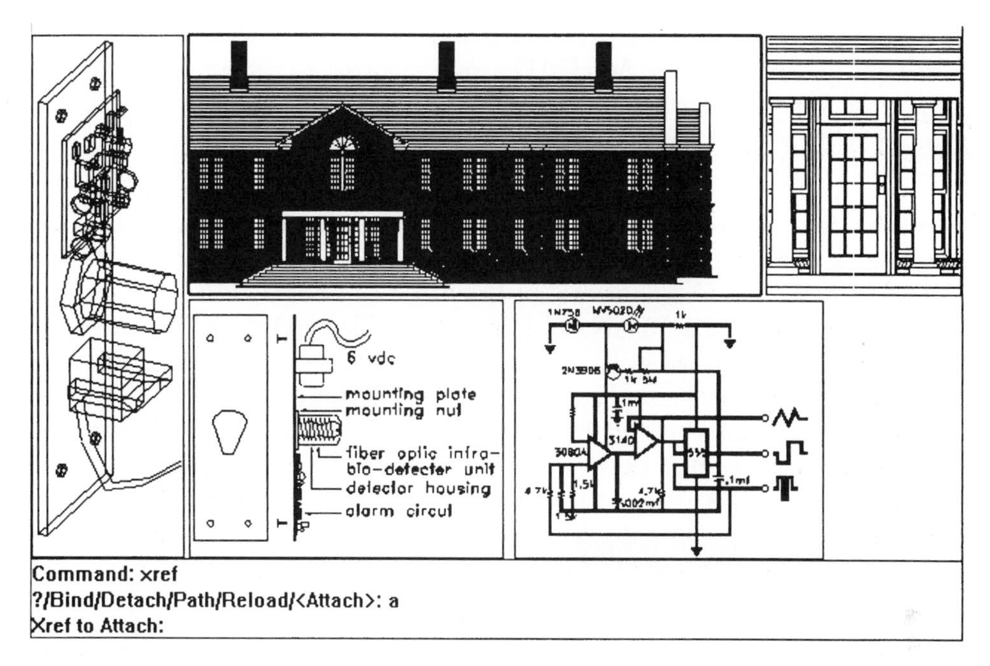

Figure 11–1 *This drawing appears to be a single drawing but references four other drawings*

The XREF command, when combined with the networking capability of AutoCAD, gives the project manager powerful features to cope with the problems of file management. The project manager instantaneously sees the work of the departments and designers working on aspects of the contract. If necessary, you can overlay a drawing where appropriate, track the progress, and maintain document integrity. At the same time, departments need not lose control over individual designs and details.

If you need to make changes to an attached external reference file while you are in the host drawing, you can do so by the REFEDIT command. AutoCAD 2004 also allows you to open an attached external reference in a separate window. You do not need to browse to and open the xref file. With the XOPEN command, the external reference opens immediately in a new window. You can make necessary changes and save the changes. Immediately the changes will be reflected in the host drawing.

In AutoCAD you can control the display of the external reference file by means of clipping, so you can display only a specific section of the reference file.

EXTERNAL REFERENCES AND DEPENDENT SYMBOLS

The symbols that are carried into a drawing by an external reference are called dependent symbols, because they depend on the external file, not on the current drawing, for their characteristics. The symbols have arbitrary names and include blocks, layers, linetypes, text styles, and dimension styles.

When you attach an external reference drawing, AutoCAD automatically renames the xref's dependent symbols. AutoCAD forms a temporary name for each symbol by combining its original name with the name of the xref itself. The two names are separated by the vertical bar (|) character. Renaming the symbols prevents the xref's objects from taking on the characteristics of existing symbols in the drawing.

For example, you created a drawing called PLAN1 with layers 0, First-fl, Dim, and Text, in addition to blocks Arrow and Monument. If you attach the PLAN1 drawing as an external reference file, the layer First-fl will be renamed as PLAN1|first-fl, Dim as PLAN1|dim, and Text as PLAN1|text, as shown in Figure 11–2. Blocks Arrow and Monument will be renamed as PLAN1|Arrow and PLAN1|Monument. The only exceptions to renaming are unambiguous defaults like layer 0 and linetype continuous. The information on layer 0 from the reference file will be placed on the active layer of the current drawing when the drawing is attached as an external reference of the current drawing. It takes on the characteristics of the current drawing.

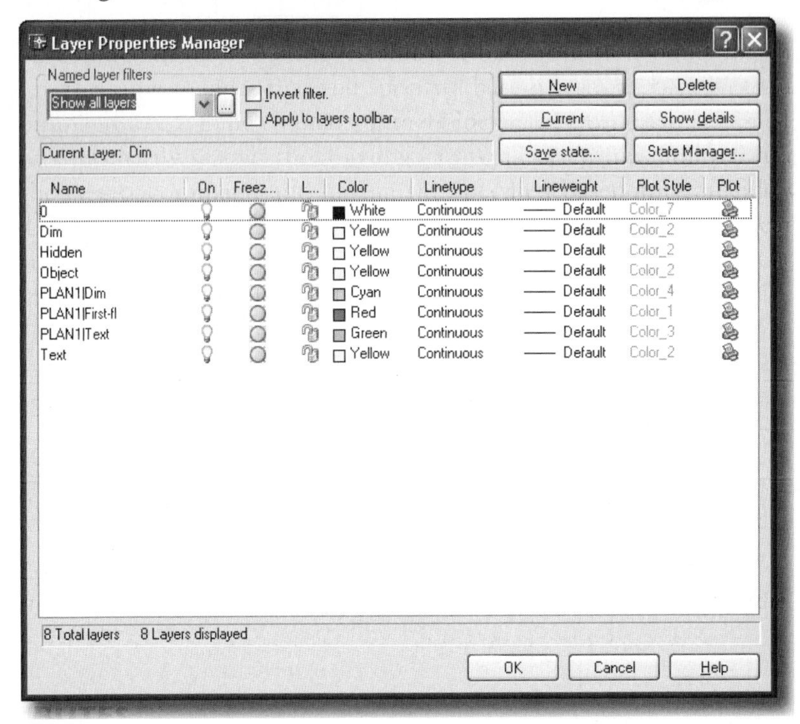

Figure 11–2 *Layer Properties Manager*

This prefixing is carried to nested xrefs. For example, if the external file PLAN1 included an xref named "Title" that has a layer Legend, it would get the symbol name PLAN1|Title|Legend if PLAN1 were attached to another drawing.

This automatic renaming of an xref's dependent symbols has two benefits:

- It allows you to see at a glance which named objects belong to which external reference file.
- It allows dependent symbols to have the same name in both the current drawing and an external reference, and coexist without any conflict.

The AutoCAD commands and dialog boxes for manipulating named objects do not let you select an xref's dependent symbols. Usually, dialog boxes display these entries in lighter text.

For example, you cannot insert a block that belongs to an external reference drawing in your current drawing, nor can you make a dependent layer the current layer and begin creating new objects.

You can control the visibility of the layers (ON/OFF, Freeze/Thaw) of an external reference drawing and, if necessary, you can change the color and linetype. When the VISRETAIN system variable is set to 0 (default), any changes you make to these settings apply only to the current drawing session. They are discarded when you end the drawing. If VISRETAIN is set to 1, then the current drawing visibility, color, and linetype for xref dependent layers take precedence. They are saved with the drawing and are preserved during xref reload operations.

There may be times when you want to make your xref data a permanent part of your current drawing. To make an xref drawing a permanent part of the current drawing, use the Bind option of the XREF command. With the Bind option, all layers and other symbols, including the data, become part of the current drawing. This is similar to inserting a drawing via the INSERT command.

If necessary, you can make dependent symbols such as layers, linetypes, text styles, and dim styles part of the current drawing by using the XBIND command instead of binding the whole drawing. This allows you to work with the symbol just as if you had defined it in the current drawing.

ATTACHING AND MANIPULATING WITH THE XREF COMMAND

The XREF command provides various options for attaching and detaching external references files.

Invoke the XREF command:

Reference toolbar	Choose the External Reference command (see Figure 11–3)
Insert menu	Choose Xref Manager...
Command: prompt	**xref** (ENTER)

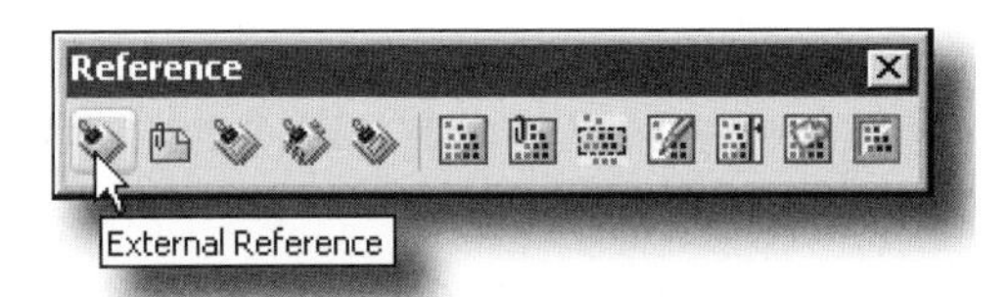

Figure 11–3 *Invoking the* EXTERNAL REFERENCE *command from the Reference toolbar*

AutoCAD displays the Xref Manager dialog box, similar to Figure 11–4.

Note: Type **-xref** and press ENTER at the "Command:" prompt for all the options available at the Command: prompt level for attaching and manipulating external reference drawings. The options are the same ones available in the Xref Manager dialog box.

AutoCAD provides two options for listing attached external reference drawings: the List View and the Tree View. By default, the List View option (see Figure 11–4) displays a list of the attached reference files and their associated data. To sort a column alphabetically, select the column heading. A second click sorts it in reverse order. To resize a column's width, select the separator between columns and drag the pointing device to the right or left.

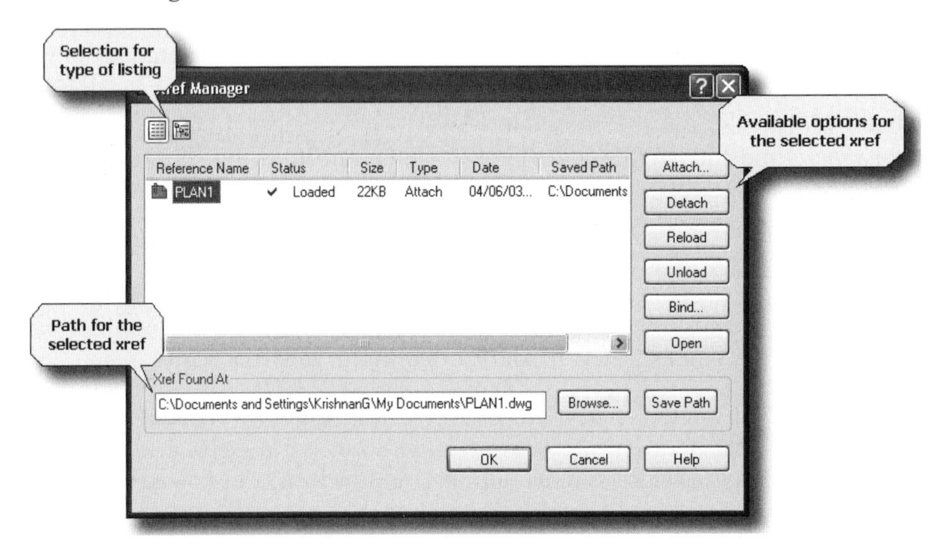

Figure 11–4 *Xref Manager dialog box*

The external reference file's **Reference Name** does not have to be the same as its original file name. To rename the external file, double-click the name or press F2. AutoCAD allows you to rename the file. The new name can contain up to 255 characters, including embedded spaces and punctuation.

The **Status** column displays the state of the external reference file, which can be Loaded, Unloaded, Unreferenced, Unresolved, Orphaned, Reload, or Not found. A detailed discussion of these states is provided later in the chapter.

The **Size** column shows the file size of the corresponding external reference drawing. The size is not displayed if the external reference is unloaded, not found, or unresolved.

The **Type** column indicates whether the external reference is an attachment or an overlay.

The **Date** column displays the last date the associated drawing was modified. The date is not displayed if the external reference is unloaded, not found, or unresolved.

The **Saved Path** column shows the saved path of the associated external reference file.

Selecting any field highlights the external file's reference name.

You can also display the information as a Tree View. To do so, click the **Tree View** button located at the top left of the dialog box, or press F4. To switch back to the List View, press F3.

In a Tree View listing, AutoCAD displays a hierarchical representation of the external reference in alphabetical order, as shown in Figure 11–5. Tree View shows the level of nesting relationship of the attached external references, whether they are attached or overlaid, whether they are loaded, unloaded, marked for reload or unload, or not found, unresolved, or unreferenced.

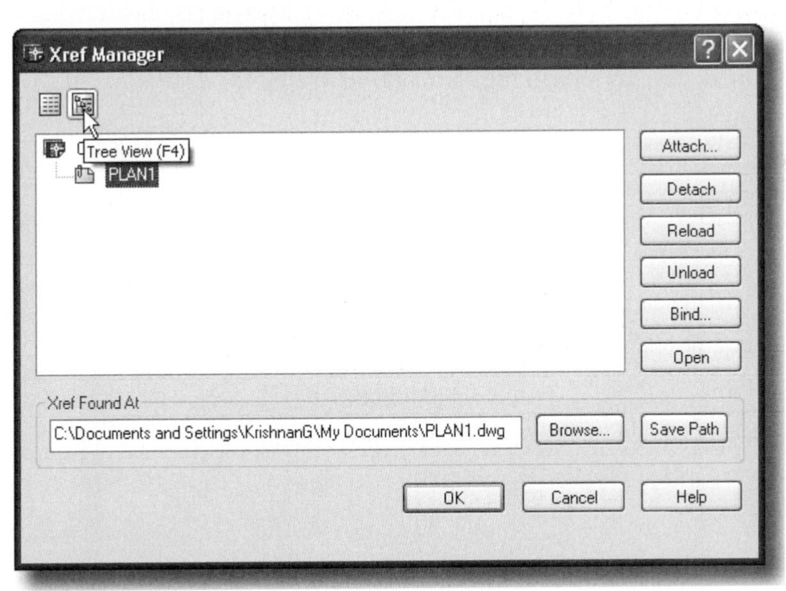

Figure 11–5 *Xref Manager dialog box with Tree View listing*

ATTACHING EXTERNAL REFERENCE DRAWINGS

The **Attach** option allows you attach a drawing as an external reference. Use it when you want to attach a new external reference file or a copy of the external reference file already attached to the current drawing. If you attach a drawing that itself contains an attached external reference, the attached external reference appears in the current drawing. If another person is currently editing the external reference drawing, the drawing attached is based on the most recently saved version.

To attach a drawing as an external reference, invoke the Attach option:

Xref Manager dialog box	Choose the Attach button
Reference toolbar	Choose the External Reference Attach command (see Figure 11–6)
Command: prompt	**xattach** (ENTER)

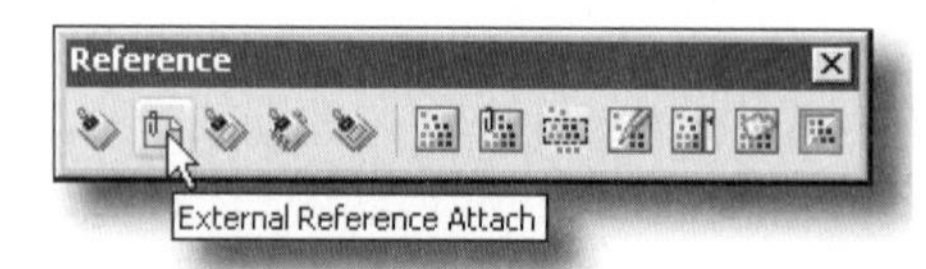

Figure 11–6 *Invoking the* EXTERNAL REFERENCE ATTACH *command from the Reference toolbar*

AutoCAD displays the Select Reference File dialog box. Select the drawing file from the appropriate directory to attach to the current drawing. AutoCAD displays External Reference dialog box, as shown in Figure 11–7.

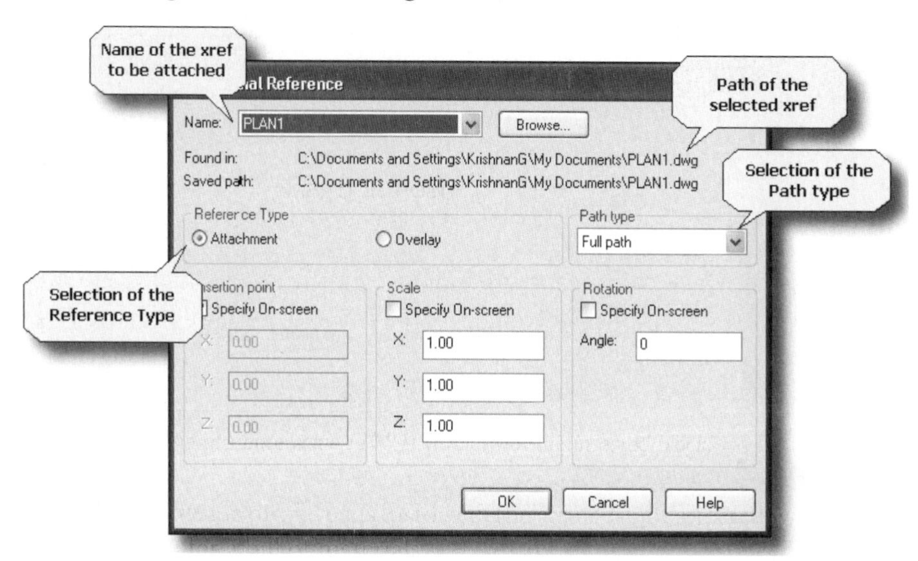

Figure 11–7 *External Reference dialog box*

Once an external reference drawing is attached to the current drawing, the external reference drawing name is added to the list box located next to the **Browse...** button. Choose the **Browse...** button to display the Select a Reference dialog box, in which you can select new xrefs for the current drawing.

When an attached external reference file name is selected from the list box, AutoCAD displays its path where the xref was found. If no path was saved for the xref or if the xref is no longer located at the specified path, AutoCAD searches for the xref in the following order:

1. Current folder of the host drawing
2. Project search paths defined on the **Files** tab in the Options dialog box and in the PROJECTNAME system variable
3. Support search paths defined on the **Files** tab in the Options dialog box
4. Start-in folder specified in the Windows® application shortcut

AutoCAD displays saved path, if any, that is used to locate the xref. This path can be an absolute (fully specified) path, a relative (partially specified) path, or no path.

In the **Reference Type** section of the dialog box, select one of the two available options: **Attachment** or **Overlay**. When you attach an external reference in the **Attachment** mode, the external reference will be included in the drawing when the drawing itself is attached as an external reference to another drawing. If you attach an external reference in the **Overlay** mode, in contrast, an overlaid external reference is not included in a drawing when the drawing itself is attached as an external reference or overlaid external reference to another drawing.

For example, PLAN-A drawing is attached as an overlaid external reference to PLAN-B. Then PLAN-B is attached as an external reference to PLAN-C. PLAN-A is not seen in PLAN-C because it is overlaid in PLAN-B, not attached as an external reference. But, if PLAN-A is attached as an external reference in PLAN-B, and in turn PLAN-B is attached in PLAN-C, both PLAN-A and PLAN-B will be seen in PLAN-C.

So, the only behavioral difference between overlays and attachments is how nested references are handled. Overlaid external references are designed for data sharing. If necessary, you can change the status from **Attachment** to **Overlay**, or vice versa, by double-clicking the field in the **Type** column in the Xref Manager dialog box or by selecting the appropriate radio button in the **Reference Type** section of the External Reference dialog box.

The **Path type** option menu specifies whether the saved path to the xref is set to **No path**, **Full path** or **Relative path**. When you set the path type to **No path**, AutoCAD first looks for the xref in the folder of the host drawing. This option is useful when the xref files are in the same folder as the host drawing. Instead, if you set the path type to **Full path**, AutoCAD saves the xref's precise location to the host drawing. This option is the most precise but the least flexible. If you move a project folder, AutoCAD cannot resolve any xrefs that are attached with a full path. And if you set the path type to **Relative path**, AutoCAD saves the xref's location relative to the host drawing. If you move a project folder, AutoCAD can resolve xref's attached with a relative path, as long as the xref's locative relative the host drawing has not changed. You must save the current drawing before you can set the path type to **Relative path**.

In the **Insertion point** section of the External Reference dialog box, you can specify the insertion point of the external reference.

In the **Scale** section of the External Reference dialog box, you can specify the insertion point, X scale factor, Y scale factor, and Z scale factor of the external reference.

In the **Rotation** section of the External Reference dialog box, you can specify the rotation angle of the external reference.

The **Insertion point**, **Scale**, and **Rotation** features are similar to those in the insertion of a block, explained in Chapter 10.

Choose the **OK** button to attach the selected external reference drawing to the current drawing.

Note: Once a drawing is attached as a reference file, the Manage Xref's icon is displayed in the status bar that allows you to open the Reference Manager.

DETACHING EXTERNAL REFERENCE DRAWINGS

The **Detach** option in the Xref Manager dialog box allows you detach one or more external reference drawings from the current drawing. Only the external reference drawings attached or overlaid directly to the current drawing can be detached. You cannot detach an external reference drawing referenced by another external reference drawing. If the external reference is currently being displayed as part of the current drawing, it disappears when you detach it.

To detach an external reference drawing from the current drawing, first select the drawing name from the displayed list in the Xref Manager dialog box, and then choose the **Detach** button. AutoCAD detaches the selected external reference drawing(s) from the current drawing. You can select several files at once by using the standard Windows methods (holding down SHIFT or CTRL while selecting).

RELOADING EXTERNAL REFERENCE DRAWINGS

The **Reload** option allows you to update one or more external reference drawings any time while the current drawing is in AutoCAD. When you open a drawing, it automatically reloads any external references attached. The Reload option has been provided to reread the external drawing from the external drawing file whenever it is desirable to do so from within AutoCAD. The Reload option is helpful in a network environment to get the latest version of the reference drawing while you are in an AutoCAD session.

To reload external reference drawing(s), first select the drawing name(s) from the displayed list in the Xref Manager dialog box, and then choose the **Reload** button. AutoCAD reloads the selected external reference drawing(s). You can select several files at once by using the standard Windows methods (holding down SHIFT or CTRL while selecting).

UNLOADING EXTERNAL REFERENCE DRAWINGS

The **Unload** option allows you to unload one or more external reference drawings from the current drawing. Unlike the Detach option, the Unload option merely suppresses the display and regeneration of the external reference definition, to help current session editing and improve performance. This option can also be useful when a series of external reference drawings needs to be viewed during a project on an as-needed basis. Rather than have the referenced files displayed at all times, you can reload the drawing when you require the information.

To unload external reference drawing(s), first select the drawing name(s) from the displayed list in the Xref Manager dialog box, and then choose the **Unload** button. AutoCAD turns off the display of the selected external reference drawing(s). You can select several files at once by using the standard Windows methods (holding down SHIFT or CTRL while selecting).

The results of **Unload** and **Reload** take effect when you close the dialog box.

BINDING EXTERNAL REFERENCE DRAWINGS

The **Bind** option allows you to make your external reference drawing data a permanent part of the current drawing. To bind external reference drawing(s), first select the drawing name(s) from the displayed list in the Xref Manager dialog box, and then choose the **Bind...** button. AutoCAD displays the Bind Xrefs dialog box shown in Figure 11–8. Select one of the two available bind types: **Bind** or **Insert.** You can select several files at once by using the standard Windows methods (holding down SHIFT or CTRL while selecting).

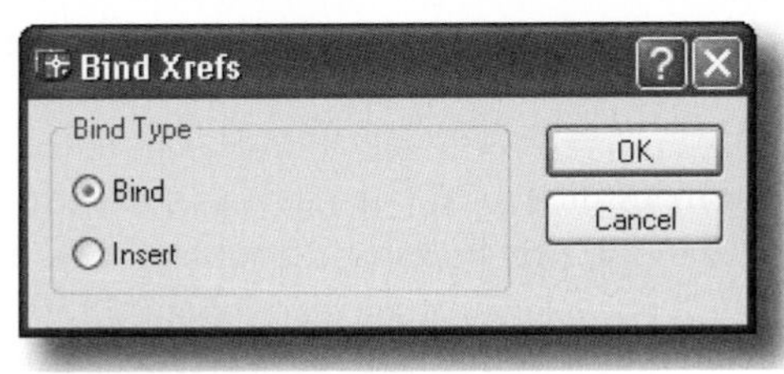

Figure 11–8 *Bind Xrefs dialog box*

With **Bind**, the external reference drawing becomes an ordinary block in your current drawing. It also adds the dependent symbols to your drawing, letting you use them as you would any other named objects. In the process of binding, AutoCAD renames the dependent symbols. The vertical bar symbol (|) is replaced with three new characters: a $, a number, and another $. The number is assigned by AutoCAD to ensure that the named object will have a unique name.

For example, if you bind an external reference drawing named PLAN1 that has a dependent layer PLAN1|FIRST-FL, AutoCAD will try to rename the layer to PLAN1$0$FIRST-FL. If there is already a layer by that name in the current drawing, then AutoCAD tries to rename the layer to PLAN1$1$FIRST-FL, incrementing the number until there is no duplicate.

If you do not want to bind the entire external reference drawing, but only specific dependent symbols, such as a layer, linetype, block, dimension style, or text style, then you can use the XBIND command, explained later in this chapter, in the section on "Adding Dependent Symbols to the Current Drawing."

With **Insert**, the external reference drawing is inserted in the current drawing just like inserting a drawing with the INSERT command. AutoCAD adds the dependent symbols to the current drawing by stripping off the external reference drawing name.

For example, if you insert an external reference drawing named PLAN1 that has a dependent layer PLAN1|FIRST-FL, AutoCAD will rename the layer to FIRST-FL. If there is already a layer by that name in the current drawing, then the layer FIRST-FL would assume the properties of the layer in the current drawing.

OPENING THE EXTERNAL REFERENCE

If you need to make changes to attached external reference file while you are in the host drawing, choose the **Open** button, and AutoCAD opens the selected xref for editing in a new window. The new window is displayed after the Xref Manager is closed. You can also open an external reference for editing by invoking the XOPEN command.

CHANGING THE PATH

To change to a different path or file name for the currently selected external reference file, choose the **Browse...** button in the Xref Manager dialog box. AutoCAD displays the Select New Path dialog box, in which you can specify a different path or file name.

SAVING THE PATH

To save the path as it appears in the **Saved Path** field of the currently selected external reference file (and displayed in the **Xref Found At** text box), select the **Save Path** button. AutoCAD saves the path of the currently selected external reference file.

After making the necessary changes in the Xref Manager dialog box, choose the **OK** button to keep the changes and close the dialog box.

ADDING DEPENDENT SYMBOLS TO THE CURRENT DRAWING

The XBIND command lets you permanently add a selected subset of external reference-dependent symbols to your current drawing. The dependent symbols include the block, layer, linetype, dimension style, and text style. Once the dependent symbol is added to the current drawing, it behaves as if it were created in the current drawing and saved with the drawing when you close the drawing session. While adding the dependent symbol to the current drawing, AutoCAD removes the vertical bar symbol (|) from each dependent symbol's name, replacing it with three new characters: a $, a number, and another $ symbol.

For example, you might want to use a block that is defined in an external reference. Instead of binding the entire external reference with the Bind option of the XREF command, it is advisable to use the XBIND command. With the XBIND command, the block and the layers associated with the block will be added to the current drawing. If the block's definition contains a reference to an external reference, AutoCAD binds that xref and all its dependent symbols as well. After binding the necessary dependent symbols, you can detach the external reference file.

Invoke the XBIND command:

Reference toolbar	Choose the External Reference Bind command (see Figure 11–9)
Modify menu	Choose Object > External Reference > Bind
Command: prompt	**xbind** (ENTER)

Figure 11–9 *Invoking the* EXTERNAL REFERENCE BIND *command from the Reference toolbar*

AutoCAD displays the Xbind dialog box, similar to the Figure 11–10.

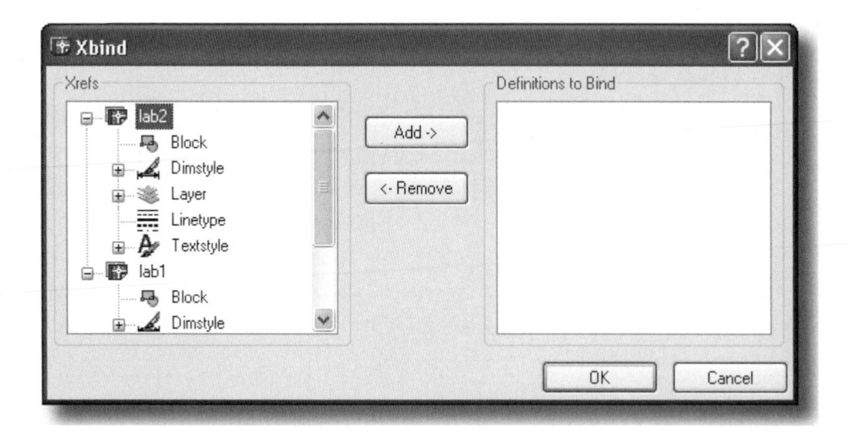

Figure 11–10 *Xbind dialog box*

Note: To invoke the XBIND command at the "Command:" prompt level, type **-xbind** and press ENTER at the "Command:" prompt. The available options are the same ones as in the Xbind dialog box.

On the left side of the Xbind dialog box AutoCAD lists the external reference files currently attached to the current drawing. Double-click on the name of the external reference file, and AutoCAD expands the list by listing the dependent symbols. Select the dependent symbol from the list and choose the **Add->** button. AutoCAD moves the selected dependent symbol into the **Definitions to Bind** list. If necessary, return it to the external reference dependent list from the **Definitions to Bind** list by choosing the **<-Remove** button after selecting the appropriate dependent symbol.

Choose the **OK** button to bind the definitions to the current drawing.

CONTROLLING THE DISPLAY OF EXTERNAL REFERENCES

The XCLIP command allows you to control the display of unwanted information by clipping the external reference drawings and blocks. Clipping does not edit or change the external reference or block, it just prevents part of the object from being displayed. The defined clipping boundary can be visible or hidden. You can also define the front and back clipping planes.

The clipping boundary is created coincident with the polyline. Valid boundaries are 2D polylines with straight or spline-curved segments. Polylines with arc segments, or fit-curved polylines, can be used as the definition of the clip boundary, but the clip boundary will be created as a straight segment representation of that polyline. If the polyline has arcs, the clip boundary is created as if it had been decurved prior to being used as a clip boundary. An open polyline is treated as if it were closed.

The XCLIP command can be applied to one or more external references or blocks. If you set the clip boundary to OFF, the entire external reference or block is displayed. If you subsequently set the clip boundary to ON, the clipped drawing is displayed again. If necessary, you can delete the clipping boundary; AutoCAD redisplays the entire external reference or block. In addition, AutoCAD also allows you to generate a polyline from the clipping boundary.

Invoke the XCLIP command:

Reference toolbar	Choose the External Reference Clip command (see Figure 11–11)
Modify menu	Choose Clip > Xref
Command: prompt	**xclip** (ENTER)

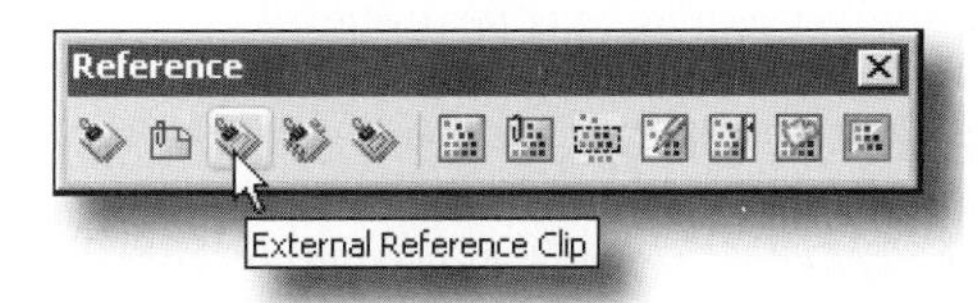

Figure 11–11 *Invoking the* EXTERNAL REFERENCE CLIP *command from the Reference toolbar*

AutoCAD prompts:

Command: **xclip**
Select objects: *(select one or more external references and/or blocks to be included in the clipping and press* ENTER *to complete the selection)*
Enter clipping option [ON/OFF/Clipdepth/Delete/generate Polyline/New boundary] <New> *(right-click and select one of the available options from the shortcut menu)*

CREATING A NEW CLIPPING BOUNDARY

The New option (default) allows you to define a rectangular or polygonal clip boundary or generates a polygonal clipping boundary from a polyline. AutoCAD prompts:

Specify clipping boundary:
[Select polyline/Polygonal/Rectangular] <Rectangular>: *(right-click and select one of the available options from the shortcut menu)*

The Rectangular option (default) allows you to define a rectangular boundary by specifying the opposite corners of a window. The clipping boundary is applied in the current UCS and is independent of the current view.

The Select polyline option defines the boundary by using a selected polyline. The polyline can be open or closed, and can be made of straight-line segments, but cannot intersect itself.

The Polygonal option defines a polygonal boundary by specifying points for the vertices of a polygon.

Once the clipping boundary is defined, AutoCAD displays only the portion of the drawing that is within the clipping boundary and then exits the command.

If you already have a clipping boundary of the selected external reference drawing, and you invoke the New boundary option, then AutoCAD prompts:

Delete old boundary(s)? [Yes/No] <Yes>: *(select one of the two available options)*

If you choose Yes, the entire reference file is redrawn and the command continues; if you choose No, the command sequence is terminated.

Note: The display of the boundary border is controlled by the XCLIPFRAME system variable. If it is set to 1 (ON), then AutoCAD displays the boundary border; if it is set to 0 (OFF), then AutoCAD does not display the boundary border.

CONTROLLING THE DISPLAY OF THE CLIPPED BOUNDARY

The ON/OFF option controls the display of the clipped boundary. The OFF option displays all of the geometry of the external reference or block, ignoring the clipping boundary. The ON option displays the clipped portion of the external reference or block only.

SETTING THE FRONT AND BACK CLIPPING PLANES

The Clipdepth option sets the front and back clipping planes on an external reference or block. Objects outside the volume defined by the boundary and the specified depth are not displayed.

DELETING THE CLIPPING BOUNDARY

The Delete option removes the clipping boundary for the selected external reference or block. To turn off the clipping boundary temporarily, use the OFF option explained earlier. The Delete option erases the clipping boundary and the clipdepth and displays the entire reference file.

Note: The ERASE command cannot be used to delete clipping boundaries.

GENERATING A POLYLINE

AutoCAD draws a polyline coincident with the clipping boundary. The polyline assumes the current layer, linetype, and color settings. When you delete the clipping boundary, AutoCAD deletes the polyline. If you need to keep a copy of the polyline, then invoke the generate Polyline option. AutoCAD makes a copy of the clipping boundary. You can use the PEDIT command to modify the generated polyline, and then redefine the clipping boundary with the new polyline. To see the entire external reference while redefining the boundary, use the OFF option to turn off the clipping boundary.

EDITING REFERENCE FILES/XREF EDIT CONTROL

You can edit block references and external references while working in a drawing session by means of the REFEDIT command. This is referred to as in-place reference editing. If you select a reference for editing and it has attached xrefs or block definitions, the nested references and the reference are displayed and available for selection in the Reference Edit dialog box.

Note: You can edit only one reference at a time. Block references inserted with the MINSERT command cannot be edited. You cannot edit a reference file if it is use by someone else.

You can also display the attribute definitions for editing if the block reference contains attributes. The attributes become visible and their definitions can be edited along with the reference geometry. Attributes of the original reference remain unchanged when the changes are saved back to the block reference. Only subsequent insertions of the block will be affected by the changes.

As mentioned earlier, you can also edit external reference by opening in a separate window with the XOPEN command.

Invoke the REFEDIT:

Modify menu	Choose Xref and Block Editing > Edit Reference In-Place
Command: prompt	**refedit** (ENTER)

AutoCAD prompts:

Command: **refedit**
Select reference: *(select external references and/or blocks to be included in the editing)*

AutoCAD displays the Reference Edit dialog box as shown in Figure 11–12.

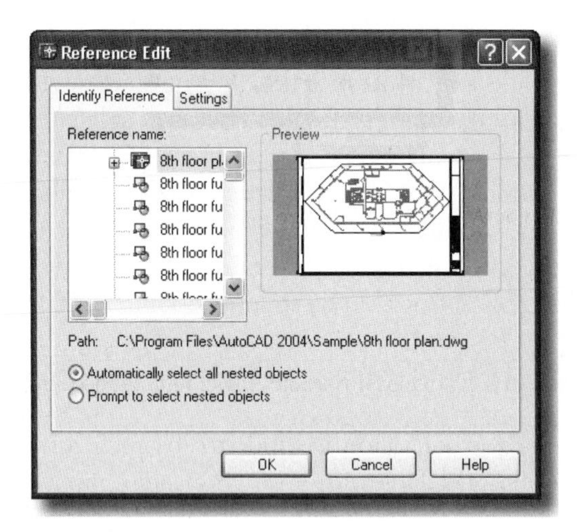

Figure 11–12 *Reference Edit dialog box (Identify Reference tab)*

The **Identify Reference** tab provides visual aids as shown in Figure 11–12 for identifying the reference to edit and controls how the reference is selected. If you select an object that is part of one or more nested references, the nested references are displayed in the dialog box. Objects selected that belong to any nested references cause all the references to become candidates for editing. Select the specific reference you want to edit by choosing the name of the reference in the **Reference name** list box of the Reference Edit dialog box. This will lock the reference file to prevent other users from opening the file. Only one reference can be edited in place at a time. The path of the selected reference is displayed at the bottom of the dialog box. If the selected reference is a block, no path is displayed.

The **Preview** section of the dialog box displays a preview image of the currently selected reference. The preview image displays the reference as it was last saved in the drawing. The reference preview image is not updated when changes are saved back to the reference.

Choosing the **Automatically select all nested objects** radio button controls whether nested objects are included automatically in the reference editing session. If this option is chosen, all the objects in the selected reference will be automatically included in the reference editing session (becoming part of the working set). Instead if the **Prompt to select nested objects** radio button is chosen, then nested objects must be selected individually in the reference editing session. If this option is chosen, after you close the Reference Edit dialog box and enter the reference edit state, AutoCAD prompts you to select the specific objects in the reference that you want to edit.

The **Settings** tab as shown in Figure 11–13 provides options for editing references. Selecting the **Create unique layer, style, and block names** check box controls whether layers and other named objects extracted from the reference are uniquely altered. If this option is set to ON, named objects in xrefs are altered (names are prefixed with $#$), similar to the way they are altered when you bind xrefs. If it is set to OFF, the names of layers and other named objects remain the same as in the reference drawing. Named objects that are not altered to make them unique assume the properties of those in the current host drawing that share the same name.

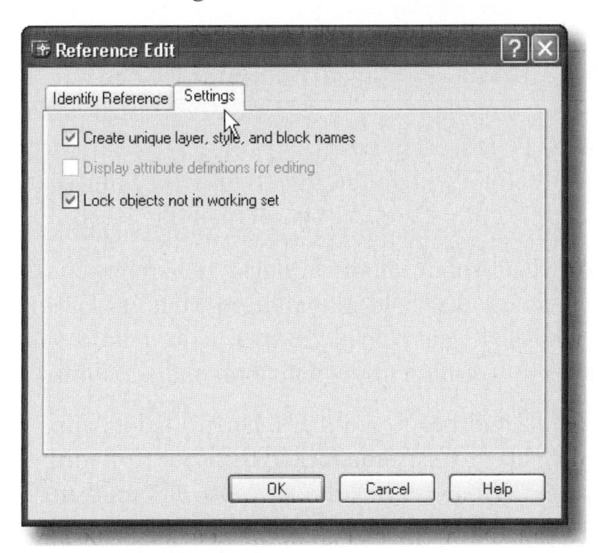

Figure 11–13 *Reference Edit dialog box (Settings tab)*

The **Display attribute definitions for editing** check box controls whether all variable attribute definitions in block references are extracted and displayed during reference editing. If this option is set to ON, the attributes (except constant attributes) are made invisible, and the attribute definitions are available for editing along with the selected reference geometry. When changes are saved back to the block reference, the attributes of the original reference remain unchanged. The new or altered attribute definitions affect only subsequent insertions of the block; the attributes in existing block instances are not affected. Xrefs and block references without definitions are not affected by this option.

The **Lock objects not in working set** check box locks all objects not in the working set. If this option is set to ON, it will prevent you from accidentally selecting and editing objects in the host drawing while in a reference editing state. The behavior of locked objects is similar to objects on a locked layer. If you try to edit locked objects, they are filtered from the selection set.

Choose the **OK** button to close the Reference Edit dialog box. AutoCAD prompts to select objects if the **Prompt to select nested objects** option is selected.

The objects you choose are temporarily extracted for modification in the current drawing and become the *working set.* The working set objects stand out so they can be distinguished from other objects. All other objects not selected appear faded. You can now perform modifications on the working set objects.

Note: Make sure **Reference Edit fading intensity** is set to appropriate setting in the Options dialog box (**Display** tab).

ADDING/REMOVING OBJECTS FROM THE WORKING SET

If a new object is created while editing a reference, it is usually added to the working set automatically. However, if making changes to objects outside the working set causes a new object to be created, it will not be added to the working set.

Objects removed from the working set are added to the host drawing and removed from the reference when the changes are saved back. Objects created or removed are automatically added to or deleted from the working set. You can tell whether an object is in the working set or not by the way it is displayed on the screen; a faded object is not in the working set. When a reference is being edited, the Refedit toolbar is displayed as shown in Figure 11–14.

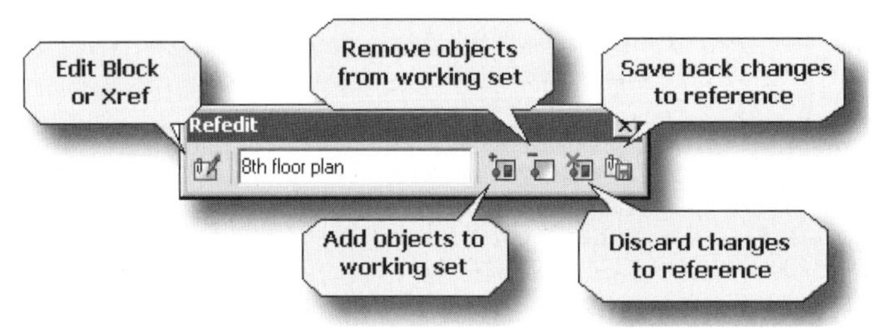

Figure 11–14 *The Refedit toolbar*

To add objects to the working set, invoke **Add objects to the working set** from the Refedit toolbar and select objects to be added. You can only select items when the type of space (Model or Paper) is in effect that was in effect when the REFEDIT command was initiated. To remove objects from the working set, invoke **Remove objects from the working set** from the Refedit toolbar and select objects to be removed. As mentioned earlier, if you remove objects from the working set and save changes, the objects are removed from the reference and added to the current drawing. Any changes you make to objects in the current drawing (not in the xref or block) are not discarded. Once the modifications are complete, invoke **Save back changes to reference** to save the changes to the reference file and to discard the changes, invoke **Discard changes to reference** from the Refedit toolbar.

MANAGING EXTERNAL REFERENCES

Several tools are available to help in the management and tracking of external references.

One of the tracking mechanisms is an external ASCII log file that is maintained on each drawing that contains external references. This file, which AutoCAD generates and maintains automatically, has the same name as the current drawing and a file extension .XLG. You can examine the file with any text editor and/or print it. The log file registers each attach, bind, detach, and reload of each external reference for the current drawing. AutoCAD writes a title block to the log file that contains the name of the current drawing, the date and time, and the operation being performed. Once a log file has been created for a drawing, AutoCAD continues to append to it. The log file is always placed in the same folder as the current drawing. The log file is maintained only if the XRECTL system variable is set to 1. The default setting for XRECTL is 0.

External references are also reported in response to the ? option of the -XREF command and the BLOCK command. Because of the external reference feature, the contents of a drawing may now be stored in multiple drawing files. This means that new backup procedures are required to handle drawings linked in external reference partnerships. Three possible solutions are:

1. Make the external reference drawing a permanent part of the current drawing prior to archiving with the Bind option of the XREF command.
2. Modify the current drawing's path to the external reference drawing so that they are both stored in the same folder, and then archive them together.
3. Archive the folder location of the external reference drawing with the drawing which references it. Tape backup machines do this automatically.

In AutoCAD, a combination of demand loading and the saving of drawings with indexes helps you increase the performance of drawing with external references. In conjunction with the XLOADCTL and INDEXCTL system variables, demand loading provides a method of displaying only those parts of the referenced drawing that are necessary.

The XLOADCTL system variable controls whether demand loading is set to ON or OFF and whether it opens the original drawing or a copy. If XLOADCTL is set to 0, then AutoCAD turns off demand loading, and the entire reference file is loaded. If XLOADCTL is set to 1, then AutoCAD turns on the demand loading, and the reference file is kept open. AutoCAD loads only the objects that are necessary to display on the current drawing. AutoCAD places a lock on all reference drawings that are set for demand loading. Other users can open those reference drawings, but they cannot save changes to them. If XLOADCTL is set to 2, then AutoCAD turns on demand loading and a copy of the reference file is opened. AutoCAD makes a temporary copy of the externally referenced file and demand-loads the temporary file. Other users are allowed to edit the original drawing. When you disable demand loading, AutoCAD reads in the entire reference drawing regardless of layer visibility or clip instances.

The INDEXCTL system variable determines whether layer, spatial, or layer & spatial indexes are created when a drawing file is saved. Using layer and spatial indexes increases performance when AutoCAD is demand-loading external references. If INDEXCTL is set to 0, then indexes are not created; if INDEXCTL is set to 1, a layer index is created. The layer index maintains a list of objects that are on specific layers, and with demand loading it determines which objects need to be read in and displayed. If INDEXCTL is set to 2, then a spatial index is created. The spatial index organizes lists of objects based on their location in three-dimensional space, and it determines which objects lie within the clip boundary and reads only those objects into the current session. If INDEXTCTL is set to 3, both layer and spatial indexes are created and saved with the drawing. If you intend to take full advantage of demand loading, then INDEXCTL should be set to 3.

Note: If the drawing you are working on is not going to be referenced by another drawing, it is recommended that you set the INDEXCTL to 0 (OFF).

AutoCAD provides another system variable, VISRETAIN, to control the visibility of layers in the external reference drawing. If the VISRETAIN system variable is set to 0 (OFF), any changes you make to settings of the external reference drawing's layers, such as ON/OFF, Freeze/Thaw, Color, and Linetype, apply to the current drawing session only. If VISRETAIN is set to 1 (ON), then any changes you make to settings of the external reference drawing's layers take precedence over the external reference layer definition.

Another tool that you can use to manage external references is Autodesk Reference Manager. It provides tools to list referenced files in selected drawings and to modify the saved reference paths without opening the drawing files in AutoCAD. With Reference Manager, drawings with unresolved references can be easily identified and fixed.

Reference Manager is a stand-alone application that you can access from the Autodesk program group under Programs in the Start menu (Windows).

IMAGES

The IMAGE command provides various options for attaching and detaching a raster or bit-mapped bi-tonal, 8-bit gray, 8 bit color, or 24-bit color image file into the drawing. The image formats that can be inserted into AutoCAD include BMP, TIFF, RLE, JPG, PCX, FLIC, GIF, IG4, IG5, RIC, PCT, CALSI, and TGA. More than one image can be displayed in any viewport, and the number and size of images is not limited.

Invoke the IMAGE command:

Reference toolbar	Choose the IMAGE command (see Figure 11–15)
Insert menu	Choose Raster Image
Command: prompt	**image** (ENTER)

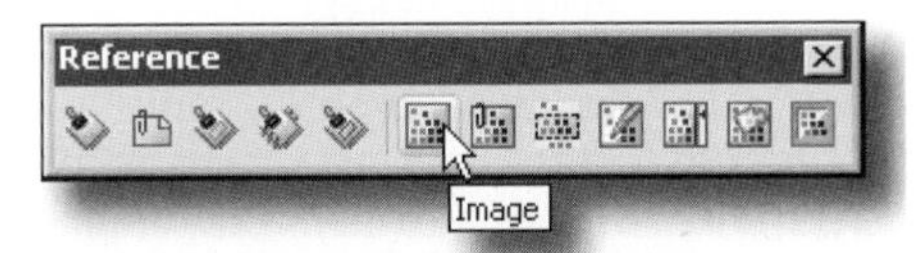

Figure 11–15 *Invoking the* IMAGE *command from the Reference toolbar*

AutoCAD displays the Image Manager dialog box, similar to Figure 11–16.

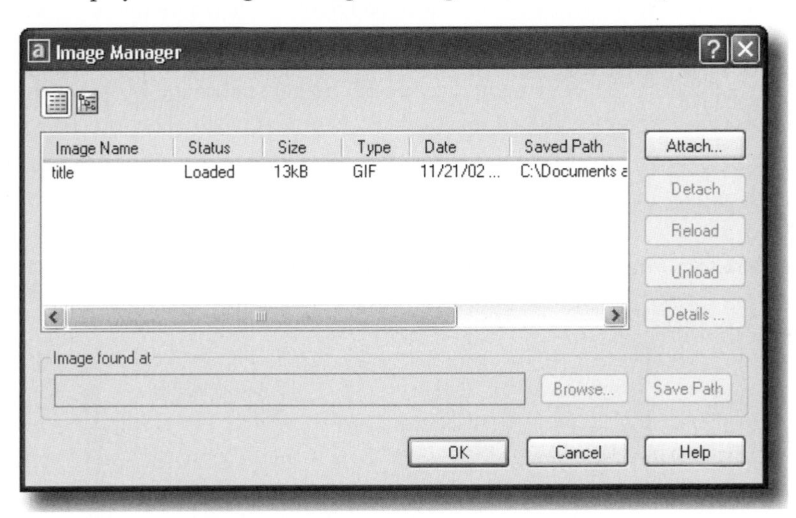

Figure 11–16 *Image Manager dialog box*

The Image Manager dialog box looks a lot like the Xref Manager dialog box. In the Image Manager dialog box, AutoCAD lists the images attached to the current drawing. The information provided in the list box of the Image Manager dialog box, such as **Image name, Status, Size, Type, Date,** and **Saved Path,** is similar to the information provided in the Xref Manager dialog box. You can switch between **List View** and **Tree View,** as in the Xref Manager dialog box, by clicking the two buttons at the top left of the Image Manager dialog box.

ATTACHING AN IMAGE TO THE CURRENT DRAWING

The **Attach** option allows you to attach an image object to the current drawing. Invoke the **Attach** option:

Image Manager dialog box	Choose the **Attach** button
Reference toolbar	Choose the Image Attach command (see Figure 11–17)
Command: prompt	**imageattach** (ENTER)

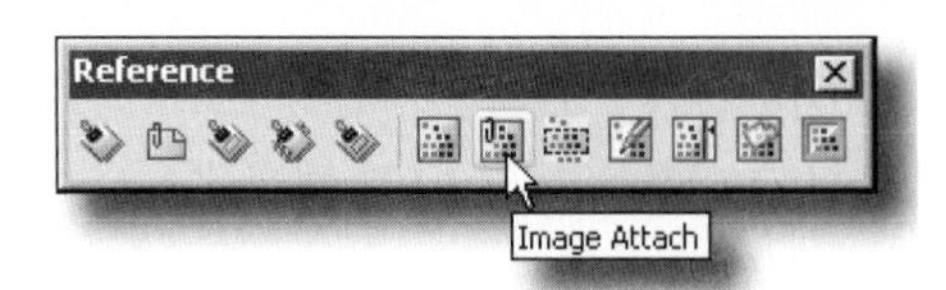

Figure 11–17 *Invoking the* IMAGE ATTACH *command from the Reference toolbar*

AutoCAD displays the Select Image dialog box. Select the image file from the appropriate directory and choose the **Open** button. AutoCAD displays the Image dialog box, similar to Figure 11–18.

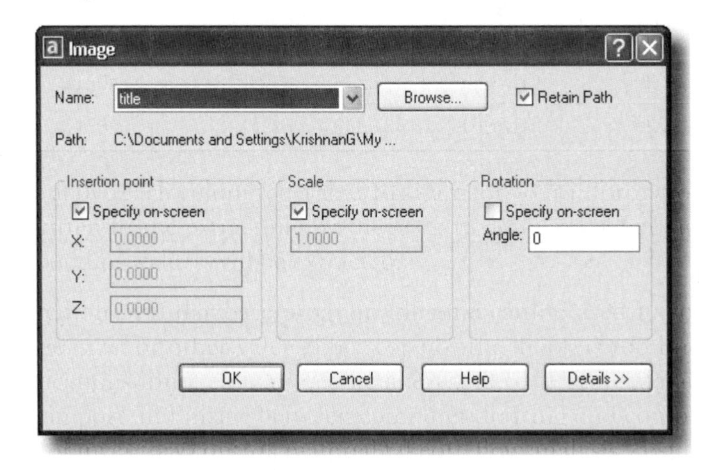

Figure 11–18 *Image dialog box*

To select a different image choose the **Browse...** button to display the Select Image File dialog box. Select the appropriate image file to attach to the current drawing.

Once an image file is attached to the current drawing, the image file name is added to the list box located next to the **Browse...** button. When an attached image file name is selected from the list box, its path is displayed below.

The **Retain Path** check box determines whether or not the full path to the image file is saved. If **Retain Path** is set to ON, then the image file path is saved in the drawing database; if it is set to OFF, the name of the image file is saved without a path in the database. AutoCAD searches for the image file in the AutoCAD Support File Search Path and in the paths associated with the **Files** tab of the Options dialog box.

In the **Insertion point** section of the Image dialog box, you can specify the insertion point of the external reference.

In the **Scale** section of the Image dialog box, you can specify X scale factor, Y scale factor, and Z scale factor of the external reference.

In the **Rotation** section of the Image dialog box, you can specify the rotation angle of the external reference.

The **Insertion point**, **Scale**, and **Rotation** features are similar to those in the insertion of a block, explained in Chapter 10.

Choose the **Details>>** button to display the image information for the selected image file. This information includes image resolution in horizontal and vertical units, image size by width and height in pixels, and image size by width and height in the current selected units.

DETACHING AN IMAGE FROM THE CURRENT DRAWING

The **Detach** option in the Image Manager dialog box removes the selected image definitions from the drawing database and erases all the associated image objects from the drawing and from the display. To detach an image file from the current drawing, first select the image file name from the displayed list in the Image Manager dialog box, and then select the **Detach** button. AutoCAD detaches the selected image file from the current drawing.

RELOADING AN IMAGE INTO THE CURRENT DRAWING

The **Reload** option loads the most recent version of an image. To reload the image file to the current drawing, first select the image file name from the displayed list in the Image Manager dialog box, and then select the **Reload** button. AutoCAD reloads the selected image file into the current drawing.

UNLOADING AN IMAGE FROM THE CURRENT DRAWING

The **Unload** option unloads the image data from working memory without erasing the image objects from the drawing. It is highly recommended that you unload images that are no longer needed for editing. By unloading the images, you can improve performance by reducing the memory requirement for AutoCAD. To unload the image file from the current drawing, first select the image file name from the displayed list in the Image Manager dialog box, and then select the **Unload** button. AutoCAD unloads the selected image file from the current drawing.

DETAILS OF THE IMAGE FILE

The Details option provides detailed information about the selected image, including the image name, saved path, active path, file creation date and time, file size and type, color, color depth, width and height in pixels, resolution, default size in units, and a preview image. To display the detailed information of the selected image file, first select the image file name from the displayed list in the Image Manager dialog box, then select the **Details...** button. AutoCAD displays the detailed information of the selected image file in the Image File Details dialog box, similar to the Figure 11–19.

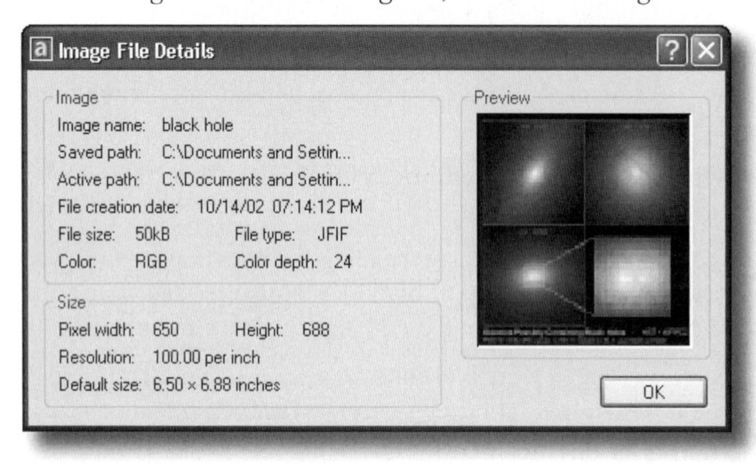

Figure 11–19 *Image File Details dialog box*

CHANGING THE PATH

To change to a different path or file name for the currently selected image file, choose the **Browse...** button in the Image Manager dialog box. AutoCAD displays the Image dialog box, in which you can select a different path or file name.

SAVING THE PATH

To save the path as it appears in the **Saved Path** field of the currently selected image file (and displayed in the **Image Found At** text box), select the **Save Path** button. AutoCAD saves the path of the currently selected image file.

After making the necessary changes in the Image Manager dialog box, choose the **OK** button to keep the changes and close the dialog box.

CONTROLLING THE DISPLAY OF THE IMAGE OBJECTS

The IMAGECLIP command allows you to control the display of unwanted information by clipping the image object; this is similar to the use of the XCLIP command for external references and blocks.

Invoke the IMAGECLIP command:

Reference toolbar	Choose the IMAGE CLIP command (see Figure 11–20)
Modify menu	Choose Clip > Image
Command: prompt	**imageclip** (ENTER)

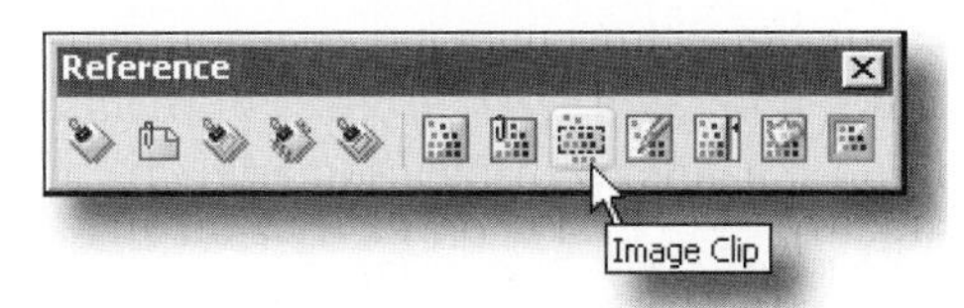

Figure 11–20 *Invoke the IMAGE CLIP command from the Reference toolbar*

AutoCAD prompts:

Command: **imageclip**
Select image to clip: *(select the image to clip)*
Enter image clipping option [ON/OFF/Delete/New boundary] <New>: *(right-click and select one of the available options from the shortcut menu)*

Creating a New Clipping Boundary

The New boundary option (default) allows you to define a rectangular or polygonal clip boundary. AutoCAD prompts:

Enter clipping type [Polygonal/Rectangular] <Rectangular>: *(right-click and select one of the available options from the shortcut menu)*

The Rectangular option (default) allows you to define a rectangular boundary by specifying the opposite corners of a window. The rectangle is always drawn parallel to the edges of the image.

The Polygonal option defines a polygonal boundary by specifying points for the vertices of a polygon.

If you already have a clipping boundary of the selected image, and you invoke the New boundary option, then AutoCAD prompts:

Delete old boundary(s)? [Yes/No] <Yes>: *(select one of the two available options)*

If you choose Yes, the entire reference file is redrawn and the command continues; if you choose No, the command sequence is terminated.

Once the clipping boundary is defined, AutoCAD displays only the portion of the image that is within the clipping boundary and then exits the command.

Note: The display of the boundary border is controlled by the IMAGEFRAME system variable. If it is set to 1 (ON), then AutoCAD displays the boundary border; if it is set to 0 (OFF), then AutoCAD does not display the boundary border.

Controlling the Display of the Clipped Boundary

The ON/OFF option controls the display of the clipped boundary. The OFF option displays all of the image, ignoring the clipping boundary. The ON option displays the clipped portion of the image only.

ADJUSTING THE IMAGE SETTINGS

The IMAGEADJUST command controls the brightness, contrast, and fade values of the selected image. Invoke the IMAGEADJUST command:

Reference toolbar	Choose the Image Adjust command (see Figure 11–21)
Modify menu	Choose Object > Image > Adjust
Command: prompt	**imageadjust** (ENTER)

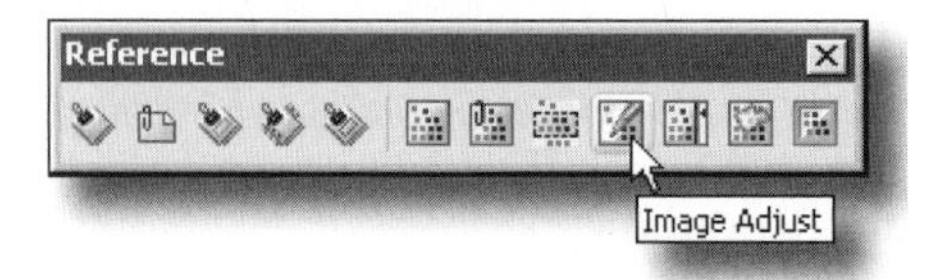

Figure 11–21 *Invoking the* IMAGE ADJUST *command from the Reference toolbar*

AutoCAD prompts:

Command: **imageadjust**
Select image to adjust: *(select the images, and press* ENTER *to complete selection)*

AutoCAD displays the Image Adjust dialog box, similar to Figure 11–22.

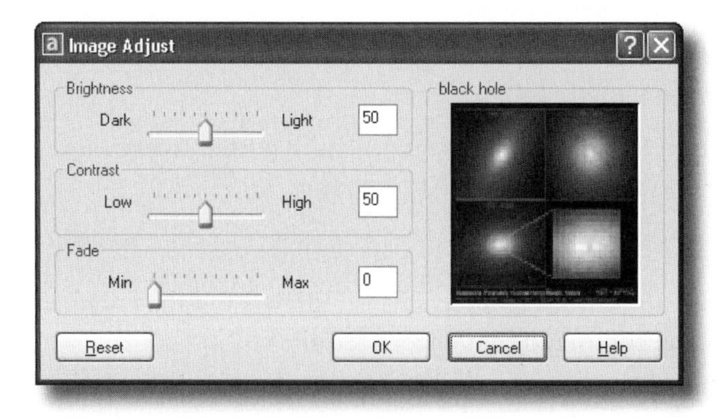

Figure 11–22 *Image Adjust dialog box*

You can adjust the **Brightness**, **Contrast**, and **Fade** within the range of 0 to 100.

Select the **Reset** button to reset values for the brightness, contrast, and fade parameters to the default settings of 50, 50, and 0, respectively.

ADJUSTING THE DISPLAY QUALITY OF IMAGES

The IMAGEQUALITY command controls the display quality of images. The quality setting affects display performance. A high-quality image takes longer to display. Changing the setting updates the display immediately without causing a regeneration. Images are always plotted using a high-quality display.

Invoke the IMAGEQUALITY command:

Reference toolbar	Choose the Image Quality command (see Figure 11–23)
Modify menu	Choose Object > Image > Quality
Command: prompt	**imagequality** (ENTER)

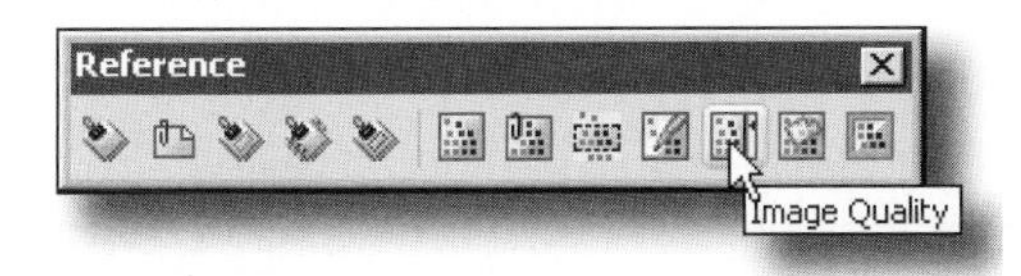

Figure 11–23 *Invoking the* IMAGE QUALITY *command from the Reference toolbar*

AutoCAD prompts:

Command: **imagequality**
Enter image quality setting [High/Draft] <current>: *(right-click and select one of the available options from the shortcut menu)*

The High option produces a high-quality image on screen, and the Draft option produces a lower-quality image on screen.

CONTROLLING THE TRANSPARENCY OF AN IMAGE

The TRANSPARENCY command controls whether the background pixels in an image are transparent or opaque. Invoke the TRANSPARENCY command:

Reference toolbar	Choose the Image Transparency command (see Figure 11–24)
Modify menu	Choose Object > Image > Transparency
Command: prompt	**transparency** (ENTER)

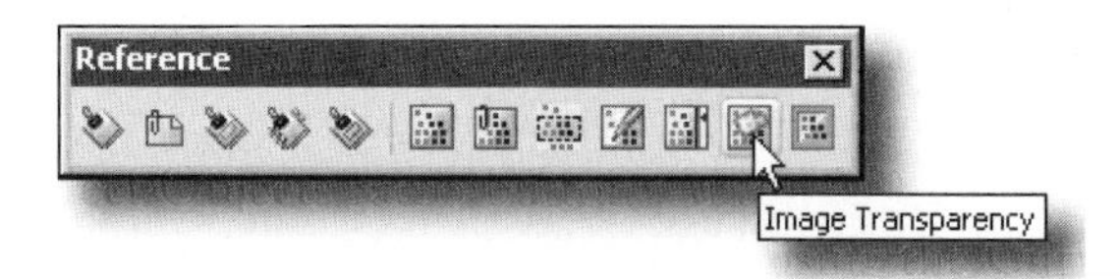

Figure 11–24 *Invoking the* IMAGE TRANSPARENCY *command from the Reference toolbar*

AutoCAD prompts:

Command: **transparency**
Select image: *(select the images, and press* ENTER *to complete selection)*
Enter transparency mode [ON/OFF] <current>: *(select one of the two available options)*

The ON option turns transparency on so that objects beneath the image are visible. The OFF option turns transparency off so that objects beneath the image are not visible.

CONTROLLING THE FRAME OF AN IMAGE

The IMAGEFRAME command controls whether image frames are displayed or hidden from view.

Invoke the IMAGEFRAME command:

Reference toolbar	Choose the Image Frame command (see Figure 11–25)
Modify menu	Choose Object > Image > Frame
Command: prompt	**imageframe** (ENTER)

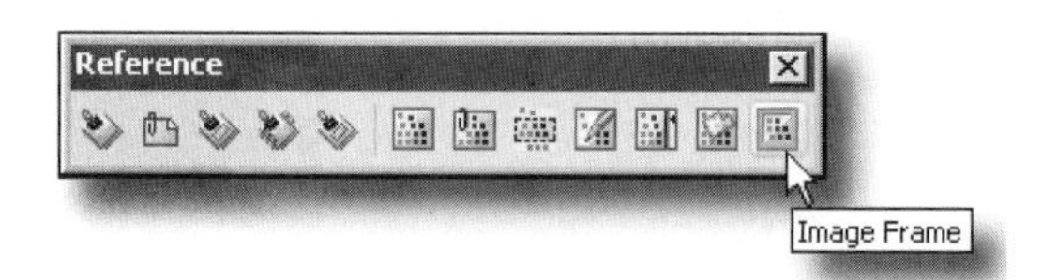

Figure 11–25 *Invoking the* IMAGE FRAME *command from the Reference toolbar*

AutoCAD prompts:

Command: **imageframe**
Select image: *(select the images, and press* ENTER *to complete selection)*
Enter image frame setting [ON/OFF] <current>: *(right-click and select one of the available options from the shortcut menu)*

The ON option turns Imageframe on so that the frames around images are visible. The OFF option turns Imageframe off so that frames around images are not visible.

Open the Exercise Manual PDF file for Chapter 11 on the accompanying CD for project and discipline specific exercises.

If you have the accompanying Exercise Manual, refer to Chapter 11 for project and discipline specific exercises.

REVIEW QUESTIONS

1. If an externally referenced drawing called "FLOOR.DWG" contains a block called "TABLE" and is permanently bound to the current drawing, the new name of the block is:

   a. FLOOR0TABLE

   b. FLOOR|TABLE

   c. FLOOR$0TABLE

   d. FLOOR_TABLE

   e. FLOOR$|$TABLE

2. If an externally referenced drawing called "FLOOR.DWG" contains a block called "TABLE," the name of the block is listed as:

   a. FLOOR0TABLE

   b. FLOOR|TABLE

   c. FLOOR$0TABLE

   d. FLOOR_TABLE

   e. FLOOR$|$TABLE

3. The maximum number of files which can be externally referenced into a drawing is:

   a. 32

   b. 1,024

   c. 8,000

   d. 32,000

   e. only limited by memory

4. If you want to retain, from one drawing session to another, any changes you make to the color or visibility of layers in an externally referenced file, the system variable that controls this is:

   a. XREFRET

   b. RETXREF

   c. XREFLAYER

   d. VISRETAIN

   e. these changes cannot be saved from one session to another

5. XREFs are converted to blocks if you detach them.

   a. True b. False

6. When detaching XREFs from your drawing, it is acceptable to use wild cards to specify which XREF should be detached.

   a. True b. False

7. Overlaying XREFs rather than attaching them causes AutoCAD to display the file as a bit map image, rather than a vector based image.

   a. True b. False

8. To make a reference file a permanent part of the current drawing database, use the XREF command with the:

   a. Attach option
   b. Bind option
   c. ? option
   d. Reload option
   e. Path option

9. The XREF command is invoked from the toolbar:

   a. Draw
   b. Modify
   c. External Reference
   d. Any of the above
   e. None of the above

10. The Attach option of the XREF command is used to:

    a. bind the external drawing to the current drawing
    b. attach a new external reference file to the current drawing
    c. reload an external reference drawing
    d. all of the above

11. The following are the dependent symbols that can be made a permanent part of your current drawing, except:

    a. Blocks
    b. Dimstyles
    c. Text Styles
    d. Linetypes
    e. Grid and Snap

12. Including an image file in a drawing will incorporate the image similar to the way a drawing file is merged by using:

    a. INSERT

    b. XREF

    c. WBLOCK

13. Which of the following is not a valid file type to use with the IMAGE command?

    a. BMP
    b. TIF
    c. WMF
    d. JPG
    e. GIF

14. Which of the following parameters can be adjusted on a bitmapped image?

    a. Brightness
    b. Contrast
    c. Fade
    d. all of the above
    e. none of the above

AutoCAD DesignCenter

INTRODUCTION

AutoCAD DesignCenter makes it much easier to manage content within your drawing. Content includes blocks, external references, layers, raster images, linetypes, layouts, text styles, dimension styles, and custom content created by third-party applications. You can now manage content between your drawing and other sources such as other drawings, whether currently open, stored on any drive, or even elsewhere on a network or somewhere on the Internet. The AutoCAD DesignCenter provides a program window with a specialized drawing file-handling section. It allows you to drag and drop content and images into your current drawing or attach a drawing as an external reference. AutoCAD 2004 has added a page to the DesignCenter called DC Online that gives immediate and direct access to thousands of symbols, manufacturers' product information, and content aggregators' sites. Content in the AutoCAD DesignCenter can be dragged into a tool palette for use in the current drawing.

After completing this chapter, you will be able to do the following:

- Open, undock, move, resize, dock, and close AutoCAD DesignCenter
- Locate drawings, files, and their content in a manner similar to Windows Explorer
- Use the DesignCenter content area and Tree View
- Preview images, drawings, content, and their written descriptions
- Customize and use Autodesk *Favorites* folder
- Manage blocks, layers, xrefs, layouts, dimstyles, textstyles, and raster images
- Manage web-based content and custom content from third-party applications
- Create shortcuts to drawings, folders, drives, the network, and Internet locations
- Browse sources for content and drag and drop (copy) into drawings
- Access symbols and information directly over the Internet

CONTENT

As mentioned above, content includes block definitions, external references, layer names and compositions, raster images, linetypes, layouts, text styles, dimension styles, and custom content created by third-party applications. For example, a layer of one name may have the same properties as a layer of another name. But, because its name is part of its composition, it can be identified as a unique layer. Also, two items of content can have the same name if they are not the same type of content. For example, you can name both a text style and a dimension style "architectural".

Note: You can only view the name of an item of content in the content area along with a raster image if it is a block or image and a description if one has been written. The item itself still resides in its container. Through the DesignCenter, you can drag and drop copies of the item's definition into your current drawing, but you cannot edit the item itself from the DesignCenter.

CONTENT TYPE

Content types are only types or categories and not the items themselves. Lines, circles, and other objects are not included in what are considered content types, although a block may be made up of such objects. To view a particular item of content you must first select the name of its content type (under a drawing name) in the Tree View or double-click the name of its content type in the content area (See Figure 12–6 for identifying the Tree View pane and the content area). For example, if you were trying to locate an xref named "1st floor plan", you could search through drawing names and select the content type named "xref" under each drawing name. Under each drawing name there will be a content type named "xref", but under each content type named "xref" there may or may not be any xrefs listed.

CONTAINER

The primary container is the drawing. It contains the blocks, images, linetypes, and other definitions that are most commonly sought to add to your current drawing. A folder can be considered a container because it contains files, and a drawing is a file. An image can also be a file in a folder. An image is normally not a container like a drawing. Drawings and images can be dragged and dropped into your current drawing from the content area. So, a drawing can be both content and a container.

Figure 12–1 shows an example of a content (block named Conference Seat), content type (Block), and Container (drawing named *8th floor furniture*).

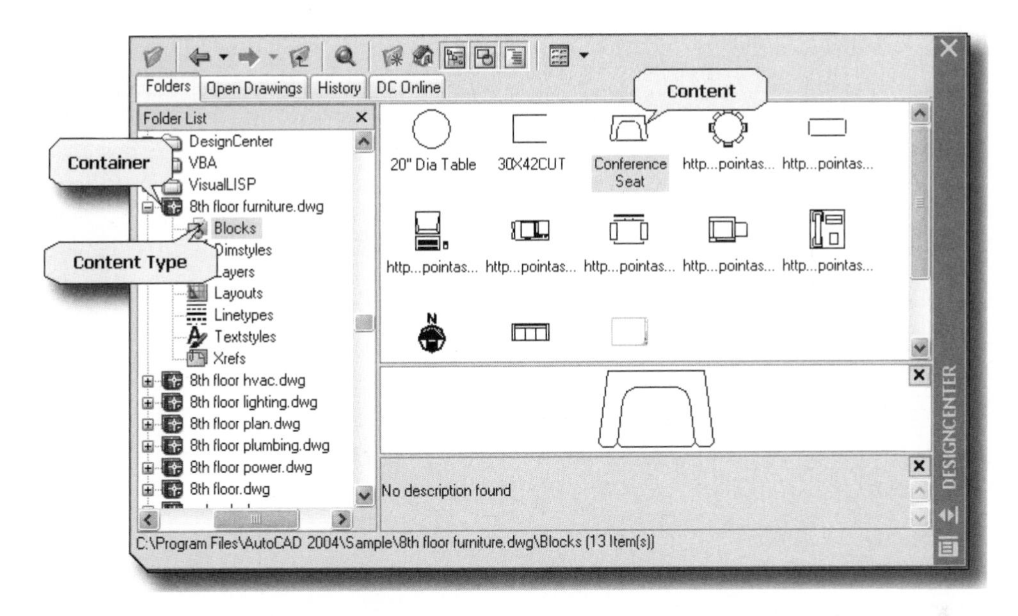

Figure 12–1 *An example showing a content, content type, and container from the DesignCenter window*

OPENING THE DesignCenter WINDOW

The AutoCAD DesignCenter is a window rather than a dialog box. It is like calling up a special program that runs along with AutoCAD and expedites file managing and drawing content handling tasks.

Invoke the AutoCAD DesignCenter command:

Standard toolbar	Choose the DESIGNCENTER command (see Figure 12–2)
Tools menu	Choose DesignCenter
Command: prompt	**adcenter** (ENTER)

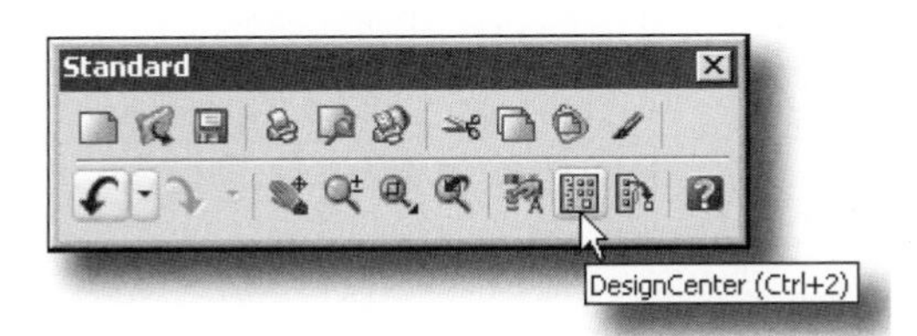

Figure 12–2 *Invoking the AutoCAD DesignCenter from the Standard toolbar*

AutoCAD displays the AutoCAD DesignCenter window, similar to Figure 12–3.

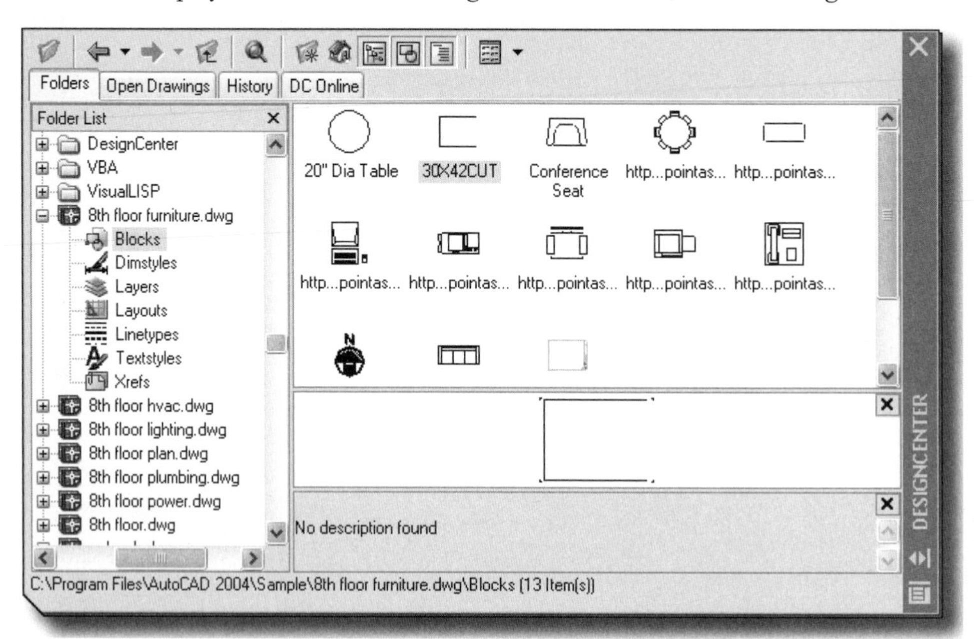

Figure 12–3 *AutoCAD DesignCenter window*

POSITIONING THE DesignCenter WINDOW

DOCKED POSITION

The default position of the DesignCenter is docked at the left side of the drawing area. This is where it will be located when you invoke the ADCENTER command after AutoCAD is started for the first time, as shown in Figure 12–4. However, once the DesignCenter has been repositioned and the drawing session is closed with the DesignCenter window open, the next time you open AutoCAD it will be at its relocated position.

FLOATING (UNDOCKED) POSITION

You can undock the DesignCenter by double-clicking on its title bar or border. Or you can drag the window into the drawing area until its drag preview image clears its present position, as shown in Figure 12–5.

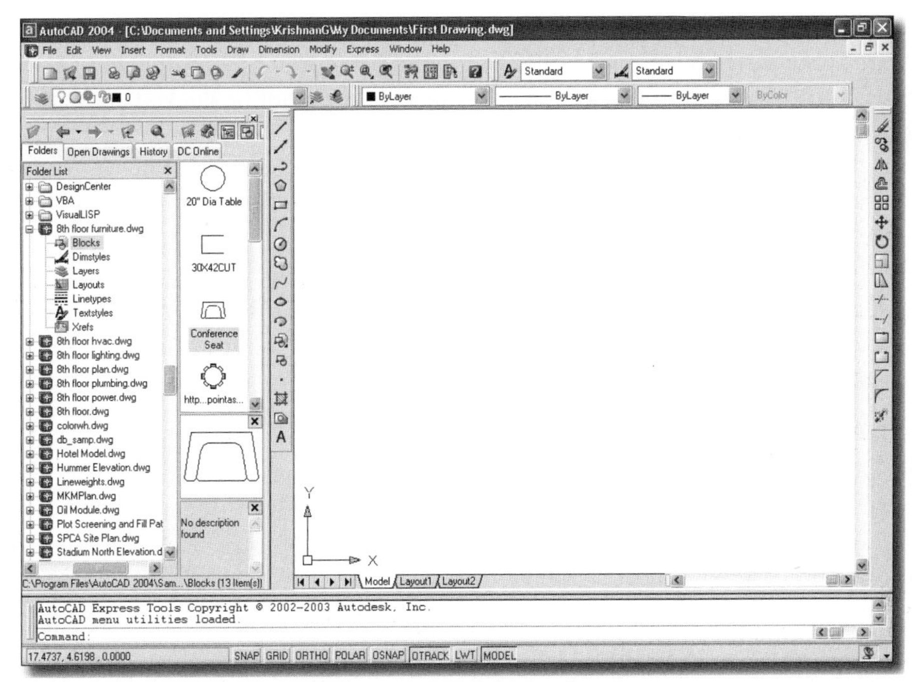

Figure 12–4 *DesignCenter docked on the left side of the drawing area*

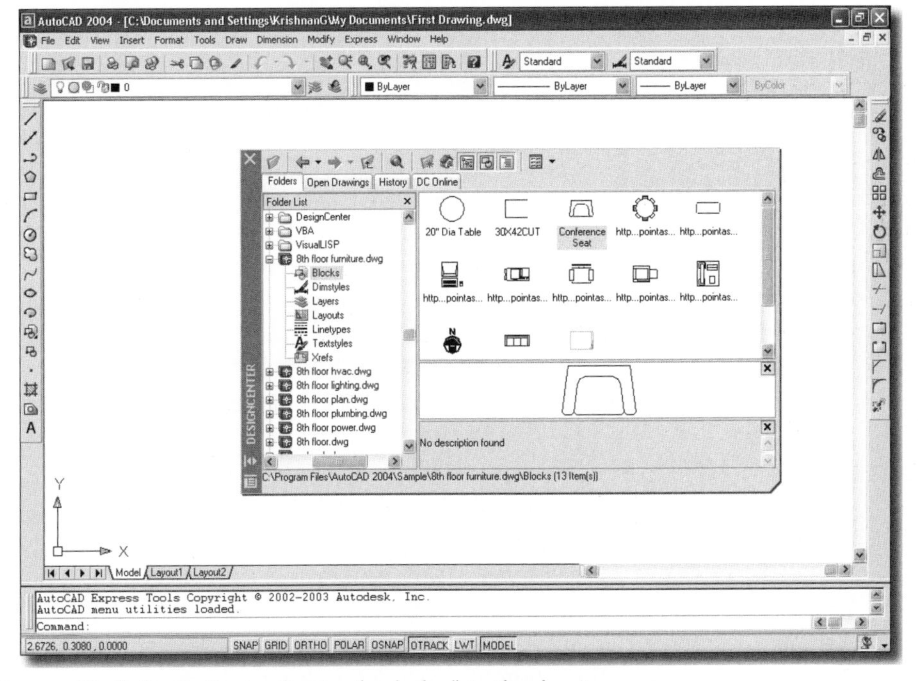

Figure 12–5 *DesignCenter floating (undocked) in the drawing area*

REDOCKING

To redock the DesignCenter, double-click the title bar or border or drag the Design-Center to the left or right side of the drawing area. It cannot be docked at the top or bottom of the screen. When undocked, the DesignCenter can be resized by holding down the pick button over an edge or corner of the window and, when the cursor becomes a double arrow, dragging the edge or corner with the pointing device.

PARTS OF THE DesignCenter WINDOW

The DesignCenter has four tabs: **Folders**, **Open Drawings**, **History** and **DC Online**. The buttons on the toolbar for the **DC Online** tab are different from those for the other three tabs.

FOLDERS

The DesignCenter, when the **Folders** tab is displayed, consists of five major areas: toolbar, content area, Tree View pane, Preview pane, and Description pane. The content area and toolbar are always displayed as shown in Figure 12–6. The Tree View, Preview, and Description panes can be turned off and on when desired.

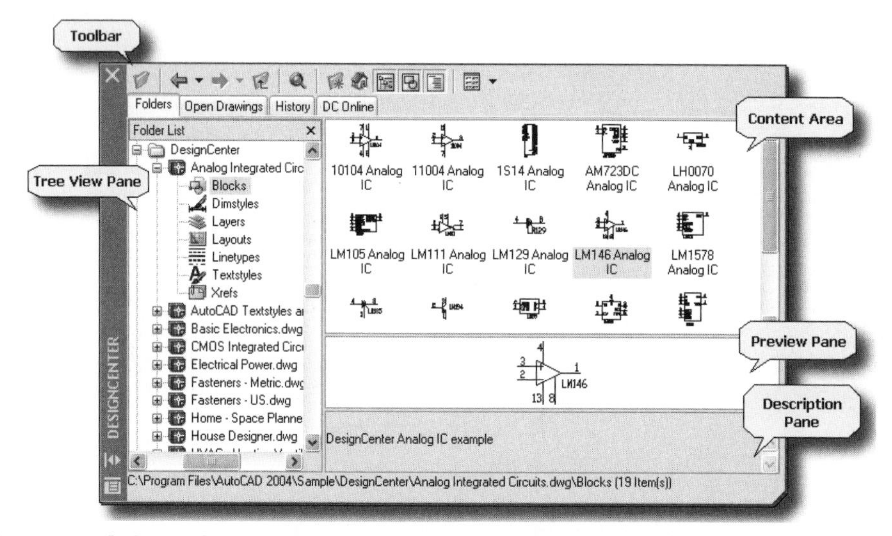

Figure 12–6 *Parts of the DesignCenter window with the Folders tab displayed*

TOOLBAR

The toolbar contains buttons designed especially to help you select and manage the type of content desired within the DesignCenter window, as shown in Figure 12–6.

CONTENT AREA

The content area (right pane) is AutoCAD DesignCenter's primary area for displaying the names and icons representing content, as shown in Figure 12–6.

TREE VIEW

The Tree View or navigation pane is an optional area on the left side of the DesignCenter window for displaying files such as drawings and images and their locations. Choose the Tree View button on the toolbar, as shown in Figure 12–7, to display the Tree View.

Figure 12–7 *Invoking the Tree View from the DesignCenter toolbar*

The Tree View also displays content types. The Tree View displays multiple levels and operates in a manner similar to the Windows Explorer. Figure 12–8 shows the DesignCenter window with the Tree View window display on and off within the **Folders** tab.

Note: The Tree View can be used to navigate up and down through the hierarchy of networks, drives, folders, files, and content type. But, you cannot drag and drop to, from, or within the Tree View.

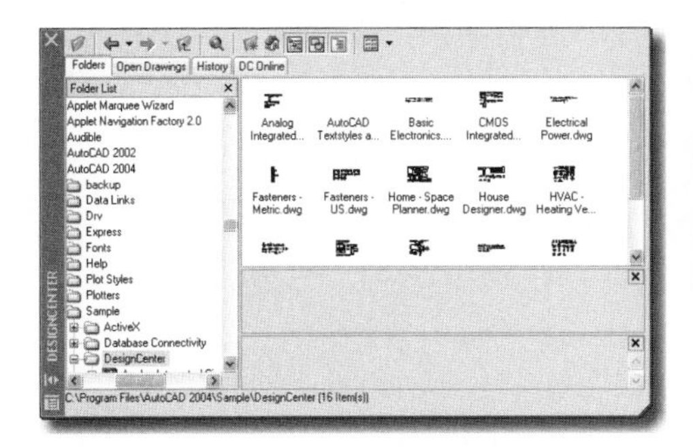

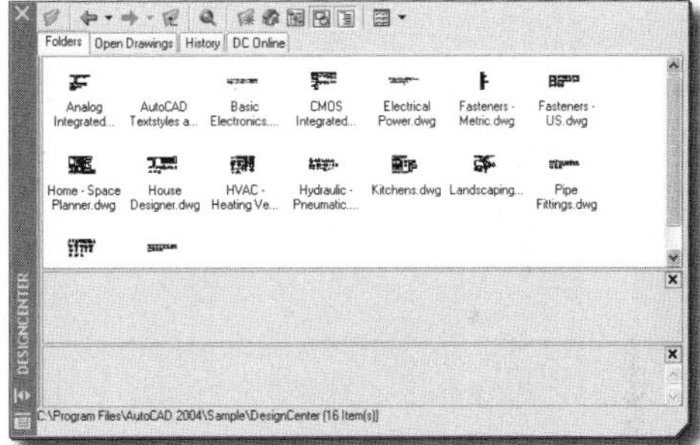

Figure 12–8 *AutoCAD DesignCenter window with Tree View window display on (left) and off (right)*

PREVIEW

The Preview pane displays a raster image of the selected item of content if the item is a drawing, block, or image, as shown in Figure 12–6.

DESCRIPTION

The Description pane displays a written description of the item of content selected on the content area if a description has been written, as shown in Figure 12–6.

TOOLBAR BUTTONS

The buttons on the toolbar are used to manage the DesignCenter panes and what is being displayed in them, as shown in Figure 12–9.

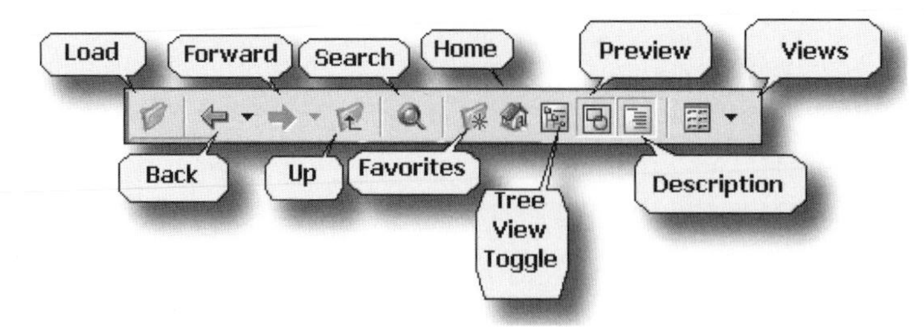

Figure 12–9 *AutoCAD DesignCenter showing toolbar buttons*

LOAD

The **Load** button causes the Load dialog box to be displayed, as shown in Figure 12–10. It is similar to the Windows File Manager dialog box. From it you can select a drawing file whose contents will be loaded into the content area.

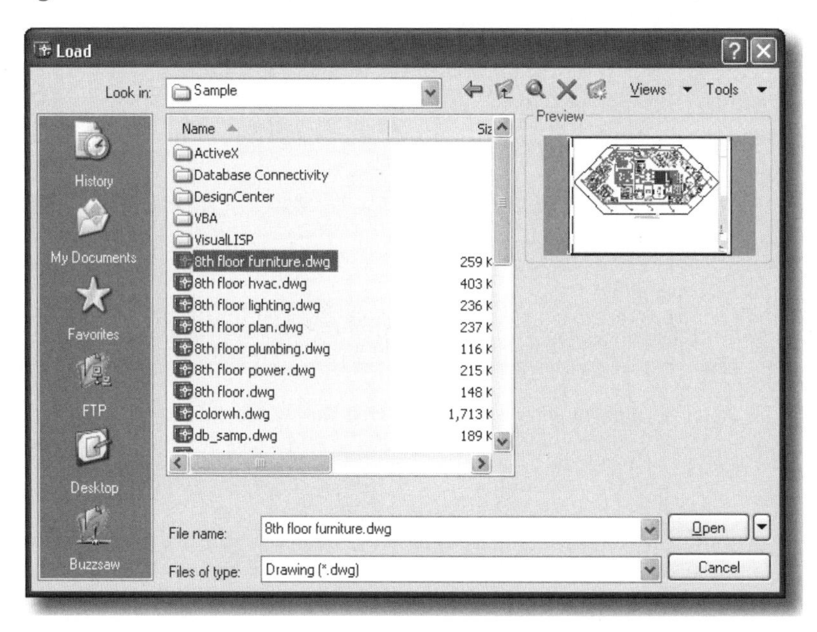

Figure 12–10 *Load dialog box*

AutoCAD displays a bitmap image in the **Preview** section. In addition, there is a window on the left side of the dialog box displaying quick access icons to folders on your computer:

Desktop and *My Documents*; icons for **History** (recently opened drawings) and **Favorites**; and locations on the Internet: **Buzzsaw.com, RedSpark.com**, and **FTP**.

Buttons to the right of the **Look in:** list box are **Back to <the last folder>, Up one level, Search the Web, Delete, Create New Folder, Views** and **Tools**. The **Back to <the last folder>, Up one level, Search the Web, Delete**, and **Create New Folder** are options that are similar to most file handling dialog boxes for reaching the location of the drawing you wish to open.

Views

The **Views** button displays a pulldown menu with the options of **list** or **details**, to determine how folders and files are displayed, and a preview option that opens a Preview window to show a thumbnail sketch of drawing selected.

Tools

The **Tools** button displays a pulldown menu with the options of **Find, Locate, Add/ Modify FTP Locations**, and **Add to Favorites**.

Note: This Find dialog box is similar to the Search dialog box described in this chapter. This Find dialog box is used primarily to locate drawing files. The Search dialog box is described in the section on "Searching for Content".

Choosing the **Find** option causes the Find dialog box to be displayed. Various drives and folders are searched using search criteria. The Find dialog box combines the usual Windows file/path search of files by name and location in the **Name & Location** tab and by date ranges in the **Date Modified** tab. The **Name & Location** tab, shown in Figure 12–11, permits you to search a specific path or paths that meet the search criteria.

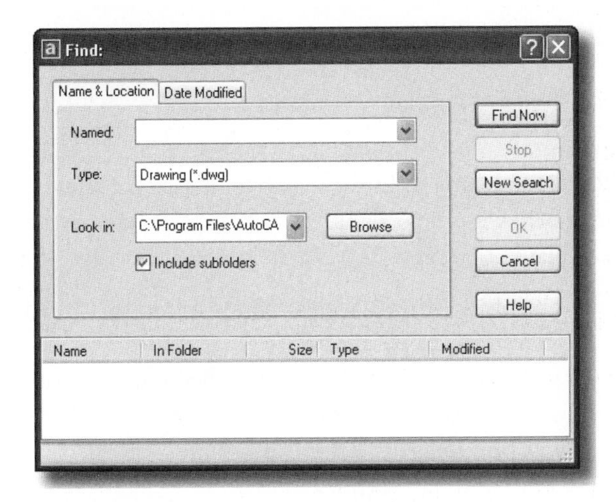

Figure 12–11 *Name & Location tab of the Find dialog box*

Named: permits you to read a list of files that have been searched. Here is where you can enter a name to search.

Type: permits you to specify the file type.

Look in: lets you specify the drive/path to search.

The **Browse** button displays the tree form of drives and folders available for searching.

The search will include subfolders if the **Include subfolders** check box is selected.

The **Date Modified** tab, shown in Figure 12–12, permits you specify dates and/or date ranges as criteria to search for drawings.

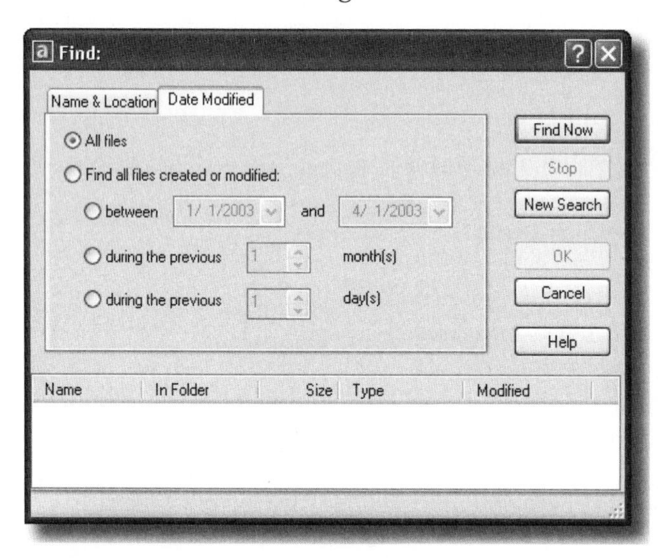

Figure 12–12 *Date Modified tab of the Find dialog box*

The **All Files** selection causes the search to display the names of all files meeting the criteria of the Name & Location tab parameters selected. If the **Named:** text box is blank, then AutoCAD will include drawings of all names.

The options below the **Find all files created or modified:** radio button include options for date ranges that are **between** (text box for date) **and** (text box for date), **during the previous** (text box for number of months) **months**, and **during the previous** (text box for number of days) **days.** The **between** (text box for date) **and** (text box for date) option has pulldown calendars to allow you to select the month and day with the mouse.

The **Find Now** button starts the search.

The **Stop** button will stop the search.

The **New Search** button clears the **Named:** text box for a new search.

Once the search for drawings that match the name and criteria has caused one or more drawings to be listed in the list box at the bottom of the dialog box, you can highlight onem and it will be opened for editing when you choose the **OK** button.

The **Cancel** button closes the Find dialog box.

NAVIGATION

The **Back** and **Forward** arrows when selected, cause the contents in the Tree View and content area to scroll backwards and forwards respectively through the history of viewing activity. The **Up** button causes the Tree View to display contents of the parent folder.

SEARCHING FOR CONTENT

The Search feature provides a means to locate containers, content type, and content in a manner similar to the Windows Find feature. Choosing the **Search** button causes the Search dialog box to be displayed as shown in Figure 12–13.

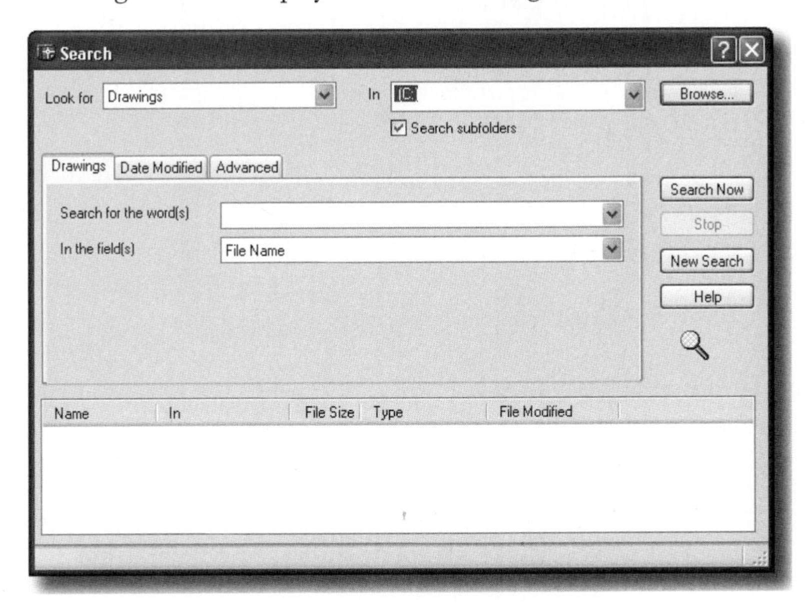

Figure 12–13 *Search dialog box with the Drawings tab displayed*

Note: This Search dialog box is similar to the Find dialog box described earlier in this chapter. This Search dialog box is used primarily to locate content in containers in addition to drawing files. The Find dialog box is described in the section on "Load" and is used mainly for locating drawing files.

From the **Look for** list box in the Search dialog box, select the type of content you wish to find. Available options include **Blocks, Dimstyles, Drawings, Drawings and Blocks, Layers, Layouts, Linetypes, Textstyles**, and **Xrefs**.

From the **In** list box, select the location for searching. Available options include My Computer, local hard drives (C:) (and any others), 3½ Floppy (A:), and any other drives.

The **Browse...** button causes the Browse For Folder dialog box to be displayed, which has a file manager window as shown in Figure 12–14. Here you can specify a path by stepping through the levels to the location which you wish to search. When the final location is highlighted, select **OK**, and the path to this location is displayed in the **In** list box.

Figure 12–14 *Browse For Folder dialog box*

The number of tabs that will be displayed in the Search dialog box depends on the type of content selected in the **Look for** list box. Each of the content types will have a tab that corresponds to the type of content selected. If **Drawings** is selected, there are two additional tabs: **Date Modified** and **Advanced**.

On the **Drawings** tab, the **Search for the word(s)** text box lets you enter the word(s), such as a drawing name or author, to determine what to search for. The **In the field(s)** text box lets you select a field type to search. Fields include **File Name, Title, Subject, Author**, and **Keywords**.

On the **Date Modified** tab (available only when **Drawings** is selected in the **Look for** text box of the Search dialog box) you can specify a search of drawing files by the date they were modified, as shown in Figure 12–15. The search will include all files that comply with other filters if the **All files** radio button is selected. If the **Find all files created or modified:** radio button is selected, you can limit the search to the range of dates specified by selecting one of the following secondary radio buttons. The **between** and **and** radio button lets you limit the search to drawings modified between the dates entered. The **during the previous months** radio button lets you limit the search to drawings modified during the number of previous months entered. The **during the previous days** radio button lets you limit the search to drawings modified during the number of previous days entered.

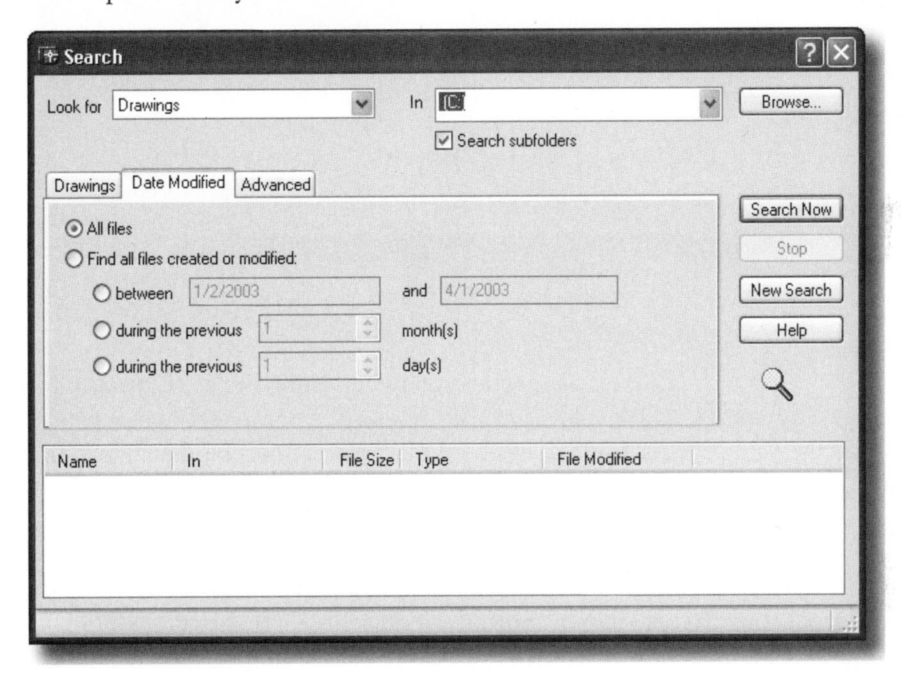

Figure 12–15 *Display of the Date Modified tab in the Search dialog box*

On the **Advanced** tab (available only when **Drawings** is selected in the **Look For** text box of the Search dialog box) you can specify a search of files by additional parameters as shown in Figure 12–16. In the **Containing:** text box you can specify a file by one of four options: **Block Name, Block and Drawing Description, Attribute Tag,** and **Attribute Value.** The search will then be limited to items containing the text entered in the **Containing text:** text box. The **Size is:** text boxes let you limit the search to drawings that are **At Least** or **At Most** as many KBs (size of the drawing file) as the number entered into the second text box.

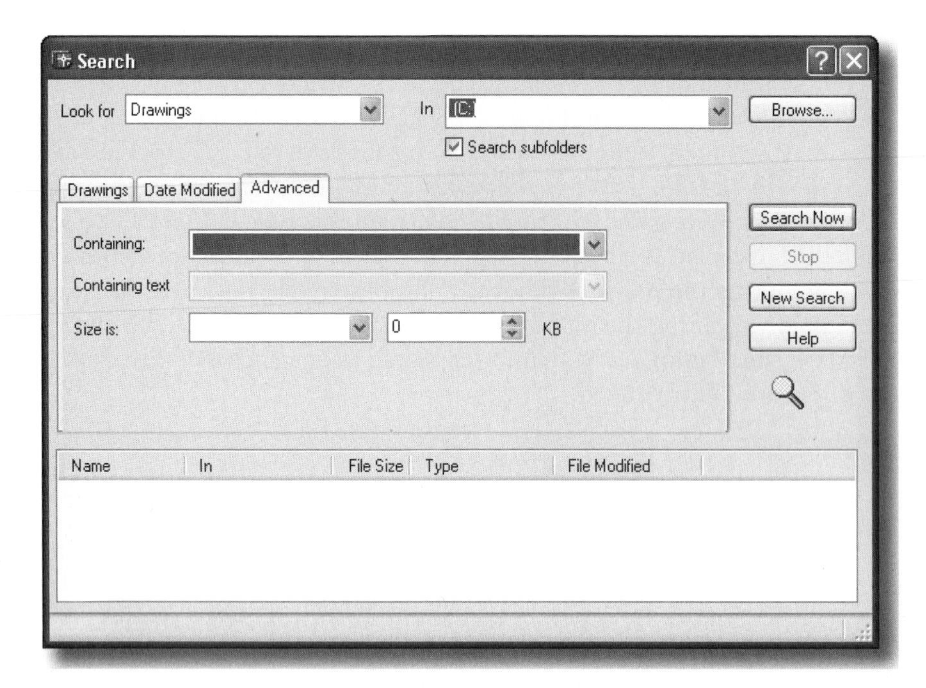

Figure 12–16 *Display of the Advanced tab in the Search dialog box*

All of the other tabs (**Blocks**, **Dimstyles**, **Drawings and Blocks**, **Layers**, **Layouts**, **Linetypes**, **Textstyles**, and **Xrefs**) have a **Search for the name** text box in which you can enter the name of the item you wish to find. This can be a block name, dimstyle, or one of the other content types.

To the right of the tab section are four buttons: **Search Now**, **Stop**, **New Search**, and **Help**.

Once the parameters have been specified, such as the path, content type, and date modified, selecting the **Search Now** button initiates the search. Select the **Stop** button to terminate the search. Selecting the **New Search** button lets you specify new parameters for another search. Selecting the **Help** button causes the AutoCAD Command Reference (Help) dialog box to be displayed.

Each time a search is performed, the text entered in the **Search for the name** text box or the **Search for the word(s)** text box is saved. If you wish to repeat the same search again, select the down arrow next to the text box and select the text associated with the search you wish to repeat.

FAVORITES

The **Favorites** button causes the Autodesk Favorites feature to display icons in the content area representing shortcuts to frequently used files. These are files for which you have previously set up shortcut icons to make them quickly accessible. The Tree View displays and highlights the subfolder *Favorites* of the folder Autodesk, as shown in Figure 12–17.

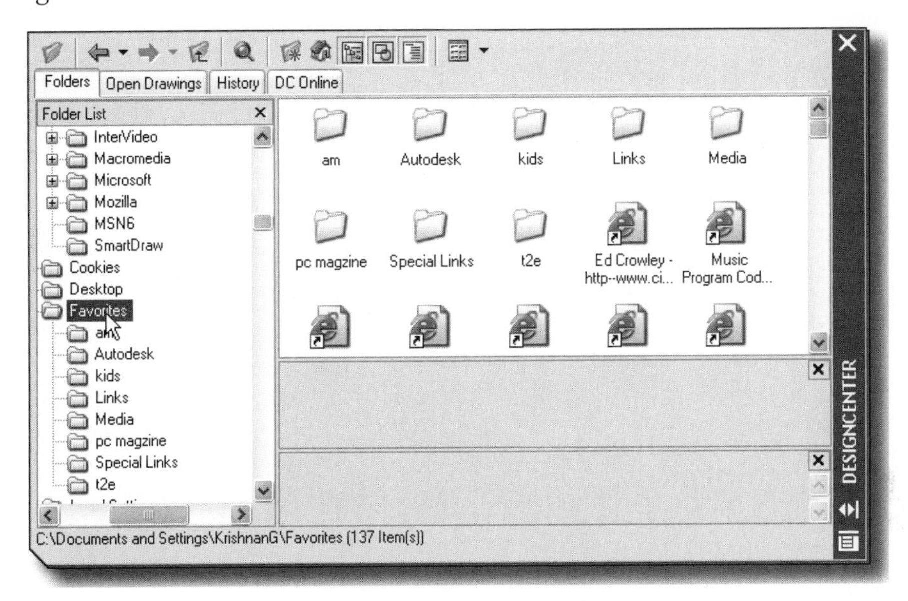

Figure 12–17 *The Tree View with the Autodesk Favorites folder selected*

Note: Only the shortcuts are in the *Favorites* folder. The files themselves remain in their original locations

A drawing or image file or a folder can be added to *Favorites* by right-clicking on the file name or folder name in the content area or Tree View and choosing Add To Favorites from the shortcut menu. To view and organize items in *Favorites*, choose Organize Favorites from the shortcut menu. An Autodesk window is displayed for managing items in *Favorites*.

When the *Favorites* folder is highlighted in the Tree View, the content area displays the icons, which are shortcuts to other files or folders. Double-click on the icon for the desired folder or file.

After you have double-clicked the desired icon in the content area, the selected file or folder will be displayed in the Tree View and it will be highlighted.

And, in turn, the content area will display the contents of the highlighted folder or file. Figure 12–18 shows the selection of the *Favorites* folder and corresponding selection.

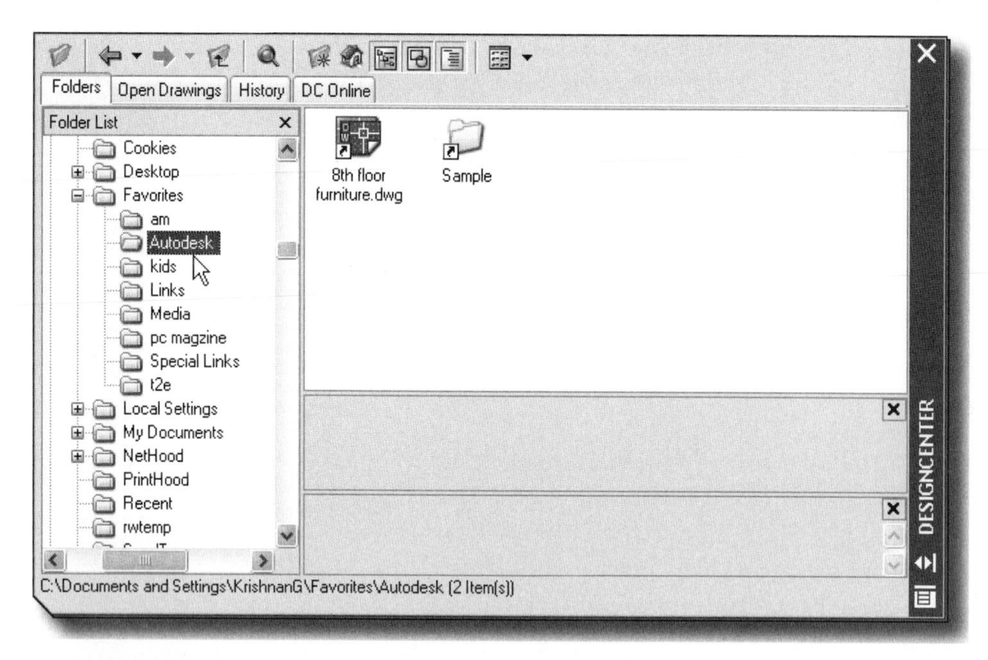

Figure 12–18 *Selection of a subfolder in the Favorites folder*

TREE VIEW, PREVIEW, AND DESCRIPTION

The buttons for the Tree View, Preview, and Description simply cause their respective panes to be displayed or not displayed, depending on their current status.

HOME

The **Home** button causes the Tree View to display the drives, folders, subfolders, and files that are located on your computer's desktop. The hierarchy is expanded, and the *DesignCenter* folder is highlighted with its contents displayed in the content area , as shown in Figure 12–19. This is the type of view you normally see when you open Windows Explorer.

From there you can step through the path(s) necessary to the desired location. Selection of the **Folders** tab is used whenever you want to return from Open Drawings tree view and History view (explained later this section) to display the drives, folders, subfolders, and files that are located on your computer's desktop.

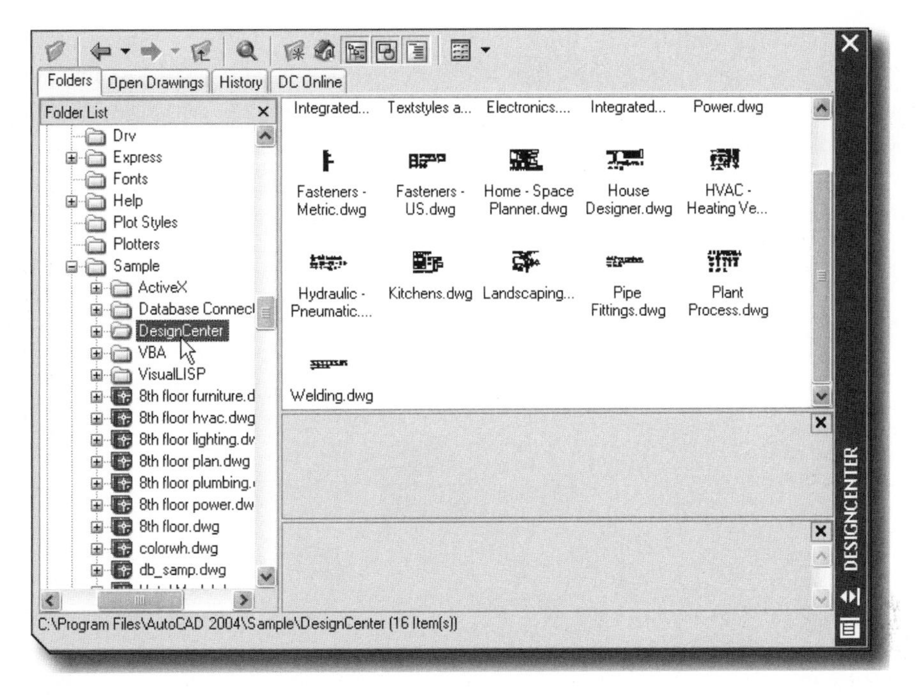

Figure 12–19 *DesignCenter displaying the Desktop with the DesignCenter folder highlighted*

VIEWING CONTENT

The names of content type, such as blocks, linetypes, linestyles, and so on, and containers such as drawings, image files, folders, drives, networks, and Internet locations, can be viewed in the Tree view as well as the content area. Names of content can be viewed in the content area but not in the Tree View pane. Either pane can be used to move up and down through the path from the drive to the item of content. It is usually quicker to navigate the path in the Tree View because of its ability to display multiple levels of hierarchy. When the container (a folder for drawings/images and a drawing/content type for blocks, images, and other items) appears in the Tree View, select it and then view the name of the items of content (if any) in the content area.

USING THE TREE VIEW

As mentioned earlier, in the Tree View you can view the content type and container, but not the actual content. You can use the Tree View to display the icon for an item of content in the content area. Just doing this will not, however, cause a raster image to be displayed in the Preview panel, as shown in using the content area in this section.

The folder named *Sample* that comes with the AutoCAD program contains numerous drawings which will be used in this section as examples. First, select the Up button as many times as necessary to display the level (in the Tree View pane) that includes the branch leading through the folder named *AutoCAD2004* (or the folder in which AutoCAD has been installed) to the sub-folder named *Sample*. This, again, might require going all the way back to a particular drive. Then click the plus sign or double-click the folder name or drive name to display the folders within a folder. Figure 12–20 shows the folder named *Sample* highlighted in the Tree View. Note that the list of content folders is displayed in the content area.

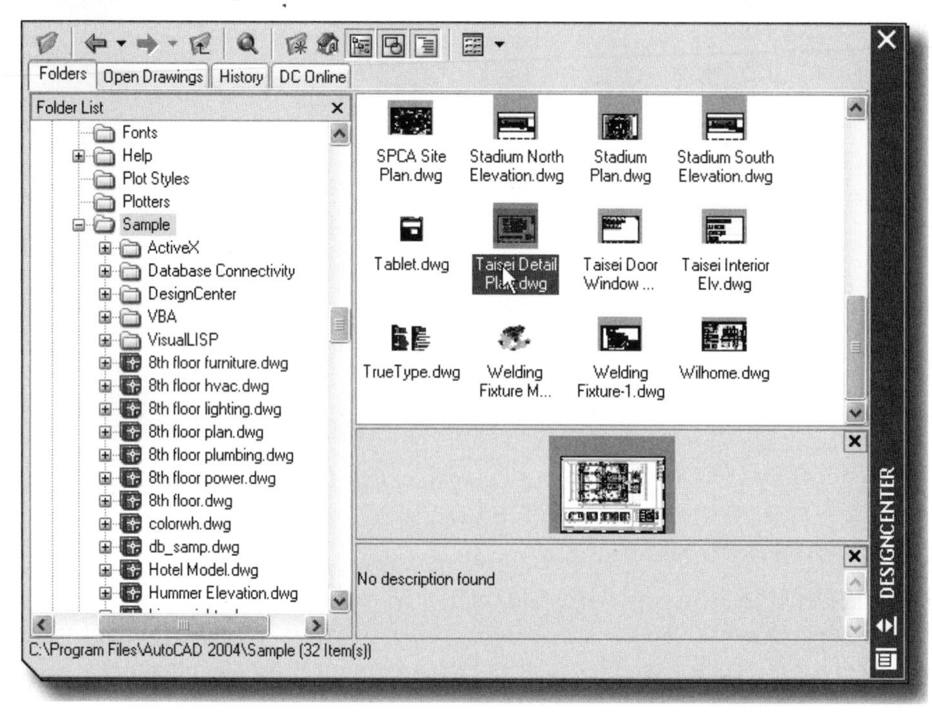

Figure 12–20 *The Tree View showing the drawing named Taisei Detail Plan in the folder named Sample*

Select the Taisei Detail Plan in the Tree View. Figure 12–21 shows the drawing named Taisei Detail Plan highlighted in the Tree View. Note that the list of content types is displayed in the content area.

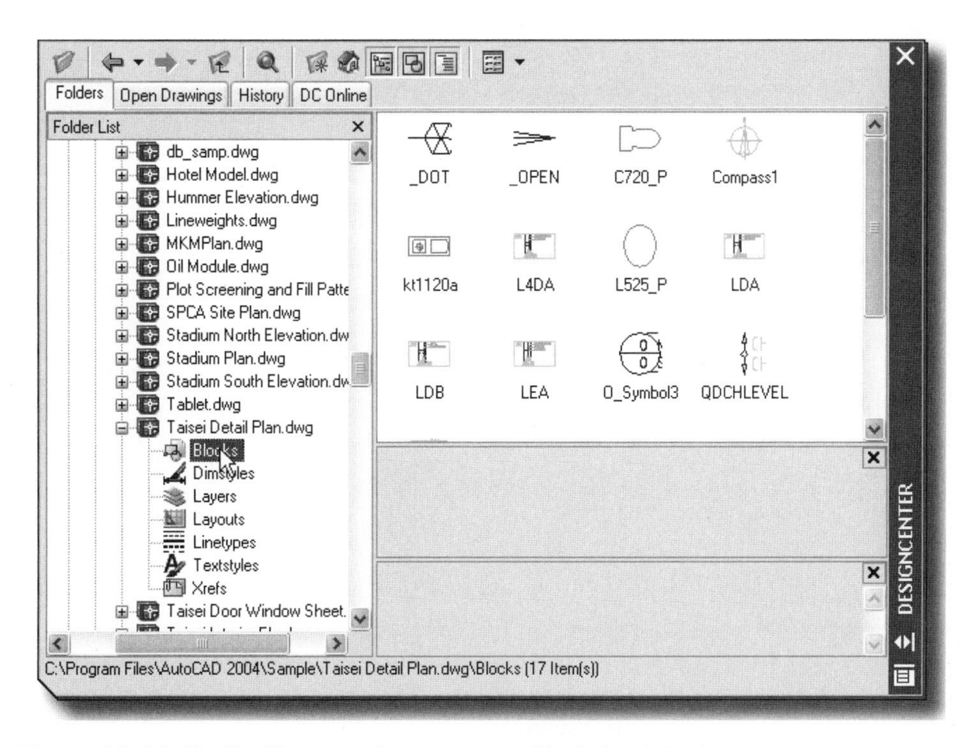

Figure 12–21 *The Tree View with the content type Blocks highlighted*

Next, double-click on the drawing named "Taisei Detail Plan" in the Tree View or select the box with the plus in it to the left of the drawing name. The list of content types will be listed below the drawing name, as well as in the content area. To display the names of individual items of content, blocks for example, select the content type Blocks in the Tree View, as shown in Figure 12–22.

When one of the block icons displayed in the content area is chosen, AutoCAD displays the corresponding preview image and description, if any, in the Preview pane and Description pane, respectively. Figure 12–23 shows the selection of a block named C720 P, and its corresponding preview image and description are shown in the Preview pane and in the Description pane.

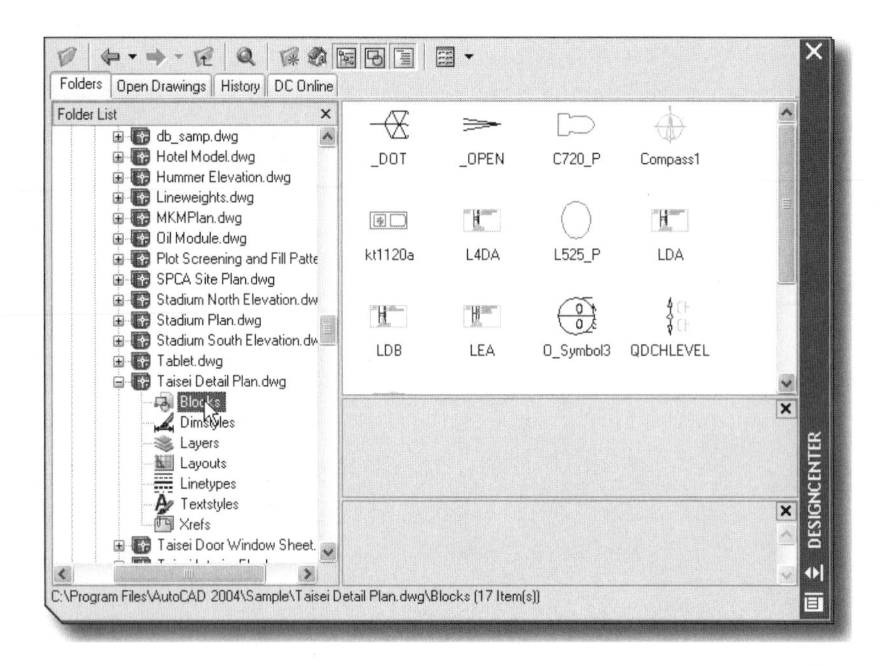

Figure 12–22 *An example of the Tree View displaying container and content type and the content area displaying content*

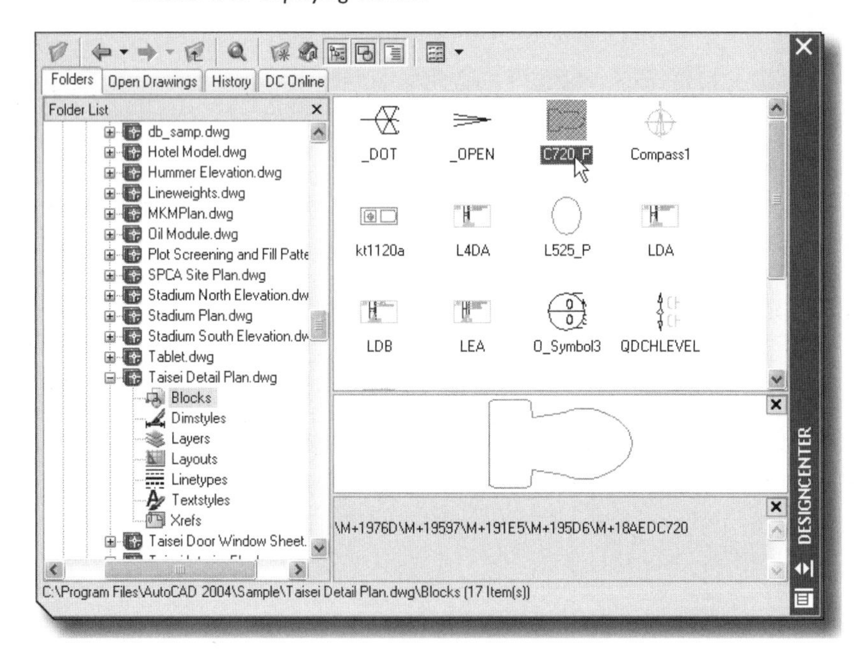

Figure 12–23 *Selection of a block named **UB1216J** and the corresponding preview image and description are shown in the **Preview** pane and in the Description pane*

USING THE CONTENT AREA

As mentioned earlier, the content area is used for displaying content. It can also display containers and content type. A container is a source of content such as a drawing. In the hierarchy of what can be displayed, containers are one step higher than content types. Content types are one step higher than content. From drawing files and other containers, the next step up is a folder. Then, progressing up through the folders (if there is a hierarchy of folders), next is the drive, which may lead to a network, a Web site on the Internet, a floppy diskette, or a CD-ROM.

Note: The content area displays only items of the same level that are members of a single container one level above them. The content area does not display more than one level at a time. For example, content types such as blocks, xrefs, and layers are members of one drawing. If there are any blocks in the drawing, then they are listed when you have selected that drawing's content type called Blocks. Several drawings may be members of a particular folder. Folders are members of one drive or folders may actually be subfolders of a folder one level up. Remember, in the content area, only one level of the hierarchy is displayed and all items shown are members of the same component/container one level above. For displaying more than one level at a time, use the Tree View.

You can navigate up and down through the hierarchy of drives, folders, drawings, content types, and content in the content area. It is not as easy as in the Tree View, however, because only one level is displayed at a time. If the content is a drawing, block, or image, you can view a raster image of it in the Preview pane. Simply select the item in the content area. Remember that you can enlarge the Preview pane for viewing, even if only temporarily. As mentioned earlier, you can double-click the title bar to dock or undock the DesignCenter.

For example, there is a block named "UB1216J_F" in the drawing named "*Taisei Detail Plan*" in the folder named *Sample*. To get to the block named "UB1216J_F" in the content area, as described for the example in the Tree View above, select the **Up** button as many times as necessary to display the level that includes the branch leading to the drawing named "*Taisei Detail Plan*". This might require going all the way back to a particular drive. Then double-click the sequence of drive and folders on the path to the drawing named "*Taisei Detail Plan*" in the content area, and then double-click on "*Taisei Detail Plan*" icon. When you double-click the icon for "*Taisei Detail Plan*", the icons for content types will be displayed, including blocks, dimstyles, layers, layouts, linetypes, textstyles, and xrefs, as shown in Figure 12–24.

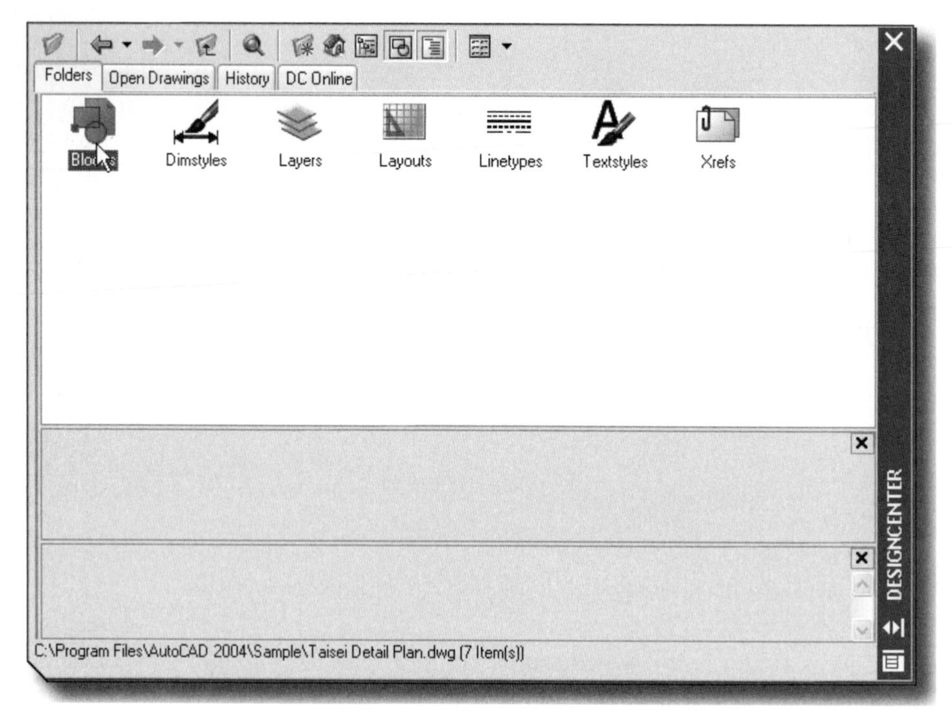

Figure 12–24 *The Content Area with the display of the content types*

AutoCAD also displays the path to the selected drawing at the bottom of the Design-Center window, as shown in Figure 12–24. Select the Blocks icon, and AutoCAD displays the icons (when the display is set to Large icons) representing all the blocks in the selected drawing. Choose one of the block icons (do not double-click), and AutoCAD displays corresponding raster images and descriptions in the Preview pane and Description pane respectively. Figure 12–25 shows the selection of the block named "UB1216J_F" and the corresponding display of its raster image and description. You can enlarge the Preview pane to get a bigger picture. Note, however, that it is a raster image, and sharp details are not available in close-up views.

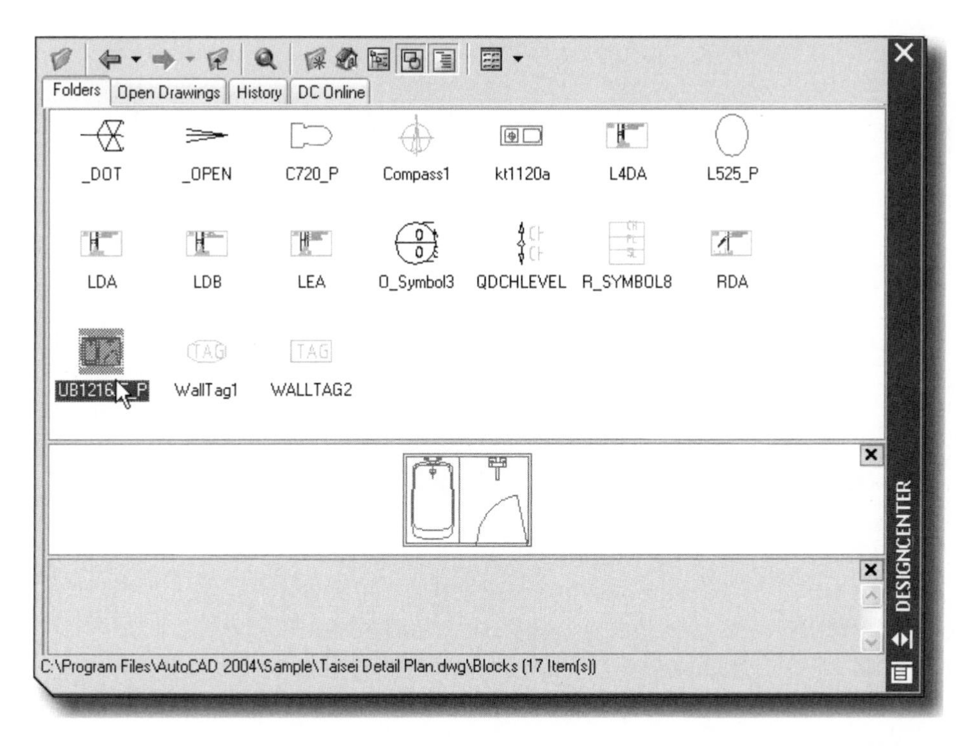

Figure 12–25 *The Preview pane displaying the raster image of the block named "**UB1216J_F**"*

VIEWING IMAGES

If the item of content is a bitmap image, its name can be displayed in the content area by selecting the name of its container in the Tree View. For example, there is a sub-folder named *Sample* in the folder named *AutoCAD 2004*. In the folder named *Database Connectivity* is another folder named *CAO*. When *CAO* is highlighted in the Tree View, you can choose the bitmap file named *dbcm_query.bmp* in the content area, and the icon resembling a question mark is displayed in the Preview pane, as shown in Figure 12–26. Also, there is a description that can be viewed in the Description pane.

All the images in the subfolder have a file extension of .BMP. In another sub-folder named *Textures* whose path is *MyComputer/Local Disk /Documents and Settings/< User Name>/Local Settings/ Application Data/Autodesk/AutoCAD 2004/ R16.0/enu/Textures*, there are files with the extension .TGA. You can view these files in the same manner as the .BMP file in the previous example, as shown in Figure 12–27.

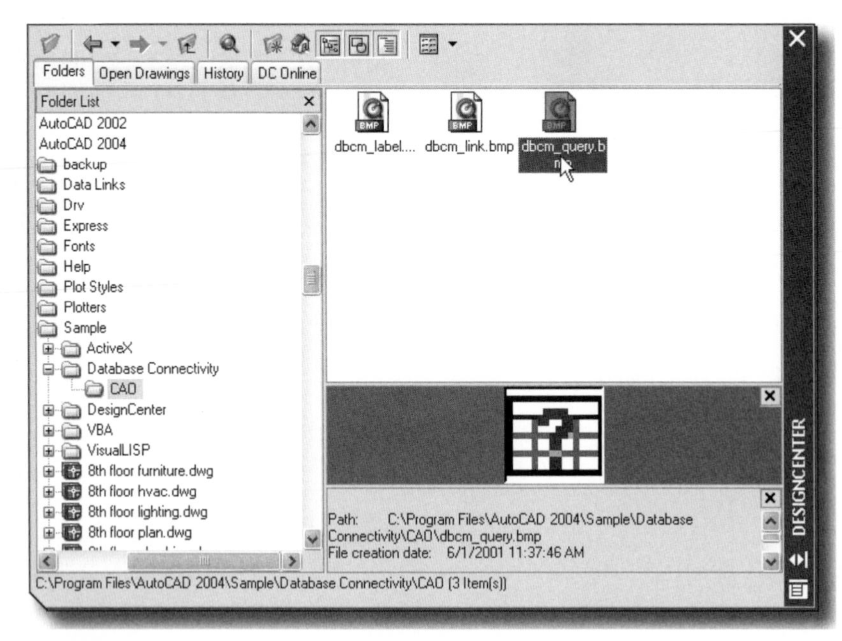

Figure 12–26 *Display of the image named "dbcm_query.bmp" with its raster image in the Preview pane and its corresponding description in the Description pane*

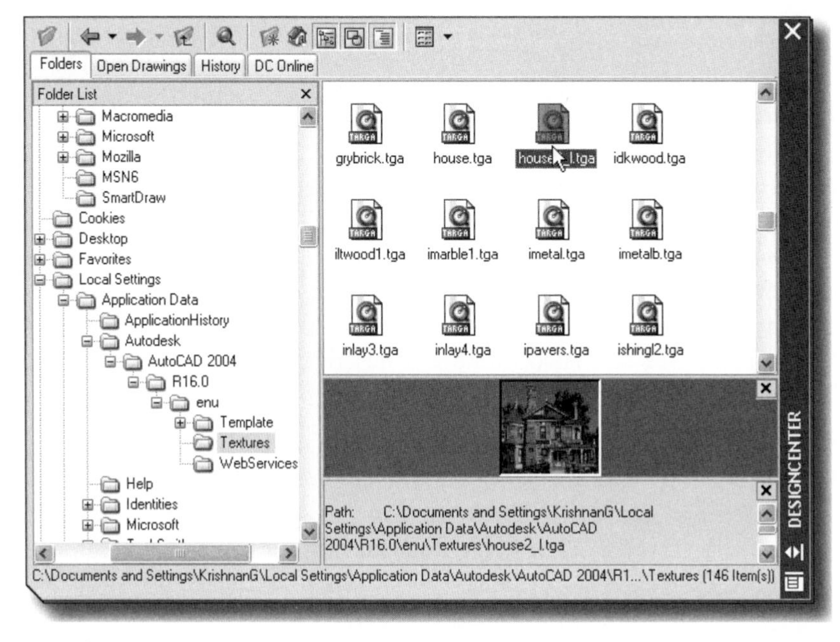

Figure 12–27 *The image named "House_2l.tga" with its raster image in the Preview pane and corresponding description in the Description pane*

Note: When you have selected the name of a folder in the Tree View, if there are any files in that folder that are viewable, their names will appear in the content area. Some folders may have other files in them, but if they are not viewable, their names will not appear.

LOAD

As described in the section on "Toolbar Buttons," the **Load** button causes the Load dialog box to be displayed, from which you can select a drawing file whose contents will be loaded onto the content area. You can also load a drawing or folder from the Find dialog box by dragging and dropping it into the content area or Tree View pane. Or you can right-click and select LOAD INTO CONTENT AREA from the shortcut menu.

Note: The Load feature does not load the item selected into your drawing. It only loads its icon into the content area, as shown in the next section on adding content to drawings.

REFRESHING THE CONTENT AREA AND TREE VIEW

When file manipulations are made to the content area or Tree View, they do not always show up immediately on the screen. If this happens, you can right-click in one of the panes and choose REFRESH from the shortcut menu. The views will be updated.

THE OPEN DRAWINGS TAB

The **Open Drawings** tab of the DesignCenter window displays a list in the Tree View of drawings that are currently open, as shown in Figure 12–28. When you select one of the drawings, its content types will be displayed in the content area. Or you can double-click on the file name, and it will display the content types at one level below the name you double-clicked on. Selecting one of the content types causes content of that type (if any exists in the drawing) to be listed in the content area.

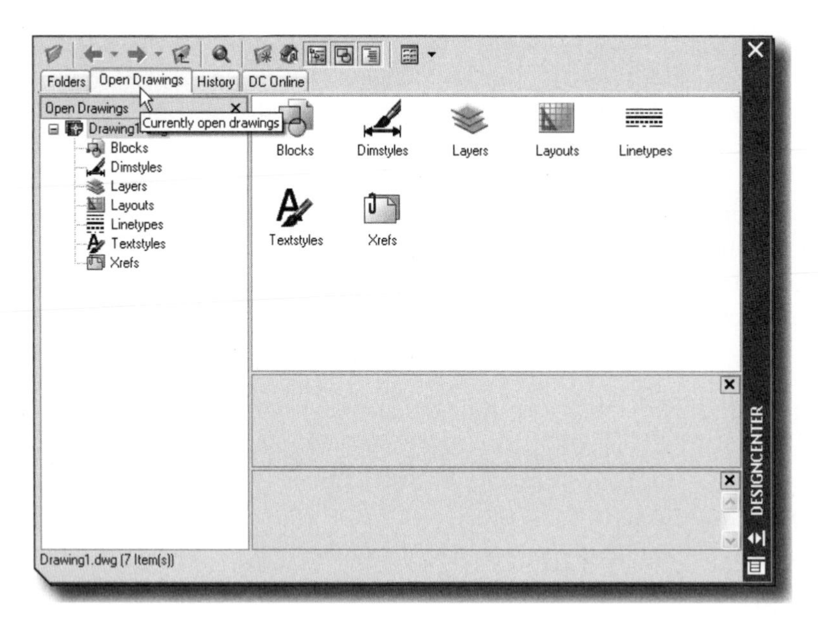

Figure 12–28 *The DesignCenter window with the Open Drawings tab displayed*

THE HISTORY TAB

The **History** tab of the DesignCenter window displays the last 20 items accessed through AutoCAD DesignCenter, including their paths, as shown in Figure 12–29.

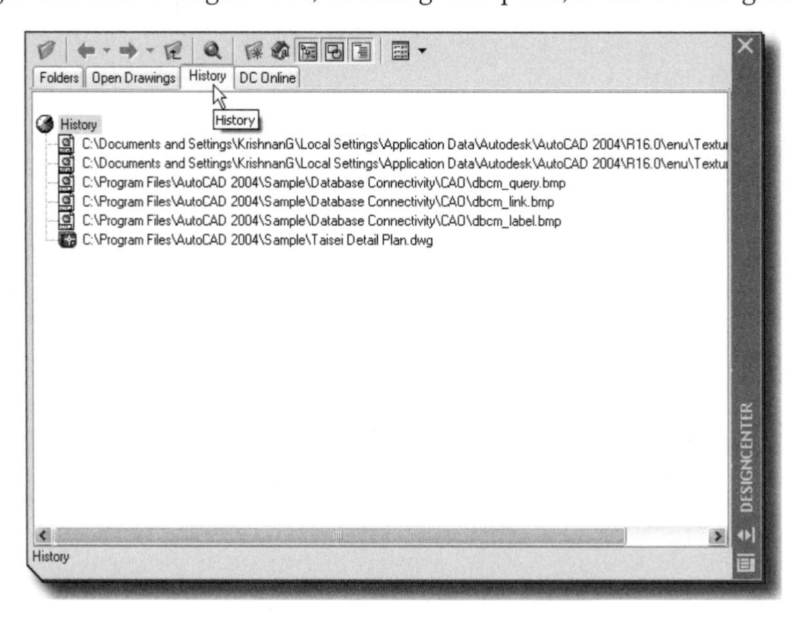

Figure 12–29 *The DesignCenter window with the History tab displayed*

THE DC ONLINE TAB

The **DC Online** tab of the DesignCenter window, when selected, causes AutoCAD to log on to the DesignCenter Online on the Internet, provided your computer is Internet-ready (see Figure 12–30).

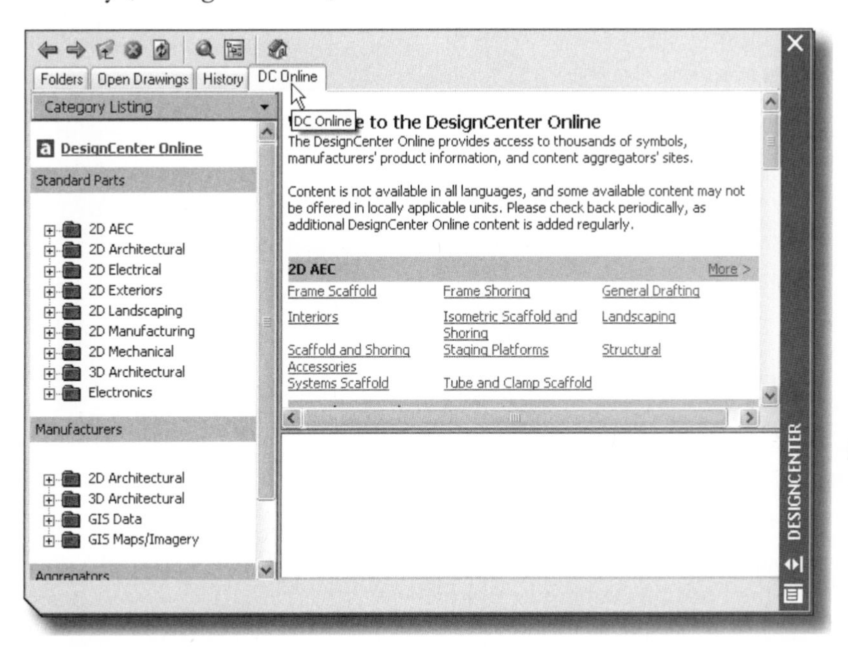

Figure 12–30 *The DesignCenter window with the DC Online tab displayed and Category Listing selected.*

There are two panes in the **DC Online** tab of the DesignCenter window. The left pane has four views: Category Listing, Search, Settings and Collections.

Category Listing

In the Category Listing view there are three categories: Standard Parts, Manufacturers, and Aggregators (see Figure 12–30). The Standard Parts category includes groups of drawings and images that can be used in architectural and engineering design disciplines such as architecture, landscaping, mechanical and GIS. For example, you can select the box to the left of the **2D Architectural** group and then select the box to the left of the **Landscaping** sub-group when expanded. This expands to the list of types of content. If you select **Tables** from this list, the right pane will display thumbnail sketches of drawings or images that are available for download (see Figure 12–31). When you select an image, additional links are displayed in the lower half of the right pane along with a larger sketch of the content.

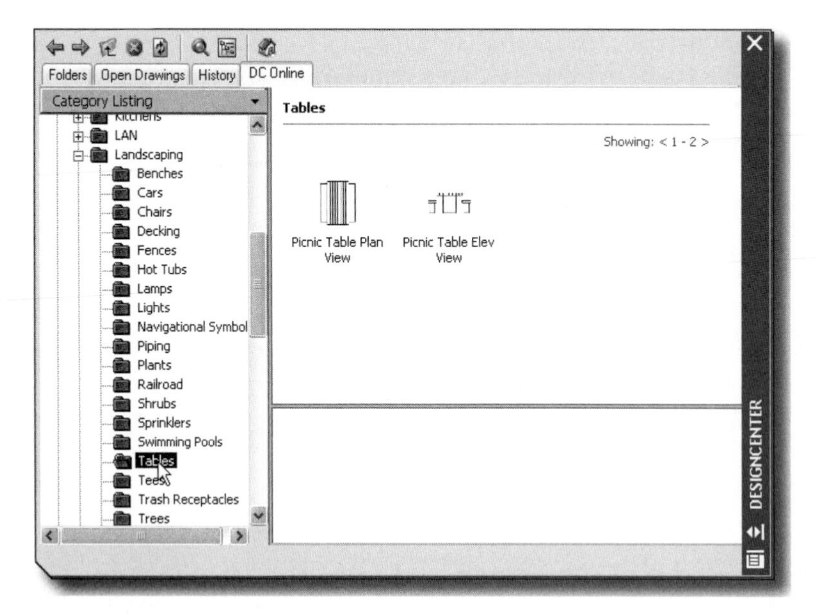

Figure 12–31 *Category Listings with 2D Architectural/Landscaping/Tables groups displayed*

Another group of content related to tables can be found under the **2D Architectural** group by selecting the **Furniture** subgroup and then selecting **Tables** (see Figure 12–32).

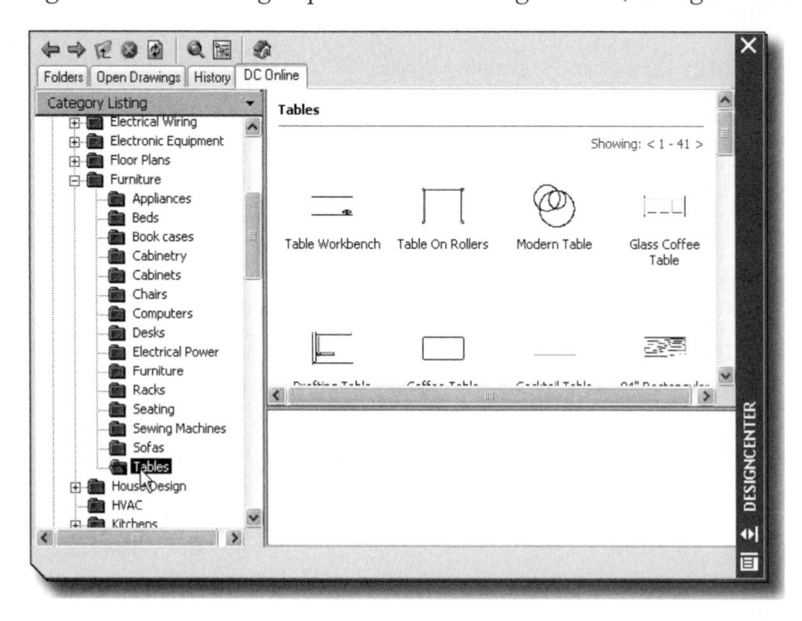

Figure 12–32 *Category Listings with 2D Architectural/Furniture/Tables groups displayed*

Note: If you are looking for drawings, images, or links to drawings and images of a particular type of content (such as tables in the example above), there is usually more than one group or path to a group that includes such drawings, images or links. See the explanation of using the Search view in this section.

The Manufacturers category includes groups of Web sites or Internet addresses of manufacturers of products used in architectural and engineering construction. Through these Web sites, drawings and images can be downloaded where available. For example, in the Manufacturers group (similar to the selection in the Category Listing group), you can select the box to the left of the **2D Architectural** group and then select the box to the left of the **Landscaping** subgroup when expanded. This expands to the list of types of content. If you select **Outdoor Furnishing** from this list, the right pane will display one or more Internet addresses of Web sites from which you can access drawings, images, or other content data that are available for download (see Figure 12–33).

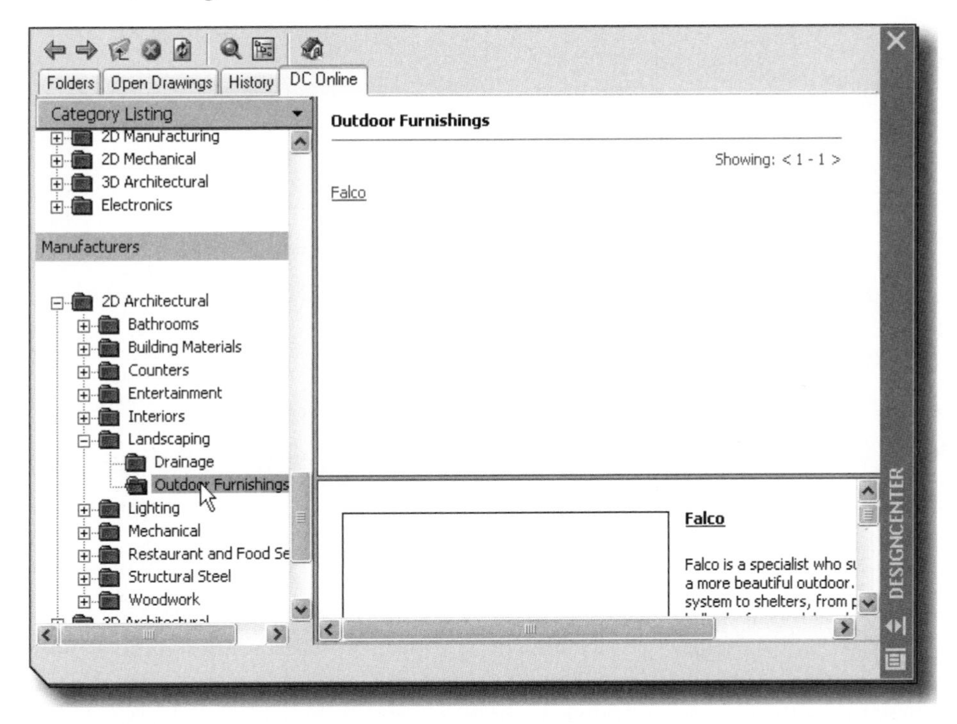

Figure 12–33 *Manufacturers with 2D Architectural/Landscaping/Outdoor Furnishings groups displayed*

The Aggregators category contains lists of libraries compiled by commercial catalog providers. Like the items in the Manufacturers category, you can access Web sites that contain or lead to drawings and blocks for use in architectural and engineering design and drafting. For example, in the Aggregators group (similar to the selection in the Category Listing and Manufacturers group), you can select the **AEC Aggregators** from this list, and the right pane will display one or more Internet addresses of Web sites from which you can access drawings, images, or other content data that are available for download (see Figure 12–34).

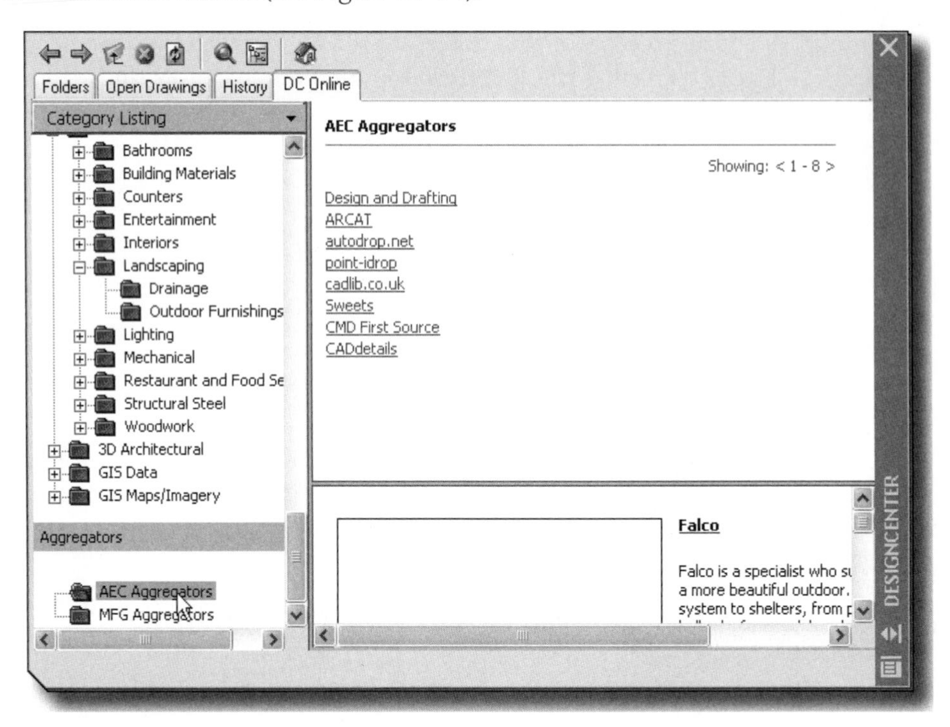

Figure 12–34 *Aggregators groups displayed*

Search

In the Search view there is a text box in which you can enter words or combinations of characters to tell AutoCAD what type of content to search for. You can display details for how to use Boolean and multiple-word search strings by selecting the **Need Help?** link. Figure 12–35 shows an example of entering **table** in the text box and selecting the **Search** button. The right pane shows the result of the search. Selecting one of the content in the right pane causes additional links to be displayed in the lower half of the right pane along with a larger sketch of the content.

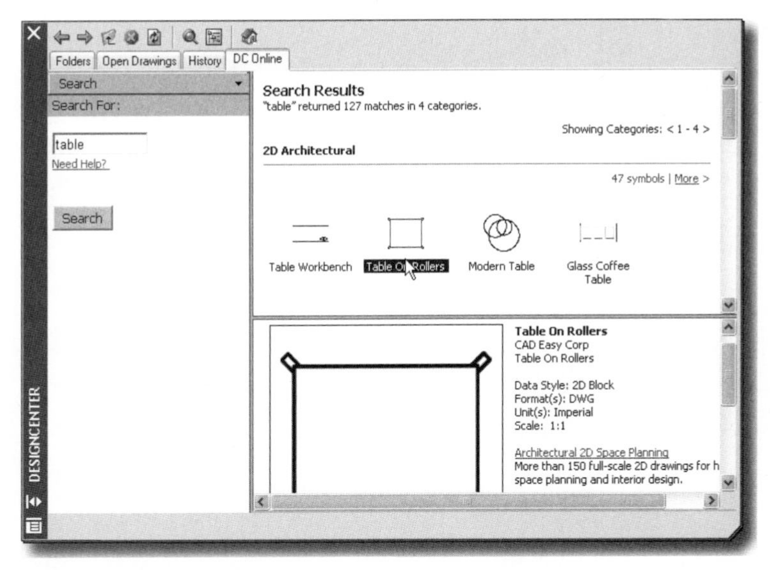

Figure 12–35 *The DesignCenter window with the DC Online tab displayed and Search view selected*

Settings

In the Settings view there are two text boxes: **Number of Categories per page:**, and **Number of Items per page:**. As shown in Figure 12–36, you can select from 5, 10, and 20 in the Number of Categories per page and from 50, 100, and 200 in the Number of Items per page text boxes to specify the **Max Search** numbers for Categories and Items respectively.

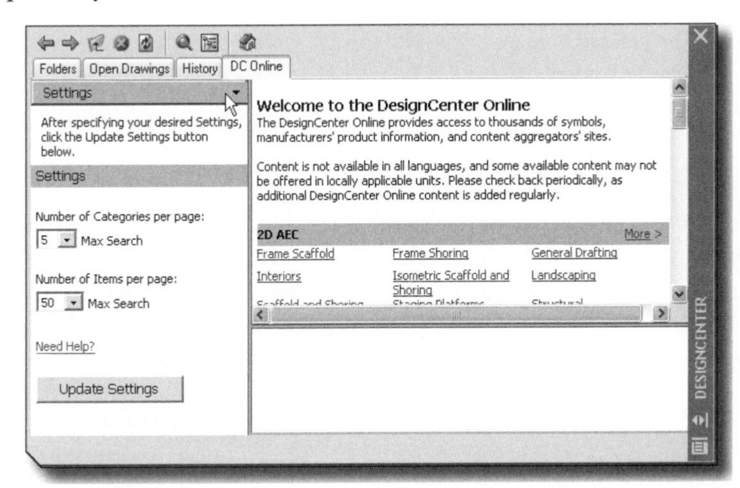

Figure 12–36 *The DesignCenter window with the DC Online tab displayed and Settings view selected*

Collections

In the Collections view there is a list of collections for each of the three categories with a check box beside each collection (see Figure 12–37). Place a check in a particular collection's check box that you wish to be displayed in the Category Listing view. Once you have selected/deselected the desired collections, select the **Update Collections** button, and AutoCAD will return to the Category Listing view displaying the list of collections whose check boxes have checks in them.

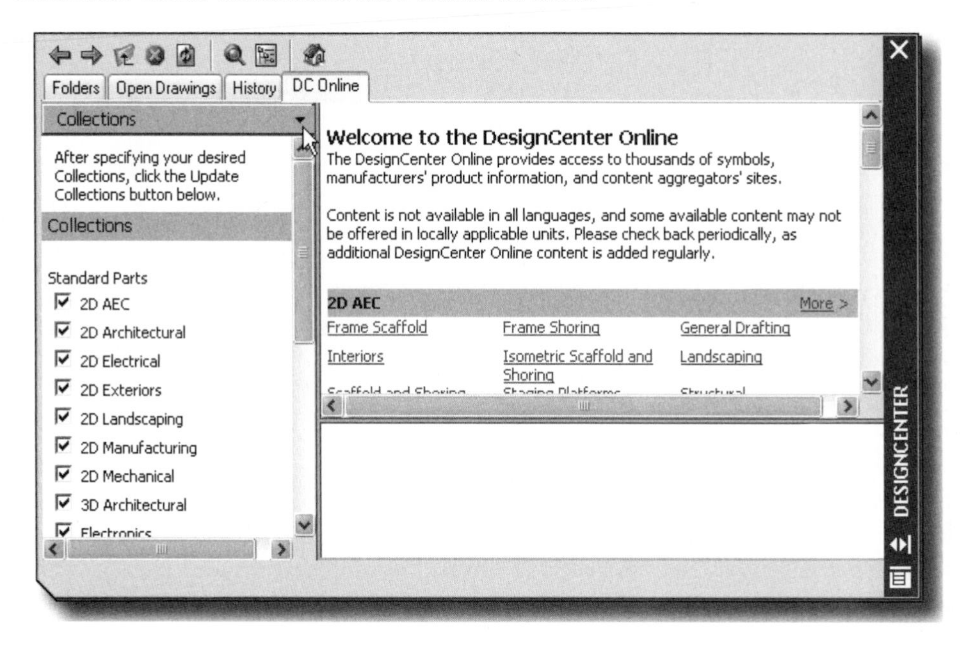

Figure 12–37 *The DesignCenter window with the DC Online tab displayed and Collections view selected*

ADDING CONTENT TO DRAWINGS

As discussed in the introduction to this chapter, blocks, external references, layers, raster images, linetypes, layouts, text styles, dimension styles, and custom content created by third-party applications are the content types that can be added to the current drawing session by using the AutoCAD DesignCenter. Content in the AutoCAD DesignCenter can be dragged into a tool palette for use in the current drawing.

LAYERS, LINETYPES, TEXT STYLES, AND DIMENSION STYLES

A definition of a layer or linetype or a style created for text or dimensions can be dragged into the current drawing from the content area. It will become part of the current drawing as though it had been created in that drawing.

BLOCKS

A block definition can be inserted into the current drawing by dragging its icon from the content area into the drawing area. You cannot, however, do this while a command is active. It must be done at the command prompt. Only one block definition at a time can be inserted from the content area. Figure 12–38 shows a Block in the content area of the AutoCAD DesignCenter being dragged into the tool palette of the current drawing.

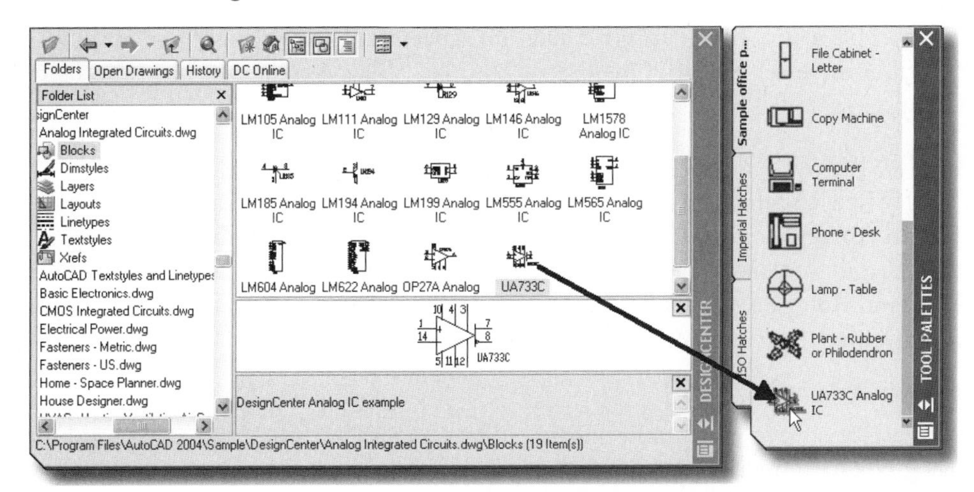

Figure 12–38 *The block named **UA733C Analog IC** being dragged from the DesignCenter window into the tool palette of the current drawing.*

Note: If you double-click a block and it has nested blocks, the hierarchy is flattened.

There are two methods of inserting blocks from the AutoCAD DesignCenter. One method uses *Autoscaling*, which scales the block reference as needed based on the comparison of the units in the source drawing to the units in the target drawing. Another method is to use the Insert dialog box to specify the insertion point, scale, and rotation.

Note: When dragging and dropping using the automatic scaling method, the dimension values inside the blocks will not be true.

When you drag a block definition from the content area or the Find dialog box into your drawing, you can release the button on the pointing device (drop) when the block is at the desired location. This is useful when the desired location can be specified with a running object snap mode in effect. The block will be inserted with the default scale and rotation.

To invoke the Insert dialog box, double-click the block icon or right-click the block definition in the content area or Find dialog box and then select **Insert Block** from the shortcut menu. In the Insert dialog box, specify the **Insertion point**, **Scale**, and **Rotation**, or select **Specify On-screen**. You can select **Explode** to have the block definition exploded on insertion.

RASTER IMAGES

A raster image such as a digital photo, print screen capture saved in a paint program as a bitmap, or a company logo can be copied into the current drawing by dragging its icon from the content area into the drawing area. Then specify the **Insertion point**, **Scale**, and **Rotation**. Or you can right-click the image icon and choose ATTACH IMAGE from the shortcut menu.

EXTERNAL REFERENCES

To attach an xref to the current drawing, drag its icon from the content area or Find dialog box with the right button on the pointing device into the drawing area. Then release the button and select ATTACH from the shortcut menu. The Attach Xref dialog box is displayed, from which you can choose between **Attachment** or **Overlay** as the **Reference Type** option. From the Attach Xref dialog box, specify the **Insertion point**, **Scale**, and **Rotation**, or select **Specify On-screen**

NAME CONFLICTS

When you copy, insert, or attach content into a drawing that already has an item of the content type with the same name, AutoCAD will display a warning. The item is not added to the drawing. If the item is a block or xref, AutoCAD checks to determine if the name is already listed in the database. The warning "Duplicate definition of [object][name] ignored" is displayed. If the xref exists in the drawing, the warning "Xref [name] has already been defined. Using existing definition" is displayed. If the item with the duplicate name is a layer, linetype, or other item that is not a block or xref, the warning "Add [object] operation performed. Duplicate definitions will be ignored" is displayed.

OPENING DRAWINGS

To open a drawing being displayed in the content area, right-click its icon and choose **Open In Window** from the shortcut menu. You can also drag and drop the icon in the drawing area. Be sure to drop the icon in an area that is clear of another drawing. If necessary, resize or minimize any other drawing(s) first.

EXITING AutoCAD DesignCenter

You can exit AutoCAD DesignCenter by either selecting the X in the upper right corner of the DesignCenter Window or entering **adcclose** at the AutoCAD Command: prompt.

REVIEW QUESTIONS

1. What command is used to invoke the AutoCAD DesignCenter?

   a. DSNCEN

   b. DGNCEN

   c. ADCENTER

   d. DGNCTR

2. Which of the following is considered to be drawing content types?

   a. blocks

   b. layers

   c. linetypes

   d. all the above

3. AutoCAD allows items to be directly edited from the DesignCenter.

   a. True

   b. False

4. Which of the following are not drawing content types?

   a. lines

   b. circles

   c. arcs

   d. all of the above

5. Within the DesignCenter a drawing is considered to be a ______________?

   a. content

   b. container

   c. folder

   d. a & b

   e. a & c

6. Invoking the AutoCAD DesignCenter command opens the DesignCenter dialog box.

   a. True

   b. False

7. The default position for the DesignCenter is in the lower right corner.

   a. True

   b. False

8. Which of the following areas of the DesignCenter displays the names and icons representing content?

   a. Toolbar

   b. Content area

   c. Tree

   d. Preview pane

   e. Description pane

9. Large Icons, Small Icons, List and Details are four optional modes of displaying content using which of the following buttons?

   a. LOAD

   b. FIND

   c. UP

   d. VIEWS

10. Within the Tree View pane which button can display the previous twenty items and their paths accessed through the AutoCAD DesignCenter?

    a. Open Drawings

    b. Desktop

    c. History

    d. None of the above

Utility Commands

INTRODUCTION

After completing this chapter, you will be able to do the following:

- Create, customize, and use tool palettes
- Partial Load drawings
- Manage drawing properties
- Use the geometric calculator to perform geometric calculations
- Manage named objects
- Delete unused named objects
- Use the utility display commands
- Use object properties
- Use X,Y,Z filters
- Use the SHELL command
- Set up a drawing by means of the MVSETUP utility
- Use the Layer Translator
- Use the TIME and AUDIT commands
- Customize AutoCAD settings
- Export and Import data
- Use object linking and embedding
- Understand CAD Standards
- Use script and slide commands

TOOL PALETTES

AutoCAD 2004 introduces a new feature called tool palettes. Tool palettes are created to make it quicker and easier to insert blocks, draw hatch patterns and implement custom tools developed by a third party. Blocks and hatch patterns are the primary tools that are managed with tool palettes.

Tool palettes are separate tabbed areas within the Tool Palettes window. This allows blocks and hatch patterns of similar usage and type to be grouped in their own tool palette. For example, one tool palette can be named Sample Office Project, with blocks representing office furniture's and fixtures as shown in Figure 13–2.

Figure 13–2 shows the default Tool Palettes window that comes with AutoCAD 2004 with other tabs attached: Sample Office Project, Imperial Hatches, and ISO Hatches, which all contain icons representing blocks, hatch patterns or both.

The TOOLPALETTES command causes the Tool Palettes window to be displayed.

Invoke the TOOLPALETTES command:

Standard toolbar	Choose the Tool Palettes command (see Figure 13–1)
Tools menu	Choose Tool Palettes Window
Command: prompt	**toolpalettes** (ENTER)

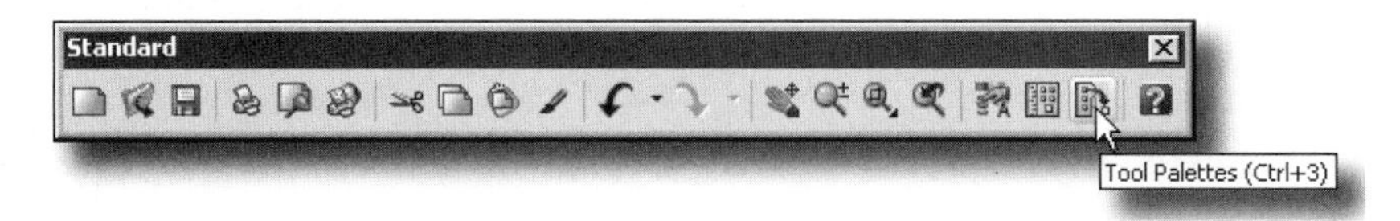

Figure 13–1 *Invoking the* TOOL PALETTES *command from the Standard toolbar*

AutoCAD displays the Tool Palettes window as shown in Figure 13–2.

Note: You can open or close Tool Palettes window with the CTRL+3 combination keystrokes.

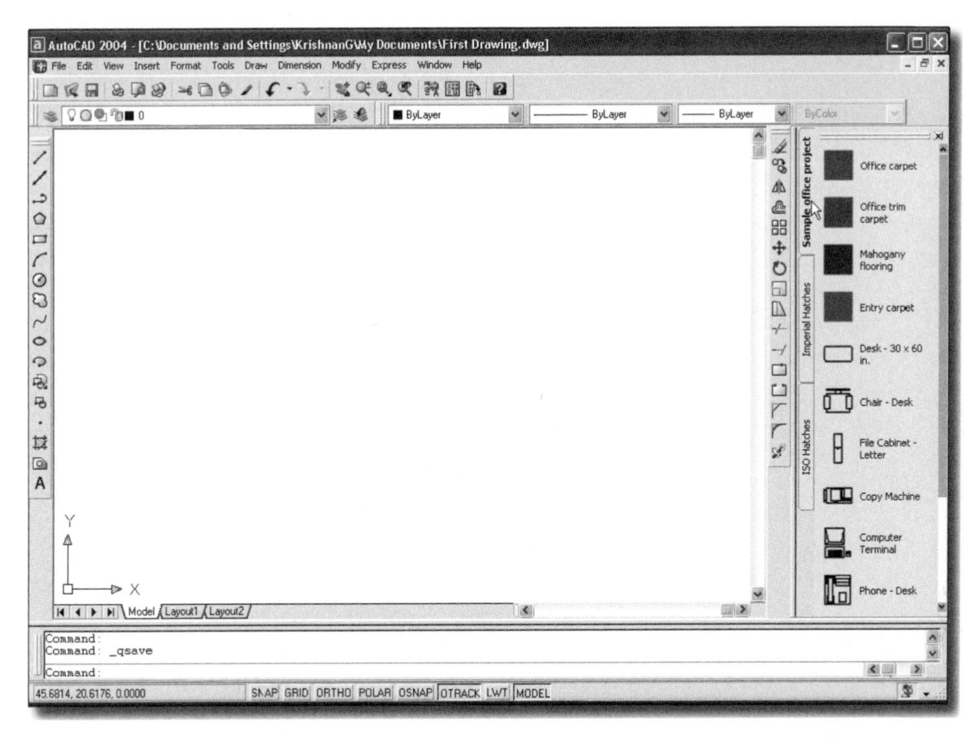

Figure 13–2 *The Tool Palettes window in the docked position with the Sample Office Project tab displayed*

DOCKING AND UNDOCKING THE TOOL PALETTES WINDOW

The default position for the Tool Palettes window is docked on the right side of the screen. Its position can be changed by placing the cursor over the double line bar at the top of the window and either double-clicking or dragging the window into the screen area (or across to a docking position on the left side of the screen). Double-clicking causes the Tool Palettes window to become undocked and to float in the drawing area as shown in in Figure 13–3. When the Tool Palettes window is undocked, it can be docked by double-clicking in the title bar (which may be on the left or right side of the window) or by placing the cursor over the title bar and dragging the window all the way to the side you wish to dock it.

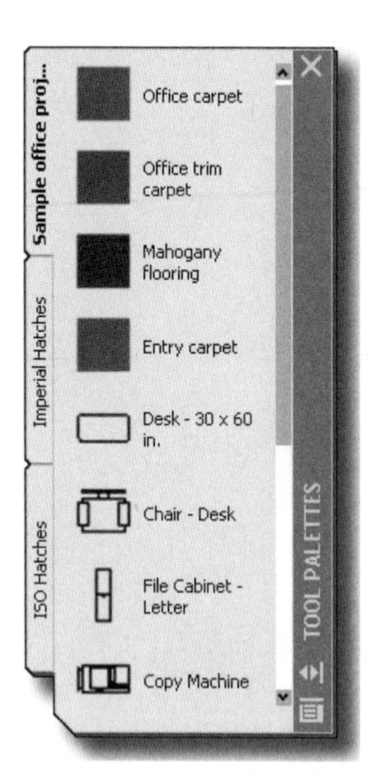

Figure 13–3 *The Tool Palettes window in the floating position*

TOOL PALETTES WINDOW SHORTCUT MENUS

A shortcut menu is displayed when you right-click in the Tool Palettes window as shown in Figure 13–4.

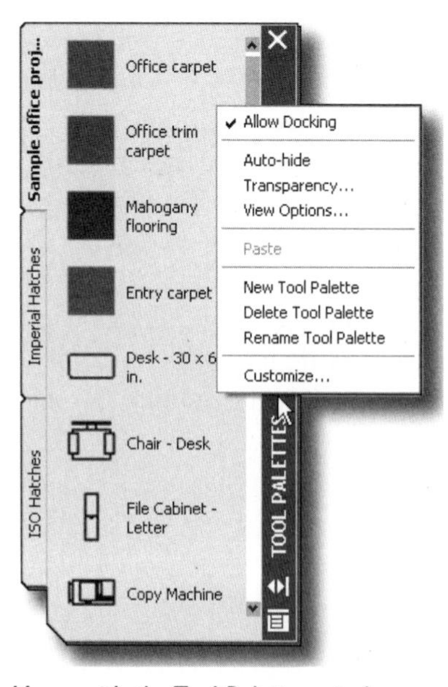

Figure 13–4 *The Shortcut Menu with the Tool Palettes window*

The **Allow Docking** option, when checked, allows you to drag the Tool Palettes window to one side of the screen and dock it. When not checked, the window cannot be docked.

The **Auto-Hide** option, when checked, causes the Tool Palettes window (only when floating) to be hidden, except for the title bar, when the cursor is not over the title bar or the window. To display the window, move the cursor over the title bar.

The **Transparency** option, when selected, causes the Transparency dialog box to be displayed as shown in Figure 13–5. When the indicator on the **Less-More** slider bar is to the left (Less), the Tool Palettes window (only when floating) is opaque. The closer the indicator is to the right (More), the more transparent the window will become. The **Turn off window transparency** check box, when checked, prevents the Tool Palettes window from becoming transparent.

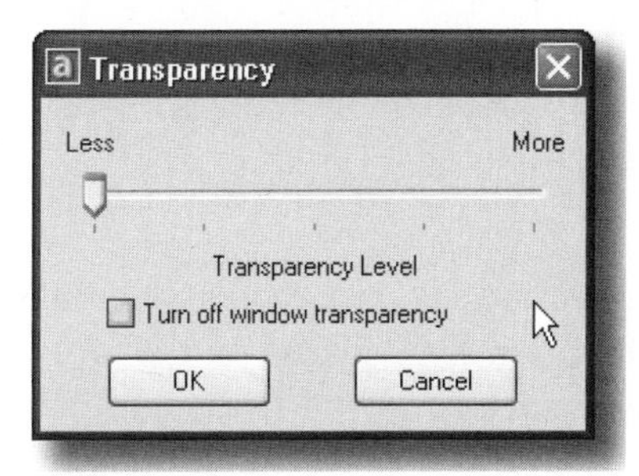

Figure 13–5 *The Transparency dialog box*

The **View Options** option, when selected, causes the View Options dialog box to be displayed as shown in Figure 13–6. In the dialog box, the **Image size:** slider bar allows you to change the size of the images. Under the **View style:** section are three radio buttons. The **Icon only** radio button, when selected, causes the block/hatch pattern shortcut to be displayed as an image only without text. The **Icon with text** radio button, when selected, causes the block/hatch pattern shortcut to be displayed as an image with the descriptive text below it. The **List view** radio button, when selected, causes the block/hatch pattern shortcut to be displayed as an image with the descriptive text to its right, allowing for a more compressed listing of the symbols when used with a small image. The **Apply to:** list box allows you to choose whether the changes are applied to the Current Tool Palette or to All Tool Palettes. To exit the View Options dialog box and accept the changes, choose the **OK** button. To exit without accepting the changes, choose the **Cancel** button.

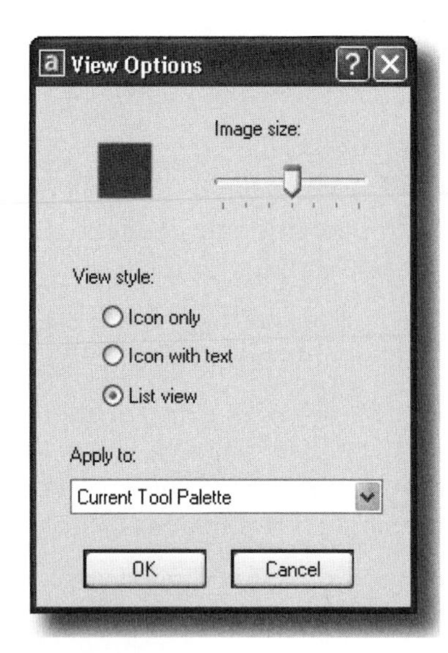

Figure 13–6 *The View Options dialog box*

The **Paste** option, when active and selected, causes a block/hatch pattern shortcut that has been copied to the Clipboard to be placed on the tool palette. If there is nothing on the Clipboard, or if whatever is on the Clipboard is not a block/hatch pattern shortcut, then the paste option is not active and cannot be selected.

The **New Tool Palette** option, when selected, causes a new tool palette to be created. The new tool palette will be empty, its name will temporarily be New Tool Palette, and there will be a text box displayed with the name "New Tool Palette" highlighted that you can change by overwriting or you can accept by pressing enter.

The **Delete Tool Palette** option, when selected, causes the Confirm Tool Palette Deletion dialog box to be displayed, stating "Are you sure you want to remove the 'New Tool Palette' tool palette? This deletion is permanent unless you first export it to an external file." You can accept the deletion by choosing the **OK** button or reject it by choosing the **Cancel** button.

The **Rename Tool Palette** option, when selected, causes a text box to be displayed with the existing name of the tool palette highlighted that you can change by overwriting or you can retain by pressing enter.

The **Customize** option, when selected, causes the Customize dialog box to be displayed with the Tool Palettes tab selected as shown in Figure 13–7.

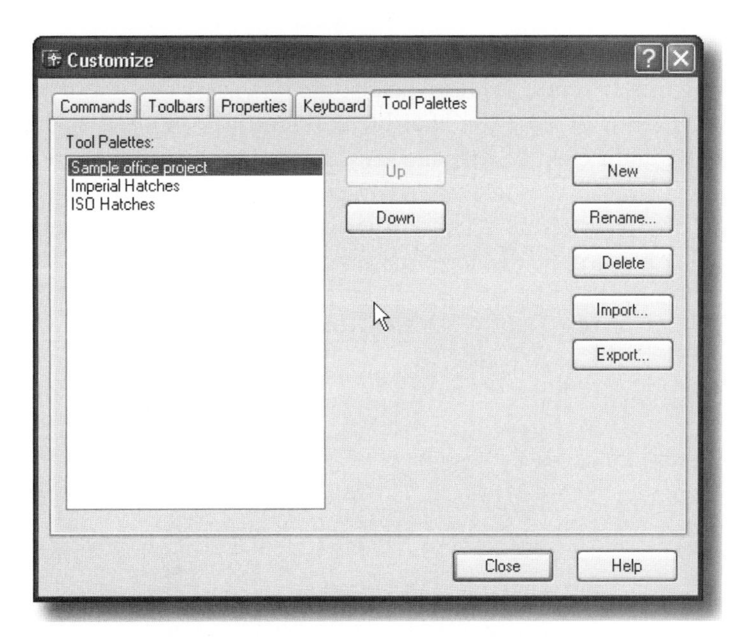

Figure 13–7 *The Customize dialog box with the Tool Palettes tab selected*

The **Up** and **Down** buttons in the Customize dialog box cause the highlighted tab to be moved toward the top or bottom of the screen respectively in relationship to other tabs on the side of the Tool Palettes window. The **New, Rename,** and **Delete** buttons operate in the same manner as the **New Tool Palette, Rename Tool Palette,** and the **Delete Tool Palette** options described above. The **Import** button, when selected, causes the Import Tool Palette dialog box to be displayed, from which a Tool Palette file can be selected and added to the Tool Palettes window. The **Export** button, when selected, causes the Export Tool Palette dialog box to be displayed, which allows you to save the highlighted tool palette as a file with an extension of XTP to a specified location.

INSERT BLOCKS/HATCH PATTERNS FROM A TOOL PALETTE

To insert a block from a tool palette, simply place the cursor on the block symbol in the tool palette, press the pick button and drag the symbol into the drawing area. The block will be inserted at the point where the cursor is located when the pick button is released. This procedure is best implemented by using the appropriate OSNAP mode. Another method of inserting a block from a tool palette is to select the block symbol in the tool palette and then select a point in the drawing area for the insertion point.

To draw a hatch pattern that is a tool in a tool palette, place the cursor on the hatch pattern symbol in the tool palette, press the pick button and drag the symbol into the boundary to receive the hatch pattern and release the pick button. Another method of drawing a hatch pattern that is a tool in a tool palette is to select the hatch pattern symbol in the tool palette and then select a point within a boundary in the drawing area.

BLOCK TOOL PROPERTIES

Blocks whose symbols appear in a tool palette are not, as a rule, blocks defined in the current drawing. Usually, they reside as block definitions in another drawing or in some cases they might even be drawing files. As a tool in a tool palette, a block has tool properties. To access the block/drawing tool properties, right-click on the tool's symbol in the tool palette, and select **Properties** from the shortcut menu. The Tool Properties dialog box will be displayed as shown is Figure 13–8.

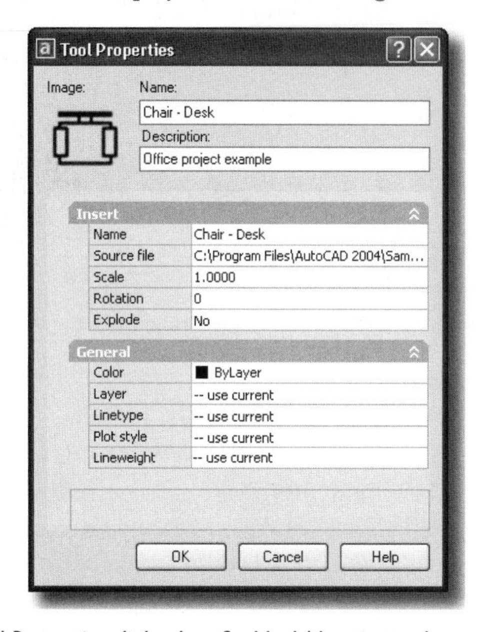

Figure 13–8 *The Tool Properties dialog box for block/drawing tools*

In addition to the properties of name, scale, rotation, layer, color and linetype, an important property of a tool representing a block is its source file.

The Insert properties of a block/drawing tool in a tool palette include **Name**, **Source file**, **Scale**, **Rotation**, and **Explode**. The General properties include **Color**, **Layer**, **Linetype**, and **Plot style**.

Source file

The **Source file** property edit box lists the path to the drawing file where the block that is a tool in the tool palette resides. Or, if the tool is a drawing file, the edit box lists the path to it.

Note: Any change to the path/file name in the ***Source file*** property text box will prevent the block/drawing from being inserted as anticipated when that tool is selected for insertion.

Scale

The **Scale** factor is applied equally to all coordinates when the block is inserted.

Rotation

The angle specified for the **Rotation** property is applied to the block about its insertion point.

Explode

If the **Explode** list box is set to **No**, the block will be inserted as an unexploded block. If it is set to **Yes**, the separate parts that make up the block will be drawn as separate entities, that is, as a block that is inserted and exploded.

Color

The **Color** property options include **ByLayer, ByBlock, use current**, or choice of assigned standard or custom color which can be specified in the connected list box.

Layer

The **Layer** property options include **use current** or one of the layers in the drawing as listed in the connected list box.

Linetype

The **Linetype** property options include **ByLayer, ByBlock, Continuous, use current** or one of the linetypes in the drawing as listed in the connected list box.

Plot style

The **Plot style** property options include **ByLayer, ByBlock, use current** or one of the plot styles in the drawing as listed in the Select Plot Style dialog box that is displayed when the **Other** option is selected.

PATTERN TOOL PROPERTIES

Hatch patterns whose symbols appear in a tool palette have tool properties. The Pattern properties of a hatch pattern tool in a tool palette include **Type, Pattern Name, Angle, Scale, Spacing**, and **ISO pen width**, and **Double**. The General properties include **Color, Layer, Linetype, Plot style**, and **Lineweight**. To access the pattern tool properties, right-click on the tool's symbol in the tool palette, and select PROPERTIES from the shortcut menu. The Tool Properties dialog box will be displayed as shown in Figure 13–9.

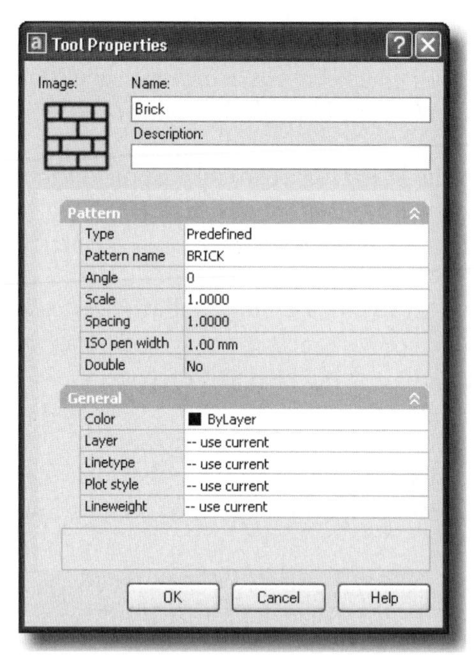

Figure 13–9 *The Tool Properties dialog box for pattern tools*

Type

The **Type** property options are available from the **Hatch Pattern Type** dialog box that is displayed when the ellipsis is selected in the **Type** list of the Tool Properties dialog box. The pattern types include **User-defined, Predefined** from one of the hatch patterns listed in the **Pattern** list box, or **Custom** from one of the hatch patterns in the drawing as listed in the **Custom Pattern** list box.

Pattern name

The **Pattern name** property options are available in the Hatch Pattern Palette dialog box that is displayed when the ellipsis is selected in the **Pattern name** list of the Tool Properties dialog box. The pattern names (which are the patterns themselves) are included under the **ANSI, ISO**, and **Other Predefined** tabs, **User-defined, Predefined** from one of the hatch patterns listed in the **Pattern** list box, or **Custom** from one of the hatch patterns in the drawing as listed in the **Custom Pattern** list box.

Angle

The angle specified for the **Angle** property is applied to the pattern.

Scale

The **Scale** factor is applied equally to both coordinates when the hatch pattern is drawn. This option is not changeable for a **User-defined** pattern type.

Spacing

The **Spacing** property specifies the distance between the lines of the hatch pattern. This option is available only for a **User-defined** pattern type.

ISO pen width

The **ISO pen width** property edit box specifies the plotting pen that will determine the widths of the pattern lines if you select a linetype whose name begins with ACAD_ISO.

Double

If the **Double** list box is set to **No**, the hatch pattern will be drawn with one set of parallel lines at the spacing specified. If it is set to **Yes**, the hatch pattern will be drawn with two sets of lines, perpendicular to each other at the spacing specified. This option is available only for a **User-defined** pattern type.

Color

The **Color** property options include **ByLayer**, **ByBlock**, **use current**, or choice of assigned standard or custom color, which can be specified in the connected list box.

Layer

The **Layer** property options include **use current** or one of the layers in the drawing as listed in the connected list box.

Linetype

The **Linetype** property options include **ByLayer**, **ByBlock**, **Continuous**, **use current** or one of the linetypes in the drawing as listed in the connected list box.

Plot style

The **Plot style** property options include **ByLayer**, **ByBlock**, **use current** or one of the Plot styles in the drawing as listed in the Select Plot Style dialog box that is displayed when the **Other** option is selected.

Lineweight

The **Lineweight** property options include **ByLayer**, **ByBlock**, **Default**, **use current** or one of the lineweights as listed in the connected list box.

CREATING & POPULATING TOOL PALETTES

A new tool palette can be created from the Tool Palette shortcut menu by selecting the **New Tool Palette** option as described earlier in this section.

You can also create a new tool palette from the Tree view or Content area of the DesignCenter. From the Tree view or Content area highlight an item and then right-click to display the shortcut menu. If the selected item is a folder, one of the available options in the shortcut menu is **Create Tool Palette Of Blocks** as shown in Figure 13–10. If there are no blocks in the folder, a message will be displayed stating "Folder does not contain any drawing files". If the selected item is a drawing, then one of the available options in the shortcut menu is **Create Tool Palette**. If the drawing does not contain any blocks, a message will be displayed stating "Drawing does not contain any block definitions".

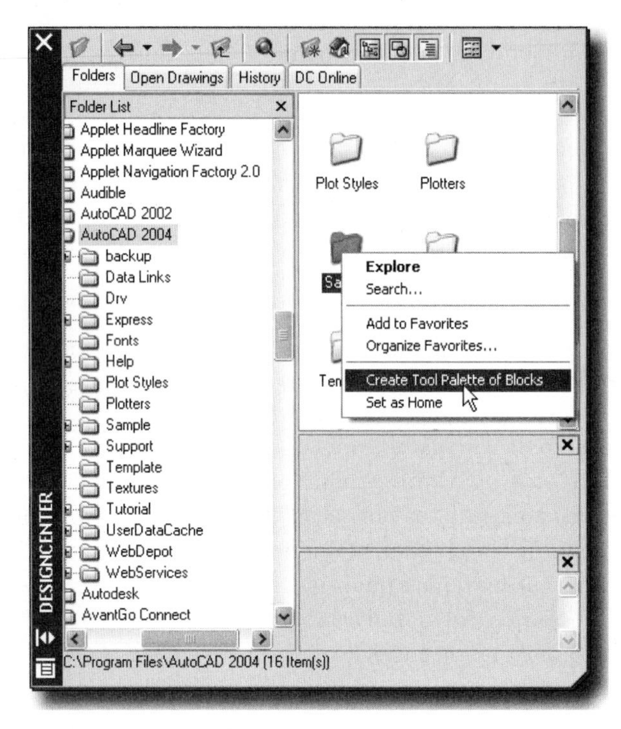

Figure 13–10 *Shortcut menu – Design Center content area*

When one of the options to create a tool palette is selected, AutoCAD creates a new tool palette, which will be populated with the drawings from the selected folder or blocks from the selected drawing.

From the Content area, in addition to creating tool palettes by right-clicking a folder or drawing, you can also right-click on a block and select the **Create Tool Palette** option from the shortcut menu. In this case the new tool palette will contain only the selected block. In this case you will be prompted to name the tool palette.

You can also drag and drop a block from a drawing or a drawing from the Content area of the DesignCenter on to the one of the existing palettes.

PARTIAL LOAD

The **Partial Open** command, described in Chapter 1, allows you to work with just part of a drawing by loading geometry only from specific views or layers. When a drawing is partially open, specified and named objects are loaded. Named objects include blocks, layers, dimension styles, linetypes, layouts, text styles, viewport configurations, UCSs, and views. With the **Partial Open** command you can load additional information into the current drawing as long as the drawing is partially open.

The **Partial Load** command allows you to load additional geometry into a current partially-loaded drawing by loading geometry only from specific views or layers.

Invoke the **Partial Load** command:

Files menu	Choose Partial Load
Command: prompt	**partiaload** (ENTER)

AutoCAD displays the Partial Load dialog box as shown in Figure 13–11.

Note: If you invoke the PARTIALOAD command in a drawing that has not been partially opened AutoCAD will respond with the message "Command not allowed unless the drawing has been partially opened."

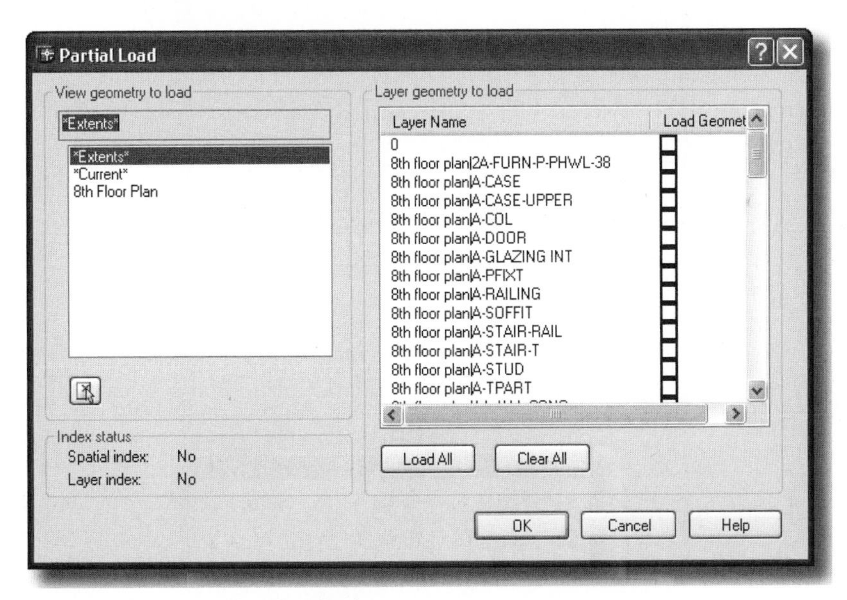

Figure 13–11 *The Partial Load dialog box*

In the Partial Load dialog box, select a view. The default view is Extents. Only geometry from model space views that are saved in the current drawing can be loaded. Select layer(s). No layer geometry is loaded into the drawing if you do not select any layers to load. However, all drawing layers will exist in the drawing. Even if the geometry from a view is specified to load, and no layer is specified to load, no geometry is loaded. When the view(s) and layer(s) have been specified, choose **OK** to partially load the geometry from the specified layers.

DRAWING PROPERTIES

Drawing properties allow you to keep track of your drawings by having properties assigned to them. The properties you assign will identify the drawing by its title, author, subject, and keywords for the model or other data. Hyperlink addresses or paths can be stored along with ten custom properties.

Invoke the Drawing Properties command:

Files menu	Choose Drawing Properties
Command: prompt	**dwgprops** (ENTER)

AutoCAD displays the Drawing Properties dialog box as shown in Figures 13–12.

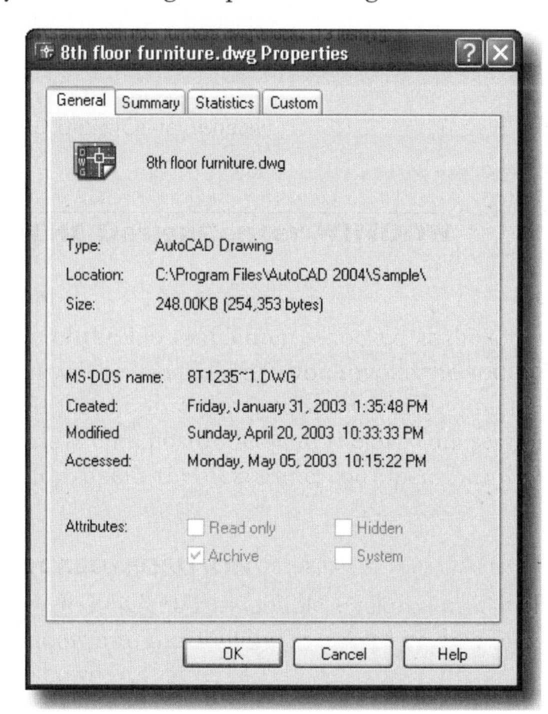

Figure 13–12 *The Drawing Properties dialog box with the General tab displayed*

The Drawing Properties dialog box has four tabs: **General, Summary, Statistics,** and **Custom.**

The **General** tab (see Figure 13–12) displays the drawing type, location, size, and other information. These come from the operating system. These fields are read-only. However, the attributes options are made available by the operating system if you access file properties through Windows Explorer®.

The **Summary** tab (see Figure 13–13) lets you enter a title, subject, author, keywords, comments, and a hyperlink base for the drawing. Keywords for drawings sharing a common property will help in your search. For a hyperlink base, you can specify a path to a folder on a network drive or an Internet address.

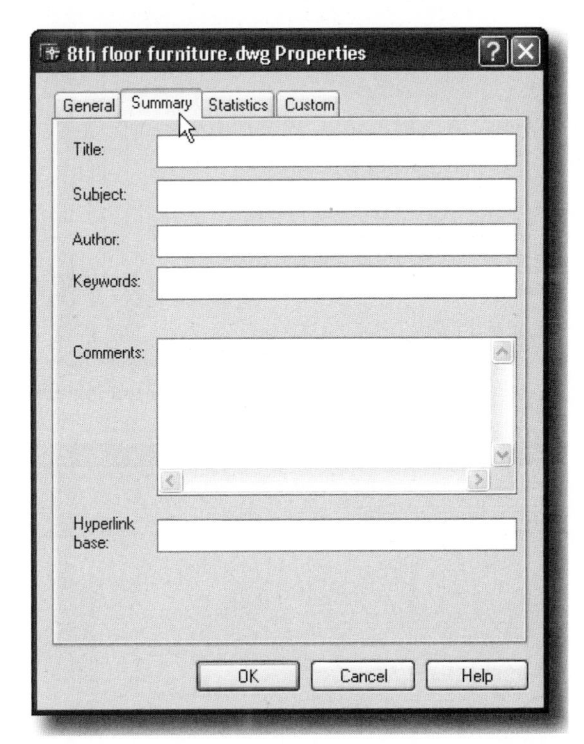

Figure 13–13 *The Drawing Properties dialog box with the Summary tab displayed*

The **Statistics** tab (see Figure 13–14) displays information such as the dates files were created and last modified. You can search for all files created at a certain time.

The **Custom** tab (see Figure 13–15) lets you specify up to ten custom properties. Enter the names of the custom fields in the left column, and the value for each custom field in the right column.

Note: Properties entered in the Drawing Properties dialog box are not associated with the drawing until you save the drawing.

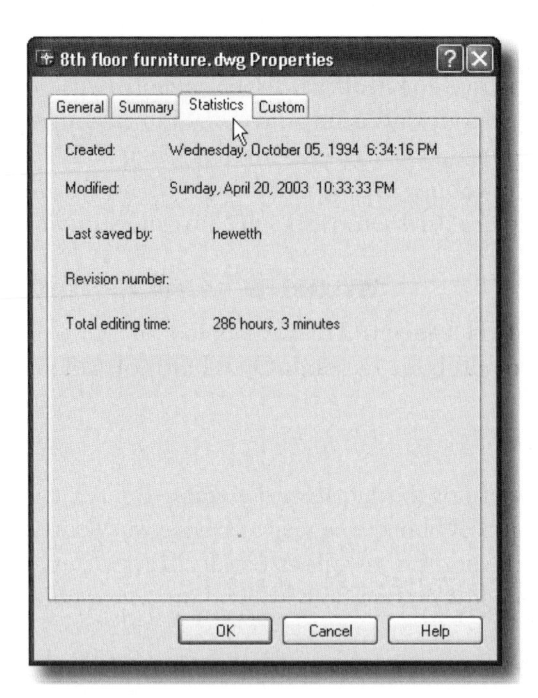

Figure 13–14 *The Drawing Properties dialog box with the Statistics tab displayed*

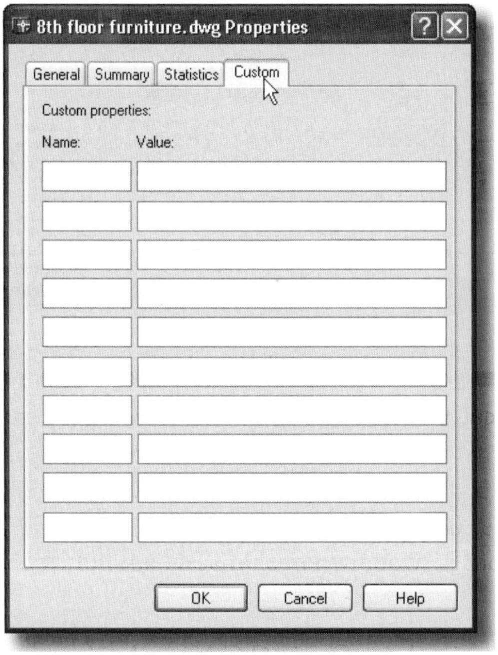

8th floor furniture.dwg Properties
General
Summary
Statistics
Custom
Custom properties:
Name:
Value:
OK
Cancel
Help

Figure 13–15 *The Drawing Properties dialog box with the Custom tab displayed*

THE GEOMETRIC CALCULATOR (CAL)

The CAL command is an on-line calculator that evaluates real, integer, and vector expressions. The CAL command, when entered at the "Command:" prompt (or **'cal** transparently when prompted for input) switches you to the calculator mode. When the "Command:" prompt is preceded by double angle brackets (>>), AutoCAD is awaiting input, which must be in the acceptable calculator format. For example, you can solve a simple algebraic equation by entering the following:

```
Command: cal
Initializing...
>>Expression: A=pi*4^2 (ENTER)
```

producing the following response:

```
50.2655
```

This sequence might be used to derive the area of a circle with a radius of 4. The variable A and the equal sign are not required to arrive at the solution. Entering pi*4^2 will suffice. However, the A= serves to give a user-defined name (the letter A in this case) to a variable so that the result of the equation can be set as the value of the named variable for later use in AutoLISP or in another calculator expression. In the built-in calculator, "pi" has been assigned a preset value of approximately 3.141592654, which is the ratio of a circle's circumference to its diameter. The "*" and the "^" are the symbols for multiply and exponent, respectively, as is noted later in the listing of arithmetic operators you may use in calculator expressions.

You may also enter the built-in calculator transparently (while in the middle of another command), as shown here during the LINE command and in Figure 13–16:

```
Command: line
Specify first point: 'cal
>>Expression: (mid+end)/2
>>Select entity for MID snap: (select one line)
>>Select entity for END snap: (select other line)
Specify next point or [Undo]: (ENTER)
```

This expression causes AutoCAD to pause while the user selects points by object snapping to a midpoint and an endpoint. Then AutoCAD uses the midpoint between those two points as the starting point of the line to be drawn.

Responses to the ">>Expression:" and other calculator (with leading >>'s) prompts vary according to the desired function.

In addition to using CAL (and 'CAL transparently) while in AutoCAD, you can use CAL as an operator within an AutoLISP expression, followed by a calculator expression enclosed in double quotes.

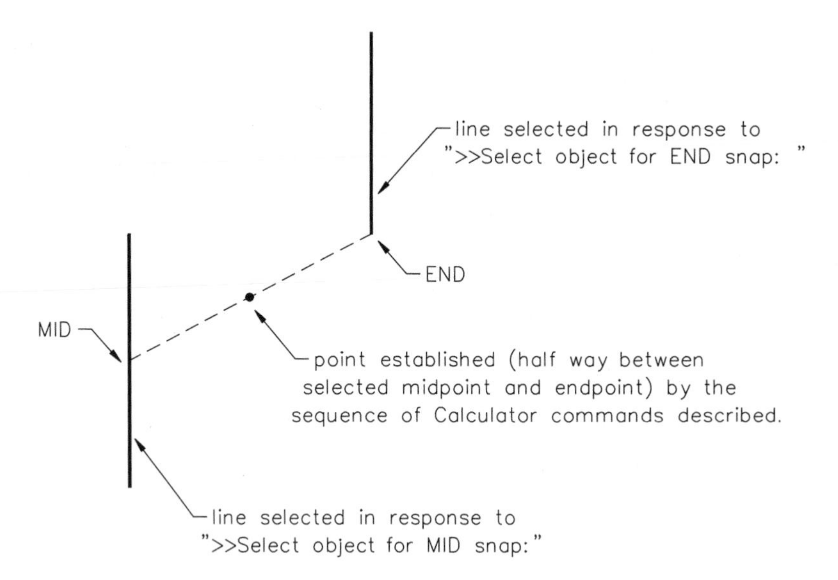

Figure 13–16 *Using the* CAL *command transparently during the* LINE *command*

CAL can be considered an AutoLISP subroutine with an argument of a calculator expression-string. In the following example, CAL finds the midpoint between two other midpoints of parallel and equal-length lines 1 and 2, as shown in Figure 13–17.

> Command: **(setq pt1 (cal "(mid+mid)/2"))**
> >>Select entity for MID snap: *(select one line)*
> >>Select entity for MID snap: *(select other line)*
> (4.5 5.0 0.0)

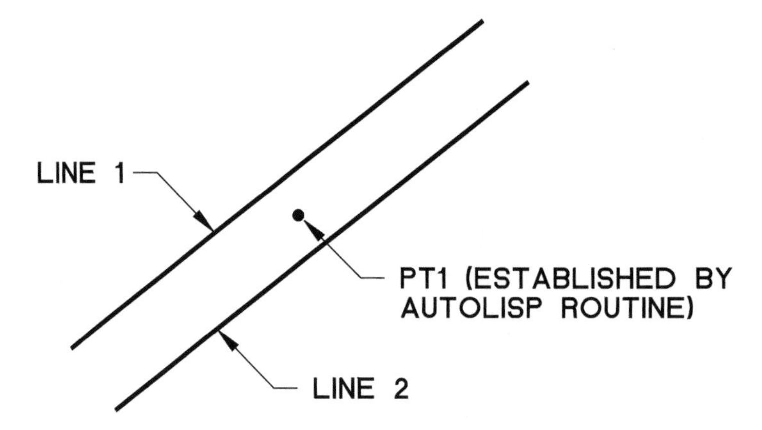

Figure 13–17 CAL finds the midpoint between two other midpoints parallel and equal-length lines 1 and 2

Things you can do directly or interactively (with AutoLISP) through this programmable calculator include the following:

- Use expressions similar to common algebra to solve equations.
- Name and store values directly (or indirectly through expressions) to variables for later use in expressions, by AutoLISP or in AutoCAD commands.
- Derive information about the existing geometry, or use available data to generate or analyze geometric constructs such as points, vectors, components (*X*, *X*, and *Z* coordinates) of a point, coordinate systems, and axes.
- Apply Object Snap modes.

NOTATIONS AND FORMAT

Calculator expressions are entered in a format similar to algebra, known as infix notation. Operational precedences are as follows:

1. Left to right for operators of equal precedence
2. Parentheses, from innermost out
3. Exponents first; multiplication and division second; addition and subtraction last

Numeric operators that can be used include the following:

+	add
-	subtract
*	multiply
/	divide
^	raise to a power
[ ]	signifies expression

Vector operators that can be used include the following:

+	add vectors
-	subtract vectors
*	multiply a vector by a real number or scalar product of vectors
/	divide a vector by a real number
&	vector product of vectors
[ ]	signifies expression

Point/vector operators and symbols that can be used include the following:

[ ]	encloses point or vector
,	separates coordinate values
<	precedes angle
@	precedes distance
*	denotes WCS (overrides UCS)

Architectural distances can be entered as either feet'-inches" or feet'inches".

Angles can be entered in degrees/minutes/seconds as degrees d, minutes', seconds", radians r, or gradians g. Degrees must be entered as 0d if the angle in d/m/s is less than one degree.

Points and vectors are entered by enclosing the coordinate values in square brackets, separated by commas. Zero values can be omitted as long as the commas are there. For example, the following values are valid point/vectors:

[,,]	equals	[0,0,0]
[8,9,]	equals	[8,9,0]
[,,7]	equals	[0,0,7]

Coordinate systems can be entered in the following formats:

System	Format	Example
polar	=[dist<angle]	[7<45]
cylindrical	=[dist<angle,z]	[3<0.57r,1.0]
spherical	=[dist<angle1,angle2]	[7<30<60]
relative	=[@x,y,z]	[@1.0,2.5,3.75]
WCS	=[*x,y,z]	[*5,6,20]

Conversion of point/vectors between the User Coordinate System and the World Coordinate System can be achieved by the following functions:

w2u converts from the WCS to the UCS

u2w converts from the UCS to the WCS

A point/vector expression can include arithmetic operators, as the following example shows:

```
[4+3<45]
[3<3.14/6,0.5*2]
```

Other point/vector operators include the following:

sin, asin	sine and arcsine of angle
cos, acos	cosine and arccosine of angle
tang, atan	tangent and arctangent of angle
ln	natural log of the number
log	base-10 log of the number
exp	natural exponent of the number
exp10	base-10 exponent of the number
sqr, sqrt	square and square root
abs	absolute value
round	number rounded to nearest integer
trunc	integer with decimal portion removed
r2d	converts angle in radians to degrees
d2r	converts angle in degrees to radians
pi	quotient of circumference divided by diameter
vec	obtains the vector between two points
vec1	obtains a unit vector between two points

Coordinate values of point/vectors can be obtained singly and in pairs by using one of the following operators:

xyof(point)	returns new point with *X* and *Y* coordinates of point
xzof(point)	returns new point with *X* and *Z* coordinates of point
yzof(point)	returns new point with *Y* and *Z* coordinates of point
xof(point)	returns new point with *X* coordinate of point

yof(point)	returns new point with *Y* coordinate of point
zof(point)	returns new point with *Z* coordinate of point
rxof(point)	returns real that is *X* coordinate of point
ryof(point)	returns real that is *Y* coordinate of point
rzof(point)	returns real that is *Z* coordinate of point

Cur

The CUR operator pauses and prompts the user to pick a point with the cursor, using that point in the expression.

The @ symbol causes the last point to be used in the expression. This is similar to the relative coordinate method of specifying points.

Object Snap

Using the first three letters of any of the Object Snap modes in an expression causes AutoCAD to pause and prompt the user for a point subject to the Object Snap mode specified.

Finding Points

You can obtain a point on a line by entering the two endpoints and either a distance or a scalar value, as follows:

pld(pt1,pt2,distance) returns a point that is the given distance from pt1 in the direction of pt2 from pt1.

pld([1,2,3], [5,5,3],3.5) obtains pt3, as shown in Figure 13–18.

plt(pt1,pt2,proportion) returns a point that is in the direction of pt2 from pt1, and its distance from pt1 will be the product of the proportion multiplied by the distance between pt1 and pt2.

plt(pt1,pt2, 0.5) obtains pt 3, as shown in Figure 13–19.

plt(pt1, pt2, 7/8) obtains pt3, as shown in Figure 13–20.

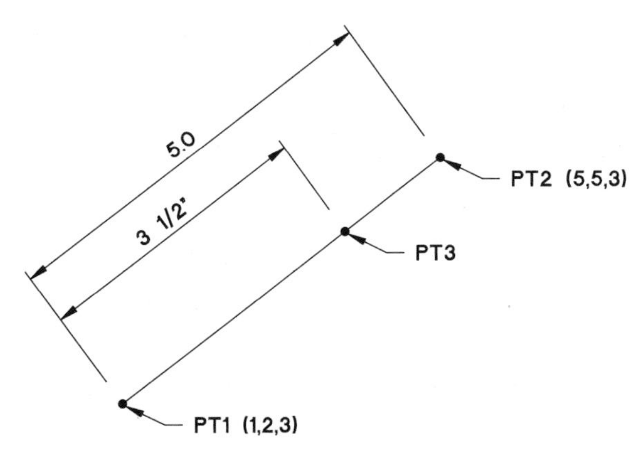

Figure 13–18 *Obtaining a point on a line by entering the two endpoints*

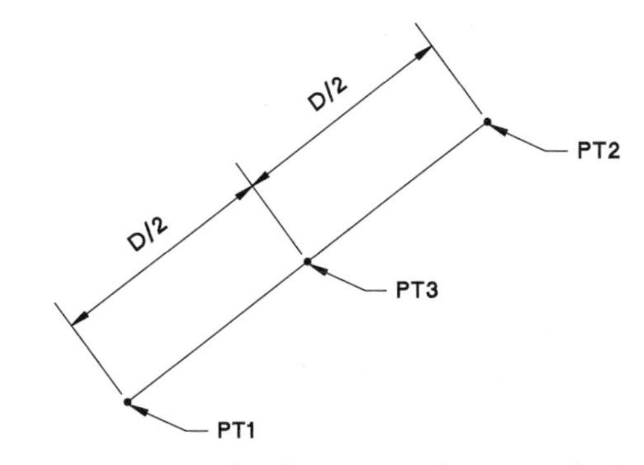

Figure 13–19 *Obtaining a point on a line by entering the two endpoints and a distance*

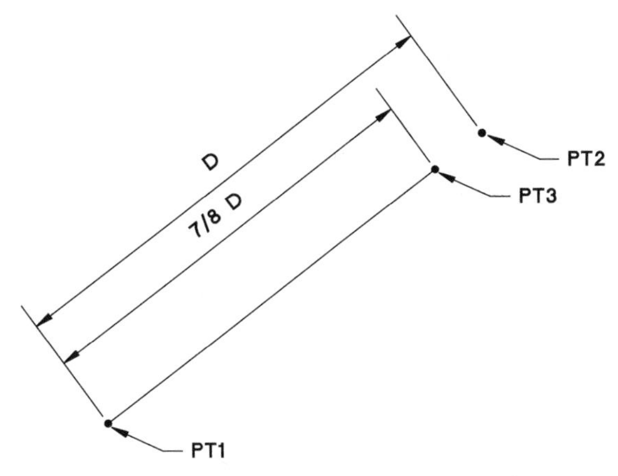

Figure 13–20 *Obtaining a point on a line by entering the two endpoints and a scalar value*

ADDITIONAL GEOMETRIC FUNCTIONS

Special points, vectors, distances, radii, and angles can be derived by using the calculator functions described in this section.

Rotation

Rotation of a point about an axis can be achieved with the ROT command.

rot(p,origin,angle) returns the point *p* rotated *ang* angle about an axis through the origin in the *y* direction, as shown in Figure 13–21.

rot(p,Axpt1,Axpt2,ang) returns the point *p* rotated *ang* angle about an axis through Axpt1 and Axpt2, as shown in Figure 13–22.

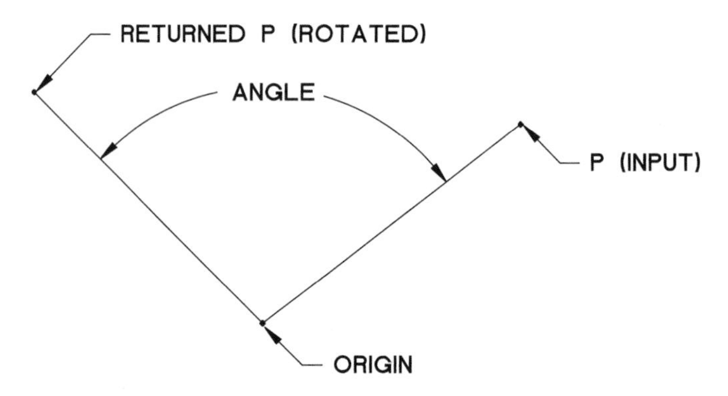

Figure 13–21 *Rotation of a point about an axis through the origin in the Y direction*

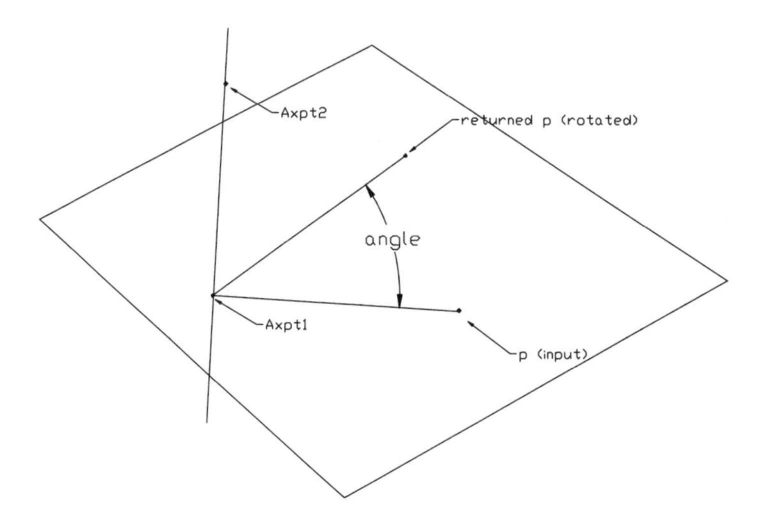

Figure 13–22 *Rotation of a point about an axis through Axpt1 and Axpt2*

Intersection Point

Ill(pt1,pt2, pt3,pt4) returns the intersection point between two lines (p1,p2) and (p3,p4). AutoCAD considers all points three-dimensional.

ilp(p1,p2,p3,p4,p5) returns the intersection point between a line (p1,p2) and a plane passing through three points (p3,p4,p5).

Distance

dpl(pt1,pt2,pt3) returns the distance (shortest) between point pt1 and a line through pt2 and pt3.

dpp(pt1,pt2,pt3,pt4) returns the distance (shortest) between point pt1 and a plane defined by the three points pt2, pt3, and pt4.

Radius

RAD pauses for the user to select an arc, circle, or *2D* polyline arc segment, and returns its radius.

Angle

ang(v) returns the angle between the *X* axis and a line defined by the vector v.

ang(pt1,pt2) returns the angle between the *X* axis and a line defined by points pt1 and pt2.

ang(apex,pt1,pt2) returns the angle between lines defined by apex to pt1 and apex to pt2.

ang(apex,pt1,pt2,pt3) returns the angle between lines defined by apex to pt1 and apex to pt2 and measured counterclockwise about the axis defined by apex to pt3.

Normal Vector

nor returns the unit normal vector (*3D*) of a selected arc, circle, or polyline arc segment.

nor(v) returns the unit normal vector (*2D*) to the vector v.

nor(pt1,pt2) returns the unit normal vector (*2D*) to the line through pt1 and pt2.

nor(pt1,pt2,pt3) returns the unit normal vector (*3D*) to the plane defined by the three points pt1, pt2, and pt3.

SHORTCUT OPERATIONS

Following are the shortened versions of calculator operators:

dee	=	dist(end,end)
ille	=	ill(end,end,end,end)
mee	=	(end+end)/2
nee	=	nor(end,end)
vee	=	vec(end,end)
vee1	=	vec1(end,end)

MANAGING NAMED OBJECTS

The RENAME command allows you to change the names of blocks, dimension styles, layers, linetypes, text styles, views, User Coordinate Systems, or viewport configurations.

Invoke the RENAME command:

Format menu	Choose Rename
Command: prompt	**rename** (ENTER)

AutoCAD displays the Rename dialog box, shown in Figure 13–23.

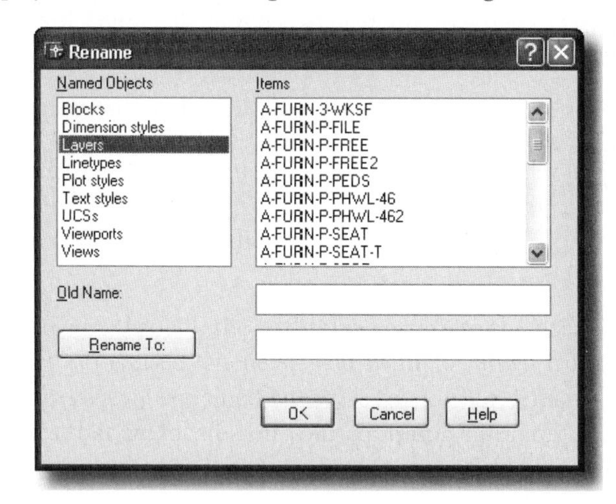

Figure 13–23 *Rename dialog box*

Note: Except for the layer named 0 and the linetype named Continuous, you can change the name of any of the named objects.

In the **Named Objects** list box, select the object name you want to change. The **Items** list box displays the names of all objects that can be renamed. To change the object's name, pick the name in the Items list box or enter it into the **Old Name:** text box. Enter the new name in the **Rename To:** text box, and select the **Rename To:** button to update the object's name in the Items list box. To close the dialog box, choose the **OK** button.

DELETING UNUSED NAMED OBJECTS

The PURGE command is used to selectively delete any unused named objects.

Invoke the PURGE command:

File menu	Choose Drawing Utilities > Purge
Command: prompt	**purge** (ENTER)

AutoCAD prompts:

Command: **purge**

AutoCAD displays the Purge dialog box as shown in Figure 13–24.

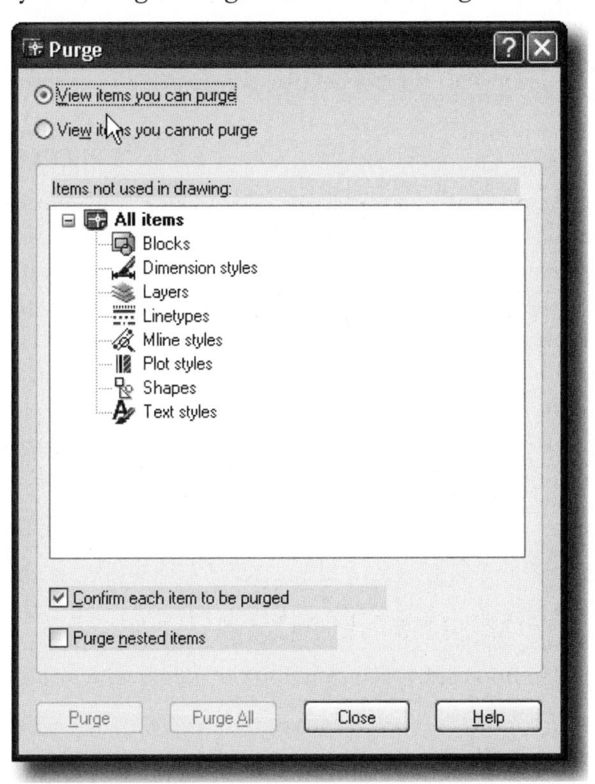

Figure 13–24 *Purge dialog box with the Items not used in drawing: section displayed*

If the **View items you can purge** radio button is selected, the **Items not used in drawing:** text window lists categories of named items, under which are listed the individual named items that have been defined in the drawing but are not currently being used. For example, if a layer has been defined but has nothing drawn on it and it is not the current layer, it can be purged. Or, if the drawing contains a block definition but the block has not been inserted, then it can be purged. Objects such as lines, circles, and other basic unnamed drawing elements cannot be purged.

If the **Confirm each item to be purged** check box is checked, then AutoCAD will display the **Confirm Purge** dialog box and ask you to reply by selecting a **Yes** or **No** button before continuing.

If the **Purge nested items** check box is checked, then AutoCAD will purge nested items within any item selected.

Once an individual named item in the list has been selected or a category of items has been selected that has items that can be purged, the **Purge** button at the bottom of the dialog box becomes operational. The **Purge All** button causes all items that can be purged to be purged.

If the **View items you cannot purge** radio button is selected, the **Items currently used in drawing:** text window lists categories of named items, under which are listed the individual named items that have been defined in the drawing but are currently being used, as shown in Figure 13–25. For example, if a layer has something drawn on it or it is the current layer, it cannot be purged. Or, if a block has been inserted in the drawing, then it cannot be purged. The text area under the list of items informs you why a selected item cannot be purged. For example, if the Standard Text Style is selected, the message "The default text style, STANDARD, cannot be purged" will be displayed.

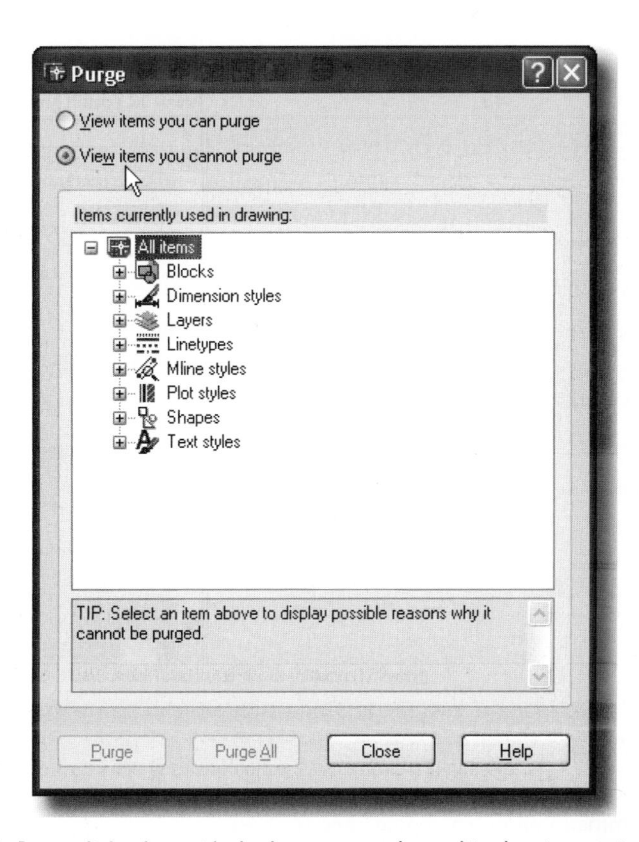

Figure 13–25 *Purge dialog box with the Items currently used in drawing: section displayed*

If from the Command: prompt you enter **–purge**, AutoCAD prompts:

> Enter type of unused objects to purge
> [Blocks/Dimstyles/LAyers/LTypes/Plotstyles/SHapes/textSTyles/Mlinestyles/All]:

If you choose the All option, AutoCAD prompts:

> Enter type of unused objects to purge [Blocks/Dimstyles/LAyers/LTypes/Plotstyles/SHapes/textSTyles/Mlinestyles/All]: *(right-click and select one of the available options to purge from the shortcut menu)*

If you have chosen an option from the submenu other than All, or if you specify one of the available options at the command prompt, AutoCAD prompts:

> Enter name(s) to purge <*>: *(specify the name(s) of the object(s) to purge, or with wild-card characters specify the objects to purge)*
> Verify each name to be purged? [Yes/No] <Y>: *(press* ENTER *for AutoCAD to prompt for verification of each symbol name displayed before it is purged, or enter n, for AutoCAD to purge without verification)*

Depending on the response, AutoCAD prompts for verification of each symbol name before it is purged.

The PURGE command removes only one level of reference. For instance, if a block has nested blocks, the PURGE command removes the outer block definition only. To remove the second-, third-, or deeper-level blocks within blocks, you must repeat the PURGE command until there are no referenced objects. You can use the PURGE command at any time during a drawing session.

Individual shapes are part of a *.SHX* file. They cannot be renamed, but references to those that are not being used can be purged. Views, User Coordinate Systems, and viewport configurations cannot be purged, but the commands that manage them provide options to delete those that are not being used.

Note: The PURGE command cannot be used in a partially opened drawing.

COMMAND MODIFIER—MULTIPLE

MULTIPLE is not a command, but when used with another AutoCAD command, it causes automatic recalling of that command when it is completed. You must press ESC to terminate this repeating process. Here is an example of using this modifier to cause automatic repeating of the ARC command:

Command: **multiple** (ENTER)
Enter command name to repeat: **arc** (ENTER)

You can use the MULTIPLE command modifier with any of the draw, modify, and inquiry commands. PLOT, however, will ignore the MULTIPLE command modifier.

UTILITY DISPLAY COMMANDS

The utility display commands include VIEW, REGENAUTO, DRAGMODE, and BLIPMODE.

SAVING VIEWS

The VIEW command allows you to give a name to the display in the current viewport and have it saved as a view. You can recall a view later by using the VIEW command and responding with the name of the view desired. This is useful for moving back quickly to needed areas in the drawing without having to resort to zoom and pan.

Invoke the VIEW command:

View menu	Choose Named Views...
Command: prompt	**view** (ENTER)

AutoCAD displays the View dialog box, shown in Figure 13–26.

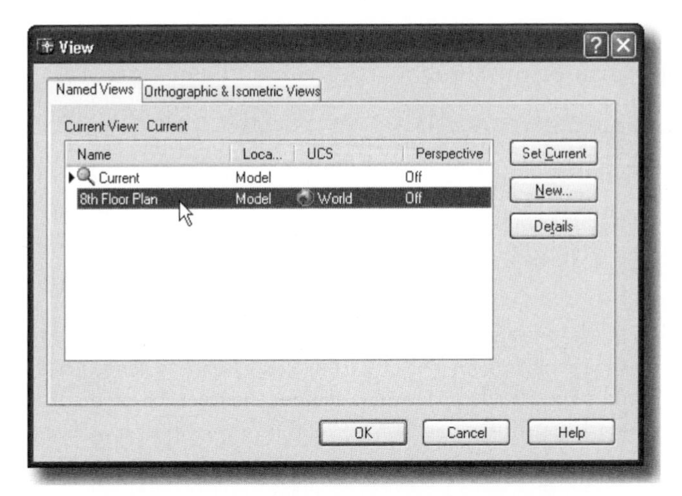

Figure 13–26 *View dialog box with the Named Views tab displayed*

There are two tabs in the View dialog box: **Named Views** and **Orthographic & Isometric Views.** In the **Named Views** tab, AutoCAD lists any saved view(s) in the Views list box. You can right-click on a view name and select one of the options in the shortcut menu as shown in Figure 13–27. Options include **Set Current, Rename, Delete,** and **Details....** The **Set Current** option makes the highlighted view the current view. The RENAME option highlights the view name for editing. It can be edited in the same manner as you would a file name after clicking it once and then once again a second or so after the first click. The **Delete** option deletes the highlighted option. The DETAILS option causes a dialog box to be displayed, described later in this section.

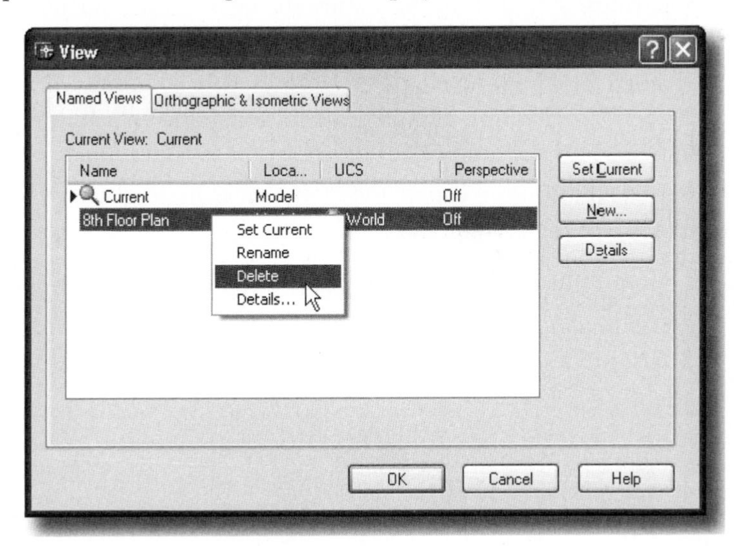

Figure 13–27 *View dialog box after right-clicking on a named view in the Current View list box*

To create a new view, choose the **New...** button. AutoCAD displays the New View dialog box as shown in Figure 13–28.

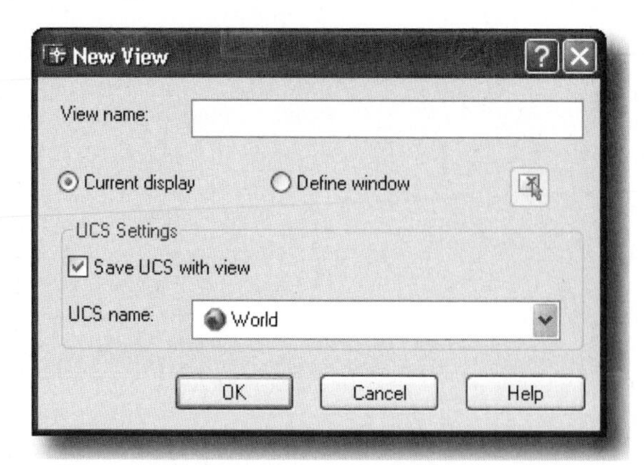

Figure 13–28 *New View dialog box*

In the **New View** dialog box, you can enter a name for the new view in the **View name:** text box. There are two radio buttons for specifying the area for the new view: **Current display** and **Define window**. If the **Current display** radio button is selected, the view will include the current display as the new view. If the **Define window** radio button is selected, the **Define New Window** button will be highlighted. Select the **Define New Window** button, and you will be prompt to specify diagonally opposite corners of a window that will define the area for the new view.

From the **UCS Settings** section of the New View dialog box, the available Coordinate Systems (User or World) can be selected in the **UCS name:** list box. You can select the **Save UCS with view** check box and have the coordinate system that is displayed in the **USC name:** list box saved with the new view. Select **OK** to close the New View dialog box.

Selecting the **Set Current** button causes the view that is highlighted in the Current View list box to become the current view.

Selecting the **Details** button causes the View Details dialog box to be displayed, as shown in Figure 13–29.

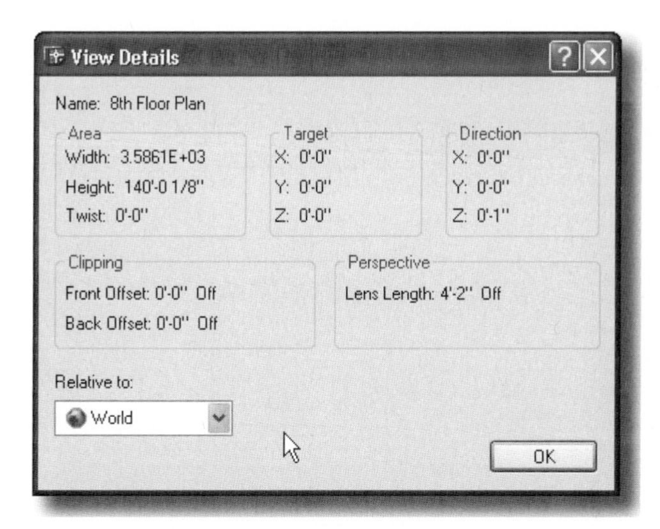

Figure 13–29 *View Details dialog box*

The View Details dialog box shows data for the view's **Area, Target, Direction, Clipping** and **Perspective** and which Coordinate system the view is **Relative to.**

The **Orthographic & Isometric Views** tab presents additional options as shown in Figure 13–30. In the **Current View**: list box of the **Orthographic & Isometric Views** tab are listed six orthographic views (Top, Bottom, Front, Back, Left, and Right) and four isometric views (Southwest, Southeast, Northeast, and Northwest). Selecting one of the standard orthographic or isometric views and then selecting the **Set Current** button and the **OK** button will cause the view to be taken from the direction specified by the view name. For example, if you are viewing the front view of a building and select Top, your view will be from the traditional plan view direction. It will not, of course, automatically cut away the roof and ceiling (if the roof and ceiling are part of the *3D* drawing) and give you the horizontal section that you expect to see in a plan view.

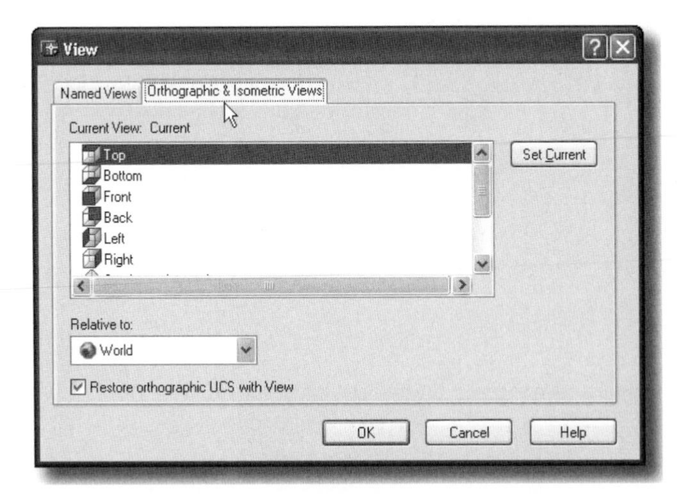

Figure 13–30 *View dialog box with the Orthographic & Isometric Views tab displayed*

CONTROLLING THE REGENERATION

The REGENAUTO command controls automatic regeneration. When the REGENAUTO is set to ON, AutoCAD drawings regenerate automatically. When it is set to OFF, then you may have to regenerate the drawing manually to see the current status of the drawing.

Invoke the REGENAUTO command:

Command: prompt	**regenauto** (ENTER)

AutoCAD prompts:

Command: **regenauto**
Enter mode [ON/OFF] <ON>: *(select one of the two available options)*

When you set REGENAUTO to OFF, what you see on the screen may not always represent the current state of the drawing. When changes are made by certain commands, the display will be updated only after you invoke the REGEN command. But waiting time can be avoided as long as you are aware of the status of the display. Turning the REGENAUTO setting back to ON will cause a regeneration. If a command should require regeneration while REGENAUTO is set to OFF, you will be prompted:

About to regen, proceed? <Y>

Responding with No will abort the command.

Regeneration during a transparent command will be delayed until a regeneration is performed after that transparent command. The following message will appear:

REGEN QUEUED

CONTROLLING THE DRAWING OF OBJECTS

The DRAGMODE command controls the way dragged objects are displayed. Certain draw and modify commands display highlighted dynamic (cursor-following) representations of the objects being drawn or edited. This can slow down the drawing process if the objects are very complex. Setting DRAGMODE to OFF turns off dragging.

Invoke the DRAGMODE command:

Command: prompt	**dragmode** (ENTER)

AutoCAD prompts:

Command: **dragmode**
Enter new value [ON/OFF/Auto] <Auto>: *(right-click and select one of the three available options from the shortcut menu)*

When DRAGMODE is set to OFF, all calls for dragging are ignored. Setting DRAGMODE to ON allows dragging by use of the DRAG command modifier. Setting DRAGMODE to Auto (default) causes dragging wherever possible.

When DRAGMODE is set to ON the DRAG command modifier can be used wherever dragging is permitted. For example, during the MOVE prompt you can use the following:

Command: **move**
Select objects: *(select the objects)*
Specify base point or displacement: *(specify the base point)*
Specify second point of displacement or
<use first point as displacement>: **drag**

At this point the selected objects follow the cursor movement.

CONTROLLING THE DISPLAY OF MARKER BLIPS

The BLIPMODE command controls the display of marker blips. When BLIPMODE is set to ON, a small cross mark is displayed when points on the screen are specified with the cursor or by entering their coordinates. After you edit for a while, the drawing can become cluttered with these blips. They have no effect other than visual reference and can be removed at any time by using the REDRAW, REGEN, ZOOM, or PAN commands. Any other command requiring regeneration causes the blips to be removed. When BLIPMODE is set to OFF, the blips marks are not displaced.

Invoke the BLIPMODE command:

Command: prompt	**blipmode** (ENTER)

AutoCAD prompts:

Command: **blipmode**
Enter mode [ON/OFF] <OFF>: *(right-click and select one of the two available options from the shortcut menu)*

CHANGING THE DISPLAY ORDER OF OBJECTS

The DRAWORDER command allows you to change the display order of objects as well as images. This will ensure proper display and plotting output when two or more objects overlay one another. For instance, when a raster image is attached over an existing object, AutoCAD obscures them from view. With the help of the DRAWORDER command, you can make the existing object display over the raster image.

Invoke the DRAWORDER command:

Modify II toolbar	Choose the DRAWORDER command (see Figure 13–31)
Tools menu	Choose Display Order
Command: prompt	**draworder** (ENTER)

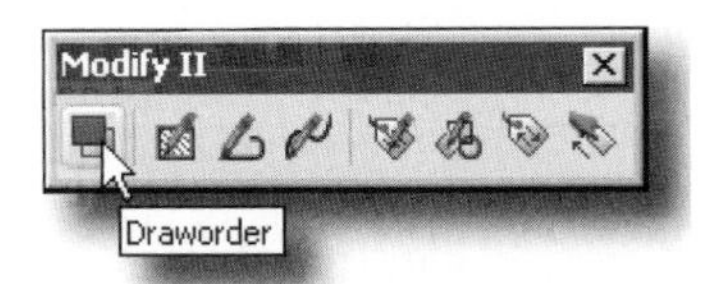

Figure 13–31 *Invoking the DRAWORDER command from the Modify II toolbar*

AutoCAD prompts:

Command: **draworder**
Select objects: *(select the objects for which you want to change the display order, and press ENTER to complete object selection)*
Enter object ordering option [Above object/Under object/Front/Back] <Back>: *(right-click and select one of the available options from the shortcut menu)*
Select reference object: *(select the reference object for changing the order of display)*

When multiple objects are selected for reordering, the relative display order of the objects selected is maintained.

Above Object

The Above object option moves the selected object(s) above a specified reference object.

Under Object

The Under object option moves selected object(s) below a specified reference object.

Front

The Front option moves selected object(s) to the front of the drawing order.

Back

The Back option moves selected object(s) to the back of the drawing order.

Note: The DRAWORDER command terminates when selected object(s) are reordered. The command does not continue to prompt for additional objects to reorder.

OBJECT PROPERTIES

There are three important properties that control the appearance of objects: color, linetype, and lineweight. You can specify the color, linetype, and lineweight for the objects to be drawn with the help of the LAYER command, as explained in Chapter 3. You can do the same thing by means of the COLOR, LINETYPE, and LINEWEIGHT commands.

COLOR COMMAND

The COLOR command allows you to specify a color for the objects to be drawn, separate from the layer color.

Invoke the COLOR control command:

Properties toolbar	Choose Select color... from the option menu (see Figure 13–32)
Format menu	Choose Color...
Command: prompt	**color** (ENTER)

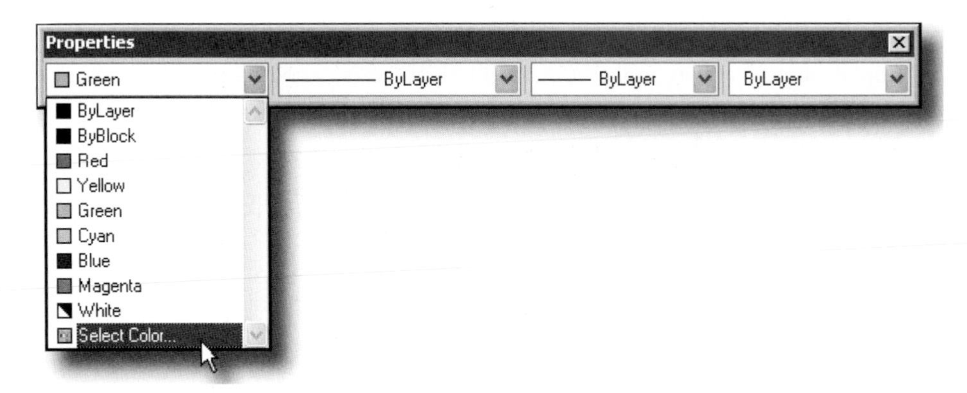

Figure 13–32 *Selecting Select color... from the option menu in the Properties toolbar*

AutoCAD displays the Select Color dialog box as shown in Figure 13–33. From the color option menu in the Properties toolbar or the text box on the Index Color tab of the Select Color dialog box, the color may be entered as a standard name (red, green, cyan, yellow, magenta, blue, white, or green) or by the number code (1 through 255). You can also select one of the colors from the chart or one of the color bars below the chart in the Index tab of the Select color dialog box. Its name or number will be displayed in the text box and it will become the current color. If you reply with a standard name or number code, this becomes the current color. All new objects you create are drawn with this color, regardless of which layer is current, until you again set the color to BYLAYER or BYBLOCK. BYLAYER causes the objects drawn to assume the color of the layer on which it is drawn. BYBLOCK causes objects to be drawn in white until selected for inclusion in a block definition. Subsequent insertion of a block that contains objects drawn under the BYBLOCK option causes those objects to assume the color of the current setting of the COLOR command. The default is set to BYLAYER. You can use the PROPERTIES command to change the color of the existing objects. AutoCAD 2004 introduces new color capabilities. In the Select Color dialog box, two additional new tabs are provided: **True Color** (see Figure 13–34) and **Color Books** (see Figure 13–35). Instead of choosing from 256 standard colors, you can also choose colors from the True Color graphic interface located on the **True Color** tab with its controls for Hue, Saturation, Luminance, and Color Model or from standard **Color Books** (such as Pantone) located on the **Color Books** tab. **True Color** and **Color Books** options make it easier to match colors in your drawing with colors of actual materials.

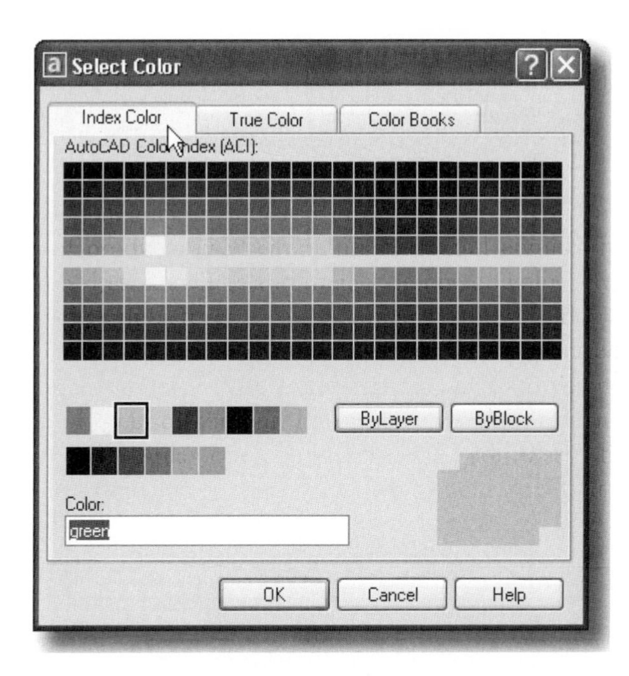

Figure 13–33 *Select Color dialog box with Index Color tab displayed*

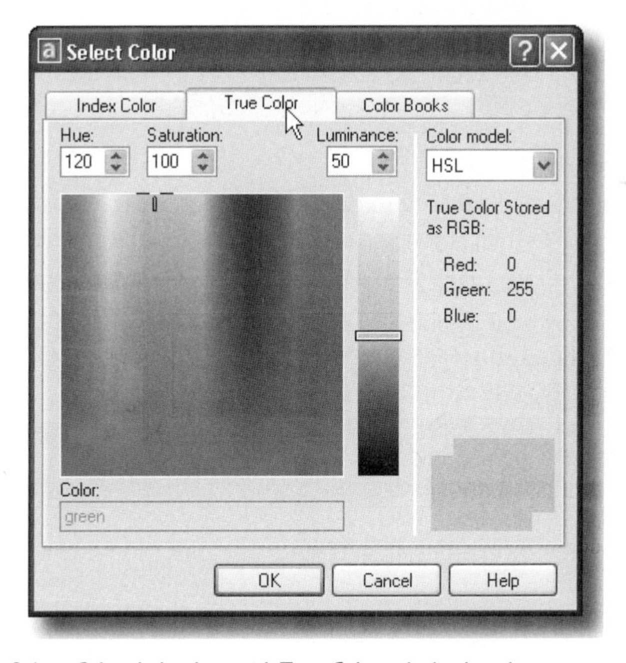

Figure 13–34 *Select Color dialog box with True Color tab displayed*

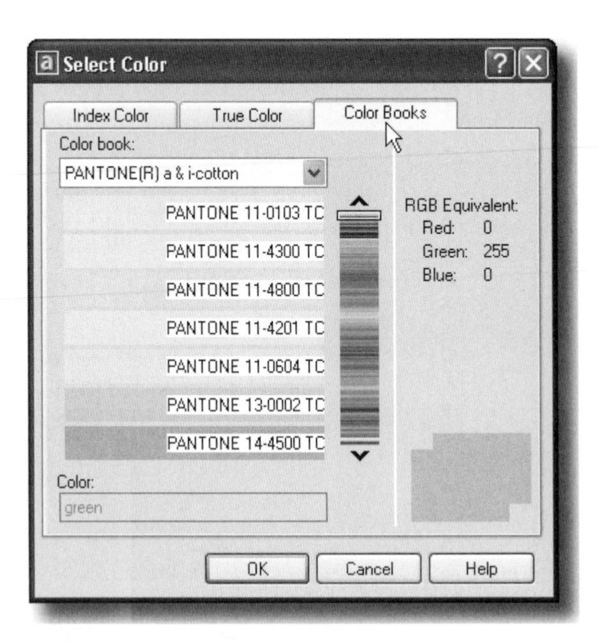

Figure 13–35 *Select Color dialog box with Color Books tab displayed*

Note: As noted in Chapter 3, the options to specify colors by both layer and the COLOR command can cause confusion in a large drawing, especially one containing blocks and nested blocks. You are advised not to mix the two methods of specifying colors in the same drawing.

LINETYPE COMMAND

The LINETYPE command allows you to draw lines with different dash/dot/space combinations. It is used to load linetype definitions from a library or to let you create custom linetypes.

A linetype must exist in a library file and be loaded before you can apply it to an object or layer. Standard linetypes are in the library file called ACAD.LIN and are not loaded with the LAYER command. You must load the linetype before you assign it to a specific layer.

Linetypes are combinations of dashes, dots, and spaces. Customized linetypes permit "out of line" objects in a linetype such as circles, wavy lines, blocks, and skew segments.

Dash, dot, and space combinations eventually repeat themselves. For example, a six-unit-long dash, followed by a dot between two one-unit-long spaces, repeats itself according to the overall length of the line drawn and the LTSCALE setting.

Lines with dashes (not all dots) usually have dashes at both ends. AutoCAD automatically adjusts the lengths of end dashes to reach the endpoints of the adjoining line. Intermediate dashes will be the lengths specified in the definition. If the overall length of the line is not long enough to permit the breaks, the line is drawn continuous.

There is no guarantee that any segments of the line fall at a particular location. For example, when placing a centerline through circle centers, you cannot be sure that the short dashes will be centered on the circle centers as most conventions call for. To achieve this effect, the short and long dashes have to be created by either drawing them individually or by breaking a continuous line to create the spaces between the dashes. This also creates multiple in-line lines instead of one line of a particular linetype. Or you can use the DIMENSION command Center option to place the desired mark.

Individual linetype names and definitions are stored in one or more files whose extension is .LIN. The same name may be defined differently in two different files. Selecting the desired one requires proper responses to the prompts in the Load option of the LINETYPE command. If you redefine a linetype, loading it with the LINETYPE command will cause objects drawn on layers assigned to that linetype to assume the new definition.

Mastering the use of linetypes involves using the LAYER command, the LINETYPE command, the LTSCALE command, and knowing which files contain the linetype definition(s) desired. Also, with the LINETYPE command you can define custom linetypes.

Invoke the LINETYPE command:

Properties toolbar	Choose Other... from Linetype Control (see Figure 13–36)
Format menu	Choose Linetype...
Command: prompt	**linetype** (ENTER)

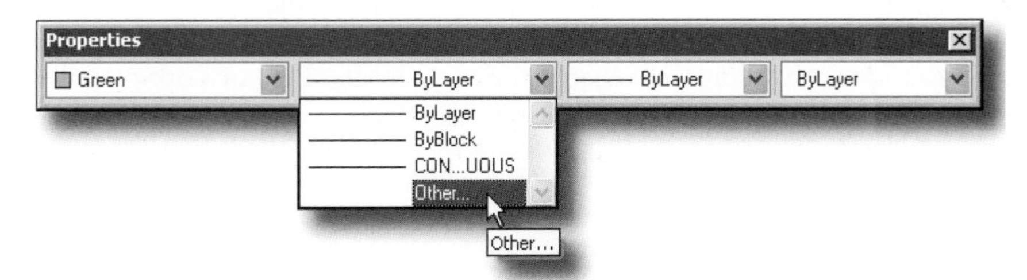

Figure 13–36 *Invoking the* LINETYPE *command from the Properties toolbar*

AutoCAD displays the Linetype Manager dialog box, similar to Figure 13–37.

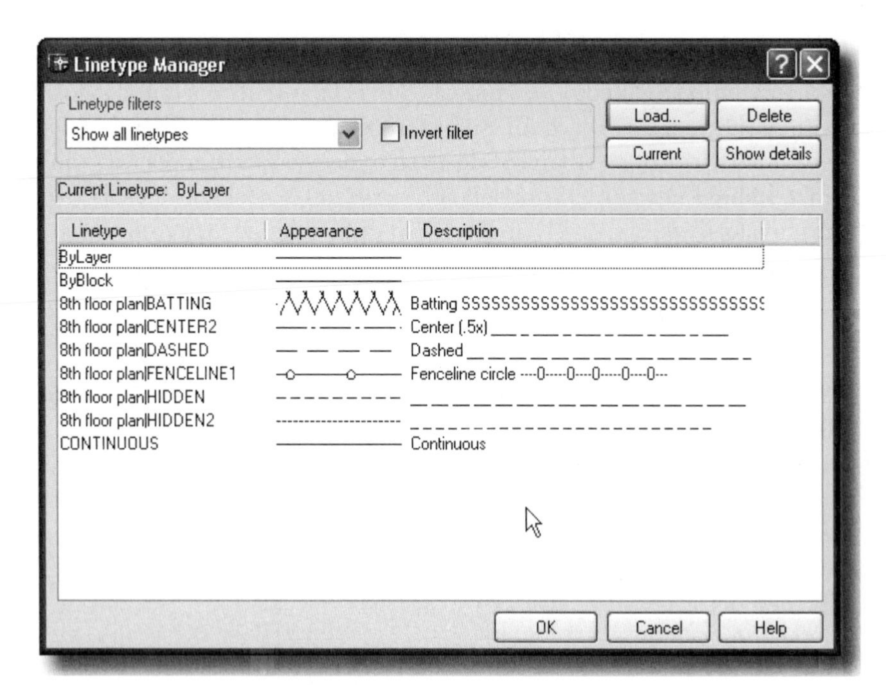

Figure 13–37 *Linetype Manager dialog box*

AutoCAD lists the available linetypes for the current drawing and displays the current linetype setting next to the **Current** button. By default, it is set to ByLayer. To change the current linetype setting, select the appropriate linetype from the Linetype list box and select the **Current** button. All new objects you create will be drawn with the selected linetype, irrespective of the layer you are working with, until you again set the linetype to ByLayer or ByBlock. ByLayer causes the object drawn to assume the linetypes of the layer on which it is drawn. ByBlock causes objects to be drawn in Continuous linetype until selected for inclusion in a block definition. Subsequent insertion of a block that contains objects drawn under the ByBlock option will cause those objects to assume the linetype of the block.

To load a linetype explicitly into your drawing, choose the **Load...** button. AutoCAD displays the Load or Reload Linetypes dialog box, as shown in Figure 13–38.

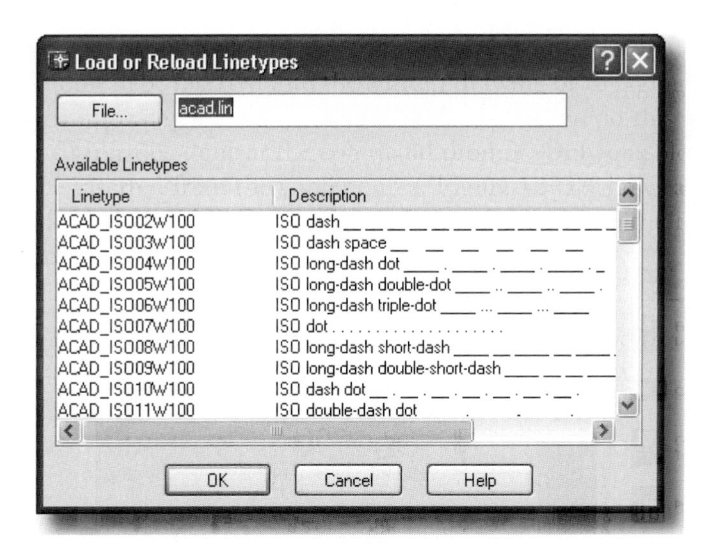

Figure 13–38 *Load or Reload Linetypes dialog box*

By default, AutoCAD lists the available linetypes from the ACAD.LIN file. Select the linetype to load from the **Available Linetypes** list box, and choose the **OK** button.

If you need to load linetypes from a different file, choose the **File** button in the Load or Reload Linetypes dialog box. AutoCAD displays the Select Linetype File dialog box. Select the appropriate linetype file and choose the **OK** button. In turn, AutoCAD lists the available linetypes from the selected linetype file in the Load or Reload Linetypes dialog box. Select the appropriate linetype to load from the **Available Linetypes** list box, and choose the **OK** button.

To delete a linetype that is currently loaded in the drawing, first select the linetype from the list box in the Linetype Manager dialog box, and then select the **Delete** button. You can delete only linetypes that are not referenced in the current drawing. You cannot delete linetype Continuous, ByLayer, or ByBlock. Deleting is the same as using the PURGE command to purge unused linetypes from the current drawing.

To display additional information of a specific linetype, first select the linetype from the Linetype list box in the Linetype Manager dialog box, and then choose the **Show details** button. AutoCAD displays an extension of the dialog box, listing additional settings, as shown in Figure 13–39.

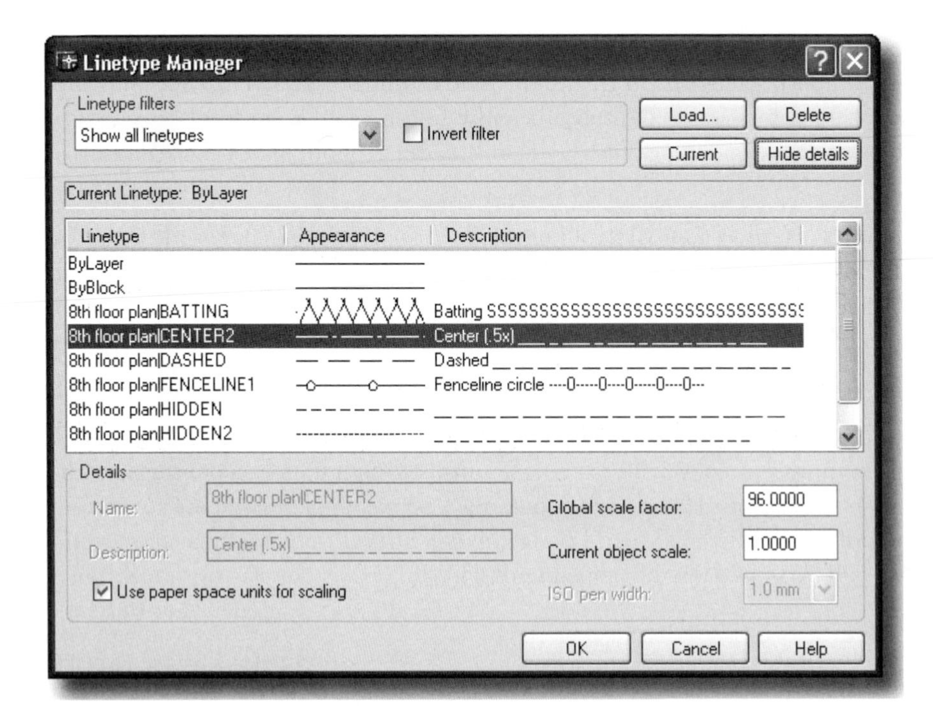

Figure 13–39 *Extended Linetype Manager dialog box*

The **Name:** and **Description:** edit fields display the selected linetype name and description, respectively.

The **Global scale factor:** text field displays the current setting of the LTSCALE factor. The **Current object scale:** text field displays the current setting of the CELTSCALE factor. If necessary, you can change the values of the LTSCALE and CELTSCALE system variables.

The **ISO pen width:** box sets the linetype scale to one of a list of standard ISO values. The resulting scale is the global scale factor multiplied by the object's scale factor.

Setting the **Use paper space units for scaling** check box to ON scales linetypes in paper space and model space identically.

After making the necessary changes, click the **OK** button to keep the changes and close the Linetype Manager dialog box.

Note: As noted in Chapter 3, the options to specify linetypes by both the LAYER and the LINETYPE commands can cause confusion in a large drawing, especially one containing blocks and nested blocks. You are advised not to mix the two methods of specifying linetypes in the same drawing.

LINEWEIGHT COMMAND

The LINEWEIGHT command allows you to specify a lineweight for the objects to be drawn, separate from the assigned lineweight for the layer.

Invoke the LINEWEIGHT command:

Properties toolbar	Select one of the lineweights available from the option menu (see Figure 13–40)
Format menu	Choose Lineweight...
Command: prompt	**lineweight** (ENTER)

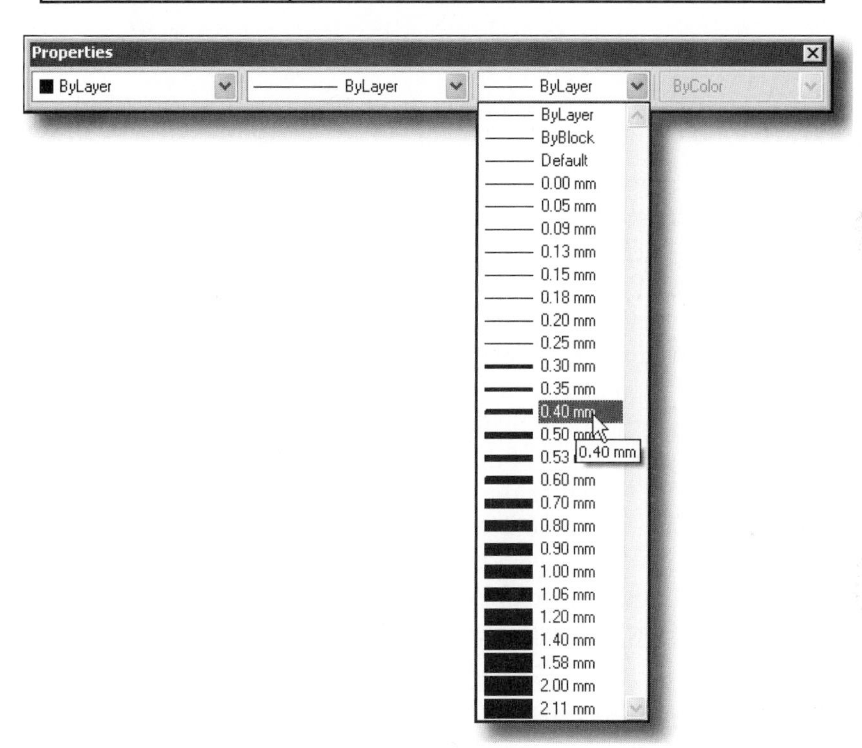

Figure 13–40 *Selecting one of the available lineweights from the option menu in the Properties toolbar*

If you select **Lineweight** frpm the **Format** menu or type **lineweight** at the Command: prompt, AutoCAD displays the Lineweight Settings dialog box, as shown in Figure 13–41.

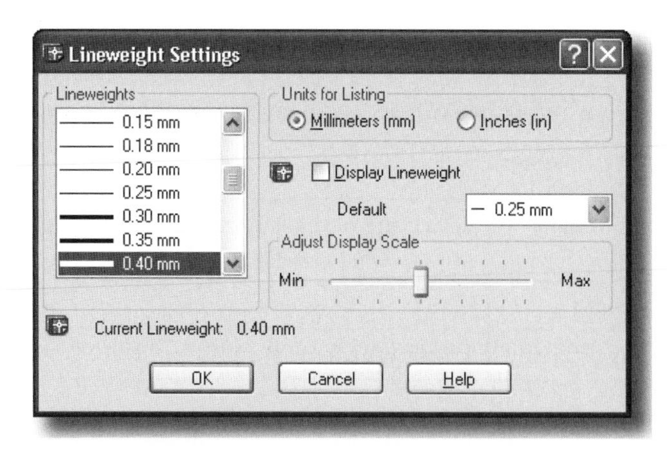

Figure 13–41 *Lineweight Settings dialog box*

AutoCAD displays current lineweight at the bottom of the dialog box. To change the current lineweight, select one of the available lineweights from the **Lineweights** list box. Bylayer is the default. All new objects you create are drawn with the current lineweight, regardless of which layer is current, until you again set the lineweight to Bylayer, Byblock, or Default. Bylayer causes the objects drawn to assume the lineweight of the layer on which it is drawn. Byblock causes objects to be drawn in default until selected for inclusion in a block definition. Subsequent insertion of a block that contains objects drawn under the Byblock option causes those objects to assume the Lineweight of the current setting of the LINEWEIGHT command. The **Default** selection causes the objects to be drawn to the default value as set by the LWDEFAULT system variable and defaults to a value of 0.01 inches or 0.25 mm. You can also set the default value from the **Default** option menu located in the right side of the dialog box. The lineweight value of 0 plots at the thinnest lineweight available on the specified plotting device and is displayed at one pixel wide in model space. You can use the PROPERTIES command to change the lineweight of the existing objects.

The **Units for Listing** section specifies whether lineweights are displayed in millimeters or inches. You can also set **Units for Listing** by using the LWUNITS system variable.

The **Display Lineweight** toggle controls whether lineweights are displayed in the current drawing. If it is set to ON, lineweights are displayed in model space and paper space. AutoCAD regeneration time increases with lineweights that are represented by more than one pixel. If it is set to OFF, AutoCAD performance improves. Performance slows down when working with lineweights set to ON in a drawing.

The **Adjust Display Scale** controls the display scale of lineweights on the Model tab. On the Model tab, lineweights are displayed in pixels. Lineweights are displayed using a pixel width in proportion to the real-world unit value at which they plot. If you are using a high-resolution monitor, you can adjust the lineweight display scale to better display different lineweight widths. The Lineweight list reflects the current display scale. Objects with lineweight that are displayed with a width of more than one pixel may increase AutoCAD regeneration time. If you want to optimize AutoCAD performance when working in the Model tab, set the lineweight display scale to the minimum value or turn off lineweight display altogether.

Choose the **OK** button to close the dialog box and keep the changes in the settings.

Note: As noted in Chapter 3, the options to specify lineweights by both the LAYER and the LINEWEIGHT commands can cause confusion in a large drawing, especially one containing blocks and nested blocks. You are advised not to mix the two methods of specifying lineweights in the same drawing.

X, Y, AND Z FILTERS—AN ENHANCEMENT TO OBJECT SNAP

The AutoCAD filters feature allows you to establish a *2D* point by specifying the individual (*X* and *Y*) coordinates one at a time in separate steps. In the case of a *3D* point you can specify the individual (*X*, *Y*, and *Z*) coordinates in three steps. Or you can specify one of the three coordinate values in one step and a point in another step, from which AutoCAD extracts the other two coordinate values for use in the point being established.

The filters feature is used when you being prompted to establish a point, as in the starting point of a line, the center of a circle, drawing a node with the POINT command, or specifying a base point or second point in displacement for the MOVE or COPY command, to mention just a few.

Note: During the application of the filters feature there are steps where you can input either single coordinate values or points, and there are steps where you can input only points. It is necessary to understand these restrictions and options and when one type of input is more desirable than the other.

When selecting points during the use of filters, you need to know which coordinates of the specified point are going to be used in the point being established. It is also essential to know how to combine Object Snap modes with those steps that use point input.

The filters feature is actually an enhancement to either the object snap or the @ (last point) feature. The AutoCAD ability to establish a point by snapping to a point on an existing object is one of the most powerful features in AutoCAD, and being able to have AutoCAD snap to such an existing point and then filter out selected coordinates for use in establishing a new point adds to that power. Therefore, in most cases you will not use the filters feature if it is practical to enter in all of the coordinates from the keyboard, because entering in all the coordinates can be done in a single step. The filters feature is a multistep process, and each step might include substeps, one to specify the coordinate(s) to be filtered out and another to designate the Object Snap mode involved.

FILTERS WITH @

When AutoCAD is prompting for a point, the filters feature is initiated by entering a period followed by the letter designation for the coordinate(s) to be filtered out. For example, if you draw a point starting at (0,0) and use the relative polar coordinate response @3<45 to determine the endpoint, you can use filters to establish another point whose *X* coordinate is the same as the *X* coordinate of the end of the line just drawn. It works for *Y* and *Z* coordinates and combinations of *XY*, *XZ*, and *YZ* coordinates also. The following command sequence shows how to apply filter to a line that needs to be started at a point whose *X* coordinate is the same as that of the end of the previous line and whose *Y* coordinate is 1.25. The line will be drawn horizontally 3 units long. The sequence is as follows:

Command: **line**
Specify first point: **0,0**
Specify next point or [Undo]: **@3<45**
Specify next point or [Undo]: (ENTER)

Command: **line** (or ENTER)
Specify first point: **.x**
of **@**
(need YZ): **0,1.25**
Specify next point or [Undo]: **@3<0**

Entering .x initiates the filters feature. AutoCAD then prompts you to specify a point from which it can extract the *X* coordinate. The @ (last point) does this. The new line has a starting point whose *X* coordinate is the same as that of the last point drawn. By using the filters feature to extract the *X* coordinate, that starting point will be on an imaginary vertical line through the point specified by @ in response to the "of" prompt.

When you initiate filters with a single coordinate (.x in our example) and respond with a point (@), the prompt that follows asks for a point also. From it (the second point specified) AutoCAD extracts the other two coordinates for the new point.

Even though the prompt is for "*YZ*," the point may be specified in *2D* format as 0,1.25 (the *X* and *Y* coordinates), from which AutoCAD takes the second value as the needed *Y* coordinate. The *Z* coordinate is assumed to be the elevation of the current coordinate system.

You can use the two-coordinate response to initiate filters. Then specify a point, and all that AutoCAD requires is a single value for the final coordinate. An example of this follows.

Command: **line**
Specify first point: **0,0**
Specify next point or [Undo]: **@3<45**
Specify next point or [Undo]: (ENTER)

Command: **line** (or ENTER)
Specify first point: **.xz**
of @
(need Y): **1.25**
Specify next point or [Undo]: **@3<0**

You can also specify a point in response to the "(need Y):" prompt:

(need Y): **0,1.25** *(or pick a point on the screen)*

In this case, AutoCAD uses the *Y* coordinate of the point specified as the *Y* coordinate of the new point.

Remember, it is an individual coordinate in *2D* (one or two coordinates in *3D*) of an existing point that you wish AutoCAD to extract and use for the new point. In most cases you will be object snapping to a point for the response. Otherwise, if you knew the value of the coordinate needed, you would probably enter it at the keyboard.

FILTERS WITH OBJECT SNAP

Without filters, an Object Snap (Osnap) mode establishes a new point to coincide with one on an existing object. With filters, an Osnap mode establishes selected coordinates of a new point to coincide with corresponding coordinates of one on an existing object.

Extracting one or more coordinate values to be applied to corresponding coordinate values of a point that you are being prompted to establish is shown in the following example.

In Figure 13–42, a 2.75" by 7.1875" rectangle has a 0.875"-diameter hole in its center. A board drafter would determine the center of a square or rectangle by drawing diagonals and centering the circle at their intersection. AutoCAD drafters (without filters) could do the same, or they might draw orthogonal lines from the midpoint of a horizontal line and from the midpoint of one of the vertical lines to establish a centering intersection. The following command sequence shows steps in drawing a rectangle with a circle in the center using filters.

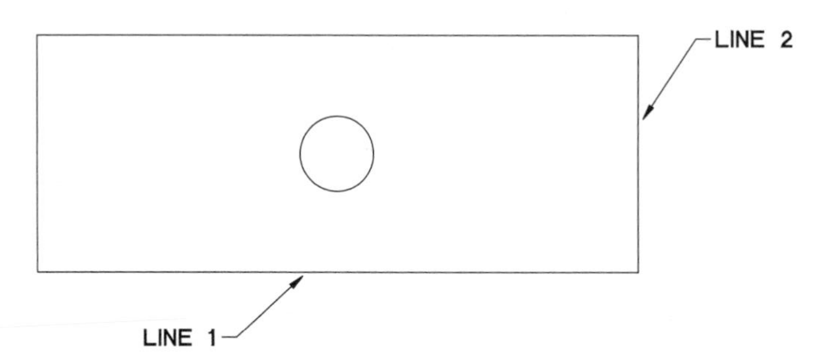

Figure 13–42 *Extracting coordinate values to be applied to corresponding coordinate values*

Command: **line**
Specify first point: *(select a point)*
Specify next point or [Undo]: **@2.75<90**
Specify next point or [Undo]: **@7.1875<0**
Specify next point or [Close/Undo]: **@2.75<270**
Specify next point or [Close/Undo]: **c**

Command: **circle**
Specify center point for circle or [3P/2P/Ttr (tan tan radius)]: **.x**
of **mid**
of (select line 1)
(need YZ): **mid**
of (select line 2)
Specify radius of circle or [Diameter]: **d**
Specify diameter of circle <previous diameter >: **.875**

SHELL COMMAND

The SHELL command allows you to execute operating system programs without leaving AutoCAD. You can execute any operating system program as long as there is sufficient memory to execute.

Invoke the SHELL command:

Command: prompt	**shell** (ENTER)

AutoCAD prompts:

Command: **shell**
OS Command: (invoke one of the available operating system's utility programs)

When the utility program is finished, AutoCAD takes you back to the "Command:" prompt. If you need to execute more than one operating system program, give a null response to the "OS command:" prompt. AutoCAD responds with the appropriate operating system prompt. You can now enter as many operating system commands as you wish. When you are finished, you may return to AutoCAD by entering EXIT. It will take you back to the "Command:" prompt.

If there is not enough free memory for the SHELL command, the following message will appear:

Shell error: insufficient memory for command.

Where there is insufficient memory, you can execute the SH command instead of the SHELL command. SH requires less memory than the SHELL command and can be used to access internal DOS commands, such as DIR, COPY, and TYPE. If the need arises, you can adjust the amount of memory required for the SH and SHELL commands by modifying the ACAD.PGP file. For additional information, see Chapter 18 on "Customizing AutoCAD."

In the Windows 95 (and later), and Windows NT 4.xx operating systems, instead of invoking the SHELL command, you can switch between programs by selecting the appropriate program from the task bar. Depending on the memory available in the computer, you can open multiple programs and switch between the programs by selecting appropriate open programs from the task bar.

Note: Do not delete the AutoCAD lock files or temporary files created for the current drawing when you are at the operating system prompt.

SETTING UP A DRAWING

The MVSETUP command is used to control and set up the view(s) of a drawing, including the choice of standard plotted sheet sizes with a border, the scale for plotting on the selected sheet size, and multiple viewports. MVSETUP is an AutoLISP routine that can be customized to insert any type of border and title block.

Options and associated prompts depend on whether the TILEMODE system variable is set to ON (1) or OFF (0). When TILEMODE is set to ON, **Tiled Viewports** is enabled. When TILEMODE is set to OFF, the **Floating Viewports** menu item is enabled. Other paper space–related drawing setup options are available.

Invoke the MVSETUP command:

Command: prompt	**mvsetup** (ENTER)

AutoCAD prompts depend whether you are working on Model tab or Layout tab. If you are working on Model tab, then AutoCAD prompts:

Enable paper space? [No/Yes] <Y>: *(press* ENTER *to enable paper space, and AutoCAD changes the* TILEMODE *setting to 0, or enter n and press* ENTER *to stay in the* TILEMODE *setting of 1)*

If you are working in Layout tab, then AutoCAD prompts:

Enter an option [Align/Create/Scale viewports/Options/Title block/Undo]: *(right-click and select one of the available options from the shortcut menu)*

The following is the procedure for setting up the drawing when you are working on Model tab. AutoCAD prompts:

Enable paper space? [No/Yes] <Y>: **n** *(*ENTER*)*
Enter units type [Scientific/Decimal/Engineering/Architectural/Metric]: *(select a unit type. Depending on the units selected, AutoCAD lists the available scales. Select one of the available scales, or you can even specify a custom scale factor)*
Enter the scale factor: *(specify a scale factor)*
Enter the paper width: *(specify the paper width at which the drawing will be plotted)*
Enter the paper height: *(specify the paper height at which the drawing will be plotted)*

AutoCAD sets up the appropriate limits to allow you to draw to full scale and also draws a bounding box enclosing the limits. Draw the drawing to full scale; when you are ready to plot, specify the scale mentioned earlier to plot the drawing.

Following is the procedure for setting up the drawing when you are working in the Layout tab. AutoCAD prompts:

Enter an option [Align/Create/Scale viewports/Options/Title block/Undo]:

The options are explained here in the order that is logical to complete the drawing setup.

Title Block

The Title block option allows you to select an appropriate title block. AutoCAD prompts:

Enter title block option [Delete objects/Origin/Undo/Insert] <Insert>: *(right-click and select one of the available options from the shortcut menu)*

The Insert option (default) allows you to insert one of the available standard title blocks. AutoCAD lists the available title blocks and prompts you to select one of the available title blocks as follows:

0. None
1. ISO A4 Size(mm)
2. ISO A3 Size(mm)

3. ISO A2 Size(mm)
4. ISO A1 Size(mm)
5. ISO A0 Size(mm)
6. ANSI-V Size(in)
7. ANSI-A Size(in)
8. ANSI-B Size(in)
9. ANSI-C Size(in)
10. ANSI-D Size(in)
11. ANSI-E Size(in)
12. Arch/Engineering (24 x 36in)
13. Generic D size Sheet (24 x 36in)

Enter number of title block to load or [Add/Delete/Redisplay]: *(select one of the available title blocks and press* ENTER, *and AutoCAD inserts a border and title block, as shown in Figure 13–43, or enter an option)*

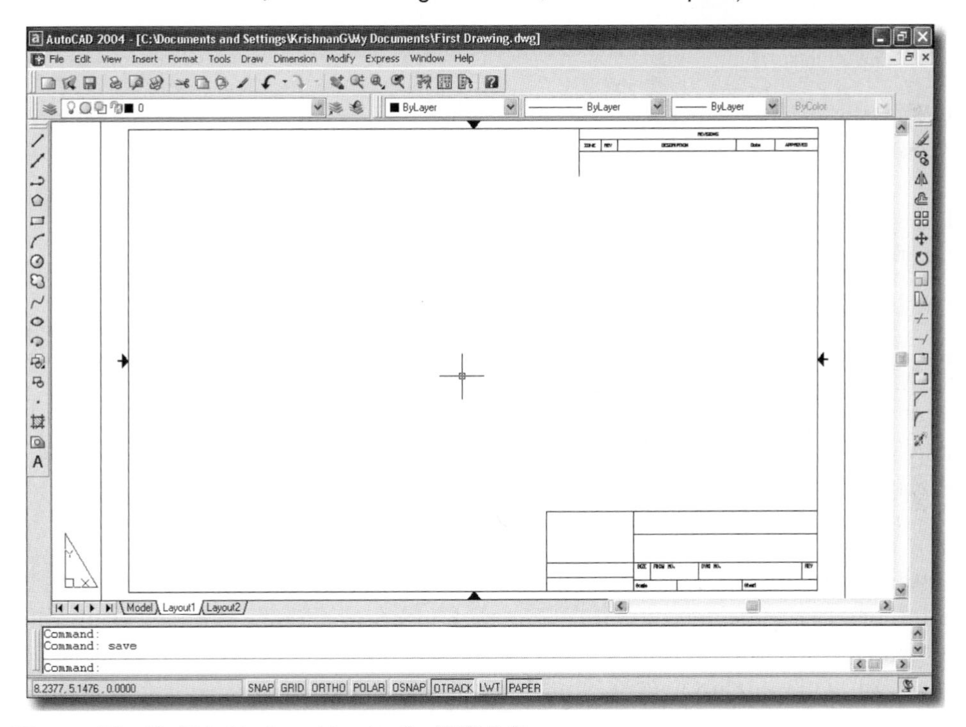

Figure 13–43 *Title block and border for ANSI-B Size*

The Add option allows you to add a title block drawing to the available list.

The Delete option allows you to delete an entry from the available list.

The Redisplay option redisplays the list of title block options.

Create

The Create option allows you to establish viewports. AutoCAD prompts:

> Enter option [Delete objects/Create viewports/Undo] <Create>: *(press* ENTER *to create viewports, and AutoCAD lists the available viewport layout options, or enter an option)*

The Create viewports option (default) allows you to create multiple viewports to one of the standard layouts. AutoCAD prompts:

> Available Mview viewport layout options:
>
> 0: None
> 1: Single
> 2: Std. Engineering
> 3: Array of Viewports
>
> Enter layout number to load or [Redisplay]: *(select one of the available layouts)*

The **None** selection creates no viewports. The **1** selection creates a single viewport whose size is determined during subsequent prompts responses. The **2** selection creates four viewports with preset viewing angles by dividing a specified area into quadrants. The size is determined by responses to subsequent prompts. The **3** selection creates a matrix of viewports along the *X* and *Y* axes.

The Delete objects option deletes the existing viewports.

The Undo option reverses operations performed in the current MVSETUP session.

Scale Viewports

The Scale viewports option adjusts the scale factor of the objects displayed in the viewports. The scale factor is specified as a ratio of paper space to model space. For example, 1:48 is 1 paper space unit for 48 model space units (scale for 1/4" = 1'0").

Options

The Options option lets you establish several different environment settings that are associated with your layout. AutoCAD prompts:

> Enter an option [Layer/LImits/Units/Xref] <exit>: *(enter an option)*

The Layer option permits you to specify a layer for placing the title block.

The LImits option permits you to specify whether or not to reset the limits to drawing extents after the title block has been inserted.

The Units option permits you to specify whether sizes and point locations will be translated to inch or millimeter paper units.

The Xref option permits you to specify whether the title block is to be inserted or externally referenced.

Align

The Align option causes AutoCAD to pan the view in one viewport so that it aligns with a basepoint in another viewport. Whichever viewport the other point moves to becomes the active viewport. AutoCAD prompts:

```
Enter an option [Angled/Horizontal/Vertical alignment/Rotate view/Undo]:
```

The Angled option causes AutoCAD to pan the view in a viewport in a specified direction.

The Horizontal option causes AutoCAD to pan the view in one viewport, aligning it horizontally with a basepoint in another viewport.

The Vertical alignment option causes AutoCAD to pan the view in one viewport, aligning it vertically with a basepoint in another viewport.

The Rotate view option causes AutoCAD to rotate the view in a viewport around a basepoint.

Undo

The Undo option causes AutoCAD to undo the results of the current MVSETUP command.

LAYER TRANSLATOR

The Layer Translator is used to make selected layers in the current drawing match layers in another drawing or in a CAD standards file.

Invoke the LAYTRANS command:

Tools menu	Cad Standards > Layer Translator...
Command: prompt	**laytrans** (ENTER)

AutoCAD displays the Layer Translator dialog box, as shown in Figure 13–44.

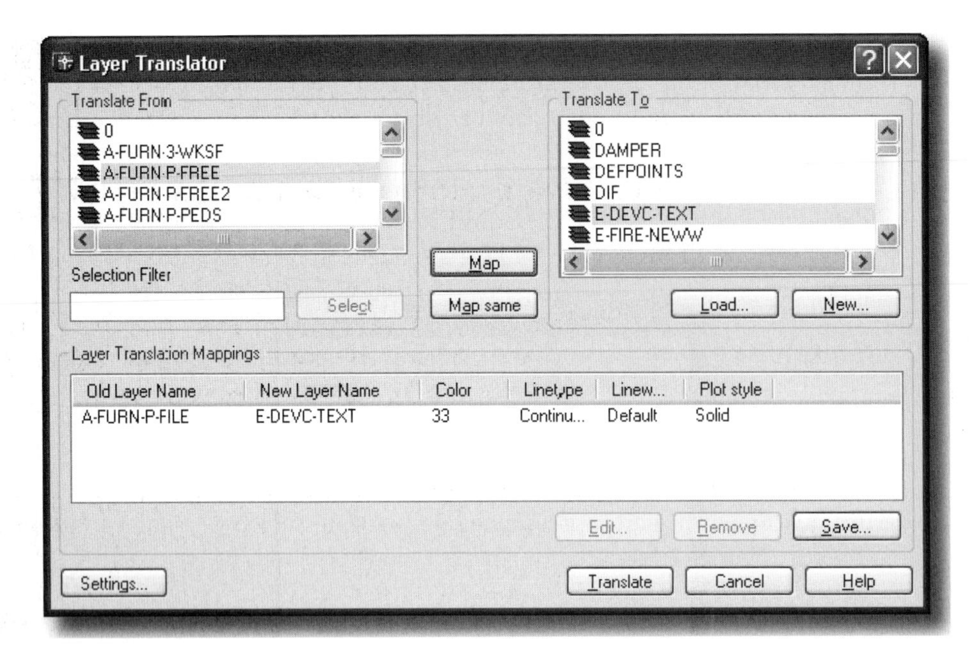

Figure 13–44 *Layer Translator dialog box*

Translate From

In the **Translate From** text box you can select the layer(s) to be changed. Or you can supply a selection filter. A dark colored icon indicates that the layer is referenced in the drawing; a white icon indicates the layer is unreferenced. You can delete unreferenced layers from the drawing by right-clicking in the **Translate From** list and choosing PURGE LAYERS.

Selection Filter

In addition to any layers previously selected, you can specify a naming pattern to filter layers to be selected in the **Translate From** list. You can include wildcards.

Select

The **Select** button causes the layers specified in **Selection Filter** to be selected.

Map

The **Map** button causes the layers selected in **Translate From** to be mapped to the layer selected in **Translate To**.

Map same

The **Map same** button causes all layers that have the same name in both lists to be mapped from **Translate From** to **Translate To**.

Translate To

The layers you can translate the current drawing's layers to are listed in the **Translate To** text box.

Load

The **Load** button lets you load layers in the **Translate To** list from a drawing, drawing template, or standards file that you specify. If the specified file contains saved layer mappings, those mappings are applied to the layers in the **Translate From** list and are displayed in **Layer Translation Mappings**. You can load layers from more than one file. If you load a file that contains layers of the same name as layers already loaded, the original layers are retained and the duplicate layers are ignored. Similarly, if you load a file containing mappings that duplicate mappings already loaded, the original mappings are retained and the duplicates are ignored.

New

The **New…** button lets you define a new layer to be shown in the **Translate To** list for translation. If you select a **Translate To** layer before choosing **New…**, the selected layer's properties are used as defaults for the new layer. You cannot create a new layer with the same name as an existing layer.

Layer Translation Mapping

In the **Layer Translation Mapping** section, each translated layer is listed by categories for **Old Layer Name, New Layer Name, Color, Linetype, Lineweight**, and **Plot Style**.

Edit

The **Edit** button causes the Edit Layer dialog box to be displayed, similar to Figure 13–45.

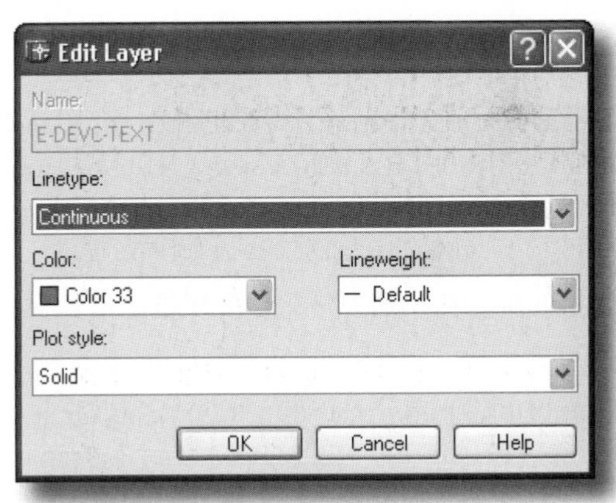

Figure 13–45 *Edit Layer dialog box*

The Edit Layer dialog box has text boxes for **Linetype, Color, Lineweight**, and **Plot style.** Those that are not inactive can not be changed.

Remove

The **Remove** button is used to remove the selected translation mapping from the **Layer Translation Mappings** list.

Save

The **Save** button is used to save the current layer translation mappings to a file for later use.

Settings

The **Settings** button causes the Settings dialog box to be displayed, similar to Figure 13–46.

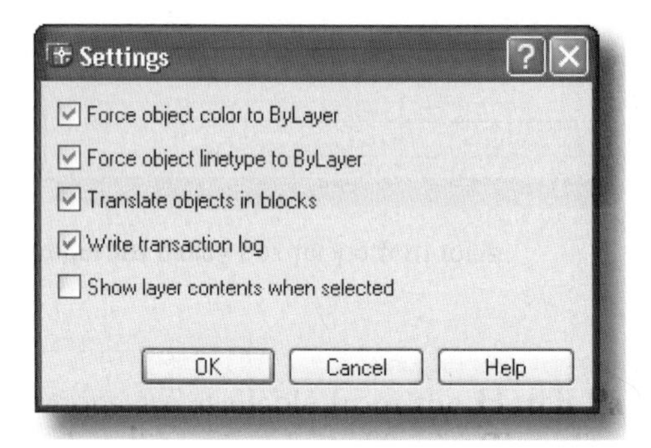

Figure 13–46 *Settings dialog box*

The **Force object color to ByLayer** check box, when checked, causes every object translated to take on the color assigned to its layer. If it is not checked, every object retains its original color.

The **Force object linetype to ByLayer** check box, when checked, causes every object translated to take on the linetype assigned to its layer. If it is not checked, every object retains its original linetype.

The **Translate objects in blocks** check box, when checked, causes objects nested within blocks (including nested blocks) to be translated. If it is not checked, nested objects in blocks are not translated.

When the **Write transaction log** option is checked, a log file is created detailing the results of the translation. The log file is assigned the same name as the translated drawing, with a *.log* file name extension and is created in the same folder. If it is not checked, no log file is created.

When checked, the **Show layer contents when selected** option, causes only the layers selected in the Layer Translator dialog box to be displayed in the drawing area. If it is not checked, all layers in the drawing are displayed.

Translate

The **Translate** button starts translation of the layers you have mapped. If you have not saved the current layer translation mappings, you are prompted to save the mappings before translation begins.

TIME COMMAND

The TIME command displays the current time and date related to your current drawing session. In addition, you can find out how long you have been working in AutoCAD. This command uses the clock in your computer to keep track of the time functions and displays to the nearest millisecond using 24-hour military format.

Invoke the TIME command:

Tools menu	Choose Inquiry > Time
Command: prompt	**time** (ENTER)

The following listing is displayed in the text screen followed by a prompt:

```
Current time: Sunday, April 25, 1999 at 9:46:10:110 PM
Times for this drawing:
Created: Sunday, April 25, 1999 at 9:26:29:370 PM
Last updated: Sunday, April 25, 1999 at 9:26:29:370 PM
Total editing time: 0 days 00:19:40.790
Elapsed timer (on): 0 days 00:19:40.790
Next automatic save in: 0 days 01:40:24.455
Enter option [Display/ON/OFF/Reset]:
```

The first line gives today's date and time.

The third line gives the date and time the current drawing was initially created. The drawing time starts when you initially begin a new drawing. If the drawing was created by means of the WBLOCK command, it is the date and time the command was executed that is displayed here.

The fourth line gives the date and time the drawing was last updated. Initially set to the drawing creation time, this is updated each time you use the END or SAVE command.

The fifth line gives the length of time you are in AutoCAD. This timer is continuously updated by AutoCAD while you are in the program, excluding plotting and printer plot time. This timer cannot be stopped or reset.

The sixth line provides information about the stopwatch timer. You can turn this timer ON or OFF and reset it to zero. This timer is independent of other functions.

The seventh line provides information about when the next automatic save will take place.

Display Option

The Display option redisplays the time functions, with updated times.

ON Option

The ON option turns the stopwatch timer to ON, if it is OFF. By default it is ON.

OFF Option

The OFF option turns the stopwatch timer to OFF and displays the accumulated time.

Reset Option

The Reset option resets the stopwatch timer to zero.

To exit the TIME command at the prompt, give a null response or press ESC.

AUDIT COMMAND

The AUDIT command serves as a diagnostic tool to correct any errors or defects in the database of the current drawing. AutoCAD generates an extensive report of the problems, and for every error detected AutoCAD recommends action to correct it.

Invoke the AUDIT command:

File menu	Choose Drawing Utilities >Audit
Command: prompt	**audit** (ENTER)

AutoCAD prompts:

Command: **audit**
Fix any errors detected? [Yes/No] <N>: *(specify y for yes or n for no)*

If you respond with Y or Yes, AutoCAD will fix all the errors detected and display an audit report with detailed information about the errors detected and fixing them. If you answer with N or No, AutoCAD will just display a report and will not fix any errors.

In addition, AutoCAD creates an ASCII report file (AUDITCLT system variable should be set to ON) describing the problems and the actions taken. It will save the file in the current directory, using the current drawing's name with the file extension .ADT.

You can use any ASCII editor to display the report file on the screen or print it on the printer, respectively.

Note: If a drawing contains errors that the AUDIT command cannot fix, open the drawing with the RECOVER command to retrieve the drawing and correct its errors.

OBJECT LINKING AND EMBEDDING (OLE)

Object linking and embedding (OLE) is a Microsoft Windows feature that combines various application data into one compound document. AutoCAD has client as well as server capabilities. As a client, AutoCAD permits you to have objects from other Windows applications either embedded in or linked to your drawing.

When an object is inserted into an AutoCAD drawing from an application that supports OLE, the object can maintain a connection with its source file. If you insert an embedded object into AutoCAD (client), it is no longer associated with the source (server). If necessary, you can edit the embedded data from inside the AutoCAD drawing by using the original application. But at the same time, this editing does not change the original file.

If, instead, you insert an object as a linked object into AutoCAD (client), the object remains associated with its source (server). When you edit a linked object in AutoCAD by using the original application, the original file changes as well as the object inserted into AutoCAD.

Linked or embedded objects appear on the screen in AutoCAD and can be printed or plotted using Windows system drivers.

Let's look at an example of object linking between an AutoCAD (server) drawing and Microsoft Word (client). Figure 13–47 shows a drawing of a desk, a computer, and a chair, that contains various attribute values. We are going to link this drawing to a Microsoft Word document.

From the **Edit** menu select **Copy**, and AutoCAD prompts you to select objects. Select the computer, the table, and the chair, and press ENTER. This will copy the selection to the Windows clipboard. Minimize the AutoCAD program.

Instead of selecting specific objects, you can copy the current view into the Windows clipboard by invoking the COPYLINK command from the **Edit** menu.

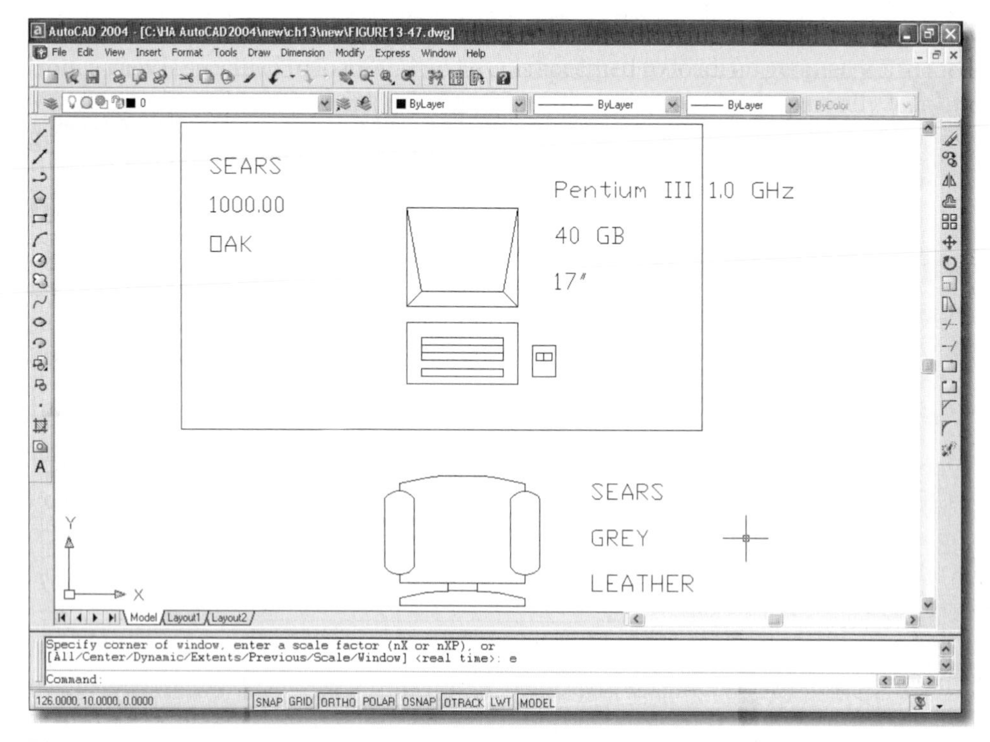

Figure 13–47 *Drawing of a desk, a computer, and a chair with attribute values*

Open the Microsoft Word program from the desktop by double-clicking the Word program icon. The Microsoft Word program is displayed, as shown in Figure 13–48.

From the **Edit** menu in Microsoft Word, select **Paste Special...**, and Word displays the Paste Special dialog box shown in Figure 13–49. Select the **Paste Link** button. This will insert the AutoCAD drawing object into Microsoft Word, as shown in Figure 13–50.

Minimize the Word program and maximize AutoCAD or if the AutoCAD program is not open, double-click the drawing image in the Word document, which will launch the AutoCAD program with the image drawing open. Edit the values of the attributes in the computer block to Pentium IV 2.5 GHz, 120.0 GB, 21", which represent a Pentium IV computer with a 120.0 GB hard drive and a 21" monitor.

Switch back to Word, and from the **Edit** menu select **Links**. Word will display the Links dialog box shown in Figure 13–51.

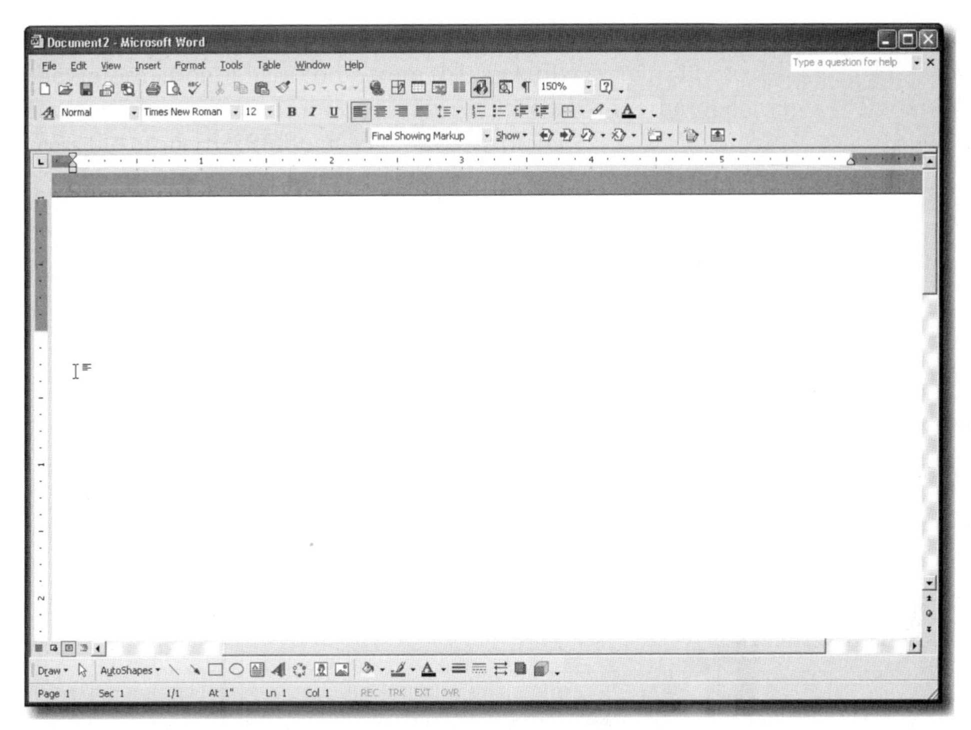

Figure 13–48 *Microsoft Word program*

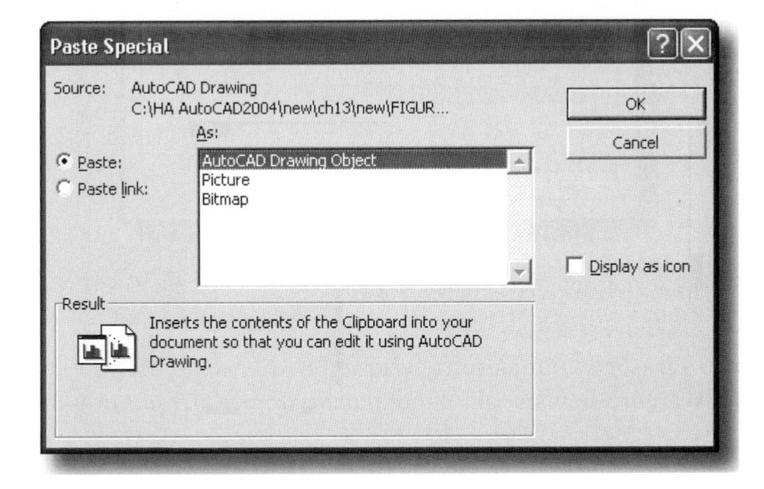

Figure 13–49 *Paste Special dialog box*

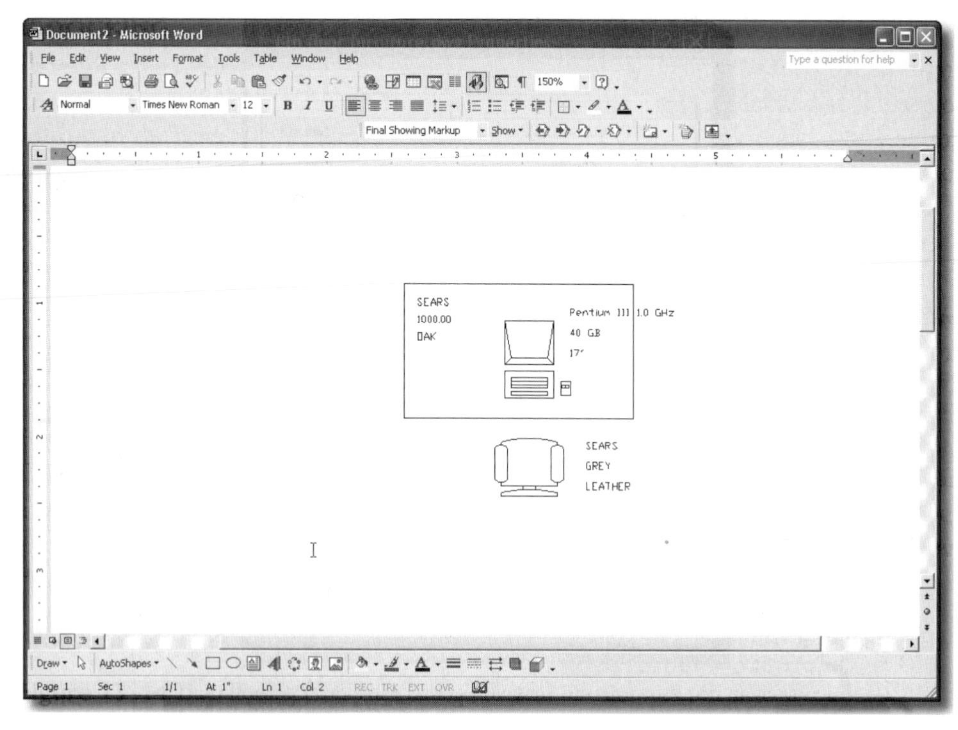

Figure 13–50 *Microsoft Word document with the AutoCAD drawing*

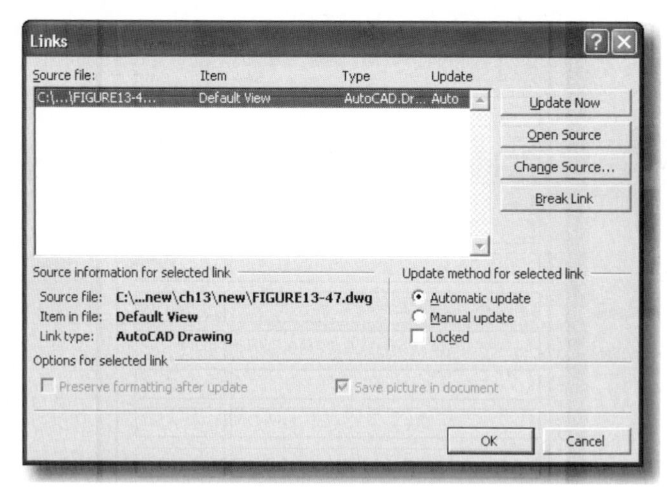

Figure 13–51 *Links dialog box*

Select the **Update Now** button and then choose the **OK** button. The image in the Word document is updated, as shown in Figure 13–52.

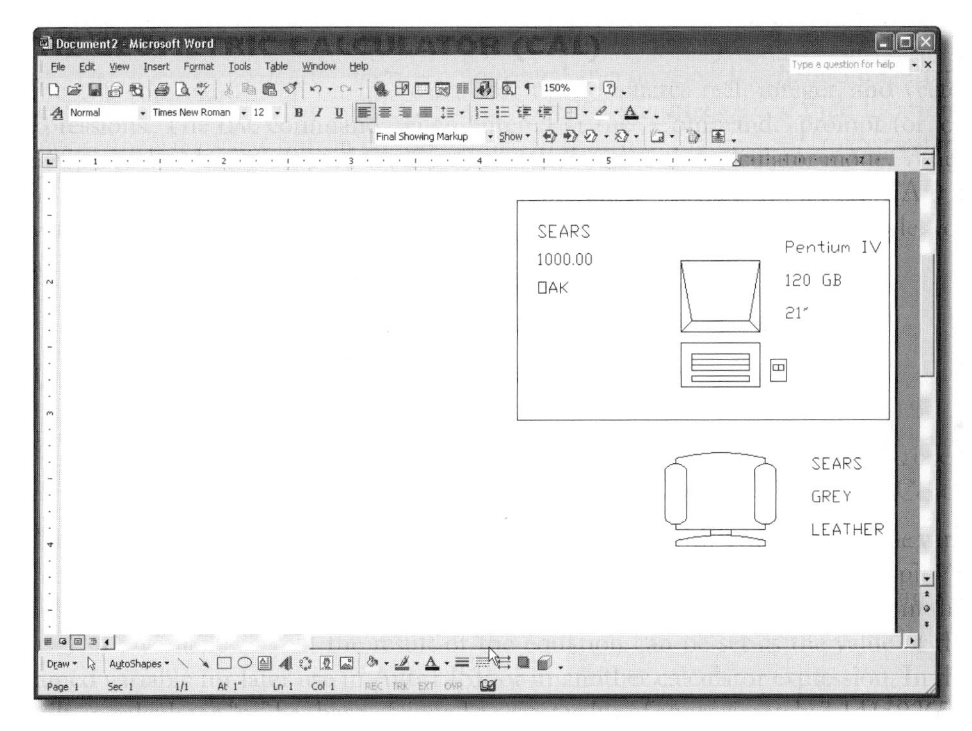

Figure 13–52 *AutoCAD drawing updated in Word as the client*

In our example, AutoCAD is the server and Microsoft Word is the client.

Conversely, you can place a linked object in AutoCAD, where AutoCAD is the client and another application is the server. Let's look an example in which AutoCAD is the client and Excel is the server.

Start the Excel program and create a spreadsheet. Copy the contents into the Windows clipboard.

From the **Edit** menu in AutoCAD, select **Paste Special.** AutoCAD displays the Paste Special dialog box. Select the **Paste Link** radio button and Microsoft Excel Worksheet from the list box, and click the **OK** button.

The Excel spreadsheet will be linked to the drawing, as shown in Figure 13–53. AutoCAD is now the client and Microsoft Excel is the server.

To edit the spreadsheet, double-click anywhere on the spreadsheet, which in turn will launch Excel with the spreadsheet document open. Any changes made to the spreadsheet will be reflected in the drawing. Figure 13–54 shows the changes that were made in the spreadsheet.

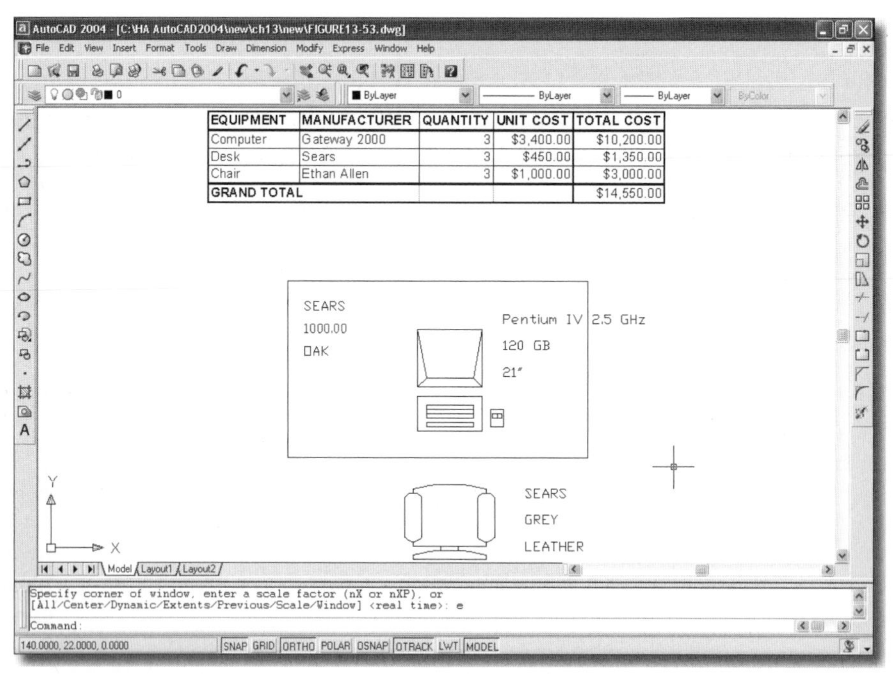

Figure 13–53 *An Excel spreadsheet in the AutoCAD drawing*

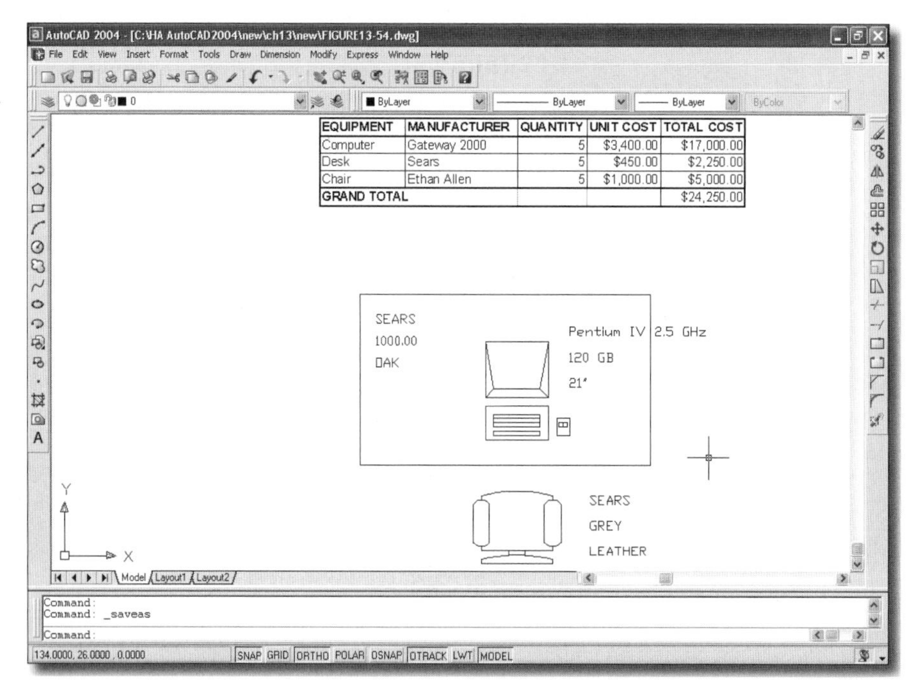

Figure 13–54 *AutoCAD drawing showing the changes made in the Excel spreadsheet*

Here is another example in which an AutoCAD drawing is the client, for both embedding from Word and linking from Excel.

Figure 13–55 shows both an AutoCAD screen and an Excel spreadsheet, in which the spreadsheet is being used for area calculations. The AREA cells are formulas that calculate the product of the corresponding WIDTH and LENGTH cells. In turn, the TOTAL cell is the sum of the AREA cells.

Figure 13–56 shows the same AutoCAD screen and a Word document that was used for typing the GENERAL NOTES, which were in turn embedded into the AutoCAD drawing.

Figure 13–57 shows how the cell value in the WIDTH for ROOM1 has been changed, resulting in changes in the AREA and TOTAL cells. Because this object (the spreadsheet consisting of four columns and seven rows, including the title) was paste linked into the AutoCAD drawing, the linked object automatically reflects the changes.

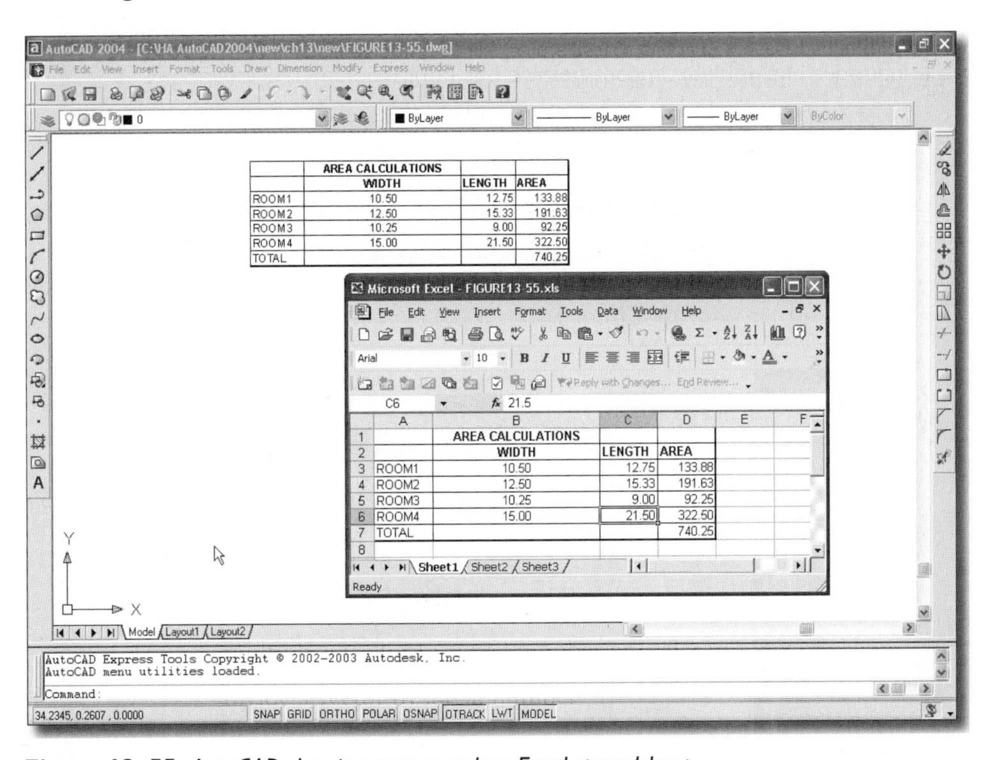

Figure 13–55 *AutoCAD drawing screen and an Excel spreadsheet*

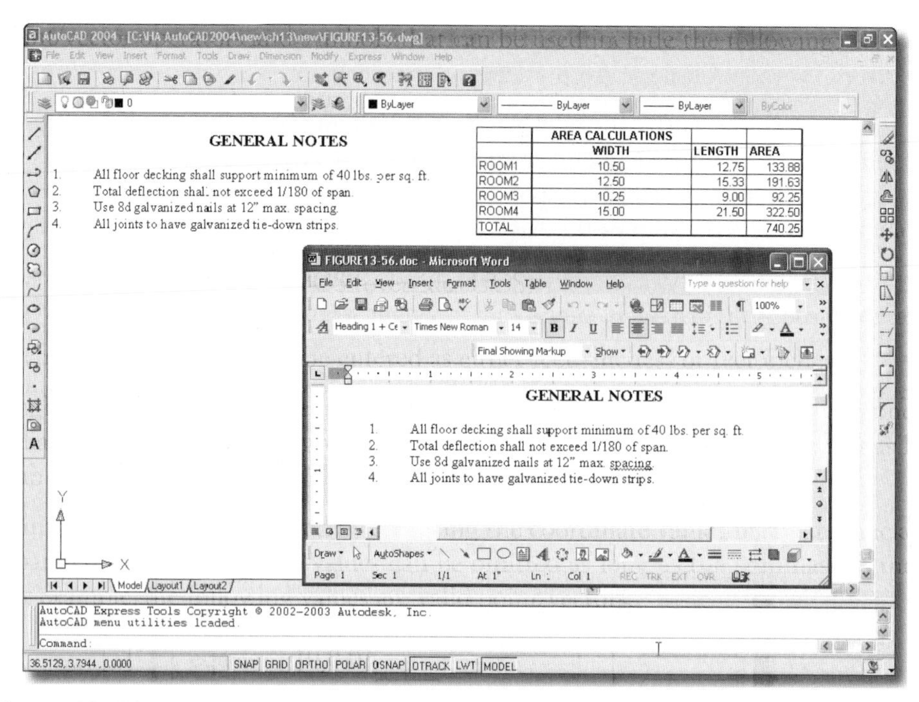

Figure 13–56 *AutoCAD drawing and a Word document*

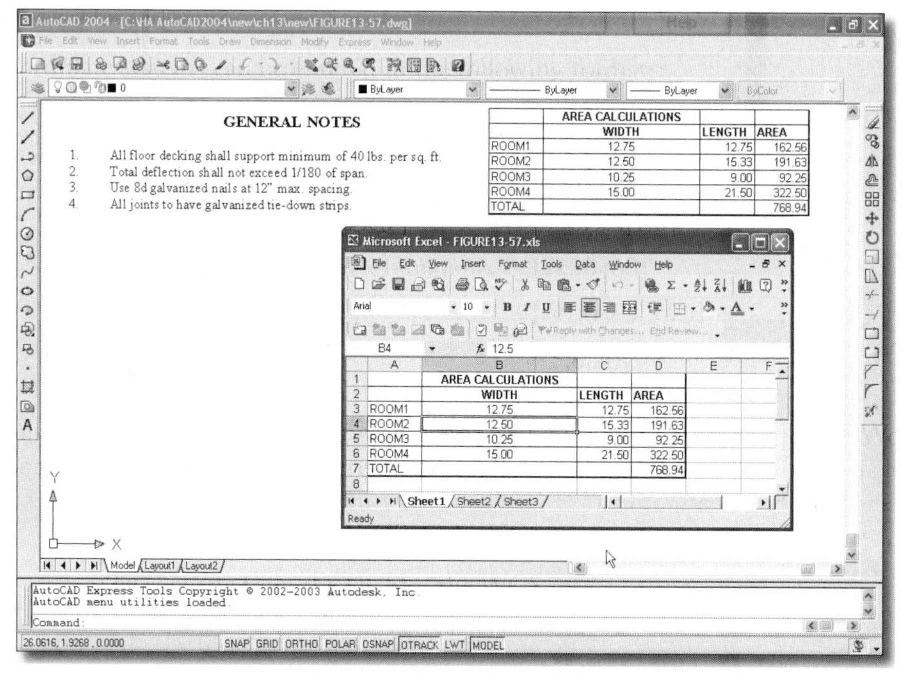

Figure 13–57 *Changes shown in the spreadsheet and AutoCAD screen*

AutoCAD displays the OLE Properties dialog box as shown in Figure 13–58 whenever you insert an OLE object into a drawing.

Figure 13–58 *OLE Properties dialog box*

If necessary, you can change the size of the object by changing the height and width in drawing units, or enter a percentage of the current height or width in the **Scale** section. You can also change the text size and OLE plot quality. If you need to change the OLE properties after pasting the objects into the current drawing, first select the OLE object, and from the shortcut menu select PROPERTIES or at the Command: prompt invoke the OLESCALE command. AutoCAD displays the OLE Properties dialog box.

SECURITY, PASSWORDS, AND ENCRYPTION

Electronic drawing files, like their paper counterparts, often need to have the information they contain protected from unauthorized viewing. AutoCAD provides password and encryption capabilities to achieve this. Also, it might be necessary to determine that the person who last edited and saved the drawing is the person who was supposed to edit and save it. To accomplish this, AutoCAD allows the use of Digital Signatures.

PASSWORDS

AutoCAD password protection makes it possible to prevent a drawing file from being opened without first entering the pre-assigned password.

To assign a password to a drawing, invoke the SECURITYOPTIONS command:

Tools menu	Choose Options, select Open and Save tab and choose Security Options... button
Command: prompt	**securityoptions** (ENTER)

AutoCAD displays the Security Options dialog box, similar to Figure 13–59.

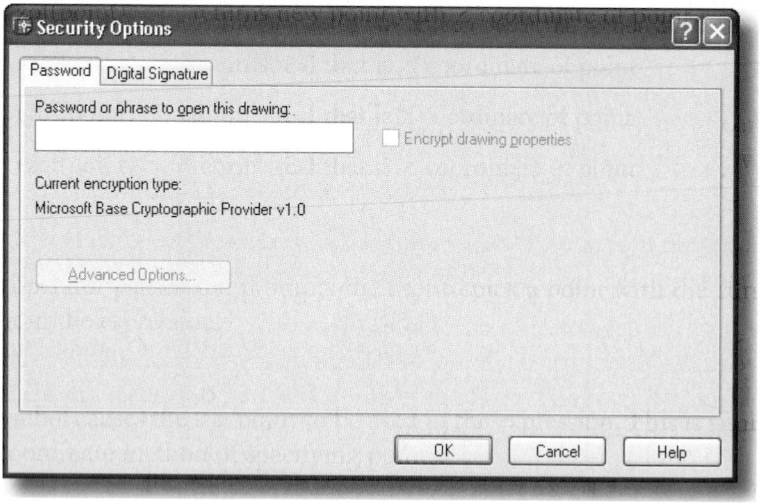

Figure 13–59 *Security Options dialog box with the Password tab displayed*

On the **Password** tab of the Security Options dialog box, enter a password in the **Password or phrase to open this drawing:** text box. This prevents the drawing from being opened without first entering the password specified. Passwords can be a single word or a phrase and are not case-sensitive.

To view data in a password-protected drawing, you open the drawing in a standard way and enter the password in the **Enter password to open drawing** text box in the Password dialog box, as shown in Figure 13–60. Unless the title, author, subject, keywords, or other drawing's properties, were encrypted when the password was attached, you can view the properties in the Properties dialog box in Windows Explorer.

Figure 13–60 *Password dialog box*

ENCRYPTION

You can encrypt drawing properties, such as the title, author, subject, and keywords, thus requiring a password to view the properties and thumbnail preview of the drawing. If you decide to specify an encryption type and key length, you can select them from the ones available on your computer. On the **Password** tab of the Security Options dialog box, after you have entered a password in the **Password or phrase to open this drawing:** text box, check the **Encrypt drawing properties** check box. Under the **Password or phrase to open this drawing:** text box AutoCAD displays the current encryption type. To change the encryption type, select the **Advanced Options...** button. AutoCAD displays the Advanced Options dialog box as shown in Figure 13–61. The note in the upper windows warns "Note: Encryption providers vary depending on operating system and country. Before changing the encryption provider you should confirm that the intended recipient of this drawing has a computer with the encryption provider you choose." In the **Choose an encryption provider:** text box you can select one of the encryption providers listed. From the **Choose a key length:** text box you can choose a key length. The higher the key length, the higher the protection.

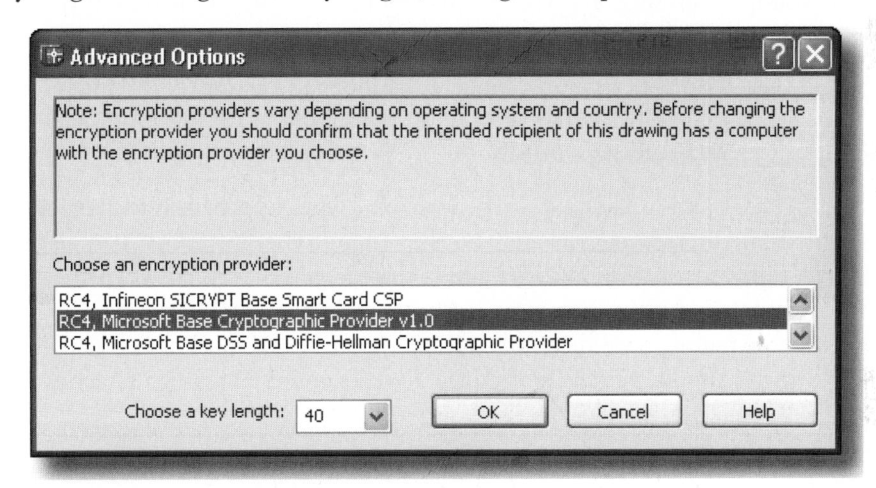

Figure 13–61 *Advanced Options dialog box*

DIGITAL SIGNATURE

AutoCAD provides a means to sign the drawing file electronically. This means that it is possible to verify that a drawing has had a distinct and unique digital signature attached to it when it was last saved. Along with this positive electronic identification, you can also apply a time stamp and comments.

To attach a digital signature to a drawing, you must first obtain a digital ID. This can be done by contacting a certificate authority through a search engine in your Internet browser, using the term "digital certificate." Once a digital ID has been established on your computer, invoke the SECURITYOPTIONS command:

Tools menu	Choose Options, select Open and Save tab and choose Security Options... button
Command: prompt	**securityoptions** (ENTER)

AutoCAD displays the Security Options dialog box. Select the **Digital Signature** tab, and AutoCAD displays various options available for Digital Signature.

AutoCAD displays a list of digital IDs that you can use to sign files. Includes information about the organization or individual to whom the digital ID was issued, the digital ID vendor who issued the digital ID, and when the digital ID expires. Select one of the available digital IDs and set the **Attach digital signature after saving drawing** check box to ON. From the **Signature information** section of the **Digital Signature** tab you can select time stamp to be attached with the digital ID from the **Get time stamp from:** text box. You can also add comments to the digital ID in the **Comment**: text box.

A digital ID has a name, expiration date, serial number and certain certifying information. The certificate authority that you obtain the digital ID from can provide Low, Medium, and High levels of security.

From the digital signature feature, you can determine whether the file was changed since it was signed, whether the signers are who they claim to be, and if they can be traced. The digital signature is considered invalid if the file was corrupted when the digital signature was attached, it was corrupted in transit, or if the digital signature is no longer valid. In order to maintain validity of the digital signature you must not add a password to the drawing or modify or save it after the digital signature has been attached.

Note: The digital signature status is displayed when you open a drawing if the SIGWARN system variable is set to ON. If it is set to OFF the signature status is displayed only if the signature is invalid.

In the **Open and Save** tab of the Options dialog box, if the **Display Digital Signature Information** check box is checked, then when you open a drawing that has a digital signature attached, the Digital Signature Contents dialog box is displayed, providing information on the status of the drawing and the signer. In the **Other Fields** list, you can obtain information about the issuer, beginning and expiration dates and serial number of the digital signature.

CUSTOM SETTINGS WITH THE OPTIONS DIALOG BOX

The **Options** dialog box allows you to customize the AutoCAD settings. AutoCAD allows you to save and restore a set of custom preferences called a profile. A profile can include preference settings that are not saved in the drawing, with the exception of pointer and printer driver settings. By default, AutoCAD stores your current settings in a profile named <Unnamed Profile>.

To open the Options dialog box, invoke the OPTIONS command:

Tools menu	Choose Options
Command: prompt	**options** (ENTER)

AutoCAD displays the Options dialog box, similar to Figure 13–62. From the Options dialog box, the user can control various aspects of the AutoCAD environment. The Options dialog box has nine tabs; to make changes to any of the sections, select the corresponding tab from the top of the Options dialog box.

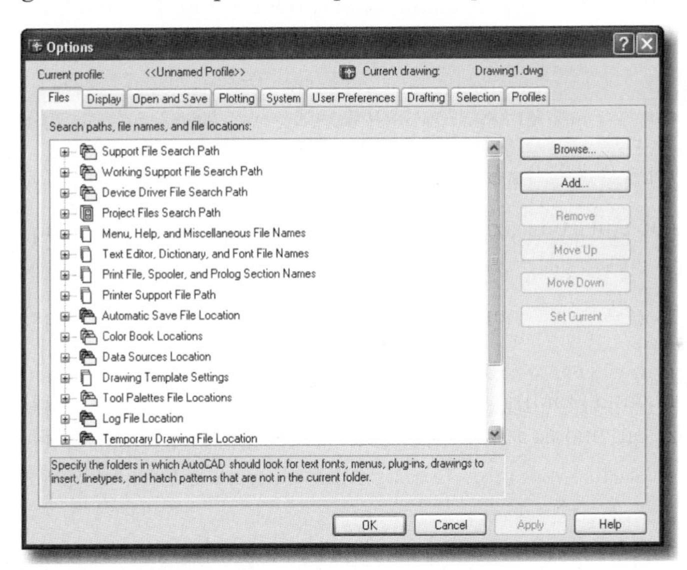

Figure 13–62 *Options dialog box (with the Files tab selected)*

FILES

The **Files** tab of the Options dialog box, shown in Figure 13–62, specifies the directory in which AutoCAD searches for support files, driver files, project files, template drawing file location, temporary drawing file location, temporary external reference file location, and texture maps. It also specifies the location of menu, help, log, text editor, and dictionary files.

When you choose the **Browse...** button, AutoCAD displays the Browse for Folder or Select a File dialog box, depending on what you selected from the list.

When you choose the **Add** button, AutoCAD adds a search path for the selected folder.

When you choose the **Remove** button, AutoCAD removes the selected search path or file.

When you select the **Move Up** button, AutoCAD moves the selected search path above the preceding search path.

When you select the **Move Down** button, AutoCAD moves the selected search path below the following search path.

When you select the **Set Current** button, AutoCAD makes the selected project or spelling dictionary current.

DISPLAY

The **Display** tab of the Options dialog box, shown in Figure 13–63, controls preferences that relate to AutoCAD performance.

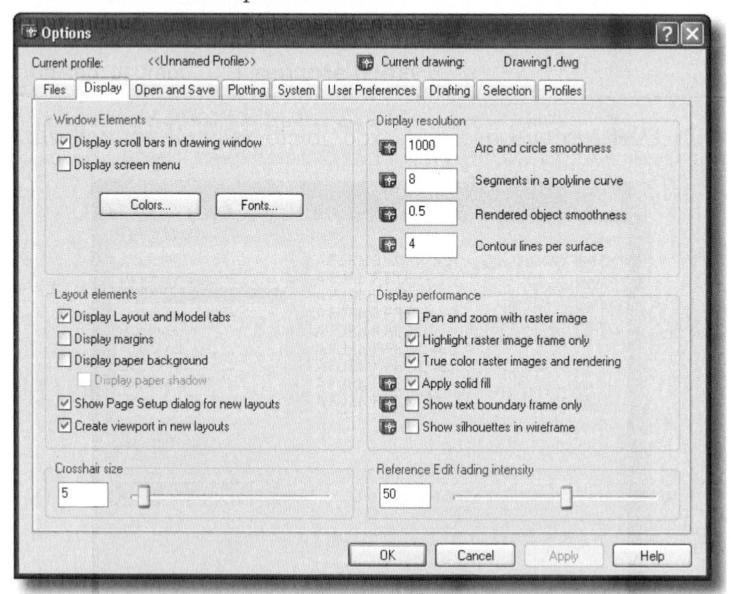

Figure 13–63 *Options dialog box with the Display tab selected*

Window Elements

The **Window Elements** section controls the parameters of the AutoCAD drawing window. The **Display scroll bars in drawing window** check box (set to OFF by default) specifies whether to display scroll bars at the bottom and right sides of the drawing window. The **Display screen menu** check box (set to OFF by default) specifies whether to display the screen menu on the right side of the drawing window. Choose the **Colors...** button, and AutoCAD displays the AutoCAD Window Colors dialog box, which can be used to set the colors for drawing area, screen menu, text window, and command line. Choose the **Fonts...** button, and AutoCAD displays the Graphics Window Font dialog box, which can be used to specify the font AutoCAD uses for the screen menu and command line and in the text window.

Display Resolution

The **Display resolution** section lets you set the resolution in the following text boxes:

- Arc and circle smoothness
- Segments in a polyline curve
- Rendered object smoothness
- Contour lines per surface

Layout Elements

The **Layout elements** section has check boxes to toggle on and off the following:

- Display Layout And Model tabs
- Display margins
- Display paper background
- Display paper shadow
- Show page setup dialog for new layouts
- Create viewport in new layouts

Display Performance

The **Display performance** section has check boxes to toggle on and off the following:

- Pan and zoom with raster image
- Highlight raster image frame only
- True color raster images and rendering
- Apply solid fill
- Show text boundary frame only
- Show silhouettes in wireframe

OPEN AND SAVE

The **Open And Save** tab of the Options dialog box, shown in Figure 13–64, lets you determine formats and parameters for drawings, external references, and ObjectARX applications as they are opened or saved.

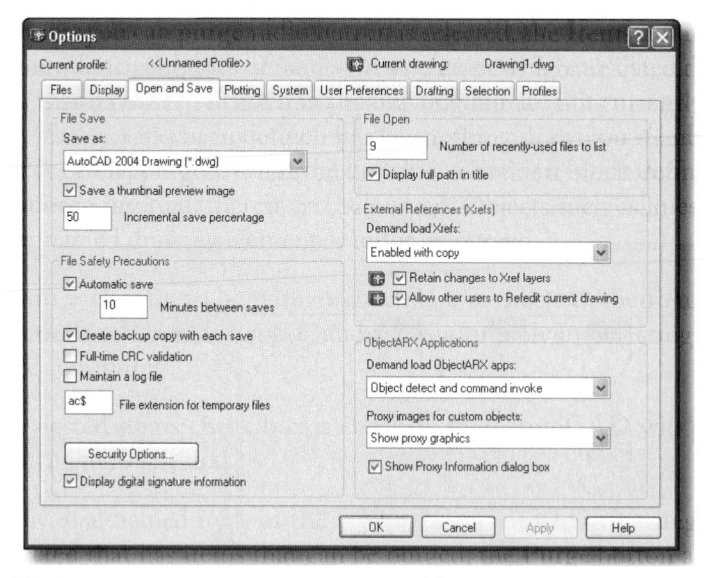

Figure 13–64 *Options dialog box with the Open And Save tab selected*

File Save

In the **File Save** section, the **Save As:** text box lets you select a default save format when you invoke the SAVEAS command. Choose from the following formats:

- AutoCAD 2004 Drawing (*.dwg)
- AutoCAD 2000/LT2000 Drawing (*.dwg)
- AutoCAD Drawing Template File (*.dwt)
- AutoCAD 2004 DXF (*.dxf)
- AutoCAD R12/LT2 DXF (*.dxf)

File Open

The **File Open** section shows the number of recently used files to list and has a check box that causes AutoCAD to display the full path in the title when checked.

External References (Xrefs)

The **External References (Xrefs)** section has a text box from which you can select **Disabled, Enabled,** or **Enabled With Copy. Disabled** means that demand loading is not on in the current drawing. Someone else can open and edit an xref file except as it is being read into the current drawing. **Enabled** means that demand loading is ON in the current drawing. No one else can edit an xref file while the current drawing is open. However, someone else can reference the xref file. **Enabled with Copy** means that demand loading is ON in the current drawing. An xref file can still be opened and edited by someone else. AutoCAD only uses a copy of the xref file, treating it as a completely separate file from the original xref.

The **Retain changes to Xref layers** check box controls the properties of layers in xrefs to be saved as they are changed in the current drawing for reloading later. Another check box entitled **Allow other users to Refedit current drawing** allows the current drawing to be edited while being referenced by other drawing(s).

File Safety Precautions

The **File Safety Precautions** section helps to detect errors and avoid losing data. The **Automatic save** check box with the **Minutes between saves** text box allow you to determine if and for what intervals periodic automatic saves will be performed. The **Create backup copy with each save** check box allows you to determine whether a backup copy is created when you save the drawing. **The Full-time CRC validation** check box allows you to determine whether a cyclic redundancy check (CRC) is performed when an object is read into the drawing. Cyclic redundancy check is a mechanism for error-checking. If you suspect a hardware problem or AutoCAD error is causing your drawings to be corrupted, set this check box on. The **Maintain a log file** check box allows you to determine whether the contents of the text window are written to a log file. Use the **Files** tab in the Options dialog box to specify the name and location of the log file. The **File extension for temporary files** text box allows you to specify an extension for temporary files on a network. The default extension is .AC$. AutoCAD 2004 has added the **Security Options...** button and **Display digital signature information** check box, which are described in the previous section on "Security, Passwords, and Encryption" in this chapter.

ObjectARX Applications

The **ObjectARX Applications** section controls parameters for AutoCAD Runtime Extension applications and proxy graphics. The **Demand load ObjectARX Apps** text box let you select when and if a third-party application is demand-loaded when a drawing has custom objects that were created in that application. The **Disable load on demand** option turns off demand-loading. The **Custom object detect** option selection demand-loads the source application when you open a drawing that contains custom objects. It does not demand-load the application when you invoke one of the application's commands. The **Command invoke** option selection demand-loads the source application when you invoke the application's command. This setting does not demand-load the application when you open a drawing that contains custom objects. The **Object detect and command invoke** option selection demand-loads the source application when you open a drawing that contains custom objects or when you invoke one of the application's commands.

The **Proxy images for custom objects** text box controls how custom objects in the drawings are displayed. The **Do not show proxy graphics** option, when selected, causes custom objects in drawings not to be displayed. The **Show proxy graphics** option, when selected, causes custom objects in drawings to be displayed. The **Show proxy bounding box** option, when selected, causes a box to be displayed in place of custom objects in drawings. The **Show Proxy Information dialog box** check box, when checked, causes a warning to be displayed when you open a drawing that contains custom objects.

PLOTTING

The **Plotting** tab of the Options dialog box, shown in Figure 13–65, lets you select the parameters for plotting your drawing.

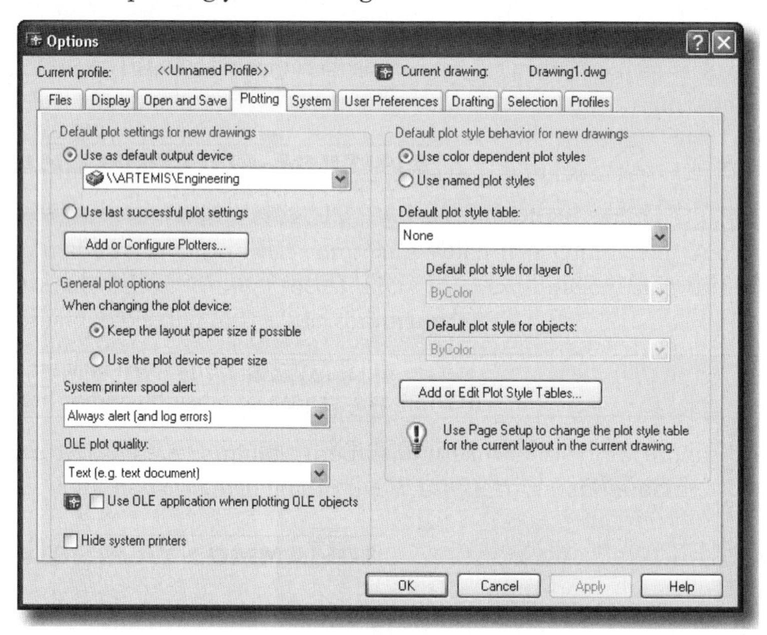

Figure 13–65 *Options dialog box (with the Plotting tab selected)*

Default Plot Settings For New Drawings

The **Default plot settings for new drawings** section determines plotting parameters for new drawings. It will also determine settings for drawings created in earlier releases of AutoCAD prior to AutoCAD 2000 that have never been saved in AutoCAD 2000 format. The **Use as default output device** text box determines the default output device for new drawings and for drawings created in an earlier release of AutoCAD that have never been saved in AutoCAD 2000 format. The list displays all plotter configuration files (PC3) that are found in the plotter configuration search path. It also displays all system printers configured in the system. The **Use last successful plot settings** check box, when checked, uses the settings of the last successful plot for the current settings. The **Add or Configure Plotters...** buttons lets you add or configure a plotter from the Plotters program window. See the chapter on plotting for adding and configuring plotters.

General Plot Options

The **General plot options** section lets you set general parameters such as paper size, system printer alert parameters, and OLE objects. Under **When changing the plot device:**, the **Keep the Layout Paper Size If Possible** radio button, when selected, applies the paper size in the **Layout Settings** tab in the Page Setup dialog box provided the selected output device is able to plot to this paper size. If it cannot, AutoCAD displays a warning message and uses the paper size specified either in the plotter configuration file (PC3) or in the default system settings if the output device is a system printer. The **Use the plot device paper size** radio button applies the paper size in either the plotter configuration file (PC3) or in the default system settings if the output device is a system printer.

The **System printer spool alert** list box selection determines if a warning will be displayed if the plotted drawing is spooled through a system printer because of an input or output port conflict. The **Always alert (and log errors)** option, when selected, displays a warning and always logs an error when the plotted drawing spools through a system printer. The **Alert first time only (and log errors)** option, when selected, displays a warning once and always logs an error when the plotted drawing spools through a system printer. The **Never alert (and log first error)** option, when selected, will never display a warning and logs only the first error when the plotted drawing spools through a system printer. The **Never alert (do not log errors)** option, when selected, never displays a warning or logs an error when the plotted drawing spools through a system printer. There is a checkbox that causes AutoCAD to **Hide system printers** when checked.

The **OLE plot quality** list box selection determines plotted OLE objects' quality. Options include Line Art, Text, Graphics, Photograph, and High Quality Photograph.

The **Use OLE application when plotting OLE objects** check box starts the application that creates the OLE object when you plot a drawing with OLE objects. This will help optimize quality of OLE objects.

Default Plot Style Behavior

The **Default plot style behavior for new drawings** section is used to set up parameters for plot styles in all drawings. A plot style is a set of parameters defined in a plot style table and used when the drawing is plotted. See the Chapter 8 on plotting . The **Use color dependent plot styles** radio button, when selected, disables the plot style list on the Object Properties toolbar. This is the default setting. The **Use named plot styles** radio button, when selected, enables the list.

When the **Use color dependent plot styles** option is selected, color-dependent plot styles are used in both new drawings and drawings created in earlier versions of AutoCAD. Color-dependent plot styles utilize numbers from the AutoCAD color index to create a plot style table. This is a file with a .CTB file extension. Each color is identified by a number ranging from 1 to 255 or a name. You can assign color numbers to individual pens on a pen plotter. This can allocate specific property settings in the plotted drawing. If this option is selected, a plot style is created for each color setting.

To change the default plot style behavior for a drawing, select the **Use named plot styles** option before opening or creating a drawing. Changing the default settings using the Options dialog box affects only new drawings or drawings created in a release of AutoCAD prior to ACAD2000 and that have never been saved in AutoCAD 2000 format. This setting is saved with the drawing. Once a drawing is saved with **Use color dependent plot styles** as the default, you can change the default to **Use named plot styles** with a migration utility. But, once a drawing is saved with **Use named plot styles** as the default, it cannot be changed to **Use color dependent plot styles**.

The **Default plot style table** list box selection allows you to determine which plot style table will be attached to new drawings. A plot style table is a file. It has a file extension of .CTB or .STB and it defines plot styles. If you are using color-dependent plot styles, this option lists the value of None along with all color dependent plot style tables found in the search path. This option lists all named plot styles tables when named plot styles are being used.

The **Default plot style for layer 0** list box selection determines the plot style for Layer 0 for new drawings or drawings created in a release of AutoCAD prior to ACAD2000 and have never been saved in AutoCAD 2000 format. The Normal style and any other style defined in the currently loaded plot style table are listed.

The **Default plot style for objects** list box selection determines the default plot style assigned for new objects. It includes Normal, BYBLOCK, and BYLAYER, and any other style defined in the currently loaded plot style table.

The **Add or Edit Plot Style Tables...** button causes the Plot Styles program window to be displayed for managing (creating and editing) plot style tables. See the Chapter 8 on plotting.

SYSTEM

The **System** tab of the Options dialog box, shown in Figure 13–66, has sections for managing the 3D graphics display, pointing devices, dbConnect options, and general options.

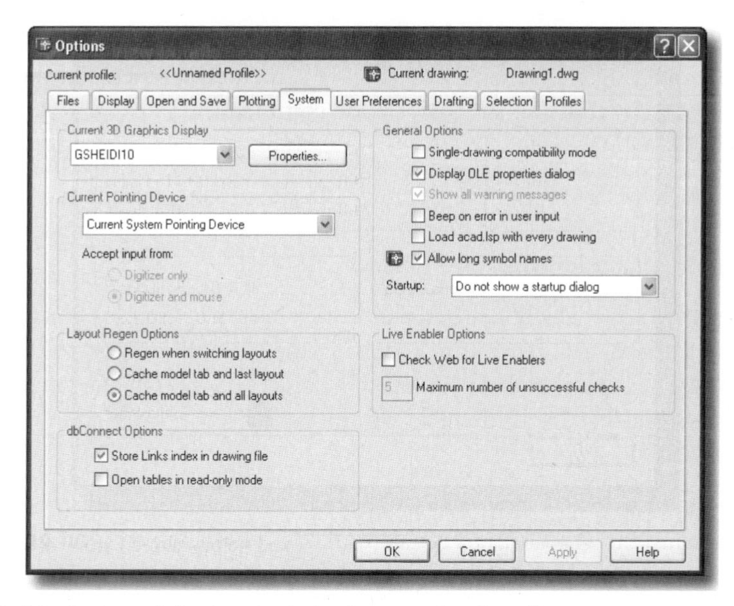

Figure 13–66 *Options dialog box (with the System tab selected)*

Current 3D Graphics Display

The **Current 3D Graphics Display** section text box allows you to specify the current display from the list. The **Properties…** button causes the 3d Graphics System Configuration dialog box to be displayed.

In the **3d Graphics System Configuration** dialog box, the **Adaptive Degradation** section allows you to determine to what degree the view can degrade to maintain the speed of the view manipulation. For example, selecting Wireframe will allow the display to speed up. The **Flat shaded** check box, when checked, allows the display to change to flat shaded. The **Wireframe** check box, when checked, allows the display to change to wireframe. The **Bounding box** check box, when checked, allows the display to change to a bounding box which will display in place of objects. The **Maintain speed FPS** text box lets you specify the display speed in frames per second.

The **Dynamic Tessellation** section allows you to determine the smoothness of the objects in a drawing. The **Dynamic tessellation** check box, when checked, determines that tessellations will be drawn. The **Surface tessellation** and **Curve tessellation** slide bars let you set the tessellation from **Low detail (less memory)** to **High detail (more memory)** for surface and curves respectively. The **Number of tessellations to cache** text box allows you to specify whether the number of tessellations in the cache is 1 or more. When the number is 1, all viewports have the same tessellation. Increasing the number of tessellations can reduce the occurrence of regenerations when more than one viewport is being used.

The **Render options** section allows you to change settings for the display of lights, materials, textures, and transparency in 3D views, including objects in the 3D Orbit view and objects shaded using the SHADEMODE command.

The **Enable lights** check box, when checked, illuminates objects and attached materials in 3D views by lights that are defined using the LIGHT command. If the check box is not checked, or if the LIGHT command has not been used for the drawing, then AutoCAD uses default lighting for 3D views

The **Enable materials** check box, when checked, causes materials for objects that have materials in 3D views that were attached using the RMAT command to be displayed. AutoCAD uses the default global material if the object has no attached material. If the check box is not checked, or if the RMAT command has not been used in the drawing, then no materials are displayed.

The **Enable textures** check box, when checked, shows textures attached to objects in 3D views using the RMAT and SETUV commands. The **Enable materials** check box must also be checked for textures to be visible.

The **Transparency** section allows you to adjust the transparency quality of images. This affects the time to redraw the screen. The **Low quality** setting is a screen-door effect that achieves transparency without increasing redraw time. The **Medium quality** setting which is blending, makes image quality better. The **High quality** setting, which is blending and extra processing, produces an image without visual artifacts, but it increases redraw time. The **Enable materials** check box must also be checked for transparency to be turned on.

The **Geometry** section determines how AutoCAD displays isolines in 3D and whether AutoCAD displays back faces in 3D.

The **Isolines always on top** check box, when checked, causes isolines to be displayed for front and back faces in all shade modes except hidden. When this check box is checked, AutoCAD displays the isolines for the back faces on top of the shading for the objects in 3D. When this check box is not checked, AutoCAD hides the isolines for the back faces.

The **Discard back faces** check box, when checked, causes back faces to not be drawn.

The **Acceleration** section determines whether software or hardware acceleration is used in 3D. The **Software** radio button, when selected, uses the software graphics system to perform all of the drawing tasks in 3D. The **Hardware** radio button, when selected, uses the hardware graphics card to perform most of the drawing tasks in 3D. This speeds up the drawing time. The **Use geometry acceleration (single precision)** check box, when checked, causes geometry acceleration to be used. The **Enable anti-aliasing lines** check box, when checked, causes anti-aliasing lines to be used. Lines will appear smoother.

Current Pointing Device

The **Current Pointing Device** section of the **System** tab determines the parameter for the pointing device(s) being used. The text box lists available pointing device drivers from which to choose. The **Current system pointing device** option causes the system pointing device to be the current pointing device. The Wintab Compatible Digitizer ADI 4.2 by Autodesk option causes the **Wintab compatible digitizer** to be the current pointing device. The **Accept input from:** options determines if the digitizer only is active or both the mouse and digitizer are active. These are determined by the **Digitizer only** and the **Digitizer and mouse** radio buttons.

Layout Regen Options

The **Layout Regen Options** sections of the **System** tab lets you specify how AutoCAD updates the display list in the Model and layout tabs. The display list for each tab is updated either by regenerating the drawing when you switch to that tab or by saving the display list to memory and regenerating only the modified objects when you switch to that tab.

The **Regen when switching layouts** radio button, when selected, causes AutoCAD to regenerate the drawing each time you switch tabs.

The **Cache model tab and last layout** radio button, when selected, saves the display list to memory for the Model tab and the last layout made current. When checked, it supresses regenerations when you switch between the two tabs. Regenerations for all other layouts still occur when you switch to those tabs.

The **Cache model tab and all layouts** radio button, when selected, causes AutoCAD to regenerate the drawing the first time you switch to each tab. For the remainder of the drawing session, when you switch to those tabs, the display list is saved to memory and regenerations are suppressed.

dbConnect Options

The **dbConnect Options** section of the **System** tab allows you to manage the options associated with database connectivity.

The **Store links index in drawing file** check box, when checked, causes AutoCAD to store the database index within the drawing file. When this option is checked, performance during link selection operations is enhanced. When this option is not checked, the drawing file size is decreased and the opening process is enhanced for drawings with database information.

The **Open tables in read-only mode** check box, when checked, causes AutoCAD to open database tables in Read-only mode within the drawing file.

General Options

The **General Options** section of the **System** tab has check boxes relating to a variety of options for system parameter settings. The **Single-drawing compatibility mode** check box, when checked, causes AutoCAD to open only one drawing at a time (Single-drawing Interface or SDI). Otherwise AutoCAD can open multiple drawing sessions (Multi-drawing Interface or MDI). The **Display OLE properties dialog** check box, when checked, causes the OLE Properties dialog box to be displayed when inserting OLE objects into AutoCAD drawings. The **Show all warning messages** check box, when checked, causes all dialog boxes that include a **Don't Display This Warning Again** option to be displayed. Dialog boxes with warning options will be displayed regardless of previous settings specific to each dialog box. The **Beep on error in user input** check box, when checked, causes an alarm beep when AutoCAD detects an invalid entry. The **Load acad.lsp with every drawing** check box, when checked, causes AutoCAD to load the acad.lsp file into every drawing. The **Allow long symbol names** check box, when checked, allows you to use up to 255 characters for named objects.

From the **Startup:** text box, you can select one of three possible initial views that might appear when starting AutoCAD: **Show TODAY startup dialog**, **Show the traditional startup dialog**, and **Do not show a startup dialog**. These options are explained in Chapter 1 under "STARTING AutoCAD".

Live Enabler Options

The **Live Enabler Options** section of the **System** tab allows you to specify how AutoCAD checks for Object Enablers. Using Object Enablers, you can display and use custom objects in AutoCAD drawings even when the ObjectARX application that created them is unavailable.

The **Check Web for Live Enablers** check box, when checked, causes AutoCAD to check for Object Enablers on the Autodesk Web site.

The **Maximum number of unsuccessful checks** text box allows you to specify the number of times AutoCAD will continue to check for Object Enablers after unsuccessful attempts.

USER PREFERENCES

The User Preferences tab of the Options dialog box, shown in Figure 13–67, controls options that optimize the way you work in AutoCAD.

Figure 13–67 *Options dialog box (with the User Preferences tab selected)*

Windows Standard Behavior

The **Windows Standard Behavior** section of the **User Preferences** tab lets you apply Windows techniques and methods in AutoCAD. The **Windows standard accelerator keys** check box, when checked, arranges for Windows standards to be applied in interpreting keyboard accelerators (for example, CTRL+C equals COPYCLIP). Otherwise, AutoCAD standards will be applied for keyboard accelerators (for example, CTRL+C equals Cancel; CTRL+V toggles among the viewports). The **Shortcut menus in drawing area** check box, when checked, causes shortcut menus to be displayed when the pointing device is right-clicked. Otherwise, right-clicking is the same as pressing ENTER. The **Right-click customization** button causes the Right-Click Customization dialog box to be displayed.

The Right-Click Customization dialog box determines whether right-clicking in the drawing area displays a shortcut menu or is the same as pressing ENTER. This will allow you to have a right-click invoke ENTER while a command is active. You can also disable the following Command shortcut menus options:

The **Default Mode** section of the Right-Click Customization dialog box determines the effect of right-clicking when no objects are selected. The **Repeat last command** radio button causes right-clicking to be the same as pressing ENTER. The **Shortcut menu** radio button causes right-clicking to display a shortcut menu when applicable.

The **Edit Mode** section of the Right-Click Customization dialog box determines the effect of right-clicking when one or more objects are selected. The **Repeat last command** radio button causes right-clicking to be the same as pressing ENTER. The **Shortcut menu** radio button causes right-clicking to display the **Edit** shortcut menu.

The **Command Mode** section of the Right-Click Customization dialog box determines the effect of right-clicking when a command is in progress. The **Enter** radio button causes right-clicking to be the same as pressing ENTER when a command is in progress. The **Shortcut menu: always enabled** radio button causes right-clicking to display the Command shortcut menu. The **Shortcut menu: enabled when command options are present** radio button causes right-clicking to display the Command shortcut menu to be displayed only when options are currently available from the command line. Otherwise, right-clicking is the same as pressing ENTER.

Drag-and-drop scale

The **Drag-and-drop scale** section of the **User Preferences** tab allows you to control the default scale for dragging objects into a drawing using i-drop or DesignCenter. The **Source content units:** text box allows you to set the units AutoCAD uses for an object being inserted into the current drawing when no insert units are specified with the INSUNITS system variable. The **Target drawing units:** text box allows you to set the units AutoCAD uses in the current drawing when no insert units are specified with the INSUNITS system variable. (INSUNITSDEFTARGET system variable). Each of these allow you to choose from the following units: Inches, Feet, Miles, Millimeters, Centimeters, Meters, Kilometers, Microinches, Mills, Yards, Angstroms, Nanometers, Microns, Decimeters, Decameters, Hectometers, Gigameters, Astronomical Units, Light Years, and Parsecs If Unspecified-Unitless is selected, the object is not scaled when inserted.

Hyperlink

The **Hyperlink** section of the **User Preferences** tab determines display property settings of hyperlinks. The **Display hyperlink cursor and shortcut menu** check box, when checked, causes the hyperlink cursor and shortcut menu to be displayed when the cursor is over an object that contains a hyperlink. The **Display hyperlink tooltip** check box, when checked, causes the hyperlink tooltip to be displayed when the cursor is over an object that contains a hyperlink.

Priority for Coordinate Data Entry

The **Priority for Coordinate Data Entry** section of the **User Preferences** tab determines how input of coordinate data affects AutoCAD's actions. The **Running object snap** radio button, when selected, causes running object snaps to be used at all times instead of specific coordinates. The **Keyboard entry** radio button, when selected, causes the coordinates that you enter to be used at all times and overrides running object snaps. The **Keyboard entry except scripts** radio button, when selected, causes the specific coordinates that you enter to be used rather than running object snaps, except in scripts.

Object Sorting Methods

The **Object Sorting Methods** section of the User Preferences tab controls the order in which objects are sorted. The **Object selectio**n check box, when checked, causes AutoCAD to sort selectable objects from those created first to those created last. The latest object will be selected when two objects are chosen at the same time. Otherwise the selection is random. The **Object snap** check box, when it is checked and a selection is made using Object Snap, causes AutoCAD to sort selectable objects from those created first to those created last. The latest object will be selected when two objects are chosen at the same time. The **Regens** check box, when checked, causes AutoCAD to sort objects from those created first to those created last when redrawing objects during a REGEN or REGENALL. The **Plotting** check box, when checked, causes AutoCAD to sort objects from those created first to those created last when redrawing objects when plotting.

Associative Dimensioning

The **Make new dimensions associative** check box, when checked, causes new dimensions to be drawn as associative dimensions and will be associated with the objects being dimensioned.

Hidden Line Settings

The **Hidden Line Settings...** button causes the Hidden Line Settings dialog box to be displayed. See Figure 13–68. This allows you to change the display properties of hidden lines. These settings are in effect only when the HIDE command is used or when the Hidden option of the SHADEMODE command is used.

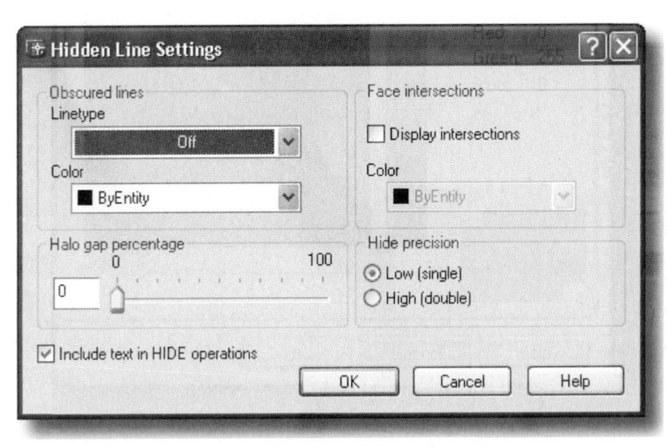

Figure 13–68 *Hidden Line Settings dialog box*

The **Obscured lines** section of the Hidden Line Settings dialog box allows you to specify the linetype and color of obscured lines which are lines that are made visible by changing its color and linetype. The **Linetype** list box allows you to select from a list of linetypes or select **Off**. The **Color** list box allows you to select from available colors.

In the **Halo gap percentage** section, you specify the distance to shorten a haloed line at the point where it will be hidden. Moving the slider bar specifies the distance as a percentage of one inch. It is not affected by the zoom level.

The **Include text in** HIDE **operations** check box, when checked, causes text objects created by the TEXT, DTEXT, or MTEXT command to be included during a HIDE command.

In the **Face intersections** section, the **Display intersections** check box, when checked, causes intersection polylines to be displayed. The **Color** check box/text box, when checked, allows you specify the color of intersection polylines.

The **Hide precision** section allows you to select between the **Low (single)** radio button for low accuracy (and low memory usage) and the **High (double)** radio button for high accuracy in the method of creating hides and shades.

Lineweight Settings

The **Lineweight Settings…** button causes the Lineweight Settings dialog box to be displayed. This allows you to set lineweight options.

DRAFTING

The **Drafting** tab of the Options dialog box, shown in Figure 13–69, lets you customize drafting options in AutoCAD. In this tab are sections for **AutoSnap Settings, AutoSnap Marker Size, AutoTrack Settings, Alignment Point Acquisition**, and **Aperture size**. The options available in these sections are explained in Chapter 3 in the section on Drafting Settings.

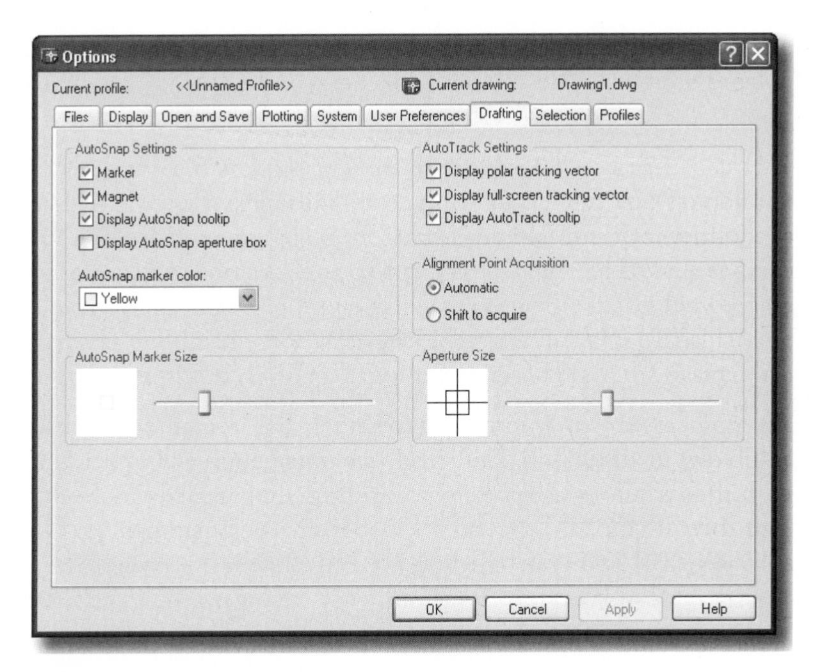

Figure 13–69 *Options dialog box with the Drafting tab selected*

SELECTION

The **Selection** tab of the Options dialog box, shown in Figure 13–70, lets you customize selection options in AutoCAD. In this tab are sections for **Selection Modes**, **Pickbox Size**, **Grips**, and **Grip Size**. The options available for the Selection Modes are explained in Chapter 5 in the section on Object Selection Modes. The options available for Grips are explained in Chapter 6 in the section on Editing With Grips.

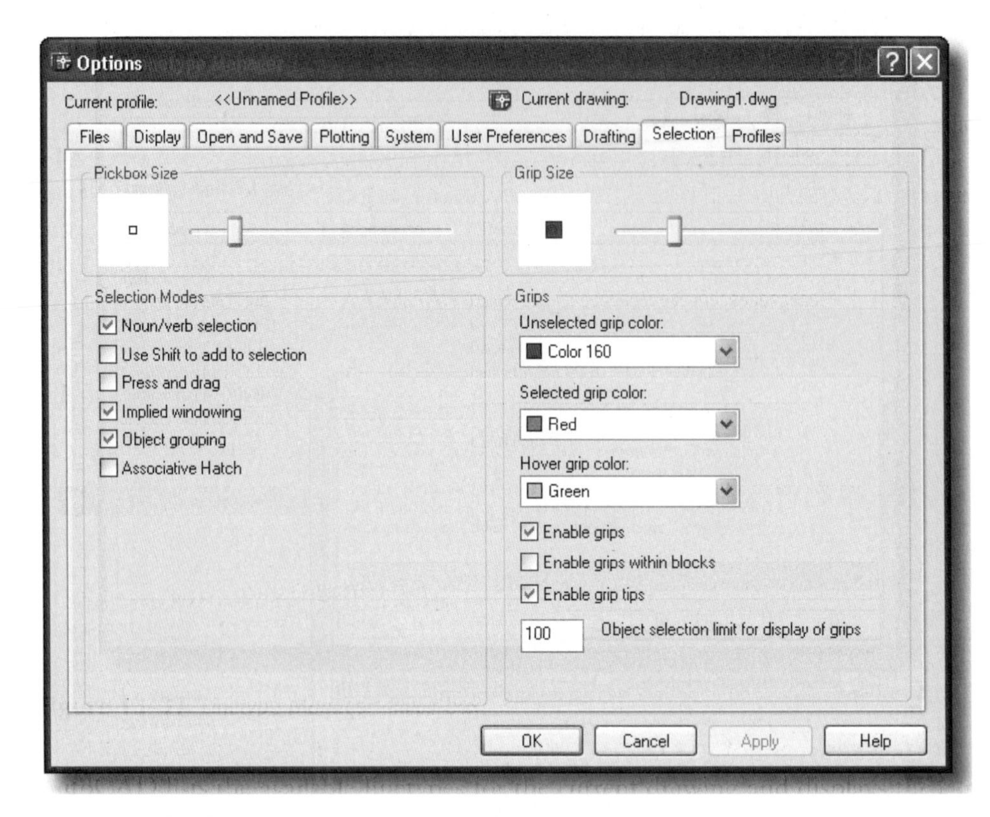

Figure 13–70 *Options dialog box with the Selection tab selected*

PROFILES

The **Profiles** tab of the Options dialog box, shown in Figure 13–71, lets you manage profiles. A profile is a named and saved group of environment settings. This profile can be restored as a group when desired. AutoCAD stores your current options in a profile named Unnamed Profile. AutoCAD displays the current profile name, as well as the current drawing name, in the Options dialog box. The profile data is saved in the system registry and can be written to a text file (an ARG file). AutoCAD organizes essential data and maintains changes in the registry as necessary.

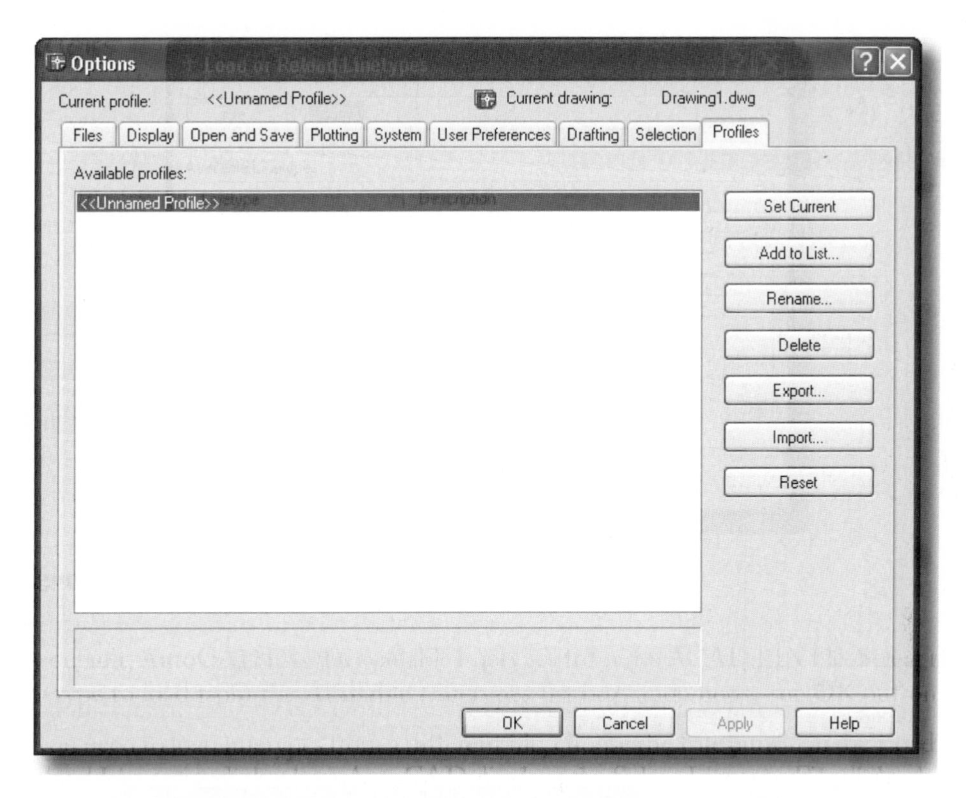

Figure 13–71 *Options dialog box with the Profiles tab selected*

A profile can be exported to or imported from different computers. If changes have been made to your current profile during an AutoCAD session and you want to save them in the ARG file, the profile must be exported. After the profile with the current profile name has been exported, AutoCAD updates the ARG file with the new settings. Then the profile can be imported again into AutoCAD, thus updating your profile settings.

The **Set Current** button makes the profile that is highlighted in the **Available profiles** list box the current profile. Choosing the **Add To List...** button lets you name and save the current environment settings as a profile. The **Rename...** button lets you rename the highlighted profile. The **Delete** button lets you delete the highlighted profile. The **Export...** button causes the Export Profiles dialog box to be displayed. This is a file manager dialog box in which the highlighted profile can be saved to the path you specify. The **Import...** button causes the Import Profiles dialog box to be displayed. This is a file manager dialog box in which you can select a profile from a saved path to be imported. The **Reset** button causes the highlighted profile to be reset. The default profile name is listed in the description pane at the bottom of the dialog box.

After making all the necessary changes, click the **Apply** button to apply the changes. Then choose the **OK** button to close the Options dialog box.

SAVING OBJECTS IN OTHER FILE FORMATS (EXPORTING)

The EXPORT command allows you to save a selected object in other file formats, such as .BMP, .DXF, .DWF, .SAT, .3DS, and .WMF. Invoke the EXPORT command:

File menu	Choose Export
Command: prompt	**export** (ENTER)

AutoCAD displays the Export Data dialog box, similar to Figure 13–72. In the **Files of type:** list box, select the format type in which you wish to export objects. Enter the file name in the **File name:** edit box. Select the **Save** button, and AutoCAD prompts:

Select objects: *(select the objects to export, and press* ENTER *to complete object selection)*

AutoCAD exports the selected objects in the specified file format using the specified file name.

Table 13–1 lists the format types available in AutoCAD for exporting the current drawing.

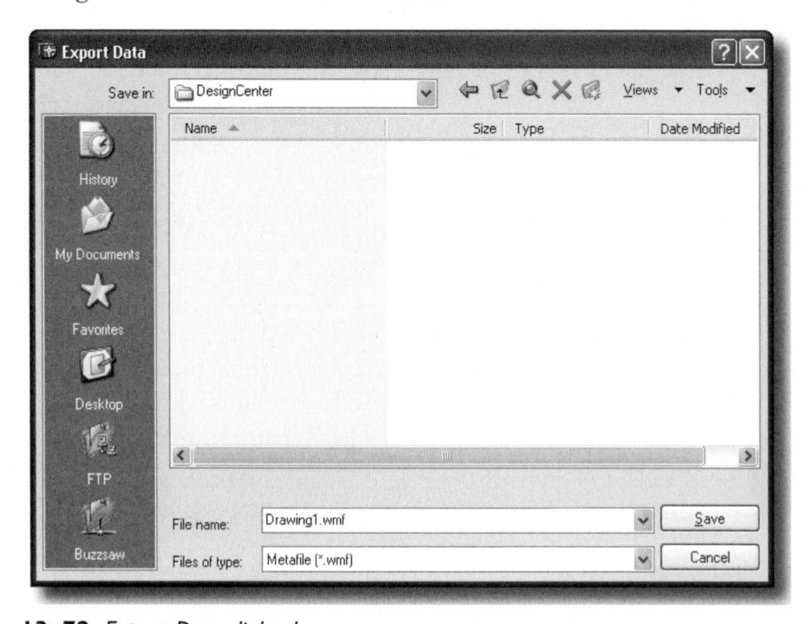

Figure 13–72 *Export Data dialog box*

Table 13–1 Exportable Format Types

Format Type	Description
3DS	3D Studio file
BMP	Device-independent bitmap file
DWG	AutoCAD 2000 drawing file (same as invoking the WBLOCK command)
DXX	AutoCAD attribute extract DXF file (same as invoking the ATTEXT command)
EPS	Encapsulated PostScript file
SAT	ACIS solid-object file
STL	Solid object stereo-lithography file
WMF	Windows metafile

IMPORTING VARIOUS FILE FORMATS

The IMPORT command allows you to import various file formats, such as .3DS, .DXF, .EPS, .SAT, and WMF, into AutoCAD. Invoke the IMPORT command:

Insert toolbar	Select IMPORT command (see Figure 13–73)
Command: prompt	**import** (ENTER)

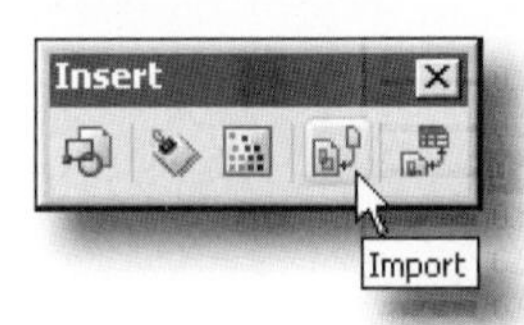

Figure 13–73 *Invoking the* IMPORT *command from the Insert toolbar*

AutoCAD displays the Import File dialog box, similar to Figure 13–74. In the **Files of Type:** list box, select the format type you wish to import into AutoCAD. Select the file from the appropriate directory from the list box and click the **Open** button. AutoCAD imports the file into the AutoCAD drawing.

Table 13–2 lists the format types available to import into AutoCAD.

Table 13–2 Format types available to import into AutoCAD

Format Type	Description
3DS format	Imports a 3D Studio file
DXF format	Imports drawing interchange file
EPS format	Imports Encapsulated PostScript file
SAT format	Imports ACIS solid object file
WMF format	Imports Windows Metafile

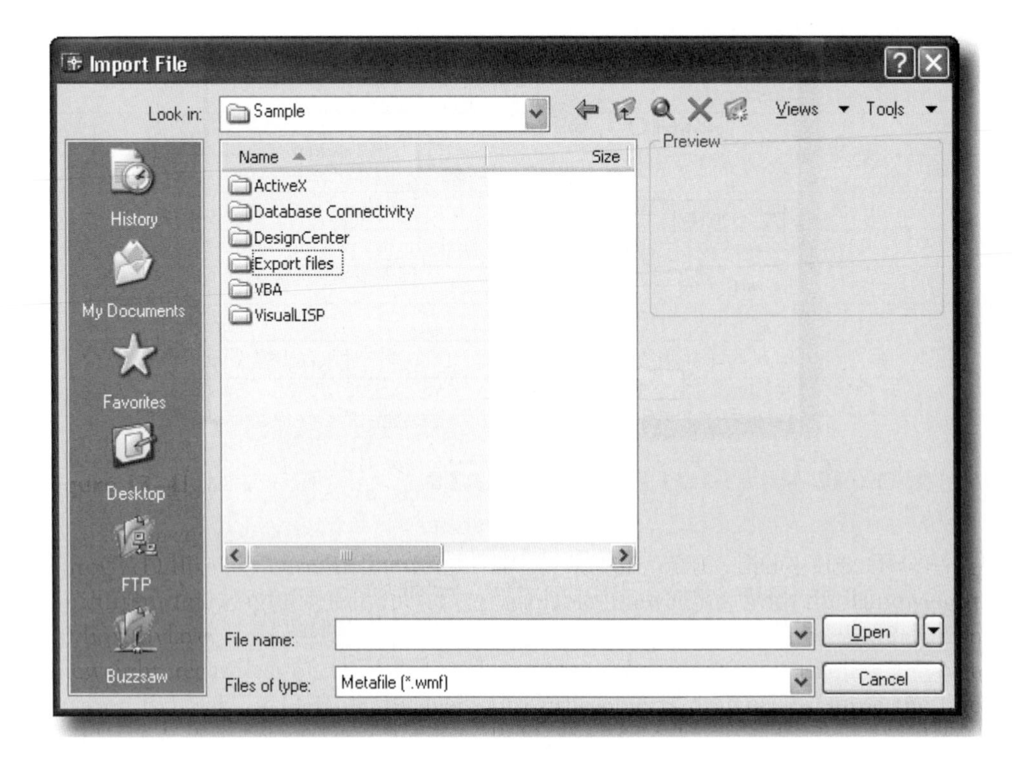

Figure 13–74 *Import File dialog box*

STANDARDS

AutoCAD has a feature that allows you to verify that the layers, dimension styles, linetypes, and text styles of the drawing you are working in conform to an accepted standard, such as a company, trade, or client standard. To utilize this feature, your drawing must have some standard drawing(s) with which it is associated.

You can create a standards file from an existing drawing or you can create a new drawing and save it as a standards file with an extension of .DWS. Open an existing drawing from which you want to create a standards file, invoke the SAVEAS command, and enter a name for the standards file in the Save Drawing As dialog box. Select AutoCAD 2004 Drawing Standards (*.dws) from the **Files of type** list and then choose the **Save** button. You can also create a new drawing and set appropriate standards for layers, text styles, dimension styles and linetypes. Invoke the SAVEAS command, enter a name for the standards file in the Save Drawing As dialog box,select AutoCAD 2004 Drawing Standards (*.dws) from the **Files of type** list and then click the **Save** button.

The STANDARDS command lets you obtain information about the standards files that are associated with the current drawing.

Invoke the STANDARDS command:

Tools menu	Select Cad Standards > Configure
Status bar tray	Select Associated Standards File icon>Configure Standards
Command: prompt	**standards** (ENTER)

AutoCAD displays the Configure Standards dialog box as shown in Figure 13-75.

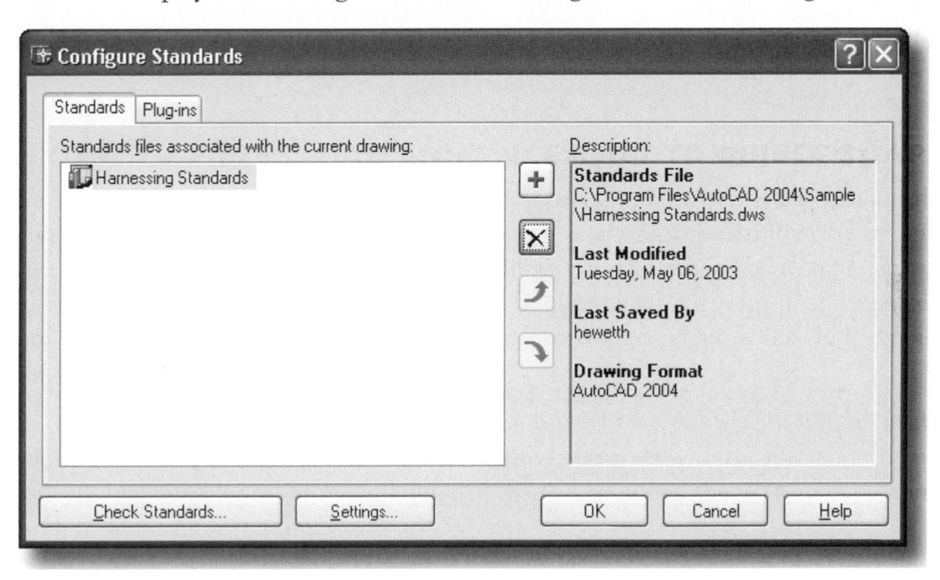

Figure 13–75 *Configure Standards dialog box with the Standards tab selected*

If a drawing has an associated standards file, there will be an Associated Standards File(s) icon on the status bar tray at the bottom right corner of the drawing area.

The **Standards files associated with the current drawing** section lists all standards (DWS) files that are associated with the current drawing. To add a standards file, choose **Add Standards File** button or press the F3 function key. AutoCAD displays the Select Standards File dialog box. Select a standards file from an appropriate folder. To remove a standards file from the current drawing, choose **Remove Standards File** button or press the DELETE key on your keyboard. If conflicts arise between multiple standards in the list (for example, if two standards specify layers of the same name but with different properties), the standard that appears first in the list takes precedence. To change the position of a standards file in the list, select it and choose the **Move Up** or **Move Down** button.

In the **Plug-ins** tab of the Configure Standards dialog box (see Figure 13-76) the **Plug-ins used when checking standards** section lists the standards plug-ins that are installed on the current system. For the CAD Standards Extension, a standards plug-in is installed for each of the named objects for which standards can be defined (layers, dimension styles, linetypes, and text styles). The **Description** section has descriptions of the **Purpose**, **Version**, and **Publisher** of the plug-in that is highlighted in the **Plug-ins used when checking standards** section.

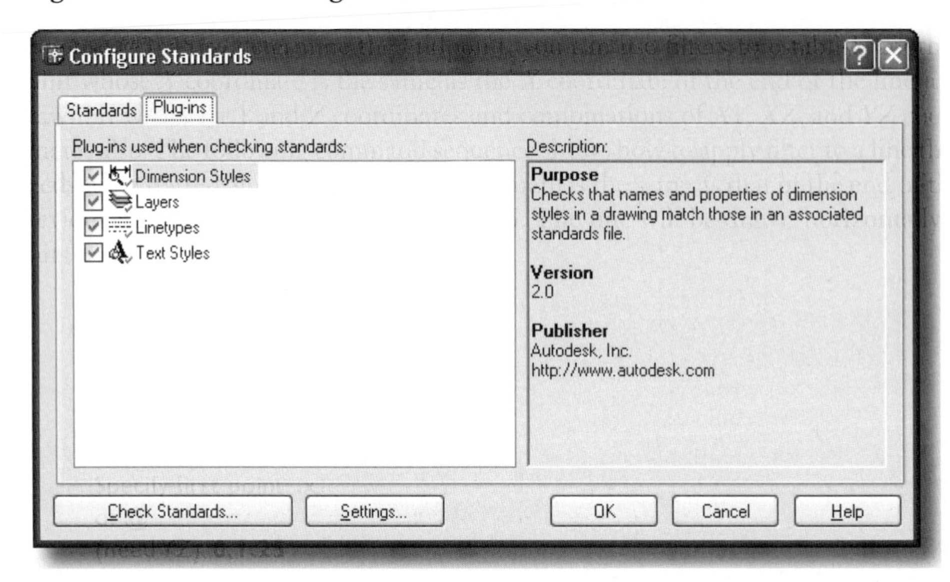

Figure 13–76 *Configure Standards dialog box with the Plug-ins tab selected*

The CHECKSTANDARDS command analyzes the current drawing for standards violations.

Invoke the CHECKSTANDARDS command:

Tools menu	Select Cad Standards > Check (see Figure 13–77)
Command: prompt	**checkstandards** (ENTER)

AutoCAD displays the Check Standards dialog box.

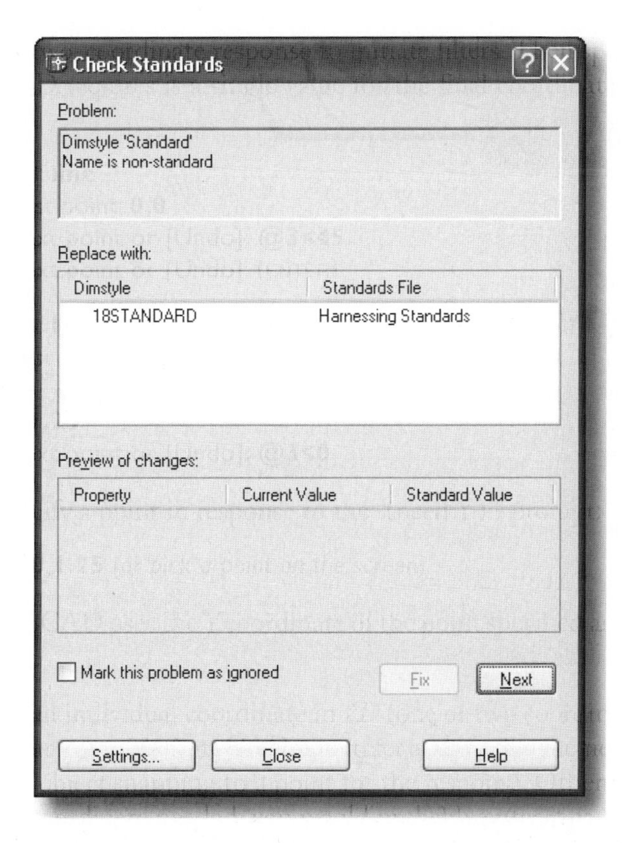

Figure 13–77 *Check Standards dialog box*

The Check Standards dialog box has sections titled **Problem**, **Replace with**, and **Preview of changes.**

Problem

In the **Problem** section there is a description of a nonstandard object in the current drawing. To fix a problem, select a replacement from the **Replace with** list and then choose the **Fix** button or press the F4 function key.

Replace With

The **Replace with** section lists possible replacements for the current standards violation. If a recommended fix is available, it is preceded by a check mark.

Preview of Changes

The **Preview of changes** section indicates the properties of the nonstandard AutoCAD object that will be changed if the fix currently selected in the **Replace with** list is applied.

Choose the **Close** button to close the Check Standards dialog box without applying a fix to the standards violation currently displayed in the **Problem** section.

In AutoCAD 2004 you can use CAD Standards tools to check for violations as you work. You are immediately alerted whenever you create a non-standard named object.

SLIDES AND SCRIPTS

Slides are quickly-viewable, non-editable views of a drawing or parts of a drawing. There are two primary uses for slides. One is to have a quick and ready picture to display symbols, objects, or written data for informational purposes only. The other very useful application of slides is to be able to display a series of pictures, organized in a prearranged sequence for a timed slide show. This is a very useful tool for demonstrations to clients or in a showroom. This feature supplements the time-consuming calling up of views required when using the ZOOM, PAN, or other display commands. The "slide show" is implemented through the SCRIPT command (described later in this chapter).

It should be noted that a slide merely masks the current display. Any cursor movement or editor functions employed while a slide is being displayed affects the current drawing under the slide and not the slide itself.

MAKING A SLIDE

The current display can be made into a slide with the MSLIDE command. The current viewport becomes the slide while you are working in model space. The entire display, including all viewports, becomes the slide when using MSLIDE while you are working in paper space. The MSLIDE command takes a picture of the current display and stores it in a file, so be sure it is the correct view.

Invoke the MSLIDE command:

Command: prompt	**mslide** (ENTER)

The Create Slide File dialog box appears, as shown in Figure 13–78. The default is the drawing name, which you can use as the slide file name by pressing ENTER. Or you can type any other name, as long as you are within the limitations of the operating system file-naming conventions. AutoCAD automatically appends the extension .SLD. Only objects that are visible in the screen drawing area (or in the current viewport when in model space) are made into the slide.

If you plan to show the slide on different systems, you should use a full-screen view with a high-resolution display for creating the slide.

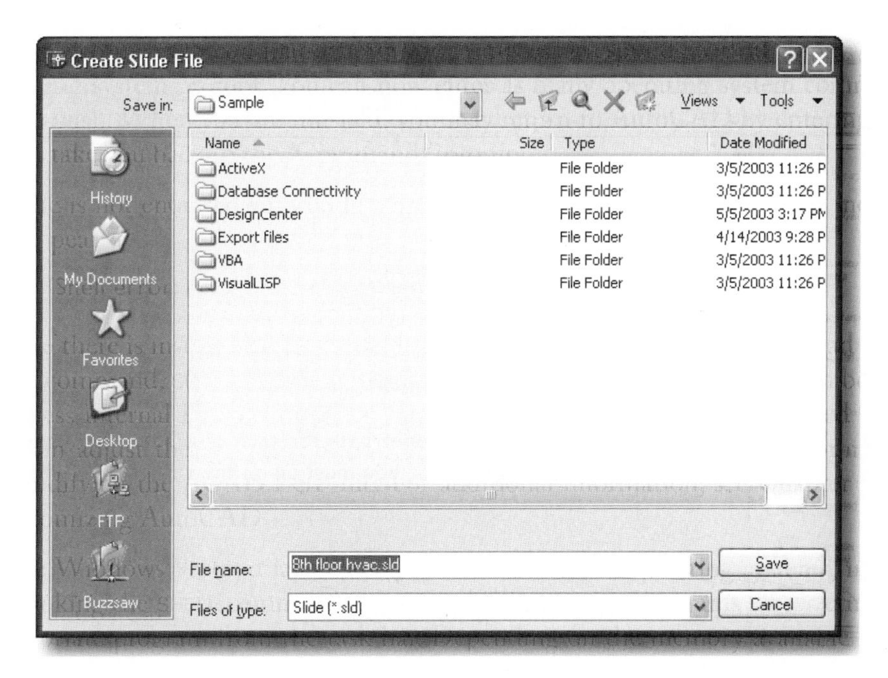

Figure 13–78 *Create Slide File dialog box*

SLIDE LIBRARIES

The SLIDELIB program provided with AutoCAD is used to create a slide library file in conjunction with a list of slide file names. Each slide file name must correspond to an actual slide whose extension is .SLD. The list of file names is in a separate file, written in ASCII format, with each slide file name on a separate line by itself. The file names in the list may or may not include the .SLD extension, but they must include the path (drive and/or folder) for the SLIDELIB command to access the proper file for inclusion in the library.

A file named SLDS001 could read as follows:

pic_1

A:pic_2.sld

c:\dwgs\pic_3.sld

b:\other\pic_6

Each line represents a slide name, some with extensions, some without. Each one has a different path. When the SLIDELIB command is used to address this file, all the slides listed will be included in the library you specify. The paths are not included when used with the SLIDELIB program. If there were a PIC_2.SLD on another path, it would not be included or it would conflict with the one specified for its particular library file. This allows for control of which slides will be used if the same slide file name exists in several drive/directory locations. The SLIDELIB program appends .SLB to the library file it creates from the slide list.

The preceding listing file can be used to create a library file named ALLPICS.SLB using the form:

slidelib *library* [< *slidelist*]

From the DOS window enter the following:

slidelib allpics [< slds001]

Note: In the DOS window, the proper path for each of the three elements of the function must either be entered or you must be on the same folder/directory as the element(s).

VIEWING A SLIDE

The VSLIDE command displays a slide in the current viewport.

Invoke the VSLIDE command:

Command: prompt	**vslide** (ENTER)

The Select Slide File dialog box appears. This is similar to most file-managing dialog boxes. Select the slide to display in the current viewport. If you have stored the slide in a library file and are calling it up for view, you can use the library file keyname followed by the slide name in parentheses. For example, if a slide named PIC_1 (actually stored under the filename pic_1.sld) is in a library file called ALLPICS.SLB, you can use the following format:

Command: **vslide**
Slide file <default>: **allpics(pic_1)**

Note: The library file lists only the address of each slide. It can be on a path or directory other than that of the slide(s) listed in it. It is necessary to specify the path to the library (.SLB file), but it is not necessary to specify the path in front of the slide name. For example, the ALLPICS.SLB file might be on the working directory, while the slide PIC_1 is on drive F in the directory SL1, and the slide PIC_2 is on drive F in the directory SL2. The sequences to call them into view would be as follows:

Command: **vslide**
Slide file <default>: **allpics(pic_1)**

Command: **vslide**
Slide file <default>: **allpics(pic_2)**

If the ALLPICS.SLB file is on drive F in the subdirectory SLB, the sequences are as follows:

Command: **vslide**
Slide file <default>: **f:\slb\allpics(pic_1)**

Command: **vslide**
Slide file <default>: **f:\slb\allpics(pic_2)**

SCRIPTS

Of the many means available to enhance AutoCAD through customization, scripts are among the easiest to create. Scripts are similar to the macros that can be created to enhance word processing programs. They permit you to combine a sequence of commands and data into one or two entries. Creating a script, like most enhancements to AutoCAD, requires that you use a text editor to write the script file (with the extension .SCR), which contains the instructions and data for the SCRIPT command to follow.

Because script files are written for use at a later time, you must anticipate the conditions under which they will be used. Therefore, familiarity with sequences of prompts that will occur and the types of responses required is necessary to have the script function properly. Writing a script is a simple form of programming.

The Script Text

A script text file must be written in ASCII format. That is, it must have no embedded print codes or control characters that are automatically written in files when created with a word processor in the document mode. If you are not using one of the available line editors, be sure you are in the nondocument, programmer, or ASCII mode of your word processor when creating or saving the file. Save the script with the file extension .SCR.

Each command can occupy a separate line, or you can combine several command/data responses on one line. Each space between commands and data is read as an ENTER, just as pressing the SPACEBAR is read while in AutoCAD. The end of a line of text is also the same as an ENTER.

Spaces and Ends of Lines in Script Files

The following script file contains several commands and data. The commands are GRID, LINE, CIRCLE, LINE, and CIRCLE again. The data are the response ON, coordinates (such as 0,0 and 5,5), and distances, such as radius 3.

```
GRID ON LINE 0,0 5,5  CIRCLE 3,3 3
LINE 0,5 5,0
CIRCLE 5,2.5 1
```

The first line includes the GRID, LINE, and CIRCLE commands, and their responses. Note the two spaces after 5,5; these are required to simulate the double ENTER. Not obvious is the extra space following the 5,0 response in the second LINE command. This extra space and the invisible CR-LF (carriage-return, linefeed) code that ends every line in a text file combine to simulate pressing the SPACEBAR twice. This is necessary, again, to exit the LINE command.

Some text editors automatically remove blank spaces at the end of text lines. To guard against that, an alternative is to have a blank line indicate the second ENTER, as follows:

```
GRID ON LINE 0,0 5,5 CIRCLE 3,3 3
LINE 0,5 5,0

CIRCLE 5,2.5 1
```

Invoke the SCRIPT command:

Tools menu	Choose Run Script
Command: prompt	**script** (ENTER)

AutoCAD displays the Select Script File dialog box. Select the appropriate script file from the list box and choose the **Open** button. AutoCAD executes the command sequence from the script file.

Changing Block Definitions with a Script File

Using a script to perform a repetitive task is illustrated in the following example. This application also offers some insight on changing the objects in an inserted block with attributes without affecting the attribute values.

Figure 13–79 shows a group of drawings all of which utilize a common block with attribute values in one insertion that are different from the attribute values of those in other drawings. In this case, the border/title block is a block named BRDR. It was originally drawn with the short lines around and outside of the main border line. It was discovered that these lines interfered with the rollers on the plotter and needed to be removed. The BRDR block definition is shown in Figure 13–80. Remember, the inserted block has different attribute values in each drawing, such as drawing number, date, and title.

You want to change objects in the block but maintain the attribute values as they are. There are two approaches to redefinition. One is to find a clear place in the drawing and insert the block with an asterisk (*). This is the same as inserting and exploding the block. Then you make the necessary changes in the objects and make the revised group into a block with the same block name. This redefines all insertions of blocks with that same name in the drawing. In this case there is only one insertion. You must be attentive to how any changes to attributes might affect the already inserted block of that name.

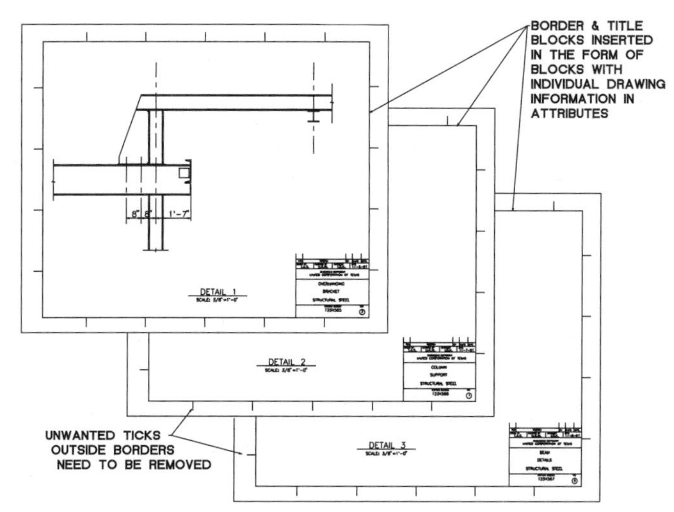

Figure 13–79 *Drawings utilizing common block and attribute values that are different from those of other drawings*

The second method is to use the WBLOCK command to place a copy of the block in a file with the same name. This makes a new and separate drawing of the block. Then you exit the current drawing, call up the newly created drawing, make the required changes in the objects, and end the drawing that was created by the WBLOCK command. Then you re-enter the drawing in which the block objects need to be changed. You now use the INSERT command and respond with "blockname=" and have the block redefined without losing the attribute values. For example, if the block name is BRDR, the sequence would be as follows:

Command: **insert**
Enter block name or [?]: **brdr=**
Block "b" already exists. Redefine it? [Yes/No] <N>: **y**
Block BRDR redefined
Regenerating drawing
Specify insertion point or [Scale/X/Y/Z/Rotate/PScale/PX/PY/PZ/PRotate]:
(press ENTER *to cancel the command sequence)*

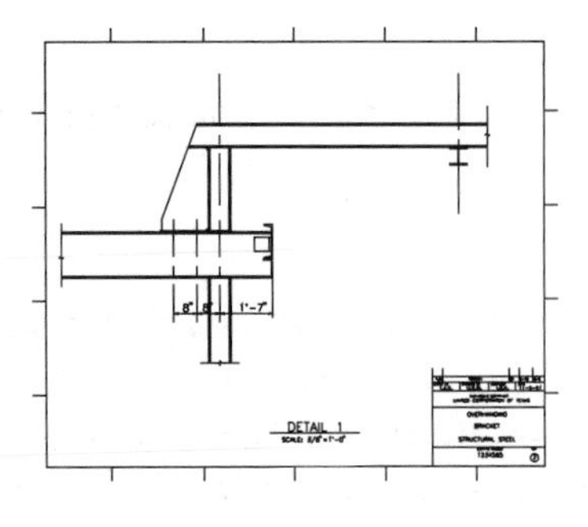
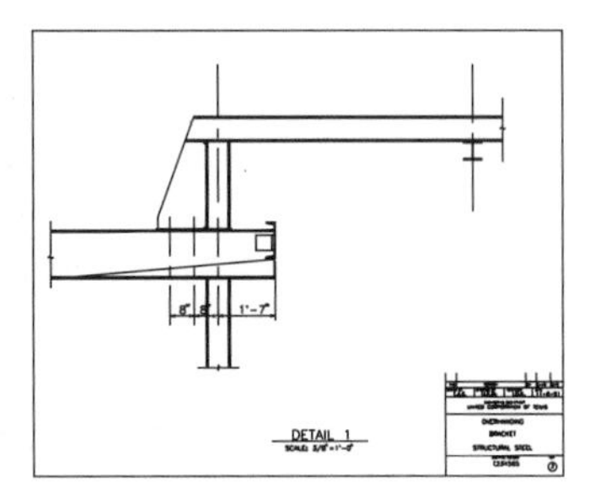

Figure 13–80 *A block definition*

The key to this sequence is the equals sign (=) following the block name. This causes AutoCAD to change the definition of the block named BRDR to be that of the drawing named BRDR, although it maintains the attribute values as long as attribute definitions remain unchanged.

If the preceding procedure must be repeated many times, this is where a script file can be employed to automate the process. In the following example, we show how to apply the script to a drawing named PLAN_1. The sequence included an ENTER as it was described to be used while in AutoCAD. This expedited the operation by not actually having the block inserted, but only having its definition brought into the drawing. Because an ENTER during the running of a script file causes the script to terminate, those keystrokes cannot be in the middle of a script; besides, invoking in a SCRIPT command requires using the AutoLISP function "(command)." The script can be written in a file (called BRDRCHNG.SCR for this example) in ASCII format as follows:

```
INSERT
BRDR=
0,0       (the 0,0 is followed by six spaces)
ERASE L (the L is followed by one space)
REDRAW
```

Now, from the "Command:" prompt, you can apply the script to drawing PLAN_1 by responding as follows:

```
Command: script
```

There are several important aspects of this sequence:

Lines 1–2 This is where you might enter if you were not in the SCRIPT command and have the definition of the block named BRDR take on that of the drawing BRDR without actually having to continue with the insertion in the drawing.

Line 3 In this case, 0,0 as the insertion point is arbitrary because the inserted block is going to be erased anyway. Special attention is given to the six spaces following the insertion point. These are the same as pressing SPACEBAR or ENTER six times in response to the "X-scale," "Y-scale," and "Rotation angle" prompts, and the number of spaces that follow must correspond to the number of attributes that require responses for values. In this example there were three attributes. Again, the fact that the responses are null is immaterial, because this insertion will not be kept.

Line 4 The ERASE L is self-explanatory, but do not forget that after the "L" you must have another space (press SPACEBAR again) to terminate the object selection process and complete the ERASE command.

Line 5 The REDRAW command is not really required except to show the user for a second time that the changes have been made before the script ends.

UTILITY COMMANDS FOR A SCRIPT

Following are the utility commands that may be used within a script file.

Specifying the Delay

The DELAY command causes the script to pause for the number of milliseconds that have been specified by the command. To set the delay in the script for five seconds, you would write:

```
DELAY 5000
```

Resuming the Script

The RESUME command causes the script to resume running after the user has pressed either or to interrupt the script. It may be entered as follows:

```
Command: resume
```

Graphscr and Textscr

The GRAPHSCR and TEXTSCR commands are used to flip or toggle the screen to the graphics or text mode, respectively, during the running of the script. Each is simply entered as a command in the script, as follows:

Command: **graphscr**

or

Command: **textscr**

These two screen toggle commands can be used transparently by preceding them with an apostrophe.

Repeating the Script

The RSCRIPT command, when placed at the end of a script, causes the script to repeat itself. With this feature you can have a slide show run continuously until terminated by ESC key.

A repeating demonstration can be set up to show some sequences of commands and responses as follows:

```
GRID ON
LIMITS 0,0 24,24
ZOOM A
CIRCLE 12,12 4
DELAY 2000
COPY L M 12,12 18,12 12,18 6,12 12,6 (an extra space at the end)
DELAY 5000
ERASE W 0,0 24,24 (an extra space at the end)
DELAY 2000
LIMITS 0,0 12,9
ZOOM A

GRID OFF
TEXT 1,1 .5 0 THAT'S ALL FOLKS!
ERASE L (an extra space at the end)
RSCRIPT
```

This script file utilizes the DELAY and RSCRIPT subcommands. Note the extra spaces where continuation of some actions must be terminated.

The SCRIPT command can be used to show a series of slides, as in the following sequence:

```
VSLIDE SLD_A
VSLIDE *SLD_B
DELAY 5000
```

```
VSLIDE
VSLIDE *SLD_C
DELAY 5000
VSLIDE
DELAY 10000
RSCRIPT
```

This script uses the asterisk (*) before the slide name prior to the delay. This causes AutoCAD to load the slide, ready for viewing. Otherwise, there would be a blank screen between slides while the next one is being loaded. The RSCRIPT command repeats the slide show.

REVIEW QUESTIONS

1. All of the following can be renamed using the RENAME command, except:

   a. a current drawing name

   b. named views within the current drawing

   c. block names within the current drawing

   d. text style names within the current drawing

2. The PURGE command can be used:

   a. after an editing session

   b. at the beginning of the editing session

   c. at any time during the editing session

   d. a and b only

3. To calculate the product of 4 times 21.5 times pi and store it for use with later AutoCAD commands, you should respond to the ">>Expression" prompt of the CAL command with:

   a. A=* 4 21.5 pi

   b. A= 4 * 21.5 * 3.14

   c. A= 4 * 21.5 * pi

   d. A= (* 4 21.5 pi)

   e. none of the above

4. To use the product from the previous question to respond to a scale factor in the SCALE command you would type:

   a. %A

   b. ^A

   c. !A

   d. A

   e. (A)

5. The most common setting for the current drawing color is:

   a. Red

   b. Bylayer

   c. Byblock

   d. White

   e. none of the above

6. Which of the following are not valid color names in AutoCAD?

   a. Brown

   b. Red

   c. Yellow

   d. Magenta

   e. none of the above (i.e. all are valid)

7. To change the background color for the graphics drawing area, you should use:

   a. COLOR
   b. BGCOLOR
   c. PREFERENCES
   d. SETTINGS
   e. CONFIG

8. AutoCAD release 14 can be used to edit a drawing saved as an ACAD2004 drawing.

   a. True
   b. False

9. ACAD2004 can be used to edit a drawing that is saved in an AutoCAD release 14 drawing.

   a. True
   b. False

10. All the following items can be purged from a drawing file, except:

   a. Text styles
   b. Blocks
   c. System variables
   d. Views
   e. Linetypes

11. The following can be deleted with the PURGE command, except:

   a. blocks not referenced in the current drawing
   b. linetypes that are not being used in the current drawing
   c. layer 0, if it is not being use
   d. none of the above (i.e. all can be purged)

12. With OLE (object linking and embedding), you can copy or move information from one application to another while retaining the ability to edit the information in the original application.

   a. True
   b. False

13. If the TIME command is not turned off during lunch break, the TIME command will:

   a. include the lunch break time
   b. exclude the lunch break time
   c. turn itself off after 10 minutes of inactivity
   d. automatically subtract one hour for lunch
   e. none of the above

14. The VIEW command:

    a. serves a purpose similar to the PAN command

    b. will restore previously saved views of your drawing

    c. is normally used on very small drawings

    d. none of the above

15. When using a .X point filter, AutoCAD will request that you complete the point selection process by entering:

    a. An *X* coordinate

    b. A *Y* and *Z* coordinate

    c. An OSNAP mode

    d. Nothing, .X is not a valid filter

16. The geometric calculator function MEE will:

    a. calculate the midpoint of a selected line

    b. calculate the midpoint between the endpoints of any two objects

    c. return the node name for the computer you are working on

    d. average a string of numbers

    e. none of the above

17. To load a script file called "SAMPLE.SCR", use:

    a. SCRIPT, then SAMPLE

    b. LOAD, then SAMPLE

    c. LOAD, SCRIPT, then SAMPLE

    d. none of the above

18. The AutoCAD command used for viewing a slide is:

    a. VSLIDE

    b. VIEWSLIDE

    c. SSLD

    d. SLIDE

    e. MSLIDE

19. If a REDRAW is performed while viewing a slide:

    a. the command will be ignored

    b. the current slide will be deleted

    c. the current drawing will be displayed

    d. AutoCAD will load the drawing the slide was created from

    e. none of the above

20. A script file is identified by the following extension:

a. SCR d. SPT

b. BAK e. none of the above

c. DWK

21. A slide file is identified by the following extension:

a. SLD d. SLU

b. SCR e. none of the above

c. SLE

22. Slides can be removed from the display with the command:

a. ZOOM ALL d. both a and b

b. REGEN e. none of the above

c. OOPS

23. The SCRIPT command cannot be used to:

a. insert blocks

b. create layers

c. place text

d. create another script file

e. none of the above (i.e. all are possible)

24. To create a slide library, you must:

a. use the AutoCAD command SLIDELIB

b. add slides to the library one at a time

c. create an ASCII text file listing all slides to be included in the library

d. have all slide files in a common directory

25. To cause a script file to execute in an infinite loop, you should place what command at the end of the file?

a. REPEAT d. BEGIN

b. RSCRIPT e. none of the above

c. GOTO:START

26. If AutoCAD is executing in an infinite script file loop, how can you terminate the loop?

a. press BACKSPACE d. press F1

b. press CTRL + C e. none of the above

c. press ALT + C

27. What command permits a user to work with just a portion of a drawing by loading geometry from specific views or layer?

    a. pload c. partialopen

    b. partopn d. pltopn

28. What command permits additional geometry to be loaded into the current partially loaded drawing?

    a. PLOAD c. PARTADD

    b. PARTIAL LOAD d. ADDGEO

29. Can the PURGE command be used in partially open drawings?.

    a. Yes

    b. No

30. What term can be added to draw, modify, or inquiry commands which will cause them to automatically repeat?

    a. redo d. multiple

    b. repeat e. no such option

    c. return

31. When one of more objects overlay each other which command is used to control their order of display?

    a. DWGORDER c. VIEWORDER

    b. DRAWORDER d. ARRANGE

32. Which command permits you to execute operating system programs without exiting AutoCAD 2004?

    a. RUN c. OPSYS

    b. EXOPRG d. SHELL

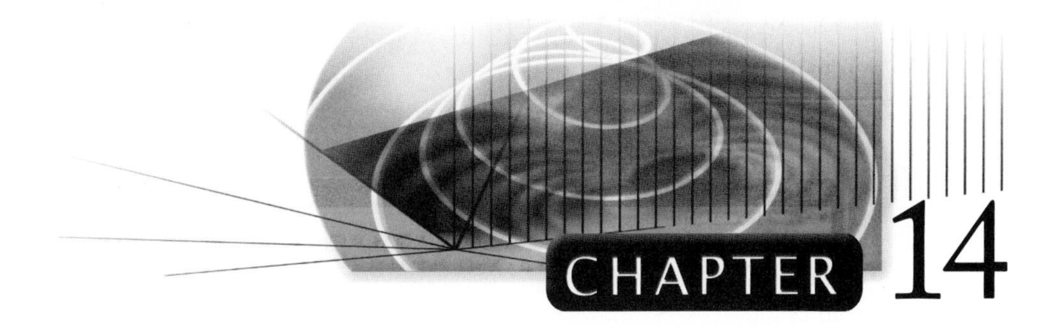

Internet Utilities

INTRODUCTION

The Internet is the most important way to convey digital information around the world. You are probably already familiar with the best-known uses of the Internet: e-mail (electronic mail) and surfing the Web (short for "World Wide Web"). E-mail lets users exchange messages and data at very low cost. The Web brings together text, graphics, audio, and video in an easy-to-use format. Other uses of the Internet include FTP (file transfer protocol, for effortless binary-file transfer), Gopher (presents data in a structured, subdirectory-like format), and Usenet, a collection of more than 100,000 news groups.

AutoCAD allows you to interact with the Internet in several ways. You can launch a Web browser from within AutoCAD. AutoCAD can create DWF (short for "Design Web Format") files for viewing drawings in two-dimensional format on Web pages. AutoCAD can open and insert drawings from, and save drawings to the Internet. AutoCAD 2004 transforms AutoCAD into the Internet design platform, delivering enhanced Internet collaboration and communication capabilities to users.

After completing this chapter, you will be able to do the following:

- Launch the default Web browser
- Use Communication Center
- Open drawings from the Internet
- Save drawings to the Internet
- Create and use hyperlinks
- Create and view DWF files
- Etransmit
- Publish to Web
- Use Meet Now

LAUNCHING THE DEFAULT WEB BROWSER

The BROWSER command lets you start a Web browser from within AutoCAD. By default, the BROWSER command uses the Web browser program that is registered in your computer's Windows operating system. AutoCAD prompts for the URL (short for "uniform resource locator"). The URL is the Web site address, such as http://www.autodesk.com. (More about URLs in the next section.) The BROWSER command can be used in scripts, toolbar or menu macros, and AutoLISP routines to access the Internet automatically.

Invoke the BROWSER command:

Web toolbar	Choose the BROWSE THE WEB command (see Figure 14–1)
Command: prompt	**browser** (ENTER)

Figure 14–1 *Invoking the* BROWSE THE WEB *command from the Web toolbar*

AutoCAD prompts:

Command: **browser**
Enter Web location (URL) <default location>: *(specify a new location, or press* ENTER *to accept the default location)*

You can specify a default URL in the Options dialog box. Select the **Files** tab, choose **Menu, Help**, and **Miscellaneous File Name**, select **Default Internet Location** (see Figure 14–2) and specify the default URL Web location. When you invoke the BROWSER command and press ENTER to accept the default location, AutoCAD launches the Web browser, which contacts the Web site. Figure 14–3 shows Internet Explorer and the Autodesk Web site.

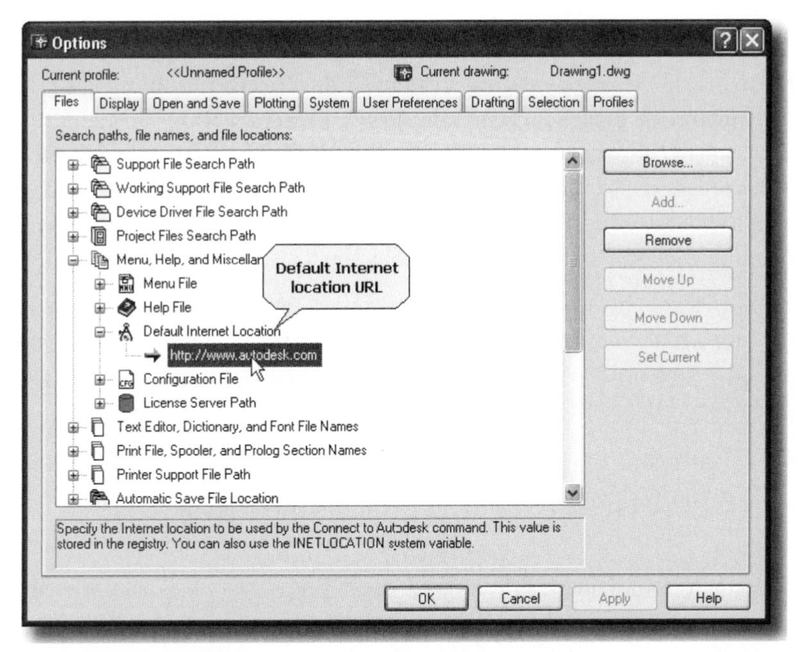

Figure 14–2 *Options dialog box with Files tab selected*

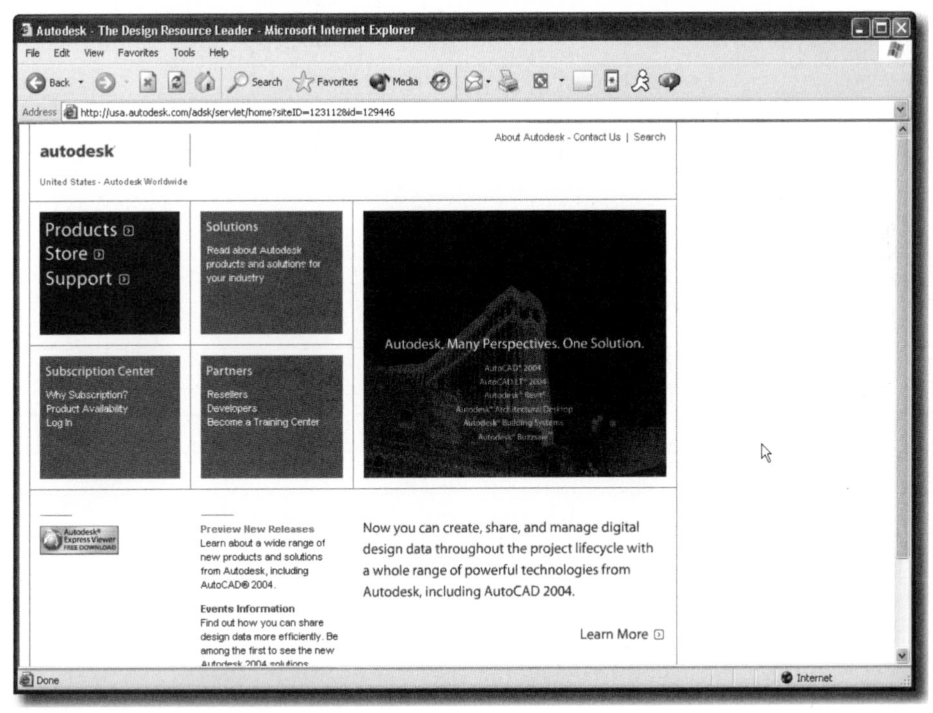

Figure 14–3 *Internet Explorer displaying the Autodesk Web site*

UNIFORM RESOURCE LOCATOR

As mentioned earlier, the file-naming system of the Internet is known as URL, short for "uniform resource locator." The URL system allows you to find any resource (a file) on the Internet. Resources include text files, Web pages, program files, audio, or movie clips—in short, anything you might find on your own computer. The primary difference is that these resources are located on somebody else's computer. A typical URL looks like the examples shown in Table 14–1.

The "http://" prefix is not required. Most of today's Web browsers automatically add in this routing prefix, which saves you a few keystrokes. URLs can access several different kinds of resources—such as Web sites, e-mail, or news groups—but always take on the following format:

```
scheme://netloc
```

The "scheme" accesses the specific resource on the Internet, including those listed in Table 14–2.

The "://" characters indicate a network address. Table 14–3 lists the formats recommended for specifying URL-style file names with the BROWSE command. *Servername* is the name of the server, such as "www.autodesk.com." The *pathname* is the same name as a subdirectory or folder name. The *drive:* is the driver letter, such as C: or D:. A local file is a file located on your computer. The *localhost* is the name of the network host computer.

Table 14–1 Example URLs

Example	Meaning
http://www.autodesk.com	Autodesk primary Web site
http://data.autodesk.com	Autodesk data publishing Web site
news://adesknews.autodesk.com	Autodesk news server
ftp://ftp.autodesk.com	Autodesk FTP server

Table 14–2 URL Prefix Meanings

Scheme Prefix	Meaning
file://	Files on your computer's hard drive or local network
ftp://	File transfer protocol (downloading files)
http://	Hypertext transfer protocol (Web sites)
mailto://	Electronic mail (e-mail)
news://	Usenet news (news groups)
telnet://	Telnet protocol
gopher://	Gopher protocol

Table 14–3 Recommended Formats for Specifying URL-Style File Names

Drawing Location	Template URL
Web site	http://servername/pathname/filename
FTP site	ftp://servername/pathname/filename
Local file	file:///drive:/pathname/filename file:///drive:/pathname/filename file://\\ localPC\pathname\filename file:////localPC/pathname/filename
Network file	file://localhost/drive:/pathname/filename file://localhost/drive:/pathname/filename

COMMUNICATION CENTER

Communication Center resides in the Tray section (lower right corner of the AutoCAD window) and can be customized to offer you just the right amount of information. Here you can choose to be notified about such things as maintenance patches, product support information, subscription information and extension announcements, articles and tips. Maintenance patches include any program updates and fixes to the existing product. Product support information includes ground-breaking news from the Product Support team at Autodesk. Subscription information and extension announcements include announcements and subscription program news if you are an Autodesk subscription member. If new information becomes available, a bubble announcement is displayed as shown in Figure 14–4. If it is selected, the Communication Center window is displayed as shown in Figure 14–5.

Figure 14–4 *Bubble announcement from the Communication Center*

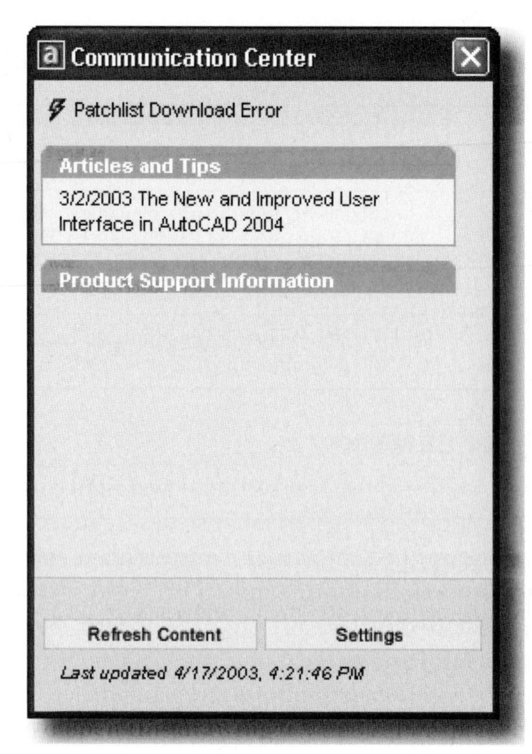

Figure 14–5 *Communication Center window*

Choose one of the available topics from the list for detailed information.

Communication Center is an interactive feature that must be connected to the Internet to deliver content and information. Each time Communication Center is connected, it sends information to Autodesk so that the correct information can be returned. All information is sent anonymously to maintain your privacy. The information that will be sent to Autodesk includes Product Name, Product Release Number, Product Language and Country.

To customize Communication Center options, choose **Settings** in the Communication Center window. AutoCAD displays the Configuration Settings dialog box as shown in Figure 14–6.

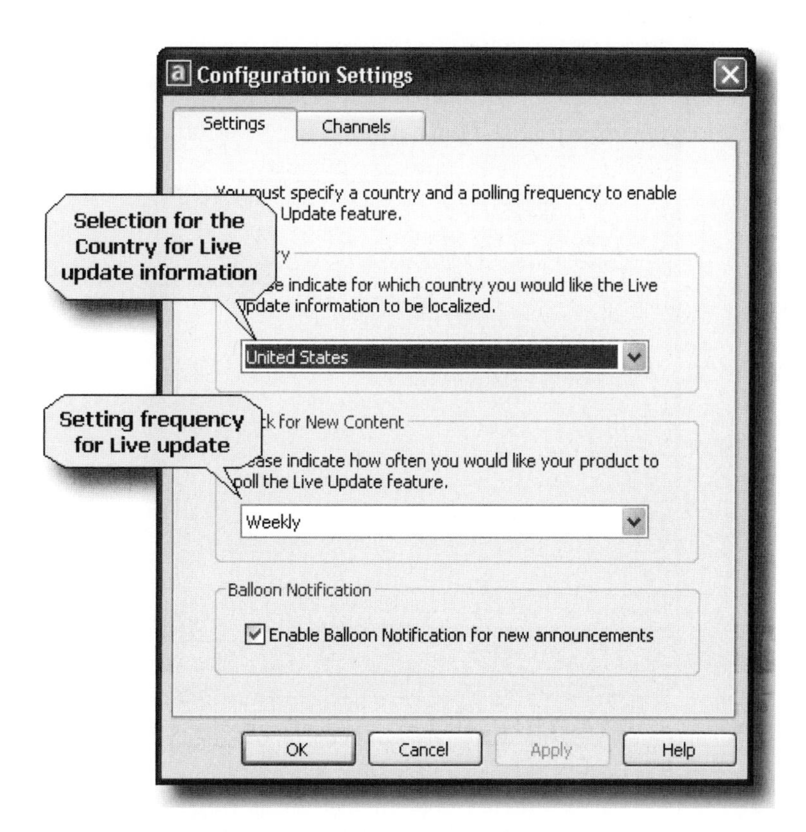

Figure 14–6 *Configuration Settings dialog box with Settings tab selected*

In the **Settings** tab of the dialog box, choose the **Country** where the software is registered, so that Communication Center can provide information that is designed specifically for your location. Choose how often you want Communication Center to synchronize with Autodesk servers to update the software. Set the **Balloon Notification** toggle switch to on or off. When it is set to ON. Communication Center balloon messages are displayed above the status bar when a new announcement is received. If the balloon notification is disabled in the tray settings, the Balloon Notification setting in the Communication Center is ignored.

In the **Channels** tab of the dialog box as shown in Figure 14–7 specify the information that you want displayed in Communication Center.

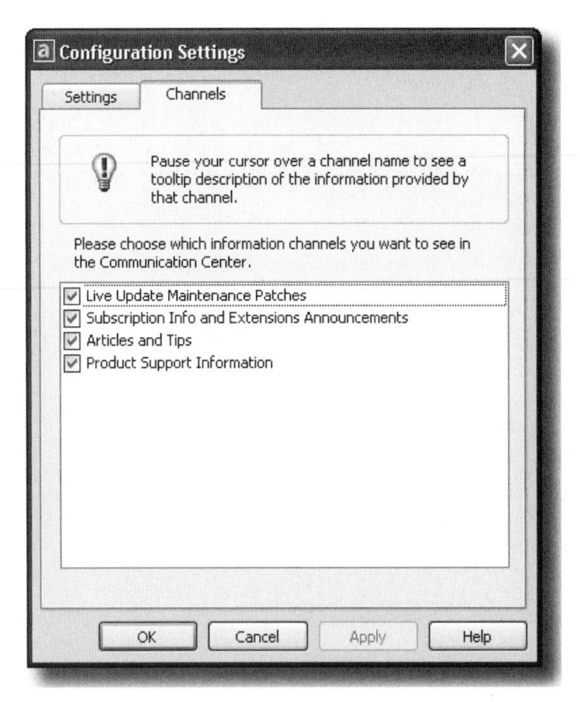

Figure 14–7 *Configuration Settings dialog box with Channels tab selected*

In the Configuration Settings dialog box, choose the settings and options that you want to use and then choose **Apply**. Choose **OK** to close the Configuration Settings dialog box and then close the Communication Center window.

OPENING AND SAVING DRAWINGS FROM THE INTERNET

AutoCAD allows you to open and save drawing files from the Internet or an intranet. You can also attach externally referenced drawings stored on the Internet/intranet to drawings stored locally on your system. Whenever you open a drawing file from an Internet or intranet location, it is first downloaded into your computer and opened in the AutoCAD drawing area. Then you can edit the drawing and save it, either locally or back to the Internet or intranet location for which you have appropriate access privileges.

To open an AutoCAD drawing from an Internet/intranet location, invoke the OPEN command:

Standard toolbar	Choose the OPEN command (see Figure 14–8)
File menu	Choose Open
Command: prompt	**open** (ENTER)

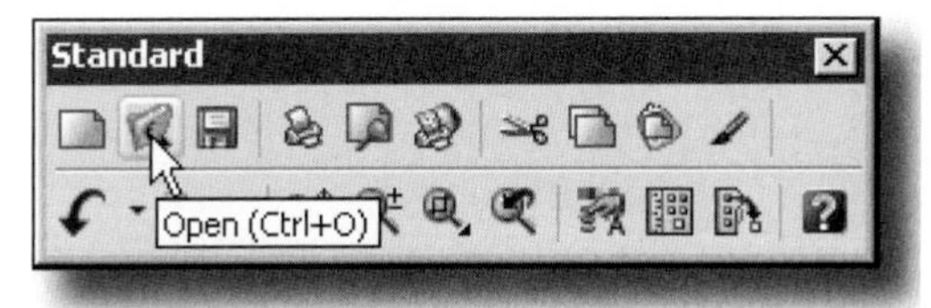

Figure 14–8 *Invoking the* OPEN *command from the Standard toolbar*

AutoCAD displays the Select File dialog box, as shown in Figure 14–9.

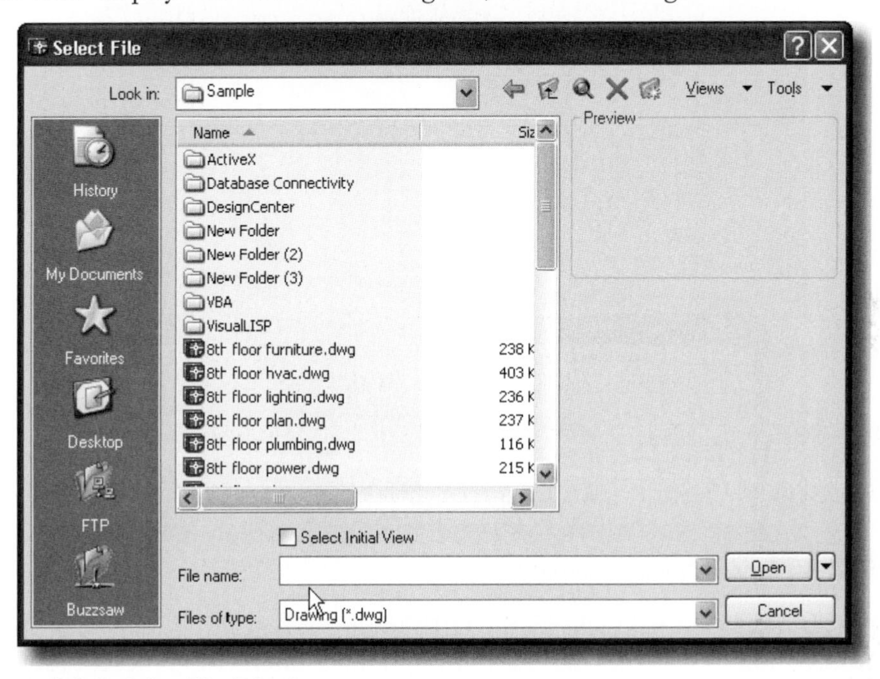

Figure 14–9 *Select File dialog box*

In the **File name:** text field, specify the URL to the file you wish to open, and choose the **Open** button to open the drawing from the specified Internet/intranet location. Be sure to specify the transfer protocol, such as ftp:// or http://, and the file extension, such as DWG or DWT. You can also choose the **Search the Web** button to open the Browse the Web dialog box. From there you can navigate to the Internet location where the file is stored. You can also access the Buzzsaw.com and FTP locations by selecting the appropriate tabs provided on the left side of the dialog box.

To save an AutoCAD drawing to an Internet/intranet location, invoke the SAVEAS command:

File menu	Choose Save As
Command: prompt	**saveas** (ENTER)

AutoCAD displays the Save Drawing As dialog box, as shown in Figure 14–10.

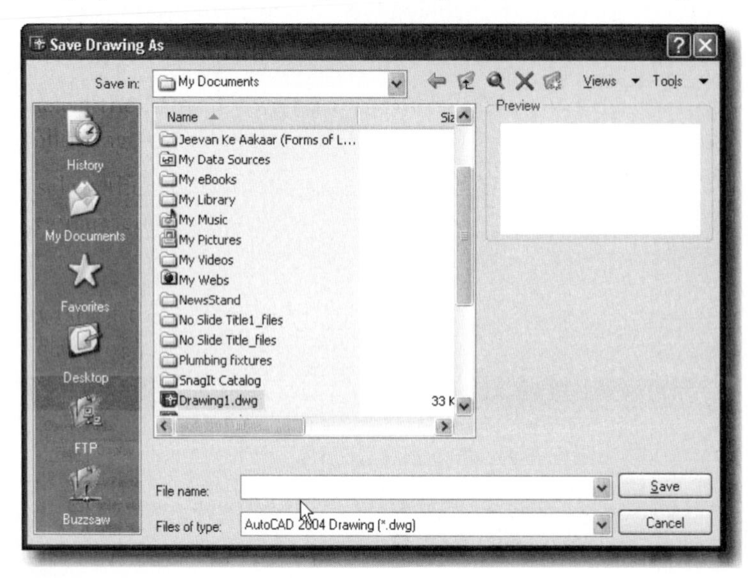

Figure 14–10 *Save Drawing As dialog box*

In the **File name:** text field, specify the URL to the file you wish to save and choose the **Save** button to save the drawing to the specified Internet/intranet location. Be sure to specify the transfer protocol and file extension (such as DWG or DWT). You can also choose the **Search the Web** button to open the Browse the Web dialog box. From there you can navigate to the Internet location where the file is to be saved.

Note: You can save the drawing to a specific Internet location by using the FTP protocol only.

To attach an xref to a drawing stored on the Internet/Intranet location, invoke the XATTACH command:

Reference toolbar	Choose the External Reference Attach command (see Figure 14–11)
Insert menu	Choose External Reference...
Command: prompt	**xattach** (ENTER)

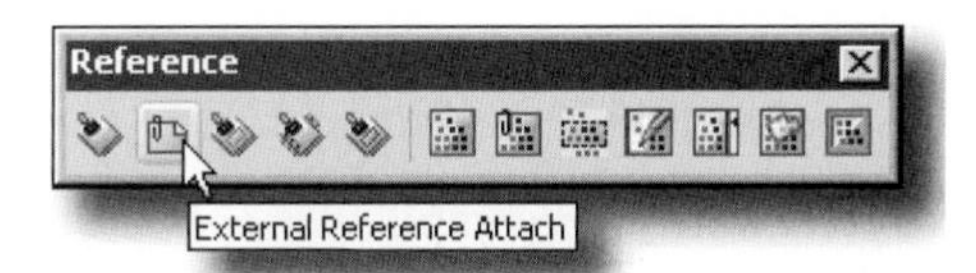

Figure 14–11 *Invoking the* EXTERNAL REFERENCE ATTACH *command from the Reference toolbar*

AutoCAD displays the Select Reference File dialog box, as shown in Figure 14–12.

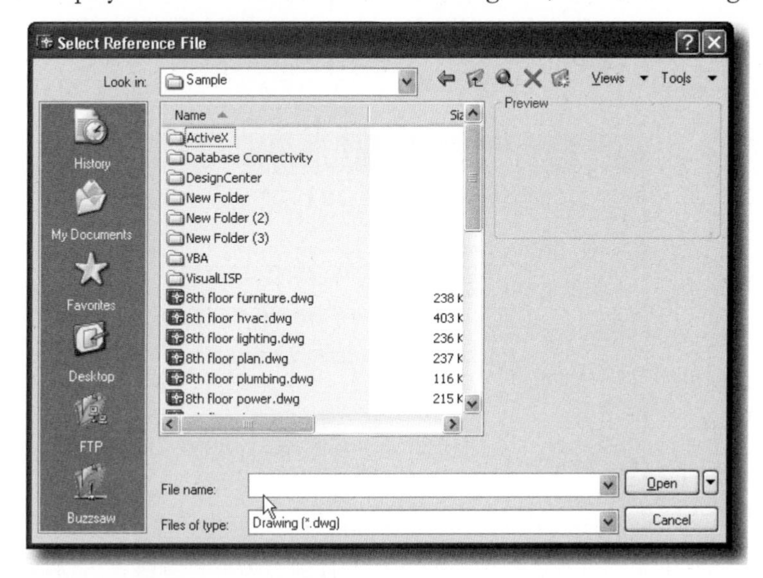

Figure 14–12 *Select Reference File dialog box*

In the **File name:** text field, specify the URL to the file you wish to open and choose the **Open** button to attach the drawing as a reference file from the specified Internet/intranet location. Be sure to specify the transfer protocol, such as ftp:// or http://, in the URL. You can also choose the **Search the Web** button to open the Browse the Web dialog box. From there you can navigate to the Internet location where the file is stored. You can also access the Buzzsaw.com and FTP locations by selecting the appropriate tabs provided on the left side of the dialog box.

WORKING WITH HYPERLINKS

AutoCAD allows you to create hyperlinks that provide jumps to associated files. Hyperlinks provide a simple and powerful way to quickly associate a variety of documents with an AutoCAD drawing. For example, you can create a hyperlink that opens another drawing file from the local drive or network drive, or from an Internet Web site. You can also specify a named location to jump to within a file, such as a view name in an AutoCAD drawing, or a bookmark in a word processing program.

You can also attach a URL to jump to a specific Web site. You can attach hyperlinks to any graphical object in an AutoCAD drawing.

AutoCAD allows you to create both *absolute* and *relative* hyperlinks in your AutoCAD drawings. Absolute hyperlinks store the full path to a file location, whereas relative hyperlinks store a partial path to a file location, relative to a default URL or name of the directory you specify using the HYPERLINKBASE system variable.

You can also specify the relative path for a drawing in the Drawing properties dialog box (**Summary** tab) shown in Figure 14–13. The Properties dialog box is opened from the **File** menu.

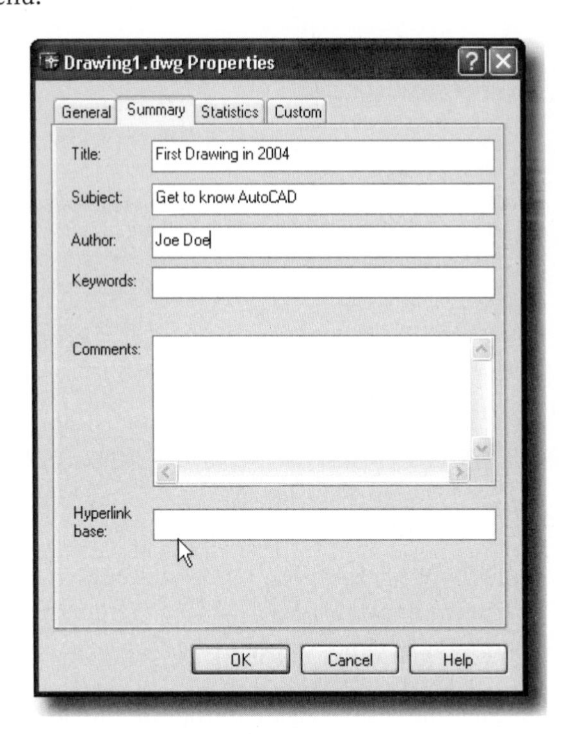

Figure 14–13 *Drawing Properties dialog box (Summary tab)*

Whenever you attach a hyperlink to an object, AutoCAD provides a cursor feedback when you position the cursor over the object. To activate the hyperlink, first select the object (make sure the PICKFIRST system variable is set to 1). Right-click to display the shortcut menu and activate the link from the **Hyperlink** sub-menu, as shown in Figure 14–14.

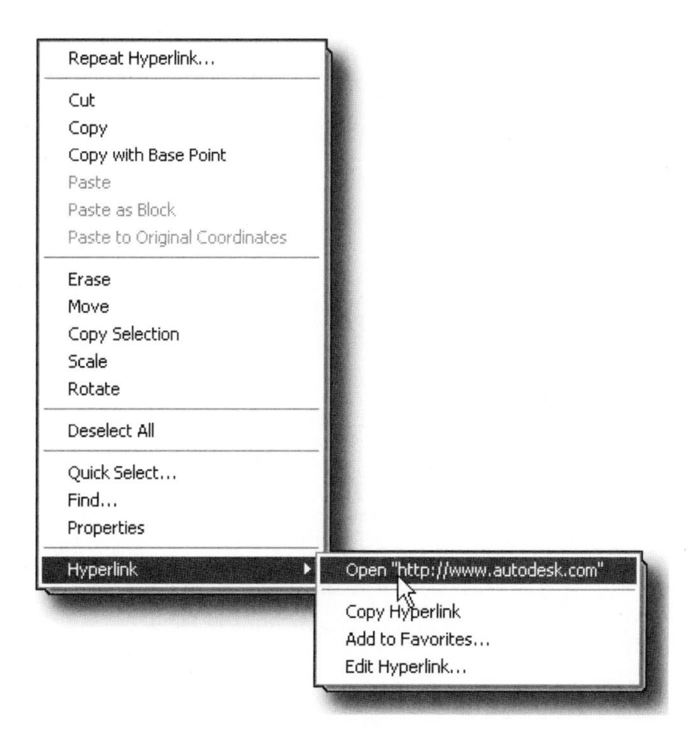

Figure 14–14 *Hyperlink sub-menu on the shortcut menu*

When you create a hyperlink that points to an AutoCAD drawing template (DWT) file, AutoCAD creates a new drawing file based on the selected template rather than opening the actual template when you activate the hyperlink. With this method there is no risk of accidentally overwriting the original template.

When you create a hyperlink that points to an AutoCAD named view and activate the hyperlink, the named view that was created in the model space is restored in the **Model** tab, and the named view that was created in paper space is restored in the **Layout** tab.

CREATING A HYPERLINK

To create an hyperlink, invoke the HYPERLINK command:

Insert menu	Choose Hyperlink
Command: prompt	**hyperlink** (ENTER)

AutoCAD prompts:

Command: **hyperlink**
Select objects: *(select one or more objects to attach the hyperlink and press* ENTER*)*

AutoCAD displays the Insert Hyperlink dialog box, as shown in Figure 14–15.

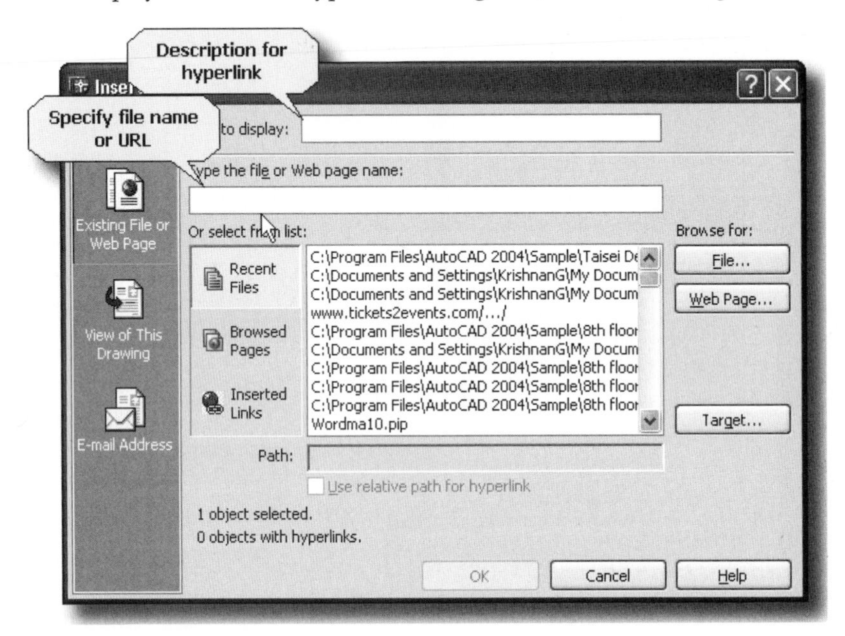

Figure 14–15 *Insert Hyperlink dialog box*

Specify a description for the hyperlink in the **Text to Display**: text field. This is useful when the file name or URL is not helpful in identifying the contents of the linked file. Specify the URL or path with the name of the file that you wish to have associated with the selected objects in the **Type the File or Web Page Name:** text field. Or choose one of the Browse for buttons: **File...**, **Web Page...**, or **Target....** Choosing the **File...** button opens the Browse the Web – Select Hyperlink dialog box (standard file selection dialog box). Use the dialog box to navigate to the file that you want associated with the hyperlink. Choosing the **Web Page...** button opens the AutoCAD browser. Use the browser to navigate to a Web page to associate with the hyperlink. Choosing the **Target...** button opens the Select Place in Document dialog box, in which you specify a link to a named location in a drawing. The named location that you select is the initial view that is restored when the hyperlink is executed. You can also select the path, with the name of the file or URL, from the list box categorized from **Recent Files**, **Browsed Pages**, and **Inserted Links**.

The **Path:** text field displays the path to the file associated with the hyperlink. If **Use relative path for hyperlink** is selected, only the file name is listed. If **Use relative path for hyperlink** is cleared, the full path and the file name are listed.

The **Use relative path for hyperlink** check box toggles the use of a relative path for the current drawing. If this option is selected, the full path to the linked file is not stored with the hyperlink. AutoCAD sets the relative path to the value specified by the HYPERLINKBASE system variable or, if this variable is not set, the current drawing path. If this option is not selected, the full path to the associated file is stored with the hyperlink.

Choosing the **Existing File or Web Page** tab located on the left side of the Insert Hyperlink dialog box creates a hyperlink to an existing file or Web page. Choosing the **View of This Drawing** tab allows you to select a named view in the current drawing to link to. Choosing the **E-mail Address** tab specifies an email address to link to. When the hyperlink is executed, a new email is created using the default system email program.

Choose the **OK** button to create the hyperlink to the selected objects and close the Insert Hyperlink dialog box.

EDITING A HYPERLINK

To edit a hyperlink, invoke the HYPERLINK command:

Insert menu	Choose Hyperlink...
Command: prompt	**hyperlink** (ENTER)

AutoCAD prompts:

Command: **hyperlink**
Select objects: *(select one or more objects to edit the hyperlink and press* ENTER*)*

AutoCAD displays the Edit Hyperlink dialog box, similar to Figure 14–16. Make necessary changes in the **Text to display:**, and/or **Type the file or Web page name:** text fields. Choose the **OK** button to accept the changes and close the Edit Hyperlink dialog box.

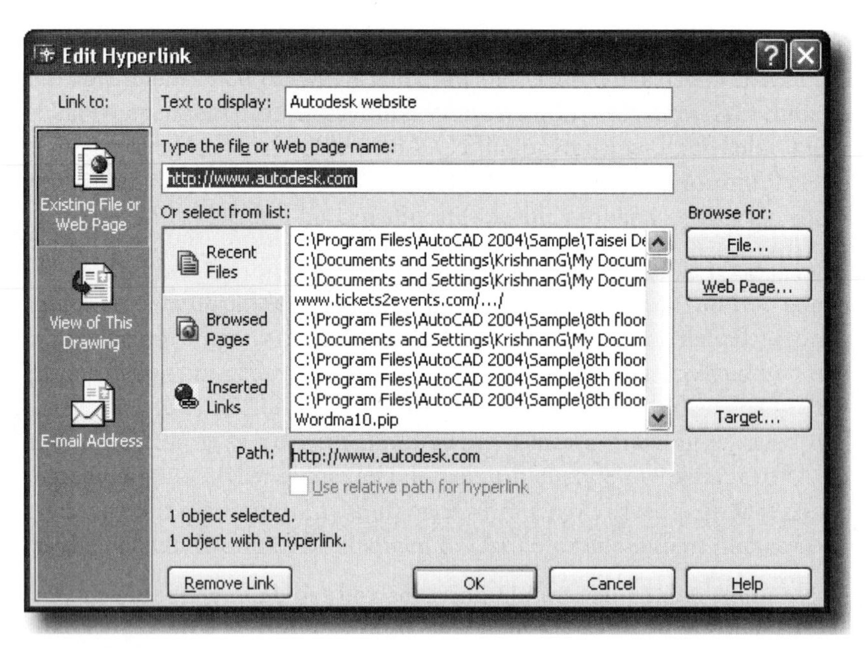

Figure 14–16 *Edit Hyperlink dialog box*

REMOVING THE HYPERLINK

To remove the hyperlink, invoke the HYPERLINK command:

Insert menu	Choose Hyperlink...
Command: prompt	**hyperlink** (ENTER)

AutoCAD prompts:

Command: **hyperlink**
Select objects: *(select one or more objects from which to remove the hyperlink and press* ENTER*)*

AutoCAD displays the Edit Hyperlink dialog box, as shown in Figure 14–17.

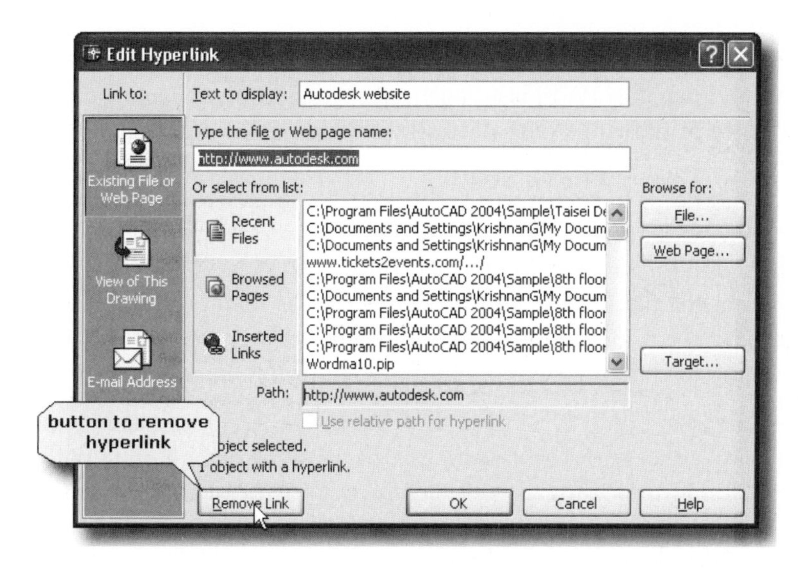

Figure 14–17 *Choosing the Remove Link button from the Edit Hyperlink dialog box*

Click the **Remove Link** button. AutoCAD removes the hyperlink from the selected objects and closes the Edit Hyperlink dialog box.

HYPERLINKS IN BLOCKS

AutoCAD allows you to attach hyperlinks to blocks, including nested objects contained within blocks. If the blocks contain any relative hyperlinks, the relative hyperlinks adopt the relative base path of the current drawing when you insert them.

Whenever you attach a hyperlink to a block reference, AutoCAD provides a cursor feedback when you position the cursor over the inserted block. To activate the hyperlink, first select the block reference (make sure the PICKFIRST system variable is set to 1). Right-click to display the shortcut menu and activate the hyperlink associated with the currently selected block element from the **Hyperlink** sub-menu.

Whenever you include objects that have hyperlinks in a block definition, you can activate the hyperlink from any of the block references. If you attach a hyperlink to a block reference, then you will have the choice of activating the block hyperlink or selected object hyperlink. To remove or edit hyperlinks of the objects within the block, you must explode the block reference, and then proceed with the removing or editing of hyperlinks. You can remove or edit the hyperlink that was attached to a block reference without exploding the block.

Note: When a hyperlink is attached to a block reference that already contains an object with a hyperlink, the cursor feedback for that block will only show the block hyperlink. The object hyperlink can still be accessed through the **Hyperlink** sub-menu as previously described.

DESIGN WEB FORMAT

Design Web Format™ (DWF) is an open, secure file format developed by Autodesk for the transfer of drawings over networks, including the Internet. DWF files are highly compressed, so they are much smaller (more than 50 percent smaller than DWG file) and faster to transmit, enabling the communication of rich design data, without the overhead associated with typical heavy CAD drawings. DWF files are not a replacement for native CAD formats such as DWGs and don't allow editing of the data within the file. The sole purpose of DWF is to allow designers, engineers, developers, and their colleagues to communicate design information and intent to anyone needing to view, review, or print design information.

The latest release of DWF is DWF 6. DWF 6 has been re-designed to enable users to build a complex set of design documents, pages, or layouts in one DWF file. This allows engineering professionals to build the equivalent of a drawing set within a single DWF file for distribution and communication to the many people in the various professions working on a project.

To view DWF files, you need Autodesk Express Viewer, which is a free, lightweight, high-performance application. Autodesk Express Viewer enables users to view DWF files using the advanced graphics systems found in many of the Autodesk design products. Autodesk Express Viewer has a very flexible printing system that enables users to print to scale, fit to page, tile, or selectively print a variety of layouts from the new DWF 6 publishing format. Autodesk Express Viewer is available as both a stand-alone application or as a control that is used within the Miscrosoft Internet Explorer browser, providing a simple, easy to use, and common Windows interface for even the novice user to master viewing and printing of design data. Check the www.autodesk.com Web site for the latest version of Autodesk Express Viewer. Always use the latest version of the application to view the DWF files.

USING ePLOT TO CREATE DWF FILES

AutoCAD's ePlot feature allows you to create DWF files that can be opened, viewed, and plotted with Autodesk Express Viewer, which is available as a free download from the www.autodesk.com Web site.

Express Viewer supports real-time panning and zooming as well as control over the display of layers, named views, and embedded hyperlinks. With ePlot, you can specify a variety of settings, such as pen assignments, rotation, and paper size, all of which control the appearance of plotted DWF files. With ePlot you can also create DWF files that have rendered images and multiple viewports displayed in **Layout** tab.

Note: By default, AutoCAD plots all objects with a lineweight of 0.06 inches, even if you haven't specified lineweight values in the Layer Properties Manager. If you want to plot without any lineweight, clear the **Plot Object Lineweights** option from the **Plot Settings** tab in the Plot dialog box.

AutoCAD provides a preconfigured plotter driver file called DWF6 ePlot to create DWF files. It generates electronic drawing files that are optimized for either printing or viewing. The files you create are stored in Design Web Format (DWF). DWF files can be opened, viewed, and plotted by anyone using Autodesk Express Viewer.

DWF files are created in a vector-based format (except for inserted raster image content) and are typically compressed. Compressed DWF files can be opened and transmitted much faster than AutoCAD drawing files. Their vector-based format ensures that precision is maintained.

DWF files are an ideal way to share AutoCAD drawing files with others who don't have AutoCAD. Because the Autodesk Express Viewer interface is easy to use, even those with no CAD knowledge can easily view and navigate a DWF file. Autodesk Express Viewer is installed with AutoCAD 2004, or you can get as a free download from www.autodesk.com. For more information about using Autodesk Express Viewer, see the Autodesk Express Viewer online Help.To create a .DWF file, invoke the plot command:

Standard toolbar	Choose the PLOT command
File menu	Choose Plot
Command: prompt	**plot** (ENTER)

AutoCAD displays the Plot dialog box similar to Figure 14–18.

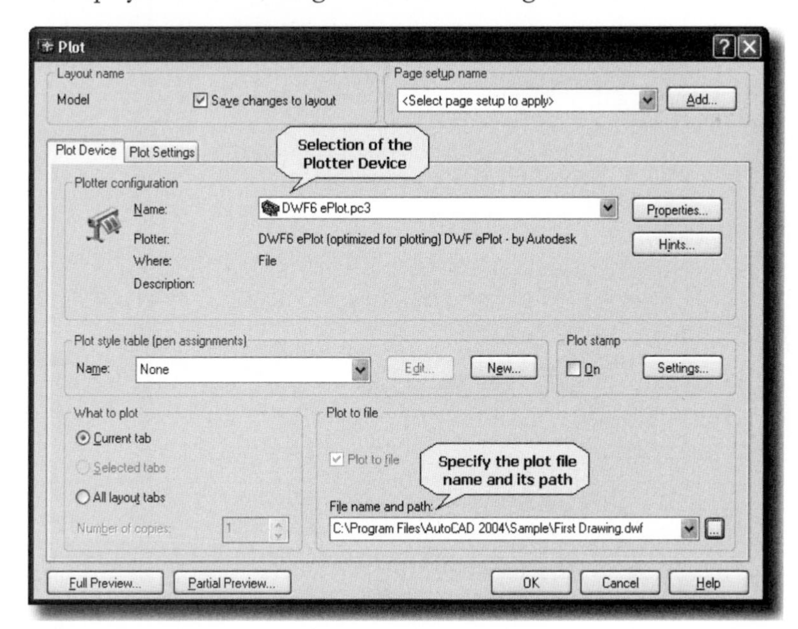

Figure 14–18 *Plot dialog box*

In the **Plot Device** tab (Figure 14–18), select DWF6 ePlot.pc3 from the **Name:** list box (Plotter configuration section) to create a DWF file. Specify a file name and location or network folder to plot the file in the **File name and path:** text field (**Plot to file** section). You can only plot DWF files to the Internet using the FTP protocol.

Make necessary changes to paper size and paper units, **Drawing orientation**, **Plot area**, **Plot scale**, **Plot offset**, **Plot options**, and choose the **OK** button to create the DWF file.

Setting Custom Properties to Create DWF files

AutoCAD allows you to fine-tune custom plotting properties such as resolution, file compression, background color, inclusion of paper boundary, and so on, to create DWF files.

To modify custom plotting properties, open the Plot dialog box and choose the **Plot Device** tab. Be sure to select an DWF6 ePlot plotting device and then choose the **Properties** button. AutoCAD displays the Plotter Configuration Editor dialog box, as shown in Figure 14–19.

Choose the **Device and Document Settings** tab, and then select **Custom Properties** from the tree window (see Figure 14–19). Choose the **Custom Properties...** button. AutoCAD displays the DWF6 ePlot Properties dialog box, as shown in Figure 14–20.

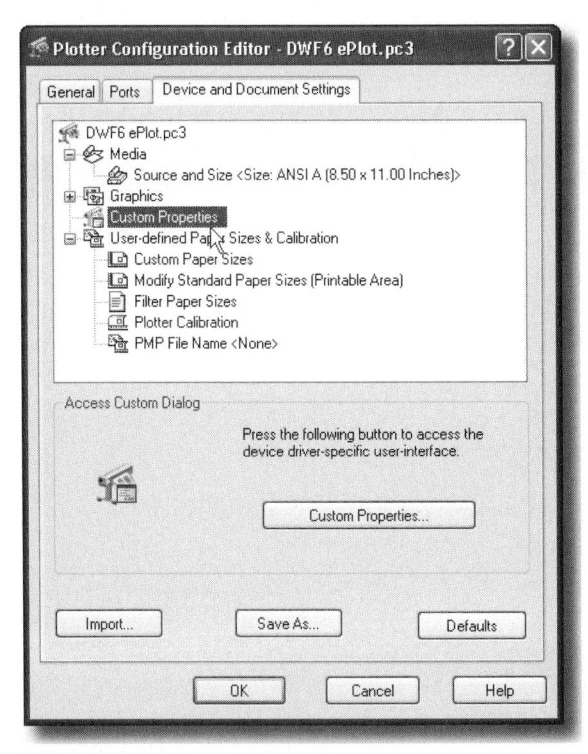

Figure 14–19 *Plotter Configuration Editor dialog box*

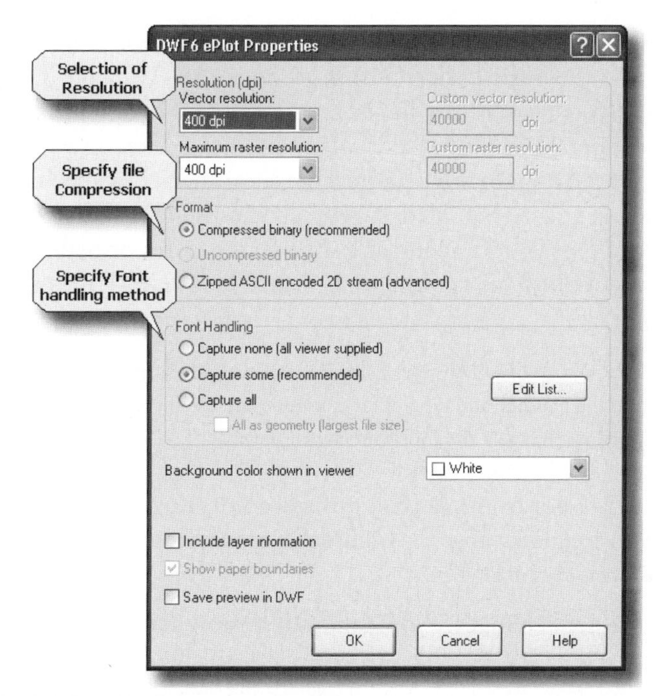

Figure 14–20 *DWF6 ePlot Properties dialog box*

In the **Resolution** section, select appropriate vector and raster resolution from the **Vector Resolution** pulldown menu and **Maximum raster resolution** pulldown menu respectively. The higher the resolution of the DWF file, the greater the precision and larger the file size. If there are large numbers of objects in the drawing, then it is recommended that you create a DWF file with high resolution. When you create DWF files intended for plotting, select a resolution to match the output of your plotter or printer.

Specify a file compression option from the **Format** section. By default, DWF files are created in a compressed binary format. If necessary, you can create DWF files as uncompressed binary format or ASCII format (uncompressed text) files. You do not lose any data by compressing the DWF files. It is the recommended format.

The **Font Handling** section specifies the inclusion and handling of fonts in DWF files. **Capture none (all viewer supplied)** selection specifies that no fonts will be included in the DWF file. In order for the fonts used in the source drawing for the DWF file to be visible in the DWF file, the fonts must be present on the DWF viewer's system. If the fonts used to create the DWF file are not present on the viewer's system, other fonts will be substituted. **Capture some (recommended)** selection specifies that fonts used in the source drawing for the DWF file that are selected in the Available True Type Fonts dialog box will be included in the DWF file. The selected fonts do not need to be available on the DWF viewer's system in order for them to appear in the DWF file. Choose the **Edit List...** button to open the Available True Type Fonts dialog box, where you can edit the list of fonts eligible for capture in the DWF file. **Capture all** selection specifies that all fonts used in the drawing will be included in the DWF file. The **All as geometry (largest file size)** check box specifies that all fonts used in the drawing will be included as geometry in the DWF file. If you select this option, you should plot your drawing at a scale factor of 1:1 or better to ensure good quality in the output file. This option is only available for DWF files created with the DWF6 ePlot model.

Note: The size of a DWF file can be affected by the font-handling settings, the amount of text, and the number and type of fonts used in the file. If the size of your DWF file seems too large, try changing the font-handling settings.

Specify background color from the **Background color shown in viewer** option menu. In addition, specify toggle settings for **Include layer information**, **Show paper boundaries**, and **Save preview in DWF.**

After making necessary changes, choose the **OK** button to close the DWF Properties dialog box. Choose the **OK** button to save the settings and close the Plotter Configuration Editor dialog box. AutoCAD displays the changes to the Changes to a Printer Configuration File dialog box, as shown in Figure 14–21

Figure 14–21 *Changes to a Printer Configuration File dialog box*

There are two available options. If you want to apply the changes in the settings to the current plot, then select the **Apply changes to the current plot only** radio button. If you want to save the settings and apply them to all plots, select the **Save changes to the following file** radio button and specify the name of the file in the edit field. Choose the **OK** button to close the Changes to a Printer Configuration File dialog box.

Choose the **OK** button to create the DWF file and close the Plot dialog box.

VIEWING DWF FILES

AutoCAD itself cannot display DWF files, nor can DWF files be converted back to DWG format without using file translation software from a third-party vendor. In order to view a DWF file, you need Autodesk Express Viewer, Volo View Express™ or Volo™ View.

Figure 14–22 shows an example of DWF displayed in Autodesk Express Viewer.

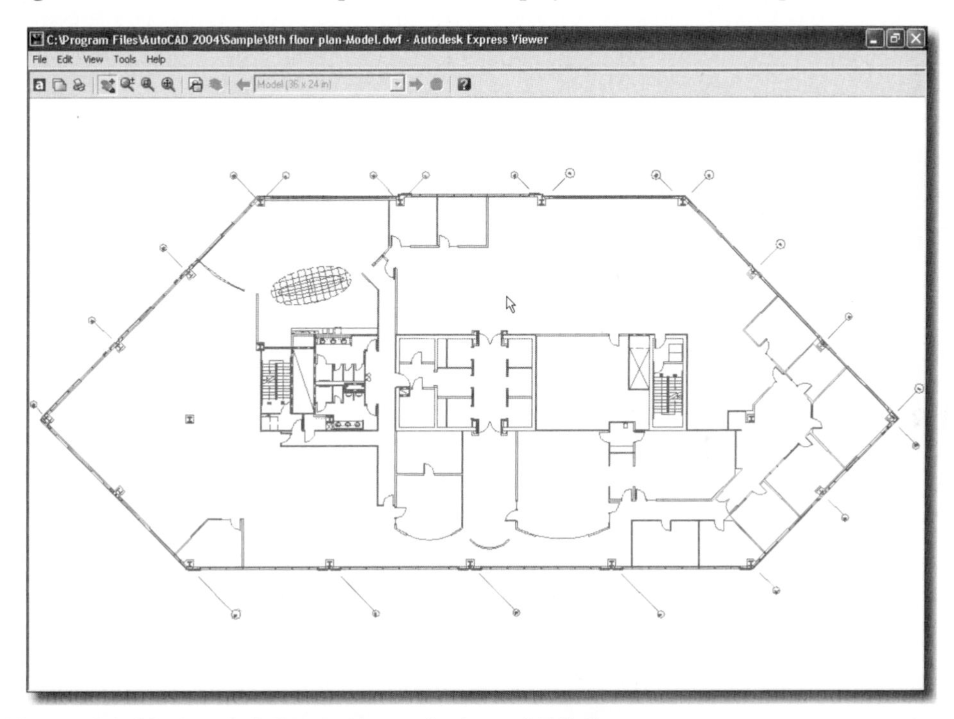

Figure 14–22 *Autodesk Express Viewer displaying DWF file*

Volo View Express also lets you view and print the basic design drawing file formats, such as DWF, DWG, etc., independently of AutoCAD® software. You don't need to install or know how to use AutoCAD in order to access, mark up, or accurately print computer-aided designs. It enables you to quickly load and open drawings, and then pan and zoom between different drawing layers. Volo View Express has expanded editing capabilities that save time during the revision process. Along with real-time pan and zoom, there's a zoom extents and window and also includes a dynamic 3D Orbit rotation feature similar to that of AutoCAD. You can also shade 3D models, display AutoCAD lineweights, and mark up redline designs.

Volo View has all the capabilities of Volo View Express and, in addition, you can print to scale and Full Active Shapes markup. Autodesk Express Viewer and Volo View Express are available as free downloads from the www.autodesk.com Web site, and Volo View is available for a nominal price.

PUBLISHING AUTOCAD DRAWINGS TO THE WEB

You use the Publish to Web Wizard to easily and seamlessly publish AutoCAD drawings to the Web. Drawings are published in HTML format using three predefined image types: DWF, JPEG and PNG. The DWF image type translates and publishes specified layouts into DWF, which are easily viewed with either Autodesk Express Viewer or Volo View Express. With the JPEG and PNG image types, you specify a drawing perspective, and AutoCAD translates the specified layout into a JPEG or PNG raster image. Anyone with a standard browser can view JPEG or PNG content. To move your content to the Web or a company's intranet, just specify the server location and configuration, and the content uploads automatically. Once it's posted, updating it is simple and fast.

To publish AutoCAD drawings, invoke the PUBLISHTOWEB command:

File menu	Choose Publish to Web
Command: prompt	**publishtoweb** (ENTER)

AutoCAD displays the introductory text of the Publish to Web - Begin page as shown in Figure 14–23. Choose one of the following two radio buttons:

- **Create New Web Page** – Allows you to create a new Web page.
- **Edit Existing Web Page** – Allows you to edit an existing page by adding or removing Web pages.

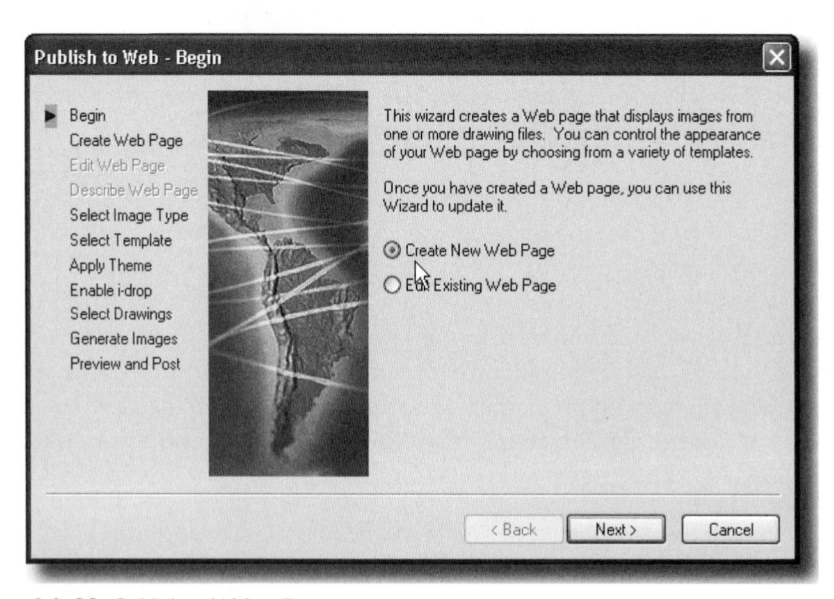

Figure 14–23 *Publish to Web – Begin*

Choose the **Next >** button, and AutoCAD displays the Publish to Web – Create Web Page page, shown in Figure 14–24. Specify the name of your Web page in the first text field and specify the parent directory in your file system where the Web page folder will be created. If necessary, provide a description to appear on your Web page in the last text field in the wizard.

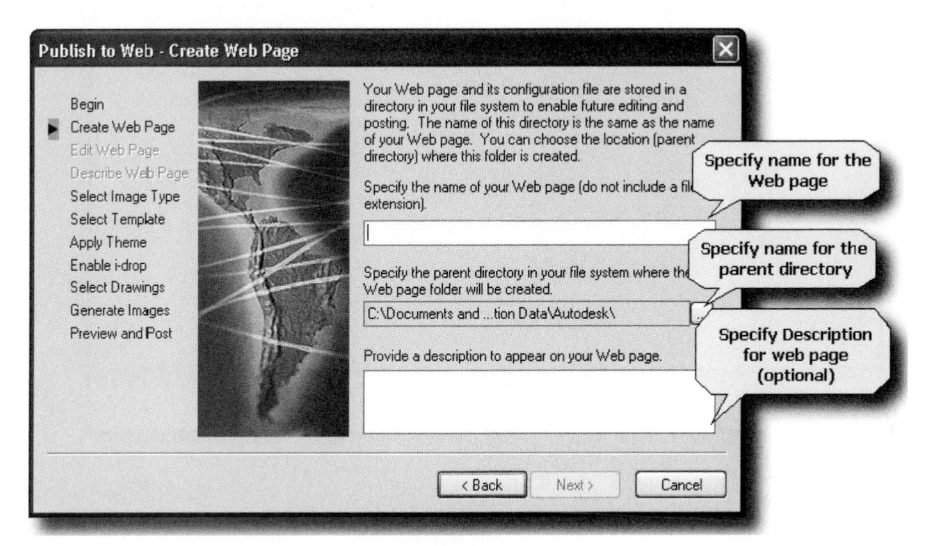

Figure 14–24 *Publish to Web – Create Web Page*

Choose the **Next >** button, and AutoCAD displays the Publish to Web – Select Image Type page shown in Figure 14–25. Choose one of the following three image types:

- DWF Image type – The DWF template translates and publishes specified DWG files in DWF; they can then be viewed with Autodesk Express Viewer, Volo, or Volo View Express. DWF files are inserted into your completed Web page in a size that is optimized to display well with most browser settings.
- JPEG Image type – JPEG files are raster-based representations of AutoCAD drawing files. JPEGs are one of the most common formats used on the Web. Due to the compression mechanism used, this format is not suitable for large drawings or drawings that have a lot of text.
- PNG Image type – PNG (Portable Network Graphics) are raster-based representation of drawing files. Most browsers now support the PNG image type, making more suitable than JPEG for creating images of AutoCAD drawings.

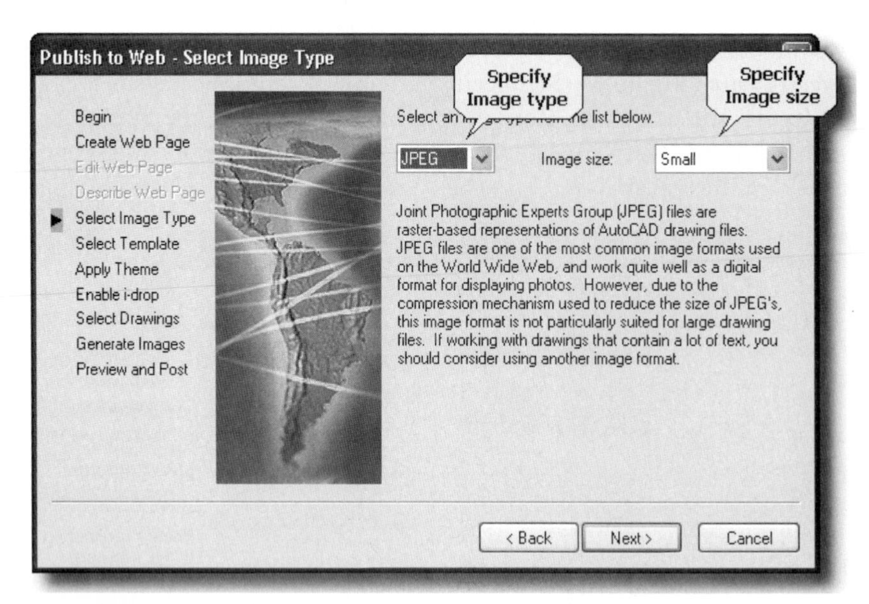

Figure 14–25 *Publish to Web –Select Image Type*

Choose the **Next** > button, AutoCAD displays the Publish to Web – Select Template page shown in Figure 14–26. Select one of the available templates. The **Preview** pane demonstrates how the selected template will affect the layout of drawing images in your Web page.

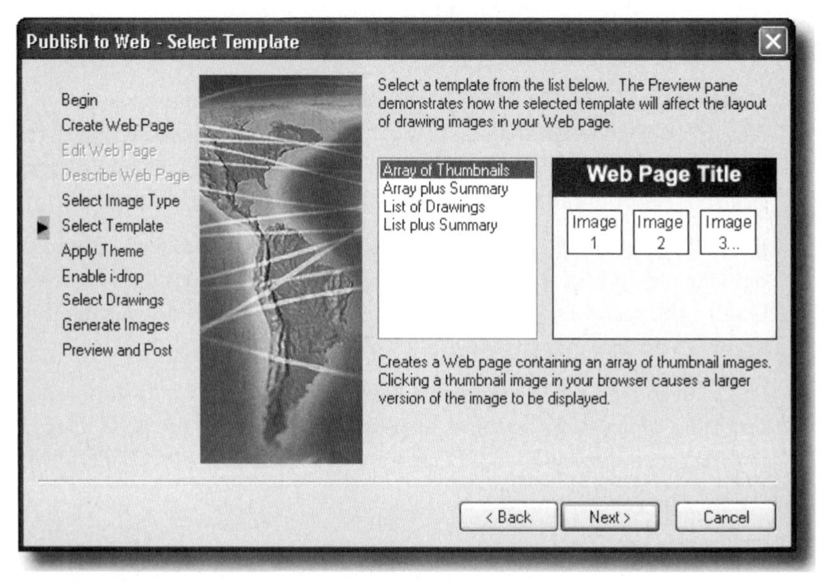

Figure 14–26 *Publish to Web – Select Template*

Choose the **Next >** button, and AutoCAD displays the Publish to Web – Apply Theme page, shown in Figure 14–27. Choose one of the available themes. Themes are preset elements (fonts and colors) that control the appearance of various elements of your completed Web page. The **Preview** pane demonstrates how the selected theme will display the layout of your Web page. The available themes are: Autumn Fields, Classic, Cloudy sky, Dusky Maze, Ocean Waves, Rainy Day, and Supper Club.

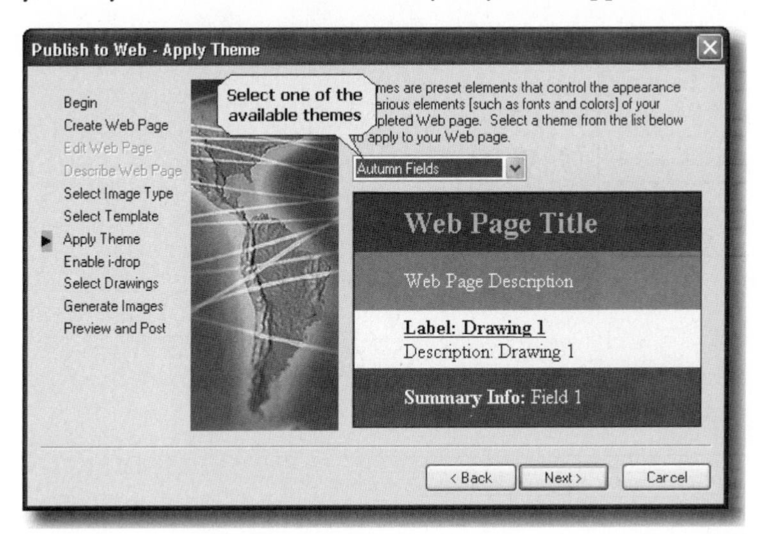

Figure 14–27 *Publish to Web – Apply the Theme*

Choose the **Next >** button, and AutoCAD displays the Publish to Web – Enable i-drop page, shown in Figure 14–28.

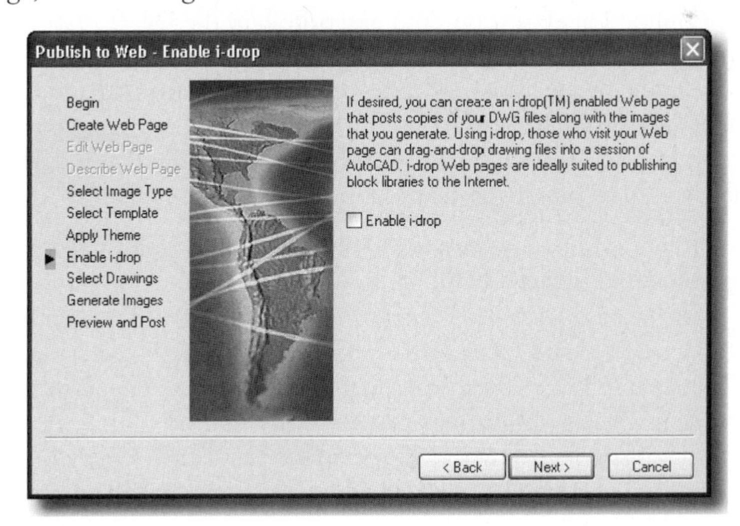

Figure 14–28 *Publish to Web – Enable i-drop*

If desired, you can create an i-drop enabled Web page that posts copies of your DWG files along with the images. Using i-drop, visitors to your Web site can drag and drop drawing files into a session of AutoCAD.

Choose the **Next** > button, and AutoCAD displays the Publish to Web – Select Drawings page, shown in Figure 14–29.

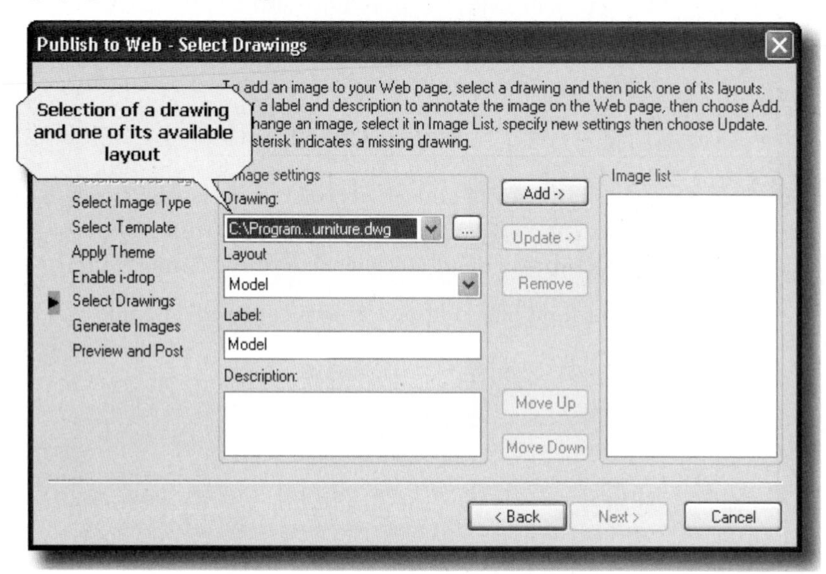

Figure 14–29 *Publish to Web – Select Drawings*

Select a drawing and then choose one of its layouts from the **Layout** pull-down menu. Specify a label in the **Label:** text box and description in the **Description:** text box to annotate the selected image on the Web page. Choose the **Add** button to add the selected image to the Image list box. If necessary, you can change the properties of the selected image and choose the **Update** button to apply the changes or you can remove it from the selection by choosing the **Remove** button. The **Move Up** and **Move Down** buttons allow you to rearrange the selected images.

Choose the **Next** > button, and AutoCAD displays the Publish to Web – Generate Images page, shown in Figure 14–30.

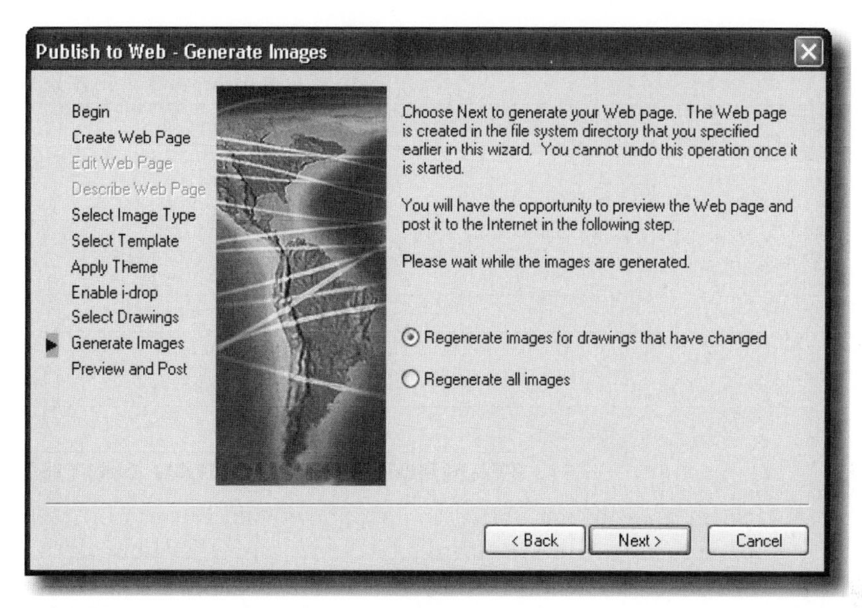

Figure 14–30 *Publish to Web – Generate Images*

Choose one of the following two radio buttons:

- **Regenerate images for drawings that have changed** – Generates the images for all the selected drawings that have changed.
- **Regenerate all images** – Generates the images for all the selected drawings.

AutoCAD creates the Web pages and stores them in the file directory that you specified earlier in the wizard. You cannot undo this operation once it is started.

Choose the **Next** > button, and AutoCAD displays the Publish to Web –Preview and Post page, shown in Figure 14–31. To preview the Web pages, choose the **Preview** button. AutoCAD opens the default browser (Internet Explorer or Netscape Navigator) and displays the Web pages with appropriate links. To post the Web pages, close the browser, and choose the **Post Now** button. AutoCAD displays the Posting Web File Handling dialog box. Choose the URL where you want to post it and choose the **Save** button. If necessary, choose the **Send Email** button to create and send an email message that includes a hyperlinked URL to its location.

Choose the **Finish** button to close the Publish to Web Wizard.

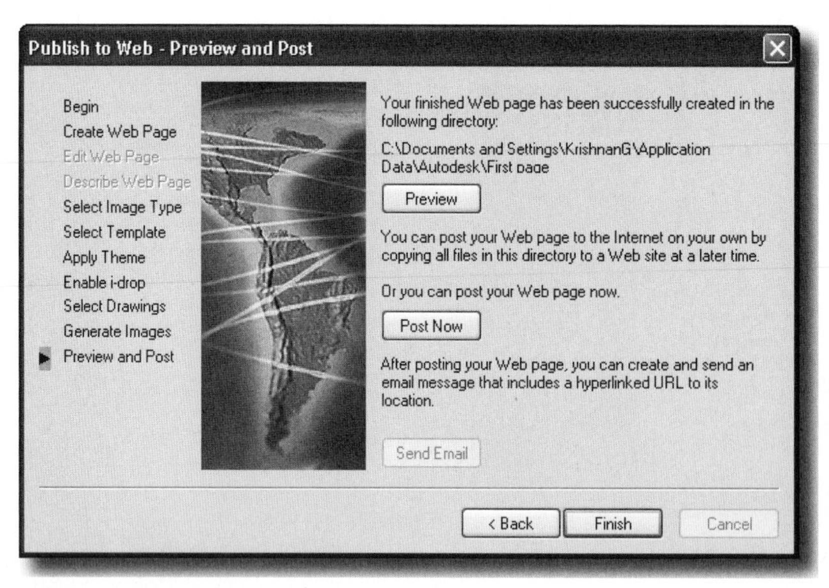

Figure 14–31 *Publish to Web – Preview and Post*

DESIGN PUBLISHER

Design Publisher allows you to assemble a collection of drawings and plot directly to paper or publish to a DWF (Design Web Format) file. AutoCAD allows you to publish your drawing sets as either a single multi-sheet DWF format file or multiple single-sheet DWF format files, or to plot to designated plotter in the page setup. You can publish to devices (plotters or files) specified in the page setups for each layout. With Design Publisher, you have the flexibility to create electronic or paper drawing sets for distribution. The recipients can then view or plot your drawing sets.

You can customize your drawing set for a specific user, and you can add and remove sheets in a drawing set as a project evolves. Design Publisher allows you to publish directly to paper or to an intermediate electronic format that can be distributed using email, FTP sites, project Web sites, or CD. You can open DWF files with Autodesk Express Viewer, Volo View Express or Volo View.

To create multi-sheet drawing sets for publishing to a single multi-sheet DWF file, a plotting device, or a plot file, invoke the PUBLISH command:

Standard toolbar	Choose the PUBLISH command (see Figure 14–32)
File menu	Choose Publish…
Command: prompt	**publish** (ENTER)

Figure 14–32 *Invoking the* PUBLISH *command from the Standard toolbar*

AutoCAD displays the Publish Drawing Sheets dialog box similar to Figure 14–33.

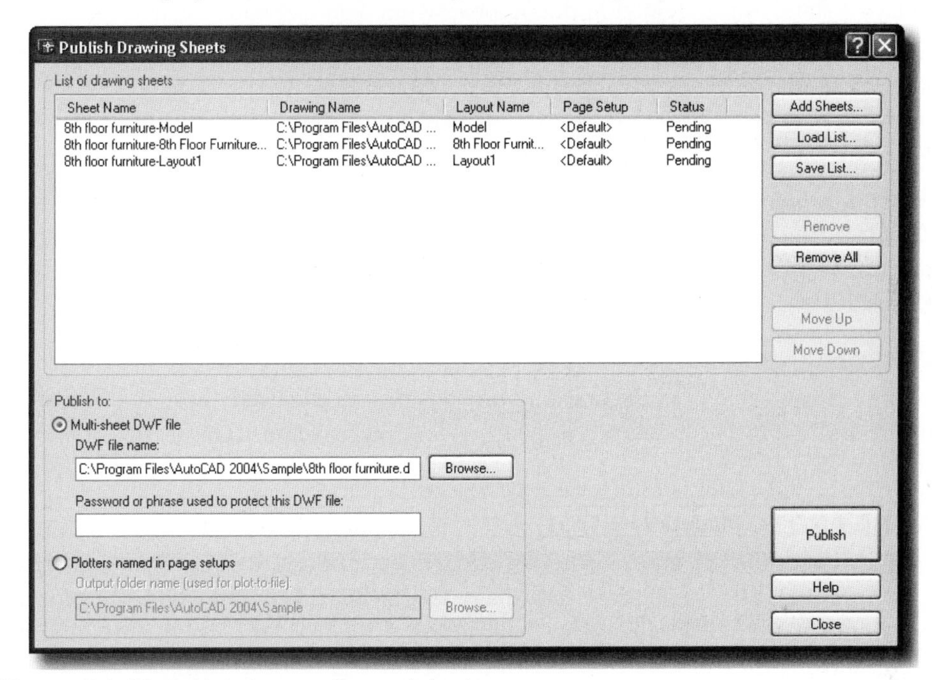

Figure 14–33 *Publish Drawing Sheets dialog box*

The **List of drawing sheets** section lists the drawing sheets to be included for publishing.

The **Sheet Name** column displays the combined name of drawing and the layout name separated with a dash (-). If necessary, you can rename from the RENAME option available from the shortcut menu. Drawing sheet names must be unique within a single DWF file.

The **Drawing Name** column displays the name of the selected drawing file. Displays the full path name only if the DISPLAY DRAWING FILE PATH NAMES setting on the shortcut menu is set to on.

The **Layout Name** column displays layout name of the selected drawing. Model tab is included only if the INCLUDE MODEL WHEN ADDING SHEETS is set to ON.

The **Page Setup** column displays the named page setup for the selected layout. You can change the page setup by clicking CHANGE PAGE SETUP on the shortcut menu. Only Model tab page setups can be applied to Model tab sheets, and only paper space page setups can be applied to paper space layouts.

Status column displays a message as each drawing sheet is published.

ADDING AND REMOVING SHEETS

To add sheets to the existing selection, choose the **Add Sheets...** button, AutoCAD displays a standard file selection dialog box, where you can add sheets to the list of drawing sheets. The layout names from those files are extracted, and one sheet is added to the list of drawing sheets for each layout. New drawing sheets are always appended to the end of the current list. You can drag selections into the list of drawing sheets.

To remove sheets, choose the **Remove** button, and AutoCAD deletes the currently selected drawing sheet from the list of sheets. To remove all sheets, choose the **Remove All** button, and AutoCAD deletes all the sheets from the list. If the current list of drawing sheets has not been saved, AutoCAD allows you to first save the new list before all drawing sheets are removed.

REARRANGING THE SHEETS

To move the selected drawing sheet up one position in the list, choose the **Move Up** button. and to move the selected drawing sheet down one position in the list choose the **Move Down** button.

SAVING AND LOADING THE LIST

To save the current selection of the drawing sheets, choose the **Save List...** button, and AutoCAD displays the Save List dialog box, where you can save the current list of drawings as a DSD (Drawing Set Descriptions) file. DSD files are used to describe lists of drawing files and selected lists of layouts within those drawing files.

To load a saved list to the current selection, choose the **Load List...** button, and AutoCAD displays a standard file selection dialog box. You can select a DSD file or a BP3 (Batch Plot) file to load. AutoCAD displays the Replace or Append dialog box if a list of drawing sheets is present in the Publish Drawing Sheets dialog box. You can either replace the existing list of drawing sheets with the new sheets or append the new sheets to the current list.

SELECTION OF THE MODE TO PUBLISH

AutoCAD provides two modes to publish: a single multi-sheet DWF format file or plotting individual layouts to a designed plotter in page setup.

Choose the **Multi-sheet DWF file** radio button to generate a single multi-sheet DWF file. Specify the name of the single multi-sheet DWF file in the **DWF File name** text box and/or choose the **Browse...** button to open the Select DWF file dialog box, where you can select a target folder for your DWF file. If necessary, specify the password in the **Password or phrase used to protect to this DWF file:** text box to protect the DWF file being created. DWF passwords are case sensitive. The password or phrase can be made up of letters, numbers, punctuation, or non-ASCII characters.

Note: If you lose or forget the password, it cannot be recovered. Keep a list of passwords and their corresponding DWF file names in a safe place.

When you choose the **Publish** button to publish, AutoCAD prompts to confirm the password that you entered in the Publish Drawing Sheets dialog box. If the two passwords do not match, you must click **Publish** again to re-enter the correct password.

Choose the **Plotters named in page setups** radio button to plot to output devices given for each drawing sheet in the page setups. Specify the name of the folder in the **Output folder name (used for plot-to-file)** text box to save the single-sheet plot files and/or choose the **Browse…** button to open a Select Folder dialog box, where you can select a target folder for the generated DWF file.

PUBLISH

Choose the **Publish** button to begin the publishing operation, which creates a single-sheet DWF file or a single multi-sheet DWF file, or plots to a device or file depending on the radio button selected in the **Publish to** area.

If a drawing sheet fails to plot, Design Publisher continues plotting the remaining sheets in the drawing set. A log file is created that contains detailed information, including any errors or warnings encountered during the publishing process.

You can stop publishing after a sheet has finished plotting. If you stop publishing a multi-sheet DWF file before it is complete, no output file is generated. After publishing is complete, the Status field is updated to show the results.

Choose the **Close** button to close Publish Drawing Sheets dialog box.

eTRANSMIT UTILITY

The eTransmit utility allows you to select and bundle together the drawing file and its related files. You can create a transmittal set of files as a compressed self-extracting executable file, as a compressed zip file or as a set of uncompressed files in a new or existing folder. You can include all the reference files attached to the drawing file, word files, spreadsheet, etc. to be part of the bundle. It is easier to transmit by e-mail one single compressed file consisting of a drawing file and several related files.

To create a transmittal set of a drawing and related files, invoke the ETRANSMIT command:

File menu	Choose eTransmit
Command: prompt	**etransmit** (ENTER)

AutoCAD displays the Create Transmittal dialog box, shown in Figure 14–34.

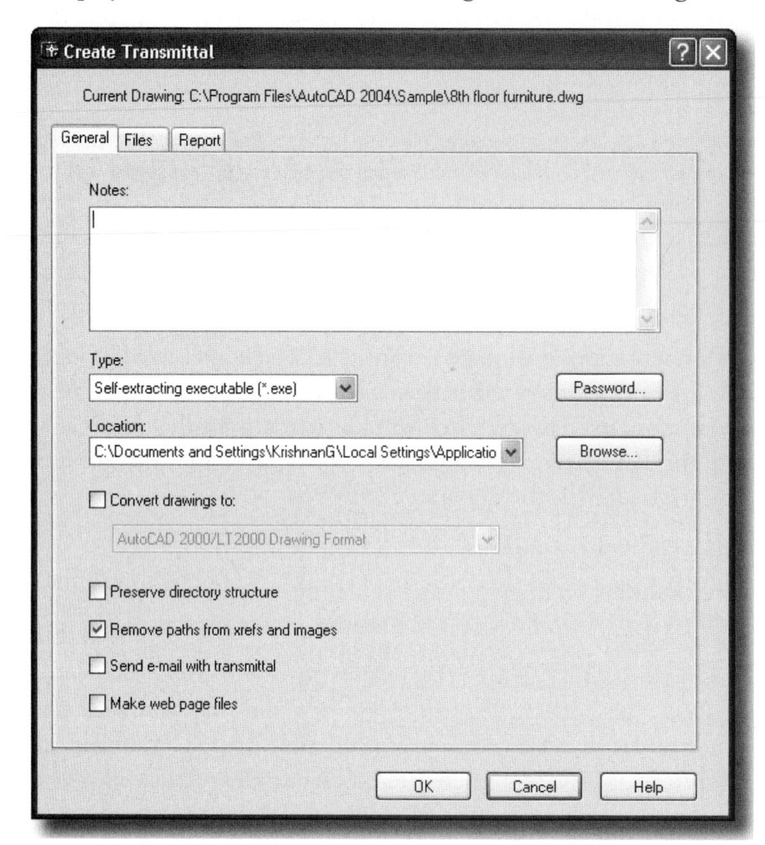

Figure 14–34 *Create Transmittal dialog box*

GENERAL SETTINGS

On the **General** tab, you can select the type, location and related settings of transmittal sets. Enter transmittal notes in the **Notes:** text box. The notes are included in the transmittal report. You can specify a template of default notes to be included with all your transmittal sets by creating an ASCII text file called *etransmit.txt.* This file must be saved to a location specified by the **Support File Search Path** option on the **Files** tab of the Options dialog box.

Select one of the three available transmittal type from the **Type:** list box as shown in Figure 14–35. The **Folder (set of files)** selection creates a transmittal set of uncompressed files in a new or existing folder. The **Self-extracting executable (*.exe)** selection creates a transmittal set of files as a compressed, self-extracting executable file. Double-clicking the resulting EXE file decompresses the transmittal set and restores the files. The **Zip (*.zip)** selection creates a transmittal set of files as a compressed zip file. To restore the files, you need a decompression utility such as the shareware applications, PKZIP or WinZip. Choose the **Password...** button to specify a password for your transmittal set. AutoCAD displays the eTransmit – Set Password dialog box, shown in Figure 14–36.

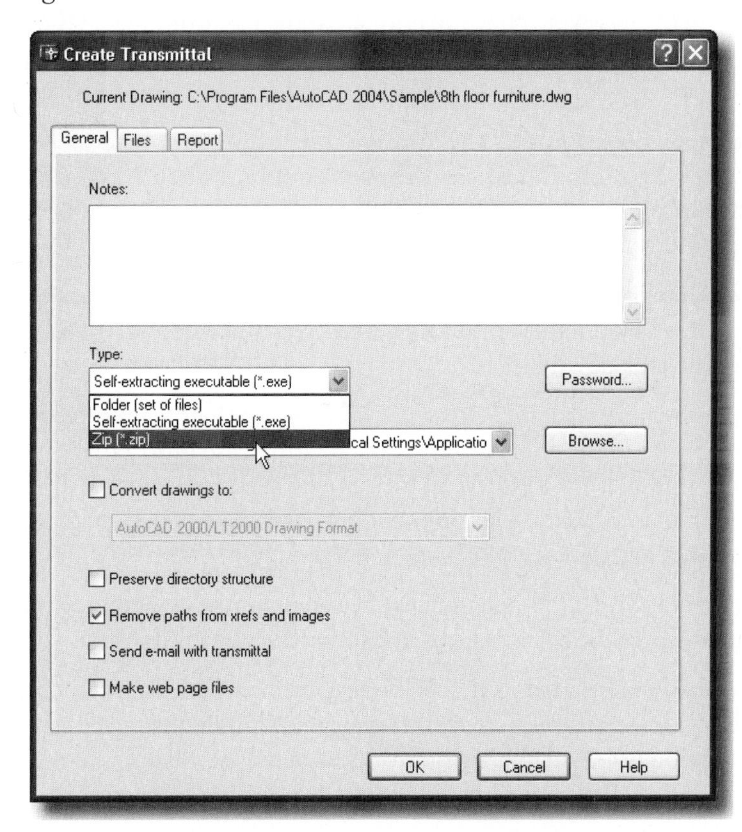

Figure 14–35 *Create Transmittal dialog box with General Tab selected*

Figure 14–36 *e-Transmit – Set Password dialog box*

Specify an optional password in the **Password for compressed transmittal:** text field for the transmittal set. When others attempt to open the transmittal set, they will need to provide this password to access the files. Password protection cannot be applied to folder transmittal sets. Retype the password in the **Password confirmation:** text field to confirm the password that you entered in the password field. Choose the **OK** button to close the dialog box. If the two passwords do not match, you are prompted to re-enter them.

The **Location:** text box specifies the location in which the transmittal set is created. To specify a new location, choose the **Browse...** button. AutoCAD opens a standard file selection dialog box, in which you navigate to the location where you want to create the transmittal set.

The **Convert drawings to:** check box controls the file format of all drawings included in a transmittal set. When this option is selected, you can select an AutoCAD drawing format from the pull-down list.

The **Preserve directory structure** check box controls whether to preserve the directory structure of all files in the transmittal set, facilitating ease of installation on another system. If this option is set to OFF, all files are installed to the target directory when the transmittal set is installed. This option is not available if you're saving a transmittal set to an Internet location.

The **Remove paths from xrefs and images** check box controls whether to remove paths from any cross-referenced drawings or images in the transmittal set.

The **Send e-mail with transmittal** check box controls whether to launch the default email application when the transmittal set is created. This makes it convenient to send an email notifying others of the new transmittal set.

The **Make web page files** check box controls whether to generate a Web page that includes a link to the transmittal set.

FILES SELECTION

On the **Files** tab (see Figure 14–37), AutoCAD lists the files to be included in the transmittal set. By default, all files associated with the current drawing (such as related xrefs, plot styles, and fonts) are listed. You can add additional files to the transmittal set or remove existing files. Related files that are referenced by URLs are not included in the transmittal set. The two buttons in the upper left allow you to toggle between list view and tree view. All files to be included in the transmittal set will have a check mark next to the file name. To remove a file from the transmittal set, click in the check box. Right-click in the file list to display a shortcut menu, from which you can clear all check marks or apply check marks to all files. To add files to the selection list, choose the **Add file...** button. AutoCAD opens a standard file selection dialog box, in which you select additional files to include in the transmittal set. The **Include fonts** check box controls whether to include the current drawing's associated font files (*.txt* and *.shx*) with the transmittal set. True Type fonts are not included in the transmittal set, as they are proprietary to the local computer on which they are installed. If any required True Type fonts are not present on the computer to which the transmittal set is copied, the font specified by the FONTALT system variable is substituted.

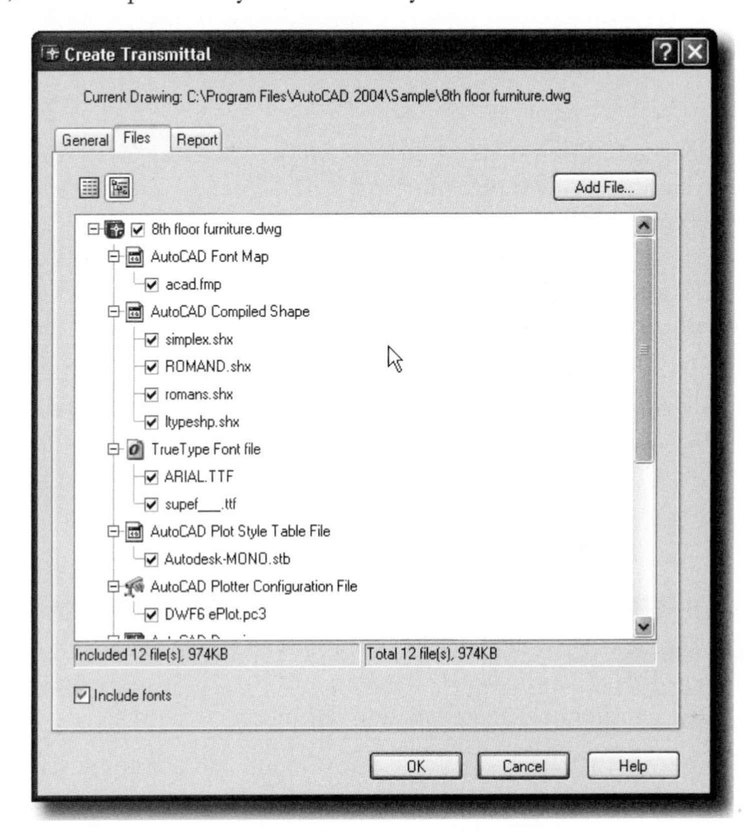

Figure 14–37 *Create Transmittal dialog box with Files tab selected*

SAVING THE TRANSMITTAL SET REPORT

On the **Report** tab, AutoCAD displays report information that is included with the transmittal set (see Figure 14–38). It includes any transmittal notes you entered on the **General** tab, as well as,distribution notes automatically generated by AutoCAD that detail what steps must be taken for the transmittal set to work properly. If you have created a text file of default notes, the notes are also included in the report.

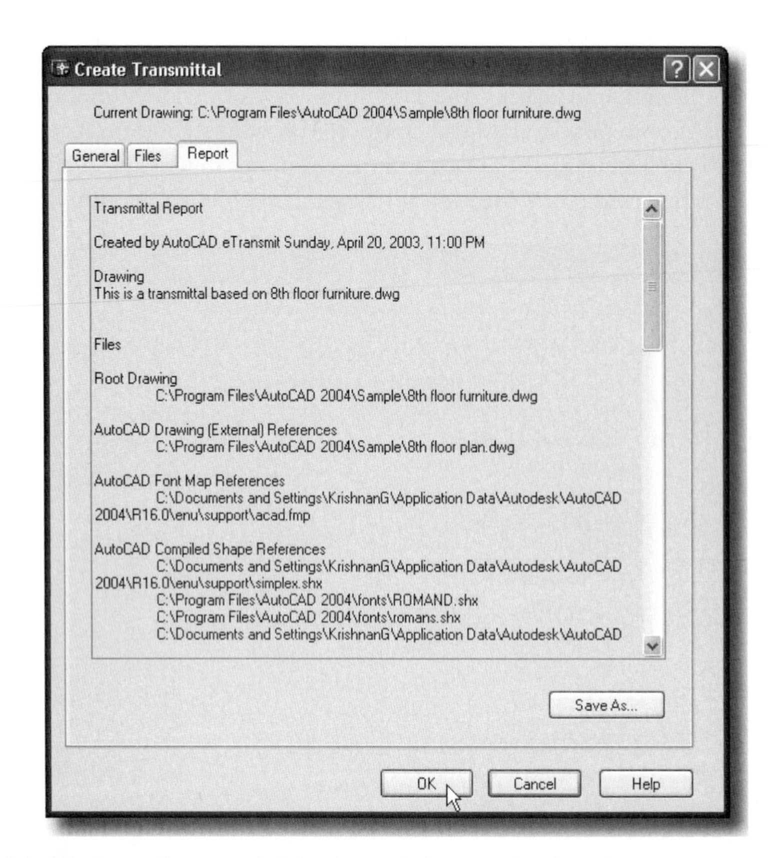

Figure 14–38 *Create Transmittal dialog box with Report tab selected*

Choose the **Save As...** button to open a file save dialog box. Here, you specify the location in which to save the report file. Note that a report file is automatically included with all transmittal sets that you generate; by choosing **Save As,** you save an additional copy of the report file, for archival purposes.

Choose the **OK** button to create the transmittal set and close the Create Transmittal dialog box.

REVIEW QUESTIONS

1. The command to invoke the Internet browser in AutoCAD is:

   a. INTERNET
   b. BROWSER
   c. HTTP
   d. WWW
   e. none of the above

2. DWF is short for:

   a. DraWing Format.
   b. Design Web Format.
   c. DXF Web Format.

3. The purpose of DWF files is to view:

   a. *2D* drawings on the Internet.
   b. *3D* drawings on the Internet.
   c. *3D* drawings in another CAD system.
   d. all of the above.

4. URL is short for:

   a. Union Region Lengthen.
   b. Earl.
   c. Useful Resource Line.
   d. Uniform Resource Locator.

5. Which of the following URLs are valid:

   a. www.autodesk.com
   b. http://www.autodesk.com
   c. all of the above.
   d. none of the above.

6. What is the purpose of a URL?

   a. to access files on computers, networks, and the Internet.
   b. it is a universal file-naming system for the Internet.
   c. to create a link to another file.
   d. all of the above.

7. FTP is short for:

   a. Forwarding Transfer Protocol

   b. File Transfer Protocol

   c. File Transference Protocol

   d. File Transfer Partition

8. URLs are used in an AutoCAD drawing to browse the Internet.

   a. True   b. False

9. The purpose of URLs is to let you create ____________ between files.

   a. backups

   b. links

   c. copies

   d. partitions

10. You can attach a URL to rays and xlines.

    a. True   b. False

11. Compression in the DWF file causes it to take ________ time to transmit over the Internet.

    a. more

    b. less

    c. all of the above

    d. none of the above

12. A "plug-in" lets a Web browser:

    a. plug into the Internet

    b. display a file format

    c. log in to the Internet

    d. display a URL

13. A Web browser can view DWG drawing files over the Internet.

    a. True   b. False

14. Which AutoCAD command is used to start a Web browser?

    a. WEBSTART

    b. WEBDWG

    c. BROWSERS

    d. LAUNCH

15. Before a URL can be accessed you must always type the prefix "http://".

    a. True   b. False

16. Which AutoCAD command is used to open an AutoCAD drawing file from an Internet site?

    a. LAUNCH

    b. OPEN

    c. START

    d. LAUNCHFILE

17. Once an Internet AutoCAD drawing file has been modified it can only be saved back to the Internet site.

    a. True   b. False

18. AutoCAD creates hyperlinks as being _____________.

    a. Absolute

    b. Relative

    c. Polar

    d. Only a or b

19. Hyperlinks can only be attached to text objects within a drawing.

    a. True   b. False

20. The system variable PICKFIRST must be set to a value of 1 before a hyperlink can be attached.

    a. True   b. False

21. DWF files are compressed to as much as _______ of the original .DWG file size.

    a. 1/8

    b. 1/4

    c. 1/2

    d. equal to

AutoCAD 3D

INTRODUCTION

After completing this chapter, you will be able to do the following:

- Define a User Coordinate System
- View in *3D* using the VPOINT, DVIEW, and 3DORBIT commands
- Create *3D* objects
- Use the REGION command
- Use the 3DPOLY and 3DFACE commands
- Create meshes
- Edit in *3D* using the ALIGN, ROTATE3D, MIRROR3D, 3DARRAY, EXTEND, and TRIM commands
- Create solid shapes: solid box, solid cone, solid cylinder, solid sphere, solid torus, and solid wedge
- Create solids from existing *2D* objects and regions
- Create solids by means of revolution
- Create composite solids
- Edit *3D* solids using the CHAMFER, FILLET, SECTION, SLICE, and INTERFERE commands, in addition to Face editing, body editing, and edge editing
- Obtain the mass properties of a solid
- Place a multiview in paper space
- Generate Views in viewports
- Generate profiles

WHAT IS *3D*?

In *two-dimensional* drawings you have been working in a single plane with two axes, *X* and *Y*. In *three-dimensional* drawings you work with the *Z* axis, in addition to the *X* and *Y* axes, as shown in Figure 15–1. Plan views, sections, and elevations represent only two dimensions. Isometric, perspective, and axonometric drawings, on the other hand, represent all three dimensions. For example, to create three views of a cube, the cube is simply drawn as a square with thickness and then viewed along each of the three axes. Drawing in this manner is referred to as extruded *2D*. Only objects that are extrudable can be drawn by this method. Other views are achieved by rotating the viewpoint or the object, just as if you were physically holding the cube. You can get an isometric or perspective view by simply changing the viewpoint.

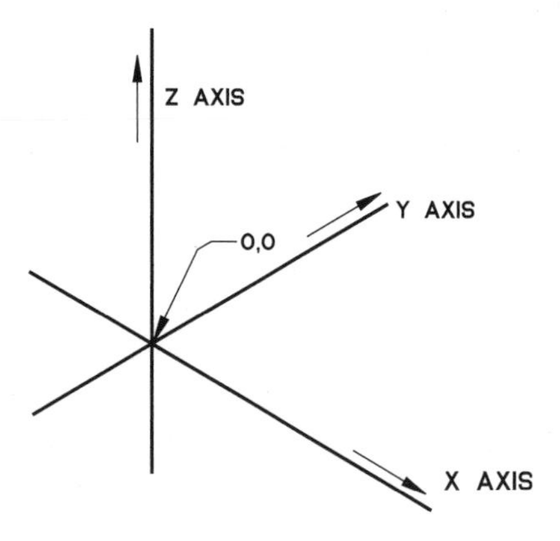

Figure 15–1 *The X, Y, and Z axes for a 3D drawing*

Whether you realize it or not, all the drawings you have done in previous chapters were created by AutoCAD in true *3D*. What this means is that every line, circle, or arc that you have drawn, even if you think you have drawn it in *2D*, is really stored with three coordinates. By default, AutoCAD stores the *Z* value as your current elevation with a thickness of zero. What you think of now as *2D* is really only one of an infinite number of views of your drawing in *3D* space.

Drawing objects in *3D* provides three major advantages:

- An object is drawn once and then can be viewed and plotted from any angle (viewpoint).
- A *3D* object holds mathematical information that can be used in engineering analysis such as finite element analysis and computer numerical control (CNC) machinery.
- Shading and rendering enhances the visualization of an object.

There are two major limitations to working in *3D*:

1. Whenever you want to input *3D* coordinates whose *Z* coordinate is different from the current construction plane's elevation, you have to use the keyboard instead of your pointing device. One exception is to Object Snap to an object not in the current construction plane. The input device (mouse or digitizer) can supply AutoCAD with only two of the three coordinates at a time. Three-

dimensional-input devices exist, but there is no practical support by AutoCAD for them at this time. So you are limited to using the keyboard.

2. Determining where you are in relationship to an object in *3D* space is difficult.

COORDINATE SYSTEMS

AutoCAD provides two types of coordinate systems. One is a single fixed coordinate system called the World Coordinate System, and the other is an infinite set of user-defined coordinate systems available through the User Coordinate System.

The World Coordinate System (WCS) is fixed and cannot be changed. In this system (when viewing the origin from 0,0,1), the *X* axis starts at the point 0,0,0, and values increase as the point moves to the operator's right; the *Y* axis starts at 0,0,0, and values increase as the point moves to the top of the screen; and finally, the *Z* axis starts at the 0,0,0 point, and values get larger as it comes toward the user. All drawings from previous chapters are created with reference to the WCS. The WCS is still the basic system in virtually all *2D* AutoCAD drawings. However, because of the difficulty in calculating *3D* points, the WCS is not suited for many *3D* applications.

The User Coordinate System (UCS) allows you to change the location and orientation of the *X*, *Y*, and *Z* axes to reduce the number of calculations needed to create *3D* objects. The UCS command lets you redefine the origin of your drawing and establish the positive *X* and the positive *Y* axes. New users think of a coordinate system simply as the direction of positive *X* and positive *Y*. But once the directions of *X* and *Y* are defined, the direction of *Z* will be defined as well. Thus, the user has only to be concerned with *X* and *Y*. As a result, when you are drawing in *2D*, you are also somewhere in *3D* space. For example, if a sloped roof of a house is drawn in detail using the WCS, each endpoint of every object on the inclined roof plane must be calculated. On the other hand, if the UCS is set to the same plane as the roof, each object can be drawn as if it were in the plan view. You can define any number of UCSs within the fixed WCS and save them, assigning each a user-determined name. But at any given time, only one coordinate system is current and all coordinate input and display is relative to it. This is unlike a rotated snap grid in which direction and coordinates are based on the WCS. If multiple viewports are active, AutoCAD allows you to assign a different UCS to each viewport. Each UCS can have a different origin and orientation for various construction requirements.

RIGHT-HAND RULE

The directions of the *X*, *Y*, and *Z* axes change when the UCS is altered; hence, the positive rotation direction of the axes may become difficult to determine. The right-hand rule helps in determining the rotation direction when changing the UCS or using commands that require object rotation.

To remember the orientation of the axes, do the following:

1. Hold your right hand with the thumb, forefinger, and middle finger pointing at right angles to each other, as shown in Figure 15–2.
2. Consider the thumb to be pointing in the positive direction of the *X* axis.
3. The forefinger points in the positive direction of the *Y* axis.
4. The middle finger points in the positive direction of the *Z* axis.

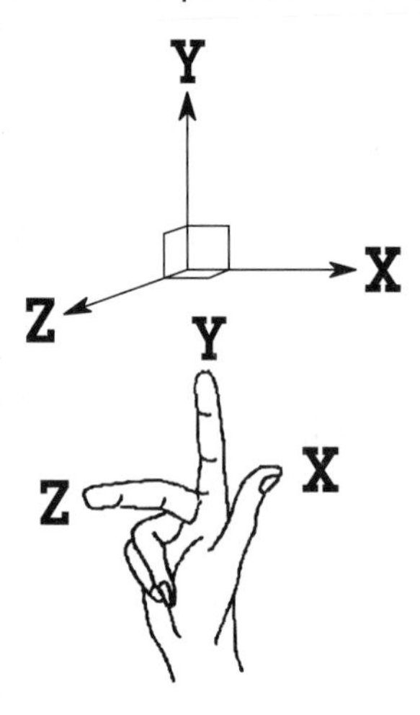

Figure 15–2 *The correct hand position when using the right-hand rule*

SETTING THE DISPLAY OF THE UCS ICON

The UCS icon provides a visual reminder of how the UCS axes are oriented, where the current UCS origin is, and the viewing direction relative to the UCS *XY* plane. AutoCAD provides two methods of displaying icons: 2D UCS style and 3D UCS style and displays different coordinate system icon in paper space and in model space, as shown in Figure 15–3.

The *X* and *Y* axis directions are indicated with arrows labeled appropriately, and the *Z* axis is indicated by the placement of the icon and in both cases (2D UCS style and 3D UCS style), a plus sign (+) appears at the base of the icon when it is positioned at the origin of the current UCS. With the 2D UCS icon a W is displayed in the *Y* axis arrow and with the 3D UCS icon, a square is displayed in the *XY* plane at the origin when the UCS is the same as the world coordinate system. When looking straight up or down on the *Z* plane, the icon seems flat. When viewed at any other angle, the icon looks skewed. In the case of 2D UCS style, the orientation of the *Z* axis is indicated further by the presence or absence of a box at the base of the arrows that create the icon. If the box is visible, you are looking down on (from the positive *Z* side of) the *XY* plane; if the box is not present, the bottom of the *XY* plane is being viewed (from the negative *Z* side of it). With the 3D UCS icon, the *Z* axis is solid when viewed from above the *XY* plane and dashed when viewed from below the *XY* plane. See Figure 15–4 for different orientations of the UCS icon.

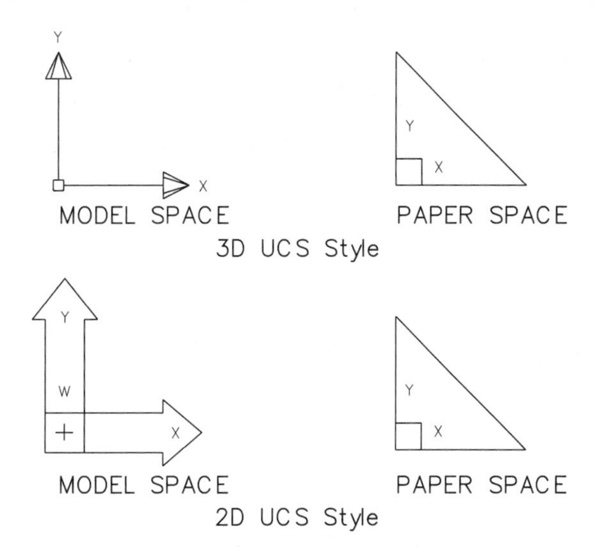

Figure 15–3 *The UCS icon for model space and for paper space*

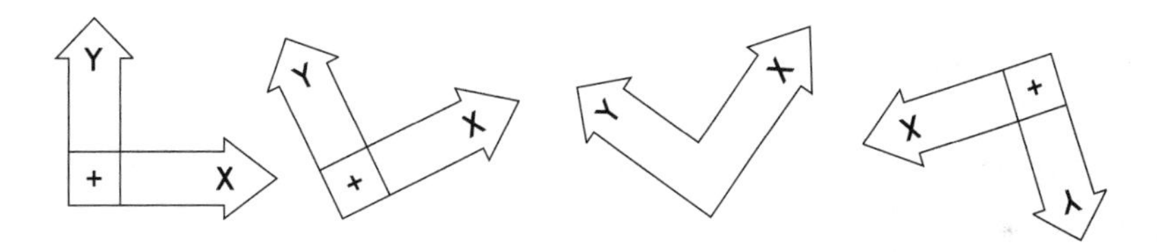

Figure 15–4 *The UCS icon in different orientations*

Note: If the viewing angle comes within 1 degree of the XY plane, the 2D UCS style icon will change to a "broken pencil," as shown in Figure 15–5. When this icon is showing in a view, it is recommended that you avoid trying to use the cursor to specify points in that view, because the results may be unpredictable. The 3D UCS icon does not use a broken pencil icon.

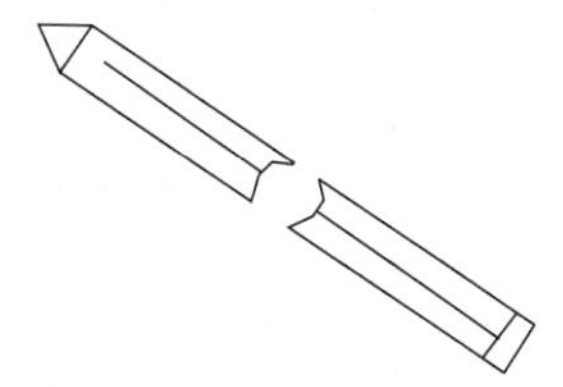

Figure 15–5 *The UCS icon becomes a broken pencil when the viewing angle is within 1 degree of the XY plane*

The display and placement on the origin of the UCS icon is handled by the UCSICON command. Invoke the UCSICON command:

Command: prompt	**ucsicon** (ENTER)

AutoCAD prompts:

Command: **ucsicon** (ENTER)
Enter an option [ON/OFF/All/Noorigin/Origin/Properties]<default>: *(select one of the available options or right-click for shortcut menu and select one of the available options)*

ON

The ON option allows you to set the icon to ON if it is OFF in the current viewport.

OFF

The OFF option allows you to set the icon to OFF if it is ON in the current viewport.

Noorigin

The Noorigin option (default setting) tells AutoCAD to display the icon at the lower left corner of the viewport, regardless of the location of the UCS origin. This is like parking the icon in the lower left corner.

Origin

The Origin option forces the icon to be displayed at the origin of the current coordinate system.

Note: If the origin is off screen, the icon is displayed at the lower left corner of the viewport.

All

The All option determines whether the options that follow affect all of the viewports or just the current active viewport. This option is issued before each and every option if you want to affect all viewports. For example, to turn ON the icon in all the viewports and display the icon on the origin, the following sequence of prompts is displayed:

Command: **ucsicon**
Enter an option [ON/OFF/All/Noorigin/Origin] <default>: **a**
Enter an option [ON/OFF/Noorigin/Origin]: **on**
Command: (ENTER)
Enter an option [ON/OFF/All/Noorigin/Origin]: **a**
Enter an option [ON/OFF/Noorigin/Origin]: **or**

Properties

The Properties option displays the UCS Icon dialog box as shown in Figure 15-6, in which you can control the style, visibility, and location of the UCS icon.

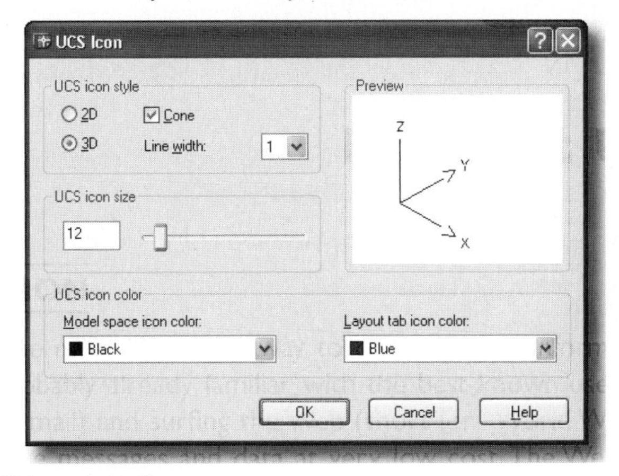

Figure 15–6 *UCS Icon dialog box*

The **UCS icon style** specifies display of either the 2D or the 3D UCS icon and its appearance. The **2D** selection displays a 2D icon without a representation of the Z axis, and the **3D** selection displays a 3D icon. Setting the **Cone** toggle ON displays a 3D cone arrowheads for the X and Y axes (available only when 3D UCS icon is selected). If **Cone** toggle is set to OFF, 2D arrowheads are displayed instead. The **Line width** controls the line width of the UCS icon, and you can set to one of the three available selections: 1, 2, or 3 pixels (available only when 3D UCS icon is selected).

The **Preview** section displays a preview of the selected UCS icon style in model space.

The **UCS icon size** section controls the size of the UCS icon as a percentage of viewport size. The default value is 12, and the valid range is from 5 to 95.

Note: Size of the UCS icon is proportional to the size of the viewport in which it is displayed.

The **UCS icon color** section allows you to select the color of the UCS icon in model space viewports and in the Layout tab.

Choose the **OK** button to save the settings and close the UCS Icon dialog box.

Note: You can also open UCS Icon dialog box from **View** menu (**Display > UCS icon > Properties...**)

DEFINING A NEW UCS

The UCS is the key to almost all 3D operations in AutoCAD, as mentioned earlier. Many commands in AutoCAD are traditionally thought of as 2D commands. They are effective in *3D* because they are always relative to the current UCS. For example, the ROTATE command rotates in only the *X* and *Y* directions; if you want to rotate an object in the *Z* direction, change your UCS so that *X* or *Y* is now in the direction of what was previously *Z*. Then you could use the ROTATE command.

The UCS command lets you redefine the origin in your drawing. Broadly, you can define the origin by five methods:

- Specify a data point for an origin, specify a new *XY* plane by providing three data points, or provide a direction for the *Z* axis.
- Define an origin relative to the orientation of an existing object.
- Define an origin by selecting a face.
- Define an origin by aligning with the current viewing direction.
- Define an origin by rotating the current UCS around one of its axes.

Invoke the UCS command:

UCS toolbar	Choose the UCS command (see Figure 15–7)
Tools menu	Choose New UCS
Command: prompt	**ucs** (ENTER)

Figure 15–7 *Invoking the UCS command from the UCS toolbar*

AutoCAD prompts:

Command: **ucs**
Enter an option [New/Move/orthoGraphic/Prev/Restore/Save/Del/Apply/?/World]<World>: *(select the new option or right-click for the shortcut menu and select New option)*

Specify origin of new UCS or [ZAxis/3point/OBject/Face/View/X/Y/Z] <0,0,0>: *(specify a point to define a new origin or select one of the available options to define a new origin or right-click for the shortcut menu and select one of the available options)*

Origin

The new data point defines a new UCS by shifting the origin of the current UCS, leaving the directions of the *X*, *Y*, and *Z* axes unchanged. You can also invoke the ORIGIN UCS command:

UCS toolbar	Choose the Origin UCS command (see Figure 15–8)
Tools menu	Choose New UCS > Origin

Figure 15–8 *Invoking the* ORIGIN UCS *command from the UCS toolbar*

AutoCAD prompts:

Specify new Origin point (0,0,0): *(Specify a new origin point relative to the origin of the current UCS, as shown in Figure 15–9)*

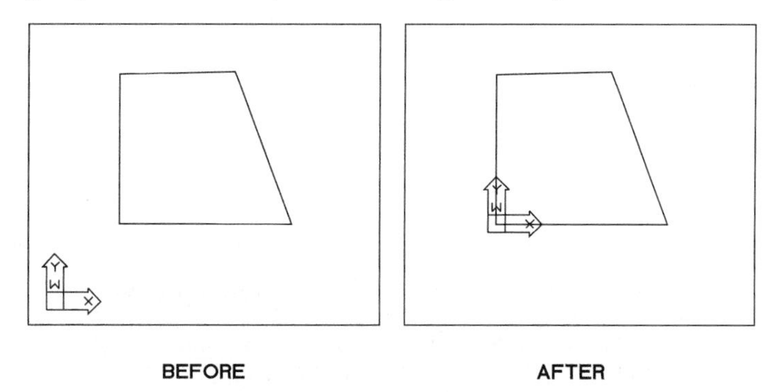

Figure 15–9 *Specifying a new origin point relative to the origin of the current UCS*

ZAxis

The ZAxis option allows you to define an origin by specifying a point and the direction for the *Z* axis. AutoCAD arbitrarily, but consistently, sets the direction of the *X* and *Y* axes in relation to the given *Z* axis. You can also invoke the Z AXIS UCS command:

UCS toolbar	Choose the Z Axis Vector UCS command (see Figure 15–10)
Tools menu	Choose New UCS > Z Axis Vector

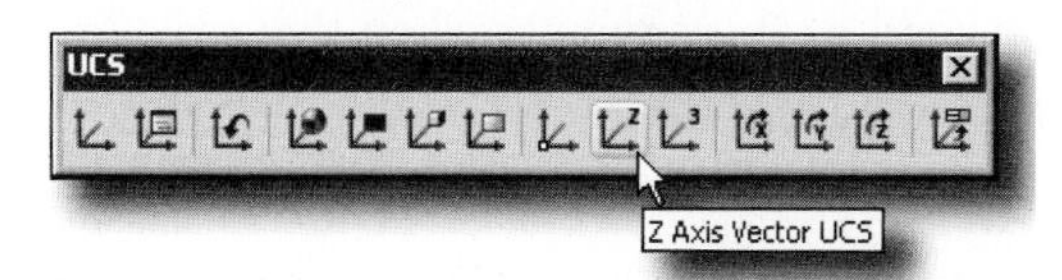

Figure 15–10 *Invoking the* Z AXIS VECTOR UCS *command from the UCS toolbar*

AutoCAD prompts:

> Specify new Origin point (0,0,0): *(specify a point to define a new origin point)*
> Specify point on positive portion of the Z axis <default>: *(specify the direction for the Z axis)*

If you give a null response to the second prompt, the *Z* axis of the new coordinate system will be parallel to the previous one. This is similar to using the Origin option.

3point

The 3point option is the easiest and most often used option for controlling the orientation of the UCS. This option allows the user to select three points to define the origin and the directions of the positive *X* and *Y* axes. The origin point acts as a base for the UCS rotation, and when a point is selected to define the direction of the positive *X* axis, the direction of the *Y* axis is limited because it is always perpendicular to the *X* axis. When the *X* and *Y* axes are defined, the *Z* axis is automatically placed perpendicular to the *XY* plane. You can also invoke the 3 POINT UCS command:

UCS toolbar	Choose the 3 Point UCS command (see Figure 15–11)
Tools menu	Choose New UCS > 3 point

Figure 15–11 *Invoking the 3 POINT UCS command from the UCS toolbar*

AutoCAD prompts:

> Specify new Origin point (0,0,0): *(specify the origin point)*
> Specify Point on positive portion of the X-axis <default>: *(specify a point for the positive X axis)*
> Specify Point on positive-Y portion of the UCS XY plane <default>: *(specify a point for the positive Y axis)*

Specify a data point and the direction for the positive *X* and *Y* axes, as shown in Figure 15–12. The points must not form a straight line. If you give a null response to the first prompt, the new UCS will have the same origin as the previous UCS. If you give a null response to the second or third prompt, then that axis direction will be parallel to the corresponding axis in the previous UCS.

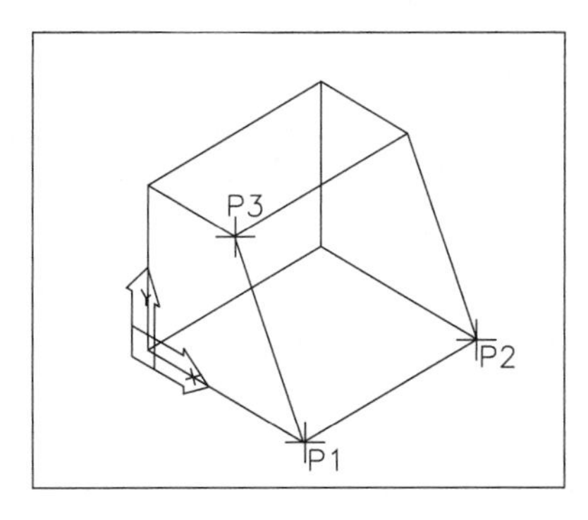

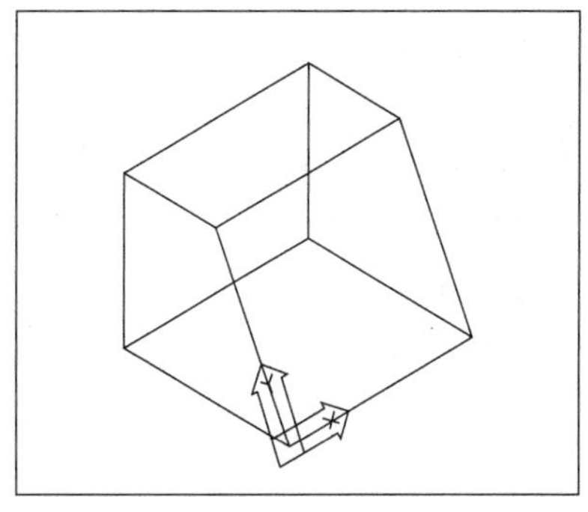

Figure 15–12 *Specifying a data point and the direction for the positive X and Y axes using the 3point option*

OBject

The OBject option lets you define a new coordinate system by selecting an object. The actual orientation of the UCS depends on how the object was created. When the object is selected, the UCS origin is placed at the first point used to create the object (in the case of a line, it will be the closest endpoint; for a circle, it will be the center point of the circle); the *X* axis is determined by the direction from the origin to the second point used to define the object. And the *Z* axis direction is placed perpendicular to the *XY* plane in which the object sits. Table 15–1 lists the location of the origin and its *X* axis for different types of objects.

Table 15–1 Location of the Origin and Its X Axis for Different Types of Objects

Object	Method of UCS Determination
Line	The endpoint nearest the specified point becomes the new UCS origin. The new *X* axis is chosen so that the line lies in the *XZ* plane of the new UCS.
Circle	The circle's center becomes the new UCS origin, and the *X* axis passes through the point specified.
Arc	The arc's center becomes the new UCS origin, and the *X* axis passes through the endpoint of the arc closest to the pick point.
2D polyline	The polyline's start point becomes the new UCS origin, with the *X* axis extending from the start point to the next vertex.
Solid	The first point of the solid determines the new UCS origin, and the *X* axis lies along the line between the first two points.
Dimension	The new UCS origin is the middle point of the dimension text, and the direction of the *X* axis is parallel to the *X* axis of the UCS in effect when the dimension was drawn.

You can also invoke the OBJECT UCS command:

UCS toolbar	Choose the Object UCS command (see Figure 15–13)
Tools menu	Choose New UCS > Object

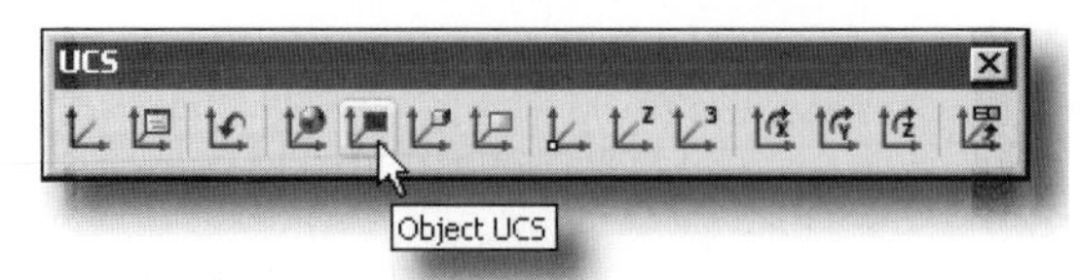

Figure 15–13 *Invoking the* OBJECT UCS *command from the UCS toolbar*

AutoCAD prompts:

> Select object to align UCS: *(identify an object to define a new coordinate system)*

Identify an object to define a new coordinate system, as shown in Figure 15–14.

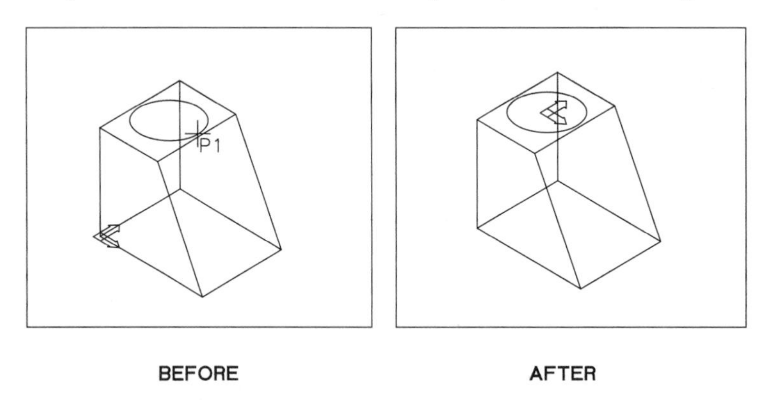

Figure 15–14 *Defining a new coordinate system by the* OBJECT UCS *command*

Face

The Face option aligns the UCS to the selected face of a solid object. To select a face, click within the boundary of the face or on the edge of the face. The face is highlighted and the UCS *X* axis is aligned with the closest edge of the first face found. You can also invoke the FACE UCS command:

UCS toolbar	Choose the Face UCS command (see Figure 15–15)
Tools menu	Choose New UCS > Face

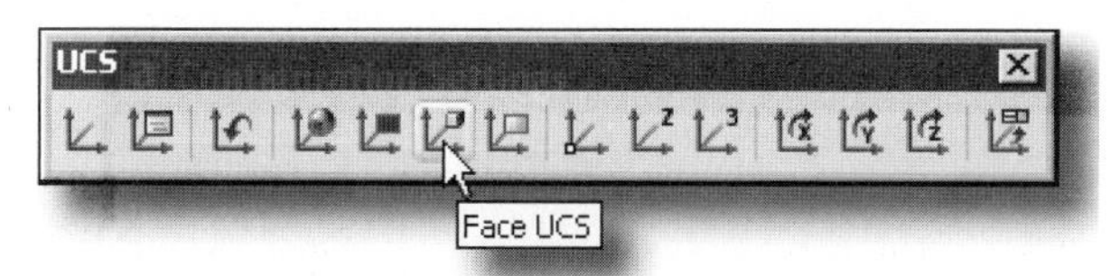

Figure 15–15 *Invoking the* FACE UCS *command from the UCS toolbar*

AutoCAD prompts:

> Select face of solid object: *(select the face of the solid object)*
> Enter an option [Next/Xflip/Yflip] <accept>: *(select one of the available options or right-click for the shortcut menu and select one of the available options)*

The **Next** option locates the UCS on either the adjacent face or the back face of the selected edge.

The **Xflip** option rotates the UCS 180 degrees around the *X* axis.

The **Yflip** option rotates the UCS 180 degrees around the *Y* axis.

The **Accept** option accepts the location. The prompt repeats until you accept a location.

View

The View option places the *XY* plane parallel to the screen, and makes the *Z* axis perpendicular. The UCS origin remains unchanged. This method is used mainly for labeling text, which should be aligned with the screen rather than with objects. You can also invoke the VIEW UCS command:

UCS toolbar	Choose the View UCS command (see Figure 15–16)
Tools menu	Choose New UCS > View

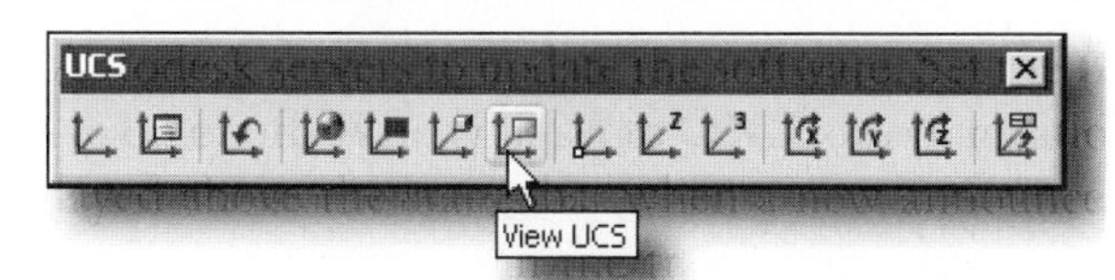

Figure 15–16 *Invoking the* VIEW UCS *command from the UCS toolbar*

X/Y/Z Rotation

The X/Y/Z rotation option lets you define a new coordinate system by rotating the *X*, *Y*, and *Z* axes independently of each other. You can show AutoCAD the desired angle by specifying two points, or you can enter the rotation angle from the keyboard. In either case, the new angle is specified relative to the *X* axis of the current UCS. See Figures 15–17, 15–18, and 15–19 for examples of rotating the UCS around the *X*, *Y*, and *Z* axes, respectively.

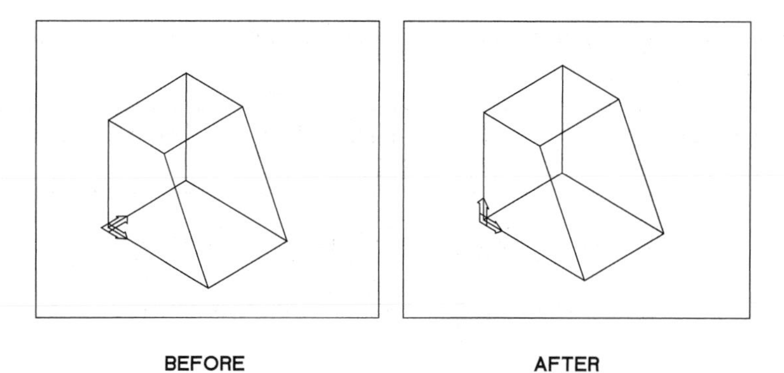

Figure 15–17 *Example of rotating the UCS around the X axis*

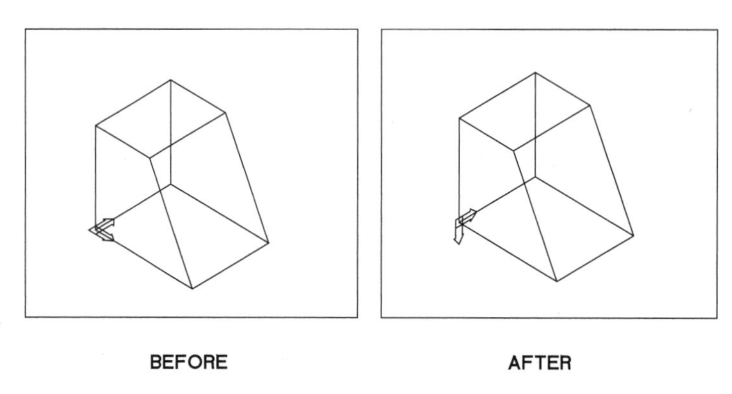

Figure 15–18 *Example of rotating the UCS around the Y axis*

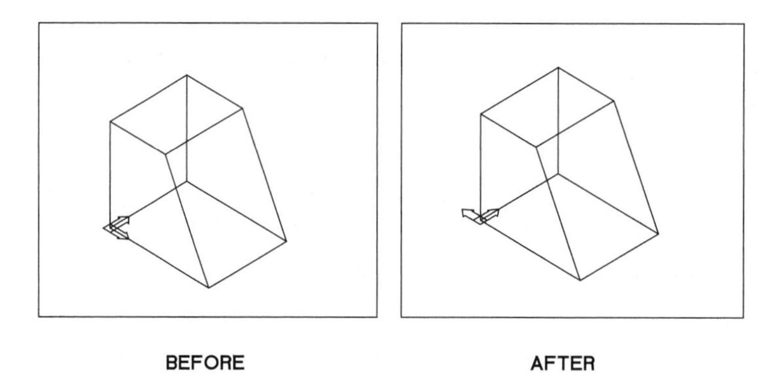

Figure 15–19 *Example of rotating the UCS around the Z axis*

Previous

The Previous option is similar to the Previous option of the ZOOM command. AutoCAD saves the last 10 coordinate systems in both model space and paper space. You can step back through them by using repeated Previous options. You can also invoke the UCS PREVIOUS command:

UCS toolbar	Choose the UCS Previous command (see Figure 15–20)

Figure 15–20 *Invoking the* UCS PREVIOUS *command from the UCS toolbar*

Restore

The Restore option allows you to restore any previously saved UCS.

Note: You can also restore a previously saved UCS by invoking the DDUCS command, which in turn displays a dialog box listing the previously saved UCS.

Save

The Save option allows you to save the current UCS under a user-defined name.

Delete

The Delete option allows you to delete any saved UCS.

?

The ? option lists the name of the UCS you specify, the origin, and the *X*, *Y*, and *Z* axes for each saved coordinate system, relative to the current UCS. To list all the UCS names, accept the default, or you can specify wild cards.

World

The World option returns the drawing to the WCS. You can also invoke the WORLD UCS command:

UCS toolbar	Choose the World UCS command (see Figure 15–21)
Tools menu	Choose New UCS > World

Figure 15–21 *Invoking the* WORLD UCS *command from the UCS toolbar*

SELECTING A PREDEFINED ORTHOGRAPHIC UCS

In addition to various methods explained earlier for defining a new UCS, AutoCAD's UCS dialog box allows you to change the UCS to one of the available standard orthographic settings. In addition, the UCS dialog box lists saved user coordinate systems and allows you to modify UCS icon settings and UCS settings saved with a viewport. Invoke the UCSMAN command to open UCS dialog box:

UCS toolbar	Choose the Display UCS Dialog command (see Figure 15–22)
Tools menu	Choose Orthographic UCS > Preset...
Command: prompt	**ucsman** (ENTER)

Figure 15–22 *Invoking the* DISPLAY UCS DIALOG *command from the UCS toolbar*

AutoCAD displays the UCS dialog box with the Orthographic UCSs tab selected, as shown in Figure 15–23.

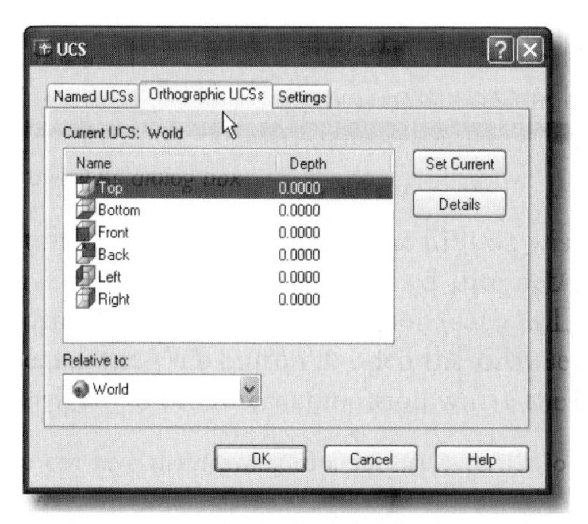

Figure 15–23 *UCS dialog box with the Orthographic UCSs tab selected*

AutoCAD displays the name of the current UCS view at the top of the dialog box. If the UCS setting has not been saved and named, the current UCS reads UNNAMED. AutoCAD lists the standard orthographic coordinate systems in the current drawing. The orthographic coordinate systems are defined relative to the UCS specified in the **Relative to:** list box. By default it is set to World coordinate system. To set the UCS to one of the orthographic coordinate systems, select one of the six listed names and choose the **Set Current** button, double-click on the listed name, or right-click and select Set Current from the shortcut menu. The **Depth** field lists the distance between the orthographic coordinate system and the parallel plane passing through the origin of the UCS base setting (stored in the UCSBASE system variable). To change the depth, double-click on the Depth field, and AutoCAD displays Orthographic UCS Depth dialog box. Make the necessary changes to the depth and choose the **OK** button to close the dialog box.

To set the UCS to one of the saved UCSs, select the **Named UCSs** tab, and AutoCAD lists the saved UCS in the current drawing, similar to Figure 15–24. Select one of the listed UCS saved names and choose the **Set Current** button, double-click on the listed name, or right-click and select SET CURRENT from the shortcut menu.

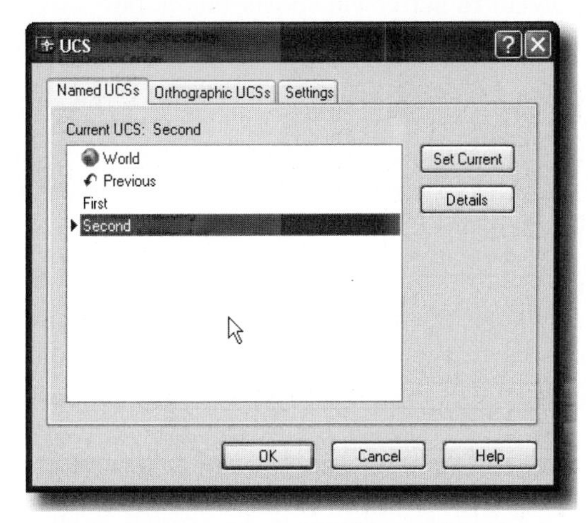

Figure 15–24 *UCS dialog box with the Named UCSs tab selected*

The Settings tab of the UCS dialog box displays and allows you to modify UCS icon settings and UCS settings saved with a viewport, as shown in Figure 15–25.

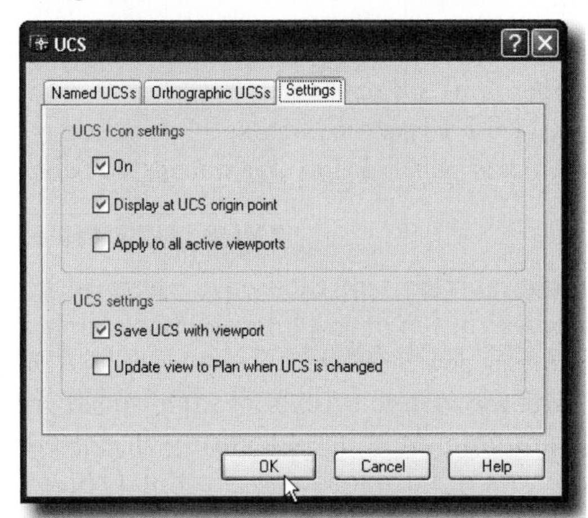

Figure 15–25 *UCS dialog box with the Settings tab selected*

The **UCS icon settings** section of the dialog box specifies the UCS settings for the current viewport.

The **On** toggle button controls the display of the UCS icon in the current viewport.

The **Display at UCS origin point** toggle button controls the display of the UCS icon at the origin of the current coordinate system for the current viewport. If this option is set to off, or if the origin of the coordinate system is not visible in the viewport, the UCS icon is displayed at the lower left corner of the viewport.

The **Apply all to active viewports** toggle button applies the UCS icon settings to all active viewports in the current drawing.

The **UCS settings** section specifies the UCS settings for the current viewport.

The **Save UCS with viewport** toggle button saves the coordinate system setting with the viewport. If this option is set to off, the viewport reflects the UCS of the viewport that is current.

The **Update view to Plan when UCS is changed** toggle button restores plan view when the coordinate system in the viewport changes. Plan view is restored when the dialog box is closed and the selected UCS setting is restored.

Choose the **OK** button to close the UCS dialog box and accept the changes.

You can also set the current UCS to one of the saved UCS or to one of the standard orthographic UCS from the UCS II toolbar drop down menu, as shown in Figure 15–26.

Figure 15–26 *UCS II toolbar*

VIEWING A DRAWING FROM PLAN VIEW

The PLAN command provides a convenient means of viewing a drawing from plan view. The definition of a plan means that you are at positive *Z* and looking perpendicularly down on the *XY* plane, with *X* to the right and *Y* pointing up. You can select the plan view of the current UCS, a previously saved UCS, or the WCS.

Invoke the PLAN command:

View menu	Choose *3D* views > Plan view
Command: prompt	**plan** (ENTER)

AutoCAD prompts:

> Command: **plan**
> Enter an option [Current ucs>/Ucs/World] <current>: *(select one of the available options or right-click for the shortcut menu and select one of the available options)*

Current ucs

The Current ucs option displays the plan view of the current UCS. This is the default option.

Ucs

The Ucs option displays a plan view of a previously saved UCS. When you select this option, AutoCAD prompts for a name of the UCS.

World

The World option displays the plan view of the World Coordinate System.

VIEWING IN *3D*

Until now, you have been working on the plan view, or the *XY* plane. You have been looking down at the plan view from a positive distance along the *Z* axis. The direction from which you view your drawing or model is called the viewpoint. You can view a drawing from any point in model space. From your selected viewpoint, you can add objects, modify existing objects, or suppress the hidden lines from the drawing.

The VPOINT, DVIEW, and 3DORBIT commands are used to control viewing of a model from any point in model space.

VIEWING A MODEL BY MEANS OF THE VPOINT COMMAND

To view a model in *3D*, you may have to change the viewpoint. The location of the viewpoint can be controlled by means of the VPOINT command. The default viewpoint is 0,0,1; that is, you are looking at the model from 0,0,1 (on the positive *Z* axis above the model) to 0,0,0 (the origin).

Invoke the VPOINT command:

View menu	Choose *3D* Views > Viewpoint
Command: prompt	**vpoint** (ENTER)

AutoCAD prompts:

Command: **vpoint**
Specify a view point or [Rotate] <display compass and tripod>: *(select one of the available options or right-click for the shortcut menu and select one of the available options)*

The default method requires you to enter *X*, *Y*, and *Z* coordinates from the keyboard. These coordinates establish the viewpoint. From this viewpoint, you will be looking at the model in space toward the model's origin. For example, a 1,–1,1 setting gives you a –45-degree angle projected in the *XY* plane and 35.264-degree angle above the *XY* plane (top, right, and front views); looking at the model origin (0,0,0). You can set the viewpoint to any *X*, *Y*, *Z* location. Table 15–2 lets you experiment with the rotation of *3D* objects.

Table 15–2 Various Viewpoint Settings for Rotating *3D* objects

Viewpoint Setting	Displayed View(s)	Viewpoint Setting	Displayed View(s)
0,0,1	Top	–1,–1,1	Top, Front, Left side
0,0,–1	Bottom	1,1,1	Top, Rear, Right side
0,–1,0	Front	–1,1,1	Top, Rear, Left side
0,1,0	Rear	1,–1,–1	Bottom, Front, Right side
1,0,01	Right side	–1,–1,–1	Bottom, Front, Left side
–1,0,0	Left side	1,1,–1	Bottom, Rear, Right side
1,–1,1	Top, Front, Right side	–1,1,–1	Bottom, Rear, Left side

If instead of entering coordinates you give a null response (press ENTER or the SPACEBAR), a compass and axes tripod appear on the screen, as shown in Figure 15–27. The compass, in the upper right of the screen, is a *2D* representation of a globe. The center point of the circle represents the north pole (0,0,1), the inner circle represents the equator, and the outer circle represents the south pole (0,0,–1), as shown in Figure 15–28. A small cross is displayed on the compass. You can move the cross with your pointing device. If the cross is in the inner circle, you are above the equator looking down on your model. If the cross is in the outer circle, you are looking from beneath your drawing, or from the Southern Hemisphere. Move the cross, and the axes tripod rotates to conform to the viewpoint indicated on the compass. When you achieve the desired viewpoint, press the pick button on your pointing device or press ENTER. The drawing regenerates to reflect the new viewpoint position.

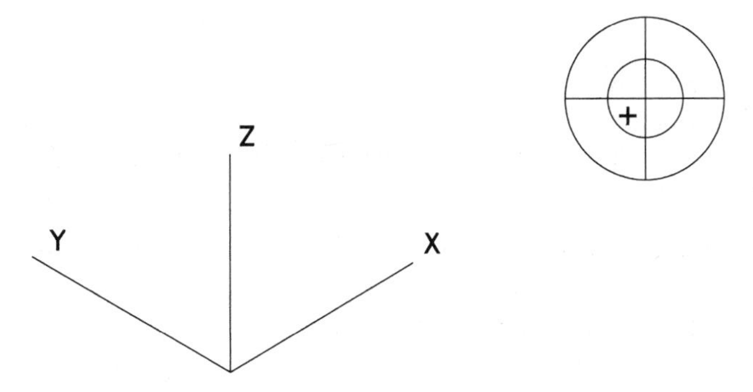

Figure 15–27 *The* VPOINT *command's compass and axes tripod*

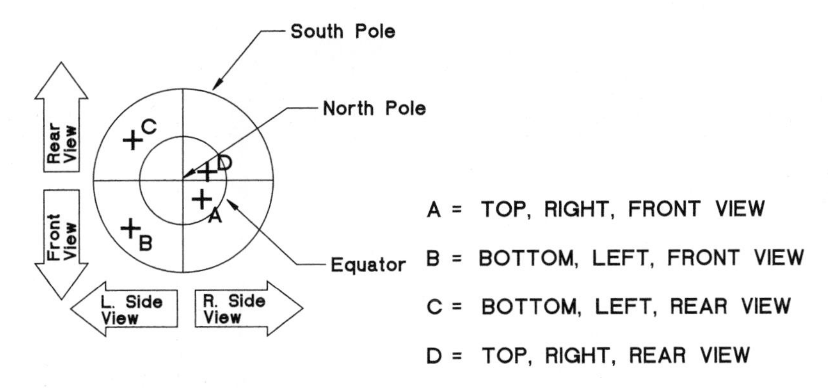

Figure 15–28 *Components of the* VPOINT *command's compass and its poles*

Rotate Option

The Rotate option allows you to specify the location of the viewpoint in terms of two angles. The first angle determines the rotation in the *XY* plane from the *X* axis (0 degrees) clockwise or counterclockwise. The second angle determines the angle from the *XY* plane up or down. When you select the Rotate option, AutoCAD prompts:

> Enter angle in X-Y plane from X axis <current>: *(specify the angle in the* XY *plane from the* X *axis)*
> Enter angle from X-Y plane <current>: *(specify the angle from the* XY *plane)*

AutoCAD regenerates to reflect the new viewpoint position.

Viewpoint Presets

AutoCAD provides Viewpoint Presets dialog box when you invoke the DDVPOINT command. The dialog box lets you set a *3D* viewing direction by specifying an angle from the *X* axis and an angle from the *XY* plane. This is similar to using the Rotate option of the VPOINT command.

Invoke the DDVPOINT command:

View menu	Choose *3D* Views > Viewpoint Presets...
Command: prompt	**ddvpoint** (ENTER)

AutoCAD displays the Viewpoint Presets dialog box, similar to the one shown in Figure 15–29. Specify viewing angles from the image tile, or enter their values in the text boxes. You specify the view direction relative to the current UCS or the WCS, and the viewing angles are updated accordingly. The new angle is indicated by the white arm; the current viewing angle is indicated by the red arm. By selecting the **Set to Plan View** button, you can set the viewing angles to display the plan relative to the selected coordinate system.

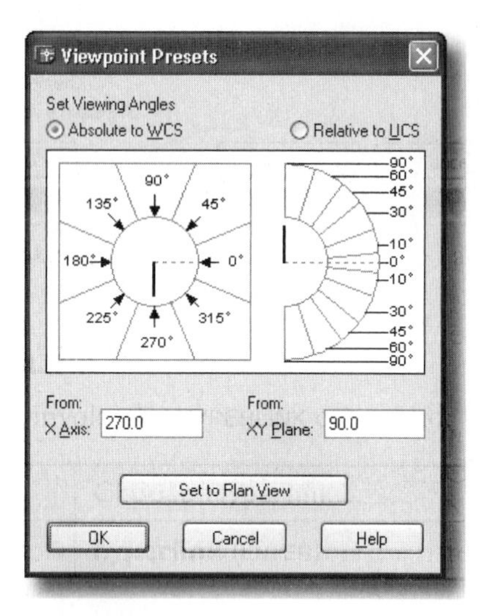

Figure 15–29 *Viewpoint Presets dialog box*

Note: By default, AutoCAD always places the model to your current viewpoint position in reference to the WCS, not the current UCS. If necessary, you can change the WORLDVIEW system variable from 1 (default) to 0; AutoCAD then places the model in reference to the UCS for your current viewpoint position. It is recommended that you keep the WORLDVIEW set to 1 (default). Regardless of the WORLDVIEW setting, you are always looking through your viewpoint to the WCS origin.

VIEWING A MODEL BY MEANS OF THE DVIEW COMMAND

The DVIEW command is an enhanced VPOINT command. Here you visually move an object around on the screen, dynamically viewing selected objects as the view changes. The DVIEW command provides either parallel or perspective views, whereas the VPOINT command provides only parallel views. In the case of a parallel view, parallel lines always remain parallel, whereas in perspective view, parallel lines converge from your view to a vanishing point. Figures 15–30 and 15–31 show parallel and perspective views, respectively, of a model. The viewing direction is the same in each case.

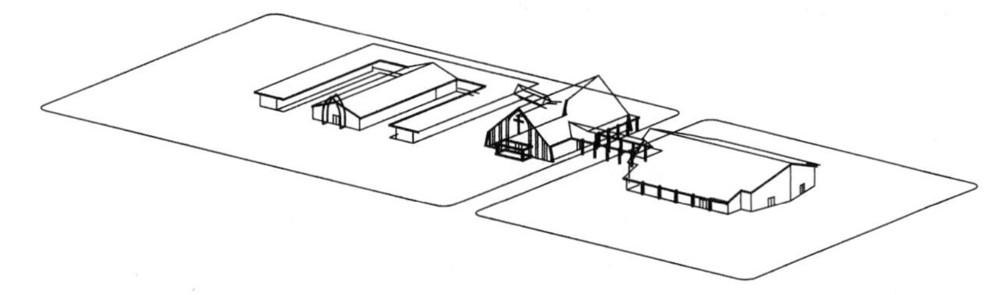

Figure 15–30 *The* DVIEW *command displaying a model with parallel projection*

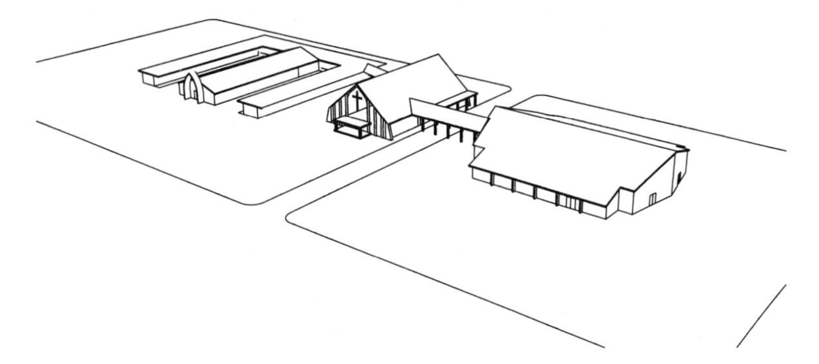

Figure 15–31 *The* DVIEW *command displaying the same model with perspective projection*

Invoke the DVIEW command:

Command: prompt	**dview** (ENTER)

AutoCAD prompts:

Command: **dview**
Select objects or (use DVIEWBLOCK): *(select objects or press* ENTER*)*

All or any part of the objects in the drawing can be selected for viewing during the DVIEW command process. But once you exit the DVIEW command, all objects in the drawing are represented in the new view created. If your drawing is too large to display quickly in the DVIEW display, small portions can be selected and used to orient the entire drawing. The purpose of this is to save time on slower machines and still give dynamic rotation so that you can quickly and effortlessly adjust the view of your object before you begin working with it.

If you give a null response (press ENTER or SPACEBAR) to the "Select objects:" prompt, AutoCAD provides you with a picture of a *3D* house. Whatever you do to the *3D* house under DVIEW will be done to your current drawing when you exit the DVIEW command.

Each time you exit the DVIEW command, AutoCAD performs an unconditional regeneration. No matter what the current setting for REGENAUTO is, AutoCAD automatically performs a regeneration.

After selecting the objects, AutoCAD prompts with the following options:

> [CAmera/TArget/Distance/POints/PAn/Zoom/TWist/CLip/Hide/Off/Undo]: *(select one of the available options or right-click for the shortcut menu and select one of the available options)*

CAmera

The CAmera option is one of the six options that adjust what is seen in the view. With the CAmera option, the drawing is stationary while the camera moves. It can move up and down (above or below) or it can move around the target to the left or right (clockwise or counterclockwise). When you are moving the camera, the target is fixed.

When you select the CAmera option, AutoCAD prompts:

> Specify camera location, or enter angle from *X-Y* plane [or Toggle (angle in)] <current>:

You can specify the amount of rotation you want by positioning the graphics cursor in the graphics area. When you move the cursor, you will see the object begin to rotate dynamically, and the AutoCAD status line displays a continuous readout of the new angle. Move the cursor to the desired angle and then press the pick button. Or you could type the desired angle from the keyboard. Either way, you have selected an angle of view above or below the target.

Next, AutoCAD prompts for the desired rotation angle of the camera around the target:

> Specify camera location, or enter angle in *X-Y* plane from *X* axis or [Toggle (angle from)] <current>:

You can move the camera 180 degrees clockwise and 180 degrees counterclockwise around the target. Specify the angle by using the cursor to specify a point on the screen. Or you could type in the desired angle from the keyboard. Either way, you have selected an angle of view around the target clockwise or counterclockwise.

The **Toggle angle** option allows you to move between two angle input modes.

AutoCAD takes you back to the 11-option prompt of the DVIEW command. When the new angle of view is correct, you exit the command sequence by giving a null response (press ENTER or the SPACEBAR). This takes you back to the "Command:" prompt.

When you exit DVIEW, your entire drawing will rotate to the same angle of view as the few objects that you selected.

The following command sequence shows an example of using the CAmera option of the DVIEW command.

Command: **dview**
Select objects or <use DIVEWBLOCK>: *(select the objects)*
Enter option [CAmera/TArget/Distance/POints/PAn/Zoom/TWist/CLip/Hide/Off/Undo]: **ca**
Specify camera location, or enter angle from *X-Y* plane or [Toggle (angle in)] <current>: **45**
Specify camera location, or enter angle in the *X-Y* plane from *X* axis or [Toggle (angle from)] <current>: **45**
Enter option [CAmera/TArget/Distance/POints/PAn/Zoom/TWist/CLip/Hide/Off/Undo]: (ENTER)

TArget

The TArget option is similar to the CAmera option, but in this case the target is rotated around the camera. The camera remains stationary except for maintaining its lens on the target point. The prompts are similar to those for the CAmera option. There may seem to be no difference between the CAmera and TArget options, but there is a difference in the actual angle of view. For instance, if you elevate the camera 75 degrees above the target, you are then looking at the target from the top down. On the other hand, if you raise the target 75 degrees above the camera, you are then looking at the target from the bottom up. The angles of view are reversed. The real difference comes when you are typing in the angles rather than visually picking them.

The following command sequence shows an example of using the TArget option of the DVIEW command.

Command: **dview**
Select objects or <use DVIEWBLOCK>: *(select the objects)*
Enter option [CAmera/TArget/Distance/POints/PAn/Zoom/TWist/CLip/Hide/Off/Undo]: **ta**
Specify camera location, or enter angle from X-Y plane or [Toggle (angle in)] <current>: **75**
Specify camera location, or enter angle in the *X-Y* plane from *X* axis or [Toggle (angle from)] <current>: **75**
Enter option [CAmera/TArget/Distance/POints/PAn/Zoom/TWist/CLip/Hide/Off/Undo]: (ENTER)

Distance

The Distance option creates a perspective projection from the current view. The only information required for this option is the distance from the camera to the target point. Once AutoCAD knows the distance, it will apply the correct perspective. When perspective viewing is on, a box icon appears on the screen, as shown in Figure 15–32, in place of the UCS icon. Some commands (like ZOOM and PAN) will not work while perspective is on. You turn on the perspective just for visual purposes or for plotting.

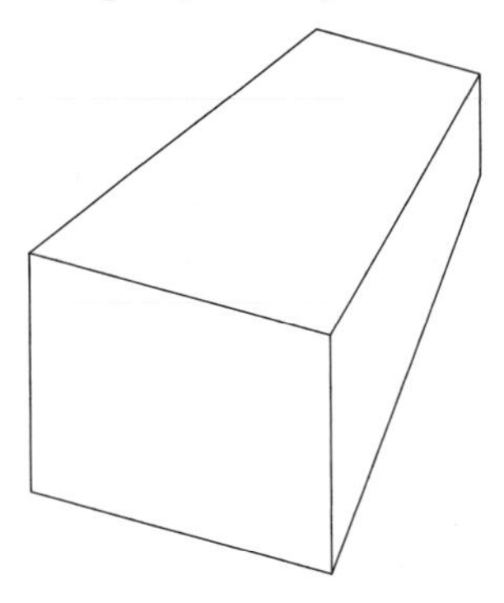

Figure 15–32 *The perspective box icon appears when perspective mode is on*

When you select the Distance option, AutoCAD prompts:

Specify a new camera—target distance <current>:

In addition to the prompt, you also see a horizontal bar at the top of the screen. The bar goes from 0x to 16x. These are factor distances times your current distance from the object. Moving the slider cursor toward the right increases the distance between the target and the camera. Moving the slider cursor toward the left reduces the distance between the target and the camera. The current distance is represented by 1x. For instance, moving the slider cursor to 3x makes the new distance three times the previous distance. Or you could also type the desired distance in the current linear units from the keyboard.

The following command sequence shows an example of using the Distance option of the DVIEW command.

Command: **dview**
Select objects or <use DVIEWBLOCK>: *(select the objects)*
Enter option [CAmera/TArget/Distance/POints/PAn/Zoom/TWist/CLip/Hide/Off/Undo]: **d**
Specify a new camera—target distance <current>: **75**
Enter option [CAmera/TArget/Distance/POints/PAn/Zoom/TWist/CLip/Hide/Off/Undo]: (ENTER)

Off

The Off option turns off the perspective view. The following command sequence shows an example of using the Off option of the DVIEW command.

Command: **dview**
Select objects or <use DVIEWBLOCK>: *(select the objects)*
Enter option [CAmera/TArget/Distance/POints/PAn/Zoom/TWist/CLip/
Hide/Off/Undo]: **o**
Enter option [CAmera/TArget/Distance/POints/PAn/Zoom/TWist/CLip/
Hide/Off/Undo]: (ENTER)

Note: To turn the perspective on again, select the Distance option and press ENTER for all the defaults. There is no option called On for turning on the perspective view.

POints

The POints option establishes the location of the camera as well as the target points. This gives AutoCAD the basic information needed to create the view. The location of the camera and target points must be specified in a parallel projection. If perspective is set to ON, AutoCAD temporarily turns it off while you specify the new location for camera and target points, and then redisplays the image in perspective.

When you select the POints option, AutoCAD prompts:

Specify target point: *(specify the target location)*
Specify camera point: *(specify the camera location)*

After the locations are defined, the screen shows the new view immediately.

The following command sequence shows an example of using the POints option of the DVIEW command.

Command: **dview**
Select objects or <use DVIEWBLOCK>: *(select the objects)*
Enter option [CAmera/TArget/Distance/POints/PAn/Zoom/TWist/CLip/
Hide/Off/Undo]: **po**
Specify target point: *(specify a point)*
Specify camera point: *(specify a point)*
Enter option [CAmera/TArget/Distance/POints/PAn/Zoom/TWist/CLip/
Hide/Off/Undo]: (ENTER)

PAn

The PAn option allows you to view a different location of the model by specifying the pan distance and direction. This option is similar to the regular PAN command. The following command sequence shows an example of using the PAn option of the DVIEW command.

Command: **dview**
Select objects or <use DVIEWBLOCK>: *(select the objects)*
Enter option [CAmera/TArget/Distance/POints/PAn/Zoom/TWist/CLip/
Hide/Off/Undo]: **pa**
Specify displacement base point: *(specify a point)*

```
Specify second point: (specify a point)
Enter option [CAmera/TArget/Distance/POints/PAn/Zoom/TWist/CLip/
   Hide/Off/Undo]: (ENTER)
```

Zoom

The Zoom option lets you zoom in on a portion of the model. This option is similar to the regular AutoCAD ZOOM CENTER command, with the center point lying at the center of the current viewport. This option is controlled by a scale factor value.

When you select the Zoom option, AutoCAD prompts:

```
Specify zoom scale factor <current>: (specify the zoom scale factor)
```

In addition to the prompt, you see a horizontal bar at the top of the screen. The slider bar lets you specify a zoom scale factor, with 1x being the current zoom level. Any value greater than 1 increases the size of the objects in the view; any decimal value less than 1 decreases the size.

Note: When the perspective is set to ON by the Distance option, the Zoom option prompts for a lens size rather than a zoom factor, but the effect is similar. The larger the lens size, the closer the object.

TWist

The TWist option rotates or twists the view. It allows you to rotate the image around the line of sight at a given angle from zero, with zero being to the right. The angle is measured counterclockwise.

The following command sequence shows an example of using the TWist option of the DVIEW command.

```
Command: dview
Select objects or <use DVIEWBLOCK>: (select the objects)
Enter option [CAmera/TArget/Distance/POints/PAn/Zoom/TWist/CLip/
   Hide/Off/Undo]: tw
Specify view twist angle <current>: (select a point)
Enter option [CAmera/TArget/Distance/POints/PAn/Zoom/TWist/CLip/
   Hide/Off/Undo]: (ENTER)
```

CLip

The CLip option hides portions of the object in view so that the interior of the object can be seen or parts of the complex object can be more clearly identified.

The CLip option has three suboptions: Back, Front, and Off.

The Back suboption eliminates all parts of the object in view that are located beyond the designated point along the line of sight.

The Front suboption eliminates all parts of the object in view that are located between the camera and the front clipping plane.

The Off suboption turns off front and back clipping.

The following command sequence shows an example of using the CLip option of the DVIEW command.

Command: **dview**
Select objects or <use DVIEWBLOCK>: *(select the objects)*
Enter option [CAmera/TArget/Distance/POints/PAn/Zoom/TWist/CLip/Hide/Off/Undo]: **cl**
Enter clipping option [Back/Front/Off] <current>: **b**
Specify distance from target or [On/Off] <current>: *(specify the distance, or turn on and off the previously defined clipping plane)*
Enter option [CAmera/TArget/Distance/POints/PAn/Zoom/TWist/CLip/Hide/Off/Undo]: (ENTER)

Hide

The Hide option is similar to the regular AutoCAD HIDE command.

Undo

The Undo option will undo the last DVIEW operation. You use it to step back through multiple DVIEW operations.

To exit the Undo option in the DVIEW command, press ENTER without selecting any of the available options, and AutoCAD returns you to the "Command:" prompt. It is the default option of the DVIEW command.

USING 3DORBIT

The 3DORBIT command allows you to view the model interactively in the current viewport. By using the pointing device you can manipulate the view of the model. You can view the entire model or any object in your model from different points around it.

Invoke the 3DORBIT command:

3D orbit toolbar	Choose the 3D Orbit command (see Figure 15–33)
View menu	Choose *3D* Orbit
Command: prompt	**3dorbit** (ENTER)

Figure 15–33 *Invoking the 3DORBIT command from the 3D Orbit toolbar*

AutoCAD prompts:

```
Command: 3dorbit
Press esc or enter to exit, or right-click to display shortcut-menu.
```

AutoCAD displays an arcball (as shown in Figure 15–34) a circle divided into four quadrants by smaller circles. While 3DORBIT is active, the point (target) that you are viewing stays stationary and the camera moves around the target. By default, the center of the arcball is the target point.

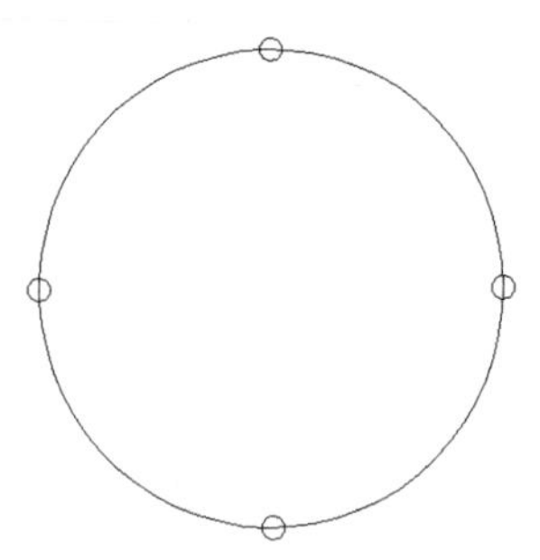

Figure 15–34 *Display of the arcball with a 3D model*

Click and drag the cursor to rotate the view. When you move your cursor over different parts of the arcball, the cursor icon changes.

When you move the cursor inside the arcball a small sphere encircled by two lines is displayed similar to the first icon as shown in Figure 15–35. By clicking and dragging, you can manipulate the view freely. You can drag horizontally, vertically, and diagonally.

When you move the cursor outside the arcball a circular arrow around a small sphere is displayed similar to the second icon as shown in Figure 15–35. Clicking outside the arcball and dragging the cursor around the arcball moves the view around an axis that extends through the center of the arcball, perpendicular to the screen.

When you move the cursor over one of the smaller circles on the left or right of the arcball a horizontal ellipse around a small sphere is displayed similar to the third icon as shown in Figure 15–35. Clicking and dragging from either of these points rotates the view around the vertical or *Y* axis that extends through the center of the arcball. The *Y* axis is represented on the cursor by a vertical line.

When you move the cursor over one of the smaller circles on the top or bottom of the arcball a vertical ellipse around a small sphere is displayed similar to the fourth icon as shown in Figure 15–35. Clicking and dragging from either of these points rotates the view around the horizontal or *X* axis that extends through the center of the arcball. The *X* axis is represented on the cursor by a horizontal line.

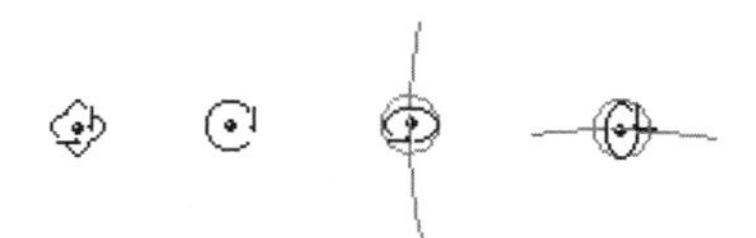

Figure 15–35 *Various cursor icons displayed when the 3DORBIT command is invoked*

Panning and Zooming in the *3D* Orbit View

You can pan and zoom while you are viewing the model with 3DORBIT command. It works similar to the REALTIME PAN and REALTIME ZOOM commands (see Chapter 3 for a detailed explanation). Select pan and zoom options from the shortcut menu while 3DORBIT is active, from the *3D* Orbit toolbar, or enter **3dpan** and **3dzoom** at the Command: prompt to invoke the pan and zoom options of the 3DORBIT command, respectively.

Using Projection Options in the *3D* Orbit View

AutoCAD allows you to display a perspective or a parallel projection of the view while 3DORBIT is active. The **perspective** view changes the view so that lines parallel to each other converge at a point. Objects appear to recede into the distance while parts of the objects appear larger and closer to you. The shapes are somewhat distorted when the object is very close. This view correlates most closely to what your eye sees. The **parallel** view changes the view so that two parallel lines never converge at a single point. The shapes in the drawing always remain the same and do not appear distorted when they are closer. You can select perspective and parallel options from the shortcut menu while 3DORBIT is active.

Shading Objects in the *3D* Orbit View

AutoCAD allows you to view shaded view while 3DORBIT is active. You can change the way objects are shaded using the different shading modes. The available modes are Wireframe, hidden, Flat Shaded, Gourand Shaded, Flat Shaded – Edges On, Gourand Shaded – Edges On. You can select one of the available shading options from the shortcut menu while 3DORBIT is active. See Chapter 16 for a detailed explanation of various shading options.

Note: Shading is applied even after exiting the 3DORBIT command. Use SHADEMODE to change the shading mode when 3DORBIT is not active.

Adjusting Clipping Planes in the *3D* Orbit View

AutoCAD allows you to set clipping planes for the objects in *3D* orbit view. Objects or parts of the objects that are beyond a clipping plane cannot be seen in the view. Open the Adjust Clipping Planes window by selecting ADJUST CLIPPING PLANES from the shortcut menu while 3DORBIT is active, from the *3D* Orbit toolbar, or type **3dclip** at the Command: prompt. In the Adjust Clipping Planes window, there are two clipping planes, front and back. The front and back clipping planes are represented as lines at the top and bottom of the Adjust Clipping Planes window. You can adjust the lines with your pointing device and you can use toolbar buttons or the options from the shortcut menu to choose the clipping plane that you want to adjust.

If clipping planes are ON when you exit the *3D* orbit view, they remain on in the *2D* and *3D* view. The only way to turn off the clipping planes is to remove the check mark from the FRONT CLIPPING ON or BACK CLIPPING ON in the shortcut menu while 3DORBIT is active.

Using Continuous Orbit in the *3D* Orbit View

AutoCAD allows you to display the model in a continuous motion. To start the continuous motion, first select the CONTINUOUS ORBIT option from the shortcut menu when 3DORBIT is active. Then click and drag in the direction that you want the continuous orbit to move. Then release the pick button. The orbit continues to move in the direction that you indicated with your pointing device. The speed of the orbit rotation is determined by the speed with which you move the pointing device. If necessary, you can change the direction by clicking and dragging in a new direction. To stop continuous orbit, choose PAN, ZOOM, ORBIT, or ADJUST CLIPPING PLANES from the shortcut menu.

Resetting to Preset Views in the *3D* Orbit View

AutoCAD allows you to reset the view to the view that was current when your first entered the *3D* orbit view or set to one of the preset views, such as top, front, side, isometric, and so on. You can choose the reset view or one of the preset views from the shortcut view when 3DORBIT is active.

To exit 3DORBIT command, press ESC or ENTER.

WORKING WITH MULTIPLE VIEWPORTS IN 3D

As mentioned in Chapter 3, AutoCAD allows you to set multiple viewports to provide different views of your model. For example, you might set up viewports that display top, front, right side, and isometric views. You can do so with the help of the VPOINT, DVIEW, or 3DORBIT commands. To facilitate editing objects in different views, you can define a different UCS for each view. Each time you make a viewport current, you can begin drawing using the same UCS you used the last time that viewport was current.

The UCSVP system variable controls the setting for saving the UCS in the current viewport.

When UCSVP is set to 1 (default setting) in a viewport, the UCS last used in that viewport is saved with the viewport and is restored when the viewport is made current again.

When UCSVP is set to 0 in a viewport, its UCS is always the same as the UCS in the current viewport.

For example, you might set up four viewports: top view, front view, right side view, and isometric view. If you set the UCSVP system variable to 0 in the isometric viewport and to 1 in the top view, front view, and right side view, when you make the front viewport current, the isometric viewport's UCS reflects the UCS front viewport. Likewise, making the top viewport current switches the isometric viewport's UCS to match that of the top viewport.

CREATING *3D* OBJECTS

As mentioned earlier, there are several advantages to drawing objects in *3D*, including viewing the model at any angle, automatic generation of standard and auxiliary *2D* views, rendering and hidden-line removal, interference checking, and engineering analysis.

AutoCAD supports three types of *3D* modeling: wireframe, surface, and solid.

The wireframe model consists of only points, lines, and curves that describe the edges of the object. In AutoCAD you can create a wireframe model by positioning *2D* (planar) objects anywhere in *3D* space. In addition, AutoCAD provides additional commands, such as 3DPOLY, for creating a wireframe model.

The surface model is more sophisticated than the wireframe model. It defines not only the edges of a *3D* object but also its surfaces. The AutoCAD surface modeler defines faceted surfaces by using a polygonal mesh. It is possible to create a mesh to a flat or curved surface by locating the boundaries or edges of the surface.

Solid modeling is the easiest type of *3D* modeling. Solids are the unambiguous and informationally complete representation of the shape of a physical object. Fundamentally, solid modeling differs from wireframe or surface modeling in two ways:

- The information is more complete in the solid model.
- The method of construction of the model itself is inherently straightforward.

In wireframe or surface modeling, objects are created by positioning lines or surfaces in *3D* space. In solid modeling, you build the model as you would with building blocks; from beginning to end, you think, draw, and communicate in *3D*. One of the main benefits of solid modeling is its ability to be analyzed. You can calculate the mass properties of a solid object, such as its mass, center of gravity, surface area, and moments of inertia.

Each modeling type uses a different method for constructing *3D* models, and the use of each editing method varies among model types. It is recommended that you not mix modeling methods. It is possible in AutoCAD to convert between model types from solids to surfaces and from surfaces to wireframe; however, you cannot convert from wireframe to surfaces or surfaces to solids.

2D DRAW COMMANDS IN *3D* SPACE

You can use most of the draw commands discussed in previous chapters with a *Z* coordinate value. But *2D* objects such as polylines, circles, arcs, and solids are constrained to the *XY* plane of the current UCS. For these objects, the *Z* value is accepted only for the first coordinate to set the elevation of the *2D* object above or below the current plane. When you specify a point by using an Object Snap mode, it assumes the *Z* value of the point to which you snapped.

SETTING ELEVATION AND THICKNESS

You can create new objects by first setting up a default elevation (*Z* value). Subsequently, all the objects drawn assume the current elevation as the *Z* value whenever a *3D* point is expected but you supply only the *X* and *Y* values. The current elevation is maintained separately in model space and paper space.

Similarly, you create new objects with extrusion thickness by presetting a value for the thickness. Subsequently, all the objects drawn, such as lines, polylines, arcs, circles, and solids, assume the current thickness and extrude in their *Z* direction. For example, you can draw a cylinder by drawing a circle with preset thickness, or you can draw a cube simply by drawing a square with preset thickness.

Note: Thickness can be positive or negative. Thickness is in the Z axis direction for 2D objects. For 3D objects that can accept thickness, it is always relative to the current UCS. They will appear oblique if they do not lie in or parallel the current UCS. If thickness is added to a line drawn directly in the Z direction, the line appears to extend beyond its endpoint in the positive or negative thickness direction. Text and dimensions ignore the thickness setting.

To set the default elevation and thickness, invoke the ELEV command:

Command: prompt	**elev** (ENTER)

AutoCAD prompts:

Command: **elev**
Specify new default elevation <current>: *(specify the elevation, or press* ENTER *to accept the current elevation setting)*
Specify new default thickness <current>: *(specify the thickness, or press* ENTER *to accept the current thickness setting)*

For example, the following are the command sequences to draw a six-sided polygon at zero elevation with a radius of 2.5 units and a height of 4.5 units, and to place a cylinder at the center of the polygon with a radius of 1.0 unit at an elevation of 2.0 units with a height of 7.5 units, as shown in Figure 15–36.

Command: **elev**
New current elevation <0.0000>: (ENTER)
New current thickness <0.0000>: **4.5**

Command: **polygon** *(draw a polygon with a radius of 2.5 units)*
Command: **elev**
New current elevation <0.0000>: **2.0**
New current thickness <0.0000>: **7.5**
Command: **circle** *(draw a circle with a radius of 1.0 unit)*

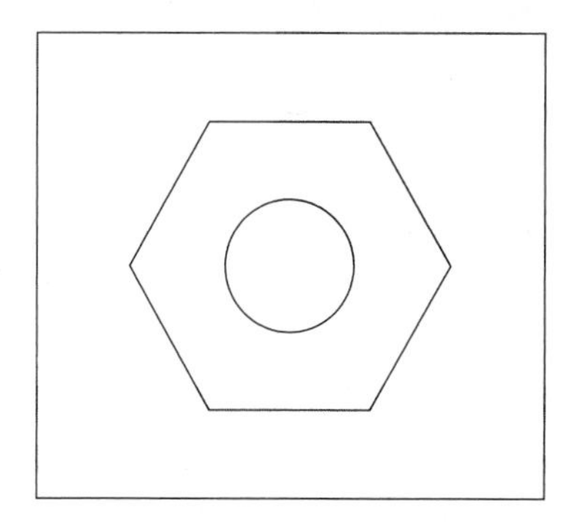

MODEL (PLAN VIEW)

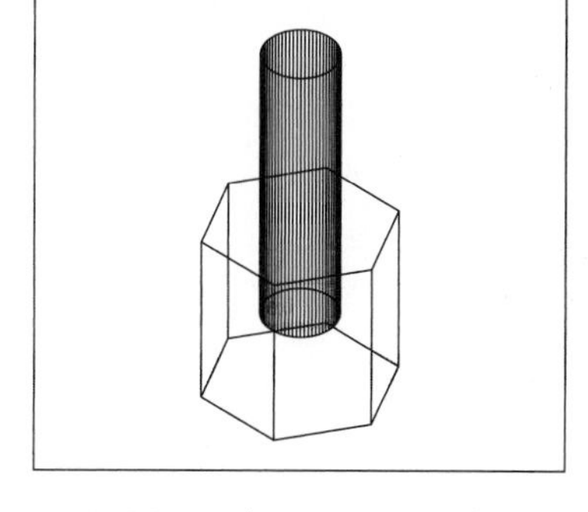

MODEL (VPOINT 1,-1,1)

Figure 15–36 *Specifying a new current elevation and thickness*

You can change the thickness and elevation (*Z* coordinate) of the existing objects by invoking the PROPERTIES command.

CREATING A REGION OBJECT

Regions are *2D* areas you create from closed shapes or loops. The REGION command creates a region object from a selection set of objects. Closed polylines, lines, curves, circular arcs, circles, elliptical arcs, ellipses, and splines are valid selections. Once you create a region, you can extrude it with the EXTRUDE command to make a *3D* solid. You can also create a composite region with the UNION, SUBTRACTION, and INTERSECTION commands. If necessary, you can hatch a region with the BHATCH command.

AutoCAD converts closed *2D* and planar *3D* polylines in the selection set to separate regions and then converts polylines, lines, and curves that form closed planar loops. If more than two curves share an endpoint, the resultant region might be arbitrary. Each object retains its layer, linetype, and color. AutoCAD deletes the original objects after converting them to regions and, by default, does not hatch the regions.

Invoke the REGION command:

Draw toolbar	Choose the REGION command (see Figure 15–37)
Draw menu	Choose Region
Command: prompt	**region** (ENTER)

Figure 15–37 *Invoking the* REGION *command from the Draw toolbar*

AutoCAD prompts:

Command: **region**
Select objects: *(select the objects, and press* ENTER *to complete object selection)*

DRAWING *3D* POLYLINES

The 3DPOLY command draws polylines with independent *X*, *Y*, and *Z* axis coordinates using the continuous linetype. The 3DPOLY command works similar to the PLINE command, with a few exceptions. Unlike the PLINE command, 3DPOLY draws only straight-line segments without variable width. Editing a *3D* polyline with the PEDIT command is similar to editing a *2D* polyline, except for some options. *3D* polylines cannot be joined, curve-fit with arc segments, or given a width or tangent.

Invoke the 3DPOLY command:

Draw menu	Choose the 3D Polyline command
Command: prompt	**3dpoly** (ENTER)

AutoCAD prompts:

Command: **3dpoly**
Specify start point of polyline: *(specify a point)*
Specify endpoint of line or [Undo]: *(specify a point or select an option)*
Specify endpoint of line or [Undo]: *(specify a point or select an option)*
Specify endpoint of line or [Close/Undo]: *(specify a point or select one of the available options)*

The available options are similar to those for the PLINE command, described earlier in Chapter 4.

CREATING *3D* FACES

When you create a *3D* model, it is often necessary to have solid surfaces for hiding and shading. These surfaces are created with the 3DFACE command. The 3DFACE command creates a solid surface, and the command sequence is similar to that for the SOLID command. Unlike the SOLID command, you can give differing *Z* coordinates for the corner points of a face, forming a section of a plane in space. Unlike the SOLID command, a 3DFACE is drawn from corner to corner clockwise or counterclockwise around the object (and it does not draw a "bow tie"). A *3D* face is a plane defined by either three or four points used to represent a surface. It provides a means of controlling which edges of a *3D* face will be visible. You can describe complex, *3D* polygons using multiple *3D* faces, and you can tell AutoCAD which edges you want to be drawn. If you have an object with curved surfaces, then the 3DFACE command is not suitable. One of the mesh commands is more appropriate, as explained later in the chapter.

Invoke the 3DFACE command:

Surfaces toolbar	Choose the 3DFACE command (see Figure 15–38)
Draw menu	Choose Surfaces > *3D* Face
Command: prompt	**3dface** (ENTER)

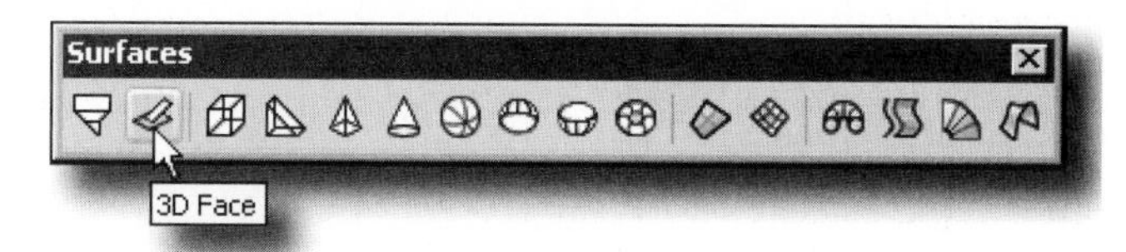

Figure 15–38 *Invoking the 3DFACE command from the Surfaces toolbar*

AutoCAD prompts:

Command: **3dface**
Specify first point or [Invisible]: *(specify the first point or enter i)*

Specify the first point, and AutoCAD prompts you for the second, third, and fourth points in sequence. Then AutoCAD closes the face from the fourth point to the first point and prompts for the third point. If you give a null response to the prompt for the third point, AutoCAD closes the *3D* face with four sides, terminates the command, and takes you to the "Command:" prompt.

If you want to draw additional faces in one command sequence, the last two points of the first face become the first two points for the second face. And the last two points of the second face become the first two points of the third face, and so on. You have to be very careful in drawing several faces in one command sequence, since AutoCAD does not have an Undo option that works inside the 3DFACE command. A single mistake can cause the entire face to be redrawn. For this reason, it is a good idea to draw *3D* faces one at a time.

For example, the following command sequence demonstrates the placement of *3D* faces, as shown in Figure 15–39.

Command: **3dface**
Specify first point or [Invisible]: *(select point A1)*
Specify second point or [Invisible]: *(select point A2)*
Specify third point or [Invisible] <exit>: *(select point A3)*
Specify fourth point or [Invisible] <create three-sided face>: *(select point A4)*
Specify third point or [Invisible] <exit>: *(select point A5)*
Specify fourth point or [Invisible] <create three-sided face>: *(select point A6)*
Specify third point or [Invisible] <exit>: *(select point A7)*
Specify fourth point or [Invisible] <create three-sided face>: *(select point A8)*
Specify third point or [Invisible] <exit>: *(select point A1)*
Specify fourth point or [Invisible]<create three-sided face>: *(select point A2)*
Specify third point or [Invisible] <exit>: (ENTER)

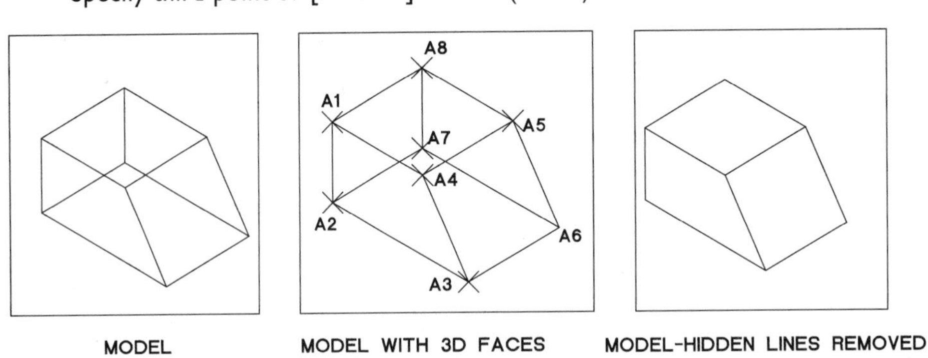

Figure 15–39 *Drawing a 3D object with the* 3DFACE *command*

The surface created, as shown in Figure 15–39, required four faces to cover it. Some of the edges are overlapping, which is not acceptable when viewing the object. The 3DFACE command allows face edges to be "invisible." To create an invisible edge, the letter **i** must be entered at the prompt for the first point of the edge to be invisible, and then the point can be entered.

The following command sequence shows the placement of *3D* faces for invisible edges, as shown in Figure 15–39.

Command: **3dface**
Specify first point or [Invisible]: *(select point A1)*
Specify second point or [Invisible]: *(select point A8)*
Specify third point or [Invisible] <exit>: *(select point A5)*
Specify fourth point or [Invisible] <create three-sided face>: *(select point A4)*
Specify third point or [Invisible] <exit>: *(select point A3)*
Specify fourth point or [Invisible] <create three-sided face>: *(select point A6)*
Specify third point or [Invisible] <exit>: *(select point A7)*
Specify fourth point or [Invisible] <create three-sided face>: **i** (ENTER)
Specify fourth point or [Invisible]<create three-sided face>: *(select point A8)*
Specify third point or [Invisible] <exit>: *(select point A1)*

Specify fourth point or [Invisible]<create three-sided face>: *(select point A2)*
Specify third point or [Invisible] <exit>: *(select point A3)*
Specify fourth point or [Invisible] <create three-sided face>: *(select point A4)*
Specify third point or [Invisible] <exit>: (ENTER)

3DFACE commands ignore thickness. The SPLFRAME system variable controls the display of invisible edges in *3D* faces. If SPLFRAME is set to a nonzero value, all invisible edges of *3D* faces are displayed.

Controlling the Visibility of a *3D* Face

The EDGE command allows you to change the visibility of *3D* face edges. You can selectively set the edges to ON/OFF.

Invoke the EDGE command:

Surfaces toolbar	Choose the EDGE command (see Figure 15–40)
Draw menu	Choose Surfaces > Edge
Command: prompt	**edge** (ENTER)

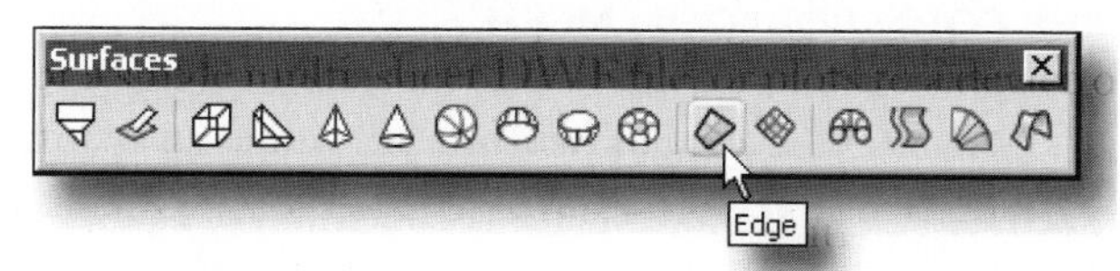

Figure 15–40 *Invoking the* EDGE *command from the Surfaces toolbar*

AutoCAD prompts:

Command: **edge**
Specify edge of 3dface to toggle visibility or [Display]: *(select the edges and press* ENTER *to complete selection, or select the Display option)*

The Display option highlights invisible edges of *3D* faces so you can change the visibility of the edges. AutoCAD prompts:

Enter selection method for display of hidden edges [Select/All] <All>:

The default option displays all the invisible edges. Once the edges are displayed, AutoCAD allows you to change the status of the visibility.

The Select option allows you to selectively identify hidden edges to be displayed. Then, if necessary, you can change the status of the visibility.

Note: All edges are visible regardless of the visibility setting if the SPLFRAME system variable is set to 1 (ON). If necessary, you can also use the PROPERTIES command to modify 3D faces.

CREATING MESHES

A *3D* mesh is a single object. It defines a flat surface or approximates a curved one by placing multiple *3D* faces on the surface of an object. It is a series of lines consisting of columns and rows. AutoCAD lets you determine the spacing between rows (M) and columns (N).

It is possible to create a mesh for a flat or curved surface by locating the boundaries or edges of the surface. Surfaces created in this fashion are called geometry-generated surfaces. Their size and shape depend on the boundaries used to define them and on the specific formula (or command) used to determine the location of the vertices between the boundaries. AutoCAD provides four different commands to create geometry-generated surfaces: RULESURF, REVSURF, TABSURF, and EDGESURF. The differences between these types of meshes depend on the types of objects connecting the surfaces. In addition, AutoCAD provides two additional commands for creating polygon mesh: 3DMESH and PFACE. The key to using meshes effectively is to understand the purpose and requirement of each type of mesh and to select the appropriate one for the given condition.

CREATING A FREE-FORM POLYGON MESH

You can define a free-form *3D* polygon mesh by means of the 3DMESH command. Initially, it prompts you for the number of rows and columns, in terms of mesh M and mesh N, respectively. Then it prompts for the location of each vertex in the mesh. The product of M x N gives the number of vertices for the mesh.

Invoke the 3DMESH command:

Surfaces toolbar	Choose the 3DMESH command (see Figure 15–41)
Draw menu	Choose Surfaces > *3D* Mesh
Command: prompt	**3dmesh** (ENTER)

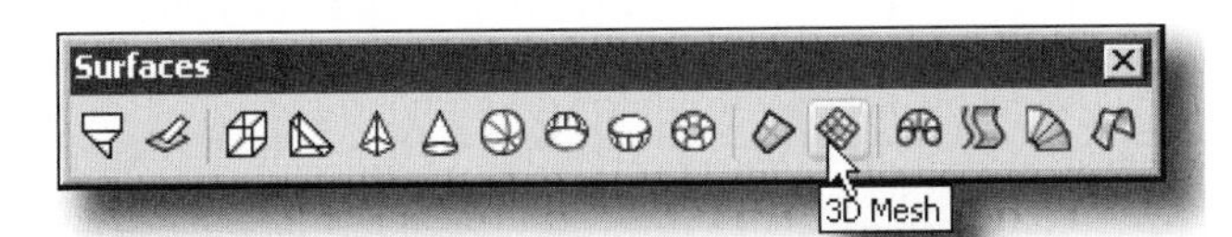

Figure 15–41 *Invoking the* 3DMESH *command from the Surfaces toolbar*

AutoCAD prompts:

Command: **3dmesh**
Enter size of mesh in M direction: *(specify an integer value between 2 and 256)*
Enter size of mesh in N direction: *(specify an integer value between 2 and 256)*

The points for each vertex must be entered separately, and the M value can be considered the number of lines that will be connected by faces, while the N value is the number of points each line consists of. Vertices may be specified as *2D* or *3D* points, and may be any distance from each other.

The following command sequence creates a simple 5 x 4 polygon mesh. The mesh is created between the first point of the first line, the first point of the second line, and so on, as shown in Figure 15–42.

Command: **3dmesh**
Enter size of mesh in M direction: **5**
Enter size of mesh in N direction: **4**
Specify location for vertex (0,0): *(select point A1)*
Specify location for vertex (0,1): *(select point A2)*
Specify location for vertex (0,2): *(select point A3)*
Specify location for vertex (0,3): *(select point A4)*
Specify location for vertex (1,0): *(select point B1)*
Specify location for vertex (1,1): *(select point B2)*
Specify location for vertex (1,2): *(select point B3)*
Specify location for vertex (1,3): *(select point B4)*
Specify location for vertex (2,0): *(select point C1)*
Specify location for vertex (2,1): *(select point C2)*
Specify location for vertex (2,2): *(select point C3)*
Specify location for vertex (2,3): *(select point C4)*
Specify location for vertex (3,0): *(select point D1)*
Specify location for vertex (3,1): *(select point D2)*
Specify location for vertex (3,2): *(select point D3)*
Specify location for vertex (3,3): *(select point D4)*
Specify location for vertex (4,0): *(select point E1)*
Specify location for vertex (4,1): *(select point E2)*
Specify location for vertex (4,2): *(select point E3)*
Specify location for vertex (4,3): *(select point E4)*

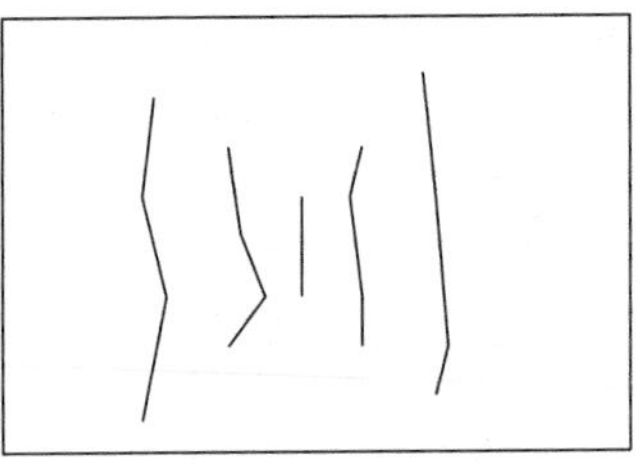

LINE DIAGRAM

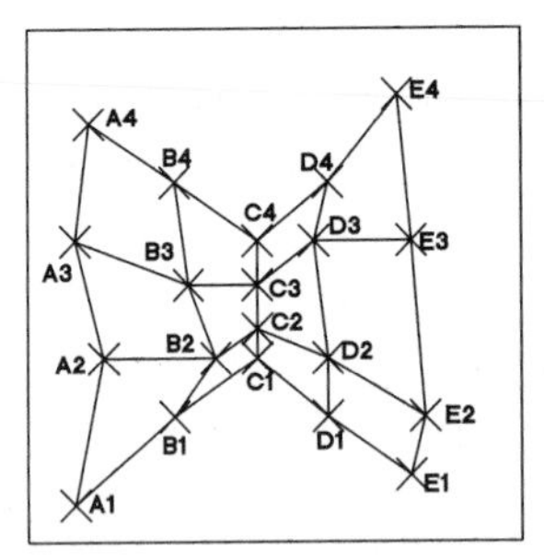

LINE DIAGRAM WITH 3D FACES

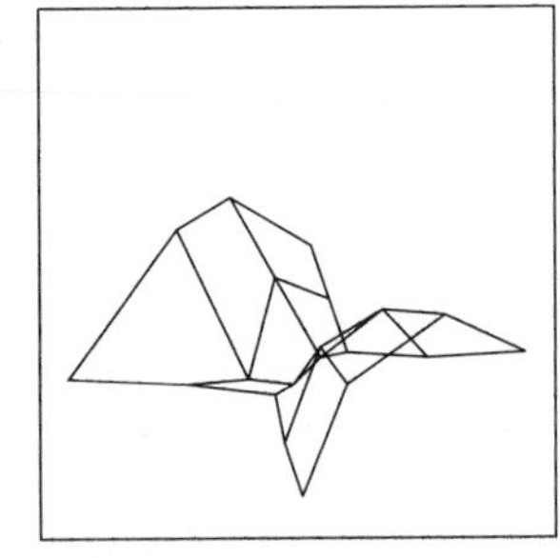

MODEL AT VPOINT 1,−1,1

Figure 15–42 *Creating a 3D mesh*

Note: Specifying 3D mesh of any size can be time-consuming and tedious. It is preferable to use one of the commands for geometry-generated surfaces, such as RULESURF, REVSURF, TABSURF, or EDGESURF. The 3DMESH command is designed primarily for AutoLISP and ADS applications.

CREATING A *3D* POLYFACE MESH

The PFACE command allows you to construct a mesh of any topology you desire. This command is similar to the 3DFACE command, but it creates surfaces with invisible interior divisions. You can specify any number of vertices and *3D* faces, unlike the other meshes. Producing this kind of mesh lets you conveniently avoid creating many unrelated *3D* faces with the same vertices.

AutoCAD first prompts you to pick all the vertex points, and then you can create the faces by entering the vertex numbers that define their edges.

Invoke the PFACE command:

Command: prompt	**pface** (ENTER)

AutoCAD prompts:

Command: **pface**
Specify location for vertex 1: *(specify a point)*

One by one, specify all the vertices used in the mesh, keeping track of the vertex numbers shown in the prompts. You can specify the vertices as *2D* or *3D* points and place them at any distance from one another. Enter a null response (press ENTER) after specifying all the vertices, and AutoCAD prompts for a vertex number that has to be assigned to each face. You can define any number of vertices for each face, and enter a null response (press ENTER). AutoCAD prompts for the next face. After all the vertex numbers for all the faces are defined, enter a null response (press ENTER), and AutoCAD draws the mesh.

The following command sequence creates a simple polyface for a given six-sided polygon, with a circle of 1" radius drawn at the center of the polygon at a depth of –2, as shown in Figure 15–43.

Command: **pface**
Specify location for vertex 1: *(select point A1)*
Specify location for vertex 2 or <define faces>: *(select point A2)*
Specify location for vertex 3 or <define faces>: *(select point A3)*
Specify location for vertex 4 or <define faces>: *(select point A4)*
Specify location for vertex 5 or <define faces>: *(select point A5)*
Specify location for vertex 6 or <define faces>: *(select point A6)*
Specify location for vertex 7 or <define faces>: (ENTER)
Face 1, Vertex 1:
Enter a vertex number or [Color/Layer]: **1**
Face 1, Vertex 2:
Enter a vertex number or [Color/Layer] <new face>: **2**
Face 1, Vertex 3:
Enter a vertex number or [Color/Layer] <new face>: **1**
Face 1, Vertex 4:
Enter a vertex number or [Color/Layer] <new face>: **2**
Face 1, Vertex 5:
Enter a vertex number or [Color/Layer] <new face>: **1**
Face 1, Vertex 6:
Enter a vertex number or [Color/Layer] <new face>: **2**
Face 1, Vertex 7:
Enter a vertex number or [Color/Layer] <new face>: (ENTER)
Face 2, Vertex 1:
Enter a vertex number or [Color/Layer]: (ENTER)

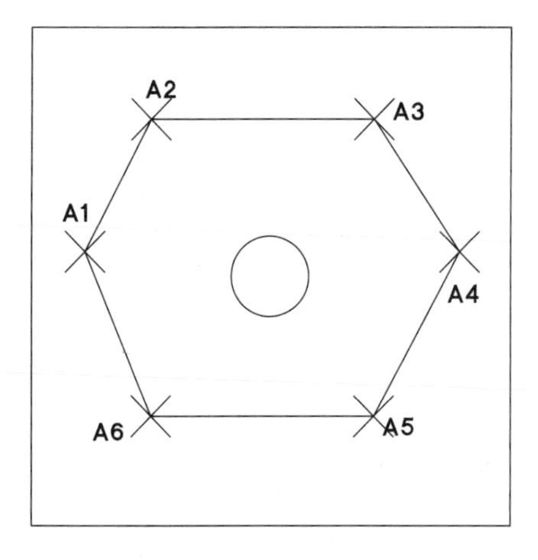

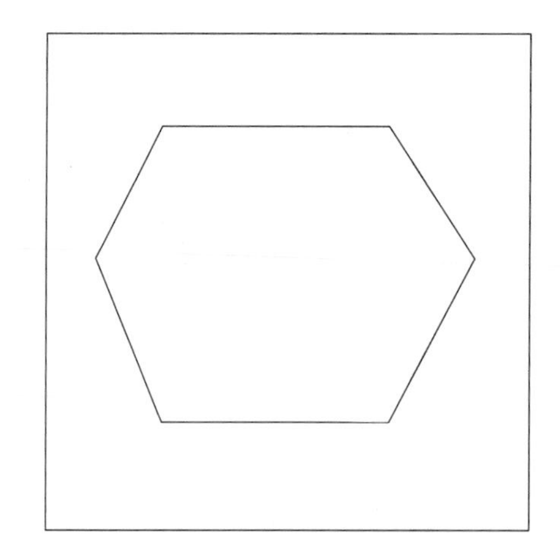

Figure 15–43 *Creating a polyface for a given six-sided polygon with a circle at the center*

If necessary, you can make an edge of the polyface mesh invisible by entering a negative number for the beginning vertex of the edge. By default, the faces are drawn on the current layer and with the current color. However, you can create the faces in layers and colors different from the original object. You can assign a layer or color by responding to the "Enter a vertex number or [Color/Layer]:" prompt with L for layer or C for color. AutoCAD then prompts for the name of the layer or color, as appropriate. It will continue with the prompts for vertex numbers. The layer or color you enter is used for the face you are currently defining and for any subsequent faces created.

Note: Specifying the layer or color within the PFACE command does not change object properties for subsequent commands. Specifying PFACE of any size can be time-consuming and tedious. It is preferable to use one of the commands for geometry-generated surfaces, such as RULESURF, REVSURF, TABSURF, or EDGESURF. The PFACE command is designed primarily for AutoLISP and ADS applications.

CREATING A RULED SURFACE BETWEEN TWO OBJECTS

The RULESURF command creates a polygon mesh between two objects. The two objects can be lines, points, arcs, circles, *2D* polylines, or *3D* polylines. If one object is open, such as a line or an arc, the other must be open too. If one is closed, such as a circle, so must the other be. A point can be used as one object, regardless of whether the other object is open or closed. But only one of the objects can be a point.

RULESURF creates an M x N mesh, with the value of mesh M a constant 2. The value of mesh N can be changed depending on the required number of faces. This can be done with the help of the SURFTAB1 system variable. By default, SURFTAB1 is set to 6.

The following command sequence shows how to change the value of SURFTAB1 from 6 to 20:

Command: **surftab1**
Enter new value for SURFTAB1 <6>: **20**

Invoke the RULESURF command:

Surfaces toolbar	Choose the RULESURF command (see Figure 15–44)
Draw menu	Choose Surfaces > Ruled Surface
Command: prompt	**rulesurf** (ENTER)

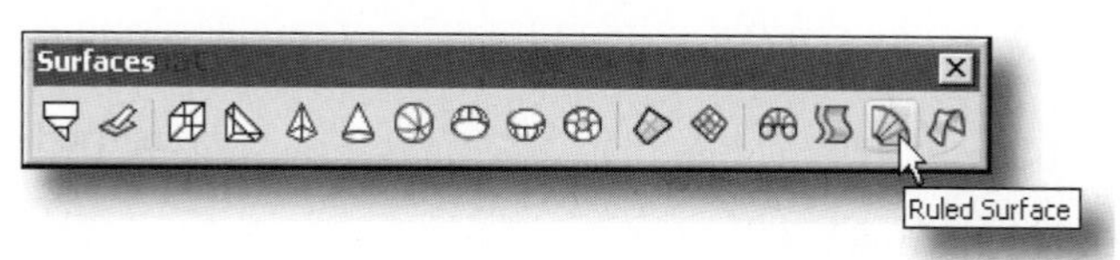

Figure 15–44 *Invoking the* RULESURF *command from the Surfaces toolbar*

AutoCAD prompts:

Command: **rulesurf**
Current wire frame density: SURFTAB1=<current>
Select first defining curve: *(select the first defining curve)*
Select second defining curve: *(select the second defining curve)*

Identify the two objects to which a mesh has to be created. See Figure 15–45, in which an arc (A1–A2) and a line (A3–A4) were selected and a mesh was created with SURFTAB1 set to 15. Two lines (B1–B2 and B3–B4) were selected and a mesh was created with SURFTAB1 set to 20. A cone was created by drawing a circle at an elevation of 0 and a point (C1) at an elevation of 5, followed by the application of the RULESURF command with SURFTAB1 set to 20.

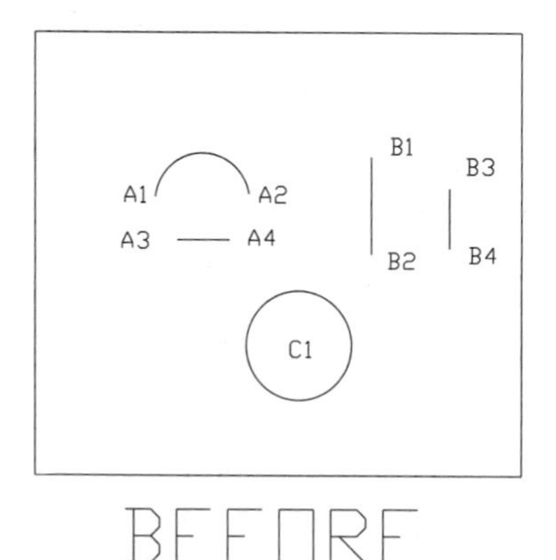

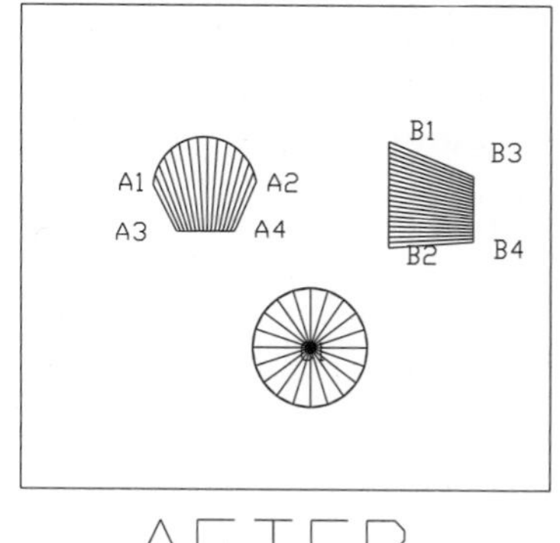

Figure 15–45 *Creating ruled surfaces with the* RULESURF *command*

Note: When you identify the two objects, make sure to select on the same side of the objects, left or right. If you pick the left side of one of the sides and the right side of the other, you will get a bow-tie effect.

CREATING A TABULATED SURFACE

The TABSURF command creates a surface extrusion from an object with a length and direction determined by the direction vector. The object is called the defining curve and can be a line, arc, circle, *2D* polyline, or *3D* polyline. The direction vector can be a line or open polyline. The endpoint of the direction vector nearest the specified point will be swept along the path curve, describing the surface. Once the mesh is created, the direction vector can be deleted. The number of intervals along the path curve is controlled by the SURFTAB1 system variable, similar to the RULESURF command. By default, SURFTAB1 is set to 6.

Invoke the TABSURF command:

Surfaces toolbar	Choose the TABSURF command (see Figure 15–46)
Draw menu	Choose Surfaces > Tabulated Surface
Command: prompt	**tabsurf** (ENTER)

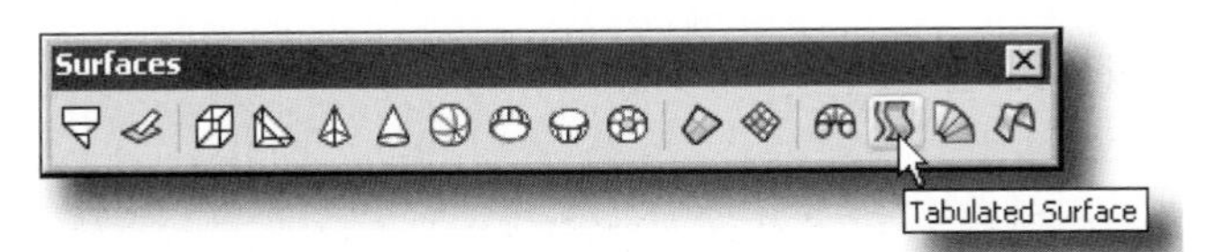

Figure 15–46 *Invoking the* TABSURF *command from the Surfaces toolbar*

AutoCAD prompts:

Command: **tabsurf**
Select object for path curve: *(select the path curve)*
Select object for direction vector: *(select the direction vector)*

The location at which the direction vector is selected determines the direction of the constructed mesh. The mesh is created in the direction from the selection point to the nearest endpoint of the direction vector. In Figure 15–47, a mesh was created with SURFTAB1 set to 16 by identifying a polyline as the path curve and the line as the direction vector.

Note: The length of the *3D* mesh is the same as that of the direction vector.

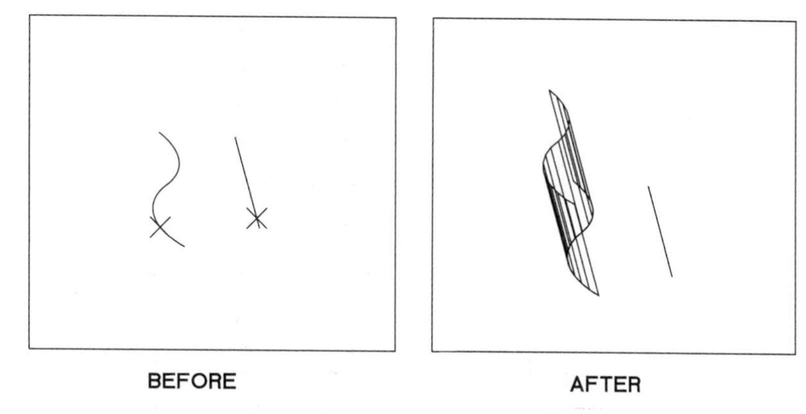

Figure 15–47 *Creating a tabulated surface with the* TABSURF *command*

CREATING A REVOLVED SURFACE

The REVSURF command creates a *3D* mesh that follows the path defined by a path curve and is rotated around a centerline. The object used to define the path curve may be an arc, circle, line, *2D* polyline, or *3D* polyline. Complex shapes consisting of lines, arcs, or polylines can be joined into one object using the PEDIT command, and then you can create a single rotated mesh instead of several individual meshes.

The centerline can be a line or polyline that defines the axis around which the faces are constructed. The centerline can be of any length and at any orientation. If necessary, you can erase the centerline after the construction of the mesh. Thus it is recommended that you make the axis longer than the path curve so that it is easy to erase after the rotation.

In the case of REVSURF, both the mesh M size as well as mesh N are controlled by the SURFTAB1 and SURFTAB2 system variables, respectively. The SURFTAB1 value determines how many faces are placed around the rotation axis and can be an integer value between 3 and 1024. The SURFTAB2 determines how many faces are used to simulate the curves created by arcs or circles in the path curve. By default, SURFTAB1 and SURFTAB2 are set to 6.

The following command sequence shows how to change the value of SURFTAB1 from 6 to 20 and that of SURFTAB2 from 6 to 15:

Command: **surftab1**
Enter new value for SURFTAB1 <6>: **20**

Command: **surftab2**
Enter new value for SURFTAB1 <6>: **15**

Invoke the REVSURF command:

Surfaces toolbar	Choose the REVSURF command (see Figure 15–48)
Draw menu	Choose Surfaces > Revolved Surface
Command: prompt	**revsurf** (ENTER)

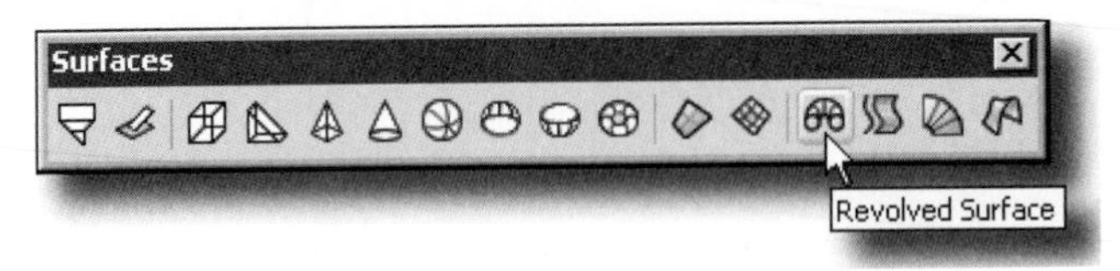

Figure 15–48 *Invoking the* REVSURF *command from the Surfaces toolbar*

AutoCAD prompts:

Command: **revsurf**
Current wire frame density: SURFTAB1=<current> SURFTAB2=<current>
Select object to revolve: *(select the path curve)*
Select object that defines the axis of revolution: *(select the axis of revolution)*
Specify Start angle<0>: *(specify the start angle, or press* ENTER *to accept the default angle)*
Specify Included angle (+=ccw,-=cw)<360>: *(specify the included angle, or press* ENTER *to accept the default)*

For the "Specify Start angle:" prompt, it does not matter if you are going to rotate the curve 360 degrees (full circle). If you want to rotate the curve only a certain angle, you must provide the start angle in reference to three o'clock (the default) and then indicate the angle of rotation in the counterclockwise (positive) or clockwise (negative) direction. See Figure 15–49, in which a mesh was created with SURFTAB1 set to 16 and SURFTAB2 set to 12, by identifying a closed polyline as the path curve and the vertical line as the axis of revolution, and then rotated 360 degrees.

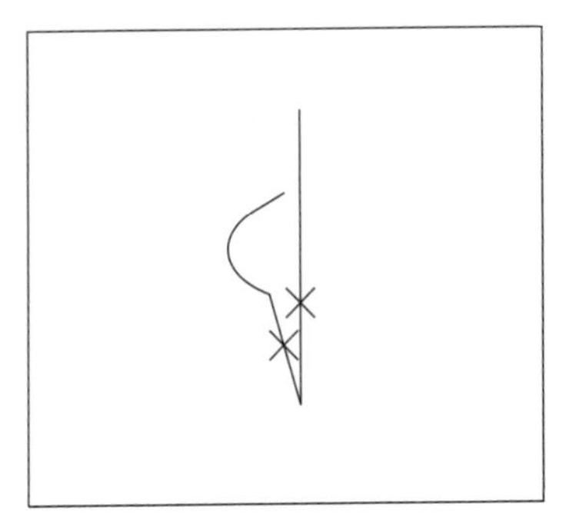

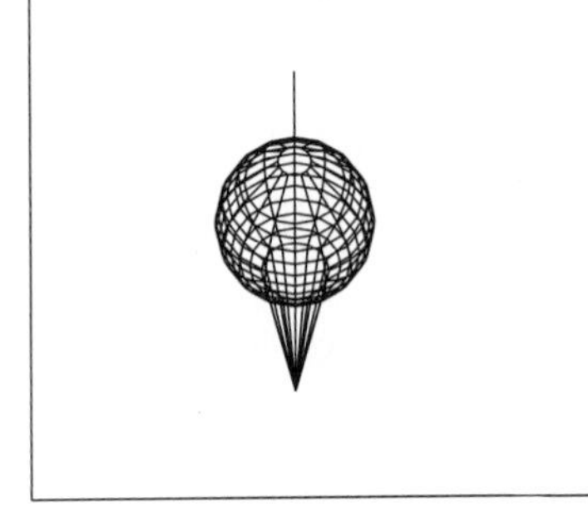

BEFORE AFTER

Figure 15–49 *Creating a meshed surface with the* REVSURF *command*

CREATING AN EDGE SURFACE WITH FOUR ADJOINING SIDES

The EDGESURF command allows a mesh to be created with four adjoining sides defining its boundaries. The only requirement for EDGESURF is that the mesh has exactly four sides. The sides can be lines, arcs, or any combination of polylines and polyarcs. Each side must join the adjacent one to create a closed boundary.

In EDGESURF, both the mesh M size and mesh N can be controlled by the SURFTAB1 and SURFTAB2 system variables, respectively, just as in REVSURF.

Invoke the EDGESURF command:

Surfaces toolbar	Choose the EDGESURF command (see Figure 15–50)
Draw menu	Choose Surfaces > Edge Surface
Command: prompt	**edgesurf** (ENTER)

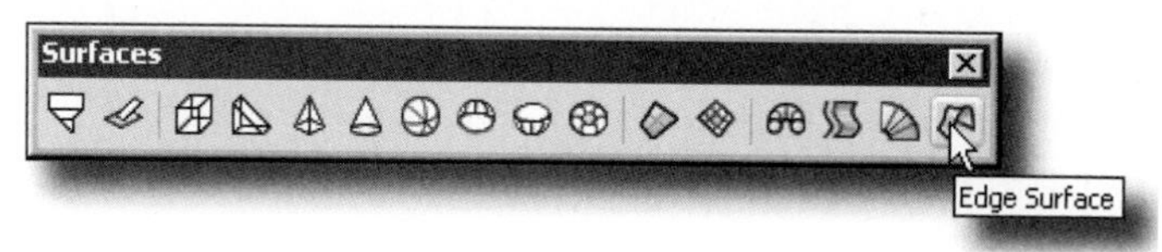

Figure 15–50 *Invoking the* EDGESURF *command from the Surfaces toolbar*

AutoCAD prompts:

Command: **edgesurf**
Current wire frame density: SURFTAB1=<current> SURFTAB2=<current>
Select edge 1 for surface edge: *(select the first edge)*
Select edge 2 for surface edge: *(select the second edge)*
Select edge 3 for surface edge: *(select the third edge)*
Select edge 4 for surface edge: *(select the fourth edge)*

When picking four sides, you must be consistent in picking the beginning of each polyline group. If you pick the beginning of one side and the end of another, the final mesh will cross and look strange. See Figure 15–51, in which a mesh was created with SURFTAB1 set to 25 and SURFTAB2 set to 20 by identifying four sides.

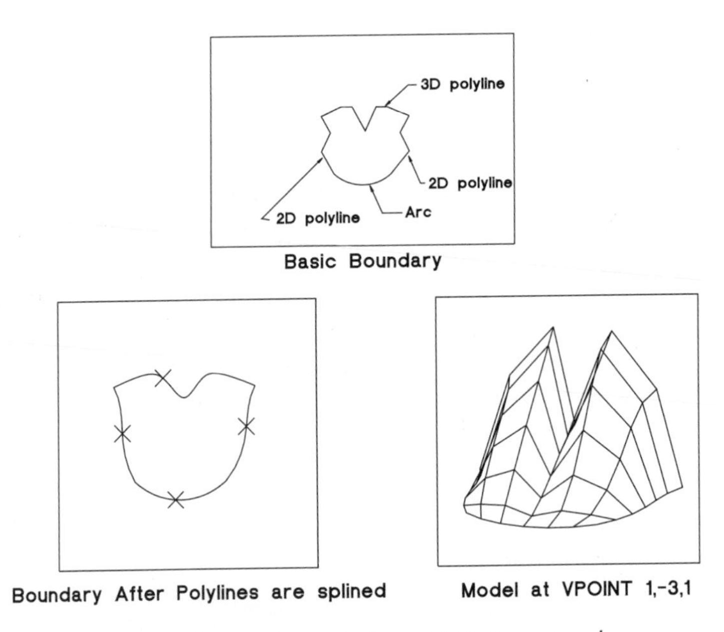

Figure 15–51 *Creating a meshed surface with the* EDGESURF *command*

EDITING POLYMESH SURFACES

As with blocks, polylines, hatch, and dimensioning, you can explode a mesh. When you explode a mesh it separates into individual *3D* faces. Meshes can also be altered by invoking the PEDIT command, similar to editing polylines using the PEDIT command. Most of the options under the PEDIT command can be applied to meshes, except giving width to the edges of the polymesh. For a detailed explanation of the PEDIT command, refer to Chapter 5.

EDITING IN *3D*

This section describes how to perform various *3D* editing operations, such as aligning, rotating, mirroring, arraying, extending, and trimming.

ALIGNING OBJECTS

The ALIGN command allows you to translate and rotate objects in *3D* space regardless of the position of the current UCS. Three source points and three destination points define the move. The ALIGN command lets you select the objects to move, and then subsequently prompts for three source points and three destination points.

Invoke the ALIGN command:

Modify menu	Choose *3D* Operation > Align
Command: prompt	**align** (ENTER)

AutoCAD prompts:

Command: **align**
Select objects: *(select the objects)*

Select the objects to move and press ENTER; AutoCAD then prompts for three source points and three destination points. Temporary lines are displayed between the matching pairs of source and destination points. If you enter all six points, the move consists of a translation and two rotations based on the six points. The translation moves the 1st source point to the 1st destination point. The first rotation aligns the line defined by the 1st and 2nd source points with the line defined by the 1st and 2nd destination points. The second rotation aligns the plane defined by the three source points with the plane defined by the three destination points.

If instead of entering three pairs of points you enter two pairs of points, the transformation reduces to a translation from the 1st source point to the 1st destination point and a rotation such that the line passing through the two source points aligns with the line passing through the two destination points. The transformation occurs in either *2D* or *3D*, depending on your response to the following prompt:

<2d> or 3d transformation:

If you enter 2d or press ENTER, the rotation is performed in the *XY* plane of the current UCS. If you enter 3d, the rotation is in the plane defined by the two destination points and the 2nd source point.

If you enter only one pair of points, the transformation reduces to a simple translation from the source to the destination point. This is similar to using the AutoCAD regular MOVE command without the dynamic dragging.

ROTATING OBJECTS ABOUT A *3D* OBJECT

The ROTATE3D command lets you rotate an object about an arbitrary *3D* axis.

Invoke the ROTATE3D command:

Modify menu	Choose *3D* Operation > Rotate 3D
Command: prompt	**rotate3d** (ENTER)

AutoCAD prompts:

Command: **rotate3d**
Select objects: *(select objects)*

AutoCAD lists the options for selecting the axis of rotation:

Specify first point on axis or define axis by [Object/Last/View/Xaxis/Yaxis/Zaxis/ 2points]: *(specify a point or select one of the available options)*

2points

The 2points option prompts you for two points. The axis of rotation is the line that passes through the two points, and the positive direction is from the first to the second point.

Object

The Object option lets you select an object and then derives the axis of rotation based on the type of object selected. Valid objects include line, circle, arc, and pline.

Last

The Last option specifies the last-used axis. If there is no last axis, a message to that effect is displayed and the axis selection prompt is redisplayed.

View

The View option prompts you to select a point. The axis of rotation is perpendicular to the view direction and passes through the selected point. The positive axis direction is toward the viewer.

X/Y/Zaxis

The X/Y/Z axis option prompts you to select a point. The axis of rotation is parallel to the standard axis of the current UCS and passes through the selected point.

Once you have selected the axis of rotation, AutoCAD prompts:

> Specify by Rotation angle or [Reference]: *(specify the rotation angle, or enter* ***r****, for reference)*

AutoCAD rotates the selected object(s) to the specified rotation angle. The Reference option allows you to specify the current orientation as the reference angle or to show AutoCAD the angle by pointing to the two endpoints of a line to be rotated and then specifying the desired new rotation. AutoCAD automatically calculates the rotation angle and rotates the selected object appropriately.

See Figure 15–52 for an example of rotating a cylinder around the Z axis.

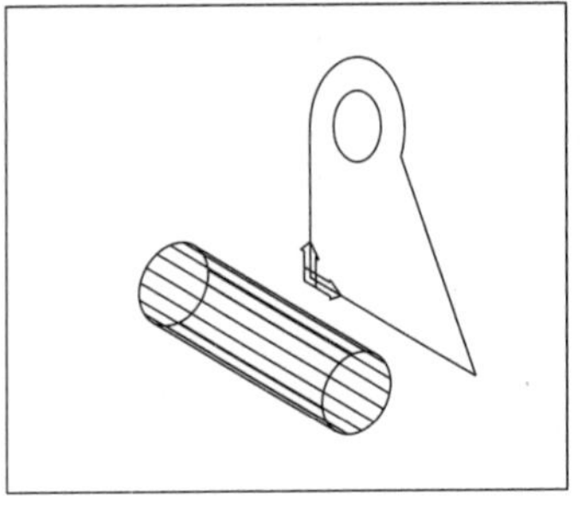

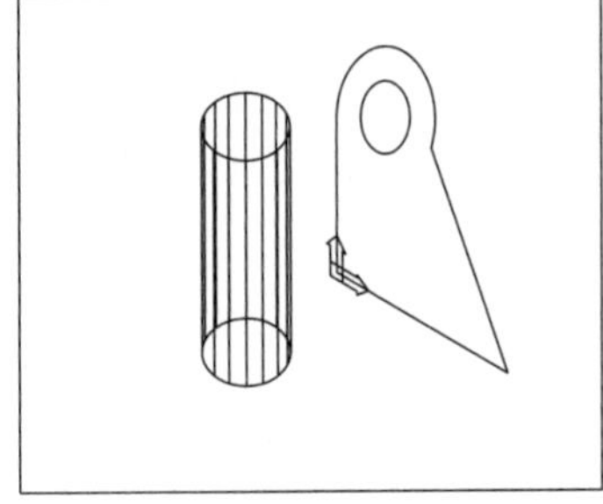

BEFORE AFTER

Figure 15–52 *Rotating a cylinder about the Z axis with the* ROTATE3D *command*

MIRRORING ABOUT A PLANE

The MIRROR3D command lets you mirror a selected object about a plane.

Invoke the MIRROR3D command:

Modify menu	Choose *3D* Operation > Mirror 3D
Command: prompt	***mirror3D*** (ENTER)

AutoCAD prompts:

Command: **mirror3d**
Select objects: *(select the objects)*

Select the objects to mirror and press ENTER. AutoCAD then lists the options for selecting the mirroring plane:

Specify first point of mirror plane [3 points] or [Object/Last/Zaxis/View/XY/YZ/XZ/3points] <3points>: *(select one of the options to specify the mirroring plane)*
Delete source objects? <No>: *(enter* ***y*** *to delete the objects or* ***n*** *not to delete the objects)*

3points

The 3points option prompts you for three points. The mirroring plane is the plane that passes through the three selected points.

Object

The Plane by Object option lets you select an object, and the mirroring plane is aligned with the plane of the object selected. Valid objects include: circle, arc, and pline.

Last

The Last option specifies the last used plane. If there is no last plane, a message is displayed to that effect and the plane selection prompt is redisplayed.

Zaxis

The Zaxis option prompts you to select two points. The mirroring plane is the plane specified by a point on the plane and a point on the plane's normal (perpendicular to the plane).

View

The View option prompts you to select a point. The mirroring plane is created perpendicular to the view direction and passes through the selected point.

XY/YZ/XZ

The XY/YZ/XZ option prompts you to select a point. The mirroring plane is created parallel to the standard plane of the current UCS and passes through the selected point.

See Figure 15–53 for an example of mirroring a cylinder aligned with the plane of the object selected (pline).

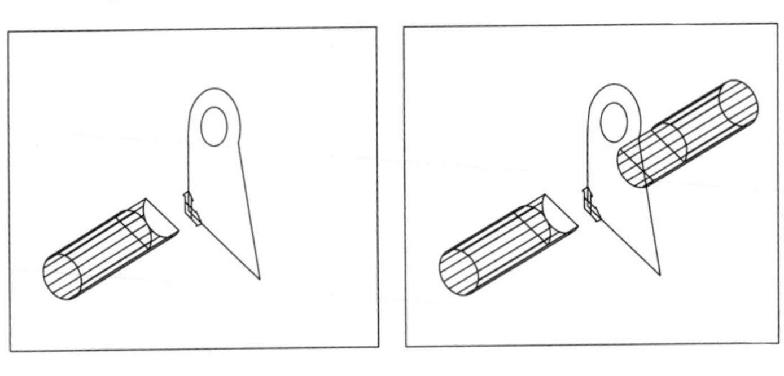

Figure 15–53 *Mirroring a cylinder about a polyline object*

CREATING A *3D* ARRAY

The 3DARRAY command is used to make multiple copies of selected objects in either rectangular or polar array in *3D*. In the rectangular array, specify the number of columns (*X* direction), the number of rows (*Y* direction), the number of levels (*Z* direction), and the spacing between columns, rows, and levels. In the polar array, specify the number of items to array, the angle that the arrayed objects are to fill, the start point and endpoint of the axis about which the objects are to be rotated, and whether or not the objects are rotated about the center of the group.

Invoke the 3DARRAY command:

Modify menu	Choose *3D* Operation > 3D Array
Command: prompt	**3darray** (ENTER)

AutoCAD prompts:

Command: **3Darray**
Select objects: *(select the objects to array)*
Enter type of array [Rectangular/Polar] (R): *(select one of the two available options)*

Rectangular Array

To generate a rectangular array, enter R (for rectangular array) and AutoCAD prompts:

Enter the number of rows(—)<1>: *(specify the number of rows or press* ENTER*)*
Enter the number of columns(|||)<1>: *(specify the number of columns or press* ENTER*)*

Number of levels(...)<1>: *(specify the number of levels or press* ENTER*)*
Specify the distance between rows(—)<1>: *(specify a distance)*
Specify the distance between columns(—)<1>: *(specify a distance)*
Specify the distance between levels(—)<1>: *(specify a distance)*

Any combination of whole numbers of columns, rows, and levels may be entered. AutoCAD includes the original object in the number you enter. An array must have at least two columns, two rows, or two levels. Specifying one row requires that more than one column be specified, or vice versa. Specifying one level creates a *2D* array. Column, row, and level spacing can be different from one another. They can be entered separately when prompted, or you can select two points and let AutoCAD measure the spacing. Positive values for spacing generate the array along the positive *X*, *Y*, and *Z* axes. Negative values generate the array along the negative *X*, *Y*, and *Z* axes.

Polar Array

To generate a polar array, enter **p** (for polar array) and AutoCAD prompts:

Enter the number of items in the array: *(specify the number of items in the array; include the original object)*
Specify the angle to fill (+=ccw, -=cw) <360>: *(specify an angle, or press* ENTER *for 360 degrees)*
Rotate arrayed objects? [Yes/No] <Y>: *(enter **y** to rotate the objects as they are copied, or enter **n** not to rotate the objects as they are copied)*
Specify center point of array: *(specify a point)*
Specify second point on axis of rotation: *(specify a point for the axis of rotation)*

EXTENDING AND TRIMMING IN 3D

AutoCAD allows you to extend an object by means of the EXTEND command (explained in Chapter 4) to any object in *3D* space or to trim an object to any other *3D* space by means of the TRIM command (explained in Chapter 4), regardless of whether the objects are on the same plane or parallel to the cutting or boundary edges. Before you select an object to extend or trim in *3D* space, specify one of the three available projection modes: None, UCS, or View. The **None** option specifies no projection. AutoCAD extends/trims only objects that intersect with the boundary/cutting edge in *3D* space. The **UCS** option specifies projection onto the *XY* plane of the current UCS. AutoCAD extends/trims objects that do not intersect with the boundary/cutting objects in *3D* space. The **View** option specifies projection along the current view direction. the PROJMODE System variable allows you to set one of the available projection modes. You can also set the projection mode by selecting the Project option available in the EXTEND command.

In addition to specifying the projection mode, you have to specify one of the two available options for the edge. The edge determines whether the object is extended/trimmed to another object's implied edge or only to an object that actually intersects it in *3D* space. The available options are Extend and No Extend. The Extend option extends the boundary/cutting object/edge along its natural path to intersect another object or its implied edge in *3D* space. The No Extend option specifies that the object is extended/trimmed only to a boundary/cutting object/edge that actually intersects it in *3D* space. The EXTEDGE system variable allows you to set one of the available modes. You can also set the Edge by selecting the Edge option available in the EXTEND command.

CREATING SOLID SHAPES

As mentioned earlier, solids are the most informationally complete and least ambiguous of the modeling types. It is easier to edit a complex solid shape than to edit wireframes and meshes.

You create solids from one of the basic solid shapes: box, cone, cylinder, sphere, torus, or wedge. The user-defined solids can be created by extruding or revolving *2D* objects and regions to define a *3D* solid. In addition, you can create more complex solid shapes by combining solids together by performing a Boolean operation—union, subtraction, or intersection.

Solids can be further modified by filleting and chamfering their edges. AutoCAD provides commands for slicing a solid into two pieces or obtaining a *2D* cross-section of a solid.

Like meshes, solids are displayed as a wireframe until you hide, shade, or render them. AutoCAD provides commands to analyze solids for their mass properties (volume, moments of inertia, center of gravity, etc.). AutoCAD allows you to export data about a solid object to applications such as NC (numerical control) milling and EXTEDGE FEM (finite element method) analysis. If necessary, you can use the AutoCAD EXPLODE command to explode solids into mesh and wireframe objects.

Note: The ISOLINES system variable controls the number of tessellation lines used to visualize curved portions of the wireframe. The default value for ISOLINES is set to 4.

CREATING A SOLID BOX

The BOX command creates a solid box or cube. The base of the box is defined parallel to the current UCS by default. The solid box can be drawn by one of two options: by providing a center point or a starting corner of the box.

Invoke the BOX command:

Solids toolbar	Choose the BOX command (see Figure 15–54)
Draw menu	Choose Solids > Box
Command: prompt	**box** (ENTER)

Figure 15–54 *Invoking the* BOX *command from the Solids toolbar*

AutoCAD prompts:

Command: **box**
Specify corner of box or [CEnter]<0,0,0>: *(specify a point, or type* **c** *for the Center option)*

First, by default, you are prompted for the starting corner of the box. Once you provide the starting corner, the box's dimensions can be entered in one of three ways.

The default option lets you create a box by locating the opposite corner of its base rectangle first, and then its height. The following command sequence defines a box, as shown in Figure 15–55, using the default option:

Command: **box**
Specify corner of box or [CEnter]<0,0,0>: **3,3**
Specify corner or [Cube/Length]: **7,7**
Specify height: **4**

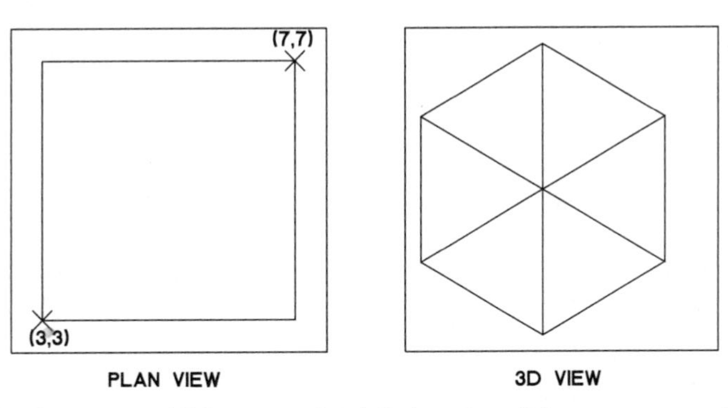

Figure 15–55 *Creating a solid box using the default option of the* BOX *command*

The **Cube** option allows you to create a box in which all edges are of equal length. The following command sequence defines a box using the Cube option:

Command: **box**
Specify corner of box or [CEnter]<0,0,0>: **3,3**
Specify corner or [Cube/Length]: **c**
Specify length: **3**

The **Length** option lets you create a box by defining its length, width, and height. The following command sequence defines a box using the Length option:

Command: **box**
Specify corner of box or [CEnter]<0,0,0>: **3,3**
Specify corner or [Cube/Length]: **l**
Specify length: **3**
Specify width: **4**
Specify height: **3**

CEnter

The CEnter option allows you to create a box by locating its center point. Once you locate the center point, a line rubberbands from this point to help you visualize the size of the rectangle. Then AutoCAD prompts you to define the size of the box by entering one of the following options:

Specify corner or [Cube/Length]: *(select one of the available options)*

Note: Once you create a box you cannot stretch it or change its size. However, you can extrude the faces of a box with the SOLIDEDIT command.

CREATING A SOLID CONE

The CONE command creates a cone, either round or elliptical. By default, the base of the cone is parallel to the current UCS. Solid cones are symmetrical and come to a point along the *Z* axis. The solid cone can be drawn two ways: by providing a center point for a circular base or by selecting the elliptical option to draw the base of the cone as an elliptical shape.

Invoke the CONE command:

Solids toolbar	Choose the CONE command (see Figure 15–56)
Draw menu	Choose Solids > Cone
Command: prompt	**cone** (ENTER)

Figure 15–56 *Invoking the CONE command from the Solids toolbar*

AutoCAD prompts:

Command: **cone**
Current wire frame density ISOLINES=4

Specify center point for base of cone or [Elliptical] <0,0,0>: *(specify a point, or type* **e** *for the Elliptical option)*

By default, AutoCAD prompts you for the center point of the base of the cone and assumes the base to be a circle. Subsequently, you are prompted for the radius (or enter **D**, for diameter). Enter the appropriate value and then it prompts for the apex/height of the cone. The height of the cone is the default option, and it allows you to set the height of the cone, not the orientation. The base of the cone is parallel to the current base plane. The Apex option, in contrast, prompts you for a point. In turn, it sets the height and orientation of the cone. For example, the following command sequence lists the steps in drawing a cone, as shown in Figure 15–57, using the default option:

Command: **cone**
Current wire frame density ISOLINES=4

Specify center point for base of cone or [Elliptical] <0,0,0>: **5,5**
Specify radius base of cone or [Diameter]: **3**
Specify height of cone or [Apex]: **4**

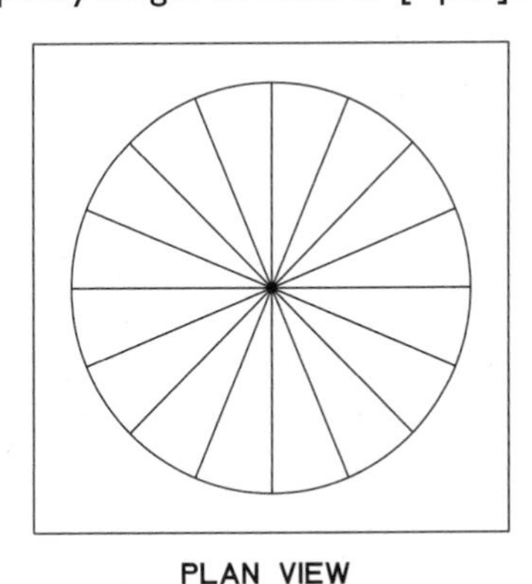

PLAN VIEW

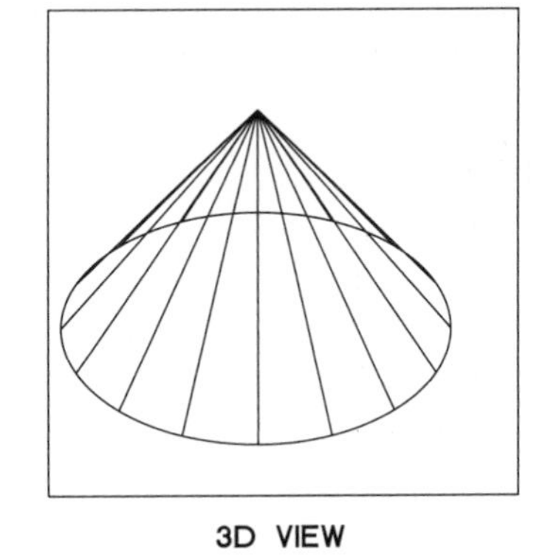

3D VIEW

Figure 15–57 *Creating a solid cone using the default option of the* CONE *command*

Elliptical

Selecting the Elliptical option indicates that the base of the cone is an ellipse. The prompts are identical to the regular AutoCAD ELLIPSE command. For example, the following command shows steps in drawing a cone using the Elliptical option:

Command: **cone**
Current wire frame density ISOLINES=4

Specify center point for base of cone or [Elliptical] <0,0,0>: **e**
Specify axis endpoint of ellipse for base of cone or [Center]: **3,3**
Specify second axis endpoint of ellipse for base of cone: **6,6**
Specify length of other axis for base of cone: **5,7**
Specify height of cone or [Apex]: **4**

CREATING A SOLID CYLINDER

The CYLINDER command creates a cylinder of equal diameter on each end and similar to an extruded circle or an ellipse. The solid cylinder can be by means of one of two options: by providing a center point for a circular base, or by selecting the elliptical option to draw the base of the cylinder as an elliptical shape.

Invoke the CYLINDER command:

Solids toolbar	Choose the CYLINDER command (see Figure 15–58)
Draw menu	Choose Solids > Cylinder
Command: prompt	**cylinder** (ENTER)

Figure 15–58 *Invoking the* CYLINDER *command from the Solids toolbar*

AutoCAD prompts:

Command: **cylinder**
Current wire frame density ISOLINES=4
Specify center point for base of cylinder or [Elliptical] <0,0,0>: *(specify a point, or type* **e** *for the Elliptical option)*

The prompts are identical to those for a cone. For example, the following command sequence lists the steps in drawing a cylinder, as shown in Figure 15–59, using the default option:

Command: **cylinder**
Current wire frame density ISOLINES=4
Specify center point for base of cylinder or [Elliptical] <0,0,0>: **5,5**
Specify radius for base of cylinder or [Diameter]: **3**
Specify height of cylinder or [Center of other end]: **4**

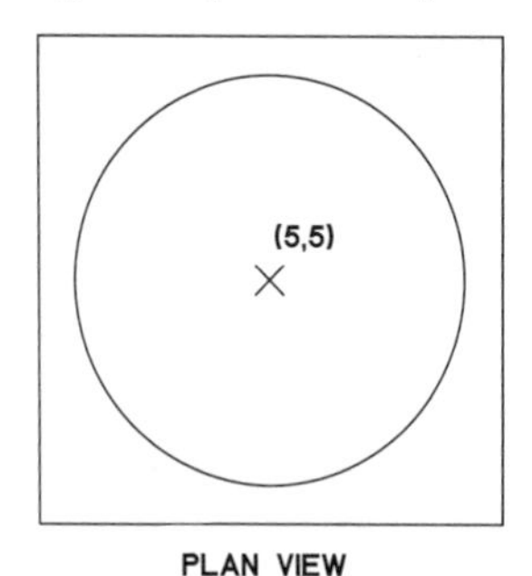

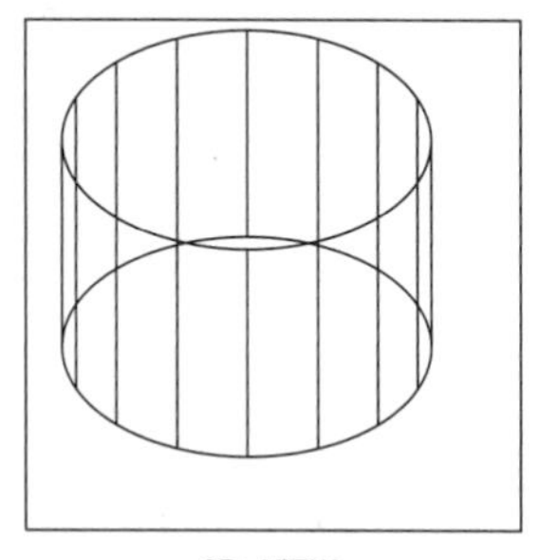

PLAN VIEW   3D VIEW

Figure 15–59 *Creating a solid cylinder using the default option of the* CYLINDER *command*

CREATING A SOLID SPHERE

The SPHERE command creates a *3D* body in which all surface points are equidistant from the center. The sphere is drawn in such a way that its central axis is coincident with the *Z* axis of the current UCS.

Invoke the SPHERE command:

Solids toolbar	Choose the SPHERE command (see Figure 15–60)
Draw menu	Choose Solids > Sphere
Command: prompt	**sphere** (ENTER)

Figure 15–60 *Invoking the* SPHERE *command from the Solids toolbar*

AutoCAD prompts:

Command: **sphere**
Current wire frame density ISOLINES=4
Specify center of sphere <0,0,0>: *(specify a point)*

First, AutoCAD prompts for the center point of the sphere; then you can provide the radius or diameter to define a sphere.

For example, the following command sequence shows steps in drawing a sphere, as in Figure 15–61:

Command: **sphere**
Current wire frame density ISOLINES=4
Specify center of sphere <0,0,0>: **5,5**
Specify radius of sphere or [Diameter]: **3**

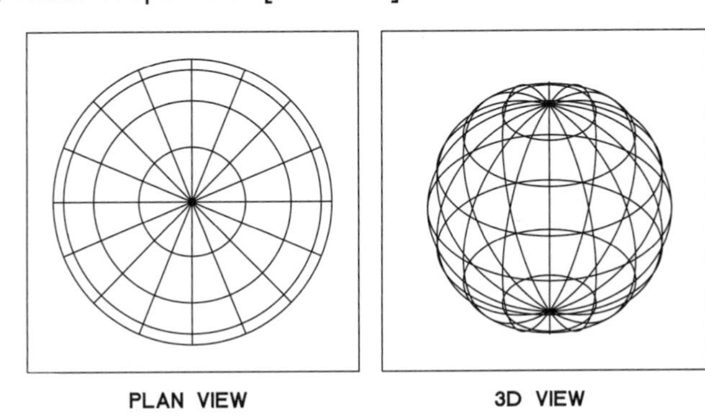

Figure 15–61 *Creating a solid sphere using the default option of the* SPHERE *command*

CREATING A SOLID TORUS

The TORUS command creates a solid with a donut-like shape. If a torus were a wheel, the center point would be the hub. The torus is created lying parallel to and bisected by the *XY* plane of the current UCS.

Invoke the TORUS command:

Solids toolbar	Choose the TORUS command (see Figure 15–62)
Draw menu	Choose Solids > Torus
Command: prompt	**torus** (ENTER)

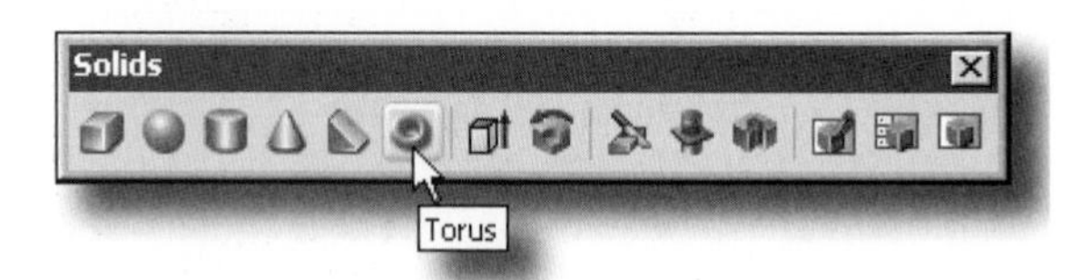

Figure 15–62 *Invoking the* TORUS *command from the Solids toolbar*

AutoCAD prompts:

Command: **torus**
Current wire frame density ISOLINES=4

Specify center of torus<0,0,0>: *(specify a point)*

AutoCAD prompts for the center point of the torus and then subsequently for the diameter or radius of the torus and the diameter or radius of the tube, as shown in Figure 15–63. You can also draw a torus without a center hole if the radius of the tube is defined as greater than the radius of the torus. A negative torus radius would create a football-shaped solid.

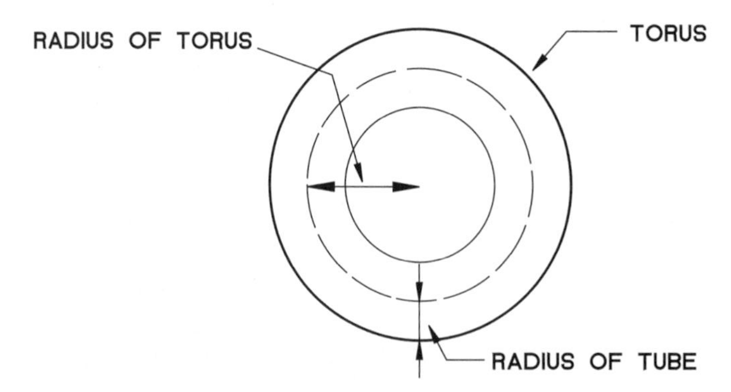

Figure 15–63 *Creating a solid torus with a center hole using the* TORUS *command*

For example, the following command sequence lists the steps in drawing a torus, as shown in Figure 15–64:

Command: **torus**
Current wire frame density ISOLINES=4

Specify center of torus<0,0,0>: **5,5**
Specify radius of torus or [Diameter]: **3**
Specify radius of tube or [Diameter]: **0.5**

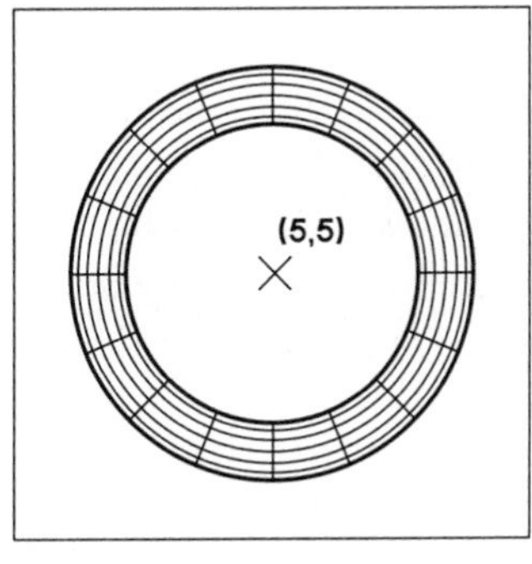

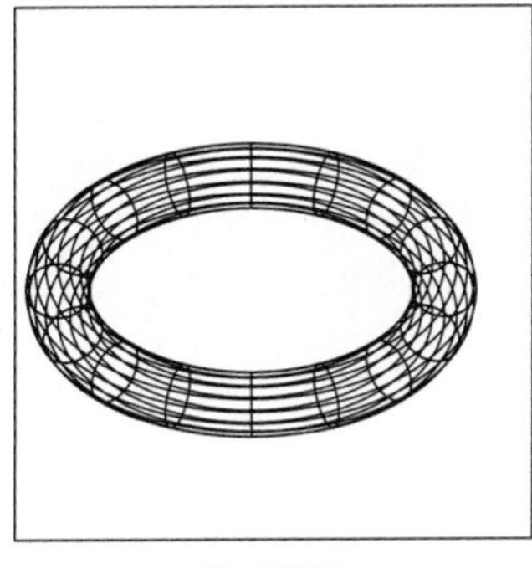

Figure 15–64 *Creating a torus by specifying the baseplane and central axis direction using the* TORUS *command*

CREATING A SOLID WEDGE

The WEDGE command creates a solid like a box that has been a cut in half diagonally along one face. The face of the wedge is always drawn parallel to the current UCS, with the sloped face tapering along the *Z* axis. The solid wedge can be drawn by one of two options: by providing a center point of the base or by providing starting corner of the box.

Invoke the WEDGE command:

Solids toolbar	Choose the WEDGE command (see Figure 15–65)
Draw menu	Choose Solids > Wedge
Command: prompt	**wedge** (ENTER)

Figure 15–65 *Invoking the* WEDGE *command from the Solids toolbar*

AutoCAD prompts:

Command: **wedge**
Specify first corner of wedge or [CEnter] <0,0,0>: *(specify a point, or type* **c** *for the corner of the wedge)*

First, by default you are prompted for the starting corner of the box. Once you provide the starting corner, AutoCAD prompts:

Specify corner or [Cube/Length]: *(specify corner of the wedge or select one of the available options)*

The wedge dimensions can be specified by using one of the three options. The Specify corner option lets you create a wedge by locating first the opposite corner of its base rectangle and then its height. The Cube option allows you to create a wedge in which all edges are of equal length. The Length option lets you create a box by defining its length, width, and height.

CEnter

The CEnter option allows you to create a wedge by first locating its center point. Once you locate the center point, a line rubber-bands from this point to help you visualize the size of the rectangle. Then AutoCAD prompts you to define the size of the box by entering one of the following options:

Specify corner or [Cube/Length]: *(specify corner of the wedge or select one of the available options)*

CREATING SOLIDS FROM EXISTING 2D OBJECTS

The EXTRUDE command creates a unique solid by extruding circles, closed polylines, polygons, ellipses, closed splines, donuts, and regions. Because a polyline can have virtually any shape, the EXTRUDE command allows you to create irregular shapes. In addition, AutoCAD allows you to taper the sides of the extrusion.

Note: A polyline must contain at least 3 but not more than 500 vertices and none of the segments can cross each other. See Figure 15–66 for examples of shapes that cannot be extruded. If the polyline has width, AutoCAD ignores the width and extrudes from the center of the polyline path. If a selected object has thickness, AutoCAD ignores the thickness.

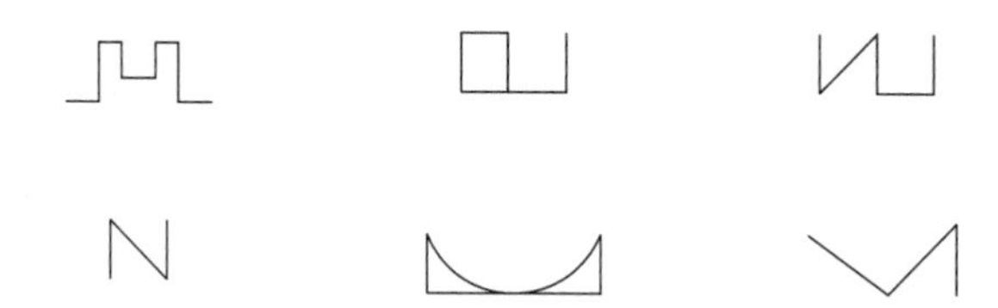

EXAMPLES OF SHAPES THAT CANNOT BE EXTRUDED (SHOWN IN PLAN VIEW)

Figure 15–66 *Shapes (shown in plan view) that cannot be extruded using the* EXTRUDE *command*

Invoke the EXTRUDE command:

Solids toolbar	Choose the EXTRUDE command (see Figure 15–67)
Draw menu	Choose Solids > Extrude
Command: prompt	**extrude** (ENTER)

Figure 15–67 *Invoking the* EXTRUDE *command from the Solids toolbar*

AutoCAD prompts:

Command: **extrude**
Current wire frame density ISOLINES=4

Select objects: *(select the objects to extrude and press* ENTER *to complete the selection)*

Specify Height of Extrusion or [Path]: *(specify height of extrusion or select path option)*

Height of Extrusion

The Height of Extrusion option (default) allows you to specify the distance for extrusion. Specifying a positive value extrudes the objects along the positive *Z* axis of the current UCS, and a negative value extrudes along the negative *Z* axis.

Path

The Path option allows you to select the extrusion path based on a specified curve object. All the profiles of the selected object are extruded along the chosen path to create solids. Lines, circles, arcs, ellipses, elliptical arcs, polylines, or splines can be paths. The path should not lie on the same plane as the profile, nor should it have areas of high curvature. The extruded solid starts from the plane of the profile and ends on a plane perpendicular to the path's endpoint. One of the endpoints of the path should be on the plane of the profile. Otherwise, AutoCAD moves the path to the center of the profile.

Once you specify the Height of Extrusion and path appropriately, AutoCAD prompts:

Specify angle of taper for extrusion <0>: *(specify the angle)*

Specify an angle between –90 and +90 degrees, or press ENTER or SPACEBAR to accept the default value of 0 degrees. If you specify 0 degrees as the taper angle, AutoCAD extrudes a *2D* object perpendicular to its *2D* plane, as shown in Figure 15–68. Positive angles taper in from the base object; negative angles taper out.

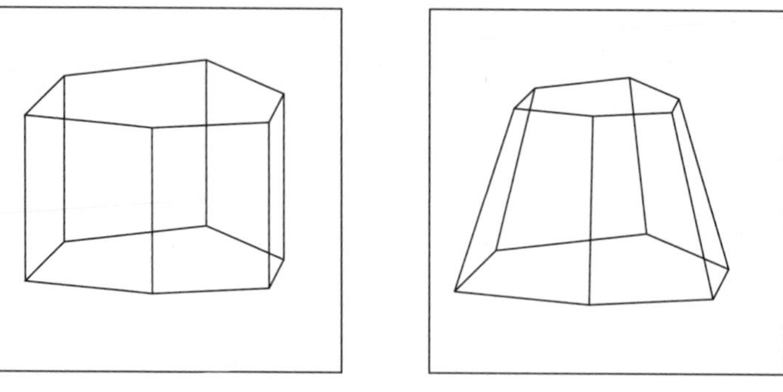

Figure 15–68 *Creating a solid with the* EXTRUDE *command with 0 degrees and with 15 degrees of taper angle*

Note: It is possible for a large taper angle or a long extrusion height to cause the object, or portions of the object, to taper to a point before reaching the extrusion height.

CREATING SOLIDS BY MEANS OF REVOLUTION

The REVOLVE command creates a unique solid by revolving or sweeping a closed polyline, polygon, circle, ellipse, closed spline, donut, and region. Polylines that have crossing or self-intersecting segments cannot be revolved. The REVOLVE command is similar to the REVSURF command. The REVSURF command creates a surface of revolution, whereas REVOLVE creates a solid of revolution. The REVOLVE command provides several options for defining the axis of revolution.

Invoke the REVOLVE command:

Solids toolbar	Choose the REVOLVE command (see Figure 15–69)
Draw menu	Choose Solids > Revolve
Command: prompt	**revolve** (ENTER)

Figure 15–69 *Invoking the* REVOLVE *command from the Solids toolbar*

AutoCAD prompts:

> Current wire frame density ISOLINES=4
> Command: **revolve**
> Select objects: *(select the objects to revolve and press* ENTER *to complete the selection)*
> Specify start point for axis of revolution –or define axis by [Object/X (axis)/Y(axis)]: *(specify start point for axis of revolution or select one of the available options)*

Start Point of Axis

The Start point of axis option (default) allows you to specify two points for the start point and the endpoint of the axis, and the positive direction of rotation is based on the right-hand rule.

Object

The Object option allows you select an existing line or single polyline segment that defines the axis about which to revolve the object. The positive axis direction is from the closest to the farthest endpoint of this line.

X axis

The *X* axis option uses the positive *X* axis of the current UCS as the axis of the revolution.

Y axis

The *Y* axis option uses the positive *Y* axis of the current UCS as the axis of the revolution.

Once you specify the axis of revolution, AutoCAD prompts:

> Specify angle of revolution <full circle>: *(Specify the angle for revolution)*

The default is for a full circle. You can specify any angle between 0 and 360 degrees.

CREATING COMPOSITE SOLIDS

As mentioned earlier in this chapter, you can create a new composite solid or region by combining two or more solids or regions via Boolean operations. Although the term Boolean implies that only two objects can be operated upon at once, AutoCAD lets you select many solid objects in a single Boolean command. There are three basic Boolean operations that can be performed in AutoCAD:

Union

Subtraction

Intersection

The UNION, SUBTRACT, and INTERSECT commands let you select both the solids and regions in a single use of the commands, but solids are combined with solids, and regions combined only with regions. Also, in the case of regions you can make composite regions only with those that lie in the same plane. This means that a single command creates a maximum of one composite solid, but might create many composite regions.

UNION OPERATION

Union is the process of creating a new composite object from one or more original objects. The union operation joins the original solids or regions in such a way that there is no duplication of volume. Therefore, the total resulting volume can be equal to or less than the sum of the volumes in the original solids or regions. The UNION command performs the union operation.

Invoke the UNION command:

Solids Editing toolbar	Choose the UNION command (see Figure 15–70)
Modify menu	Choose Solids Editing > Union
Command: prompt	**union** (ENTER)

Figure 15–70 *Invoking the* UNION *command from the Solids Editing toolbar*

AutoCAD prompts:

Command: **union**
Select objects: *(select the objects to make one composite object and press* ENTER *to complete the selection)*

You can select more than two objects at once. The objects (solids or regions) can be overlapping, adjacent, or nonadjacent.

For example, the following command sequence shows steps in creating a composite solid by joining two cylinders, as shown in Figure 15–71.

Command: **union**
Select objects: *(select cylinders A and B and press* ENTER*)*

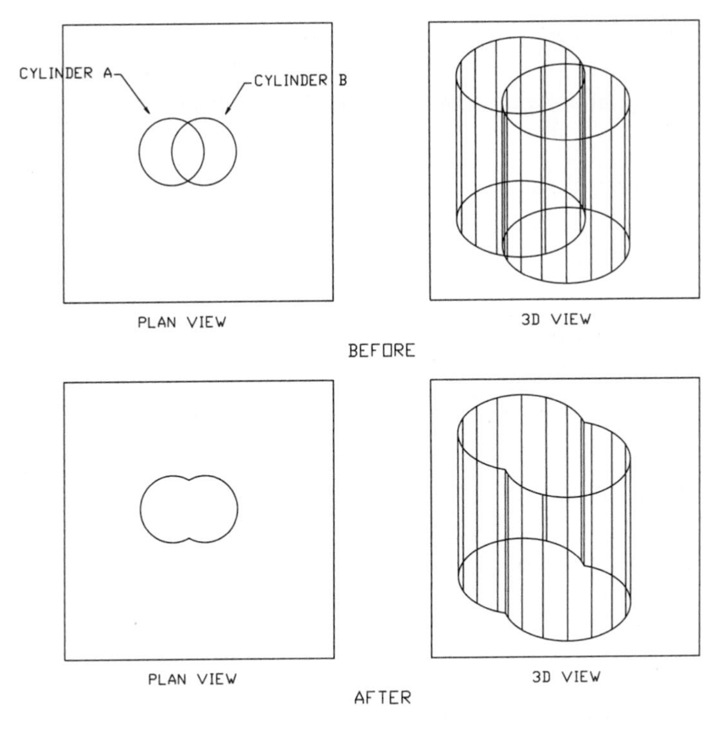

Figure 15–71 *Creating a composite solid by joining two cylinders using the* UNION *command*

SUBTRACTION OPERATION

Subtraction is the process of forming a new composite object by starting with one object and removing from it any volume that it has in common with a second object. In the case of solids, they are created by subtracting the volume of one set of solids from another set. If the entire volume of the second solid is contained in the first solid, then what is left is the first solid minus the volume of the second solid. However, if only part of the volume of the second solid is contained within the first solid, then only the part that is duplicated in the two solids is subtracted. Similarly, in the case of regions, they are created by subtracting the common area of one set of existing regions from another set. The SUBTRACT command performs the subtraction operation.

Invoke the SUBTRACT command:

Solids Editing toolbar	Choose the SUBTRACT command (see Figure 15–72)
Modify menu	Choose Solids Editing > Subtract
Command: prompt	**subtract** (ENTER)

Figure 15–72 *Invoking the* SUBTRACT *command from the Solids Editing toolbar*

AutoCAD prompts:

Command: **subtract**
Select solids and regions to subtract from...
Select objects: *(select the objects from which you will subtract other objects and press* ENTER*)*

You can select one or more objects as source objects. If you select more than one, they are automatically joined. After selecting the source objects, press ENTER or the SPACEBAR, and AutoCAD prompts you to select the objects to subtract from the source object.

Select solids and regions to subtract...
Select objects: *(select the objects to subtract and press* ENTER*)*

If necessary, you can select one or more objects to subtract from the source object. If you select several, they are automatically joined before they are subtracted from the source object.

Note: Objects that are neither solids nor regions are ignored.

For example, the following command sequence shows steps in creating a composite solid by subtracting cylinder B from A, as shown in Figure 15–73.

Command: **subtract**
Select objects: *(select cylinder A and press* ENTER *or the* SPACEBAR*)*
Objects to subtract from them...
Select objects: *(select cylinder B and press* ENTER *or the* SPACEBAR*)*

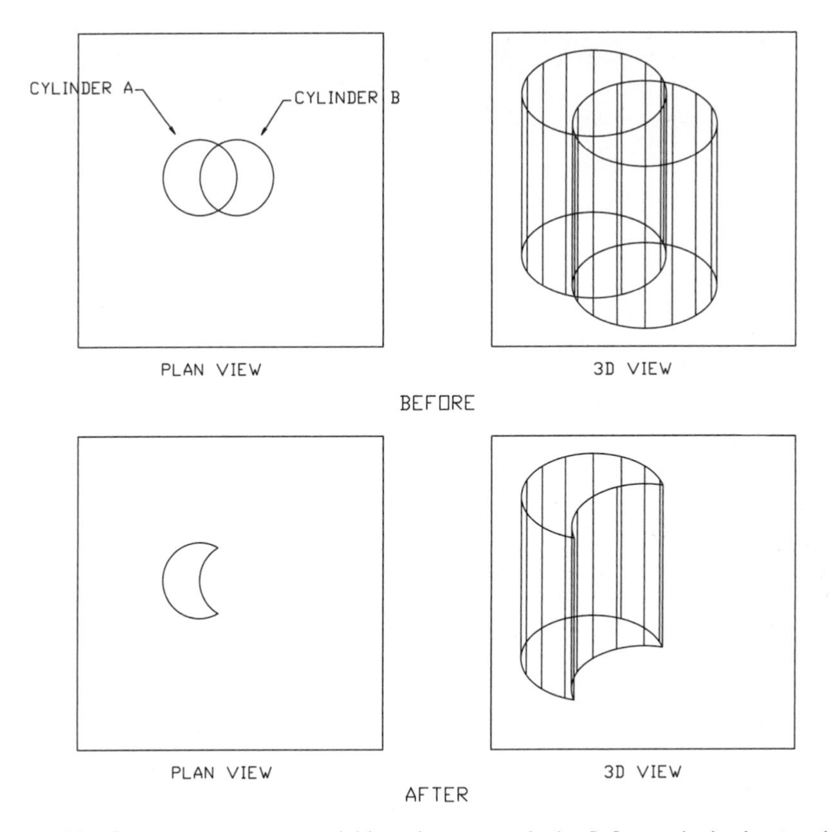

Figure 15–73 *Creating a composite solid by subtracting cylinder B from cylinder A using the* SUBTRACT *command*

INTERSECTION OPERATION

Intersection is the process of forming a composite object from only the volume that is common to two or more original objects. In the case of solids, you can create a new composite solid by calculating the common volume of two or more existing solids. Whereas in the case of regions, it is done by calculating the overlapping area of two or more existing regions. The INTERSECT command performs the intersection operation.

Invoke the INTERSECT command:

Solids Editing toolbar	Choose the INTERSECT command (see Figure 15–74)
Modify menu	Choose Solids Editing > Intersect
Command: prompt	**intersect** (ENTER)

Figure 15–74 *Invoking the* INTERSECT *command from the Solids Editing toolbar*

AutoCAD prompts:

Command: **intersect**
Select objects: *(select the objects for intersection and press* ENTER *to complete the selection)*

For example, the following command sequence shows the steps in creating a composite solid by intersecting cylinder A with cylinder B, as shown in Figure 15–75:

Command: **intersect**
Select objects: *(select cylinders A and B and press* ENTER *or the* SPACEBAR*)*

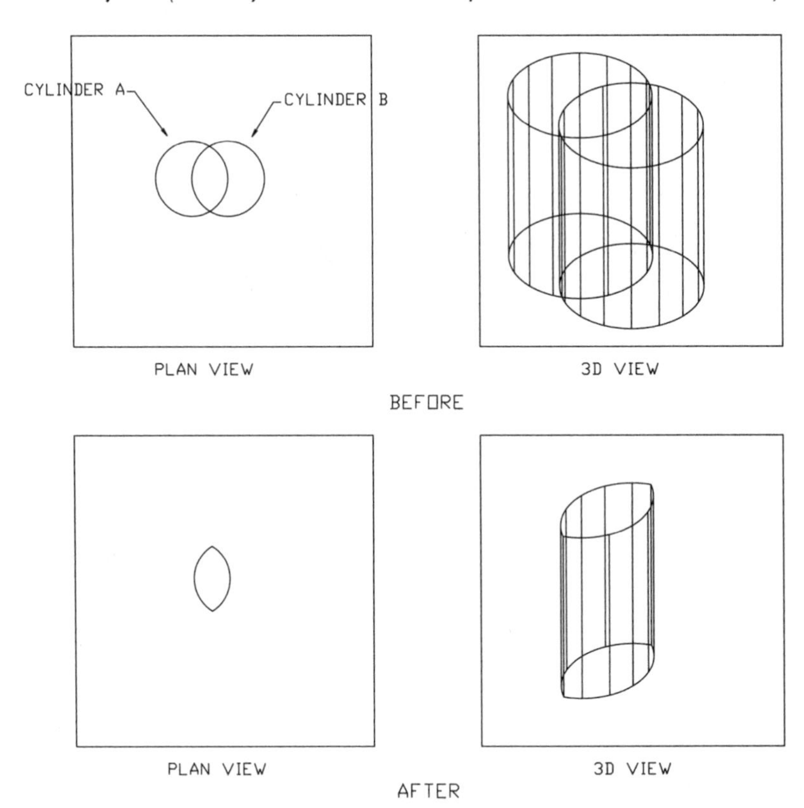

Figure 15–75 *Creating a composite solid by intersecting cylinder A from cylinder B using the* INTERSECT *command*

EDITING *3D* SOLIDS

AutoCAD makes the work of creating solids a little easier by providing editing tools, including chamfering or filleting the edges, creating a cross-section through a solid, creating a new solid by cutting the existing solid and removing a specified side, and creating a composite solid from the interference of two or more solids. In addition, AutoCAD provides additional editing tools such as extrude faces, move faces, offset faces, delete faces, rotate faces, taper faces, color faces, copy faces, color and copy edges, imprint, clean, separate solids, shell and check. If necessary, you can always use the AutoCAD modify and construct commands, such as MOVE, COPY, ROTATE, SCALE, and ARRAY to edit solids.

CHAMFERING SOLIDS

The CHAMFER command (explained in Chapter 4) can also be used to bevel the edges of an existing solid object.

Invoke the CHAMFER command:

Modify toolbar	Choose the CHAMFER command (see Figure 15–76)
Modify menu	Choose Chamfer
Command: prompt	**chamfer** (ENTER)

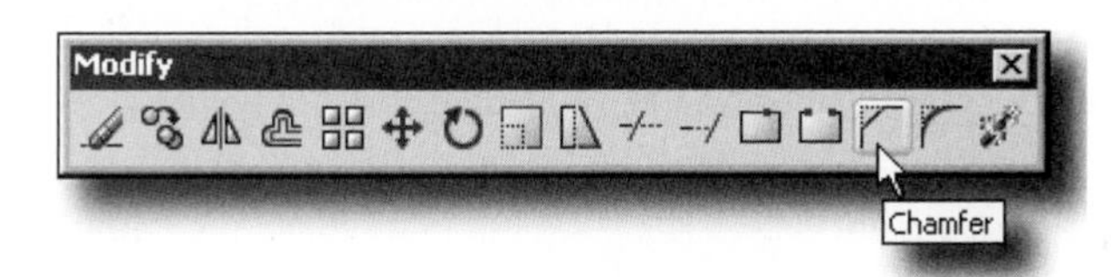

Figure 15–76 *Invoke the* CHAMFER *command from the Modify toolbar*

AutoCAD prompts:

Command: **chamfer**
Select first line or [Polyline/Distances/Angle/Trim/Method]: *(select an edge on a 3D solid)*

If you pick an edge that is common to two surfaces, AutoCAD highlights one of the surfaces and prompts:

Base surface selection

Enter surface selection option [Next/OK (current)] <OK>:

If this is the surface you want, press ENTER or SPACEBAR to accept it. If it is not, enter **N** (for next) to highlight the adjoining surface and then press ENTER or the SPACEBAR. AutoCAD prompts:

Specify base surface chamfer distance <default>: *(specify a distance, or press the* ENTER *key or* SPACEBAR *to accept the default)*
Specify other surface chamfer distance <default>: *(specify a distance, or press the* ENTER *key or* SPACEBAR *to accept the default)*

Once you provide the chamfer distances, AutoCAD prompts:

Select an edge or [Loop]: Select the edges of the highlighted surface you want chamfered, and then press ENTER or the SPACEBAR.

The **Loop** option allows you to select one of the edges on the base surface, and AutoCAD automatically selects all edges on the base surface for chamfering.

The following command sequence draws a chamfer for a solid object, as shown in Figure 15–77.

Command: **chamfer**
Select first line or [Polyline/Distances/Angle/Trim/Method]: *(select the edge)*
Enter surface selection option [Next/OK (current)] <OK>: (ENTER)
Specify base surface chamfer distance <default>: **0.25**
Specify other surface chamfer distance <default>: **0.5**
Select an edge or [Loop]: *(select the first edge)*
Loop/<Select edge>: *(select the second edge)*

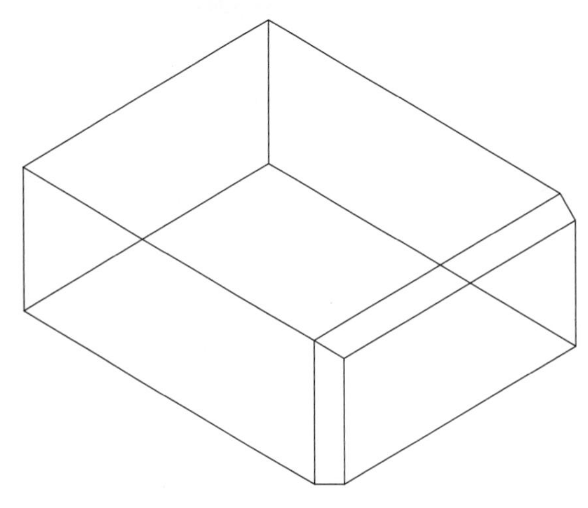

Figure 15–77 *Example of chamfering a solid surface*

FILLETING SOLIDS

The FILLET command (explained in Chapter 4) can also be used to round the edge of an existing solid object.

Invoke the FILLET command:

Modify toolbar	Choose the FILLET command (see Figure 15–78)
Modify menu	Choose Fillet
Command: prompt	**fillet** (ENTER)

Figure 15–78 *Invoking the FILLET command from the Modify toolbar*

AutoCAD prompts:

Command: **fillet**
Select first object or [Polyline/Radius/Trim]: *(select an edge in a 3D solid)*

If necessary, you can select multiple edges; but you must select the edges individually after specifying the radius for the fillet. AutoCAD prompts:

Enter fillet radius <default>: *(specify radius for fillet, or press ENTER or SPACEBAR to accept the default)*
Select an edge or [Chain/Radius]: *(select addition edges and press ENTER to complete the selection)*

The following command sequence draws a fillet for a solid object, as shown in Figure 15–79.

Command: **fillet**
Select first object or [Polyline/Radius/Trim]: *(select the solid edge)*
Enter fillet radius <default>: **0.5**
Select an edge or [Chain/Radius]: (ENTER)

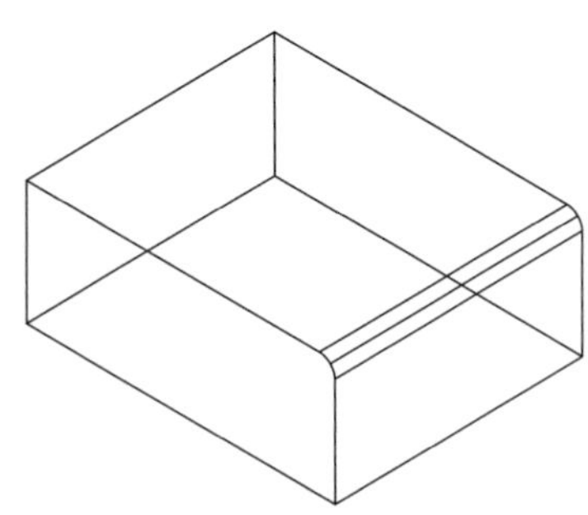

Figure 15–79 *Example of filleting a solid surface*

SECTIONING SOLIDS

The SECTION command creates a cross-section of one or more solids. The cross-section is created as one or more regions. The region is created on the current layer and is inserted at the location of the cross-section. If necessary, you can use the MOVE command to move the cross-section.

Invoke the SECTION command:

Solids toolbar	Choose the SECTION command (see Figure 15–80)
Draw menu	Choose Solids > Section
Command: prompt	**section** (ENTER)

Figure 15–80 *Invoking the* SECTION *command from the Solids toolbar*

AutoCAD prompts:

Command: **section**
Select objects: *(select the objects from which you want the cross-section to be generated)*
Specify first point on Section plane by [Object/Zaxis/View/XY/YZ/ZX/3points] <3points>: *(specify one of the three points to define a plane or select one of the available options)*

3points

The 3points option (default) allows you to define a section plane by locating three points. The first point is the origin, the second point determines the positive direction of the *X* axis for the section plane, and the third point determines the positive *Y* axis of the section plane. This option is similar to the 3point option of the AutoCAD UCS command.

Object

The Object option aligns the sectioning plane with a circle, ellipse, circular or elliptical arc, *2D* spline, or *2D* polyline segment.

Zaxis

The Zaxis option defines the section plane by locating its origin point and a point on the *Z* axis (normal) to the plane.

View

The View option aligns the section plan with the viewing plane of the current viewport. Specifying a point defines the location of the sectioning plane.

XY

The XY option aligns the sectioning plane with the *XY* plane of the current UCS. Specifying a point defines the location of the sectioning plane.

YZ

The YZ option aligns the sectioning plane with the *XY* plane of the current UCS. Specifying a point defines the location of the sectioning plane.

ZX

The ZX option aligns the sectioning plane with the *XY* plane of the current UCS. Specifying a point defines the location of the sectioning plane.

Figure 15–81 shows a hatched cross-section produced with the SECTION command.

Note: The section must be hatched using hatching techniques of Chapter 9.

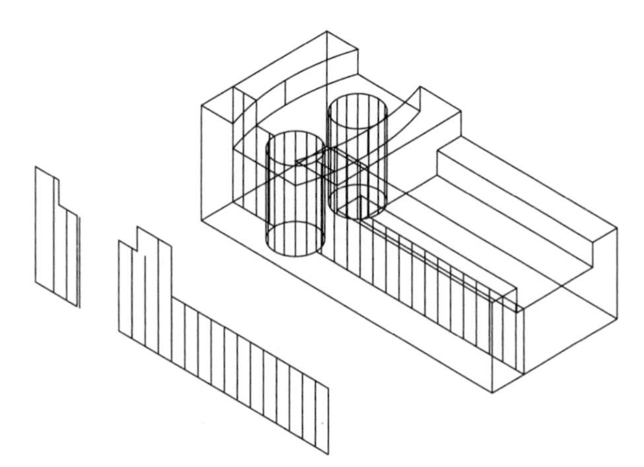

Figure 15–81 *Creating a 2D hatched cross-section using the* SECTION *command*

SLICING SOLIDS

The SLICE command allows you to create a new solid by cutting the existing solid and removing a specified portion. If necessary, you can retain both portions of the sliced solid(s) or just the portion you specify. The sliced solids retain the layer and color of the original solids.

Invoke the SLICE command:

Solids toolbar	Choose the SLICE command (see Figure 15–82)
Draw menu	Choose Solids > Slice
Command: prompt	**slice** (ENTER)

Figure 15–82 *Invoking the* SLICE *command from the Solids toolbar*

AutoCAD prompts:

Command: **slice**
Select objects: *(select the objects to create a new solid by slicing)*

After selecting the objects, press ENTER or SPACEBAR. AutoCAD prompts you to define the slice plane:

Specify first point on slicing plane by [Object/Zaxis/View/XY/YZ/ZX/ 3points] <3points>: *(specify one of the three points to define a plane or select one of the available options)*

The options are the same as those for the SECTION command explained earlier in this chapter.

After defining the slicing plane, AutoCAD prompts you to indicate which part of the cut solid is to be retained:

Specify a Point on desired side of the plane or [Keep Both sides]:

The default option allows you to select with your pointing device the side of the slice that has to be retained in your drawing.

The **Both sides** option allows you to retain both portions of the sliced solids.

Figure 15–83 shows two parts of a solid model that have been cut using the SLICE command and moved apart using the MOVE command.

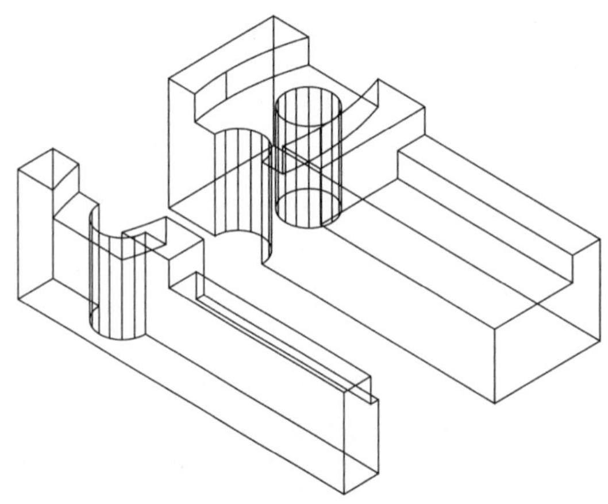

Figure 15–83 *Cutting a solid model into two parts using the* SLICE *command*

SOLID INTERFERENCE

The INTERFERE command checks the interference between two or more solids and creates a composite solid from their common volume.

There are two ways to determine the interference between solids:

- Select two sets of solids. AutoCAD determines the interference between the first and second sets of solids.
- Select one set of solids instead of two. AutoCAD determines the interference between all of the solids in the set. They are checked against each other.

Invoke the INTERFERE command:

Solids toolbar	Choose the INTERFERE command (see Figure 15–84)
Draw menu	Choose Solids > Interfere
Command: prompt	**interfere** (ENTER)

Figure 15–84 *Invoking the* INTERFERE *command from the Solids toolbar*

AutoCAD prompts:

Command: **interfere**
Select the first set of solids: *(select the first set of solids and press* ENTER *or* SPACEBAR*)*
Select the second set of solids: *(select the second set of solids or press* ENTER*)*

The second selection set is optional. Press ENTER or SPACEBAR if you do not want to define the second selection set. If the same solid is included in both the selection sets, it is considered part of the first selection set and ignored in the second selection set. AutoCAD highlights all interfering solids and prompts:

Create interference solids? [Yes/No] <N>:

Entering **y** creates and highlights a new solid on the current layer that is the intersection of the interfering solids. If there are more than two interfering solids, AutoCAD prompts:

Highlight pairs of interfering solids? [Yes/No] <N>:

If you specify **y** for yes, and if there is more than one interfering pair, AutoCAD prompts:

Enter an option [Next pair/exit[<Next>: *(Specify* **x** *or* **n***)*

Pressing ENTER cycles through the interfering pairs of solids, and AutoCAD highlights each interfering pair of solids. Enter **x** to complete the command sequence.

EDITING FACES OF *3D* SOLIDS

AutoCAD allows you to edit solid objects by extruding faces, copying faces, offsetting faces, moving faces, rotating faces, tapering faces, coloring faces, and deleting faces.

Extruding Faces

AutoCAD allows you to extrude selected faces of a *3D* solid object to a specified height or along a path. Specifying a positive value extrudes the selected face in its positive direction (usually outward); and a negative value extrudes in the negative direction (usually inward). Tapering the selected face with a positive angle tapers the face inward, and a negative angle tapers the face outward. Tapering the selected face to 0 degrees extrudes the face perpendicular to its plane.

Face extrusion along a path is based on a path curve, such as lines, circles, arcs, ellipses, elliptical arcs, polylines, or splines.

Invoke the Extrude Faces option of the SOLIDEDIT command to extrude selected faces:

Solids Editing toolbar	Choose Extrude Faces (see Figure 15–85)
Modify menu	Choose Solids Editing > Extrude faces
Command: prompt	**solidedit** (ENTER)

Figure 15–85 *Invoking the* EXTRUDE FACES *option of the* SOLIDEDIT *command from the Solids Editing toolbar*

AutoCAD prompts:

Command: **solidedit**
Enter a solids editing option [Face/Edge/Body/Undo/eXit] <eXit>: **f**
Enter a face editing option [Extrude/Move/Rotate/Offset/Taper/Delete/Copy/coLor/Undo/eXit] <eXit>: **e**
Select faces or [Undo/Remove]: *(select faces and press* ENTER *to complete the selection)*
Specify height of extrusion or [Path]: *(specify height of extrusion or select path option to extrude along a path)*
Specify angle of taper for extrusion <0>: *(press* ENTER *to accept the default angle or specify the angle for taper for extrusion)*

Enter a face editing option [Extrude/Move/Rotate/Offset/Taper/Delete/Copy/coLor/Undo/eXit] <eXit>: *(select the eXit option to exit the face editing)*
Enter a solids editing option [Face/Edge/Body/Undo/eXit] <eXit>: *(select the eXit option to exit solids editing)*

Figure 15–86 shows an example of a solid model in which one of the faces is extruded by a positive value with a 15 degree tapered angle with the Extrude Faces of the SOLIDEDIT command.

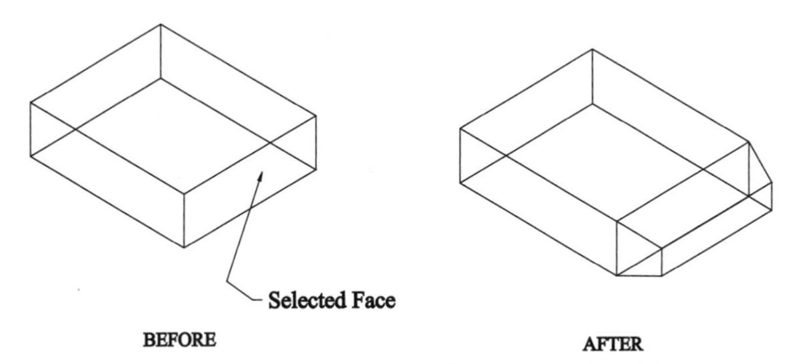

Figure 15–86 *An example in extruding one of the faces in a solid model.*

Copying Faces

The Copy Faces option of the SOLIDEDIT command allows you to copy selected faces of a *3D* solid object. AutoCAD copies selected faces as regions or bodies. Prompts are similar to the regular COPY command.

Invoke the Copy Faces option of the SOLIDEDIT command to copy selected faces:

Solids Editing toolbar	Choose Copy Faces (see Figure 15–87)
Modify menu	Choose Solids Editing > Copy faces
Command: prompt	**solidedit** (ENTER)

Figure 15–87 *Invoking the Copy Faces option of the* SOLIDEDIT *command from the Solids Editing toolbar*

AutoCAD prompts:

Command: **solidedit**
Enter a solids editing option [Face/Edge/Body/Undo/eXit] <eXit>: **f**

Enter a face editing option [Extrude/Move/Rotate/Offset/Taper/Delete/Copy/coLor/Undo/eXit] <eXit>: **c**
Select faces or [Undo/Remove]: *(select faces and press* ENTER *to complete the selection)*
Specify a base point or displacement: *(specify a base point)*
Specify a second point of displacement: *(specify a second point of displacement or press* ENTER *to consider the original selection point as a base point)*
Enter a face editing option [Extrude/Move/Rotate/Offset/Taper/Delete/Copy/coLor/Undo/eXit] <eXit>: *(select the eXit option to exit the face editing)*
Enter a solids editing option [Face/Edge/Body/Undo/eXit] <eXit>: *(select the eXit option to exit solids editing)*

Figure 15–88 shows an example of a solid model in which one of the faces is copied by the Copy Faces of the SOLIDEDIT command.

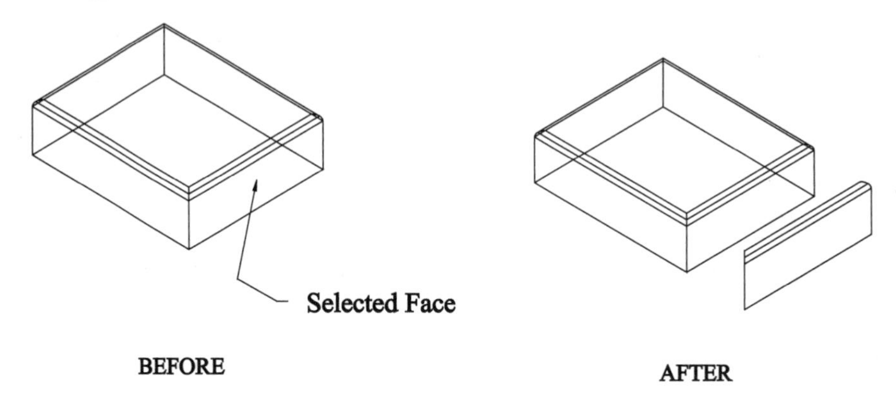

Figure 15–88 *An example of copying one of the faces in a solid model.*

Offsetting Faces

AutoCAD allows you to uniformly offset selected faces of a *3D* solid object by a specified distance. New faces are created by offsetting existing ones inside or outside at a specified distance from their original positions. Specifying a positive value increases the size or volume of the solid; a negative value decreases the size or volume of the solid.

Invoke the Offset Faces option of the SOLIDEDIT command to offset selected faces:

Solids Editing toolbar	Choose Offset Faces (see Figure 15–89)
Modify menu	Choose Solids Editing > Offset faces
Command: prompt	**solidedit** (ENTER)

Figure 15–89 *Invoking the Offset Faces option of the* SOLIDEDIT *command from the Solids Editing toolbar*

AutoCAD prompts:

Command: **solidedit**
Enter a solids editing option [Face/Edge/Body/Undo/eXit] <eXit>: **f**
Enter a face editing option [Extrude/Move/Rotate/Offset/Taper/Delete/Copy/coLor/Undo/eXit] <eXit>: **o**
Select faces or [Undo/Remove]: *(select faces and press* ENTER *to complete the selection)*
Specify the offset distance: *(specify the offset distance)*
Enter a face editing option [Extrude/Move/Rotate/Offset/Taper/Delete/Copy/coLor/Undo/eXit] <eXit>: *(select the eXit option to exit the face editing)*
Enter a solids editing option [Face/Edge/Body/Undo/eXit] <eXit>: *(select the eXit option to exit solids editing)*

Figure 15–90 shows an example of a solid model in which one of the faces is offset (positive value) by the Offset Faces option of the SOLIDEDIT command.

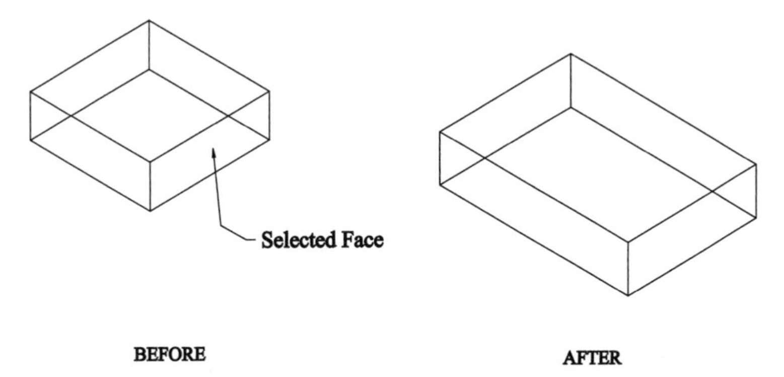

Figure 15–90 *An example of offsetting one of the faces (positive value) in a solid model.*

Moving Faces

The Move Faces option of the SOLIDEDIT command allows you to move selected faces of a *3D* solid object. You can move holes from one location to another location in a *3D* solid. Prompts are similar to the regular MOVE command.

Invoke the Move Faces option of the SOLIDEDIT command to move selected faces:

Solids Editing toolbar	Choose Move Faces (see Figure 15–91)
Modify menu	Choose Solids Editing > Move faces
Command: prompt	**solidedit** (ENTER)

Figure 15–91 *Invoking the Move Faces option of the* SOLIDEDIT *command from the Solids Editing toolbar*

AutoCAD prompts:

Command: **solidedit**
Enter a solids editing option [Face/Edge/Body/Undo/eXit] <eXit>: **f**
Enter a face editing option [Extrude/Move/Rotate/Offset/Taper/Delete/Copy/coLor/Undo/eXit] <eXit>: **m**
Select faces or [Undo/Remove]: *(select faces and press* ENTER *to complete the selection)*
Specify a base point or displacement: *(specify a base point)*
Specify a second point of displacement: *(specify a second point of displacement or press* ENTER *to consider the original selection point as a base point)*
Enter a face editing option [Extrude/Move/Rotate/Offset/Taper/Delete/Copy/coLor/Undo/eXit] <eXit>: *(select the eXit option to exit the face editing)*
Enter a solids editing option [Face/Edge/Body/Undo/eXit] <eXit>: *(select the eXit option to exit solids editing)*

Figure 15–92 shows an example of a solid model in which an elliptical cylinder is moved by the Move Faces option of the SOLIDEDIT command.

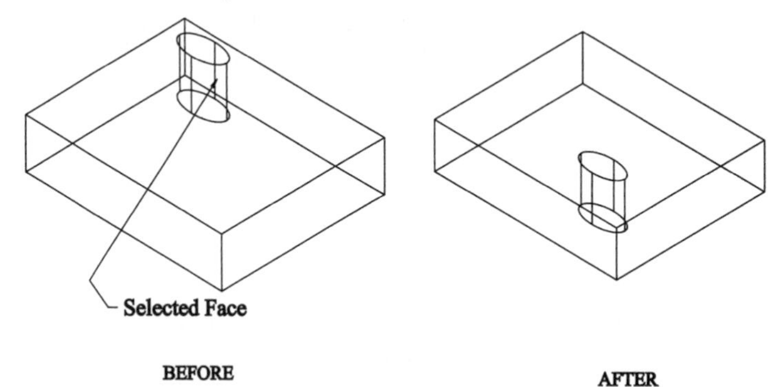

Figure 15–92 *An example of moving the elliptical cylinder in a solid model.*

Rotating Faces

The Rotating Faces option of the SOLIDEDIT command allows you to rotate selected faces of a *3D* solid object by choosing a base point to relative or absolute angle. All *3D* faces rotate about a specified axis.

Invoke the Rotating Faces option of the SOLIDEDIT command to rotate selected faces:

Solids Editing toolbar	Choose Rotate Faces (see Figure 15–93)
Modify menu	Choose Solids Editing > Rotate faces
Command: prompt	**solidedit** (ENTER)

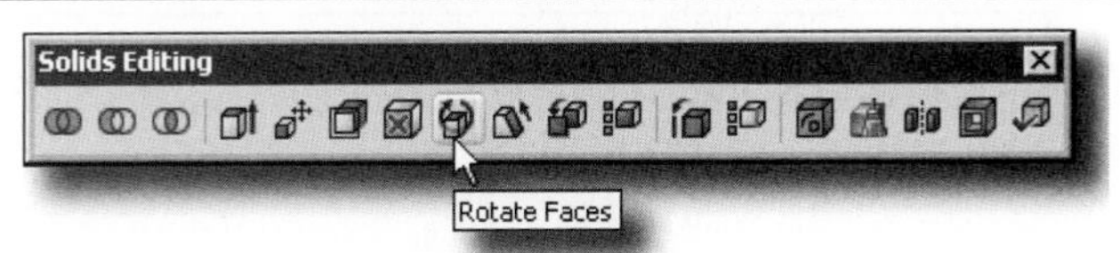

Figure 15–93 *Invoking the Rotating Faces option of the* SOLIDEDITS *command from the Solids Editing toolbar*

AutoCAD prompts:

Command: **solidedit**
Enter a solids editing option [Face/Edge/Body/Undo/eXit] <eXit>: **f**
Enter a face editing option [Extrude/Move/Rotate/Offset/Taper/Delete/ Copy/coLor/Undo/eXit] <eXit>: **r**
Select faces or [Undo/Remove]: *(select faces and press* ENTER*to complete the selection)*
Specify an axis point or [Axis by object/View/Xaxis/Yaxis/Zaxis] <2points>: *(specify an axis point or select one of the available options)*
Enter a face editing option [Extrude/Move/Rotate/Offset/Taper/Delete/ Copy/coLor/Undo/eXit] <eXit>: *(select the eXit option to exit the face editing)*
Enter a solids editing option [Face/Edge/Body/Undo/eXit] <eXit>: *(select the eXit option to exit solids editing)*

Figure 15–94 shows an example of a solid model in which elliptical cylinder is rotated by the Rotated Faces option of the SOLIDEDIT command.

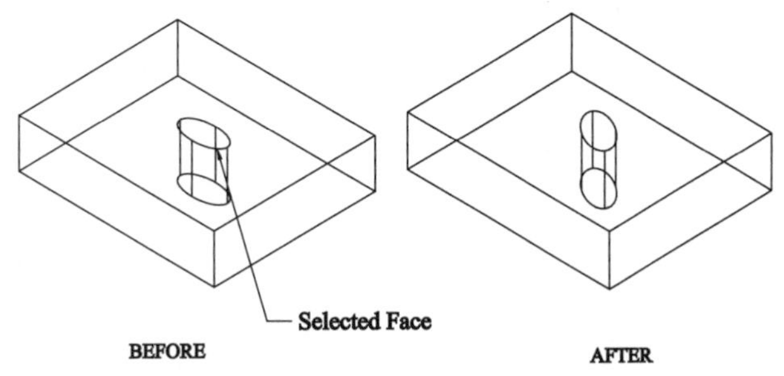

Figure 15–94 *An example of rotating elliptical cylinder of a solid model.*

Tapering Faces

The SOLIDEDIT command allows you to taper selected faces of a *3D* solid object with a draft angle along a vector direction. Tapering the selected face with a positive angle tapers the face inward, and a negative angle tapers the face outward.

Invoke the Taper Faces option of the SOLIDEDIT command to taper selected faces:

Solids Editing toolbar	Choose Taper Faces (see Figure 15–95)
Modify menu	Choose Solids Editing > Taper faces
Command: prompt	**solidedit** (ENTER)

Figure 15–95 *Invoking the Taper Faces option of the* SOLIDEDIT *command from the Solids Editing toolbar*

AutoCAD prompts:

Command: **solidedit**
Enter a solids editing option [Face/Edge/Body/Undo/eXit] <eXit>: **f**
Enter a face editing option [Extrude/Move/Rotate/Offset/Taper/Delete/Copy/coLor/Undo/eXit] <eXit>: **t**
Select faces or [Undo/Remove]: *(select faces and press* ENTER *to complete the selection)*
Specify the base point: *(specify the base point)*
Specify another point along the axis of tapering: *(specify a point to define the axis of tapering)*
Specify the taper angle: *(specify the taper angle and press* ENTER *to continue)*
Enter a face editing option [Extrude/Move/Rotate/Offset/Taper/Delete/Copy/coLor/Undo/eXit] <eXit>: *(select the eXit option to exit the face editing)*
Enter a solids editing option [Face/Edge/Body/Undo/eXit] <eXit>: *(select the eXit option to exit solids editing)*

Figure 15–96 shows an example of a solid model in which the cylinder tapered angle is changed by the Taper Faces option of the SOLIDEDIT command.

Coloring Faces

The Color Faces option of the SOLIDEDIT command allows you to change color of selected faces of a *3D* solid object. You can choose a color from the Select Color dialog box. Setting a color on a face overrides the color setting for the layer on which the solid object resides.

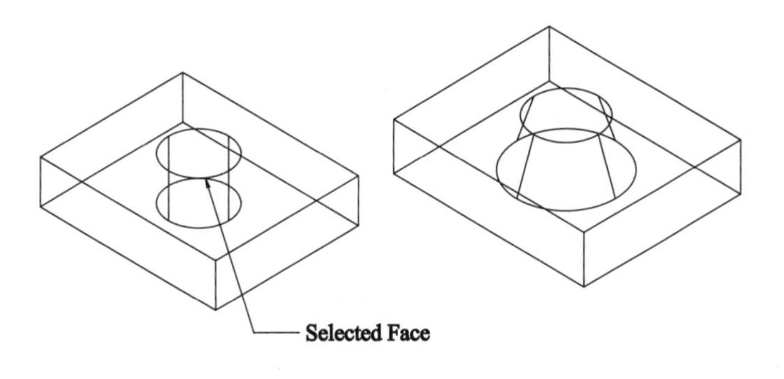

Figure 15–96 *An example of change in a tapered angle of the cylinder in a solid model.*

Invoke the Color Faces option of the SOLIDEDIT command to change the color of selected faces:

Solids Editing toolbar	Choose Color Faces (see Figure 15–97)
Modify menu	Choose Solids Editing > Color faces
Command: prompt	**solidedit** (ENTER)

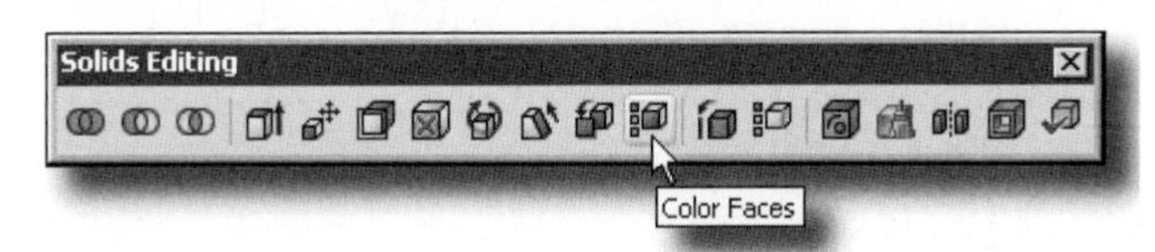

Figure 15–97 *Invoking the Color Faces option of the* SOLIDEDIT *command from the Solids Editing toolbar*

AutoCAD prompts:

Command: **solidedit**
Enter a solids editing option [Face/Edge/Body/Undo/eXit] <eXit>: **f**
Enter a face editing option [Extrude/Move/Rotate/Offset/Taper/Delete/Copy/coLor/Undo/eXit] <eXit>: **l**
Select faces or [Undo/Remove]: *(select faces and press* ENTER *to complete the selection)*

AutoCAD displays the Select Color dialog box. Select the color to change for selected faces and choose the **OK** button. AutoCAD prompts:

Enter a face editing option [Extrude/Move/Rotate/Offset/Taper/Delete/Copy/coLor/Undo/eXit] <eXit>: *(select the eXit option to exit the face editing)*
Enter a solids editing option [Face/Edge/Body/Undo/eXit] <eXit>: *(select the eXit option to exit solids editing)*

AutoCAD changes the color of the selected faces of the *3D* solid model.

Deleting Faces

The Delete Faces option of the SOLIDEDIT command allows you to delete selected faces, holes, and fillets of a *3D* solid object.

Invoke the Delete Faces option of the SOLIDEDIT command to delete selected faces:

Solids Editing toolbar	Choose Delete Faces (see Figure 15–98)
Modify menu	Choose Solids Editing > Delete faces
Command: prompt	**solidedit** (ENTER)

Figure 15–98 *Invoking the Delete Faces option of the* SOLIDEDIT *command from the Solids Editing toolbar*

AutoCAD prompts:

Command: **solidedit**
Enter a solids editing option [Face/Edge/Body/Undo/eXit] <eXit>: **f**
Enter a face editing option [Extrude/Move/Rotate/Offset/Taper/Delete/Copy/coLor/Undo/eXit] <eXit>: **d**
Select faces or [Undo/Remove]: *(select faces and press* ENTER *to complete the selection)*

Note: If the object is a simple box, AutoCAD responds with: Modeling Operation Error – Gap cannot be filled. (This informs us that there is no way to join the opposite sides to fill the gap.

Enter a face editing option [Extrude/Move/Rotate/Offset/Taper/Delete/Copy/coLor/Undo/eXit] <eXit>: *(select the eXit option to exit the face editing)*
Enter a solids editing option [Face/Edge/Body/Undo/eXit] <eXit>: *(select the eXit option to exit solids editing)*

AutoCAD deletes the selected faces of the *3D* solid model.

EDITING EDGES OF *3D* SOLIDS

AutoCAD allows you to copy individual edges and change color of edges on a *3D* solid object. The edges are copied as lines, arcs, circles, ellipses or spline objects.

Copying Edges

The Copy Edges option of the SOLIDEDIT command allows you to copy selected edges of a *3D* solid object. AutoCAD copies selected edges as lines, arcs, circles, ellipses, or splines. Prompts are similar to the regular COPY command.

Invoke the Copy Edges option of the SOLIDEDIT command to copy selected edges:

Solids Editing toolbar	Choose Copy Edges (see Figure 15–99)
Modify menu	Choose Solids Editing > Copy edges
Command: prompt	**solidedit** (ENTER)

Figure 15–99 *Invoking the Copy Edges option of the* SOLIDEDIT *command from the Solids Editing toolbar*

AutoCAD prompts:

Command: **solidedit**
Enter a solids editing option [Face/Edge/Body/Undo/eXit] <eXit>: **e**
Enter an edge editing option [Copy/coLor/Undo/eXit] <eXit>: **c**
Select edges or [Undo/Remove]: *(select edges and press* ENTER *to complete the selection)*
Specify a base point or displacement: *(specify a base point)*
Specify a second point of displacement: *(specify a second point of displacement or press* ENTER *to consider the original selection point as a base point)*
Enter an edge editing option [Copy/coLor/Undo/eXit] <eXit>: *(select the eXit option to exit the edge editing)*
Enter a solids editing option [Face/Edge/Body/Undo/eXit] <eXit>: *(select the eXit option to exit solids editing)*

Figure 15–100 shows an example of a solid model in which the edges (right side of the model) are copied by the Copy Edges option of the SOLIDEDIT command.

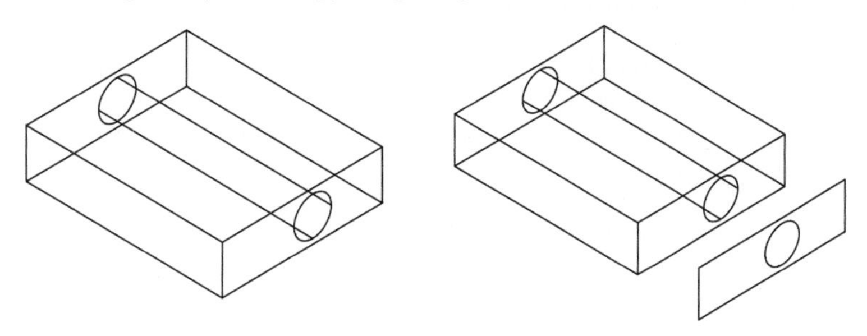

Figure 15–100 *An example of copying edges of a solid model.*

Coloring Edges

The Color Edges option of the SOLIDEDIT command allows you to change color of selected edges of a *3D* solid object. You can choose a color from the Select Color dialog box. Setting a color on an edge overrides the color setting for the layer on which the solid object resides.

Invoke the Color Edges option of the SOLIDEDIT command to change the color of selected edges:

Solids Editing toolbar	Choose Color Edges (see Figure 15–101)
Modify menu	Choose Solids Editing > Color edges
Command: prompt	**solidedit** (ENTER)

Figure 15–101 *Invoking the Color Edges option of the* SOLIDEDIT *command from the Solids Editing toolbar*

AutoCAD prompts:

Command: **solidedit**
Enter a solids editing option [Face/Edge/Body/Undo/eXit] <eXit>: **e**
Enter an edge editing option [Copy/coLor/Undo/eXit] <eXit>: **l**
Select edges or [Undo/Remove]: *(select faces and press* ENTER *to complete the selection)*

AutoCAD displays the Select Color dialog box. Select the color to change for selected faces and choose the OK button. AutoCAD prompts:

Enter an edge editing option [Copy/coLor/Undo/eXit] <eXit>: *(select the eXit option to exit the edge editing)*
Enter a solids editing option [Face/Edge/Body/Undo/eXit] <eXit>: *(select the eXit option to exit solids editing)*

AutoCAD changes the color of the selected edges of the *3D* solid model.

IMPRINTING SOLIDS

AutoCAD allows you to have an imprint of an object on the selected solid. The object to be imprinted must intersect one or more faces on the selected solid in order for imprinting to be successful. Imprinting is limited to the following objects: arcs, circles, lines, *2D* and *3D* polylines, ellipses, splines, regions, bodies, and *3D* solids.

Invoke the Imprint option of the SOLIDEDIT command to have an imprint of an object on the selected object:

Solids Editing toolbar	Choose IMPRINT (see Figure 15–102)
Modify menu	Choose Solids Editing > Imprint
Command: prompt	**solidedit** (ENTER)

Figure 15–102 *Invoking the Imprint option of the* SOLIDEDIT *command from the Solids Editing toolbar*

AutoCAD prompts:

Command: **solidedit**
Enter a solids editing option [Face/Edge/Body/Undo/eXit] <eXit>: **b**
Enter a body editing option [Imprint/seParate solids/Shell/cLean/Check/Undo/eXit] <eXit>: **i**
Select a 3D solid: *(select a 3D solid object)*
Select an object to imprint: *(select an object to imprint)*
Delete the source object <N>: *(press* ENTER *to delete the source objects or enter* **y** *to keep the source object)*
Select an object to imprint: *(select another object to imprint or press* ENTER *to complete the selection)*
Enter a body editing option [Imprint/seParate solids/Shell/cLean/Check/Undo/eXit] <eXit>: *(select the eXit option to exit body editing)*
Enter a solids editing option [Face/Edge/Body/Undo/eXit] <eXit>: *(select the eXit option to exit solids editing)*

AutoCAD creates imprint of the selected object.

SEPARATING SOLIDS

AutoCAD separates solids from a composite solid. But it cannot separate solids if the composite *3D* solid object shares a common area or volume. After separation of the *3D* solid, the individual solids retain the layers and colors of the original.

Invoke the Separate option of the SOLIDEDIT command to separate solids:

Solids Editing toolbar	Choose SEPARATE (see Figure 15–103)
Modify menu	Choose Solids Editing > Separate
Command: prompt	**solidedit** (ENTER)

Figure 15–103 *Invoking the Separate option of the* SOLIDEDIT *command from the Solids Editing toolbar*

AutoCAD prompts:

Command: **solidedit**
Enter a solids editing option [Face/Edge/Body/Undo/eXit] <eXit>: **b**
Enter a body editing option [Imprint/seParate solids/Shell/cLean/Check/Undo/eXit] <eXit>: **p**
Select a *3D* solid: *(select a 3D solid object and press* ENTER *to complete the selection)*
Enter a body editing option [Imprint/seParate solids/Shell/cLean/Check/Undo/eXit] <eXit>: *(select the eXit option to exit body editing)*
Enter a solids editing option [Face/Edge/Body/Undo/eXit] <eXit>: *(select the eXit option to exit solids editing)*

AutoCAD separates the selected composite solid.

SHELLING SOLIDS

AutoCAD creates a shell or a hollow thin wall with a specified thickness from the selected *3D* solid object. AutoCAD creates new faces by offsetting existing ones inside or outside their original positions. AutoCAD treats continuously tangent faces as single faces when offsetting. A positive offset value creates a shell in the positive face direction; a negative value creates a shell in the negative face direction.

Invoke the Shell option of the SOLIDEDIT command to create a shell:

Solids Editing toolbar	Choose SHELL (see Figure 15–104)
Modify menu	Choose Solids Editing > Shell
Command: prompt	**solidedit** (ENTER)

Figure 15–104 *Invoking the Shell option of the* SOLIDEDIT *command from the Solids Editing toolbar*

AutoCAD prompts:

Command: **solidedit**
Enter a solids editing option [Face/Edge/Body/Undo/eXit] <eXit>: **b**

Enter a body editing option [Imprint/seParate solids/Shell/cLean/Check/
Undo/eXit] <eXit>: **s**
Select a *3D* solid: *(select a 3D solid object)*
Remove faces or [Undo/Add/ALL]: *(select faces to be excluded from shelling and press* ENTER *to complete the selection)*
Specify the shell offset value: *(specify the shell offset value)*
Enter a body editing option [Imprint/seParate solids/Shell/cLean/Check/
Undo/eXit] <eXit>: *(select the eXit option to exit body editing)*
Enter a solids editing option [Face/Edge/Body/Undo/eXit] <eXit>: *(select the eXit option to exit solids editing)*

AutoCAD creates a shell with the specified thickness from the *3D* solid object.

CLEANING SOLIDS

AutoCAD allows you to remove edges or vertices if they share the same surface or vertex definition on either side of the edge or vertex. All redundant edges, imprinted as well as used, on the selected *3D* solid object are deleted.

Invoke the Clean option of the SOLIDEDIT command to clean solids:

Solids Editing toolbar	Choose CLEAN (see Figure 15–105)
Modify menu	Choose Solids Editing > Clean
Command: prompt	**solidedit** (ENTER)

Figure 15–105 *Invoking the Clean option of the* SOLIDEDIT *command from the Solids Editing toolbar*

AutoCAD prompts:

Command: **solidedit**
Enter a solids editing option [Face/Edge/Body/Undo/eXit] <eXit>: **b**
Enter a body editing option [Imprint/seParate solids/Shell/cLean/Check/
Undo/eXit] <eXit>: **l**
Select a *3D* solid: *(select a 3D solid object and press* ENTER *to complete the selection)*
Enter a body editing option [Imprint/seParate solids/Shell/cLean/Check/
Undo/eXit] <eXit>: *(select the eXit option to exit body editing)*
Enter a solids editing option [Face/Edge/Body/Undo/eXit] <eXit>: *(select the eXit option to exit solids editing)*

AutoCAD removes the selected edges or vertices of the selected *3D* model.

CHECKING SOLIDS

AutoCAD checks to see if the selected solid object is a valid *3D* solid object. With a *3D* solid model, you can modify the object without incurring ACIS failure error messages. If the selected solid *3D* model is not valid, you cannot edit the object.

Invoke the Check option of the SOLIDEDIT command to clean solids:

Solids Editing toolbar	Choose CHECK (see Figure 15–106)
Modify menu	Choose Solids Editing > Check
Command: prompt	**solidedit** (ENTER)

Figure 15–106 *Invoking the Check option of the* SOLIDEDIT *command from the Solids Editing toolbar*

AutoCAD prompts:

Command: **solidedit**
Enter a solids editing option [Face/Edge/Body/Undo/eXit] <eXit>: **b**
Enter a body editing option [Imprint/seParate solids/Shell/cLean/Check/ Undo/eXit] <eXit>: **c**
Select a 3D solid: *(select a 3D solid object and press* ENTER *to complete the selection)*
Enter a body editing option [Imprint/seParate solids/Shell/cLean/Check/ Undo/eXit] <eXit>: *(select the eXit option to exit body editing)*
Enter a solids editing option [Face/Edge/Body/Undo/eXit] <eXit>: *(select the eXit option to exit solids editing)*

AutoCAD checks the solid *3D* model and displays with appropriate information about the selected solid.

MASS PROPERTIES OF A SOLID

The MASSPROP command calculates and displays the mass properties of selected solids and regions. The mass properties displayed for solids are mass, volume, bounding box, centroid, moments of inertia, products of inertia, radii of gyration, and principal moments with corresponding principal directions. The mass properties are calculated based on the current UCS.

Invoke the MASSPROP command:

Inquiry toolbar	Choose the MASSPROP command (see Figure 15–107)
Tools menu	Choose Inquiry > Region/Mass Properties
Command: prompt	**massprop** (ENTER)

Figure 15–107 *Invoking the* MASSPROP *command from the Inquiry toolbar*

AutoCAD prompts:

> Command: **massprop**
> Select objects: *(select the objects and press* ENTER *to complete the selection)*

The MASSPROP command displays the object mass properties of the selected objects in the text window, as shown in Figure 15–108. Then, AutoCAD prompts:

> Write analysis to a file? [Yes/No] <N>: *(type* **y** *to create a file or* **n** *for not to create a file)*

```
AutoCAD Text Window
Edit
Mass:                    105.66
Volume:                  105.66
Bounding box:         X: 0.00  --  8.00
                      Y: 0.00  --  7.00
                      Z: -3.00  --  2.00
Centroid:             X: 3.55
                      Y: 3.50
                      Z: -1.25
Moments of inertia:   X: 1989.68
                      Y: 2177.49
                      Z: 3640.82
Products of inertia: XY: 1312.68
                     YZ: -461.91
                     ZX: -600.08
Radii of gyration:    X: 4.34
                      Y: 4.54
                      Z: 5.87
Press ENTER to continue:
Principal moments and X-Y-Z directions about centroid:
                      I: 497.04 along [0.97 0.00 -0.25]
                      J: 681.40 along [0.00 1.00 0.00]
                      K: 1048.65 along [0.25 0.00 0.97]

Write to a file ? <N>:
```

Figure 15–108 *Mass properties listing*

If you enter y, AutoCAD prompts for a file name and saves the file in ASCII format.

Note: You can also use the AutoCAD LIST and AREA commands to obtain information about coordinates and areas of solid(s).

HIDING OBJECTS

The HIDE command hides objects (or displays them in different colors) that are behind other objects in the current viewport. Complex models are difficult to read in wireframe form, and benefit in clarity when the model is displayed with hidden lines removed. HIDE considers circles, solids, traces, wide polyline segments, *3D* faces, polygon meshes, and the extruded edges of objects with a thickness to be opaque surfaces hiding objects that lie behind them. The HIDE command remains active only until the next time the display is regenerated. Depending on the complexity of the model, hiding may take from a few seconds to even several minutes.

Invoke the HIDE command:

Render toolbar	Choose the HIDE command (see Figure 15–109)
View menu	Choose Hide
Command: prompt	**hide** (ENTER)

Figure 15–109 *Invoking the HIDE command from the Render toolbar*

AutoCAD prompts:

Command: **hide**

There are no prompts to be answered. The current viewport goes blank for a period of time, depending on the complexity of the model, and is then redrawn with hidden lines removed temporarily. Hidden-line removal is lost during plotting unless you specify that AutoCAD remove hidden lines in the plotting configuration.

PLACING MULTIVIEWS IN PAPER SPACE

The SOLVIEW command creates untiled viewports (Layout) using orthographic projection to lay out orthographic views and sectional views. View-specific information is saved with each viewport as you create it. The information that is saved is used by the SOLDRAW command, which does the final generation of the drawing view. The SOLVIEW command automatically creates a set of layers that the SOLDRAW command uses to place the visible lines, hidden lines, and section hatching for each view. In addition, the SOLVIEW command creates a specific layer for dimensions that are visible in individual viewports. The SOLVIEW command applies the following conventions in naming the layers:

Layer	Name
For visible lines	VIEW NAME-VIS
For hidden lines	VIEW NAME-HID
For dimensions	VIEW NAME-DIM
For sections	VIEW NAME-HAT6

The viewport objects are drawn on the VPORTS layer. The SOLVIEW command creates the VPORTS layer if it does not already exist. All the layers created by SOLVIEW are assigned the color white and the linetype continuous. The stored information is deleted and updated when you run the SOLDRAW command, so do not draw any objects on these layers.

Invoke the SOLVIEW command:

Solids toolbar	Choose Setup View (see Figure 15–110)
Draw menu	Choose Solids > Setup > View
Command: prompt	**solview** (ENTER)

Figure 15–110 *Invoking the SOLVIEW command by selecting SETUP view from the Solids toolbar*

AutoCAD prompts:

Command: **solview**
Enter an option [Ucs/Ortho/Auxiliary/Section]: *(select one of the available options)*

Ucs

The Ucs option creates a profile view relative to a User Coordinate System. AutoCAD prompts:

Enter an option [Named/World/?/Current] <Current>: *(select one of the available options)*

The Named option uses the *XY* plane of a named UCS to create a profile view. After prompting for the name of the view, AutoCAD prompts as follows:

Enter view scale <1.0>: *(specify the scale factor for the view to be displayed)*
Specify View center: *(specify the center location for the viewport to be drawn in the paper space and press* ENTER*)*
Specify first corner of viewport: *(specify a point for one corner of the viewport)*
Specify opposite corner of viewport: *(specify a point for the opposite corner of the viewport)*
Enter view name: *(specify a name for the newly created view)*

The World option uses the *XY* plane of the WCS to create a profile view. The prompts are the same as just described for the Named option.

The ? option lists the names of existing User Coordinate Systems. After the list is displayed, press any key to return to the first prompt.

The Current option, which is the default option, uses the *XY* plane of the current UCS to create a profile view. The prompts are the same as explained earlier for the Named option.

If no untiled viewports exist in your drawing, the UCS option allows you to create an initial viewport from which other views can be created.

Ortho

The Ortho option creates a folded orthographic view from an existing view. AutoCAD prompts:

Specify side of viewport to project: *(select the one of the edges of a viewport)*
Specify view center: *(specify the center location for the viewport to be drawn in the paper space and press* ENTER*)*
Specify first corner of viewport: *(specify a point for one corner of the viewport)*
Specify opposite corner of viewport: *(specify a point for the opposite corner of the* viewport)
Enter view name: *(specify a name for the newly created view)*

Auxiliary

The Auxiliary option creates an auxiliary view from an existing view. An auxiliary view is one that is projected onto a plane perpendicular to one of the orthographic views and inclined in the adjacent view. AutoCAD prompts:

Specify first point of inclined plane: *(specify a point)*
Specify second point of incline plane: *(specify a point)*
Specify side to view from: *(specify a point that determines the side from which you will view the plane)*
Specify view center: *(specify the center location for the viewport to be drawn in the paper space and press* ENTER*)*
Specify first corner of viewport: *(specify a point for one corner of the viewport)*
Specify opposite corner of viewport: *(specify a point for the opposite corner of the viewport)*
Enter view name: *(specify a name for the newly created view)*

Section

The Section option creates a sectional view of solids with crosshatching. AutoCAD prompts:

Specify first point of cutting plane: *(specify a point to define the first point of the cutting plane)*
Specify second point of cutting plane: *(specify a point to define the second point of the cutting plane)*
Specify side to view from: *(specify a point to define the side of the cutting plane from which to view)*
Enter view scale: *(specify the scale factor for the view to be displayed)*
Specify view center: *(specify the center location for the viewport to be drawn in the paper space and press* ENTER*)*
Specify first corner of viewport: *(specify a point for one corner of the viewport)*
Specify opposite corner of viewport: *(specify a point for the opposite corner of the viewport)*
Enter view name: *(specify a name for the newly created view)*

To exit the command sequence, press ENTER.

GENERATING VIEWS IN VIEWPORTS

The SOLDRAW command generates sections and profiles in viewports that have been created with the SOLVIEW command. Visible and hidden lines representing the silhouette and edges of solids in the viewports are created and then projected to a plane perpendicular to the viewing direction. AutoCAD deletes any existing profiles and sections in the selected viewports, and new ones are generated. In addition, AutoCAD freezes all the layers in each viewport, except those required to display the profile or section.

Invoke the SOLDRAW command:

Solids toolbar	Choose Setup Drawing (see Figure 15–111)
Draw menu	Choose Solids > Setup > Drawing
Command: prompt	**soldraw** (ENTER)

Figure 15–111 *Invoking the* SOLDRAW *command by selecting* SETUP DRAWING *from the Solids toolbar*

AutoCAD prompts:

Command: **soldraw**
Select viewports to draw...
Select objects: *(select the viewports to be drawn)*

AutoCAD generate views in the selected viewports.

GENERATING PROFILES

The SOLPROF command creates a profile image of a solid, including all of its edges, according to the view in the current viewport. The profile image is created from lines, circles, arcs, and/or polylines. SOLPROF will not give correct results in perspective view; it is designed for parallel projections only.

The SOLPROF command will work only when you are working in layout tab and you are in model space.

Invoke the SOLPROF command:

Solids toolbar	Choose Setup Profile (see Figure 15–112)
Draw menu	Choose Solids > Setup > Profile
Command: prompt	**solprof** (ENTER)

Figure 15–112 *Invoking the* SOLPROF *command by selecting* SETUP PROFILE *command from the Solids toolbar*

AutoCAD prompts:

Command: **solprof**
Select objects: *(select one or more objects)*

Select one or more solids and press ENTER. The next prompt lets you decide the placement of hidden lines of the profile on a separate layer:

Display hidden profile lines on separate layer? [Yes/No] <Y>:

Enter **Y** or **N**. If you answer **Y** (default option), two block inserts are created—one for the visible lines in the same linetype as the original and the other for hidden lines in the hidden linetype. The visible lines are placed on a layer whose name is PV-(viewport handle of the current viewport). The hidden lines are placed on a layer whose name is PH-(viewport handle or the current viewport). If these layers do not exist, AutoCAD will create them. For example, if you create a profile in a viewport whose handle is 6, then visible lines will be placed on a layer PV-6 and hidden lines on layer PH-6. To control the visibility of the layers, you can turn the appropriate layers on and off.

The next prompt determines whether *2D* or *3D* entities are used to represent the visible and hidden lines of the profile:

Project profile lines onto a profile? [Yes/No] <Y>:

Enter **Y** or **N**. If you answer **Y** (default option), AutoCAD creates the visible and hidden lines of the profile with *2D* AutoCAD entities; **N** creates the visible and hidden lines of the profile with *3D* AutoCAD entities.

Finally, AutoCAD asks if you want tangential edges deleted. A tangential edge is an imaginary edge at which two facets meet and are tangent. In most of the drafting applications, the tangential edges are not shown. The prompt sequence is as follows for deleting the tangential edges:

Delete tangential edges? [Yes/No] <Y>:

Enter **Y** to delete the tangential edges and **N** to retain them.

Open the Exercise Manual PDF file for Chapter 15 on the accompanying CD for project and discipline specific exercises.

If you have the accompanying Exercise Manual, refer to Chapter 15 for project and discipline specific exercises.

REVIEW QUESTIONS

1. What type of coordinate system does AutoCAD use?
   a. Lagrangian System
   b. Right-hand system
   c. Left-hand system
   d. Maxwellian system
2. If the origin of the current coordinate system is off screen, then the UCSICON, if it is set to Origin, will:
   a. appear in the lower left corner of the screen
   b. attempt to be at the origin, and will therefore not be on screen
   c. appear at the center of the screen
   d. force AutoCAD to perform a ZOOM so that it will be on screen
3. Which of the following UCS options will perform a translation only on the coordinate system?
   a. Rotate
   b. Z-axis
   c. Origin
   d. View
   e. none of the above
4. Which of the following UCS options will perform only a rotation of the coordinate system?
   a. Rotate
   b. Z-axis
   c. Origin
   d. View
   e. none of the above

5. Which of the following UCS options will perform both a rotation and a translation of the coordinate system?

   a. Rotate
   b. *Z*-axis
   c. Origin
   d. View
   e. none of the above

6. The command used to generate a perspective view of a *3D* model is:

   a. VPOINT
   b. PERSPECT
   c. VPORT
   d. DVIEW
   e. none of the above

7. The VPOINT command will allow you to specify the viewing direction in all of the following methods *except*:

   a. 2 points
   b. 2 angles
   c. 1 point
   d. dynamically dragging the *X*, *Y*, and *Z* axes
   e. none of the above (that is, all are valid)

8. Objects can be assigned a negative thickness.

   a. True
   b. False

9. Regions:

   a. are *3D* objects
   b. can be created from any group of objects
   c. can be used in conjunction with Boolean operations
   d. cannot be crosshatched

10. Which of the following mesh commands is primarily used in a user-defined function, rather than from the keyboard?

    a. 3DMESH
    b. RULESURF
    c. TABSURF
    d. EDGEMESH
    e. 3DFACE

11. The PFACE command:

    a. generates a mesh

    b. requests the vertexes of the object

    c. can create an opaque (will hide) flat polygon

    d. all of the above

12. To align two objects in *3D* space, invoke the command called:

    a. ALIGN

    b. ROTATE

    c. 3DROTATE

    d. TRANSFORM

    e. 3DALIGN

13. *3D* arrays are limited to rectangular patterns (that is, rows, columns, and layers).

    a. True b. False

14. The UNION command will allow you to select more than two solid objects concurrently.

    a. True b. False

15. Which of the following is not a standard solid primitive shape used by AutoCAD?

    a. SPHERE

    b. BOX

    c. DOME

    d. WEDGE

    e. CONE

16. Fillets on solid objects are limited to planar edges.

    a. True b. False

17. A command which will split a solid object into two solids is:

    a. TRIM

    b. SLICE

    c. CUT

    d. SPLIT

    e. DIVIDE

18. To display a more realistic view of a *3D* object, use:
    a. HIDE
    b. VIEW
    c. DISPLAY3D
    d. MAKE3D
    e. VIEWEDIT
19. To force invisible edges of *3D* faces to display, what system variable should be set to 1?
    a. 3DEGE
    b. 3DFRAME
    c. SPLFRAME
    d. EDGEFACE
    e. EDGEVIEW
20. Advantages of solid modeling include:
    a. creation of objects which are manufacturable
    b. interfaces with Computer Aid Manufacturing
    c. analysis of physical properties of objects
    d. all of the above
21. The SECTION command:
    a. creates a region
    b. creates a polyline
    c. crosshatches the area where a plane intersects a solid
    d. is an alias for the HATCHcommand
    e. will give you a choice of either a or b
22. Which AutoCAD command permits a 3D model to be viewed interactively in the current viewport?
    a. 3DVIEW
    b. 3DPAN
    c. 3DROTATE
    d. 3DORBIT

23. When the cursor is positioned inside the arcball of the 3DORBIT command the 3D object can be viewed ___________?

    a. horizontally

    b. vertically

    c. diagonally

    d. all of the above

24. When the cursor is positioned outside the arcball of the 3DORBIT command the 3D object can be viewed ___________?

    a. horizontally

    b. vertically

    c. diagonally

    d. all of the above

    e. both a or b

25. Which of the following is not an option in the 3DORBIT command?

    a. pan

    b. zoom

    c. tilt

    d. none of the above

26. AutoCAD allows you to display a perspective or a parallel projection of the view while 3DORBIT is active.

    a. True

    b. False

27. AutoCAD allows you to view shaded view while 3DORBIT is active.

    a. True

    b. False

28. Which of the following AutoCAD commands allows you to set clipping planes for the objects in 3D orbit view?

    a. CLIPLANE

    b. 3DCLIP

    c. 3DFACED

    d. CLIPVIEW

29. The Continuous Orbit option of 3DORBIT permits the 3D model to be set into motion, even in shaded mode?

    a. True   b. False

30. AutoCAD allows you to edit solid objects by __________.

    a. extruding faces

    b. copying faces

    c. offsetting faces

    d. moving faces

    e. all the above

31. When offsetting a face of a 3D solid a positive value will increase the volume of the solid.

    a. True   b. False

32. A hole in a 3D solid model can be relocated using the Move Faces option within the SOLIDEDIT command.

    a. True   b. False

33. Colors of individual edges of a 3D solid can be changed using SOLIDEDIT?

    a. True   b. False

34. When an edge is copied using the SOLIDEDIT command it is copied as an __________.

    a. Arc

    b. Spline

    c. Circle

    d. Line

    e. all of the above

35. When a composite 3D solid is separated into individual solids each solid element will retain the layer and color settings of the original solid.

    a. True   b. False

36. AutoCAD refers to the removal of specified edges or vertices as __________.

    a. Extraction

    b. Cleaning

    c. Removing

    d. Erasing

Rendering

INTRODUCTION

Shading or rendering turns your three-dimensional model into a realistic (eye-catching) image. The AutoCAD SHADEMODE command allows you to produce quick, shaded models. However, the AutoCAD RENDER command gives you more control over the appearance of the final image. You can add lights and control lighting in your drawing and define the reflective qualities of surfaces in the drawing, making objects dull or shiny. You can create the rendered image of your *3D* model entirely within AutoCAD.

After completing this chapter, you will be able to do the following:

- Render a *3D* model
- Create and modify lighting—ambient light, distant light, point light, and spotlight
- Create and modify a scene
- Create and modify a material
- Save and replay an image

SHADING A MODEL

The AutoCAD SHADEMODE command lets you produce a shaded picture of the *3D* model in the current Viewport. It also provides various shading and wireframe options, and you can even edit shaded objects without regenerating the drawing. You have little control over lighting, and AutoCAD uses a single light that is logically placed just over your right shoulder in the current viewport.

Invoke the SHADEMODE command:

View menu	Choose Shade
Command: prompt	**shademode** (ENTER)

AutoCAD prompts:

Command: **shademode**
Enter option [2D wireframe/3D wireframe/Hidden/Flat/Gouraud/fLat+edges/gOuraud+edges] <Flat+Edges>: *(right-click and select one of the available options from the shortcut menu)*

2D Wireframe

The 2D wireframe option displays the objects using lines and curves to represent the boundaries. Raster and OLE objects, linetypes, and lineweights are visible.

3D Wireframe

The 3D wireframe option displays the objects using lines and curves to represent the boundaries. Raster and OLE objects, linetypes, and lineweights are visible. AutoCAD also displays a shaded *3D* user coordinate system (UCS) icon.

Hidden

The Hidden option displays the objects using the *3D* wireframe representation and hides the lines representing the back faces. Text objects are ignored unless it is given a thickness. The thickness can be as large a value as you want or as small as 0.000001 units (or any value other than 0).

Flat

The Flat option displays the objects shaded between the polygon faces. The objects appear flatter and less smooth than Gouraud shaded objects. Materials that have been applied to objects are shown.

Gouraud

The Gouraud option displays the shaded objects and smooths the edges between polygon faces, giving the objects a smooth, realistic appearance. Materials that have been applied to objects are shown.

Flat with edges on

The Flat+edges option displays the objects as a combination of Flat and Wireframe options. The objects are flat shaded with the wireframe showing through.

Gouraud with edges on

The Gouraud+edges option displays the objects as a combination of Gouraud and Wireframe options. The objects are Gouraud shaded with the wireframe showing through.

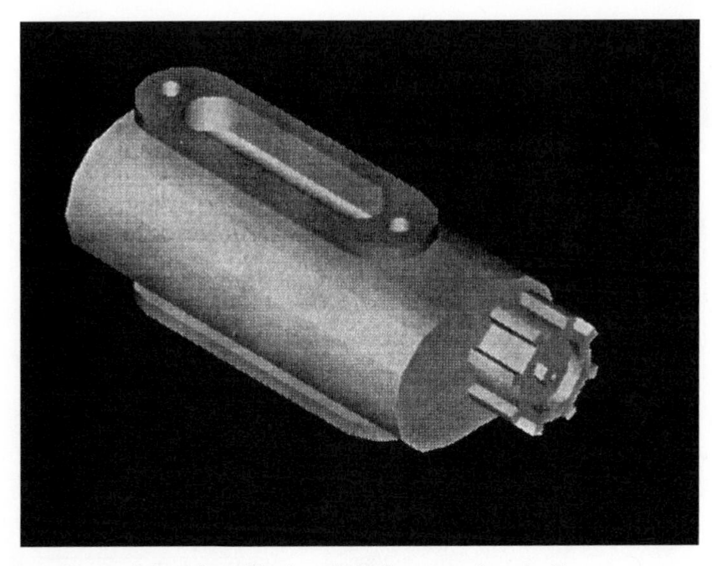

Figure 16–1 *A 3D model rendered by AutoCAD (courtesy Autodesk)*

RENDERING A MODEL

The AutoCAD render facility allows you to create realistic models from your AutoCAD drawings. With the AutoCAD RENDER command you can adjust lighting factors, material finishes, and camera placement. All these options give you a great deal of flexibility. If you wish to take rendering "to the max," you may wish to purchase 3D Studio Max or 3D VIZ to gain the greatest flexibility and photorealism.

The tools available for rendering allow you to adjust the type and quality of the rendering, set up lights and scenes, and save and replay images. But you can always use the RENDER command without any other AutoCAD render setup. By default, RENDER uses the current view if no scene or selection is specified. If there are no lights specified, the RENDER command assumes a default over-the-shoulder distant light source with an intensity of 1.

Invoke the RENDER command:

Render toolbar	Choose the RENDER command (see Figure 16–2)
View menu	Choose Render > Render...
Command: prompt	**render** (ENTER)

Figure 16–2 *Invoking the RENDER command from the Render toolbar*

AutoCAD displays the Render dialog box, as shown in Figure 16–3.

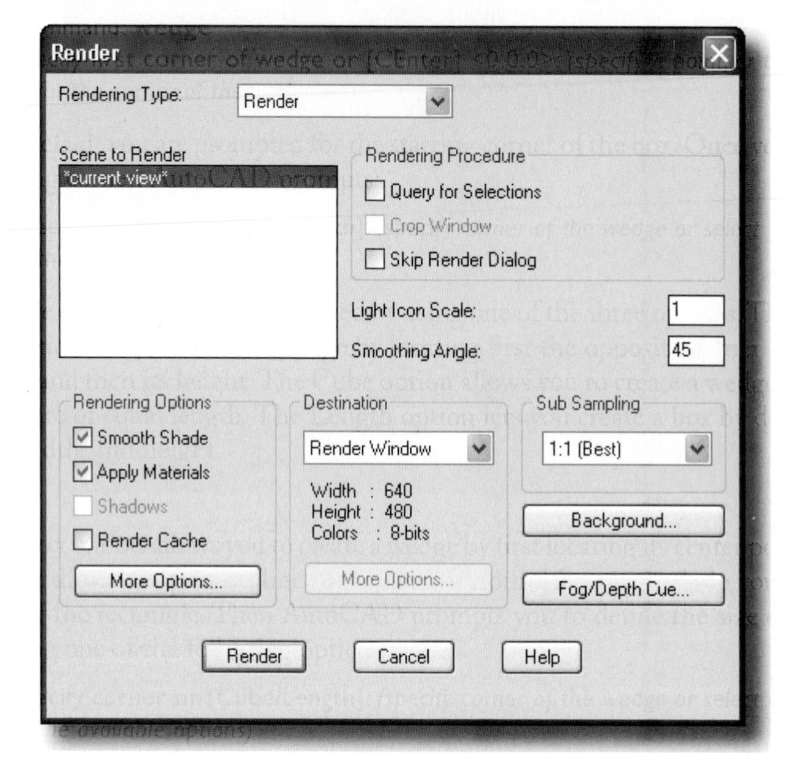

Figure 16–3 *Render dialog box*

The **Rendering Type:** list box lists the available rendering types, including AutoCAD Render, Photo Real, and Photo Raytrace. The default type is set to AutoCAD Render.

The **Scene to Render** list box lists the scenes available in the current drawing, including the current view, from which you can select the scene/view for rendering.

The **Rendering Procedure** section of the dialog box sets the default value for rendering. Three settings are available:

The **Query for Selections** toggle controls whether or not to display a prompt to select objects to render.

The **Crop Window** toggle controls the prompts for whether or not to pick an area on the screen for rendering.

The **Skip Render Dialog** toggle controls whether or not to skip the Render dialog box and render the current view.

The **Light Icon Scale:** text field controls the size of the light blocks inserted in the drawing. The default is set to a scale factor of 1.0. If necessary, you can change the value to some other real number to rescale the blocks. Overhead, Direct, and Sh_spot are the blocks affected by the scale factor.

The **Smoothing Angle:** text field sets the angle at which AutoCAD interprets an edge. The default value is 45 degrees. Angles greater than 45 degrees are considered edges, and those less than 45 degrees as smoothed.

The **Rendering Options** section of the dialog box controls the rendering display. Four main settings plus additional options are available:

> The **Smooth Shade** option yields a smooth appearance. Depending on the object's surfaces and the direction of the screen's lights, smooth shading adds a cleaner, more realistic appearance to a rendering.
>
> The **Apply Materials** option applies surface materials you have defined and attaches them by color or to specific object(s). If the **Apply Materials** option is not selected, then all objects in the drawing assume the color, ambient, reflection, transparency, refraction, bump map, and roughness attribute values defined for the *GLOBAL* material.
>
> The **Shadows** option generates shadows when selected. This option is applicable only when **Photo Real** or **Photo Raytrace** rendering is selected.
>
> The **Render Cache** option specifies that rendering information be written to a cache file on the hard disk. As long as the drawing or view is unchanged, the cached file is used for subsequent renderings, eliminating the need for AutoCAD to retessellate.
>
> For fine-tuning the rendering quality, choose the **More Options...** button. AutoCAD displays the Windows Render Options dialog box. The options available vary depending on whether you have selected **Render, Photo Real, Photo Raytrace**, or a third-party application as your rendering type.

The **Destination** section of the dialog box controls the image output setting. Three options are available:

> The **Viewport** option allows AutoCAD to render to the current viewport.
>
> The **File** option renders to a file. When the **File** option is selected, you must choose the **More Options...** button to select the file type, set the colors in the output file, and set the postscript options, if necessary.
>
> The **Render Window** option renders the model in a window, as shown in Figure 16–4.

The OPEN command (available from the File menu) in the Render window allows you to open three types of files: bitmap *(.BMP)*, .dib, .rle *(.CLP)* files. The SAVE command (available from the File menu) in the Render window allows you to save an image to a bitmap file. If necessary, you can also use the PRINT command (available from the File menu) to print the image. The COPY command (available from the Edit menu) can be used to copy an image from the active render window to the clipboard. The OPTIONS command (available from the File menu) displays the Windows Render Options dialog box, as shown in Figure 16–5, from which you can select aspect ratios (Size in Pixels) and resolutions (Color Depth) for bitmap images.

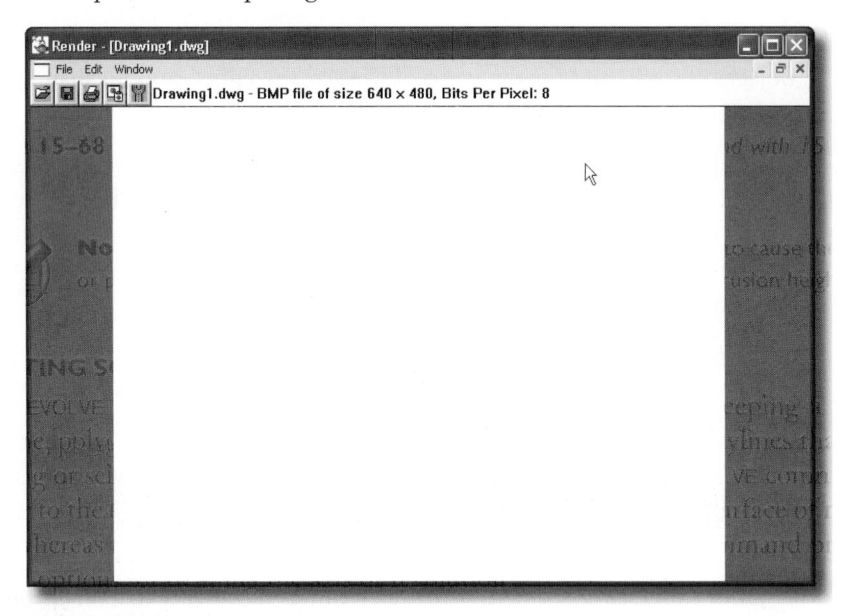

Figure 16–4 *Render window*

The **Sub Sampling** section of the Render dialog box controls the rendering time and image quality without abandoning effects such as shadows by rendering a fraction of all pixels. The available ratios include 1:1 (for best quality) to 8:1 (fastest).

The **Background...** button allows you to set the background for your scene. Choose the **Background...** button to display the Background dialog box, as shown in Figure 16–6. Following are the four options available to select the type of background for rendering:

The **Solid** option selects a one-color background. Use the controls in the **Colors** section to specify the color.

The **Gradient** option lets you specify a two- or three-color gradient background. Use the Colors section controls and the Horizon, Height, and Rotation controls on the lower right of the dialog box to define the gradient.

The **Image** option enables you to use a bitmap *(.BMP)* file for the background. Manipulate the controls in the **Image** and **Environment** sections to define the bitmap.

The **Merge** option enables you to use the current AutoCAD image as the background.

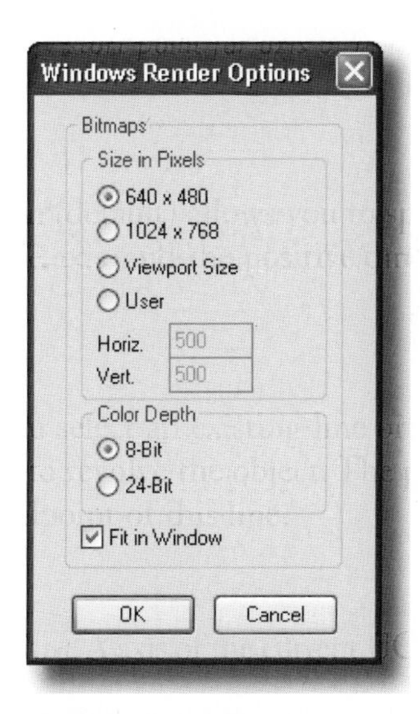

Figure 16–5 *Windows Render Options dialog box*

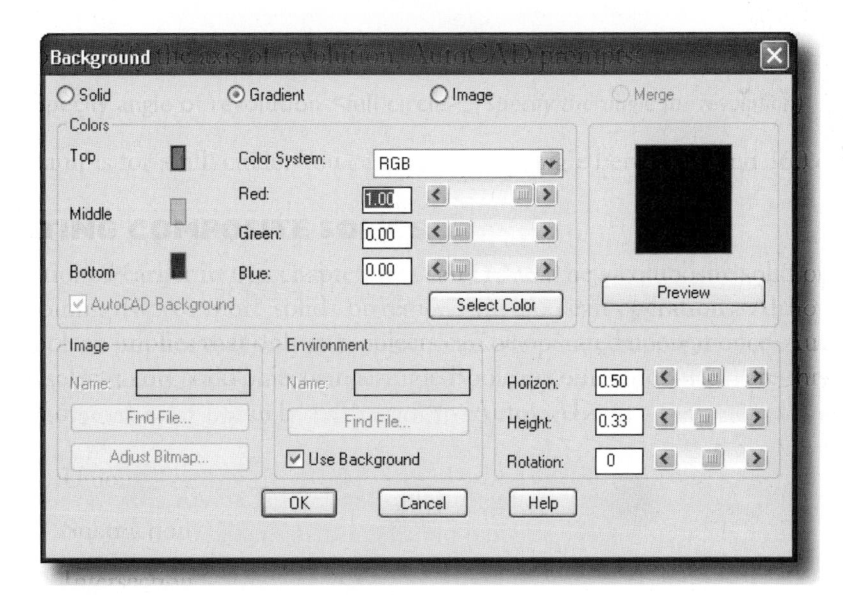

Figure 16–6 *Background dialog box*

The **Fog/Depth Cue...** button in the Render dialog box provides settings for visual cues for the apparent distance of objects. Choose the **Fog/Depth Cue...** button to display the Fog/Depth Cue dialog box, as shown in Figure 16–7.

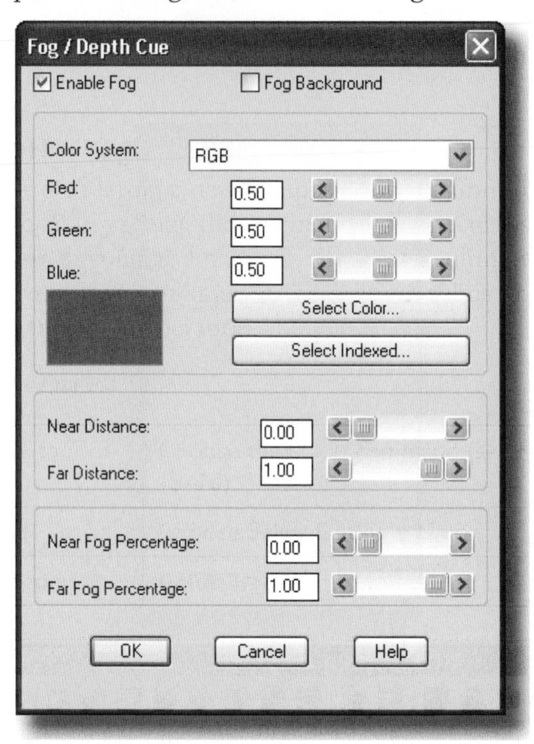

Figure 16–7 *Fog/Depth Cue dialog box*

The **Enable Fog** toggle sets the fog to ON and OFF without affecting the settings in the dialog box. The **Fog Background** toggle controls whether or not to apply the fog to the background as well as to the geometry. The **color controls** section of the dialog box controls whether AutoCAD uses the red-green-blue color system or the hue-lightness-saturation color system. The **Near Distance** and **Far Distance** controls define where the fog starts and ends. Each value is a percentage of the distance from the camera to the back clipping plane. The **Near Fog Percentage** and **Far Fog Percentage** controls define the percentage of fog at the near and far distances, ranging from 0% fog to 100% fog.

After making all the necessary changes in the Render dialog box, choose the **Render** button. AutoCAD renders the selected objects in the selected destination window.

SETTING UP LIGHTS

AutoCAD's render facility gives you great control over four types of lights in your renderings:

> **Ambient light** can be thought of as background light that is constant and distributed equally among all objects.
>
> **Distant light** gives off a fairly straight beam of light that radiates in one direction. Another property of distant light is that its brilliance remains constant, so an object close to the light receives as much light as a distant object.
>
> **Point light** can be thought of as a ball of light. A point light radiates beams of light in all directions. Point lights also have more natural characteristics. Their brilliance may be diminished as the light moves away from its source. An object that is near a point light appears brighter; an object that is farther away will appear darker.
>
> **Spotlight** is very much like the kind of spotlight you might be accustomed to seeing at a theater or auditorium. Spotlights produce a cone of light toward a target that you specify.

Invoke the LIGHTS command:

Render toolbar	Choose the LIGHTS command (see Figure 16–8)
View menu	Choose Render > Light...
Command: prompt	**light** (ENTER)

Figure 16–8 *Invoking the* LIGHTS *command from the Render toolbar*

AutoCAD displays the Lights dialog box, as shown in Figure 16–9.

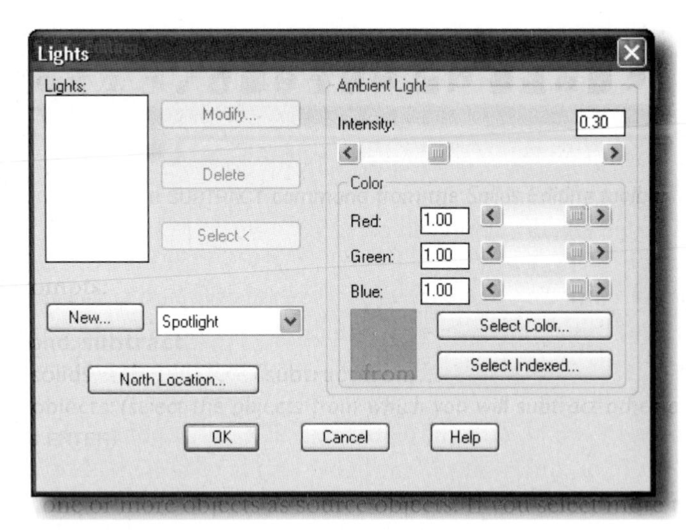

Figure 16–9 *Lights dialog box*

CREATING A NEW LIGHT

In the Lights dialog box, select one of the three available light types from the list box located to the right of the **New...** button. Then choose the **New...** button. Depending on the type of light selected, AutoCAD displays the New Point Light dialog box (shown in Figure 16–10), New Distant Light dialog box (shown in Figure 16–11), or the New Spotlight dialog box (shown in Figure 16–12).

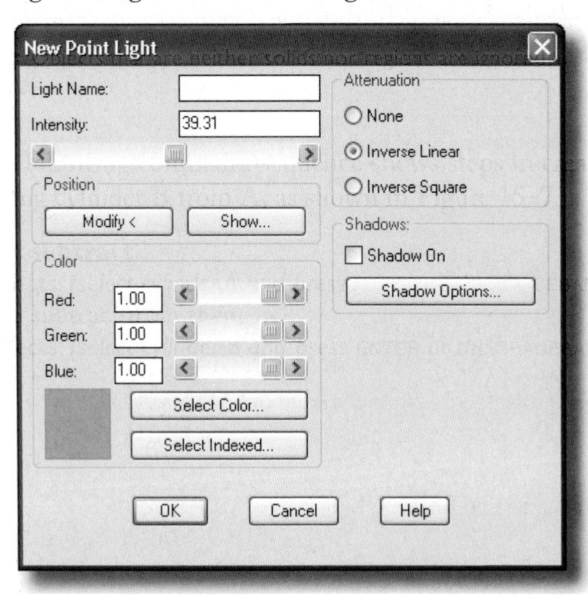

Figure 16–10 *New Point Light dialog box*

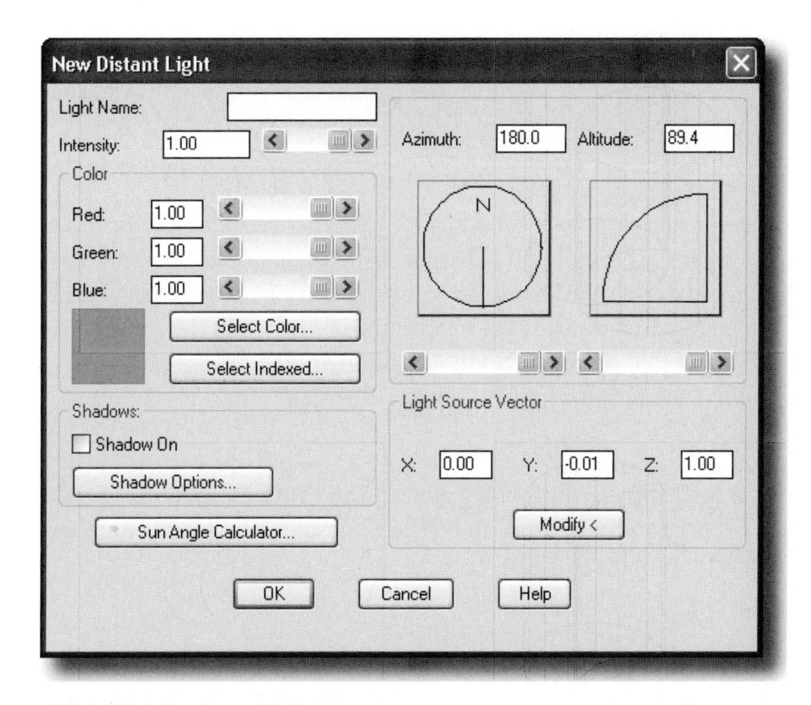

Figure 16–11 *New Distant Light dialog box*

Figure 16–12 *New spotlight dialog box*

Light Name

Specify the name of the light in the **Light Name:** text box. The name must be eight or fewer characters.

Intensity

The slider bar located below the **Intensity:** text box changes the brightness of the light, with 0 turning the light off. Distant light intensity values may range from 0 to 1. Point lights have a more complex intensity setting. This setting can be any real number. The factors that control the maximum intensity are the extents of the drawing and the current rate of falloff. Spotlight intensity factors are the same as for point lights except that the falloff is always inverse linear.

Position

The **Modify** < and **Show...** buttons located in the **Position** section let you modify or look at the *X,Y,Z* coordinate location of the light and its target. (available only with point lights and spot lights).

Color

The **Color** section controls the current color of the light. To set the color you can either adjust the Red, Green, and Blue slider bars or choose the **Select Color...** or **Select Indexed...** button to choose true color or index color respectively.

Attenuation

The setting of the attenuation controls how light diminishes over distance. The attenuation applies to both point light as well as spotlight. Select one of the three radio buttons. Selection of **None** sets no attenuation. Objects far from the point light are as bright as objects close to the light. Selection of **Inverse Linear** falloff decreases the light intensity linearly as the distance increases. For example, an object 10 units away from the light source will be 1/10 as illuminated as an object adjacent to the light source. Selection of **Inverse Square** falloff decreases the light intensity by the inverse of the squared distance. An object 10 units away from the light source receives 1/100 the amount of light of an item adjacent to the light source. This function provides rapid falloff so that a point light can be more localized.

Azimuth and Altitude

The **Azimuth and Altitude** edit boxes in the New Distant Light dialog box specify the position of the distant light by using the site-based coordinates. Azimuth can be set at any value between −180 and 180. Altitude can be set at any value between 0 and 90.

Light Source Vector

The **Light Source Vector** section in the New Distant Light dialog box displays the coordinates of the light vector that result from the light position you set using Azimuth and Altitude. You can also enter values directly in the edit boxes. AutoCAD updates the corresponding Azimuth and Altitude values.

Hotspot and Falloff

The **Hotspot and Falloff** edit fields in the New Spotlight dialog box specify the angle that defines the brightness cone of light and the full cone of light, respectively.

Once you set the appropriate values in the dialog box, choose the **OK** button to create a new light and close the dialog box. AutoCAD lists the name of the newly created light in the **Lights:** list box of the Lights dialog box.

AMBIENT LIGHT

The Ambient Light slider bar allows you to adjust the intensity of the ambient (background) light from a value of 0 (off) to 1 (bright). The **Color** section controls the current color of the ambient light. To set the color you can adjust either the Red, Green, or Blue slider bars or choose the **Select Color...** button or **Select Indexed...** button. Choose the **OK** button to accept the changes and close the Lights dialog box.

MODIFYING A LIGHT

In the Lights dialog box, first select the light name to modify from the **Lights:** list box. Then choose the **Modify...** button. AutoCAD displays the appropriate Modify dialog box, depending on the light type selected. Make the necessary changes, and choose the **OK** button to accept the changes. You cannot change a light type. For example, you cannot make a point light a distant light. But you can delete the point light and insert a new distant light in the same location.

DELETING A LIGHT

First, select the light name to delete from the **Lights:** list box. Then choose the **Delete** button. AutoCAD deletes the selected light.

SELECTING A LIGHT

The **Select <** button allows you to select a light from the screen. AutoCAD temporarily dismisses the dialog box while you specify a light on screen using the pointing device. The Lights dialog box returns, with the selected light highlighted in the **Lights:** list.

SETTING UP A SCENE

Inserting and adjusting lights allows you to render images from an unlimited number of viewpoints. If necessary, you can save a certain combination of lights and a particular view as a scene, which you can recall at any time. A scene represents a particular view of the drawing together with one or more lights. Making a scene avoids re-creating a particular set of conditions every time you need to render that image. The VPOINT and DVIEW commands are used to control viewing of a model from any point in model space, and the LIGHT command allows you to add one or more lights to the model or modify lights. The SCENE command allows you to save a scene. You can have an unlimited number of scenes in a drawing.

Note: The VIEW command allows you to save a view but doesn't save the lights, whereas the scene can include both the view and the light positions.

Invoke the SCENES command:

Render toolbar	Choose the SCENES command (see Figure 16–13)
View menu	Choose Render > Scene...
Command: prompt	**scene** (ENTER)

Figure 16–13 *Invoking the* SCENES *command from the Render toolbar*

AutoCAD displays the Scenes dialog box, as shown in Figure 16–14.

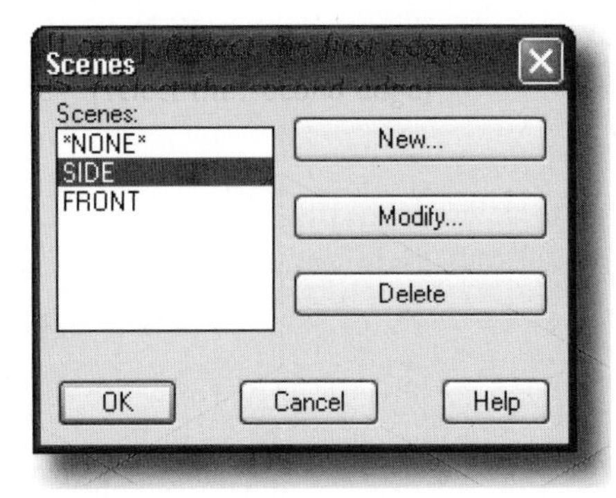

Figure 16–14 *Scenes dialog box*

AutoCAD lists the scenes available in the current drawing in the **Scenes:** list box.

CREATING A NEW SCENE

The **New...** button in the Scenes dialog box allows you to add a new scene to the current drawing. When you choose the **New...** button, AutoCAD displays the New Scene dialog box, as shown in Figure 16–15. (A similar dialog box is displayed when you select the **Modify...** button, except that dialog box is called Modify Scene.)

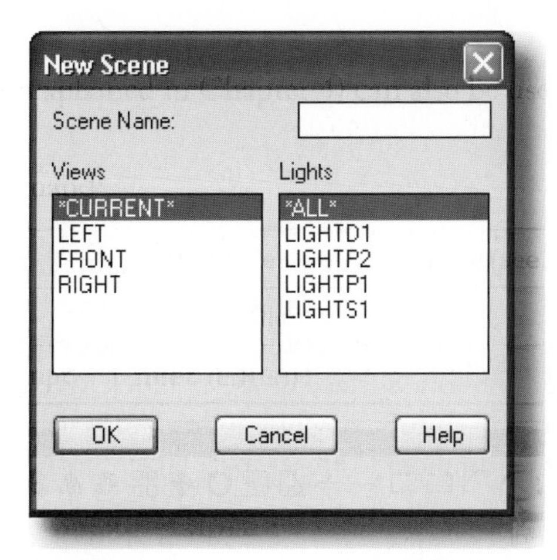

Figure 16–15 *New Scene dialog box*

Scene Name

Specify the name of the scene in the **Scene Name:** text box, which may be up to eight characters long.

Views

The **Views** list box displays the list of the views in the current drawing. *CURRENT* is the current view in the active viewport. The active view in the current scene is highlighted. Selecting another view makes it the new view of the scene. You can have only one view in a scene.

Lights

The **Lights** list box displays the list of lights in the current drawing. *ALL* represents all the lights in the drawing. When you select the *ALL* option, all the lights in the drawing are added to the scene. The lights in the current scene are highlighted. Holding down the CTRL key while selecting a non-highlighted light adds that light to the scene. Selecting a highlighted light deselects that light and removes it from the scene.

Note: You can create a scene with no lights, so that in this case the only lighting in the scene is ambient light.

Choose the **OK** button to create a new scene.

MODIFYING AN EXISTING SCENE

The **Modify...** button in the Scenes dialog box allows you to modify the selected scene. AutoCAD displays the Modify Scene dialog box and allows you to add or delete views and lights or change the name of the scene.

DELETING A SCENE

The **Delete** button in the Scenes dialog box deletes the selected scene from the current drawing.

MATERIALS

The RMAT command gives you the power to modify the light reflection characteristics of the objects you will render. By modifying these characteristics, you make objects appear rough or shiny. These finish characteristics are stored in the drawing via surface property blocks. The drawing contains one surface property block for each finish you create, an attribute from the name, and AutoCAD color index (ACI) if assigned. You can modify materials by manipulating ambient, diffuse, specular, and roughness factors.

Invoke the RMAT command:

Render toolbar	Choose the MATERIALS command (see Figure 16–16)
View menu	Choose Render > Materials...
Command: prompt	**rmat** (ENTER)

Figure 16–16 *Invoking the MATERIALS command from the Render toolbar*

AutoCAD displays the Materials dialog box, as shown in Figure 16–17.

The **Materials:** list box lists the available materials. The default for objects with no other material attached is *GLOBAL*.

Materials Library

Choosing the **Materials Library...** button displays the Materials Library dialog box, shown in Figure 16–18, from which you can select a material.

The Materials Library dialog box allows you to import a predefined material from an .MLI materials library into the current drawing. The **Current Drawing:** list box in the dialog box lists the materials currently in the drawing. The **Current Library** list box lists the materials available in the library file. If necessary, you can preview a sample of the material selected in the list. The sample can be applied to a sphere or square. You can preview only one material at a time. To import materials from the **Current Library** list box into the current drawing, first select the materials you want to import from the **Current Library** list box and then click the **<-Import** button. AutoCAD adds the selected materials to the **Current Drawing:** list box. Choose **OK** button to close the Materials Library dialog box, and AutoCAD returns control to the Materials dialog box.

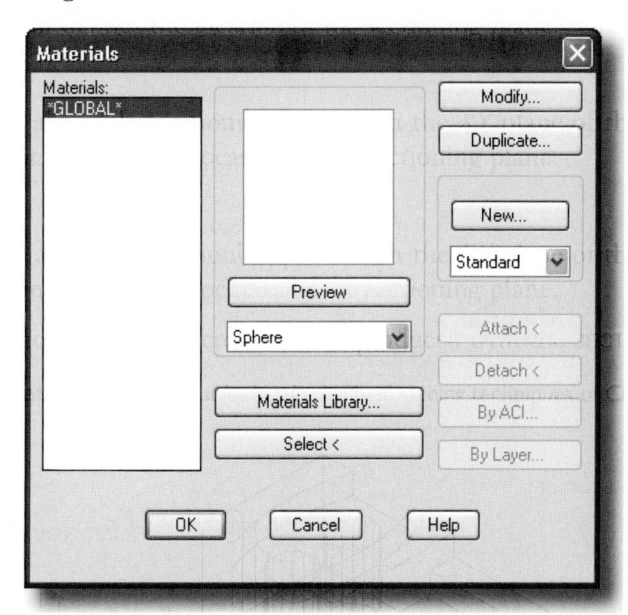

Figure 16–17 *Materials dialog box*

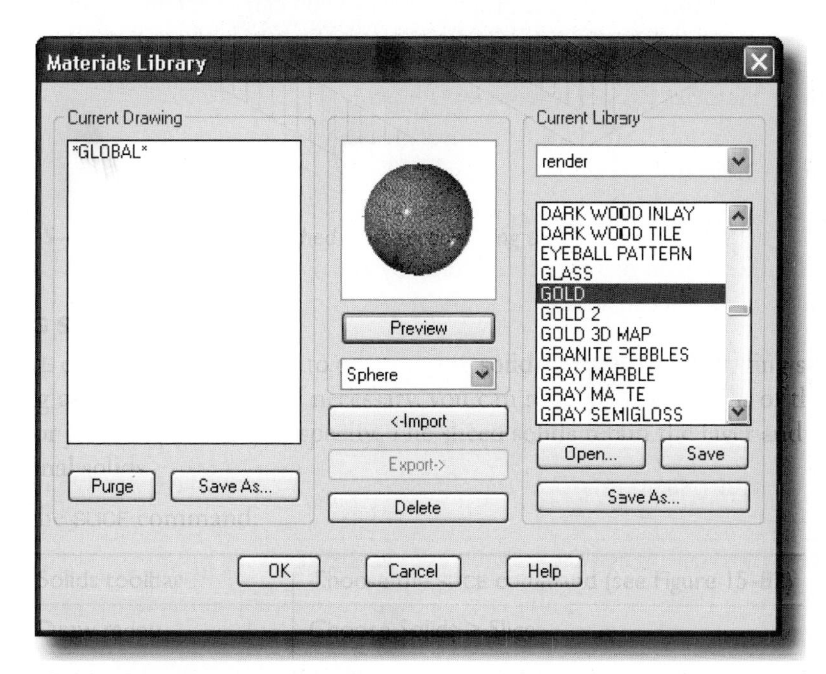

Figure 16–18 *Materials Library dialog box*

Selecting Objects

The **Select <** button allows you to select an object to display the attached material. AutoCAD temporarily removes the Materials dialog box and displays the graphics area so you can select an object and display the attached material. After you select the object, the Materials dialog box reappears highlighting the material applied to the object, with the method of attachment displayed at the bottom of the dialog box.

Modify Material

The **Modify...** button allows you to modify a material by displaying the Modify Standard Material dialog box.

Creating Duplicate Material

The **Duplicate...** button duplicates a selected material and displays the New Standard Material dialog box. Make the necessary changes, and save the material with a new material name.

Creating a New Material

The **New...** button allows you to create a new material by displaying the New Standard Material dialog box.

Attaching a Material Property to an Object

The **Attach <** button displays the graphics area so you can select an object and attach the current material to it.

Detaching Material Property from an Object

The **Detach <** button displays the graphics area so you can select an object and detach the material from it.

Attaching AutoCAD Color Index

The **By ACI...** button displays the Attach by AutoCAD Color Index (ACI) dialog box, from which you can select the available ACI to attach to a material.

Attaching Material Property to a Layer

The **By Layer...** button displays the Attach by Layer dialog box, from which you can select a layer by which to attach a material.

Choose the **OK** button to accept the changes and close the Materials dialog box.

SETTING PREFERENCES FOR RENDERING

The Rendering Preferences dialog box allows you to establish the default settings for rendering. Invoke the RPREF command:

Render toolbar	Choose the RENDER PREFERENCES command (see Figure 16–19)
View menu	Choose Render > Preferences...
Command: prompt	**rpref** (ENTER)

Figure 16–19 *Invoking the* RENDER PREFERENCES *command from the Render toolbar*

AutoCAD displays the Rendering Preferences dialog box, as shown in Figure 16–20. The available settings in the Rendering Preferences dialog box are same as in the Render dialog box. For a detailed explanation of the available settings, refer to the earlier section on "Rendering a Model."

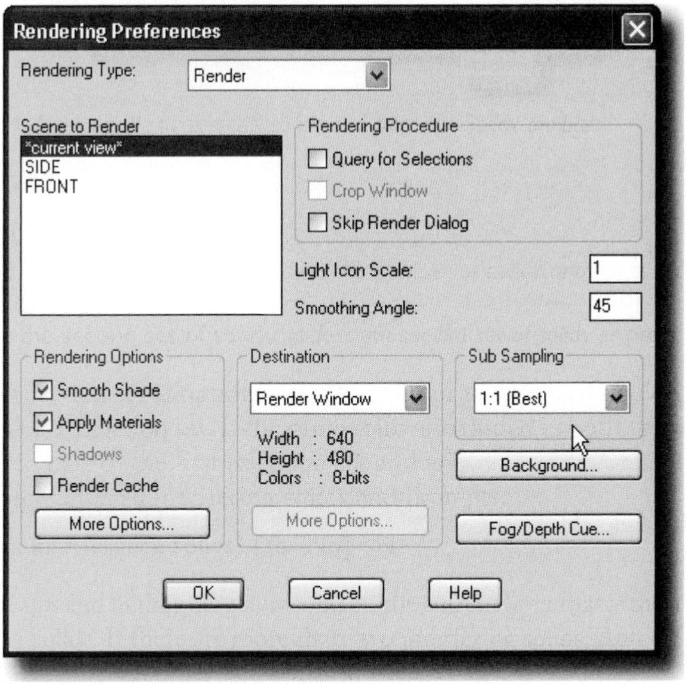

Figure 16–20 *Rendering Preferences dialog box*

SAVING AN IMAGE

You can save the contents of the frame buffer to a BMP, TIFF, or TGA file format by invoking the SAVEIMG command.

Invoke the SAVEIMG command:

Tools menu	Choose Display Image > Save...
Command: prompt	**saveimg** (ENTER)

AutoCAD displays the Save Image dialog box, as shown in Figure 16–21.

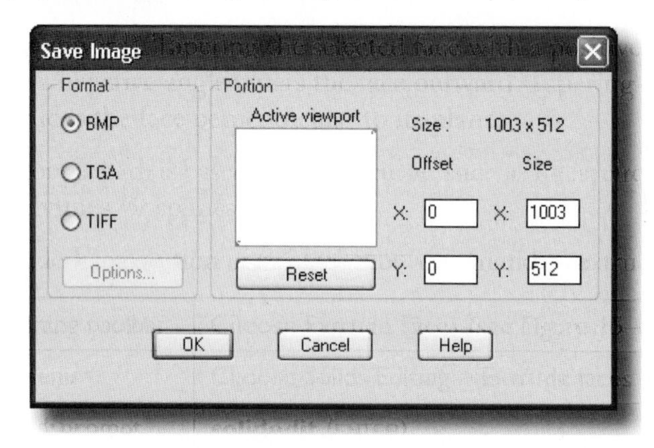

Figure 16–21 *Save Image dialog box*

Format

Choose the file format for the output image. You can save an image in any one of the three industry standard file formats: BMP, TGA, or TIFF.

Portion

The Portion section specifies the portion of the image to be rendered.

Choose the **OK** button to save the image. AutoCAD displays the Image File dialog box. Select the appropriate directory in which to save the file, and type the file name in the **File name:** text field. Choose the **Save** button to save the image to the given file name.

VIEWING AN IMAGE

The REPLAY command allows you to load a BMP, TIFF, or TGA file into the frame buffer for viewing, converting, or using as a background for a composite rendering.

Invoke the REPLAY command:

Tools menu	Choose Display Image > View...
Command: prompt	**replay** (ENTER)

AutoCAD displays the Replay File dialog box. Select a file or enter a file name. After you select an image file, choose the OPEN button, and an Image Specifications dialog box is displayed, as shown in Figure 16–22.

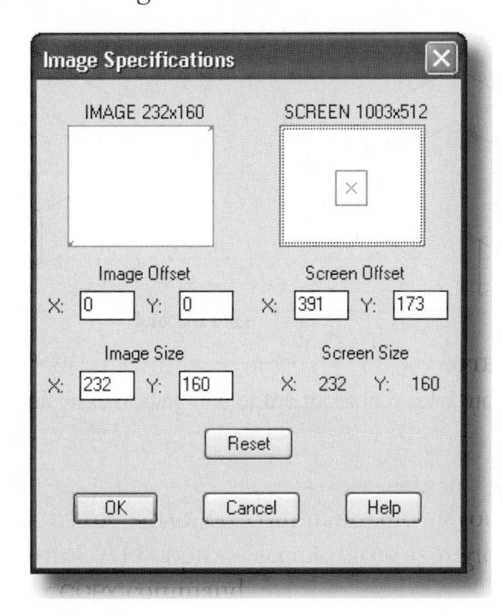

Figure 16–22 *Image Specifications dialog box*

Image

The image tile in the Image Specifications dialog box lets you select a smaller part of the image you want to display. The default size of the image in the image tile reflects the entire display size in pixel measurement, with offset set to 0,0, the lower left corner of the image. To resize the image, specify two points, one for the lower left corner and the other for the upper right corner of the image in the image tile. AutoCAD automatically draws a box to mark the bounds of the reduced image area and updates the values in the *X* and *Y* Image Size coordinates.

Screen

The screen tile lets you adjust the offset location of the selected, sized image in relation to your screen. The tile displays the size of your screen or your current viewport. To change the offset, select a point in this tile to offset the center of the image to that point. AutoCAD automatically redraws the image size boundaries to mark the new offsets and updates the values in the *X* and *Y* Screen Offset coordinates.

Reset

Resets the size and offset values to the original values.

Choose the **OK** button to close the Image Specifications dialog box.

STATISTICS

The STATISTICS command gives detailed information on your last rendering. This can be useful for diagnosing problems with your drawing.

Invoke the STATISTICS command:

Render toolbar	Choose the STATISTICS command (see Figure 16–23)
View menu	Choose Render > Statistics...
Command: prompt	**stats** (ENTER)

Figure 16–23 *Invoking the* STATISTICS *command from the Render toolbar*

AutoCAD displays the Statistics dialog box. If necessary, you can save the information in the Statistics dialog box to a file.

REVIEW QUESTIONS

1. Text is ignored by the SHADEMODE command.

   a. True

   b. False

2. Approximately where is the light source for the SHADE command?

   a. pure ambient light (i.e. all around)

   b. just behind your right shoulder

   c. directly behind the object

   d. in the upper right corner of the screen

3. It is possible to make a slide of a shaded image.

   a. True

   b. False

4. If you save your drawing while a shaded view is displayed, the preview of the drawing will appear shaded.

   a. True

   b. False

5. Which of the following is not a valid type of light you can add to a drawing for the RENDER command?

   a. point

   b. spot

   c. distant

   d. ambient

   e. none of the above (i.e. all are valid)

6. It is possible to make a slide of a rendered image.

   a. True

   b. False

7. If there is a single point light source in your drawing, objects will be darker if you set the attenuation for that light to:

   a. none

   b. inverse linear

   c. inverse square

   d. both b and c

8. Distant light sources do not attenuate (i.e. get dimmer with distance).

   a. True

   b. False

9. The SCENE command operates much like the VIEW command, but it also saves all the lights you have added to the drawing.

   a. True

   b. False

10. The maximum number of scenes that can be saved in a drawing is:

    a. 8

    b. 64

    c. 256

    d. 32,000

    e. unlimited (except by memory)

11. Material finishes can be assigned based on the color of the object in AutoCAD.

    a. True

    b. False

12. Which of the following file types is not available to the SAVEIMG command:

    a. BMP

    b. GIF

    c. TGA

    d. TIF

13. By default all lights emit white light.

    a. True

    b. False

14. The smoothing angle is the minimum angle:

    a. between faces which AutoCAD will ignore

    b. between the viewing angle and the face which AutoCAD will display the face

    c. between the incident light and the face which AutoCAD will still reflect the light

    d. none of the above

15. Both 2D and 3D models can be rendered using the SHADEMODE command?

 a. True

 b. False

16. The image option allows which type of files to be used as a background image?

 a. .GIF

 b. .BMP

 c. .JPG

 d. .TIF

17. Which AutoCAD command is used to control the position from which a 3D model is viewed in model space?

 a. VPORTS

 b. VIEWRES

 c. RENDER

 d. VPOINT

18. Which AutoCAD command is used to adjust the light reflection characteristics of an object?

 a. REFL

 b. LGTRFL

 c. LIGHTREF

 d. RMAT

19. Which of the following depicts the point at which a light's value fades to zero?

 a. fade

 b. highlight

 c. falloff

 d. diminish

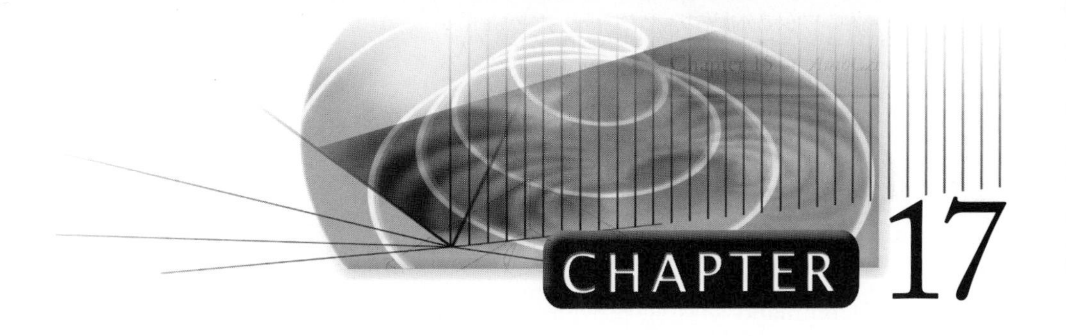

Customizing AutoCAD

INTRODUCTION

Off the shelf, AutoCAD is extremely powerful. But, like many popular engineering and business software programs, it does not automatically do all things for all users. It does—probably better than any software available—permit users to make changes and additions to the core program to suit individual needs and applications. Word processors offer a feature by which you can save a combination of many keystrokes and invoke them at any time with just one or a combination of two keystrokes. This is known as a *macro*. Spreadsheet and database management programs (as well as other types of programs) have their own library of user functions that can also be combined and saved as user-named, custom-designed commands. These programs also allow you to create and save standard blank forms for later use, to be filled out as needed. Using these features to make your copy of a generic program unique and more powerful for your particular application is known as *customizing*.

Customizing AutoCAD can include several facets requiring various skill levels. The topics in this chapter include creating command aliases, linetypes, shapes, and hatch patterns; menu customization; and command macros. A text editor that lets you save a file in ASCII or plain text format such as Microsoft Windows NOTEPAD.EXE is needed in order to customize AutoCAD menus. A macro is a sequence of commands executed by a single user selection from a custom menu. Macros can be written that help maintain certain standards in design and drafting conventions, allowing the user to concentrate on the project at hand.

After completing this chapter, you will be able to do the following:

- Create command aliases
- Create, modify, and understand the AutoCAD menu structure
- Create custom linetypes, hatch patterns, and fonts

EXTERNAL COMMANDS AND ALIASES

AutoCAD allows you to run certain programs, including internal and external DOS commands. Windows brings true multitasking to AutoCAD by making it possible to use AutoCAD simultaneously with word processors, database and spreadsheet programs, and many other applications. You can also include aliases in the ACAD.PGP file for regular AutoCAD commands. An alias is nothing but a nickname.

Command aliasing provides an alternate keystroke for invoking a command, not options of the command. The ZOOM command has a Windows option; however, you cannot define an alias "ZW" for ZOOM WINDOW. There is an alias for ZOOM, the letter "Z". Prior to modifying the ACAD.PGP file, it is recommended that you make a backup copy of the file, such as XACAD.PGP, so that if you make a mistake, you can restore the original version. Following is an extract from the ACAD.PGP file.

```
;

;

; AutoCAD Program Parameters File For AutoCAD 2004
; External Command and Command Alias Definitions

; Copyright (C) 1997-2002 by Autodesk, Inc.

; Each time you open a new or existing drawing, AutoCAD searches
; the support path and reads the first acad.pgp file that it finds.

; -- External Commands --
; While AutoCAD is running, you can invoke other programs or utilities
; such Windows system commands, utilities, and applications.
; You define external commands by specifying a command name to be used
; from the AutoCAD command prompt and an executable command string
; that is passed to the operating system.

; -- Command Aliases --
; You can abbreviate frequently used AutoCAD commands by defining
; aliases for them in the command alias section of acad.pgp.
; You can create a command alias for any AutoCAD command,
; device driver command, or external command.

; Recommendation: back up this file before editing it.

; External command format:
; <Command name>,[<DOS request>],<Bit flag>,[*]<Prompt>,

; The bits of the bit flag have the following meanings:
```

```
;  Bit 1: if set, don't wait for the application to finish
;  Bit 2: if set, run the application minimized
;  Bit 4: if set, run the application "hidden"
;  Bit 8: if set, put the argument string in quotes
;
;  Fill the "bit flag" field with the sum of the desired bits.
;  Bits 2 and 4 are mutually exclusive; if both are specified, only
;  the 2 bit is used. The most useful values are likely to be 0
;  (start the application and wait for it to finish), 1 (start the
;  application and don't wait), 3 (minimize and don't wait), and 5
;  (hide and don't wait). Values of 2 and 4 should normally be avoided,
;  as they make AutoCAD unavailable until the application has completed.
;
;  Bit 8 allows commands like DEL to work properly with filenames that
;  have spaces such as "long filename.dwg".  Note that this will interfere
;  with passing space delimited lists of file names to these same commands.
;  If you prefer multiplefile support to using long file names, turn off
;  the "8" bit in those commands.

;  Examples of external commands for command windows

CATALOG,    DIR /W,        8,File specification: ,
DEL,        DEL,           8,File to delete: ,
DIR,        DIR,           8,File specification: ,
EDIT,       START EDIT,    9,File to edit: ,
SH,         ,              1,*OS Command: ,
SHELL,      ,              1,*OS Command: ,
START,      START,         1,*Application to start: ,
TYPE,       TYPE,          8,File to list: ,

; Examples of external commands for Windows
; See also the (STARTAPP) AutoLISP function for an alternative method.

EXPLORER,   START EXPLORER, 1,,
NOTEPAD,    START NOTEPAD,  1,*File to edit: ,
PBRUSH,     START PBRUSH,   1,,


; Command alias format:
;   <Alias>,*<Full command name>
```

```
;  The following are guidelines for creating new command aliases.
;  1. An alias should reduce a command by at least two characters.
;      Commands with a control key equivalent, status bar button,
;      or function key do not require a command alias.
;      Examples: Control N, O, P, and S for New, Open, Print, Save.
;  2. Try the first character of the command, then try the first two,
;      then the first three.
;  3. Once an alias is defined, add suffixes for related aliases:
;      Examples: R for Redraw, RA for Redrawall, L for Line, LT for
;      Linetype.
;  4. Use a hyphen to differentiate between command line and dialog
;      box commands.
;      Example: B for Block, -B for -Block.
;
; Exceptions to the rules include AA for Area, T for Mtext, X for
    Explode.

;  -- Sample aliases for AutoCAD commands --
;  These examples include most frequently used commands.

3A,          *3DARRAY
3DO,         *3DORBIT
3F,          *3DFACE
3P,          *3DPOLY
A,           *ARC
ADC,         *ADCENTER
AA,          *AREA
AL,          *ALIGN
AP,          *APPLOAD
AR,          *ARRAY
-AR,         *-ARRAY
ATT,         *ATTDEF
-ATT,        *-ATTDEF
ATE,         *ATTEDIT
-ATE,        *-ATTEDIT
ATTE,        *-ATTEDIT
B,           *BLOCK
-B,          *-BLOCK
BH,          *BHATCH
BO,          *BOUNDARY
```

```
-BO,            *-BOUNDARY
BR,             *BREAK
C,              *CIRCLE
CH,             *PROPERTIES
-CH,            *CHANGE
CHA,            *CHAMFER
CHK,            *CHECKSTANDARDS
COL,            *COLOR
COLOUR,         *COLOR
CO,             *COPY
CP,             *COPY
D,              *DIMSTYLE
DAL,            *DIMALIGNED
DAN,            *DIMANGULAR
DBA,            *DIMBASELINE
DBC,            *DBCONNECT
DC,             *ADCENTER
DCE,            *DIMCENTER
DCENTER,        *ADCENTER
DCO,            *DIMCONTINUE
DDA,            *DIMDISASSOCIATE
DDI,            *DIMDIAMETER
DED,            *DIMEDIT
DI,             *DIST
DIV,            *DIVIDE
DLI,            *DIMLINEAR
DO,             *DONUT
DOR,            *DIMORDINATE
DOV,            *DIMOVERRIDE
DR,             *DRAWORDER
DRA,            *DIMRADIUS
DRE,            *DIMREASSOCIATE
DS,             *DSETTINGS
DST,            *DIMSTYLE
DT,             *TEXT
DV,             *DVIEW
E,              *ERASE
ED,             *DDEDIT
EL,             *ELLIPSE
EX,             *EXTEND
```

```
EXIT,          *QUIT
EXP,           *EXPORT
EXT,           *EXTRUDE
F,             *FILLET
FI,            *FILTER
G,             *GROUP
-G,            *-GROUP
GR,            *DDGRIPS
H,             *BHATCH
-H,            *HATCH
HE,            *HATCHEDIT
HI,            *HIDE
I,             *INSERT
-I,            *-INSERT
IAD,           *IMAGEADJUST
IAT,           *IMAGEATTACH
ICL,           *IMAGECLIP
IM,            *IMAGE
-IM,           *-IMAGE
IMP,           *IMPORT
IN,            *INTERSECT
INF,           *INTERFERE
IO,            *INSERTOBJ
L,             *LINE
LA,            *LAYER
-LA,           *-LAYER
LE,            *QLEADER
LEN,           *LENGTHEN
LI,            *LIST
LINEWEIGHT,    *LWEIGHT
LO,            *-LAYOUT
LS,            *LIST
LT,            *LINETYPE
-LT,           *-LINETYPE
LTYPE,         *LINETYPE
-LTYPE,        *-LINETYPE
LTS,           *LTSCALE
LW,            *LWEIGHT
M,             *MOVE
MA,            *MATCHPROP
```

```
ME,                 *MEASURE
MI,                 *MIRROR
ML,                 *MLINE
MO,                 *PROPERTIES
MS,                 *MSPACE
MT,                 *MTEXT
MV,                 *MVIEW
O,                  *OFFSET
OP,                 *OPTIONS
ORBIT,              *3DORBIT
OS,                 *OSNAP
-OS,                *-OSNAP
P,                  *PAN
-P,                 *-PAN
PA,                 *PASTESPEC
PARTIALOPEN,        *-PARTIALOPEN
PE,                 *PEDIT
PL,                 *PLINE
PO,                 *POINT
POL,                *POLYGON
PR,                 *PROPERTIES
PRCLOSE,            *PROPERTIESCLOSE
PROPS,              *PROPERTIES
PRE,                *PREVIEW
PRINT,              *PLOT
PS,                 *PSPACE
PTW,                *PUBLISHTOWEB
PU,                 *PURGE
-PU,                *-PURGE
R,                  *REDRAW
RA,                 *REDRAWALL
RE,                 *REGEN
REA,                *REGENALL
REC,                *RECTANG
REG,                *REGION
REN,                *RENAME
-REN,               *-RENAME
REV,                *REVOLVE
RO,                 *ROTATE
RPR,                *RPREF
```

```
RR,          *RENDER
S,           *STRETCH
SC,          *SCALE
SCR,         *SCRIPT
SE,          *DSETTINGS
SEC,         *SECTION
SET,         *SETVAR
SHA,         *SHADEMODE
SL,          *SLICE
SN,          *SNAP
SO,          *SOLID
SP,          *SPELL
SPL,         *SPLINE
SPE,         *SPLINEDIT
ST,          *STYLE
STA,         *STANDARDS
SU,          *SUBTRACT
T,           *MTEXT
-T,          *-MTEXT
TA,          *TABLET
TH,          *THICKNESS
TI,          *TILEMODE
TO,          *TOOLBAR
TOL,         *TOLERANCE
TOR,         *TORUS
TP,          *TOOLPALETTES
TR,          *TRIM
UC,          *UCSMAN
UN,          *UNITS
-UN,         *-UNITS
UNI,         *UNION
V,           *VIEW
-V,          *-VIEW
VP,          *DDVPOINT
-VP,         *VPOINT
W,           *WBLOCK
-W,          *-WBLOCK
WE,          *WEDGE
X,           *EXPLODE
XA,          *XATTACH
```

```
XB,             *XBIND
-XB,            *-XBIND
XC,             *XCLIP
XL,             *XLINE
XR,             *XREF
-XR,            *-XREF
Z,              *ZOOM

; The following are alternative aliases and aliases as supplied
;  in AutoCAD Release 13.

AV,             *DSVIEWER
CP,             *COPY
DIMALI,         *DIMALIGNED
DIMANG,         *DIMANGULAR
DIMBASE,        *DIMBASELINE
DIMCONT,        *DIMCONTINUE
DIMDIA,         *DIMDIAMETER
DIMED,          *DIMEDIT
DIMTED,         *DIMTEDIT
DIMLIN,         *DIMLINEAR
DIMORD,         *DIMORDINATE
DIMRAD,         *DIMRADIUS
DIMSTY,         *DIMSTYLE
DIMOVER,        *DIMOVERRIDE
LEAD,           *LEADER
TM,             *TILEMODE

; Aliases for Hyperlink/URL Release 14 compatibility
SAVEURL,        *SAVE
OPENURL,        *OPEN
INSERTURL,      *INSERT

; Aliases for commands discontinued in AutoCAD 2000:
AAD,            *DBCONNECT
AEX,            *DBCONNECT
ALI,            *DBCONNECT
ASQ,            *DBCONNECT
ARO,            *DBCONNECT
ASE,            *DBCONNECT
```

```
DDATTDEF,          *ATTDEF
DDATTEXT,          *ATTEXT
DDCHPROP,          *PROPERTIES
DDCOLOR,           *COLOR
DDLMODES,          *LAYER
DDLTYPE,           *LINETYPE
DDMODIFY,          *PROPERTIES
DDOSNAP,           *OSNAP
DDUCS,             *UCS

; Aliases for commands discontinued in AutoCAD 2004:
ACADBLOCKDIALOG,   *BLOCK
ACADWBLOCKDIALOG,  *WBLOCK
ADCENTER,          *ADCENTER
BMAKE,             *BLOCK
BMOD,              *BLOCK
BPOLY,             *BOUNDARY
CONTENT,           *ADCENTER
DDATTE,            *ATTEDIT
DDIM,              *DIMSTYLE
DDINSERT,          *INSERT
DDPLOTSTAMP,       *PLOTSTAMP
DDRMODES,          *DSETTINGS
DDSTYLE,           *STYLE
DDUCS,             *UCSMAN
DDUCSP,            *UCSMAN
DDUNITS,           *UNITS
DDVIEW,            *VIEW
DIMHORIZONTAL,     *DIMLINEAR
DIMROTATED,        *DIMLINEAR
DIMVERTICAL,       *DIMLINEAR
DOUGHNUT,          *DONUT
DTEXT,             *TEXT
DWFOUT,            *PLOT
DXFIN,             *OPEN
DXFOUT,            *SAVEAS
PAINTER,           *MATCHPROP
PREFERENCES,       *OPTIONS
RECTANGLE,         *RECTANG
SHADE,             *SHADEMODE
VIEWPORTS,         *VPORTS
```

The External Commands are defined at the top of the ACAD.PGP file.

The format for a command line is as follows:

```
<Command name>,<executable>,flags,[*]<Prompt>,  <Return code>
```

An example of lines in an ACAD.PGP file for specifying external commands is as follows:

```
SH,,0, *OS Command: ,0
SHELL,,0, *OS Command: ,0
TYPE,TYPE, 0, File to type: ,0
CATALOG,DIR/W,0, File specification: ,0
DEL,DEL, 0, File to erase: ,0
DIR,DIR, 0, File specification: ,0
EDIT,, 0, File to edit: ,0
```

The command name (to be entered at the "Command:" prompt) should not be the same as an AutoCAD command and should be in uppercase characters.

The executable string is sent to the operating system as the name of a command. It can contain parameters and switches.

The **flags** is a required bitcoded parameter. The combination of active instructions is the sum of the integer values.

0 Start the application and wait for it to finish.

1 Don't wait for the application to finish.

2 Run the application minimized.

4 Run the application "hidden."

8 Put the argument string in quotes.

Bit values 2 and 4 are mutually exclusive; if both are specified only the 2 bit is used. Using values 2 or 4 without value 1 should be avoided, because AutoCAD becomes unavailable until the application has completed.

Bit value 8 lets commands such as **del** to work properly with file names that have spaces embedded in them, eliminating the possibility of passing a space-delimited list of file names to such commands. Do not use the bit value 8 if you wish to have multiple file support.

The prompt (optional) is used to inform the operator if additional input is necessary. If the prompt is preceded by an asterisk (*), then the user response may contain spaces. The response must be terminated by pressing ENTER. Otherwise, pressing SPACEBAR or ENTER will terminate the response.

The return code is a bit-coded specification. The number you specify will represent one or more of the bitcodes. For example, if you specify 3, then bitcodes 1 and 2 will be in effect. The values are as follows:

0: Return to text screen.

1: Load DXB file. This causes a file named $cmd.dxb to be loaded into the drawing at the end of the command.

2: Construct block from DXB file. This causes the response to the prompt to become the name of a block to be added to the drawing, consisting of objects in the $cmd.dxb file written by the file command. This code must be used in conjunction with bitcode 1. This may not be used to redefine a previously defined block.

The format to define an alias is as follows:

```
<Alias>,*<Full command name>
```

The abbreviation preceding the comma is the character or characters to be entered at the "Command:" prompt. The asterisk (*) must precede the command you wish to be invoked. It may be a standard AutoCAD command name, a custom command name that has been defined in and loaded with AutoLISP or ADS, or a display or machine driver command name. Aliases cannot be used in scripts. You can prefix a command in an alias with the hyphen that causes a command line version to be used instead of a dialog box, as shown here:

```
BH, *-BHATCH
```

See Appendix H for list of available aliases from the default ACAD.PGP file.

CUSTOMIZING MENUS

When you launch the standard version of AutoCAD 2004, you are presented with the standard set of menus and toolbars. Selecting an item from a menu or toolbar might execute a command, an AutoLISP routine, or a macro, or cause another menu to be displayed. Menus are user definable and are created/edited using text editors. Menu files also define the functionality and appearance of the menu area. If you perform an application-specific task on a regular basis that requires multiple steps to accomplish this task, you can place this in a menu macro and have AutoCAD complete all the required procedures in a single step while pausing for input if necessary. Menu macros are similar to script files (files ending with *.SCR). Script files are also capable of executing many commands in sequence but have no decision-making capability and cannot pause for interactive user input.

AutoCAD 2004 supports the following kinds of menus:

- Pull-down and shortcut menus
- Screen menus
- Image tile menus
- Pointing device menus
- Tablet menus
- Toolbars
- Keyboard accelerators
- Help strings and tool tips
- Menu groups

MENU FILE TYPES

Prior to AutoCAD Release 13 for Windows, there were two types of menu files: ASCII text files ending with the file extension .MNU, and a compiled version of the same file with the extension .MNX. The ACAD.MNU file is provided with the AutoCAD program, and if, when you launch AutoCAD, the ACAD.MNX file is not present or any changes were made to the ACAD.MNU file, AutoCAD automatically compiles and creates a new ACAD.MNX. If you created your own custom menu—for example, CUSTOM.MNU—then upon loading the menu, AutoCAD automatically creates CUSTOM.MNX. The ACAD.MNU and ACAD.MNX files remained applicable in the DOS and UNIX versions of AutoCAD Release 14. Because of the additional functionality introduced in the Windows version of AutoCAD Release 14, a new menu scheme was introduced.

The following table lists the menu files used by AutoCAD for current releases:

Menu File Type	Description
.MNU	Template menu file, ASCII text
.MNC	Compiled menu file. This binary file contains the command strings and menu syntax that defines the functionality and appearance of the menu.
.MNR	Menu resource file. This binary file contains the bitmaps used by the menu.
.MNS	Source menu file (AutoCAD generated)
.MNT	Menu resource file. This file is generated when the MNR file is unavailable, for example, read-only.
.MNL	Menu LISP file. This file contains AutoLISP expressions that are used by the menu file and are loaded into memory when a menu file with the same file name is loaded.

AutoCAD 2004 comes with the ACAD.MNU file. This is located in the "ACAD2002\SUPPORT" folder, assuming AutoCAD is installed on the C: drive and Documents and Settings\<user name>\Application Data\Autodesk\R16.0\enu\ Support is the subdirectory. If your location is different, substitute the drive and path location for your situation. The .MNS file is similar in structure to the .MNU file and is created automatically and dynamically updated by AutoCAD when you add new toolbars. The .MNC file compiles when you reload AutoCAD or dynamically update the menu. The .MNR file is created automatically when .MNS is completed and when buttons on the toolbars are created, changed, or added. The .MNR menu resource file is used for storing bitmaps (*.BMP) for the icons and is a binary file.

When you add toolbar information to the menu, the *.MNS file is the file that is updated. The toolbars are denoted in the ***TOOLBARS major section. If you delete the *.MNS file, you will lose the newly created toolbars, because AutoCAD creates a new *.MNS file based on the *.MNU file. If you want to add the toolbars to your *.MNU file, copy the ***TOOLBARS section from your *.MNS file and paste it into your *.MNU file. If you delete your *.MNS file, AutoCAD rebuilds it based on the information found in the *.MNU file. AutoCAD has another file that is constantly updated, called the ACAD.INI file. Windows applications contain many *.INI files. This is where the initialization and configuration is kept for Windows applications. When you move toolbars and dock menus, the information is read to the ACAD.INI file so that when you launch AutoCAD, the menus are located in their new positions.

MENU FILE STRUCTURE

Menus are divided into sections relating to specific menu areas. Menu sections can contain submenus that you can reference and display as needed. The command strings and macro syntax that define the result of a menu selection are called menu macros. The Windows Notepad editor program can only open a file size less than 64K in memory. To view the contents of the ACAD.MNU file, use the Windows WordPad program, which is located in the Accessories group. Prior to opening the ACAD.MNU file, copy it and save it as XACAD.MNU. Then if something should happen, you can restore the original. It is important that files remain as text files without any print or formatting codes.

Following are the major sections in menu files for DOS and Windows versions (major sections are denoted by ***):

> ***BUTTONSn_Pointing device button menu, where n is either 1 or 2.
>
> ***BUTTONS1_The normal menu used by a mouse or tablet cursor (puck).
>
> ***BUTTONS2_Holding down SHIFT and pressing the right mouse button (or on a three-button mouse, pressing the middle button) activates the menu.

***AUXn Auxiliary device button, where n is either 1 or 2.

***POPn Pulldown and cursor menus, where n is a number from 0 to 16 (*Note:* 0 is used only for cursor menus).

***POP0 The cursor menu, which follows the crosshairs. Pressing SHIFT and the left mouse button (or pressing button 3 on a digitizing puck) activates this menu.

***POP1 through ***POP16 The pulldown menus. The AutoCAD regular menu file does not use all the available pulldown menus.

***SCREEN Screen menu area, which slowly and surely is disappearing from AutoCAD for Windows. It may not be supported in future versions.

***TABLETn Tablet menu area, where n is a number from 1 to 4.

***TABLET1 through ***TABLET4 The four menu areas of a digitizing tablet.

***HELPSTRINGS Text that is displayed in the status bar when a pulldown or shortcut menu item is highlighted, or when the cursor is over a toolbar button

***ACCELERATORS Accelerator key definitions

***TOOLBARS Toolbar menus are the most flexible and easily customized method of interface now available for the AutoCAD user. Groups of buttons can be placed around or on the screen in changeable proportions in toolbars. These buttons can be used to invoke commands, macros, or user-defined programs that can make drawing easier and more accurate.

You can use the Find feature of Windows WordPad to locate the major sections.

Menu system does not have to use all of the preceding major menu sections. AutoCAD for Windows has additional major sections explained later in the chapter.

Following is the menu code taken from the ACAD.MNU file.

```
***POP1
**FILE
ID_MnFile     [&File]
ID_New        [&New...\tCtrl+N]^C^C_new
ID_Open       [&Open...\tCtrl+O]^C^C_open
ID_FILE_CLOSE [&Close]
ID_PartialOp  [$(if,$(eq,$(getvar,fullopen),0),,~)Pa&rtial Load]^C^C_
    partiaload
              [—]
ID_Save       [&Save\tCtrl+S]^C^C_qsave
```

```
ID_Saveas     [Save &As...]^C^C_saveas
ID_Export     [&Export...]^C^C_export
              [--]
ID_PlotSetup [Pa&ge Setup...]^C^C_pagesetup
ID_PlotMgr    [Plotter &Manager...]^C^C_plottermanager
ID_PlotStyMgr [Plot St&yle Manager...]^C^C_stylesmanager
ID_Preview    [Plot Pre&view]^C^C_preview
ID_Print      [&Plot...\tCtrl+P]^C^C_plot
              [--]
ID_MnDrawing [->Drawing &Utilities]
ID_Audit        [&Audit]^C^C_audit
ID_Recover      [&Recover...]^C^C_recover
                [--]
ID_MnPurge      [->&Purge]
ID_PurgeAll       [&All]^C^C_purge _a
                  [--]
ID_PurgeLay       [&Layers]^C^C_purge _la
ID_PurgeLin       [Li&netypes]^C^C_purge _lt
ID_PurgeTxt       [&Text Styles]^C^C_purge _st
ID_PurgeDim       [&Dimension Styles]^C^C_purge _d
ID_PurgeMln       [&Multiline Styles]^C^C_purge _m
ID_PurgeBlk       [&Blocks]^C^C_purge _b
ID_PurgePlt       [&Plot Styles]^C^C_purge _p
ID_PurgeShp       [<-<-&Shapes]^C^C_purge _sh
ID_SendMail   [Sen&d...]
ID_Props      [Drawing Propert&ies...]^C^C_dwgprops
              [--]
ID_MRU        [Drawing History]
              [--]
ID_APP_EXIT   [E&xit]
```

POP1 is the major section, which is the File (first) menu on the menu bar of AutoCAD. All the words that begin with ID_ are menu tag names. Labels are enclosed in square brackets []. The ampersand (&) is how you underscore the following letter, which allows menu selections from the keyboard, with the ALT + "letter" combination common to all Windows programs. The label (or letters between the brackets) is what appears in the pulldown menu. The first label in the menu is what appears on the menu bar as "File"; all other labels appear on the menu itself. This is only for POP menus 1 to 16 and IMAGE menus. The text following the closing bracket is the command macro that AutoCAD will execute.

Here is the syntax:

```
[&Open...\tCtrl+O]^C^C_open
```

The ^C^C is the cancel command, which is followed by the open command. The symbol ^ before the C is a caret, which is SHIFT + 6 on the keyboard. This combination of characters executes the ESC sequence to cancel any previous command. Note that there are two of them. This is because some AutoCAD commands require the user to press ESC twice before canceling a command. After the cancellation, the OPEN command will execute. This is an example of a simple macro.

MENU MACRO SYNTAX

Following is a partial list of the codes you will encounter in menu macros:

Syntax	Description
***	Denotes major sections of the menu
**	Denotes subsections located between major sections
[ ]	Menu label
;	Semicolon; equivalent to pressing ENTER on the keyboard
^M	Caret M; equivalent to pressing ENTER on the keyboard
^I	Caret I; equivalent to pressing TAB on the keyboard
space	A space character; equivalent to pressing ENTER on the keyboard
\	Pause for user input
'	Issue a command transparently while in another command
*	Repeat a command until user cancels
+	Allows the long macros to be continued onto the next line

The following example macro will create a layer called "EL_OFFEQ" (Electrical Office Equipment), assign the color "RED", and make it the current layer:

```
[EL_OFFEQ]^C^C-LAYER;M;EL-OFFEQ;C;RED;;;
```

This is the equivalent of typing the -LAYER command, selecting the Make option, typing "EL-OFFEQ" as the desired layer name, selecting the Color option to assign the color Red, and finally, pressing ENTER three times to exit the command. Selecting this macro will execute all of this in one operation. Use the Make option of the -LAYER command in case the layer does not yet exist. If the layer does exist it will become the current layer. Note that there is no space after the ^C^C.

EXAMPLE OF CREATING A MENU FILE

Let's go through the sequence of steps in creating a menu file incorporating all the facets of menu sections. In order to create the menu file, first create the blocks with attributes shown in Figure 17–1, and save the drawing in your current working directory.

Figure 17–1 shows three blocks: a computer, a table, and a chair. Create the objects on layer 0 and the attribute definitions on layer ATTRIB. Make layer 0 "white" and layer ATTRIB "yellow." Define the attributes as visible, and verify. Following is the list of attribute tags with the corresponding block names.

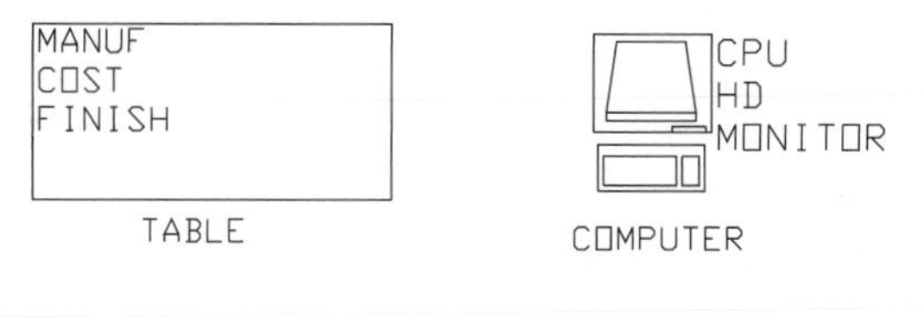

Figure 17–1 *Drawing showing the three blocks required to complete the menu exercise*

Block Name	Attribute Flag
COMPUTER	CPU HD MONITOR
TABLE	MANUF COST FINISH
CHAIR	MANUF COLOR MATL

Save the blocks as separate drawing files, COMP.DWG, CHAIR.DWG, and TABLE1.DWG, by using the WBLOCK command.

Create a menu system file called TEST.MNU that will insert various items of furniture into a drawing using either a standard set of values or a custom set of values. The purpose of the menu system is to insert the blocks with a single pick from a menu that will automatically create the standard layer for insertion of the block, allow for custom values (if the Custom option is selected), and set layer 0 as the current layer.

Invoke the AutoCAD program from the Windows file manager. Begin a new drawing, and launch the Windows Notepad editor from the Accessories group. Type the following code, and save the file as TEST.MNU in your current working directory.

```
***POP1
**Furniture
[&Furniture]
[->&Standard Furniture]
 [&Computer]^C^C^CATTREQ;0;-LAYER;M;EL_OFFEQ;COLOR;1;;;+
-INSERT;COMP;\;;\ATTREQ;1;-LAYER;SET;0;;
 [&Table]^C^C^CATTREQ;0;-LAYER;M;FR_OFF;COLOR;3;;;+
```

```
-INSERT;TABLE1;\;;\ATTREQ;1;-LAYER;SET;0;;
 [<-C&hair]^C^C^CATTREQ;0;-LAYER;M;FR_OFF;COLOR;3;;;+
-INSERT;CHAIR;\;;\ATTREQ;1;-LAYER;SET;0;;
[->Custo&m Furniture]
 [Com&puter]^C^C^CATTDIA;0;-LAYER;M;EL_OFFEQ;COLOR;1;;;+
-INSERT;COMP;\;;\\\\\\\-LAYER;S;0;;ATTREQ;1
 [Tab&le]^C^C^CATTDIA;0;-LAYER;M;FR_OFF;COLOR;3;;;+
-INSERT;TABLE1;\;;\\\\\\\-LAYER;S;0;;ATTREQ;1
 [<-Chai&r]^C^C^CATTDIA;0;-LAYER;M;FR_OFF;COLOR;3;;;+
-INSERT;CHAIR;\;;\\\\\\\-LAYER;S;0;;ATTREQ;1
[—]
[Layer EL_OFFEQ]^C^C^C-LAYER;M;EL_OFFEQ;COLOR;1;;;
[Layer FR_OFF]^C^C^C-LAYER;M;FR_OFF;COLOR;3;;;
[—]
[Layer 0]^C^C^C-LAYER;SET;0;;
```

Go back to the AutoCAD program, and load the TEST.MNU menu file by typing **MENU** at the "Command:" prompt and then pressing ENTER or SPACEBAR. The Select Menu File dialog box appears, as shown in Figure 17–2.

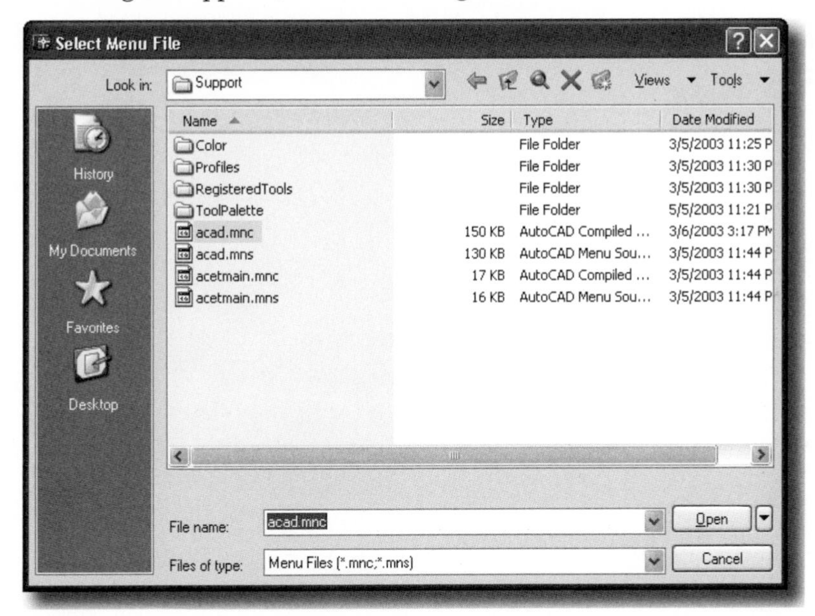

Figure 17–2 *Select Menu File dialog box*

Select the TEST.MNU file from the appropriate drive and directory, and choose the **Open** button. TEST.MNU will replace ACAD.MNU. The right mouse button on your tablet puck/mouse (or ENTER) will not respond to any action, because in the

TEST.MNU file the ***BUTTONS1 major section is not defined. Only the pulldown menu **Furniture** is displayed, as shown in Figure 17–3.

To test the menu, select the **Computer** option from the **Standard Furniture** cascading menu. AutoCAD prompts for an insertion point and rotation angle. From the **Custom Furniture** cascading menu, select the **Computer** option. AutoCAD prompts for the insertion point and rotation angle, and in addition prompts for three attribute values. If you come across an error message, go back to the TEST.MNU file, check the syntax, and make the necessary changes. Reload the TEST.MNU file. Be sure you enter the full name and extension when you specify TEST.MNU so that the source file TEST.MNS and compiled file TEST.MNC are updated with the latest changes or corrections. Invoke the commands again, and make sure the menu is working properly. Let's examine some of the unique features of the TEST.MNU file line by line.

```
***POP1
```

Indicates that this is a pull-down menu located at position 1.

```
[&Furniture]
```

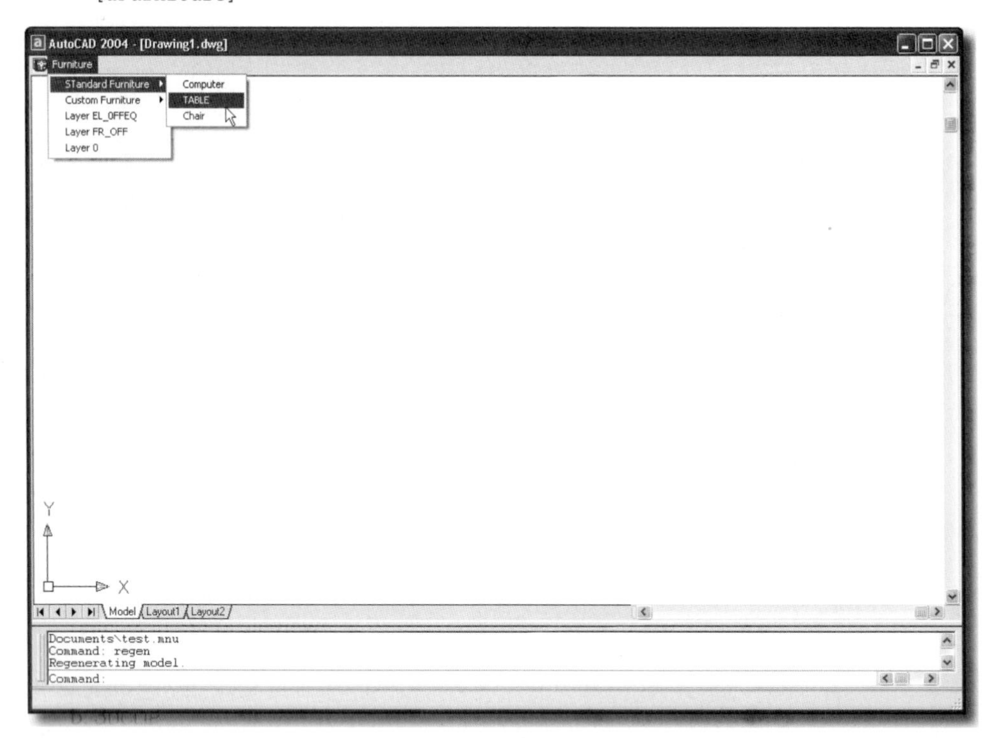

Figure 17–3 *Pull-down menu Furniture*

This is the title that will appear on the menu bar. The ampersand before the F will underline the letter F, allowing the menu to be selected using the ALT + F combination.

```
[->&Standard Furniture]
```

This is the beginning of the cascading menu, as noted by the "->." The letter S will be underscored and "Standard Furniture" will appear as the first item in the "Furniture" pulldown menu.

```
[&Computer]^C^C^CATTREQ;0;-LAYER;M;EL_OFFEQ;COLOR;1;;;+
-INSERT;COMP;\;;\ATTREQ;1;-LAYER;SET;0;;
```

This is the first item available in the cascaded menu. The ^C's will cancel any previous command. It is advisable to use three ^C combinations, because some commands require more than two exits to terminate. Following the ^C is ATTREQ, a setvar variable whose function is to control attribute requests. If it has a value of 0, the block will be inserted without prompting for attributes using standard values, which, in this case, is what we want. Note the use of the semicolon, which is equivalent to pressing ENTER. Next, the -LAYER command is invoked, creating an "EL_OFFEQ" layer and setting the color to "RED" (1 is the code number for the color RED), followed by three returns to terminate the LAYER command. The hyphen prefix causes the prompts to be invoked at the Command: prompt instead of using the dialog box. The use of the "+" symbol at the end of the line allows the code to continue onto the next line. There are no spaces between the final return ";" and the "+". The macro continues by invoking the -INSERT command and inserting a block called COMP. The first "\" backslash character after the block name causes a pause, which allows the user to pick an insertion point. The two semicolons (;;) that follow the backslash character accept the default scale factors. The next backslash character is for the user to pick a rotation angle. The ATTREQ setvar variable is reset to the default value of 1. The final command restores layer 0 as the current layer.

```
[->Custo&m Furniture]
[Com&puter]^C^C^CATTDIA;0;-LAYER;M;EL_OFFEQ;COLOR;1;;;+
INSERT;COMP;\;;\\\\-LAYER;S;0;;ATTDIA;1
```

This code is for a custom **Computer** option from the **Custom Furniture** cascading menu. In the Custom menu, the intent is to allow the operator to insert a block and at the same time allow the user to input nonstandard values for the attributes of the block. ATTDIA (Attribute dialog box), which is a setvar variable, is set to 0. This will disable the Attribute dialog box from popping up on the screen. Instead, the user can input attribute values from the "Command:" prompt. The LAYER command creates the "EL_OFFEQ" layer, whose color is set to "RED." The "+" is used to allow the continuation of the command onto the next line. The INSERT command is invoked, and the COMP block is inserted, pausing for the user to pick an insertion point.

The macro is followed by two semicolons, ";;", which accept the default scale factor. The following seven backslash characters, "\\\\\\\", causes a pause for the operator to pick the rotation angle of the block and input the three attribute values for the block followed by the three verifications of the attribute values. The final command is to restore layer 0 as the current layer.

Note: It is considered good programming practice to reset the system variables to their default values.

```
[Layer EL_OFFEQ]^C^C^C-LAYER;M;EL_OFFEQ;COLOR;1;;;
[Layer FR_OFF]^C^C^C-LAYER;M;FR_OFF;COLOR;3;;;
[—]
[Layer 0]^C^C^C-LAYER;SET;0;;
```

This macro creates new layers if they do not already exist and assigns the appropriate colors. The [—] code draws a line (separator) on the pulldown menu to group similar elements visually.

MENUGROUPS AND PARTIAL MENU LOADING

One of the problems with the TEST.MNU file menu is that in order to use it, you have to replace the AutoCAD default menu file (ACAD.MNU) even though TEST.MNU is incomplete, containing only one major section. One way around this problem is to rename the major section in TEST.MNU from ***POP1 to ***POP11 and to append this menu to the AutoCAD default ACAD.MNU file. Then you can use all the available standard menu commands in addition to the custom menu. The ACAD.MNU file that comes standard with AutoCAD uses ***POP1 to ***POP10. This has been the traditional approach, which causes rather large menu files just to use the additional functionality. If AutoCAD changes the default menu file, the user will sometimes be forced to make necessary changes to additional functionality to accommodate the ACAD.MNU file changes. Windows Notepad can only edit a file that is less than 64K in size. You would have to use some other editor to append the ***POP11 to the copy of the standard AutoCAD menu. But AutoCAD provides the ability for partial loading of menus so you can mix and match the functionality of various menus. Changes in one menu will not affect the other menus. AutoCAD achieves this new functionality with the addition of a new major section group called ***MENUGROUP. Each menu file can only have one major section called MENUGROUP.

The ***MENUGROUP= major section is how AutoCAD tracks which menus are loaded and referenced. A MENUGROUP string definition can be up to 32 alphanumeric characters (spaces and punctuation marks are not allowed). The ***MENUGROUP=

label must precede all menu definitions that use the name-tag mechanism, which will be covered later. Open the TEST.MNU file in the Notepad editor, make the following changes, and save the file as TEST.MNU to demonstrate partial menu loading.

```
//This is a test menu file and it demonstrates the basic new
//functionality of the new menu name tag syntax.
***MENUGROUP=test
***POP1
**Furniture
ID_Furn [&Furniture]
ID_Furns [->&Standard Furniture]
ID_Comps [&Computer]^C^C^CATTREQ;0;-LAYER;M;EL_OFFEQ;COLOR;1;;;+
-INSERT;COMP;\;;\ATTREQ;1;-LAYER;SET;0;;
ID_Tables [&Table]^C^C^CATTREQ;0;-LAYER;M;FR_OFF;COLOR;3;;;+
-INSERT;TABLE1;\;;\ATTREQ;1;-LAYER;SET;0;;
ID_Chairs [<-C&hair]^C^C^CATTREQ;0;-LAYER;M;FR_OFF;COLOR;3;;;+
-INSERT;CHAIR;\;;\ATTREQ;1;-LAYER;SET;0;;
ID_Furnc [->Custo&m Furniture]
ID_Compc [Com&uter]^C^C^CATTDIA;0;-LAYER;M;EL_OFFEQ;COLOR;1;;;+
-INSERT;COMP;\;;\\\\-LAYER;S;0;;ATTDIA;1
ID_Tablec   [Tab&le]^C^C^CATTDIA;0;-LAYER;M;FR_OFF;COLOR;3;;;+
-INSERT;TABLE1;\;;\\\\-LAYER;S;0;;ATTDIA;1
ID_Chairc   [<-Chai&r]^C^C^CATTDIA;0;-LAYER;M;FR_OFF;COLOR;3;;;+
-INSERT;CHAIR;\;;\\\\-LAYER;S;0;;ATTDIA;1
[—]
[Layer EL_OFFEQ]^C^C^C-LAYER;M;EL_OFFEQ;COLOR;1;;;
[Layer FR_OFF]^C^C^C-LAYER;M;FR_OFF;COLOR;3;;;
[—]
ID_Lyr0 [Layer 0]^C^C^C-LAYER;SET;0;;
```

The text that follows the // is considered comment lines by the AutoCAD program and is ignored in menu file compilation.

The name tags before the labels in the menu begin with "ID_". This can be any string; however, AutoCAD recommends using "ID_" as part of the string. The other requirement is that name tags be unique. Details on how name tags are used are provided later in the chapter. To partially load the menus, invoke **Customize Menus** from the **Tools** menu. AutoCAD displays the Menu Customization dialog box, as shown in Figure 17–4.

Choose the **Browse...** button, and AutoCAD displays the Select Menu file dialog box. Select the TEST.MNU file from the appropriate drive and directory, and choose the **Open** button to close the dialog box. AutoCAD displays the name of the menu file selected in the **File Name:** edit box. Choose the **Load** button. TEST.MNU will be added to the **Menu Groups:** list box. The functionality of the TEST.MNU menu is added to the AutoCAD menu. If the TEST.MNU file had toolbars, then they would have appeared on the screen. To display the TEST.MNU group as part of the AutoCAD menu bar, first select TEST.MNU from the **Menu Groups:** list box. Then choose the **Menu Bar** tab, as shown in Figure 17–5. AutoCAD displays the **Menu Bar** tab, as shown in Figure 17–6.

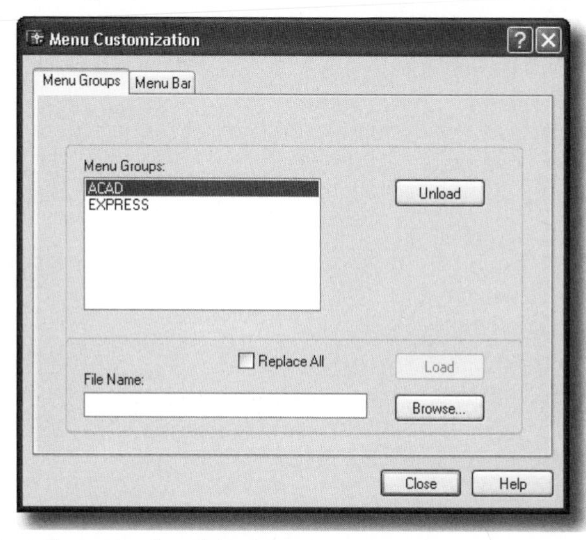

Figure 17–4 *Menu Customization dialog box*

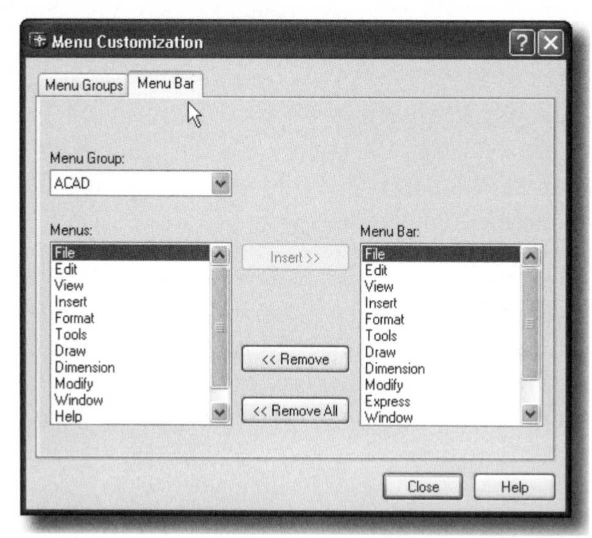

Figure 17–5 *Choosing the Menu Bar tab in the Menu Customization dialog box*

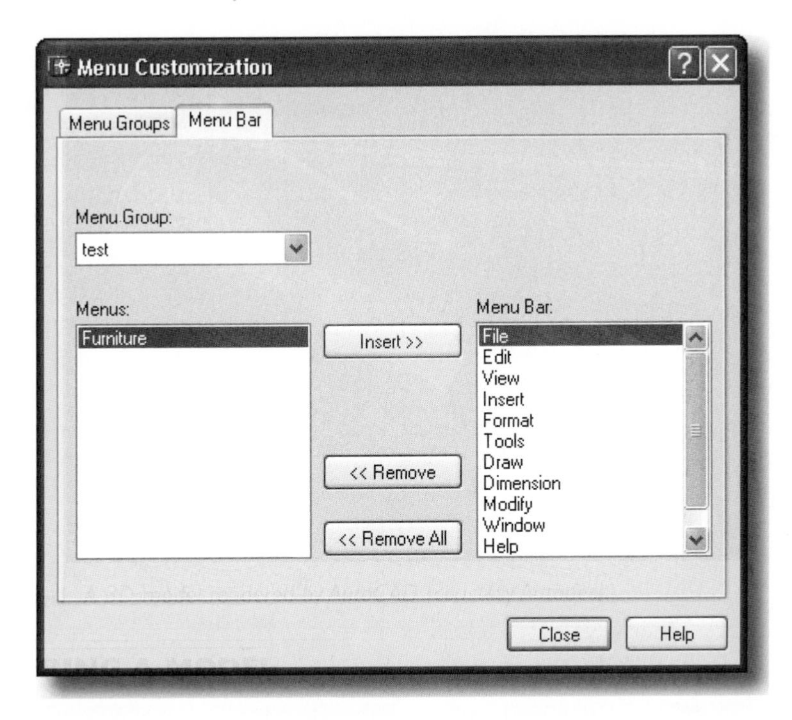

Figure 17–6 *Menu Bar tab of the Menu Customization dialog box*

The **Menu Bar** tab of the dialog box shows the Furniture item menu on the left side and the AutoCAD menu bar on the right side. Select Furniture on the left side and Help on the right side, and then choose the **Insert>>** button. AutoCAD inserts the Furniture item before the Help item on the right side of the dialog box. Choose the **Close** button to accept the configuration. AutoCAD adds the **Furniture** menu item to the standard AutoCAD menu bar between the **Tools** and **Help** menus, complete with its own functionality, as shown in Figure 17–7.

Partial menu loading gives you the ability to blend menus together and achieve the functionality as if the menu system were one menu. Test the menu items from the AutoCAD menu bar and from the **Furniture** menu. The system works as if dealing with one menu. To unload the TEST.MNU menugroup, open the Menu Customization dialog box (**Menu Groups** tab), select the "test" menu group from the **Menu Groups:** list box, and then choose the **Unload** button.

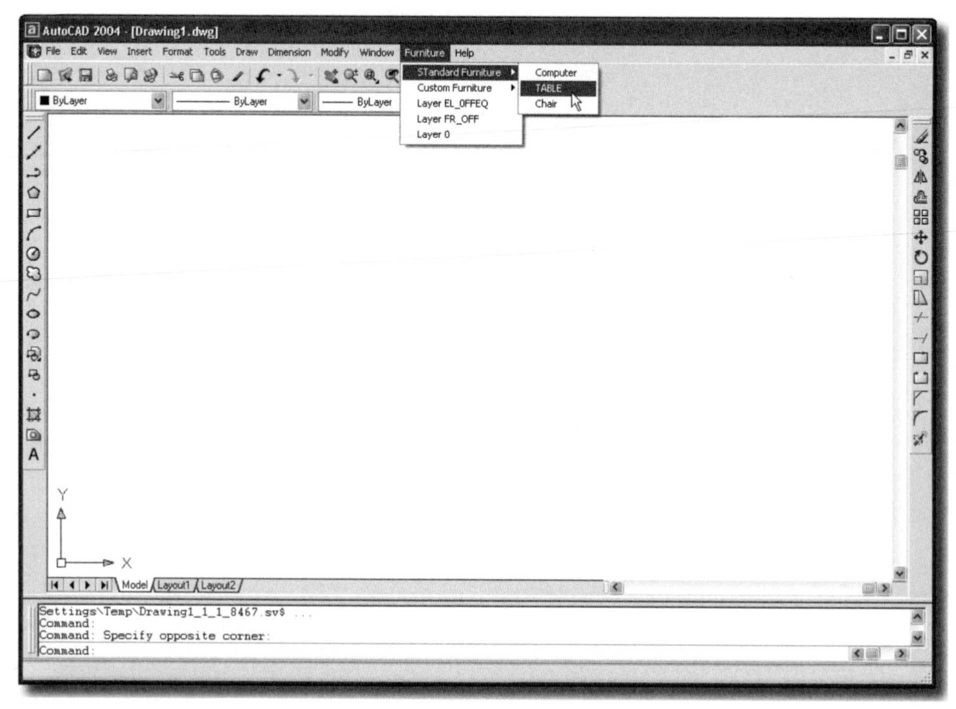

Figure 17–7 *Furniture menu item as part of the AutoCAD standard menu bar*

HELPSTRINGS

In the menu file, you can also include help strings that will be displayed in the status bar whenever a command is selected from the pull-down menu. Open the TEST.MNU file in the Notepad editor and add the ***HELPSTRINGS section, as shown below. Save the file as TEST.MNU to demonstrate the display of help strings.

```
//This is a test menu file and it demonstrates the basic new
//functionality of the new menu name tag syntax.
***MENUGROUP=test
***POP1
**Furniture
ID_Furn [&Furniture]
ID_Furns [->&Standard Furniture]
ID_Comps [&Computer]^C^C^CATTREQ;0;-LAYER;M;EL_OFFEQ;COLOR;1;;;+
INSERT;COMP;\;;\ATTREQ;1;-LAYER;SET;0;;
ID_Tables [&Table]^C^C^CATTREQ;0;-LAYER;M;FR_OFF;COLOR;3;;;+
INSERT;TABLE1;\;;\ATTREQ;1;-LAYER;SET;0;;
ID_Chairs [<-C&hair]^C^C^CATTREQ;0;-LAYER;M;FR_OFF;COLOR;3;;;+
INSERT;CHAIR;\;;\ATTREQ;1;-LAYER;SET;0;;
ID_Furnc [->Custo&m Furniture]
```

```
ID_Compc [Com&uter]^C^C^CATTDIA;0;-LAYER;M;EL_OFFEQ;COLOR;1;;;+
INSERT;COMP;\;;\\\\-LAYER;S;0;;ATTDIA;1
ID_Tablec [Tab&le]^C^C^CATTDIA;0;-LAYER;M;FR_OFF;COLOR;3;;;+
INSERT;TABLE1;\;;\\\\-LAYER;S;0;;ATTDIA;1
ID_Chairc   [<-Chai&r]^C^C^CATTDIA;0;-LAYER;M;FR_OFF;COLOR;3;;;+
INSERT;CHAIR;\;;\\\\-LAYER;S;0;;ATTDIA;1
[—]
[Layer EL_OFFEQ]^C^C^C-LAYER;M;EL_OFFEQ;COLOR;1;;;
[Layer FR_OFF]^C^C^C-LAYER;M;FR_OFF;COLOR;3;;;
[—]
ID_Lyr0 [Layer 0]^C^C^C-LAYER;SET;0;;
***HELPSTRINGS
ID_Comps [Standard Computer]
ID_Tables [Standard Table]
ID_Chairs [Standard Chair]
ID_Compc [Custom Computer]
ID_Tablec [Custom Table]
ID_Chairc [Custom Chair]
```

The text inside the square brackets next to the tag name will appear on the status bar when the command is invoked from the menu bar. For example, selecting the **Standard Computer** option from the **Custom Furniture** menu causes the text inside the square brackets to appear on the status bar.

To load the menu with changes, select **Customize Menu** from the **Tools** menu. From the **Menu Group** unload the TEST.MNU file if it is already loaded, and then reload the updated file with changes. Select the **Browse...** button, and AutoCAD displays the Select Menu file dialog box. Select the TEST.MNU file from the appropriate directory, and click the **Open** button to return to the Customize Menu dialog box. AutoCAD displays the name of the menu file selected in the **Menu Groups:** edit box. Select TEST in the edit box and then select the **Load** button. A warning is displayed that states "Loading of a template menu file (MNU file) overwrites and redefines the menu source file (MNS file), which results in the loss of any toolbar customization changes that have been made." You are then asked, "Continue loading MNU file?" Select **Yes**. TEST.MNU will be added to the **Menu Groups:** list box. The functionality of the TEST.MNU menu is added to the AutoCAD menu.

To display the TEST.MNU group as part of the AutoCAD menu bar, first select TEST.MNU from the **Menu Groups:** list box. Then choose the **Menu Bar** tab. In the **Menu Group:** pulldown text box, select TEST. The **Menus:** text box on the left side shows the Furniture menu and the default AutoCAD list of menu bar items in the **Menu Bar:** text box on the right side. Select the Furniture item on the left side and Help on the right side, and then select the **Insert>>** button. AutoCAD inserts the

Furniture item before the Help item on the right side of the dialog box. Choose the **Close** button to accept the configuration. AutoCAD adds the **Furniture** menu item to the standard AutoCAD menu bar between the **Tools** and **Help** menus, complete with its own functionality. Test the **Furniture** menu to see whether the help string appears on the status bar.

CUSTOMIZING MENU CONTENT, TOOLBARS, SHORTCUT KEYS, AND BUTTONS

It has been pointed out previously in this book that there is almost always more that one way to invoke any AutoCAD command. For example, the LINE command can be invoked from the **Draw** pulldown menu, the Draw toolbar, the traditional (rarely used) screen menu, or entering **line** or **l** (for a shortcut key) at the keyboard and pressing ENTER. Or if LINE were the last command used, you can repeat it by pressing ENTER or you can call up a shortcut menu by right-clicking in the Command window and then choosing LINE from the Recent Commands list.

The Customize dialog box allows you to add or delete commands from menus, toolbars, and shortcut keys, therefore, enhancing these methods of entering commands to suit your needs. The Customize dialog box has four tabs: **Commands, Toolbars, Properties, Keyboard,** and **Tool Palettes**.

Commands

On the **Commands** tab, under the **Categories** section, is a list of types of commands that coincide with the items on the Menu bar at the top of the screen, as shown in Figure 17–8. When you select one of the items in the list, such as Edit, the **Commands** section of the **Commands** tab displays the commands available on the Edit pulldown menu. The list for the Edit category includes Undo, Redo, Cut to Clipboard and other edit commands.

Note: Clear item on the pulldown menu is called Erase in the Commands section of the dialog box. And it is the same as the Del(ete) key on the keyboard. None of these will send the removed objects to the Windows Clipboard as the CUT command does.

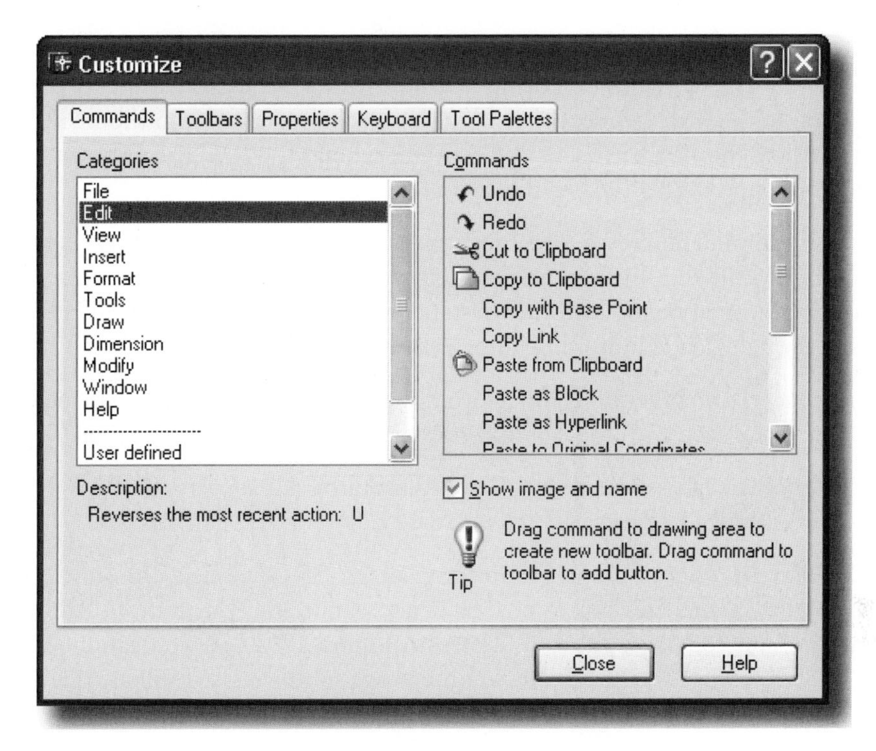

Figure 17–8 *Customize dialog box with Commands tab displayed*

The **Description**: section describes the command that is highlighted in the **Commands** section. If the **Show image and name** check box is checked the image icon associated with the highlighted command is displayed in the **Commands** section along with the name of the command. If the **Show image and name** check box is not checked, only the image icon is displayed. As noted beside the Tip, you can drag a command from the **Commands** section to the screen, and AutoCAD will create a new toolbar named Toolbar1 (or Toolbar*n* with *n* being the last number created) comprised of that command. You can drag other commands to the newly created toolbar (or even other existing toolbars) and have those commands added to the toolbar.

Toolbars

On the **Toolbars** tab is a list of standard AutoCAD toolbars and custom created toolbars (if there are any). When you check one of the toolbars' checkboxes in the list, AutoCAD displays that toolbar on the screen. You can dock the toolbar by dragging it (by its title bar) to one of the edges of the screen.

The easiest way to create toolbars while you are working in AutoCAD is to let AutoCAD add the code automatically to the *.MNS file. If you want to add this information to the *.MNU file, copy and paste this information into the *.MNU file and then delete the *.MNS file. AutoCAD will rebuild the *.MNS file based on the information found in the *.MNU file.

In this section, you will create three new toolbars, as shown in Figure 17–9, as part of the TEST.MNU file: one for custom and standard computers, one for custom and standard tables, and one for custom and standard chairs. Make sure the TEST menu file is still partially loaded into the AutoCAD menu.

To create a new toolbar, select **Toolbars** from the **View** menu. AutoCAD displays the Customize dialog box with the **Toolbars** tab displayed, as shown in Figure 17–10.

From the **Toolbars** tab, select the **New…** button. AutoCAD displays the New Toolbar dialog box, as shown in Figure 17–11.

From the **Save toolbar in menu group:** drop-down list box, select the "test" group. In the **Toolbar Name:** edit box, type Chairs and choose the **OK** button. AutoCAD displays the new toolbar Chairs, as shown in Figure 17–12. If you cannot see the newly created toolbar on the screen, you may have to move the Toolbars dialog box out of the way to find it.

To add buttons to the empty Chairs toolbar, choose the **Commands** tab in the Customize dialog box. Then highlight **User defined** in the **Categories** list box and from the **Commands** text box highlight **User defined button** as shown in Figure 17–13. Hold the pick button down while the cursor is over User defined button and drag it to the newly created toolbar. It will create a new button with a blank image.

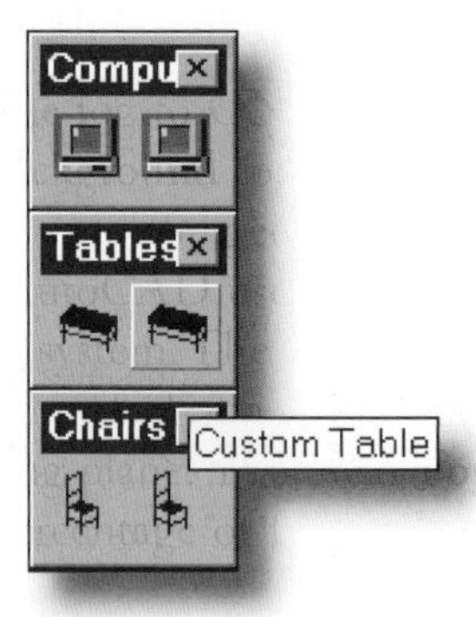

Figure 17–9 *Toolbars for a custom menu*

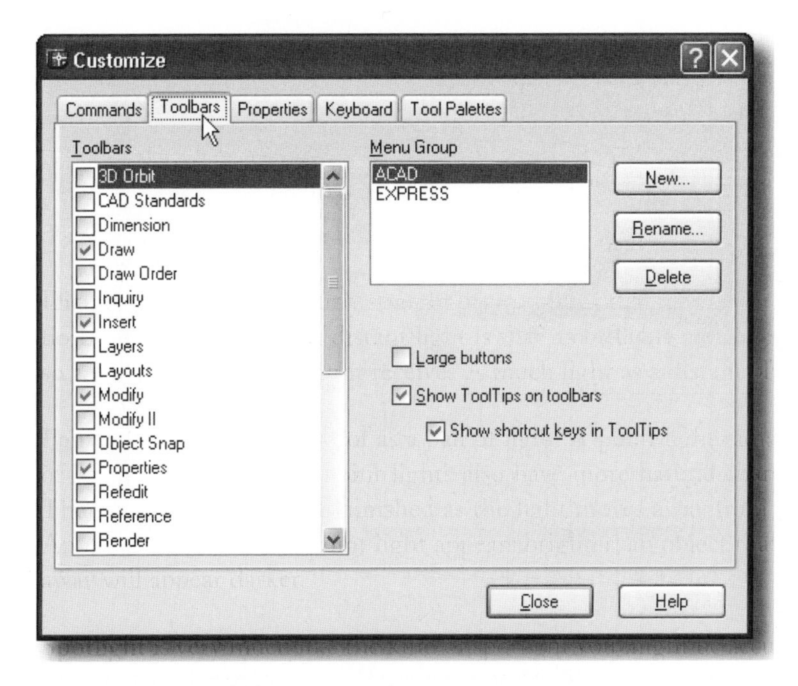

Figure 17–10 *Customize dialog box with Toolbars tab displayed*

Figure 17–11 *New Toolbar dialog box*

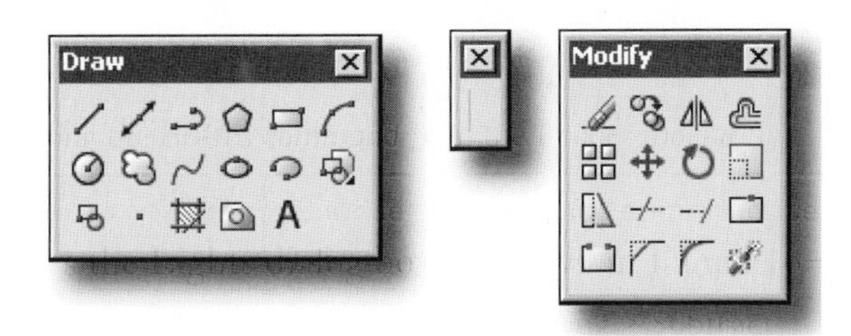

Figure 17–12 *Chairs toolbar*

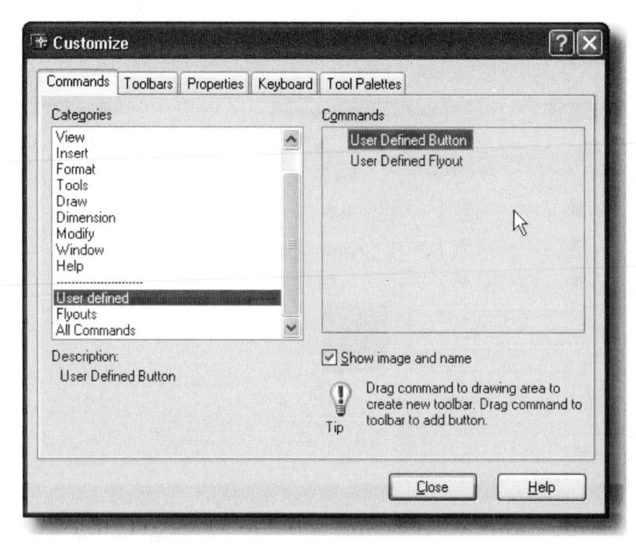

Figure 17–13 *Adding a new button to a toolbar*

On the **Properties** tab of the Customize dialog box, you are given a Tip to select a toolbar item to view or modify its properties. When you select the new button on the new toolbar the **Properties** tab will become a **Button Properties** tab with the **Name:** and **Description:** text boxes displaying "User Defined Button" as shown in Figure 17–14. If you had selected another (existing) button, for example the button for the LINE command, you would have been shown the button properties for that button. Or if you had selected a button that had a flyout menu (one with a small black arrow in the lower right corner), then the tab would become a **Flyout Properties** tab and you would have been shown the properties of that flyout menu.

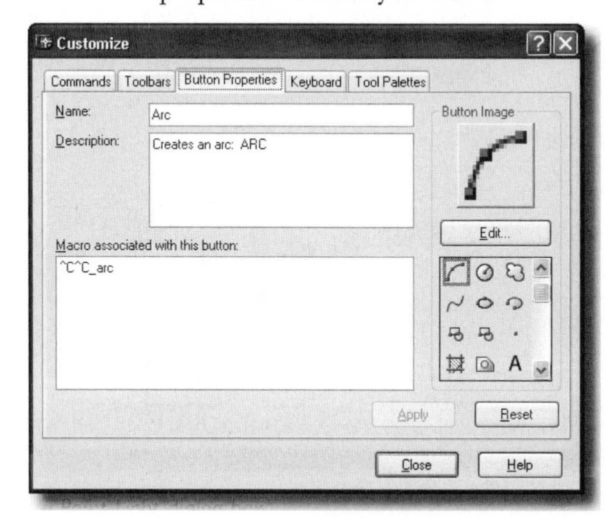

Figure 17–14 *Button Properties tab of the Customize dialog box*

Type Custom Chair in the **Name:** edit box. This is the tooltip that will be displayed when the mouse pointer passes over the button. In the **Help:** edit box, type "Custom Office Furniture". This will be the message displayed on the status bar when the mouse pointer passes over the button. In the **Macro:** edit section, delete the existing code and replace it with the following code taken from the TEST.MNU file created earlier for Custom Chair:

```
^C^C^CATTDIA;0;-LAYER;M;FR_OFF;COLOR;3;;;-INSERT;CHAIR;\;;\\\\
    -LAYER;S;0;;ATTDIA;1
```

Do not add "+" in the macro. Although this will work for the POP menus, it will not work in the **Button Properties** tab. Remember that spaces and semicolons are considered the ENTER key by AutoCAD. You may now wish to edit the Button Icon by choosing the **Edit...** button in the **Button Properties** tab. AutoCAD displays the Button Editor dialog box shown Figure 17–16.

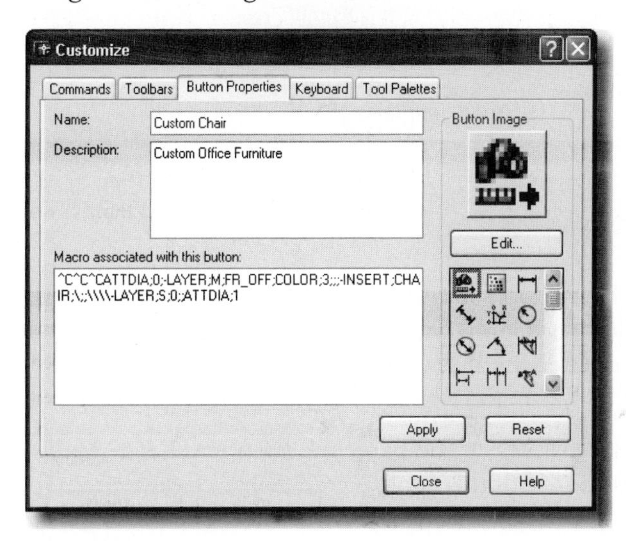

Figure 17–15 *Button Editor dialog box*

Make the necessary changes to the icon by using the tools provided in Button Editor dialog box. Select the **Save** or **Save As...** button to save the changes. This will allow you to save the icon as a bitmap (files with the *.BMP extension). Choose the **Close** button to return to the **Button Properties** tab of the Customize dialog box. For the changes to take effect, select the **Apply** button on the **Button Properties** tab. Once again AutoCAD will save the changes to the TEST.MNS file.

Repeat the procedure to add another button to the Chairs toolbar, adding a chair as standard furniture. Make sure to copy the appropriate code from the TEST.MNU file. In addition, create two more toolbars: one called Tables and the other called Computers, as shown in Figure 17–16.

Let's look at an example of creating a flyout toolbar called Office Furniture, as shown in Figure 17–17. In order to create a flyout toolbar, we need to reference existing toolbars.

Figure 17–16 *Table and Computer toolbars*

Figure 17–17 *Flyout toolbar*

The Office Furniture toolbar will include three flyout icons, namely, Computers, Chairs, and Tables. The Computers flyout will include macros for Standard Computer and Custom Computer. Similarly, the Chairs and Tables flyouts will include Standard Chair and Custom Chair and Standard Table and Custom Table, respectively.

To create a new toolbar, select **Toolbars** from the **View** menu. AutoCAD displays the Customize dialog box with the **Toolbars** tab displayed. From the **Toolbars** tab select the **New...** button. AutoCAD displays the New Toolbar dialog box.

From the **Save toolbar in menu group:** drop-down list box select the "test" group. In the **Toolbar Name:** edit box, type Office Furniture and select the **OK** button. AutoCAD displays the new Office Furniture toolbar. If you cannot see the newly created toolbar on the screen, you may have to move the Toolbar dialog box out of the way to find it.

To add buttons to the empty Office Furniture toolbar, click the **Commands** tab in the Customize dialog box. Then highlight **User defined** in the **Categories** list box and from the **Commands** text box highlight **User Defined Flyout.**

Select the highlighted **User Defined Flyout** and drag it to the Office Furniture toolbar. Place two additional flyout buttons in the Office Furniture toolbar. Move your mouse pointer to the Office Furniture toolbar and right-click on the left button. A message appears warning you to associate another toolbar with this flyout as shown in Figure 17–18. Choose the **OK** button. The **Properties** tab of the Customize dialog box becomes the **Flyout Properties** tab, as shown in Figure 17–19.

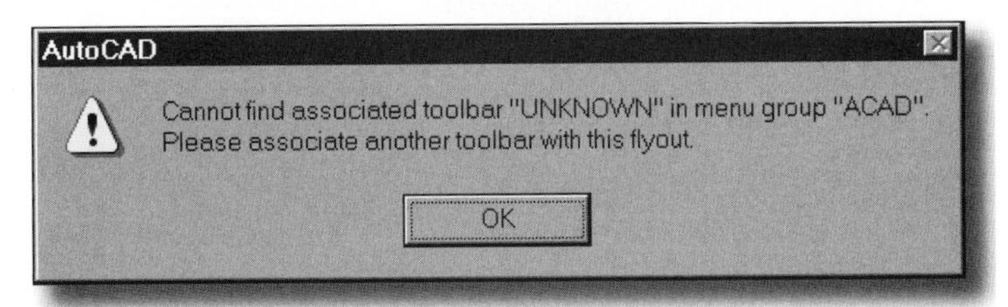

Figure 17–18 *Warning displayed when new flyout button selected*

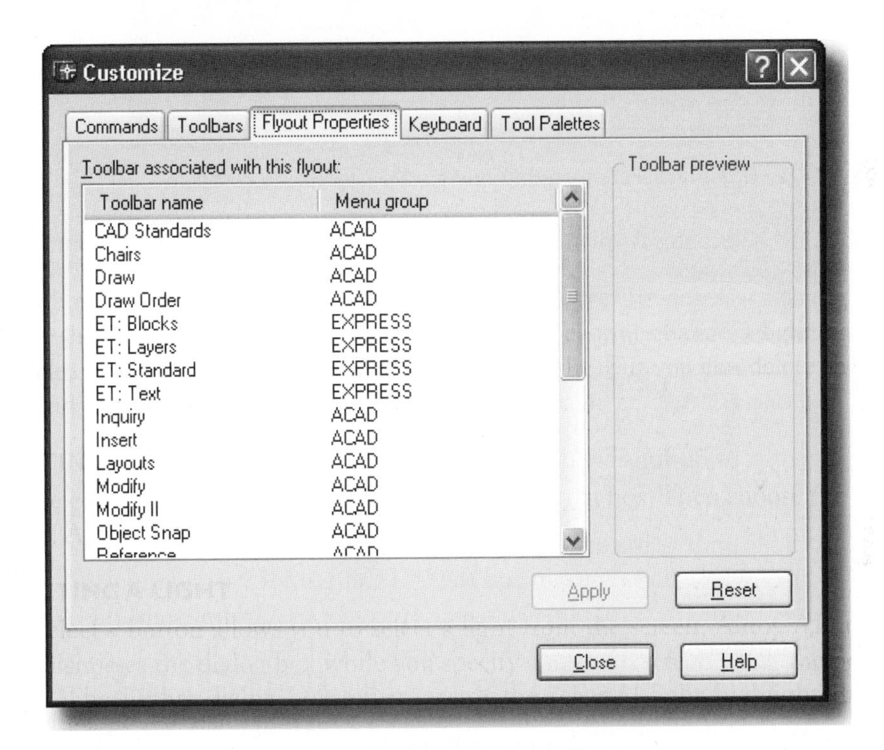

Figure 17–19 *Flyout Properties tab of the Customize dialog box*

Note: Highlighting **Chair** in the textbox may or may not immediately make it available to associate with the new toolbar. In that case, highlight another toolbar in the text box and then go back and highlight the Chair toolbox again. When it is available to associate with the new toolbar, its icon(s) will appear in the **Toolbar Preview** section of the **Flyout Properties** tab as shown in Figure 17–20.

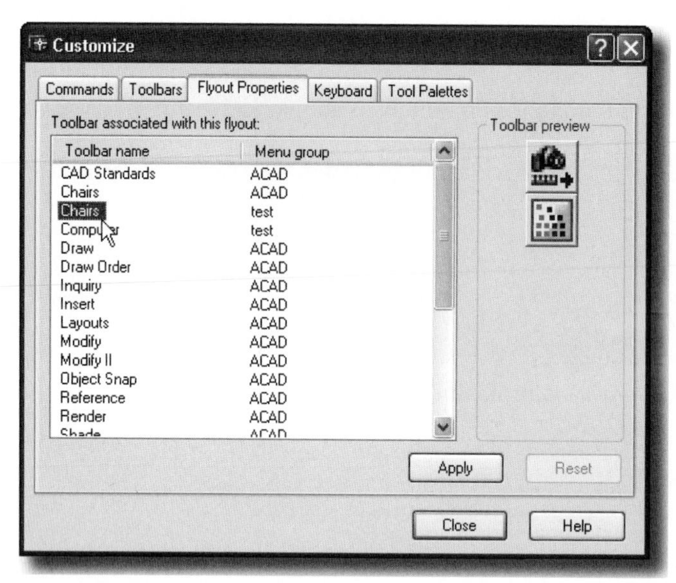

Figure 17–20 *Flyout Properties tab with Chair toolbar available to associate*

You may now select the **Apply** button and have the Chair toolbar as a flyout on the Office Furniture toolbar.

Note: Flyout buttons on toolbars are, in fact, toolbars themselves. They are simply condensed into a single button with the button representing the last function used being displayed.

Once all the flyout buttons are defined, close the Customize dialog box.

When you create the toolbars, AutoCAD adds all descriptions of the toolbars to the TEST.MNS file under a major section called ***TOOLBARS. AutoCAD creates and maintains the *.MNS files. You can add the ***TOOLBARS section to your *.MNU template file. From the Program Manager Accessories Group, start the Notepad editor program and open the TEST.MNS file. From the TEST.MNS file, select the ***TOOLBARS major section and paste it into the TEST.MNU file. You can have multiple copies of the Notepad editor open at the same time. Following is the line-by-line listing of the code of the TEST.MNU file after button bars are added.

```
//This is a test menu file and it demonstrates the basic new
//functionality of the new menu name tag syntax.
***MENUGROUP=test
***POP1
**Furniture
ID_Furn [&Furniture]
ID_Furns [->&Standard Furniture]
```

```
ID_Comps     [&Computer]^C^C^CATTREQ;0;-LAYER;M;EL_OFFEQ;COLOR;1;;;+
INSERT;COMP;\;;\ATTREQ;1;-LAYER;SET;0;;
ID_Tables    [&Table]^C^C^CATTREQ;0;-LAYER;M;FR_OFF;COLOR;3;;;+
INSERT;TABLE1;\;;\ATTREQ;1;-LAYER;SET;0;;
ID_Chairs    [<-C&hair]^C^C^CATTREQ;0;-LAYER;M;FR_OFF;COLOR;3;;;+
INSERT;CHAIR;\;;\ATTREQ;1;-LAYER;SET;0;;
ID_Furnc [->Custo&m Furniture]
ID_Compc     [Com&uter]^C^C^CATTDIA;0;-LAYER;M;EL_OFFEQ;COLOR;1;;;+
INSERT;COMP;\;;\\\\-LAYER;S;0;;ATTDIA;1
ID_Tablec    [Tab&le]^C^C^CATTDIA;0;-LAYER;M;FR_OFF;COLOR;3;;;+
INSERT;TABLE1;\;;\\\\-LAYER;S;0;;ATTDIA;1
ID_Chairc    [<-Chai&r]^C^C^CATTDIA;0;-LAYER;M;FR_OFF;COLOR;3;;;+
INSERT;CHAIR;\;;\\\\-LAYER;S;0;;ATTDIA;1
[—]
[Layer EL_OFFEQ]^C^C^C-LAYER;M;EL_OFFEQ;COLOR;1;;;
[Layer FR_OFF]^C^C^C-LAYER;M;FR_OFF;COLOR;3;;;
[—]
ID_Lyr0 [Layer 0]^C^C^C-LAYER;SET;0;;
***HELPSTRINGS
ID_Furn [Office Furniture]
ID_Furns [Standard Office Furniture]
ID_Comps [Standard Computer]
ID_Tables [Standard Table]
ID_Chairs [Standard Chair]
ID_Furnc [Custom Office Furniture]
ID_Compc [Custom Computer]
ID_Tablec [Custom Table]
ID_Chairc [Custom Chair]
***TOOLBARS
**COMPUTER
ID_Computer [_Toolbar("Computer", _Floating, _Show, 200, 100,0)]
ID_Layers       [_Button("Standard Computer", ICON2580.bmp,
          ICON_32_-LAYERS)]^C^C^CATTREQ;0;
     -LAYER;M;EL_OFFEQ;COLOR;1;;;INSERT;COMP;\;;
      \ATTREQ;1;-LAYER;SET;0;;
ID_Layers_0     [_Button("Custom Computer", ICON7216.bmp, ICON_32_-LAYER
     S)]^C^C^CATTDIA;0;-LAYER;M;EL_OFFEQ;COLOR;1;;;INSERT;COMP;\;;
      \\\\-LAYER;S;0;;ATTDIA;1
**TABLES
ID_Tables_0  [_Toolbar("Tables", _Floating, _Show, 200, 200, 0)]
ID_StandardComputer [_Button("Standard Table", ICON1629.bmp,
     ICON499.bmp)]^C^C^CATTREQ;0;-LAYER;M;FR_OFF;COLOR;3;;;INSERT;TABLE1;\;;
      \ATTREQ;1;-LAYER;SET;0;;
ID_CustomComputer [_Button("Custom Table", ICON5498.bmp,
```

```
    ICON8877.bmp)]^C^C^CATTDIA;0;-LAYER;M;FR_OFF;COLOR;3;;;INSERT;TABLE1;\;;
     \\\\-LAYER;S;0;;ATTDIA;1
**CHAIRS
ID_Chairs    [_Toolbar("Chairs", _Floating, _Show, 200, 300, 0)]
ID_StandardTable [_Button("Standard Chair", ICON2854.bmp, ICON5060.bmp)
    ]^C^C^CATTREQ;0;-LAYER;M;FR_OFF;COLOR;3;;;INSERT;CHAIR;\;;
     \ATTREQ;1;-LAYER;SET;0;;
ID_CustomTable [_Button("Custom Chair", ICON665.bmp,
    ICON4474.bmp)]^C^C^CATTDIA;0;-LAYER;M;FR_OFF;COLOR;3;;;INSERT;CHAI
    R;\;;
     \\\\-LAYER;S;0;;ATTDIA;1
**OFFICE_FURNITURE
**TB_OFFICE_FURNITURE
                [_Toolbar("Office Furniture", _Floating, _Show, 200,
    400, 0)]
ID_             [_Flyout("Computer", ICON8617.bmp, ICON_32_BLANK,
    _OtherIcon, test.COMPUTER)]
ID__0           [_Flyout("Tables", ICON3090.bmp, ICON_32_BLANK,
    _OtherIcon, test.TABLES)]
ID__1           [_Flyout("Chairs", ICON9328.bmp, ICON_32_BLANK,
    _OtherIcon, test.CHAIRS)]
```

Make sure you save the TEST.MNU file. If you deleted the TEST.MNS file without copying the ***TOOLBARS section to the TEST.MNU file, you will lose the newly created toolbars. During the process of adding new toolbars, it is always recommended to copy the ***TOOLBARS sections to the *.MNU file. Remember, if there is no *.MNS file, AutoCAD creates it for you based on the *.MNU template file.

You can manually insert the ***TOOLBARS section of code into your TEST.MNU file, but let AutoCAD do the job and then copy it into the *.MNU template file. Let's examine, line by line, some of the unique features of the code in the ***TOOLBARS section of the TEST.MNU file.

```
***TOOLBARS
```

Defines the major section for toolbars.

```
**COMPUTER
```

Menu subsection.

```
ID_Computer [_Toolbar("Computer", _Floating, _Show, 200, 100, 0)]
    ID_Computer
```

Tag name by which the item on the menu is referenced.

```
        _Toolbar
```

Signifies that the menu is dealing with a toolbar definition.

```
"Computer"
```

Name that appears above the toolbar.

```
_Floating
```

Indicates that the toolbar is floating and not docked. Possible values are _Top, _Bottom, _Left, _Right, or _Floating. The underscore character in front of the keywords is not required; however, it is used for the international versions of AutoCAD. The keywords are not case-sensitive.

```
_Show
```

Visibility keyword, available options include _Show or _Hide.

```
200, 100, 0
```

The first number is the distance in pixels from the left edge of the screen. The second number is the distance in pixels from the top edge. The final number is the number of rows in the toolbar.

```
ID_Layers          [_Button("Standard
      Computer",ICON2580.bmp,ICON_32_-LAYERS)]macro
      ID_Layers
```

Tag name by which this item on the menu is referenced.

```
_Button
```

Signifies that the menu is dealing with a button definition.

```
"Standard Computer"
```

This is the tooltip text that is displayed when the mouse pointer passes over the button.

```
ICON2580.bmp
```

ID of the small-icon (16 x 16) bitmap.

```
ICON_32_-LAYERS
```

ID of the large-icon (32 x 32) bitmap.

```
macro
```

Command sequence required to complete the macro.

Flyout Section of the Menu

```
**TB_OFFICE_FURNITURE
      [_Toolbar("Office Furniture", _Floating, _Show, 200, 400, 0)]
ID_   [_Flyout("Computer", ICON8617.bmp, ICON_32_BLANK, _OtherIcon,
    test.COMPUTER)]
```

```
ID__0 [_Flyout("Tables", ICON3090.bmp, ICON_32_BLANK, _OtherIcon,
    test.TABLES)]
ID__1 [_Flyout("Chairs", ICON9328.bmp, ICON_32_BLANK, _OtherIcon,
    test.CHAIRS)]
[_Toolbar("Office Furniture", _Floating, _Show, 200, 400, 0)]
```

The toolbar menu item is the same as for the toolbar description explained earlier.

```
ID_   [_Flyout("Computer", ICON8617.bmp, ICON_32_BLANK, _OtherIcon,
    test.COMPUTER)]
```

`ID_`

Tag name by which the item on the menu is referenced.

`Flyout`

Signifies that the menu is dealing with a flyout definition.

`"Computer"`

Tooltip text that is displayed when the mouse pointer passes over the button.

`ICON8617.bmp`

ID of the small-icon (16 x 16) bitmap.

`ICON_32_BLANK`

ID of the large-icon (32 x 32) bitmap.

`_OtherIcon`

Has one of two possible values: _OtherIcon or _OwnIcon. The OtherIcon value allows the button to switch icons and displays the last icon selected. The OwnIcon value will not permit this switching.

ACCELERATOR KEYS

AutoCAD supports user-defined accelerator keys.

On the **Keyboard** tab of the Customize dialog box, select "test" from the **Menu Group** text box and then select Chair Toolbar from the **Categories** text box. See Figure 17–21. The commands Custom Chair and Standard Chair are listed in the **Commands** text. Select Standard Chair and then click in the **Press new shortcut key** text box. Then, while holding down the CTRL and SHIFT keys, press **1**. This will cause "Ctrl+Shift+1" to be displayed in the **Press new shortcut key** text box. Then select the Assign button, and this sequence of accelerator keys will be assigned to the Standard Chair macro that has been written for that button (Standard Chair) in the Chair toolbar.

Note: If the selected command already has a sequence of accelerator keys assigned to it, they will be displayed in the **Current Keys** text box. If the combination of accelerator keys that you have entered into the **Press new shortcut key** text box is already assigned to another command, the message below the **Press new shortcut key** text box will tell you so. Otherwise the message will read "Currently assigned to: [unassigned]."

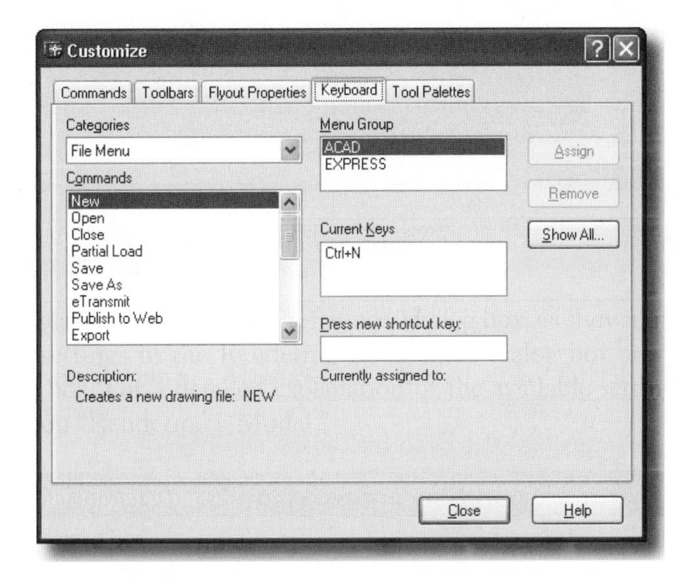

Figure 17–21 *Keyboard tab of the Customize dialog box*

Once the keystrokes have been satisfactorily assigned to the command, select the **Close** button. Now, pressing the newly assigned combination of accelerator keys (Ctrl+Shift+1) while at the Command: prompt will activate the macro that has been written to insert the Standard Chair from the Chair toolbar that is part of the Test menu.

Apart from including macros, you can also include AutoLISP code in the menus (refer to Chapter 18) or have the menu macros load and run AutoLISP applications.

DIESEL

As mentioned earlier, a macro is a string of commands that AutoCAD executes when the menu item is selected. You can also add AutoLISP in a macro and call an AutoLISP routine from a macro. AutoCAD provides a macro language alternative to AutoLISP, called DIESEL. DIESEL is an acronym for Direct Interpretively Evaluated String Expression Language. The subject of DIESEL is beyond the scope of this book, but let's go through an example to see the application of DIESEL.

DIESEL allows you to customize the AutoCAD status line through the use of the MODEMACRO setvar variable, and it also allows modification of the appearance of pull-down menus. DIESEL, though a macro language and similar in style to AutoLISP, does not have the power and flexibility of AutoLISP.

The simplest use of DIESEL is through the MODEMACRO setvar variable. At the AutoCAD "Command": prompt, type **MODEMACRO** and press ENTER or SPACEBAR. AutoCAD prompts:

New value for MODEMACRO, or . for none <"">: **Captain CAD**

Type in a text string as shown in this example and press ENTER. AutoCAD displays the text string in the status bar, as shown in Figure 17–22.

Figure 17–22 *Display of the text string in the status bar*

To return to the default value, type **MODEMACRO** at the "Command:" prompt and press ENTER or SPACEBAR. Type in a single period to the prompt and press ENTER or SPACEBAR. The status bar returns to the default settings.

As another example, a string with arguments is provided in response to the MODEMACRO prompt:

New value for MODEMACRO, or . for none <"">:-LAYER = $(getvar,clayer)

AutoCAD gets the current layer name and displays it in the status bar, as shown in Figure 17–23. DIESEL expressions can also be placed as part of the menus.

Figure 17–23 *Display of the current layer name in the status bar*

CUSTOMIZING TABLET MENUS

The previous descriptions and examples of menu customization can be applied to customizing the tablet menu area. One problem not encountered in a tablet menu is that of the display. The brackets that enclose non-active items will not be visible except to someone who is reading the filename.MNU file from which the filename.MNX file was compiled.

The main concern in customizing the tablet part of the menu is in placing the programming lines in the right order and in the right area so they will correspond to the preprinted overlay that will be configured on the tablet for use with the menu.

A menu may have up to four tablet areas. They will have the headings ***TABLET1, ***TABLET2, ***TABLET3, and ***TABLET4. The first program line following a heading will correspond to the configured overlay's upper-left column/row rectangle. Subsequent program lines will correspond the rectangle to the right in the same row as its predecessor until the end of the row is reached. Then the next program line will be on the extreme left rectangle of the next row. For example, a tablet menu area with 12 program lines might be in any one of the 6 following arrangements:

1	2	3	4	5	6	7	8	9	10	11	12

1	2	3	4	5	6
7	8	9	10	11	12

1	2	3	4
5	6	7	8
9	10	11	12

1	2	3
4	5	6
7	8	9
10	11	12

1	2
3	4
5	6
7	8
9	10
11	12

1
2
3
4
5
6
7
8
9
10
11
12

If there are more rectangles specified in the tablet configuration than there are program lines, the extras will be non-active when picked. If there is an excess of program lines in the menu, they will, of course, not be accessible.

DIALOG CONTROL LANGUAGE

AutoCAD Release 12 introduced programmable dialog boxes, and this feature is continued in AutoCAD 2004. Programming dialog boxes requires a thorough knowledge of AutoLISP and/or C/C++. The description of the dialog box is a text file known as a DCL (Dialog Control Language) file. The DCL file is a description of the various parts of the dialog and of the elements the dialog contains. The following is an example of a DCL file.

```
cesdoor : dialog
{
   label = "CESCO Doors";
   : row
   {
      : radio_cluster
      {
         key = "thick";
         : boxed_radio_column
         {
            label = "Thickness";
            : radio_button
            {
               key = "t1";
               label = "1\"";
               value = "1";
            }
            : radio_button
            {
               key = "t2";
               label = "2\"";
            }
            : radio_button
            {
               key = "t4";
               label = "4\"";
            }
         }//boxed radio column - Thickness
      }//radio cluster - Thick
      : radio_cluster
```

```
      {
         key = "hinge";
         : boxed_radio_column
         {
            label = "Hinge";
            spacer;
            : radio_button
            {
               key = "rt";
               label = "Right";
            }
            : radio_button
            {
               key = "lt";
               label = "Left";
               value = "1";
            }
            spacer;
         }//boxed radio column - Hinge location
      }//radio_cluster - Hinge
   }// row - top
   : row
   {
      : radio_cluster
      {
         key = "opng";
         : boxed_radio_column
         {
            label = "Open";
            width = 9;
            : radio_button
            {
               key = "in";
               label = "In";
            }
            : radio_button
            {
               key = "out";
               label = "Out";
               value = "1";
            }
         }//boxed radio column - Opening
```

```
        }//radio_cluster - opng
        : radio_cluster
        {
           key = "hand";
           : boxed_radio_column
           {
              label = "Handles";
              : radio_button
              {
                 key = "hndl2";
                 label = "2";
                 value = "1";
              }
              : radio_button
              {
                 key = "hndl3";
                 label = "3";
              }
           }//boxed radio column - Number of handles
        }//radio_cluster - hand
     }//row - 2nd row
     spacer_1;
     ok_cancel;
}
```

When AutoCAD is loaded into memory, it automatically loads two files: ACAD.DCL and BASE.DCL. These files contain various attributes on which you can build your dialog boxes. Each element (:radio_button, for example) contains a key (key = "hndl3"). These keys are the names to which AutoLISP and C/C++ refer in order to retrieve values from them and act upon them. Figure 17–24 shows a sample dialog box.

Figure 17–24 *Sample dialog box*

Selecting the OK button causes AutoLISP or C/C++ to read the various keys to determine which values the user selected and pass the results to the calling program. When writing programs that utilize dialog boxes, a certain portion of the code is dedicated to reading and responding to the dialog box (apart from the DCL file describing the dialog box itself). This provides the user with a more detailed view of all the inputs and can be used to limit the user inputs to the program. Dialog boxes can utilize radio buttons, toggle boxes, horizontal and vertical sliders, list boxes, edit boxes, buttons, image tiles, and more to enhance the front end to any program requiring user input.

A complete discussion of programming dialog boxes is beyond the scope of this book. Refer to AutoCAD's Customization Guide for more details.

REVIEW QUESTIONS – A

1. What extension must a file have when you are writing a custom menu?
2. How many menus may be active simultaneously?
3. What AutoCAD command sequence will activate a menu written as a file named *XYZ.MNU*?
4. Under the ***BUTTONS menu, which program line corresponds to the second button on the mouse/puck?
5. Placing three blank spaces in a menu program line is equivalent to what action from the keyboard?
6. What is the purpose of the caret (^) in a menu program line?
7. What is the purpose of the "^C" at the beginning of a menu program line?
8. Create a menu file to include a program line that will draw a rectangle whose opposite corners are points with coordinates 1,1 and 5,2.
9. Create a menu file to include a program line that will draw a circle of diameter 2 units and whose center point has the coordinates 3,3.
10. Create a menu file to include a program line that will array the circle in Exercise 9 in a polar array every 30 degrees, with the center of the array at a point whose coordinates are 4,3.
11. Create a menu file to include a program line that will permit a line-arc-line continuation starting at point 1,1, with user picks for the second, third, and fourth points.
12. Write lines that set UNITS as follows:

    a. Decimal to four-place display

    b. Architectural to 1/8" display

    c. Each of the above to include a mechanism to return to the graph screen

CUSTOMIZING HATCH PATTERNS

Certain concepts about hatch patterns should be understood before learning to create one.

1. Hatch patterns are made up of lines or line segment/space combinations. There are no circles or arcs available in hatch patterns like there are in shapes and fonts (which will be covered next).
2. A hatch pattern may be one or more series of repeated parallel lines or repeating dot or line segment/space combinations. That is, each line in one so-called family is like every other line in that same family. And each line has the same offset and stagger (if it is a segment/space combination), relative to its adjacent sibling, as every other line.
3. One hatch pattern can contain multiple families of lines. One family of lines may or may not be parallel to other families. With properly specified basepoints, offsets, staggers, segment/space combinations, lengths, and relative angles, you can create a hatch pattern from multiple families of segment/space combinations that will display repeated closed polygons.
4. Each family of lines is drawn with offsets and staggers based on its own specified basepoint and angle.
5. All families of lines in a particular hatch pattern will be located (basepoint), rotated, and scaled as a group. These factors (location, angle of rotation, and scale factor) are determined when the hatch pattern is loaded by the HATCH command and used to fill a closed polygon in a drawing.
6. The pattern usually can be achieved by different ways of specifying parameters.

Hatch patterns are created by including their definition in a file whose extension is .PAT. This can be done by using a text editor such as Notepad or a word processor in the non-document (or programmer) mode, which will save the text in ASCII format. Your hatch pattern definition can also be added to the ACAD.PAT file. You can also create a new file specifically for a pattern.

Each pattern definition has one header line giving the pattern name/description and a separate specification line describing each family of lines in the pattern.

The header line has the following format:

```
*pattern-name[,description]
```

The pattern name will be the name for which you will be prompted when using the HATCH command. The description is optional and is there so that someone reading the .PAT file can identify the pattern. The description has no effect, nor will it be displayed while using the HATCH command. The leading asterisk denotes the beginning of a hatch pattern.

The format for a line family is as follows:

```
angle, x-origin, y-origin, delta-x, delta-y [,dash-1, dash-2...]
```

The brackets "[]" denote optional segment/space specifications used for noncontinuous-line families. Note also that any text following a semicolon (;) is for comment only and will be ignored. In all definitions, the angle, origins, and deltas are mandatory (even if their values are zero).

An example of continuous lines that are rotated at 30 degrees and separated by 0.25 units is shown in Figure 17–25:

```
*P30, 30 degree continuous
30, 0,0, 0,.25
```

The 30 specifies the angle.

The first and second zero specify the coordinates of the origin.

The third zero, though required, is meaningless for continuous lines.

The 0.25 specifies the distance between lines.

A pattern of continuous lines crossing at 60 degrees to each other could be written as follows (see Figure 17–26):

```
*PX60,x-ing @ 60
30, 0,0, 0,.25
330, 0,0, 0,.25
```

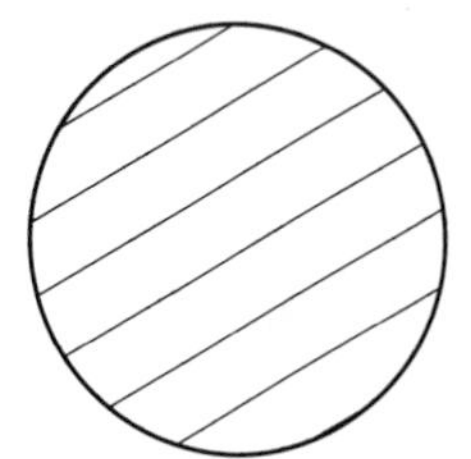

Figure 17–25 *A hatch pattern with continuous lines rotated at 30 degrees and separated by 0.25 units*

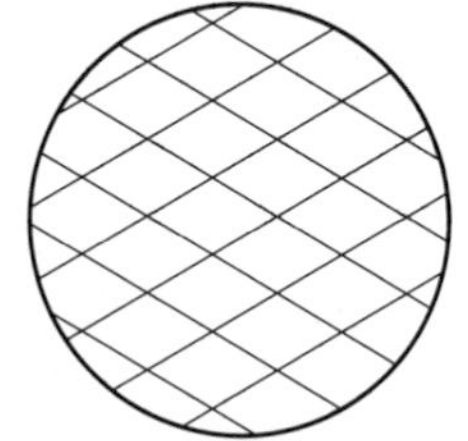

Figure 17–26 *A hatch pattern with continuous lines crossing at 60 degrees to each other and separated by 0.25 units*

A pattern of lines crossing at 90 degrees but having different offsets is as follows (see Figure 17–27):

```
*PX90, x-ing @ 90 w/ 2:1 rectangles
0, 0,0, 0,.25
90, 0,0, 0,.5
```

Note the effect of the delta-Y. It is the amount of offset between lines in one family. Hatch patterns with continuous lines do not require a value (other than zero) for delta-*X*. Orthogonal continuous lines also do not require values for the *X* origin unless they are used in a pattern that includes broken lines.

To illustrate the use of a value for the *Y* origin, two parallel families of lines can be written to define a hatch pattern for steel as follows (see Figure 17–28):

```
*steel
45, 0,0, 0,1
45, 0,.25, 0,1
```

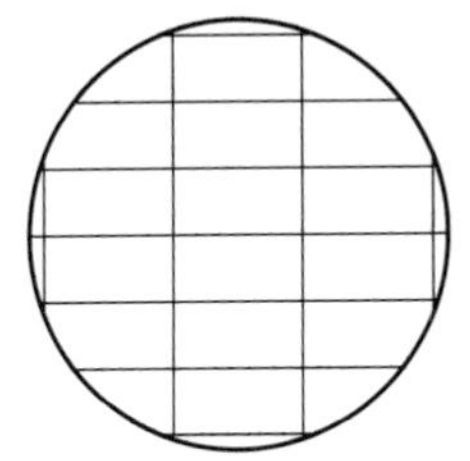

Figure 17–27 *A hatch pattern with lines crossing at 90 degrees and having different offsets*

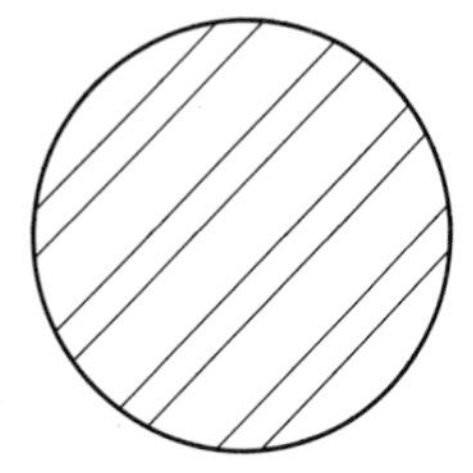

Figure 17–28 *Defining a hatch pattern for steel*

Three concepts are worthy of note in this example.

- If the families were not parallel, then specifying origins other than zero would serve no purpose.
- Parallel families of lines should have the same delta-*Y* offsets. Different offsets would serve little purpose.
- Most important, the delta-*Y* is at a right angle to the angle of rotation, but the *Y* origin is in the *Y* direction of the coordinate system. The steel pattern as written in the example would fill a polygon, as shown in Figure 17–29. Note the dimensions when used with no changes to the scale factor of 1.0 or a rotation angle of zero.

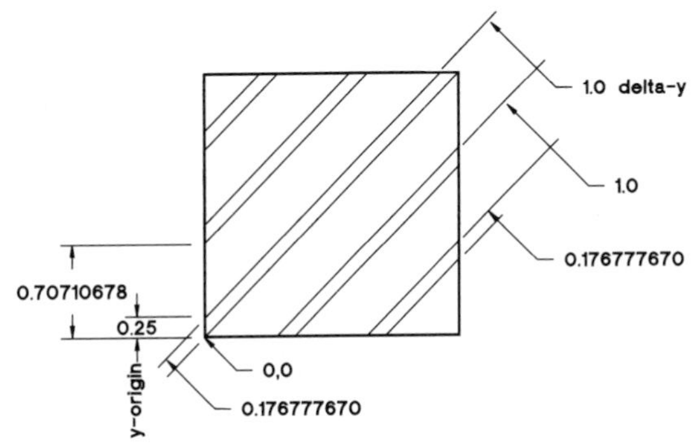

Figure 17–29 *The steel hatch pattern with the delta-Y at a right angle to the angle of rotation and the Y origin in the Y direction of the coordinate system*

CUSTOM HATCH PATTERNS AND TRIGONOMETRY

The dimensions in the hatch pattern resulting from a 0.25 value for the delta-*Y* of the second line-family definition may not be what you expected; see Figure 17–30. If you wished to have a 0.25 separation between the two line families (see Figure 17–31), then you must either know enough trigonometry/geometry to predict accurate results or else put an additional burden on the user to reply to prompts with the correct responses to achieve those results. For example, you could write the definition as follows:

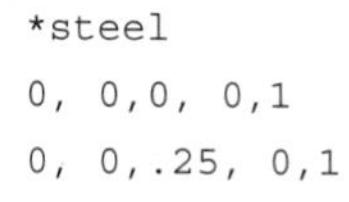

```
*steel
0, 0,0, 0,1
0, 0,.25, 0,1
```

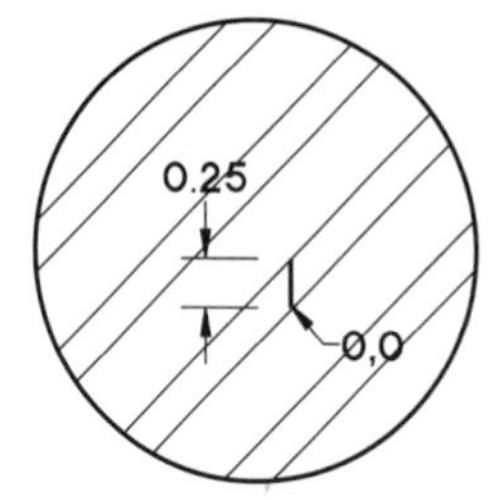

Figure 17–30 *Hatch pattern dimensions resulting from a 0.25 value for the delta Y of the second line-family definition*

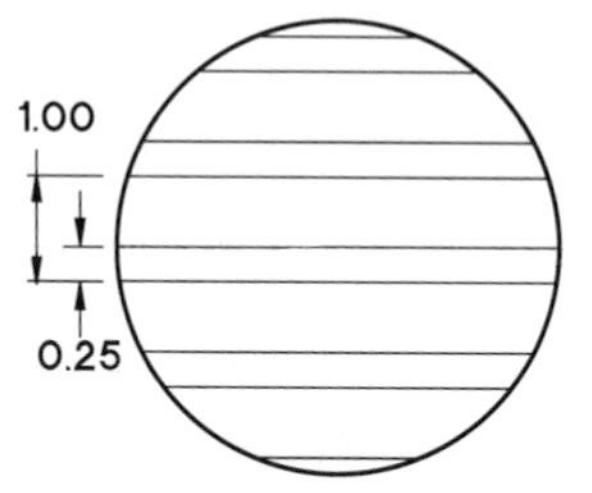

Figure 17–31 *Steel hatch pattern defined with a 0.25 separation between the two line families*

In order to use this pattern as shown, the user will have to specify a 45-degree rotation when using it. This will maintain the ratio of 1 to .25 between the offset (delta-*Y*) and the spacing between families (*Y* origin). However, if you wish to avoid this inconvenience to the user, but still wish to have the families separated by .25, you can write the definition as follows:

```
*steel
45, 0,0, 0,1
45, 0,.353553391, 0,1
```

The value for the *Y* origin of .353553391 was obtained by dividing .25 by the sine (or cosine) of 45 degrees, which is .70710678. The *X* origin and *Y* origin specify the coordinates of a point. Therefore, setting the origins of any family of continuous lines merely tells AutoCAD that the line must pass through that point. See Figure 17–32 for the trigonometry used.

For families of lines that have segment/space distances, the point determined by the origins can tell AutoCAD not only that the line passes through that point, but that one of the segments will begin at that point. A dashed pattern can be written as follows (see Figure 17–33):

```
*dashed
    0, 0,0, 0,.25, .25,-.25
```

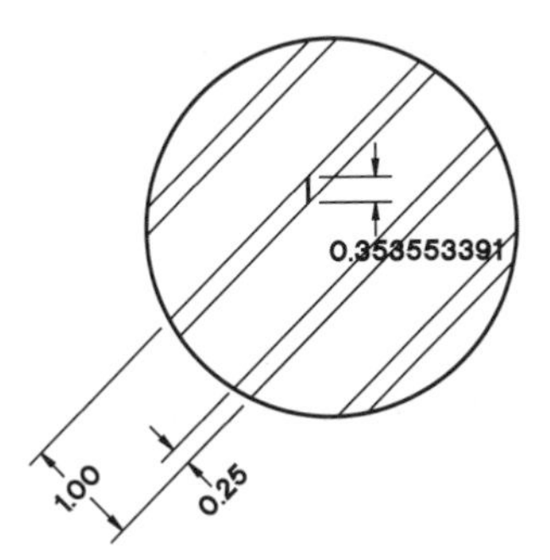

Figure 17–32 *Steel hatch pattern defined with a 45 degree rotation to maintain a 1:25 offset ratio*

Note that the value of the *X* origin is zero, thus causing the dashes of one line to line up with the dashes of other lines. Staggers can be produced by giving a value to the *X* origin as follows (see Figure 17–34):

```
*dashstagger
0, 0,0, .25,.25, .25,-.25
```

In a manner similar to defining linetypes, you can cause lines in a family to have several lengths of segments and spaces (see Figure 17–35).

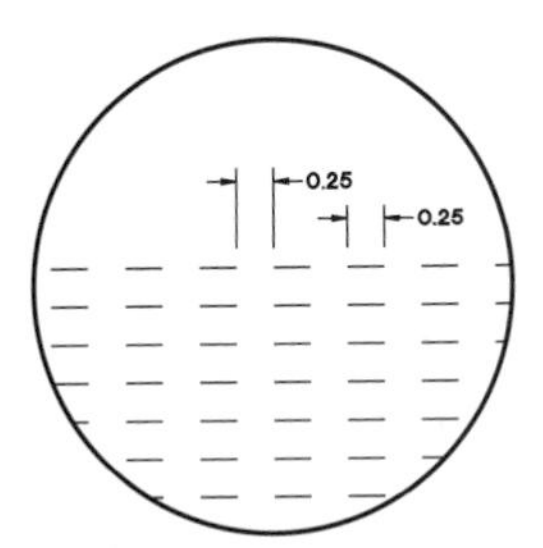

Figure 17–33 *Writing a dashed pattern*

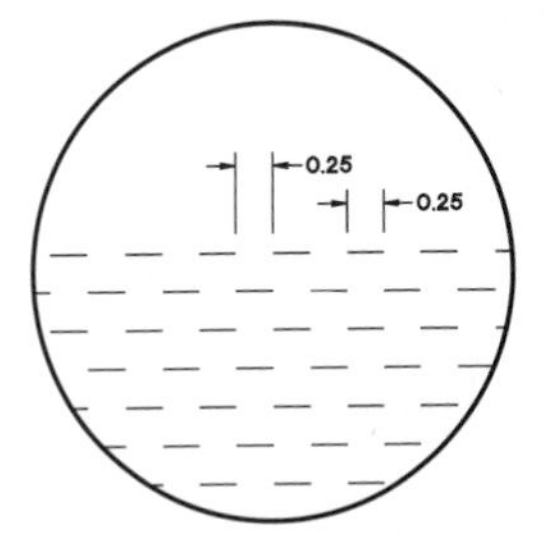

Figure 17–34 *Writing a staggered dashed pattern*

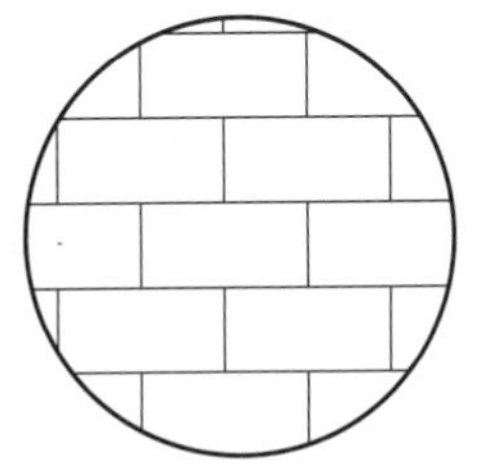

Figure 17–35 *A pattern with several lengths of segments and spaces*

```
*simple
0, 0,0, 0,.5
90, 0,0, 0,1, .5,-.5
```

A similar, but more complex hatch pattern could be written as follows (see Figure 17–36):

```
*complex
45, 0,0, 0,.5
-45, 0,0, 0,1.414213562, 0,1.41421356
```

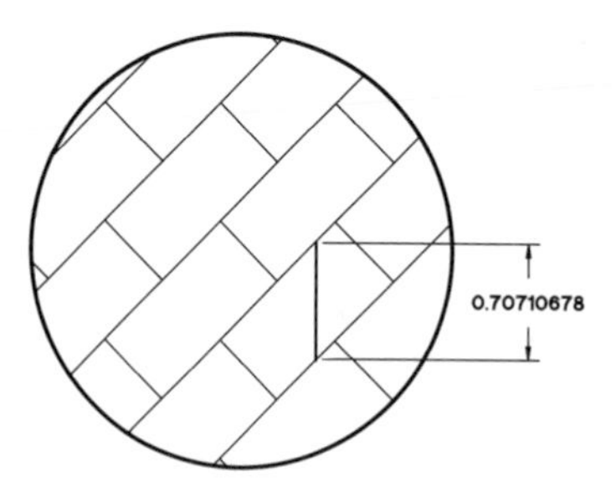

Figure 17–36 *A pattern with more complex hatch patterns*

REPEATING CLOSED POLYGONS

Creating hatch patterns with closed polygons requires planning. For example, a pattern of 45/90/45-degree triangles, as shown in Figure 17–37, should be started by first extending the lines, as shown in part a of Figure 17–37. Extend the construction lines through points of the object parallel to other lines of the object. Note the grid that emerges when you use the lines and distances obtained to determine the pattern.

It is also helpful to sketch construction lines that are perpendicular to the object lines. This will assist you in specifying segment/space values. In the example, two of the lines are perpendicular to one another, thus making this easier. Figures 17–38 through 17–41 illustrate potential patterns of triangles. Once the pattern is selected, the grid, and some knowledge of trigonometry, will assist you in specifying all of the values in the definition for each line family.

For pattern PA the horizontal line families can be written as follows (see Figure 17–42):

```
0, 0,0, 0,1, 1,-1
```

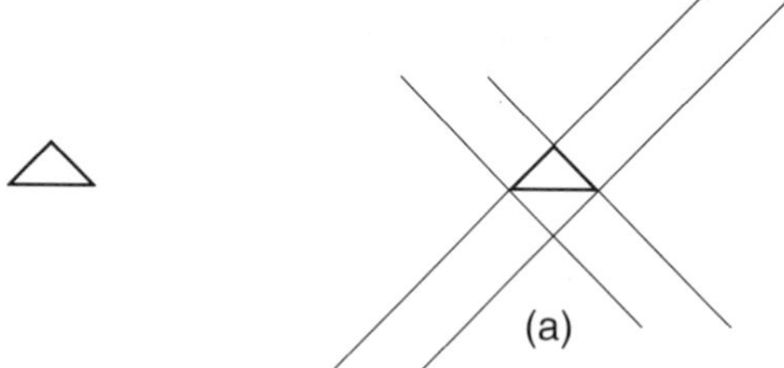

Figure 17–37 *Closed polygons*

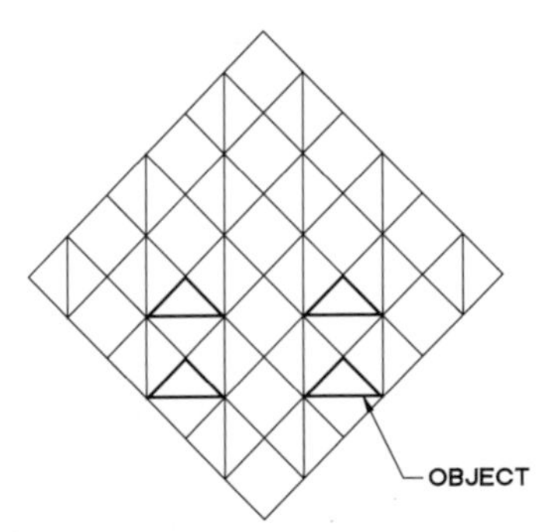

Figure 17–38 *Creating triangular hatch patterns—method #1*

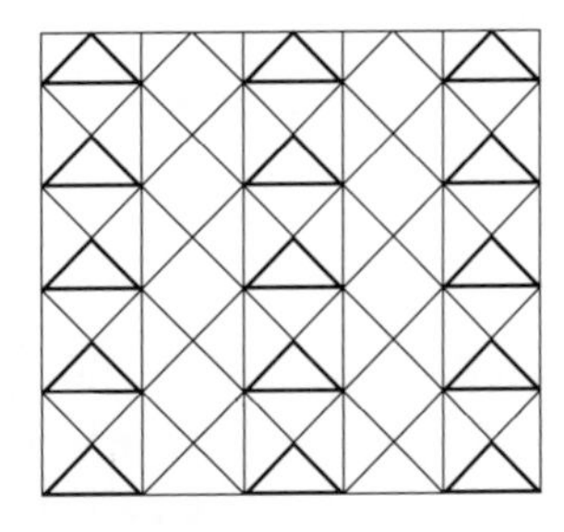

Figure 17–39 *Creating triangular hatch patterns—method #2*

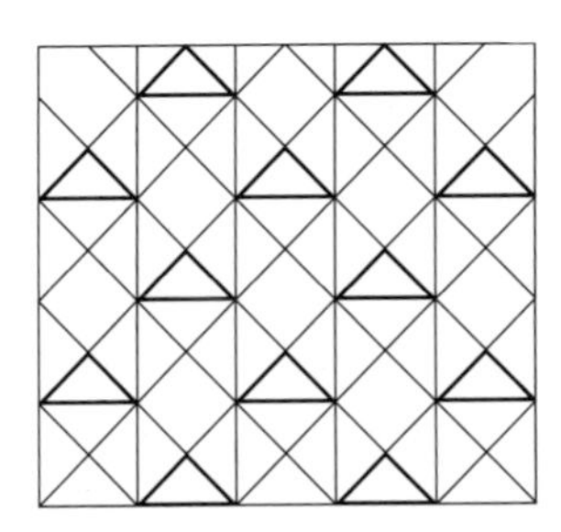

Figure 17–40 *Creating triangular hatch patterns—method #3*

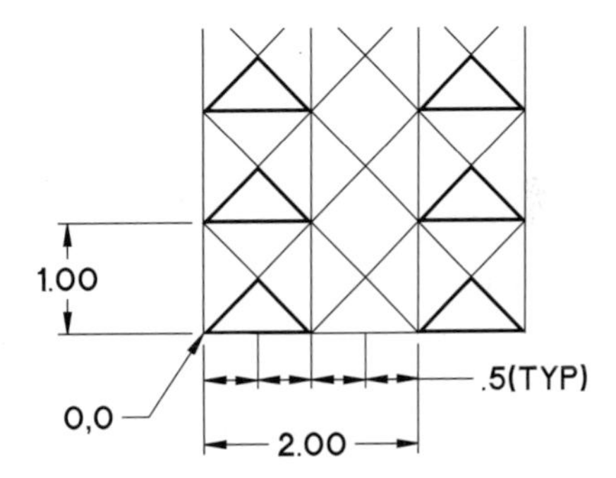

Figure 17–41 *Creating triangular hatch patterns—method #4*

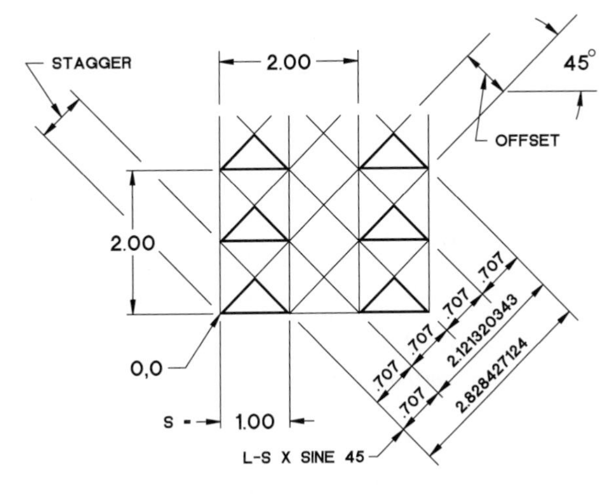

Figure 17–42 *Triangular patterns*

The specifications for the 45-degree family of lines can be determined by using the following trigonometry:

```
sin 45 degrees = 0.70710678
S = 1
L = S times sin 45 degrees
L = 1 times sin 45 degrees = 0.70710678
```

Note that the trigonometry function is applied to the hypotenuse of the right triangle. In the example, the hypotenuse is 1 unit. A different value would simply produce a proportional result; i.e., a hypotenuse of .5 would produce L = S x 0.70710678 = 0.353553391. The specifications for the 45-degree family of lines could be written as follows:

```
45, 0,0 0.70710678,0.70710678, 0.70710678,-2.121320343
angle,origin,offset,  stagger,            segment,     space
```

For the 135-degree family of lines, the offset, stagger, segment, and space have the same values (absolute) as for the 45-degree family. Only the angle, the *X* origin, and the sign (+ or –) of the offset or stagger may need to be changed.

The 135-degree family of lines could be written as follows:

```
135, 1,0,  -0.70710678,-0.70710678, 0.70710678,-2.121320343
```

Putting the three families of lines together under a header could be written as follows, and as shown in Figure 17–43.

```
*PA,45/90/45 triangles stacked
0, 0,0, 0,1, 1,-1
45, 0,0 0, 0.70710678,0.70710678, 0.70710678,-2.121320343
135, 1,0,  -0.70710678,0.70710678, 0.70710678,-2.121320343
```

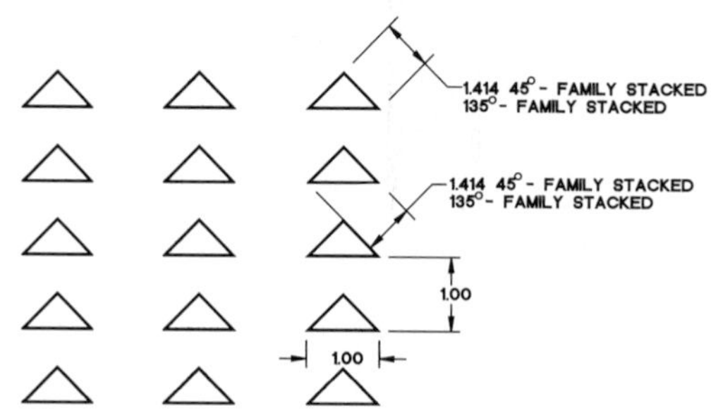

Figure 17–43 *Families of triangular patterns*

In the preceding statement, "could be written" tells you that there may be other ways to write the definitions. As an exercise, write the descriptions using 225 degrees instead of 45 degrees and 315 instead of 135 for the second and third families, respectively.

As a hint, you determine the origin values of each family of lines from the standard coordinate system. But to visualize the offset and stagger, orient the layout grid so that the rotation angle coincides with the zero angle of the coordinate system. Then the signs and the values of delta-*X* and delta-*Y* will be easier to establish along the standard plus for right/up and negatives for left/down directions. Examples of two hatch patterns, PB and HONEYCOMB, follow.

The PB pattern can be written as follows, and as shown in Figure 17–44.

```
*PB, 45/90/45 triangle staggered
0, 0,0, 1,1, 1,-1
45, 0,0, 0,1.414213562, 0.70710678,-0.70710678
135, 1,0, 0,1.414213562, 0.70710678,-0.70710678
```

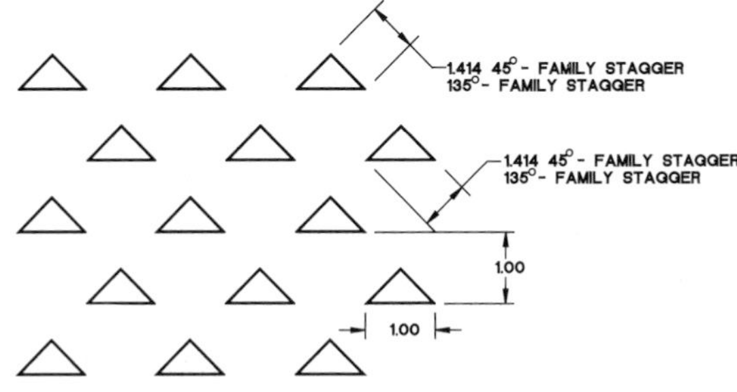

Figure 17–44 *Example of the pattern*

Note that this alignment simplifies the definitions of the second and third families of lines over the PA pattern.

The Honeycomb pattern can be written as follows, and as shown in Figure 17–45.

```
*HONEYCOMB
90, 0,0, 0.866025399,0.5, 0.577350264,-1.154700538
330, 0,0, 0.866025399,0.5, 0.577350264,-1.154700538
30, 0.5, -0.288675135, 0.866025399, 0.5, 0.577350264,-1.154700538
```

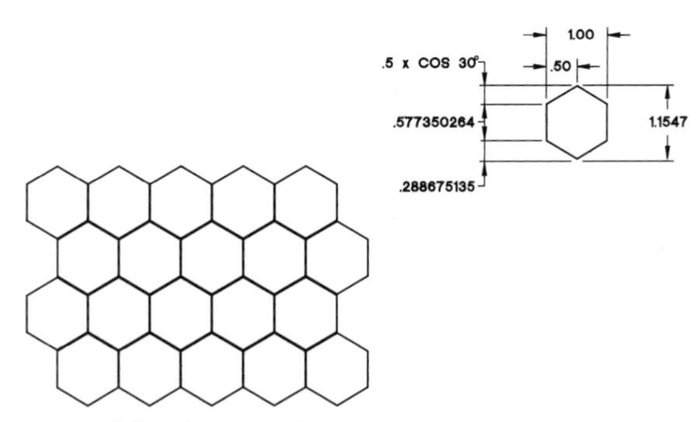

Figure 17–45 *Example of the Honeycomb pattern*

CUSTOMIZING SHAPES AND TEXT FONTS

Shapes and fonts are written in the same manner, and both are stored in files with the .SHP file extension. The .SHP files must be compiled into .SHX files. This section covers how to create and save .SHP files and how to compile .SHP files into .SHX files. To compile an .SHP file into an .SHX file for shapes or fonts, enter:

Command: **compile**

From the Select Shape File dialog box, select the file to be compiled. If the file has errors they will be reported; otherwise, you will be prompted:

Compilation successful
Output file name .SHX contains nnn bytes

The main difference between shapes and fonts is in the commands used to place them in a drawing. Shapes are drawn by using the SHAPE command, and fonts are drawn using commands that insert text, such as TEXT or DIM. Whether or not an object in an .SHP/.SHX file can be used with the SHAPE command or as a font character is partly determined by whether its shape name is written in uppercase or lowercase (explained below).

Each shape or character in a font in an .SHP or .SHX file is made up of simplified objects. These objects are simplified lines, arcs, and circles. The reason they are referred to as simplified is because in specifying their directions and distances, you cannot use decimals or architectural units. You must use only integers or integer fractions. For example, if the line distance needs to be equal to 1 divided by the square root of 2 (or .7071068), the fraction 70 divided by 99 (which equals .707070707) is as close as you can get. Rather than call the simplified lines and arcs "objects," we will refer to them as "primitives."

Individual shapes (and font characters) are written and stored in ASCII format. .SHP/.SHX files may contain up to 255 SHAPE-CHARACTERS. Each SHAPE-CHARACTER definition has a header line, as follows:

```
*shape number, defbytes, shapename
```

The codes that describe the SHAPE-CHARACTER may take up one or more lines following the header. Most of the simple shapes can be written on one or two lines. The meaning of each item in the header is as follows:

The shape number may be from 1 to 255 with no duplications within one file.

Defbytes is the number of bytes used to define the individual SHAPE-CHARACTER, including the required zero that signals the end of a definition. The maximum allowable bytes in a SHAPE-CHARACTER definition is 2,000.

Defbytes (the bitcodes) in the definition are separated by commas. You may enclose pairs of bitcodes within parentheses for clarity of intent, but this does not affect the definition.

The *shapename* should be in uppercase if it is to be used by the SHAPE command. Like a block name is used in the BLOCK command, you enter the shapename when prompted to do so during the SHAPE command. If the shape is a character in a font file, you may make any or all of the shapename characters lowercase, thereby causing the name to be ignored when compiled and stored in memory. It will serve for reference only in the .SHP file for someone reading that file.

PEN MOVEMENT DISTANCES AND DIRECTIONS

The specifications for pen movement distances and directions (whether the pen is up or down) for drawing the primitives that will make up a SHAPE-CHARACTER are written in bitcodes. Each bitcode is considered one defbyte. Codes 0 through 16 are not DISTANCE-DIRECTION codes, but special instructions-to-AutoCAD codes (to be explained shortly, after the DISTANCE-DIRECTION codes discussion).

DISTANCE-DIRECTION codes have three characters. They begin with a zero. The second character specifies distance. More specifically, it specifies vector length, which may be affected by a scale factor. Vector length and scale factor combine to determine actual distances. The third character specifies direction. There are 16 standard directions available through use of the DISTANCE-DIRECTION bitcode (or defbyte). Vectors 1 unit in length are shown in the 16 standard directions in Figure 17–46.

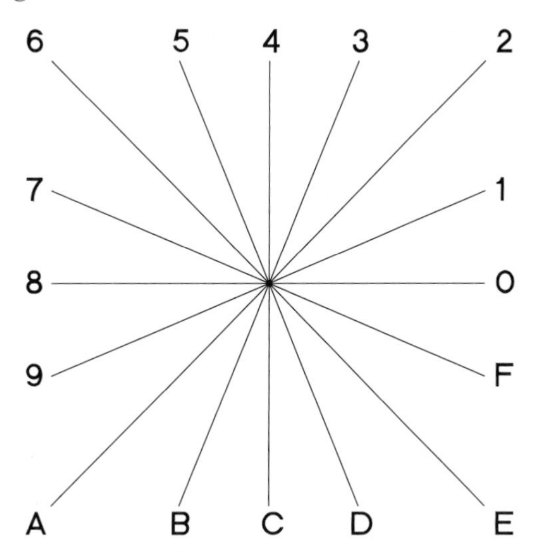

Figure 17–46 *DISTANCE-DIRECTION bitcodes*

Directions 0, 4, 8, and C are equivalent to the default 0, 90, 180, and 270 degrees, respectively. Directions 2, 6, A, and E are 45, 135, 225, and 315 degrees, respectively. But the odd-numbered direction codes are not increments of 22.5 degrees, as you might think. They are directions that coincide with a line whose delta-*X* and delta-*Y* ratio are 1 unit to 2 units. For example, the direction specified by code 1 is equivalent to drawing a line from 0,0 to 1,.5. This equates to approximately 26.56505118 degrees (or the arctangent of 0.5). The direction specified by code 3 equates to 63.434494882 degrees (or the arctangent of 2) and is the same as drawing a line from 0,0 to .5,1.

Distances specified will be measured on the nearest horizontal or vertical axis. For example, 1 unit in the 1 direction specifies a vector that will project 1 unit on the horizontal axis. Three units in the D direction will project 3 units on the vertical axis (downward). So the vector specified as 1 unit in the 1 direction will actually be 1.118033989 units long at an angle of 26.65606118 degrees, and the vector specified as 3 units in the D direction will be 3.354101967 units long at an angle of 296.5650512 degrees. See Figure 17–47 for examples of specifying direction.

To illustrate the codes specifying the DISTANCE-DIRECTION vector, the following example is a definition for a shape called "oddity" that will draw the shape shown in Figure 17–48.

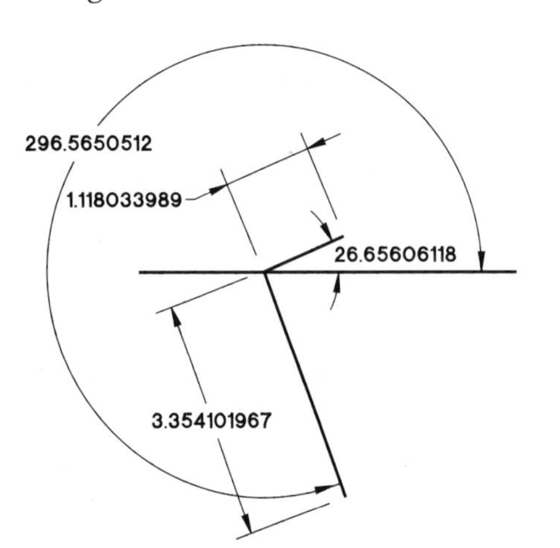

Figure 17–47 *Specified directions*

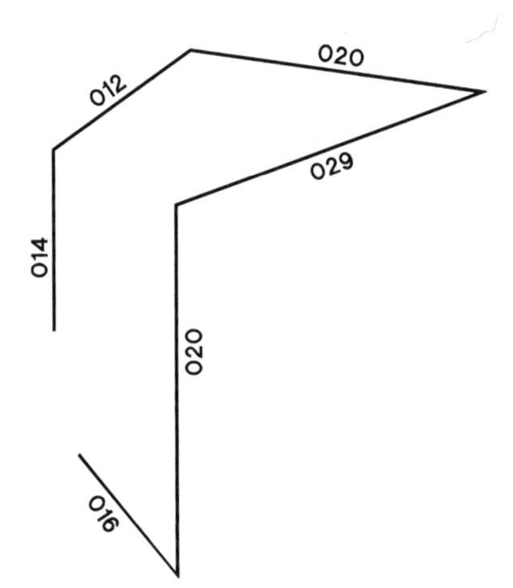

Figure 17–48 *Distance-direction vector specifying code*

```
*200,7,ODDITY
014,012,020,029,02C,016,0
```

To draw the shape named "oddity", you would first load the shape file that contains the definition and then use the SHAPE command as follows:

Command: **shape**
Name (or ?): **oddity**
Starting point: *(specify a point)*
Height <default>: *(specify a scale factor)*
Rotation angle <default>: *(specify a rotation angle)*

Note: An alternative to the standard DISTANCE-DIRECTION codes is to use codes 8 and 9 to move the pen by paired (delta-*X*,delta-*Y*) ordinate displacements.This is explained in the following section on "Special Codes."

SPECIAL CODES

Special codes can be written in decimal or hexadecimal. You can specify a special code as 0 through 16 or as 000 through 00E. A three-character defbyte with two leading zeros will be interpreted as a hexadecimal special code. A code 10 is a special code in decimal. However, 010 is equivalent to decimal 16. But more important, it will be interpreted by AutoCAD as a DISTANCE-DIRECTION code with a vector length of 1 and a direction of 0. The hexadecimal equivalent to 10 is 00A. The code functions are as follows.

Code 0: End of Shape The end of each separate shape definition must be marked with the code 0.

Codes 1 and 2: Pen Up and Down The "PEN DOWN" (or DRAW) mode is on at the beginning of each shape. Code 2 turns the DRAW mode off or lifts the pen. This permits moving the pen without drawing. Code 1 turns the DRAW mode on.

Note the relationship between the insertion point specified during the SHAPE command and where you wish the object and its primitives to be located. If you wish for AutoCAD to begin drawing a primitive in the shape at a point remote from the insertion point, then you must lift the pen with a code 2 and move the pen (with the proper codes) and then lower the pen with a code 1. Movement of the pen (directed by other codes) after a "PEN DOWN" code 1 is what causes AutoCAD to draw primitives in a shape.

Codes 3 and 4: Scale Factors Individual (and groups of) primitives within a shape can be increased or decreased in size by integer factors as follows: Code 3 tells AutoCAD to divide the subsequent vectors by the number that immediately follows the code 3. Code 4 tells AutoCAD to multiply the subsequent vectors by the number that immediately follows the code 4.

Scale factors are cumulative. The advantage of this is that you can specify a scale factor that is the quotient of two integers. A two-thirds scale factor can be achieved by a code 4 followed by a factor of 2 followed by a code 3 followed by a factor of 3. But the effects of scale factor codes must be reversed when they are no longer needed. They do not go away by themselves. Therefore, at the end of the definition (or when you wish to return to normal or other scaling within the definition), the scale factor must be countered. For example, when you wish to return to the normal scale from a two-thirds scale, you must use code 3 followed by a factor of 3 followed by a code 4 followed by a factor of 2. There is no law that states you must always return to normal from a scaled mode. You can, with codes 3 and 4 and the correct factors, change from a two-thirds scale to a one-third scale for drawing additional primitives within the shape. You should *always,* however, return to the normal scale at the end of the definition. A scale factor in effect at the end of one shape will carry over to the next shape.

Codes 5 and 6: Saving and Recalling Locations Each location in a SHAPE definition is specified relative to a previous location. However, once the pen is at a particular location, you can store that location for later use within that SHAPE definition before moving on. This is handy when an object has several primitives starting or ending at the same location. For example, a wheel with spokes would be easier to define by using code 5 to store the center location, drawing a spoke, and then using code 6 to return to the center.

Storing and recalling locations are known as *pushing* and *popping* them, respectively, in a stack. The stack storage is limited to four locations at any one time. The order in which they are popped is the reverse of the order in which they were pushed. Every location pushed must be popped.

More pushes than pops will result in the following error message:

```
Position stack overflow in shape nnn
```

More pops than pushes will result in the following error message:

```
Position stack underflow in shape nnn
```

Code 7: Subshape One shape in an .SHP/.SHX file can be included in the definition of another shape in the same file by using the code 7 followed by the inserted shape's number.

Codes 8 and 9: *X-Y* Displacements Normal vector lengths range from 1 to 15 and can be drawn in one of the 16 standard directions unless you use a code 8 or code 9 to specify *X-Y* displacements. A code 8 tells AutoCAD to use the next two bytes as the *X* and *Y* displacements, respectively. For example, 8, (7,–8) tells AutoCAD to move the pen a distance that is 7 in the *X* direction and *8* in the Y direction. The parentheses are optional, for viewing effects only. After the displacement bytes, specifications revert to normal.

Code 9 Code 9 tells AutoCAD to use all following pairs of bytes as *X-Y* displacements until terminated by a pair of zeros. For example; 9,(7,–8),(14,9), (–17,3),(0,0) tells AutoCAD to use the three pairs of values for displacements for the current mode and then revert to normal after the (0,0) pair.

Code 00A: Octant Arc Code 00A (or 10) tells AutoCAD to use the next two bytes to define an arc. It is referred to as an octant (an increment of 45 degrees) arc. Octant arcs start and end on octant boundaries. Figure 17–49 shows the code numbers for the octants. The specification is written in the following format:

```
10, radius. (-)OSC
```

The radius may range from 1 to 255. The second byte begins with zero and specifies the direction by its sign (clockwise if negative, counterclockwise otherwise), the starting octant (S), and the number of octants it spans (C), which may be written as 0 to 7, with 0 being 8 (a full circle). Figure 17–50 shows an arc drawn with the following codes:

```
10,(2,-043)
```

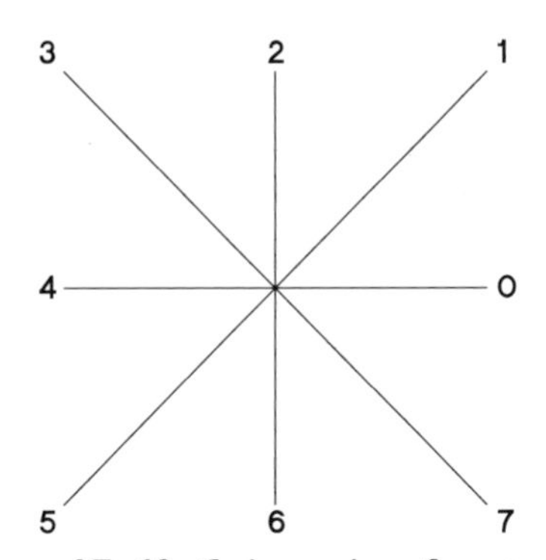

Figure 17–49 *Code numbers for octants*

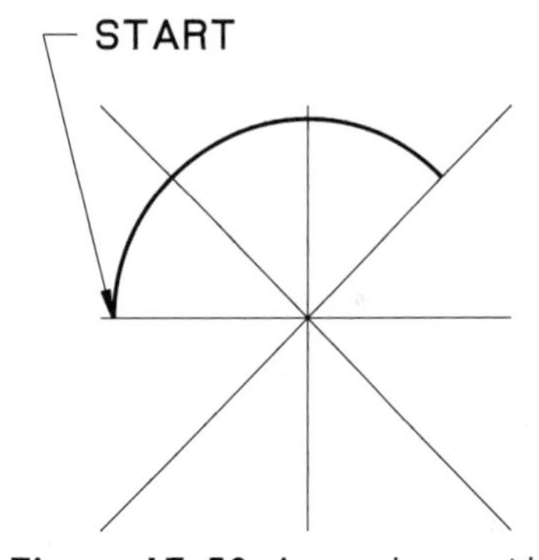

Figure 17–50 *An arc drawn with code 10,(2,–043)*

The arc has a radius of 2, begins at octant arc 4, and turns 135 degrees (3 octants) clockwise.

Code 00B: Fractional Arc Code 00B (11) can be used to specify an arc that begins and ends at points other than the octants. The definition is written as follows:

```
11,start-offset, end-offset, high-radius, low-radius, (-)OSC
```

Start and end offsets specify how far from an octant the arc starts and ends. The high-radius, if not zero, specifies a radius greater than 255. The low-radius is specified in the same manner as the radius in a code 10 arc, as are the starting octant and octants covering specifications in the last byte. The presence of the negative also signifies a clockwise direction.

The units of offset from an octant are a fraction of 1 degree times 45 divided by 256, or approximately .17578125 degrees. For example, if you wish to specify the starting value near 60 degrees, the equation would be:

```
offset = (60-45)*(256/45) = 85.333333
```

So the specification value would be 85.

To end the arc at 102 degrees, the equation would be:

```
offset = (102-90)*(256/45) = 68.2666667
```

So the specification value would be 68.

To draw an arc with a radius of 2 that starts near 60 degrees and ends near 102 degrees, the specifications would be as follows:

```
11,(85,68,0,2,012)
```

The last byte (012) specifies the starting octant to be 1 (45 degrees) and the ending octant to be 2 (90 degrees).

Codes 00C and 00D: Bulge-Specified Arc Codes 00C and 00D (12 and 13) are used to specify arcs in a different manner from octant codes. Codes 00C and 00D call out bulge factors to be applied to a vector displacement. The effect of using code 00C or 00D involves specifying the endpoints of a flexible line by the *X-Y* displacement method and then specifying the bulge. The bulge determines the distance from the straight line between the endpoints and the extreme point on the arc. The bulge can range from −127 to 127. The maximum/minimum values (127 or −127) define a 180-degree arc (half circle). Smaller values define proportionately smaller-degree arcs. That is, an arc specified some value, say x, will be x times 180 divided by 127 degrees. A bulge value of zero will define a straight line.

Code 00C Precedes a single-bulge-defined arcs; 00D precedes multiple arcs. This is similar to the way codes 008 and 009 work on *X-Y* displacement lines. Code 00D, like 009, must be terminated by a 0,0 byte pair. You can specify a series of bulge arcs and lines without exiting the code 00D by using the zero bulge value for the lines.

Code 00E: Flag Vertical Text Command Code 00E (14) is only for dual-orientation text font descriptions, where a font might be used in either horizontal or vertical orientations. When code 00E is encountered in the SHAPE definition, the next code will be ignored if the text is horizontal.

TEXT FONTS

Text fonts are special SHAPE files written for use with AutoCAD TEXT drawing commands. The shape numbers should correspond to ASCII codes for characters. Table 17–1 shows the ASCII codes. Codes 1 through 31 are reserved for special control characters. Only code 10 (line feed) is used in AutoCAD. In order to be used as a font, the file must include a special shape number, 0, to describe the font. Its format is as follows:

```
*0,4,fontname
above, below, modes, 0
```

Note: Codes 1 to 31 are for control characters, only one of which is used in AutoCAD text fonts.

"Above" specifies the number of vector lengths that uppercase letters extend above the baseline, and "below" specifies the number of vector lengths that lowercase letters extend below the baseline. A modes byte value of zero (0) defines a horizontal (normal) mode, and a value of two (2) defines dual-orientation (horizontal or vertical). A value of 2 must be present in order for the special code 00E (14) to operate.

Standard AutoCAD fonts include special shape numbers 127, 128, and 129 for the degrees symbol, plus/minus symbol, and diameter dimensioning symbol, respectively.

The definition of a character from the TXT.SHP file is as follows:

```
*65,21,uca
2,14,8,(-2,-6),1,024,043,04D,02C,2,047,1,040,2,02E,14,8,(-4,
  -3),0
```

Note that the number 65 corresponds to the ASCII character that is an uppercase "A." The name "uca" (for uppercase a) is in lowercase to avoid taking up memory. As an exercise, you can follow the defbytes to see how the character is drawn. The given character definition starts by lifting the pen. A font containing the alphanumeric characters must take into consideration the spaces between characters. This is done by having similar starting and stopping points based on each character's particular width.

Table 17–1 *ASCII Codes for Text Fonts*

Code	Character	Code	Character	Code	Character
32	space	64	@	96	‘ left apostrophe
33	!	65	A	97	a
34	“ double quote	66	B	98	b
35	#	67	C	99	c
36	$	68	D	100	d
37	%	69	E	101	e
38	&	70	F	102	f
39	‘ apostrophe	71	G	103	g
40	(	72	H	104	h
41	)	73	I	105	i
42	*	74	J	106	j
43	+	75	K	107	k
44	, comma	76	L	108	l
45	- hyphen	77	M	109	m
46	. period	78	N	110	n
47	/	79	O	111	o
48	0	80	P	112	p
49	1	81	Q	113	q
50	2	82	R	114	r
51	3	83	S	115	s
52	4	84	T	116	t
53	5	85	U	117	u
54	6	86	V	118	v
55	7	87	W	119	w
56	8	88	X	120	x
57	9	89	Y	121	y
58	: colon	90	Z	122	z
59	; semicolon	91	[	123	{
60	<	92	\ backslash	124	| vertical bar
61	=	93	]	125	}
62	>	94	^ caret	126	~ tilde
63	?	95	_ underscore		

CUSTOM LINETYPES

Linetype definitions are stored in files with an .LIN extension. Approximately 40 standard linetype definitions are stored for use in the acad.lin file. The definitions are in ASCII format and can be edited, or you can add new ones of your own by using either a text editor in the non-document mode or the Create option of the LINETYPE command. Or you can save new or existing linetype definitions in another filename.lin file.

Simple linetypes consist of series of dashes, dots, and spaces. Their definitions are considered the in-line pen-up/pen-down type. Complex linetypes have repeating "out-of-line" objects, such as text and shapes, along with the optional in-line dashes, dots, and spaces. These are used in mapping/surveying drawings for such things as topography lines, fences, utilities, and many other descriptive lines. Instrumentation/control drawings also use many lines with repeating shapes to indicate graphically the purpose of each line.

Each linetype definition in a file comprises two lines. The first line must begin with an asterisk, followed by the linetype name and an optional description, in the following format:

```
*ltname,description
```

The second line gives the alignment and description by using proper codes and symbols, in the following format:

```
alignment,patdesc-1,patdesc-2,...
```

A simple linetype definition for two dashes and a dot, called DDD, could be written as follows:

```
*DDD,___ ___ . ___ ___ . ___ ___ .
A,.75,-.5,.75,-.5,0,-.5
```

The linetype name is DDD. A graphic description of underscores, spaces, and periods follow. The dashes are given as .75 in length (positive for pen down) separated by spaces –.5 in length (negative for pen up), with the 0 specifying a dot. No character other than the A should be entered for the alignment; it is the only one applicable at this time. This type of alignment causes the lines to begin and end with dashes (except for linetypes with dots only).

The complex linetype definitions include a descriptor (enclosed in square brackets) in addition to the alignment and dash/dot/space specification. A shape descriptor will include the shape name, shape file, and optional transform specification, as follows:

```
[shapename,filename,transform]
```

A text descriptor will include the actual text string (in quotes), the text style, and optional transform specification, as follows:

```
["string",textstyle,transform]
```

Transform specifications (if included) can be one or more of the following:

A=##	absolute rotation
R=##	relative rotation
S=##	scale

```
X=##     X offset
Y=##     Y offset
```

The ## for rotation is in decimal degrees (plus or minus); for scale and offset it is in decimal units.

The following example of an embedded shape in a line for an instrument air line (with repeating circles) could be written as follows:

```
*INSTRAIR, ___ [CIRC] ____ [CIRC] ____
A,2.0,-.5,[CIRC,ctrls.shx],-.5
```

If the ctrls.shx file contains a proper shape description of the desired circle, it will be repeated in the broken line (with spaces on each side) when applied as the INSTRAIR linetype. If the scale of the circle needed to be doubled in order to have the proper appearance, it could be written as follows:

```
*INSTRAIR, ___ [CIRC] ____ [CIRC] ____
A,2.0,-.5,[CIRC,ctrls.shx,S=2],-.5
```

The following example of an embedded text string in a line for a storm sewer (with repeating SS's) could be written as follows:

```
*STRMSWR,____ SS ____ SS ____
A,3.0,-1.0,["SS",simplex,S=1,R=0,X=0,Y=-0.125],-1.0
```

EXPRESS TOOLS

In addition to the customization tools, AutoCAD has a set of tools under the category of Express tools that will make your job easier in using AutoCAD. You can install AutoCAD Express Tools from the install program provided in the AutoCAD 2004 CD. Before you install the Express tools, AutoCAD must already be installed on your system.

See Appendix I for a list of commands available with a brief description.

CUSTOMIZING AND PROGRAMMING LANGUAGE

There are various other topics with respect to the customization of AutoCAD, some of which are beyond the scope of this book. AutoCAD provides a programming language called AutoLISP (refer to Chapter 18). AutoLISP is a structured programming language similar in a number of ways to other programming languages. A compiled language (AutoCAD supports a number of these, too) is first converted into object code (this is called compiling) or machine language and then linked with various other compiled object code modules (this is called linking) to form an executable file. C/C++ and ARx (AutoCAD Runtime Extension) are examples of compiled programming languages.

REVIEW QUESTIONS – B

1. The purpose of the ACAD.PGP file is to:
   a. allow other programs to be accessed while editing a drawing
   b. enable shape files to be complied
   c. store system configurations
   d. serve as a "file manager" for system variables
   e. none of the above

2. The standard AutoCAD screen menu:
   a. is stored in a file named ACAD.MNU
   b. can be viewed using the DOS TYPE command
   c. contains the screen menu items found in the AutoCAD screen menus
   d. all of the above

3. In an AutoCAD menu file, a semicolon contained in a menu item will tell the computer to:
   a. prompt the user for input
   b. press ENTER
   c. press CTRL
   d. ignore the rest of the line as a comment
   e. none of the above

4. AutoCAD menu files are stored with what type of file extension?
   a. DWG
   b. DXF
   c. MNU
   d. MEN
   e. none of the above

5. When developing screen menus, the information you would like to see displayed in the screen menus should be:
   a. typed in uppercase letters only
   b. enclosed with square brackets "[]"
   c. longer than four characters, but shorter than ten characters
   d. all of the above

6. How many characters between brackets will display in a screen menu?
   a. 2
   b. 4
   c. 6
   d. 8
   e. 10

7. What symbology signifies the heading of a menu device such as a digitizer or table area?

a. ****
b. ***
c. **S
d. none of the above

8. What is the purpose of the backslash "\" in a menu line?

a. to pause for user input
b. to terminate a command
c. to press E
d. none of the above

9. What file defines external commands and their parameters?

a. ACAD.DWK
b. ACAD.EXT
c. ACAD.PGP
d. ACAD.LSP
e. ACAD.CMD

10. The system variable used in conjunction with DIESEL to modify the contents of the status line is:

a. STATUSLINE
b. MACRO
c. MODEMACRO
d. MODESTATUS

11. AutoCAD allows you to create your own icons for use in toolbar menus.

a. True
b. False

12. To specify a cascade menu on a pulldown menu, you should use:

a. \
b. >
c. ->
d. #>

13. Command aliases are stored in what file?

a. ACAD.INI
b. ACAD.ALS
c. ACAD.PGP
d. ACAD.MNU
e. none of the above

14. If you define a help string for a menu entry, it will display:

a. on the status line
b. as a small label by the icon
c. if you invoke the help command
d. at the command prompt
e. none of the above

15. When creating a partial menu, the one entry which will be included is:

a. PARTIAL
b. MENUGROUP
c. SECTION
d. MENU
e. ENTRYNUMBER

16. Which of the following would not be a valid line in a definition of a custom crosshatching pattern?

 a. 0,0,1,0,1
 b. 90,1,0,1,0
 c. 180,0,0,1,1,–1,1
 d. 270,1,2,0,1,1,–1,0,–1
 e. 360,1,2,1,–1

17. The first line of a custom crosshatching definition always begins with:

 a. *
 b. **
 c. ***
 d. !
 e. !!

18. To define a custom linetype with three elements—a 1 unit dash, a 0.5 unit dash, and a dot, all separated by 0.25 unit spaces—what would the definition look like?

 a. A,1,0.25,0.5,0.25,0
 b. A,1,.25,0.5,0.25,0,0.25
 c. A,1,–0.25,0.5,–0.25,0
 d. A,1,–0.25,0.5,–0.25,0,–0.25
 e. A,1,0.5,0,–0.25

19. The bit code which specifies a direction of 12 o'clock in a shape file is:

 a. 0
 b. 4
 c. 8
 d. 12
 e. C

20. When specifying the name of a shape character, in order to conserve memory you should:

 a. use upper-case
 b. use lower-case
 c. preface the name with an *
 d. use a short name
 e. it does not matter, all shapes require the same amount of memory regardless of their names

21. A macro permits the user to save and invoke a series of keystrokes with a combination of only one or two key strokes?

 a. True
 b. False

22. Which of the following can be customized in AutoCAD?

 a. aliases
 b. linetypes
 c. shapes
 d. hatch patterns
 e. all of the above

23. Command aliases can include command options?

    a. True

    b. False

24. Which of the following is the correct format to define an alias using *L* for the LINE command?

    a. Line = L

    b. L = <line>

    c. L, *line

    d. L = *line

25. Which of the following is not a valid AutoCAD menu type?

    a. Pull-down

    b. Screen

    c. Tablet

    d. Tool tip

    e. all of the above

26. When writing a script file command aliases cannot be used.

    a. True

    b. False

27. Which of the following is not a valid AutoCAD menu file extension?

    a. .mnu

    b. .mno

    c. .mns

    d. .mnc

    e. none of the above

28. How many pulldown menus does AutoCAD have available to use?

    a. 12

    b. 16

    c. 24

    d. 32

    e. unlimited

29. Which of the following characters, when included in a macro will repeat the command until the user cancels?

    a. :

    b. ;

    c. +

    d. *

    e. >

30. Which of the following characters, when placed in a macro will underscore the letter that follows and permits menu selection from the keyboard by combining the ALT key and a specified key?

    a. ~

    b. &

    c. #

    d. @

    e. none of the above

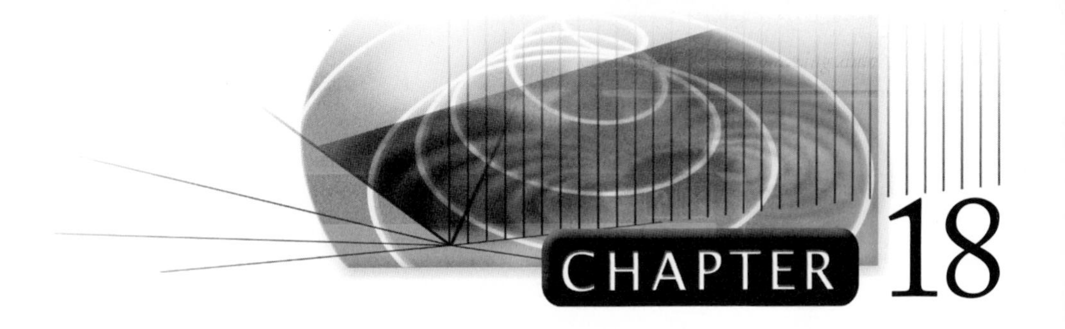

The Tablet and Digitizing

INTRODUCTION

The AutoCAD program consists of several major components in the form of program files, along with many other supporting files. One major part of the program, the menu files (in the forms of *filename*.MNU, *filename*.MNR, *filename*.MNS, *filename*.MNL, and *filename*.MNC), determine how various devices work with the program. The devices controlled by the menu include the buttons on the pointing device (mouse or tablet puck), the toolbars, pulldown menus, function box keys (not commonly used), and tablet. This chapter covers configuring and using the tablet menu.

First, using the tablet part of any menu requires that the AutoCAD program be installed and configured for the particular make and model of digitizing tablet that is properly connected to the computer.

Note: The installation configuration permits using a mouse as well as a digitizing tablet. But because a mouse and a tablet are both pointing devices, you must choose one or the other at any one time. This pre-startup configuration is not to be confused with configuring the installed tablet with the TABLET command while in AutoCAD.

Second, a preprinted template (paperlike overlay to place on the digitizer surface) with up to four rectangular menu areas that has properly arranged columns and rows of pick areas within each menu area should be used. Each pick area corresponds to a command or line of programming in the *filename*.MNU menu file in effect. It is also necessary to set aside a screen area on the tablet if you wish to control the screen cursor with the tablet's puck, as shown in Figure 18–1.

After completing this chapter, you will be able to do the following:

- Configure the tablet menu
- Calibrate the tablet for digitizing

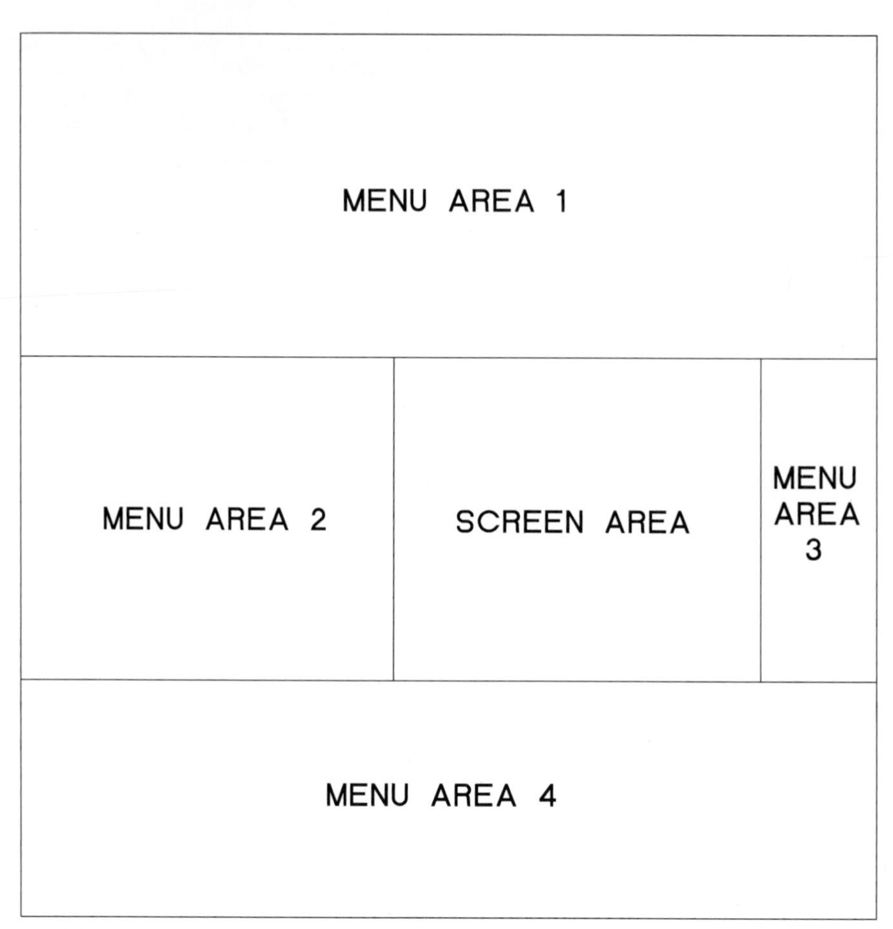

Figure 18–1 *The four screen areas of the overlay*

TABLET OPERATION

When a *filename*.MNU file, a digitizing tablet, and an overlay have been installed and properly set up to work together, you can use the digitizer's attached puck on the tablet surface to achieve the following results.

Normal operation The most common use of a tablet is to allow the user to move the puck and press the pick button while pointing to one of the various commands or symbols on a preprinted overlay, and be able to invoke that particular command or initiate a program (perhaps written by that user) that will draw the object(s) that that symbol represents. In addition, when the puck is moved within the overlay's designated screen area, the screen cursor will mimic the puck movement, thereby permitting the user to specify points or select objects on the screen.

Mouse movement Some tablets have an option that causes the puck to emulate mouse-type movement rather than absolute movement. Mouse emulation means that if you pick the puck up off of the tablet surface and move it to another place in the screen area, the screen cursor does not move. The cursor moves only with puck movement while it is on the tablet and in the screen area. Absolute (normal) tablet-puck operation means that, once configured, each point in the tablet's screen area corresponds to only one point on the screen. So, while in the normal mode, if you pick up the puck and put it down at another point in the tablet screen area, the cursor will immediately move to the screen's corresponding point.

Paper copying By switching Tablet mode to ON, you can cause points on the tablet to correspond to drawing coordinates rather than to screen pixel locations as it does in the normal or mouse operations just described. This allows you to fix a drawing (like a map) on the tablet surface, select two points on the map, and specify their coordinate locations on the map, after which the puck movement around the map will cause screen cursor movement to correspond to the same coordinates in the computer-generated drawing. Options and precautions for using this feature (referred to as *digitizing*) are discussed later in this chapter.

TABLET CONFIGURATION

The intended procedure is to have a preprinted template (the overlay) arranged on a sheet that can be fixed to the tablet. Tablet menu areas can then be configured to coincide with the template. Although you could try to configure a bare tablet, it would be difficult to select the required points for rectangular menu areas and also impractical to try and place a template on the tablet after it was configured in such a manner. However, if a tablet has been configured for one template, you can use another template in the same location without reconfiguring as long as the areas are the same. One benefit of this is being able to change from one set of icons/symbols or commands to another set without having to configure again. However, a change in the menu must be made in order to accommodate changes in the template, even if the configuration is the same.

The ACAD.MNC (compiled from ACAD.MNU) menu file supports a multi-button pointing device and the tablet overlay that is provided with the AutoCAD program package. That overlay is approximately 11" x 11" and has four areas for selecting icon/commands and a screen area, as shown in Figure 18–1. The pointing device (the puck furnished with every tablet) usually has three or more buttons (the menu supports up to a 10-button puck or mouse), and the cursor movement on the screen mimics the puck's movement in the tablet's configured screen area.

CUSTOM MENUS

Chapter 17 describes how to customize a menu file. Most of the explanations and examples refer to the screen menu, primarily because of its complexity. The same principles of customizing the tablet portion of a menu can be applied.

TABLET COMMAND

The TABLET command is used to switch between digitizing paper drawings and normal command/icon/screen area selections on a configured overlay. The TABLET command is also used to calibrate a paper drawing for digitizing or to configure the overlay to suit the current menu.

Invoke the TABLET command:

Tools menu	Choose Tablet
Command: prompt	**tablet** (ENTER)

AutoCAD prompts:

Command: **tablet**
Enter an option [ON/OFF/CAL/CFG]: *(right-click and select one of the available options from the shortcut menu)*

TABLET CONFIGURATION

The CFG option is used to set up the individual tablet menu areas and the screen pointing area. At this time a preprinted overlay should have been fixed to the tablet. Its menu areas should suit the menu you wish to use. The sequence of prompts is as follows:

Command: **tablet**
Option (ON/OFF/CAL/CFG): **cfg**
Enter number of tablet menus desired (0-4) <0>:

Select the number of individual menu areas desired (with a limit of 4). The next prompt asks:

Digitize upper left corner of menu area n:
Digitize lower left corner of menu area n:
Digitize lower right corner of menu area n:

The "n" refers to tablet menu areas of the corresponding tablet number in the menu. If the three corners you digitize do not form a right angle (90 degrees), you will be prompted to try again. Individual areas may be skewed on the tablet and with respect to each other, but such an arrangement usually does not provide the most efficient use of total tablet space. Tablet areas should not overlap.

The next prompts are:

> Enter the number of columns for menu area n:
> Enter the number of rows for menu area n:

Enter the numbers from the keyboard. The area will be subdivided into equal rectangles determined by the row and column values you have entered. If the values you enter do not correspond to the overlay row/column values, the results will be unpredictable when you try to use the tablet. Remember also that the overlay must suit the menu being used. After the menus areas have been specified, AutoCAD prompts:

Do you want to respecify the Fixed Screen Pointing Area? [Yes/No] <N>:

If you respond with **y**, AutoCAD prompts:

> Digitize lower left corner of Fixed Screen pointing area: *(specify a point with the pick button)*

> Digitize upper right corner of Fixed Screen pointing area: *(specify a point with the pick button)*

AutoCAD then prompts:

> Do you want to specify the Floating Screen Pointing area? [Yes/No] <N>:

If you respond with **y**, AutoCAD prompts:

> Do you want the Floating Screen Area to be the same size as the Fixed Screen Pointing Area? [Yes/No] <Y>:

If you respond with **n**, AutoCAD prompts:

> Digitize lower-left corner of the Floating Screen pointing area: *(specify a point with the pick button)*

> Digitize upper-right corner of the Floating Screen pointing area: *(specify a point with the pick button)*

In order to toggle the Floating Screen Area ON and OFF, press F12.

AutoCAD prompts:

> The F12 Key will toggle the Floating Screen Area ON and OFF. Would you also like to specify a button to toggle the Floating Screen Area? [Yes/No] <N>:

If you respond with y, AutoCAD prompts:

```
Press any non-pick button on the digitizer puck that you wish to designate
   as the toggle for the Floating Screen Area.
```

The standard AutoCAD overlay is installed as follows:

```
Command: tablet
Option (ON/OFF/CAL/CFG): cfg
Enter number of tablet menus desired (0-4) <default>: 4
Do you want to realign tablet menu areas? <N>: y (if required)
```

At this time, digitize areas 1 through 4, as shown in Figure 18–2. The values for columns and rows must be entered as follows:

MENU AREA	COLUMN	ROW
1	25	9
2	11	9
3	9	7
4	25	7

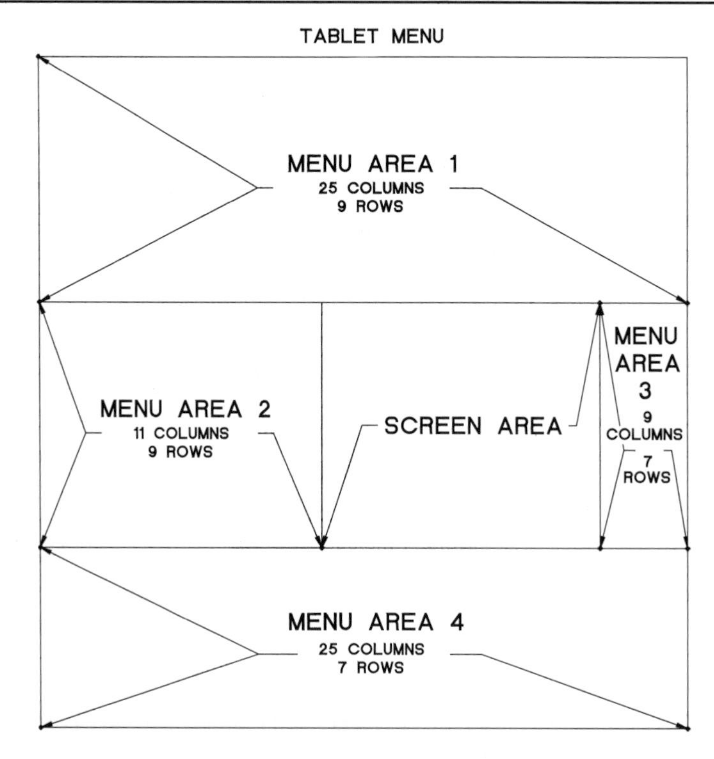

Figure 18–2 *Digitizing the screen areas of the tablet menu*

These values are for the ACAD.MNC menu. Other menus may vary.

ON/OFF OPTION

The default setting of the Tablet mode is OFF. The OFF setting does not incapacitate the tablet as you might think; rather, it means that you are not going to use the tablet for digitizing (making copies of paper drawings). With the Tablet mode set to OFF you can use the tablet to select command/icons in the areas programmed accordingly. You may also use the puck in the screen area of the tablet to control the screen cursor.

In order to digitize paper drawings, you must respond to the prompt as follows:

> Command: **tablet**
> Option (ON/OFF/CAL/CFG): **on**

Most systems have a toggle key to switch the Tablet mode ON and OFF. With many PCs, the toggle is either F10 or CTRL + T.

CALIBRATION

If the tablet has been calibrated already, the last calibration coordinates will still be in effect. If not, or if you wish to change the calibration (necessary when you move the paper drawing on the tablet), you can respond as follows:

> Option (ON/OFF/CAL/CFG): **cal**
> Digitize point #1: *(digitize the first known point)*

The point you select on the paper drawing must be one whose coordinates you know. The next prompt asks you to enter the actual paper drawing coordinates of the point you just digitized:

> Enter coordinates point for #1: *(enter those known coordinates)*

You are then prompted to digitize and specify coordinates for the second known point:

> Digitize point for #2: *(digitize the second known point)*
> Enter coordinates point #2: *(enter those known coordinates)*
> Digitize point #3 (or press ENTER to end): *(digitize the third known point or press* ENTER*)*

An example of a drawing that might be digitized is a map, as shown in Figure 18–3.

If, for example, the map in Figure 18–3 has been printed on an 11" x 17" sheet and you wish to digitize it on a 12" x 12" digitizer, you can overlay and digitize on one-half of the map at a time. You may use the coordinates 10560,2640 and 7920,5280 for two calibrating points. But because *X* coordinates increase toward the left, you must consider them as negative values in order to make them increase to the right. Therefore, in calibrating the map, you may use coordinates –10560,2640 and –7920,5280 to calibrate the first half and coordinates –7920,2640 and –5280,5280 for the second half.

The points on the paper should be selected so that the *X* values increase toward the right and the *Y* values increase upward.

Once calibration has been initiated in a particular space (model or paper), turning on the Tablet mode must be done while in that particular space.

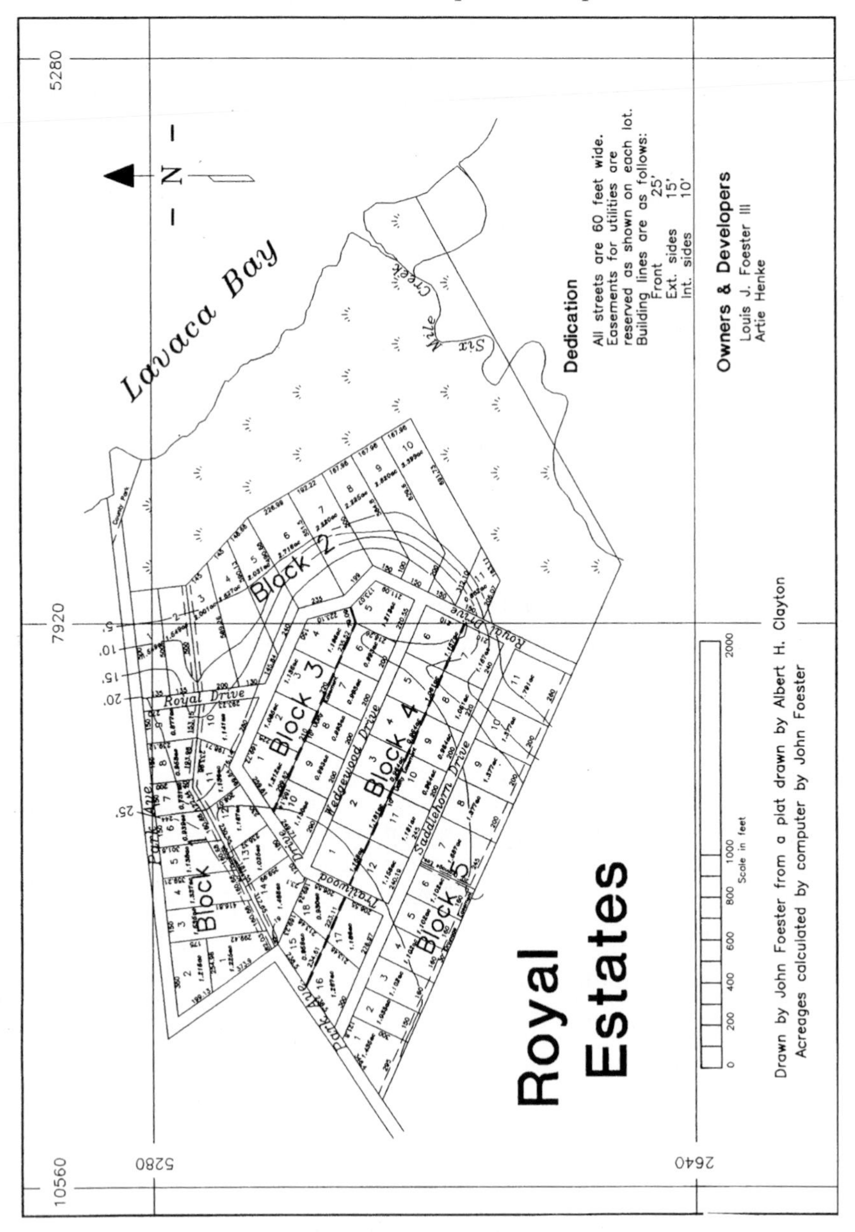

Figure 18–3 *An example of a drawing digitized as a map*

TRANSFORMATION OPTIONS

Tablet calibration can be done by one of several transformation methods. These include Orthogonal, Affine, Projective, and Multiple-Point. The method you choose may depend on the condition of the map/drawing that is to be digitized and the desired accuracy.

If three or more points are entered, AutoCAD uses each of the three transformation types (Orthogonal, Affine, and Projective) to compute the transformation to determine which best fits the calibration points. Entering more than four points can cause computing the best-fitting projective transformation to take a long time. The process can be canceled by pressing ESC. A table is displayed when the computations are complete showing the number of calibration points with a column for each transformation type.

If no failures of projection transformation have occurred, you are prompted to choose a transformation type:

> Enter transformation type [Orthogonal/Affine/Projective/Repeat table]
> <Repeat>: *(Enter an option or press* ENTER)

This prompt includes only transformation types for which the outcome was Success, Exact, or Canceled.

Orthogonal

This option involves two points. It results in uniform scaling and rotation. The translation is arbitrary. Orthogonal (two-point) translation is most suitable for tracing paper drawings that are dimensionally and rotationally accurate (right angles are not skew). It is advisable to use this option for long, narrow applications (a pipeline, for example).

Affine

This option involves three points and can be applied to a paper drawing with scale factors, rotation, and right angle representations that are not to an acceptable accuracy in two dimensions. This can be applied to drawings with parallel lines that are represented parallel, but with an *X* direction scale/*Y* direction scale differential that is out of tolerance. Right angles may not be represented by right angles.

Whether or not you should use the Affine option depends on whether or not those lines that should be parallel are represented by parallel lines on the paper drawing. You can check the report displayed (in the form of a table) when you have digitized at least three points. If the RMS error (described in this section) is small, then the Affine option should be acceptable.

Projective

This option involves four points, to simulate a translation comparable to a perspective in which points from one plane converge while passing through another plane (skew to the first) to one point of view. This option is applicable to copying paper sheets with irregularities that differ from one area to another (known as rubber-sheeting) and parallel lines that are not always represented as parallel. However, lines do project as lines.

Repeat Table

Redisplays the computed table, which rates the transformation types.

THE CALIBRATION TABLE

If you use three or more points during calibration, AutoCAD computes the range or error (if any) and reports the results and displays information designed to help you determine if the paper copying process is acceptable. An example of a calibration table follows:

Four Calibration Points

Transformation type:	Orthogonal	Affine	Projective
Outcome of fit:	Success	Success	Exact
RMS Error:	143'-6.2"	73'-4.1"	
Standard deviation:	62'-2.7"	1'-7.8"	
Largest residual:	193'-1"	74'-1"	
At point:	2	3	
Second largest residual:	177'-9"	7'-7"	
At point:	3	2	

OUTCOME OF FIT

Cancelled Cancelled occurs only with projective transformations. It indicates the fit has been cancelled.

Exact The number of points used was exactly correct and the transformation defined from them was valid.

Failure Points selected, though the correct number, were probably collinear or coincidental.

Impossible Insufficient points were selected for the transformation type under which "Impossible" is reported.

Success The transformation defined was valid and more points were used than were required.

RMS Error

If the transformation is reported as a "Success," then the RMS (root mean square) error is reported. This is the square root of the average of the squares of the distances (called residuals) of each selected point from their respective targets.

Standard Deviation

This reports the standard deviation of all residuals (the distance each point misses its target).

Point(s)/Residual(s)

The two points whose residuals (see "RMS Error" and "Standard Deviation") are largest and second largest are reported along with their respective residual values.

TABLET MODE AND SKETCHING

Sketching while in Tablet mode operates similarly to sketching with a mouse or on a tablet with the Tablet mode off. The difference is that the entire tablet surface is used for digitizing while the Tablet mode is on, making the maximum area available for tracing but making the pulldown menus inaccessible.

EDITING SKETCHES

Once sketched lines have been recorded and the SKETCH command has been terminated, you can use regular editing commands (like COPY, MOVE, ERASE) to edit the individual line segments or sketched polylines (discussed next) just as though they had been drawn by the LINE or PLINE command. In the case of sketched polylines, the PEDIT command can be used for editing.

SKETCHING IN POLYLINES

You can cause AutoCAD to make the created sketch segments into polylines instead of lines by setting the SKPOLY system variable to a nonzero value.

LINETYPES IN SKETCHING

You should use the Continuous linetype while sketching, whether with regular lines or polylines.

REVIEW QUESTIONS

1. What filename extension is used to store the menu command entries for a tablet menu?

   a. .TAB
   b. .MNU
   c. .CMD
   d. .PGP
   e. .INI

2. How many different tablet menu areas can be specified on a digitizing pad?

   a. 1
   b. 2
   c. 3
   d. 4
   e. 5

3. When aligning a tablet menu with the digitizing pad, how many points are required?

   a. 1
   b. 2
   c. 3
   d. 4
   e. 5

4. To toggle the digitizing pad between menu functions and paper copying, you press:

   a. F10
   b. F11
   c. F12
   d. CTRL + O
   e. none of the above

5. The minimum number of points required to calibrate a digitizing pad using an Affine calibration is:

   a. 1
   b. 2
   c. 3
   d. 4
   e. 5

6. If you have an isometric drawing on paper and wish to digitize an orthographic view of the top, what type of calibration would you use?

   a. 2 point (Orthogonal)
   b. 3 point (Affine)
   c. 4 point (Projective)
   d. cannot be done

7. If you have photograph of a building and wish to digitize the front elevation, what type of calibration would you use?

   a. 2 point (Orthogonal)

   b. 3 point (Affine)

   c. 4 point (Projective)

   d. cannot be done

8. What does RMS stand for in RMS error?

   a. Real Measure Statistic

   b. Root Mean Square

   c. Radical Motion Setting

   d. ReMainder Sum

   e. none of the above

9. If you select three points in a straight line to calibrate the digitizing pad, what "Outcome of Fit" will AutoCAD report for an Affine fit:

   a. Canceled

   b. Exact

   c. Failure

   d. Impossible

   e. Success

10. When using the SKETCH command, what system variable determines if lines segments are polylines?

    a. SKLINE

    b. SKPOLY

    c. SKTYPE

    d. LINESEG

    e. POLYGEN

11. What command will allow you to generate freehand lines when digitizing a paper drawing?

    a. DIGITIZE

    b. TABLET

    c. SKETCH

    d. FREEHAND

    e. DRAW

12. The option of the TABLET command that allows for the configuration of a tablet menu is:

    a. ON

    b. OFF

    c. CAL

    d. CFG

    e. MENU

13. The RMS error will always be lower (or equal) for a projective fit versus an orthogonal fit when 6 points are selected.

    a. True
    b. False

14. AutoCAD's default menu template has how many menu areas?

    a. 1
    b. 2
    c. 3
    d. 4
    e. 5

15. To toggle the floating screen area on and off press which of the following?

    a. F10
    b. F11
    c. F12
    d. CTRL + F
    e. none of the above

16. The three corners that are selected to define the tablet menu area must form a _______ degree angle.

    a. 30
    b. 45
    c. 60
    d. 90

17. Which of the following is used to toggle the tablet mode on and off?

    a. CTRL + B
    b. CTRL + M
    c. CTRL + T
    d. none of the above

18. Which of the following is not a tablet transformation method?

    a. Orthogonal
    b. Affine
    c. Polygonal
    d. Projective
    e. Multi-Point

19. The Projective transformation option allows copying paper with irregularities from one area to another. This type of copying is known as _______________.

    a. Rubber banding
    b. Rubber sheeting
    c. Rubber stamping
    d. Applique

APPENDIX a

Hardware Requirements

INTRODUCTION

The configuration of your CAD system is a combination of the hardware and software you have assembled to create your system. There are countless PC configurations available on the market. The goal for a new computer user should be to assemble a PC workstation that does not block future software and hardware upgrades. This section lists the essential hardware required to run AutoCAD 2004.

When you install AutoCAD 2004, it automatically configures itself to the various drivers already set up for the Windows operating system. The Options dialog box can be used to custom configure your installation, as explained in Chapter 13.

RECOMMENDED CONFIGURATION

The minimum configuration required for AutoCAD 2004 includes the following:

1. Intel® Pentium® III or later, with 800 MHz or faster processor, or compatible
2. Microsoft® Windows® XP (Professional, Home Edition, or Tablet PC Edition), Windows 2000, or Windows NT® 4.0 (SP6a or later). Microsoft® Internet Explorer 6.0 or higher
3. 256 MB RAM (recommended), 300 MB free disk space for installation
4. 1024x768 VGA with true color (minimum)
5. Peripherals: CD ROM drive, mouse/pointing device
   Optional Peripherals: Printer or plotter, digitizer

WINDOWS: THE OPERATING ENVIRONMENT

Windows NT 4.0, Windows 2000, or Windows XP is the name of the operating environment. The operating system provides the graphical user interface and device independence, and lets you run more than one program at a time.

The graphical user interface (or GUI, for short) consists of the windows, icons, scroll bars, menu bar, button bar, and other graphical elements you see on the Windows screen. The GUI lets you use the mouse to control most aspects of a software program, including selecting commands, resizing windows, clicking on icons, and drawing.

Device independence means that all software applications running under Windows use the same device drivers, such as for the display, the mouse, and the printer. AutoCAD uses the Windows 2000 Professional and Windows XP device driver.

The multitasking and task switching feature of Windows 2000 Professional and Windows XP operating system lets you be more productive. Multitasking means that Windows can run more than one program at a time, unlike DOS, which runs just one program at a time. Task switching means that you can quickly switch from task to task (or from one program to another).

INPUT DEVICES

AutoCAD supports several input device configurations. Data may be entered via the keyboard, the mouse, or a digitizing tablet.

KEYBOARD

The keyboard is one of the primary input methods. It can be used to enter commands and responses.

MOUSE

The mouse is used with the keyboard as a tracking device to move the crosshairs on the screen. This method is fast and far surpasses the keyboard for positioning the screen crosshairs.

The mouse is equipped with two or more buttons. The left button is the pick button; the other buttons can be programmed to perform any of the AutoCAD commands.

DIGITIZER

The digitizer is another means of input supported by AutoCAD, but it is not well supported by Windows. Hence, Autodesk provides a "mole" mode that lets the mouse and digitizer work at the same time in Windows while AutoCAD is running.

The digitizing tablet is a flat, sensitized electromechanical device that can recognize the location of the tablet cursor. The tablet cursor moves over the surface of the digitizing tablet, causing a corresponding movement of the screen cursor.

Another use of the digitizing tablet is to overlay a tablet menu, configured to enter commands when they are selected by placing the crosshairs of the puck at the corresponding digitizer coordinates. This process automates any AutoCAD command options that are represented on the menu, and eliminates the use of the keyboard except to enter an occasional command value.

The tablet cursor (puck) may have from 4 to 16 buttons. Except for the first button, the current menu can be programmed to cause the remaining buttons to perform any of the regular AutoCAD commands. The first button is always the pick button.

PLOTTERS AND PRINTERS

AutoCAD supports several types of output devices for producing hard copies of drawings. The most common devices are the thermal plotter, pen plotter, inkjet printer, and laser printer.

THERMAL PLOTTERS

The thermal plotter heats up the drawing paper, changing the color to black to create a hard copy of the drawing. The resolution of the output, which is measured in dots per inch (dpi), may range from a low-quality (100 dpi) product to medium-quality (400 dpi) product.

PEN PLOTTERS

The pen plotter draws continuous lines on the paper. The drawing pen is driven by vector commands that correspond to the *X* and *Y* coordinates on the drawing paper. Drawing media vary between mylar, vellum, and bond. The pens used in such plotters also vary in size and quality, thus affecting the drawing resolution. Resolution is typically between the equivalent of 1000 and 2000 dpi.

Pen plotters come in all shapes and sizes, and over the years pen plotters have improved in the areas of speed, accuracy, and price. The pen plotter is an established favorite for those who demand clear line and shape definition. Although prices vary on the different types of pen plotters, it is possible to get a good desktop pen plotter in the area of $2,000. Contact your dealer to find out if a particular model is supported.

INKJET PRINTERS

AutoCAD supports the ability to generate a hard copy of your drawing with an inkjet printer, either in black and white or in color. Inkjet printers form characters and graphics by squirting ink onto standard paper. Inkjet printers provide the most inexpensive means for color printing. They are generally used to provide rough draft hard copies for graphics and text data. They are an excellent and cost-effective way to produce check plots in A-, B-, and C-size.

LASER PRINTERS

Laser printers are the newest printing device on the CAD market. They deliver a low-cost, medium- to high-resolution hard copy that can be driven from a personal computer. Laser printers boast resolutions from 300 dpi to 1200 dpi and printing speeds of about 4 to 12 pages per minute (ppm) for text and 1 to 2 pages per minute for graphics.

As with any raster device, laser printers require a vector-to-raster conversion to generate most graphics. AutoCAD can plot design files on laser printers that support the HP LaserJet PCL and PostScript graphic output languages. The cost of laser printers ranges from under $1,000 to over $12,000, yet the increase in resolution makes the price a good investment for many serious CAD users.

APPENDIX b

Listing of AutoCAD Commands and Dimensioning Variables

Command Aliases	Explanation	Options	Toolbar	Pull-Down Menu
'About	Displays a dialog box with the AutoCAD version and serial numbers, a scrolling window with the text of the *acad.msg* file, and other information			Help
Acisin	Imports an ACIS file			Insert
Acisout	Exports AutoCAD solid objects to an ACIS file			Export
Adcclose	Closes AutoCAD DesignCenter			TYPE IN
Adcenter	Manages content		Standard	
Adcnavigate	Directs the Desktop in AutoCAD DesignCenter to the file name, directory location, or network path you specify			TYPE IN
Align	Moves and rotates objects to align with other objects			Modify
Ameconvert	Converts AME solid models to AutoCAD solid objects			TYPE IN
Aperture	Regulates the size of the object snap target box	Select (1–50 pixels) to increase or reduce size of box		TYPE IN
Appload	Loads AuotLISP, ADS, and ARX applications			Tools
Arc (A)	Draws an arc of any size	A Included angle C Center point D Starting direction E Endpoint L Length of chord R Radius S Start point ENTER Continues arc from endpoint of line or arc	Draw	Draw
Area	Calculates the area of a polygon, pline, or circle	A Sets add mode S Sets subtract mode O Calculates area of the circles, ellipses, splines, polylines, polygons, regions, and solids	Object Properties	Tools

Command Aliases	Explanation	Options	Toolbar	Pull-Down Menu
Array	Copies selected objects in circular or rectangular pattern	P Polar (circular) arrays about a center point R Rectangular arrays objects in horizontal rows and vertical columns	Modify I	Modify
Arx	Loads, unloads, acquaints you with ARX applications			
Asexxx	Related to Structured Query Launguage applications			
Assist	Opens the Active Assistance window		Standard	Help
Attdef	Creates an attribute definition that assigns (tags) textual information to a block	I Regulates visibility C Regulates constant/variable mode V Regulates verify mode P Regulates preset mode		TYPE IN
Attdisp	Regulates the visibility of attributes in the drawing	ON Makes all attribute tags visible OFF Makes all attributes invisible N Normal visibility set individually		View
Attedit	Permits the editing of attributes			Modify
Attext	Extracts attribute information from a drawing	C CDF comma-delimited format D DXF format S SDF format E Select objects		TYPE IN
Attredef	Redefines a block and updates associated attributes			TYPE IN
Attsync	Updates all instances of specified block with current attributes defined for the block	? Lists all blocks in drawing Name Lets you enter block name Select Lets you select on screen	Modify II	TYPE IN
Audit	Performs drawing integrity check while in AutoCAD	Y Fixes errors encountered N Reports, but does not fix, errors		File
Background	Sets up the background for your scene	Solid Selects one color background Gradient Specifies a two- or three-color gradient background Image Uses a bitmap file for the background Merge Uses the current AutoCAD® image as the background Colors Sets color for a solid or gradient background Preview Displays a preview of the current Background settings Image Specifies the image file name including BMP, JPG, PCX, TGA, & TIFF	Render	

Command Aliases	Explanation	Options		Toolbar	Pull-Down Menu
		Environment	Defines an environment for creating reflection and refraction effects on objects with reflective, raytraced materials		
		Horizon	Represents the percentage of unrotated height		
		Height	Represents a percentage of the second color in a three-color gradient. Use the box or scroll bar to set the value		
		Rotation	Sets an angle at which you can rotate a gradient background		
Base	Defines the origin point for insertion of one drawing into another				Draw
Battmann	Displays Block Attribute Manager			Modify II	Modify
Bhatch	Fills an automatically defined boundary with a hatch pattern through the use of dialog boxes; also allows previewing and repeated adjustments without starting over each time			Draw	Draw
Blipmode	Turns blip markers on and off				TYPE IN
Block	Makes a compound object from one or more objects	?	Lists names of defined blocks a group of objects	Draw	Draw
Blockicon	Generates preview images for blocks created with Release 14 or earlier				TYPE IN
Bmpout	Exports selected objects to bitmap format				
Boundary	Creates a polyline of a closed boundary				Draw
Box	Creates a 3D solid box			Solids	Draw
Break	Breaks out (erases) part of an object or splits it into parts	F	Allows you to reselect the first point again	Modify I	Modify
Browser	Launches the Web browser				
Cal	Evaluates mathematical and geometric expressions				TYPE IN
Camera	Sets a different camera and target location			View	
Chamfer	Makes a chamfer at the intersection of two lines	D	Sets chamfer distance	Modify I	Modify
		P	Chamfers all intersections of a pline figure		
		A	Sets the chamfer distances using a chamfer distance for the first line and an angle for the second		
		T	Controls whether AutoCAD trims the selected edges to the chamfer line endpoints		
		M	Controls whether AutoCAD uses two distances or a distance and an angle to create the chamfer		

Command Aliases	Explanation	Options	Toolbar	Pull-Down Menu
Change	Makes changes in the location, size, orientation, and other properties of selected objects; is very helpful for editing text	P Changes properties of objects E Elevation C Color LA Layer LT Linetype T Thickness LW Lineweight PL Plotstyle		TYPE IN
Checkstandards	Checks for standards violations		Tools	Tools
Chprop	Makes changes in the properties of selected objects	C Color LA Layer LT Linetype T Thickness LW Lineweight PL Plotstyle		TYPE IN
Circle (C)	Draws a circle of any size; default is center point and radius	2P Two endpoints on diameter 3P Three points on circle. D Enters circle diameter TTR Tangent, Tangent, Radius R Enter radius	Draw	Draw
Close	Closes the current drawing			File
Closeall	Closes all open drawings			TYPE IN
Color	Sets color for objects by name or number; also sets color to be by block or layer BYBLOCK BYLAYER	number Sets color by number name Sets a color by name Retains color of block Uses color of layer 1. Red 2. Yellow 3. Green 4. Cyan 5. Blue 6. Magenta 7. White		TYPE IN
Compile	Compiles shape and font files			TYPE IN
Cone	Creates a 3D solid cone		Solids	Solids
Convert	Converts associative hatches and 2D polylines to optimized format			
Convertctb	Converts color-dependent plot style table			TYPE IN
Convertpstyles	Converts drawing to color dependent plot styles			TYPE IN
Copy (CP)	Makes one or more copies of selected objects	M Makes more than one copy of the selected object	Modify I	Modify
Copybase	Copies objects with a specified base point			Edit
Copyclip	Copies selected objects to the Windows clipboard			
Copyhist	Text in command line history is copied to the clipboard			
Copylink	Current view is copied to the clipboard for OLE applications			
Customize	Customizes toolbars, buttons, and shortcut keys			TYPE IN
Cutclip	Cuts and copies selected objects from the drawing to the clipboard			
Cylinder	Creates a 3D solid cylinder		Solids	Solids

Command Aliases	Explanation	Options	Toolbar	Pull-Down Menu
Dbcclose	Closes the dbConnect Manager		Standard	
Dblclkedit	Controls double-click behavior			TYPE IN
Dbconnect	Provides an AutoCAD interface to external database tables		Standard	
Dblist	Makes a listing of every object in the drawing database		Inquiry	Tools
Ddedit	Edits text and attribute definitions		Modify II	
'Ddptype	Specifies point object display mode and sizes			
Ddvpoint	Sets the 3D viewing direction			View
Delay	Sets the timing for a sequence of command used in a script file			TYPE IN
Detachurl	Removes Hyperlinks from the selected objects			TYPE IN
Dim	Accesses the dimensioning mode			TYPE IN
Dimxx	See the end of this appendix for Dimensioning Commands			
Dist	Determines the distance between two points		Inquiry	Tool
Divide	Places markers along selected objects, dividing them into a specified number of parts	B Sets a specified block as a marker		Draw
Doughnut (Donut)	Draws a solid circle or a ring with a specified inside and outside diameter		Draw	Draw
Dragmode	Allows control of the dynamic specification (dragging) feature for all appropriate commands	ON Honors drag requests when applicable OFF Ignores drag requests A Sets Auto mode: drags whenever possible		TYPE IN
Draworder	Changes the display order of images and objects			
Dsettings	Specifies settings for Snap mode, grid, and polar and object snap tracking			Tools
Dsviewer	Opens the Aerial View window		View (Aerial View)	
Dview (DV)	Defines parallel or visual perspective views dynamically	CA Selects the camera angle relative to the target CL Sets front and back clipping planes D Sets camera-to-target distance, turns on perspective H Removes hidden lines on the selection set OFF Turns perspective off PA Pans the drawing across the screen PO Specifies the camera and target points TA Rotates the target point about the camera TW Twists the view around your line of sight U Undoes a Dview subcommand X Exits the Dview command Z Zooms in/out, or sets lens length		View

Command Aliases	Explanation	Options	Toolbar	Pull-Down Menu
Dwgprops	Sets and displays the properties of the current drawing		File	
Dxbin	Inserts specially coded binary files into a drawing			File
Eattedit	Edits attributes in a block reference		Modify II	Modify
Eattext	Exports attribute to external file		Modify II	Tools
Edge	Changes the visibility of 3D face edges		Surfaces	Draw
Edgesurf	Constructs a 3D polygon mesh approximating a Coons surface patch (a bicubic surface interpolated between four adjoining edges)		Surfaces	Draw
Elev	Sets the elevation and extrusion thickness for entities to be drawn in 3D drawings			TYPE IN
Ellipse	Draws ellipses using any of several methods	C Selects center point R Selects rotation rather than second axis I Draws isometric circle in current isoplane A Creates an elliptical arc	Draw	Draw
Erase (E)	Deletes objects from the drawing		Modify I	Modify
Etransmit	Creates a set with a transmittal drawing and associated files			TYPE IN
Explode	Changes a block or polyline back into its original objects		Modify I	Modify
Export	Saves objects to other file formats			
Extend	Extends a line, arc, or polyline to meet another object	U Undoes last extension P Specifies the Projection mode AutoCAD uses when extending objects E Extends the object to another object's implied edge, or only to an object that actually intersects it in 3D space	Modify I	Modify
Extrude	Creates unique solid primitives by extruding existing 2D objects		Solids	Draw
Fill	Determines if solids, traces, and wide polylines are automatically filled	ON Solids, traces, and wide polylines filled OFF Solids, traces, and wide polylines outlined		TYPE IN
Fillet	Constructs an arc of specified radius between two lines, arcs, or circles	P Fillets an entire polyline; sets fillet radius R Sets fillet radius T Controls whether AutoCAD trims the selected edges to the fillet arc endpoints	Modify I	Modify
Filter	Creates lists to select objects based on properties			TYPE IN

Command Aliases	Explanation	Options	Toolbar	Pull-Down Menu
Find	Finds, replaces, selects, or zooms to specified text		Stanadard	
Fog	Provides visual cues for the apparent distance of objects		Render	
Gotourl	Opens URL associated with selected object			TYPE IN
Graphscr F2	Flips to the graphics display on single-screen systems; used in command scripts and menus			TYPE IN
Grid F7 On/Off toggle	Displays a grid of dots, at desired spacing, on the screen	ON Turns grid on OFF Turns grid off S Locks grid spacing to snap resolution A Sets grid aspect (differing *X–Y* spacings) number Sets grid spacing (0 = use snap spacing) number X Sets spacing to multiple of snap spacing	Status bar	
Group	Creates a named selection set of objects		Standard	Tools
Hatch	Creates crosshatching and patternfilling	name Uses hatch pattern name from library file U Uses simple user-defined hatch pattern ? Lists selected names of available hatch patterns NAME and U can be followed by a comma and a hatch style from the following list: I Ignores internal structure N Normal style: turns hatch lines off and on when internal structure is encountered O Hatches outermost portion only		TYPE IN
Hatchedit	Modifies an existing associative hatch block		Modify II	Modify
'Help or '?	Displays a list of valid commands and data entry options or obtains help for a specific command or prompt	To get a set of Help modes, use ESC and F2 for flip screen	Standard	Help
Hide	Regenerates a 3D visualization with hidden lines removed		Render	View
HIsettings	Displays Hidden Line Settings dialog box to set the display properties of hidden lines			TYPE IN
Hyperlink	Attaches a hyperlink to a graphical object or modifies an existing hyperlink		Standard	
Hyperlinkoptions	Controls the visibility of the hyperlink cursor and the display of hyperlink tooltips			TYPE IN
Id	Displays the coordinates of a point selected on the drawing			Tools

Command Aliases	Explanation	Options		Toolbar	Pull-Down Menu
Imagexxx	Commands used for modifying and displaying images to the clipboard				
Import	Imports various file formats into AutoCAD				
Insert	Inserts a copy of a block or Wblock complete drawing into the current drawing	fname	Loads fname as block	Draw	Draw
		fname=f	Creates block fname from file f		
		*name	Retains individual part objects		
		C	(as reply to *X* scale prompt) Specifies scale via two points (Corner specification of scale)		
		XYZ	(as reply to *X* scale prompt) Readies Insert for *X*, *Y*, and *Z* scales		
		~	Displays a File dialog box		
		?	Lists names of defined blocks		
Insertobj	Inserts embedded or linked objects				
Interfere	Finds the interference of two or more solids and creates a composite solid from their common volume			Solids	Solids
Intersect	Creates composite solids or regions from the intersection of two or more solids or regions			Modify II	Modify
Isoplane CTRL + E	Changes the location of the isometric crosshairs to left, right, and top plane	L	Left plane		TYPE IN
		R	Right plane		
		T	Top plane		
		ENTER	Toggle to next plane		
Jpgout	Displays the Create Raster File dialog box, creates a JPEG file from selected objects				TYPE IN
Justifytext	Changes text justification point			Text	Modify
Layer (LA)	Allows for the creation of drawing layers and the assigning of color and linetype properties	C	Sets layers to color selected	Object Properties	Format
		F	Freezes layers		
		LT	Sets specified layers to linetype		
		M	Makes a layer the current layer, creating it if necessary		
		N	Creates new layers		
		ON	Turns on layers		
		OFF	Turns off layers		
		S	Sets current layer to existing layer		
		T	Thaws layers		
		?	Lists specified layers and their associated colors, linetypes, and visibility		
		L	Lock		
		U	Unlock		
		LW	Changes the lineweight associated with a layer		
		PS	Sets the plot style assigned to a layer		
Layerp	Undoes last changes made to layer settings				TYPE IN
Layerpmode	Toggles layer change tracking on and off				TYPE IN
Layout	Creates a new layout and renames, copies, saves, or deletes an existing layout	Copy	Copies a layout	Layout	
		Delete	Deletes a layout		

Command Aliases	Explanation	Options	Toolbar	Pull-Down Menu
		New — Creates a new layout tab Template — Creates a new template based on an existing layout in a template (DWT) or drawing (DWG) file Rename — Renames a layout Save — Saves a layout Set — Makes a layout current ?-List Layouts — Lists all the layouts defined in the drawing		
Layoutwizard	Starts the Layout wizard, in which you can designate page and plot settings for a new layout			TYPE IN
Laytrans	Changes layers to specified layer standards		CAD Standards	Tools
Leader	Draws a line from an object to, and including an annotation			
Lengthen	Lengthens an object		Modify I	Modify
Light	Manages lights and lighting effects		Render	View
Limits	Sets up the drawing size	2 points — Sets lower left/upper right drawing limits ON — Enables limits checking OFF — Disables limits checking		Format
Line (L)	Draws straight lines of any length	ENTER — (as reply to "From point:") Starts at end of previous line or arc C — (as reply to "To point:") Closes polygon U — (as reply to "To point:") Undoes segment	Draw	Draw
Linetype	Defines, loads, and sets the linetype	? — Lists a linetype library C — Creates a linetype definition L — Loads a linetype definition S — Sets current object linetype; *set suboptions:* name — Sets object linetype name BYBLOCK — Sets floating object linetype BYLAYER — Uses layer's linetype for objects ? — Lists specified loaded linetypes		TYPE IN
List	Provides database information for objects that are selected		Inquiry	Tools
Load	Loads a file of user-defined shapes to be used with the SHAPE command	? — Lists the names of loaded shape files		TYPE IN
Logfileoff	Closes the log file opened by LOGFILEON			TYPE IN
Logfileon	Writes the text window contents to a file			TYPE IN

Command Aliases	Explanation	Options		Toolbar	Pull-Down Menu
Lsedit	Edits a landscape object			Render	
Lslib	Maintains libraries of landscape objects			Render	
Lsnew	Adds realistic landscape items, such as trees and bushes, to your drawings			Render	
Ltscale	Regulates the scale factor to be applied to all linetypes within the drawing				TYPE IN
Lweight	Sets the current lineweight, lineweight display options, and lineweight units				Format
Massprop	Calculates and displays the mass properties of regions or solids			Inquiry	Tools
Matchprop	Causes properties of one object to be assigned to selected objects				
Matlib	Imports and exports materials to and from a library of materials				Tools
Measure	Inserts markers at measured distances along a selected object	B	Uses specified block as marker	Inquiry	Tools
Menu	Loads a menu into the menu areas (screen, pull-down, tablet, and button)				Tools
Menuload	Loads partial menu files				TYPE IN
Menuunload	Unloads partial menu files				TYPE IN
Minsert	Inserts multiple copies of a block in a rectangular array	fname	Loads fname and forms a rectangular array of the resulting block	Draw	Draw
		fname=f	Creates block fname from file f and forms a rectangular array		
		?	Lists names of defined blocks		
		C	(as reply to *X* scale prompt) Specifies scale via two points (Corner specification of scale)		
		XYZ	(as reply to *X* scale prompt) Readies Multiple Insert for *X*,*Y*, and *Z* scales		
		~	Displays a File dialog box		
		S	Sets the scale factor for the *X*, *Y*, and *Z* axes		
		PD	Sets the scale factor for the *X*, *Y*, and *Z* axes to control the display of the block as it is dragged into position		
		PR	Sets the rotation angle of the block as it is dragged into position		
Mirror	Reflects selected objects about a user-specified axis, vertical, horizontal, or inclined			Modify I	Modify
Mirror3D	Creates a mirror image copy of objects about a plane				Modify
Mledit	Edits multiple parallel lines			Modify II	Modify

Command Aliases	Explanation	Options		Toolbar	Pull-Down Menu
Mline	Draws multiple parallel lines				
Mlstyle	Defines a style for multiple parallel lines				Format
Model	Switches from a layout tab to the Model tab and makes it current				TYPE IN
Move (M)	Moves selected objects to another location in the drawing			Modify I	Modify
Mredo	Reverses action of multiple UNDO commands			Standard	
Mslide	Creates a slide of what is displayed on the screen				TYPE IN
Mspace (MS)	Switches to model space from paper space			Status bar	View
Mtext	Creates paragraph text			Draw	Draw
Multiple	Allows the next command to repeat until canceled				TYPE IN
Mview	Sets up and controls viewports	ON	Turns selected viewport(s) on; causes model to be regenerated in the selected viewport(s)		View
		OFF	Turns selected viewport(s) off; causes model not to be displayed in the selected viewport(s)		
		Hideplot	Causes hidden lines to be removed in selected viewport(s) during paper space plotting		
		Fit	Creates a single viewport to fit the current paper space view		
		2	Creates two viewports in specified area or to fit the current paper space view		
		4	Creates four equal viewports in specified area or to fit the current paper space view		
		Restore	Translates viewport configurations saved with the VPORTS command into individual viewport objects in paper space		
		<point>	Creates a new viewport within the area specified by two points		
		O	Specifies a closed polyline, ellipse, spline, region, or circle to convert into a viewport		
		P	Creates an irregularly shaped viewport using specified points		
		3	Creates three viewports in specified area or to fit the current paper space view		

Command Aliases	Explanation	Options	Toolbar	Pull-Down Menu
Mvsetup	Sets up the specifications of a drawing			TYPE IN
New	Creates a new drawing		Standard	File
Offset	Reproduces curves or lines parallel to the one selected	number Specifies offset distance T Through: allows specification of a point through which the offset curve is to pass	Modify I	Modify
Olescale	Displays the OLE Properties dialog box			TYPE IN
Olelinks	Cancels, changes, updates OLE links			
Oops	Recalls last set of objects previously erased			TYPE IN
Open	Opens an existing drawing		Standard	File
Options	Customizes the AutoCAD settings			Tools
Ortho F8	Restricts the cursor to vertical or horizontal use	ON Forces cursor to horizontal or vertical use OFF Does not constrain cursor movement	Status bar	
Osnap	Allows for selection of precise points on existing objects	CEN Center of arc or circle END Closest endpoint of arc or line INS Insertion point of text/block/shape INT Intersection of line/arc/circle MID Midpoint of arc or line NEA Nearest point of arc/circle/line/point NOD Node (point) NON None (off) PER Perpendicular to arc/line/circle QUA Quadrant point of arc or circle QUI Quick mode (first find, not closest) TAN Tangent to arc or circle EXT Snaps to the extension point of an object APP Apparent Intersection includes two separate snap modes: PAR Snaps to an extension in parallel with an object	Object Snap	Tools
Pagesetup	Specifies the layout page, plotting device, paper size, and settings for each new layout		Layout	
'Pan (P)	Moves the display window		Standard	View
Partiaload	Loads additional geometry into a partially opened drawing			File

Command Aliases	Explanation	Options		Toolbar	Pull-Down Menu
Partialopen	Loads geometry from a selected view or layer into a drawing				TYPE IN
Pasteblock	Pastes a copied block into a new drawing				Edit
Pasteclip	Inserts clipboard data				
Pasteorig	Pastes a copied object in a new drawing using the coordinates from the original drawing				Edit
Pastespec	Specifies fomat of data imported from the clipboard				
Pcinwizard	Displays a wizard to import PCP and PC2 configuration file plot settings into the Model tab or current layout				Tools
Pedit (2D)	Permits editing of 2D polylines	C	Closes to start point	Modify II	Modify
		D	Decurves, or returns a spline curve to its control frame		
		F	Fits curve to polyline		
		J	Joins to polyline		
		O	Opens a closed polyline		
		S	Uses the polyline vertices as the frame for a spline curve (type set by SPLINETYPE)		
		U	Undoes one editing operation		
		W	Sets uniform width for polyline		
		X	Exits PEDIT command during vertex editing		
		E	Edits vertices during vertex editing		
		B	Sets first vertex for Break		
		G	Go (performs Break or Straighten operation)		
		I	Inserts new vertex after current one		
		M	Moves current vertex		
		N	Makes next vertex current		
		P	Makes previous vertex current		
		R	Regenerates the polyline		
		S	Sets first vertex for Straighten		
		T	Sets tangent direction for current vertex		
		W	Sets new width for following segment		
		X	Exits vertex editing, or cancels Break/Straighten		
		L	Generates the linetype in a continuous pattern through the vertices of the polyline		

Command Aliases	Explanation	Options	Toolbar	Pull-Down Menu
Pedit (3D)	Allows editing of 3D polylines	C Closes to start point. D Decurves, or returns a spline curve to its control frame O Opens a closed polyline S Uses the polyline vertices as the frame for a spline curve (type set by SPLINETYPE) U Undoes one editing operation X Exits PEDIT command E Edits vertices *During vertex editing:* B Sets first vertex for Break G Go (performs Break or Straighten operation) I Inserts new vertex after current one M Moves current vertex N Makes next vertex current P Makes previous vertex current R Regenerates the polyline S Sets first vertex for Straighten X Exits vertex editing, or cancels Break/Straighten	Modify II	Modify
Pedit (mesh)	Allows editing of 3D polygon meshes	D Desmoothes-restores original mesh M Opens (or closes) the mesh in the *M* direction N Opens (or closes) the mesh in the *N* direction S Fits a smooth surface as defined by SURFTYPE U Undoes one editing operation X Exits PEDIT command E Edits mesh vertices during vertex editing D Moves down to previous vertex in *M* direction L Moves left to previous vertex in *N* direction M Repositions the marked vertex N Moves to next vertex P Moves to previous vertex R Moves right to next vertex in *N* direction RE Redisplays the polygon mesh U Moves up to next vertex in *M* direction X Exits vertex editing during vertex editing	Modify II	Modify
Pface	Creates a 3D mesh of arbitrary complexity and surface characteristics			TYPE IN

Command Aliases	Explanation	Options	Toolbar	Pull-Down Menu
Plan	Puts the display in plan view (Vpoint 0,0,1) relative to either the current UCS, a specified UCS, or the WCS	C Establishes a plan view of the current UCS U Establishes a plan view of the specified UCS W Establishes a plan view of the WCS		View
Pline (PL)	Draws 2D polylines	H Sets new half-width U Undoes previous segment W Sets new line width ENTER Exits PLINE command C Closes with straight segment L Segment length (continues previous segment) A Switches to arc mode *In arc mode:* A Included angle CE Center point CL Closes with arc segment D Starting direction L Chord length, or switches to line mode R Radius S Second point of three-point arc	Draw	Draw
Plot (Print)	Plots a drawing to a plotting device or a file		Standard	File
Plotstamp	Places a plot stamp on a drawing corner and logs it to a file			TYPE IN
Plotstyle	Sets the current plot style for new objects, or the assigned plot style for selected objects			TYPE IN
Plottermanager	Displays the Plotter Manager, where you can launch the Add-a-Plotter wizard and the Plotter configuration editor			File
Pngout	Displays the Create Raster file dialog box, creates a Portable Network Graphics file from selected objects			TYPE IN
Point	Draws a single point on the drawing		Draw	Draw
Polygon	Creates regular polygons with the specified number of sides indicated	E Specifies polygon by showing one edge C Circumscribes around circle I Inscribes within circle	Draw	Draw
Preview	Displays plotted view of drawing			
Properties	Controls properties of existing objects		Standard	
Propertiesclose	Closes the Properties window			TYPE IN
Psetupin	Imports a user-defined page setup into a new drawing layout			TYPE IN
Pspace (PS)	Switches to paper space		Status bar	View

Command Aliases	Explanation	Options	Toolbar	Pull-Down Menu
Publish	Displays the Publish Drawing Sheets dialog box to begin publishing the current drawing sheets to DWF file or plotter		Standard	
Publishtoweb	Creates HTML pages including images of drawings			TYPE IN
Purge	Removes unused Blocks, text styles, layers, linetypes, and dimension styles from the drawing	A Purges all unused named objects B Purges unused blocks D Purges unused dimstyles LA Purges unused layers SH Purges unused shape files ST Purges unused text styles LT Purges linetypes		File
Qdim	Quickly creates a dimension	Continuous Creates a series of continued dimensions Staggered Creates a series of staggered dimensions Baseline Creates a series of baseline dimensions Ordinate Creates a series of ordinate dimensions Radius Creates a series of radius dimensions Diameter Creates a series of diameter dimensions Datum Point Sets a new datum point for baseline and ordinate dimensions Edit Edits a series of dimensions	Dimension	
Qleader	Quickly creates a leader and leader annotation			Dimension
Qnew	Starts a new drawing from the current default drawing template file		Standard	
Qsave	Saves the drawing without requesting a file name		Standard	File
Qselect	Quickly creates selection sets based on filtering criteria			Tools
Qtext	Enables text objects to be identified without drawing the test detail	ON Quick text mode on. OFF Quick text mode off		TYPE IN
Quit	Exit AutoCAD			File
Ray	Creates a semi-infinite line		Draw	
Recover	Attempts to recover damaged or corrupted drawings			File
Rectang	Draws a rectangular polyline		Draw	Draw
Redefine	Restores a built-in command deleted by UNDEFINE			TYPE IN
Redo	Reverses the previous command if it was U or Undo		Standard	Edit
'Redraw (R)	Refreshes or cleans up the current viewport		Standard	View

Command Aliases	Explanation	Options	Toolbar	Pull-Down Menu
'Redrawall	Redraws all viewports		Standard	View
Refclose	Saves back or discards changes made during in-place editing of a reference (an xref or a block)			Modify
Refedit	Selects a reference for editing		Refedit	
Refset	Adds or removes objects from a working set during in-place editing of a reference (an xref or a block)	Add Adds objects to the working set Remove Removes objects from the working set		Modify
Regen	Regenerates the current viewport			TYPE IN
Regenall	Regenerates all viewports			TYPE IN
Regenauto	Controls automatic regeneration performed by other commands	ON Allows automatic regenerations OFF Prevents automatic regenerations		TYPE IN
Region	Creates a region object from a selection set of existing objects		Draw	Draw
Reinit	Allows the I/O ports, digitizer, display, plotter, and *PGP* file to be reinitialized			TYPE IN
Rename	Changes the names associated with text styles, layers, linetypes, blocks, views, UCSs, viewport configurations, and dimension styles	B Renames block. D Renames dimension style LA Renames layer LT Renames linetype S Renames text style U Renames UCS VI Renames view VP Renames viewport configuration		TYPE IN
Render	Creates a realistically shaded image of a 3D wireframe or solid model		Render	View
Rendscr	Redisplays the last rendering created with the RENDER command			TYPE IN
Replay	Displays a GIF, TGA, or TIFF image			Tools
'Resume	Resumes an interrupted command script			TYPE IN
Revcloud	Creates a revision cloud			Draw
Revolve	Creates a solid by revolving a 2D object about an axis		Solids	Draw
Revsurf	Creates a 3D polygon mesh approximating a surface of revolution, by rotating a curve around a selected axis		Surfaces	Draw
Rmat	Manages rendering materials		Render	View
Rmlin	Inserts markups from an RML file into a drawing		Insert	
Rotate	Rotates existing objects to the angle selected	R Rotates with respect to reference angles	Modify I	Modify
Rotate3D	Moves objects about a 3D axis			
Rpref	Sets rendering preferences		Render	View
Rscript	Restarts a command script from the beginning			TYPE IN
Rulesurf	Creates a 3D polygon mesh approximating a ruled surface between two curves		Surfaces	Draw

Command Aliases	Explanation	Options		Toolbar	Pull-Down Menu
Save	Updates the current drawing file without exiting the Drawing Editor			Standard	File
Saveas	Same as SAVE, but also renames the current drawing				File
Saveimg	Saves a rendered image to a file				Tools
Scale	Changes the size of existing objects to the selected scale factor	R	Resizes with respect to reference size	Modify I	Modify
Scaletext	Changes size of text objects			Text	Modify
Scene	Manages scenes in model space			Render	View
Script	Executes a command script				Tools
Section	Uses the intersection of a plane and solids to create a region			Section	Modify
Securityoptions	Lets you add security settings to drawing				TYPE IN
Select	Groups objects into a selection set for use in subsequent commands				TYPE IN
Setidrophandler	Displays the Set Default i-drop Content Type dialog box where you can set default type of i-drop content for current application				TYPE IN
Setuv	Maps materials onto objects			Render	
'Setvar	Allows you to display or change the value of system variables	?	Lists specified system variables		TYPE IN
Shademode	Shades the objects in the current viewport	2D Wireframe	Displays the objects using lines and curves to represent the boundaries		View
		3D Wireframe	Displays the objects using lines and curves to represent the boundaries		
		Hidden	Displays the objects using 3D wireframe representation and hides lines representing back faces		
		Flat Shaded	Shades the objects between the polygon faces		
		Gouraud Shaded	Shades the objects and smooths the edges between polygon faces		
		Flat Shaded, Edges On	Combines the Flat Shaded and Wireframe options		
		Gouraud Shaded, Edges On	Combines the Gouraud Shaded and Wireframe options		
Shape	Draws predefined shapes	?	Lists available shape names		TYPE IN
Shell	Allows access to other programs while running AutoCAD				TYPE IN

Command Aliases	Explanation	Options	Toolbar	Pull-Down Menu
Showmat	Lists material type and method of attachment for the selected object			
Sigvalidate	Displays the Validate Digital Signatures dialog box			TYPE IN
Sketch	Allows freehand sketching	C Connect: restarts sketch at endpoint E Erases (backs up over) temporary lines P Raises/lowers sketching pen Q Discards temporary lines, remains in SKETCH R Records temporary lines, remains in SKETCH X Records temporary lines, exits SKETCH; draws line to current point	Draw	Draw
Slice	Slices a set of solids with a plane		Solids	Draw
Snap F9	Allows for precision alignment of points	number Sets snap resolution ON Aligns designated points OFF Does not align designated points A Sets aspect (differing *X–Y* spacing) R Rotates snap grid S Selects style, standard or isometric	Status bar	
Soldraw	Generates sections and profiles in viewports created with SOLVIEW			
Solid	Creates filled-in polygons		Draw	Draw
Solidedit	Edits faces and edges of 3D solid objects			Modify
Solprof	Creates a profile image of a 3D solid			
Solview	Creates floating viewports for 3D solid objects			
Spacetrans	Converts lengths between model and paper spaces		Text	
Spell	Checks the spelling in a drawing		Standard	Tools
Sphere	Creates a 3D solid sphere		Solids	Draw
Spline	Creates a quadratic or cubic spline (NURBS) curve		Draw	Draw
Splinedit	Edits a spline object		Modify II	Modify
Standards	Manages association of standards files with AutoCAD drawings		CAD Standards	Tools
Stats	Displays rendering statistics		Render	View
Status	Displays drawing setup			Tools
Stlout	Stores a solid in ASCII or binary file			File
Stretch	Allows you to move a portion of a drawing while retaining connections to other parts of the drawing		Modify I	Modify

Command Aliases	Explanation	Options		Toolbar	Pull-Down Menu
Style	Sets up named text styles, with various combinations of font, mirroring, obliquing, and horizontal scaling	?	Lists specified currently defined text style		Format
Stylesmanager	Displays the Plot Style Manager				File
Subtract	Creates a composite region or solid by subtracting the area of one set of regions from another and subtracting the volume of one set of solids from another			Modify II	Modify
Syswindows	Arranges windows				
Tablet	Allows for configuration of a tablet menu or digitizing of an existing drawing	ON	Turns tablet mode on		Tools
		OFF	Turns tablet mode off		
		CAL	Calibrates tablet for use in the current space		
Tabsurf	Creates a polygon mesh approximating a general tabulated suface defined by a path and a direction vector			Surfaces	Draw
Text	Enters text on the drawing	J	Prompts for justification options	Draw	Draw
		S	Lists or selects text style		
		A	Aligns text between two points, with style-specified width factor; AutoCAD computes appropriate height		
		C	Centers text horizontally		
		F	Fits text between two points, with specified height; AutoCAD computes an appropriate width factor		
		M	centers text horizontally and vertically		
		R	Right-justifies text		
		BL	Bottom left		
		BC	Bottom center		
		BR	Bottom right		
		ML	Middle left		
		MC	Middle center		
		MR	Middle right		
		TL	Top left		
		TC	Top center		
		TR	Top right		
'Textscr F2	Flips to the text display on singlescreen systems; used in command scripts and menus				TYPE IN
Tifout	Displays the Create Raster File dialog box Creates a TIFF file from selected objects				TYPE IN
Time	Indicates total elapsed time for each drawing	D	Displays current times		Tools
		ON	Starts user elapsed timer		
		OFF	Stops user elapsed timer		
		R	Resets user elapsed timer		
Tolerance	Creates geometric tolerances			Dimension	Draw

Command Aliases	Explanation	Options		Toolbar	Pull-Down Menu
Toolpalettes	Opens Tool Palettes window				Tools
Toolpalettesclose	Closes the Tool Palettes window				Tools
Toolbar	Customizes, hides, and displays toolbars				
Torus	Creates a donut-shaped solid			Solids	Draw
Trace	Creates solid lines of specified width			Draw	Draw
Transparency	Determines transparency of opacity of bitonal image background				
Traysettings	Displays the Tray Settings dialog box				TYPE IN
Treestat	Displays information on the drawing's current spatial index, such as the number and depth of nodes in the drawing's database; use this information with the TREEDEPTH system variable setting to fine-tune performances for large drawings				TYPE IN
Trim	Deletes portions of selected entities that cross a selected boundary edge	U	Undoes last trim operation	Modify I	Modify
U	Reverses the effect of the previous command			Standard	Edit
UCS	Defines or modifies the current User Coordinate System	D	Deletes one or more saved coordinate systems	Standard	View
		E	Sets a UCS with the same extrusion direction as that of the selected object		
		O	Shifts the origin of the current coordinate system		
		P	Restores the previous UCS		
		R	Restores a previously saved UCS		
		S	Saves the current UCS		
		V	Establishes a new UCS whose *Z* Axis is parallel to the current viewing direction		
		W	Sets the current UCS equal to the WCS		
		X	Rotates the current UCS around the *X* axis		
		Y	Rotates the current UCS around the *Y* axis		
		Z	Rotates the current UCS around the *Z* axis		
		ZA	Defines a UCS using an origin point and a point on the positive portion of the *Z* axis		
		3	Defines a UCS using an origin point, a point on the positive portion of the *X* axis, and a point on the positive *Y* portion of the *X* plane		
		?	Lists specified saved coordinate systems		

Command Aliases	Explanation	Options	Toolbar	Pull-Down Menu
Ucsicon	Controls visibility and placement of the UCS icon, which indicates the origin and orientation of the current UCS; the options normally affect only the current viewport	A Changes settings in all active viewports N Displays the icon at the lower-left corner of the viewport OR Displays the icon at the origin of the current UCS if possible ON Enables the coordinate system icon		Tools
Ucsman	Manages defined user coordinate systems		UCS	
Undefine	Deletes the definition of a built-in AutoCAD command			TYPE IN
Undo	Reverses the effect of multiple commands, and provides control over the Undo facility	number Undoes the number most recent commands A Auto: controls treatment of menu items as Undo groups B Back: undoes back to previous Undo mark C Control: enables/disables the Undo mark E End: terminates an Undo group G Group: begins sequence to be treated as one command M Mark: places marker in Undo file (for back)	Standard	Edit
Union	Creates a composite region or solid		Modify II	Modify
Units	Selects coordinate and angle display formats and precision			Format
Vbaide	Displays the Visual Basic Editor			Tools
Vbaload	Loads a global VBA project into the current AutoCAD session			Tools
Vbaman	Loads, unloads, saves, creates, embeds, and extracts VBA projects			Tools
Vbarun	Runs a VBA macro			Tools
Vbastmt	Executes a VBA statement on the AutoCAD command line			TYPE IN
Vbaunload	Unloads a global VBA project			TYPE IN
'View	Saves the current graphics display and space as a named view, or restores a saved view and space to the display	D Deletes named view R Restores named view to screen S Saves current display as named view W Saves specified window as named view ? Lists specified named views		View
Viewports or Vports	Divides the AutoCAD graphics display into multiple viewports, each of which can contain a different view of the current drawing	D Deletes a saved viewport configuration J Joins (merges) two viewports R Restores a saved viewport configuration		View

Command Aliases	Explanation	Options	Toolbar	Pull-Down Menu
		S Saves the current viewport configuration S1 Displays a single viewport filling the entire graphics area 2 Divides the current viewport into viewports 3 Divides the current viewport into three viewports 4 Divides the current viewport into four viewports ? Lists the current and saved viewport configurations		
Viewres	Adjusts the precision and speed of circle and arc drawing on the monitor			Format
Vlisp	Displays the Visual LISP interactive development environment (IDE)			Tools
Vpclip	Clips viewport objects	Object Specifies an object to act as a clipping boundary Polygonal Draws a clipping boundary Delete Deletes the clipping boundary of a selected viewport		TYPE IN
Vplayer	Sets viewport visibility for new and existing layers	? Lists layers frozen in a selected viewport Freeze Freezes specified layers in selected viewport(s) Thaw Thaws specified layers in selected viewport(s) Reset Resets specified layers to their default visibility Newfz Creates new layers that are frozen in all viewports Vpvisdfit Sets the default viewport visibility for existing layers		Type In
Vpoint	Selects the viewpoint for a 3D visualization	R Selects viewpoint via two rotation angles ENTER Selects viewpoint via compass and axes tripod x,y,z Specifies viewpoint		View
Vslide	Displays a previously created slide file	file Views slide *file Preloads next Vslide you will view		Tools
Wblock	Creates a block as a separate drawing	name Writes specified block definition = Block name same as file name * Writes entire drawing ENTER Writes selected objects		File

Command Aliases	Explanation	Options	Toolbar	Pull-Down Menu
Wedge	Creates a 3D solid with a tapered sloping face		Solids	Draw
Whohas	Displays ownership information for opened drawing files			TYPE IN
Wmxxx	Controls windows metafiles			
Xattach	Attaches an external reference			
Xbind	Permanently adds a selected subset of an external reference's dependent symbols to your drawing	Block — Adds a Block. Dimstyle — Adds a dimstyle Layer — Adds a layer Ltype — Adds a linetype Style — Adds a style	Reference	Modify
Xclip	Defines and external reference			
Xline	Creates an infinite line		Draw	Draw
Xopen	Opens xref to which selected objects belongs			TYPE IN
Xplode	Breaks a compound object into its component objects			TYPE IN
Xref	Allows you to work with other AutoCAD drawings without adding them permanently to your drawing and without altering their contents	Attach — Attaches a new Xref or a copy of an Xref that you have already attached Bind — Makes an Xref a permanent part of your drawing Detach — Removes an Xref from your drawing Path — Allows you to view and edit the file name AutoCAD uses when loading a particular Xref Reload — Updates one or more contents Xrefs at any time, without leaving and re-entering the Drawing Editor ? — Lists Xrefs in your drawing and the drawing associated with each one	Reference	Insert
'Zoom (Z)	Enlarges or reduces the display area of a drawing	number — Multiplier from original scale numberX — Multiplier from current scale number XP — Scale relative to paper space A — All C — Center D — Dynamic Pan Zoom E — Extents ("drawing uses") L — Lower left corner P — Previous V — Virtual screen maximum W — Window	Standard	View
3D	Creates three-dimensional polygon mesh objects			Draw

Command Aliases	Explanation	Options		Toolbar	Pull-Down Menu
3Darray	Creates a three-dimensional array				Modify
3Dclip	Invokes the interactive 3D view and opens the Adjust Clipping Planes window				
3Dconfig	Sets 3D configuration from command propt				TYPE IN
3Dcorbit	Invokes the interactive 3D view and enables you to set the objects in the 3D view into continuous motion			Orbit	
3Ddistance	Invokes the interactive 3D view and makes objects appear closer or farther away			Orbit	
3Dface	Draws 3D plane sections	I	Makes the following edge invisible	Surfaces	Draw
3Dmesh	Defines a 3D polygon mesh (by specifying its size in terms of *M* and *N*) and the location of each vertex in the mesh			Surfaces	Draw
3Dorbit	Controls the interactive viewing of objects in 3D			Standard	
3Dorbitctr	Specifies the center of rotation with pointing device				TYPE IN
3Dpan	Invokes the interactive 3D view and enables you to drag the view horizontally and vertically			Orbit	
3Dpoly	Creates a 3D polyline	C U ENTER	Closes the Polyline back to the first point Undoes (deletes) the last segment entered Exits 3Dpoly command	Draw	Draw
3Dsin	Imports a 3D Studio (3DS) file				Insert
3Dsout	Exports to a 3D Studio (3DS) file				TYPE IN
3Dswivel	Invokes the interactive 3D view and simulates the effect of turning the camera			3D Orbit	
3Dzoom	Invokes the interactive 3D view so you can zoom in and out on the view			3D Orbit	

DIMENSIONING COMMANDS

Command	Explanation
Dimaligned	Aligns dimension parallel with objects
Dimangular	Draws an arc to show the angle between two nonparallel lines or three specified points
Dimbaseline	Continues a linear dimension from the baseline (first extension line) of the previous or selected dimension
Dimcenter	Draws a circle/arc center mark or centerlines
Dimcontinue	Continues a linear dimension from the second extension line of the previous dimension
Dimdiameter	Dimensions the diameter of a circle or arc
Dimdisassociate	Disassociates dimension with object(s)
Dimedit	Edits dimensions
Dimlinear	Creates linear dimensions
Dimordinate	Creates ordinate point associative dimensions
Dimoverride	Overrides a subset of the dimension variable settings associated with selected dimension objects
Dimradius	Dimensions the radius of a circle or arc, with an optional center mark or centerlines
Dimreassociate	Reassociates dimension with object(s)
Dimstyle	Displays the Dimension Style Manager window
Dimtedit	Allows repositioning and rotation of text items in an associative dimension without affecting other dimension subentities
Leader	Draws a line with an arrowhead placement of dimension text

DIMENSIONING VARIABLES

Name	Description	Type	Default
DIMADEC	Controls number of places of precision displayed for angular dimension text		
DIMALT	Alternate units	Switch	Off
DIMALTD	Alternate units decimal places	Integer	2
DIMALTF	Alternate units scale factor	Scale	25.4
DIMALTRND	Alternate units rounding value		0.0000
DIMALTTD	Alternate units tolerance value	Integer	2
DIMALTTZ	Toggles suppression of zeros for tolerance values	Integer	0
DIMALTU	Sets unit format for alternate units	Integer	2
DIMALTZ	Toggles suppression of zeros for alternate values	Integer	0
DIMAPOST	Alternate units text suffix	String	None
DIMASO	Controls the associativity of dimesion objects	String	On
DIMASSOC	Obsolete dimension variable (used in AutoCAD 2002 and earlier)		
DIMASZ	Arrow size.	Distance	0.18
DIMATFIT	Arrow and text fit		3
DIMAUNIT	Angle format	Integer	0
DIMAZIN	Zero suppression in angles	Integer	0
DIMBLK	Arrow block	String	None
DIMBLK1	Separate arrow block 1	String	None
DIMBLK2	Separate arrow block 2	String	None
DIMCEN	Center mark size	Distance	0.09
DIMCLRD	Dimension line color	Color number	BYBLOCK
DIMCLRE	Extension line color	Color number	BYBLOCK
DIMCLRT	Dimension text color	Color number	BYBLOCK
DIMDEC	Decimal place for tolerance values	Integer	4
DIMDLE	Dimension Line extension	Distance	0.0
DIMDLI	Dimension line increment	Distance	0.38

Name	Description	Type	Default
DIMDSEP	Decimal separator		.
DIMEXE	Extension line extension	Distance	0.18
DIMEXO	Extension line offset	Distance	0.0625
DIMFIT	Preserves integrity of scripts	Integer	3
DIMFRAC	Fraction format		0
DIMGAP	Dimension line gap	Distance	0.09
DIMJUST	Controls horizontal text position	Imteger	0
DIMLDRBLK	Leader arrowhead block name		ClosedFilled
DIMLFAC	Length factor	Scale	1.0
DIMLIM	Limits dimensioning	Switch	Off
DIMLUNIT	Linear unit format		2
DIMLWD	Dimension line and leader lineweight		-2
DIMLWE	Extension line lineweight		-2
DIMPOST	Dimension text suffix	String	None
DIMRND	Rounding value	Scaled distance	0.0
DIMSAH	Separate arrow blocks	Switch	Off
DIMSCALE	Dimension feature scale factor	Switch	1.0
DIMSD1	Suppresses first dimension line	Switch	Off
DIMSD2	Suppresses second dimension line	Switch	Off
DIMSE1	Suppresses extension line 1	Switch	Off
DIMSE2	Suppresses extension line 2	Switch	Off
DIMSHO	Lists dimension styles	Switch	On
DIMSOXD	Suppresses outside dimension lines	Switch	Off
DIMSTYLE	Shows current dimension style	String	
DIMTAD	Text above dimension line	Switch	Off
DIMTDEC	Tolerance values	Integer	4
DIMTFAC	Tolerance text scale factor	Scale	1.0
DIMTIH	Text inside horizontal	Switch	On
DIMTIX	Text inside extension lines	Switch	Off

Name	Description	Type	Default
DIMTM	Minus tolerance value	Scaled distance	0.0
DIMTMOVE	Text movement		0
DIMTOFL	Text outside, force line inside	Switch	Off
DIMTOH	Text outside horizontal	Switch	On
DIMTOL	Tolerance dimensioning	Switch	Off
DIMTOLJ	Tolerance dimensioning justification	Integer	1
DIMTP	Plus tolerance value	Scaled distance	0.0
DIMTSZ	Tick size	Distance	0.0
DIMTVP	Text vertical position	Scale	0.0
DIMTXSTY	Dimension text style	String	Standard
DIMTXT	Text size	Distance	0.18
DIMTZIN	Controls suppression of zeros in tolerance values		
DIMUNIT	Sets unit format	Integer	2
DIMUPT	User-positioned text	Switch	Off
DIMZIN	Zero suppression	Integer	0

The following table lists the Dimension commands as entered at the "Command:" prompt and their equivalent command entered at the "Dim:" prompt.

Dimension Commands entered at the "Command:" prompt	Equivalent Dimension Commands entered at the "Dim:" prompt
DIMALIGNED	ALIGNED
DIMANGULAR	ANGULAR
DIMBASELINE	BASELINE
DIMCENTER	CENTER
DIMCONTINUE	CONTINUE
DIMDIAMETER	DIAMETER
DIMEDIT Home	HOMETEXT
DIMLINEAR Horizontal	HORIZONTAL
LEADER	LEADER
DIMEDIT New	NEWTEXT
DIMEDIT Oblique	OBLIQUE
DIMORDINATE	ORDINATE
DIMOVERRIDE	OVERRIDE
DIMRADIUS	RADIUS
- DIMSTYLE Restore	RESTORE
DIMLINEAR Rotated	ROTATED
- DIMSTYLE Save	SAVE
- DIMSTYLE Status	STATUS
DIMTEDIT	TEDIT
DIMEDIT Rotate	TROTATE
- DIMSTYLE Apply	UPDATE
- DIMSTYLE Variables	VARIABLES
DIMLINEAR Vertical	VERTICAL

OBJECT SELECTION

Object Selection Option	Meaning
point	Selects one object that crosses the small pick box. If no object crosses the pick box and Auto mode has been selected, this designated point is taken as the first corner of a Crossing or Window box
Multiple	Allows selection of multiple objects using a single search of the drawing. The search is not performed until you give a null response to the "Select objects:" prompt
Window	Selects all objects that lie entirely within a window
WPolygon	Selects objects that lie entirely within a polygon shaped selection area
Crossing	Selects all objects that lie within *or cross* a window
CPolygon	Selects all objects that lie within and crossing a polygon-shaped selection area
Fence	Selects all objects that cross a selection fence line
BOX	Prompts for two points. If the second point is to the right of the first point, selects all objects inside the box (like "Window"); otherwise, selects all objects within or crossing the box (like "Crossing")
AUto	Accepts a point, which can select an object using the small pick box; if the point you pick is in an empty area, it is taken as the first corner of a BOX (see above)
ALL	Selects all entities in the drawing except entities on frozen or locked layers
Last	Selects the most recently drawn object that is currently visible
Previous	Selects the previous selection set
Add	Establishes Add mode to add following objects to the selection set
Remove	Sets Remove mode to remove following objects from the selection set
SIngle	Sets single selection mode; as soon as one object (or one group of objects via Window/Crossing box) is selected, the selection set is considered complete and the editing command uses it without further user interaction
Undo	Undoes (removes objects last added)

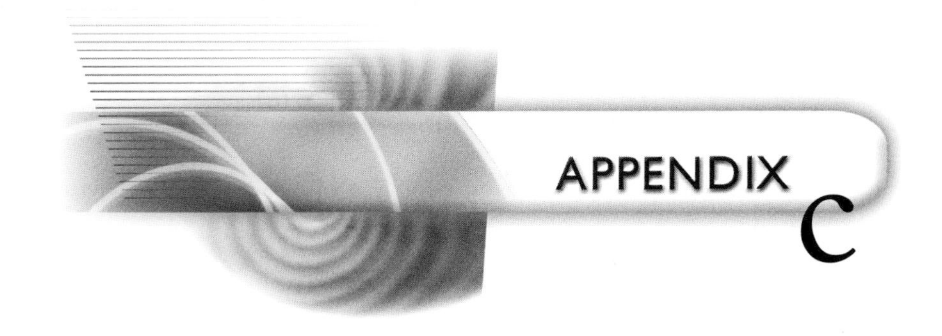

APPENDIX C

AutoCAD Toolbars

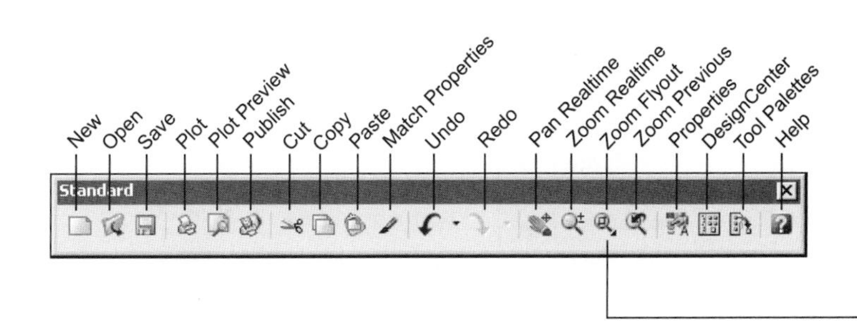

Object Snap

- Temporary Tracking Point
- Snap From
- Snap to Endpoint
- Snap to Midpoint
- Snap to Intersection
- Snap to Apparent Intersection
- Snap to Extension
- Snap to Center
- Snap to Quadrant
- Snap to Tangent
- Snap to Perpendicular
- Snap to Parallel
- Snap to Insert
- Snap to Node
- Snap to Nearest
- Snap to None
- Object Snap Settings

UCS

- UCS
- Display UCS Dialog
- UCS Previous
- World UCS
- Object UCS
- Face UCS
- View UCS
- Origin UCS
- Z Axis Vector UCS
- 3-Point UCS
- X Axis Rotate UCS
- Y Axis Rotate UCS
- Z Axis Rotate UCS
- Apply UCS

View

- Named Views
- Top View
- Bottom View
- Left View
- Right View
- Front View
- Back View
- SW Isometric View
- SE Isometric View
- NE Isometric View
- NW Isometric View
- Camera

Zoom

- Zoom Window
- Zoom Dynamic
- Zoom Scale
- Zoom Center
- Zoom In
- Zoom Out
- Zoom All
- Zoom Extents

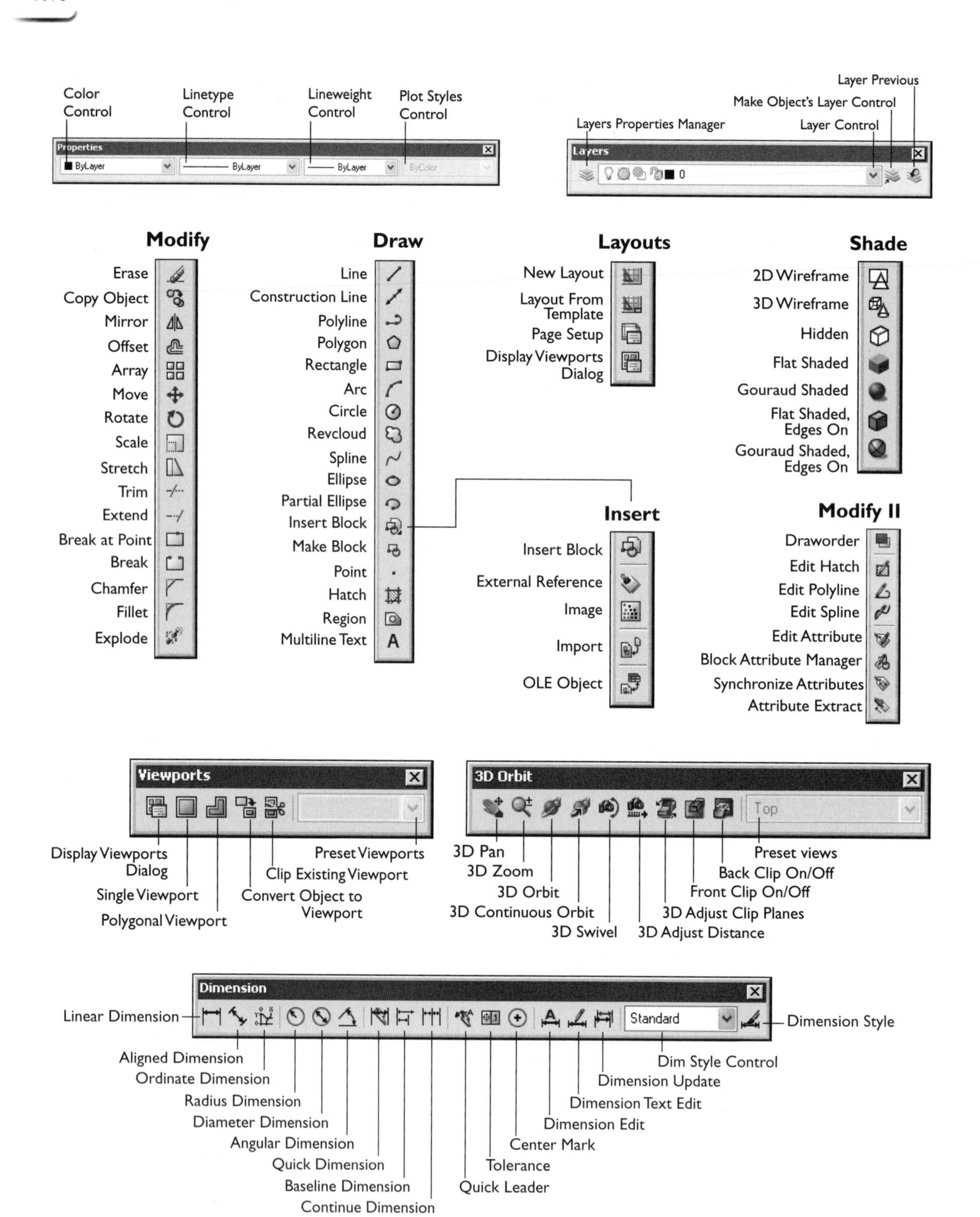

Color Control
Linetype Control
Lineweight Control
Plot Styles Control
Properties
ByLayer
ByLayer
ByLayer
ByColor
Layer Previous
Make Object's Layer Control
Layers Properties Manager
Layer Control
Layers
0
Modify
Erase
Copy Object
Mirror
Offset
Array
Move
Rotate
Scale
Stretch
Trim
Extend
Break at Point
Break
Chamfer
Fillet
Explode
Draw
Line
Construction Line
Polyline
Polygon
Rectangle
Arc
Circle
Revcloud
Spline
Ellipse
Partial Ellipse
Insert Block
Make Block
Point
Hatch
Region
Multiline Text
A
Layouts
New Layout
Layout From Template
Page Setup
Display Viewports Dialog
Shade
2D Wireframe
3D Wireframe
Hidden
Flat Shaded
Gouraud Shaded
Flat Shaded, Edges On
Gouraud Shaded, Edges On
Insert
Insert Block
External Reference
Image
Import
OLE Object
Modify II
Draworder
Edit Hatch
Edit Polyline
Edit Spline
Edit Attribute
Block Attribute Manager
Synchronize Attributes
Attribute Extract
Viewports
Display Viewports Dialog
Single Viewport
Polygonal Viewport
Convert Object to Viewport
Clip Existing Viewport
Preset Viewports
3D Orbit
Top
3D Pan
3D Zoom
3D Orbit
3D Continuous Orbit
3D Swivel
3D Adjust Distance
3D Adjust Clip Planes
Front Clip On/Off
Back Clip On/Off
Preset views
Dimension
Standard
Linear Dimension
Aligned Dimension
Ordinate Dimension
Radius Dimension
Diameter Dimension
Angular Dimension
Quick Dimension
Baseline Dimension
Continue Dimension
Quick Leader
Tolerance
Center Mark
Dimension Edit
Dimension Text Edit
Dimension Update
Dim Style Control
Dimension Style

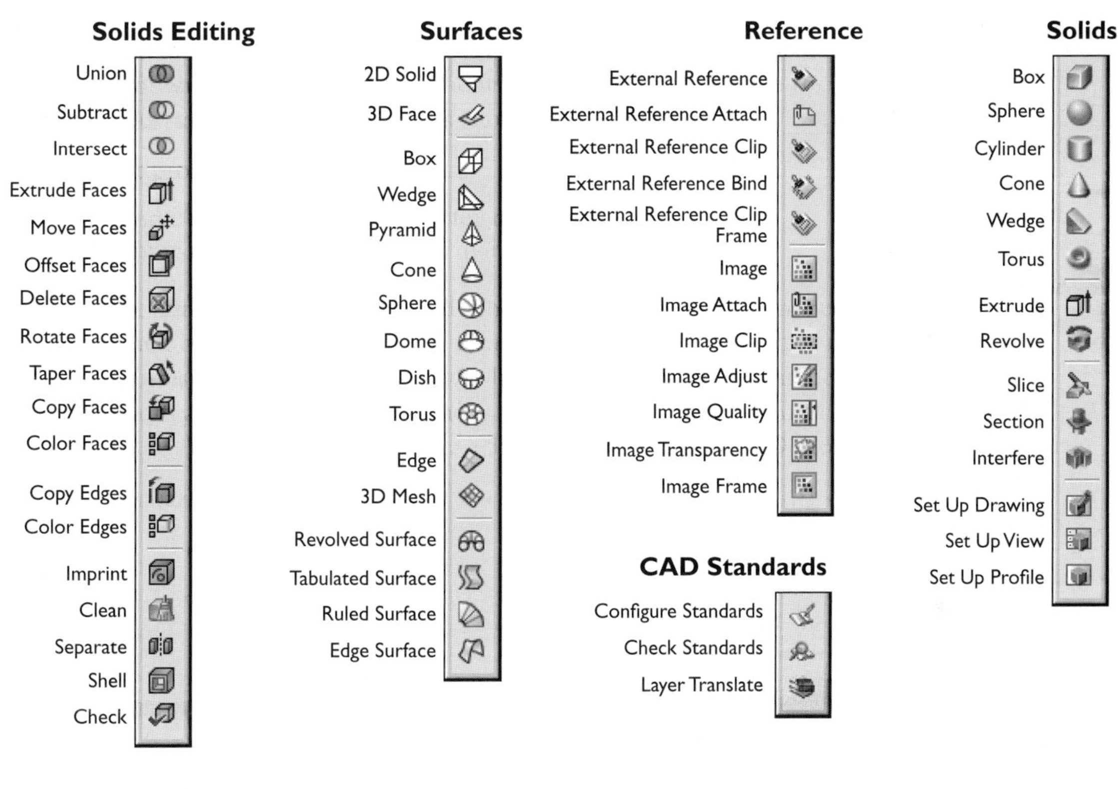
Solids Editing
Union
Subtract
Intersect
Extrude Faces
Move Faces
Offset Faces
Delete Faces
Rotate Faces
Taper Faces
Copy Faces
Color Faces
Copy Edges
Color Edges
Imprint
Clean
Separate
Shell
Check
Surfaces
2D Solid
3D Face
Box
Wedge
Pyramid
Cone
Sphere
Dome
Dish
Torus
Edge
3D Mesh
Revolved Surface
Tabulated Surface
Ruled Surface
Edge Surface
Reference
External Reference
External Reference Attach
External Reference Clip
External Reference Bind
External Reference Clip Frame
Image
Image Attach
Image Clip
Image Adjust
Image Quality
Image Transparency
Image Frame
CAD Standards
Configure Standards
Check Standards
Layer Translate
Solids
Box
Sphere
Cylinder
Cone
Wedge
Torus
Extrude
Revolve
Slice
Section
Interfere
Set Up Drawing
Set Up View
Set Up Profile

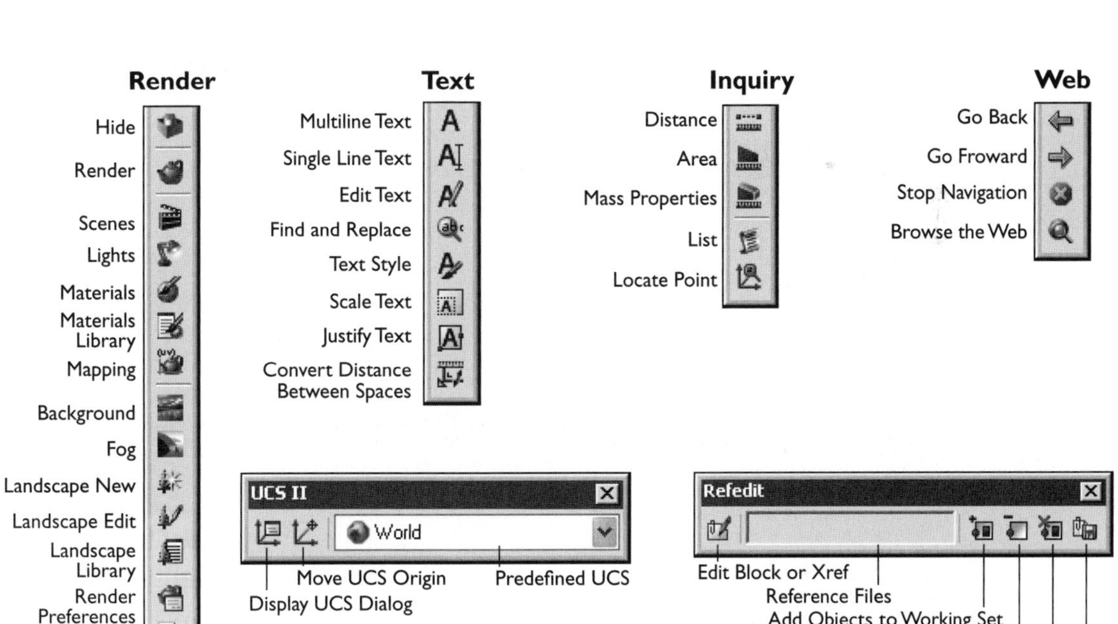
Render
Hide
Render
Scenes
Lights
Materials
Materials Library
Mapping
Background
Fog
Landscape New
Landscape Edit
Landscape Library
Render Preferences
Statistics
Text
Multiline Text
Single Line Text
Edit Text
Find and Replace
Text Style
Scale Text
Justify Text
Convert Distance Between Spaces
Inquiry
Distance
Area
Mass Properties
List
Locate Point
Web
Go Back
Go Froward
Stop Navigation
Browse the Web

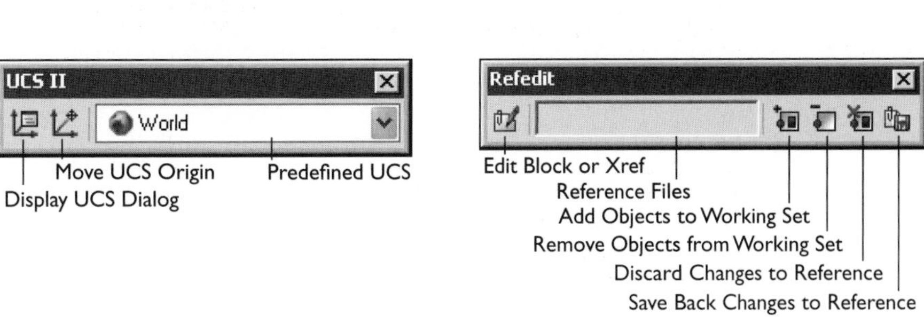
UCS II
World
Move UCS Origin
Predefined UCS
Display UCS Dialog
Refedit
Edit Block or Xref
Reference Files
Add Objects to Working Set
Remove Objects from Working Set
Discard Changes to Reference
Save Back Changes to Reference

System Variables

This is a complete listing of AutoCAD system variables. Each variable has an associated type: integer, real, point, or text string. These variables can be examined and changed (unless read-only) by means of the SETVAR command and AutoLISP (getvar and setvar) functions. Many of the system variables are saved across editing sessions; as indicated in the table, some are saved in the drawing itself, while others are saved in the AutoCAD general configuration file, *ACAD.CFG*

Variable	Default Setting	Type	Saved In	Explanation
ACADLSPASDOC	0	Integer	Registry	Controls whether AutoCAD loads the acad.lsp file into every drawing or just the first drawing opened in an AutoCAD session
ACADPREFIX	" "	String	Read-only	The directory path, if any specified by the ACAD environment variable, with path separators appended if necessary (read-only)
ACADVER		String		This is the AutoCAD version number, which can have values only like "15.0" (read-only). Note that this differs from the DXF file $ACADVER header variable, which contains the drawing database level number.
ACISOUTVER	40	Integer		Controls the ACIS version of SAT files created using the ACISOUTcommand
ADCSTATE	Varies	Integer	Not Saved	Determines if DesignCenter is active or not 0 - DesignCenter not active 1 - DesignCenter active
AFLAGS	0	Integer		Attribute flags bit-code for ATTDEF command (sum of the following): 0 = No Attribute Mode Selected 1 = Invisible 2 = Constant 4 = Verify 8 = Preset
ANGBASE	0	Real	Drawing	Angle 0 direction (with respect to the current UCS)
ANGDIR	0	Integer	Drawing	1= clockwise angles, 0 = counterclockwise (with respect to the current UCS)
APBOX	0	Integer	Registry	Turns Autosnap aperture box on or off 0 - Aperture box not displayed 1 - Apetrure box not displayed

Variable	Default Setting	Type	Saved In	Explanation
APERTURE	10	Integer	Config	Object Snap target height, in pixels (default value = 10)
APBOX				Sets AutoSnap aperture box to ON or OFF
APERTURE	10	Integer	Registry	Sets object snap target height, in pixels
AREA		Real		True area computed by Area, List, or Dblist (read-only)
ATTDIA	0	Integer	Registry	1 causes the INSERT command to use a dialog box for entry of attribute values; 0 to issue prompts
ATTMODE	1	Integer	Drawing	Attribute display mode (0 = OFF, 1 = normal, 2 = ON)
ATTREQ	1	Integer	Registry	0 assumes defaults for the values of all attributes during insert of blocks; 1 enables prompts (or dialog box) for attribute values, as selected by ATTDIA
AUDITCTL	0	Integer	Config	Controls whether an *.adt* log file (audit report file) is created 0 = Disables (or prevents) the writing of *adt* log files 1 = Enables the writing of *.adt* log files by the AUDIT command
AUNITS	0	Integer	Drawing	Angular units mode (0 = decimal degrees, 1 = degrees/minutes/seconds, 2 = grads, 3 = radians, 4 = surveyor's units)
AUPREC	0	Integer	Drawing	Angular units decimal places
AUTOSNAP	63	Integer	Registry	Controls the display of the AutoSnap marker and SnapTips and sets the AutoSnap magnet to ON or OFF
BACKZ	0.0000	Real	Drawing	Back clipping plane offset for the current viewport, in drawing units. Meaningful only if the back clipping bit in VIEWMODE is on. The distance of the back clipping plane from the camera point can be found by subtracting BACKZ from the camera-to-target distance (read-only)
BINDTYPE	0	Integer		Controls how xref names are handled when binding xrefs or editing xrefs in-place
BLIPMODE	0	Integer	Registry	Marker blips ON if 1, OFF if 0
CDATE		Real		Calendar date/time (read-only)
CECOLOR	"BY-LAYER"	String	Drawing	Current object color (read-only)
CELTSCALE	1	Real	Drawing	Sets the current global linetype scale for objects
CELTYPE	"BY-LAYER"	String	Drawing	Current object linetype (read-only)
CELWEIGHT	-1	Integer		Sets the lineweight of new objects
CHAMFERA	0.5000	Real	Drawing	First chamfer distance
CHAMFERB	1.0000	Real	Drawing	Second chamfer distance
CHAMFERC	1.0000	Real	Drawing	Sets the chamfer length
CHAMFERD	0.0000	Real	Drawing	Sets the chamfer angle
CHAMMODE	0	Integer		Sets the input method by which AutoCAD creates chamfers 0 = Requires two chamfer distances 1 = Requires one chamfer length and an angle
CIRCLERAD	0.0000	Real		Sets the default circle radius; to specify no default, enter 0 (zero)
CLAYER	"0"	String	Drawing	Sets the current layer (read-only)

Variable	Default Setting	Type	Saved In	Explanation
CMDACTIVE		Integer		Bitcode that indicates whether an ordinary command, transparent command, script, or dialog box is active (read-only). It is the sum of the following: 1 = Ordinary command is active 2 = Ordinary command and a transparent command are active 4 = Script is active 8 = Dialog box is active 16= Autolist active (only visible to an object arcs-defined)
CMDECHO	1	Integer		When the AutoLISP (command) function is used, prompts and input are echoed if this variable is 1 but not if it is 0
CMDNAMES		String		Displays in English the name of the command (and transparent command) that is currently active, for example; LINE'ZOOM indicates that the ZOOM command is being used transparently during the LINE command
CMLJUST	0	Integer	Drawing	Specifies multiline justification 0 = Top 1 = Middle 2 = Bottom
CMSCALE	1.0000	Real	Drawing	Controls the overall width of a multiline
CMSTYLE	"Standard"	String	Drawing	Sets the name of the multiline style that AutoCAD uses to draw the multiline
COMPASS	0	Integer		Controls whether the 3D compass is on or off in the current viewport
COORDS	1	Integer	Registry	If 0, coordinate display is updated on point picks only; if 1, display of absolute coordinates is continuously updated; if 2, distance and angle from last point are displayed when a distance or angle is requested
CPLOTSTYLE	"ByLayer"	Drawing		Controls the current plot style for new objects
CPROFILE		String	Registry	Stores the name of the current profile (read-only)
CTAB		String	Drawing	(Read-Only) Returns the name of the current (model or layout) tab in the drawing. Provides a means for the user to determine which tab is active
CURSORSIZE				Sets the size of the crosshairs as a percentage of screen size
CVPORT	2	Integer	Drawing	The identification number of the current viewport
DATE		Real		Julian date/time (read-only)
DBCSTATE	0	Integer	Drawing	Stores active or nonactive state of dbConnect Manager 0 - dbConnect Mgr not displayed 1 - dbConnect Mgr displayed
DBMOD		Integer		Bitcode that indicates the drawing modification status (read-only); it is the sum of the following: 1 = Entity database modified 2 = Symbol table modified 4 = Database variable modified 8 = Window modified 16 = View modified
DCTCUST		String	Registry	Displays the current custom spelling dictionary path and file name

Variable	Default Setting	Type	Saved In	Explanation
DCTMAIN		String	Registry	Displays the current main spelling dictionary file name
DEFLPLSTYLE	""	String	Registry	Specifies the default plot style for new layers
DEFPLSTYLE	"ByLayer"		Registry	Specifies the default plot style for new objects
DELOBJ	1	Integer	Registry	Controls whether objects used to create other objects are retained or deleted from the drawing database 0 = Objects are retained 1 = Objects are deleted
DEMANDLOAD	3	Integer	Registry	Determines demand loading of a third-party application if a drawing contains custom objects created in that application
DIASTAT		Integer		Dialog box exit status: if 0, the most recent dialog box was exited via CANCEL; if 1, the most recent dialog box was exited by pressing OK (read-only)
DIMxxx		Assorted	Drawing	All the dimensioning variables are also accessible as system variables (see Dimensioning Variables, Appendix B)
DISPSILH	0	Integer	Drawing	Sets display of silhouette curves of body objects in wireframe mode
DISTANCE		Real		Distance computed by DIST command (read-only)
DONUTID	0.5000	Real		Default donut inside diameter; can be zero
DONUTOD	1.0000	Real		Default donut outside diameter; must be nonzero. If DONUTID is larger than DONUTOD, the two values are swapped by the next command
DRAGMODE	2	Integer	Registry	0 = no dragging, 1 = on if requested, 2 = auto
DRAGP1	10	Integer	Registry	Regeneration-drag input sampling rate
DRAGP2	25	Integer	Registry	Fast-drag input sampling rate
DWGCHECK	0	Integer	Registry	Determines whether a drawing was last edited by a product other than AutoCAD
DWGCODEPAGE		String	Drawing	Drawing code page: This variable is set to the system code page when a new drawing is created, but otherwise AutoCAD doesn't maintain it. It should reflect the code page of the drawing and you can set it to any of the values used by the SYSCODEPAGE system variable or to "undefined." It is saved in the header DWGNAME String Drawing name as entered by the user; if the user specified a drive/directory prefix, it is included as well (read-only)
DWGPREFIX		String		Drive/directory prefix for drawing (read-only)
DWGTITLED		Integer		Bitcode that indicates whether the current drawing has been named (read-only) 0 = The drawing hasn't been named 1 = The drawing has been named
EDGEMODE	0	Integer	Registry	Sets how TRIM and EXTEND determine cutting and boundary edges
ELEVATION	0.0000	Real	Drawing	Current 3D elevation, relative to the current UCS for the current space
ERRNO	0	Integer	Not Saved	Shows number of error code when AutoLisp function calls an error

Variable	Default Setting	Type	Saved In	Explanation
EXPERT	0	Integer		Controls the issuance of certain "are you sure?" prompts: 0 = Issues all prompts normally 1 = Suppresses "About to regen, proceed?" and "Really want to turn the current layer off?" 2 = Suppresses the preceding prompts and Block's "Block already defined. Redefine it?" and Save/Wblock's "A drawing with this name already exists. Overwrite it?" 3 = Suppresses the preceding prompts and those issued by linetype if you try to load a linetype that is already loaded or create a new linetype in a file that already defines it 4 = Suppresses the preceding prompts and those issued by "Ucs Save" and "Vports Save" if the name you supply already exists 5 = Suppresses the preceding prompts and those issued by "Dim Save" and "Dim Override" if the dimension style name you supply already exists (the entries are redefined) When a prompt is suppressed, EXPERT, the operation in question, is performed as though you had responded Y to the prompt. In the future, values greater than 5 may be used to suppress additional safety prompts. The setting of EXPERT can affect scripts, menu macros, AutoLISP, and the command functions. The default value is 0.
EXPLMODE	0	Integer		Determines whether EXPLODE supports nonuniformly scaled (NUS) blocks
EXTMAX		3D point	Drawing	Upper right drawing uses extents. Expands outward as new objects are drawn; shrinks only by ZOOM All or ZOOM Extents. Reported in World coordinates for the current space (read-only)
EXTMIN		3D point	Drawing	Lower left drawing uses extents. Expands outward as new objects are drawn; shrinks only by ZOOM All or ZOOM Extents. Reported in World coordinates for the current space (read-only)
EXTNAMES	1	Integer	Drawing	Sets the parameters for named object names (such as linetypes and layers) stored in symbol tables
FACETRATIO	0	Integer		Controls the aspect ratio of faceting for cylindrical and conic ACIS solids
FACETRES	0.5	Real	Drawing	Adjusts smoothness of shaded and hidden-line-removed objects
FILEDIA	1	Integer	Registry	1 = Use file dialog boxes if possible; 0 = do not use File dialog boxes unless requested via ~ (tilde)
FILLETRAD	0.5000	Real	Drawing	Fillet radius
FILLMODE	1	Integer	Drawing	Fill mode ON if 1, OFF if 0
FONTALT	"Simplex.shp "	String	Registry	Specifies alternate font
FONTMAP	"acad.fmp"	String	Registry	Specifies font mapping file

Variable	Default Setting	Type	Saved In	Explanation
FRONTZ	0.0000	Real	Drawing	Front clipping plane offset for the current viewport, in drawing units. Meaningful only if the front clipping bit in VIEWMODE is ON and the "Front clip not at eye" bit is also ON. The distance of the front clipping bit from the camera point can be found by subtracting FRONTZ from the camera-to-target distance (read-only)
FULLOPEN		Integer		Indicates whether the current drawing is partially open (read-only)
GFANG	0	Integer	Not Saved	Specifies gradient fill angle
GFCLR1	"RGB 000, 000,255"	String	Not saved	Specifies color for one-color gradient fill or first color for two color gradient fill
GFCLR2	"RGB, 255, 255, 153"	String	Not saved	Specifies second color for a two-color gradient fill
GFCLRLUM	1.0000	Real	Not saved	Makes color a tint (mixed with white) or a shade (mixed with black) in a one color gradient fill
GFCLRSTATE	1	Integer	Not saved	Specifies whether gradient fill uses one or two colors 0 - Two-color gradient fill 1 - One-color gradient fill
GFNAME	1	Integer	Not saved	Specifies pattern of gradient fill. Values as follows: 1 - Linear 2 - Cylindrical 3 - Inverted cylindrical 4 - Spherical 5 - Inverted spherical 6 - Hemispherical 7 - Inverted hemispherical 8 - Curved 9 - Inverted Curved
GFSHIFT	0	Integer	Not saved	Specifies if pattern in gradient fill is centered or shifted up and to the left 0 - Centered 1 - Shifted up and to left
GRIDMODE	0	Integer	Drawing	1 = Grid on for current viewport, X and Y
GRIDUNIT	0.5000,0.5000	2D point	Drawing	Grid spacing for current viewport, X and Y
GRIPBLOCK	0	Integer	Registry	Controls the assignment of grips in blocks 0 = Assigns grip only to the insertion point of the block 1 = Assigns grips to entities within the block
GRIPCOLOR	5	Integer (1–255)	Registry	Color of nonselected grips; drawn as a box outline
GRIPHOT	1	Integer (1–255)	Registry	Color of selected grips; drawn as a filled box
GRIPHOVER	3	Integer	Registry	Controls fill color of grip
GRIPOBJLIMIT	100	Integer	Registry	Suppresses number of grips displayed when initial selection set contains more objects than number specified
GRIPS	1	Integer	Registry	Allows the use of selection set grips for the Stretch, Move, Rotate, Scale, and Mirror modes 0 = Disables grips 1 = Enables grips

Variable	Default Setting	Type	Saved In	Explanation
GRIPTIPS	1	Integer	Registry	Controls display of grips when cursor hovers on custom objects that support grip tips 0 - Grip tips not displayed 1 - Grip tips displayed
GRIPSIZE	3	Integer (1–255)	Registry	The size in pixels of the box drawn to display the grip
HALOGAP	0	Integer	Drawing	Specifies the distance to shorten a haloed line
HANDLES	1	Integer	Drawing	If 0, entity handles are disabled; if 1, handles are on (read-only)
HIDEPRECISION	0	Integer		Controls the accuracy of hides and shades
HIDETEXT	On	Switch	Drawing	Determines whether text objects created by TEXT, DTEXT, or MTEXT are processed during HIDE
HIGHLIGHT	1	Integer		Object selection highlighting ON if 1, OFF if 0
HPANG	0.0000	Real		Default hatch pattern angle
HPASSOC	1	Integer	Registry	Controls whether hatch patterns and gradient fills are associative 0 - Patterns/fills not associated 1 - Patterns/fills associated
HPBOUND	1			Controls BHATCH and BOUNDARY object types
HPDOUBLE	0	Integer		Default hatch pattern doubling for "U" user defined patterns 0 = Disables doubling 1 = Enables doubling
HPNAME	" ANSI31"	String		Default hatch pattern name. Up to 34 characters no spaces allowed. Returns " " if there is no default. Enter . (period) to set no default
HPSCALE	1.0000	Real		Default hatch pattern scale factor; must be nonzero
HPSPACE	1.0000	Real		Default hatch pattern line spacing for "U" user defined simple patterns; must be nonzero
HYPERLINKBASE	""	String	Drawing	Specifies the path used for all relative hyperlinks in the drawing
IMAGEHLT	0	Integer	Registry	Controls whether the entire raster image or only the raster image frame is hightlighted
INDEXCTL	0	Integer	Drawing	Controls whether layer and spatial indexes are created and saved in drawing files
INETLOCATION	www.acad.com/acaduser			Saves the Browser location used by the Internet
INSBASE	0.0000, 0.0000, 0.0000	3D point	Drawing	Insertion basepoint (set by BASE command) expressed in UCS coordinates for the current space
INSNAME	" "	String		Default block name for DDINSERT or INSERT. The name must conform to symbol-naming conventions. Returns " " if there is no default. Enter . (period) to set no default.
INSUNITS	0	Integer	Registry	When you drag a block from AutoCAD DesignCenter, specifies a drawing units value

Variable	Default Setting	Type	Saved In	Explanation
INSUNITSDEFSOURCE	0	Integer	Registry	Sets source content units value
INSUNITSDEFTARGET	0	Integer	Registry	Sets target drawing units value
INTERSECTIONCOLOR	257	Integer	Drawing	Specifies color of intersection polyline
INTERSECTIONDISPLAY	Off	Switch	Drawing	Specifies display of intersection polylines 0 - intersection plines not displayed 1 - intersection plines displayed
ISAVEBAK	1	Integer	Registry	Optimizes the speed of periodic saves, especially for large drawings in Windows
ISAVEPERCENT	50	Integer	Registry	Specifies the amount of wasted space tolerated in a drawing file
ISOLINES	4	Integer	Drawing	Specifies the number of iso lines first surface on the object
LASTANGLE	0	Real		The end angle of the last arc entered, relative to the XY plane of the current UCS for the current space (read-only)
LASTPOINT	0.0000, 0.0000, 0.0000,	3D point		The last point entered, expressed in UCS coordinates for the current space; referenced by @ during keyboard entry
LASTPROMPT	" "	String		Saves the last string echoed to the command line (read-only)
LAYOUTREGENCTL		Integer		Specifies how the display list is updated in the Model tab and layout tabs
LENSLENGTH		Real	Drawing	Length of the lens (in millimeters) used in perspective viewing, for current viewport (read-only)
LIMCHECK	0	Integer	Drawing	Limits checking for the current space: ON if 1, OFF if 0
LIMMAX	12,000, 9,000	2D point	Drawing	Upper right drawing limits for the current space, expressed in World coordinates
LIMMIN	0.0000, 0.0000,	2D point	Drawing	Lower left drawing limits for the current space, expressed in World coordinates
LISPRINT				Detrmines whether names and values of AutoLISP-defined functions and variables are preserved when you open a new drawing
LOCALE	"enu"	String		Displays the ISO language code of the current AutoCAD version
LOCALROOTPREFIX	"pathname"	String	Registry	Stores full path to the root folder where local customizable files installed
LOGFILEMODE	0	Integer	Registry	Determines whether the contents of the text window are written to a log file
LOGFILENAME				Determines the path for the log file (read-only)
LOGFILEPATH		String	Registry	Specifies the path for the log files for all drawings in a session
LOGINNAME		String		Displays the user's name as configured or input when AutoCAD is loaded (read-only)
LTSCALE	1.000	Real	Drawing	Sets global linetype scale factor
LUNITS	2	Integer	Drawing	Linear units units mode (1 = scientific, 2 = decimal, 3 = engineering, 4 = architectural, 5 = fractional)

Variable	Default Setting	Type	Saved In	Explanation
LUPREC	4	Integer	Drawing	Linear units decimal places
LWDEFAULT	25	Enum	Registry	Sets the value for the default linweight
LWDISPLAY	0	Integer	Drawing	Controls whether the lineweight is displayed in the Model or Layout tab
LWUNITS	1	Integer	Registry	Controls whether lineweight units are displayed in inches or millimeters
MAXACTVP	64	Integer	Drawing	Maximum number of viewports to regenerate at one time (read-only)
MAXSORT	200	Integer	Config	Maximum number of symbol/file names tó be sorted by listing commands; if the total number of items exceeds this number, then none of the items are sorted (default value is 200)
MBUTTONPAN	1	Integer	Registry	Controls the behavior of the third button or wheel on the pointing device
MEASUREINIT		Integer		Controls which hatch pattern and linetype files an existing drawing uses when it's opened
MEASUREMENT	0	Integer	Drawing	Sets drawing units as English or metric
MENUCTL	1	Integer	Config	Controls the page switching of the screen menu 0 = Screen menu doesn't switch pages in response to keyboard command entry 1 = Screen menu switches pages in response to keyboard command entry
MENUECHO	0	Integer		Menu echo/prompt control bits (sum of the following): 1 = Suppresses echo of menu items (^P in a menu item toggles echoing) 2 = Suppresses printing of system prompts during menu 4 = Disables ^P toggle of menu echoing 8 = Displays input/output strings The default value is 0 (all menu items and style prompts are displayed)
MENUNAME	"Acad"	Integer	Drawing	The name of the currently loaded menu file; includes a drive/path prefix if you entered it (read-only)
MIRRTEXT	1	Integer	Drawing	Mirror reflects text if nonzero, retains text direction if 0
MODEMACRO		String		Allows you to display a text string in the status line, such as the name of the current drawing, time/date stamp, or special modes. You can use MODEMACRO to display a simple string of text, or use special text strings written in the DIESEL macro language to have AutoCAD evaluate the macro from time to time and base the status line on user-selected conditions
MTEXTED	" Internal"	String	Config	Sets the name of the program to use for editing mtext objects
MTEXTFIXED	0	Integer	Registry	Controls apperence of Multiline Text Editor 0 - Editor/text drawing size/location 1 - Editor/text size/location fixed
MTJIGSTRING	"abc"	String	Registry	Sets content of sample text at cursor location when MTEXTcommand is started

Variable	Default Setting	Type	Saved In	Explanation
MYDOCUMENTSPREFIX	"pathname"	String	Registry	Stores full path to the My Documents folder for user currently logged-on
NOMUTT	0	Short		Suppresses the message display (muttering) when it wouldn't normally be suppressed
OBSCUREDCOLOR	0	Integer	Drawing	Specifies the color of obscured lines
OBSCUREDLTYPE	0	Integer	Drawing	Specifies the linetype of obscured lines
OFFSETDIST	1.0000	Real		Sets the default offset distance; if you enter a negative value, it defaults to Through mode
OFFSETGAPTYPE	0	Integer	Registry	Controls how to offset polylines when a gap is created as a result of offsetting the individual polyline segments
OLEHIDE	0	Integer	Registry	Controls the display of OLE objects in AutoCAD
OLEQUALITY	1	Integer	Registry	Controls the default quality level for embedded OLE objects
OLESTARTUP	0	Integer	Drawing	Controls whether the source application of an embedded OLE object loads when plotting
ORTHOMODE	0	Integer	Drawing	Ortho mode ON if 1, OFF if 0
OSMODE	0	Integer	Registry	Object Snap modes bitcode (sum of the following): 1 = Endpoint 128 = Perpendicular 2 = Midpoint 256 = Tangent 4 = Center 512 = Nearest 8 = Node 1024 = Quick 16 = Quadrant 2048 = Appint 32 = Intersection 4096 = Extension 64 = Insertion 8192 = Parallel
OSNAPCOORD	2	Integer	Registry	Controls whether coordinates entered on the command line override running object snaps
PALETTEOPAQUE	0	Integer	Registry	Controls whether windows can be opaque. Transparency is: 0 - turned on by user 1 - turned off by user 2 - unavailable though turned on by user 3 - unavailable and turned off by user
PAPERUPDATE	0	Integer	Registry	Controls the display of a warning dialog when attempting to print a layout with a paper size different from the paper size specified by the default for the plotter configuration file
PDMODE	0	Integer	Drawing	Point entity display mode
PDSIZE	0.0000	Real	Drawing	Point entity display size
PEDITACCEPT	0	Integer	Registry	Suppresses display of prompt "Object Selected Is Not a Polyline" 0 - turned on by user 1 - prompt suppressed
PELLIPSE	0	Integer	Drawing	Controls type of ellipse created 0 - turned on by user 1 - turned off by user
PERIMETER		Real		Perimeter computed by Area, List, or Dblist (read-only)

Variable	Default Setting	Type	Saved In	Explanation
PFACEMAX	4	Integer		Maximum number of vertices per face (read-only)
PICKADD	1	Integer	Config	Controls additive selection of objects 0 = Disables PICKADD. The most recently selected objects, either by an individual pick or windowing, become the selection set. Previously selected objects are removed from the selection set. You can add more objects to the selection set, however, by holding down SHIFT while selecting 1 = Enables PICKADD. Each object you select, either individually or by windowing, is added to the current selection set. To remove objects from the selection set, hold down SHIFT while selecting
PICKAUTO	1	Integer	Config	Controls automatic windowing when the "Select objects:" prompt appears 0 = Disables PICKAUTO 1 = Allows you to draw a selection window (both window and crossing window) automatically at the "Select objects:" prompt
PICKBOX	3	Integer	Config	Object selection target height, in pixels
PICKDRAG	0	Integer	Config	Controls the method of drawing a selection window 0 = You draw the selection window by clicking the mouse at one corner and then at the other corner 1 = You draw the selection window by clicking at one corner, holding down the mouse button, dragging, and releasing the mouse button at the other corner
PICKFIRST	1	Integer	Config	Controls the method of object selection so that you can select objects first and then use an edit/inquiry command 0 = Disables PICKFIRST 1 = Enables PICKFIRST
PICKSTYLE	1	Integer	Registry	Controls group selection and associative hatch selection
PLATFORM		String		Read-only message that indicates which version of AutoCAD is in use
PLINEGEN	1	Integer	Drawing	Sets the linetype pattern generation around the vertices of a 2D polyline. When set to 1, PLINEGEN causes the netype to be generated in a continuous pattern around the vertices of the polyline. When set to 0, polylines are generated with the linetype to start and end with a dash at each vertex. PLINEGEN doesn't apply to polylines with tapered segments
PLINETYPE	2	Integer	Registry	Determines whether AutoCAD uses optimized 2D patterns
PLINEWID	0.0000	Real	Drawing	Default polyline width; it can be zero
PLOTROTMODE	1	Integer	Registry	Controls orientation of plots
PLQUIET	0	Integer	Registry	Controls the display of optional dialog boxes and nonfatal errors for batch plotting and scripts
POLARADDANG	null	String	Registry	Contains user-defined polar angles
POLARANG	90.0000	Real	Registry	Sets the polar angle increment
POLARDIST	0.000	Real	Registry	Sets the snap increment when the SNAPSTYL system variable is set to 1 (polar snap)

Variable	Default Setting	Type	Saved In	Explanation
POLARMODE	1	Integer	Registry	Controls settings for polar and object snap tracking
POLYSIDES	4	Integer		Default number of sides for the POLYGON command; the range is 3–1024
POPUPS	1	Integer		1 if the currently configured display driver supports dialog boxes, the menu bar, pull-down menus, and icon menus; 0 if these advanced user interface features are not available (read-only)
PRODUCT	"AutoCAD"			Returns the product name (read-only)
PROGRAM	"acad"			Returns the program name (read-only)
PROJECTNAME	""	String	Drawing	Saves the current project name
PROJMODE	1	Integer	Config	Sets the current Projection mode for Trim or Extend operations
PROXYGRAPHICS	1	Integer	Drawing	Determines whether images of proxy objects are saved in the drawing
PROXYNOTICE	1	Integer	Registry	Displays a notice when you open a drawing containing custom objects created by an application that is not present
PROXYSHOW	1	Integer	Registry	Controls the display of proxy objects in a drawing
PROXYWEBSEARCH		Integer		Specifies how AutoCAD checks for Object Enablers
PSLTSCALE	1	Integer	Drawing	Controls paper space linetype scaling 0 = No special linetype scaling 1 = Viewport scaling governs linetype scaling
PSTYLEMODE	0	Integer	Drawing	Indicates whether the current drawing is in a Color-Dependent or Named Plot Style mode (read-only)
PSTYLEPOLICY	1	Integer	Registry	Controls whether an object's color property is associated with its plot style
PSVPSCALE	0	Real		Sets the view scale factor for all newly created viewports
PUCSBASE	""	String	Drawing	Stores the name of the UCS that defines the origin and orientation of orthographic UCS settings in paper space only
QTEXTMODE	0	Integer	Drawing	Quick text mode ON if 1, OFF if 0
RASTERPREVIEW	1	Integer	Drawing	Controls whether drawing preview images are saved with the drawing
REFEDITNAME	""	String		Indicates whether a drawing is in a reference-editing state and stores the reference file name (read-only)
REGENMODE	1	Integer	Drawing	Regenauto ON if 1, OFF if 0
RE-INIT		Integer		Reinitializes the I/O ports, digitizer, display, plotter, and *acad.pgp* file using the following bit codes. To specify more than one reinitialization, enter the sum of their values, for example, 3 to specify both digitizer port (1) and plotter port (2) reinitialization 1 = Digitizer port reinitialization 2 = Plotter port reinitialization 4 = Digitizer reinitialization 8 = Display reinitialization 16 = PGP file reinitialization (reload)
REMEMBERFOLDERS		Integer		Controls the default path for the Look In or Save In option in standard file selection dialog boxes

Variable	Default Setting	Type	Saved In	Explanation
REPORTERROR	1	Integer	Registry	Controls sending of error report 0 - Error Report msg not displayed 1 - Error Report msg displayed
ROAMABLEROOTPREFIX	"pathname"	String	Registry	Stores root folder full path where roamable customizable files were installed
RTDISPLAY	1	Integer	Registry	Controls the display of raster images during Realtime ZOOM
SAVEFILE	""	String	Config	Current auto-save file name (read-only)
SAVEFILEPATH	"C\TEMP\"	String	Registry	Specifies the path to the directory for all automatic save files for the AutoCAD session.
SAVENAME		String		The file name you save the drawing to (read-only)
SAVETIME	120	Integer	Config	Automatic save interval, in minutes (or 0 to disable automatic saves). The SAVETIME timer starts as soon as you make a change to a drawing, and is reset and restarts by a manual SAVE, SAVEAS, or QSAVE. The current drawing is saved to auto.sv$
SCREENBOXES		Integer		The number of boxes in the screen menu area of the graphics area. If the screen menu is disabled configured off), SCREENBOXES is zero. On platforms that permit the AutoCAD graphics window to be resized or the screen menu to be econfigured during an editing session, the value of this variable might change during the editing session (read-only)
SCREENMODE		Integer		A (read-only) bit code indicating the graphics/text state of the AutoCAD display. It is the sum of the following bit values: 0 = Text screen is displayed 1 = Graphics mode is displayed 2 = Dual-screen display configuration
SCREENSIZE		2D point		Current viewpoint size in pixels, X and Y (read-only)
SDI	0	Integer	Registry	Controls whether AutoCAD runs in single- or multiple-document interface
SHADEDGE	3	Integer	Drawing	0 = Faces shaded, edges not highlighted 1 = Faces shaded, edges drawn in background color 2 = Faces not filled, edges in object color 3 = Faces in entity color, edges in background color
SHADEDIF	70	Integer	Drawing	Ratio of ambient to diffuse light (in percentage of ambient light)
SHORTCUTMENU	11	Integer	Registry	Controls whether Default, Edit, and Command mode shortcut menus are available in the drawing area
SHPNAME	" "	String		Default shape name; must conform to symbol-naming conventions. If no default is set, it returns a " ". Enter . (period) to set no default.
SIGWARN	1	Integer	Registry	Controls display of warning when file with attached digital signature opened 0 - Warning not displayed 1 - Warning displayed
SKETCHINC	0.1000	Real	Drawing	Sketch record increment

Variable	Default Setting	Type	Saved In	Explanation
SKPOLY	0	Integer	Drawing	Sketch generates lines if 0, polylines if 1
SNAPANG	0	Real	Drawing	Snap/Grid rotation angle (UCS-relative) for the current viewport
SNAPBASE	0.0000, 0.0000,	2D point	Drawing	Snap/Grid origin point for the current viewport (in UCS XY coordinates)
SNAPISOPAIR	0	Integer	Drawing	Current isometric plane (0 = left, 1 = top, 2 = right) for the current viewport
SNAPMODE	0	Integer	Drawing	1 = Snap on for current viewport; 0 = Snap off
SNAPSTYL	0	Integer	Drawing	Snap style for current viewport (0 = standard, 1 = isometric)
SNAPTYPE	0	Integer	Registry	Sets the snap style for the current viewport
SNAPUNIT	0.5000, 0.5000,	2D point	Drawing	Snap spacing for current viewport, X and Y
SOLIDCHECK	1	Integer		Turns the solid validation on and off for the current AutoCAD session
SORTENTS	96	Integer	Drawing	Controls the display of objects sort order perations using the following codes. To select more than one, enter the sum of their codes; for example, enter 3 to specify codes 1 and 2 The default, 96, specifies sort operations for plotting and PostScript output 0 = Disables SORTENTS 1 = Sort for object selection 2 = Sort for object snap 4 = Sort for redraws 8 = Sort for MSLIDE slide creation 16 = Sort for regenerations 32 = Sort for plotting 64 = Sort for PostScript output
SPLFRAME	0	Integer	Drawing	If = 1: – the control polygon for spline-fit polylines is to be displayed – only the defining mesh of a surface-fit polygon mesh is displayed (the fit surface is not displayed) – invisible edges of 3D faces are displayed If = 0: – does not display the control polygon for spline-fit polylines – displays the fit surface of a polygon mesh, not the defining mesh – does not display the invisible edges of 3D faces
SPLINESEGS	8	Integer	Drawing	The number of line segments to be generated for each spline patch
SPLINETYPE	6	Integer	Drawing	Type of spline curve to be generated by PEDIT Spline. The valid values are: 5 = Quadratic B-spline 6 = Cubic B-spline
SURFTAB1	6	Integer	Drawing	Number of tabulations to be generated for Rulesurf and Tabsurf; also mesh density in the M direction for Resurf and Edgesurf

Variable	Default Setting	Type	Saved In	Explanation
STANDARDSVIOLATION	2	Integer	Registry	Controls notification of standards violation 0 - Notification turned off 1 - Alert is displayed 2 - Displays icon in status bar
STARTUP	0	Integer	Registry	Displays Create New Drawing dialog box at startup or use of NEW command 0 - Displays dialog box 1 - Displays Create New Dwg diabox
SURFTAB2	6	Integer	Drawing	Mesh density in the N direction for Revsurf and Edgesurf
SURFTYPE	6	Integer	Drawing	Type of surface fitting to be performed by PEDIT Smooth. The valid values are: 5 = Quadratic B-spline surface 6 = Cubic B-spline surface 8 = Bezier surface
SURFU	6	Integer	Drawing	Surface density in the M direction
SURFV	6	Integer	Drawing	Surface density in the N direction
SYSCODEPAGE		String	Drawing	Indicates the system code page specified in acad.xmf (read-only)
TABMODE	0	Integer		Controls the use of tablet mode 0 = Disables tablet mode 1 = Enables tablet mode
TARGET	0.0000, 0.0000, 0.0000,	3D point	Drawing	Location (in UCS coordinates) of the target (look-at) point for the current viewport (read-only)
TDCREATE		Real	Drawing	Time and date of drawing creation (read-only)
TDINDWG		Real	Drawing	Total editing time (read-only)
TDUCREATE		Real	Drawing	Stores the universal time and date the drawing was created
TDUPDATE		Real	Drawing	Time and date of last update/save (read-only)
TDSURTIMER		Real	Drawing	User elapsed timer (read-only)
TDUUPDATE		Real	Drawing	Stores the universal time and date of the last update/save
TEMPPREFIX	" "	String		Contains the directory name (if any) configured for placement of temporary files, with a path separator appended if necessary (read-only)
TEXTEVAL	0	Integer		If = 0, all responses to prompts for text strings and attribute values are taken literally. If = 1, text starting with "(" or "!" is evaluated as an AutoLISP expression, as for nontextual input. Note: The DTEXT command takes all input literally, regardless of the setting of TEXTEVAL
TEXTFILL	1	Integer	Registry	Controls the filling of Bitstream, TrueType, and Adobe Type 1 fonts
TEXTQLTY	50	Integer		Sets the resolution of Bitstream, TrueType, and Adobe Type 1 fonts

Variable	Default Setting	Type	Saved In	Explanation
TEXTSIZE	0.2000	Real	Drawing	The default height for new text objects drawn with the current text style (meaningless if the style has a fixed height)
TEXTSTYLE	"STANDARD"	String	Drawing	Contains the name of the current text style (read-only)
THICKNESS	0.0000	Real	Drawing	Current 3D thickness
TILEMODE	1	Integer	Drawing	1 = Release 10 compatibility mode (uses Vports) 0 = Enables paper space and viewport entities (uses MVIEW)
TOOLTIPS	1	Integer	Registry	Controls the display of Tool Tips
TPSTATE	Varies	Integer	Not saved	Controls if Tool Palettes window is active or not 0 - Tools Palettes window not active 1 - Tools Palettes window active
TRACEWID	0.0500	Real	Drawing	Default trace width
TRACKPATH	0	Integer	Registry	Controls the display of polar and object snap tracking alignment paths
TRAYICONS	1	Integer	Registry	Controls if tray is displayed in the status bar 0 - Tray not displayed 1 - Tray displayed
TRAYNOTIFY	1	Integer	Registry	Controls if service notification is displayed in the status bar 0 - Notifications not displayed 1 - Notifications displayed
TRAYTIMEOUT	5	Integer	Registry	Controls time service notification is displayed
TREEDEPTH	3020	Integer	Drawing	A 4-digit (maximum) code that specifies the number of times the tree-structured spatial index may divide into branches, hence affecting the speed in which AutoCAD searches the database before completing an action. The first two digits refer to the depth of the model space nodes, and the second two digits refer to the depth of paper space nodes. Use a positive setting for 3D drawings and a negative setting for 2D drawings
TREEMAX	10000000	Integer	Registry	Limits memory consumption during drawing regeneration
TRIMMODE	1	Integer	Registry	Controls whether AutoCAD trims selected edges for chamfers and fillets
TSPACEFAC	1	Real		Controls the multiline text line spacing distance measured as a factor of text height
TSPACETYPE	1	Integer		Controls the type of line spacing used in multiline text
TSTACKALIGN	1	Integer	Drawing	Controls the vertical alignment of stacked text
TSTACKSIZE	70	Integer	Drawing	Controls the percentage of stacked text fraction height relative to selected text's current height
UCSAXISANG	90	Integer	Registry	Stores the default angle when rotating the UCS around one of its axes using the X, Y, or Z options of the UCS command
UCSBASE	"World"	String	Drawing	Stores the name of the UCS that defines the origin and orientation of orthographic UCS settings
UCSFOLLOW	0	Integer	Drawing	The setting is maintained separately for both spaces and can be accessed in either space, but the setting is ignored while in paper space (it is always treated as if set to 0)

Variable	Default Setting	Type	Saved In	Explanation
UCSICON	3	Integer	Drawing	The coordinate system icon bitcode for the current viewport (sum of the following): 1 = On — icon display enabled 2 = Origin — if icon display is enabled, the icon floats to the UCS origin if possible 3 = On and displayed at origin
UCSNAME	" "	String	Drawing	Name of the current coordinate system for the current space; returns a null string if the current UCS is unnamed (read-only)
UCSORG	0.0000, 0.0000, 0.0000,	3D point	Drawing	The origin point of the current coordinate system for the current space; this value is always returned in World coordinates (read-only)
UCSORTHO	1	Integer	Registry	Determines whether the related orthographicUCS setting is restored automatically when an orthographic view is restored
UCSVIEW	1	Integer	Registry	Determines whether the current UCS is saved with a named view
UCSVP	1	Integer	Drawing	Determines whether the UCS in active viewports remains fixed or changes to reflect the UCS of the currently active viewport
UCSXDIR	1.0000, 0.0000, 0.0000,	3D point	Drawing	The X direction of the current UCS for the current space (read-only)
UCSYDIR	0.0000, 1.0000, 0.0000,	3D point	Drawing	The Y direction of the current UCS for the current space (read-only)
UNDOCTL	1	Integer		A (read-only) code indicating the state of the UNDO feature; it is the sum of the following values: 0 = undo is turned off 1 = Set if UNDO is enabled 2 = Set if only one command can be undone 4 = Set if Auto-group mode is enabled 8 = Set if a group is currently active
UNDOMARKS		Integer	Not Saved	Stores number of UNDO marks
UNITMODE	0	Integer	Drawing	0 = Displays fractional, feet and inches, and surveyor's angles as previously 1 = Displays fractional, feet and inches, and surveyor's angles in input format
USERI1-5	0	Integer	Drawing	Saves and recalls integer values
USERR1-5	0.0000	Real	Drawing	Saves and recalls real numbers
USERS1-5	" "	String		Saves and recalls text string data
VIEWCTR		3D point	Drawing	Center of view in current viewport, expressed in UCS coordinates (read-only)
VIEWDIR	3D Vector	3D point	Drawing	The current viewport's viewing direction expressed in World coordinates; this describes the camera point as a 3D offset from the TARGET point (read-only)

Variable	Default Setting	Type	Saved In	Explanation
VIEWMODE	0	Integer	Drawing	Viewing mode bitcode for the current viewport (read-only); the value is the sum of the following: 0 = turned off 1 = Perspective view active 2 = Front clipping on 4 = Back clipping on 8 = UCS follow mode 16 = Front clip not at eye. If On, the front clip distance (FRONTZ) determines the front clipping plane. If Off, FRONZ is ignored and the front clipping is set to pass through the camera point (i.e., vectors behind the camera are not displayed). This flag is ignored if the front clipping bit (2) is off
VIEWSIZE		Real	Drawing	Height of view in current viewport, expressed in drawing units (read-only)
VIEWTWIST	0	Real	Drawing	View twist angle for the current viewport (read-only)
VISRETAIN	1	Integer	Drawing	If = 0, the current drawing's On/Off, Freeze/Thaw, color, and linetype settings for Xref-dependent layers take precedence over the Xref's layer definition; if = 1, these settings don't take precedence
VSMAX		3D point		The upper right corner of the current viewport's virtual screen, expressed in UCS coordinates (read-only)
VSMIN		3D point		The lower left corner of the current viewport's virtual screen, expressed in UCS coordinates (read-only)
WHIPARC	0	Integer	Registry	Controls whether the display of circles and arcs is smooth
WHIPTHREAD		Integer		Controls whether to use an additional processor
WMFBKGND		Integer		Controls whether the background display of AutoCAD objects is transparent in other applications
WMFFOREGND		Integer		Controls the assignment of the foreground color of AutoCAD objects in other applications
WORLDUCS	1	Integer		If = 1, the current UCS is the same as the WCS; if = 0, it is not (read-only)
WORLDVIEW	1	Integer	Drawing	Dview and Vpoint command input is relative to the current UCS. If this variable is set to 1, the current UCS is changed to the WCS for the duration of a DVIEW or VPOINT command. Default value = 1
WRITESTAT	1	Integer	Read-only	Indicates whether a drawing file is read-only or can be written to, for developers who need to determine write status through AutoLISP
XCLIPFRAME	0	Integer	Drawing	Sets visibilty of Xref clipping boundaries
XEDIT	1	Integer	Drawing	Controls whether the current drawing can be edited in-place when being referenced by another drawing
XFADECTL	50	Integer	Registry	Controls the fading intensity for references being edited in-place

Variable	Default Setting	Type	Saved In	Explanation
XLOADCTL	1	Integer	Registry	Turns XREF demand loading on/off
XLOADPATH				Creates a path for storing temporary copies of demand-loaded Xref files
XREFCTL	0	Integer	Config	Controls whether *.xlg* files (external reference log files) are written 0 = Xref log (*.xlg*) files not written 1 = Xref log (*.xlg*) files written
ZOOMFACTOR	10	Integer	Registry	Controls the incremental change in zoom with each IntelliMouse wheel action, whether forward or backward

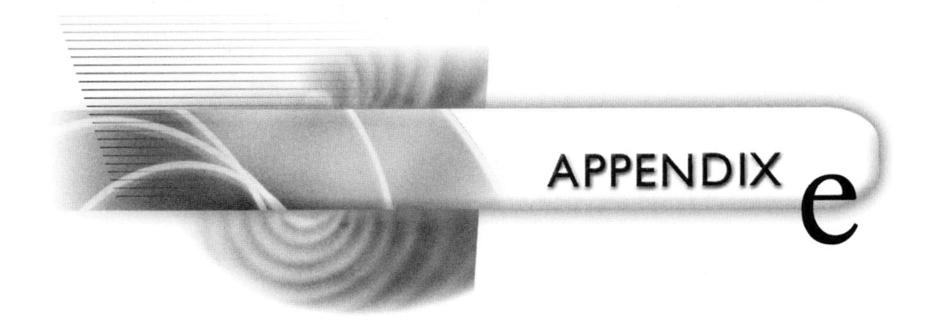

Hatch and Fill Patterns

INTRODUCTION

AutoCAD supports two types of hatch patterns: vector patterns and PostScript fill patterns. Vector patterns are made of straight lines and dots; they are defined in the Acad.Pat pattern file. You can create custom hatch patterns or purchase patterns created by third-party vendors. You place a hatch pattern with the HATCH and BHATCH commands.

PostScript fill patterns are made via PostScript PDL (page description language); they are defined in the Acad.Psf file. To create a custom fill pattern, you need to know PostScript programming. You place a fill pattern with the PSFILL command.

Over 60 hatch patterns and 12 PostScript fills are supplied with the AutoCAD package. Some are shown on the following pages.

HATCH PATTERNS

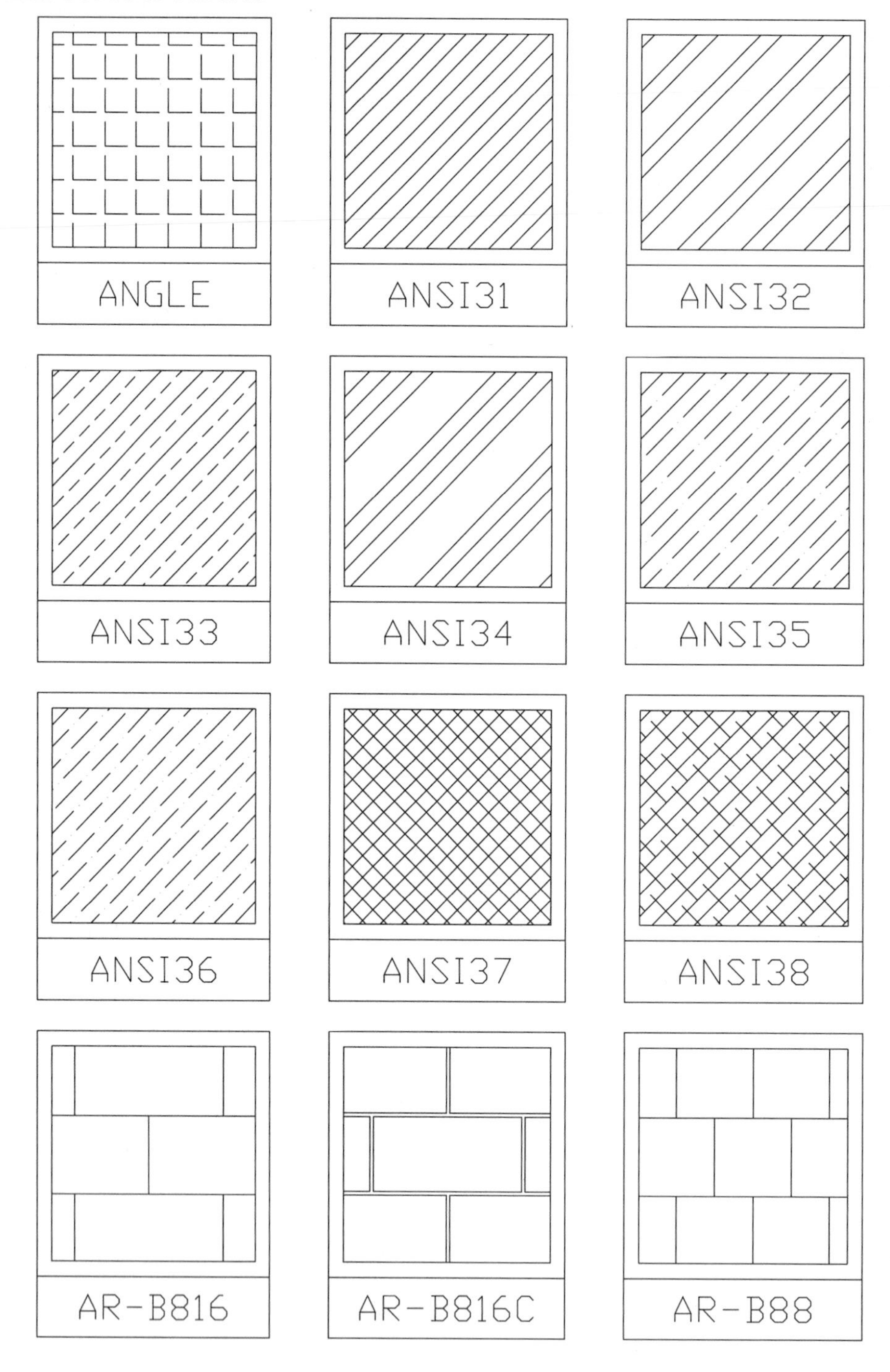

AR-BRELM
AR-BRSTD
AR-CONC
AR-HBONE
AR-PARQ1
AR-RROOF
AR-RSHKE
AR-SAND
BOX
BRASS
BRICK
BRSTONE

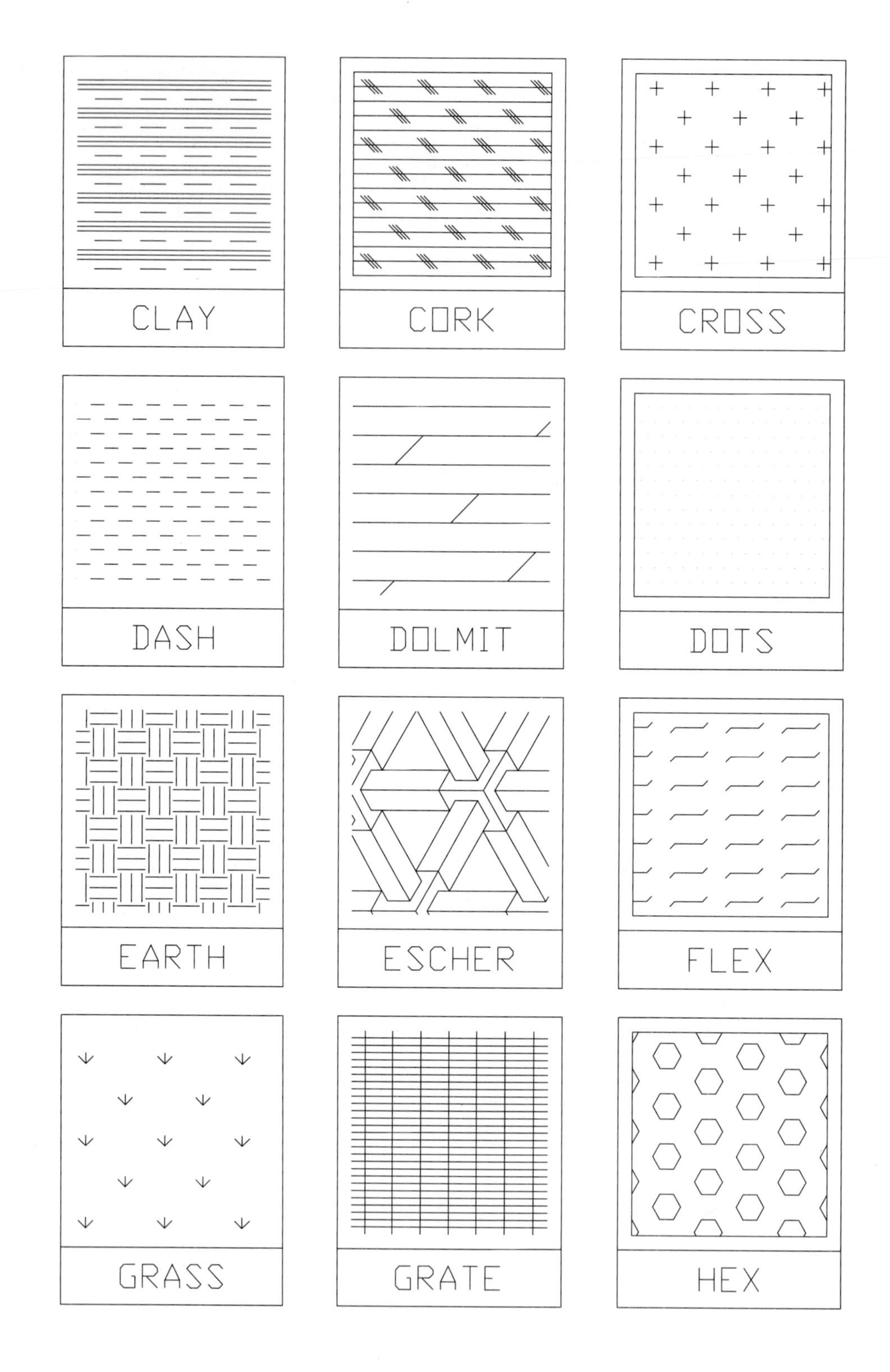
CLAY
CORK
CROSS
DASH
DOLMIT
DOTS
EARTH
ESCHER
FLEX
GRASS
GRATE
HEX

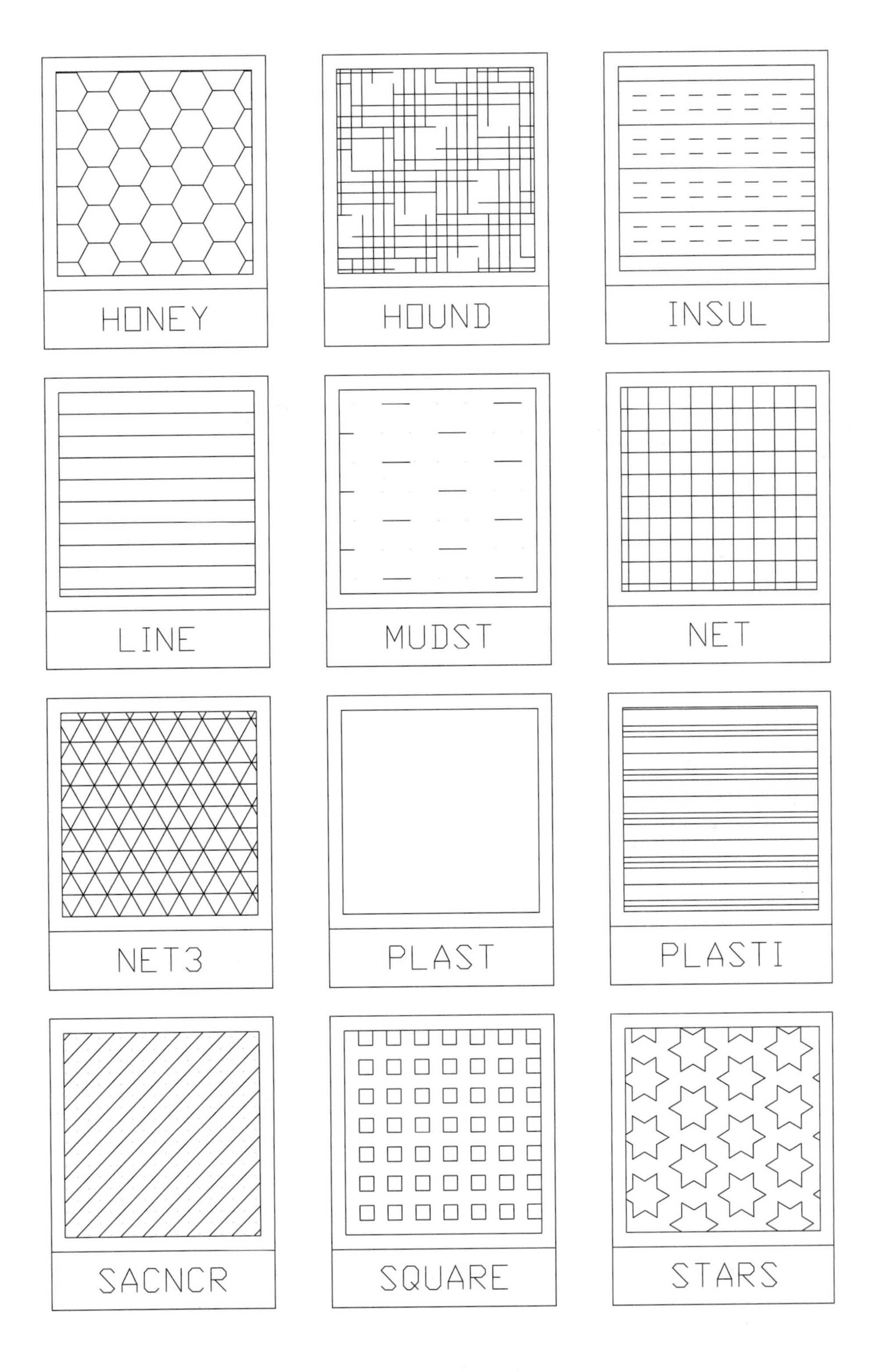
HONEY
HOUND
INSUL
LINE
MUDST
NET
NET3
PLAST
PLASTI
SACNCR
SQUARE
STARS

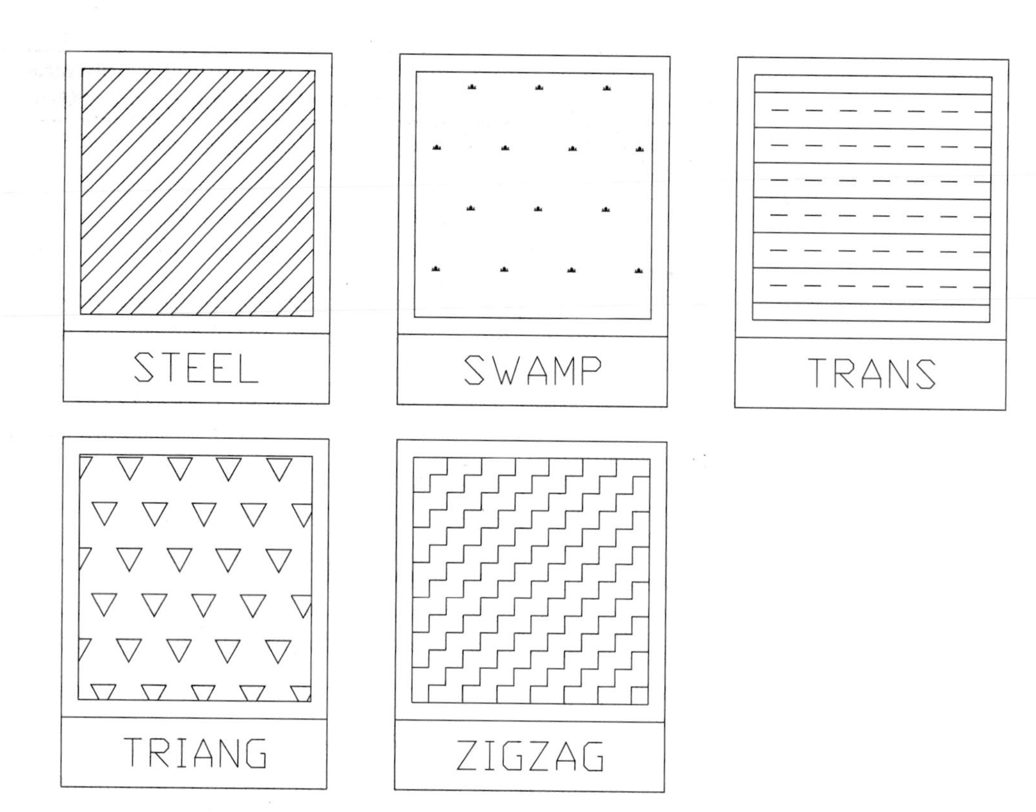
STEEL
SWAMP
TRANS
TRIANG
ZIGZAG

POSTSCRIPT FILL PATERNS

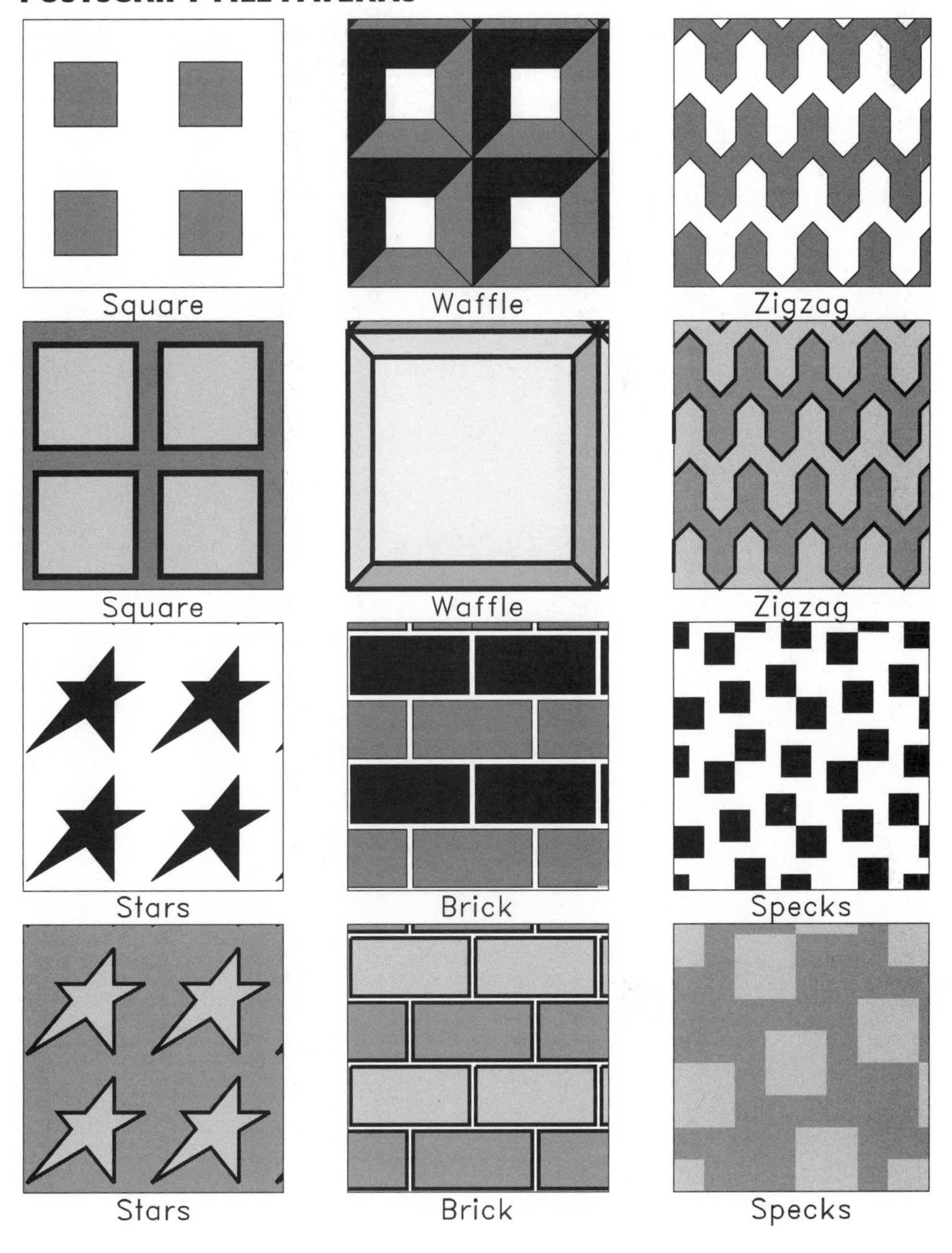

Fonts

INTRODUCTION

AutoCAD works with two types of text fonts: the original SHX-format font files and PFB PostScript font files. The AutoCAD package includes 17 SHX text fonts, five SHX symbol fonts, and 16 PFB text fonts. Some of the available fonts are shown on the following pages.

USING SHX AND PFB FILES

Other SHX font files are available from third-party developers. In addition, AutoCAD can use PostScript fonts from any source, many of which are included free with other software packages. Postscript fonts are usually stored in the \Psfonts folder.

AutoCAD does not store text fonts in the drawing file. Instead, the DWG file references SHX font definition files stored elsewhere on the hard drive. Thus, if you receive a drawing from another AutoCAD system, you might have to tell AutoCAD where to find the font files on your system.

PostScript fonts placed in an AutoCAD drawing have two anomilies: the fonts are unfilled and they are drawn 30% too small. To compensate for the reduced size, specify a text height 50% larger.

STANDARD TEXT FONTS

FAST FONTS

TXT ABCDEFGHIJKLMNOPQRSTUVWXYZ 1234567890

MONOTXT ABCDEFGHIJKLMNOPQRSTUVWXYZ 1234567890

SIMPLEX FONTS

ROMANS ABCDEFGHIJKLMNOPQRSTUVWXYZ 1234567890

SCRIPTS *ABCDEFGHIJKLMNOPQRSTUVWXYZ* 1234567890

GREEKS ΑΒΧΔΕΦΓΗΙϑΚΛΜΝΟΠΘΡΣΤϒ∇ΩΞΨΖ 1234567890

DUPLEX FONTS

ROMAND ABCDEFGHIJKLMNOPQRSTUVWXYZ 1234567890

COMPLEX FONTS

ROMANC ABCDEFGHIJKLMNOPQRSTUVWXYZ 1234567890

ITALICC *ABCDEFGHIJKLMNOPQRSTUVWXYZ 1234567890*

SCRIPTC *ABCDEFGHIJKLMNOPQRSTUVWXYZ 1234567890*

GREEKC ΑΒΧΔΕΦΓΗΙϑΚΛΜΝΟΠΘΡΣΤϒ∇ΩΞΨΖ 1234567890

TRIPLEX FONTS

ROMANT **ABCDEFGHIJKLMNOPQRSTUVWXYZ 1234567890**

ITALICT ***ABCDEFGHIJKLMNOPQRSTUVWXYZ 1234567890***

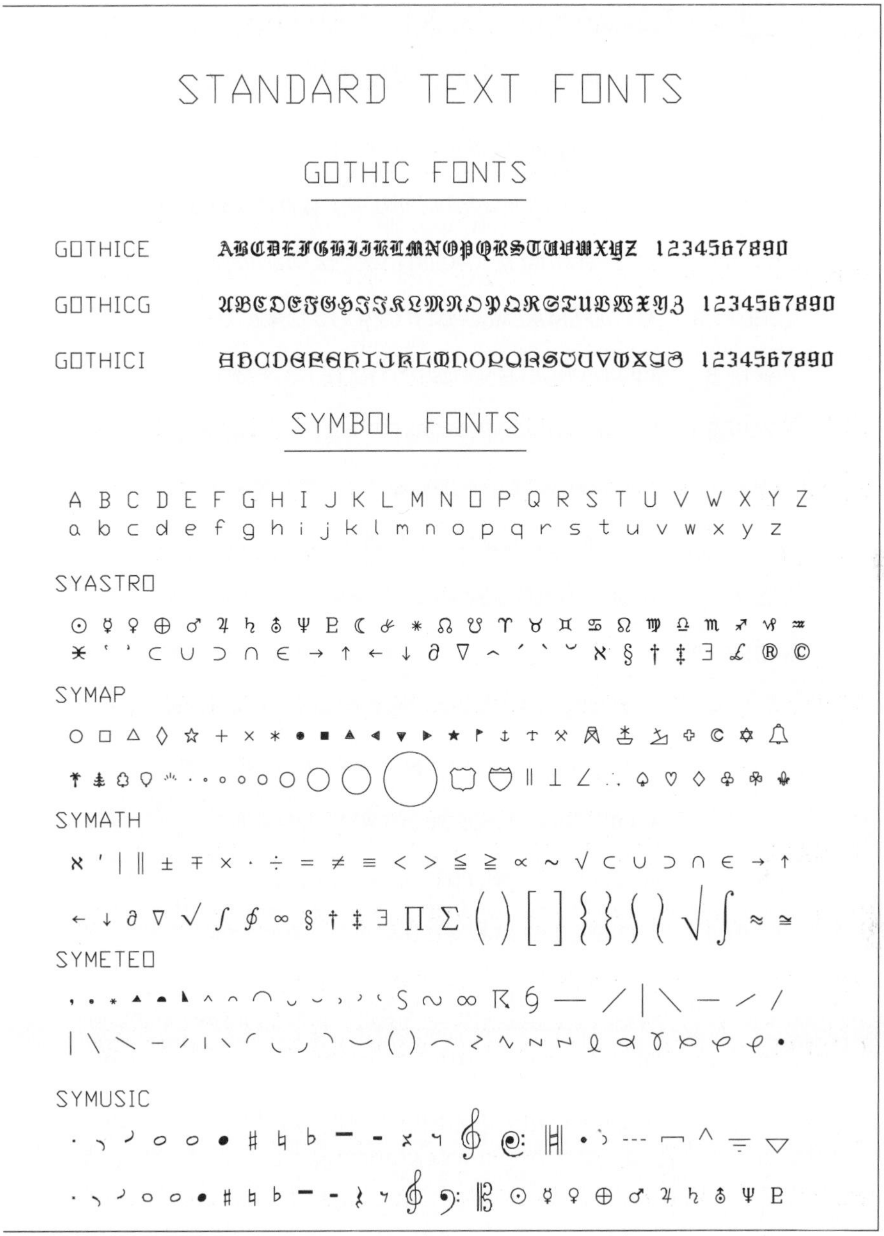

STANDARD TEXT FONTS

GOTHIC FONTS

GOTHICE ABCDEFGHIJKLMNOPQRSTUVWXYZ 1234567890

GOTHICG ABCDEFGHIJKLMNOPQRSTUVWXYZ 1234567890

GOTHICI ABCDEFGHIJKLMNOPQRSTUVWXYZ 1234567890

SYMBOL FONTS

A B C D E F G H I J K L M N O P Q R S T U V W X Y Z
a b c d e f g h i j k l m n o p q r s t u v w x y z

SYASTRO

SYMAP

SYMATH

SYMETEO

SYMUSIC

STANDARD TEXT FONTS

POSTSCRIPT FONTS

CIBT.PFB	ABCDEFGHIJKLMNOPQRSTUVWXYZ 1234567890
COBT.PFB	*ABCDEFGHIJKLMNOPQRSTUVWXYZ 1234567890*
EUR.PFB	**ABCDEFGHIJKLMNOPQRSTUVWXYZ 1234567890**
EURO.PFB	***ABCDEFGHIJKLMNOPQRSTUVWXYZ 1234567890***
PAR.PFB	**ABCDEFGHIJKLMNOPQRSTUVWXYZ 1234567890**
ROM.PFB	**ABCDEFGHIJKLMNOPQRSTUVWXYZ 1234567890**
ROMB.PFB	**ABCDEFGHIJKLMNOPQRSTUVWXYZ 1234567890**
ROMI.PFB	***ABCDEFGHIJKLMNOPQRSTUVWXYZ 1234567890***
SAS.PFB	**ABCDEFGHIJKLMNOPQRSTUVWXYZ 1234567890**
SASB.PFB	**ABCDEFGHIJKLMNOPQRSTUVWXYZ 1234567890**
SASBO.PFB	***ABCDEFGHIJKLMNOPQRSTUVWXYZ 1234567890***
SASO.PFB	***ABCDEFGHIJKLMNOPQRSTUVWXYZ 1234567890***
SUF.PFB	**ABCDEFGHIJKLMNOPQRSTUVWXYZ 1234567890**
TE.PFB	**ABCDEFGHIJKLMNOPQRSTUVWXYZ 1234567890**
TEB.PFB	**ABCDEFGHIJKLMNOPQRSTUVWXYZ 1234567890**

CYRILLIC FONTS

CYRILLIC	АБВГДЕЖЗИЙКЛМНОПРСТУФХЦЧШЩ 1234567890
CYRILTLC	АБЧДЕФГХИЩКЛМНОПЦРСТУВШЖЙЗ 1234567890

Linetypes and Lineweights

INTRODUCTION

Linetypes are defined by the ACAD.LIN file. In addition to the continuous linetype, the AutoCAD program comes with the 45 linetypes. Some of them are listed on the next page. You can add custom linetypes to the ACAD.LIN file.

Before you can use a linetype in a drawing, it must be loaded by means of the LINETYPE command. Set the linetype scale with the LTSCALE command; set independent linetype scaling in paper space with the PSLTSCALE system variable; control the generation of linetype along a polyline with the PLINEGEN system variable. The following tables list the standard lineweights and linetypes provided by AutoCAD.

AutoCAD Standard Lineweights

mm	inch		ISO
0.00			
0.05	.002		
0.09	.003		
0.13	.005		
0.15	.006		
0.18	.007		X
0.20	.008		
0.25	.010		X
0.30	.012		
0.35	.014		X
0.40	.016		
0.50	.020		X
0.53	.021		
0.60	.024		
0.70	.028		X
0.80	.031		
0.90	.035		
1.00	.039		X
1.06	.042		
1.20	.047		
1.40	.056		X
1.58	.062		
2.00	.078		X
2.11	.083		

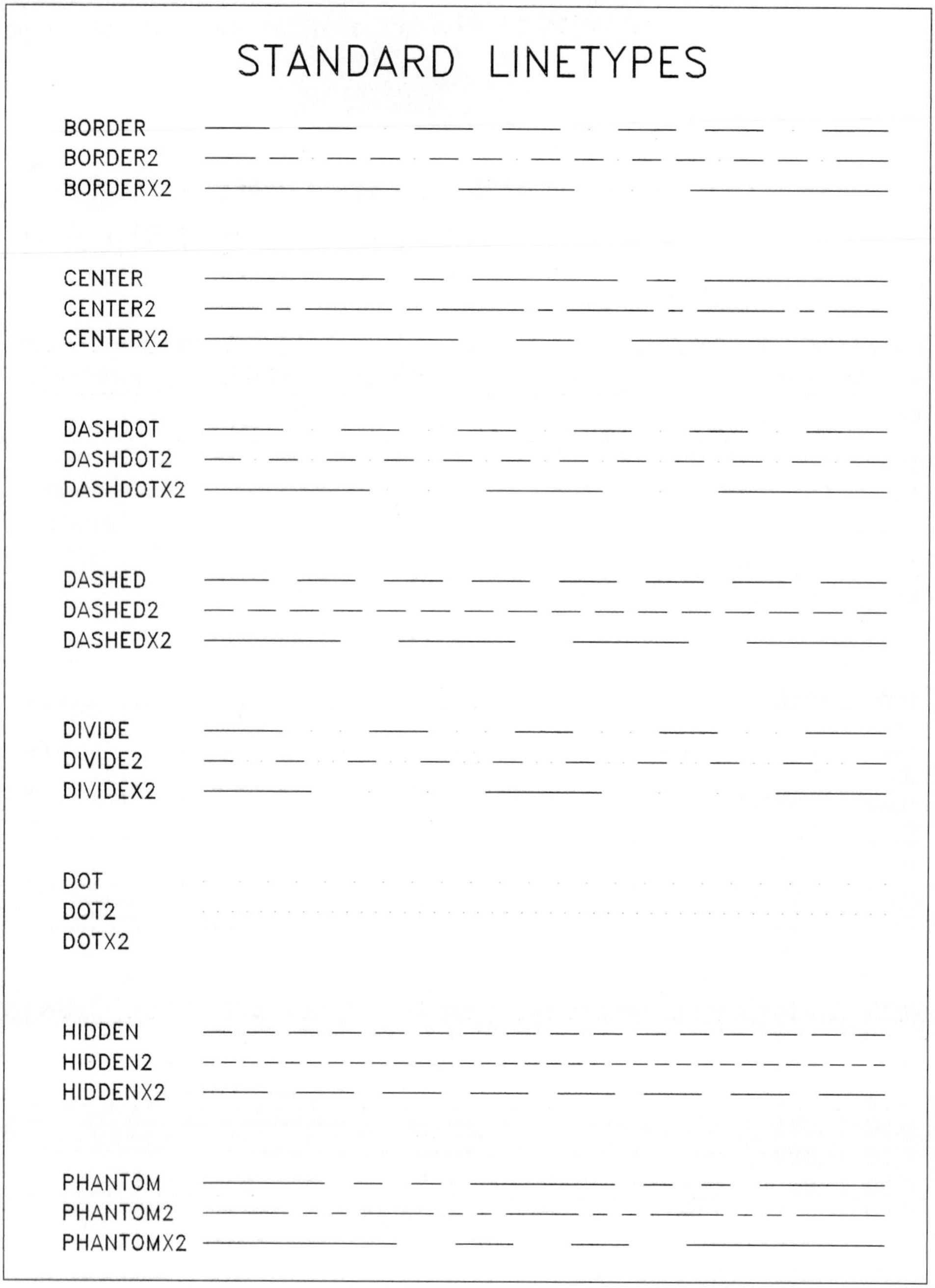

STANDARD LINETYPES

BORDER
BORDER2
BORDERX2

CENTER
CENTER2
CENTERX2

DASHDOT
DASHDOT2
DASHDOTX2

DASHED
DASHED2
DASHEDX2

DIVIDE
DIVIDE2
DIVIDEX2

DOT
DOT2
DOTX2

HIDDEN
HIDDEN2
HIDDENX2

PHANTOM
PHANTOM2
PHANTOMX2

APPENDIX h

Command Aliases

ALIAS NAME	COMMAND NAME
3A	3DARRAY
3DO	3DORBIT
3F	3DFACE
3P	3DPOLY
A	ARC
ADC	ADCENTER
AA	AREA
AL	ALIGN
AP	APPLOAD
AR	ARRAY
ATT	ATTDEF
-ATT	-ATTDEF
ATE	ATTEDIT
-ATE	-ATTEDIT
ATTE	-ATTEDIT
B	BLOCK
-B	-BLOCK
BH	BHATCH
BO	BOUNDARY
-BO	-BOUNDARY
BR	BREAK
C	CIRCLE
CH	PROPERTIES

ALIAS NAME	COMMAND NAME
-CH	CHANGE
CHA	CHAMFER
COL	COLOR
COLOUR	COLOR
CO	COPY
D	DIMSTYLE
DAL	DIMALIGNED
DAN	DIMANGULAR
DBA	DIMBASELINE
DBC	DBCONNECT
DCE	DIMCENTER
DCO	DIMCONTINUE
DDI	DIMDIAMETER
DED	DIMEDIT
DI	DIST
DIV	DIVIDE
DLI	DIMLINEAR
DO	DONUT
DOR	DIMORDINATE
DOV	DIMOVERRIDE
DR	DRAWORDER
DRA	DIMRADIUS
DS	DSETTINGS
DST	DIMSTYLE
DT	DTEXT
DV	DVIEW
E	ERASE
ED	DDEDIT
EL	ELLIPSE
EX	EXTEND
EXIT	QUIT

ALIAS NAME	COMMAND NAME
EXP	EXPORT
EXT	EXTRUDE
F	FILLET
FI	FILTER
G	GROUP
-G	-GROUP
GR	DDGRIPS
H	BHATCH
-H	HATCH
HE	HATCHEDIT
HI	HIDE
I	INSERT
-I	-INSERT
IAD	IMAGEADJUST
IAT	IMAGEATTACH
ICL	IMAGECLIP
IM	IMAGE
-IM	-IMAGE
IMP	IMPORT
IN	INTERSECT
INF	INTERFERE
IO	INSERTOBJ
L	LINE
LA	LAYER
-LA	-LAYER
LE	QLEADER
LEN	LENGTHEN
LI	LIST
LINEWEIGHT	LWEIGHT
LO	-LAYOUT
LS	LIST

ALIAS NAME	COMMAND NAME
LT	LINETYPE
-LT	-LINETYPE
LTYPE	LINETYPE
-LTYPE	-LINETYPE
LTS	LTSCALE
LW	LWEIGHT
M	MOVE
MA	MATCHPROP
ME	MEASURE
MI	MIRROR
ML	MLINE
MO	PROPERTIES
MS	MSPACE
MT	MTEXT
MV	MVIEW
O	OFFSET
OP	OPTIONS
ORBIT	3DORBIT
OS	OSNAP
-OS	-OSNAP
P	PAN
-P	-PAN
PA	PASTESPEC
PARTIALOPEN	-PARTIALOPEN
PE	PEDIT
PL	PLINE
PO	POINT
POL	POLYGON
PR	OPTIONS
PRCLOSE	PROPERTIESCLOSE
PROPS	PROPERTIES

ALIAS NAME	COMMAND NAME
PRE	PREVIEW
PRINT	PLOT
PS	PSPACE
PU	PURGE
R	REDRAW
RA	REDRAWALL
RE	REGEN
REA	REGENALL
REC	RECTANGLE
REG	REGION
REN	RENAME
-REN	-RENAME
REV	REVOLVE
RM	DDRMODES
RO	ROTATE
RPR	RPREF
RR	RENDER
S	STRETCH
SC	SCALE
SCR	SCRIPT
SE	DSETTINGS
SEC	SECTION
SET	SETVAR
SHA	SHADE
SL	SLICE
SN	SNAP
SO	SOLID
SP	SPELL
SPL	SPLINE
SPE	SPLINEDIT
ST	STYLE

ALIAS NAME	COMMAND NAME
SU	SUBTRACT
T	MTEXT
-T	-MTEXT
TA	TABLET
TH	THICKNESS
TI	TILEMODE
TO	TOOLBAR
TOL	TOLERANCE
TOR	TORUS
TR	TRIM
UC	DDUCS
UCP	DDUCSP
UN	UNITS
-UN	-UNITS
UNI	UNION
V	VIEW
-V	-VIEW
VP	DDVPOINT
-VP	VPOINT
W	WBLOCK
-W	-WBLOCK
WE	WEDGE
X	EXPLODE
XA	XATTACH
XB	XBIND
-XB	-XBIND
XC	XCLIP
XL	XLINE
XR	XREF
-XR	-XREF
Z	ZOOM

APPENDIX i

Express Tools

The following commands are included with the Express Tools package:

Command	Description
AliasEdit	Edits the aliases stored in *acad.pgp*.
AlignSpace	Aligns model space objects, whether in different viewports or with objects in paper space.
ArcText	Places text along an arc.
AttIn	Imports attribute data.
AttOut	Quickly extracts attributes in tab-delimited format.
BExtend	Extends open objects to objects in blocks and xrefs.
BlockReplace	Replaces all inserts of one block with another.
BlockToXref	Convert blocks to xrefs.
BreakLine	Creates the break-line symbol.
BScale	Scales blocks from their insertion points.
BTrim	Trims to objects nested in blocks and external references.
Burst	Explodes blocks, converts attributes to text.
ChSpace	Moves objects between model and paper space.
ChUrls	**DdEdit**-like editor for hyperlinks (URL addresses).
ClipIt	Adds arcs, circles, and polylines to the **XClip** command.
CopyM	**Copy** command with repeat, divide, measure, and array options.
CopyToLayer	Copies objects to other layers.

DimEx	Exports dimension styles to an ASCII file.
DimIm	Imports dimension style files created with **DimEx**.
EtBug	Sends bug reports to Autodesk.
ExOffset	Adds options to the **Offset** command.
ExPlan	Adds options to the **Plan** command.
FS	Selects objects that touch the selected object.
FullScreen	Toggles between full-screen and regular window.
GetSel	Selects objects based on layer and type.
ImageApp	Specifies the external image editor.
ImageEdit	Launches the image editor to edit selected images.
LayCur	Changes the layer of selected objects to the current layer.
LayDel	Deletes layers from drawings — permanently.
LayFrz	Freezes the layers of selected objects.
LayIso	Isolates layers of selected objects (all other layers are frozen).
LayLck	Locks the layers of selected objects.
LayMch	Changes the layer of selected objects to that of a selected object.
LayMrg	Merges two layers; removes the first layer from the drawing.
LayOff	Turns off layers of selected objects.
LayOn	Turns on all layers.
LayoutMerge	Places objects from layouts onto one layout.
LayThw	Thaws all layers.
LayUlk	Unlocks layer of selected object.
LayVpi	Isolates object's layer in viewport.
LayWalk	Isolates each layer in sequential order.
LMan	Saves and restores layer settings.
Lsp	AutoLISP function searching utility.
LspSurf	LISP file viewer.

MkLtype	Creates linetypes from selected objects.
MkShape	Creates shapes from selected objects.
MoCoRo	Moves, copies, rotates, and scales objects.
MoveBak	Moves *.bak* files to specified folders.
MStretch	Stretches with multiple selection windows.
NCopy	Copies objects nested inside blocks and xrefs.
Plt2Dwg	Imports HPGL files into the drawings.
Propulate	Updates, lists, and clears drawing properties.
PsBScale	Sets and updates the scale of blocks relative to paper space.
PsTScale	Sets text height relative to paper space.
QlAttach	Associate leaders to annotation objects.
QlAttach	Set Associates leaders with annotations.
QlDetach	Set Dissassociates leaders from annotations.
QQuit	Closes all drawings, and then exits AutoCAD.
ReDir	Changes paths for xrefs, images, shapes, and fonts.
RepUrls	Replaces hyperlinks.
Revert	Closes the drawing, and re-opens the original.
RText	Inserts and edits remote text objects.
RtUcs	Changes UCSs in real time.
SaveAll	Saves all drawings.
ShowUrls	Lists URLs in a dialog box.
Shp2blk	Converts from a shape definition to a block definition.
SuperHatch	Uses images, blocks, external references, or wipeouts as hatch patterns.
SysvDlg	Launches an editor for system variables.
TCase	Changes text between Sentence, lower, UPPER, Title, and toggle CASE.
TCircle	Surrounds text and multiline text with circles, slots, and rectangles.

TCount	Prefixes text with sequential numbers.
TextFit	Fits text between points.
TextMask	Places masks behind selected text.
TextUnmask	Removes masks from behind text.
TFrames	Toggles the frames surrounding images and wipeouts.
TJust	Justifies text created with the **MText** and **AttDef** commands.
TOrient	Re-orients text, multiline text, and block attributes.
TScale	Scales text, multiline text, attributes, and attribute definitions.
TSpaceInvaders	Finds and selects text with overlapping objects.
Txt2Mtxt	Converts single-line to multiline text.
TxtExp	Explodes selected text into polylines.
VpScale	Lists the scale of selected viewports.
VpSynch	Synchronizes viewports with a master viewport.
XData	Attaches xdata to objects.
XdList	Lists xdata attached to objects.
XList	Displays properties of objects nested in blocks and xref.

INDEX

Note: AutoCAD keyed commands appear in SMALL CAPS. AutoLISP functions appear in (parentheses). Page numbers with "CD" indicate material from Chapter 19 found on the CD-ROM.

SYMBOLS

A

C

D

E

G

I

J

K

L

M

N

O

P

Q

R

S

T

U

V

W

X

Z

LICENSE AGREEMENT FOR AUTODESK PRESS

A Thomson Learning Company

Educational Software/Data

You the customer, and Autodesk Press incur certain benefits, rights, and obligations to each other when you open this package and use the software/data it contains. BE SURE YOU READ THE LICENSE AGREEMENT CAREFULLY, SINCE BY USING THE SOFTWARE/DATA YOU INDICATE YOU HAVE READ, UNDERSTOOD, AND ACCEPTED THE TERMS OF THIS AGREEMENT.

Your rights:

1. You enjoy a non-exclusive license to use the enclosed software/data on a single microcomputer that is not part of a network or multi-machine system in consideration for payment of the required license fee, (which may be included in the purchase price of an accompanying print component), or receipt of this software/data, and your acceptance of the terms and conditions of this agreement.
2. You own the media on which the software/data is recorded, but you acknowledge that you do not own the software/data recorded on them. You also acknowledge that the software/data is furnished "as is," and contains copyrighted and/or proprietary and confidential information of Autodesk Press or its licensors.
3. If you do not accept the terms of this license agreement you may return the media within 30 days. However, you may not use the software during this period.

There are limitations on your rights:

1. You may not copy or print the software/data for any reason whatsoever, except to install it on a hard drive on a single microcomputer and to make one archival copy, unless copying or printing is expressly permitted in writing or statements recorded on the diskette(s).
2. You may not revise, translate, convert, disassemble or otherwise reverse engineer the software/data except that you may add to or rearrange any data recorded on the media as part of the normal use of the software/data.
3. You may not sell, license, lease, rent, loan, or otherwise distribute or network the software/data except that you may give the software/data to a student or and instructor for use at school or, temporarily at home.

Should you fail to abide by the Copyright Law of the United States as it applies to this software/data your license to use it will become invalid. You agree to erase or otherwise destroy the software/data immediately after receiving note of Autodesk Press' termination of this agreement for violation of its provisions.

Autodesk Press gives you a LIMITED WARRANTY covering the enclosed software/data. The LIMITED WARRANTY can be found in this product and/or the instructor's manual that accompanies it.

This license is the entire agreement between you and Autodesk Press interpreted and enforced under New York law.

Limited Warranty

Autodesk Press warrants to the original licensee/ purchaser of this copy of microcomputer software/ data and the media on which it is recorded that the media will be free from defects in material and workmanship for ninety (90) days from the date of original purchase. All implied warranties are limited in duration to this ninety (90) day period. THEREAFTER, ANY IMPLIED WARRANTIES, INCLUDING IMPLIED WARRANTIES OF MERCHANTABILITY AND FITNESS FOR A PARTICULAR PURPOSE ARE EXCLUDED. THIS WARRANTY IS IN LIEU OF ALL OTHER WARRANTIES, WHETHER ORAL OR WRITTEN, EXPRESSED OR IMPLIED.

If you believe the media is defective, please return it during the ninety day period to the address shown below. A defective diskette will be replaced without charge provided that it has not been subjected to misuse or damage.

This warranty does not extend to the software or information recorded on the media. The software and information are provided "AS IS." Any statements made about the utility of the software or information are not to be considered as express or implied warranties. Delmar will not be liable for incidental or consequential damages of any kind incurred by you, the consumer, or any other user.

Some states do not allow the exclusion or limitation of incidental or consequential damages, or limitations on he duration of implied warranties, so the above limitation or exclusion may not apply to you. This warranty ves you specific legal rights, and you may also have other rights which vary from state to state. Address all rrespondence to:

todeskPress
cutive Woods
xwell Drive
n Park, NY 12065
y, NY 12212-5015